7th International Conference on Modern Practice in Stress and Vibration Analysis 2009

Journal of Physics: Conference Series Volume 181

Cambridge, United Kingdom
8-10 September 2009

ISBN: 978-1-61738-303-8

Printed from e-media with permission by:

Curran Associates, Inc.
57 Morehouse Lane
Red Hook, NY 12571

Some format issues inherent in the e-media version may also appear in this print version.

Printed by Curran Associates, Inc. (2010)

For permission requests, please contact the Institute of Physics
at the address below.

Institute of Physics
Dirac House, Temple Back
Bristol BS1 6BE UK

Phone: 44 1 17 929 7481
Fax: 44 1 17 920 0979

techtracking@iop.org

Additional copies of this publication are available from:

Curran Associates, Inc.
57 Morehouse Lane
Red Hook, NY 12571 USA
Phone: 845-758-0400
Fax: 845-758-2634
Email: curran@proceedings.com
Web: www.proceedings.com

TABLE OF CONTENTS

KEYNOTE PAPERS

SYSTEM IDENTIFICATION

CONTACT MECHANICS

BIOMEDICAL APPLICATIONS

ROTOR DYNAMICS

MANUFACTURING PROCESSES 1

STRUCTURAL INTEGRITY ASSESSMENT 1

MANUFACTURING PROCESSES 2

STRUCTURAL INTEGRITY ASSESSMENT 2

COMPOSITE STRUCTURES AND MATERIALS

CIVIL ENGINEERING APPLICATIONS 1

ADVANCED MODELLING TECHNIQUES 1

CIVIL ENGINEERING APPLICATIONS 2

FRACTURE AND DAMAGE 1

IMPACT, DYNAMIC LOADING AND DAMPING 1

FRACTURE AND DAMAGE 2

NOVEL SENSOR TECHNOLOGY 1

NOVEL SENSOR TECHNOLOGY 2

IMPACT, DYNAMIC LOADING AND DAMPING 2

STRUCTURAL HEALTH MONITORING

ADVANCED MODELLING TECHNIQUES 2

Preface

The proceedings contain the papers presented at the 7th International Conference on Modern Practice in Stress and Vibration Analysis. The collection of papers represents the range of activities that are carried out to understand the functionality of engineering systems and structures through stress/strain based evaluation and dynamic response. The scope is broad and covers theoretical studies, modelling and experimental evaluations. Many of the papers cover integration techniques and approaches to better understanding of system performance and failure. All of the papers have been peer reviewed by at least two experts and represent the state of the art of research in this area.

The conference is the seventh in the series, following on from previous conferences in Bath, Glasgow, Nottingham, Dublin, Sheffield and Liverpool. Although based in the British Isles the conference has a truly international flavour with offerings from 22 countries. The conference is organised by the Institute of Physics Applied Mechanics Group (formerly the Stress and Vibration Group). It incorporates activities associated with the British Society for Strain Measurement including the Measurements Lecture, the EMex Exhibition and the Young Stress Analyst Competition.

The organising committee is grateful to the support of all of the authors, the scientific committee and keynote speakers who played significant role in the review process, to John Edwards who was instrumental in managing the paper review and submission process, Dawn Stewart and Claire Garland of the Institute of Physics for organising the conference, social programme and registration and Biana Gale of the British Society for Strain Measurement for organising the Exhibition. The organising committee is also grateful to the sponsors of the conference for their kind support and to the co-sponsors for distributing information on the conference.

Professor Janice Barton
Professor of Experimental Mechanics
University of Southampton
School of Engineering Sciences
Conference Chairman

Journal of Physics: Conference Series **181** (2009) 011001 doi:10.1088/1742-6596/181/1/011001

Organising Committee

Professor J M Barton, University of Southampton, UK (Conference Chair)
Professor K M Holford, Cardiff University, UK
Professor P S Keogh, University of Bath, UK
Professor A W Lees, Swansea University, UK
Dr C Mares, Brunel University, UK
Dr S Quinn, University of Southampton, UK
Professor V Silberschmidt, Loughborough University, UK
Dr C E Truman, University of Bristol, UK

International Scientific Committee

Professor A Ajovalasit, University of Palermo, Italy
Professor J R Barber, University of Michigan, USA
Professor A A Becker, University of Nottingham, UK
Dr C C Berggreen, DTU, Denmark
Dr R L Burguete, Airbus UK Ltd, UK
Professor M Cartmell, University of Glasgow, UK
Dr P R Cunningham, Loughborough University, UK
Professor J Eberhardsteiner, Vienna University of Technology, Austria
Professor M I Friswell, University of Bristol, UK
Dr A Ferreira, University of Porto, Portugal
Professor S D Garvey, University of Nottingham, UK
Professor E E Gdoutos, Democritus University of Thrace, Greece
Professor M Gilchrist, University College Dublin, Ireland
Professor M Grediac, Université Blaise Pascal, France
Professor J M Huntley, Loughborough University, UK
Professor D Inman, Virginia Tech, USA
Professor P Jacquot, EPFL, Switzerland
Dr I A Jones, University of Nottingham, UK
Dr S Kaczmarczyk, University of Northampton, UK
Dr A Lavrov, SINTEF Petroleum Research, Norway
Dr J Lord, National Physical Laboratory, UK
Professor M Lucas, University of Glasgow, UK
Professor Nuno Maia, Instituto Superior Técnico, Portugal
Professor A V Manzhirov, Russian Academy of Sciences, Russia
Dr R A W Mines, University of Liverpool, UK
Professor J E Mottershead, University of Liverpool, UK
Professor D Nowell, University of Oxford, UK
Professor W M Ostachowicz, PASci, Gdansk, Poland
Dr H Ouyang, University of Liverpool, UK
Dr S Patsias, Rolls-Royce, UK
Professor J E T Penny, Aston University, UK

Journal of Physics: Conference Series **181** (2009) 011001 doi:10.1088/1742-6596/181/1/011001

Dr E Petrov, Imperial College London, UK
Professor F Pierron, ENSAM Chalons en Champagne, France
Professor K Ramesh, IIT Madras, India
Professor R E Rowlands, University of Wisconsin, USA
Dr M N Sahinkaya, University of Bath, UK
Dr R Scott, University of Durham, UK
Professor C Surace, Polytechnic of Turin, Italy
Mr P Tatum, AWE, UK
Professor O T Thomsen, Aalborg University, Denmark
Dr R A Tomlinson, University of Sheffield, UK
Dr D L Tunnicliffe, BAE SYSTEMS, UK
Professor A Vautrin, ENSM, France
Professor M Wiercigroch, University of Aberdeen, UK
Professor P Withers, University of Manchester, UK
Professor K Worden, University of Sheffield, UK
Professor T Yokoyama, Okayama University of Science, Japan
Professor D Zenkert, KTH, Sweden

Journal of Physics: Conference Series **181** (2009) 011001 doi:10.1088/1742-6596/181/1/011001

Sponsors

AWE	
Flir Infra-red	
Instron	
LaVision	
Limess	
Veqter	
Vishay	

Journal of Physics: Conference Series **181** (2009) 011001 doi:10.1088/1742-6596/181/1/011001

Co-Sponsors

Association Française de Mécanique	
British Society for Strain Measurement	
Engineering Integrity Society	
European Society for Experimental Mechanics	
Emerald	
Forum for Applied Mechanics	
Institution of Mechanical Engineers	
Japanese Society for Experimental Mechanics	
NAFEMS	
Society for Experimental Mechanics	

All papers published in this volume of *Journal of Physics: Conference Series* have been peer reviewed through processes administered by the proceedings Editors. Reviews were conducted by expert referees to the professional and scientific standards expected of a proceedings journal published by IOP Publishing.

Vibration and stress analysis in the presence of structural uncertainty

R.S. Langley

Department of Engineering, University of Cambridge, Cambridge CB2 1PZ, UK.

RSL21@eng.cam.ac.uk

Abstract. At medium to high frequencies the dynamic response of a built-up engineering system, such as an automobile, can be sensitive to small random manufacturing imperfections. Ideally the statistics of the system response in the presence of these uncertainties should be computed at the design stage, but in practice this is an extremely difficult task. In this paper a brief review of the methods available for the analysis of systems with uncertainty is presented, and attention is then focused on two particular "non-parametric" methods: statistical energy analysis (SEA), and the hybrid method. The main governing equations are presented, and a number of example applications are considered, ranging from academic benchmark studies to industrial design studies.

1. Introduction

There has been much recent interest in predicting the response of structures that have uncertain or random properties, and this is particularly true in the field of vibro-acoustics. In predicting stress, vibration, and interior noise a broad frequency range is of interest and the response of an engineering structure such as a car or an aeroplane can be highly sensitive to small manufacturing variations, particularly at higher frequencies. Ideally this sensitivity should be identified and quantified at the design stage, so that the performance of the complete set of produced articles can be predicted, rather than just the performance of a nominally perfect design. Perhaps the most direct way to consider the effect of system uncertainty on the response is to adopt a *parametric model*. This approach consists of (i) identifying the uncertainties in the key physical properties (or "parameters") of the system, and (ii) propagating this uncertainty through the governing equations of motion to predict the uncertainty in the response. However both parts of this procedure can present severe practical difficulties. With regard to issue (i), the parametric model would ideally consist of the detailed statistical distribution of the system properties (material properties, dimensions, rigidity of joints, etc), but it can be extremely difficult to acquire this data, or even to identify which of the system properties are statistically significant. This problem can be overcome to some extent by adopting a less precise parametric model of the system uncertainties, consisting of perhaps bounds [1], or fuzzy-number descriptions [2], but the difficulty of propagating this uncertainty through the equations of motion, issue (ii), still remains. Typical industrial structures are described by finite element models that may have thousands or millions of degrees of freedom, and so the propagation of parametric uncertainty using Monte Carlo simulations, or more analytical means, can be highly computationally intensive. It is clear that parametric modelling is a very challenging research area, and progress is being made by the introduction of innovate Monte Carlo methods [3] or alternative propagation methods [2].

Despite the foregoing discussion, the fact that engineering systems can be large and complex is not necessarily to the disadvantage of the analyst. In fact, very complex systems can sometimes be easier to analyse than less complex systems. Of course, care is needed in the use of the term "complex" in this context; here, by "more complex" the intention is to say "more random". To consider the response of a single plate for example, the introduction of a small degree of randomness will cause the natural frequencies, the mode shapes, and the plate response to be random, and it can be challenging to work out precisely the statistics of these quantities. The results obtained will be dependent on the statistics of the random input parameters, and thus a very detailed analytical model will be required. If, on the other hand, the plate is subjected to a high degree of randomness, then it is known that the natural frequencies (excluding the lower frequencies) approach a standard distribution described by the Gaussian Orthogonal Ensemble (GOE) [4-6], regardless of the statistical details of the input parameters. In this case the prediction of the natural frequency statistics is unnecessary (since they are known) and the statistics of the response can be calculated in a relatively straight forward way [7]. This opens up the possibility of *non-parametric* models of uncertainty for complex systems, i.e. models which avoid the need to describe the detailed statistical distributions of the underlying physical parameters. One of the earliest non-parametric modelling approaches developed in engineering is known as statistical energy analysis (SEA) [8], in which the governing equations are based on the flow of vibrational energy between regions of the system known as "subsystems". By adopting a single (energy) degree of freedom for each subsystem, relatively efficient models of the system can be derived, and the mean energy response can be computed without the need for parametric modelling, large finite element models, or Monte Carlo simulations. The statistical assumptions employed in SEA are valid only at high frequencies, and significant current research effort is being directed at the development alternative, more general, non-parametric approaches. Much of this work is based on novel applications of random matrix theory (for example, [9]), or, as described below, the combination of SEA with other approaches.

Over the past decade, much research into the dynamic analysis of uncertain engineering systems has been motivated by the "mid-frequency" problem, where neither the finite element method (FE) nor statistical energy analysis (SEA) provide a satisfactory solution. At mid-frequencies the system is too complex for FE to apply (too random, with too many degrees of freedom), and not complex enough for SEA apply (not random enough for the averaging required by SEA to apply, so that very high response variances arise which cannot be predicted by standard SEA). The obvious approach is therefore to try and combine FE and SEA into a single model, so that advantage can be taken of the strengths of each method. This type of hybrid model has been proposed by many researchers, and the pioneering work of Soize [10] on fuzzy structure theory highlighted many general underlying principles. An early attempt to produce a hybrid model consisted of partitioning the response of a system component into a "long wavelength" contribution and a "short wavelength" contribution [11]. The long wavelength part was then included in an FE model, and the short wavelength part in an SEA model. A computational difficulty with this approach for general systems is that the coupling between the FE and SEA parts of the model require the evaluation of some rather difficult integrals. An alternative method was then developed in which the response of the component is partitioned not into different wavelength components, but into a "direct field" and a "reverberant field" [12]. One advantage of this approach is that the coupling between the SEA and FE parts of the model can be affected simply and efficiently through a "diffuse field reciprocity relation" [13,14]. Also, the method can be combined with aspects of random matrix theory to predict not only the mean, but also the variance of the system response [15].

In this paper attention is focused on two analysis methods: SEA and the hybrid FE-SEA approach. The basic concepts of SEA are described in Section 2, while the hybrid FE-SEA method is described in Section 3. An outline derivation of the key equations is given in each case, although the cited references provide a much more rigorous discussion of the various approximations involved. A number of example applications is given for each method.

2. Statistical Energy Analysis (SEA)

2.1. The ensemble mean response

In SEA a complex system is represented as an assembly of subsystems, and the aim is to find the ensemble mean of the vibrational energy of each subsystem. The governing equations, which arise from a consideration of power balance, have the form (see for example [8])

$$\omega\eta_j E_j + \sum_k \omega\eta_{jk} n_j (E_j / n_j - E_k / n_k) = P_{\text{in},j} \quad \text{or} \quad \mathbf{C}\hat{\mathbf{E}} = \mathbf{P}_{\text{in}} \tag{1a,b}$$

where, for subsystem j, E_j is the ensemble averaged vibrational energy (which may also be averaged over a frequency band), η_j is the loss factor, n_j is the modal density, and $P_{\text{in},j}$ is the power input from external sources. The term η_{jk} is the coupling loss factor between subsystems j and k, and ω is the vibration frequency. In Eq. (1b), $\hat{E}_j = E_j / n_j$, and the entries of the matrix $\mathbf{C}$ and the vector $\mathbf{P}_{\text{in}}$ can readily be deduced from Eq. (1a). The assumptions that underpin Eq. (1) have been discussed, for example, by Mace [16], Lyon and DeJong [17], and Fahy [18].

2.2. The ensemble variance

In reference [19] an expression has been derived for the variance of the energy of the jth subsystem within the context of SEA. The result has the form

$$\text{Var}(E_j) = \sum_k (C_{jk}^{-1})^2 \text{Var}(P_{\text{ran},k}) + \sum_k \sum_{s \neq k} [(C_{jk}^{-1} - C_{js}^{-1})\hat{E}_s]^2 \text{Var}(C_{\text{ran},ks}), \tag{2}$$

where C_{jk}^{-1} represents the jkth entry of the inverse of the matrix $\mathbf{C}$ that appears in Eq. (1b), and the two "Var" terms are given by

$$\text{Var}(P_{\text{ran},k}) = P_{\text{in},k}^2 r^2(\alpha_k, m_k', B_k'), \qquad \text{Var}(C_{\text{ran},ks}) = C_{ks}^2 r^2(\alpha_{ks}, m_k', B_k'), \tag{3,4}$$

$$m_k' = \omega\eta_k' \, n_k, \qquad \eta_k' = 1/(\omega n_k C_{kk}^{-1}), \qquad B_k' = \frac{\Delta}{\omega\eta_k'}, \tag{5-7}$$

where Δ is the frequency bandwidth over which the energy is averaged (Δ=0 for a purely harmonic analysis). The function r that appears in Eqs (3) and (4) is given by [7,20]

$$r^2(\alpha, m, B) = \frac{\alpha - 1}{\pi m}\left(\frac{1}{B^2}\right)\left\{2B\tan^{-1}(B) - \ln\left(1 + B^2\right)\right\} + \frac{1}{(\pi m)^2}\left(\frac{1}{B^2}\right)\ln\left(1 + B^2\right). \tag{8}$$

This is an approximate result arising from simplifying certain integrals that appear in the underlying derivations. A more exact result is available for the case B=0:

$$r^2(\alpha, m, 0) = \frac{1}{\pi m}\left\{\alpha - 1 + \frac{1}{2\pi m}\left[1 - e^{-2\pi m}\right] + \text{E}_1(\pi m)\left[\cosh(\pi m) - \frac{1}{\pi m}\sinh(\pi m)\right]\right\}. \tag{9}$$

Equation (2) can readily be applied as a post-processing step at the end of a standard SEA analysis. The only "non-SEA" parameters that appear in Eqs (2)-(9) are the coefficients α_k and α_{ks}, and these are readily found for various types of power input (α_k) and various types of coupling between subsystems (α_{ks}). The derivation of Eqs. (2)-(9) is fully described in reference [19] and will not be repeated here, other than to state that the main assumptions behind the equations are: (i) each subsystem exhibits statistical overlap, and the statistics of the mode shapes and natural frequencies are described by the Gaussian Orthogonal Ensemble (GOE), and (ii) the subsystems are weakly coupled, as normally required in SEA.

2.3. Example applications

An example of the application of equations (1-9) is shown in Figure 1. The results concern two plates that are coupled along a common edge. Plate 1 is subject to a point load, and Monte Carlo simulations have been performed on an ensemble of systems generated by adding small masses in random locations. The irregular curves in the figure represent the Monte Carlo results, while the smooth curves are the predictions arising from the foregoing theory. Results are shown for the mean and relative variance of the bending energy in each plate, and good agreement with theory is demonstrated. It can be noted that while the driven plate has the largest mean energy, the relative variance is greater for the second plate. This is because the response of the second plate is affected by the randomness in both plates, while the first plate response is little affected by the second plate. This type of behaviour is further illustrated in Figure 2 for a three plate system: the relative variance increases with increasing distance from the excited plate.

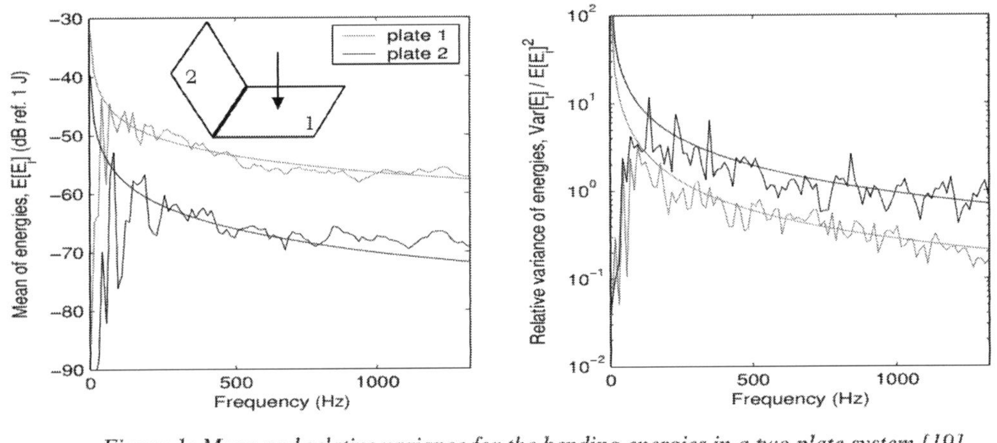

Figure 1: Mean and relative variance for the bending energies in a two plate system [19].

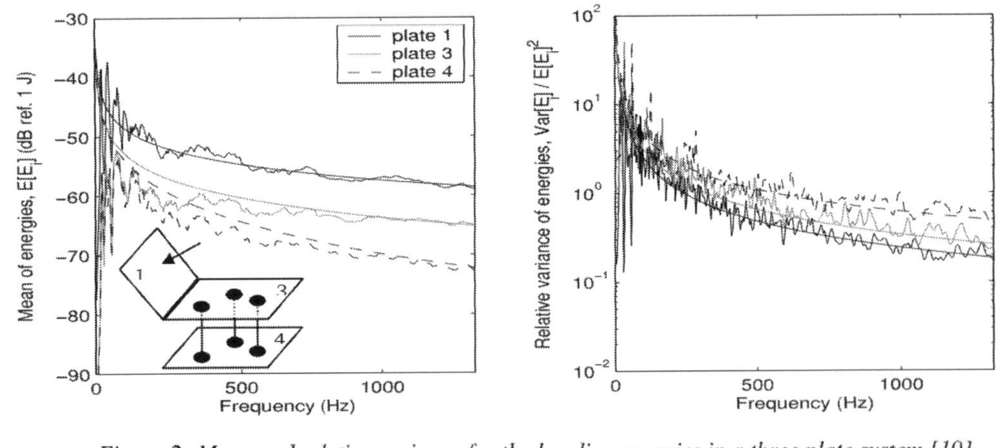

Figure 2: Mean and relative variance for the bending energies in a three plate system [19].

3. The hybrid method

3.1. Introduction and key concepts

As explained in the previous section, SEA is based on modelling a structure as a collection of subsystems. It is assumed that each subsystem has random properties across the ensemble, and that the subsystem natural frequencies are sufficiently random for the tenets of SEA to apply. In many practical structures this condition is not met: in an aircraft structure for example, the skin panels may well have highly random modes, but the frames do not, being modally sparse and relatively deterministic. Similarly, parts of an automotive structure, such as the door pillars and side rails, are stiff and modally sparse, while the door panels and roof have many modes. In such cases it would be advantageous to combine the finite element method (or some other deterministic modelling method) with SEA, and this is achieved through the hybrid method described in the following sections.

By way of background, it can be noted that the literature on mid and high frequency vibration is extensive, and there have been many efforts to combine deterministic and statistical analysis methods into a single model. Beylaev and Palmov [21] were among the first researchers to include locally averaged statistical behaviour into a global deterministic model of a system. This type of approach was developed much further by the pioneering work of Soize [10] on fuzzy structure theory, and Langley and Bremner [11] developed a hybrid FE-SEA method based on these ideas combined with a wavelength partitioning scheme. A hybrid FE-SEA method ideally combines the low frequency performance of the FE method with the high frequency performance of SEA to produce a robust method that can be applied across the whole frequency range. However, the coupling of FE and SEA in a single model is difficult because the methods differ in two ways: (i) FE is based on dynamic equilibrium while SEA is based on the conservation of energy flow, and (ii) FE is a deterministic method while SEA is inherently statistical. More recently Shorter and Langley [12] have developed a new method of effecting this coupling, which is based on wave concepts rather then the modal type of approach employed in reference [11]. At the heart of the method is a reciprocity result [13] regarding the forces exerted at the boundaries of an SEA subsystem; this centres on the concept of a "direct field" dynamic stiffness matrix, which is explained in what follows.

As discussed above, in the mid-frequency range some components of a complex structure (for example thin panels) display short wavelength vibrations and are sensitive to the effects of random uncertainties, while others (for example beams) show little variation in their dynamic properties and are essentially deterministic. In the hybrid method proposed by Shorter and Langley [12], the deterministic components are modelled by using the finite element method, while the random components are modelled as SEA subsystems. A key feature of the method is the concept of a "direct field" or "power absorbing" dynamic stiffness matrix associated with each SEA subsystem. Consider for example a thin plate that is excited at the boundaries. The excitation generates waves that propagate through the plate and are reflected repeatedly at the boundaries; the total dynamic stiffness matrix of the plate, phrased in terms of the edge degrees of freedom, has contributions from all of these reflections. Suppose now that the response is viewed in two parts:

1) The contribution from the initial generated waves, prior to any boundary reflections. This can be called the "direct field".

2) The contribution from waves produced on the first and all subsequent reflections. This can be called the "reverberant field".

The direct field dynamic stiffness matrix can be defined as that resulting from the presence of the direct field waves – this matrix corresponds to "power absorbing" behaviour, in the sense that the direct field waves all propagate energy away from the boundaries. Such a matrix can be found analytically for each of the subsystems by a variety of methods. For example, consider a thin plate with four straight boundaries. If this plate is part of a larger FE model, then the degrees of freedom of the boundary can be described in terms of the nodal degrees of freedom of the FE mesh. To employ the hybrid method we need to find the direct field dynamic stiffness matrix associated with the edge degrees of freedom of the plate. This can be done by considering each straight boundary in turn, and taking it to form a segment of the edge of a semi-infinite plate. Motion of the boundary will generate waves into the semi-infinite plate, and for a given boundary motion the generated waves can be found by Fourier transform techniques. Calculation of the boundary forces associated with the generated waves then allows the dynamic stiffness matrix to be constructed: i.e., strictly the dynamic stiffness matrix of a segment of the edge of a semi-infinite plate, when the motion of the segment is described by FE nodal degrees of freedom. This is our required "direct field" dynamic stiffness matrix for the plate edge, and repeating the process for each of the plate edges will give the total direct field dynamic stiffness matrix for the plate subsystem. It is important to note that the direct field dynamic stiffness matrix can also be viewed as the ensemble average of the full dynamic stiffness matrix when averaged over random boundary reflections. Given the concept of the direct field dynamic stiffness matrix, the hybrid FE-SEA equations can now be described.

3.2. The hybrid equations

The starting point for the hybrid method is to identify those parts of the system response that will be described by SEA subsystems. The remaining part of the system (which can be considered to be the "deterministic" part) is then modelled by using the FE method. For example, it might be decided that the bending motions of the panels of a car have a short wavelength of deformation and will be described using SEA subsystems. The bending degrees of freedom of these panels will then be omitted form the FE model of the system, at all points other than the panel boundaries. The relevant "direct field" dynamic stiffness matrix is then added to the FE model at the panel boundaries, and this augmented FE model is then used in the subsequent analysis. If the degrees of freedom of the deterministic part are labelled **q**, then the governing equations of motion (for harmonic vibration of frequency ω say) will have the form

$$\mathbf{D}_{tot}\mathbf{q} = \mathbf{f} + \sum_k \mathbf{f}_{rev}^{(k)}, \qquad \mathbf{D}_{tot} = \mathbf{D}_d + \sum_k \mathbf{D}_{dir}^{(k)} . \tag{10,11}$$

The summation is over the number of SEA subsystems in the model, and $\mathbf{D}_{dir}^{(k)}$ represents the direct field dynamic stiffness matrix associated with subsystem k. Furthermore, $\mathbf{D}_d$ is the dynamic stiffness matrix given by the finite element model of the deterministic part of the system, **f** is the set of external forces applied to this part of the system, and $\mathbf{f}_{rev}^{(k)}$ represents the force arising from the reverberant field in subsystem k, which is not accounted for in $\mathbf{D}_{dir}^{(k)}$. The matrix $\mathbf{D}_{tot}$ is the dynamic stiffness matrix of the FE model (excluding the SEA subsystem degrees of freedom), when augmented by the direct field dynamic stiffness matrix of each SEA subsystem. It should be noted that equations (10) and (11) are exact – all that has been done is to split the forces arising from the SEA subsystems into a direct field part, which is accounted for by $\mathbf{D}_{dir}^{(k)}$, and a reverberant part which is carried to the right hand side of equation (10). The following result (Shorter and Langley [13]) is central to the development of the hybrid method

$$\mathbf{S}_{ff}^{(k),rev} \equiv \mathrm{E}\left[\mathbf{f}_{rev}^{(k)}\mathbf{f}_{rev}^{(k)*T}\right] = \left(\frac{4E_k}{\omega\pi n_k}\right)\mathrm{Im}\left\{\mathbf{D}_{dir}^{(k)}\right\} \tag{12}$$

Here E_k and n_k are respectively the (ensemble average) vibrational energy and the modal density of the kth subsystem. Equation (12) implies that the cross-spectral matrix of the force exerted by the reverberant field is proportional to the resistive part of the direct field dynamic stiffness matrix, which is a form of diffuse field reciprocity statement. Now from equation (10), the response **q** can be expanded in the form

$$\mathbf{q} = \mathbf{q}_d + \sum_k \mathbf{q}^{(k)}, \quad \mathbf{q}_d = \mathbf{D}_{tot}^{-1}\mathbf{f}, \quad \mathbf{q}^{(k)} = \mathbf{D}_{tot}^{-1}\mathbf{f}_{rev}^{(k)} \,. \tag{13}$$

Now the time averaged power input to the direct field of subsystem j can be written as

$$P_{in,j} = (\omega/2)\,\mathrm{Im}\{\mathbf{q}^{*\mathrm{T}}\mathbf{D}_{dir}^{(j)}\mathbf{q}\} = (\omega/2)\sum_{rs}\mathrm{Im}\{D_{dir,rs}^{(j)}\}S_{qq,rs}\,, \tag{14}$$

where it has been noted that the dynamic stiffness matrix is symmetric. If the various contributions $\mathbf{q}^{(k)}$ that appear in equation (13) are taken to be uncorrelated and of zero mean, then equations (12)-(14) yield

$$P_{in,j} = P_{in,j}^{ext} + \sum_k \omega\eta_{jk} n_j (E_k / n_k) \tag{15}$$

where

$$P_{in,j}^{ext} = (\omega/2)\sum_{rs}\mathrm{Im}\{D_{dir,rs}^{(j)}\}\left(\mathbf{D}_{tot}^{-1}\mathbf{S}_{ff}\mathbf{D}_{tot}^{-1*\mathrm{T}}\right)_{rs}, \tag{16}$$

$$\omega\eta_{jk} n_j = (2/\pi)\sum_{rs}\mathrm{Im}\{D_{dir,rs}^{(j)}\}\left(\mathbf{D}_{tot}^{-1}\,\mathrm{Im}\{\mathbf{D}_{dir}^{(k)}\}\mathbf{D}_{tot}^{-1*\mathrm{T}}\right)_{rs}. \tag{17}$$

Given that the dynamic stiffness matrices are symmetric, it is readily shown from equation (17) that reciprocity holds, in the sense that $\eta_{jk} n_j = \eta_{kj} n_k$. As will be shown in what follows, the terms η_{jk} are equivalent to the coupling loss factors that appear in SEA.

The power output from the reverberant field in subsystem j can be written as the sum of: (i) the power dissipated through damping, (ii) the power transferred to the other subsystems, and (iii) the power dissipated in the deterministic system due to the response $\mathbf{q}^{(j)}$. Thus

$$P_{out,j} = \omega\eta_j E_j + \sum_k \omega\eta_{kj} n_k (E_j / n_j) + \omega\eta_{d,j} E_j, \tag{18}$$

where

$$\omega\eta_{d,j} = (\omega/2E_j)\,\mathrm{Im}\{\mathbf{q}^{(j)*\mathrm{T}}\mathbf{D}_d\mathbf{q}^{(j)}\} = \left(\frac{2}{\pi n_j}\right)\sum_{rs}\mathrm{Im}\{D_{d,rs}\}\left(\mathbf{D}_{tot}^{-1}\,\mathrm{Im}\{\mathbf{D}_{dir}^{(j)}\}\mathbf{D}_{tot}^{-1*\mathrm{T}}\right)_{rs}. \tag{19}$$

Equations (21) and (25) then lead to the following energy balance equation for subsystem j

$$\omega(\eta_j + \eta_{d,j})E_j + \sum_k \omega\eta_{jk} n_j (E_j / n_j - E_k / n_k) = P_{in,j}^{ext}\,. \tag{20}$$

Furthermore, the cross-spectral matrix of the response **q** can be derived from equations (12) and (13), which yields

Journal of Physics: Conference Series **181** (2009) 012001 doi:10.1088/1742-6596/181/1/012001

$$\mathbf{S}_{qq} = \mathbf{D}_{tot}^{-1}\left[\mathbf{S}_{ff} + \sum_{k}\left(\frac{4E_k}{\omega\pi n_k}\right)\mathrm{Im}\left\{\mathbf{D}_{dir}^{(k)}\right\}\right]\mathbf{D}_{tot}^{-1*\mathrm{T}}. \quad (27)$$

Equations (20) and (21) form the two main equations of the hybrid method. It is clear that these equations couple FE and SEA methodologies: equation (20) has precisely the form of SEA, but the coupling loss factors η_{jk} and loss factors $\eta_{d,j}$ are calculated by using the FE model (augmented by the direct field dynamic stiffness matrices) via equations (17) and (19); furthermore, equation (21) has the form of a standard deterministic FE analysis, but additional forces arise from the reverberant energies in the subsystems. If no SEA subsystems are included then the method becomes purely FE; on the other hand, if only the junctions between the SEA subsystems are modelled by FE, then the method becomes purely SEA, with a novel method of computing the coupling loss factors.

3.3. Academic benchmark example

The first example of the application of the hybrid method is taken from reference [22]. Two thin plates are coupled via a complex beam, as shown in Figure 3. Two models of the system are considered: (i) a detailed finite element model, (ii) a hybrid model in which the beam and the in-plane motion of the plates is modeled by FE, while the bending motion of the plates is modeled using two SEA subsystems (one for each plate). The detailed finite element model has been randomized by adding small masses to the plates in random locations, and an ensemble of 200 structures has been created in this way. The detailed finite element model has 12000 degrees of freedom, while the hybrid model has 3500 finite element degrees of freedom and two SEA subsystems.

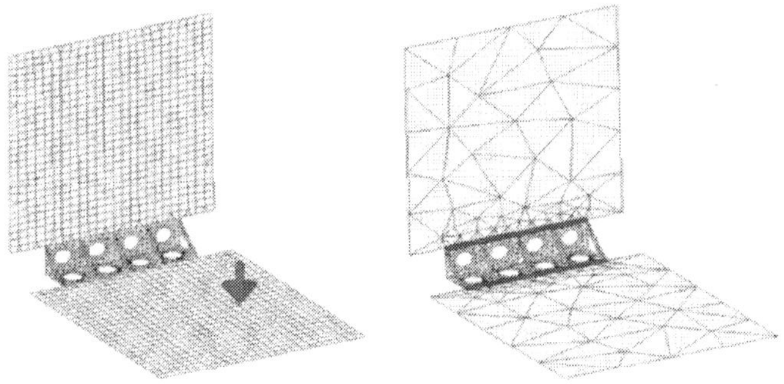

Figure 3: Two plates coupled by a beam. Left hand diagram – detailed FE model; right hand diagram – FE model used as part of the hybrid model.

The lower panel is subjected to rain-on-the roof excitation. It has been found that the hybrid method requires around 1/100 of the computer time needed to perform 20 Monte Carlo simulations using the detailed model. The predicted response of the system is shown in Figure 4, where the ensemble average yielded by the detailed by the detailed FE model is compared with the results yielded by the hybrid method. It can be seen that the hybrid method yields very good response predictions; the driven plate is fairly insensitive to the dynamics of the coupling beam, while the non-driven plate shows peaks corresponding to the beam resonances. Other benchmark example applications of the method are given in references [12] and [22], and variance predictions using an extended version of the method are presented in reference [15].

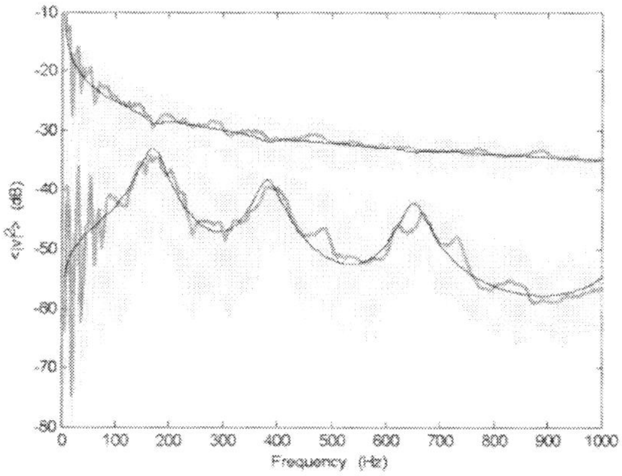

Figure 4: mean squared velocity response of the two plates. Upper curves – driven plate, lower curves - non-driven plate. For each plate the cloud of results corresponds to 200 Monte Carlo realizations, the irregular dark line is the ensemble average, the smooth dark line is the hybrid prediction.

3.4. Industrial example

The hybrid method has been implemented in the software package VA One [23], and this has been used on many industrial problems. Example applications can be found in references [24-26], covering automotive, aerospace, and rail vehicles. The application of the method to an automotive structure is shown in Figure 5 [27]. Stiff elements of the car, such as the rails, engine supports, and shock towers, are modelled using FE, whereas flexible components such as the roof, windows, doors, and the interior acoustic cavity, are represented by SEA subsystems. The use of SEA subsystems removes a huge number of degrees of freedom from the finite element model, since the flexible components are very modally dense, and require a very fine mesh if they are to be modelled using the FE method.

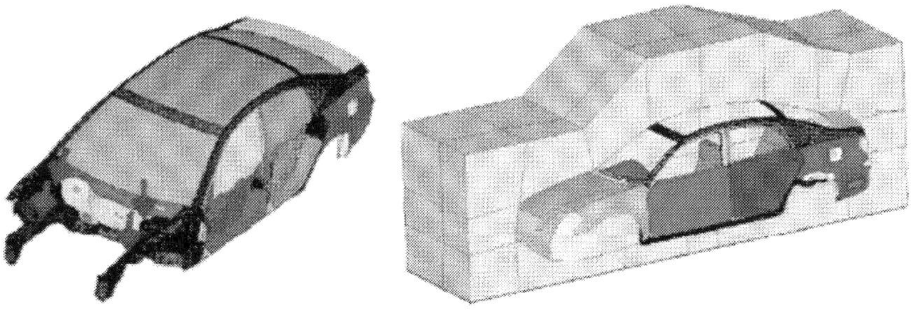

Figure 5: VA One model of an automotive structure

Typical results for the interior noise level are shown in Figure 6. The results for trimmed and untrimmed versions of the vehicle are included in the graphs. The numerical results are in general within 3 to 5 dB of the experimental results, correctly accounting for the effect of noise control treatments applied in the trimmed version. The results for the trimmed version are actually in better agreement with experiment than for the bared version, and this can be traced to better characterization of the interior cavity damping; in the absence of trim, the cavity damping is very low, and difficult to estimate. Full details of the hybrid model for this case can be found in reference [27].

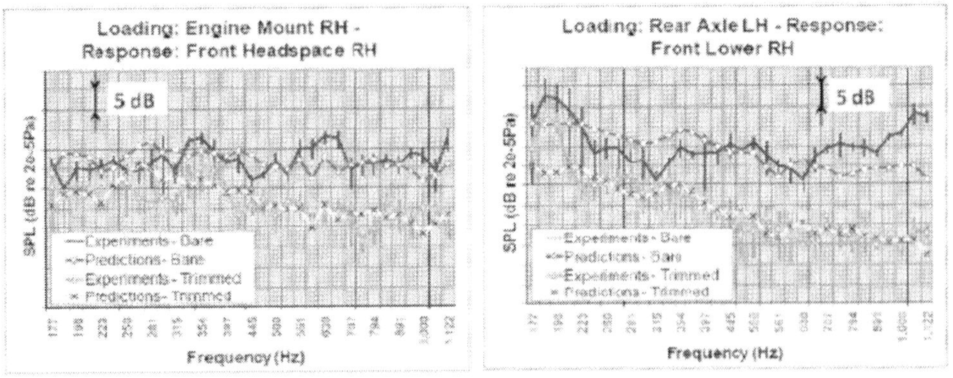

Figure 6: Velocity response at the rear window for engine mount excitation. The predicted results have been derived using the hybrid method.

4. Conclusions

This paper has consider various method for the dynamic analysis of structures in the presence of manufacturing uncertainties. Parametric and non-parametric approaches have been briefly reviewed in the introduction, and then a detailed presentation of two non-parametric methods has been given: statistical energy analysis (SEA) and the hybrid method. Each of these methods can be used to estimate the mean and variance of stress, vibration, and noise levels within a complex system, with SEA being suited to high-frequencies and the hybrid method to mid-frequencies. At low frequencies the statistics of the response of a system can be very sensitive to the statistics of the uncertain input parameters, and parametric methods are more suitable than the two methods described here. Work is continuing on combining the hybrid method with parametric models of uncertainty within the FE component of the method.

References

[1] Elishakoff I. (1995) Essay on uncertainties in elastic and viscoelastic structures: from A.M. Freudenthal's criticisms to modern convex modelling. Computers and Structures 56:871-895.

[2] Moens D,Vandepitte D. (2005) A fuzzy finite element procedure for the calculation of uncertain frequency-response functions of damped structures: Part I – procedure. J. Sound Vib. 288:431-462.

[3] Schuëller G.I. (2006) Developments in stochastic structural mechanics. Archive of Applied Mechanics 75:755-773.

[4] Mehta M.L. (1991) Random Matrices. Academic Press, San Diego.

[5] Weaver, R.L. (1989) Spectral statistics in elastodynamics. J. Acou. Soc. Am. 85:1005-1013.

[6] Lobkis, O.I., Weaver, R.L., Rozhkov, I. (2000) Power variances and decay curvature in a reverberant system. J. Sound Vib. 237: 281-302.

[7] Langley R.S., Brown, A.W.M. (2004) The ensemble statistics of the energy of a random system subjected to harmonic excitation. J. Sound Vib. 275: 823-846.

[8] Lyon, R.H. (1975) Statistical Energy Analysis, Theory and Application. Boston: MIT Press.

[9] Soize, C. (2005) A comprehensive overview of a non-parametric probabilistic approach of model uncertainties for predictive models in structural dynamics. J. Sound Vib. 288: 623-652.

[10] Soize C. (1993) A model and numerical method in the medium frequency range for vibroacoustic predictions using the theory of structural fuzzy. J. Acou. Soc. Am. 94: 849-

865.
[11] Langley R.S., Bremner P. (1999) A hybrid method for the vibration analysis of complex structural-acoustic systems. J. Acou. Soc. Am. 105:1657-1671.
[12] Shorter P.J., Langley R.S. (2005a) Vibro-acoustic analysis of complex systems. J. Sound Vib. 288: 669-700.
[13] Shorter P.J., Langley, R.S. (2005b) On the reciprocity relationship between direct field radiation and diffuse reverberant loading. J. Acou. Soc. Am. 117:85-95.
[14] Langley R.S. (2007) On the diffuse field reciprocity relationship and vibrational energy variance in a random subsystem at high frequencies. J. Acou. Soc. Am. 121:913-921.
[15] Langley R.S, Cotoni V. (2007) Response variance prediction for uncertain vibro-acoustic systems using a hybrid deterministic-statistical method. J. Acou. Soc. Am. 122: 3445-3463.
[16] Mace, B.R. (1994). On the statistical energy analysis hypothesis of coupling power proportionality and some implications for its failure. J. Sound Vib. 178: 95-112.
[17] Lyon, R.H., DeJong, R.G. (1995). Theory and Application of Statistical Energy Analysis. Butterworth-Heinemann, Boston.
[18] Fahy, F.J. (1994). Statistical energy analysis: a critical overview. Philosophical Transactions of the Royal Society of London, Series A 346:431-447.
[19] Langley, R.S. and Cotoni, V. (2004) Response variance prediction in the statistical energy analysis of built-up systems. J. Acou. Soc. Am. 115: 706-718.
[20] Langley, R.S., and Brown, A.W.M. (2004) The ensemble statistics of the band averaged energy of a random system. J. Sound Vib. 275: 847-857.
[21] Belyaev, A.K., V.A. Palmov, V.A. (1986) Integral theories of random vibration of complex structures, in *Random Vibrations – Status and Recent Developments,* edited by I. Elishakoff and R.H. Lyon, Elsevier, Amsterdam.
[22] Cotoni, V., Shorter, P.J., Langley, R.S. (2007) Numerical and experimental validation of a hybrid finite element-statistical energy analysis method. J. Acou. Soc. Am. 122: 288-299.
[23] ESI Group (2005) VA One user's manual. San Diego, USA.
[24] Cordioli, J.A., Cotoni, V. (2009) Review of some industrial applications of the hybrid FE-SEA method to vibro-acoustic prediction and analysis. NOVEM 2009, Oxford, 5-8 April.
[25] Orrenius, U., Cotoni, V., Wareing, A. (2009) Analysis of sound transmission through periodic structures typical for railway carbodies and aircraft fuselages. NOVEM 2009, Oxford, 5-8, April.
[26] Shorter, P.J. (2009) Modelling noise and vibration in complex systems. NOVEM 2009, Oxford, 5-8 April.
[27] Charpentier, A., Sreedhar, P., Cordolio, J.A., Fukui, K. (2008) Modelling process and validation of hybrid FE-SEA method to structure-borne noise path in trimmed automotive vehicle. SEA technical paper 2008-36-0574.

Journal of Physics: Conference Series **181** (2009) 012002 doi:10.1088/1742-6596/181/1/012002

Experimental Studies on the Performance of Novel Layered Materials under Highly Dynamic Loads

Arun Shukla

Simon Ostrach Professor

Dynamic Photomechanics Laboratory

Department of Mechanical Engineering and Applied Mechanics

University of Rhode Island, Kingston, RI 02881, USA

shuklaa@egr.uri.edu

Abstract. This paper focuses on the experimental observations of the performance of different layered composite material systems subjected to blast loadings. These material systems include layered composites and sandwich composite materials. A controlled blast loading of pre-defined pressure magnitude and rise time were obtained using a shock tube apparatus. Rectangular plate elements of the desired material system were subjected to such a controlled blast loading and the effects of the blast loading on these elements were studied using optical and residual strength measurements. A high speed imaging technique was utilized to study the damage modes and mechanisms in real time. It was observed that layering of a conventional composite material with a soft visco-elastic polymer provided better blast resistance and sandwiching the polymer greatly enhanced its survivability under extreme air blast conditions. Aside from layering the conventional composite material with a soft visco-elastic polymer, it was observed that layering or grading the core can successfully mitigate the impact damage and thus improve the overall blast resistance as well. In addition to these, three dimensional (3D) woven skin and core reinforcements were introduced in the conventional sandwich composites and their effects on the blast resistance were studied experimentally. It was observed that these reinforcements also enhance the blast resistance of conventional sandwich composites by changing the mechanism of failure initiation and propagation in these sandwich structures.

Journal of Physics: Conference Series **181** (2009) 012002 doi:10.1088/1742-6596/181/1/012002

1. Introduction

Composite materials have replaced metals in various engineering applications due to their numerous advantages, such as high strength/weight ratio, low cost, better stealth properties, etc. They are widely used in the defense industry (applications like naval ships, warplanes, and armored vehicles), aerospace industry (applications like wing construction, tail construction and fiber composite bodies) and transportation industry (lightweight bodies for buses, trains and trucks). In recent times, threats to such defense and civilian structures in the form of terrorist attacks have added to the significance of comprehensive understanding of material and structural damage behavior under blast loading, in order to design and fabricate these structures to withstand such loadings. Experimental and instrumentation difficulties associated with blast loadings have limited the extent of basic research in this area, but with growing concerns on safety and human lives involved, the significance of such a research cannot be understated.

Blast loading is a complex and interesting problem that has been studied from the early 20th century. When a structure is loaded with a blast wave, there are two main events occurring. There is a transfer of energy from the medium that the blast wave is traveling through to the structure, as well as an internal response of the structural material to the shock wave that is propagating in it. The internal response might lead to blast mitigation and hence good resistance to the blast wave or initiate several different failure mechanisms in the material which might lead to ultimate failure of the structure itself.

G I Taylor's work on the pressure and impulse of underwater explosion waves on freestanding plates is considered as one of the pioneering works on studying impulse transfer to plates subjected to blast loading conditions. His work revealed that the density of the material and medium in which the blast wave propagates has the maximum effect on the impulse transfer. Given this, there are many research efforts that try to use the same for mitigation of blast wave by modifying the material into several different forms like inhomogeneous, layered, anisotropic, etc. The current paper focuses mainly on novel layered sandwich composites subjected to air blast loading conditions.

Sandwich composites are highly desired for energy absorbing structural elements. While in service, structures constructed of sandwich composites may experience impact events that compromise the integrity of the structure. In recent times, the response of sandwich structures to such dynamic loadings has been an area of interest [1]. The overall dynamic response of the sandwich is dependent, among several other factors, on the construction of the skin, compressive modulus and failure strength of the core, and the interface strength between the skin and core. Due to the inherent nature of its construction, the strength of the sandwich is often limited by the strength of the core material. The thickness of the skin increased beyond certain limits, also negates the weight beneficiary advantage of sandwich construction. In lieu of these considerations and to facilitate better performance under transient conditions, several design modifications have been sought after. These include better choice of core material, introduction of soft inter layers (e.g. polyurea layer) between the core and the skin, and discretely layering or grading the core.

Since the properties of graded/layered core structures can be designed and controlled, they show great potential to be an effective core material for absorbing the blast energy and improving the overall blast resistance of sandwich structures. The numerical investigation by N.A. Apetre [2] has shown that a reasonable core design can effectively reduce the shear forces and strains within the structures. Consequently, they can mitigate or completely prevent impact damage on sandwich composites. Y. Li [3] examined the impact response of layered and graded metal-ceramic structures numerically. He found that the choice of gradation has a great significance on the impact applications and the particular design can exhibit better energy dissipation properties. Also, in recent times, Z-directional pins have been utilized [4] to modify the core in sandwich material and its high strain rate impact response has been studied. Failure modes of carbon fiber based sandwich beams reinforced with similar Z-directional pins under three-point bending [5] were studied under quasi-static loading conditions.

2. Material system

The experimental study involved the evaluation of two different major material systems, namely layered composites and sandwich composites. The layered composites material system consisted of polyurea layered materials only, while the sandwich composites material system consisted of polyurea sandwich composites, sandwich composites with 3D woven skins, core reinforced sandwich composites and sandwich composites with a stepwise graded core. A brief description on the construction and characteristic properties are given below.

2.1 Layered composites

2.1.1 *Polyurea layered materials.* Polyurea is a cross-linked amorphous isocyanate monomer or prepolymer and polyamine curative. Layered materials were fabricated by casting polyurea over E-glass vinyl ester composite (EVE) plates. The thickness of each component of the layered materials was 6.35 mm (0.25 in). Further details on the properties and fabrication of these layered plates can be found in [6].

2.2 Sandwich composites

2.2.1 *Polyurea sandwich composites.* Similar to the fabrication of layered materials, sandwich composites were fabricated out of the E-glass vinyl ester composite and Polyurea material combinations. Two different sandwich configurations were manufactured. Whereas one had a soft core (PU) sandwiched between two hard skins (EVE), the other had a hard core (EVE) sandwiched between two soft skins (PU). The sandwich composites along with their dimensions are given below,

(1) 3.18 mm (0.125") EVE + 6.35 mm (0.25") PU + 3.18 mm (0.125") EVE.
(2) 3.18 mm (0.125") PU + 6.35 mm (0.25") EVE + 3.18 mm (0.125") PU.

Based on the experimental observations, the beneficial effects of the sandwich construction was predominant only in the sandwich with soft core configuration. Hence the results and discussions focus on that particular class of the sandwich alone.

2.2.2 *Sandwich composites with 3D Woven skin.* Three-dimensional orthogonal woven composites, hereafter referred to as 3D composites are different from traditional 2D laminates in that they are able to sustain loads in three linearly independent directions. Fiber preforms constructed in a three dimensional fashion not only contain warp and weft fibers in plane of the lamina, but also a certain fraction in the through thickness, or z-direction. The composite skin materials utilized in this study were made by VARTM method from 3.25 kg/m^2 (96 oz/yd^2) areal density 3WEAVE® E-glass glass roving preforms using Dow Derakane 8084 epoxy-vinyl ester resin. This fabric construction results in ~49% warp-, ~49% fill- and ~2% Z-fiber content per preform volume. The total fiber volume fraction in a single-layer composite made with such preform was estimated as ~49%.

The core material used in the sandwich construction was made of Dow Chemical TRYMER™ 200L Polyisocyanurate foam, which is a cellular polymer. From the manufacturer's data, the density of the foam is 32 kg/m^3. This foam is not an isotropic material; besides, its tensile and compressive properties are markedly different. Specifically, from the foam manufacturer data the following tensile, compressive and shear moduli are given in the direction parallel to rise: 8.27, 5.17 and 1.79 MPa, respectively.

For constructing the sandwich composite panels, the first two layers of the front skin (to be exposed to shock wave) were made of the standard 0.81 kg/m^2 (24 oz/yd2) areal density 2-D E-glass plain weave. Those were followed by three layers of 3WEAVE® fabric and subsequently by a layer of above mentioned foam core. The back skin preform contained (counting from the back surface of the core) two layers of the same 3WEAVE® fabric followed by two layers of the same E-glass plain weave. The front skin was intentionally made one fabric layer thicker than the back one, to enhance its strength. The average front skin thickness was approximately 9.1 mm (.36 in) and the average back skin thickness was approximately 6.91mm (.27 in) respectively.

2.2.3 *Core reinforced sandwich composites.* The work on fabricating the core reinforced sandwich composites, designated as TRANSONITE®, was performed by Martin Marietta Composites. This sandwich type is produced in accordance with patented technology and provides an alternative for structural and non-structural flat panel applications. In addition to all the characteristics explained in the previous section on sandwich composites, the TRANSONITE panels have core reinforcement in the form of stitches introduced in their through thickness direction, as shown in Fig. 1 (without the core).To begin the pultrusion process, various layers of fabrics are pulled together to produce the desired thicknesses of top and bottom skin preforms. Then a foam layer is inserted between the top and bottom layers of the fabric forming a sandwich and subsequently the stitching is performed. Once the resin is infused, the sandwich is drawn through a heated die forming a cured composite sandwich. Finally, the panel is cut to the appropriate dimensions.

Fig 1 TRANSONITE composite sandwich with removed foam; through thickness "stitches", impregnated with resin, can be seen. [Ref: Martin Marietta Composites website www.martinmarietta.com]

2.2.4 *Sandwich composite with a stepwise graded core.* The sandwich composites were manufactured using E-Glass Vinyl Ester (EVE) composites and Corecell™ A-series Styrene acrylonitrile (SAN) foams. The woven roving E-glass fibers of the skin material were placed in a quasi-isotropic layout [0/45/90/-45]$_s$. The fibers were made of the 610 g/m^2 (18 oz/yd^2) areal density plain weave. The resin system used was Ashland Derakane Momentum 8084 and the front skin and the back skin consisted of identical layup and materials.

The core materials were manufactured by Gurit SP Technologies and are specifically designed for marine sandwich composite applications. Three types of Corecell™ A-series foam were used and they were A300, A500, and A800. Important material properties of the three foams include the nominal density and the compressive modulus. From the manufacturers data, the nominal densities for the A300, A500, and A800 foams are 58.5 kg/m^3, 92 kg/m^3 and 150 kg/m^3, while the compressive modulus is 32 MPa, 64 MPa, and 117 MPa respectively. Based on the density and compression modulus, the wave impedances (WI) of the three types of foams show the following relationship.

$$WI_{A800} > WI_{A500} > WI_{A300}$$

The VARTM-fabricated panels were produced from a plain weave E-glass fabric type. Fabric lay-up was an 8-ply balanced/symmetric quasi-isotropic layout [0/45/90/-45]$_s$. For the core, each layer of foam was 12.7 mm (.5 in) respectively. The overall dimensions for the samples were 102 mm (4 in) wide, 254 mm (10 in) long, and 48 mm (1.9 in) thick. The foam core itself was 38 mm (1.5 in) thick, while the skin thickness was 5 mm (.2 in). The average areal density of the samples was 19.02 kg/m^2.

3. Experimental setup

3.1 Shock tube

A shock tube, shown in Fig. 2, was used for providing blast loading in a laboratory environment. In its simplest form a shock tube consists of a long rigid cylinder, divided into a high-pressure driver section and a low pressure driven section, which are separated by a diaphragm. The tube is operated by pressurizing the high-pressure section with helium gas until the pressure difference across the diaphragm reaches a critical value and it ruptures. This rapid release of gas creates a shock wave that travels down the tube to impart air blast loading on a specimen. The intensity of blast loading can be controlled to a repeatable extent by varying the thickness and number of diaphragms used. Initial calibrations for quantification of these input parameters were performed. Shock pressure and velocity measurements were done in real time to provide accurate input parameters for the study. A dynamic pressure sensor is mounted at the muzzle section of the shock tube and graphite rods are used as break circuit initiators to measure the shock velocity. Thin mylar sheets (10 mil) are used as diaphragm material. The driver pressure and hence the shock pressure obtained from the burst of these diaphragms is controlled by the number of plies of mylar sheets used. Complete description of the shock tube and its calibration can be found in [7].

Fig 2 The shock tube facility to produce controlled blast loading

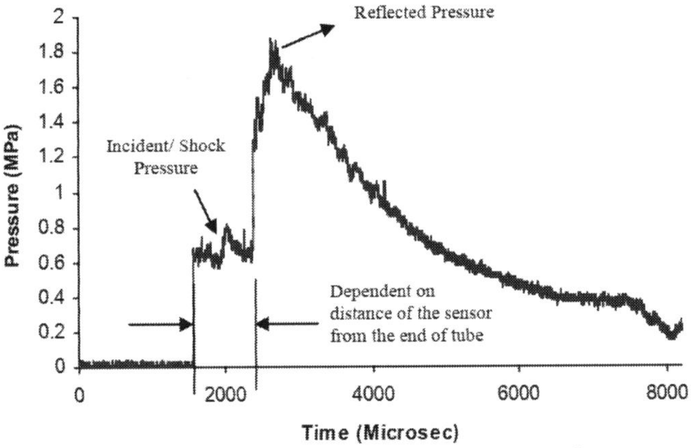

Fig 3 Typical shape of the pressure – time profiles obtained in the shock tube.

The shock tube utilized in the present study has an overall length of 8 m and is divided into a 1.82 m driver section, 3.65 m driven section and a final 2.53 m muzzle section. The diameter of the driver and driven section are 0.15m. The final diameter of the muzzle section that is in contact with the specimen is 0.07 m. Mylar diaphragms are ruptured due to pressure differential created between the driver and driven section, which develops and drives a shock wave down the length of the shock tube. A typical pressure profile obtained at the sensor location by using 3 plies of 10 mil mylar is shown in Fig 3.

3.2 High speed imaging

In the experiments reported here, the specimens are placed under simply supported boundary conditions with a span between the supports of approximately 152 mm. The dynamic loading was applied over a central circular area 76mm in diameter. One of the free ends of the simply supported plate was viewed from the side using an IMACON high-speed camera. The camera is capable of taking 16 photographs at framing rates as high as 200 million frames/second, with exposure times as low as 5 ns. Typical blast loading events are in the order of 2-6 milliseconds and the deformation of the plate is recorded using the camera over the response period of the plate. Post analysis of these images provided the macroscopic deformation mechanisms and quantitative deformation-time history of the centre point of the plate during the deformation.

4. Result and discussion

4.1 Blast resistance of layered composites

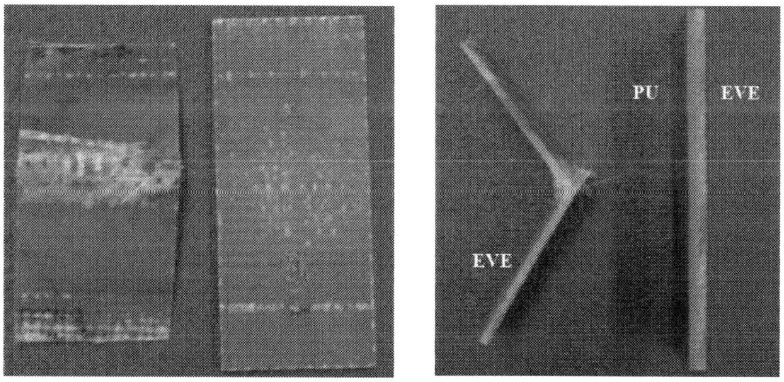

0.62 MPa 0.62 MPa 0.62 MPa 0.62 MPa

Fig. 4 Damage in E-Glass/Vinyl ester composite and layered plates under blast loading of same intensity

The layering of polyurea on to laminated composites provided enhanced blast resistance. This enhancement can be attributed to two main reasons: (1) weight (inertia) addition leading to increased impulse required for damage initiation (2) Energy dissipation in the polyurea material. Since weight addition is not a sought after method to enhance blast resistance, given the requirement of light weight structures, the beneficial factor can be justified only if the energy dissipation mechanisms in the layered structure are superior to laminated composites. It requires a systematic experimental and analytical approach to completely understand these mechanisms. It is equally important to optimize the amount of layering required and the orientation of the layering with respect to blast loading direction. When experiments were performed using Split Hopkinson bar apparatus, it was found that the orientation of polyurea layer with respect to the applied loading had no significant effect on the dynamic response of the layered sample. But due to the presence of dynamic equilibrium in these experiments, a significant effect is not expected either. So to understand the effect of orientation, two different series of experiments were conducted in the blast loading setup. In the first series, polyurea side of the layered composite faced the blast loading and this series is referred as PU/EVE composites. In the second series, the E-glass vinyl ester side of the layered composite faced the blast loading and hence referred as EVE/PU composites. A comparative case of damage caused in the laminated composite and a layered composite, PU/EVE, under the same level of blast loading is shown in Fig 4.

4.1.1 *PU/EVE layered material.* Whereas panels of plain woven composite failed at an incident shock pressure of 0.62 MPa, the PU/EVE layered composite required 0.75 MPa of incident shock pressure to fail. Fig. 5 provides a comparison of damage in the laminated and layered composites with reference to the orientation of the polyurea layer. Damage progression in PU/EVE layered composite was very similar to the damage progression in EVE but the pressure required to induce the same level of damage was higher in case of the layered composite.

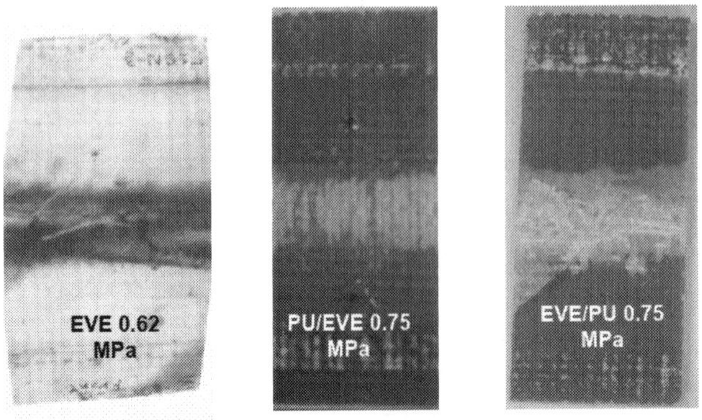

Fig. 5. Post blast view of damage in laminated and layered composites. The EVE composite side is shown in the layered materials

4.1.2 *EVE/PU layered material.* In the reverse case (EVE/PU), when EVE was on the strike face, the weaker compressive strength (compared to the tensile strength) of EVE attributed to extensive damages observed in the plate, particularly on the strike face which comprised of glass composite material. The strike face of EVE/PU material had higher delamination area shown as bright white regions in Fig. 5.

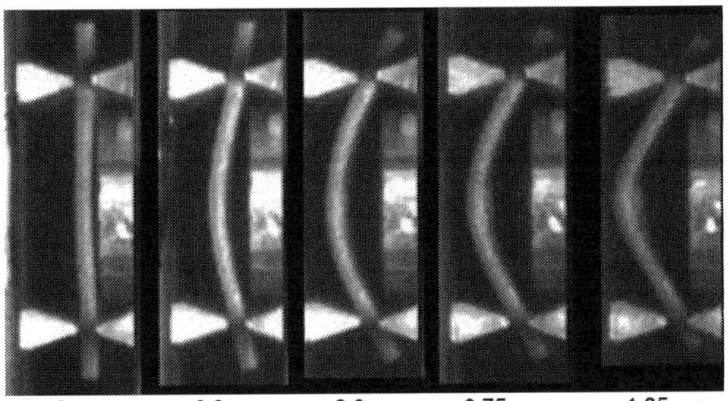

Fig. 6 Typical real-time deformation event of a EVE / PU layered composite plate, when subjected to blast load (0.75 MPa)

The qualitative difference of high speed images of bending in laminated and layered composites due to the applied blast loading is negligible. Quantitatively, the deflection of center point of the plate was calculated from the high-speed images for all the cases of laminated and layered composites. Fig. 6 shows a typical sequence of bending in layered materials when subjected to blast loading.

4.2 Blast resistance of sandwich composites

Four broad classifications of sandwich composites were evaluated for blast resistance. The first type was based on glass fiber composite and polyurea material system, the second type was based on 3D composites and polymer foam (Corecell) material system. The third type was based on glass fiber composites and stitched polymer foam material system. The fourth type was based on E-glass fiber composites and a stepwise graded corecell foam core material system

4.2.1 *Polyurea based sandwich composites.* As in the case of layered construction, two different cases of sandwich construction were evaluated for blast resistance. The construction details of these sandwich materials were discussed in the Materials section. Results are provided only for the case of sandwich with polyurea layer acting as the core with skin materials comprising of E-glass fiber composite material. This sandwich composite system showed minimal damage as shown in Fig. 7 under increasing blast loading intensities.

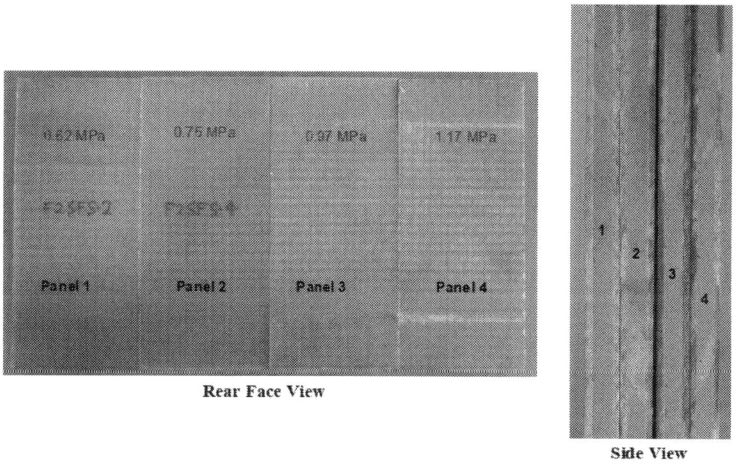

Fig. 7 Four different EVE/PU/EVE sandwich composite plates under blast loading of varying intensities (Shock pressure shown in inset)

These minimal damages in the panels were visualized on the strike face predominantly and there is no evident external damage on the rear face. On a macroscopic scale, no damage was observed in the EVE/PU/EVE sandwich system despite the fact that these sandwich panels were subjected to 85% higher pressure than the plain woven composite and 33% higher pressure than the PU/EVE layered material. Fig. 7 also shows the side view of the sandwich panels after being subjected to increasing intensities of blast loadings. There was no visible damage or deformation induced in them due to the blast loads. The damage behavior of soft-core (EVE/PU/ EVE) composites was different from that of the hard-core sandwich (PU/EVE/PU). Under similar magnitude of loading (1.17 MPa input pressure), PU/EVE/PU panel showed signs of failure as wrinkles on the strike face and shear failure on the composite core.

4.2.2 *Sandwich composites with 3D skin and Polymer foam core.* In addition to polyurea based sandwich composites, several design modifications were sought after to reinforce lightweight sandwich composite structures against blast loading. Two major design changes that were incorporated were: (1) use of 3 dimensional skin materials and (2) use of reinforced core materials; by introducing stitches in the transverse direction of the panel. Rectangular sandwich composite plates 300mm x 102mm were subjected to identical blast loading of given high intensity, with an input pressure in the range of 1.2-1.4 MPa and reflected pressure in between 5.5-5.8 MPa. Post-mortem evaluation was conducted on the shock wave tested sandwich composite. In case of Sandwich 1, the core disintegrated completely when tested and was lost beyond retrieval. The front skin showed severely fractured fibers in the central region (where the dynamic pressure was applied); it was also fractured into two layers there. The back skin completely delaminated along the interface between two 3-D woven composite layers and separated into two pieces having thicknesses 2.92 mm and 3.86 mm.

Contrary to the above, post mortem analysis of Sandwich 2 (low stitching density) showed only localized delaminations in the skins accompanied by the foam damage in the core. The damage in this case was confined to the mid-section of the front skin; minimal visual damage is observed in the back skin. The observed residual deformation of this sandwich sample was relatively small. Separation of the core chunks from the vertical stitch composite bars were seen in this sample. Buckling of the stitch bars themselves was observed under close-up visual examination. The location of such local buckling was identified by bright white shear-like spots on the side view of the sandwiches. Such bright white spots were detected closer to either one of the skins rather than in the mid-region of the core. It was also observed that micro-delaminations appeared within the skin of this sandwich specimen. Post - mortem analysis of Sandwich 3 (higher stitching density) showed minimal damage in the skins and core and practically no global permanent deformation. Both front and back skins showed no signs of fiber breakage or other failure types. Only small cracks parallel to the stitches were found in the core, and small micro-delamination cracks are present in the skins. Overall, it can be concluded that this sandwich had survived the shock loading significantly better than Sandwich 2 and did not lose its overall integrity. This improvement can be attributed to the higher stitching density and to the marginal increase in the thickness of the skin (which replaces the corresponding foam material in the core).

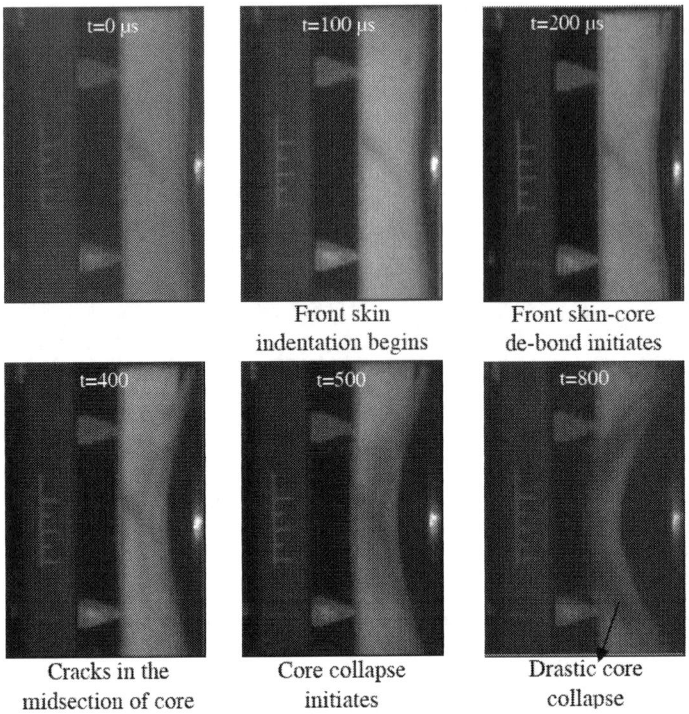

Fig. 8 Sandwich composite with 3D Woven skin when subjected to blast load (1.2 MPa Incident)

The dynamic transient behavior of all the above sandwich specimens under blast loading was recorded using a high speed digital camera and analyzed. The real time observation of shock loading of Sandwich 1 is shown in Fig. 8. The measurements showed that at 500 μ sec the central deflection magnitude was about 5.5 times higher for the front skin than for the back skin. It was also noted that initial separation at one of the sandwich edges between the front skin and the core was seen as early as at 200 μ sec, which indicates a high stress concentration at the front skin-core interface along the sample edges. The indentation failure of the front skin was observed from 200 μ sec. The onset of core failure was observed at 400 μ sec, and complete collapse of the core was seen at 500 μ sec. The conclusion is that for Sandwich 1 all important deformation, damage initiation and total failure events had occurred within 0.5 m sec time from the arrival of the blast wave.

The first principal failure mechanism evident in Sandwich 1 was progressive damage of the core, which started with some dispersed damage near the front skin and simultaneous formation of a large inclined crack in the central region of the core. These two damage zones gradually extended and coalesce at approximately 400 μ sec, which was followed by the formation of a much larger damage zone at 500 μ sec followed by a rapid crushing of the core after that. These images revealed that Sandwich 1 does not provide sufficient transient load transfer from the front skin to the back skin and, as a result, it cannot withstand the applied dynamic pressure.

Analysis of such corresponding real time observations for Sandwich 2 and Sandwich 3 revealed that the dynamic pressure was much better transferred from the front to the back of the sandwich, and we see well synchronized local bending of both skins. The core was gradually damaged as the skin bending progresses. Quantitative measurements of deflections showed that the difference between central deflection values of the front and back skins was relatively small. Also, no separation between the front skin and the core was seen in the high-speed camera frames, thus the aforementioned high stress concentration for Sandwich 1 at the skin-core interface had been reduced by core stitching. Failure mode of this stitched sandwich can be defined as a combination of core shear and face failure types. In case of Sandwich 2, the images(up to 100 μ sec) do not reveal any macroscopic damage to the sandwich. At 200 μ sec initial cracks are seen in the core and start in the mid-section of the core in the direction parallel to the loading plane. The transition of "hinge point" takes place between time instants 300 and 800 μ sec. At 900 μ sec, the cracks in the core became more pronounced and the "hinging" of the whole sandwich (including the span and overhang) along a center line was observed.

The deformation process of Sandwich 3 was very similar to the one seen for Sandwich 2. Though with a closer examination, two significant differences in the real time deformation were observed. The first difference was that the deflection values at 500 μ sec were 30% smaller at the center of the specimen in this case than for Sandwich 2. This can be attributed to the two factors: (a) more dense core stitching and (b) resin layers compensating for reduced thickness of the foam core material in Sandwich 3. Both factors provide additional stiffening to the sandwich. The second difference is in the deformation and failure progression. It was observed that at 100 μ sec there were no visible cracks in the core and the overhangs of the simply supported sandwich were straight, whereas the span portion was hinged about the support lines. The hinge point moves from the support line to the center of the sandwich as the time progresses, which was seen between 200 and 400 μ sec. Further on, at 700 μ sec the transition was completed and the whole sandwich was hinged about the center.

In a contrast to the results for Sandwich 1, the real time deformation sequences for Sandwich 3 indicated that the reinforced core damage does not include any visible macro-cracks. The through-thickness stitching of the sandwich before resin infusion makes a remarkable positive effect on the transient deformation, damage initiation and progression processes in the studied sandwich materials. The major failure modes observed in this study under the transient loading conditions can be broadly classified as front skin indentation, core shear failure and face failure [8]. Unlike in respective static loading case, where given load type usually corresponds to a well-defined, dominating single failure mode, in the case of highly transient dynamic loading (as in this study) the initiation and development of more than one different failure modes can be observed simultaneously.

4.2.3 *Sandwich composites with stepwise graded foam cores.* Two different types of graded foam core sandwich specimens were studied with the core gradation based on the densities of the respective foams. Configuration 1 consisted of a sequence of core gradation as A300/A500/ A800 (low / medium / high density), and configuration 2 consisted of a sequence of core gradation as A500/A300/A800 (medium / low / high density). With these configurations it should be noted that the first core layer is the one first subjected to the shock wave loading.

Both types of rectangular sandwich panels were subjected to an identical blast loading with an incident peak pressure of approximately 1 MPa and a shock wave velocity of approximately 1030 m/s.

The real time observations of the transient behavior of configuration 1 can be seen in Fig. 9. For configuration 1 core compression in the A300 foam is very prominent and visible.At the time when this compression is first observed, there is no obvious compression in the other two core layers. Due to the compression of the foam, the high dynamic pressure applied to the front skin is substantially weakened when it reaches the back skin. The onset of core failure, where core cracking begins, is also observed as well as the delamination of the front skin from the core. Even though this occurs, complete core collapse and failure is not observed in this configuration.

In configuration 2, A500 is the first core layer subjected to the shock wave and the core gradation begins with the foam of middle density, next the foam of least density, and then the highest density foam. Unlike configuration 1, central core compression is not as prominent in this sandwich and the delamination occurs between the core layers of foam (A500/A300) and not between the front skin and core. This delamination is due to the fact that the A300 is weaker foam in comparison to the A500 foam.

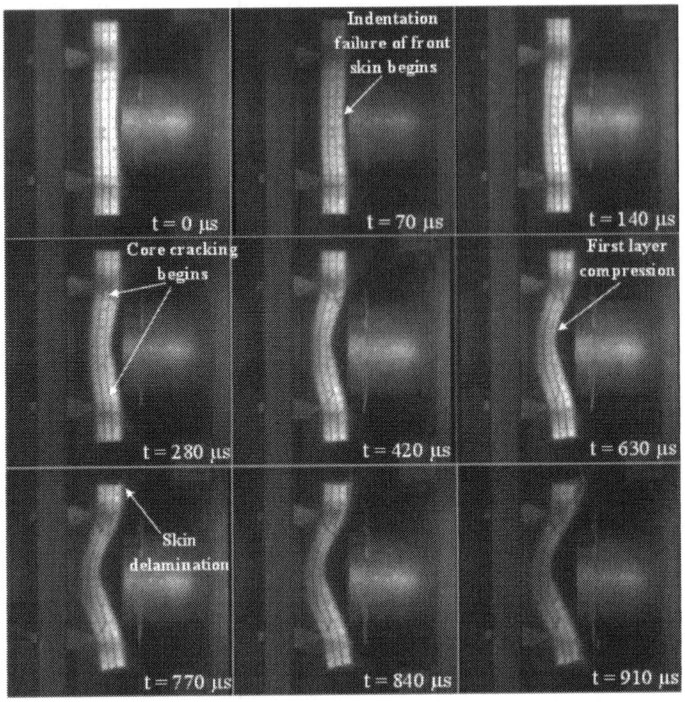

Fig. 9 Real time observation of A300/A500/A800 under shock loading

The major failure mechanism in configuration 2 is progressive damage of the core and the sandwich, which starts at the back skin. This crack became a large inclined crack and propagated through the core from the back skin to the front skin. By t = 490 μsec the crack had extended completely through the core and delamination between the A300 and A500 foam was very prominent. Also at this time, many cracks were visible in the core which was followed by a rapid crushing of the core after that and catastrophic failure of the sandwich. This showed that configuration 2 could not withstand the applied shock wave pressure which had a peak value of ~4.83MPa (710 psi).

Contrary to the case of configuration 2 the real time deformation sequences seen in Fig. 9 for configuration 1 indicated that the core lay-up improved the overall performance of the structure. The onset of core failure took twice as long to be visible in this configuration as opposed to configuration 2 and no complete core collapse was evident. Even though delamination did occur, it was between the face sheet and foam core only. Overall configuration 1 out-performed configuration 2, and this was directly related to the location of the A300 foam.

The damage patterns in the graded sandwich composites after the shock event occurred were visually examined and recorded using a high resolution digital camera as shown in Fig. 10. When configuration 1 was subjected to the highly transient loading, the damage was confined to the area where the supports were located in the shock tube and core cracking is visible in these two areas. Delamination was visible between the front skin and the foam core, as well as the back skin and the foam core. The core compression can be seen clearly and distinctively in the A300 foam.Unlike the damage visible in configuration 1, configuration 2 suffered catastrophic damage. The core of the sandwich disintegrated and the front skin (blast side) of the sandwich fractured into two pieces at the midsection. The back skin showed heavy fiber delamination in the central region as well.

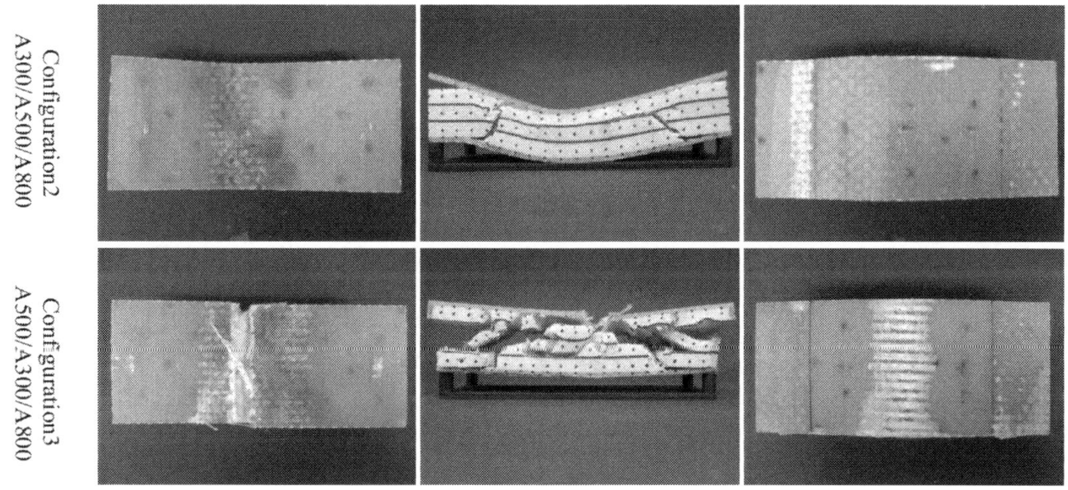

(a) Front face sheet (blast side) (b) Foam core (c) Back face sheet

Figure10. Visual examination of sandwich composites after being subjected to high intensity blast load

4. Summary

This paper focused on experimental evidence of damage mechanisms in novel layered composite material systems. Design modifications to strengthen the skin and core of sandwich composite were incorporated based on experimental study on individual components, namely the skin and the core of the sandwich composites. These modifications were observed to be beneficial to the blast resistance but they are obtained at reasonable addition to the bulk weight of the sandwich structure.

It was observed that, layering of glass fiber composites with a soft layer provides better blast resistance. This enhancement of blast resistance is more pronounced when the softer material faces the blast. It is experimentally observed that, of the different possible material constructions using polyurea and glass fiber composites, sandwich materials made by sandwiching a soft layer (PU) in between woven composite skins (EVE) had the best blast resistant properties

The sandwich composites with through-thickness stitch reinforcement in the core exhibit much lower dynamic deformation, delayed damage initiation and higher dynamic damage tolerance with minimal visual residual damage. Also by grading the core foams from lowest density to highest density, the gradation of the sandwich can help mitigate the blast damage and thus improve the overall performance of the specimen. Hence these materials are highly beneficial in constructing structures that are designed to withstand moderate to severe blast loadings.

5. Acknowledgements

The financial support and encouragement provided by Dr. Yapa Rajapakse, under Office of Naval Research Grant No.N00014-04-1-0268 is gratefully acknowledged. The support provided by the Department of Homeland Security (DHS) under Cooperative Agreement No. 2008-ST-061-ED0002 is also acknowledged. The author thanks Dr. Alex Bogdanovich of 3TEX Inc, for several inputs and discussions during the preparation of the manuscripts on 3D composites. The author acknowledges Dr. Kunigal Shivakumar of NC A&T University, and Martin Marietta Composites for supplying composites and sandwich materials used in this study. Author also thanks Dr. Stephen Nolet and TPI Composites for providing the facility for creating the stepwise graded sandwich composites used in the study. Following graduate students Arjun Tekular, Nate Gardner, Matt Jackson and Erheng Wang contributed to these projects.

References

[1] Tagarielli VL, Deshpande VS, Fleck NA (2007) The dynamic response of composite sandwich beams to transverse impact, Int J Solids Struc, 44(7-8): 2442-2457.

[2] Apetre NA, Sankar BV, Ambur DR (2006) Low-velocity impact response of sandwich beams with functionally graded core, Int J Solids Struc, 43(9): 2479-2496.

[3] Li Y, Ramesh KT, Chin ESC (2001) Dynamic characterization of layered and graded structures under impulsive loading, Int J Solids Struc, 38(34-35): 6045-6061.

[4] Vaidya UK, Nelson S, Sinn B, Mathew B (2001) Processing and high strain rate impact response of multi-functional sandwich composites, Comp Struc, 52: 429-440.

[5] Rice MC, Fleischer CA, Zupan M (2006) Study on the collapse of pin-reinforced foam sandwich cores, Exp Mech, 46(2): 197-204.

[6] Tekalur SA, Shukla A, Shivakumar K (2008) Blast resistance of polyurea based layered composite materials, Comp Struc, 84(3): 271-281.

[7] LeBlanc J, Shukla A, Rousseau C, Bogdanovich AE (2007) Shock loading of three-dimensional woven composite materials, Comp Struc, 79(3):344-355.

[8] Tekalur SA, Bogdanovich A and Shukla A 92008) Shock loading response of sandwich panels with 3-D woven E-glass composite skins and stitched foam core Composite Science and Technology, 84(3):199-209. 2008.

Journal of Physics: Conference Series **181** (2009) 012003 doi:10.1088/1742-6596/181/1/012003

Analysis Methods for the Practical Application of Fracture Mechanics

R A Ainsworth

British Energy, Barnwood, Gloucester GL4 3RS, UK.

bob.ainsworth@british-energy.com

Abstract. Advances in numerical methods have enabled detailed fracture mechanics calculations to be employed in practice. However, more often simplified analysis methods are applied as these often enable adequate margins to be demonstrated. This paper describes detailed and approximate methods for low temperature fracture and creep fracture assessments. Comparisons of the two approaches are given demonstrating that high accuracy can be obtained with simplified methods even for complex applications.

1. Introduction

When a crack is found or postulated in a component, it is necessary to estimate relevant fracture mechanics parameters in order to determine if the component is fit-for-service [1]. When the material response is elastic, estimation of the stress intensity factor, the linear elastic fracture mechanics parameter, is straightforward and comparison with the corresponding fracture toughness enables an assessment of fitness-for-service to be made. For elastic-plastic or creep material response, non-linear fracture mechanics is required and estimation of the parameters controlling fracture at the crack tip is more complex. However, methods for estimating such parameters from detailed finite element analyses have been developed so that non-linear fracture mechanics methods can now be applied in practice [2, 3].

The detailed finite element analysis methods and the associated materials data requirements are expensive and time consuming to use. Therefore, simplified approximate methods have been developed for practical application of inelastic fracture mechanics to defect assessment of plant components [4, 5]. Often, these approximate methods are sufficient to demonstrate adequate margins on limiting defect size or limiting load.

This paper first sets out the background to the fracture mechanics methods. Then the detailed finite element approach and the simplified approaches for estimating the corresponding fracture mechanics parameters are presented. Finally, axi-symmetric and three-dimensional (3-D) finite element (FE) analysis results are presented to illustrate some aspects of component response and to compare the finite element and simplified approaches. Both elastic-plastic and creep fracture approaches are addressed and attention is focussed on combined mechanical and thermal loading.

2. Fracture Mechanics Methods

There are three general classes of fracture mechanics methods. The first class is energy methods as pioneered by Griffiths for brittle fracture but subsequently applied more widely [6]. The second class is the use of parameters which describe the stress and strain (or strain rate) near the crack tip and which are therefore assumed to control fracture [5, 7, 8]. Finally, micromechanical models of brittle, ductile or creep fracture can be used with detailed stress analysis to make a direct prediction of crack tip response [9, 10].

This section concentrates on the second class, the use of parameters which describe conditions in the region of the crack tip. First, Section 2.1 describes single-parameter fracture mechanics, the most commonly applied approach. Then Section 2.2 briefly describes two-parameter methods.

2.1. Single Parameter Fracture Mechanics

For a linear elastic material under Mode I loading, the stress tensor σ_{ij} in polar co-ordinates (r,θ) centred at the crack tip may be written as

$$\sigma_{ij} = \frac{K}{(2\pi r)^{1/2}} g_{ij}(\theta) + T\delta_{1i}\delta_{1j} + O(r^{1/2}) \tag{1}$$

as $r \to 0$. Here, g_{ij} are angular functions of θ, δ_{ij} is Kronecker's delta and K is the stress intensity factor; the second-order T stress term can be regarded as the stress parallel to the crack flanks and is discussed further in Section 2.2. This second–order term is generally neglected and elastic fracture is described by the single parameter K which controls both the stress and strain fields near the crack tip.

For non-linear elastic materials in which strain, ε, is related to stress, σ, by

$$\varepsilon = \alpha\varepsilon_Y(\sigma/\sigma_Y)^n \tag{2}$$

where $\alpha\ \varepsilon_Y, \sigma_Y$ and n are constants, the stress field near the crack tip as $r \to 0$ may be written as

$$\frac{\sigma_{ij}}{\sigma_Y} = \left(\frac{J}{\alpha\sigma_Y\varepsilon_Y I_n r}\right)^{1/(n+1)} \tilde{\sigma}_{ij}(\theta, n) + Q\delta_{ij}, \quad |\theta| < \frac{\pi}{2} \tag{3}$$

where I_n and $\tilde{\sigma}_{ij}$ are functions of n as determined by the asymptotic analyses of Hutchinson [7], Rice and Rosengren [8]. Again, if the second-order term Q is neglected, fracture is controlled by the single parameter J.

The analogy between widespread plasticity and steady state creep [5] means that if creep strain rate, $\dot{\varepsilon}^c$, is related to stress by

$$\dot{\varepsilon}^c = \dot{\varepsilon}_0(\sigma/\sigma_0)^n \tag{4}$$

where $\dot{\varepsilon}_0, \sigma_0$ and n are constants, the first-order term in the stress field at the crack tip is given in a similar manner to eqn (3) by

$$\frac{\sigma_{ij}}{\sigma_0} = \left(\frac{C^*}{\sigma_0\dot{\varepsilon}_0 I_n r}\right)^{1/(n+1)} \tilde{\sigma}_{ij}(\theta, n) \tag{5}$$

where C^* is the steady state creep parameter. On initial loading, the crack tip fields are given by eqn (1), for elastic response, or eqn (3) for elastic-plastic response. Subsequently, there is a transient period during which stress redistribution occurs until the steady state conditions of eqn (5) are reached. During this transient period, the stress and strain rate fields are described by eqn (5) but with C^* replaced by the transient parameter C(t) where $C(t) \geq C^*$.

Either detailed numerical methods or simplified methods are used to evaluate the single parameters K, J, C^* and C(t) as functions of the applied loadings, material properties and crack size and shape. These are then related to corresponding material properties, which are obtained from test specimens that control fracture or crack growth. For linear material response, K is directly proportional to the applied load and independent of the material properties and there are many solutions available for evaluation of K [1]. Evaluation of J, C^* and C(t) is more complex and methods for estimating these parameters are set out in Sections 3 and 4.

2.2. Two Parameter Fracture Mechanics

It has been found that material fracture response is generally not simply defined by the first-order parameters discussed in Section 2.1. Instead fracture toughness and creep crack growth have been found to be influenced by the nature of the loading, crack size and shape. These influences may be conveniently described in terms of the second-order parameters T and Q in eqns (1, 3). Conservative fracture mechanics assessments may be made by neglecting these second-order effects and obtaining materials data from deeply cracked bend specimens in which T, Q>0 tending to minimise the fracture resistance, or maximise the creep crack growth rates. These conditions are referred to as "high constraint". For so-called low constraint conditions, such as shallow cracks and tension loading, for which T and Q are significantly negative and reduce the crack tip stress fields from the high constraint fields, the fracture resistance can be significantly higher. Although these second-order effects are not considered in detail in this paper, they are briefly discussed here.

Although T is an elastic parameter, it has been argued that it can be used to quantify crack tip constraint under elastic-plastic conditions [11]. As T can be evaluated by elastic analysis, it is relatively straightforward to determine compared to Q. Therefore, there are a number of solutions available in the literature. A compendium of these solutions has been compiled in [12] for two-dimensional test specimen geometries, some axisymmetric geometries and for surface cracked plates under tension and bending. A more comprehensive collation for 1-D cracks has been produced by Fett [13]. This includes tabulations of weight functions for a number of cases, enabling the T-stress to be evaluated for non-linear thermal or residual stresses as well as uniform tension or linear bending stress distributions.

Unlike eqn (1) for elastic response, eqn (3) is not a strict asymptotic analysis of the higher-order terms for the non-linear problem. Instead, the hydrostatic term, Q, is an approximation to the collective behaviour of a number of higher-order terms in the forward sector, $|\theta| < \pi/2$, ahead of the crack [14]. In practice, at least in two-dimensional problems, the Q stress varies slightly with distance from the crack tip and for application purposes has been evaluated at $r/(J/\sigma_Y) = 2$. Although, some solutions for Q are available in the literature [15, 16], these are of more restricted application than corresponding solutions for T.

3. Finite Element Approaches

Finite element analysis may be used to aid fracture mechanics assessments in a variety of ways. These include:

- Linear elastic analysis of the uncracked structure to determine the stresses at the position where a crack is to be assessed, for input to a weight function solution to calculate K or T for a range of crack sizes.
- Linear elastic analysis of the cracked structure to determine K or T for the crack size of interest.
- Inelastic analysis of the cracked structure to determine J, C(t) or Q for the loading, material and crack size of interest.

This last use of finite element analysis is examined in more detail here for evaluation of J or C(t). It can be seen from eqns (3, 5) that in principle these parameters can be determined from the amplitude of the stress fields at the crack tip. However, this requires material behaviour to be described by the simple forms of eqns (2, 4) and requires very fine finite element meshes at the crack tip to capture the asymptotic behaviour. Therefore, more generally, the crack tip parameters are obtained from contour integrals. For example, Rice [17] showed that a path-independent integral could be used to define J so that it was not necessary to determine the crack tip fields with high accuracy very close to the crack tip. However, this early definition of J did not allow for thermal or residual stresses in the structure and was limited to proportional loading. Extensions to determine J under more general loading conditions have been developed [2, 18] and to demonstrate that the resulting stress fields are consistent with eqn (3). Thus, it is now possible to perform detailed finite element analysis to quantify the crack tip stress field as a function of the loading applied to the structure. This is illustrated in Section 5.

For time-dependent creep calculations, finite element values C(t) can be extracted as a function of time t from a contour integral defined by

$$C(t) = \int_{\Gamma\to 0} \left[\dot{W}dy - T_i \left(\frac{\partial \dot{u}_i}{\partial x} \right) ds \right] \tag{6}$$

where T_i are components of the traction vector, $\dot{u}_i$ are the displacement rate vector components, and $\dot{W}$ is strain energy rate density. At small times, a transient creep condition occurs, during which the stress redistributes within the cracked body. Under this condition, the C(t)-integral results exhibit significant path dependence and according to the definition in eqn (6), C(t) values should be extracted close to the crack tip. Therefore, care is needed in selecting a suitably refined mesh at the crack tip and highly refined meshes have been used to obtain the results presented later in Section 5. At longer times, when the stress has re-distributed within the cracked body, a steady state condition is achieved, where path-independence occurs and eqn (6) can be replaced by:

$$C^* = \int_{\Gamma} \left[\dot{W}dy - T_i \left(\frac{\partial \dot{u}_i}{\partial x} \right) ds \right] \tag{7}$$

which defines the steady state stress field in eqn (5). This integral is path independent and therefore does not need to be evaluated close to the crack tip.

4. Simplified Methods

While the detailed numerical methods outlined in Section 3 can be followed, the results are dependent on the particular loading, crack size and material properties input to the finite element analysis. Thus, where it is required to determine margins on loadings or limiting defect sizes the methods are time consuming and expensive to use. Therefore, simplified methods which can provide more rapid assessments are attractive. Here, simplified methods for estimating J and C(t) for combined mechanical and thermal stresses are presented. Comparisons between these simplified methods and application of the detailed numerical approaches of Section 3 are then given in Section 5.

For, the general loading case considered here is to necessary to quantify the effects of combined mechanical and thermal loading applied at time t=0 on the crack tip stress field as characterized by the elastic-plastic J (Section 4.1). Then, it is necessary to characterise the subsequent variations of C(t) with time t(>0) under transient creep conditions (Section 4.2).

4.1 Estimation of J

Reference stress methods [4] were developed a number of years ago to estimate, under mechanical loading only, the elastic-plastic value of J for a cracked component, denoted J^p, from

$$\frac{J^p}{J_e^p} = \frac{E\varepsilon_{ref}^p}{\sigma_{ref}^p} + \frac{1}{2} L_r^2 \frac{\sigma_{ref}^p}{E\varepsilon_{ref}^p} \tag{8}$$

where the reference stress, σ^p_{ref}, and the parameter L_r are defined by the plastic limit load of the cracked component of interest, P_L:

$$L_r = \frac{\sigma^p_{ref}}{\sigma_y} = \frac{P}{P_L} \tag{9}$$

In eqn (8), the reference strain ε^p_{ref} is defined as the true strain at the reference stress, determined from true stress-strain data of the material, and J^p_e denotes the elastically calculated value of J due to primary stress, defined by the stress intensity factor K due to primary stress, K^p:

$$J^p_e = (K^p)^2 / E' \tag{10}$$

where $E' = E/(1-\nu^2)$ for plane strain, $E' = E$ for plane stress; and E and ν denote Young's modulus and Poisson's ratio, respectively.

When the cracked component is subject to combined primary and secondary stress, the elastic J for combined primary and secondary stresses, J^{p+s}_e can be obtained by linear superposition:

$$J^{p+s}_e = (K^p + K^s)^2 / E' \tag{11}$$

where K^s is the stress intensity factor for the secondary stresses. If eqn (11) was used to replace J^p_e in eqn (8), the resulting estimate of J could significantly overestimate the actual J, due to mechanical stress relief of the secondary stresses. Therefore, a multiplying factor V has been introduced [19] to estimate J due to combined primary and secondary stresses, J^{p+s}, from

$$J^{p+s} = \frac{\left(K^p + V \cdot K^s\right)^2}{E'} \left[\frac{E\varepsilon^p_{ref}}{\sigma^p_{ref}} + \frac{1}{2} L_r^2 \frac{\sigma^p_{ref}}{E\varepsilon^p_{ref}} \right] \tag{12}$$

It can be seen that the value of V=1 indicates a full effect of the secondary stress on the estimated J, whereas V=0 corresponds to the secondary stress being completely relaxed and not contributing to the elastic-plastic value of J. Methods for estimating V as a function of the magnitudes of the primary and secondary loadings are given in [19].

The above elastic-plastic approach has been presented in terms of an estimate of J. An alternative and fully equivalent approach [1] is to present the analysis in terms of a Failure Assessment Diagram (FAD) [20]. Such approaches are not described in detail here but it is worth noting that these are very convenient methods for practical fracture mechanics assessment.

4.2 Estimation of C(t)

Reference stress methods [4] estimate the steady state creep parameter from

$$C^* = \sigma^p_{ref}(a)\dot{\varepsilon}^c_{ref}[\sigma^p_{ref}(a), \varepsilon^c_{ref}](K^p / \sigma^p_{ref})^2 \tag{13}$$

where σ^{p}_{ref} is the reference stress for the primary loading as in Section 4.1 and $\dot{\varepsilon}^{c}_{ref}$ is the creep strain rate at the current reference stress and creep strain, ε^{c}_{ref}. Eqn (13) is analogous to eqn (8) apart from a small-scale yielding term in eqn (8).

For combined primary and secondary loading, if the initial response on loading is elastic, an initial reference stress, σ^{0}_{ref}, can be defined by

$$\sigma^{0}_{ref} = \sigma^{p}_{ref}(K^{p} + K^{s})/K^{p} \tag{14}$$

The reference stress may also be defined from the stress resultants for combined loading using an equation similar to eqn (9). This is not discussed in detail here but is particularly suitable for loadings with significant stresses not in the plane of the defect. The combined reference stress of eqn (14) may relax due to both creep straining and crack growth and the rate of change is given by

$$\frac{\dot{\sigma}_{ref}}{\sigma_{ref}} + \left[\frac{K^{s}}{(K^{p}+K^{s})}\left\{\frac{\partial K^{p}/\partial(a/w)}{K^{p}} - \frac{\partial K^{s}/\partial(a/w)}{K^{s}}\right\} - \frac{\partial\sigma^{p}_{ref}/\partial(a/w)}{\sigma^{p}_{ref}}\right]\frac{\dot{a}}{w} + \frac{E\dot{\varepsilon}^{c}_{ref}}{Z\sigma_{ref}} = 0 \tag{15}$$

for a crack of depth a in section width w, where Z is the elastic follow-up factor and the creep strain rate is calculated at the current combined reference stress. Evaluation of the elastic follow-up factor is discussed later in Section 5.

Equations (14, 15) enable the initial reference stress and its relaxation to be determined. Omitting algebraic details, the resulting value of the transient crack tip characterising parameter, generalised to the case of combined primary and secondary loading involving stress relaxation due to both creep and crack growth and plasticity on initial loading is

$$\frac{C(t)}{C^{*}} = \left(\frac{\sigma_{ref}\dot{\varepsilon}^{c}_{ref}}{\sigma^{p}_{ref}\dot{\varepsilon}^{c}_{ref,p}}\right)\left[\frac{(\varepsilon_{ref}/\varepsilon^{0}_{ref})^{n+1}}{(\varepsilon_{ref}/\varepsilon^{0}_{ref})^{n+1} - (\sigma^{0}_{ref}/E\varepsilon^{0}_{ref})}\right] \tag{16}$$

where $\dot{\varepsilon}^{c}_{ref}$ and $\dot{\varepsilon}^{c}_{ref,p}$ are the creep strain rates at σ_{ref} and σ^{p}_{ref}, respectively, ε_{ref} is the total strain at σ_{ref}, C^{*} refers to the value evaluated for the primary loading only, and ε^{0}_{ref} is the total elastic-plastic strain corresponding to the initial value of the total reference stress σ^{0}_{ref}. For pure primary loading, it is straightforward to evaluate the strain terms in eqn (16) as the reference stress is well defined from the limit load expression of eqn (9). For more general loading, the initial strain term may be obtained from an estimate of the initial value of J, J_0, using the method discussed in Section 4.1.

5. Results

In this section the approaches of Sections 3 and 4 are applied to a cylindrical geometry subjected to a variety of combined mechanical and thermal stresses. The geometry and loadings are first presented in Section 5.1. Results for J are presented in Section 5.2 and results for C(t) in Section 5.3.

5.1 Geometry and Loading

Finite element analyses have been performed for circumferentially cracked cylinders under combined mechanical and thermal loading. The pipe has mean radius, r, and thickness, t, with r/t=10. Circumferential through-wall and part-through surface cracks are considered, as shown in Fig. 1. The circumferential through-wall crack is characterized by the relative crack length, θ/π, where θ denotes the half crack angle (Fig. 1b). The circumferential part-through surface cracks have a rectangular shape, i.e., constant depth, characterized by the relative crack depth, a/t.

The mechanical loading was applied as an axial tension N. Three different types of temperature variation were applied to generate thermal stresses. The first, termed a "radial temperature gradient", has a temperature gradient through the pipe wall, as shown in Fig. 2a, and produces through-wall bending stresses. The second type has a gradient along the longitudinal direction to produce global bending stresses, as shown in Fig. 2b, and is termed an "axial temperature gradient". The final type has a temperature gradient across the cross-section, as depicted in Fig. 2c. This type of thermal loading produces global bending stresses, and is termed a "sectional temperature gradient". The magnitude of the temperature difference, ΔT_{max}, was adjusted to produce different values of the thermal stress intensity factor.

Plastic strains were described by eqn (2) with n=5 or 10. Creep strain rates were defined by eqn (4), again with n=5 or 10. For elastic-plastic-creep analyses, the value of n was taken to be the same for both plasticity and creep.

5.2 Results for elastic-plastic response

Rather than presenting results for J, it is more convenient for the combined loading cases examined to present the results in terms of the parameter V in eqn (12). Combining eqns (8, 12) leads to

$$V = \left(\sqrt{\frac{J_e^p}{J_e^s}}\right)\left(\sqrt{\frac{J^{p+s}}{J^p}} - 1\right) \tag{17}$$

where J_e^s is the elastic value of J when only the secondary stress is applied. In this case, the elastic-plastic J, J^s, can be written in terms of an initial value V_0 of V as

$$J^s = (V_o \cdot K^s)^2 / E' \tag{18}$$

Then eqn (17) becomes

$$\frac{V}{V_o} = \sqrt{\frac{J_e^p}{J^s}}\left(\sqrt{\frac{J^{p+s}}{J^p}} - 1\right) \tag{19}$$

The finite element results for the various J estimates in eqn (19) have been used to express V/V_0 for the three types of thermal loading as functions of the level of axial stress and some results are shown in Figs 3-5. More comprehensive results which show similar trends are given in [21]. It is noteworthy that the results, when presented in the form of Figs 3-5, are insensitive to crack size and the type of thermal loading. Note that the normalising load, N_{OR}, in these figures is slightly different from the limit load in eqn (9) and has been chosen to ensure that eqn (8) is accurate for axial load acting alone. Thus, the comparisons in Figs 3-5 address the influence of combined loading on J. Also shown in these figures are simplified estimates of V/V_0 given in the R6 defect assessment procedure [20]. It can be seen that these are generally conservative. An alternative less conservative estimate [21]

$$V/V_0 = \left[\frac{E\varepsilon_{ref}^p}{\sigma_{ref}^p} + \frac{1}{2}L_r^2\frac{\sigma_{ref}^p}{E\varepsilon_{ref}^p}\right]^{-1/2} \tag{20}$$

is also shown in Figs 3-5. This is similar to the equation used to estimate J in eqn (8) and it can be seen that this enables accurate but not overly conservative estimates of V, and hence J (see eqn (19)) for combined primary and secondary loading. The results in Figs 3-5 have been presented for different magnitudes of the secondary loading as expressed by the ratio

$$\beta = K^s L_r / K^p \tag{21}$$

with Fig 3 having a relatively small secondary stress, Fig 4 corresponding to a large radial temperature gradient and Fig 5 corresponding to a modest sectional temperature gradient. It is apparent that the results are similar in all cases when normalised as presented in the figures, confirming that approximate estimation methods can be used.

5.3 Results for creep response

Following the initial loading which gave the results for J reported in Section 5.2, the finite element analyses were continued with creep relaxation of the secondary stresses occurring. The resulting values of C(t) evaluated according to eqn (6) fell rapidly with time with the rate of relaxation dependent on the magnitude of both the primary and secondary loadings, as quantified by L_r, β of eqns (9, 21). The rate of relaxation is governed by the elastic follow-up factor Z in eqn (15). From observation of the finite element results [22], it was found that this could be represented by a time normalised by a redistribution time dependent on the magnitudes of the primary and secondary loadings. The normalised time is:

$$\tau^{p+s} = t / t_{red}^{p+s} \tag{22}$$

where the redistribution time is defined by the ratio of the initial value of J for the combined loading to the steady state value of C^* for the mechanical loading:

$$t_{red}^{p+s} = J_0^{p+s} / C^* \quad \text{for} \quad J_0^{p+s} < 10 J_0^p \tag{23}$$

At higher values of secondary loading the redistribution time was found to saturate so that eqn (23) was replaced by

$$t_{red}^{p+s} = 10 J_0^p / C^* \quad \text{for} \quad J_0^{p+s} \geq 10 J_0^p \tag{24}$$

Some finite element results are shown in Figs 6-8 for selected cases with initial elastic response; more comprehensive results including plasticity on initial loading show similar trends and are given in [22]. It can be seen that the results when normalised by the time of eqn (22) are insensitive to the magnitudes of the primary and secondary loadings, although there is some sensitivity to the magnitude of the sectional temperature gradient, as shown on Fig 8. Also shown on Figs 6-8 is a simplified estimate of eqn (16) from [22]:

$$\frac{C(t)}{C^*} = \frac{(1+\tau^{p+s})^{4.5}}{(1+\tau^{p+s})^{4.5} - \phi} \quad \text{with} \quad \phi = 1 - \frac{\alpha \varepsilon_Y \sigma_0^n}{\dot{\varepsilon}_0 \sigma_Y^n} \frac{C^*}{J_0^{p+s}} \tag{25}$$

where $\phi = 1$ corresponds to elastic response on initial loading, and $\phi \rightarrow 1$ corresponds to widespread plasticity on initial loading so that the steady state creep field is established at the crack tip; note that ϕ depends on the magnitude of the thermal stress for elastic-plastic response but the dependence is weak. It can be seen that this simplified estimate is accurate for the axial and radial temperature gradients but generally a little conservative for the sectional temperature gradient, with increasing conservatism at higher secondary stresses. As for the case of elastic-plastic response, this confirms that simplified estimates sufficient for practical applications can be made for elastic-plastic-creep crack tip response.

6. Concluding Remarks

This paper has first presented some background to fracture mechanics methods and their practical application by finite element and simplified methods. Then, results from finite element solutions for elastic-plastic J and creep C(t) for circumferentially cracked pipes under combined mechanical and thermal loads have been presented. The finite element results, when presented in normalised form, have been found to be not particularly sensitive to the crack geometry (length and depth), the loading mode (axial tension and bending), the material properties and the thermal loading type. This suggests that simplified methods for estimating this ratio can be used in practical defect assessments and such simplified estimates have been presented.

Acknowledgement

This paper is published with permission of British Energy Generation Ltd. Contributions from colleagues at British Energy and the University of Korea are gratefully acknowledged.

References

[1] U Zerbst, M Schödel, S E Webster and R A Ainsworth, Fitness-for-Service Assessment of Structures Containing Cracks, A Workbook based on the European SINTAP/FITNET Procedure, Elsevier, Oxford (2007).
[2] Y Lei, J-integal evaluation for cases involving non-proportional stressing, Engng Fract Mech **72**, 577-596 (2005).
[3] J L Bassani and F L McClintock, Creep relaxation of stress around a crack tip, Int J Solids Structures **17**, 479-492 (1981).
[4] R A Ainsworth, The assessment of defects in structures of strain hardening material, Engng Fract Mech **19**, 633-642 (1984).
[5] G A Webster and R A Ainsworth, High Temperature Component Life Assessment, Chapman & Hall, London (1994).
[6] The Energetics of Fracture, The Griffiths Centenary Meeting, Liverpool (1993); Institute of Materials Book 580 (1993).
[7] J W Hutchinson, Singular behaviour at the end of a tensile crack in a hardening material, J Mech. Phys. Solids **16**, 13-31 (1968).
[8] J R Rice and G F Rosengren, Plane strain deformation near a crack-tip in a power-law hardening material, J. Mech. Phys. Solids **16**, 1-12 (1968).
[9] F M Beremin, A local criterion for cleavage fracture of nuclear pressure vessel steel, Metall Trans **14A** , 2277–2287 (1983).
[10] A L Gurson, Continuum theory of ductile rupture by void nucleation and growth: Part I – Yield criteria and flow rules for porous ductile materials, ASME J Engng Mater Technol **99**, 2-15 (1977).
[11] C Betegón and J W Hancock, Two-parameter characterisation of elastic-plastic crack-tip fields, ASME J Applied Mechanics **58**, 104-113 (1991).
[12] A H Sherry, C C France and M R Goldthorpe, Compendium of T-stress solutions for two and three dimensional cracked geometries, Fatigue Fract Engng Mater Struct.**18**, 141-155 (1995).
[13] T Fett, T-stress Solutions and Stress Intensity Factors for 1-D cracks, VDI Verlag, Düsseldorf (2002).
[14] N P O'Dowd and C F Shih, Family of crack tip fields characterised by a triaxiality parameter: Part I – Structure of fields, J Mech Phys Solids **39**, 989-1015 (1991).
[15] R A Ainsworth, I Sattari-Far, A H Sherry, D G Hooton and I Hadley, Methods for including constraint effects within the SINTAP procedures, Engng Fract Mech **67**, 563-571 (2000).
[16] R A Ainsworth and N P O'Dowd, Constraint in the failure assessment diagram approach for

fracture assessment, ASME J Pres Ves Tech **117**, 260-267 (1995).
[17] J R Rice, A path independent integral and the approximate analysis of strain concentration by notches and cracks, ASME J Appl Mech **35**, 379-386 (1968).
[18] D W Beardsmore, J K Sharples, C J Madew and M Jackson, An analysis of the roles of J and Q in the assessment of fracture for quasi-statically extending cracks in residual stress fields, Paper PVP2009-77419, Proc ASME Pressure Vessels and Piping Division Conference, Prague (2009).
[19] R A Ainsworth, J K Sharples and S D Smith, Effects of residual stresses on fracture behaviour-experimental results and assessment methods, J Strain Analysis **35**, 307-316 (2000).
[20] Assessment of the Integrity of Structures Containing Defects, Procedure R6 Revision 4, Amendment 7, British Energy, Gloucester (2009).
[21] T K Song, Y-J Kim, K Nikbin and R A Ainsworth, Approximate J estimates for circumferential cracked pipes under primary and secondary stresses, submitted to Engng Fract Mech (2008).
[22] T K Song, Y-J Kim, K Nikbin and R A Ainsworth, Estimation of the transient creep parameter C(t) under copmbined mechanical and thermal stresses, to be submitted to Engng Fract Mech (2009).

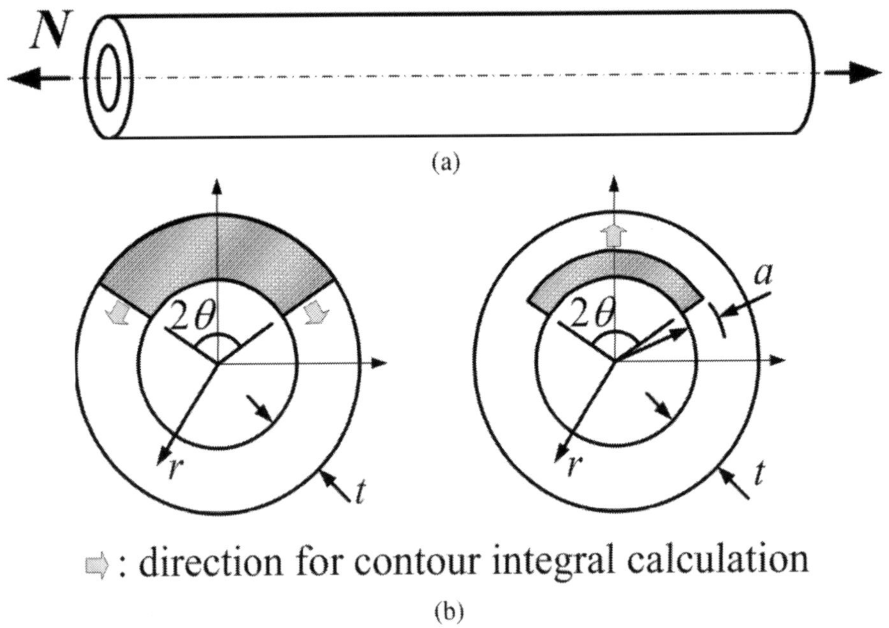

Figure 1. (a) Pipes under axial tension N, and (b) circumferential through-wall and part-through surface cracks. Directions for contour integral calculations are also shown.

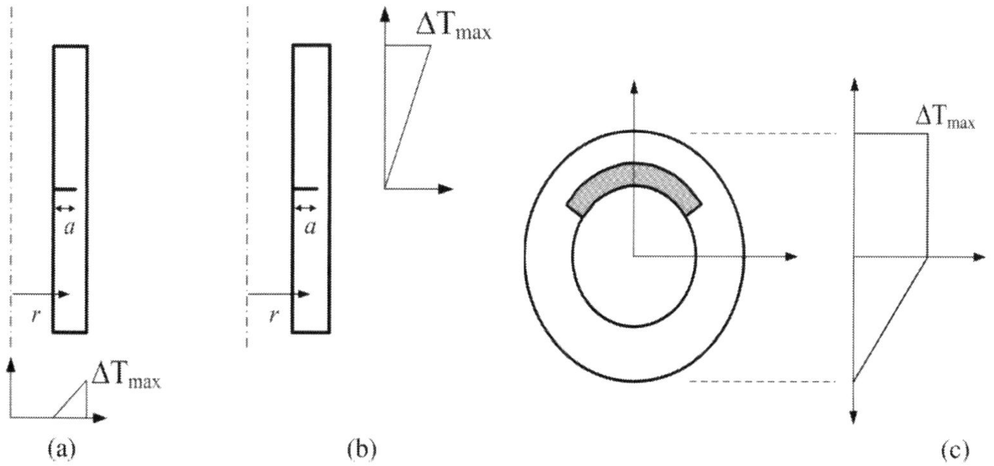

Figure 2. Thermal loading: (a) radial temperature gradient, (b) axial temperature gradient, (c) sectional temperature gradient.

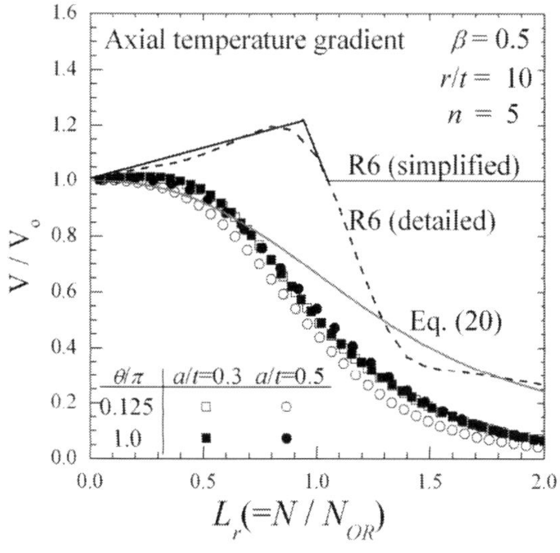

Figure 3 Elastic-plastic results for a small axial temperature gradient ($\beta = K^s L_r / K^p = 0.5$).

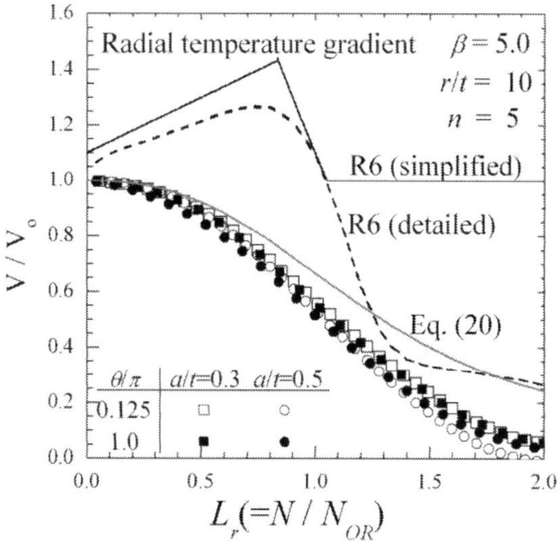

Figure 4 Elastic-plastic results for a large radial temperature gradient ($\beta = K^s L_r / K^p = 5.0$).

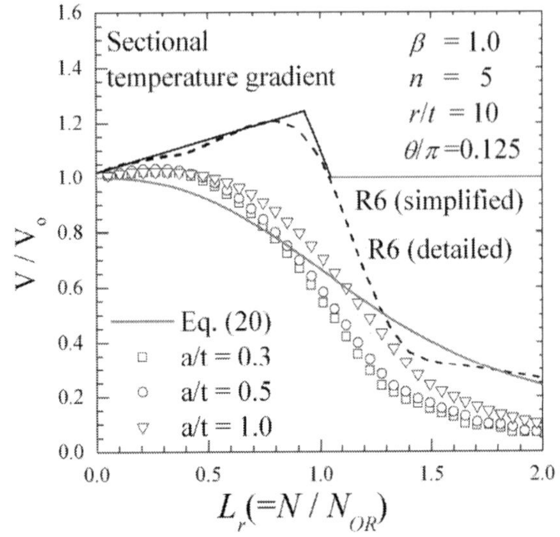

Figure 5 Elastic-plastic results for a modest sectional temperature gradient ($\beta = K^s L_r / K^p = 1.0$).

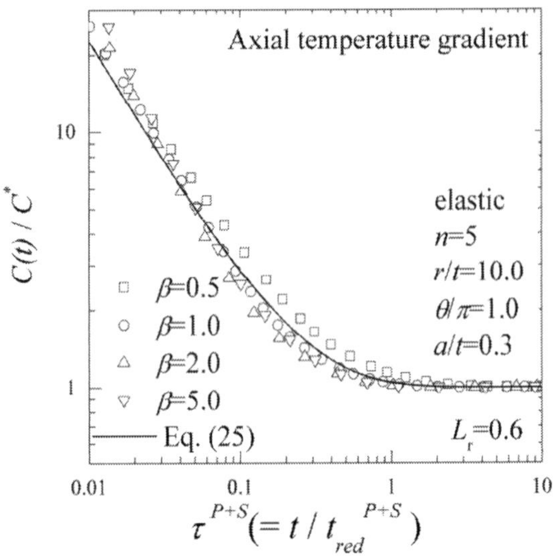

Figure 6 Creep results for the axial temperature gradient with a modest mechanical load, analysed elastically.

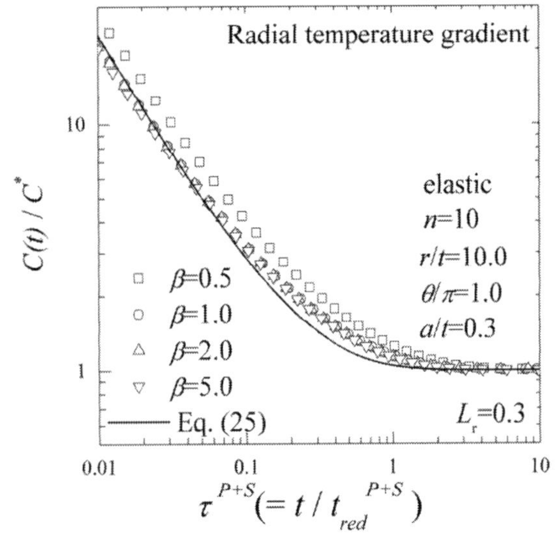

Figure 7 Creep results for the radial temperature gradient with a small mechanical load, analysed elastically.

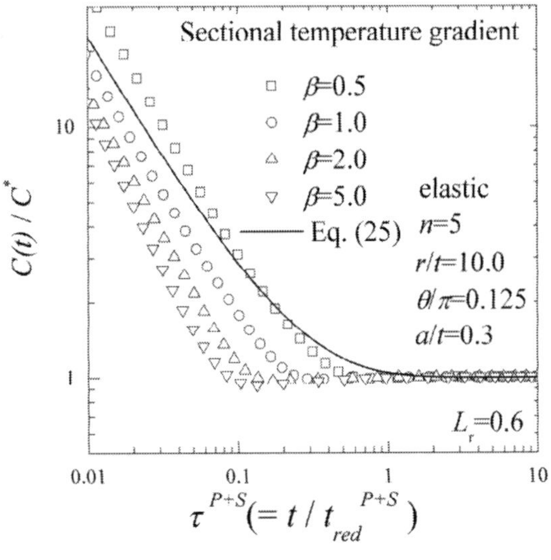

Figure 8 Creep results for the sectional temperature gradient with a modest mechanical load, analysed elastically.

Shape descriptors for mode-shape recognition and model updating

Weizhuo Wang[1], John E Mottershead[1,†] and Cristinel Mares[2]

[1] Department of Engineering, The University of Liverpool, Liverpool, UK, L69 3GH
[2] School of Engineering and Design, Brunel University, Middlesex, UK, UB8 3PH

† j.e.mottershead@liverpool.ac.uk

Abstract. The most widely used method for comparing mode shapes from finite elements and experimental measurements is the Modal Assurance Criterion (MAC), which returns a single numerical value and carries no explicit information on shape features. New techniques, based on image processing (IP) and pattern recognition (PR) are described in this paper. The Zernike moment descriptor (ZMD), Fourier descriptor (FD), and wavelet descriptor (WD), presented in this article, are the most popular shape descriptors having properties that include efficiency of expression, robustness to noise, invariance to geometric transformation and rotation, separation of local and global shape features and computational efficiency. The comparison of mode shapes is readily achieved by assembling the shape features of each mode shape into multi-dimensional shape feature vectors (SFVs) and determining the distances separating them.

1 Introduction

Accurate vibration mode shape information is available for large and complicated structures by using modern computational finite element methods and full-field optical measurement technique [1]. *Image processing* (IP) and *pattern recognition* (PR), well developed in other disciplines, enables the versatile comparison and classification of mode shapes [2, 3], which cannot be achieved by the conventional *Modal Assurance Criterion* (MAC) [4]. A set of shape features with good discriminative capability may be extracted by IP and PR methods to form a feature vector – *shape descriptor* (SD). Now the similarity/dissimilarity of the shapes can be revealed by the 'distance' of their corresponding SDs in the shape feature space according to appropriate criteria.

The moment descriptor is one of the most popular shape descriptors in image processing and pattern recognition [5] but the non-orthogonality of the moment basis gives rise to redundancy and the original shape generally becomes difficult to recover from a truncated set of moment descriptors. Teague [6] suggested using orthogonal polynomials to replace the algebraic polynomials when calculating the moment descriptor.

The *Zernike moment* (ZM) is based on a complete set of orthogonal polynomials over a circle of unit radius – Zernike polynomials. It was found later that the Zernike moment is one of the most important region-based shape descriptors because of its outstanding properties resulting from the orthogonality of the Zernike polynomials. Firstly, expressing an image as a set of mutually independent descriptors has the effect of minimising the redundancy of information. Secondly, the contribution of each order of moment to the image reconstruction can be separated, so that the process of regaining the original image is much easier than by geometric moment descriptors [7]. Rotational invariance [8, 9] is another important property of the ZM, meaning that rotating an image does not change the magnitudes of its Zernike moments. Also, the ZM is robust to noise [7] and effective, meaning that a small number of Zernike moments are usually sufficient for shape reconstruction. The ZM in vibration mode-shape recognition is described in detail by the present authors [10].

Fourier Descriptors (FDs) were originally proposed in 1960 by Cosgriff [11], and thereafter became popular among the pattern recognition community through the papers of Zahn [12], Persoon and Fu [13]. FDs are among the most popular shape representation methods for vision and pattern recognition applications. FDs refer to a class of methods, not a single method, since there are many different ways in which the FDs of a shape can be defined. The basic idea underlying this approach consists in representing the shape of interest in terms of a 1D, 2D or even 3D signal. The Fourier transform of this signal is determined and the FDs are calculated for this Fourier representation. Some properties of the FDs directly follow from the underlying theory of the Fourier transforms and series. For instance, the 0th component of the FDs obtained from the contour representation is associated with the centroid of the original shape. The invariance to geometric transformations is also a direct consequence of the properties of the Fourier series. Such properties have helped to popularize this shape representation scheme, which has been adopted in a number of applications during the last four decades.

In certain applications, local shape information is particularly significant, when certain shape features are associated with a particular portion of the object, such as a discontinuity, singularity or local frequency. *The Short-Time Fourier descriptor* (SFD) [14] and *Wavelet descriptor* (WD) [15] are two of the most powerful method for detecting local shape features, especially the WD. An important property of the wavelet descriptor is its ability to separate local features from the global shape, while avoiding the problem of a local modification affecting the whole representation. For example a partial occlusion may interfere significantly with a global shape descriptor, such as the Fourier descriptor or moment descriptor, while the wavelet descriptor of the unblocked part is not affected at all. The FD and WD in vibration mode-shape recognition are described in detail by the present authors [16].

Thus, a *shape feature vector* (SFV) can be formed by assembling the different SDs. The comparison of mode shapes is then transformed to the similarity measurement between the SFVs in the feature space. Dimensionless normalisation for the SFV is necessary before commencing the comparison to avoid the scaling effect.

2 Shape Descriptors

The theory of the various shape descriptors is described in full by the present authors [10, 16]. A brief outline of the theory is provided in this article. The general form of SDs may be expressed as,

$$\mathfrak{D} = \mathcal{T}[I(x, y)] \tag{2-1}$$

where $I(x, y)$ denotes the continuous displacement mode shape function and $\mathcal{T}[*]$ represents the transformation for extracting the shape features. More specifically, the SD can be defined by projecting the image function onto a kernel function $\mathfrak{K}(x, y)$ as

$$\mathfrak{D} = \int_{\Omega} \mathfrak{K}(x, y) I(x, y) \mathrm{d}x\mathrm{d}y \tag{2-2}$$

where Ω denotes the domain of definition. Different types of SDs (ZMs, FDs, WDs etc.) are defined by using different kernel functions.

2.1 Zernike Moment Descriptor

A set of two-dimensional complex polynomials defined over a unit circle was introduced by Zernike [17], written as

$$V_{n,m}(x, y) = V_{n,m}(\rho, \vartheta) = R_{n,m}(\rho) e^{\mathrm{i}m\vartheta} \tag{2-3}$$

where $\mathrm{i} = \sqrt{-1}$, (ρ, ϑ) is the polar coordinate, n is the non-negative integer, representing the order of the radial polynomial, m denotes the positive / negative integer subject to constraints $n-|m|$ even, $|m| \leq n$ (representing the repetition of the azimuthal angle) and $R_{n,m}$ is the radial polynomial defined as [17],

$$R_{n,m}(\rho)=\sum_{s=0}^{\frac{n-|m|}{2}}(-1)^s\frac{(n-s)!}{s!\left(\frac{n+|m|}{2}-2\right)!\left(\frac{n-|m|}{2}-2\right)!}\rho^{n-2s} \tag{2-4}$$

According to the orthogonality properties [18], the inner product of any pair of complex Zernike polynomials can be expressed as

$$\iint_{x^2+y^2\leq 1}V_{p,q}(x,y)V_{n,m}^*(x,y)\mathrm{d}x\mathrm{d}y=\frac{\pi}{n+1}\delta_{n,p}\delta_{m,q} \tag{2-5}$$

where * denotes the complex conjugate and $\delta_{n,p}$ and $\delta_{m,q}$ are Kronecker deltas. Thus, the Zernike moment descriptor can be obtained by substituting the Zernike polynomials into the kernel function in (2-2), in polar coordinates as,

$$\mathfrak{D}_{Z_{n,m}}=\frac{n+1}{\pi}\int_0^{2\pi}\int_0^1 I(\rho,\vartheta)V_{n,m}^*(\rho,\vartheta)\rho\mathrm{d}\rho\mathrm{d}\vartheta \tag{2-6}$$

The original image $I(\rho,\vartheta)$ may easily be reconstructed by virtue of the orthogonality of the ZMDs. Expression efficiency is one of the important properties of the ZMD, meaning the main shape information may be extracted by the lower order moments after truncating the higher orders of the ZMD in real applications. Rotational invariance is another significant property [9] of the ZMD, useful for recognising mode shapes of axisymmetric structures [10].

2.2 Fourier Descriptor

The FD is based on the frequency components from *Fourier transform* (FT) of the images. According to the well-known theory of the FT, the kernel function of the SD is the complex valued sinusoid,

$$\mathfrak{D}_{\mathcal{F}}(u,v)=\int_{-\infty}^{+\infty}\int_{-\infty}^{+\infty}e^{-\mathrm{i}2\pi(ux+vy)}I(x,y)\mathrm{d}x\mathrm{d}y \tag{2-7}$$

$\mathfrak{D}_{\mathcal{F}}(u,v)$ is a continuous function having the same cardinality as $I(x,y)$, and for real applications, this needs to be reduced whilst retaining as much information as possible. Generally the low frequency and higher energy components are sufficient to describe the shape. Thus, for instance, elliptical descriptors (such as *centroid, orientation, eccentricity* and *spread*) based on the FD spectrum are feasible to indicate the distribution of the frequency energy. Also, the FD spectrum can be divided into non-overlapping sub-bands. The average energy from each sub-band then becomes an element of the feature vector. The latter extraction method is easily implemented when applying the *discrete Fourier transform* (DFT),

$$\mathfrak{D}_{\mathcal{F}}(u,v)=\frac{1}{KL}\sum_{k=0}^{K-1}\sum_{\ell=0}^{L-1}e^{-\mathrm{i}2\pi\left(\frac{uk}{K}+\frac{v\ell}{L}\right)}I(k,\ell) \tag{2-8}$$

Reconstruction is straight forward by applying the inverse Fourier transform. Good approximation may be obtained by retaining a sufficient number of higher energy terms.

2.3 Wavelet Descriptor

The idea of wavelet transformation is to represent an image in terms of the superposition of wavelet with different scale levels and positions. The wavelet, having better time-frequency resolution than *short-time Fourier transform* (STFT) [19], can be expressed as

$$\psi_{b_x,b_y,a}=\frac{1}{a}\psi\left(\frac{x-b_x}{a},\frac{y-b_y}{a}\right) \tag{2-9}$$

where $a\in\mathbb{R}^+$ is the dilating scale parameter, $(b_x,b_y)\in\mathbb{R}^2$ are the translation parameters and $\psi_{b_x,b_y,a}$ is the translated and dilated version of the mother wavelet $\psi(x,y)$. The normalisation factor $1/a$ is included so that $\left\|\psi_{b_x,b_y,a}\right\|=\|\psi\|$. Depending on the applications, these parameters can be chosen as either continuous or discrete values. The definition of CWT can be expressed as an inner product of the wavelet and the image,

$$\mathfrak{D}_W(a, b_x, b_y) = \langle \psi_{b_x,b_y,a}, I(x,y) \rangle = \int_{-\infty}^{+\infty}\int_{-\infty}^{+\infty} \psi^*_{b_x,b_y,a}(x,y) I(x,y) \mathrm{d}x\mathrm{d}y \tag{2-10}$$

The mother wavelet must satisfy

$$\int_{-\infty}^{+\infty}\int_{-\infty}^{+\infty} \psi(x,y)\mathrm{d}x\mathrm{d}y = 0 \tag{2-11}$$

so that the wavelet is oscillatory with a null DC component.

Discretization of the wavelet parameters is adopted in many applications [19]. We consider the wavelet decomposition of a two-dimensional mode shape. The *two-dimensional discrete wavelet transform* (DWT2) for an image can be obtained by implementing the one-dimensional algorithm horizontally and then vertically. The outputs from each step of decomposition are the sub-images of one approximation at coarser resolution and three sub-images of detail in horizontal, vertical and diagonal directions as illustrated in Fig 1. Thus, the comparison between images can now be carried out between the sub-images at different resolutions. In additional, these coefficients can be used as the WDs and the average energy of each sub-image may be used to form the SFV. A more detailed description of the method is given by the present authors [16].

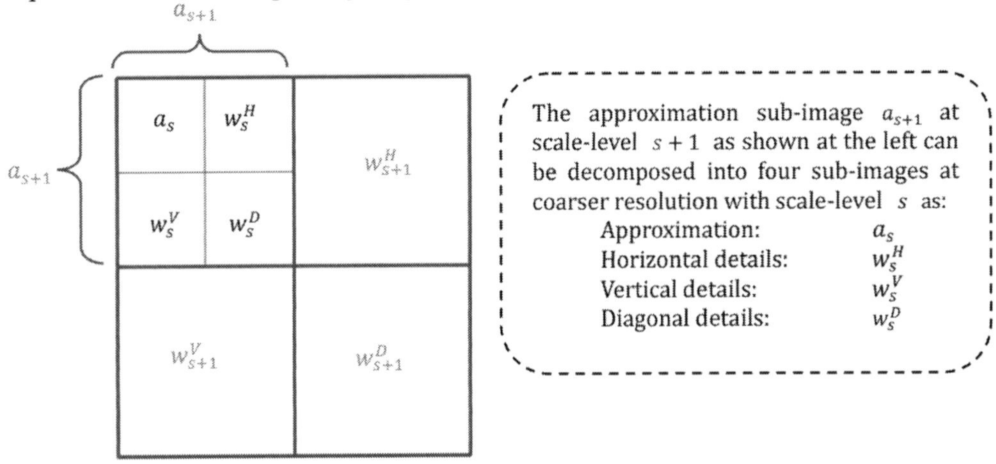

***Fig 1**: Decomposition of Sub-Images between Two Scales*

3 SD Sensitivity and Model Updating

The deterministic model updating process by the iterative sensitivity method can be written as [20],

$$\theta_{j+1} = \theta_j + \left[\boldsymbol{S}_j^T \boldsymbol{S}_j\right]^{-1} \boldsymbol{S}_j^T (z_m - z_j) \tag{3-1}$$

where θ_j and θ_{j+1} are the structural parameters of the current and next iterative steps; $\boldsymbol{S}_j$ denotes the sensitivity matrix at the current iteration; z_j is the current output and z_m is the measured output. Here, the output is considered to be the SD rather than the eigenvalues or eigenvectors in the conventional study. Thus, the sensitivity matrix, known as the SD sensitivity matrix, can be calculated by taking the derivatives of SD with respect to the structural parameters. That is,

$$S_{\mathfrak{D}} = \frac{\partial \mathfrak{D}}{\partial \theta} = \frac{\partial\left[\int_\Omega \mathfrak{K}(x,y) I(x,y,\theta)\mathrm{d}x\mathrm{d}y\right]}{\partial \theta} \tag{3-2}$$

It is noted that the continuous mode shape function can be determined by a linear combination of a set of pre-described element shape-functions $\boldsymbol{\pi}(x,y)$ with the eigenvector $\boldsymbol{\phi}(\theta)$ as [10]

$$I(x,y,\theta) = \boldsymbol{\pi}^T(x,y)\boldsymbol{\phi}(\theta) \tag{3-3}$$

Substituting (3-3) into (3-2) and applying the Leibniz integral rule yields

$$S_{\mathfrak{D}} = \left[\int_\Omega \mathfrak{K}(x,y)\boldsymbol{\pi}^T(x,y)\mathrm{d}x\mathrm{d}y\right] \frac{\boldsymbol{\phi}(\theta)}{\partial \theta} \tag{3-4}$$

Journal of Physics: Conference Series **181** (2009) 012004 doi:10.1088/1742-6596/181/1/012004

which indicates that the sensitivity of SD may be calculated analytically by multiplying together the SD of the element shape-functions and the sensitivity of eigenvector.

4 Examples

4.1 ZMD of a Circular Plate

The free vibration of a circular plate can be modelled by finite elements. Since the circular plate is a perfectly axisymmetric structure double modes are obtained. The conventional mode-shape comparison method, MAC, shows nothing about the double modes. Thus, the ZMD is applied to the first 20 modes. As illustrated in Fig 2, only a small number of the lower order ZMDs are needed to represent all the modes. Correlation of mode shapes based on the ZMD amplitudes is shown in Fig 3 where the double modes can be clearly recognised.

Furthermore, the rotational difference between any pair of double modes can be determined by a property of the ZMD [10]. The first pair of double modes 1 and 2 is shown in Fig 4. As seen from Fig 2 their ZMDs are mainly dominated by $\mathfrak{D}_{Z_{2,2}}$. The angle $\alpha = \left(\arg\left(\mathfrak{D}_{Z^{(2)}_{2,2}}\right) - \arg\left(\mathfrak{D}_{Z^{(1)}_{2,2}}\right)\right)/2 = \frac{90.1^o}{2} = 45.1^o$ matches the theoretical angular difference between double modes 1 and 2, shown in Fig 5.

The orthogonality property of the ZMDs means that they are not only excellent shape descriptors but also their sensitivities are powerful in updating finite element models. The same circular plate is now supported by an elastic foundation at the circumference. Two structural parameters, translational stiffness perpendicular to the plate (k_w) and rotational stiffness with respect to the tangent of circumference (k_a), of the elastic foundation are considered to be updated in this simulation. $k_w = 1000$ and $k_a = 2000$ were used to generate the simulated measurements. The initial values of parameters were $k_w = 700$ and $k_a = 1500$ in the first updating test and $k_w = 1500$ and $k_a = 2500$ in the second test. The ZMD sensitivities were computed using predefined values of the parameters values. The ZMDs of the mode shapes were calculated from orders 0 to 8. The updating parameters converged to the *target* in 4 and 5 iterations respectively as shown in Table 1.

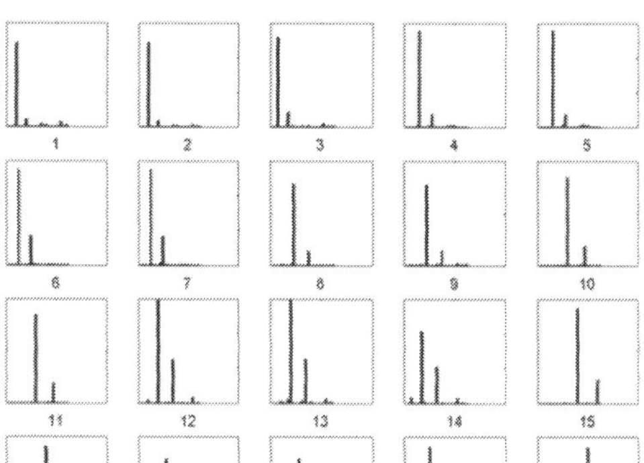

Fig 2: Amplitudes of the ZMDs

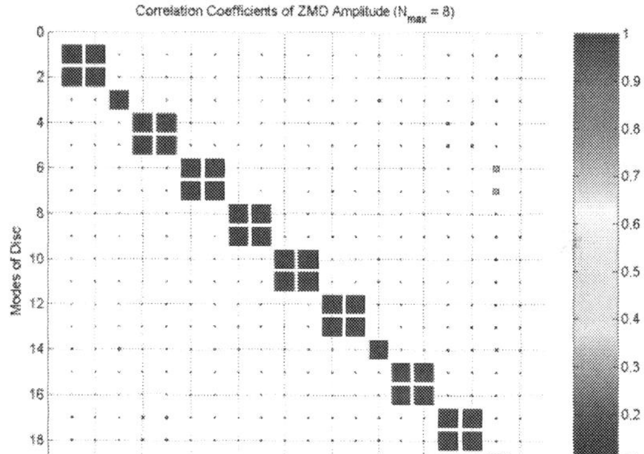

Fig 3: Correlation of the ZMD Amplitude

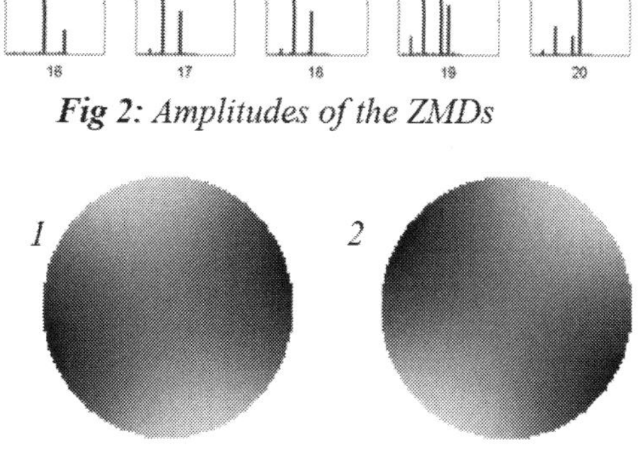

Fig 4: Mode-Shape Patterns of Mode 1 & 2

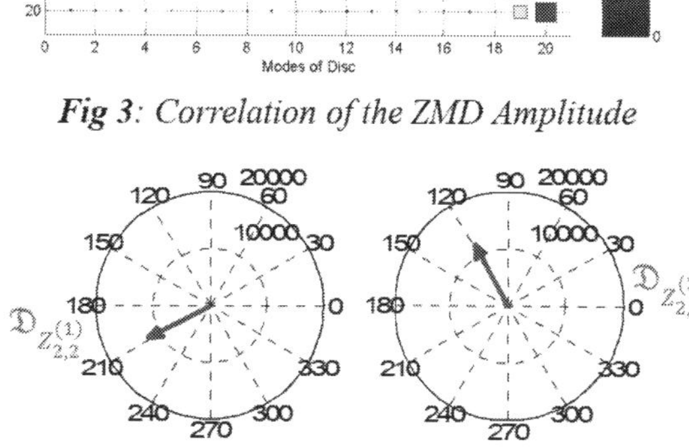

Fig 5: The $\mathfrak{D}_{Z_{2,2}}$ of Modes 1 & 2

***Table 1**: Convergence of the Updating Parameters*

	Steps	0	1	2	3	4	5	Target
Test 1	k_w	700.00	912.60	993.16	1000.02	1000.00		1000.00
	k_a	1500.00	2426.30	2031.40	1999.90	2000.00		2000.00
	Steps	**0**	**1**	**2**	**3**	**4**	**5**	**Target**
Test 2	k_w	1500.00	753.78	939.57	996.40	999.99	1000.00	1000.00
	k_a	2500.00	3119.83	2283.60	2016.91	2000.04	2000.00	2000.00

4.2 FD of a Rectangular Plate

Free vibration of a $160 \times 120 \times 2mm$ rectangular aluminium thin plate is modelled. The FDs of the mode shapes are determined by the DFT. Fig 6 shows the amplitudes of the FDs for the first 12 modes in the first quadrant. It is noted that only a few lower frequency components are significant. The elliptical descriptors of the spectrum are also presented in the figures, which describe the spectrum pattern globally. A hierarchical clustering algorithm is applied to the elliptical descriptor. Several clusters may be obtained by 'cutting' the dendrogram shown in Fig 7 at a chosen distance separating the various groups of modes. The most compact cluster ***C1*** ((1,7),10) indicates the similarity of the three spectrum patterns as shown in Fig 6. Cluster ***C2*** ((5,12),6) groups together those modes with a strongly dominant horizontal major axis of the ellipse. Also, the three compact ellipses having horizontal major axes are grouped by cluster ***C3*** (2,(8,11)). Furthermore, the similarities between FD patterns indicated by these clusters can be referred back to the similarities between mode shapes. The nodal lines of the 12 modes are shown in Fig 8. According to the closeness of mode 1, 7 and 10 (indicated by cluster ***C1***) it can be seen from their nodal lines that mode 7 has two more horizontal oscillations than mode 1 whilst mode 10 has two more vertical oscillations than mode 1. For cluster ***C2***, mode 12 has two more oscillations than mode 6 horizontally and mode 5 vertically. Similar relations between mode 2, 8 and 11 are indicated for cluster ***C3***.

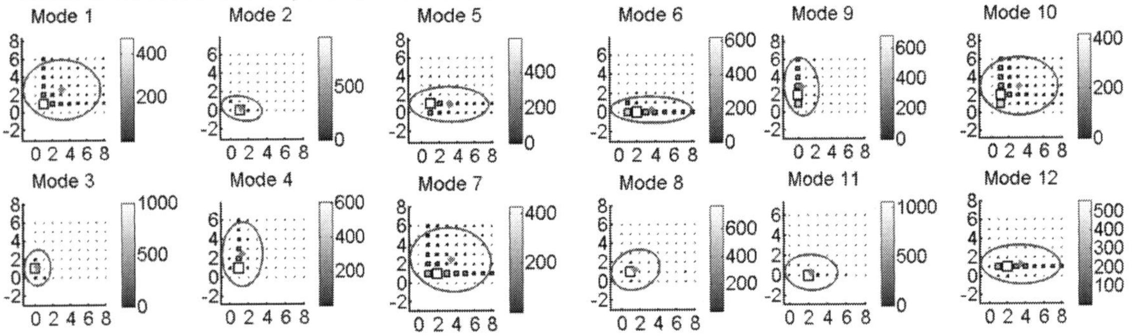

***Fig 6**: Non-Negative FD Amplitudes and Elliptical Descriptors of Mode 1-12*

4.3 WD of Rectangular Plate with a Thin Region

A rectangular plate having two regions with different thicknesses was modelled as shown in Fig 9. Four levels of decomposition by the *rbio3.1* wavelet were carried out on the mode shapes of the plate. Thirteen sub-images were obtained for every single mode-shape: 4 horizontal, 4 vertical, 4 diagonal detailed sub-images and 1 approximation sub-image. The average energy of every sub-image was calculated to determine its significance. Fig 10 shows an example of the energy percentage of each sub-image of mode 12. It is noted that the energies are mainly distributed in the higher levels of sub-images. Thus, the original mode shape pattern as shown in Fig 11 can be represented by the wavelet decomposition at level 4. The global shape information can be described by the approximation sub-image and the local shape in the thin region can be detected by the wavelet details which are shown in Fig 12.

Shape feature vector derived by the WD can be constructed by assembling the energy of each sub-

Journal of Physics: Conference Series **181** (2009) 012004 doi:10.1088/1742-6596/181/1/012004

image at different decomposition levels. Global comparison between similar structures can be implemented by the correlations of the approximations. Further details of detecting the thin region and extracting the global pattern by the WDs are described by the present authors [16].

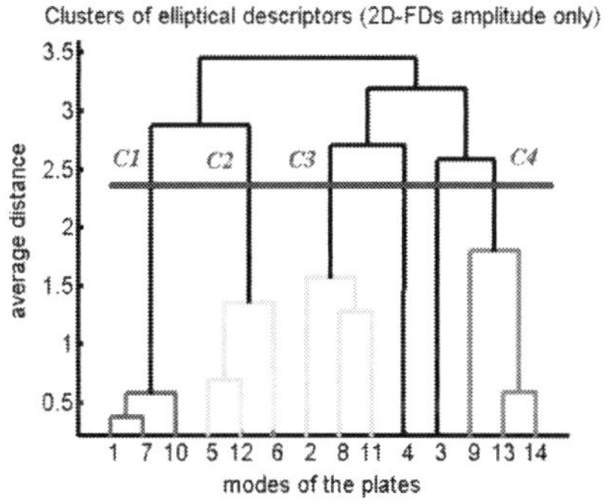

Fig 7: Dendrogram by Elliptical Descriptor

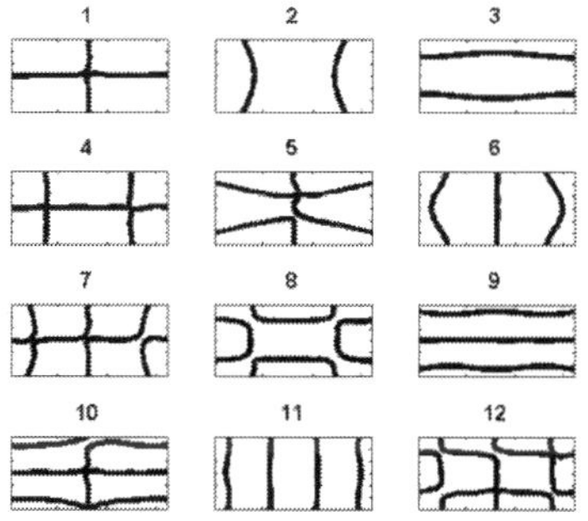

Fig 8: Nodal Lines of the Rectangular Plate

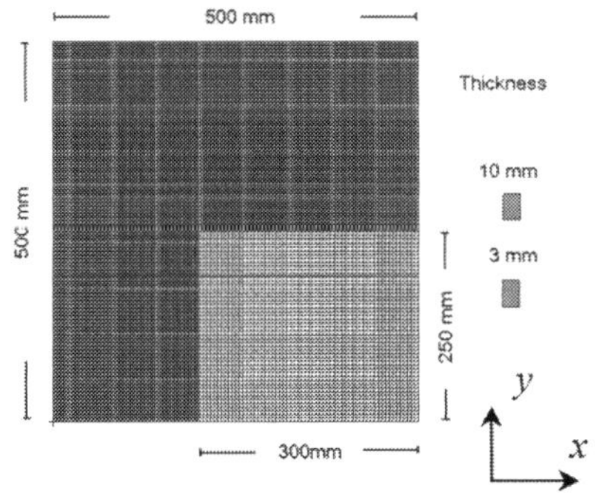

Fig 9: FE model of Plate pC

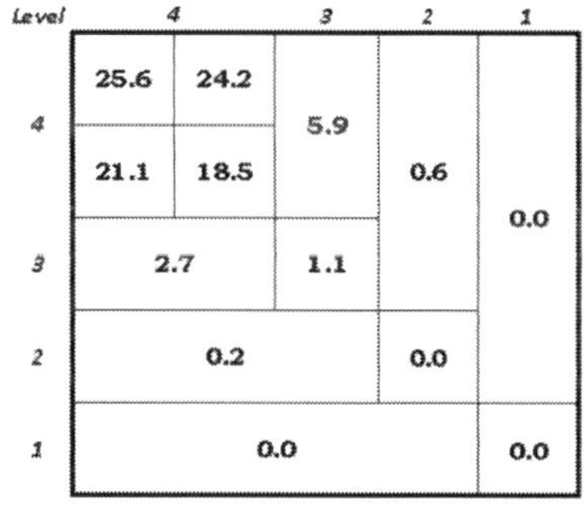

Fig 10: Percentages of Sub-Image Average Energies – Mode 12

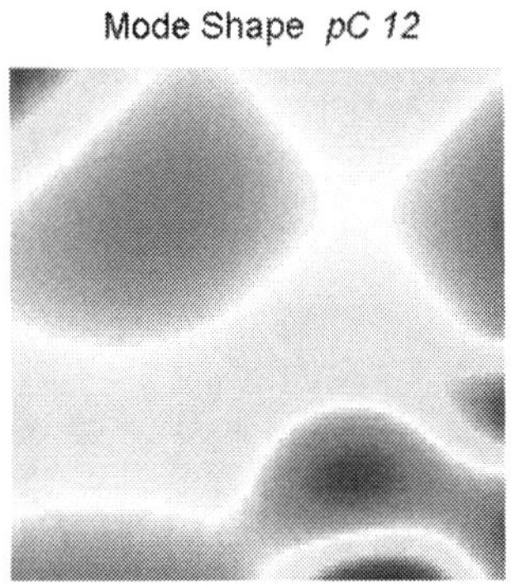

Fig 11: Mode Shape Pattern

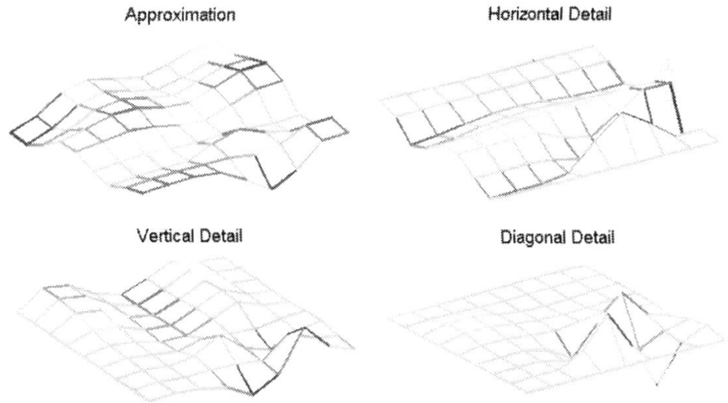

Fig 12: Decomposition Level 4

5 Conclusion

A variety of shape descriptors which demonstrate the capability of recognising subtle difference in mode shapes are presented. The SDs show the desirable properties of computational efficiency and ease of image reconstruction using a small number of SD terms. Specifically, the ZMD is powerful in discriminating circular and spherical images; the FD is more general and very effective at extracting mode-shape features by virtue of its sinusoidal kernel; the WD shows the capability of distinguishing between local and global features.

Acknowledgement

The first author wishes to acknowledge the support of an Overseas Research Studentship (ORS), a University of Liverpool Studentship, an award from the University of Liverpool Graduate Association (Hong Kong) and EC project ADVISE (Advanced Dynamic Validations using Integrated Simulation and Experimentation) Grant 218595.

References

[1] A. B. Stanbridge, M. Martarelli, and D. J. Ewins, "Measuring area vibration mode shapes with a continuous-scan LDV," *Measurement,* vol. 35, pp. 181-189, 2004.
[2] A. K. Jain, R. P. W. Duin, and J. Mao, "Statistical Pattern Recognition: A Review," *IEEE Transactions on Pattern Analysis and Machine Intelligence,* vol. 22, pp. 4-37, 2000.
[3] L. d. F. Costa and J. R. M. Cesar, *Shape Analysis and Classification Theory and Practice*: CRC Press, 2001.
[4] R. J. Allemang and D. L. Brown, "A correlation coefficient for modal vector analysis," *Proceeding of 1st International Modal Analysis Conference,* vol. 1, pp. 110-116, 1982.
[5] D. Zhang and G. Lu, "Review of shape representation and description techniques," *Pattern Recognition,* vol. 37, p. 19, 2004.
[6] M. R. Teague, "Image analysis via the general theory of moments," *Optical Society of America,* vol. 70, pp. 920-930, 1980.
[7] C. H. Teh and R. T. Chin, "On image analysis by the methods of moments," *IEEE Transactions on Pattern Analysis and Machine Intelligence,* vol. 10, pp. 496-513, 1988.
[8] R. J. Prokop and A. P. Reeves, "A survey of moment-based techniques for unoccluded object representation and recognition," *Graphical Models Image Processing,* vol. 54, pp. 438-460, 1992.
[9] A. Khotanzad and Y. H. Hong, "Invariant image recognition by zernike moments," *IEEE transaction of Pattern Analysis and Machine Intelligence,* vol. 12, pp. 489-498, 1990.
[10] W. Wang, J. E. Mottershead, and C. Mares, "Mode-shape recognition and finite element model updating using the Zernike moment descriptor," *Mechanical Systems and Signal Processing,* Submitted.
[11] R. L. Cosgriff, "Identification of Shape," Report No 820-11 of the Ohio State University Research Foundation1960.
[12] C. T. Zahn and R. Z. Roskies, "Fourier Descriptors for Plane Closed Curves," *IEEE Transactions on Computers,* vol. C21, pp. 269-281, 1972.
[13] E. Persoon and K.-S. Fu, "Shape discrimination Using Fourier Descriptors," *IEEE Transactions on Systems Man and Cybernetics,* vol. SMC-7, pp. 170-179, 1977.
[14] G. Eichmann, "Shape representation by Gabor expansion," *Hybrid Image and Signal Processing II, SPIE,* vol. 1279, pp. 86-94, 1990.
[15] S. G. Mallat, "A Theory for Multiresolution Signal Decomposition: The Wavelet Represetation," *IEEE Transactions on Pattern Analysis and Machine Intelligence,* vol. 11, pp. 674-693, July 1989.
[16] W. Wang, J. E. Mottershead, and C. Mares, "Vibration mode shape recognition using image processing," *Journal of Sound and Vibration,* submitted.
[17] F. Zernike, "Translated: diffraction theory of the cut procedure and its improved form, the phase contrast method," *Physica,* vol. 1, 1934.
[18] P. Hew, "Orthogonal functions on the unit disk having invariance in form," The University of Western Australia1996.
[19] I. Daubechies, *Ten Lectures on Wavelets*: SIAM: Society for Industrial and Applied Mathematics, 1992.
[20] M. I. Friswell and J. E. Mottershead, *Finite Element Model Updating in Structural Dynamics*: Kluwer Academic Publishers, 1995.

Journal of Physics: Conference Series **181** (2009) 012005 doi:10.1088/1742-6596/181/1/012005

Predicting shock response in uncertain structures using the Hybrid method

L H Humphry[1,a] **and R S Langley**[1,b]

[1]Department of Engineering, University of Cambridge, CB2 1PZ, U.K

E-mail: [a]lhh26@cam.ac.uk, [b]rsl21@cam.ac.uk

Abstract. Using the Hybrid method (FE + SEA) it is possible to estimate the frequency response of an uncertain structure. The current work develops the Hybrid method to allow for time domain analysis of the shock response of a structure. Problems to be overcome when taking Hybrid method results into the time domain are a) the Hybrid method frequency response has no phase information, and b) the Hybrid method frequency response is smoothed in frequency and shows no modal peaks. In this paper the first problem has been overcome, using minimum phase reconstruction. Explanation of minimum phase reconstruction and its limitations are described, and application to shock problems described.

1. Introduction

Structural vibration problems are conventionally divided into two sets: low frequency problems, where the wavelengths of vibration are long compared with the structural dimensions; and high frequency problems, where the wavelengths are short. For these problems there are well developed tools for simulating and predicting vibrations: Finite Element Analysis (FEA) [1] for low frequencies, and Statistical Energy Analysis (SEA) [2] for high frequencies.

Another class of problem exists, when a structure contains different elements that show both low and high frequency behaviour simultaneously. This is a 'mid-frequency' problem, and can be tackled using the Hybrid method [3]. The Hybrid method couples together both FEA and SEA modelled subsystems into one structural vibration model. The Hybrid method gives a frequency response result which is averaged over an ensemble of statistically similar structures, and so includes any uncertainty about the nature of a structure.

To predict the shock response of a structure a time domain result is typically sought. The nature of shock is to excite a structure at a large range of frequencies (both low and high) and so for structures with significant resonances extending over this range neither FEA nor SEA is suitable. The Hybrid method is a more suitable tool for such a problem.

Hybrid method results are frequency domain based, and must be transformed into the time domain for shock analysis. Two problems are encountered when transforming to the time domain, a) Hybrid method results lack phase information, and b) ensemble averaging has the effect of smoothing the frequency response in frequency and so removing modal peaks in the spectrum. Both of these frequency domain features have a huge effect on the time domain response.

The work contained in this paper seeks to extend the functionality of the Hybrid method to predict time domain shock responses. The first problem of phase retrieval is tackled using minimum phase reconstruction. The meaning of minimum phase is explained in Section 2, then a discussion of how minimum phase reconstruction can be carried out in practise is given in Section 3. Initial experiences of using minimum phase reconstruction for the shock response of uncertain structures are given in Section 4, and concluding remarks given in Section 5.

2. Explanation of minimum phase

For a system $F(u)$ the phase of the system is defined as $\theta(u)$, where $F(u) = |F(u)|e^{i\theta(u)}$. In the case of Hybrid method produced frequency responses, the magnitude $|F(\omega)|$ is known but the phase is not. Phase information is needed for accurate time domain realisation. In some cases phase retrieval is possible, but this is usually ambiguous. It has been shown that, except for certain special cases, there exists an infinity of functions $f_i(t)$ whose Fourier transforms $F_i(\omega)$ have the same modulus $|F(\omega)|$ on the real frequency axis [4].

2.1. Defining minimum phase

One special case where a unique relationship between the magnitude and phase of a system exists is when is a system is known to be *minimum phase.* A system is defined as minimum phase if the system and its inverse are causal and stable. In terms of the poles and zeros, for a frequency response this means that all poles *and* zeros must lie in the positive imaginary half of the complex frequency plane. A minimum phase system can be also be thought of as the 'fastest decaying' system [5]. Formally, for a causal signal $f_i(n)$ that has a given magnitude spectrum $|F(\omega)|$, the minimum phase signal $f_{mp}(n)$ is the one that maximises initial energy:

$$\sum_{n=0}^{K} |f_{mp}(n)|^2 \geq \sum_{n=0}^{K} |f_i(n)|^2, \qquad \text{for} \quad K = 0, 1, 2, \cdots, N-1. \tag{1}$$

Minimum phase systems occur frequently when studying structural vibration frequency responses. The driving point response of any physical system must be minimum phase, with one minimum phase zero between every pole. To see this, the driving point response is described by mobility — the velocity at a point resulting from a prescribed force input at the same point. This is causal and stable for all real systems. The inverse of the driving point mobility is the driving point impedance — the force that results from applying a prescribed velocity input. This is also causal and stable for all real systems. Thus, the drive point response function and its inverse are causal and stable and so both must be minimum phase functions. For non-drive point responses, the inverse function is not a simple physical function like drive point impedance, and so it cannot be assumed that the inverse is stable. Thus, the non-drive point response cannot be assumed to be minimum phase.

2.2. Phase retrieval using Hilbert Transform

As shown in Section 2, for *minimum phase* systems, there is a unique mapping between the magnitude and phase of a system's frequency response [6]. This uses a function known as the Hilbert Transform, $\mathcal{H}\{f(x)\}$, defined as

$$\mathcal{H}\{f(x)\} = \frac{1}{\pi} \int_{-\infty}^{\infty} \frac{f(\tau)}{x - \tau} dx. \tag{2}$$

The Hilbert Transform relates the transfer function phase and magnitude of a minimum phase system uniquely:

$$\arg[F(\omega)] = -\mathcal{H}\{\log(|F(\omega)|)\}, \tag{3}$$

or inversely:

$$\log(|F(\omega)|) = \log(|F(\infty)|) + \mathcal{H}\{\arg[F(\omega)]\}. \tag{4}$$

3. Using minimum phase reconstruction

For a system with magnitude frequency response $|F(\omega)|$ the phase cannot be retrieved uniquely. By assuming that the system is minimum phase the phase can be found using the Hilbert transform, this is called minimum phase reconstruction (MPR). This section details artefacts encountered when implementing an MPR algorithm, and the errors that result in the time domain when taking a minimum phase assumption for a non-minimum phase system.

3.1. Approximating the Hilbert Transform

This unique mapping given by the Hilbert transform, from phase to magnitude in minimum phase systems, is for continuous functions only. Such a function is not realisable on a digital computer, and it is to be expected that artefacts will occur as a result. For discrete systems there exists the discrete Hilbert transform (DHT) [7]. For a signal $f(k)$ with discrete Fourier transform $F(k)$, the DHT $g(k)$ is given by the inverse discrete Fourier transform of $G(k)$, where

$$G(k) = H(k) \cdot F(k), \tag{5}$$

and

$$H(k) = \begin{cases} -j, & k = 1, 2, \cdots, \frac{N}{2} - 1 \\ 0, & k = 0, \frac{N}{2} \\ +j, & k = \frac{N}{2} + 1, \frac{N}{2} + 2, \cdots, N - 1. \end{cases} \tag{6}$$

For large N the DHT approximates the continuous time Hilbert transform well. Errors in the DHT approximation of the Hilbert transform occur due to the using the DFT approximation of the Fourier transform. By using an FFT length that is long enough to give a good approximation of the Fourier transform, errors in the DHT can be reduced.

An MPR algorithm has been written in MatLab that reconstructs the minimum phase realisation of a magnitude spectrum $|F(\omega)|$, using Eq. (3) and the DHT in the frequency domain. To test the accuracy of the MPR algorithm, the MPR was validated using the magnitude spectrum of a driving point impulse response — a minimum phase signal. The spectrum $H_D(\omega)$ for the driving point response of a random system was calculated using a modal summation,

$$H_D(\omega) = \sum_{n=1}^{N} \frac{\phi_n^2}{\omega_n^2 + j\eta\omega_n\omega - \omega^2}, \tag{7}$$

where modal amplitudes ϕ_n had a Gaussian distribution, resonant frequencies ω_n had a Rayleigh distribution, and damping η was assumed constant. The DHT algorithm was validated by reconstructing the phase of $H_D(\omega)$ to give $\hat{H}_D(\omega)$.

Figure 1 shows the bode plot for a driving point response and the MPR using the DHT. It can be seen that the MPR does not reproduce the phase of original signal exactly. The magnitudes of $H_D(\omega)$ and $\hat{H}_D(\omega)$ are equal, but the phases are not quite. The MPR, $\hat{H}_D(\omega)$, shows an approximately linear phase term that increases the phase by 2π over the frequency range calculated. This error in calculated phase is due to the FFT approximation of a Fourier transform used in the calculation of the DHT. Though the phase error at high frequencies (i.e. above the Nyquist frequency) look to be large, these are all actually $2\pi - \theta_e$, where θ_e is the small phase error at the symmetric point below the Nyquist frequency.

Phase errors are worst at the Nyquist frequency, there being a π rads difference between the phase of $H_D(\omega)$ and $\hat{H}_D(\omega)$ here. As long as the magnitude of the response is low here, these errors around the Nyquist frequency do not give a significant error in the time domain. The corresponding time domain signals $h_D(t)$ and $\hat{h}_D(t)$ show only a very slight error, and only in low

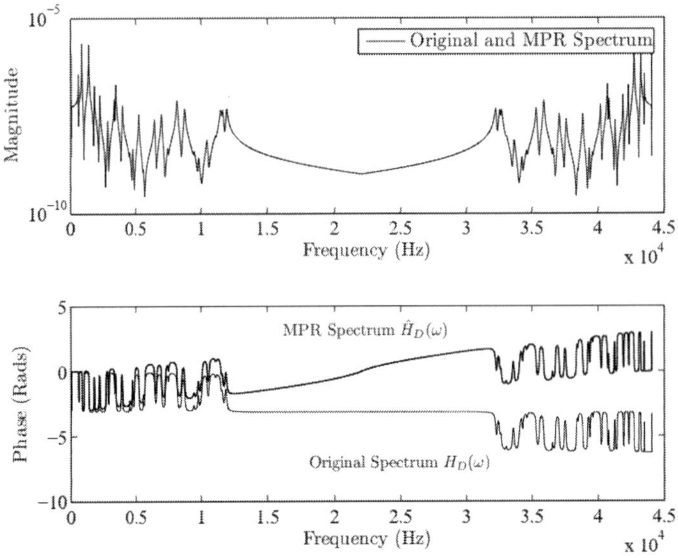

Figure 1: Bode Plot of Driving Point Response Spectrum

magnitude high frequency components which decay very rapidly. Quantitatively, the normalised L2-norm error of $h_D(t)$ and $\hat{h}_D(t)$ is given by

$$\epsilon_{L_2} = \frac{\sum_n (h_D(nT) - \hat{h}_D(nT))^2}{\sqrt{(\sum_n h_D(nT)^2)(\sum_n \hat{h}_D(nT)^2)}}, \qquad n = 0, 1, \ldots, N-1, \tag{8}$$

and has a very low value of $\epsilon_{L_2} = 2.77 \times 10^{-4}$. This error can be reduced still further by increasing the frequency window of the spectrum.

Errors can also occur for MPR when a system of low damping is used. The main effect of these errors is an increase in the decay rate of the MPR signal $\hat{h}_D(t)$. These errors are not found for spectra produced from systems with higher damping. It has been found that this error comes from the discrete approximation of $H_D(\omega)$ used in computer calculations. By increasing the frequency resolution of the discrete approximation of $H_D(\omega)$ the error seen in $\hat{h}_D(t)$ is reduced. A rule of thumb has been found for satisfactory reproduction of $\hat{h}_D(t)$: by ensuring that there are 5 frequency data points covering the half power bandwidth of every resonance of $H_D(\omega)$, no noticeable error in the time domain is observed.

Though the Hilbert transform provides a unique mapping between phase and magnitude for minimum phase signals, for work on a computer the DHT must be used, and this has been shown to produce errors. The solution has been to increase the frequency range and resolution of the spectrum being analysed. Put simply, the only way to reduce the errors that result from using the discrete Hilbert transform is to increase the size and resolution of the sampled space to better approximate the continuous space. This is a limitation of the algorithm, and cannot be avoided.

It should be noted that the use of the continuous variable ω to describe spectra here is not strictly correct for discrete spectra modelled using computers. However, effort has been made to make data lengths long enough to reduce the artefacts that occur from approximating a continuous function by a finite sampled function. By reducing these artefacts, the discrete and continuous functions become equivalent, and so the term ω (and t for time) is used where formally a discrete variable would be more appropriate.

3.2. Errors using MPR for non-minimum phase systems

Though the driving point response of a system is minimum phase, non-driving point responses typically are not. As explained in Section 2 the phase of a non-minimum phase response $H_{NMP}(\omega)$ cannot be determined uniquely from its magnitude. By assuming that the system *is* minimum phase, the MPR algorithm can be used to (incorrectly) reconstruct of the phase of $H_{NMP}(\omega)$.

Fig. 2 shows the sonogram of an impulse response of a non-minimum phase system. The system analysed is an idealised model of a periodic drill sting, and the response shown is the end to end impulse response. Fig. 2(a) shows the dispersive nature of the system, and shows echoes occurring after the initial impulse.

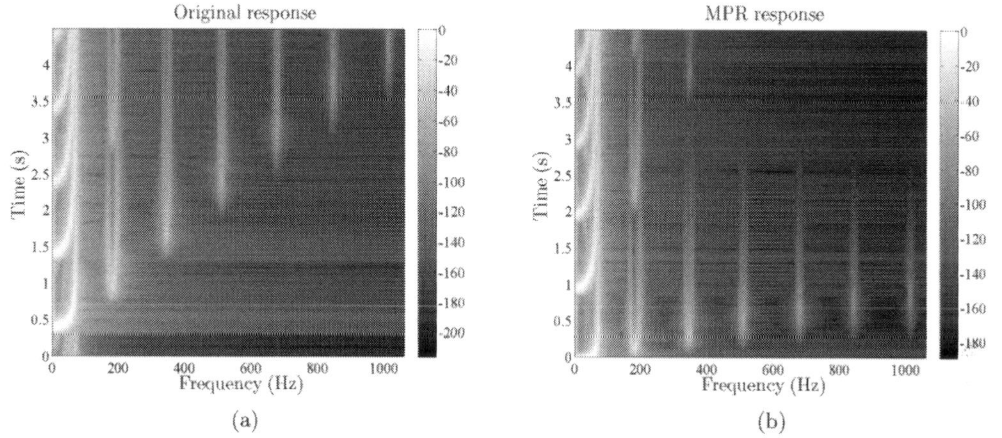

Figure 2: Sonogram of end to end impulse response: (a) original and (b) MPR

The MPR shown in Fig. 2(b) shows main three features that are of interest. Firstly, any initial delay in the impulse response is removed in the MPR. For dispersive systems it is found that different frequency components experience a different amount of time delay removal, such that all frequency components have a large response at $t = 0$. The removal of delay terms means that waves are changed in phase with one another and so response pulse shapes are changed in the time domain. Secondly, the decay of the response is not noticeably affected by MPR. And thirdly, echoes are preserved by MPR.

The lack of initial delay in MPR responses can be explained by remembering the 'fastest decay' definition of minimum phase systems given in Section 3. Clearly, any initial delay must be removed if Eq. (1) is to be satisfied.

The features of MPR seen in Fig. 2 have been observed for a range of non-minimum phase systems, including model strings, beams, acoustics ducts, and random spectra.

4. Application: peak shock prediction in uncertain structures

An uncertain structure was simulated by using a random spectrum. The MPR was found of the spectrum, and a half cosine pulse applied to both the original and MPR spectra. The time domain shock responses were obtained using the IFFT, and the magnitude of the peak response found. In this section a comparison of results given by the original non-minimum phase system and its MPR are presented.

4.1. Modelling an uncertain structure

The inertance of the structure is of interest for shock analysis, and the inertance transfer function was modelled using the modal summation

$$H_D(\omega) = \frac{a(x_2)}{F(x_1)} = \sum_{n=1}^{N} \frac{\omega^2 \phi_n(x_1)\phi_n(x_2)}{\omega_n^2 + j\eta\omega_n\omega - \omega^2}, \tag{9}$$

where x_1 defines the position of the input force, x_2 defines the listening position for the acceleration output, and loss factor η is assumed constant. Modal amplitudes ϕ_n are randomly assigned with Gaussian distribution, and resonant frequencies ω_n have a Rayleigh distribution.

The transfer function described in Eq. (9) does not decay at high frequencies, but instead has a constant trend line. For a true impulse response an infinite frequency range would be needed. To avoid this, the response for a non-idealised impulse was used.

The impulse was assumed to be a half cosine pulse, centred on $t = 0$ and with area = 1. By widening an idealised impulse in this way the high frequency components of the shock are lost. Also, this is a much more realistic model of a shock input that may occur in practise. By taking the Fourier transform of the cosine pulse the decay at high frequencies is seen:

$$H_{shock}(\omega) = \frac{\pi^2}{b^2} \frac{cos\left(\frac{\omega b}{2}\right)}{\left(\frac{\pi^2}{b^2} - \omega^2\right)} = \mathcal{F}\left\{\frac{\pi}{2b}\cos\left(\frac{\pi}{b}t\right)\right\}, \tag{10}$$

where b defines the pulse width. The power spectrum H_{shock} has a 'sinc' type structure, with rapidly decaying lobes that decrease with frequency. After the second side lobe the amplitude of the spectrum is at most 2% of the DC amplitude, and so it is assumed that after this point the effect of higher frequencies is negligible to the overall response. Thus, the rectangular windowing that must be used to create a finite length spectrum will not cause large artefacts.

Multiplying the spectra of the cosine pulse and the random structure gives the shock response of the structure. Figure 3 shows the shock response for the original system, its MPR, and for comparison the zero-phase system with the same magnitude as the original. It can be seen that the MPR response is much closer to the original than the zero-phase system. The MPR response over-predicts the value of the peak shock response, and this is expected due to the 'fastest decay' property of minimum phase systems.

4.2. Ensemble average results

Due to the random nature of the structure modelled above, the exact nature of the shock response varies depending on the random spectrum used. Using an ensemble of statistically similar structures, the peak shock response was found for original, MPR, and zero-phase spectra.

Table 1 shows the results found from an ensemble of 100 random spectra. The sampling frequency was set to 44100Hz, and an impulse duration of 160μs. Frequency resolution was varied in accordance with the rule of thumb, at least five data points covering every spectral peak. There were 100 modes in the frequency range 0–22050Hz (ie the Nyquist frequency), and a loss factor of $\eta = 0.01$ was used. 'Mean Max' refers to the ensemble mean value of the peak, 'Relative Variance Max' to the relative variance of peak response results across the ensemble. These results show that using the MPR gives a prediction of peak shock response that is comparable with the true result, and much closer in value than results predicted using a zero phase spectrum. However, the minimum phase filter over-predicts the response, by 27% in this case.

The over-prediction in the MPR peak shock response value is found for a wide range of ensembles, with different damping and modal density values. This over-prediction is typically within 50% of the original response; an acceptable error when compared with safety factors of

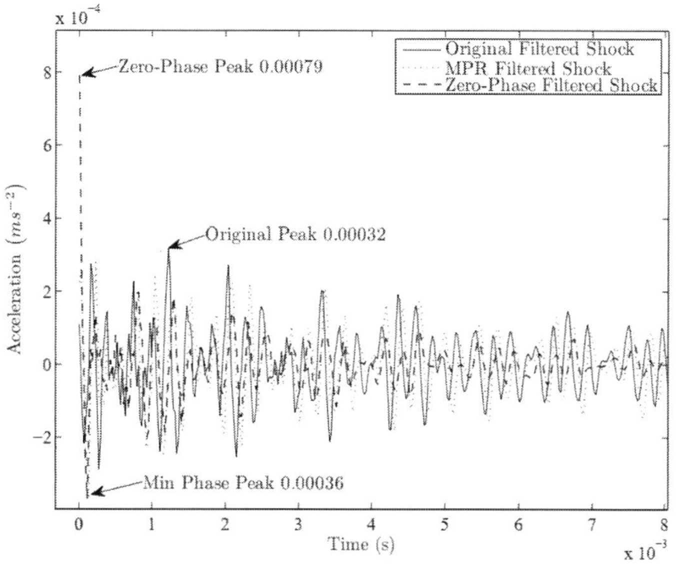

Figure 3: Impulse responses for random structure

Table 1: Peak shock response results for an ensemble of 100 random filters

	Original	Minimum Phase	Zero Phase
Mean Max	3.07E-04	3.90E-04	8.39E-04
Relative Variance Max	0.24	0.25	0.18

300-400% typically used to design against shock failure. Due to being an *over* prediction, the MPR result provides a conservative estimate for the mean peak shock response. The variance of results does not change noticeably for MPR predicted responses, and so can be used to give likelihood estimates for shock responses of uncertain structures.

4.3. Changing the correlation of modal amplitudes

The random spectra used in Sections 4.1 and 4.2 have modal amplitudes $\phi_n(x_1)$ and $\phi_n(x_2)$ that are uncorrelated, produced using the 'randn' command in MatLab. Modal amplitudes are only truly uncorrelated in a structure when the listening point x_2 is in the far field, ie $r = |x_1 - x_2| \simeq \infty$. For random structures the level of correlation between modal amplitudes at different points depends on the distance between these points, r. When $r = 0$ (driving point response), the correlation $R(r) = 1$ for all systems.

For a 2-D system the diffuse field correlation function $R(r)$ is given by:

$$R(r) = J_0(kr) \; , \tag{11}$$

where J_0 is a zeroth order Bessel function of the first kind, and k is the wave number of the system.

To allow shock responses to be calculated at a variety of listen positions, modal amplitudes $\phi_n(x_1)$ and $\phi_n(x_2)$ were chosen such that they were correlated according to Eq. (11). Again

Journal of Physics: Conference Series **181** (2009) 012005 doi:10.1088/1742-6596/181/1/012005

Table 2: Peak shock response results for an ensemble of 100 random filters

	Original	Minimum Phase	Zero Phase
Mean Max	5.13E-04	5.53E-04	9.16E-04
Relative Variance Max	0.34	0.30	0.32

an ensemble of similar spectra was used to find the peak response values for original and MPR shock responses. Results are given in Table 2 for a system where the distance r was low, meaning that the correlation between modal amplitudes was high.

When the distance between input and listening points is low it can be seen that the MPR still over-predicts the mean shock response of the ensemble, but is closer to the true result than for the uncorrelated far field case. When $r = 0$, MPR gives a very accurate prediction of the mean, and this is to be expected as at $r = 0$ a driving point response is seen.

5. Conclusions

The Hybrid method is currently not suited to shock response predictions: frequency domain results produced by the Hybrid method lack phase information and are smoothed in frequency. Using an assumption of minimum phase, the phase can be reconstructed using the Hilbert transform. A minimum phase reconstruction algorithm has been implemented, and it has been found that for accurate MPR the frequency range and resolution of the spectra being studied must both be high. Applying MPR to non-minimum phase systems produces errors in the time domain, but these are limited to a removal of any delay in waveforms. Using MPR to predict the shock response of uncertain (non-minimum phase) structures has proved very acceptable — MPR gives a mean ensemble shock response that over-predicts by less than 50% and provides a conservative estimate. MPR gives much more accurate results than using a zero phase assumption. The problem of Hybrid method frequency smoothing has not been tackled, future work must concentrate on this area to allow Hybrid method models to be used for time domain based shock prediction.

Acknowledgments

Financial support from the EPSRC gratefully acknowledged.

References

[1] Turner M J, Clough R W, Martin H C and Topp L C 1956 *Journal of the Aeronautical Sciences* **23** 805–823
[2] Lyon R H 1975 *Statistical Energy Analysis of Dynamical Systems* (M.I.T. Press)
[3] Langley R and Bremner P 1999 *Journal of the Acoustical Society of America* **105** 1657–1671
[4] Burge R, Fiddy M, Greenway A and Ross G 1976 *Proceedings of the Royal Society of London Series A-Mathematical Physical and Engineering Sciences* **350** 191–212
[5] Smith J O 2007 *Introduction to Digital Filters with Audio Applications* (`http://www.w3k.org/books/`: W3K Publishing) ISBN 978-0-9745607-1-7
[6] Oppenheim A V and Schafer R W 1975 *Digital Signal Processing* (Prentice-Hall)
[7] Cizek V 1970 *IEEE Transactions on Audio and Electroacoustics* **18** 340–343

Journal of Physics: Conference Series **181** (2009) 012006 doi:10.1088/1742-6596/181/1/012006

Damping identification in multiple degree-of-freedom systems using an energy balance approach

Marco Prandina[1], **John E Mottershead**[1] **and Elvio Bonisoli**[2]

[1] Department of Engineering, University of Liverpool, Brownlow Hill, Liverpool L69 3GH, United Kingdom

[2] Department of Production Systems, Politecnico di Torino, Corso Duca degli Abruzzi, 24 – 10129, Torino, Italy

j.e.mottershead@liv.ac.uk

Abstract. The dissipation of energy in a vibrating structure is often neglected or oversimplified in engineering design. However, there are many cases where an accurate model of damping is necessary and an effective identification is required in order to predict instabilities or unwanted vibrations of the system. The paper presents a method which balances the energy input by a sinusoidal force with the energy dissipated by damping mechanisms in order to locate and identify damping. If a complete set of measurements is available, there is no need to know or to evaluate the mass and stiffness matrices and the only data needed are the time histories of the response of the structure excited with a set of single-frequency harmonic excitations. The method has been validated by numerical simulations and tested on a real structure, a clamped aluminium beam with a magnetic eddy current dashpot.

1. Introduction

Damping in multiple degree-of-freedom systems is often modeled as linear viscous damping, i.e. forces dependent on instantaneous velocities only. This is an approximation that may not be very representative of reality, since there are many mechanisms that remove energy from a mechanical system (material damping, friction, gas pumping at interfaces, energy radiation, etc.) and often they are nonlinear. However, if the damping is light the dynamical behavior is principally determined by the relatively large elastic or inertial forces so this approach is usually chosen for its mathematical convenience and can represent a valid approximation in many cases.

The literature on the subject includes several different methods to identify the linear viscous damping matrix in multiple degree-of-freedoms systems. Some of these methods have been numerically compared by Srikantha Phani and Woodhouse [1] to bring out the merits and drawbacks of each method when dealing with some typical problems in damping identification such as modal incompleteness, noise, spatial truncation, modal overlap and computing cost. The results show that one of the drawbacks of many methods is the poor performance when spatial or modal incompleteness of the input occurs in the identification process. The authors' recent review paper [2] compares the philosophy and performances of the main strategies in the identification of linear viscous damping, namely the perturbation method [3, 4], Lancaster's formula [5, 6], the imaginary part of the inverse receptance matrix [7, 8] and the energy method [9, 10]. In [2] is shown that the last two strategies are in principle identical when dealing with linear viscous damping only and they perform better than the others when modal incompleteness is present in the measurements.

For these reasons, the method described in this paper is based on the energy strategy proposed by Liang [10] with some variations that allow the identification of damping without the need to estimate the mass and the stiffness matrices. However, the two matrices could be necessary in the case of spatial incompleteness in order to expand the measurements to all degrees of freedom.

2. Theory

The equations of motion of a n degree-of-freedom system can be written in the matrix form

$$\mathbf{M}\ddot{\mathbf{x}} + \mathbf{K}\mathbf{x} + \mathbf{D} \cdot \mathbf{f}(\mathbf{x}, \dot{\mathbf{x}}, \ddot{\mathbf{x}}) = \mathbf{g}(t) \tag{1}$$

Where $\mathbf{M} \in \Re^{n \times n}$ is the mass matrix, $\mathbf{K} \in \Re^{n \times n}$ is the stiffness matrix, $\mathbf{M} = \mathbf{M}^T$, $\mathbf{K} = \mathbf{K}^T$, $\mathbf{D} \in \Re^{n \times n}$ represents one of the possible damping matrices multiplied by $\mathbf{f}(\mathbf{x}, \dot{\mathbf{x}}, \ddot{\mathbf{x}}) \in \Re^{n \times 1}$, a function of displacements, velocities or accelerations. $\mathbf{g}(t) \in \Re^{n \times 1}$ is the excitation input vector. Premultiplying eq.(1) by $\dot{\mathbf{x}}^T$ and then integrating over time, an energy equation is obtained:

$$\int_t^{t+T_1} \dot{\mathbf{x}}^T \mathbf{M}\ddot{\mathbf{x}} dt + \int_t^{t+T_1} \dot{\mathbf{x}}^T \mathbf{K}\mathbf{x} dt + \int_t^{t+T_1} \dot{\mathbf{x}}^T \mathbf{D} \cdot \mathbf{f}(\mathbf{x}, \dot{\mathbf{x}}, \ddot{\mathbf{x}}) dt = \int_t^{t+T_1} \dot{\mathbf{x}}^T \mathbf{g}(t) dt. \tag{2}$$

In the case where the excitation force $\mathbf{g}(t)$ and the response $\mathbf{x}$ are periodic, the integration of all conservative components of eq.(2) is zero over a full cycle of periodic motion. If T is the period of the response, the sum of kinetic and potential energies over T is zero

$$\int_t^{t+T} \dot{\mathbf{x}}^T \mathbf{M}\ddot{\mathbf{x}} dt + \int_t^{t+T} \dot{\mathbf{x}}^T \mathbf{K}\mathbf{x} dt = 0 \tag{3}$$

so eq.(2) becomes

$$\int_t^{t+T} \dot{\mathbf{x}}^T \mathbf{D} \cdot \mathbf{f}(\mathbf{x}, \dot{\mathbf{x}}, \ddot{\mathbf{x}}) dt = \int_t^{t+T} \dot{\mathbf{x}}^T \mathbf{g}(t) dt. \tag{4}$$

Eq.(4) is the basis of the identification method proposed. It represents the balance between the energy dissipated by the damping mechanisms on the left hand side of the equation and the energy input to the system on the right hand side. The unknowns are represented by the coefficient matrix **D** which can be parameterized in several different ways, depending on how the damping is modeled and the number of unknowns can vary substantially depending on the parameterization chosen. Each test provides one equation and by exciting the structure with a certain number of different forces and measuring the responses, it is possible to obtain multiple versions of eq.(4) so that the number of equations is equal or bigger than the number of unknown parameters in **D**. The identification may then be carried out by solving an overdetermined inverse problem (using, for example, least squares techniques) and calculating the coefficients in **D**.

An important aspect of this method is that many of the most common damping models can be expressed in the form $\mathbf{D} \cdot \mathbf{f}(\mathbf{x}, \dot{\mathbf{x}}, \ddot{\mathbf{x}})$, such as the viscous damping or the Coulomb friction.

3. Diagonal viscous damping matrix

For a better understanding of how the method works, the simplest case of a system with diagonal viscous damping matrix is considered. In this case, the matrix **D** and the function $\mathbf{f}(\mathbf{x}, \dot{\mathbf{x}}, \ddot{\mathbf{x}})$ in eq.(4) can be written in the form:

$$\mathbf{D} = \text{diag}(c_{i,i}) \in \Re^{n \times n}, \mathbf{f}(\mathbf{x}, \dot{\mathbf{x}}, \ddot{\mathbf{x}}) = \dot{\mathbf{x}} \in \Re^{n \times 1} \tag{5}$$

where $\mathbf{D}$ is a constant $n \times n$ matrix, with positive damping coefficients $c_{i,i}$ in each ith degree of freedom of the diagonal and zero elsewhere, and $\mathbf{f}(\mathbf{x}, \dot{\mathbf{x}}, \ddot{\mathbf{x}})$ is the vector of instantaneous velocities $\dot{\mathbf{x}}$ only. Eq.(4) becomes

$$\int_t^{t+T} \dot{\mathbf{x}}^T \mathrm{diag}(c_{i,i}) \dot{\mathbf{x}} \mathrm{d}t = \int_t^{t+T} \dot{\mathbf{x}}^T \mathbf{g}(t) \mathrm{d}t \tag{6}$$

or

$$c_{1,1} \int_t^{t+T} \dot{x}_1^{\,2} \mathrm{d}t + \cdots + c_{i,i} \int_t^{t+T} \dot{x}_i^{\,2} \mathrm{d}t + \cdots + c_{n,n} \int_t^{t+T} \dot{x}_n^{\,2} \mathrm{d}t = E_a \tag{7}$$

By exciting the structure with m different excitations, multiple versions of eq.(7) are obtained and then grouped into a single matrix equation in the form:

$$\mathbf{Ac} = \mathbf{e} \tag{8}$$

where

$$\mathbf{A} = \begin{bmatrix} \int_t^{t+T} \dot{x}_1^{\,2} \mathrm{d}t_{(1)} & \cdots & \int_t^{t+T} \dot{x}_n^{\,2} \mathrm{d}t_{(1)} \\ \cdots & \cdots & \cdots \\ \int_t^{t+T} \dot{x}_1^{\,2} \mathrm{d}t_{(m)} & \cdots & \int_t^{t+T} \dot{x}_n^{\,2} \mathrm{d}t_{(m)} \end{bmatrix} \in \Re^{m \times n}, \mathbf{c} = \begin{bmatrix} c_{1,1} \\ \cdots \\ c_{n,n} \end{bmatrix} \in \Re^{n \times 1}, \mathbf{e} = \begin{bmatrix} E_{a(1)} \\ \cdots \\ E_{a(m)} \end{bmatrix} \in \Re^{m \times 1}. \tag{9}$$

If $m \geq n$, eq.(8) can be solved to obtain the individual entries of the $\mathbf{D}$ matrix by

$$\mathbf{c} = (\mathbf{A}^T \mathbf{A})^{-1} \mathbf{A}^T \mathbf{e} \tag{10}$$

Using the pseudo-inverse of $\mathbf{A}$ in an overdetermined system of equations, the solution $\mathbf{c}$ is the best solution in a least squares sense. Eq.(8) of the diagonal viscous damping matrix example can also be solved forcing the non-negative definiteness of the $\mathbf{D}$ matrix by using Non-Negative Least-Squares (NNLS) algorithms, giving more physical meaning to the results. If the measurement of velocities $\dot{x}_1, \ldots, \dot{x}_n$ is available and if it is possible to excite the structure with a sufficient number of different excitations m, the damping matrix $\mathbf{D}$ can be easily estimated.

In order to obtain the sufficient number m, different force configurations are necessary to obtain the full range of responses at a chosen frequency. Different frequencies are also necessary because the various damping terms become dominant when different modes are excited whereas varying the amplitude of the force is particularly important when trying to identify nonlinearities.

Unfortunately in real structures it is practically impossible to obtain the full vector of velocities since it is not feasible to measure the infinitely many degrees of freedom of a large structure. For this reason, in the case of spatial incompleteness, i.e. a limited number of degrees of freedom measured, a method to expand the measurements of few degrees of freedom to the whole structure is necessary. In the numerical simulations, the method described in [11] is used, based on the mode shapes of the undamped finite element model of the system. This method needs the knowledge of the stiffness and mass matrix so one of the main advantages of the identification method proposed is nullified. However, using engineering knowledge, the degrees of freedom where damping is most likely to be located can be chosen in advance and then they can be only used to identify damping.

Even if in simple cases it would be possible to measure all the degrees of freedom, the number of tests to have $m > n$ could be probably too big and expensive. In the case where $m < n$, solving eq.(8) becomes an underdetermined problem and a minimum-angle criterion is used to select the most representative degrees of freedom to locate and identify damping.

4. Location of damping using a minimum angle criterion

Once the matrix **A** of eq.(8) is obtained, either from an expansion of the available measurements or populated with selected degrees of freedom only, and $m < n$ there is still the problem of solving an underdetermined problem. There are many ways of making the problem overdetermined: reducing the number of unknowns by changing the parameterization of the damping matrix, producing extra measurements using different force configurations or reducing the size of **A** by selecting the most representative columns using a criterion that allows the location of main damping sources. The minimum-angle criterion gave good results in many simulations. This angle-based method was used in a slightly different way in a model updating example in [12]. It is known from the theory of this identification method that the vector **e** is a linear combination of some columns of the matrix **A**. These columns $\mathbf{a}_i$ correspond to the degrees of freedom where the damping matrix has non-zero elements whereas the columns corresponding to the degrees of freedom where zeros are present does not give any contribution to the total energy dissipated. An angle θ between two vectors $\mathbf{a}_i$ and **e** can be calculated using the relationship

$$\theta = \cos^{-1}\left(\frac{\mathbf{a}_i{}^T\mathbf{e}}{\sqrt{\mathbf{a}_i{}^T\mathbf{a}_i}\cdot\sqrt{\mathbf{e}^T\mathbf{e}}}\right) \tag{11}$$

Among all the columns of matrix **A**, a certain number s of columns which have the smallest angle with the energy vector **e** is selected. These columns are the most parallel to the vector **e** so they are likely to be the best representative of the vector in a one-dimensional approximation. After selecting these s columns, the concept of angle between subspaces is introduced [13]. The angle between two matrices (in this case a matrix containing some columns of **A** and a matrix containing the **e** vector only) can be calculated with the method now described: consider the two matrices $\mathbf{A}' \in \Re^{n\times s}$ and $\mathbf{e} \in \Re^{n\times 1}$, an orthogonal basis for these subspaces can be obtained by the QR algorithm

$$\mathbf{A}' = \mathbf{Q}_\mathbf{A}\mathbf{R}_\mathbf{A},\ \mathbf{e} = \mathbf{Q}_\mathbf{e}\mathbf{R}_\mathbf{e} \tag{12}$$

where $\mathbf{Q}_\mathbf{A}$ and $\mathbf{Q}_\mathbf{e}$ are orthogonal matrices of dimension $n \times s$ and $n \times 1$ respectively and $\mathbf{R}_\mathbf{A}$ and $\mathbf{R}_\mathbf{e}$ are upper triangular. If $q < m$ there will be q principal angles between the subspaces, θ_i, which are computed from the singular value decomposition of $\mathbf{Q}_\mathbf{A}{}^T\mathbf{Q}_\mathbf{e}$. Thus

$$\cos(\theta_i) = \sigma_i\left(\mathbf{Q}_\mathbf{A}{}^T\mathbf{Q}_\mathbf{e}\right) \tag{13}$$

where $\sigma_i(\,)$ indicates the ith singular value. Again, it is now possible to select a number s of subsets of two columns of **A** that have the smallest angle with **e**. These subsets will be created using combinations of one of the previously selected vectors with the other columns of **A**. This method can then be applied for bigger subsets of three columns (keeping the two selected before) and so on. When the angle between these subsets and **e** is sufficiently small, the selection can be stopped and only the selected columns will be used to solve eq.(8). In this way, matrix **A** becomes of size $n \times s$ and it can be solved with least-square techniques provided $s < m$. Other constraints can be considered in the method to obtain results with more physical meaning like, for example, the non-negative definiteness of the viscous damping matrix. The minimum-angle selection corresponds to the spatial location of damping in a real structure. By choosing the columns of **A** the method allows the location of the main damping sources in the system. However, the angle-selection itself does not guarantee the selection of the correct degrees of freedom but just one of the possible locations from an energy point of view.

5. Numerical simulation

The results obtained from a simple example on a clamped ten-element cantilever beam with four viscous dashpots (Figure 1) are shown.

5.1. Case 1

The simulated structure dimensions are $4 \times 40 \times 560$ mm, the material is aluminium and the four dashpots are located at degrees of freedom number 3, 5, 13 and 17 with values respectively equal to 0.01, 0.5, 0.1 and 1 Ns/m. The cantilever beam is excited with sinusoidal forces at frequencies close to those of the first eight natural modes. The signals from three accelerometers, contaminated with random noise and applied on degrees of freedom 7, 11 and 19 are measured. The full vector of velocities is then derived using the expansion described in [11] and the location and identification of damping is obtained by solving the energy equation using Non-Negative Least-Squares (NNLS) algorithm after selecting the columns of matrix **A** using the minimum-angle criterion.

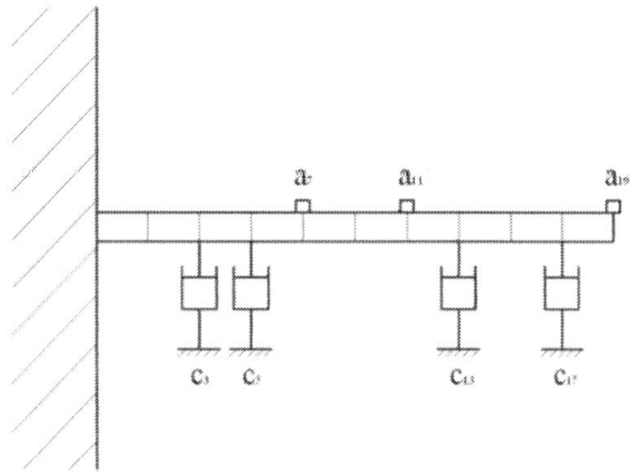

Figure 1 – Dashpots configuration and accelerometers location

In this example the viscous damping matrix is a 20×20 matrix with four positive values on the diagonal corresponding to the dashpots. The actual number of unknowns of the identification problem is 400, but since the dashpots are connected to the ground and there are no sources of damping between two elements, the number of unknowns can be reduced to 20 (the diagonal). The matrix **A** described in the previous section will be size 8×20, where each row of the matrix corresponds to one of the eight different excitations. This example was chosen in order to simulate a case where the number of excitations is smaller than the number of degrees of freedom of the structure and the measurements were taken in the wrong locations. The identified damping coefficients are summarized in Table 1. N represents the number of columns of matrix **A** selected to identify the damping, θ is the minimum angle between the selected columns and the vector **e** of input energies.

Table 1 – Located and identified damping coefficients (case 1)

N	DOF of identified dashpots				Identified damping coefficients (Ns/m)				θ
1	-	-	-	17	-	-	-	1.084	12.557
2	-	5	-	17	-	0.581	-	1.042	1.029
3	-	5	13	17	-	0.506	0.124	0.989	0.263
4	3	5	13	17	0.01	0.501	0.099	1.002	0.001
Exact	3	5	13	17	0.01	0.5	0.1	1	0

These results, considering the presence of noise and the small amount of data measured, are reasonably good. Especially the location of the main sources of damping is always correct and it can be noticed that the sources of damping are selected from the biggest to the smallest when increasing the number of dashpots to identify. This is an important aspect since even if the number of columns selected is smaller than the real number of dashpots of the structure the method can still identify the most representative dashpots and can be considered the most reasonable approximation at that level.

5.2. Case 2

In the second case the only difference is the value of the four damping coefficients, all set to 0.1 Ns/m. The results are shown in Table 2.

Table 2 – Located and identified damping coefficients (case 2)

N	DOF of identified dashpots				Identified damping coefficients (Ns/m)				θ
1	-	-	-	19	-	-	-	0.107	6.505
2	-	-	13	19	-	-	0.151	0.059	0.404
3	-	5	15	17	-	0.212	0.127	0.055	0.124
4	3	5	13	17	0.101	0.098	0.099	0.1	0.001
Exact	3	5	13	17	0.1	0.1	0.1	0.1	0

In this case, the location of the dashpots is not always correct. However, considering for example the case with N = 3, the degree of freedom 15 is close to the correct one 13 and the value of damping coefficients are similar considering the different locations. In particular, the damping coefficient of the dashpot in 5 is almost the sum of the dashpots in 3 and 5 of the original system. Considering the fact that we are using three dashpots to represent a four-dashpot system and it was assumed absolutely no knowledge about the location and values of the dashpots, these are still reasonably good results. Even the wrong solutions still hold some useful information. Moreover the identified system can be considered as an energy equivalent approximation of the original system. This can be noticed by comparing the damping ratio ζ of the first 10 modes of the identified system with the original system (Table 3).

Table 3 – Damping ratios of the first 10 modes (case 2)

Mode	Correct ζ	ζ (N = 1)	ζ (N = 2)	ζ (N = 3)
1	0.014092	0.013534	0.014096	0.014092
2	0.001496	0.002160	0.001495	0.001496
3	0.001024	0.000772	0.000894	0.001035
4	0.000338	0.000395	0.000305	0.000341
5	0.000138	0.000240	0.000149	0.000134
6	0.000190	0.000162	0.000193	0.000181
7	0.000114	0.000117	0.000118	0.000100
8	0.000057	0.000089	0.000049	0.000048
9	0.000106	0.000068	0.000073	0.000105
10	0.000085	0.000044	0.000061	0.000087

6. Experimental results

The experimental setup consists of a clamped aluminium beam (Figure 2a) with the same dimensions of the numerical simulation. A small wing is attached to the side of the beam and an eddy current dashpot (Figure 2b) consisting of two rare-earths NdFeB magnets placed above and below the wing which generates the magnetic field that creates damping. This kind of damping is a close approximation to classical viscous damping [14, 15]. There is no contact between surfaces so that friction is avoided. The damping can be varied by varying the air gap between the magnets and the wing; a gap of 2 mm is chosen and a preliminary experiment on a trolley coming down from a slope identified a viscous damping coefficient equal to 1.515 Ns/m. In the first part of the experiment, the structure without magnetic dashpot is excited with a set of 16 different excitations with frequencies close to those of the first 8 modes. The complete set of accelerations in 10 equally spaced degrees of freedom through the length of the beam is measured and an energy-equivalent viscous damping matrix is identified as the offset structural damping. Then, the experiment is repeated with the damping device attached on degree of freedom 9. The purpose of the exercise is to show that the method described above is capable of correctly locating and identifying this damping value.

Since there is no spatial incompleteness, no expansion is needed so the procedure consists in these five steps: measuring the accelerations and the input force, deriving the velocities from the accelerations, calculating the matrix **A** of integrals and the vector **e** of energies, subtracting from **e** the energies dissipated from the offset structural damping to find the effective energies dissipated by the damping device only and then solving eq.(8) for **c** using the Non-Negative Least-Squares (NNLS) solver.

a

b

Figure 2 – a) experimental setup b) eddy current magnetic dashpot.

Since the number of equations is bigger than the number of unknowns there is no need to select the columns of the matrix **A** with the minimum-angle criterion. The results of the identification are summarized in Table 4

Table 4 – Expected and identified damping coefficients

Damping coefficients	Expected (Ns/m)	Identified (Ns/m)
c_1	0	0
c_2	0	0
c_3	0	0
c_4	0	0
c_5	0	0
c_6	0	0
c_7	0	0
c_8	0	0
c_9	1.515	1.320
c_{10}	0	0.032

The main source of damping is located in the correct degree of freedom and the value is slightly smaller than the expected one with an error of about 12%. The method also identified a small source of damping at the free end of the beam that appears negligible compared to the one in degree of freedom 9. Other experiments are currently running in order to validate the method experimentally by putting more dashpots in different locations. Some Coulomb friction devices are ready to be attached to the structure to check the performances with nonlinear sources of damping too.

7. Conclusions

An energy-based method to identify damping in multi-degree-of-freedom structures is presented and a minimum-angle criterion is adopted to locate the main sources of damping. The method is intended to be used practically in real structures and to be able to identify both linear and nonlinear damping in the form $\mathbf{D} \cdot \mathbf{f}(\mathbf{x}, \dot{\mathbf{x}}, \ddot{\mathbf{x}})$. There is no need to estimate mass and stiffness matrices a priori, unless a complete set of measurements is not available. The method is validated by a numerical simulation and gives good results in the first experimental tests. Further tests are currently running to confirm the good performance of the method for both linear and nonlinear damping identification.

Acknowledgments

This work is funded by the European Union for the Marie Curie Excellence Team ECERTA under contract number MEXT-CT-2006 042383.

References

[1] Srikantha Phani, A. and Woodhouse, J., Viscous damping identification in linear vibration, *Journal of Sound and Vibration,* vol. 303, pp. 475-500, 2007.

[2] Prandina, M., Mottershead, J. E., and Bonisoli, E., An assessment of damping identification methods, *Journal of Sound and Vibration (submitted),* 2008.

[3] Lees, A. W., Use of perturbation analysis for complex modes, *Proceedings of the 17th International Modal Analysis Conference,* pp. 779-784, 1999.

[4] Adhikari, S. and Woodhouse, J., Identification of damping: Part 1, viscous damping, *Journal of Sound and Vibration,* vol. 243, pp. 43-61, 2001.

[5] Lancaster, P., Expressions for Damping Matrices in Linear Vibration Problems, *Journal of the Aerospace Sciences,* vol. 28, pp. 256-256, 1961.

[6] Pilkey, D. F., Park, G., and Inman, D. J., Damping matrix identification and experimental verification, *Proceedings of the SPIE's 6th Annual International Symposium on Smart Structures and Materials,* vol. 3672, pp. 350–357, 1999.

[7] Chen, S. Y., Ju, M. S., and Tsuei, Y. G., Estimation of mass, stiffness and damping matrices from frequency response functions, *Journal of Vibration and Acoustics-Transactions of the Asme,* vol. 118, pp. 78-82, 1996.

[8] Lee, J. H. and Kim, J., Development and validation of a new experimental method to identify damping matrices of a dynamic system, *Journal of Sound and Vibration,* vol. 246, pp. 505-524, 2001.

[9] Liang, J. W. and Feeny, B. F., Balancing energy to estimate damping parameters in forced oscillators, *Journal of Sound and Vibration,* vol. 295, pp. 988-998, 2006.

[10] Liang, J. W., Damping estimation via energy-dissipation method, *Journal of Sound and Vibration,* vol. 307, pp. 349-364, 2007.

[11] Prandina, M., Mottershead, J. E., and Bonisoli, E., Location and identification of damping parameters, in *IMAC XXVII Conference and Exposition on Structural Dynamics (accepted),* Orlando, Florida, USA 2008.

[12] Friswell, M., Mottershead, J. E., and Ahmadian, H., Combining subset selection and parameter constraints in model updating, *Transactions of the ASME,* vol. 120, pp. 854-859, 1998.

[13] Bjorck, A. and Golub, G. H., Numerical Methods for Computing Angles between Linear Subspaces, *Mathematics of Computation,* vol. 27, pp. 579-594, 1973.

[14] Nagaya, K., Kojima, H., Karube, Y., and Kibayashi, H., Braking Forces and Damping Coefficients of Eddy-Current Brakes Consisting of Cylindrical Magnets and Plate Conductors of Arbitrary Shape, *Ieee Transactions on Magnetics,* vol. 20, pp. 2136-2145, 1984.

[15] Bonisoli, E. and Vigliani, A., Passive effects of rare-earth permanent magnets on flexible conductive structures, *Mechanics Research Communications,* vol. 33, pp. 302-319, 2006.

Some issues when using Fourier analysis for the extraction of modal parameters

M. Lamb[1] and V. Rouillard[2]

Victoria University, PO Box 14428 MCMC 8001, Melbourne, Australia

[1]Matthew.Lamb@vu.edu.au, [2]Vincent.Rouillard@vu.edu.au

Abstract. It is sometimes necessary to determine the manner in which structures deteriorate with respect to time; for instance when quantifying a material's ability to withstand sustained dynamic loads. In such cases, it is well established that loss of structural integrity is reflected by variations in modal characteristics such as stiffness. This paper addresses some practical limitations of Fourier analysis with respect to temporal resolution and the uncertainties associated with extracting variations in modal parameters. The statistical analysis of numerous numerical experiments shows how techniques, such as data overlapping and zero-padding, can be used to improve the sensitivity of modal parameter extraction.

Keywords: frequency response, health monitoring, modal parameters, random vibrations.

1. Introduction

With the various parameters and spectral enhancement techniques used to obtain modal parameter estimates via Fourier analysis, it is important to evaluate their influence on the variability of the estimates obtained. This evaluation will assist analysts with the development of optimal modal parameter extraction procedures, which is particularly important in situations where modal estimates need to be obtained in the least possible amount of time. These situations include applications such as continuous structural health monitoring, which requires sensitivity in both the temporal and frequency domains.

This paper presents statistical analyses of modal parameter estimates obtained from a number of numerically simulated experiments designed to reproduce various effects of single degree-of-freedom (SDoF) physical experiments including instrument noise. Fourier analysis is used to compute estimates of the system's frequency response function (FRF) which, in turn, is used to extract the relevant modal parameters. The influence of various parameters used to compute the FRF, including those of spectral enhancement techniques such as spectral averaging, data overlapping and zero-padding, is evaluated and discussed.

2. Modal Parameter Estimation

The dynamic characteristics of all stable physical systems can be described using their frequency response function (FRF). The FRF of a system represents the complex ratio of output/response to input/excitation in the frequency domain. The most common application of FRFs is in modal analysis where they are used to extract the global properties of mechanical systems, namely their damping ratios and natural frequencies (the mode shapes of mechanical systems are also defined by the residuals of their FRF). Such modal parameter extraction requires the experimental frequency response to be fitted with an appropriate theoretical model which can, in turn, be used to extract the required modal properties. A number of modal parameter extraction techniques exist. For single degree-of-freedom (SDoF) systems, Ewins [1] recommends a technique based on fitting a circle to the Nyquist plot. Another technique uses least squares regression on the magnitude of the FRF in the region of resonance [2]. Garcia-Romeu-Martinez *et al.* [3], and later Rouillard *et al.* [4], showed that, for physical SDoF systems, the least squares regression technique is particularly robust. Based on the successful, practical, application of the least squares technique by Garcia-Romeu-Martinez *et al.*, it will be used as the basis for analysis in the remaining sections.

3. Experimental Design

The research presented in this paper addresses the behaviour of linear, viscously damped, SDoF systems in a moving base configuration [3]. The systems were analysed using their acceleration excitation and response signals.

Excitation for the SDoF systems was synthesized using a band-limited spectral function (magnitude) coupled with a uniformly-distributed phase spectrum to form a complex spectrum. The inverse Fourier transform of the complex spectrum was then taken to produce a Gaussian random signal. This process was undertaken repeatedly to produce a random signal of arbitrary duration and spectral content. This random signal was later uploaded into a simulation tool developed using the Simulink® package in Matlab® which uses incremental analysis to solve the SDoF equation of motion [5].

Post-processing was achieved using the modified, short-time Fourier transform (STFT) based Welch method described by Rouillard *et al.* [4] in conjunction with the least squares regression curve-fitting procedure, described in the same article. This technique allows for the addition of zero-padding, spectral averaging overlap (individual FRF window overlap) and changes to the analysis sub-record length (window length), all of which are often manipulated in an attempt to improve spectral estimates.

Results from the analysis of a numerical SDoF system (with its mass, stiffness and damping configured to achieve an undamped natural frequency of 20Hz and a damping ratio of 11%), along with appropriate discussions, are presented in each of the subsequent sections. In order to enable the statistical nature of the results to be analysed, a large number (50) of independent band-limited (0 to 100Hz) Gaussian excitation signals were used. The corresponding 50 independent signal pairs (excitation and response) generated were subjected to FRF analysis (including least mean squares modal parameter extraction) with variations in the parameters of interest. As well as examining the consistency (variability) of results using statistical analysis, the robustness of the potential improvements to extraneous noise, in both the excitation and response signals, was also evaluated.

4. Influence of Spectral Averaging and Sub-record Length

Spectral estimation using random signals requires some degree of averaging to minimise statistical uncertainty. However, in many applications, the time available to capture data is limited, thus restricting the amount of data available for averaging. This limited window means that a compromise needs to be made between the number of spectral averages and the analysis sub-record length. For the purposes of modal parameter extraction it is unclear which is most important. Frequency resolution, hence the sub-record length, is important as it affects the sensitivity of the frequency estimates (although it is not the limiting factor in this instance as curve-fitting is applied). However, it is not clear how much spectral averaging can be compromised.

An initial set of experiments (experiment 1) was developed to evaluate the influence of various amounts of averaging, in conjunction with varying sub-record lengths (with the combination of the two varying parameters resulting in a constant analysis window size, or temporal resolution), on modal parameter extraction. The results gathered from experiment 1 (for excitation response signal pairs containing 0% and 10% extraneous noise respectively) are presented in *Figure 1*, as the mean natural frequency extracted from the 50 signal pairs along with error bars indicating ± one standard deviation. It must be noted that during experiment 1 no spectral overlapping or zero padding was applied.

The results from experiment 1 indicate that, for the purposes of modal parameter extraction, spectral resolution, hence sub-record length, has a more significant effect on the extracted natural frequencies than reduced spectral uncertainty via spectral averaging. This is made most evident by the small variability of the results obtained using the 30second sub-record, which provides fine spectral resolution (0.033Hz) but only allows for 4 spectral averages (50% theoretical uncertainty). This finding is attributed to the fact that spectral uncertainty only creates variability in the magnitude of frequency estimates, which, due to its random nature, has a limited effect when curve-fitting using least squares regression. Conversely, coarse spectral resolution limits the amount of data available within the bandwidth appropriate for curve-fitting, thereby creating uncertainties in terms of both magnitude and spectral locality.

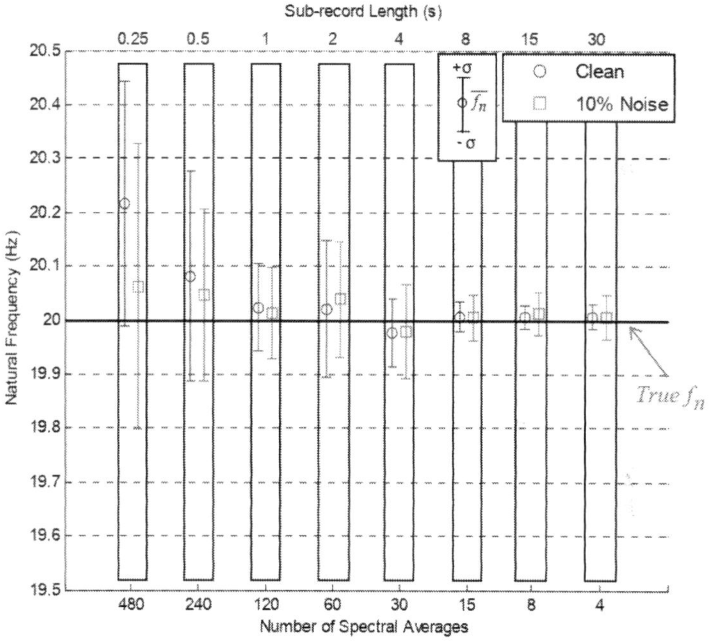

Figure 1: Experiment 1 – Spectral resolution and uncertainty.

5. Influence of Zero-padding

Given the limited temporal data available to produce a frequency domain estimate of a system with time-varying modal properties, there is a need to obtain the required frequency domain information while consuming as little as possible temporal data. This minimised consumption of temporal data can only be achieved by improving the evaluation of spectral estimates with relatively coarse frequency resolution. This improved evaluation is often attempted by artificially refining the frequency resolution of the coarse spectrum using zero-padding. However, it is uncertain whether or not zero-padding actually helps. Most agree that zero-padding does not actually improve true frequency resolution; however, the majority of authors are unclear about what it does achieve [6, 7, 8, 9].

An attempt to determine whether or not zero-padding can improve modal parameter extraction from coarse frequency spectra was made by conducting a second set of experiments (experiment 2). The signals used in the previous experiment were again analysed using various sub-record lengths (T_{sr}) with various amounts of zero-padding and spectral averaging (where the number of spectral averages was proportional to the inverse of the sub-record length, which resulted in a constant temporal resolution for each of the sub-records analysed). The results for experiments with no extraneous noise are given in *Figure 2* and those obtained from experiments containing 10% extraneous noise are given in *Figure 3*.

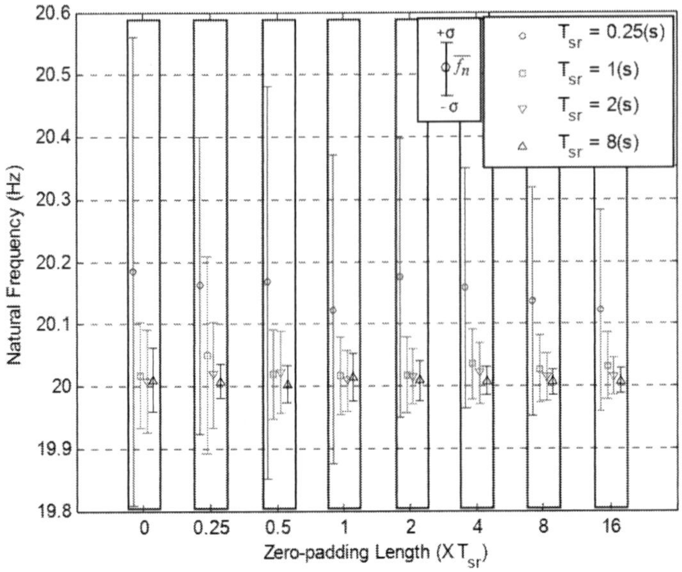

Figure 2: Experiment 2 –Influence of zero-padding on the estimation of natural frequency – uncontaminated data.

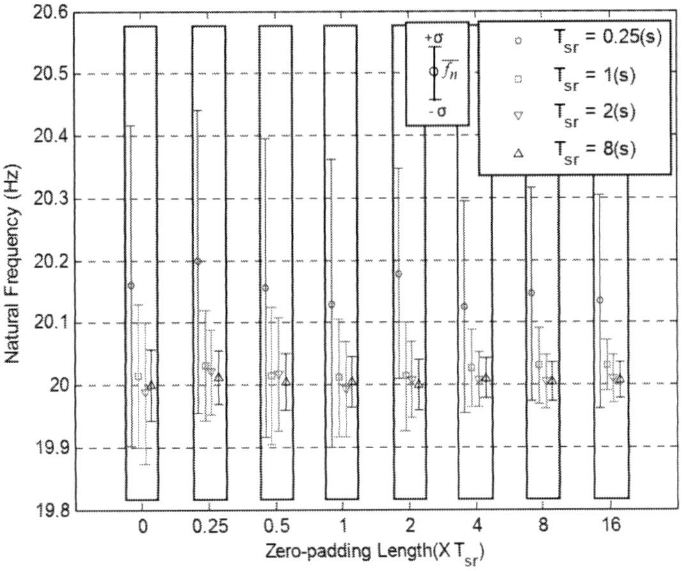

Figure 3: Experiment 2 - Influence of zero-padding on the estimation of natural frequency – data containing 10% noise.

The results obtained from experiment 2 indicate that, for signals containing as much as 10% noise, the introduction of up to 16 times zero-padding significantly reduces the variability of the frequency estimates obtained using least squares regression modal parameter extraction, irrespective of the sub-record length used. The finding that the introduction of zeros will not degrade results is significant in that it enables analysts to apply zero-padding with confidence. In saying this it must also be noted that the value of zero-padding should not be overstated and that where possible increases in spectral resolution should always be obtained by increasing the analysis sub-record length.

The improvements obtained by implementing zero-padding are attributed to the fact that it provides estimates of results at otherwise non-described frequencies, which with the implementation of averaging, effectively increases the portion of data which describes the theoretical model of the FRF, without distorting its shape. This increased portion of data improves the ability to curve-fit the FRF magnitude spectra. This suggestion follows Semmlow's [10] description of zero-padding: "Zero padding provides an interpolation of the points in the unpadded signal".

6. Influence of Overlapped Averages

It is understood that overlapped spectral averaging can be used to retrieve the statistical stability lost due to windowing. Bendat & Piersol [11] state that if 50% overlap is used with the application of the Hanning window, approximately 90% of the stability lost, due to the window, can be retrieved. Randall [12] suggests that, for the Hanning window, overlaps of greater than 50% provide no appreciable improvement in statistical uncertainty. However, according to the formulation provided by Naidu [13], the variance within time-invariant spectra, captured using the Welch method, is inversely proportional to the number of overlapped spectral averages used. This formulation suggests that maximum overlap should always be applied.

Overall, for the purposes of modal parameter extraction from time-varying systems it is uncertain if large amounts of spectral overlapping can be used to enhance results. Therefore, to gain a better understanding of the effect of overlapped spectral averaging, on modal parameter extraction, a third set of experiments (experiment 3) was undertaken. During experiment 3 varying amounts of averaging overlap (0% to 99.999%) were used in conjunction with the Tukey window [14] to analyse 50 constant parameter signal pairs. It is important to note that with the increased number of overlapped averages no temporal resolution was lost.

The results obtained from experiment 3, for a temporal resolution of 10 seconds (5 second sub-records, 2 independent averages and no zero-padding), are given in *Figure 4*. As can be seen, for signals containing as much as 10% noise, overlapped processing can generate a significant reduction in the level of variance in the extracted natural frequencies obtained using least squares regression modal parameter extraction.

Since the frequency resolution was kept constant with the addition of increased spectral overlap, the effect on the estimates must result solely from a reduction in spectral variance. Therefore, according to the formulation provided Naidu [13] the variance of the modal parameter estimates obtained should be inversely proportional to the number of overlapped spectral averages. To test this theory the variance of the natural frequency estimates, obtained from experiment 3, was plotted against the number of overlapped spectral averages and was fitted with the closest possible inverse relationship in *Figure 5*. As can be seen the reduction in the extracted natural frequency variance has a strong correlation with the fitted trend, thereby suggesting that such a relationship does in fact exist.

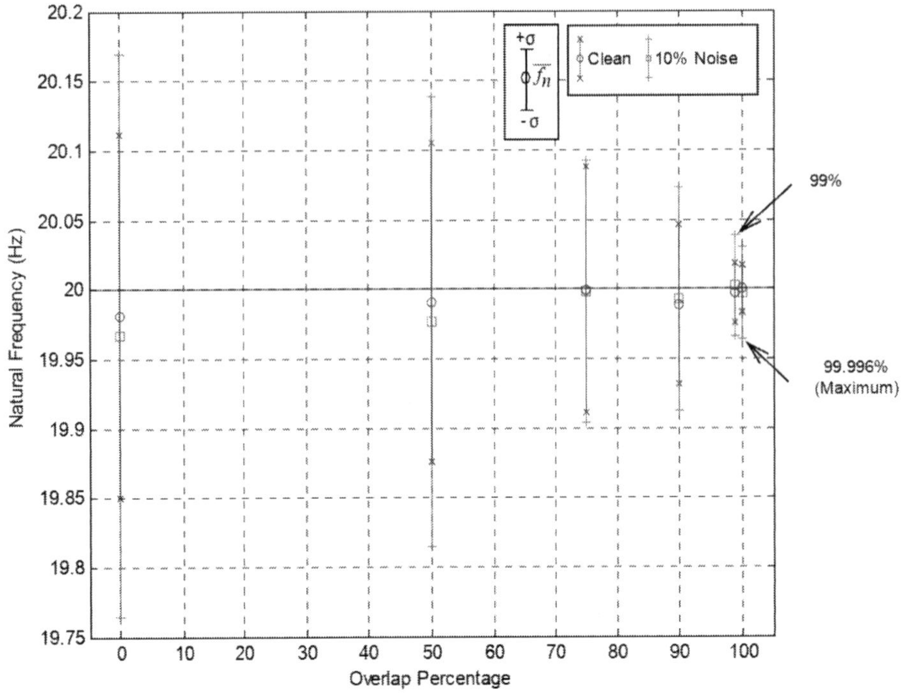

Figure 4: Experiment 3 - Influence of overlapping on the estimation of natural frequency. (Maximum represents the case where all but one sample point were overlapped)

The results in *Figure 4* and *Figure 5* clearly demonstrate that increased spectral overlapping can significantly improve modal parameter extraction results. To illustrate how these improvements are achieved the results of a typical noise-contaminated (10%) moderately high resolution FRF (0.1Hz), obtained with and without spectral overlapping are given in *Figure 6*. As can be seen, the introduction of increased overlapped averaging results in a significant reduction in the random deviation from the actual FRF, even though no additional data is analysed.

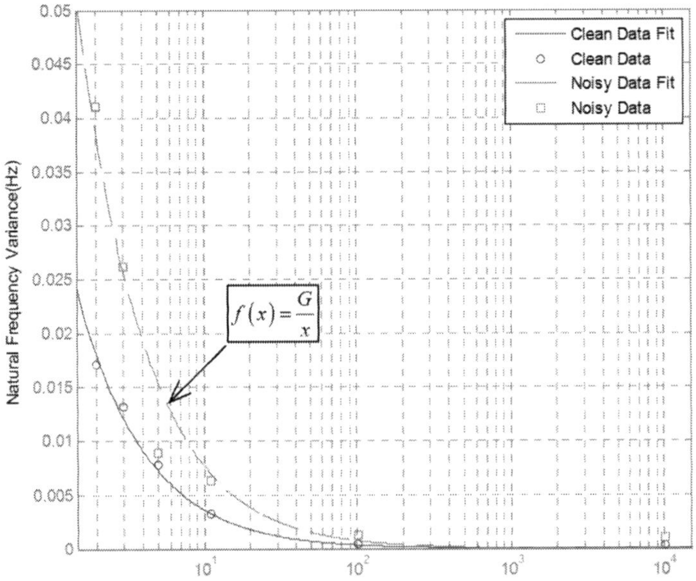

Figure 5: Experiment 3 - Influence of overlapped averaging on the estimated natural frequency variance (where x is the number of overlapped averages and G is an empirical constant).

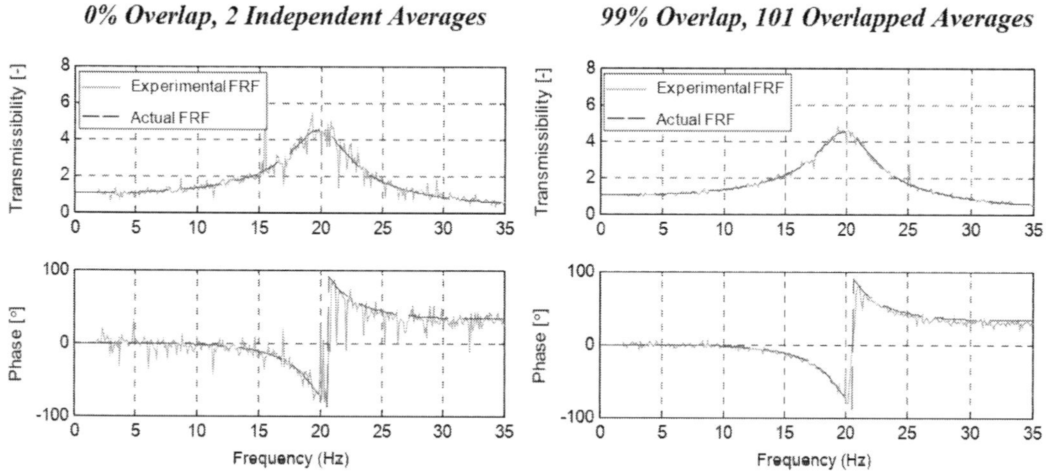

Figure 6: Experiment - 4: Enhancement of FRF estimates using overlapped averaging.

7. Conclusion

Several analysis parameters are used when computing spectral estimates. As a result of the dependence of frequency resolution on sub-record length these parameters are often manipulated in order to attempt to maintain the accuracy of results whilst minimising the amount of temporal data used. This paper discussed the effect of several common analysis parameters including spectral averaging, sub-record length (window length), overlapped averaging and zero-padding, in terms of both spectral and temporal sensitivity using a statistical evaluation. It was found that, for records containing up to 10% noise, sub-record length, hence spectral resolution, has a more pronounced influence on modal parameter extraction than the level of spectral averaging, hence spectral uncertainty. It was also found that while zero-padding may not increase the actual spectral resolution, it does allow for enhanced modal parameter extraction by reducing the variance between extracted results based on interpolated estimates at the non-described frequencies. This enhancement can be taken advantage of to maintain temporal resolution whilst minimising the level of estimation variance. Finally, it was found that while some authors suggest that large amounts of overlapped averaging provides no appreciable improvement in statistical uncertainty, for modal parameter extraction purposes, increased amounts of overlapped averaging can significantly reduce the variance of the extracted parameters. This is particularly useful as it allows for increased accuracy of spectral results without compromising temporal resolution.

8. References

[1] Ewins, D 1995, *Modal testing theory and practice*, Research Studies Press, Taunton, pp. 178-199

[2] Avitabile, P 2006, '101 ways to extract modal parameters – which one is for me?', *Experimental techniques,* (September/October 2006), Society for Experimental Mechanics

[3] Garcia-Romeu-Martinez, M.A, Rouillard, V, Sek, M & V.A Cloquell-Ballester 2007, 'Monitoring the Evolution of Fatigue in Corrugated Paperboard under Random Loads', *Proceedings of the 5th BSSM International Conference on Advances in Experimental Mechanics,* U.K. V 7-8, pp 243-250

[4] Rouillard, V, Lamb, M & Sek, M 2007, 'Determining fatigue progression in corrugated paperboard containers subjected to dynamic compression', *Proceedings of the 5th Australasian Congress on Applied Mechanics, ACAM 2007,* vol. 1, The Institute of Engineers Australia, Brisbane, pp. 331-336

[5] Rao, S 2005, *Mechanical Vibrations* SI Edition, Prentice Hall, p. 239

[6] Kay, S. 1988, 'Spectral Estimation', in J Lim & A Oppenheim (eds), *Advanced Topics in Signal Processing*, Prentice Hall, Englewood Cliffs, pp. 78

[7] Lyons, R 2004, *Understanding digital signal processing*, 2nd Edition, Prentice Hall, Upper Saddle River, p. 87

[8] Manolakis, D, Ingle, V & Kogon, S 2005, *Statistical and adaptive signal processing: Spectral estimation, signal modeling, adaptive filtering, and array processing,* Artech House, Boston, p. 202

[9] Andreas, A 2006, *Digital signal processing: Signals, systems, and filters*, McGraw-Hill Professional, New York, p. 328

[10] Semmlow, J 2005, Circuits, signals and systems for bioengineers: A Matlab based introduction, Academic Press, pp. 96-97

[11] Bendat, J & Piersol, A 2000, *Random Data: Analysis and measurement procedures*, 3rd Edition, John Wiley & Sons, New York, p. 430

[12] Randall, R.B 1987, *Frequency analysis*, 3rd Edition, Brüel & Kjaer, Denmark, K Larson & Son, pp. 174-176

[13] Naidu, P 1996, Modern spectrum analysis of time Series: Fast algorithms and error control techniques, CRC Press, pp. 173-192

[14] Harris, F 1978, 'On the use of windows for harmonic analysis with the discrete Fourier transform', *Proceedings of the IEEE*, vol. 66, no. 1, pp. 51-83

Journal of Physics: Conference Series **181** (2009) 012008 doi:10.1088/1742-6596/181/1/012008

Nonlinear parametric vibration of axially moving beams: asymptotic analysis and differential quadrature verification

Li-Qun Chen[a,b,1], **Bo Wang**[b], **Hu Ding**[b]

a Department of Mechanics, Shanghai University, Shanghai 200444, China
b Shanghai Institute of Applied Mathematics and Mechanics, Shanghai 200072, China

E-mail: lqchen@staff.shu.edu.cn

Abstract. Nonlinear parametric vibration is investigated in principal parametric resonance of an axially accelerating viscoelastic beam. The axial speed is characterized as a simple harmonic variation about the constant mean speed. The material time derivative is used in the viscoelastic constitutive relation. The transverse motion can be governed a nonlinear integro-partial-differential equation. The asymptotic analysis is performed to determine steady-state responses. It is proved that the mode uninvolved in the resonance has no effect on the steady-state response. The differential quadrature scheme is developed to verify results via the method of multiple scales. It is demonstrated that the straight equilibrium configuration becomes unstable and a stable steady-state emerges when the axial speed variation frequency is close to 2 times of any linear natural frequency. The numerical calculations validate the analytical results in the principal parametric resonance.

1. Introduction

Axially moving beams can represent many engineering devices, such as power transmission belts, band saws, and aerial cable tramways. As parametric vibration excited by the variation of the beam tension or the beam axial speed, large transverse motion of axially moving beams may occur under certain conditions. If the motion amplitude is large, the nonlinearity should be taken into account. Much research has been done on nonlinear parametric vibration of axially moving beams [1-12]. To determine the steady-state periodic response, the method of multiple scales was applied to derive the closed-form expressions of the amplitude, the existence condition and the stability conditions [2-4,8,10-12].

Addition to the method multiple scales, the method of asymptotic perturbation is also an effective approach to treat nonlinear vibration. The asymptotic approach for disturbed conservative linear systems [13-19] was developed to disturbed gyroscopic continuous systems [20,21]. In the present investigation, the asymptotic analysis will be performed to study nonlinear parametric vibration of axially moving beams.

In spite of the fact that there have been many approximate analytical investigations on nonlinear parametric vibration of axially moving beams, there are very limited researches on the topic via the direct numerical approaches. The Galerkin method is widely used in the numerical investigations [1,5-7,9]. Consequently, the results obtained via approximate analytical approaches were rarely verified numerically. Recently, a finite difference scheme was used to confirm the results via the multiscale method [11], and a differential quadrature scheme was used to confirm the results via the asymptotic stability analysis of linear parametric vibration [11]. In the present investigation, the differential

[1] To whom any correspondence should be addressed.

quadrature scheme will be applied to the nonlinear parametric vibration of axially moving beams.

2. Mathematical model

A uniform axially moving viscoelastic beam, with density ρ, cross-sectional area A, moment of inertial I and initial tension P_0, travels at time-dependent axial transport speed $\gamma(t)$ between two motionless ends separated by distance l. Consider only the bending vibration described by the transverse displacement $v(x,t)$, where t is the time and t is the axial coordinate. The viscoelastic material of the beam obeys the Kelvin model with stiffness E_0 and dynamic viscosity. Then the governing equation in the dimensionless form is [11]

$$v_{,tt}+2\gamma v_{,xt}+\dot{\gamma}v_{,x}+\left(\gamma^2-1\right)v_{,xx}+k_f^{\ 2}v_{,xxxx}+\varepsilon\alpha v_{,xxxxt}+\varepsilon\alpha\gamma v_{,xxxxx}=\varepsilon v_{,xx}\int_0^1\left[\frac{1}{2}k_1^2 v_{,x}^{\ 2}+k_2\varepsilon\left(v_{,x}v_{,xt}+\gamma v_{,x}v_{,xx}\right)\right]\mathrm{d}x, \quad (1)$$

where bookkeeping device ε is a small dimensionless parameter accounting for the fact that both the transverse displacement and the dynamic viscosity are very small, and the dimensionless variables and parameters are defined as

$$v\leftrightarrow\frac{v}{\sqrt{\varepsilon}l},\ x\leftrightarrow\frac{x}{l},\ t\leftrightarrow t\sqrt{\frac{P_0}{\rho A l^2}},\gamma(t)\leftrightarrow\gamma(t)\sqrt{\frac{\rho A}{P_0}},k_f=\sqrt{\frac{E_0 I}{P_0 l^2}},\alpha=\frac{I\eta}{\varepsilon L^3\sqrt{\rho A P_0}},k_1=\sqrt{\frac{E_0 A}{P_0 l}},k_2=\frac{\eta\sqrt{A}}{l\varepsilon\sqrt{\rho P_0}},k=\frac{K}{E_0 Il}. \quad (2)$$

Consider the axially moving beam constrained by freely rotating sleeves. Then the beam with simply supports at both ends with the boundary conditions in dimensionless form

$$v(0,t)=0,\ v_{,xx}(0,t)=0, v(1,t)=0,\ v_{,xx}(1,t)=0. \quad (3)$$

In the present investigation, the axial speed is assumed to be a small simple harmonic variation about the constant mean speed,

$$\gamma(t)=\gamma_0+\varepsilon\gamma_1\sin\omega t \quad (4)$$

where γ_0 is the mean axial speed, and $\varepsilon\gamma_1$ and ω are respectively the amplitude and the frequency of the axial speed variation, all in the dimensionless form. Substitution of equation (4) into equation1s (1) yields

$$Mv_{,tt}+Gv_{,t}+Kv=\varepsilon\left[\frac{1}{2}k_1^2 v_{,xx}\int_0^1 v_{,x}^{\ 2}\,\mathrm{d}x-2\gamma_1\sin\omega t\left(v_{,xt}-\gamma_0 v_{,xx}\right)-\omega\gamma_1\cos\omega t v_{,x}-\alpha v_{,xxxt}-\alpha\gamma_0 v_{,xxxx}\right]+O\left(\varepsilon^2\right) \quad (5)$$

where the mass, gyroscopic, and linear stiffness operators are respectively defined as

$$M=I,\ G=2\gamma_0\frac{\partial}{\partial x},\ K=\left(\gamma_0^{\ 2}-1\right)\frac{\partial^2}{\partial x^2}+v_f^{\ 2}\frac{\partial^4}{\partial x^4}. \quad (6)$$

3. Asymptotic analysis

If ε=0 in equation (5), under boundary conditions (3), the natural frequencies of the undisturbed gyroscopic continuous system

$$Mv_{,tt}+Gv_{,t}+Kv=0 \quad (7)$$

can be determined. Denote the frequencies as ω_i (i=1,2,...). If the speed variation frequency ω approaches two times of any natural frequency ω_m, principal parametric resonance may occur. A detuning parameter σ is introduced to quantify the deviation of ω from ω_m, and ω is described by

$$\omega=2\omega_m+\varepsilon\sigma\,. \quad (8)$$

To investigate the principal parametric resonance with the possible contributions of a mode not involved the resonance, the solution to equation (7) be expressed as

$$v(x,t)=\psi_m(x,\tau,\varepsilon)e^{i\omega_m t}+\psi_n(x,\tau,\varepsilon)e^{i\omega_n t}+cc, \quad (9)$$

where cc represents the complex conjugate of all preceding terms on the right hand of an equation, $\tau=\varepsilon t$ is the slowly-varying time. Functions $\psi_m(x,\tau,\varepsilon)$ and $\psi_n(x,\tau,\varepsilon)$ can be expanded in power series of ε

$$\psi_k(x,\tau,\varepsilon)=\psi_{0k}(x,\tau)+\varepsilon\psi_{1k}(x,\tau)+O\left(\varepsilon^2\right)\quad(k=m,n). \quad (10)$$

Inserting equations (8)-(10) into equations (5) and (3) and equating the coefficients of $e^{-i\omega k t}$ (k=m,n) at the order ε^0 and ε^1 in the resulting equation yield, at order ε^0

$$-\omega_k^2 M\psi_{0k} + \mathrm{i}\,\omega_k G\psi_{0k} + K\psi_{0k} = 0 \quad (k = m, n), \tag{11}$$

$$\psi_{0k}(0,\tau) = 0, \psi_{0k,xx}(0,\tau) = 0; \psi_{0k}(1,\tau) = 0, \psi_{0k,xx}(1,\tau) = 0 \quad (k = m, n), \tag{12}$$

and at order ε^1

$$\begin{aligned} &-\omega_m^2 M\psi_{1m} + \mathrm{i}\,\omega_m G\psi_{1m} + K\psi_{1m} \\ &= -2\mathrm{i}\,\omega_m \psi_{0m,\tau} - 2\gamma_0 \psi_{0m,x\tau} + \frac{1}{2}k_1^2 \bar{\psi}_{0m,xx}\int_0^1 (\psi_{0m,x})^2\,\mathrm{d}x + k_1^2 \bar{\psi}_{0n,xx}\int_0^1 \psi_{0n,x}\psi_{0m,x}\,\mathrm{d}x + k_1^2 \psi_{0n,xx}\int_0^1 \psi_{0m,x}\bar{\psi}_{0n,x}\,\mathrm{d}x \\ &+ k_1^2 \psi_{0m,xx}\int_0^1 \psi_{0n,x}\bar{\psi}_{0n,x}\,\mathrm{d}x + k_1^2 \psi_{0m,xx}\int_0^1 \psi_{0m,x}\bar{\psi}_{0m,x}\,\mathrm{d}x + \mathrm{i}\gamma_0\gamma_1 \bar{\psi}_{0m,xx}\,\mathrm{e}^{\mathrm{i}\sigma\tau} - \alpha\left(\mathrm{i}\,\omega_m \psi_{0m,xxxx} + \gamma_0 \psi_{0m,xxxxx}\right), \end{aligned} \tag{13}$$

$$\begin{aligned} &-\omega_n^2 M\psi_{1n} + \mathrm{i}\,\omega_n G\psi_{1n} + K\psi_{1n} \\ &= -2\mathrm{i}\,\omega_n \psi_{0n,\tau} - 2\gamma_0 \psi_{0n,x\tau} + \frac{1}{2}k_1^2 \bar{\psi}_{0n,xx}\int_0^1 (\psi_{0n,x})^2\,\mathrm{d}x + k_1^2 \bar{\psi}_{0m,xx}\int_0^1 \psi_{0n,x}\psi_{0m,x}\,\mathrm{d}x + k_1^2 \psi_{0m,xx}\int_0^1 \psi_{0n,x}\bar{\psi}_{0m,x}\,\mathrm{d}x \\ &+ k_1^2 \psi_{0n,xx}\int_0^1 \psi_{0m,x}\bar{\psi}_{0m,x}\,\mathrm{d}x + k_1^2 \psi_{0n,xx}\int_0^1 \psi_{0n,x}\bar{\psi}_{0n,x}\,\mathrm{d}x - \alpha\left(\mathrm{i}\,\omega_n \psi_{0n,xxxx} + \gamma_0 \psi_{0n,xxxxx}\right), \end{aligned} \tag{14}$$

$$\psi_{1k}(0,\tau) = 0, \psi_{1k,xx}(0,\tau) = 0, \psi_{1k}(1,\tau) = 0, \psi_{1k,xx}(1,\tau) = 0 \quad (k = m, n). \tag{15}$$

Under the boundary condition (12), equation (11) has the solution

$$\psi_{0k}(x,\tau) = q_k(\tau)\phi_k(x) \quad (k = m, n), \tag{16}$$

where $q_k(\tau)$ ($k=m,n$) is to be determined later and [22,23]

$$\phi_k(x) = \mathrm{e}^{\mathrm{i}\beta_{1k}x} - \frac{(\beta_{4k}^2 - \beta_{1k}^2)(\mathrm{e}^{\mathrm{i}\beta_{3k}} - \mathrm{e}^{\mathrm{i}\beta_{1k}})}{(\beta_{4k}^2 - \beta_{2k}^2)(\mathrm{e}^{\mathrm{i}\beta_{3k}} - \mathrm{e}^{\mathrm{i}\beta_{2k}})}\mathrm{e}^{\mathrm{i}\beta_{2k}x} - \frac{(\beta_{4k}^2 - \beta_{1k}^2)(\mathrm{e}^{\mathrm{i}\beta_{2k}} - \mathrm{e}^{\mathrm{i}\beta_{1k}})}{(\beta_{4k}^2 - \beta_{3k}^2)(\mathrm{e}^{\mathrm{i}\beta_{2k}} - \mathrm{e}^{\mathrm{i}\beta_{3k}})}\mathrm{e}^{\mathrm{i}\beta_{3k}x} - \left[1 - \frac{(\beta_{4k}^2 - \beta_{1k}^2)(\mathrm{e}^{\mathrm{i}\beta_{3k}} - \mathrm{e}^{\mathrm{i}\beta_{1k}})}{(\beta_{4k}^2 - \beta_{2k}^2)(\mathrm{e}^{\mathrm{i}\beta_{3k}} - \mathrm{e}^{\mathrm{i}\beta_{2k}})} - \frac{(\beta_{4k}^2 - \beta_{1k}^2)(\mathrm{e}^{\mathrm{i}\beta_{2k}} - \mathrm{e}^{\mathrm{i}\beta_{1k}})}{(\beta_{4k}^2 - \beta_{3k}^2)(\mathrm{e}^{\mathrm{i}\beta_{2k}} - \mathrm{e}^{\mathrm{i}\beta_{3k}})}\right]\mathrm{e}^{\mathrm{i}\beta_{4k}x}. \tag{17}$$

In equation (17), β_{jk} ($j=1,2,3,4$; $k=m,n$) are four roots of the following 4-order algebraic equation

$$-\omega_k^{\ 2} - 2\gamma_0\omega_k\beta_k + (\gamma_0^{\ 2} - 1)\beta_k^{\ 2} + v_f^{\ 2}\beta_k^{\ 4} = 0 \quad (k = m, n). \tag{18}$$

Substitution of equation (16) into equations (13) and (14) leads to

$$\begin{aligned} -\omega_m^2 M\psi_{1m} + \mathrm{i}\,\omega_m G\psi_{1m} + K\psi_{1m} &= -2(\mathrm{i}\,\omega_m\phi_m + \gamma_0\phi_m')\dot{q}_m - \mathrm{i}\,\omega_m\eta\phi_m^{(4)}q_m - \gamma_0\eta\phi_m^{(5)}q_m + \gamma_0\gamma_1\bar{q}_m\bar{\phi}_m''\,\mathrm{e}^{\mathrm{i}t\sigma} \\ &+ \frac{1}{2}k_1^2 q_m^2\bar{q}_m\left(2\phi_m''\int_0^1 \phi_m'\bar{\phi}_m'\,\mathrm{d}x + \bar{\phi}_m''\int_0^1 \phi_m'^{\,2}\,\mathrm{d}x\right) + k_1^2 q_n\bar{q}_n q_m\left(\phi_m''\int_0^1 \phi_n'\bar{\phi}_n'\,\mathrm{d}x + \phi_n''\int_0^1 \phi_m'\bar{\phi}_n'\,\mathrm{d}x + \bar{\phi}_n''\int_0^1 \phi_m'\phi_n'\,\mathrm{d}x\right), \end{aligned} \tag{19}$$

$$\begin{aligned} -\omega_n^2 M\psi_{1n} + \mathrm{i}\,\omega_n G\psi_{1n} + K\psi_{1n} &= -2(\mathrm{i}\,\omega_n\phi_n + \gamma_0\phi_n')\dot{q}_n - \mathrm{i}\,\omega_n\eta\phi_n^{(4)}q_n - \gamma_0\eta\phi_n^{(5)}q_n \\ &+ \frac{1}{2}k_1^2 q_n^2\bar{q}_n\left(2\phi_n''\int_0^1 \phi_n'\bar{\phi}_n'\,\mathrm{d}x + \bar{\phi}_n''\int_0^1 \phi_n'^{2}\,\mathrm{d}x\right) + k_1^2 q_m\bar{q}_m q_n\left(\phi_m''\int_0^1 \phi_n'\bar{\phi}_m'\,\mathrm{d}x + \phi_n''\int_0^1 \phi_m'\bar{\phi}_m'\,\mathrm{d}x + \bar{\phi}_m''\int_0^1 \phi_n'\phi_m'\,\mathrm{d}x\right). \end{aligned} \tag{20}$$

Equations (19) and (20) have bounded solutions only if solvability conditions are satisfied. The solvability conditions [20] demands that the right hands of the equations (19) and (20) be orthogonal to ϕ_m and ϕ_n respectively with the inner product defined by $\langle f_1, f_2\rangle = \int_0^1 f_1(x)\bar{f}_2(x)\mathrm{d}x$

for complex functions f_1 and f_2 defined on [0,1]. The orthogonality leads to

$$\dot{q}_m + (\eta\mu_m - k_1^2 C_{mn} q_n\bar{q}_n)q_m - k_1^2\kappa_m q_m^2\bar{q}_m + \mathrm{e}^{\mathrm{i}t\sigma}\gamma_1\chi_m\bar{q}_n = 0, \quad \dot{q}_n + (\eta\mu_n - k_1^2 C_{nm} q_m\bar{q}_m)q_n - k_1^2\kappa_n q_n^2\bar{q}_n = 0, \tag{21}$$

where

$$\mu_j = \frac{\mathrm{i}\,\omega_j\int_0^1 \phi_j^{(4)}\bar{\phi}_j\,\mathrm{d}x + \gamma_0\int_0^1 \phi_j^{(5)}\bar{\phi}_j\,\mathrm{d}x}{2\left(\mathrm{i}\,\omega_j\int_0^1 \phi_j\bar{\phi}_j\,\mathrm{d}x + \gamma_0\int_0^1 \phi_j'\bar{\phi}_j\,\mathrm{d}x\right)}, \kappa_j = \frac{\int_0^1 \bar{\phi}_j\bar{\phi}_j''\,\mathrm{d}x\int_0^1 \phi_j'^2\,\mathrm{d}x + 2\int_0^1 \phi_j'\bar{\phi}_j'\,\mathrm{d}x\int_0^1 \bar{\phi}_j\phi_j''\,\mathrm{d}x}{4\left(\mathrm{i}\,\omega_j\int_0^1 \bar{\phi}_j\phi_j\,\mathrm{d}x + \gamma_0\int_0^1 \bar{\phi}_j\phi_j'\,\mathrm{d}x\right)}, \chi_m = \frac{-\mathrm{i}\gamma_0\int_0^1 \bar{\phi}_m''\bar{\phi}_m\,\mathrm{d}x}{2\left(\mathrm{i}\,\omega_m\int_0^1 \phi_m\bar{\phi}_m\,\mathrm{d}x + \gamma_0\int_0^1 \phi_m'\bar{\phi}_m\,\mathrm{d}x\right)},$$

$$C_{jk} = \frac{\int_0^1 \phi_j''\bar{\phi}_j\,\mathrm{d}x\int_0^1 \bar{\phi}_k'\phi_k'\,\mathrm{d}x + \int_0^1 \phi_k''\bar{\phi}_j\,\mathrm{d}x\int_0^1 \bar{\phi}_k'\phi_j'\,\mathrm{d}x + \int_0^1 \overline{\phi_k''}\bar{\phi}_j\,\mathrm{d}x\int_0^1 \phi_k'\phi_j'\,\mathrm{d}x}{2\int_0^1 (\mathrm{i}\,\omega_j\phi_j + \gamma_0\phi_j')\bar{\phi}_j\,\mathrm{d}x} \quad (j = m, n; k = n, m). \tag{22}$$

The coefficients μ_j ($j=m,n$), χ_m, C_{mn}, C_{nm}, and κ_j ($j=m,n$) are determined by the natural frequencies and the modal function of linear system (7), which are dependent on γ_0 and k_f, and are independent of α, γ_1 and k_1. It can be numerically demonstrated that under the given boundary conditions

$$\mathrm{Re}(\mu_n)>0,\ \mathrm{Im}(\mu_n)=0;\ \mathrm{Re}(C_{jk})=0,\ \mathrm{Im}(C_{jk})>0;\mathrm{Re}(\kappa_n)=0,\ \mathrm{Im}(\kappa_n)<0. \tag{23}$$

Express the solution of equation (21) in the polar form

$$q_k(\tau)=a_k(\tau)\mathrm{e}^{\mathrm{i}\varphi_k(\tau)}\quad (k=m,n), \tag{24}$$

where real functions $a_m(\tau)$ and $\varphi_m(\tau)$ are respectively the amplitude and the phase angle of the response in the m-th principal parametric resonance, and real functions $a_n(\tau)$ and $\varphi_n(\tau)$ are possible contribution of the n-th mode to the response in the m-th principal parametric resonance. Substitution of equation (24) into equation (21) and separation of real and imaginary parts in the resulting equations give

$$\begin{aligned}\dot a_m &= \left[-\alpha\,\mathrm{Re}(\mu_m)+\gamma_1\,\mathrm{Im}(\chi_m)\sin(\sigma T_1-2\varphi_m)-\gamma_1\,\mathrm{Re}(\chi_m)\cos(\sigma T_1-2\varphi_m)\right]a_m,\\ a_m\dot\varphi_m &= -\gamma_1\left[\mathrm{Re}(\chi_m)\sin(\sigma T_1-2\varphi_m)+\mathrm{Im}(\chi_m)\cos(\sigma T_1-2\varphi_m)\right]a_m-k_1^2\,\mathrm{Im}(C_{mn})a_n^2a_m+k_1^2\,\mathrm{Im}(\kappa_m)a_m^3,\end{aligned} \tag{25}$$

$$\dot a_n=-\alpha\,\mathrm{Re}(\mu_n)a_n,\ a_n\dot\varphi_n=-k_1^2\,\mathrm{Im}(C_{nm})|A_m|^2a_n+k_1^2\,\mathrm{Im}(\kappa_n)a_n^3. \tag{26}$$

From equation (26), a_n decays to zero exponentially. Therefore, the n-th mode has actually no effect on the m-th principal parametric resonance.

Inserting a_n=0 into equation (25) leads to

$$\begin{aligned}\dot a_m &= [-\alpha\,\mathrm{Re}(\mu_m)+\gamma_1\,\mathrm{Im}(\chi_m)\sin\theta_m-\gamma_1\,\mathrm{Re}(\chi_m)\cos\theta_m]a_m,\\ a_m\dot\theta_m &= \sigma a_m+2\gamma_1[\mathrm{Re}(\chi_m)\sin\theta_m+\mathrm{Im}(\chi_m)\cos\theta_m]a_m-2k_1^2\,\mathrm{Im}(\kappa_m)a_m^3,\end{aligned} \tag{27}$$

where

$$\theta_m=\sigma T_1-2\varphi_m \tag{28}$$

Equation (27) is the same as that derived from the method of multiple scales [11].For the steady-state response, the amplitude a_m and the new phase angle θ_m in equation (27) are constant. Locating the fixed points, one can find the amplitudes of the steady-state responses. There is a fixed point at the origin (trivial zero solution), which represents the straight equilibrium configuration of the beam. In addition, there may exist nontrivial periodic solution with amplitudes defined by

$$a_{m1,2}=\frac{\sqrt{\mathrm{Im}(\kappa_m)}}{k_1}\sqrt{2\sigma\pm4\sqrt{\gamma_1^2|\chi_m|^2-\alpha^2[\mathrm{Re}(\mu_m)]^2}}, \tag{29}$$

It can be proved [8] that the larger one a_{m1} is always stable, and a_{m2} is always unstable.

From equation (9), (10), (16), and (24), the approximate expression of the stable steady-state responses in the m–th principal parametric resonance is

$$v_0(x,t)=2a_{m1}|\phi_m(x)|\cos\left[\frac{\omega}{2}t-\frac{1}{2}\arcsin\left(1-\frac{\alpha^2\,\mathrm{Re}^2(\mu_m)}{\gamma_1^2|\chi_m|^2}\right)+\arccos\frac{\mathrm{Re}\,\phi_m(x)}{|\phi_m(x)|}\right], \tag{30}$$

Hence the amplitude of the stable steady-state response is $A(x)=2a_{m1}|\phi_m(x)|$.

4. Differential quadrature verification

Stable steady-state response can be determined by solving governing equations numerically. In the following, the differential quadrature method [24,25] is applied to calculate stable steady-state responses. In numerical investigations here, let ε=1 in equation (1).

Introduce N unequally spaced sampling points as

$$x_i=\frac{1}{2}\left[1-\cos\frac{(i-1)\pi}{N-1}\right]\quad (i=1,2,\ldots,N). \tag{31}$$

The quadrature rules for the derivatives of a function at the sampling points yield

$$v_{,x}(x_i,t)=\sum_{j=1}^{N}A_{ij}^{(1)}v(x_j,t),v_{,xx}(x_i,t)=\sum_{j=1}^{N}A_{ij}^{(2)}v(x_j,t),v_{,xxxx}(x_i,t)=\sum_{j=1}^{N}A_{ij}^{(4)}v(x_j,t),v_{,xxxxx}(x_i,t)=\sum_{j=1}^{N}A_{ij}^{(5)}v(x_j,t), \tag{32}$$

where the weighting coefficients with the recurrence relationship are [24]

$$
\begin{aligned}
&A_{ij}^{(1)}=\prod_{k=1,k\neq i}^{N}\left(x_i-x_k\right)\Big/\left[\left(x_i-x_j\right)\prod_{k=1,k\neq j}^{N}\left(x_j-x_k\right)\right]\ (i,j=1,2,\cdots,N;\ j\neq i),\\
&A_{ij}^{(r)}=r\left[A_{ii}^{(r-1)}A_{ij}^{(1)}-\frac{A_{ij}^{(r-1)}}{x_i-x_j}\right]\ (r=2,\cdots,5;i,j=1,\cdots,N;\ j\neq i),A_{ii}^{(r)}=-\sum_{k=1,k\neq i}^{N}A_{ik}^{(r)}\ (r=1,\cdots,5;i=1,\cdots,N).
\end{aligned}
\tag{32}
$$

Substitution of equation (32) into equation (1) and (3) and modification of the weighting coefficient matrices to implement the boundary conditions yield

$$
\begin{aligned}
&\ddot{v}_j+\left(\sum_{k=1}^{N}\left(2\gamma A_{jk}^{(1)}+\alpha A_{jk}^{(4)}\right)\right)\dot{v}_k+\sum_{k=1}^{N}\left[\dot{\gamma}A_{jk}^{(1)}+\left(\gamma^2-1\right)A_{jk}^{(2)}+v_f^2A_{jk}^{(4)}+\alpha\gamma A_{jk}^{(5)}\right]v_k=\left(\sum_{k=1}^{N}A_{jk}^{(2)}v_k\right)\cdot\\
&\left\{\sum_{g=1}^{N}I_g\left[\frac{1}{2}k_1^2\left(\sum_{k=1}^{N}A_{gk}^{(1)}v_k\right)^2+\gamma k_2\left(\sum_{k=1}^{N}A_{gk}^{(2)}v_k\right)\left(\sum_{k=1}^{N}A_{gk}^{(1)}v_k\right)+k_2\left(\sum_{k=1}^{N}A_{gk}^{(1)}v_k\right)\left(\sum_{k=1}^{N}A_{gk}^{(1)}\dot{v}_k\right)\right]\right\},(j=2,3,\dots,N-1),\\
&v_0=v_N=0,
\end{aligned}
\tag{33}
$$

where $v_j(t)=v(x_j,t)(j=1,\dots,N)$ and weighting coefficients I_g ($g=1,2,\dots,N$) for integrals are solved from

$$
\begin{pmatrix}
1 & 1 & \cdots & 1 & 1\\
x_1 & x_2 & \cdots & x_{N-1} & x_N\\
\vdots & \vdots & & \vdots & \vdots\\
x_1^{N-2} & x_2^{N-2} & \cdots & x_{N-1}^{N-2} & x_N^{N-2}\\
x_1^{N-1} & x_2^{N-1} & \cdots & x_{N-1}^{N-1} & x_N^{N-1}
\end{pmatrix}
\begin{pmatrix} I_1\\ I_2\\ \vdots\\ I_{N-1}\\ I_N \end{pmatrix}
=
\begin{pmatrix} 1\\ 1/2\\ \vdots\\ 1/(N-1)\\ 1/N \end{pmatrix}.
\tag{34}
$$

For a set of given parameters k_f, γ_0, γ_1, ω, α, k_1, and k_2, v_j can be numerically solved from equation (33). For an odd N, $v_{(N+1)/2}$ is the beam center displacement. The stable steady-state response has constant local maximums of $v_{(N+1)/2}$, which is the amplitude of the beam center A. Changing ω and calculating corresponding A yield the curve of the amplitude via the detuning parameter.

Consider an axially moving beam with k_f=0.8 and γ_0=2.0. Then the first two natural frequencies of system (7) are ω_1=5.3692 and ω_2=30.12. Let γ_1=0.2, α=0.0001, and k_1=20. Based on the numerical solutions of equation (33), Figure 1 shows the amplitudes of the stable steady-state responses changing with the detuning parameter. The numerical results demonstrate that the lower order principal resonance has the larger stable response amplitude and the larger instability interval of the straight equilibrium. Figure 2 shows the effects of the mean axial speed based on the numerical solutions of equation (33). In figure 2, k_f=0.8, k_1=20, k_2=0, α=0.0001, γ_1=0.2, and γ_0=1.5 (dot lines) or 2.0 (solid lines). The larger mean axial speed leads to the larger stable response amplitude and the larger instability interval of the straight equilibrium. Figure 3 shows the effects of the axial speed variation amplitude based on the numerical solutions of equation (33). In figure 3, k_f=0.8, k_1=20, k_2=0, α=0.0001, γ_0=2.0, and γ_1=0.2 (dot lines) or 0.3 (solid lines). Both the stable response amplitude and the equilibrium instability interval increase with the axial speed variation amplitude. Figure 4 shows the effects of the dynamic viscosity based on the numerical solutions of equation (33). In figure 4, k_f=0.8, k_1=20, k_2=0,γ_1=0.2, γ_0=2.0, α=0.0001 (dot lines), 0.0005 (solid lines). The larger dynamic viscosity leads to the smaller stable response amplitude and the smaller instability interval of the straight equilibrium. Figure 5 shows the effects of the nonlinear coefficient k_1 based on the numerical solutions of equation (33). In figure 5, k_f=0.8, k_2=0, α=0.0001, γ_0=2.0, γ_1=0.2, and k_1=20, (dot lines) or 30 (solid lines). The stable response amplitude decreases with the nonlinear coefficient k_1, while the equilibrium instability interval is independent of k_1. Figure 6 shows the effects of the nonlinear coefficient k_2 based on the numerical solutions of equation (33). In figure 6, k_f=0.8, k_1=20, α=0.0001, γ_0=2.0, γ_1=0.2, and k_2=0, (dot lines) or 1 (solid lines). In practical cases, k_2 is usual smaller than 1. The stable response amplitude and the equilibrium instability interval are not sensitive to k_2. With respect to the mean axial speed, the axial speed variation amplitude, the dynamic viscosity, and nonlinear coefficient k_1, the numerical calculations demonstrate the qualitatively same changing tendencies as those predict by the analytical methods [8]. However, up to the first order approximation, the analytical methods cannot account for the change with nonlinear coefficient k_2.

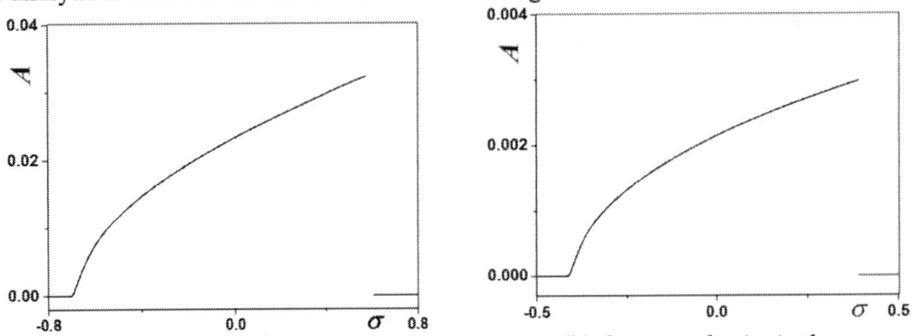

(a) the first principal parametric resonance (b) the second principal parametric resonance

Figure 1 Amplitudes of stable steady-state responses

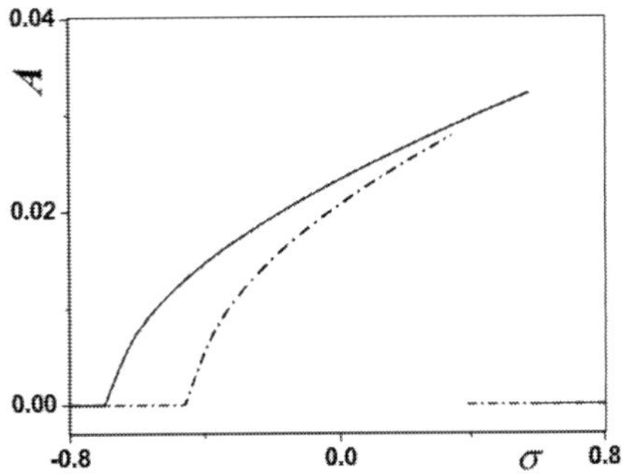

Figure 2 Effects of the mean axial speed

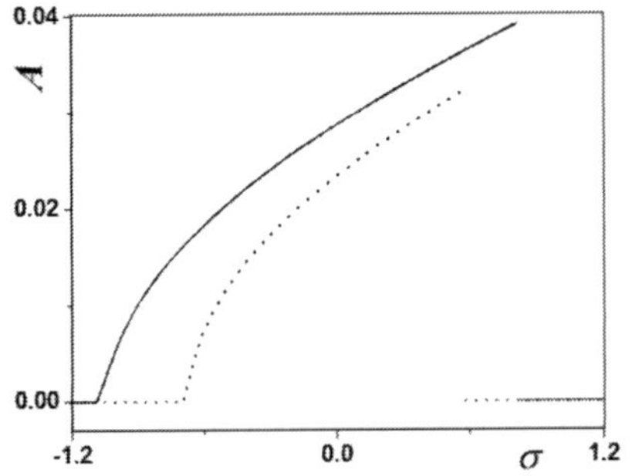

Figure 3 Effects of the axial speed variation amplitude

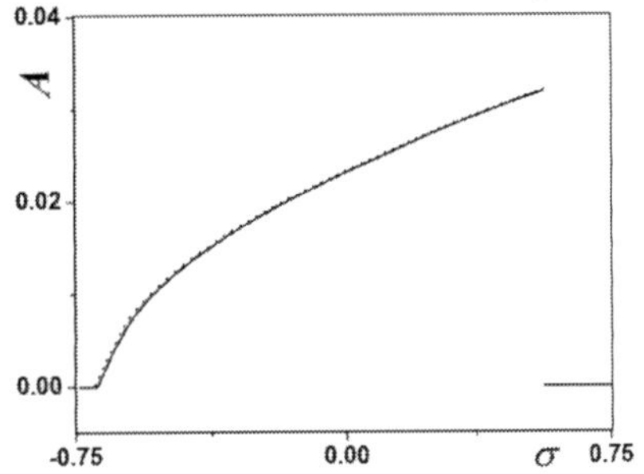

Figure 4 Effects of the dynamic viscosity

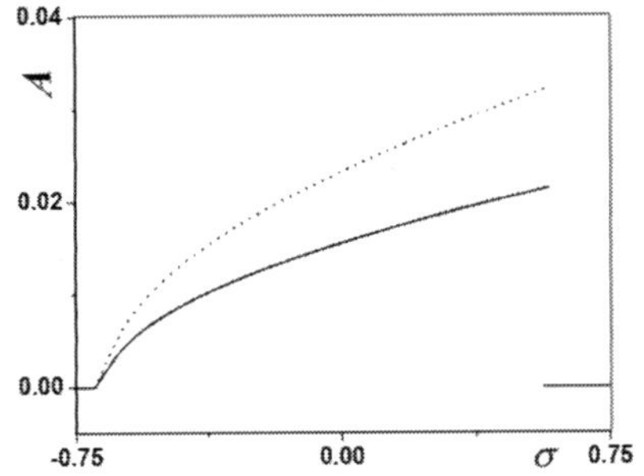

Figure 5 Effects of the nonlinear coefficient k_1

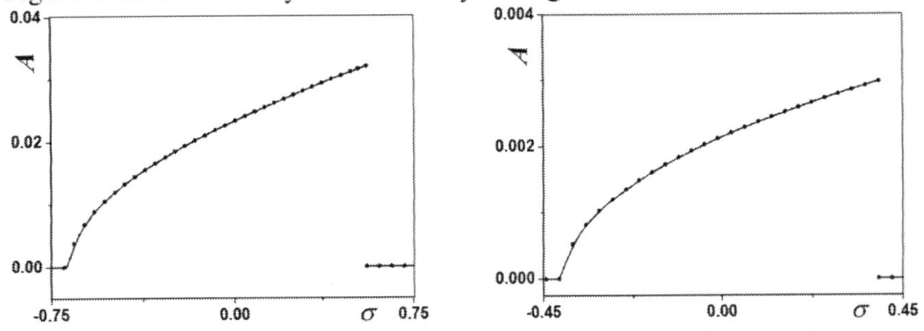

(a) the first principal parametric resonance (b) the second principal parametric resonance

Figure 6 Effects of the nonlinear coefficient k_2

Both analytical study and numerical calculations not only yield the qualitatively same results, but also they are quantitatively close. The both results are quantitatively contrasted in the case with k_f=0.8, k_1=20, k_2=0, α=0.0001, γ_0=2.0, γ_1=0.2. Figure 7 shows the comparison, in which the solid and dot lines represent the results of the asymptotic analysis method and the differential quadrature method respectively. The amplitudes from both methods are almost coincided, especially near the exact-resonance (σ=0) and in the first resonance.

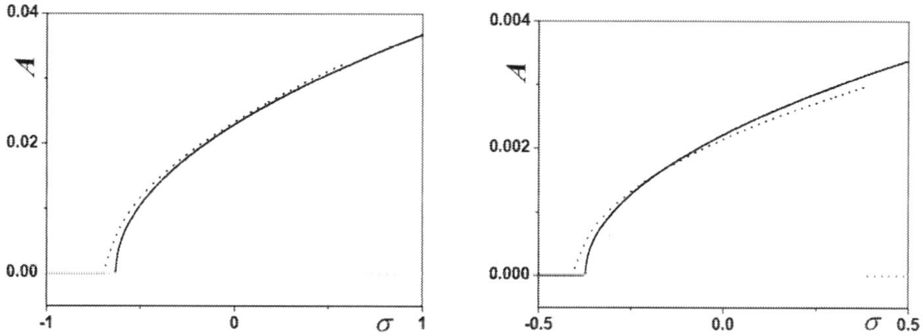

(a) the first principal parametric resonance (b) the second principal parametric resonance

Figure 7 Comparison of analytical and numerical results

5. Conclusions

This paper is devoted to steady-state response in principal parametric resonance of axially accelerating viscoelastic beams. The asymptotic analysis is carried out for a nonlinear integro-partial-differential equation to determine steady-state responses. The differential quadrature schemes are developed to verify the approximate analytical results. The investigation leads to the following conclusions. (1) Both analytical study and numerical calculations demonstrate that, when the axial speed variation frequency is close to 2 times of any linear natural frequency enough, the straight equilibrium configuration become unstable. The instability interval increases with the mean and the variation amplitude of the axial speed, decreases with the principal parametric resonance order and the dynamic viscosity, and is independent of nonlinear coefficients. Analytical study also demonstrates that the instability interval decreases with the boundary constraint stiffness. (2) Both analytical study and numerical calculations demonstrate that, when the straight equilibrium configuration begins to lose its stability, a stable steady-state response emerges. The stable steady-state response amplitude increases with the detuning parameter and the mean and the variation amplitude of the axial speed, decreases with the principal parametric resonance order, the dynamic viscosity, and the nonlinear coefficients. Therefore, in practical engineering applications, the axial speed variation frequency should be away from 2 times of any linear natural frequency in order to avoid severe transverse vibration, especially in the case of high mean translating speed and small damping and nonlinearity. (3) Analytical study demonstrates that, when the unstable straight equilibrium configuration becomes stable again, an unstable steady-state response emerges. The unstable steady-state response amplitude increases with the detuning parameter, the dynamic viscosity, and the boundary constraint stiffness, decreases with the principal parametric resonance order, the variation amplitude of the axial speed, and the nonlinear coefficient. (4) Analytical study demonstrates that the mode uninvolved in the resonance has no effect on the steady-state response.

Acknowledgments
This work was supported by the National Outstanding Young Scientists Foundation of China (Project No. 10725209) and the National Natural Science Foundation of China (Project No. 10672092).

References
[1] Ravindra, B., and Zhu, W.D., 1998, "Low dimensional chaotic response of axially accelerating

continuum in the supercritical regime," Ach. Appl. Mech., **68**, pp. 195-205.

[2] Chakraborty, G., and Mallik, A.K., 1998, "Parametrically excited nonlinear traveling beams with and without external forcing," Nonlinear Dyn., **17**, pp. 301-324.

[3] Öz, H.R., Pakdemirli, M., and Boyaci, H., 2001, "Non-linear vibrations and stability of an axially moving beam with time-dependent velocity," Int. J. Non-Linear Mech., **36**, 107-115.

[4] Parker, R. G., and Lin, Y., 2001, "Parametric instability of axially moving media subjected to multifrequency tension and speed fluctuations," ASME J. Appl. Mech., **68**, pp. 49-57.

[5] Marynowski, K., 2002, "Non-linear dynamic analysis of an axially moving viscoelastic beam," J. Theor. Appl. Mech., **40**, pp. 465-482.

[6] Marynowski, K., and Kapitaniak, T., 2002, "Kelvin-Voigt versus Bügers internal damping in modeling of axially moving viscoelastic web," Int. J. Non-Linear Mech., **37**, pp. 1147-1161.

[7] Marynowski, K., 2004, "Non-linear vibrations of an axially moving viscoelastic web with time-dependent tension," Chaos Soliton Frac., **21**, pp. 481-490.

[8] Chen, L.Q., and Yang, X.D., 2005, "Steady-state response of axially moving viscoelastic beams with pulsating speed: comparison of two nonlinear models," Int. J. Solid Struct., **42**, pp. 37-50.

[9] Marynowski, K., and Kapitaniak, T., 2007, "Zener internal damping in modelling of axially moving viscoelastic beam with time-dependent tension," Int. J. Non-linear Mech., **42**, pp. 118-131.

[10] Ghayesh, M.H., and Khadem, S.E., 2008, "Rotary inertia and temperature effects on non-nonlinear vibration, steady-state response and stability of an axially moving beam with time-dependent velocity," Int. J. Mech. Sci., **50**, pp. 389-404.

[11] Chen, L.Q., and Ding H., 2008, "Steady-state responses of axially accelerating viscoelastic beams: approximate analysis and numerical confirmation," Sci. China G: Phys., Mech. Astro., **51**, pp. 1701-1721.

[12] Ghayesh, M.H., and Balar, S., 2008, "Non-linear parametric vibration and stability of axially moving visco-elastic Rayleigh beams." Int. J. Solid Struct., **45**, pp. 6451-6467.

[13] Boertjens, G.J., and van Horssen, W.T., 1998, "On mode interactions for a weakly nonlinear beam equation," Nonlinear Dyn. **12**, pp. 23-40.

[14] Maccari, A., 1999, "The asymptotic perturbation method for nonlinear continuous systems," Nonlinear Dyn. **19**, pp. 1–18.

[15] Boertjens, G.J., and van Horssen, W.T., 2000, "An asymptotic theory for a weakly nonlinear beam equation with a quadratic perturbation," SIAM J. Appl. Math. **60**, pp. 602-632.

[16] Boertjens, G.J., and van Horssen, W.T., 2000, "On interactions of oscillation modes for a weakly non-linear undamped elastic beam," J. Sound Vib. **235**, pp. 201-217.

[17] Maccari, A., 2000, "Saddle-Node bifurcations of cycles in a relief valve," Nonlinear Dyn. **22**, pp. 225–247

[18] Andrianov, I.V., and Danishevs'kyy, V.V., 2002, "Asymptotic approach for non-linear periodical vibrations of continuous structures," J. Sound Vib. **249**, pp. 465-481.

[19] Chen, R.M., 2006, "Some nonlinear dispersive waves arising in compressible hyperelastic plates," Int. J Eng. Sci. **44**, pp. 1188-1204.

[20] Chen, L.Q., Chen, H., Lim, C.W., 2008, "Asymptotic analysis of axially accelerating viscoelastic strings," Int. J. Eng. Sci., **46**, pp. 976-985.

[21] Chen, L.Q., and Wang, B., 2009, "Stability of axially accelerating viscoelastic beams: asymptotic perturbation analysis and differential quadrature validation," Eur. J.Mech. A/Solid, in press

[22] Wickert, J.A., Mote, C.D.Jr., 1990, "Classical vibration analysis of axially moving continua," ASME J. Appl. Mech., **57**, pp. 738-744.

[23] Öz, H.R., and Pakdemirli, M., 1999, "Vibrations of an axially moving beam with time dependent velocity," Journal of Sound Vibration, **227**, pp. 239-257

[24] Bert, C. W., and Malik, M., 1996, "The differential quadrature method in computational mechanics: A review," Appl. Mech. Rev., **49**, pp. 1-28

Journal of Physics: Conference Series **181** (2009) 012009 doi:10.1088/1742-6596/181/1/012009

Non-linear friction in reciprocating hydraulic rod seals: simulation and measurement

A K Bullock, D G Tilley, D N Johnston, C R Bowen, P S Keogh

Department of Mechanical Engineering, University of Bath, Bath BA2 7AY, UK

a.k.bullock@bath.ac.uk

Abstract. Non-linear seal friction can impede the performance of hydraulic actuation systems designed for high precision positioning with favourable dynamic response. Methods for predicting seal friction are required to help develop sealing systems for this type of application. Recent simulation techniques have claimed progress, although have yet to be validated experimentally. A conventional reciprocating rod seal is analysed using established elastohydrodynamic theory and the mixed lubrication Greenwood Williamson average Reynolds model. A test rig was used to assess the accuracy of the simulation results for both instroke and outstroke. Inverse hydrodynamic theory is shown to predict a $U^{0.5}$ power law between rod speed and friction. Comparison with experimental data shows the theory to be qualitatively inaccurate and to predict friction levels an order of magnitude lower than those measured. It was not possible to model the regions very close to the inlet and outlet due to the high pressure gradients at the edges of the contact. The mixed lubrication model produces friction levels within the correct order of magnitude, although incorrectly predicts higher friction during instroke than outstroke. Previous experiments have reported higher friction during instroke than outstroke for rectangular seals, suggesting that the mixed lubrication model used could possibly be suitable for symmetric seals, although not for seal tribology in general.

Nomenclature

E	12.1×10^{6}	Seal elastic modulus (Pa)
h		Film thickness (m)
h_a		Film thickness at critical point (m)
h_m		Film thickness integration constant (m)
I		Influence coefficient
L	4.5×10^{-3}	Contact length (m)
N	10×10^{12}	Asperity density (m^{-2})
p_a	0.1×10^{6}	Gauge pressure (Pa)
p_c		Asperity contact pressure (Pa)
R	1×10^{-6}	Asperity radius (m)
U		Rod speed (ms^{-1})
x		Axial coordinate (m)
α	20×10^{-9}	Pressure-viscosity coefficient (Pa^{-1})

H	Dimensionless film thickness
H_s	Dimensionless static film thickness
H_T	Dimensionless truncated film thickness
P_c	Dimensionless asperity contact pressure p_c/p_a
P_f	Dimensionless fluid pressure p_f/p_a
P_t	Dimensionless total pressure P_f+P_c
$\hat{x}$	Dimensionless axial coordinate x/L
$\hat{\alpha}$	Dimensionless pressure-viscosity coefficient
$\hat{\sigma}$	Dimensionless roughness $\sigma_h N^{2/3} R^{1/3}$
ζ	Dimensionless rod speed $\eta_0 UL/(p_a \sigma_h^3)$

η_0	0.043	Viscosity at p_a (Pa s)
σ_h	0.4×10^{-6}	Average surface roughness height (m)
υ	0.49	Seal Poisson ratio
φ_{sc}		Shear flow factor

1. Introduction

Seal friction characteristics are important in hydraulic actuation systems designed to combine high precision positioning with favourable dynamic performance. Hydraulic seals typically have non-linear friction characteristics that can impose negative damping on the system. This non-linearity combined with slip-stick effects can impede how precisely an actuation system can follow changes in demand position. Accurate knowledge of seal friction behaviour can determine the precision and bandwidth limitations imposed on hydraulic systems by the sealing components. Simulation techniques for predicting seal friction can also be used to develop new seal concepts that allow improved precision and dynamic performance.

Previous investigations have had debatable success in obtaining agreement between simulations and experimental data. The most common approach has been to use elastohydrodynamic theory to predict the lubricant film thickness and calculate friction from the fluid shear stress. Several studies [1-4] have used the method of inverse solution of the Reynolds equation. An alternative approach [5] combining asperity contact with elastohydrodynamic theory was recently adapted for reciprocating seals in [6]. This mixed lubrication approach has yet to be definitively verified against experimental data.

Several studies [7-9] have measured seal friction experimentally. It is generally accepted that significant amounts of data scatter exist between different seals due to uncertainty in the manufactured geometry [10]. There is also relatively poor repeatability for successive cycles with a particular seal. The ISO 7986:1997 standard was recently established in attempt to reduce results variation between different studies. A definitive experimental study of seal friction to this standard was carried out by Parker Hannifin [11]. A few experimental studies [8, 12] have compared instroke friction to outstroke. Having the ability to measure friction in each direction of stroke would assist in validating simulations that predict direction-dependent friction characteristics.

The current study employs existing tribology models to predict the frictional behaviour of a standard U-cup type rod seal. An established method of the inverse solution of the Reynolds equation has been used. An alternative set of friction predictions has been obtained with the Greenwood-Williamson-Average Reynolds mixed lubrication model. A test rig has been produced and the seal friction forces measured for a range of operating conditions.

2. Methodology

2.1. Inverse hydrodynamic method

Most previous simulations of reciprocating seal tribology have used elastohydrodynamic theory. In this approach the opposing surfaces are assumed to be completely separated by a film of lubricant and the only source of friction is fluid shear stress. It is further assumed that the additional pressure required to deform the seal and create a channel for the fluid film is small compared with the pressure obtained when there is no fluid under the seal. This simplification is usually made due to the low elastic modulus of the seal material and low film thicknesses relative to the total seal depth.

To obtain the pressure distribution a finite element model (FEM) was produced for the geometry of a Parker B3 1624 rod seal (figure 1) using ANSYS 11. The fluid film between the rod and seal was not modelled at this stage. A uniform pressure of 80 bar was applied to the faces on the sealed fluid side of the seal and plane strain conditions were assumed. The pressure distribution obtained is shown in figure 2.

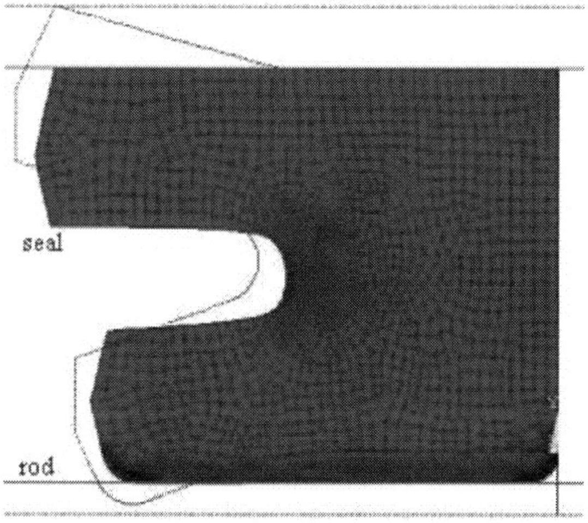

Figure 1. Finite element discretisation of the seal in order to obtain pressure distribution in the absence of a fluid film

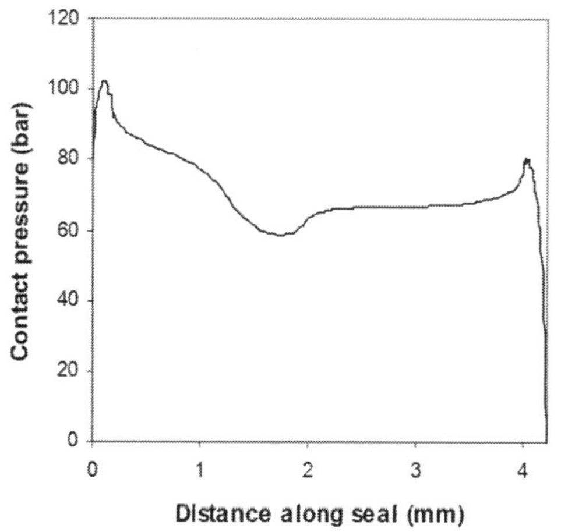

Figure 2. Contact pressure distribution obtained over the seal geometry from FEM. 80 bar sealed pressure.

Inverse hydrodynamic theory was used to calculate the film thickness from the pressure distribution. Fluid pressure and film thickness across the seal were assumed to be governed by the one-dimensional Reynolds equation with an exponential relationship between fluid pressure and viscosity:

$$\frac{d}{dx}\left(\frac{h^3}{\eta_0}e^{-\alpha p}\frac{dp}{dx}\right) = 6U\frac{dh}{dx} \tag{1}$$

A single integration produces the standard integrated form of the Reynolds equation:

$$e^{-\alpha p}\frac{dp}{dx} = 6U\eta_0\left(\frac{h-h_m}{h^3}\right) \tag{2}$$

To obtain a value for h_m equation (1) is rearranged to give

$$h^3\frac{d}{dx}\left(\frac{1}{\eta_0}e^{-\alpha p}\frac{dp}{dx}\right) - \frac{dh}{dx}\left(6U - \frac{3h^2}{\eta_0}e^{-\alpha p}\frac{dp}{dx}\right) = 0 \tag{3}$$

As the pressure distribution is known, a location where

$$\frac{d}{dx}\left(\frac{1}{\eta_0}e^{-\alpha p}\frac{dp}{dx}\right)=0 \tag{4}$$

can be identified. This point was approximately 0.09mm from the seal inlet on figure 2. At this point the local film thickness can be calculated from

$$h_a = \left(\frac{2U\eta_0 e^{-\alpha p}}{\left(\frac{dp}{dx}\right)_a} \right)^{1/2} \tag{5}$$

Using equation (5) to substitute for the pressure gradient in equation (2) allows the constant of integration in (2) to be evaluated as $h_m = \frac{2}{3}h_a$. Equation (2) is then used to calculate the film thickness across the seal. The shear stresses exerted on the rod by the fluid are then calculated. Solutions were obtained for a range of rod speeds between 0.1×10^{-3} m/s and 0.5 m/s (figures 4 and 5.)

2.2. Greenwood-Williamson-average Reynolds method

During mixed lubrication there are significant contributions to overall friction from both asperity contact and fluid viscosity. The total pressure at any point is normally assumed to be the sum of the nominal fluid and asperity contact pressures. In one approach [5] the asperity contact pressure distribution was determined from the surface mean separation using the Greenwood-Williamson (GW) contact model [13] and the fluid pressure from the solution of the Reynolds equation. This physical concept was adapted to reciprocating seals in [6] with a different numerical technique used to solve the coupled asperity and fluid equations.

For a particular film thickness distribution the fluid pressure was calculated from the average Reynolds equation using the finite difference method. A dimensionless form of the Reynolds equation was used with Patir-Cheng semi-empirical flow factors [14, 15] to account for the effects of surface roughness on flow:

$$\frac{d}{d\hat{x}}\left(\phi_x H^3 e^{-\hat{\alpha}P} \frac{dP}{d\hat{x}} \right) = 6\zeta\left(\frac{dH_T}{d\hat{x}} + \frac{d\phi_{s,c}}{d\hat{x}} \right) \tag{6}$$

A Reynolds condition cavitation model was used where a pressure gradient of zero is forced at the beginning of the cavitated region.

The GW contact model was used to calculate the nominal asperity contact pressure distribution from the film thickness. GW contact theory is based on the analytical solution of contact with parabolic asperities averaged over a Gaussian distribution of asperity heights. For a particular mean film thickness the nominal asperity contact pressure is given by

$$p_c = \frac{4}{3}\frac{1}{\left(1-\nu^2\right)} E\hat{\sigma}^{\frac{3}{2}} \frac{1}{\sqrt{2\pi}} \int_H^\infty (z-H)^{\frac{3}{2}} e^{-\frac{1}{2}z^2} dz \tag{7}$$

The additional pressure required to open the fluid channel by deflecting the seal was included in the analysis. These additional pressures were much lower than the pressures in the undeformed case (figure 2) and the final solution was relatively insensitive to these pressure additions. The relationship between pressure and seal deformation was linearised as a compliance matrix with the deflection at a particular node given by

$$H_i = H_s + \sum_{k=1}^{n} (I)_{ik} (P_t - P_{sc})_k \tag{8}$$

To obtain the compliance matrix of influence coefficients the stiffness matrix was first obtained from the FEM. Each node along the surface was deflected in sequence while the remaining surface nodes were constrained in position. The stiffness matrix produced was inverted numerically to give the compliance matrix.

In the overall solver the film thickness is iterated using the successive over-relaxation method until a film thickness profile is reached where the sum of the fluid and asperity contact pressures is equal to the sum of the undeformed pressure and the additional pressure from deflecting the seal. The final converged film thickness distribution produces fluid and asperity pressures that are compatible with the film thickness given by the undeformed and deformation pressures in equation (8). The undeformed pressure distribution was obtained from FEA and was similar to that shown in figure 2. It was found necessary to remove the curvature at the corners of the seal geometry to produce a pressure distribution that would converge with the GW-average Reynolds iteration. Allowing local pressure peaks near the inlet and outlet of the pressure distribution creates local contractions in the film thickness distribution that appear to inhibit the stability of the GW-average Reynolds coupling.

2.3. Experimental friction measurement

An experimental test rig was produced to measure the friction of a Parker B3 1624 rod seal during instroke and outstroke. Complete adherence to the ISO 7986 standard for rod seal friction measurement would have required particularly long testing programmes that were not felt to be appropriate for the current project. Additionally, the relevant ISO standard does not allow for direction-dependent friction measurement which would be useful for validating current simulation techniques. Figure 3 indicates the arrangement of the components in the test rig.

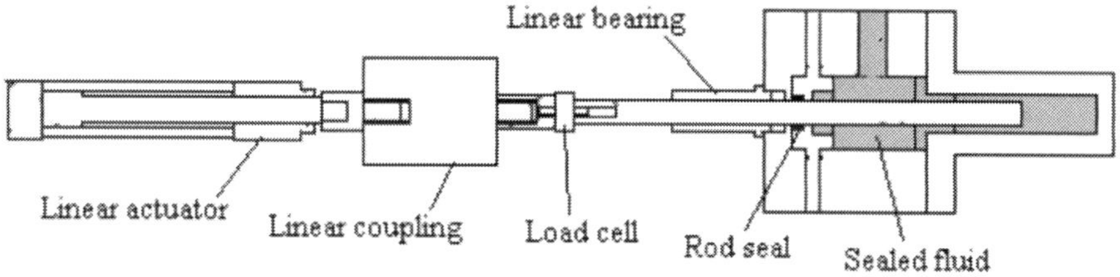

Figure 3. Test rig diagram used for measurement of rod seal friction

The linear actuator was used to extend and retract the rod through the seal housing at a constant velocity. Friction force was measured from the load cell connected in series between the rod and actuator. Hydraulic fluid was supplied to the seal housing at a pressure set by a relief valve. Data were collected during instroke and outstroke for a range of constant speeds between 0.1×10^{-3} m/s and 0.3 m/s and sealed pressures between 0 bar and 80 bar. An effective stroke length of 250 mm was used for most of the testing except at low speeds where a maximum stroke time of 15 s was set. A minimum stroke length of 5 mm was used for the lowest rod speeds.

Some post-processing of the load cell readings was required to correct for the force exerted on the end face of the rod by the pressurized fluid. The rod was retracted to a fixed position and steady state load cell readings taken for a range of sealed pressures. For all measured friction data the forces from the fluid on the rod face were interpolated from the fluid pressure measured in the housing.

3. Results

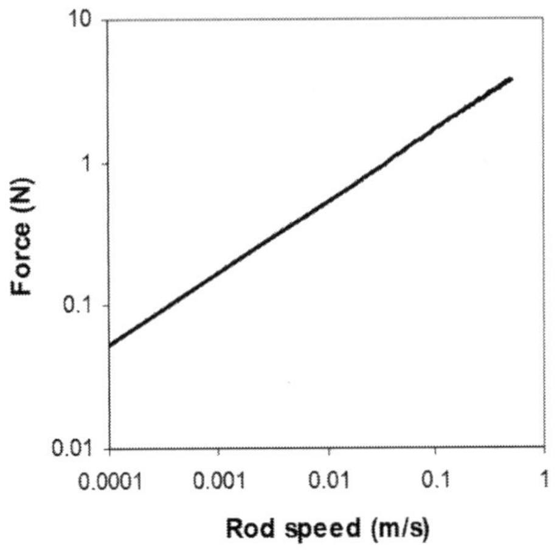

Figure 4. Outstroke friction levels from inverse EHL solutions of Reynolds equation over contact pressure distribution at 80 bar sealed pressure.

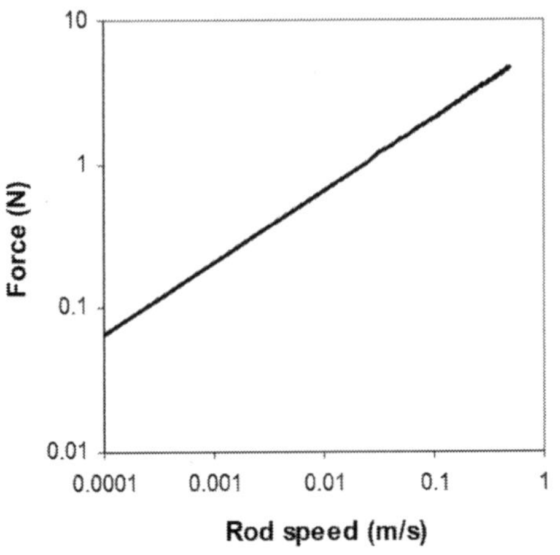

Figure 5. Instroke friction levels from inverse EHL solutions of Reynolds equation over contact pressure distribution at 80 bar sealed pressure.

Note close similarity between instroke and outstroke friction predictions from inverse EHL theory.

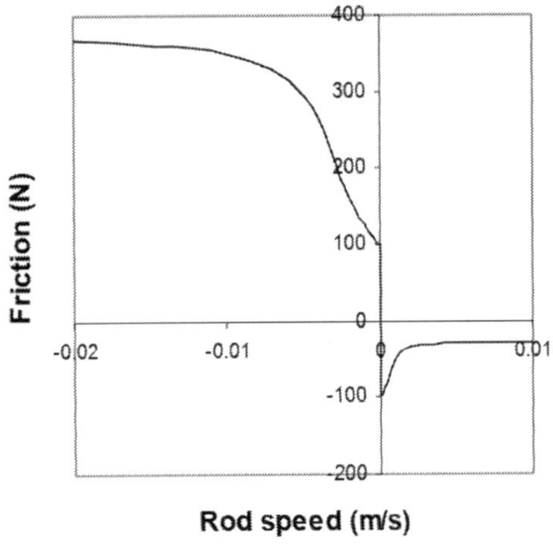

Figure 6. Seal friction levels predicted by Greenwood-Williamson-average Reynolds theory

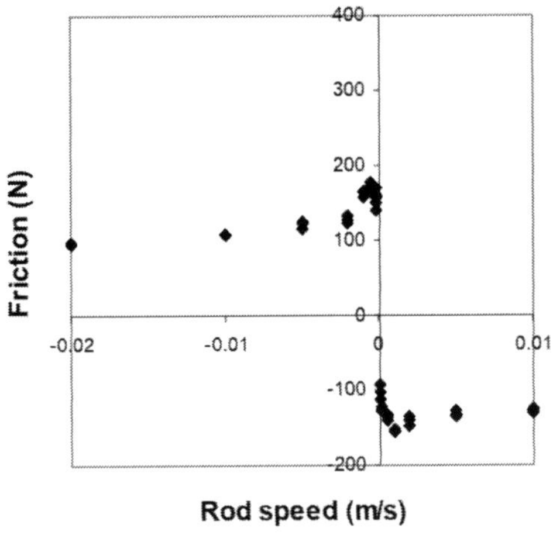

Figure 7. Seal friction levels measured from test rig during instroke and outstroke

4. Discussion

4.1. Inverse elastohydrodynamic solution

Figures 4 and 5 suggest there to be a $U^{0.5}$ power law relationship between rod speed and friction force. Previous studies such as [4] have predicted similar friction behaviour for rectangular seals when using a similar method. Inspection of equation (5) indicates the film thickness at the inflexion point in the reduced pressure distribution to be proportional to $U^{0.5}$. The fluid shear stress from the Couette term in the Reynolds equation is directly proportional to rod speed and also proportional to the reciprocal of film thickness. Therefore viscous friction is expected to be proportional to $U^{0.5}$ if the film thickness h_a is indicative of the mean film thickness.

Comparing the simulation results with the experimental measurements taken (figure 7) clearly indicates any simple relationship between rod speed and friction will not be possible. The experiments also showed the measured friction levels to be at least an order of magnitude greater than predicted by elastohydrodynamic theory. It is a known problem (noted in [12]) that the high measured friction levels suggest boundary lubrication is dominant while other experimental techniques suggest a fluid film is present.

The undeformed pressure distribution obtained from FEA (figure 2) is in close agreement with previous investigations [7] that have experimentally measured contact pressures for a U-cup type seal. Peaks in the pressure distribution near the inlet and outlet have been known about since the 1960s [16.] These peaks are caused by Hertzian stresses between parabolic surfaces.

One of the problems of incorporating the near-inlet and outlet peaks into the model is that the gradient of the Hertzian pressure profile approaches infinity as the outlet or inlet is approached. It is not physically realistic to have very high pressure gradients within the fluid. In practice it would be expected for there to be inlet and outlet regions near where the seal begins to rapidly recede from the surface and the fluid pressure distribution does not match the undeformed case. It is not possible to calculate the film thickness in very close proximity to the inlet and outlet using the undeformed pressure distribution. For the current study it has been assumed that the fluid behaviour in the inlet and outlet regions does not significantly affect the film thickness outside these regions.

4.2. GW-average Reynolds solution

Figure 6 indicates there to be significantly higher friction levels during instroke than during outstroke. The higher friction levels occur due to fluid cavitation extending from the outlet where (in the absence of fluid pressure) high asperity contact pressures are required to create the required total pressures. Coulomb friction from asperity shear contributes far more than viscous fluid shear to total friction.

There is a reasonable experimental agreement for the outstroke friction predicted by the GW-average Reynolds simulation. Although there is the obvious problem that the high friction levels simulated during instroke do not appear to occur in practice. Inclusion of the GW contact model allows friction levels of the correct order of magnitude to be predicted compared with pure elastohydrodynamic theory.

4.3. Experimental data and other experimental studies

There is a reasonable agreement with a previous study of a similar seal [7] for the speed dependence of friction. It is significant that there appears to be little direction dependence on the friction levels. Previous experimental studies [8, 12] that have compared instroke and outstroke behaviour have been concerned with rectangular profile seals. These studies showed rectangular seals to have significantly higher friction levels during instroke. It is unclear why rectangular seals should have quite different friction characteristics between instroke and outstroke while U-cup seals appear not to. The fact that the GW-average Reynolds model predicts higher friction during instroke suggests the technique may produce accurate results if applied to rectangular seals. For the current line of research into developing new seal designs and concepts it is not particularly useful to have a modelling technique that is only accurate for a narrow range of seal geometries. It is possible that the GW-average Reynolds method does not describe what physically happens in rectangular seals and may produce reasonably accurate friction predictions by coincidence.

5. Conclusions

Use of inverse hydrodynamic theory predicts a $U^{0.5}$ power law relating rod speed to friction. This relationship was not observed in experimental data. Elastohydrodynamic modelling predicts friction levels an order of magnitude lower than are observed in practice. This indicates conventional lubrication approaches to be unsuitable for friction modelling in elastomeric seals. It is not possible to include the regions immediately near the inlet and outlet with the method used. The GW-average Reynolds model was found to predict much higher friction for instroke than outstroke due to fluid cavitation occurring near the outlet during instroke. In cavitated regions the asperity contact pressure is higher, which increases friction through additional asperity shear.

Rod seal friction measured from the test rig showed similar speed-dependent behaviour to previous publications. There was little difference in measured friction between instroke and outstroke for the U-cup seal. Previous studies of rectangular seals have reported higher friction during instroke. It is possible that the GW-average Reynolds model may predict accurate results for rectangular seals, although not for reciprocating seals in general.

References

[1] Hirano F, 1967. Dynamic inverse problems in hydrodynamic lubrication. *Proc. of the Third BHRA Int. Conf. on Fluid Sealing*. BHRA: Cambridge, UK

[2] Hirano F and Kaneta M, 1971. Theoretical investigation of friction and sealing characteristics of flexible seals for reciprocating motion. *Proc. of the Fifth BHRA Int. Conf. on Fluid Sealing*.

[3] Hooke C J and O'Donoghue J P, 1972. Elastohydrodynamic lubrication of soft highly deformed contacts. *Proc. of the Inst. of Mech. Eng. Part C, Mech. Eng. Sci.* **14** 34-48

[4] Johannesson H L, 1983. Oil leakage and friction forces of reciprocating O-ring seals considering cavitation. *J. of Lubrication Technology Trans. of the ASME*. **105** 288-296

[5] Johnson K L, Greenwood J A and Poon S Y, 1972. Simple theory of asperity contact in elastohydrodynamic lubrication *Wear* **19** 91-108

[6] Salant R F, Maser N and Yang B, 2007. Numerical model of a reciprocating hydraulic rod seal. *ASME J. of Tribology* **129** 91-97

[7] Kawahara Y et al, 1981 Oil film formation of oil seals for reciprocating motion, *Proc. of the Ninth BHRA Int. Conf. on Fluid Sealing* Noordwijkerhout, Netherlands.

[8] Field J G and Nau B S, 1975. The effects of design parameters on the lubrication of reciprocating rubber seals. *Proc. of the Seventh BHRA Int. Conf. on Fluid Sealing*. BHRA: Nottingham, UK

[9] Wassink D B et al, 2001. Physically based modeling of reciprocating lip seal friction. *J. of Tribology*, **123** 404-412

[10] Nau B S 1999. A historical review of studies of polymeric seals in reciprocating hydraulic systems *Proc. of the Inst. of Mech. Eng. Part J, J. of Engineering Tribology* **213** 215-226

[11] Papatheodorou T, 2005. Influence of hard chrome plated rod surface treatments on sealing behaviour of hydraulic rod seals. *Sealing Technology April 2005*. 5-10

[12] Field J G and Nau B S, 1973. Film thickness and friction measurements during reciprocation of a rectangular section rubber seal ring. *Proc. of the Sixth BHRA Int. Conf. on Fluid Sealing*. BHRA: Munich, Germany

[13] Greenwood J A and Williamson J B P, 1966. Contact of nominally flat surfaces *Proc. of Royal Society of London, Ser. A* **295** 300-319

[14] Patir N and Cheng H S, 1978. An average flow model for determining effects of three-dimensional roughness on partial hydrodynamic lubrication. *J. of Lubrication Technology Trans. of the ASME*. **100** 12-17

[15] Patir N and Cheng H S, 1979. Application of average flow model to lubrication between rough sliding surfaces. *J. of Lubrication Technology Trans. of the ASME*. **101** 220-229

[16] Dowson D and Sayles P D, 1969. The development of elastohydrodynamic conditions in a reciprocating seal. *Proc. of the Fourth BHRA Int. Conf. on Fluid Sealing*

Plain bearing stresses due to forming and oil film pressure

A. Burke-Veliz[1], D. Wang[2], N. Wahdy[2], P.A.S. Reed[1], D. Merritt[2], S. Syngellakis[1].

[1]University of Southampton, School of Engineering Sciences, Southampton SO17 1BJ, UK.

[2]MAHLE Engine Systems UK Ltd, 2 Central park Drive, Rugby CV23 0WE, U.K.

allan.burke@itesm.mx

Abstract. This paper describes a methodology for assessing critical stress ranges arising in automotive plain bearings during engine operations. An industry-produced and run simulation program provides information on oil film pressure and overall bearing deformation during accelerated performance tests. This code performs an elasto-hydrodynamic lubrication analysis accounting for the compliance of the housing and journal. Finite element analyses of a multi-layer bearing are performed to assess the conditions responsible for possible fatigue damage over the bearing lining. The residual stresses arising from the forming and fitting process are first assessed. The stress analyses over the engine cycle show the intensity and distribution of cyclic tensile and compressive stresses in the bearing. The location of maximum stress range is found to be consistent with the damage observed in accelerated fatigue tests. Critical zones are identified in the lining for possible fatigue crack initiation and growth studies.

1. Introduction

The introduction of new materials and design processes in the automotive industry, driven by the quest for improved power efficiency, has resulted in smaller, more lightweight and durable components, which however operate under more demanding service conditions. The design of plain bearings is one case in point, because of their importance to the powertrain system and their strategic position within the engine. The bearing design process is complex and involves assessment of manufacture, assembly and service conditions. The study of the latter is especially interesting and challenging since it is currently based on elasto-hydrodynamic lubrication (EHL) analyses whereby the hydrodynamic oil film pressure is coupled with the elastic deformation of the housing and the journal. The predicted film pressure, shown schematically in Figure 1, is characterized by steep gradients, which may lead to damage over the bearing surface arising from fatigue, wear, cavitation erosion and fretting among others.

The study of fatigue damage in plain bearings has mainly involved semi-empirical approaches to the assessment of the total service life. Thus, the number of cycles to failure has been linked to the specific load, that is, the load applied through the connecting rod divided by the projected bearing area. Simplified two-dimensional [1-3] and, to a lesser extent, three-dimensional [4] finite element analyses of bearings under hydrodynamic film pressure have attempted to link the likely position of fatigue crack initiation and growth to the maximum tensile tangential stress developing in the bearing lining.

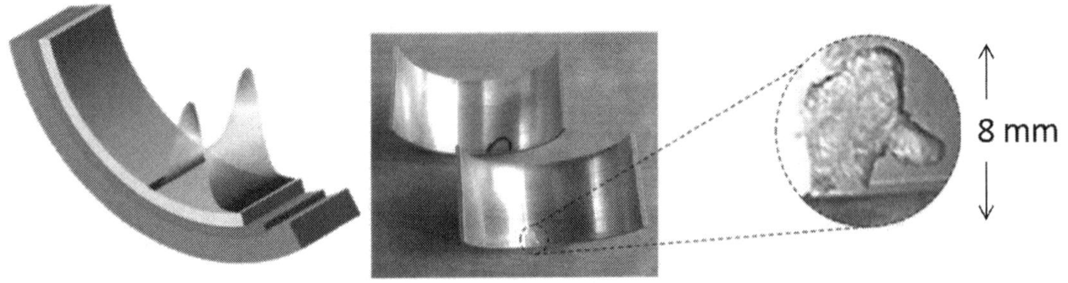

Figure 1. Tri-layer bearing with representative hydrodynamic film pressure profile (left) and fatigue damage in the lining (right).

The elastic flat strip models by Martin et al [1] and bearing shell by McCallion and Lofti [2] showed a good correlation between the fatigue crack origins in test specimens and their assessments of the maximum tangential stress. In the early 1990s, Bahai and Xu [3] predicted the stresses due to the oil film pressure through an axisymetric elasto-plastic finite element bearing shell model; they concluded that tensile deviatoric tangential stresses are present at the observed crack initiation site during the whole engine cycle while their hydrostatic counterpart dominates the overall stress state in a compressive environment. Hacifazlioglu and Karadeniz [4] studied the effects of the housing rigidity and bearing surface discontinuities on tangential tensile stresses in a three-dimensional flat strip model. Perez et al [5] used an elastic bearing shell model to assess the effects of housing rigidity as well as backing stiffness and lining thickness on tangential stresses. The influence of manufacturing and assembly processes on bearing performance was not accounted for in these studies. The incorporation of the forming process and interference fit, in particular, into an integrated analysis procedure is therefore an area where further work is necessary.

Appropriate design methodologies play a major role in the life cycle of modern products; it is also essential to produce high quality components in the shortest possible time. Therefore, tailored design strategies have been developed for automotive plain bearings in order to satisfy customer needs. They are based on analysing three major contributors to stress development: fitting of the bearing in its housing, internal forces in the engine and hydrodynamic oil film pressure. Standard, efficient design procedures in MAHLE Engine Systems rely on exact solutions and two-dimensional EHL analyses to evaluate the bearing interference fit stresses and hydrodynamic film pressure on the bearing surface, respectively.

Advanced three dimensional analyses employed by MAHLE include more sophisticated techniques that take into account the bearing stiffness, wear and surface profile among other factors. Initially, the assembly process is simulated leading to the resulting bearing deformation; this is achieved through non-linear FE contact analysis, based on the general-purpose package ABAQUS [3]. Thus, the bearing shape and clearances between the bearing and housing surfaces are estimated. Using the finite difference in-house code SABRE-EHL [6, 7], it is possible to estimate the hydro-dynamic oil film pressure and housing deformation during a complete engine cycle. At the same time, different indicators related to strength, wear, cavitation erosion and fretting are generated to assess the propensity for damage.

In this work, this design methodology was taken one step further using oil film pressure and housing deformation estimates from SABRE-EHL to generate detailed information about the stresses arising in the bearing. The analysis of manufacturing and service conditions was combined to provide a global view of the possible failure mechanisms in bearings. Based on simplifying assumptions, the forming process was simulated for an initial evaluation of a realistic distribution of residual stresses to be incorporated into the stress analysis under service conditions. The latter analysis is expected to lead to a better understanding of the influence of complex and changing film pressure and concomitant housing deformation under EHL conditions on bearing performance. Thus, it will be possible to assess the importance of particular geometric and material parameters and correlate them with experimental observation of lining detachment, a condition that is particularly harmful to bearing performance.

2. Methodology

2.1. Component modelling

SABRE-EHL carries out a coupled analysis to compute the oil film pressure throughout an engine cycle. This involves a heat balance analysis to determine the oil viscosity and material expansion leading to estimates of the clearances between bearing and journal. Extended SABRE-EHL analyses may include the input of oil feed and journal profile, account for grooves and elliptical shapes as well as solid-to-solid contact with flexible components. The latter requires the input of the elastic compliances of both the housing-bearing assembly and the journal [7]. Hence, the housing deformations are estimated during the process and, together with the oil film pressure, are later transferred to the FE stress analysis of just the bearing to assess its behaviour under service conditions.

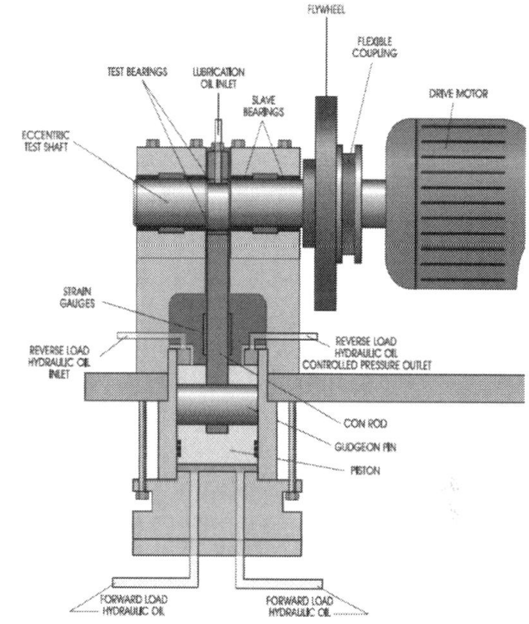

Figure 2. Sapphire test rig

The stresses and deformations experienced by the bearing are clearly dependent on the engine design and, in particular, on the specific load and housing distortion generated during the combustion cycle. Light and flexible housings promote large deflections in the bearing and may lead to premature bearing damage. The assessment of bearing architectures within MAHLE Engine Systems is carried out through an accelerated performance testing procedure using a standardised housing with enhanced stiffness. Such a test is carried out on a rig, shown schematically in Figure 2, designated as Sapphire [6]. It consists of an eccentric shaft that forces a piston to travel in the vertical direction through a connecting rod. At the same time, the rig applies some resistance to the movement of the piston due to the oil present in the chamber of a hydraulic ram. This oil is forced into another chamber and its flow is regulated by valves to obtain the desired resistance of travel or specific load applied to the bearing. The precise measurement of the load applied to the bearing is carried out through strain gauges positioned at the connecting rod. A great amount of experimental fatigue data are generated under such conditions; this makes the Sapphire rig output an ideal candidate for comparison with the results of numerical simulations. For this reason, SABRE-EHL analyses have been extensively applied to various Sapphire rig testing conditions and the output entered into subsequent stress analyses.

2.2. Bearing forming simulation

For this particular study, a tri-layer bearing architecture was analysed using a three dimensional FE bearing shell model built on ANSYS 11.0, a general purpose finite element package [8]. The analysis was based on a multi-linear elasto-plastic strain hardening and large deformation theory.

The first part of the numerical work dealt with the forming process generating a concave shell from a flat strip; previous manufacturing operations were not simulated given that thermal treatment relieves residual stresses developing prior to the forming operation. The simulation of the forming process and the estimation of residual stresses were carried out through the modelling of an initially flat, multi-layered strip placed between a punch and a die of identical semi-circular profile. Only half of the bearing was modelled by assuming symmetry with respect to the mid-plane across its width. This is consistent with the format of the output generated by the component model. The analysed tri-layer architecture consisted of a backing layer, an interlayer and lining layer whose properties and thicknesses are listed in Table 1. The length of the strip was set such that it would form a half bearing shell of 26.4 mm outer radius and a width of 29 mm.

Table 1. Material properties of layers

Layer	Material description	Thickness (mm)	Young's Modulus (GPa)	Poisson Ratio	Yield stress (MPa)	Ultimate stress (MPa)	Strain hardening exponent
Lining	Aluminium alloy	0.38	70	0.33	53	166	0.238
Inter-layer	Pure aluminium	0.04	67	0.33	37	78	0.19
Backing	Low-carbon steel alloy	1.8	198	0.3	459	512	0.058

Apart from the material non-linearity arising from elasto-plastic material behaviour, the analysis also involves geometric non linearity due to contact between the forming tools and the strip. Friction was included in contact modelling with a coefficient of friction of 0.1 [9]. For the sake of computational efficiency, both die and punch were modelled as rigid due to their stiffness being much higher than that of the flat strip. The mesh of the flat strip was built according to that used in SABRE-EHL to simplify the application of boundary conditions in later analysis and transfer of residual stress data, which are associated with the elements and their internal integration points.

Initial simulations involving the tri-layered architecture exhibited a non-convergent behaviour, which probably arose from the considerably lower stiffness and strength of the interlayer compared to those of the backing and the lining. This problem also arose in the initial attempts at stress analysis of the bearing under service conditions and can be addressed by refining the mesh and increasing the solution steps. Such measures however lead to excessive computational cost. For this reason, the interlayer was initially considered as part of the lining leaving the assessment of its structural influence on the overall stiffness of the component to a later stage.

The forming simulation was carried out in four stages. During the initial phase when the punch comes into contact with the flat strip, the latter deflected by only 2.5% of the total forming depth requiring a large number of iterations to reach a numerically stable stage. The main deformation stage was more stable but it required a large number of solution steps to complete the travel of the punch to the bottom of the die. During the load release stage, some contact between the punch and the flat strip is maintained because of the spring back force. Finally, an additional upward displacement was imposed on the bearing to overcome the frictional force keeping it attached to the die.

2.3. Stress analysis under service conditions

The residual stresses generated by the bearing forming simulation were taken into account at this stage of the analysis by transferring them into the bearing FE model. As already mentioned, the mesh used in this model was consistent with the mesh of the component model as well as that of the forming analysis to simplify the post-processing operations. An additional step was introduced at the beginning of the analysis to account for the fitting deformations that occur during assembly. Relevant modelling performed at MAHLE predicted that fitting causes a housing radial displacement of around 0.028 mm; this was applied to the outer backing surface.

Pressures and housing deformations obtained from SABRE-EHL simulating Sapphire rig operation were applied, respectively, to the lining surface and the outer backing surface of the model. The performed stress analysis corresponded to an experimental set up at 3000 rpm, a specific load of 97 MPa and oil feed temperature of 140°. The respective film pressure and housing deformation are shown in Figure 3 over half of the model; a peak pressure of around 200 MPa occurs when the crank angle is 180° with the 0° position at the centre of the top bearing shell (further from the connecting rod). A grid of 2002 points was used at intervals of 2 degrees over the engine cycle. The manipulation process of this data for the purpose of stress analysis was carried out in MATLAB 7.3, where only 36 load steps at intervals of 10 degrees were applied due to the computational cost involved.

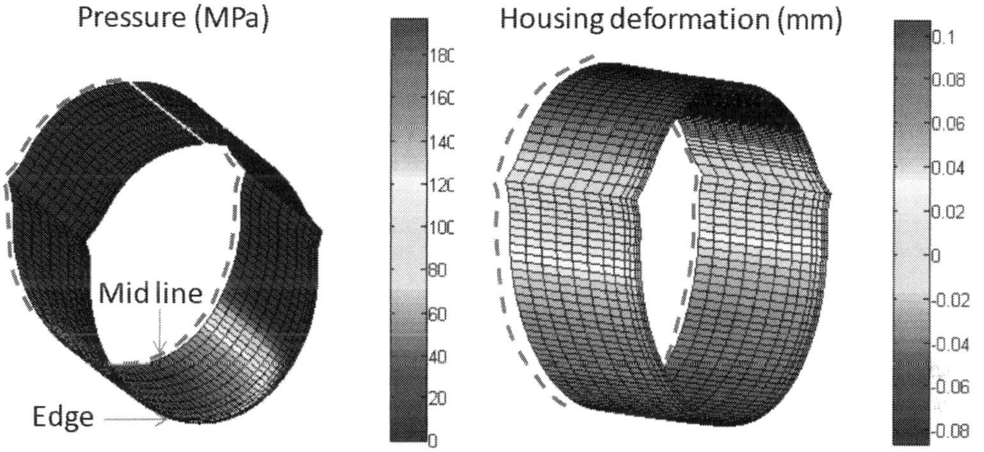

Figure 3. Film pressure and housing deformation at crank angle 180° of Sapphire rig test.

3. Results and discussion

3.1. Residual stresses

The forming analysis process took around 12 hrs on a 2.8 GHz processor with 2 GB of RAM memory. The solution showed that, at around 20% of the total travel of the punch, contact forces above 1000 N develop causing significant material yielding and deformation. The forming load increases as the bearing is driven to the bottom of the die due to the strain hardening material model and the contact area growth. The post-processing of results focused on the magnitude and distribution of the residual longitudinal stress, which becomes the tangential stress relative to the cylindrical frame of reference associated with the formed semi-circular shell bearing. As expected, the tangential stress was found to be discontinuous at the layers' interface showing high stress gradients across the layers' thickness; in contrast, it was almost constant across the component width. Its circumferential variations at the lining surface and the lining-backing interface at two key stages of the forming process are shown in Figure 4.

As indicated in Figure 4, the compressive tangential stress does not vary significantly through the thickness of the lining at end of the second stage of the simulation, that is, when the outer backing surface is in full contact with the die. As the bearing is separated from the die, elastic recovery occurs mainly driven by the reverse deformation of the backing layer, which provides most of the component stiffness. This generates tensile tangential stress in the lining; the tension at the surface is much greater than that developing at the lining-backing interface and considerably higher than the yield stress of the lining material (see Table 1); it also remains almost constant over a wide, centrally located portion of the component where the peak contact pressure occurs. As deformation progresses, an approximately constant compressive radial stress develops in that part of the bearing at a value around two times the yield stress of the lining material. The shear stresses predicted by the simulation did not appear to be significant.

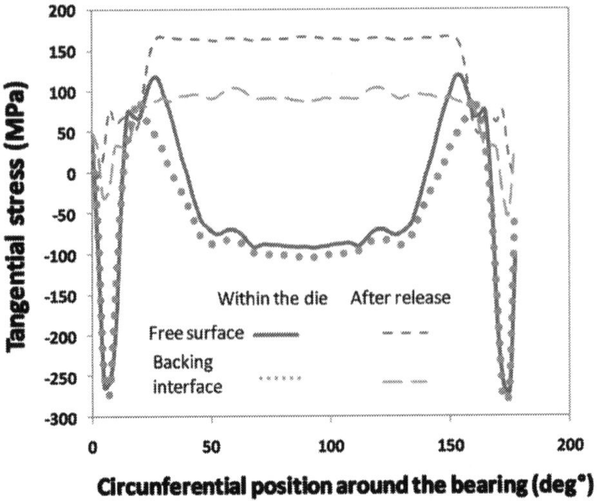

Figure 4. Circumferential variation of the tangential stress in the lining at the symmetry plane perpendicular to the axial direction.

3.2. Stresses under service conditions

The analysis of the bearing based on accelerated Sapphire test data generated by SABRE-EHL was performed on the same computing facilities as those used for the forming simulation with a typical run providing the stresses over a complete journal cycle lasting about 24 hrs. The analyses performed showed that all stresses arising from the EHL regime varied substantially during the cycle depending on the film pressure profile and the associated housing deformation. The results for the total tangential stress along the surface of the bearing at an axial section 1.8 mm from the bearing edge are shown in Figure 5. Greater deformations and stresses were predicted in the central part of the lower bearing shell on the side of the connecting rod (see Figure 2). For this reason, the angular position ranges from 120° to 240° in Figure 5(a) where the spatial stress variation has been plotted for a range of crank angles within the third simulated cycle.

For the assessment of fatigue performance, the same stress results were also plotted versus time represented by the crankshaft angular position; Figure 5(b) shows the stress cycles at four key circumferential locations. These plots reveal that there is essentially one representative crank cycle in the accelerated performance test and provide an indication of the range of mean stresses. Of equal importance is the distribution of the stress amplitudes, which were systematically evaluated over the whole lining surface and their contour map shown in Figure 6. It can be seen that they have been predicted to range from 60 to 220 MPa. However, it is worth noting by referring to Figure 5(b) that the highest amplitudes correspond to fully compressive stress cycles. This explains the high-cycle fatigue life of bearings, in spite of the extremely severe loading conditions.

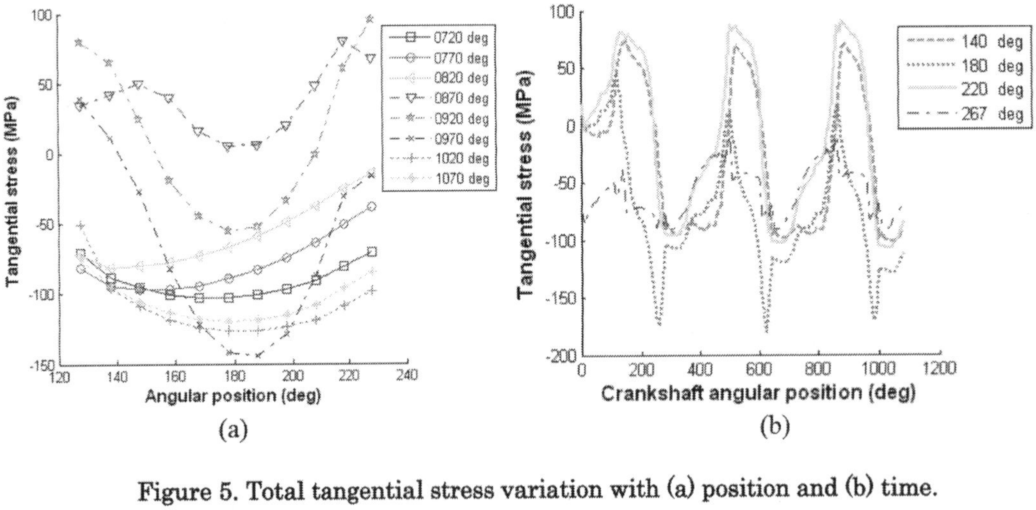

(a) (b)

Figure 5. Total tangential stress variation with (a) position and (b) time.

The stress amplitude contours in Figure 6 show that their maximum can be found around 190° in the circumferential direction and it has very small variation across the bearing width at that location. This is consistent with the position of the peak oil film pressure and the circumferential location of the damage observed during Sapphire rig tests. The latter appears closer to the edge rather than the middle of the bearing as anticipated from the numerical results, but still originates from the highest stress zone.

The high radial pressure applied to the bearing lining resulted in excessive plastic deformation and, in consequence, the appearance of high positive strains in the axial direction caused by the Poisson effect. This was also predicted by Bahai and Xu [3] and is clearly seen in the tangential plastic strain plots of Figure 7. The circumferential distributions of these strains shown in Figure 7(a) confirm that their high positive values, which correspond to the high pressure range of crank angle locations. The strain cycles shown in Figure 7(b) indicate considerable amplitude as well as positive mean value. The influence of such predictions on fatigue behaviour should be examined in conjunction with the stress cycles of Figure 5(b).

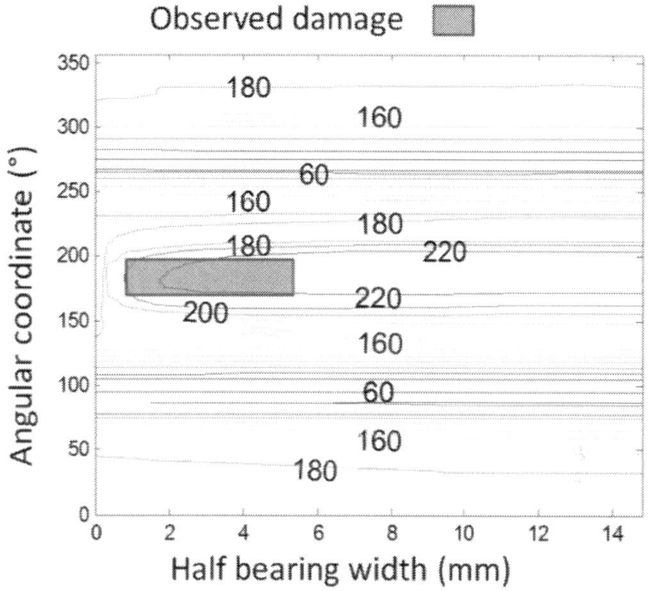

Figure 6. Stress amplitude contour map with observed damage in tests.

4. Concluding remarks

The stress analysis of a bearing under service conditions reveals that the inclusion of assembly and manufacturing conditions into the modelling leads to a better understanding of the significance of the stresses arising during the engine cycle to the bearing fatigue life. Stress estimates reveal that tensile tangential stresses exist along the lining surface despite the predominantly compressive state of stress caused by the assembly process and the oil film pressure. The latter is countered by the tensile residual stresses due to forming leading to a greater mean stress during the cyclic loading than originally thought.

As pointed out earlier, modern engines are designed to be considerably more compliant than the Sapphire rig. Under such more realistic conditions, the housing deformation is expected to play a more critical role in this problem with even greater tensile tangential stress values predicted especially when the inertial load is applied to the bearing shell opposite to the connecting rod thus causing greater deformation to the housing. There is therefore a need for further validation of the methodology and its application to bearings in real automotive engines.

The simulation of bearing response to the conditions generated during the accelerated performance test and its estimation of the stress and strain fields showed a good correlation between predictions of potential lining damage and experimental observations. The potential damage location was extended along the axial direction of the bearing showing similar stress amplitude values. The damage location along this direction could be largely influenced by the variability of the material microstructure and loading misalignments. Future work in this area will address the possible spread of damage while the component remains in service by applying already developed crack growth methodology for layered material systems to bearings subjected to such conditions.

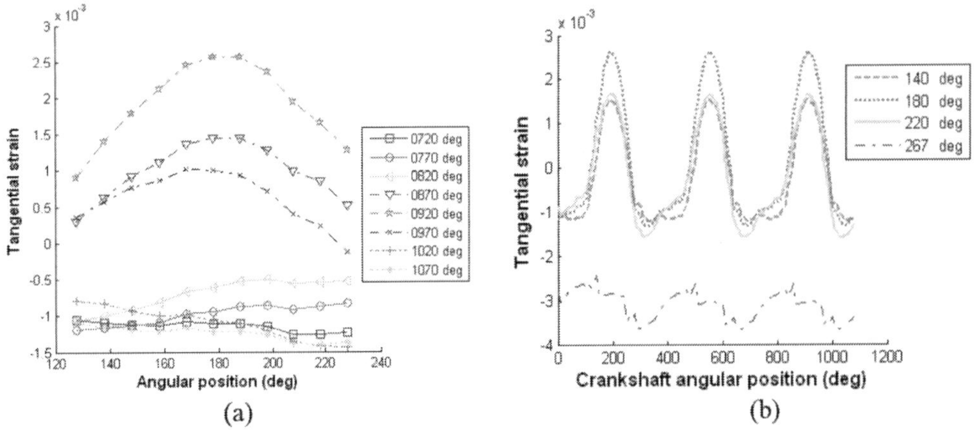

Figure 7. Tangential plastic strain variation with (a) position and (b) time.

References

[1] Martin FA, Garner DR, Adams DR. Hydrodynamic Aspects of Fatigue in Plain Bearings. J Lubr Technol-Trans ASME. 1981;103:150-156.

[2] McCallion H, Lotfi M. Tensile surface stresses and fatigue in plain journal bearings. Tribol Int. 1992;25(4):247-257.

[3] Bahai H, Xu H. Three-dimensional elastoplastic finite element and elastohydrodynamic analyses of journal bearings. Proc Inst Mech Eng Part C-J Eng Mech Eng Sci. 1997;211:143-152.

[4] Hacifazlioglu S, Karadeniz S. A parametric study of stress sources in journal bearings. Int J Mech Sci. 1996;38(8-9):1001-1015.

[5] Perez M, Thomazi C, Syngellakis S. Stress analysis of fluid-film bearings. In: Su D, editor. Proceedings of the International Conference on Gearing, Transmissions, and Mechanical Systems: Professional Engineering Publishing, 2000. p. 729-738.

[6] Xu H, Jones GJ, Aoyama S, Ushijima K, Okamoto Y, Kitahara K. Simulation of bearing wear and its Influence upon bearing performance based on elastohydrodynamic analysis. SAE 1999 World Congres. Detroit, Michigan; 1999.

[7] Mian AO, Merritt D, Wang D. The effect of crankshaft flexibility on the EHL of connecting rod bearings. SAE 2002 World Congres. Detroit, Michigan; 2002.

[8] ANSYS Release 11.0, ANSYS Academic Research. Canonsburg, PA 15317: ANSYS, Inc.; 2008.

[9] Avallone AE, Baumeister T, Sadegh A, editors. Marks' Standard Handbook for Mechanical Engineers: McGraw-Hill Professional; 2006.

Modelling Flat Spring Performance Using FEA

Blessing O Fatola, Patrick Keogh and Ben Hicks

Department of Mechanical Engineering, University of

Corresponding author bf223@bath.ac.uk

Abstract. This paper reports how the stiffness of a Flat Spring can be predicted using nonlinear Finite Element Analysis (FEA). The analysis of a Flat Spring is a nonlinear problem involving contact mechanics, geometric nonlinearity and material property nonlinearity. Research has been focused on improving the accuracy of the model by identifying and exploring the significant assumptions contributing to errors.

This paper presents results from some of the models developed using FEA software. The validation process is shown to identify where improvements can be made to the model assumptions to increase the accuracy of prediction. The goal is to achieve an accuracy level of ±10 % as the intention is to replace practical testing with FEA modelling, thereby reducing the product development time and cost. Results from the FEA models are compared with experimental results to validate the accuracy.

1. Introduction

The term Flat Spring is commonly applied to a wide range of shapes made out of flat strip materials that, when deflected by an external load, store and release energy. Flat Springs are usually manufactured from high carbon spring steel, nickel-silver, high-nickel alloys, stainless steel, and phosphor-bronze. In addition to performing as springs they are frequently used as stops, connectors, hinges, braces, and fasteners. A Flat Spring can either be in the form of a cantilever spring or in a more complex form (Fig 1). The cantilever springs are produced by bending while the complex form springs are produced by bending and forming processes.

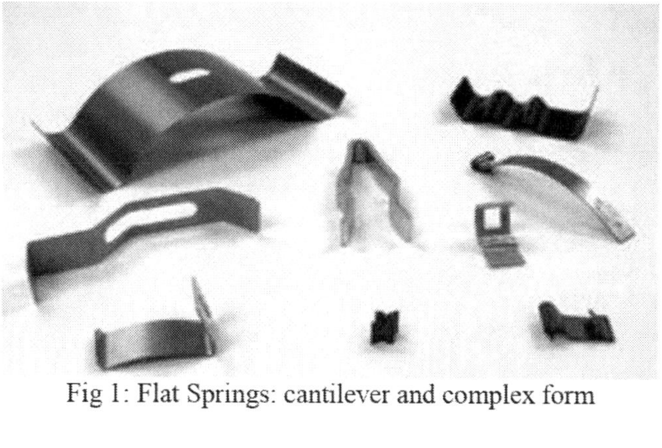

Fig 1: Flat Springs: cantilever and complex form

The stiffness of a Flat Spring depends mainly on the geometry, material properties and the load boundary conditions. Depending on the particular application, these factors can be varied to produce the desired spring stiffness.

2. Design and Performance Evaluation

Currently, Flat Spring design relies on simplified models and trial and error methods of development. These simplified models are based upon simple beam theory and curved beam theory. By using these models the performance of simple cantilever Flat Springs can be reliably predicted. However, this is not the case for complex Flat Spring geometries; hence there is a requirement to model the performance of these springs using FEA.

Fig 2: Complex formed Flat Spring geometry

The evaluation of Flat Spring performance primarily involves a compression analyses in order to obtain the stiffness of the spring and the force generated at a specified displacement. This paper focuses on creating an FEA model for spring compression of a complex form Flat Spring. The stiffness of a spring has both linear and nonlinear characteristics (Fig 4). The goal is to replace some of the initial prototype evaluation with an FEA model, thereby reducing development cost. The prototype evaluation involves iterative design, manufacture and testing using a Zwick (compression) machine (Fig 4). In order to confidently replace some of this practical process with simulation, an accuracy of ±10% is required. Flat springs are mostly designed to perform within elastic region and the boundary between the elastic and plastic range. The performance within the elastic region can be accurately predicted within an error range of ±20% using equations derived from a database of empirical results.

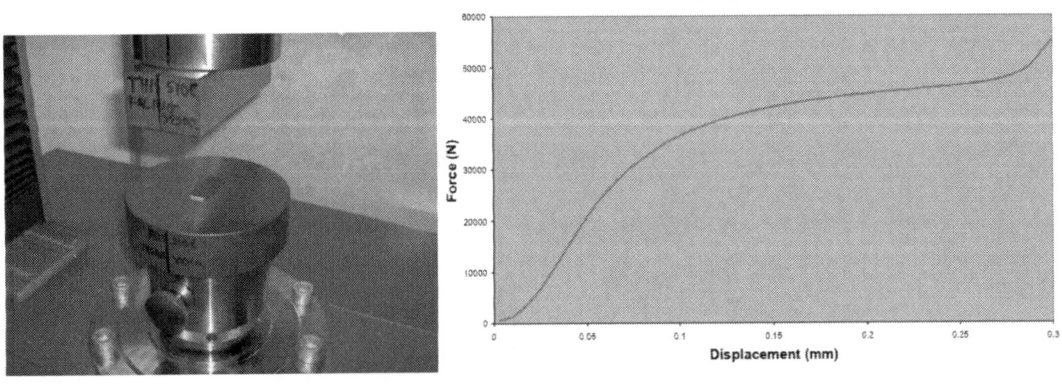

Fig 4: Flat Spring Stiffness obtained from a Zwick compression machine

3. Finite Element Model

The application of FEA usually involves making a number of assumptions and simplifications through discretisation while trying to denote a real system with a mathematical model. These simplifications are essential as they do resolve some of the challenges resulting from modelling complex nonlinear problems. The simplifications also reduce the simulation time required to solve these models. However, these assumptions can also be one of the main sources of error between the theoretical model and the real system. Errors in FEA are either due to critical assumptions/simplifications that affect the accuracy of the model or incorrect assumptions made while constructing the model.

Flat Spring compression analysis involves loading a spring between two flat rigid plates (Fig 5). The analysis can be simplified as a static nonlinear analysis. This is because the expected stresses will exceed the limit of proportionality of the material during compression. There will also be major changes in geometry and boundary conditions due to the application of load (contact) and the direction of load application changes with deformation (pressures forces). The critical model assumptions that will influence the accuracy of the model are those of the contact model, material properties and definition of the geometry. Each one of these factors will be explained in the following sections.

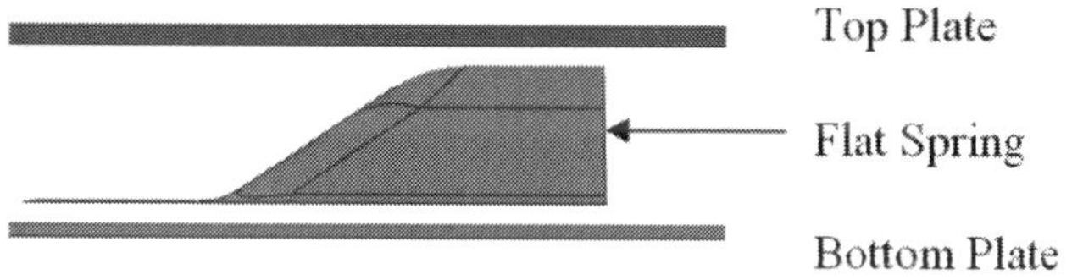

Fig 5: Flat Spring compression between two rigid plates

3.1. Contact

Contact problems are common sources of boundary nonlinearity in stress analysis as the contact region will change during the loading cycle. Using FEA, many contact problems can be solved with high accuracy. There are, however, a number of limitations to contact analysis using FEA due to the capabilities and fidelity of the numerical solvers. The numerical technique for solving contact involves identification of the points on the boundaries of the bodies in contact, then application of appropriate boundary conditions or models to prevent penetration and to simulate the interaction between the surfaces.

Contact interaction has been setup between the top plate and the top surface of the strip and the bottom plate and the bottom surface of the strip as shown in Fig 5. The model defined to govern the interaction is considered to be frictionless for tangential sliding while a contact pressure constraint is applied in the normal direction. There is no limit in the contact formulation on the magnitude of contact pressure that can be transmitted between the surfaces. A frictionless model was chosen over a classical isotropic Coulomb friction model as it is assumed that the shear force transmitted across the interface will not significantly affect the normal stiffness of the model. The specified constraints are applied when the nodes on the flat plate surface penetrates the surface of the indentation. The surfaces separate when the contact pressure between them becomes zero or negative, when the constraint is removed. In order to validate this assumption, a sensitivity test was carried out to see the effect of friction on the FEA results (Fig 6).

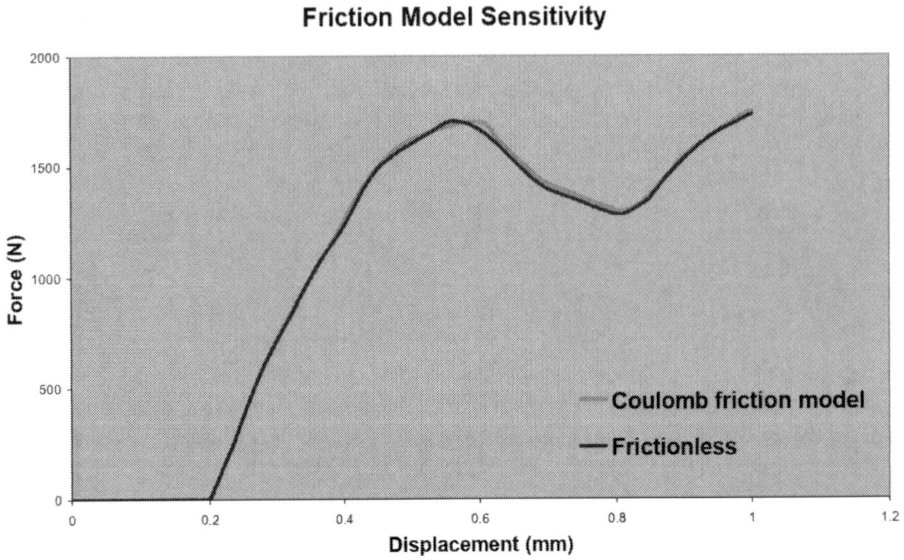

Fig 6: Friction sensitivity analysis

3.2. Material Properties

The material properties were modelled using both elastic and plastic material parameters. These were obtained from a tensile test of the spring (stainless) steel used to manufacture the spring. Within the FEA software the elastic property was modelled by the standard Young's modulus and Poisson ratio, while the plastic property was defined by its yield point and its post-yield hardening. Two points were used to define the post yield hardening, which were interpolated linearly to obtain intermediate values. The material model was assumed to be homogeneous and isotropic. This assumption ignores the residual stress in the spring, which may be generated during the forming process. The spring was modelled as a shell structure. Shell structures assume a linear stress distribution across the defined thickness.

3.3. Geometry

The geometric definitions required for the model include the 3D spring geometry and the two flat plates required to compress the spring. Initially the spring geometry was created in a CAD package using the tool dimensions and setup to predict the geometry of an indentation. A number of idealisations of the geometry were made in order to model the complex geometry of the spring (Fig 7). The mid-surface of the 3D geometry was extracted to model it a shell structure. The thickness of the spring was assumed to be uniform. The top and the bottom plate were modelled using rigid bodies - assuming infinite stiffness. To ensure consistency the critical dimensions of the spring were measured using a contact measurement system. The dimensions were measured in the x and y axes. Dimensions such as the blend radius could not be measured accurately hence they were simplified as edge radii.

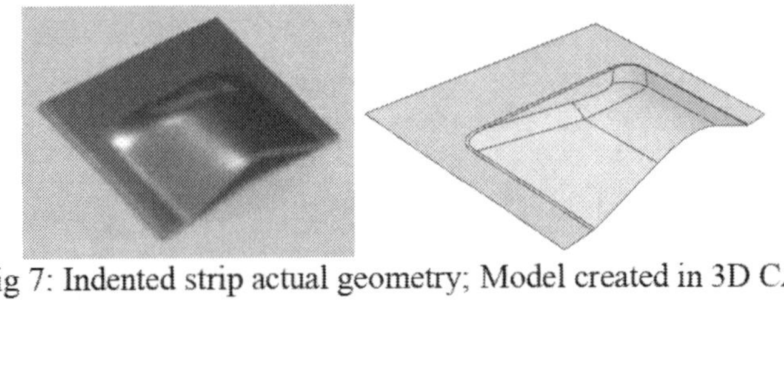

Fig 7: Indented strip actual geometry; Model created in 3D CAD

4. Analysis and Results

A total of ten benchmark springs were modelled using the approach previously described. These rings are similar in shape but are of different sizes (i.e. spring height & width). Some spring models gave results within the specified accuracy requirement while others fell outside the accuracy requirement (Fig 8). The complex flat spring with large indentation height to surface area ratio gave results well outside the 10% error bar when compare to experimental results. The main significant difference between the model and the practical setup was in the geometric idealisations made.

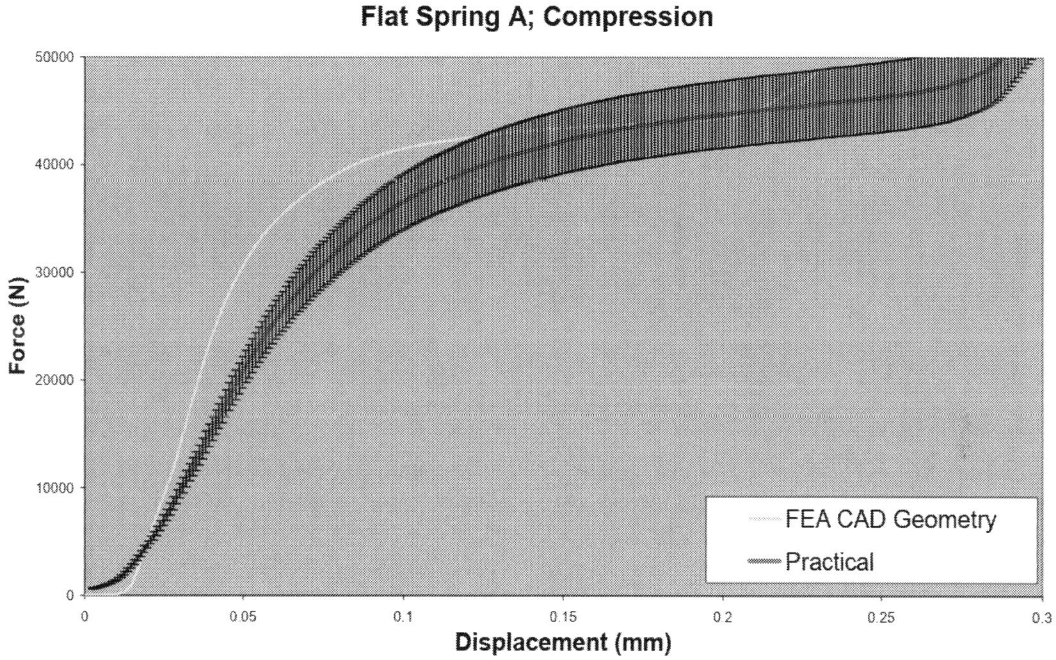

Fig 8: FEA and practical result (±10% Error range)

In the cases of the less accurate results, the FEA model reliably predicts the maximum force generated, however, it over-predicts the elastic stiffness of the strip and the inception of plastic variation of stiffness. These are critical points, especially if the FEA model is to be used for design and development purposes.

As expected the geometry of the spring is a critical factor, hence in order to achieve accurate results the geometry must be accurately modelled with minimal abstraction. In other to validate the FEA model with the practical results it is essential to ensure that the critical dimensions on the practical geometry are close to the FEA geometry. Another source of error in the CAD geometry is that it does not account for the effect of factors that influence the final spring geometry during forming or bending e.g. spring back and permanent material deformation. This will affect sheet deformation, thinning, and material flow. In the case of a Flat Spring, even if these critical factors can be accurately measured, modelling them presents another set of challenges. Many simplifications will still be made in order to produce an accurate 3D representation of an indentation. The solution to this is to model the forming process of these springs and then compress the subsequent geometry. The formed model will then account for all of the factors previously mentioned.

The forming model was created using FEA and the CAD geometry used was that of the forming blade and the forming die, which can be accurately represented (Fig 9). The Flat Strip was modelled as a shell with the thickness and material properties of stainless steel. Frictionless contact was setup between the forming blade and the strip as well as the die and the strip. The forming blade was then displaced by the distance required to achieve the required indentation height taking into consideration the spring back that occurs during the forming process.

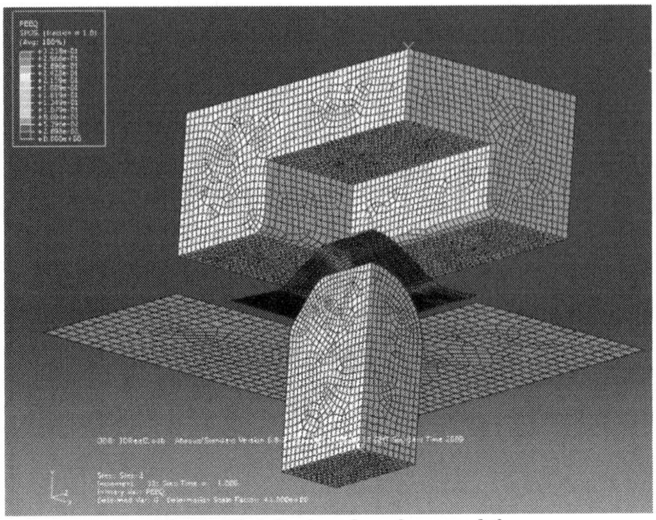

Fig 9: Flat Spring forming model

The mesh geometry created was imported into the indentation compression model to replace the CAD geometry created initially. Figure 10 shows the new spring geometry, which gives a better representation of the actual geometry for FEA. Blend radii and edge corner radii are better represented in the formed geometry. Another advantage of the formed geometry is the mesh created (Fig 10). Meshing a flat strip before forming will create a better mesh structure (i.e. no element distortion and Hex elements) when compared to the mesh created on a 3D CAD geometry (Tetrahedral elements with distortion).

Fig 10: Formed geometry (above) and CAD geometry mesh (Left)

Using the same setup from the formed model the results showed considerable improvement and fell well within the ±10 % accuracy range (Fig 11).

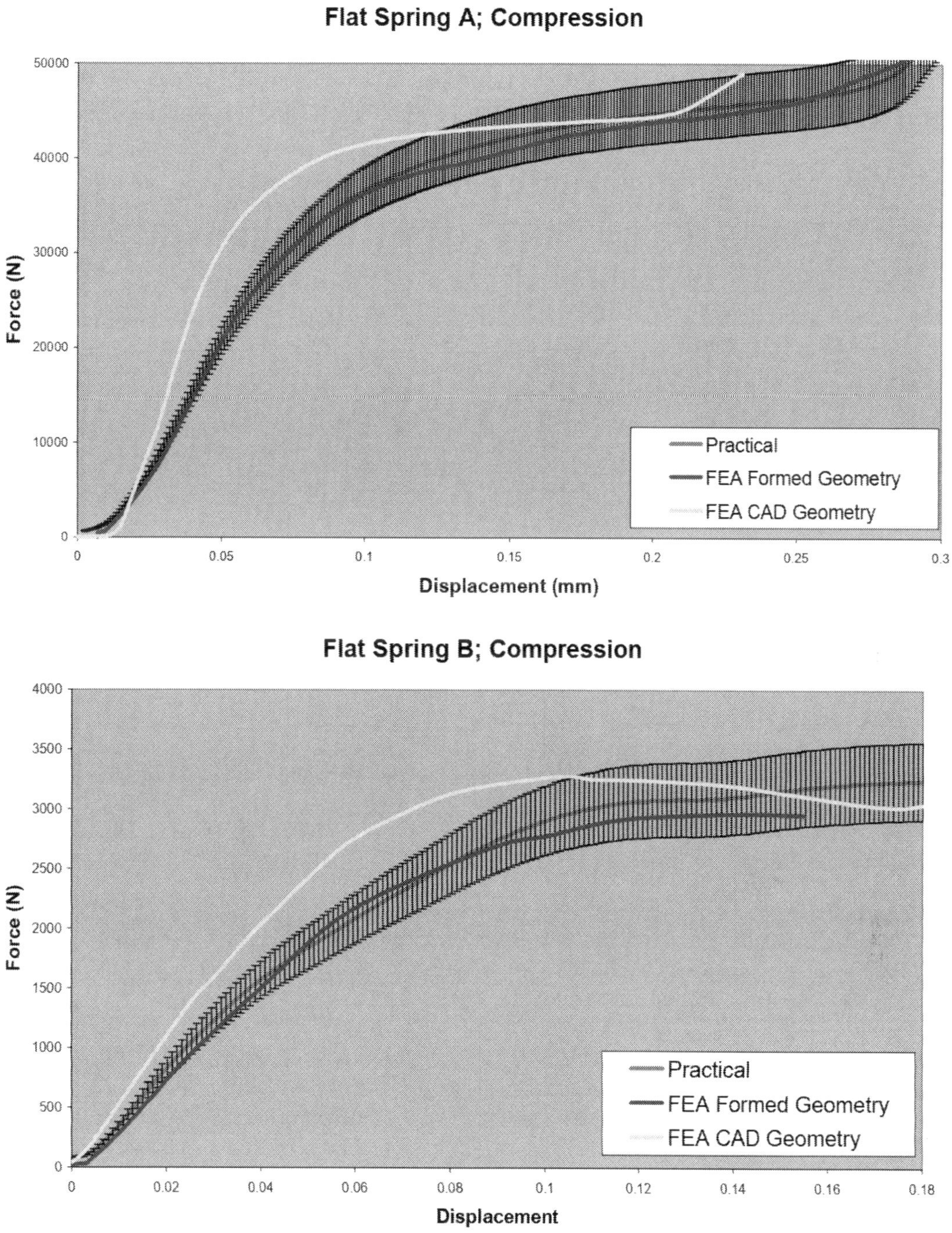

Fig 11: Practical result (±10% Error range) and FEA results

5. Discussion

Resolving the assumptions due to the geometry has improved the accuracy of the FEA model. This does not eliminate all sources of error as some Flat Spring geometries do have a different structural response to the compression load, for example some springs will buckle under the load. An understanding of the structural response is required to make the right assumptions while constructing the model. Figure 12 shows an example of a Flat Spring, which was assumed to be a static nonlinear problem. However examination of the practical ring indicated that the spring does exhibit buckling characteristics, which is not accounted for in the model. The formed model predicts the elastic nature of the spring; however, it fails to accurately predict the buckling and post buckling characteristics.

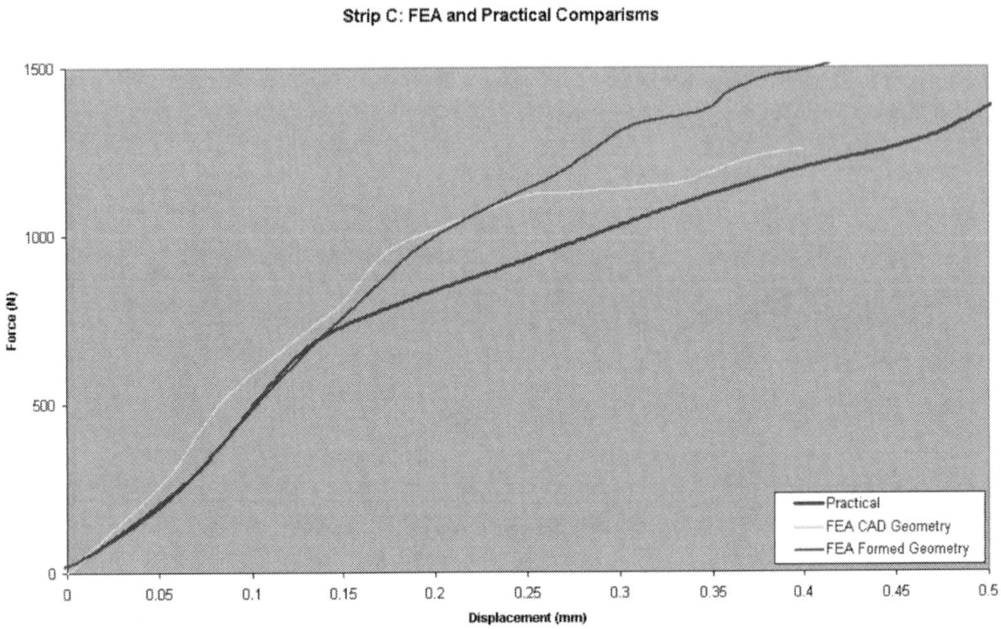

Fig 12: Flat Spring exhibiting buckling behavior under compression load

Future work is intended to use the modelling and analysis technique to develop new products and to improve exiting ones using design of experiments.

6. Conclusion

A study has been undertaking to enable the prediction of Flat Spring stiffness characteristics to be achieved with some degree of confidence. It is shown that assumptions in geometry typically of those from CAD, may lead to significant errors, these errors were overcome by performing a pre-analysis representing the forming of an indentation during manufacture. The resulting geometry was then more representative of the actual geometry and hence force/deflection prediction better matched the measured experimental results – achieving the accuracy requirement of ±10% error.

References

[1] A.M. Wahl, "Mechanical Springs", Spring Manufacturers Institute Second Edition
This reference has two entries but the second one is not numbered (it uses the 'Reference (no number)' style.
[2] P. Kurowski, "More errors that mar FEA results" Machine Design, March 21, 2002.
[3] P. Qin, G. Dentel and Mikhail Mesh; "Multi-Leaf spring and Hotchkiss Suspension CAE simulation" Abaqus Users Conference 2002
[4] Q. Feng and N.K. Prinja , "NAFEMS benchmark tests for finite element modelling of contact, gapping and Sliding" NAFEM October 2001

Contact problems for rough elastic solids being in nearly full contact or regarded as such for further elaboration

R L Salganik[1], A N Mokhel[1] and A A Fedotov[2*]

[1] Laboratory of Geomechanics, A. Ishlinsky Institute for Problems in Mechanics, Russian Academy of Sciences, prosp. Vernadskogo 101, block 1, Moscow 119526, Russian Federation
[2] Department of Physics, «MATI» - Russian State University of Aviation Technology, Orshanskaya 3, Moscow 121552, Russian Federation

E-mail: true-ten@yandex.ru

Abstract. Problems on nearly full contact between elastic bodies with nominally flat rough boundaries are considered, so that only relatively small crack-like no-contact areas are present. Such a case is of essential applied interest and has been considerably less examined than the case of comparatively small contact areas, when only relatively sparsely located "contact spots" appear. This case may really appear under sufficiently large loads and appropriate roughness and compliance conditions (e.g. when metallic and polymeric materials contact) or it may be regarded as a zero- approximation, when roughness is essentially multi-scale and a procedure of successive approximations (iterations) is applied to solve corresponding contact problems. The latter situation is mainly considered below for plane strain, and, in a lesser degree, for 3D conditions, assuming no shear interaction. Particularly, the case is considered, when there is a coarse-scale Λ and a fine-scale λ ($\lambda \ll \Lambda$) of roughness extent. In zero approximation, only coarse-scale roughness is taken into account, assuming that it is presented by a correspondingly large solitary socket. Fine-scale roughness is considered in first approximation only based on the Greenwood-Williamson model [1]. Similarly, the situations, where bending of a layer or a multi-layer structure plays dominant role are considered.

1. Introduction

The problem of taking into account the effect of surface roughness of the bodies, contacting with each other under loading, on their stress-strain state, which has long been a key problem in contact mechanics, proceeds to attract the interest from researchers, especially as more new computer-aided methods for its solving appear. An important way of solving this problem is application of asymptotic methods. These are effective in the case of comparatively (other parameters being fixed): (i) small loads, when the contact appears mainly over relatively small and sparsely located areas, or "contact spots", and (ii) large loads, when relatively small crack-like areas of absence of the contact - no-contact areas - appear. The first case is more often discussed. The second case termed as the case of nearly full contact has been studied in a considerably lesser degree, despite its significant applied interest, e. g. when dealing with contacting of metallic and polymeric materials under relatively large loads or with formation of seals. It was comparatively recently that the case was seen as deserving

* Corresponding author.

detailed study (see especially [2, 3]). On the other hand, this case may also be considered as a zero approximation, when the roughness is essentially multi-scale and a procedure of successive approximations (iterations) is applied to solve corresponding contact problem. It is the latter situation that is mainly considered here. Particularly, the case is considered, when there are only two scales of roughness extent: a coarse-scale Λ and a fine-scale λ ($\lambda \ll \Lambda$). In zero approximation, only coarse-scale roughness (scale Λ) is taken into account, so that the contact is assumed to be full. This presumes also neglecting the thickness of an equivalent layer determined by the fine-scale roughness (scale λ) proper.

Thus, the situation of nearly full contact is the basic, both, when it is of interest *per se*, and, when it is considered as allowing to find zero approximation, in solving multi-scale roughness related problems by successive approximations (iterations) method. Problems, corresponding to this situation, are considered here based on works [4, 5].

Any shear interaction between the contacting bodies is neglected.

Proceeding from the well known fact that the contact interaction under such conditions is determined only by the form of the gap between the interacting bodies, it may be assumed, without loss of generality, that one of these bodies is absolutely rigid (from here on, for brevity, - rigid).

The problem to be solved after solving the above basic problem is to find how the disturbance resulted from the coarse-scale features of roughness (scale Λ) found in zero approximation affects the phenomena of interest connected with the fine-scale roughness (scale λ). The back influence of what then happens on the stress-strain state found in zero approximation is neglected, which is admissible, since $\lambda \ll \Lambda$.

Consideration in zero approximation is performed below for the case, when the roughness of scale Λ is presented by a solitary socket elongated in a certain direction (extended defect), in the rigid body. This defect is shown in its cross section normal to the above direction in figure 1.

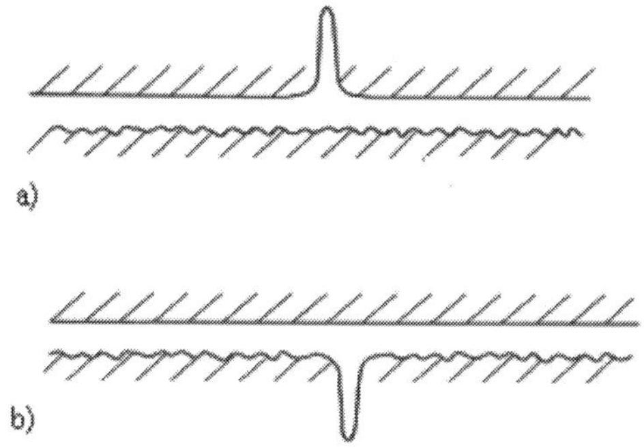

Figure 1. Two possible schemes of two-scale roughness of bodies with nominally flat boundaries. Scheme b) corresponds to an abraded surface of mild steel specimen [1], Fig. 6 (see also [6], Fig. 2).

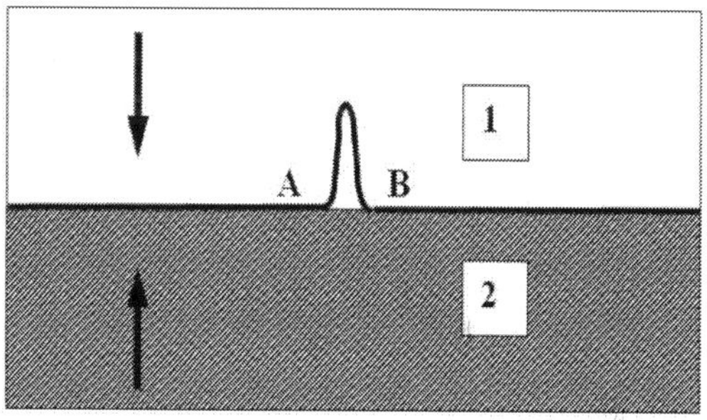

Figure 2. Scheme of squeezing of elastic (2) and rigid (1) bodies resulting in partial penetration of the elastic body into a large-scale socket in the rigid body (AB – no-contact area).

This allows consider that the plane strain conditions prevail in the cross-sections of this defect parallel to the plane of figure 1 and passing not too close to its ends. Obviously, the result obtained for this case may also be applied to the case, with a number of such sockets distributed sparsely enough to neglect their interaction. Such a situation may appear when the roughness relief contains many comparatively large sockets of the above type considered as defects, which may be present in abraded surfaces (see the example given in [1, Fig. 6]) or in road coatings. When so, one may single out the above sockets, formulate the zero approximation problem for them and solve it, using the solution of the corresponding problem for any one of these sockets, as if it were the only one present. Further, in zero approximation the effects of an equivalent layer determined by the fine-scale roughness proper are neglected. This allows to neglect the very presence of this layer and, therefore, employ for zero approximation the results obtained in [4, 5].

Fine-scale roughness is considered below in first approximation only, assuming that this may be done in frame of the Greenwood-Williamson model [1] (from here on – the GW model), the parameter values of the latter being determined, taking into account the solution of the problem at hand, obtained in zero approximation (on successful application of this model to analyze experimental results and on further developments of this model see in [6, 7, 8]).

2. Two bulk bodies contacting under plane strain conditions of loading

The consideration in this section is largely based on the results of work [9].

2.1. Zero approximation (Plane strain; for more detail on this subsection see: [4, 5])
Consider, under plane strain conditions, the problem on stress-strain state arising in a semi-infinite elastic body as a result of pressing it against a rigid body, by a pressure p uniformly distributed "at infinity", when, in the latter body, there is an elongated socket of the above type (see figure 2), a region of no contact being formed within the socket, while the contact boundary outside this region being flat. The rigid body is considered occupying the half-plane $y > 0$. The thickness of an equivalent layer determined by the fine-scale roughness is neglected based on the considerations expounded in the Introduction.

The solution of the problem at hand may be found by means of adding the solutions of two following problems: problem 1, where it is assumed that the contact takes place everywhere [10], and problem 2, where it is assumed that there is an area of no contact, or no-contact area, situated in the region, where, according to the solution of problem 1, mainly tensile stresses act.

The no-contact area is modeled by a mathematical cut assumed to be situated in an infinite homogeneous elastic body and loaded over its opposite faces by the stresses that are equal in their values and are opposite in their signs to those that are obtained, in the corresponding places, according to the solution of problem 1. The stresses "at infinity" in this infinite body with the cut are assumed vanishing. The coordinates of the ends of the above cut are found from the condition of continuity of the stresses at them. As a result, adding the solutions of the two problems yields the sought for solution, since:

(a) the opposite faces of the cut prove out to be stress free;
(b) the shear stresses on the line of the cut in the infinite body prove out to be equal to zero due to symmetry condition;
(c) the normal stresses at the ends of the cut are continuous;
(d) "at infinity", i.e. far away from the no-contact area, the stress state in the elastic body turns, in the limit, into the stress state determined by the pressure given here initially.

The basic formulae for the symmetric elongated socket of a general form, $v_b(x) = -\zeta(x)$, determining the half-length of the no-contact area, a, and the distribution of contact pressure, p_c, on the prolongation of this area, are obtained in [5, 4].

2.1.1. Example: elastic field disturbance caused by solitary socket. Consider the case, when roughness of the rigid body occupying the half-plane $y > 0$ is presented by an elongated solitary socket, whose form is $\zeta = -\zeta_0 l^2 / (x^2 + l^2)$, where ζ_0 and l are parameters of dimension of length ($\zeta_0 << l$).

The half-length of the no-contact area and contact pressure are [5]:

$$a = l\left[\left(\frac{E\zeta_0}{2(1-\nu^2)lp}\right)^{2/3} - 1\right]^{1/2}, \qquad p_c = p\frac{x\sqrt{x^2-a^2}(3l^2+2a^2+x^2)}{(x^2+l^2)^2}$$

It is convenient to present these dependences in dimensionless form by expressing in them all the quantities of dimension of length through their ratios to l and all the quantities of dimension of stress through their ratios to $E\zeta_0/2(1-\nu^2)l$ (the latter quantity is the critical pressure, which for the first time no contact area appears at). After doing so x, a and p_c turn into $\bar{x}=x/l$, $\bar{a}=a/l$ and $\bar{p}_c = p_c 2(1-\nu)l/(E\zeta_0)$ respectively. Corresponding calculated dependences are plotted in figures 3, 4.

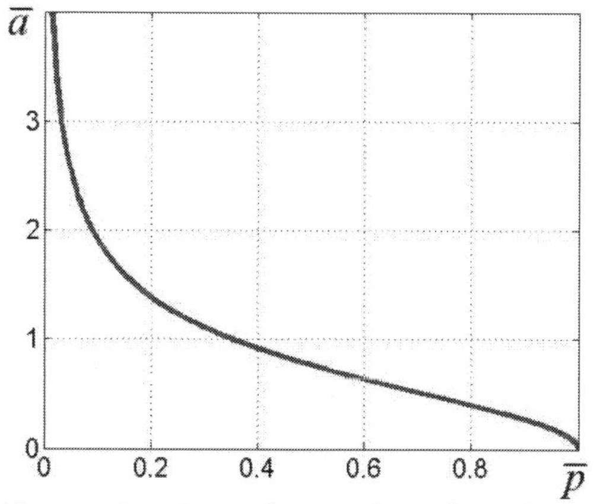

Figure 3. Dependence (in dimensionless quantities) of half-length of no-contact area $\bar{a}$ on pressure $\bar{p}$ applied remotely from the solitary socket

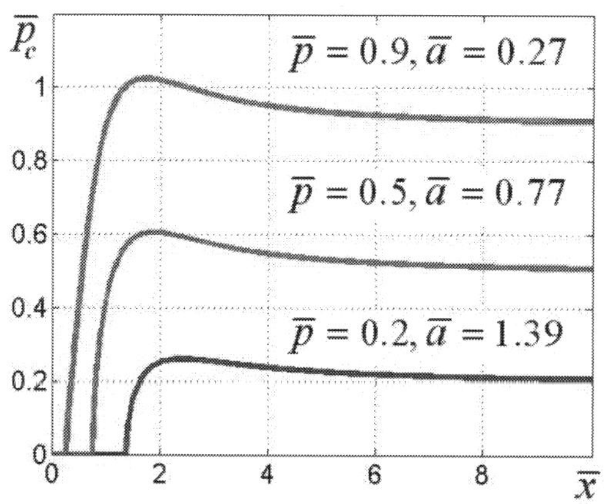

Figure 4. Distributions of contact pressure on the prolongation of the no-contact area at various levels of the pressure $\bar{p}$ applied remotely from the solitary socket

2.2. First approximation

2.2.1. Introduction. The aim of the consideration performed in this subsection is to find the effects resulting from the action of the disturbance of the stress-strain state caused by the solitary socket considered in section 2.1.2 on the fine-scale roughness characterized by a given distribution of the roughness profile heights. When solving this problem, the back influence of these effects on the above disturbance is neglected, which is admissible since $\lambda \ll \Lambda$.

These effects will be considered below on the prolongation of the no contact area considered in section 2.1.2., where the contact pressure distribution has been found. Since $\Lambda \gg \lambda$, any local value of the contact pressure at the prolongation of the no-contact area may be considered equal to the constant value of the nominal contact pressure that is the ratio of the force acting on the contacts between the bodies to the nominal area of their contact. The nominal pressure appears in the GW model, where the extent of the contact boundary is considered infinite and all the conditions are considered invariable along it. Note that, in the GW model, the separation between the interacting bodies is considered prescribed, while the force (or the corresponding pressure) squeezing them is found by calculations. In the case under study, a reverse problem appears, i.e. a need to find the separation corresponding to the value of the prescribed force or to the value of the prescribed nominal pressure acting on the interacting bodies. This nominal pressure should be equated to the local contact pressure that appears at a given point on the prolongation of the no-contact area. Thus, first a solution of the reverse problem is required to find the distribution of the separation between the interacting bodies and the distribution of the actual area of contact (realized through the micro-areas of scale λ) on the prolongation of the no-contact area. These are the quantities used to judge on the influence of the contact on its own behavior, namely – the tribological, electrical, or, possibly, some other one.

2.2.2. Basic quantities appearing in the GW model. According to the GW model [1], surface roughness is modeled by a system of spherical segments with same radius and of random height. The basic quantities needed for the model application are the total force, F, acting on the area of question, and the actual area of contact (the total area of the micro-areas of scale λ). The quantities F and A_r

Journal of Physics: Conference Series **181** (2009) 012012 doi:10.1088/1742-6596/181/1/012012

depend on the separation, b, between the interacting bodies (scale λ). As applied to the problem under consideration, instead of the dependence of F on the given separation, the dependence of the nominal pressure $p = F / A_{nom}$ on b, where A_{nom} is the nominal area, will be employed. The dependence of p on b is monotonous. Consequently, the reverse dependence of b on p exists. For the example considered in [1] (see figure 2(a) built for the following parameter values given on page 305: $\eta = N / A_{nom} = 300(mm)^{-2}$ (here N is the number of the spherical segments on the area at hand with its nominal area being A_{nom}); $\beta \cdot \sigma = 10^{-4} mm^2$ (here β is the segment radius and σ is the standard deviation of the roughness relief heights); $E_*(\beta / \sigma) = 25 Kg / mm^2$; $A_{nom} = 1cm^2 = 100mm^2$, $E_* = E / (1 - \nu^2))$ (here E and ν are the Young modulus and the Poisson ratio of the material at hand), the corresponding dependence $b(p)$ was calculated, taking also the Young modulus $E = 22000 Kg / mm^2$ and Poisson ratio $\nu = 0.3$. This dependence is presented in figure 5.

2.2.3. *Solution of problem in frame of the GW model.* According to the approach expounded in Introduction, p should be equated to p_c from section 2.1.2 under an appropriate choice of the parameter values given there, including those specifying the socket. As a result, the dependence of separation on x will be found. The dependence $b(x)$ known, the dependence $A_r(x)$ can be found. For the elongated socket of the form, $\zeta = -\zeta_0 l^2 / (x^2 + l^2)$, considered in section 2, taking the parameter values ζ_0 = 1 µm, l=100 µm ($\zeta_0 = 0{,}01 \cdot l$) and the above values of the elastic properties E and ν of the deformable body, calculations have been performed for the following values of the pressure applied remotely from the socket (i.e. "at infinity"): p_1 = 0.12 Kg / mm^2 and p_2 = 0.3 Kg / mm^2. For the socket under consideration, the following values of the no-contact area length correspond to these values: 2 mm and 1,5 mm respectively. The corresponding distributions of the contact pressure near the no-contact area are shown in figure 6. Figure 6 shows that under given value of pressure, p, the whole totality of the contact pressure values below p is realized, and further the region of elevated values of the contact pressure emerges.

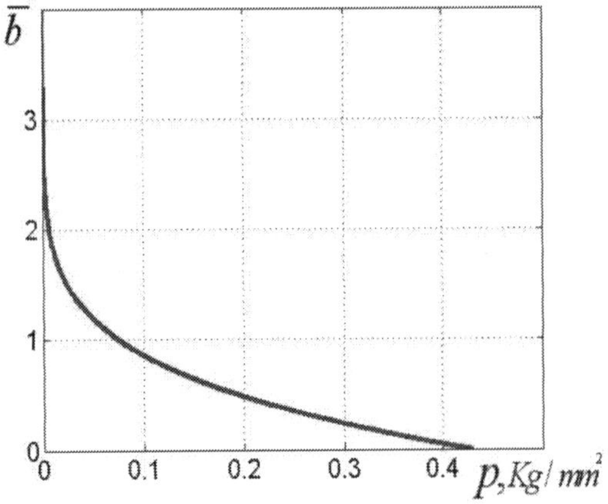

Figure 5. Dependence of dimensionless separation $\bar{b} = b / \sigma$ of interacting bodies on nominal pressure

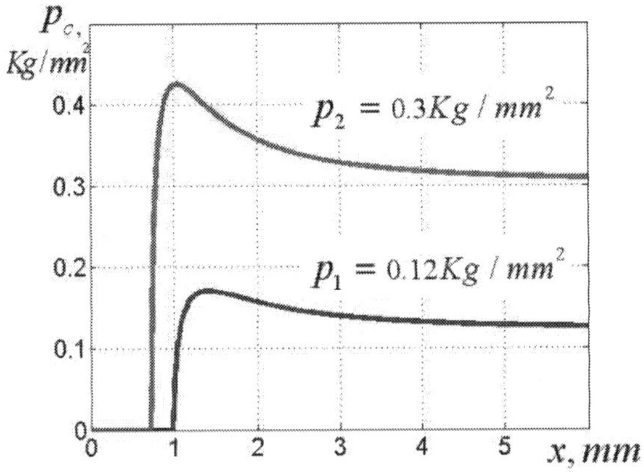

Figure 6. Distribution of contact pressure on prolongation of no-contact area for two levels of pressure applied remotely from the socket

Figures 7 and 8, show calculated dependences of the dimensionless separation $\bar{b} = b / \sigma$ and of the actual area of contact (the total area of the micro-areas of contact) A_r on the prolongation of the no-contact area, presented for two levels of the pressure applied remotely from the socket. The curves in figures 7 and 8 display gradual changes in the whole range of change of x, with the exception of narrow regions directly adjoining to the end of the no-contact area. The sizes of the regions, where $\bar{b}(x)$ and $A_r(x)$ are changed very sharply, amount to ~ (0.1÷0.2 mm). Outside these regions, $\bar{b}(x)$ and $A_r(x)$ are changed gradually enough. Approximately, the relative changes of $\bar{b}(x)$ and $A_r(x)$ on distances of ~ 0.1 mm do not exceed 10%. Note that solid vertical lines in figure 7 are asymptotes for the dependence $\bar{b}$ on x.

When solving the problem, it was assumed that, in zero approximation, all effects associated with the fine-scale roughness modeled by an equivalent layer and with the presence of this layer *per se,* can be neglected. The non-homogeneity of the stress-strain state associated with this layer and its very presence may begin to reveal themselves only in sufficiently close vicinity of the end of the no-contact area, where, however (according to the solution found in zero approximation), the contact pressure is small. Therefore, the effects resulted from the above non-homogeneity and the presence of the fine-scale roughness related equivalent layer are presumed negligible.

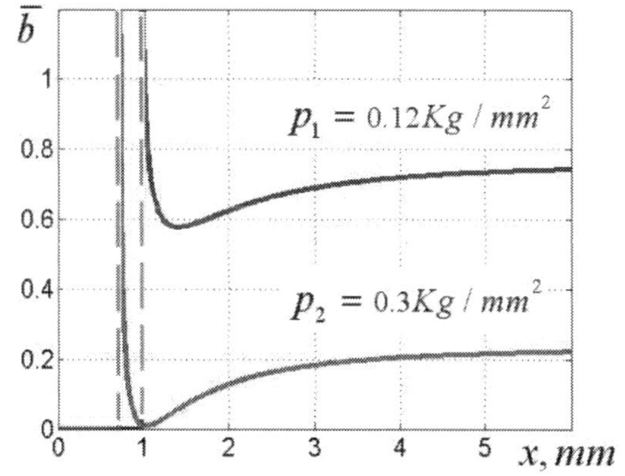

Figure 7. Distributions of dimensionless separation $\bar{b} = b / \sigma$ on prolongation of no-contact area

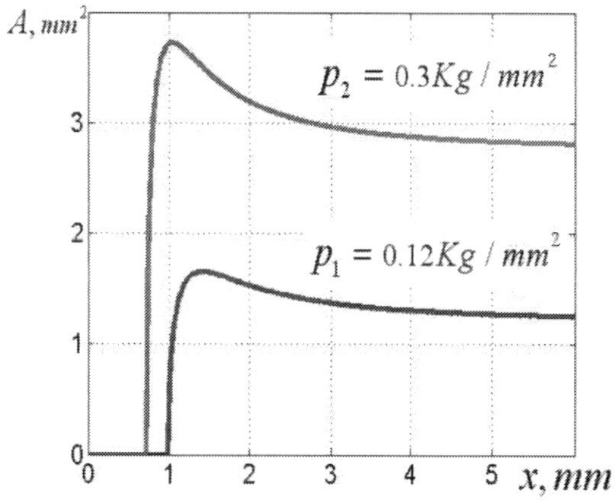

Figure 8. Distributions of actual area of contact $A = A_r(x)$ on prolongation of no-contact area

3. Generalizations

At first, zero approximation for the problems on space loading of bulk bodies and on bending and first approximation for these problems are considered.

3.1 Zero approximation

3.1.1 Two bulk bodies contacting under space conditions of loading (for more detail see [4,5]) Let the deformable body with the same properties and type of roughness of the nominally flat boundary as above contacting with the rigid body, containing a solitary socket, occupies the half-space $z > 0$ in the Cartesian coordinate system xyz and is loaded at infinity, as above, by the uniformly distributed constant pressure, p so that a no-contact area forms inside the socket and the contact boundary outside this area is flat. This case is in its essence analogous to that considered in section 2.1 and, therefore, allows an approach similar to one described in section 2.1. We presume that many sockets, and the corresponding no-contact areas, are distributed sparsely enough so that the interaction among them may be neglected and the final result is obtained by adding the elastic field disturbances caused by all the no-contact areas. At this juncture, note that in the cases when the just mentioned interaction is essential, it may be taken into account by applying the method of iterations analogous to that considered in [11] Thus, the problem is again reduced to obtaining the disturbance of the elastic field caused by a single no-contact area.

3.1.2. Bending. Layer on elastic foundation Consider a problem, where nearly full contact of the above type appears between a semi-infinite material and an infinite layer undergoing bending on an elastic foundation, the material, sufficiently far away from its boundary with the layer, being acted upon by a uniform pressure. The material is assumed to be well less rigid than the layer. A new parameter is introduced, which is (under fixed values of other parameters) the ratio of the elasticity modulus of the material to that of the layer, allowing to solve the problem by iterations method with respect to this parameter, as it has been the case, when the roughness of the layer is presented by a solitary socket of the same form, as the one used in section 2.1.1 [12].

Multi-layer structure. Consider the problem, similar to that described in section 3.1 with the exception that instead of a single layer resting on an elastic foundation there is a multi-layer structure formed by layers also working on bending. Considering conditions ensuring sufficiently gradual and smooth changes of the elastic fields, one may use a continuum approximation [13-15]. The work of the structure on bending is considered to be ensured due to: (i) mutual gliding of the forming layers possessing high enough flexural rigidity, (ii) the presence of alternating with the layers strata of negligibly small flexural rigidity (as compared with that of the layers), (iii) combined effect of what has been mentioned in (i) and (ii).

Let us assume that, the rigid body, pressing against the layered structure has a cavity, so that within the region on the boundary, $z=0$, of the layered structure, being contiguous with this cavity, there is no pressure acting on this structure. On the other hand, this structure situated at $z \geq 0$ has to be considered loaded by a uniform pressure at infinity, i.e. at $z \to +\infty$. One may readily see that, in order to find the bending-related elastic field, the corresponding problem is to be solved for the multi-layer structure at $z > 0$, assuming that within the region on plane $z > 0$, being contiguous with the cavity in the absolutely rigid material (i.e. pressing against the multi-layer structure) this structure is loaded by the constant pressure equal, with the opposite sign, to that produced at $z \to +\infty$. Outside the region at $z=0$, where the constant pressure is applied to the structure, the deflection in the structure at $z=0$ has to be equated to zero. For solutions of plane strain and axially symmetric conditions problem see [14, 15]. Corrections to these solutions and some calculation results for them are presented in [15].

3.2. First approximation

The approach to obtain this approximation remains basically the same, as in section 2.2.2. The only distinction is that the expression for the contact pressure p_c should now be that obtained for the every problem under consideration. For the problems on bending, this is true with a reservation: since in these problems the role of bending is important, taking into account the non-linearity in deformation behavior of the equivalent layer determined by the fine-scale roughness may become, to a certain extent, essential.

4. Discussion

Significance of the obtained results consists in the following. It is seen that, in the presence of two sub-macroscopic length scales of the roughness extent, with one of these, Λ, related to the solitary socket in one of the interacting bodies, and determined by the extent of the no-contact area and the other one $\lambda \ll \Lambda$, related to the micro-areas of contact in the GW mode, and determined by a typical size of these areas, appearing is relatively gradual (to the extent of smallness of λ, as compared with Λ) non-homogeneity of the stress-strain state of the material or structure, such as the multi-layer, treated approximately as a continuum. These non-homogeneous states may be considered as locally homogeneous, allowing, in first approximation, the application of the GW model formulated for strictly homogeneous conditions. In the upshot, it becomes possible to consider, based on the GW model, the development of tribological and other phenomena, which this model allows to predict provided they appear under strictly homogeneous conditions, also under essentially non-homogeneous conditions. This considerably broadens the scope of the GW model application. At the same time the proposed approach allows to considerably expand applicability of the above homogeneous models, including the just mentioned applications, by means of employing the quasi-homogeneous approximation of the same type, as that considered above.

Since for real material surfaces there is a wide variety of geometries of features of their roughness (including those of the type of sockets), note that in [4, 5] a scheme of calculation of the stress state disturbances in the vicinity of relatively large-scale features of roughness is given and an approach is developed allowing to consider an axially symmetric socket and sockets of more general forms. Methods are proposed for considering non-smoothly shaped sockets, which results in the stress concentration (including the infinite one) in the near vicinity of the no-contact area contours.

The problems of the kind considered here are important for the applications associated with (a) friction and wear, (b) sealing, where it is necessary to ensure connectedness of the region of contact, (c) conductivity, in particular electrical one, when the no-contact areas play role of screens on the way of electric current flowing normally to them (see [16], where an analogous situation has been considered, taking also into account thermo-elastic effect caused by Joule heating).

In connection with the problem of wearing, note, in particular, that the situations with relatively large sockets formed in the body interacting with a considerably more compliant one may arise in the process of wearing of the more rigid body of the two, when digging up of the wear particles occurs with formation of corresponding sockets (defects).

And finally note that the above consideration of interaction of the rigid body with a socket with the multi-layer structure described as a continuum presents attractive potentialities of its use in coping with corresponding problems in springs of vehicles under loading.

Acknowledgments
This work has been supported by the Russian Foundation for Basic Research (projects 08-01-00860 and 09-08-01148)

References

[1] Greenwood J A and Williamson J B P 1966 *Proc. Roy. Soc. London* A **295** 300-319
[2] Manners W 1998 *Proc. Roy. Soc. London A* **454** 3203–3221
[3] Persson B N J 2006 *Surf. Sci. Rep.* **61** 201-227
[4] Salganik R L, Markov V G, Mokhel A N and Fedotov A A 2006 *Preprint № 825 IPM RAS* (in Russian)
[5] Salganik R L, Mokhel A N and Fedotov A A 2007 Vestnik MAI **14**(4) 119-126 (in Russian)
[6] Greenwood J A and Wu J J 2001 *Meccanica* **36** 617–630
[7] Siniawski M T, Harris S J and Wang Q 2005 *Tribology Letters* **20**(1) 21-30
[8] Gao Y-F and Bower A F 2006 Rough Surface Plasticity and Adhesion across Length Scales *Nanomechanics of Materials and Structures* eds. T J Chuang et al. (Dodrecht: Springer) pp 277–287
[9] Mokhel A N, Salganik R L and Fedotov A A 2008 *Comp.Cont. Mech.* **1**(4) 61-68 (in Russian)
[10] Westergaard H M 1939 *ASME Journal of Applied Mechanics* **6** A49-A53
[11] Goryacheva I G 1998 *Contact Mechanics In Tribology* (Dodrecht: Kluwer Academic Publisher)
[12] Salganik R L, Zhuravlev A B, Mokhel A N and Fedotov A A 2007 *Preprint № 825 IPM RAS* (in Russian)
[13] Sonntag G 1957 *Forsch. Geb. Ingenieurwesens* **23** 3–8
[14] Salganik R L 1987 *Mech. of Solids* **22**(3) 45–43
[15] Salganik R L and Ustinov K B 2009 *Mech. of Solids* **44**(1) (in press)
[16] Salganik R L 1978 *Mech. of Solids* **13**(5) 25–32

Modeling of contact interaction for solids with inhomogeneous coatings

Kirill E. Kazakov
Ishlinsky Institute for Problems in Mechanics, Russian Academy of Sciences,
Vernadsky Ave. 101 bldg 1, Moscow, 119526 Russia
E-mail: kazakov-ke@yandex.ru

Abstract. We study the contact interaction between rigid punches and viscoelastic foundations with thin inhomogeneous coatings. We consider plane contact problems for inhomogeneous aging viscoelastic basements. We present the statements of the problems and derive their basic mixed integral equations. The exact solution in analytic form of these equations is constructed by using the Manzhirov generalized projection method. We make qualitative conclusions. We present numerical computations of model problems, including the problems in which the inhomogeneity of the coating is described by rapidly oscillating functions and discontinuous functions.

1. Statement of the problem

We assume that a viscoelastic layer with a elastic coating lies on a rigid basis. At time τ_0, the force $P(t)$ with eccentricity $e(t)$ starts to indent a smooth rigid punch of width $2a$ (Fig. 1) into the surface of such a foundation. The punch base shape is described by a function $g(x)$ ($g(x) \geqslant 0$, $\exists x_0 \in [-a, a] : g(x_0) = 0$). The coating is assumed to be thin compared with the contact area, i.e., its thickness satisfies the condition $h \ll 2a$, and its properties depend on the longitudinal coordinate, i.e., the stiffness of coating is the function of x. The lower layer of an arbitrary thickness H is made of viscoelastic materials. We denote the moment of its production by $\tau_2 \leqslant \tau_0$. We assume that the coating rigidity is less than the rigidity of the lower layer or they are of the same order of magnitude [1–3]. We consider the case of plane strain.

To derive the integral equation of the problem, we replace the punch by some normally distributed load $p(x,t) = -q(x,t)$ acting on the same region ($-a \leqslant x \leqslant a$) and equal to zero outside this region. Then the vertical displacement of the upper face of the foundation described above under the action of the normal force $q(x,t)$ can be written as [1–3]

$$u_z(x,t) = \frac{\theta q(x,t)h}{R(x)} + (\mathbf{I} - \mathbf{V}_2)\mathbf{F}\frac{2(1-\nu_2^2)q(x,t)}{\pi E_2(t-\tau_2)}, \tag{1}$$

$$\mathbf{V}_2 f(x,t) = \int_{\tau_0}^{t} K^{(2)}(t-\tau_2, \tau-\tau_2) f(x,\tau)\,d\tau, \quad K^{(2)}(t,\tau) = E_2(\tau)\frac{\partial}{\partial\tau}\Big[\frac{1}{E_2(\tau)} + C^{(2)}(t,\tau)\Big],$$

$$\mathbf{F} f(x,t) = \int_{-a}^{a} k_{\rm pl}\Big(\frac{x-\xi}{H}\Big) f(\xi,t)\,d\xi, \quad k_{\rm pl}(s) = \int_0^\infty \frac{L(u)}{u}\cos(su)\,du,$$

where ν_2 and $E_k(t)$ are the Poisson ratio and Young's modulus of the lower viscoelastic layer; $\mathbf{I}$ is the identity operator; $\mathbf{V}_2$ is the Volterra integral operator with tensile creep kernel $K^{(2)}(t,\tau)$;

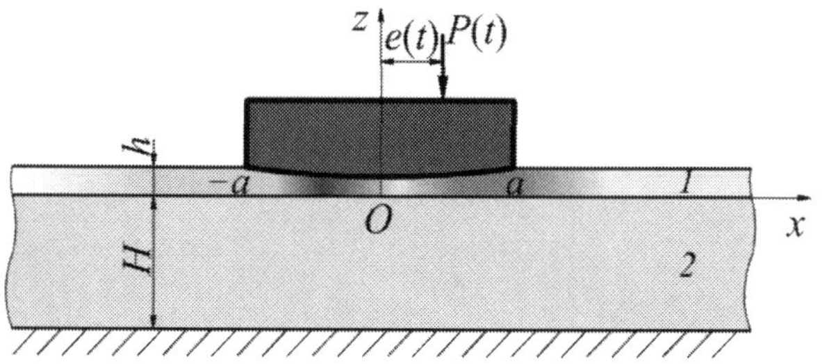

Figure 1. Plane contact problem for viscoelastic solids (*2*) with inhomogeneous coatings (*1*)

$C^{(2)}(t,\tau)$ is the tensile creep function; $R(x)$ is the stiffness of the coating; in the case of a smooth coating-layer contact, we have $R(x) = E_1(x)/[1-\nu_1^2(x)]$, and in the case of an perfect contact, $R(x) = E_1(x)[1-\nu_1(x)]/\{[1+\nu_1(x)][1-2\nu_1(x)]\}$, where $\nu_1(x)$ and $E_1(x)$ are the Poisson ratio and Young's modulus of the coating; $\mathbf{F}$ is the integral operator with the known kernel of the plane contact problem $k_{\text{pl}}[(x-\xi)/H]$ (see [3, 4]); in the case of a smooth contact between the lower layer and the rigid base,

$$L(u) = \frac{\cosh 2u - 1}{\sinh 2u + 2u},$$

and in the case of a perfect contact,

$$L(u) = \frac{2\varkappa \sinh 2u - 4u}{2\varkappa \cosh 2u + 4u^2 + 1 + \varkappa^2}, \quad \varkappa = 3 - 4\nu_2.$$

By equating the vertical displacements of the upper face of the coating with the displacement of the rigid punch and taking into account (1) and the fact that the contact interaction is conformal, we obtain the integral equation of our problem in the form

$$\frac{\theta q(x,t)h}{R(x)} + (\mathbf{I} - \mathbf{V}_2)\mathbf{F}\frac{2(1-\nu_2^2)q(x,t)}{\pi E_2(t-\tau_2)} = \delta(t) + \alpha(t)x - g(x) \quad (-a \leqslant x \leqslant a), \tag{2}$$

where $\delta(t)$ is the punch settlement and $\alpha(t)$ is its tilt angle.

We supplement Eq. (2) with the condition of the punch equilibrium on the foundation:

$$\int_{-a}^{a} q(\xi,t)\,d\xi = P(t), \quad \int_{-a}^{a} \xi q(\xi,t)\,d\xi = M(t). \tag{3}$$

Here $M(t) = e(t)P(t)$ denotes the moment of application of the force $P(t)$.

In (2) and (3), we make the change of variables by the formulas

$$x^* = \frac{x}{a}, \quad \xi^* = \frac{\xi}{a}, \quad t^* = \frac{t}{\tau_0}, \quad \tau^* = \frac{\tau}{\tau_0}, \quad \tau_2^* = \frac{\tau_2}{\tau_0}, \quad \lambda = \frac{H}{a}, \quad \delta^*(t^*) = \frac{\delta(t)}{a}, \quad \alpha^*(t^*) = \alpha(t),$$
$$g^*(x^*) = \frac{g(x)}{a}, \quad c^*(t^*) = \frac{E_2(t-\tau_2)}{E_0}, \quad m^*(x^*) = \frac{E_0}{R(x)(1-\nu_2^2)}\frac{h}{2a}, \quad e^*(t^*) = \frac{e(t)}{a},$$
$$q^*(x^*,t^*) = \frac{2(1-\nu_2^2)q(x,t)}{E_2(t-\tau_2)}, \quad P^*(t^*) = \frac{2P(t)(1-\nu_2^2)}{E_2(t-\tau_2)a}, \quad M^*(t^*) = \frac{2M(t)(1-\nu_2^2)}{E_2(t-\tau_2)a^2}, \tag{4}$$
$$\mathbf{V}_2^* f(x^*,t^*) = \int_1^{t^*} K_2(t^*,\tau^*) f(x^*,\tau^*)\,d\tau^*, \quad K_2(t^*,\tau^*) = K^{(2)}(t-\tau_2,\tau-\tau_2)\tau_0,$$
$$\mathbf{F}^* f(x^*,t^*) = \int_{-1}^{1} k_{\text{pl}}^*(x^*,\xi^*) f(\xi^*,t^*)\,d\xi^*, \quad k_{\text{pl}}^*(x^*,\xi^*) = \frac{1}{\pi}k_{\text{pl}}\Big(\frac{x-\xi}{H}\Big) = \frac{1}{\pi}k_{\text{pl}}\Big(\frac{x^*-\xi^*}{\lambda}\Big).$$

Then, omitting the asterisks, we obtain a mixed integral equation in the form

$$c(t)m(x)q(x,t) + (\mathbf{I} - \mathbf{V}_2)\mathbf{F}q(x,t) = \delta(t) + \alpha(t)x - g(x) \quad (-1 \leqslant x \leqslant 1) \tag{5}$$

with the additional conditions

$$\int_{-1}^{1} q(\xi,t)\,d\xi = P(t), \quad \int_{-1}^{1} \xi q(\xi,t)\,d\xi = M(t). \tag{6}$$

Now we divide Eq. (5) by $\sqrt{m(x)}$ and introduce the notation

$$Q(x,t) = \sqrt{m(x)}q(x,t), \quad k(x,\xi) = \frac{k_{\mathrm{pl}}(x,\xi)}{\sqrt{m(x)}\sqrt{m(\xi)}}, \quad \mathbf{A}Q(x,t) = \int_{-1}^{1} k(x,\xi)Q(\xi,t)\,d\xi.$$

Then the integral equation (5) can be reduced to the following integral equation with the Hilbert–Schmidt kernel $k(x,\xi)$ (see, e.g., [5]):

$$c(t)Q(x,t) + (\mathbf{I} - \mathbf{V}_2)\mathbf{A}Q(x,t) = \frac{\delta(t)}{\sqrt{m(x)}} + \frac{\alpha(t)x}{\sqrt{m(x)}} - \frac{g(x)}{\sqrt{m(x)}} \quad (-1 \leqslant x \leqslant 1). \tag{7}$$

The additional conditions (6) take the form

$$\int_{-1}^{1} \frac{Q(\xi,t)}{\sqrt{m(\xi)}}\,d\xi = P(t), \quad \int_{-1}^{1} \frac{Q(\xi,t)}{\sqrt{m(\xi)}}\xi\,d\xi = M(t). \tag{8}$$

In what follows, we construct the solution of the two-dimensional equation (7), which contains integral operators with constant as well as variable limits of integration, with the additional conditions (8) taken into account.

2. Solution of the problem

We seek a solution of Eq. (7) under conditions (8) in the class of functions continuous in time t in the Hilbert space $L_2[-1,1]$ (e.g., see [1, 6]). To this end, we first construct an $L_2[-1,1]$-orthonormal system of functions such that it contains $1/\sqrt{m(x)}$ and the remaining basis functions can be represented as products of functions depending on x by the weight function $1/\sqrt{m(x)}$. A system of functions satisfying the above conditions can be constructed by the formulas [7]:

$$\int_{-1}^{1} p_i(\xi)p_j(\xi)\,d\xi = \delta_{ij}, \quad p_n(x) = \frac{P_n(x)}{\sqrt{m(x)}}, \quad P_0(x) = \frac{1}{\sqrt{J_0}}, \quad J_n = \int_{-1}^{1} \frac{\xi^n}{m(\xi)}\,d\xi,$$

$$P_n(x) = \frac{1}{\sqrt{\Delta_{n-1}\Delta_n}} \begin{vmatrix} J_0 & J_1 & \cdots & J_n \\ J_1 & J_2 & \cdots & J_{n+1} \\ \vdots & \vdots & \ddots & \vdots \\ 1 & x & \cdots & x^n \end{vmatrix}, \quad \Delta_{-1} = 1, \quad \Delta_n = \begin{vmatrix} J_0 & J_1 & \cdots & J_n \\ J_1 & J_2 & \cdots & J_{n+1} \\ \vdots & \vdots & \ddots & \vdots \\ J_n & J_{n+1} & \cdots & J_{2n} \end{vmatrix}.$$

Note that if $m(x) = \mathrm{const}$, then the polynomials $p_n(x)$ are the orthonormal Legendre polynomials.

The Hilbert space $L_2[-1,1]$ can be represented as the direct sum of orthogonal subspaces $L_2[-1,1] = L_2^{(1)}[-1,1] \oplus L_2^{(2)}[-1,1]$, where $L_2^{(1)}[-1,1]$ is the Euclidean space with basis $\{p_0(x), p_1(x)\}$ and $L_2^{(2)}[-1,1]$ is the Hilbert space with basis $\{p_2(x), p_3(x), \ldots\}$. The integrand

and the right-hand side of (7) can also be represented as the algebraic sum of functions continuous in time t and ranging in $L_2^{(1)}[-1,1]$ and $L_2^{(2)}[-1,1]$, respectively, i.e.,

$$Q(x,t) = Q_1(x,t) + Q_2(x,t), \quad f(x,t) = f_1(x,t) + f_2(x,t),$$
$$Q_1(x,t) = z_0(t)p_0(x) + z_1(t)p_1(x),$$
$$f_1(x,t) = \left[\sqrt{J_0}\delta(t) + \frac{J_1}{\sqrt{J_0}}\alpha(t) - g_0\right]p_0(x) + \left[\frac{\sqrt{J_0J_2 - J_1^2}}{\sqrt{J_0}}\alpha(t) - g_1\right]p_1(x), \quad f_2(x,t) = -g_2(x),$$
$$\frac{g(x)}{\sqrt{m(x)}} = g_0p_0(x) + g_1p_1(x) + g_2(x), \quad g_0 = \int_{-1}^{1} \frac{g(\xi)}{\sqrt{m(\xi)}}p_0(\xi)\,d\xi, \quad g_1 = \int_{-1}^{1} \frac{g(\xi)}{\sqrt{m(\xi)}}p_1(\xi)\,d\xi.$$

Note that the representation for $Q(x,t)$ contains the known term $Q_1(x,t)$, which is determined by the additional conditions (8):

$$z_0(t) = \frac{P(t)}{\sqrt{J_0}}, \quad z_1(t) = \frac{J_0M(t) - J_1P(t)}{\sqrt{J_0(J_0J_2 - J_1^2)}},$$

and the term $Q_2(x,t)$ is to be found. Conversely, for the right-hand side, one should find $f_1(x,t)$, while $f_2(x,t) \equiv 0$. These peculiarities permit one to class the resulting problem as a specific case of the generalized projection problem stated and solved in [8].

Following [8], we can introduce the orthogonal projection operator, mapping the space $L_2[-1,1]$ onto $L_2^{(1)}[-1,1]$:

$$\mathbf{P}_1\phi(x,t) = \int_{-1}^{1} \phi(\xi,t)[p_0(x)p_0(\xi) + p_1(x)p_1(\xi)]\,d\xi.$$

Obviously, the orthoprojector $\mathbf{P}_2 = \mathbf{I} - \mathbf{P}_1$ maps the space $L_2[-1,1]$ onto $L_2^{(2)}[-1,1]$. In addition, the following relations hold:

$$\mathbf{P}_if(x,t) = f_i(x,t), \quad \mathbf{P}_iQ(x,t) = Q_i(x,t), \quad i = 1, 2.$$

Using [8], we apply the orthogonal projection operator $\mathbf{P}_2$ to Eq. (7). As a result, we obtain the equation for determining $Q_2(x,t)$ with a known right-hand side:

$$c(t)Q_2(x,t) + (\mathbf{I} - \mathbf{V}_2)\mathbf{P}_2\mathbf{A}Q_2(x,t) = -g_2(x) - (\mathbf{I} - \mathbf{V}_2)\mathbf{P}_2\mathbf{A}Q_1(x,t). \quad (9)$$

It is necessary to construct its solution in the form of a series in the eigenfunctions of the operator $\mathbf{P}_2\mathbf{A}$, which, as one can show using the results obtained in [8], is a compact strongly positive self-adjoint operator from $L_2^{(2)}[-1,1]$ into $L_2^{(2)}[-1,1]$. The system of eigenfunctions of such an operator is a basis in the space $L_2^{(2)}[-1,1]$, see [9]. The spectral problem for the operator $\mathbf{P}_2\mathbf{A}$ can be written in the form

$$\mathbf{P}_2\mathbf{A}\varphi_k(x) = \gamma_k\varphi_k(x),$$
$$\varphi_k(x) = \sum_{i=2}^{\infty} \varphi_i^{(k)}p_i(x), \quad k = 2, 3, \ldots, \quad k(x,\xi) = \sum_{m=0}^{\infty}\sum_{n=0}^{\infty} R_{mn}p_m(x)p_n(\xi),$$
$$R_{mn} = \int_{-1}^{1}\int_{-1}^{1} k(x,\xi)p_m(x)p_n(\xi)\,dx\,d\xi, \quad R_{nm} = R_{mn}, \quad m, n = 0, 1, \ldots,$$
$$\sum_{n=2}^{\infty} R_{mn}\varphi_n^{(k)} = \gamma_k\varphi_m^{(k)}, \quad k, m = 2, 3, \ldots$$

Journal of Physics: Conference Series **181** (2009) 012013 doi:10.1088/1742-6596/181/1/012013

We expand the functions $Q_2(x,t)$ and $g_2(x)$ with respect to the new basis functions $\varphi_k(x)$ ($k = 2, 3, \ldots$) in $L_2^{(2)}[-1,1]$, i.e., $Q_2(x,t) = \sum_{k=2}^{\infty} z_k(t)\varphi_k(x)$, $g_2(x) = \sum_{k=2}^{\infty} g_k\varphi_k(x)$, substitute this representation into (9), and see that the unknown expansion functions $z_k(t)$ ($k = 2, 3, \ldots$) can be determined by the formula

$$z_k(t) = -(\mathbf{I}+\mathbf{W}_k)\frac{g_k + (\mathbf{I}-\mathbf{V}_2)\left[z_0(t)K_k^{(0)} + z_1(t)K_k^{(1)}\right]}{c(t)+\gamma_k},$$

$$K_k^{(0)} = \sum_{n=2}^{\infty} R_{0n}\varphi_n^{(k)}, \quad K_k^{(1)} = \sum_{n=2}^{\infty} R_{1n}\varphi_n^{(k)}, \quad g_k = \int_{-1}^{1}\frac{g(\xi)}{\sqrt{m(\xi)}}\varphi_k(\xi)\,d\xi,$$

$$\mathbf{W}_k f(x,t) = \int_1^t R_k^*(t,\tau) f(x,\tau)\,d\tau, \quad k = 2, 3, \ldots,$$

where $R_k^*(t,\tau)$ ($k = 2, 3, \ldots$) is the resolvent of the kernel $K_k^*(t,\tau) = \gamma_k K_2(t,\tau)/[c(t)+\gamma_k]$.

Note that the resulting solution has the following structure

$$q(x,t) = \frac{1}{m(x)}\left[z_0(t)P_0(x) + z_1(t)P_1(x) + \ldots\right],$$

i.e., one can explicitly single out the weight function $m(x)$, and hence the coating stiffness function $R(ax)$ related to it by the change of variables (4). The formulas thus obtained permit deriving efficient analytic solutions for the layers with coatings of thickness described by complicated, in particular, rapidly oscillating functions, which can hardly be done by other well-known methods.

Determining the contact pressure under the punch, we can find the unknown punch settlement and the tilt angle of the punch. To this end, we must apply the operator $\mathbf{P}_1$ to Eq. (7):

$$\begin{aligned} &c(t)Q_1(x,t) + (\mathbf{I}-\mathbf{V}_2)\mathbf{P}_1\mathbf{A}Q(x,t) \\ &\quad = \left[\sqrt{J_0}\delta(t) + \frac{J_1}{\sqrt{J_0}}\alpha(t) - g_0\right]p_0(x) + \left[\sqrt{\frac{J_0J_2 - J_1^2}{J_0}}\alpha(t) - g_1\right]p_1(x). \end{aligned}$$

The obtained relation readily implies the formulas for the tilt angle $\alpha(t)$ and the settlement $\delta(t)$:

$$\alpha(t) = \sqrt{\frac{J_0}{J_0J_2 - J_1^2}}\left\{g_1 + c(t)z_1(t) + (\mathbf{I}-\mathbf{V}_2)\left[R_{10}z_0(t) + R_{11}z_1(t) + \sum_{k=2}^{\infty} K_k^{(1)} z_k(t)\right]\right\},$$

$$\delta(t) = \frac{1}{\sqrt{J_0}}\left\{-\alpha(t)\frac{J_1}{\sqrt{J_0}} + g_0 + c(t)z_0(t) + (\mathbf{I}-\mathbf{V}_2)\left[R_{00}z_0(t) + R_{01}z_1(t) + \sum_{k=2}^{\infty} K_k^{(0)} z_k(t)\right]\right\}.$$

We note that even in the case of centrally applied load, the variable thickness of the coating usually leads to the punch shift, which, in the case of a viscoelastic foundation, varies in the course of time.

Thus, we have constructed an analytic solution (in the form of series with any prescribed accuracy distinguishing the singularity) of the plane contact problem for a viscoelastic layer with a thin elastic inhomogeneous coating, which permits efficiently solving problems with complicated functions describing the shape of the punch base.

Journal of Physics: Conference Series **181** (2009) 012013 doi:10.1088/1742-6596/181/1/012013

3. Numerical calculations

We consider the two-layered foundation whose lower layer is made of a viscoelastic aging material and whose upper thin layer is made of an elastic material. There is an ideal contact between the layers. We assume that the variation in the modulus of elasticity of the viscoelastic material is insignificant in time, i.e., $E_2(t) \equiv E_2$. We represent the creep function of the coating material in the form [1]:

$$C(t,\tau) = (C_0 + A_0 e^{-\chi\tau})\left[1 - e^{-\psi(t-\tau)}\right].$$

We use the following values of the parameters of the elastic foundation and the viscoelastic coating [1]: $c(t) \equiv 1$, $\nu_2 = 0.1$, $C_0E_2 = 0.5522$, $A_0E_2 = 4$, $\chi\tau_0 = 0.31$, $\psi\tau_0 = 0.6$ ($\tau_0 = 10$ days, $\tau_2 = 0$ days).

We consider the case in which the load $P(t) \equiv 1$ and the eccentricity $e(t) \equiv 0$ are given. We assume that the punch base shape is described by a function $g(x) \equiv 0$ (the punch is plane) and the function $m(x)$, which is inversely proportional to the coating stiffness, is given by the formula

$$m(x) = \begin{cases} 0.18 + 0.12\cos[2\pi(x-0.3)/0.05] & \text{for } x \in [0.3, 0.35], \\ 0.3 & \text{for } x \notin [0.3, 0.35]. \end{cases} \tag{10}$$

This function models a coating with increased stiffness in the range $x \in [0.3, 0.35]$. Fig. 2 presents the distribution of contact pressure under the punch at different times ($t = 1$, $t = 10$, and the steady-state distribution as $t \to \infty$). In the course of time, the contact pressure are

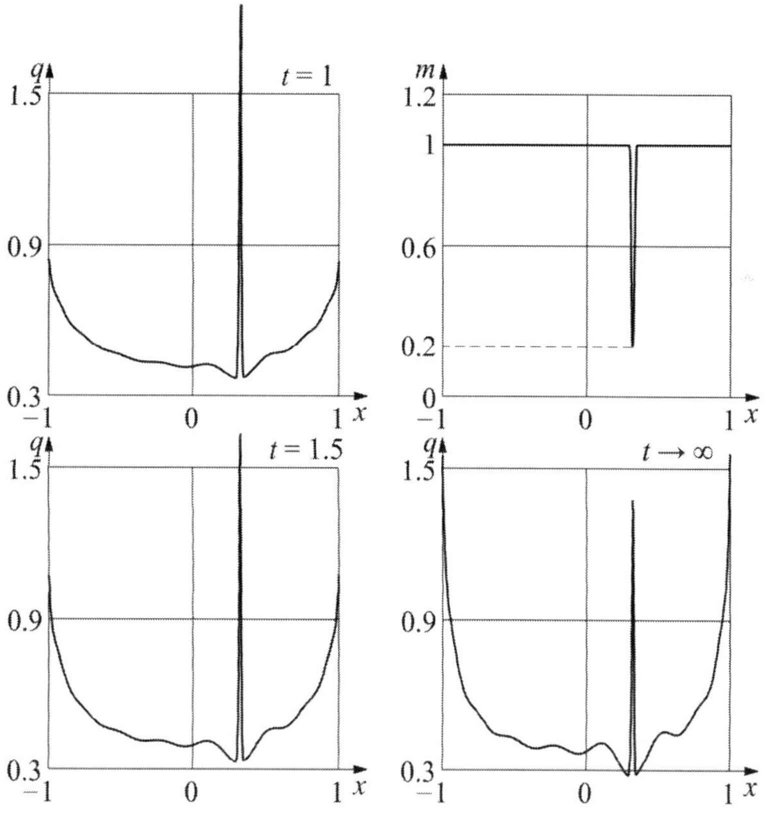

Figure 2. The distribution of contact pressure under the punch at different times ($t = 1$, $t = 1.5$, and the steady-state distribution as $t \to \infty$) for $P(t) \equiv 1$, $e(t) \equiv 0$, and (10)

redistributed, but the explicit dependence on the function $m(x)$ can be seen everywhere. As follows from the example, the obtained solutions detect all specific features (such as oscillating functions) of the surface properties.

For nonuniform and nonuniformly hardened surfaces, one often uses piecewise constant stiffness functions $R(x)$ of the form $R(x) = R_i$ for $x \in \Omega_i$ $(i = 1, 2, \ldots)$. Figure 3 shows the contact pressure distribution for the case in which the coating uses two materials whose interface goes along the punch axis:

$$m(x) = 1 \text{ for } x < 0, \quad m(x) = 2 \text{ for } x > 0. \tag{11}$$

The contact pressure experiences jumps at the points where the coating stiffness varies. We see that the stresses vary differently in different regions. In the higher-stiffness region ($x < 0$), the initial stresses on the right ($x = 0$) exceed those on the left ($x = -1$), but the diagram changes with time, and the stresses on the left become larger than those on the right. The stresses in the region with the softer coating material are more stable. Needless to say, the stiffness asymmetry results in a considerable punch skewing on the layer. The punch settlement and tilt angle curves are presented in Fig. 4.

Thus, the solution obtained permits solving contact problem for the case in which the coating stiffness is described by a discontinuous function.

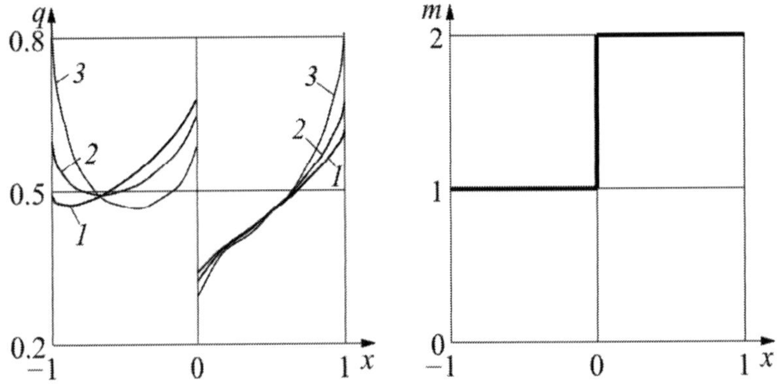

Figure 3. The distribution of contact pressure under the punch at different times (1 — $t = 1$, 2 — $t = 1.5$, 3 — the steady-state distribution as $t \to \infty$) for $P(t) \equiv 1$, $e(t) \equiv 0$, and (11)

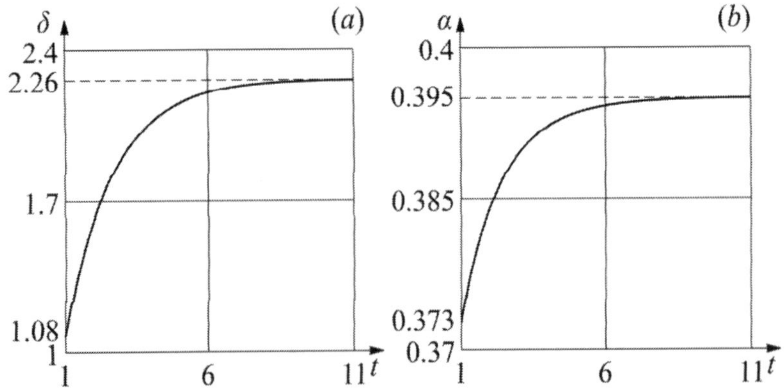

Figure 4. The punch settlement (a) and the tilt angle (b) for $P(t) \equiv 1$, $e(t) \equiv 0$, and (11)

Conclusion

We pose and solve plane contact problems for viscoelastic aging basements with elastic inhomogeneous coatings and rigid punches. The solution of the problem is obtained analytically, and, in the expressions for the contact stresses, the stiffness function of the foundation is distinguished explicitly, which allows one to perform computations for actual inhomogeneities of the coating surface, which are described by oscillating and discontinuous functions. The explicit formulas were obtained for the punch settlement and the tilt angle.

Acknowledgments

The research was financially supported by the Russian Foundation for Basic Research (projects nos. 08-01-00553-a and 08-01-91302-IND-a) and by the Department of Energetics, Mechanical Engineering, Mechanics and Control Processes of the Russian Academy of Sciences (program no. 13 OE).

References

[1] Arutyunyan N Kh and Manzhirov A V 1999 *Contact Problems in the Theory of Creep* (Erevan: Izd-vo NAN RA) [in Russian].
[2] Manzhirov A V 2001 Contact problems for inhomogeneous aging viscoelastic solids In *Mechanics of Contact Interactions*, ed I I Vorovich and V M Alexandrov (Moscow: Fizmatlit) pp 549–565 [in Russian].
[3] Alexandrov V M and Mkhitaryan S M 1983 *Contact Problems for Bodies with Thin Coatings and Interlayers* (Moscow: Nauka) [in Russian].
[4] Vorovich I I, Alexandrov V M, and Babeshko V A 1974 *Nonclassical Mixed Problems of Elasticity* (Moscow: Nauka) [in Russian].
[5] Goursat E 1927 *Cours d'Analyse Mathématique* tome III *Intégrales Infiniment Voisines; Équations aux Dérivées Partielles du Second Ordre; Équations Intégrales; Calcul des Variations* (Paris: Gauthier–Villars).
[6] Vladimirov V S 1981 *Equations of Mathematical Physics* (Moscow: Nauka) [in Russian].
[7] Szegö G 1959 *Orthogonal Polynomials* (Providence: Amer. Math. Soc).
[8] Manzhirov A V 2005 Mixed integral equations of contact mechanics and tribology In *Mixed Problems of Mechanics of Solid. Proceedings of V Russian Conf. with Intern. Participation*, ed Acad. N F Morozov (Saratov: Izd-vo Saratov Univ.) pp 221–224 [in Russian].
[9] Kolmogorov A N and Fomin S V 1999 *Elements of the Theory of Functions and Functional Analysis* (New York: Dover Publications).

Stresses in ultrasonically assisted bone cutting

K Alam[1], A V Mitrofanov[1], M Bäker[2] and V V Silberschmidt[1]

[1] Wolfson School of Mechanical and Manufacturing Engineering, Loughborough University, UK

[2] Institut für Werkstoffe, Technische Universität Braunschweig, Langer Kamp 8, 38106 Braunschweig, Germany

Email: k.alam2@lboro.ac.uk

Abstract. Bone cutting is a frequently used procedure in the orthopaedic surgery. Modern cutting techniques, such as ultrasonic assisted drilling, enable surgeons to perform precision operations in facial and spinal surgeries. Advanced understanding of the mechanics of bone cutting assisted by ultrasonic vibration is required to minimise bone fractures and to optimise the technique performance. The paper presents results of finite element simulations on ultrasonic and conventional bone cutting analysing the effects of ultrasonic vibration on cutting forces and stress distribution. The developed model is used to study the effects of cutting and vibration parameters (e.g. amplitude and frequency) on the stress distributions in the cutting region.

1. Introduction

Understanding the mechanical behaviour of bones that leads to their failure is necessary for diagnosis and prevention of bone fracture and diseases. Knee and hip implant surgeries are performed around the world and are considered the most common operations in clinical practice. A total of 300,000 knee arthroplasties are performed each year in the United States alone with the number increasing every year [1]. For years, researchers have employed mechanical testing and finite element (FE) models to study how and why bone fractures under various loading conditions [2,3,4,5]. Bone has a capability to adapt to its mechanical environment and to reconstruct functionally its structure and geometry [6,7]. One of the important mechanical factors affecting bone remodelling is the stress in bone tissue. For remodelling to occur, bone should posses some stress present in its tissue to stimulate the osteocyte over a relatively long period. This stress will be a type of residual stress that satisfies the force equilibrium condition in the bone material [8,9].

Bone is subjected to stresses in cutting processes which are the essential part of any orthopaedic surgery. Limited studies have been performed to measure the level of stresses in bone cutting. In [10], the chip generation process was observed under an optical microscope while cutting pig cortical bone – i.e. the hard outside layer of the long bone such as femur or tibia. The measured shear angle was approximately 27° for all tool rake angles, and continuous type cutting chips were generated. The shear stress was approximately 40 MPa independent of the tool rake angle. The shear stress on the shear plane was decreased with an increase in the tool rake angle and was found to depend on cutting orientation. A FE analysis of bone cutting has not been reported in literature so far due to unavailability of correct material properties for such modelling and difficulties inherent to numerical modelling of material separation.

Several modern bone cutting techniques were introduced that aim at minimizing the invasiveness and improving the safety of the operation. Ultrasonically assisted bone cutting (UAC) is one of such techniques. UAC has been previously utilized in cutting super alloys, composites, food and biomaterials [11,12]. In UAC, a high-frequency vibration (frequency 10-50 kHz, amplitude 10-30 μm) is superimposed on the movement of a cutting tool (see figure 1). Compared to conventional cutting (CC), this technique results in a decrease in the cutting force and an improvement in the surface finish [13]. Dynamics of UAC as a non-linear vibro-impact process was studied in [14] and the amplitude response of the cutting tool under loading was analysed for this cutting technique. However, the influence of UAC on the material behaviour in bone cutting has not been studied yet.

Analysis of the mechanics of cutting and its effect on stress distributions in the bone tissue would facilitate optimal prosthesis design and fabrication. This may also be helpful in reducing the risk of mechanical failures and in improving bone-implant longevity. In bone cutting, two types of stresses are necessary to be measured: (1) stresses generated during the chip formation which affect cutting mechanics and (2) residual stresses that are important for the bone-implant interface strength. The current work is aimed at investigating the level of stresses induced in bone when a hard cutting tool penetrates into it in the ultrasonically assisted/vibratory mode. Calculated stresses are compared with those generated in the conventional cutting mode. The effect of ultrasonic frequency and amplitude on material stresses is discussed. The model allows us to study the bone-tool interaction in vibratory mode, hence enabling the optimization / evaluation of the cutting procedure.

2. FE model description

The MSC.Marc general FE code [15] was used in the development of our two-dimensional FE model of bone cutting. A rectangular part of the bone was modelled with dimensions of 2 mm × 1 mm. The work piece was modelled with four-node, isoparametric, arbitrary quadrilateral elements and assuming plane-strain conditions. The cutting tool was modelled with rigid lines as the stiffness of the tool material was much greater than that of the bone.

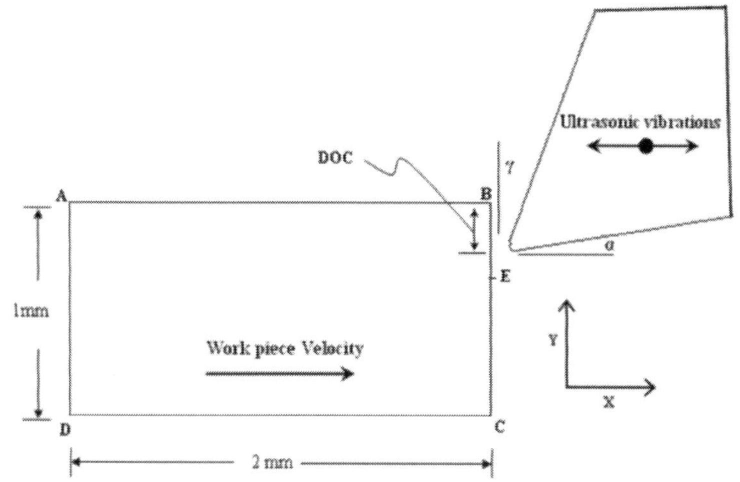

Figure 1. Scheme of the relative movement of the work piece and tool in UAC.

In simulations of UAC, the work piece was moved with a constant velocity in the X-direction, and the tool was vibrated with the required frequency and amplitude. In conventional cutting (CC), the tool position was fixed, and the workpiece was advanced with the constant velocity as before. Remeshing was applied every 10 increments in simulations to avoid convergence issues associated with highly distorted elements. The effect of remeshing frequency on magnitudes of the calculated force and stresses was negligible. Increasing the remeshing frequency significantly increased the analysis time. A contact was defined between the tool and the workpiece, where relative motion between them was allowed under friction conditions. A shear friction model frequently used for modelling of cutting processes was chosen for simulations. The software [15] used in simulations has the capability to detect the contact automatically. To check the contact stability in the iteration process, the motion of the node was controlled with regard to its penetration of a surface by a certain amount.

The developed FE model was fully thermo-mechanically coupled in order to properly reflect interconnection between thermal and mechanical processes. The FE model presupposes formation of the continuous chip, which is in a good agreement with the results of our high-speed filming experiments of bone drilling and other results reported [16,17]. All the simulations were continued until full chip formation was achieved, i.e. cutting forces, stresses and strains reached their saturated levels.

2.1. Material data

The bone tissue is a heterogeneous material with complex multi-level microstructure that is approximated in this paper with an equivalent homogeneous material (EHM). An elastic modulus of bone was measured from the nanoindentation test and incorporated into the FE model. Material separation ahead of the tool occurred when the breaking stress (obtained from the tensile test) was reached. The Johnson-Cook (JC) material model [12, 18] that accounts for the strain rate sensitivity and temperature effects was utilised for the bone tissue as it allowed an accurate approximation of our experimental data. In this model, the bone is considered as an elastic-plastic material with nonlinear strain hardening:

$$\sigma_Y = \left(A + B\varepsilon_p^n\right)\left(1 + C\ln\left(\frac{\dot{\varepsilon}_p}{\dot{\varepsilon}_o}\right)\right)\left(1 - T^{*m}\right) \tag{1}$$

where A, B, C and n are constants and $T^* = \dfrac{\left(T - T_{room}\right)}{\left(T_{melt} - T_{room}\right)}$;

ε_p and $\dot{\varepsilon}$ are the plastic strain and strain rate, respectively; T_{room} and T_{melt} are the room and melting temperatures, respectively; the term T^* was omitted in simulations at this stage of investigations. The JC constants are calculated by fitting the equation into the stress-strain graphs for different load rates obtained from our experimental tensile tests on the bone. Their magnitudes used in the simulations are given in table 1. Other material properties were taken from literature and are provided in table 2.

Table 1. Johnson-Cook parameters used in cutting simulations.

A (MPa)	B (MPa)	C	n
50	101	0.03	0.08

Table 2. Material properties of bone used in simulations

Property name	Value	Reference
Elastic modulus	20 GPa	measured
Density	2.26×10^3 kg/m^3	[19]
Thermal conductivity	0.2, 0.53 W/m K	[20, 21]
Specific heat	1.26 kJ/kg K	[19]

2.2. Boundary conditions
Constraints on the elements were not feasible in simulations with remeshing due to the software limitations. To resolve this, rigid lines AD, DC and CE (shown in figure 1) were created at the left, bottom and right edges of the work piece and were permanently glued to it. Kinematic boundary conditions were applied at these lines, which were transferred to the work piece elements. The cutting tool was vibrated harmonically in the cutting direction given by equation (2)

$$u_x = -a\cos\omega t, \quad u_y = 0, \tag{2}$$

where u is the tool displacement, $\omega = 2\pi f$ is angular frequency, f is vibration frequency and a is amplitude. For the frequency and amplitude chosen for simulations, the linear velocity of the tool was higher than the cutting velocity (i.e. the velocity of the work piece). This led to the intermittent contact condition between the tool and work piece. Analyses were run assuming dry cutting conditions, i.e. cutting without irrigation.

2.3. Cutting parameters
In simulations, various parameters, such as the ultrasonic amplitude, frequency and cutting velocity were changed to observe their effect on stresses. Cutting parameters and their magnitudes used in simulations are given in table 3.

Table 3. Cutting parameters used in simulations of the bone cutting process.

Parameters	Magnitudes, used in FEA
Cutting speed (mm/sec)	100; 150; 200; 250; 300
Friction coefficient	0.3
Tool nose radius (μm)	10
Tool rake angle, γ (deg)	15
Amplitude (μm)	10; 15; 20; 30
Frequency (kHz)	10; 20; 30

3. Results
Simulations were run from the initial contact between the cutting tool and workpiece until a state of the fully formed chip. The stress distribution in the bone material changed with the penetration depth and achieved a stable value exceeding the material yield limit when the chip was formed. The increase in stress levels beyond the yield strength of the material is due to strain hardening. The separation of the

material in the cutting process occurs in the vicinity of the cutting edge that moves along the prescribed straight line in the simulations. The material fails when the stress in it exceeds the critical value.

3.1. Stress calculations

To estimate the level of stresses induced by the tool in the bone tissue, simulation results obtained for UAC were compared to those of CC for identical cutting conditions. The simulations were performed with various cutting speeds as given in table 3; in other case studies the level of 300 mm/s was used. In CC, the stress reached almost a constant value of 140 MPa at 0.7 ms and remained stable for the remaining cut (up to full chip formation). At the initial stage of cutting, stresses were relatively small and concentrated around the tool tip in both the chip and machined surface. As the cutting process continued until the shear band was fully formed, the deformation increased and spread over a zone where shear banding takes place. The deformation zones produced in the material when the tool advanced are shown in figure 2; they are characterised by high levels of stresses (bright colours in the figure). It was noticed that the field of maximum shear stress is almost the same in primary and secondary deformation zones. The maximum equivalent stress fluctuated between 135-142 MPa and the measured shear angle was approximately 40°. The stress fluctuation can be attributed to numerical errors inherent to finite element simulations as well as to the frequent remeshing scheme.

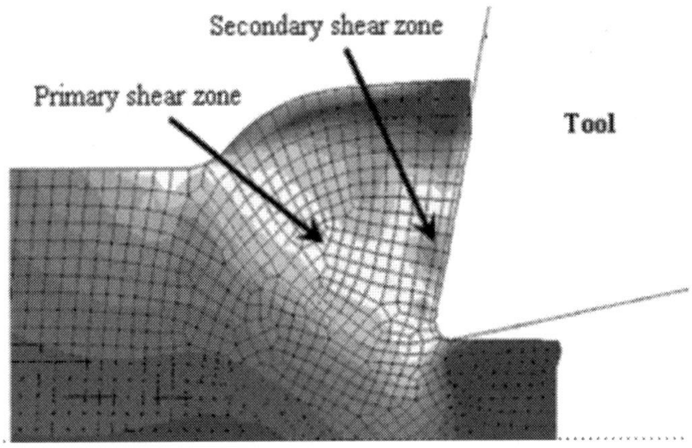

Figure 2. Deformation zones produced when tool penetrated into material in conventional cutting of bone (cutting speed 300 mm/s)

To fully understand the material's response in the interaction with the ultrasonic tool, the stress variation with tool motion is split into two stages: (a) when the tool penetrated into the material in the forward direction (b) when the tool travelled away from the chip. In UAC, during the continuous chip formation, a dynamically changing stress distribution in the process zone was observed. Figure 3 shows characteristic stages of a single cycle of vibration that took about 0.05 ms. When the tool moved forward, the stress increased with tool penetration and attained a maximum value of approximately 140 MPa. The attainment of the maximum penetration depth was characterized by the highest levels of generated stresses in the process zone and indicated the end of loading/penetration stage (figure 3a). When the tool moved in the opposite direction, the level of maximum stress dropped to an average value of 30 MPa (figure 3b). Those were the residual stresses induced in the material due to its plastic deformation. During each cycle cutting force increased in the same fashion and achieved a peak value at maximum penetration depth (end of loading) as shown in figure 4. The force dropped as the tool started moving away from the work piece and eventually vanished when separated from it. The maximum stress values were almost the same for all loading/cutting stages in UAC. As these maximum stresses were reached only during the loading part of

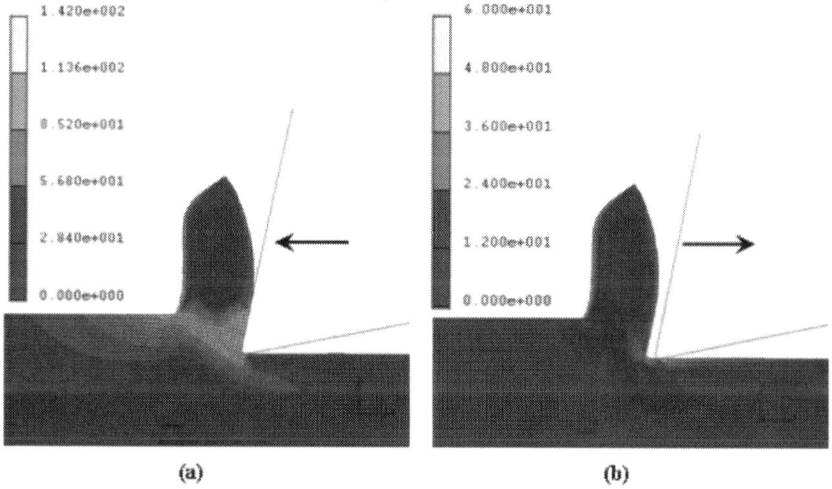

Figure 3. Distribution of von Mises (equivalent) stresses for a single cycle of ultrasonic vibration: (a) tool penetration. (b) tool moving away (f = 20 kHz and a = 20 µm). Duration of each cycle is 0.05 ms.

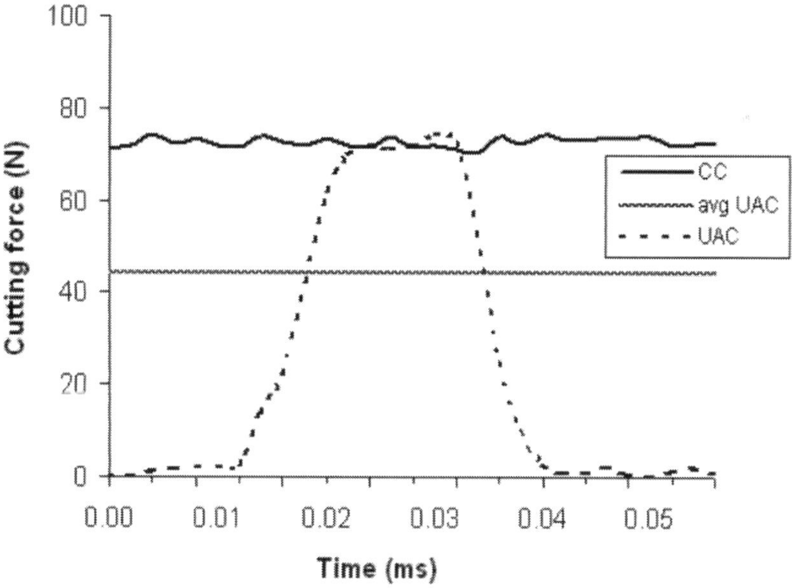

Figure 4. Comparison of calculated forces in cutting tool in simulations of CC and UAC. For UAC, f=20 kHz; and a = 20 µm.

Journal of Physics: Conference Series **181** (2009) 012014 doi:10.1088/1742-6596/181/1/012014

the ultrasonic cycle, the average over a cycle stress developed in the bone material was considerably smaller than that in CC.

Comparison of forces in CC and UAC for the duration of one cycle is shown in figure 4. The average cutting force in UAC was calculated and its magnitude was approximately 40% lower than the respective magnitude for CC. For maximum penetration of the tool in the ultrasonic cycle, the level of stress induced in the material by the tool below its cutting tip (i.e. below the cut surface) decreased with increasing depth from the cut surface. The stress level is plotted in figure 5 against the depth.

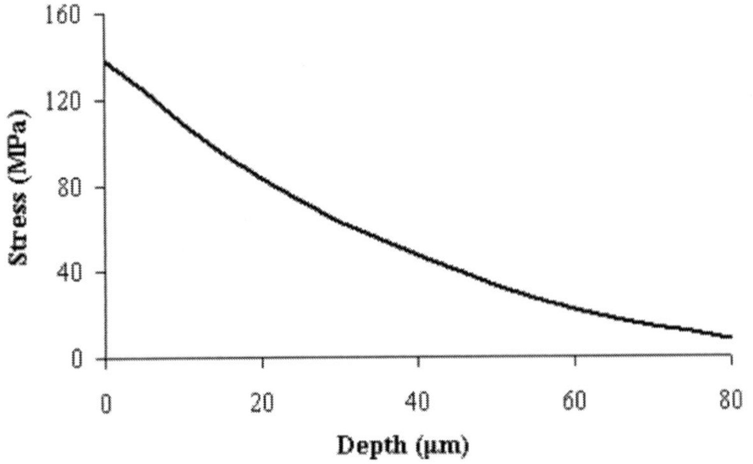

Figure 5. Variation of material stress with depth below the cut surface at the maximum tool penetration. f=20 kHz and a = 20 μm.

3.2 Effect of amplitude and frequency on stresses

To study the effect of amplitude of ultrasonic vibration on the stress developed in the bone material, the amplitude was varied between 10 and 30 μm (see table 3) while other parameters of UAC were kept constant. It was found that the amplitude of ultrasonic vibration did not influence the maximum stress magnitude. The maximum stress was achieved at the penetration depth of about 10 μm (for f = 20 kHz and a = 20 μm) and remained nearly constant until the tool got the maximum displacement in the forward direction. For a constant frequency, the only effect of increasing amplitude was the increase in duration, for which the stress induced in the material retained its maximum magnitude. Changing ultrasonic frequency resulted in the speed at which the stress state changed in the material. The variation of cutting velocity was found to have no influence on peak stresses.

4. Conclusions

A 2D FE model was developed to simulate both conventional cutting (CC) and ultrasonically assisted cutting (UAC) of the bone material. The model is fully parametric and can simulate the cutting process for various cutting parameters, such as cutting velocity, tool's vibrational frequency and amplitude. The model allowed the study of chip formation predicting distributions of stresses and cutting forces in the bone material. Stress distributions in the cutting region were studied. When subjected to ultrasonic vibration, the material experienced a varying stress state depending on the motion and location of the tool in each cycle. The average stress level in the material using UAC was lower than that produced in CC.

References

[1] Harrysso O L, Hosni Y A and Nayfeh J F 2007 *BMC. Musculoskelet. Disord.* 8 91
[2] Currey J D 1970 *J. Clin. Orthop. Relat. Res.*73 210–31
[3] Bonfield W 1987 *J. Biomech.* 20 1071–81
[4] Norman T L, Vashishth D and Burr DB 1995 *J. Biomech.* 28 309–20
[5] Zioupos P and Currey J D 1998 *Bone.* 22 57–66.
[6] Yasuda Y, Noguchi, K and Sata T 1955 *J. Bone. Joint. Surg.* 37A 1292
[7] Goodship A E, Lanyon L E and McFie H 1979 *J. Bone. Joint. Surg.* 61A 539
[8] Tadano S and Todo M 1999 *IUTAM Symposium on Synthesis in Bio Solid Mechanics*, Kluwer Academic Publications 139-59
[9] Tadano S and Okoshi T 2006 *J. Biomed. Mater. Eng.* 16(1) 11-21.
[10] Mitsuishi M 2005 *Annals CIRP* 54 41-6
[11] Lucas M, MacBeath A, McCulloch E and Cardoni A 2006 *Ultrasonics* 44 503–9.
[12] Mitrofanov A V, Babitsky V I and Silberschmidt V V 2004 *J. Mat. Proc. Tech.* 153–54 233-239
[13] Babitsky V, Kalashnikov A, Meadows and Wijesundara A 2003 *J. Mat. Proc. Tech.* 132 157-67
[14] Astashev V K and Babitsky VI 1998 *Ultrasonics* 36 89-96
[15] MSC. Marc User's Guide, Version 2007. MSC Software Corporation, Los Angeles
[16] Naohiko S, Mamoru M, Yoshinori F and Tadaaki S 2006 *J. Jap. Soc. Adv. Prod. Tech.* 24 8-13
[17] Mitsuishi M, Warisawa S and Sugita N 2004 *Annals CIRP* 53 107-12
[18] Hortig C and Svendsen B 2007 *J. Mat. Proc. Tech.* 186 66–76
[19] Huiskes J 1980 *Acta Orthop. Scandin., Suppl..* 185 44-108
[20] Davidson S R H and James D F 2000 *Med. Eng. Phys.* 22(10) 741-47
[21] Moses W M, Witthaus F W, Hogan H A and Laster W R 1995 *Exp. Therm. Fluid. Sci.* 11(1) 34-9

Journal of Physics: Conference Series **181** (2009) 012015 doi:10.1088/1742-6596/181/1/012015

Biomechanical evaluation of a corporectomy in porcine lumbar specimens using flexible polymer belts.

Beltrán-Fernández, J.A. [1a], Hernández-Gómez, L.H. [1b], Ruiz-Muñoz, Edgar[1c], González-Rebattú, A. [2d], Rodríguez-Cañizo, R.G. [3e], Urriolagoitia-Calderón, G. [1], Urriolagoitia-Sosa, G[1f], Hernández-Moreno, Hilario[4g].

[1] Escuela Superior de Ingeniería Mecánica y Eléctrica (ESIME) Unidad Zacatenco, Edificio 5, 3er Piso, 07738, D.F., México, Tel.: 52(55)57296000, ext. 54691

[2] Hospital Regional ISSSTE 1° de Octubre, Av. Instituto Politécnico Nacional, Núm. 1669, Col. Magdalena de las Salinas, 07760, México.

[3] Escuela Superior de Ingeniería Mecánica y Eléctrica (ESIME) Unidad Azcapotzalco, Av. Granjas No. 682 Col. Sta. Catarina, 02550, México.

[5] Escuela Superior de Ingeniería Mecánica y Eléctrica (ESIME) Unidad Ticomán, Edificio de Laboratorios Pesados, Av. Ticomán No. 600, Col. San José Ticomán, 07340 México D.F., México Tel. 52(55)57296000, ext. 56060, 56084.

E-mail: [a]jbeltran@ipn.mx, [b]luishector56@hotmail.com, [c]edrm_ipn@hotmail.com [d]janosclub@hotmail.com, [e]ricname@hotmail.com, [f]guiurri@hotmail.com, [g]hihernandezm@ipn.mx

Abstract. This paper presents the experimental results of a biomechanical evaluation in lumbar porcine specimens (L2-L4), instrumented with flexible polymer belts, under fatigue and tensile loading. The clinical effect called facetary arthrosis is evaluated. An experimental analysis was carried on 3 lumbar porcine specimens. In two of them, polyamide belts are fixed on the interspinous ligament from L2 to L4. Specimens are taken from pigs which are 6 month old. For the present work, the stiffness reduction of the spine and the biomechanical behaviour of the belts in conjunction with the interspinous ligament are evaluated. The purpose is to determine the failure conditions for the elements of the specimen (vertebral disk, supra and intraspinous ligament and vertebral body). Under static loading, which is the base line case, the elements of the specimen failed as a typical healthy structure. While in the fatigue combined with static loading, the element failed in different order. Additionally, the stiffness changed in accordance with the fatigue loading conditions. Because of the simplicity of this alternative technique, a high level of the structural integrity is preserved, as no holes are made on the spinous process in order to insert the fixation screws. Furthermore, there is a cost reduction.

[1] Escuela Superior de Ingeniería Mecánica y Eléctrica (ESIME) Unidad Zacatenco, Edificio 5, 3er Piso, 07738, D.F., México, Tel.: 52(55)57296000, ext. 54691

Journal of Physics: Conference Series **181** (2009) 012015 doi:10.1088/1742-6596/181/1/012015

1. 1. Introduction

One of the most important topics in human health relates to the spine. Lumbar vertebrae diseases are the main causes of back pain. There are several factors which contribute to this, and can be classified on 6 levels: (1) **Idiopathic back pain**, associated to bad postural hygiene, obesity, pregnancy and an excessive endeavour; (2) **Traumatic back pain**, which can be associated with muscular tears, vertebral fractures, vertebral luxations, column sprains, rupture of vertebral disks, traffic accidents, vertebral nipping and working accidents. Fractured vertebrae or lesions can cause bone displacement and as a consequences different injures of the spinal cord and nerves, and hence paralysis, can be developed; (3) **Emotional disrupters and tension,** as depression, economic problems, personal disorders (schedules, diets) and lack of physical activity; (4) **Stress and back pain**, excessive physical work, constipation, working tension, insomnia, anorexia and weight loss; (5) **Degenerative back pain,** caused by excessive use and wear, arthritis and espondiloarthritis, *Disk degeneracy*: disk protusion, disk hernia, *nervous degeneracy*: Sciatica, Radiculopathies, *Arthrosis*: Facctary and vertebral arthrosis. And lastly, (6) **other causes**, *congenital ones*: hypercifosis, scoliosis, the sacralisation of the fifth lumbar, the lumbarisation of the sacrum, hyperlordosis, rectifications, spondilolisis. *Espondilolistesis*: Anterolistesis, retrolistesis, spinal estenosis, bifid spine, infections, circulatory, post-surgical fibrosis, osteoporosis and tumours. [1]

Regarding this paper, Hyperlordosis is being considered, which in some cases, can also overload the facetary joint, and eventually, this can be affected by a facetary syndrome. This syndrome can also cause pain in the lower back and is referred as a pain in the legs.

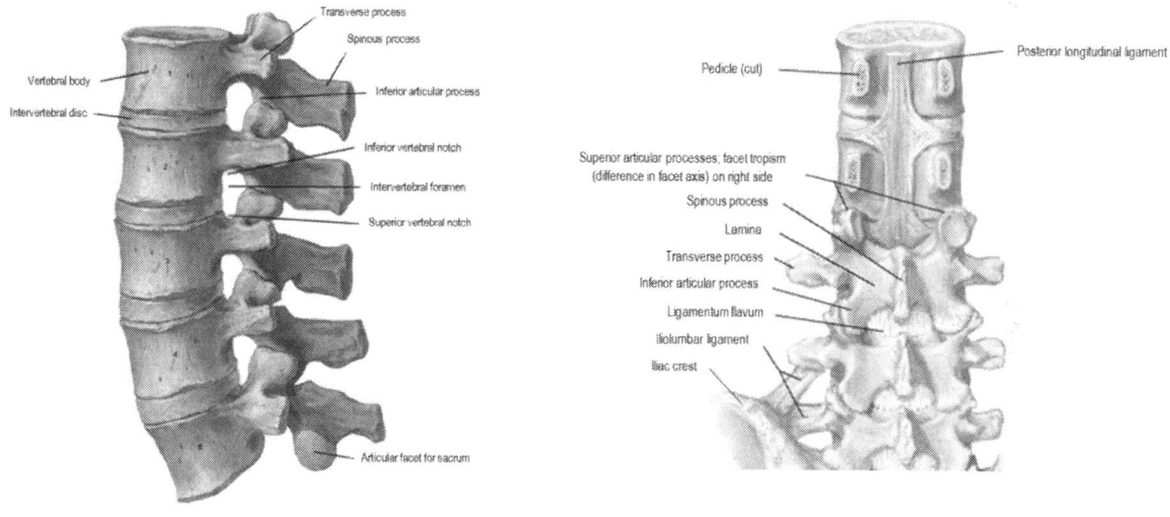

Figure 1. Lumbar vertebrae anatomy, showing main ligaments [2]

Figure 1 shows the lumbar vertebrae anatomy of a human spine. The lumbar spine is a unique bony and ligamentous structure. It can withstand excessive loads while, simultaneously, neurologic functions are protected, providing flexibility and stability. This intricate interplay provides efficient motion and constant mobility among the different anatomic components of the lumbar spine.

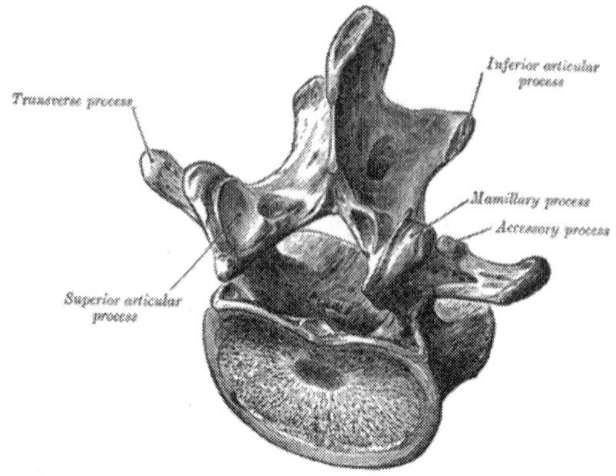

Figure 2. Lumbar vertebra [3]

The five vertebral bodies and intervertebral disks of the lumbar spine withstand significant physiological loads. The intervertebral segment of the lumbar spine consists of three characteristic articulations, the disk-vertebral body and two posterior apophyseal (facet) joints, which can resist high loads and stresses. The vertebral bodies are cylindrical masses of cancellous bone with a cortical shell. The disks consist of the annulus fibrous, the nucleus pulpous, and the cartilaginous and bony end plates of the vertebral bodies. In accordance to the figure 2, this complex bony anatomy serves as a load-bearing structure, a passive restraint to torsional strain and excessive tensioning of annulus fibrous, and as a means of protection against disk injury.

In order to establish the involved factors in the analysis of the mechanical effect, it is necessary to define the facetary arthritis as the degeneration of the cartilage that separates the joint formed by two superposed vertebral laminae [4]. Facetary arthritis is caused by normal wear of the cartilage in the facet joint. In young people, this cartilage is thick and cushions the burden borne by the joint, but it gets thinner through the time. It is normal to find a certain degree of thinning after 30 years.

According to Riew et. al [5], nearly 84% of people will experience back pain at some point in their lives. In a patient with reduced mobility, 16% of axial loading is supported by the vertebral column and is transferred to the facet joints. However, in extension, 47% can be loaded when the space between vertebral disks is limited or there is degenerative arthritis. In this way, the facet does not allow excessive movement in vertebral disks [6].

On the other hand, the most important consideration within medical treatment of the vertebral column is to preserve the human natural movement. One of the main injures which affects it is the facetary arthritis and the replacement of the vertebral disk. It produces tensile forces on the vertebral surgery and the medical term is known as: arthrodesis.

In figure 4, the pain zone is located on the inferior side of the vertebral disk, as a consequence of the excessive movement and contact between facet joints, affected by arthritis and tensile loading. Mechanical pain occurs when damaged disks and joints, which connect the vertebrae, become inflamed from excessive motion of the vertebrae. This type of pain is commonly felt in the lower back and may radiate into the buttocks and upper thighs. In order to get more knowledge of the problem, a biomechanical study is required.

Medical treatments involve the use of alternative surgical techniques. This work is focused on the evaluation of the space reduction caused by facetary arthritis testing lumbar porcine specimens instrumented with polyamide collars. This reduces the space between facet joints and contributes to the vertebral fusion. The affected lumbar zone is stabilised, and, at the same time, the use of external distracters is avoided and no holes are made on the facet joints. Corporectomy is associated to the surgical technique used by medics in order to fix polymer belts between vertebral body and ligaments. [9]

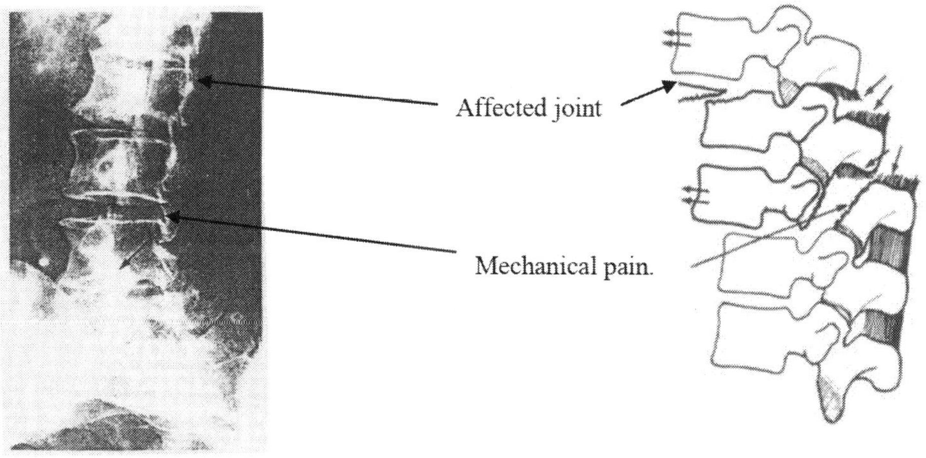

Figure 3. Facet joint movement, affecting the lumbar zone [5]

Figure 4. Facet joint movement, affecting the lumbar zone [7]

2. Materials and Methods

2.1 Testing conditions.

Three porcine lumbar specimens, which are six months old, were considered. In all cases muscles were removed and the interspinous and supraspinous ligaments were not affected. For the first specimen any collar was used. However, two polyamide 6/6 collars, fixed on L3-L4 and L4-L5 interspinous ligaments, were used on the other specimens. This constrains the movement for the spinous process. In figures 5 and 6 is shown the testing arrangement.

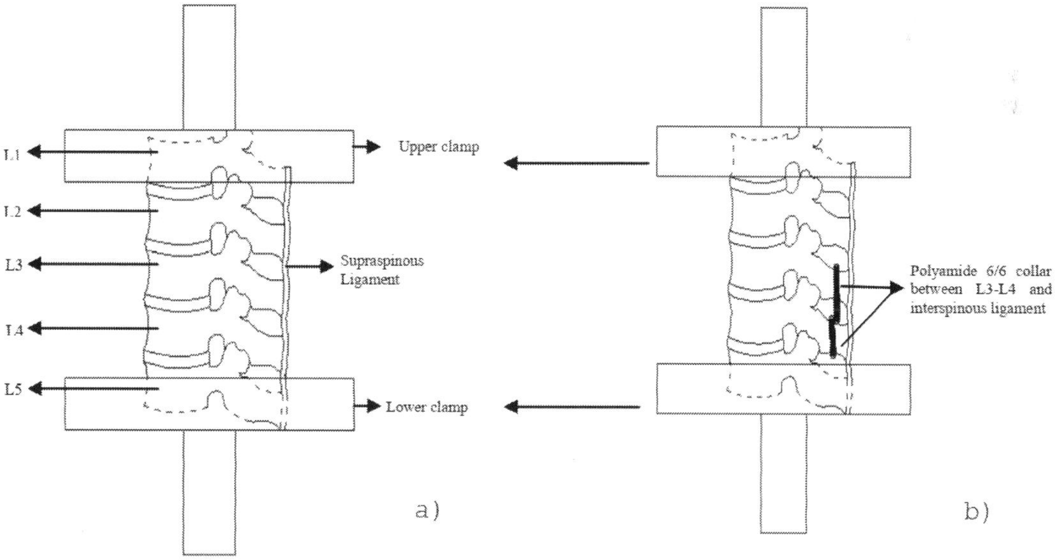

Figure 5. Lateral view. a) Intact porcine specimen, b) Instrumented porcine specimens with polyamide 6/6 collars.

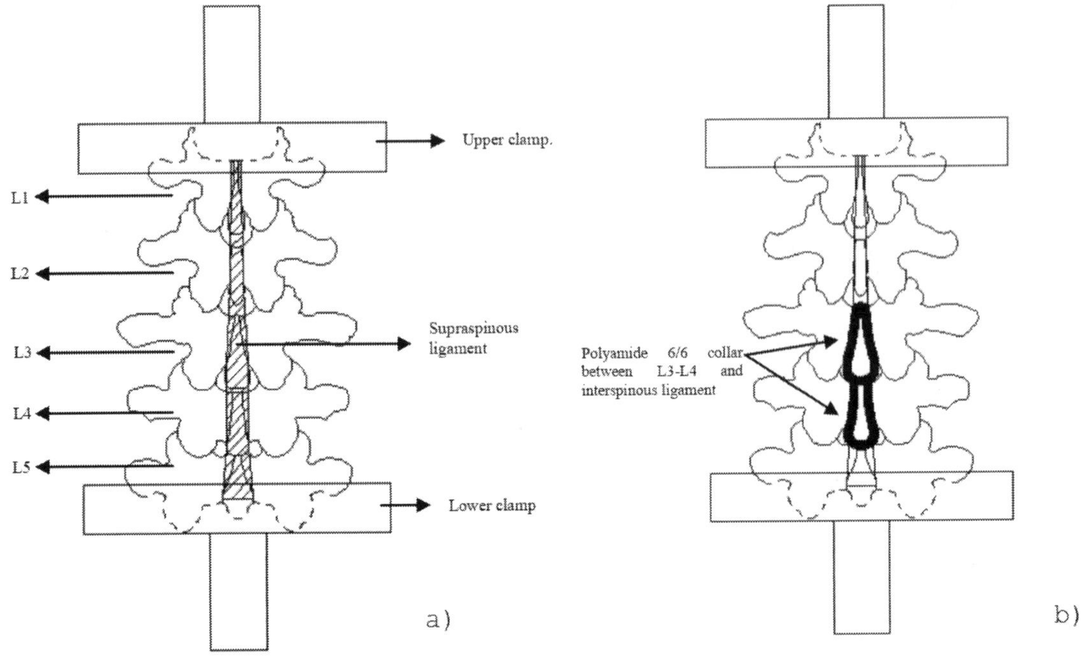

Figure 6. Posterior view. a) Intact porcine specimen, b) Instrumented porcine specimens with polyamide 6/6 collars.

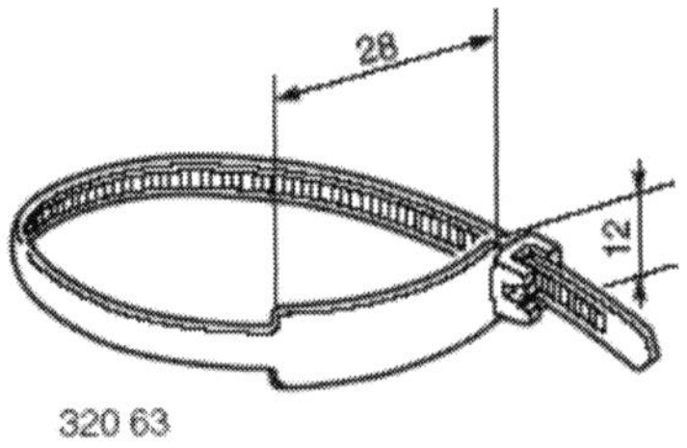

Figure 7. Polyamide 6/6 collar used within the experimental testing. (Units in mm)

2.2 Analysed specimens

The loading conditions are shown in Table1. The first case is the base line analysis, because an intact specimen without any collar was tested under tensile loading. Its purpose is to evaluate the influence of the collar on the damage process. A maximum tensile load of 3000 N was applied. Regarding the other two specimens, which were not subjected to a preliminary fatigue analysis, it was carried on a preliminary fatigue analysis. In first instance, after applying a tension load, high cycle fatigue load was applied. By contrast, the third specimen was subjected to low cycle fatigue load. The failure conditions of the vertebral bodies are developed under these loads. All the tests were performed on an INSTRON 8501 universal testing machine, and the mechanical properties of the porcine specimens and polyamide 6/6 collar are shown on table 2.

Table 1. Loading conditions

	Condition	Loading condition
Specimen 1 (figure 6-a)	Intact - removed muscle	Tensile static loading (up to 3000 N) - 1 mm/sec.
Specimen 2 (figure 6-b)	Instrumented with two Polyamide 6/6 collars	Tensile static loading - 1100 N. High cycle fatigue load - 1,000,000 cycles. Variable fatigue load from 1200 to 2000 N. Average load = 1600 N, amplitude = 400 N)
Specimen 3 (figure 6-b)	Instrumented with two Polyamide 6/6 collars	Tensile static loading - 1100 N. Low cycle fatigue load - 100,000 cycles. Variable fatigue load from 1200 to 2800 N. Inversepositive relationship fatigue R = 0.42, average stress = 2000 N, amplitude = 800 N)

Table 2. Mechanical properties of the porcine specimens and polyamide 6/6 collar.

	Description	Mechanical properties
Polyamide 6/6 collar	Legrand colring polyamide 6/6, catalogue 320 43.**[8]**	Maximum tensile strength 227 N 20000 hrs at 85 °C
Cancellous bone	Porcine specimen	Young's Modulus 466 MPa **[9]** Poisson's Ratio - 0.2
Cortical bone	Porcine specimen	Young's modulus 12 GPa Poisson's ratio - 0.2 **[10]**

Cast plaster, was used to fix the specimens on the clamps. In all the tests, the structural integrity among the clamps and the specimens was maintained.

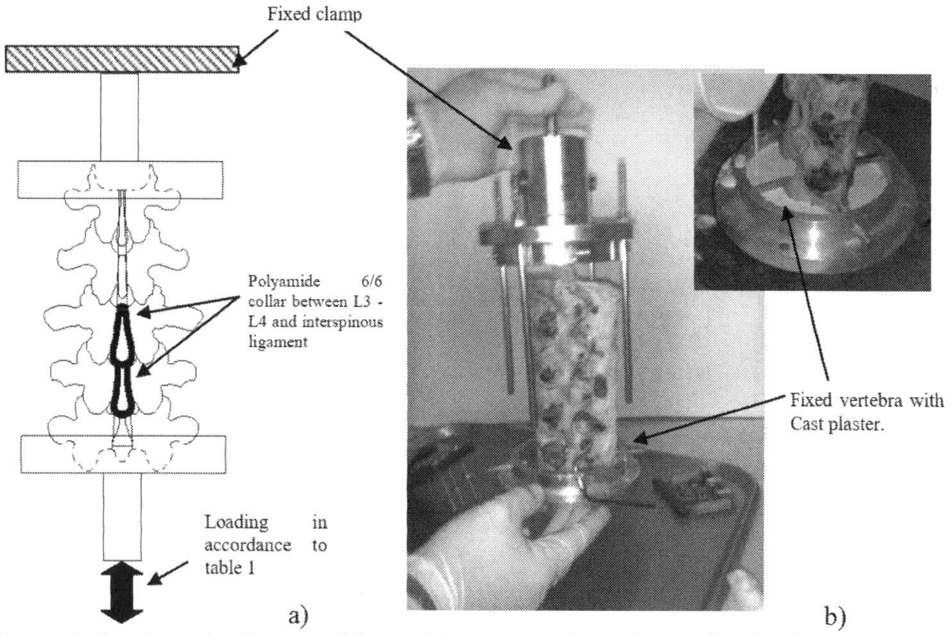

Figure 8. Specimen loading conditions. a) Instrumented specimens, b) Aluminium clamps.

3. Results

Main results of the three porcine specimens are summarized in table 3, in accordance with the loading conditions of table 1.

Table 3. Experimental results

Specimen	Condition	No. Cycles reported on the testing	Failure load
1	Intact - removed muscle – figure 6-a.	N/A	- Vertebral disk was broken at 1287 N - Intraspinous ligament was broken at 2686 N - Cortical bone (Vertebral body) was broken at 3000 N (separar cada elementos y su no de ciclos)
2	Instrumented with two Polyamide 6/6 Collars	867496	- Vertebral disk was broken at 1215 N - Intraspinous ligament was broken – 1834 N - Cortical bone (Vertebral body) was broken at 1986 N - The collar was unbroken.
3	Instrumented with two Polyamide 6/6 Collars	26895	- Vertebral disk was broken at 1386 N - Intraspinous ligament was broken at 2386 N - Cortical bone (Vertebral body) was broken at 2750 N - Collar was broken at 2580 N.

All the tests were stopped when the ligaments were broken. In particular, the collar was unbroken in the second test. On the other side, when the fatigue load was increased, the number of cycles diminished considerably and the collar was broken after the ligaments and vertebral bodies failed.

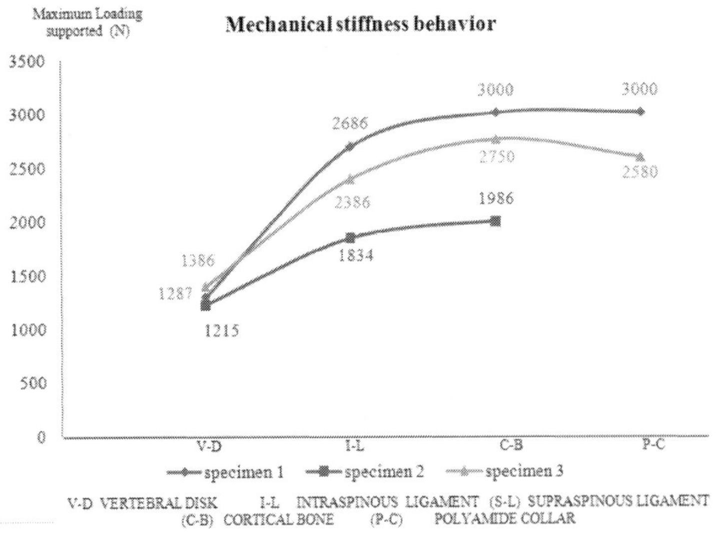

Figure 9. Mechanical Stiffness behaviour for the tested specimens.

4. Discussion

In accordance with the results shown on the Table 3 and figure 9, several aspects of the mechanical behaviour for the instrumented porcine specimens can be discussed:

The stiffness of the intact specimen is the base line for the rest of instrumented specimens on the mechanical testing. The influence of the high and low cycle fatigue loading on each test show different mechanical effects over the specimens. The results show that the stiffness is reduced as the number of fatigue cycles increases. At this respect, it can be said that the accumulated damage can be evaluated as a function of the stiffness reduction. In this way, figure 9 may give a glance of the damage introduced by fatigue loading.
Regarding the high loading conditions and low cycle fatigue, the cortical bone of the vertebral body was broken at a lower load than in the static test in which no belts were used. Nevertheless, the behaviour of the vertebral disk is similar to that experienced in static conditions.
On the other hand, as the number of fatigue cycles is reduces, the stiffness of the instrumented vertebral body increases. In this case, the polyamide collar was broken after the supraspinous and intraspinous ligaments were damaged.

5. Conclusions

Facetary arthrosis was defined as a complex syndrome which degenerate the integrity of the structure of each vertebrae **[11]**. The alternative surgical technique using polyamide collars referred in this work contributes to reduce this disease. In general terms, the polyamide collar contributed to reduce the space between damaged lumbar vertebrae. However, the resultant arrangement has a complex geometry. Accordingly, the mechanical behaviour was evaluated considering the relation between the applied load and the clamps displacement.

In all the cases, the applied loads were in the range in which failure of the vertebral components is expected. The collar was broken when static load was combined with high stress fatigue loading. According to these results, it is expected that under normal loading conditions, the integrity of this arrangement is maintained.

One point of concern relates to the damage introduced as the number of fatigue cycles is increased. This has to be analyzed in more detail in future work, as well as the effects of torsion and bending.

6. Acknowledgements

The authors kindly acknowledge the grants given by the National Council for Science and Technology of Mexico and the National Polytechnic Institute and the support given to this project by ISSSTE Hospital 1° de Octubre.

7. References

[1] López-Rodríguez MA, Varela M and Camba MA. Uses of radiofrequency for the management of benign chronic pain. A review of the literature published in the past 6 years. Rev Soc Esp. Dolor 2001; 8: 397-411.
[2] Netter, Frank H. "Atlas de Anatomía Humana". Editorial Masson. Segunda Edición. Láminas 144, 147. Canadá 2000.
[3] Grants Atlas of Anatomy. Seventh Edition
[4] Pimenta L, Bellera F, León M: Clasificación radiográfica de la degeneración facetaria lumbar: Evaluación intra e interobservador. Acta Ortop Mex 2005; 19(3): 95-98.

[5] Marqués J. "El dolor lumbar". Editorial JIMS. pp. 282-283. España 1989.
[6] Adams MA, Hutton WC: The mechanical function of the lumbar apophyseal joints. Spine 1983; 8(3): 327-330.
[7] Roca, Jaime B. "Tratamiento quirúrgico del dolor lumbar". pp. 79, 85. Editorial JIMS. España 1987.
[8] ® Legrand Catalogue 2006/07, Polyamide colrings, p.p 247-251.
[9] Beltrán-Fernández J.A, & et al, Mechanical Behavior of a Calcium Phosphate Ceramic Bone Graft used in the Rehabilitation of a C4 Human Vertebra", published by Transtech in "Applied Mechanics and Materials". 5th BSSM International Conference on Advances in Experimental Mechanics, University of Manchester, UK, 4-6 September 2007.
[10] Beltrán-Fernández J.A, & et al, Modelling of a Cervical Plate and Human Cervical Section C3 – C5 under Compression Loading Conditions Using the Finite Element Method, published by Transtech in "Applied Mechanics and Materials", Vols. 13-14-2008 p.p. 49-56. 6th BSSM International Conference on Advances in Experimental Mechanics, National Physical Laboratory, UK, 9-11 September 2008
[11] R.O. Ferracutti, A. Czerniecki, J. Paloto y N. Molinari, Analysis of the Causes of Intervertebral Disk Hernia, Univ. Nacional del Sur, Dpto. de Ingeniería, Avda. Alem N°1253, (8000) Bahía Blanca-Argentina

Journal of Physics: Conference Series **181** (2009) 012016 doi:10.1088/1742-6596/181/1/012016

Real-Time Surgical Simulation for Deformable Soft-Tissue Objects With a Tumour using Boundary Element Techniques

P. Wang [1], A. A. Becker [1], I. A. Jones [1], A.T. Glover [2], S. D. Benford [2] and M. Vloeberghs[3]

1 Department of Mechanical, Materials and Manufacturing Engineering
2 School of Computer Science and Information Technology
3 Academic Division of Child Health, School of Human Development
University of Nottingham, UK

E-mail: a.a.becker@nottingham.ac.uk

Abstract. A virtual-reality real-time simulation of surgical operations that incorporates the inclusion of a hard tumour is presented. The software is based on Boundary Element (BE) technique. A review of the BE formulation for real-time analysis of two-domain deformable objects, using the pre-solution technique, is presented. The two-domain BE software is incorporated into a surgical simulation system called VIRS to simulate the initiation of a cut on the surface of the soft tissue and extending the cut deeper until the tumour is reached.

1. Introduction

The training of neurosurgeons typically involves the stages of observing an experienced surgeon, undertaking operations under close supervision, and finally undertaking unsupervised operations to gain experience. There are currently concerns about the decreasing opportunities for the training of neurosurgeons, for reasons which include: (i) the introduction of working time directives, which limit the amount of time trainee surgeons can spend in the theatre, (ii) concerns about inexperienced surgeons performing their earliest operations on live patients, and (iii) concerns about litigation

All these are acting as drivers towards the use of virtual reality simulation to replace or supplement the initial on-the-job training of neurosurgeons. In order to address this need, the present authors and their collaborators have developed a prototype virtual reality surgical simulator system (called VIRS), in which several basic skills in neurosurgery can be simulated. Initial developments of the VIRS system have been reported elsewhere [1-5], and include the following real-time facilities:

- Pushing (prodding) and pulling (pinching)
- Initiation of a cut or separation of tissue, including multiple-cuts
- Contact with a rigid retractor to separate the cut surfaces
- Cutting deeper into the deformable organ
- Post-cutting manipulation of the deformable organ
- Gravitational deformation of the cut tissue
- 3D stereo vision

The present paper reports further developments to the prototype VIRS system to simulate the presence of hard tumours. This will initially entail probing into the brain model in order to detect (feel) the presence of a hard tumour (after possibly analysing MRI scans of the patient to visualise the tumour itself). The 'cutting' process involves separating the various structures of the brain with a probe-like device to ascertain the tumour's exact position. Once confirmed, the surgeon may cut the surrounding brain tissue to reveal the enclosed tumour, before its subsequent removal. To perform this, the VR system would also need to incorporate the facility for a two-handed operation, i.e. the use of two haptic devices that can interact independently with the brain structure.

2. Background

In order to undertake a virtual reality simulation of surgery, a real-time deformable and cuttable model of the tissue is required. An excellent review of the various deformable model techniques applied to surgery simulation is presented by Meier et al [6], concentrating on material published up to 2001; techniques used include various heuristic methods (spring-mass, deformable splines, linked volumes and mass-tensor), as well as continuum-mechanical models such as the Finite Element (FE) and Boundary Element (BE) methods. A review of deformable models was also undertaken by Montagnat et al [7], The survey in this paper is not intended to duplicate the reviews in those papers; however, a brief review of applications and developments dating from the most recent few years is included here for completeness.

Nonlinear FE calculations are very well-established in engineering applications and have also been applied, for example, to the simulation of laparoscopic surgery on the liver [8] via the explicit FE method. Other recent surgical simulation applications of the FE method include laparoscopic uterine surgery [9] and suturing [10-12], though there appears to be no publications on the use of FE for neurosurgical simulation.

In their review, Meier et al [6] identified BE analysis as being potentially one of the most promising routes to real-time deformable models for surgical simulation. However, with the exception of the VIRS project reported here, relatively little work has been done to continue the progress made in the late 1990s by James and Pai [13,14] and Monserrat et al [15], and subsequent developments in this area appear to have been restricted to the same authors [16,17].

The "pre-solution" techniques described in the present team's earlier publications [1,2] utilise the superposition principle to create look-up tables of pre-computed displacements and forces, and can be used to achieve real-time interaction with either FE or BE models of a deformable physically-based virtual object. However, the creation of new incisions or cuts is a difficult task because of the need to re-calculate the displacements and forces. Zhong et al [18] describe an adaptive FE approach which is based on pre-solutions but uses an iterative updating approach for each cutting step. Wu and Heng [19] describe an interactive FE approach where the FE mesh is partitioned into "operational" and "non-operational" regions, the operational region being capable of being subjected to cutting.

Several papers, notably those of Mahvash and Hayward [20] and of Hansen et al [21], concentrate on the interactions between the surgical instruments and the simulated tissue.

Other physically-based methods used include a meshless approach [22] which appears to be restricted to simulations which do not include cutting. Zhong et al [23] describe a new methodology based on autowaves and the propagation of energy through a deformable structure. Its application to virtual reality surgery is illustrated via a deformable kidney model. Mass-spring methods continue to be used, and one ingenious variation on this tunes the parameters defining such a model to match the behaviour of a FE model of the structure [24, 25]; however, real-time cutting was not incorporated into this system. Other methods such as spline-based techniques continue to be used, for example for simulation of the intestines [26].

3. Review of the Boundary Element Formulation

BE analysis relies upon the existence of surface integral equations to model the continuum behaviour. An integral equation can be derived to describe the effect of a point force at a an internal "load" point on the displacements (u) and tractions (t) at a "surface" point within the domain of the problem with volume V and surface S. Details of the derivation of the integral equation can be found in BE Textbooks, e.g. Brebbia et al [27] and Becker [28]).

In linear elastic BE analysis, only the surface is discretized into elements, and the displacements and tractions over each element are described by an interpolation shape function. For simplicity, constant triangular surface elements are used in this work where the displacement is assumed to be linear over each element and the elements can be represented by a node placed at the centroid of the triangle. Other more sophisticated surface elements, such as quadratic or higher -order elements, can also be used. The simplicity of the triangular surface elements means that fine meshes are needed to model awkward geometries.

By taking each node on the boundary in turn as the "load" point and performing numerical integrations over all the surface elements, a set of linear algebraic equations is obtained as follows:

$$[A]\{u\} = [B]\{t\} + \{R_G\} \tag{3}$$

where $[A]$ and $[B]$ are fully populated matrices that contain the numerical integrations and $\{R_G\}$ contains the gravitational load. Further manipulation is required in order to group the unknown tractions and displacements on one side of the equation and the known quantities on the other side, in order to yield a set of simultaneous linear equations involving a system matrix [A*], as follows:

$$[A^*]\{x\} = \{R_{bc}\} + \{R_G\} \tag{4}$$

where $[A^*]$ is the solution matrix, $\{x\}$contains all unknown variables (either displacements or tractions), and $\{R_{bc}\}$ contains all prescribed values of displacements and tractions. After solving the algebraic equations, all tractions and displacements on the surface are known.

In practice, the task of solving such an equation system is too lengthy to perform in real time, at least in terms of solving for each load case. In order to address this problem the following approach is taken: (i) A matrix of "pre-solutions" is created by applying, respectively in the x, y and z directions, a hypothetical unit displacement on all surface nodes in turn and solving the system of equations without gravitational loads, (ii) Using the principle of superposition, which is only applicable to a linear system, the relevant data from the pre-solution matrices are superimposed to arrive at the deformations resulting from any set of applied displacements along with the applied forces which are required to apply these displacements.

In brain surgery, the cutting process is performed using a special pair of forceps operating on the principle of diathermy (high-frequency electric current passing between the tips). The aim is to pinch the brain tissue to be removed but not to close the forceps completely. The material, which has been softened by the forceps, is sucked away with a suction tube. The difficulty of performing cutting in real time has already been alluded to within the literature survey, and arises from the fact that cutting (the separation of elements) alters the geometry of a body and invalidates all the pre-solutions which have been computed based on the old geometry. In references [4] and [5], the present team describe some of the approaches taken here to cutting, which are summarised only briefly in this paper. The problem can be tackled from several viewpoints:

- Because of the nature of the cutting process, it is assumed that a gap is introduced between the newly-created cut surfaces, rather than assuming that the two surfaces form a "crack" (with no gap between the separated surfaces). This avoids problems in BE analysis associated with having two coincident surfaces (and particularly the problem of singularities caused by coincident nodes).
- Adding new elements to the BE model will leave a large block of the system matrices unchanged, with the existing system matrices simply being expanded by three much smaller sub-matrices. This makes it possible to re-use the original solutions by making use of the inverse of the original [A*] matrix within the calculation of the inverse of the new (expanded) [A*] matrix. Although this was pursued for the situation where the cut took the form of a narrow crack and is reported in reference [4], it is not directly applicable to the case where cutting results in a finite-width gap because some of the original surface elements are either modified or removed. The original system matrices will therefore now contain several invalid rows and columns, making the re-use of the existing solution inappropriate. Re-calculation of the solutions using an iterative approach is therefore necessary.
- In order to allow the simulation to continue without disruption after cutting, an approximate update is performed in real-time during each cutting step by interpolating displacements from either side of

the cut so that the newly-created surfaces within the cut can have plausible deformations. However, this does not take account of the structural weakening of the organ after material has been removed; this will particularly affect the deformations of the area near the cut.

- Whilst the simulation continues uninterrupted using the approximate (interpolated) solution, the pre-solutions may be re-calculated to take account of the changed structure of the organ. For this purpose, a multi-threaded BE program was written to allow the re-calculation to take place as a background task. With current technology the updating process remains rather slow (minutes rather than seconds), and we anticipate that the structuring of the code to support this multi-threaded approach will allow for the parallelisation of certain aspects of the system, which will be investigated in due course.

4. Modelling a Hard Tumour

To model a tumour grown inside the brain tissues, two domains are defined; one for the tumour and one for the soft brain tissue, as illustrated for a simplified case in Figure 1. A tumour is usually harder than the surrounding soft-tissues and this can be modelled by assigning a greater Young's modulus to the tumour than the soft tissues. To obtain the system equations, the numerical integrations are performed separately over domain (a) and domain (b) respectively. For easier understanding, these two domains are represented separately in Figure 2. Domain (*b*) represents the tumour which has only one surface (*T*) whereas the soft tissue, domain (*a*), has two surfaces, an internal surface (*T*) and an outside surface (*S*).

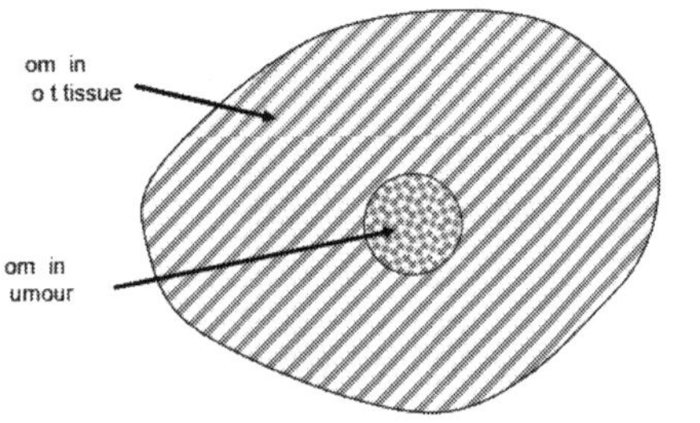

Figure 1: A hard tumour inside the soft-tissue

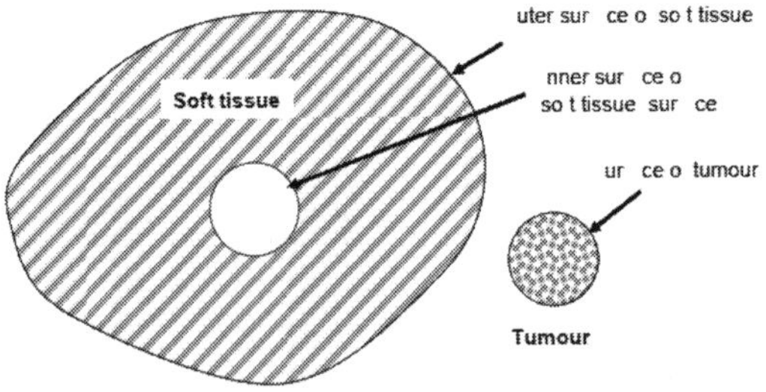

Figure 2: Surfaces of the tumour and the surrounding soft-tissue

Since the cutting process takes place only in domain (*b*) without affecting the surface (*T*) of the tumour, the numbering system for the nodes allows the tumour nodes to appear first, followed by the node numbers of the surrounding soft tissues. This is very convenient for the subsequent matrix manipulation that will be necessary after a 'surgical cut' has been initiated and propagated in the surrounding soft tissue. All new terms from the numerical integrations relating to the newly generated nodes and elements arising from the 'surgical cutting' will be grouped together in the matrix system, so that the matrix expansion due to re-meshing occurs only within the set of sub-matrix coefficients containing relevant to the surface 'S'.

Before the system equations can be solved, boundary conditions must be applied to prohibit rigid-body movement of the whole organ. The traditional boundary conditions are known (prescribed) values of tractions or displacements specified at nodes on the outer boundary, and their treatment forms standard matrix manipulations (see e.g. Becker [28]). In addition, the contact conditions between the hard tumour and the soft tissue must be satisfied. These are different to the surface boundary conditions and form the relationships between the tumour and soft tissues which are always bonded to each other. Nodes at the contact points must satisfy the equilibrium conditions (equal and opposite tractions) and compatibility (continuity of displacements with no gaps forming between the contact surfaces). Since the number of unknowns in the above equation remains equal to the number of the simultaneous equations, all the unknown displacements and tractions on boundaries of both domains can be computed.

Figure 3 shows a symmetrical half of a simplified three dimensional geometry where the hard tumour is located inside the soft tissue. For clarity, the tumour is moved out and shown separately. The tumour is designed as a half sphere and is discretized into 72 triangular surface elements in which 48 elements are on the spherical surface and 24 are on the bottom circular surface. The soft tissue domain is descretized into 616 surface elements.

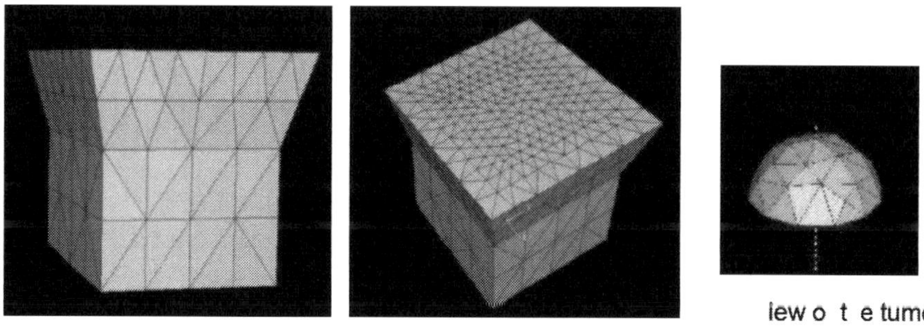

Figure 3: 3D simplified model of soft tissue surrounding a tumour

5. Pre-Solutions for the Soft Tissue and Hard Tumour

For a linear system, the superposition principle can be applied. The deformations due to the gravitational load $\{R_G\}$ can be calculated separately then added together when required. This is a constant vector for a defined model and can be solved only once and saved in an array.

The contact of a surgery tool with the surface of an organ can be considered as a non-zero displacement applied on a specific surface node of the deformable model. For real-time implementation, a look-up table or a set of pre-solutions can be created prior to the simulation by imposing a hypothetical unit displacement, respectively in the x, y and z directions, on all surface nodes in turn, solving the system of equations and storing the solutions in look-up tables. During the surgical simulation, these solutions are retrieved, scaled and added together to represent the real deformation of the model. This procedure can be easily performed in real-time.

Creating the pre-solution look-up tables involves applying a hypothetical unit displacement in turn at each node/element in order to find the displacement and traction solutions. The traction vector $\{t\}$ contains zeros except the node/element on which the unit displacement applying. The computed traction resulting from the 'unit displacement load' can be calculated directly and forms the pre-solution vector to be found.

6. Implementation

6.1 Simulation of a two-handed operation

To enable better manipulation of the organ surface while a cut is introduced, it is necessary incorporate a second haptic device, which simulates a two-handed operation. This can be accommodated using the superposition principle, provided that the behaviour of the deformable organ remains linear elastic. No further pre-solutions need to be computed, and the displacements and feedback forces caused by the actions of the two haptic devices can be obtained in real-time and saved in separate arrays to record separate results for each haptic device.

The superposition principle is used to arrive at the final resultant deformation of the organ displayed on the screen, and the feedback forces passed to the relevant haptic device. If one of the haptic devices is released from contacting the object, its relevant displacements and forces are immediately set to zero so that the deformation displayed is only caused by the other haptic device remaining in contact with the object. If both haptic devices are released, there is no deformation and the deformable object returns to its initial model. Figure 4 shows an example of the application of two haptic devices.

6.2 Simulation of self-contact

After initiating a cut, self contact may occur as the two separated surfaces may come into contact due to gravity or further manipulation of the cut. It should be emphasised that self contact is much more complex than the contact of the organ with a retractor which assumes that the retractor is a rigid non-deformable body. The main difficulty associated with the contact of two deformable bodies is that they cannot be linearly scaled. Contact problems are non-linear and the solution requires incremental and iterative approaches.

Iterations have to be performed in order to arrive at the correct contact area. In each iteration, the system matrices must be modified and re-computed to simulate the latest contact conditions between the contact elements. This means that the system equations must be solved again and the contact area updated. Although the solution time can be reduced to under 0.5 second through the use of fast iterative solvers, the iterations needed to arrive at the correct contact area require several seconds to complete. Furthermore, to improve the accuracy of the computed displacements and forces, it may become necessary to re-mesh the surface around the contact areas which would increase the overall processing time.

The BE technique has been shown to be very accurate in modelling complex three-dimensional contact problems between two deformable bodies, see, e.g. [29]. However, it has not been possible to achieve the contact computations in real-time. The real-time simulation of self-contact or contact between deformable bodies remains a challenge for further research.

6.3 Cutting to Reach a Tumour

Figure 5 shows a simplified three-dimensional example in a wire-frame form. In order to simulate a 'hard' tumour, Young's modulus of the tumour is specified as five times greater than the Young's modules of the soft tissue. During a surgical simulation, the outer surface of the soft tissue is 'virtually' deformed by one or two indenters, which will result in a small deformation of the hard tumour. With experience, the user can identify the location of the tumour by gently prodding the outer surface of the soft tissue. In practice, the position of the tumour would be determined by a pre-surgery MRI scan.

A cut can be introduced on the surface of the soft tissue by one of the two indenters, and then propagated deeper in the soft tissue until the tumour surface is reached. During the cutting process, new surfaces are automatically created either side of the cutting path. This operation can be performed in real-time using a multi-threading programming approach as discussed in reference [4]. The process of extending the cut such that the tumour surface is separated from the contact surface in real-time is planned for future development of this surgical simulation system.

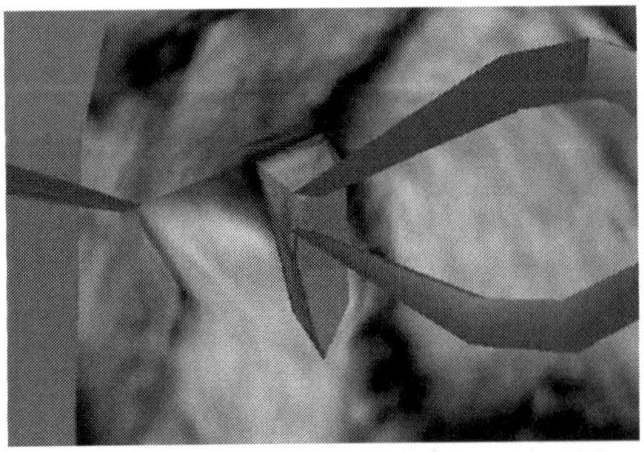

Figure 4: Example showing the interaction with two haptic devices

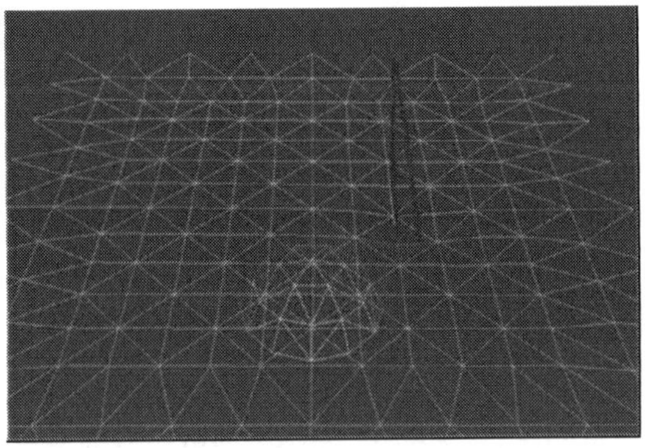

Figure 5: Wire frame model showing pinching on the surface of the soft tissue

7. Conclusions

This paper presents the development and the implementation of a two-domain surgical simulation model which incorporates a hard tumour included inside soft-tissue. Both domains are deformable, but the tumour is given a higher modulus of elasticity to distinguish it from the soft tissue. BE formulations are derived for the system matrices based on using algorithms to create pre-solution look-up tables. The two-domain BE model has been incorporated within the authors' surgical software system (VIRS) which allows continuous real-time cutting until the tumour surface is located.

Acknowledgement

This research is supported by the UK Engineering and Physical Sciences Research Council (EPSRC) via grant GR/R84030/01.

References

[1] Wang , P., Becker, A.A., Glover, A.T., Benford, S. D., Greenhalgh, C. M. , Vloeberghs, M. and Jones, I. A., "Application of the Boundary Element Method to the Simulation of Surgery including Haptic Feedback", Proceedings of the Seventh International Conference on Computational Structures Technology, Lisbon, 7-9 September 2004, B.H.V. Topping and C.A. Mota Soares, (Editors), Civil-Comp Press, Stirling, paper 100 (2004).

[2] Wang P., Becker A. A., Glover A.T., Jones I. A., Benford S. D., Vloeberghs M. and Greenhalgh C. "Designing a Virtual Reality Simulator for Neurosurgery", Proceedings of the 7th International Conference on Virtual Reality, edited by S. Richir and B. Taravel, pages 35-39, 19-22 April 2005, Laval, France, 2005.

[3] Becker, A. A., Wang, P., Jones, I. A. , Glover, A.T., Benford, S. D., Greenhalgh, C. M. and Vloeberghs, M., "Virtual reality surgical simulation- the role of computational mechanics techniques", Innovations in Engineering Computational Technology (Chapter 2), Edited by B.H.V. Topping, G. Morntero and R. Montenegro, pp. 23-39, Saxe-Coburg Publications, Stirlingshire, Scotland, 2006.

[4] Wang P., Becker A.A., Jones I. A., Glover A.T., Benford S. D., Greenhalgh C. M. and Vloeberghs M., "Virtual reality surgery simulation of cutting and retraction in neurosurgery with force-feedback" *Computer Methods & Programs in Biomedicine*, Vol. 84, pp. 11-18, 2006.

[5] Wang P., Becker A.A., Jones I. A., Glover A.T., Benford S. D., Greenhalgh C. M. and Vloeberghs M., "Virtual reality simulation of surgery with haptic feedback based on the Boundary Element Method." *J. Computers and Structures*, accepted, in print.

[6] Meier U., Lopez O., Monserrat C., Juan M.C., Alcaniz M. Real-time deformable models for surgery simulation: a survey. *Computer methods and programs in biomedicine*, Volume 77, Issue 3, March 2005, pp. 183-197.

[7] Montagnat J., Delingette H. and Ayache N., A review of deformable surfaces: topology, geometry and deformation. Image and Vision Computing, vol. 19, no. 14, pp. 1023-1040, 2001

[8] Picinbono G., Delingette H and Ayache N., Non-linear anisotropic elasticity for real-time surgery simulation. Graphical Models, 65, 2003, 305-321.

[9] Tillier Y., Paccini A., Durand-Reville M, Bay F. and Chenot J-L. Three-dimensional finite element modelling for soft tissues surgery. International Congress Series 1256 (2003), 349-355.

[10] Yoshida H., Tsutsumi S., Mizunuma M. and Yanai A., A surgical simulation system of skin sutures using a three-dimensional finite element method. Clinical Biomechanics, 16, 621-626, 2001.

[11] Berkley J., Turkiyyah G., Berg D., Ganter M and Weghorst S., Real-time finite element modelling for surgery simulation: an application to virtual suturing. IEEE Trans. On Visualization and Computer Graphics, 10(3), 314-325, 2004.

[12] Berkley, J., Turkiyyah, G., Berg, D., Ganter, M. and Weghorst, S. , Real-Time Finite Element Modeling for Surgery Simulation: An Application to Virtual Suturing IEEE Transactions on Visualization and Computer Graphics, Vol. 10, No. 3, May/June 2004.

[13] James, D.L. and Pai, D.K., ArtDefo: Accurate real time deformable objects, Proc SIGGRAPH99, pp. 65-72. See also http://www.cs.ubc.ca/~djames/deformable/

[14] James, D.L. and Pai, D.K., Pressure masks for point-like contact with elastic models. In Proceedings of the Fifth Phantom User Group Workshop, J.K. Salisbury and M.A. Srinivasan (Eds), 2000.

[15] Monserrat, C., Meier, U., Alcaniz, M., Chinesta, F and Juan, M.C., A new approach for the real-time simulation of tissue deformations in surgery simulation. Computer Methods and Programs in Biomedicine, Vol. 64 (2001), pp. 77-85.

[16] Monserrat C., Meier U., Acaniz M, Chinesta F and Juan M.C., A new approach for the real-time simulation of tissue deformations in surgery simulation. Computer Methods and Programs in Biomedicine, 64 (2001), 77-85.

[17] James D.L., Pai D.K. A Unified Treatment of Elastostatic Contact Simulation for Real Time Haptics. Haptics-e, the electronic journal of haptic research, 2(1), September 27, 2001.

[18] Zhong H., Wachowiak M.P., and Peters T.M. Adaptive finite element technique for cutting in surgical simulation Medical Imaging 2005: Visualization, Image-Guided Procedures, and Display, Proceedings of SPIE, Vol. 5744 pp. 604-611, April 2005.

[19] Wu W., Heng P.A., An improved scheme of an interactive finite element model for 3D soft-tissue cutting and deformation. Visual Comput, (online journal), 2005.

[20] Mahvash M., Hayward V., High-Fidelity Haptic Synthesis of Contact with Deformable Bodies. IEEE Computer Graphics and Applications, 48-55, March/April 2004.

[21] Hansen K.V., Brix L.,. Pedersen C.F,. Haase J.P, Larsen O.V., Modelling of interaction between a spatula and a human brain. Medical Image Analysis, 8, 23–33, 2004.

[22] De S., Kim J. Lim Y.J. Srinivasan M.A. The point collocation-based method of finite spheres (PCMFS) for real time surgery simulation. Computers and Structures, 83 1515–1525, 2005.

[23] Zhong Y., Shirinzadeh B., Alici G., Smith J., An autowave based methodology for deformable object simulation. Computer-Aided Design, 38, 740–754, 2006.

[24] Choi K-S., Sun H. and Heng P-A., An efficient and scalable deformation model for virtual reality-based medical applications. Artificial intelligence in medicine (2004), 34, 51-69.

[25] Choi K-S., Sun H. Heng P-A and Zou J., Deformable simulation using force propagation model with finite element optimization. Computers and Graphics (2004) 28 559-568.

[26] France L., Lenoir J., Angelidis A., Meseure P., Cani M.-P., Faure F., Chaillou C. A layered model of a virtual human intestine for surgery simulation, Medical Image Analysis, 9, 123–132, 2005.

[27] Brebbia, CA, Telles, JCF, Wrobel, LC. Boundary Element Techniques, Springer Verlag, Berlin, 1983.

[28] Becker, A.A. "The Boundary Element Method in Engineering", McGraw-Hill, London, 1992.

[29] Leahy, J.G. and Becker, A.A. "A quadratic boundary element formulation for three-dimensional contact problems with friction", J. Strain Analysis, 34, 235-251, 1999.

The Development of a Robotic Approach to Therapeutic Ultrasound

Z Qiu[1], C Xie[1], S Cochran[1], G Corner[2], Z Huang[1], C Song[1], S Daglish[3]

[1]School of Engineering, Physics and Mathematics, University of Dundee, Dundee, UK.

[2]Ninewells Hospital, Dundee, UK.

[3]University of Strathclyde Glasgow, Glasgow, UK

Email: z.y.huang@dundee.ac.uk

Abstract. Despite the recent wealth of research into therapy with high intensity focused ultrasound (HIFU), its clinical application still has limitations. These relate to the accuracy of the HIFU beam, the ability to visualize the target volume, and the understanding of the beam interaction with tissue. In the work reported in this paper, the output characteristics of a single channel HIFU transducer have been investigated in detail with the aid of a six-axis robot. Results show that by altering the drive parameters of the transducer clearly defined thermal or mechanical damage can be produced. The nature and patterns of damage produced by pre-programmed treatment volumes are now being investigated in tissue.

1. Introduction

To reduce suffering from traditional open surgery, in terms of trauma to the body, blood loss, long-term recovery, postoperative pain, and the risk of infection, minimally- and even non-invasive techniques are an alternative to the traditional route in the treatment of localized malignancy for most patients. Currently, regional chemotherapy and radiotherapy are regarded as the main cancer treatments. However, both of them have major side effects. Compared to these two techniques, another truly non-invasive modality, high intensity focused ultrasound (HIFU) is potentially significant as it can destroy deep-seated tumors without damaging normal tissue surrounding the tumor [1, 2], and without increasing the risk of distant metastases [3].

Just as sunlight is focused down to a spot using a magnifying glass to increase thermal intensity, HIFU is focused on to the tumour within the human body and the ultrasonic energy is converted to heat due to absorption. When the rate of heating is much greater than the rate of cooling, the temperature in the focal area can rise above a threshold of 56 ℃ in 1-2 s, approach 80 ℃ rapidly [4], which will consequently cause cell death through coagulative necrosis while the surrounding or overlying normal tissue is spared. And the tissue denaturation and coagulation occurs irreversibly [5 - 7].

Typically, the lesion caused by HIFU exposure tends to be ellipsoidal in shape, with a length of 1 - 2 cm along the beam axis and a diameter of 1 – 3 mm [8]. Compared to the whole target volume, it is required that the focal zone of HIFU beam should be scanned throughout the entire tumour, that is to say lesions are generally placed side by side and layered from the most distant part of the tumour to the nearest until the required volume has been covered. During scanning, localisation, repeatability and the accuracy of lesion placement are very important parameters which may result in unpredictable damage to normal tissue. Therefore, a robotic manipulator may be employed as a precise and accurate targeting mechanism [9, 10].

The first robot-assisted surgical application, the PUMA 560, was used in the 1980s to place a needle and obtain a stereotactic brain biopsy under CT guidance [11]. Subsequently, robotic technology has been applied in a variety of different surgical fields, such as cardiothoracic surgery, neurosurgery, radio surgery, urology and gastrointestinal surgery.

This paper reports on research into issues relating to the design, development and testing of a reliable HIFU treatment system, particularly addressing the challenge of improving the treatment accuracy. The system under investigation is intended to be deployed by an industrial robotic manipulator to allow the focus to be positioned rigidly and precisely, and to make the tool reconfigurable to meet the needs of the particular location or procedure through its reprogrammable property. It is not envisioned that the complete system could be used in the operating room but that it will provide indications of the usefulness of robotic techniques for implementation with equipment directly compatible with the operating room.

A graphical user interface written in the LabVIEW programming language (National Instruments, Newbury, England) serves as the control centre to define HIFU treatments, making it possible to control experimental parameters from one computer. This interface sets up the communication network of the HIFU system to connect each element together, including the robotic manipulator. It can be used to change and control the HIFU parameters which contribute to the lesion formation during experimental procedures, and to acquire, analyze, display and then store related data.

2. The Robot Assisted HIFU System

This paper describes the development of the robot assisted HIFU system. The RX 90 robotic manipulator applied in this system is produced by *STÄUBLI* Ltd, Switzerland. It is a modified industrial robot with six degrees of freedom (X, Y, Z, yaw, pitch, and roll) [12, 13]. Although not a clinical system, this manipulator is very stable, programmable and flexible. Its end effector can be placed in any orientation within the work envelope, which is almost a sphere, and an inner counting system is used to return the absolute position of the end effector of the robot. The default speed of this manipulator on boot up is 150 mms^{-1}, and the acceleration/deceleration is 90 mms^{-2}, with ±0.02 mm repeatability at constant temperature. Figure 1 illustrates the configuration of the HIFU system [14-17].

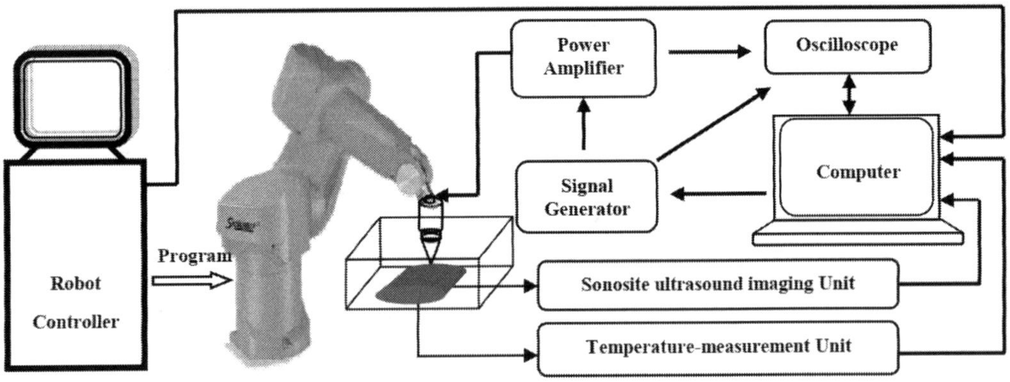

Figure 1 The configuration of the HIFU system

The HIFU treatment system comprises four co-operating modules:

1. *HIFU-generator module:* Composed of signal generator (Agilent 33220A Function / Arbitrary Waveform Generator, maximum frequency 20 MHz), power amplifier (ENI 3100LA broadband power amplifier, Gain 55 dB) and single HIFU transducer (Precision Acoustics Ltd, fixed ideal focal length 72 mm, operating frequency 1.09 MHz).
2. *Monitoring and measuring module*: Includes oscilloscope (Tektronix TDS 2024B, Serial digital oscilloscope, 200 MHz bandwidth, 2.0 GSa/s), temperature-measurement unit and ultrasound imaging unit (Sonosite 180).
3. *Robotic control module*: A *STÄUBLI* RX90 robot with CS7 robotic controller. The motion of this robotic manipulator is controlled by code in the V+ programming language.
4. *System control module*: A graphical user interface written in NI LabVIEW programming language to serve as the computer control centre.

For safety, a 20 dB attenuator was connected between the signal generator and power amplifier to reduce the ±10 V maximum output of the signal generator to below the 1 V rms maximum input amplitude of the power amplifier. The HIFU transducer was attached to the robot with a specific holder which was designed to replace the conventional robot gripper. The movement of HIFU beam was controlled with the robotic arm by running robot programs, using the V+ language.

Serial communication between the robot and a laptop computer, which serves as the system control centre to communicate with individual components, was established by an RS232 link. The signal generator, oscilloscope and the temperature-measurement unit were also connected to the same laptop via its USB 2.0 port. The key HIFU parameters such as the exposure duration were controlled by a LabVIEW program. Measurement data were collected by a special I/O functional node in LabVIEW, and then processed, displayed, and stored.

A polyacrylamide (PAA) gel phantom was made as an experimental sample for HIFU treatment to visualize the effects of the temperature rise in the focal area and the formation of a lesion. As a thermally sensitive indicator, liquid fresh egg white separated from egg yolk shortly before phantom preparation was added to the phantom material [18]. The optically transparent PAA gel phantom has similar acoustic properties to those of soft tissue, including speed of sound, density, and attenuation coefficient. The solution was mixed with 44.5% volume in volume (v/v) degassed water; 30% (v/v) egg white; 24.8% (v/v) 40% w/v acrylamide (SIGMA A4058); 0.5% (v/v) of 10% ammonium peroxodisulfate (SIGMA 215589); and 0.2% (v/v) of TEMED (VWR, 110-18-9) [19, 20]. Degassed water was used as the coupling medium as it transmits HIFU without spatial distortion.

The characterization of the focused field establishes a set of parameters of great importance to HIFU treatment. The beam profile and the focal length can be estimated through measuring the ultrasonic field [21]. The temperature rise, the key feature of HIFU exposure [22], can be recorded by medical thermography, here using an ALTAIR workstation provided by Cedip Ltd., a JADE MWIR camera, an FG9800 digital PCI frame grabber, and ALTAIR software for image acquisition and processing.

3. Acoustic Field Measurements

The measurement system was set up as shown in Figure 2(a). The HIFU transducer was fixed vertically on the end effector of the robot and a custom-made ultrasonic receiving transducer, with a diameter of 3.3 mm and a resonant operating frequency of 30 MHz, was placed at the bottom of water tank to measure the acoustic energy. The signal was amplified through the receive amplifier of a pulser/receiver (DPR 300, JSR Ultrasonics, New York, USA).

The robot was controlled by V+ program code to follow a step-by-step procedure which was a meander line in one x-y plane. Then the transducer was moved to the next plane, 1 mm higher in the +z direction, and the same motion was repeated. That is to say the robotic motion contributed a 3D matrix to measure the acoustic field. The robot stopped at each step and sent a signal to the computer to trigger the LabVIEW program, and then started recording the peak-to-peak voltage value measured by the receiver. Four data points were collected at each step and their average value was treated as the final value and stored in a Microsoft Excel file.

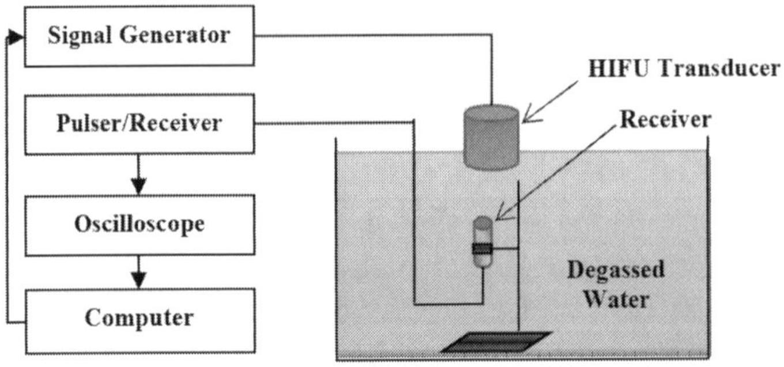

(a)

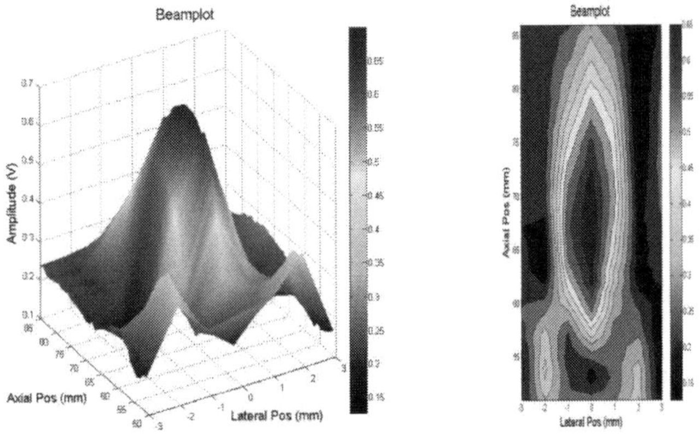

(b)

Figure 2 (a) The system used to measure the ultrasonic field. (b) Contour map of the focal zone in ultrasonic field along the axial direction

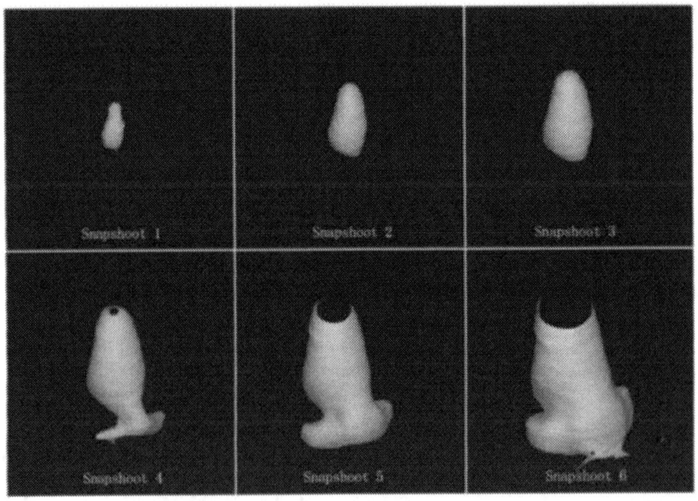

Figure 3 Snapshot frames of isosurface amplitudes within the focal zone.

In this paper, the robotic motion formed a 7 x 7 x 36 point matrix in space. There were 7 steps in the x and y directions respectively and 36 steps in z direction, with each step of 1 mm. The distance between the surface of the HIFU transducer and the receiving transducer was changed from 50 mm to 85 mm. Figure 2(b) shows the contour map of the focal zone along the axial direction, from which the maximum amplitude i.e. the focal point was obtained as at a distance of 67 mm from the HIFU transducer. Confirming the value of the experimental measurement, the ultrasonic field did not have as compact a focal zone as expected.

These data were later processed by an IDL7.0 program. Some snapshot frames are shown in Figure 3 indicating the focal zone of the acoustic field. Snapshots 1 - 6 show the isosurface amplitudes from the inside of the focused beam to the outside respectively.

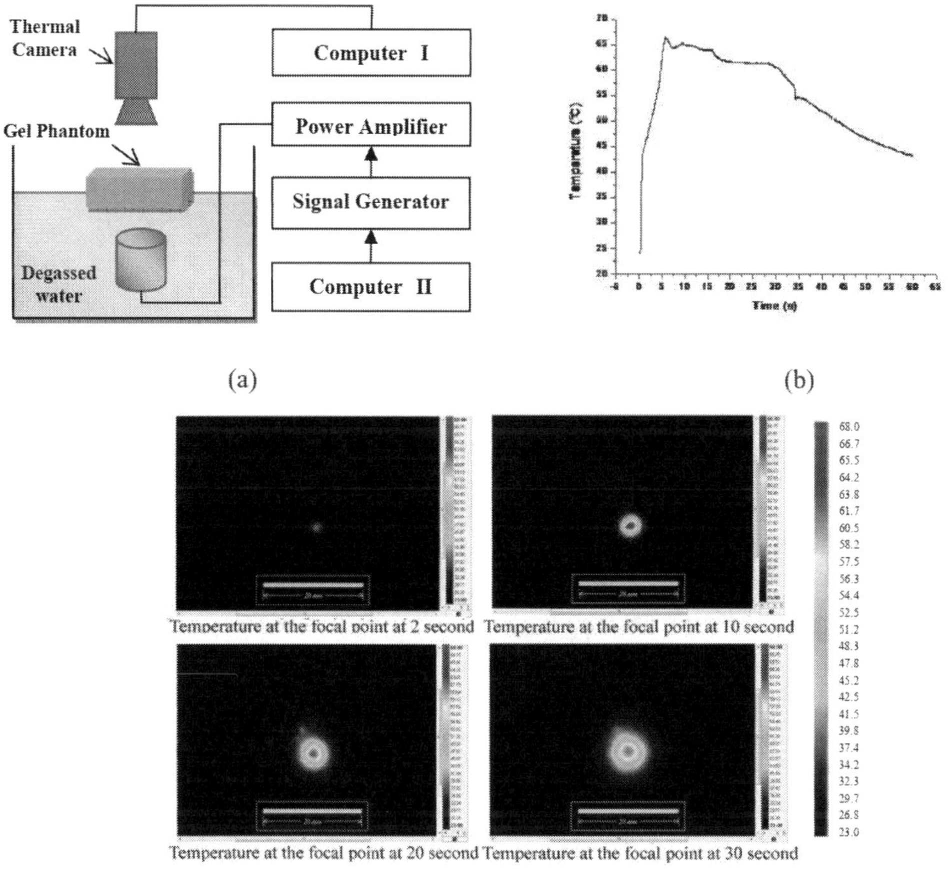

(c)

Figure 4 (a) The system to measure the temperature changes during HIFU exposure; (b) Temperature vs. time graph of 30 s exposure with an additional 30 s cooling time; (c) Medical thermography at the focal point at 2, 10, 20, and 30 s respectively.

4. Temperature Measurement

Being distinct from the main configuration of the HIFU system, the temperature measurement system is shown in Figure 4(a). The transducer was immersed in degassed water, and placed 67 mm below the upper surface of the gel phantom where the transducer focal point was acquire from just below the sample surface. The thermal camera was placed above the gel phantom with an air path to its upper surface. The temperature changes at the phantom surface were measured every 50 ms for 30 s exposure time with an additional 30 s cooling time. The power input to the HIFU transducer was approximately 70 W. Figure 4(b) is the temperature vs. time graph during the total 60 s procedure, the temperature reached 56 ℃ in about 4 seconds from an initial gel temperature of 24℃, and reached 66.49 C in about 5.7 seconds. That is in broad agreement with the study by ter Haar (2001) showed that the temperature will exceed 56 ℃ in 1-2 seconds at an intensity of 1 $kW.cm^{-2}$. Figure 4 (c) displays the images captured at 2 s, 10 s, 20 s and 30 s during this 30 s exposure time. The temperature scale is in Celsius.

5. Experiments and Results

Ex vivo experiments were carried out in the gel phantom by varying three main experimental parameters, the exposure time (ranging from 5 s to 55 s), transducer input power (30 W and 70 W) and the distances between the transducer and the phantom (ranging from 65 to 80 mm). The robotic arm holding the HIFU transducer was controlled by V+ codes to move along a specific path in space, to place lesions step-by-step and to vary the distance between the transducer and phantom. A LabVIEW program was used to set up communications within this system and to control other trial parameters. Using the HIFU system assisted by both the robot and LabVIEW, multiple series of lesions were created in one procedure, under various experimental conditions.

Figure 5 shows one series of deep-seated lesions created within phantom material by using different exposure durations ranging from 5 to 55 s. The phantom was imaged using a Sonosite 180 medical ultrasound machine. Figure 6(a), (b) and (c) are the same series of lesions from an overhead view. As can be seen, the lesions have been generated in a straight line with a measured separation of 7 mm.

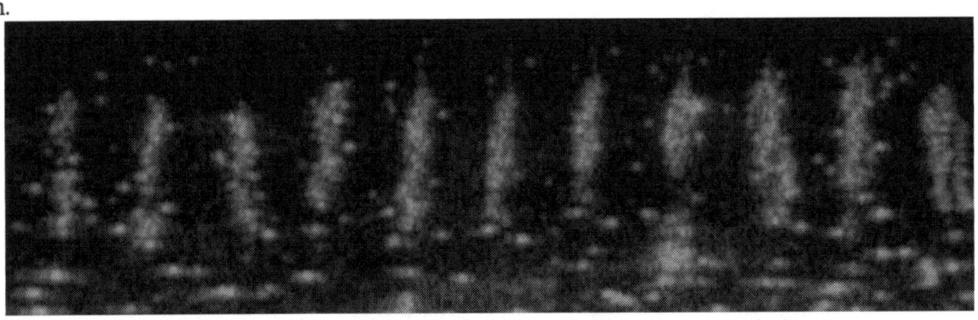

Figure 5 Ultrasound image of a series of deep-seated lesions generated with different exposure times.

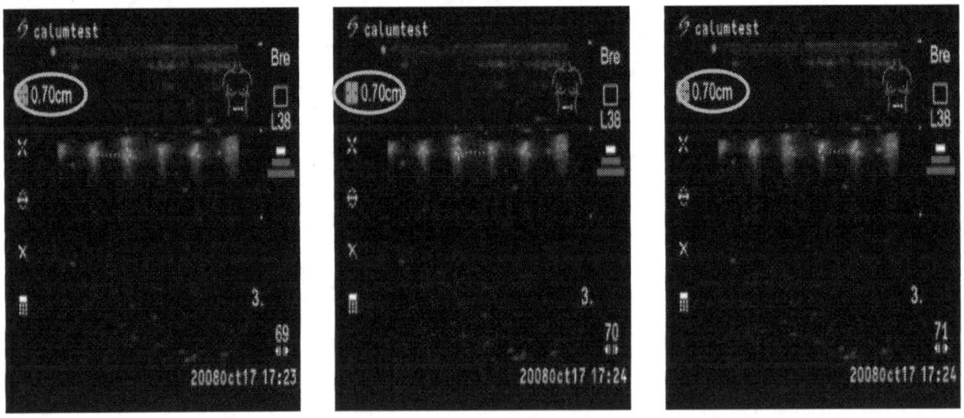

(a) (b) (c)

Figure 6 Lesions in overhead view with the measured distance between adjacent lesions

A further pre-programmed treatment was investigated to demonstrate both the flexibility of this robot assisted HIFU system and the ability to produce complex patterns. A smiling face was "drawn" on the surface of a HIFU phantom, as shown in Figure 7. The robotic arm carrying the HIFU transducer was driven to move to specific locations pre-planned in V+ code. The duration of exposure was controlled by enabling and disabling automatically the output of the signal generator providing the input to the power amplifier using LabVIEW.

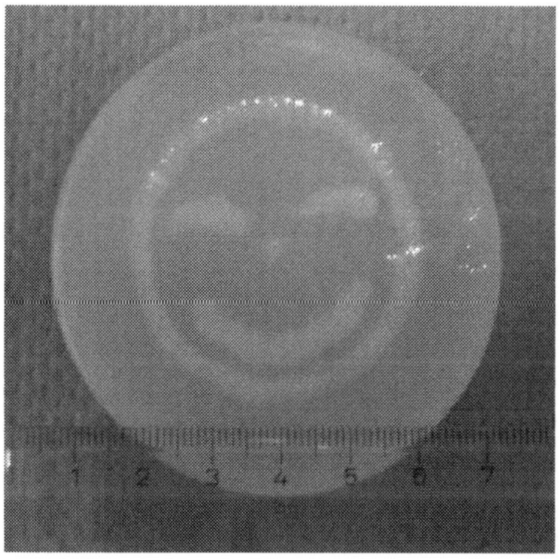

Figure 7 The smiling face lesion in PAA-egg white phantom caused by HIFU exposure

6. Conclusion

In this paper, a HIFU treatment system assisted by a modified industrial robot (*STÄUBLI* RX 90) and computer control was first described. Based on this system, HIFU beam profile and temperature increases were investigated to fully understand the principles of HIFU treatment through thermal ablation. Experiments were carried out with PAA gel phantom material to test not only the effects of HIFU treatment but also the flexibility of the robotic system and the feasibility to produce lesions in complex patterns. The main conclusion of this investigation is that HIFU thermal ablation can be successfully applied with the assistance of the robot and other hardware under computer control.

References

[1] J E Kennedy, G R Ter Haar, D Cranston (2003) "High intensity focused ultrasound: surgery of the future?" The British Journal of Radiology; 76: 590-599.

[2] Gail ter Haar, D Sinnett, and I H Rivens (1989) "High-intensity focused ultrasound—a surgical technique for the treatment of discrete liver tumours." Phys. Med. Biol. 34: 1743-1750.

[3] Yifei Xing, Xiaochun Lu, et al. (2008) "The effect of high intensity focused ultrasound treatment on metastases in a murine melanoma model" Biochemical and Biophysical Research Communications.

[4] Gail ter Haar, Clarke RL, Vaughan MG, Hill CR(1991) "Trackless surgery using focused ultrasound: technique and case report." Min Inv Ther; 1: 13-5.

[5] Gail ter Haar (2001) "Acoustic surgery" Physics Today; December: 29-34

[6] Gail ter Haar et al. (2006) Therapeutic Ultrasound Team. Annual Report.

[7] Gail ter Haar (2000) "Intervention and therapy" Ultrasound in Med. & Biol.; 26: s51-s54.

[8] Gail ter Haar (2007) "Turning up the power: High intensity focused ultrasound (HIFU) for the treatment of cancer" Ultrasound; 15(2): 73-77.

[9] R.D. Howe and Y. Matsuoka (1999) "Robotics for surgery Annu. Rev." Biomedical Engineering. 1999.01 211-240.

[10] N.G Hockstein, C.G Gourin, R.A Faust, D.J Terris (2007) "A history of robots: from science fiction to surgical robotics" J. Robotic Surg. 1: pp 113-118

[11] YS Kwoh, J Hou, EA Jonckheere, S Hayati (1988) "A robot with improved absolute positioning accuracy for CT guided stereotactic brain surgery" IEEE Trans. Biomed. Eng. 35 (2): 153-160.

[12] V+ instruction manual of STAUBLI RX 90 robot.

[13] Stirling Paatz (2008) "Anatomy of a robot" Engineering & Technology; 3, 26 Jan-8 Feb: 42-44.

[14] Sunita Chauhan, Gail ter Haar (2007) "FUSBOT: Empirical studies using a surgical robotic system for urological applications" CP911, 6th International Symposium on Therapeutic Ultrasound. pp 117-121.

[15] Fucheng Sun, Ye He, Rui Li (2007) "High intensity focused ultrasound tumor therapy system and its application" CP911, 6th International Symposium on Therapeutic Ultrasound pp. 92-97.

[16] Liangshou Wu, Viren Amin, Ron Roberts et al. (2007) "An interactive HIFU therapy Planning using simulation and visualization" CP911, 6th International Symposium on Therapeutic Ultrasound. pp. 150-156.

[17] Selvan Pather, Brian L. Davies, Roger D. Hibberd (2002) "The development of a robotic system for HIFU surgery applied to liver tumours" 7th International Conference on Control, Automation, Robotics and Vision, Volume 2, Singapore. pp. 572-577.

[18] Gabriela Wilzbach Divkovic, Marko Liebler, Klaus Braun et al. (2007) "Thermal properties and changes of acoustic parameters in an egg white phantom during heating and coagulation by high intensity focused ultrasound" Ultrasound in Med. & Biol., Vol. 33, No.6, pp. 981-986.

[19] Gabriela Wilzbach Divkovic, Jurgen W.Jenne. (2004) "Egg white phantoms for HIFU" In: ter Haar GR, Rivens I, eds. Proceedings of the 4th International Symposium on Therapeutic Ultrasound, New York: American Institute of Physics: 143–146.

[20] Kenji Takegami, Yukio Kaneko, Toshiaki Watanabe et al. (2004) "Polyacrylamide gel containing egg white as new model for irradiation experiments using focused ultrasound" Ultrasound in Med. & Biol., Vol. 30, No. 10, pp. 1419-1422.

[21] Yuebing Wang, Fangwei Ye. (2007) "The measurement of HIFU field using a reflective needle hydrophone" CP911, 6th International Symposium on Therapeutic Ultrasound. pp. 3-8.

[22] R.L. Clarke, G. R. Ter Haar. (1997) "Temperature rise recorded during lesion formation by high- intensity focused ultrasound" Ultrasound in Med. & Biol. vol. 23, No. 2, pp. 299-306.

Mathematical modeling of growth processes in nature and engineering: A variational approach

A. V. Manzhirov[1] **and S. A. Lychev**[1]

[1]Ishlinsky Institute for Problems in Mechanics of the Russian Academy of Sciences, Vernadsky Ave 101 Bldg 1, Moscow, 119526, Russian Federation

E-mail: manzh@ipmnet.ru

Abstract. We present a variational approach to the mathematical theory of accreted solids. One main point in this approach is that the operator of the accretion problem proves to be self-adjoint with respect to an appropriately modified convolution bilinear form, and it is this linear form that we use in the construction of the variational functional.

Our growing solid model can be efficiently applied to describe processes such as concreting, pyrolytic deposition, laser spraying, electrolytic deposition, polymerization, solidification of melts, crystal growth, glacier and ice cover freezing, sedimentary and volcanic rock forming, and biological tissue growth. These applications will be considered elsewhere.

1. Introduction

Deformation processes in a solid whose composition, mass, or volume varies in a piecewise continuous manner owing to *accretion*—the addition of new material to the body outer surface—were studied in [1–12]. The solid mechanics problems arising in the modeling of such processes are completely new and constitute a separate field of research, known as *mechanics of accreted solids*. Its importance is due to the fact that the manufacturing of virtually all objects considered in solid mechanics (buildings, structures, structural components, machine parts, etc.) involves accretion. (Specific examples include continuous concrete structure erection, metal solidification, spray deposition of semiconducting films, and crystal growth—one could easily continue the list.)

We consider the variational statement of the piecewise continuous accretion problem and suggest a method for solving the resulting variational problem.

The following *main stages* of deformation clearly manifest themselves in a piecewise continuous accretion process:

(i) The initial stage occurring prior to accretion.

(ii) The continuous accretion stage.

(iii) The deformation stage occurring after accretion (or growth) stops.

Stages (ii) and (iii) alternate, and each of the stages is naturally characterized by its start time and termination time. For stage (i), these are the times at which the load is applied to the solid and then growth starts; for stage (ii), the times at which growth starts and then stops; and for stage (iii), conversely, the times at which growth stops and then starts. After a number of alternations, the process is usually finished at the third stage, for which the termination time is taken to be arbitrarily large. The original solid occurring at the initial stage will also be called the *basic solid*. The rest of the solid at any given time is called the *additional solid*; it may have a complex structure and consists of the collection of solids formed on different time intervals of continuous accretion [stage (ii)]. We refer to these as sub-solids, so that the additional solid is the union of sub-solids. The domains occupied by the former and latter can be disconnected. The whole solid, which is the union of the basic and additional solids, will be called the *accreted* (or *growing*) *solid*. Note that accretion may well occur without the basic solid, starting from an infinitesimal material element.

The part of the surface where infinitesimal pieces of the material are deposited at the current time is called the *accretion* (or *growth*) *surface*. The growth surface may be disconnected in general. In particular, it can be the entire surface of the solid. Finally, the part of the surface of the original or growing solid that coincides with the growth surface at the growth start time will be called the base surface. The base surface is clearly the part of the surface of the solid on which material will be deposited during the next stage of continuous accretion. At various stages, it usually coincides with the surface between the basic solid and the additional solid or with a surface between sub-solids.

To study the stress-strain state of the accreted solid, one should know the deformation laws

• Of the basic body from the time at which the load is applied until the time at which accretion starts.

• Of the deposited material from the time at which the load is applied to this material until the time of its deposition on the accreted solid.

While the state of the original body is determined from the solution of the problem at the stage preceding accretion, the previous history of the strain tensor of the infinitesimally thin continuously deposited material is assumed to be known.

Prior to stating the problem considered in the present paper, we point out that the growth problem for a solid differs dramatically from a problem involving the removal of material. The latter is solely characterized by the fact that the domain occupied by the body is reduced, subject to the standard equations and boundary conditions.

2. Main results

The aim of this section is to obtain the variational statement of the problems under consideration. This permits us to apply direct variational methods for numerical simulation and furthermore define correct boundary and initial conditions in the above-mentioned applied initial–boundary value problems.

Let us obtain kinematic relations and use material coordinates that specify unique material particles. The positions of these particles in Euclidian space can be characterized by the spatial coordinates $\boldsymbol{\chi}(\boldsymbol{x};t)$. We denote the time variable by t. We assume that the accretion scenario is known; i.e. the accretion time $\tau^*(\boldsymbol{x})$ is prescribed for each particle $\boldsymbol{x}$. Suppose that annihilation of material particles does not occur. Thus, the function $\boldsymbol{\chi}(\boldsymbol{x};t)$ is well defined for each $\boldsymbol{x}$ in the corresponding interval $(\tau^*(\boldsymbol{x}),\infty)$. The set $\mathcal{B}$ of accreted material particles is assumed to be a simply connected domain in the topological space. The boundary $\partial\mathcal{B}$ of $\mathcal{B}$ is assumed to be regular. For simplicity, we suppose that the accreted material is in natural state at the accretion or deposition time $\tau^*(\boldsymbol{x})$. All these assumptions permit us to determine the displacements as follows:

$$\boldsymbol{u}(\boldsymbol{x};t) = \int_{\tau^*(\boldsymbol{x})}^{t} \boldsymbol{v}(\boldsymbol{x};\tau)\,d\tau, \tag{1}$$

where $\boldsymbol{v}(\boldsymbol{x};t)$ is the velocity of the particle with material coordinate $\boldsymbol{x}$,

$$\boldsymbol{v}(\boldsymbol{x};t) = \frac{\partial\boldsymbol{\chi}(\boldsymbol{x};t)}{\partial t}.$$

Note that the displacement defined in such a way is the *displacement of a particle with respect to its position at the accretion time.* Hence we have a well-defined vector field $\boldsymbol{u}$ in $\mathcal{B}$, and it is reasonable to define the deformation gradient in the interior of $\mathcal{B}$ by the formula

$$\boldsymbol{F} = \boldsymbol{\nabla} \otimes \boldsymbol{u} = \boldsymbol{\nabla} \otimes \int_{\tau^*(\boldsymbol{x})}^{t} \boldsymbol{v}(\boldsymbol{x};\tau)\, d\tau.$$

We point out that, in contrast with conventional elasticity [13], the Hamiltonian operator and the definite integral over time do not commute, because the lower integration limit $\tau^*(\boldsymbol{x})$ depends on the material variables. Indeed,

$$\boldsymbol{F} = \int_{\tau^*(\boldsymbol{x})}^{t} \boldsymbol{\nabla} \otimes \boldsymbol{v}(\boldsymbol{x};\tau)\, d\tau - \boldsymbol{\nabla}\tau^*(\boldsymbol{x}) \otimes \boldsymbol{v}(\boldsymbol{x};\tau^*(\boldsymbol{x})). \tag{2}$$

Now we introduce the velocity gradient in the classical context,

$$\boldsymbol{D} = \boldsymbol{\nabla} \otimes \boldsymbol{v}(\boldsymbol{x};\tau).$$

It is easily seen that

$$\boldsymbol{D} = \dot{\boldsymbol{F}},$$

where the dot stands for differentiation with respect to t, but at the same time it follows from Eq. (2) that

$$\boldsymbol{F} = \int_{\tau^*(\boldsymbol{x})}^{t} \boldsymbol{D}\, d\tau + \boldsymbol{\Omega}.$$

Here we have used the notation

$$\boldsymbol{\Omega} = -\boldsymbol{\nabla}\tau^*(\boldsymbol{x}) \otimes \boldsymbol{v}(\boldsymbol{x};\tau^*(\boldsymbol{x})).$$

Recall that $\boldsymbol{v}(\boldsymbol{x};\tau^*(\boldsymbol{x}))$ is the material particle velocity at the accretion time.

Consider the tensor field

$$\tilde{\boldsymbol{F}} = \int_{\tau^*(\boldsymbol{x})}^{t} \boldsymbol{D}\, d\tau,$$

similar to the deformation gradient. This tensor field can be referred to as a primitive of the velocity gradient. Obviously,

$$\boldsymbol{\nabla} \times \boldsymbol{F} = \boldsymbol{0}, \quad \boldsymbol{\nabla} \times \tilde{\boldsymbol{F}} = \boldsymbol{\nabla} \times \boldsymbol{\Omega} \neq \boldsymbol{0}.$$

This means that, in general, $\boldsymbol{F}$ satisfies the compatibility conditions, while $\tilde{\boldsymbol{F}}$ still does not. Indeed,

$$\boldsymbol{\nabla} \times \boldsymbol{\Omega} = -\boldsymbol{\nabla}\tau^*(\boldsymbol{x}) \otimes \boldsymbol{\nabla} \times \boldsymbol{v}^*(\boldsymbol{x}).$$

Therefore, if the initial velocity field $\boldsymbol{v}^*(\boldsymbol{x})$ is not potential, then the tensor field $\tilde{\boldsymbol{F}}$ does not satisfy the compatibility conditions. Of course, the difference disappears if we use the derivatives of the above-mentioned tensor fields. This leads to the statement of the problem in terms of velocities.

Now we construct the variational statement of the problems under consideration with the use of the following well-known assertion: if a linear operator $\mathcal{A}$ with domain $\mathcal{D}$ is self-adjoint with respect to a symmetric bilinear form $\langle .,. \rangle$ and if this bilinear form is nondegenerate, i.e., satisfies the condition

$$\Big(\forall \boldsymbol{u} \in \mathcal{D}\ \langle \boldsymbol{u}, \boldsymbol{v} \rangle = 0\Big) \Rightarrow \boldsymbol{v} = \boldsymbol{0},$$

then the stationary points of the quadratic functional

$$\mathcal{J}[\boldsymbol{u}] = \frac{1}{2}\langle \boldsymbol{u}, \mathcal{A}[\boldsymbol{u}]\rangle$$

defined on the same domain $\mathcal{D}$ are exactly the solutions of the equation

$$\mathcal{A}[\boldsymbol{u}] = \mathbf{0}.$$

This statement is obvious. Nevertheless, except in trivial cases, the operators that arise in continuum mechanics are not self-adjoint with respect to classical bilinear forms, which usually coincide with the standard inner products in the corresponding function spaces. The problems concerning the variational statement of dynamic elasticity by Hamilton's principle are well known. That is, the extension of variational formalism to novel problems depends on an appropriate definition of nonclassical bilinear forms. Noticeable progress in this field was obtained by Gurtin [14], Tonti [15], and Belli and Morosi [16]. They proposed the use of the convolution as a bilinear form. In particular, this permits one to obtain a well-defined variational statement for linear dynamic elasticity and thermoelasticity.

In what follows, we use a modification of the convolution bilinear form adapted to growing elastic solids. We assume that accretion of additional material takes place only on the boundary and that the classical conservation laws hold in the interior parts of the body. Thus, we can expect that the variational principles for elastic and thermoelastic growing solids and the principles recently proposed by Gurtin and Tonti differ in the boundary terms and function domains.

Consider an elastic accreted solid. It follows from the conservation of momentum that in an open neighborhood $\mathcal{N}_{\boldsymbol{x}}$ of an arbitrary interior point $\boldsymbol{x}$ one has

$$\boldsymbol{\nabla}\cdot[\boldsymbol{E}:\boldsymbol{\nabla}\otimes\boldsymbol{u}] + \rho\boldsymbol{f} - \rho\ddot{\boldsymbol{u}} = \mathbf{0},$$

where ρ is the mass density, $\boldsymbol{f}$ is the body force per unit mass, and $\boldsymbol{E}$ is the elasticity tensor. We use this notation for brevity; in conventional linear elasticity, the symmetric part of the deformation gradient is customarily used.

Consider the class $\mathscr{A}$ of smooth compactly supported functions defined in $\mathcal{N}_{\boldsymbol{x}}\times(\tau^*, t)$. Thus, the functions belonging to $\mathscr{A}$, together with their all derivatives, are zero in the complement of the set $\mathcal{N}_{\boldsymbol{x}}\times(\tau^*, t)$. On the class $\mathscr{A}$, we introduce the differential operator

$$\mathcal{A}[\boldsymbol{u}] = \boldsymbol{\nabla}\cdot[\boldsymbol{E}:\boldsymbol{\nabla}\otimes\boldsymbol{u}] - \rho\ddot{\boldsymbol{u}}$$

and the bilinear form

$$\langle \boldsymbol{u}, \boldsymbol{v}\rangle = \int_{\mathcal{N}_{\boldsymbol{x}}}\int_{\tau^*(\boldsymbol{x})}^{t} \boldsymbol{u}(\boldsymbol{x}; t+\tau^*-\tau)\cdot\boldsymbol{v}(\boldsymbol{x};\tau)\, d\tau\, dV. \tag{3}$$

A straightforward computation readily shows that this form is symmetric,

$$\forall \boldsymbol{u}, \boldsymbol{v}\in\mathscr{A} \quad \langle \boldsymbol{u}, \boldsymbol{v}\rangle = \langle \boldsymbol{v}, \boldsymbol{u}\rangle,$$

and nondegenerate. The latter assertion is a consequence of well-known functional properties of the convolution.

The Gauss divergence theorem with respect to material variables, the rule of integration by parts with respect to time t, and the assumption

$$\boldsymbol{E} = \boldsymbol{E}^{(3412)} \tag{4}$$

(where $\boldsymbol{E}^{(3412)}$ is the tensorial isomer with dyadic permutation 3412) give the following relation for arbitrary elements of $\mathscr{A}$:

$$\langle \mathcal{A}[\boldsymbol{u}], \boldsymbol{v}\rangle = \langle \boldsymbol{u}, \mathcal{A}[\boldsymbol{v}]\rangle.$$

In other words, the operator $\mathcal{A}$ is self-adjoint with respect to the bilinear form (3). Note that assumption (4) always holds in the framework of classical hyperelasticity and corresponds to the existence of an elastic potential.

Now we are in a position to construct a variational statement for the problem without body forces for functions of the class $\mathscr{A}$ as follows:

$$\mathcal{J}[\boldsymbol{u}] = \frac{1}{2}\langle \boldsymbol{u}\cdot\mathcal{A}[\boldsymbol{u}]\rangle, \quad \delta\mathcal{J}[\boldsymbol{u}] = 0. \tag{5}$$

If the tensor $\boldsymbol{E}$ is positive definite, then problem (5) has only the trivial solution. It follows from this remark that the properties of elastic bodies manufactured by accretion in the absence of body forces and with trivial initial conditions are identical to those of classical elastic bodies.

Taking into account the body forces, we obtain

$$\mathcal{J}[\boldsymbol{u}] = \frac{1}{2}\langle \boldsymbol{u}\cdot\mathcal{A}[\boldsymbol{u}]\rangle + \mathcal{R}_1,$$

where the additional operator $\mathcal{R}_1$ depends on the prescribed body force distribution $\boldsymbol{f} = \boldsymbol{f}(\boldsymbol{x}; t)$,

$$\mathcal{R}_1 = \langle \boldsymbol{u}, \rho\boldsymbol{f}\rangle.$$

Now let us expand the domain $\mathcal{N}_{\boldsymbol{x}}$ as much as possible. This possibility is limited to the current body boundary corresponding to time t. The above-mentioned expansion involves additional terms in the definition of the functional,

$$\mathcal{J}[\boldsymbol{u}] = \frac{1}{2}\langle \boldsymbol{u}\cdot\mathcal{A}[\boldsymbol{u}]\rangle + \mathcal{R}_1 + \mathcal{R}_2 + \mathcal{R}_3.$$

The term $\mathcal{R}_2$ depends on the prescribed initial velocity distribution $\boldsymbol{v}^*(\boldsymbol{x}) = \boldsymbol{v}(\boldsymbol{x}; \tau^*(\boldsymbol{x}))$,

$$\mathcal{R}_2 = -\frac{1}{2}\int_{\mathcal{B}} \boldsymbol{v}^*(\boldsymbol{x})\rho\boldsymbol{v}(\boldsymbol{x}; t)\, dV,$$

and the term $\mathcal{R}_3$ depends on prescribed traction distribution $\boldsymbol{p} = \boldsymbol{p}(\boldsymbol{x}; t)\Big|_{\partial\mathcal{B}}$ on the current body boundary,

$$\mathcal{R}_3 = \frac{1}{2}\int_{\partial\mathcal{B}}\int_{\tau^*(\boldsymbol{x})}^{t} \boldsymbol{p}(\boldsymbol{x}; t + \tau^* - \tau)\cdot\boldsymbol{v}(\boldsymbol{x}; t)\, d\tau\, dV.$$

Note that now the bilinear form is defined in the entire material region $\mathcal{B}$,

$$\langle \boldsymbol{u}, \boldsymbol{v}\rangle = \int_{\mathcal{B}}\int_{\tau^*(\boldsymbol{x})}^{t} \boldsymbol{u}(\boldsymbol{x}; t + \tau^* - \tau)\cdot\boldsymbol{v}(\boldsymbol{x}; \tau)\, d\tau\, dV,$$

and the function class $\mathscr{A}$ is expanded to the class $\mathcal{D}$ of twice differentiable functions that satisfy the above-mentioned boundary and initial conditions. Therefore, we obtain the variational principle that can be stated as follows:

$$2\,\mathcal{J}[\boldsymbol{u}] = \int_{\mathcal{B}}\int_{\tau^*(x)}^{t} \boldsymbol{u}(\boldsymbol{x}; t + \tau^* - \tau)\cdot\{\boldsymbol{\nabla}\cdot[\boldsymbol{E}(\boldsymbol{x}) : \boldsymbol{\nabla}\otimes\boldsymbol{u}(\boldsymbol{x}; \tau)] - \rho\ddot{\boldsymbol{u}}(\boldsymbol{x}; \tau) + 2\rho\boldsymbol{f}(\boldsymbol{x}; \tau)\}\, d\tau\, dV$$
$$+ \int_{\partial\mathcal{B}}\int_{\tau^*}^{t} \boldsymbol{p}(\boldsymbol{x}; t + \tau^* - \tau)\cdot\boldsymbol{u}(\boldsymbol{x}; t)\, d\tau\, dV - \int_{\mathcal{B}} \boldsymbol{v}^*(\boldsymbol{x})\cdot\rho\dot{\boldsymbol{u}}(\boldsymbol{x}; t)\, dV, \quad \boldsymbol{u}\in\mathcal{D}, \quad \delta\mathcal{J} = 0.$$

To prove the well-posedness of this variational statement, let us compute the variation of the functional $\mathcal{J}[\boldsymbol{u}]$. We have

$$2\,\delta\mathcal{J} = \langle \delta\boldsymbol{u}\cdot\mathcal{A}[\boldsymbol{u}]\rangle + \langle \boldsymbol{u}\cdot\mathcal{A}[\delta\boldsymbol{u}]\rangle + 2\,\delta\mathcal{R}_1 + \delta\mathcal{R}_2 + \delta\mathcal{R}_3.$$

It follows from the Gauss divergence theorem and the rule of integration by parts with respect to t that

$$\begin{aligned}\langle \boldsymbol{u}\cdot\mathcal{A}[\delta\boldsymbol{u}]\rangle = \langle \mathcal{A}[\boldsymbol{u}]\cdot\delta\boldsymbol{u}\rangle + \int_{\partial\mathcal{B}} \boldsymbol{n}\cdot\int_{\tau^*(\boldsymbol{x})}^{t} \Big[&\boldsymbol{u}(\boldsymbol{x};t+\tau^*-\tau)\cdot\boldsymbol{E}(\boldsymbol{x}):\boldsymbol{\nabla}\otimes\delta\boldsymbol{u}(\boldsymbol{x},\tau) \\ &-\boldsymbol{\nabla}\otimes\boldsymbol{u}(\boldsymbol{x};t+\tau^*-\tau):\boldsymbol{E}(\boldsymbol{x})\cdot\delta\boldsymbol{u}(\boldsymbol{x};\tau)\Big]\,d\tau\,dA \\ +\int_{\mathcal{B}} [\dot{\boldsymbol{u}}(\boldsymbol{x};\tau^*)\delta\boldsymbol{u}(\boldsymbol{x};t) - \dot{\boldsymbol{u}}(\boldsymbol{x};t)\delta\boldsymbol{u}(\boldsymbol{x};\tau^*) &- \boldsymbol{u}(\boldsymbol{x};\tau^*)\delta\dot{\boldsymbol{u}}(\boldsymbol{x};t) + \boldsymbol{u}(\boldsymbol{x};t)\delta\dot{\boldsymbol{u}}(\boldsymbol{x};\tau^*)]\,\rho\,dV,\end{aligned}$$

where $\boldsymbol{n}$ is the outward unit normal to the current material boundary ∂B. Moreover,

$$\delta\mathcal{R}_1 = \langle\delta\boldsymbol{u},\rho\boldsymbol{f}\rangle,\quad \delta\mathcal{R}_2 = -\int_{\mathcal{B}} \boldsymbol{v}^*(\boldsymbol{x})\rho\delta\boldsymbol{v}(\boldsymbol{x};t)\,dV,\quad \delta\mathcal{R}_3 = \int_{\partial\mathcal{B}}\int_{\tau^*(\boldsymbol{x})}^{t} \boldsymbol{p}(\boldsymbol{x};t+\tau^*-\tau)\delta\boldsymbol{v}(\boldsymbol{x};t)\,d\tau\,dA.$$

Since the traction on the boundary and the initial velocities are prescribed and the initial displacements are always zero by virtue of (1), we have

$$\boldsymbol{E}(\boldsymbol{x}):\boldsymbol{\nabla}\otimes\delta\boldsymbol{u}(\boldsymbol{x};\tau) = \boldsymbol{0},\quad \delta\boldsymbol{u}(\boldsymbol{x};\tau^*) = \delta\dot{\boldsymbol{u}}(\boldsymbol{x};\tau^*) = \boldsymbol{u}(\boldsymbol{x};\tau^*) = \boldsymbol{0}.$$

Hence, we finally obtain

$$\begin{aligned}\delta\mathcal{J} = \langle\delta\boldsymbol{u}\cdot(\mathcal{A}[\boldsymbol{u}]+\rho\boldsymbol{f})\rangle + \frac{1}{2}\int_{\mathcal{B}} [\dot{\boldsymbol{u}}^*(\boldsymbol{x},\tau^*) - \boldsymbol{v}^*(\boldsymbol{x})]\,\rho\delta\boldsymbol{v}(\boldsymbol{x};t)\,dV + \\ +\frac{1}{2}\int_{\partial\mathcal{B}}\int_{\tau^*(\boldsymbol{x})}^{t} [\boldsymbol{p}(\boldsymbol{x};t+\tau^*-\tau) - \boldsymbol{n}\cdot(\boldsymbol{\nabla}\otimes\boldsymbol{u}(\boldsymbol{x};t+\tau^*-\tau):\boldsymbol{E}(\boldsymbol{x}))]\cdot\delta\boldsymbol{v}(\boldsymbol{x};\tau)\,d\tau\,dA.\end{aligned}$$

By using the fundamental lemma of calculus of variations, we obtain the Euler–Lagrange equations

$$\mathcal{A}[\boldsymbol{u}] + \rho\boldsymbol{f} = \boldsymbol{0},\quad \text{i.e.}\quad \boldsymbol{\nabla}\cdot[\boldsymbol{E}:\nabla\otimes\boldsymbol{u}] + \rho\boldsymbol{f} - \rho\ddot{\boldsymbol{u}} = \boldsymbol{0},$$

the natural boundary conditions

$$\boldsymbol{p}(\boldsymbol{x};\tau)\Big|_{\partial\mathcal{B}} = \boldsymbol{n}\cdot(\boldsymbol{\nabla}\otimes\boldsymbol{u}(\boldsymbol{x};\tau):\boldsymbol{E}(\boldsymbol{x}))\Big|_{\partial\mathcal{B}},$$

and the natural initial conditions

$$\dot{\boldsymbol{u}}^*(\boldsymbol{x},\tau^*) = \boldsymbol{v}^*(\boldsymbol{x}).$$

Recall that the initial displacements are always zero by virtue of (1).

Certainly, we could use the velocity $\boldsymbol{v}(\boldsymbol{x};t) = \dot{\boldsymbol{u}}(\boldsymbol{x};t)$ instead of the displacement as a basic variable in the variational statement. This brings about some inessential transformations of the variational principle and the Euler–Lagrange equations; for example, the equation of motion becomes[1]

$$\boldsymbol{\nabla}\cdot\left[\boldsymbol{E}:\nabla\otimes\int_{\tau^*}^{t}\boldsymbol{v}(\tau)\,d\tau\right] + \rho\boldsymbol{f} - \rho\dot{\boldsymbol{v}} = \boldsymbol{0}.$$

[1] The following form of the equation of motion is also used:

$$\boldsymbol{\nabla}\cdot\left[\boldsymbol{E}:\int_{\tau^*}^{t}\nabla\otimes\boldsymbol{v}(\tau)\,d\tau\right] + \rho\boldsymbol{f} - \rho\dot{\boldsymbol{v}} = \boldsymbol{0}.$$

It is clear that these forms are different. But if we compute the derivatives of the left- and right-hand sides of both equations, they become identical. For this reason, the equations in terms of velocities are often considered.

The generalization of the variational principle to the nonlinear case is based on the Vainberg theorem [17]. Let $\mathcal{A}[\boldsymbol{u}]$ be a nonlinear operator. By the Vainberg theorem, the assertion that the variation of the functional

$$\mathcal{I}[\boldsymbol{u}] = \langle \boldsymbol{u}, \int_0^1 \mathcal{A}[\lambda \boldsymbol{u}]\, d\lambda \rangle$$

is zero if and only if $\boldsymbol{u}$ satisfies the equation

$$\mathcal{A}[\boldsymbol{u}] = \boldsymbol{0}$$

holds provided that the Fréchet derivative is self-adjoint.

In the nonlinear case, the momentum balance law implies that

$$\boldsymbol{\nabla}\cdot\rho\frac{\partial\Psi}{\partial(\boldsymbol{\nabla}\otimes\boldsymbol{u})} + \rho\boldsymbol{f} - \rho\ddot{\boldsymbol{u}} = \boldsymbol{0},$$

where $\Psi = \Psi(\boldsymbol{x}, \boldsymbol{\nabla}\otimes\boldsymbol{u})$ is the elastic potential. The corresponding nonlinear differential operator is

$$\mathcal{A}[\boldsymbol{u}] = \boldsymbol{\nabla}\cdot\rho\frac{\partial\Psi}{\partial(\boldsymbol{\nabla}\otimes\boldsymbol{u})} - \rho\ddot{\boldsymbol{u}}. \tag{6}$$

One can readily show that if the condition

$$\boldsymbol{E} = \boldsymbol{E}^{(3412)}, \quad \boldsymbol{E} = \rho\frac{\partial^2\Psi}{\partial(\boldsymbol{\nabla}\otimes\boldsymbol{u})(\partial\boldsymbol{\nabla}\otimes\boldsymbol{u})}$$

holds, then the Fréchet derivative of the operator (6) is self-adjoint with respect to the convolution bilinear form. Note that this condition is equivalent to the condition that Ψ is two times differentiable and is usually accepted in continuum mechanics. Thus, one can obtain the functional in the form

$$\mathcal{I}[\boldsymbol{u}] = \langle \boldsymbol{u}, \int_0^1 \mathcal{A}[\lambda \boldsymbol{u}]\, d\lambda \rangle + \mathcal{R}_1 + \mathcal{R}_2 + \mathcal{R}_3,$$

or, in expanded form,

$$\mathcal{I}[\boldsymbol{u}] = \int_{\mathcal{B}}\int_{\tau^*(\boldsymbol{x})}^{t} \boldsymbol{u}(\boldsymbol{x}; t+\tau^*-\tau)\cdot\left[\int_0^1 \boldsymbol{\nabla}\cdot\rho\frac{\partial\Psi(\boldsymbol{x}, \lambda\boldsymbol{\nabla}\otimes\boldsymbol{u}(\boldsymbol{x};\tau))}{\lambda\partial(\boldsymbol{\nabla}\otimes\boldsymbol{u})}\, d\lambda - \frac{\rho}{2}\ddot{\boldsymbol{u}}(\boldsymbol{x};\tau) + \rho\boldsymbol{f}(\boldsymbol{x};\tau)\right] d\tau\, dV$$
$$+\frac{1}{2}\int_{\partial\mathcal{B}}\int_{\tau^*(\boldsymbol{x})}^{t} \boldsymbol{p}(\boldsymbol{x}; t+\tau^*-\tau)\cdot\boldsymbol{u}(\boldsymbol{x};t)\, d\tau\, dV - \frac{1}{2}\int_{\mathcal{B}} \boldsymbol{v}^*(\boldsymbol{x})\cdot\rho\dot{\boldsymbol{u}}(\boldsymbol{x};t)\, dV,$$
$$\boldsymbol{u}\in\mathcal{D}, \quad \delta\mathcal{I} = 0.$$

3. Conclusion

Thus, we have described the variational statement of the accretion problem. The further development of the present theory is based on the refinement of the boundary and initial operators $\mathcal{R}_2$ and $\mathcal{R}_3$. That permits one to take into account specific features of the accretion process such as the initial tension, phase transitions, etc. In conclusion, note that the same procedure allows us to obtain the variational statements of dynamic problems for growing thermoelastic bodies in linear and nonlinear approaches. This topic will be considered elsewhere.

Acknowledgments

The research was financially supported by the Russian Foundation for Basic Research (under grants Nos. 08-01-00553-a, 08-01-91302-IND a, and 09-08-01194-a) and by the Branch of Power Engineering, Mechanical Engineering, Mechanics, and Control Processes of the Russian Academy of Sciences (Program No. 13 OE).

References

[1] Arutyunyan N. Kh. and Manzhirov A. V. 1999 *Contact Problems of Creep Theory* [in Russian].
[2] Arutyunyan N. Kh., Manzhirov A. V., and Naumov V. E. 1991 *Contact Problems in Mechanics of Growing Bodies* [in Russian].
[3] Manzhirov A. V. 1990 *PMM [Appl. Math. Mech.]* **54:5** 842–50.
[4] Manzhirov A. V. and Chernysh V. A. 1992 *Mech. Solids* **5** 25–37.
[5] Manzhirov A. V. 1995 *PMM [Appl. Math. Mech.]* **59:5** 836–48.
[6] Manzhirov A. V. and Mikhin M. N. 2004 *Vestnik Samarsk. Un-ta. Natural Series* **4(34)** 82–98.
[7] Manzhirov A. V. and Parshin D. A. 2006 *Mech. Solids* **1** 66–83.
[8] Manzhirov A. V. and Parshin D. A. 2006 *Mech Solids* **6** 149–66.
[9] Manzhirov A. V. and Parshin D. A. 2006 *Indo-Russian workshop "Problems in nonlinear mechanics of solids with large deformation"* **6** 71–79.
[10] Manzhirov A. V. and Kazakov K. E. 2008 *Prace IPPT - IFTR Reports* **2** 248–249.
[11] Manzhirov A. V. 2008 *XXII International Congress of Theoretical and Applied Mechanics. CD-ROM Proceedings.*
[12] Gupta N K and Manzhirov A. V. Eds 2008 *Topical Problems in Solid Mechanics.*
[13] Truesdell C. and Toupin R. A. 1960 *Handbuch der Physik. The Classical Field Theories.* Band III/1 226–858.
[14] Gurtin M. E. 1964 *Arch. Rat. Mech.* **16:1** 34–50.
[15] Tonti E. 1973 *Annali di Matematica Pura ed Applicata* **95:1** 331–59.
[16] Belli G. and Morosi C. 1974 *Meccanica* **9:4** 239–43.
[17] Vainberg M. M. 1964 *Variational Methods in the Study of Nonlinear Operators.*

Smart machines with flexible rotors

A.W.Lees

Swansea University, Singleton Park, Swansea, SA2 8PP, UK.
a.w.lees@swansea.ac.uk

Abstract. The concept of smart machinery is of significant current interest. Several technologies are relevant in this quest including magnetic bearings, shape memory alloys (SMA) and piezo-electric activation. Recently a smart bearing pedestal was proposed based on SMAs and elastomeric O-rings. However, such a device is clearly relevant only for the control of rigid rotors, for flexible rotors there is a need for some modification on the rotor itself. In this paper, rotor actuation by piezo-electric patches on the rotor is studied. A methodology is presented for the calculation of rotor behaviour and an appropriate control strategy is developed.

1. INTRODUCTION

The concept of Smart Machinery is in its infancy and has received significantly less attention than has been devoted to the related study of Smart Structures. A smart machine is one in which there is some facility for automatic diagnosis of faults coupled with a capacity to apply corrective loads to optimize machine duty until such time as corrective maintenance can be undertaken. The precise form which these facilities take will depend significantly on the duty of the machine in question. The development of smart machines is inherently more complex than the associated structural problem owing to the rotor motion and problems associated with the bearings. However, in essence, the idealized machine has three main features:

a) The facility to infer its own internal state
b) A capability to diagnose faults
c) The introduction of corrective forces in the event of faults.

The simplest case to consider is a machine with a rigid rotor, such as occurs in a range of small machines. This is a particularly simple case to consider sing the natural frequencies (or more particularly, critical speeds) are controlled entirely by the stiffness of the bearing pedestals. Lees et al [1] have shown how a controllable bearing pedestal may be designed using Shape Memory Alloys and elastomers, although it is appreciated that there are a number of possible routes to achieve this goal. Other approaches to the introduction of controllable support stiffness have been considered by Zak et al [2] and Cartmell et al [3]. Recently Maslen [4] has given a brief review of progress on Smart Machinery, but this too focuses of bearings, with particular emphasis on magnetic bearings: this is not surprising as this technology offers great promise.

However, looking further ahead one is naturally led to consider the possibility of controlling machines with flexible rotors. This is important for two reasons: in machinery generally there is a trend towards the use of flexible rotors and secondly, adequate control would enable the operation of much light machines leading to higher efficiencies and better material utilization.

2 PIEZO-ELECTRIC PATCHES

Piezo-electric devices have been used for the control of structures in aero-space applications for a number of years. There are numerous papers on this and an introductory account is given in reference [5]. Control may be exerted by either a modification of natural frequencies or, more usually, by changing the damping properties. The initial part of this study considers the possible effect of such devices on a rotor. Signals may be input by means of slip-rings.

Consider a simple, single degree of freedom rotor system. We suppose there is a vibration amplitude a, and we may assume that the deflection is of the form

$$w(x) = 4\frac{x}{L}\left(1-\frac{x}{L}\right) \tag{1}$$

Using this deflection, the extension of the neutral axis is given by

$$\Delta L = \frac{1}{2}\int_0^L \left(\frac{dw}{dx}\right)^2 dx \tag{2}$$

However, the situation under consideration here is the introduction of a piezo-electric patch, not on the neutral axis, but on the extremity of the rotor. Assuming this to be displaced by r from the neutral axis, the extension here will be rather different and will be given by

$$\Delta S \approx \frac{1}{2}\int_0^L \left(\frac{dw}{dx}+r\frac{d^2w}{dx^2}\right)^2 dx \tag{3}$$

The establishing of a numerical methodology is a relatively straightforward task. What is rather more demanding is the determination of the most effective philosophy to use. In any event it is a fairly straightforward matter to supply the requisite electrical signals to the rotor and this can be achieved either using slip-rings or telemetry systems. Whilst the voltages will be high, the currents and consequently the power will be negligible.

There are four possible approaches to use the forces arising in the piezo-electric MFC patches which may be outlined as follows:

a) Use the forces to directly change the natural frequencies. This may be achieved by using opposite sides of the rotor to modify shaft properties. As shown below, this approach appear to be limited to very small machines.
b) Apply an axial force which will modify the rotor's properties – this is akin to the common changes of natural frequency associated with axial forces.
c) By suitable scaling and timing use the force to control damping rather than the stiffness.
d) Use the patches to induce rotor bends which can counteract the unbalance.

In the following sections, this paper examines which, if any, of these approaches offer any prospects of viability given typical performance parameters of currently available patches. It is clear that the discussion here must focus on light machine, but part of the aim must be to seek what is feasible. In the calculations which follow the prototype studied has a shaft of diameter 12mm and a span between the bearings of 1metre. The bearings each have symmetrical bearing properties with a stiffness parameter of 100,000 N/m and associated damping parameter of 800 Ns/m. The single central mass gives a resonant frequency of 124 rad/sec. and the system damping coefficient is 1.8%.

Figure 1
Basic layout showing central piezo-electric patches

3 INITIAL CALCULATIONS

Figure 1 shows the basic layout for a simple rotor. Piezo-electric multi-fibre composite patches are positioned at the centre of the rotor span, but for a system involving higher modes, further patches would be required. Depending on the phasing of the applied voltages, either forces or bending moments may be applied. By using two pairs of patches around the periphery of the rotor, resultant forces/moments can be applied at any phase angle with respect to the rotor.

For the light machine considered, Figure 2 shows the response as a function of speed for two distinct cases. The first is an unbalance at the central disc of magnitude 0.0001 kg.m, whilst the second trace shows the response due to a bend of magnitude 0.1mm at the central point. Both excitations have been given zero phase for the sake of convenience. Note that although the response is synchronous in both cases, the response curve is rather different. For the unbalance case, the response for a steady shaft rotation rate Ω is given by

$$\mathbf{K}y + \mathbf{M}\ddot{y} = me\Omega^2 \exp\left(j\Omega t\right) \tag{4}$$

The response due to the bend is given by

$$\mathbf{K}y + \mathbf{C}\dot{y} + \mathbf{M}\ddot{y} = Ky_{bend} \exp\left(j\Omega t\right) \tag{5}$$

Simple calculations based on Rayleigh methods using (1) as a trial function that is is difficult to affect a substantial change in natural frequency by the application of signal to the piezo-electric patches. They can be used to impose an axial force on the rotor, but calculations reveal that the influence on critical speeds of credible forces is fairly minor.

4. FORCED RESPONSE

It is clear from both the form of equations 4 and 5, that by adjusting phase, an imbalance may be used to counteract a rotor bend, but only at some selected speed (see [6] for a more detailed discussion). This is often done in practice, but in the present study we examine the converse situation, namely inducing a bend to counteract an unbalance. Figure 3 shows the response of the system in which a bend has been imposed to exactly cancel the imbalance response.

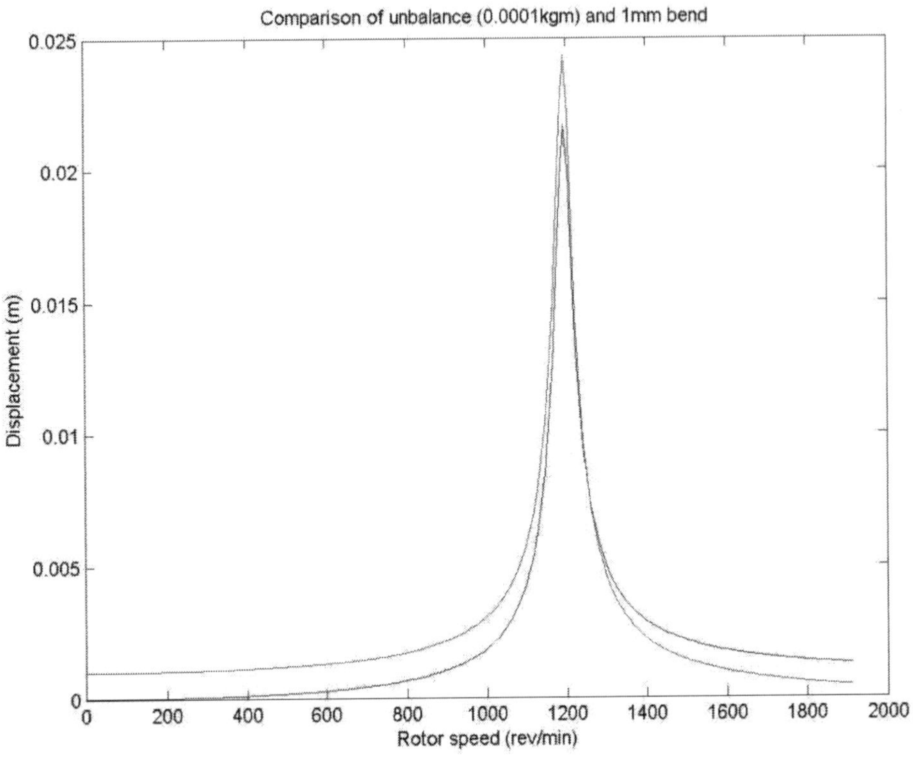

Figure 2

In the rotor fitted with a set of offset piezo-electric patches, a bend can be induced by effectively exerting bending moments about two nodal positions. In the current arrangement (shown in Figure 4) moments of P and –P were applied about neighbouring nodes and Figure 5 shows the steady state bends induce by these moments.

However, because in the case of this study, P can be time dependent, it is clear that the bend formation cannot occur instantaneously and there is inertia in the system. For the rotor, spinning at Ω, when a force $P(t)$ is applied, the development of the bend will be described by the equation

$$\mathbf{K}y_{bend} - \Omega^2 \mathbf{M}y_{bend} + \mathbf{M}\ddot{y}_{bend} = P(t) \tag{6}$$

The objective is now to solve the equations simultaneously during a machine run up or rundown. The forcing function P may be chosen. The equations are integrated using a Runge-Kutta scheme in Matlab. To do this, the equations are transformed to a state-space formulation using the variables $[y\ y_{bend}\ \dot{y}\ \dot{y}_{bend}]^T$. In this formulation, the equation of motion becomes

$$\mathbf{A}q + \mathbf{B}\dot{q} = F \tag{7}$$

To examine this in more detail we need to distinguish between the stationary and rotating frame. Let $\tilde{y}$ denote deflection in the rotating frame (i.e. stationary with respect to the rotor) then $y = \mathbf{R}\tilde{y}$ where the transformation **R** takes the diagonal block form

$$\mathbf{R} = \begin{bmatrix} r & 0 & \vdots & 0 \\ 0 & r & \vdots & 0 \\ \cdots & \cdots & \ddots & \vdots \\ 0 & 0 & 0 & r \end{bmatrix} \quad \text{and} \quad r = \begin{bmatrix} \cos\theta & \sin\theta & 0 & 0 \\ -\sin\theta & \cos\theta & 0 & 0 \\ 0 & 0 & \cos\theta & -\sin\theta \\ 0 & 0 & \sin\theta & \cos\theta \end{bmatrix} \qquad (8\ \&\ 9)$$

θ being the instantaneous position of the rotor. Clearly for the case of rotation at constant speed Ω, $\theta = \Omega t$.

The equations of motion for the complete system, whose speed varies with time, in the presence of both imbalance and activation producing a bend, can now be written as

$$\begin{aligned} &\mathbf{K}y + \mathbf{C}\dot{y} + \mathbf{M}\ddot{y} = \mathbf{KR}(t)\, y_{bend} + ma\Omega^2 e^{j\Omega t} - jma\frac{d\Omega}{dt}e^{j\Omega t} \\ &\mathbf{K}y_{bend} - \Omega^2\mathbf{M}y_{bend} + \mathbf{M}\ddot{y}_{bend} = P(t) \end{aligned} \qquad (10)$$

These equations assume shaft symmetry. They may be written in state-space form as

$$\begin{bmatrix} 0 & 0 & -K & 0 \\ 0 & 0 & 0 & -K \\ -K & -KR & -C & 0 \\ 0 & -\Omega^2 M - K & 0 & -C \end{bmatrix} \begin{Bmatrix} y \\ \tilde{y}_{bend} \\ \dot{y} \\ \dot{\tilde{y}}_{bend} \end{Bmatrix} + \begin{bmatrix} K & 0 & 0 & 0 \\ 0 & K & 0 & 0 \\ 0 & 0 & -M & 0 \\ 0 & 0 & 0 & -M \end{bmatrix} \frac{d}{dt} \begin{Bmatrix} y \\ \tilde{y}_{bend} \\ \dot{y} \\ \dot{\tilde{y}}_{bend} \end{Bmatrix} = \begin{Bmatrix} 0 \\ 0 \\ ma\Omega^2 e^{j\Omega t} + jma\dot{\Omega} e^{j\Omega t} \\ P(t) \end{Bmatrix} \qquad (11)$$

Of course, the effective phases of the deflection arising from an unbalance and a bend is somewhat involved and due recognition must be made of this. It can be shown that the combined deflection due to the two may be limited to the extent of the bend and figure 3 shows the case in which the magnitude and phase of the bend have been chosen to exactly balance at a set rotation speed.

It is easily shown that a bend and a discrete unbalance can exactly compensate at a single rotational speed, and in this case the natural frequency has been chosen as the speed at which the response is minimized. However, in the case under study here, because actuators are on the rotor, the response may, in principle be minimized at all speed simply be varying the voltages to the piezo patches. As this involves not only magnitude but phase, the control of this is non-trivial and a full discussion is beyong the scope of the present paper. However we consider now taking a fixed unbalance by varying the amplitude of Px, the applied torque between 0 and 200 Nm (forces of the corresponding magnitudes may realistically be obtained from commercially available patches, albeit with the application of considerable voltages. The phase angle between the unbalance and the bend is varied between 0 and 360 in steps of 10 degrees. The response is shown in figure 4.

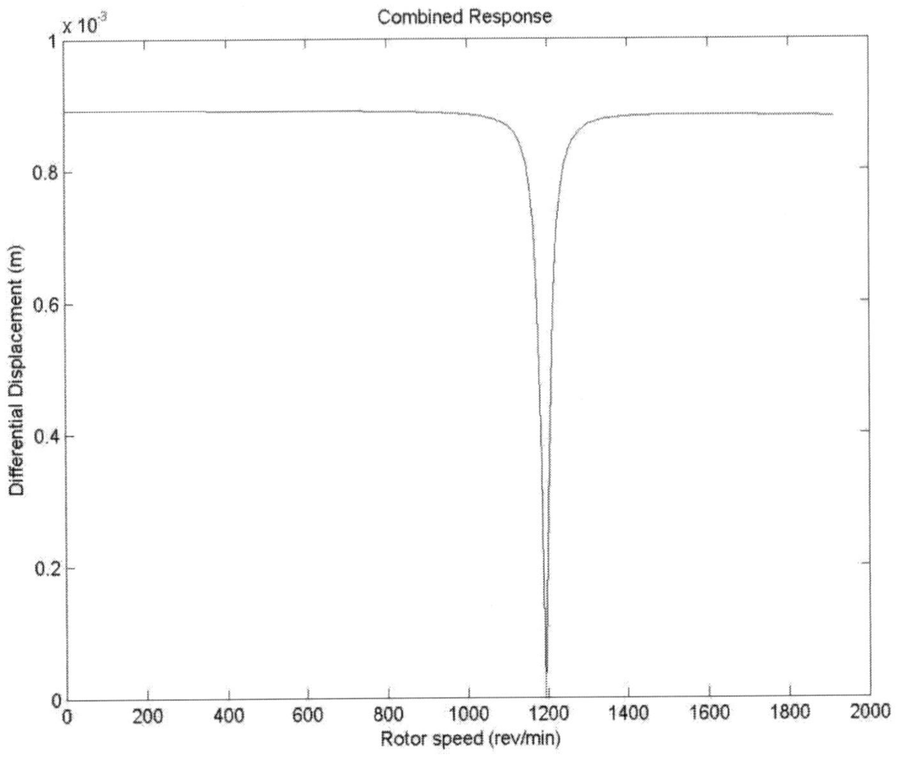

Figure 3

The point to note here is that although the unbalance used in this simulation is very high, the vibration can be reduced to negligible levels y the application of a torque of 10Nm. Since individual patches can apply 500N and more, such levels of torque are readily obtainable.

It is now important to establish the dynamic response of such a controlling parameter. The results of a preliminary calculation are shown in figure 5. The piezo-electric patches are activated after 2 seconds and a moment of 5Nm is applied to the central portion of the shaft. Note that this comfortably within the range of realistic parameters. The change in the vibration response is almost immediate but in this case, the response has not been reduced to zero. This is because the phase computation has not yet been refined. Note that a substantial change in the vibration level has resulted.

5. DISCUSSION

Interest in Mechatronics and the concept of 'Smart Machines' has developed of the past few years. But it has so far focused on the properties of bearings and support structures. Whilst this is appropriate for machines with rigid rotors, the overall trend in machinery is the enhanced use of designs involving flexible rotors. Such rotors require some controlling element to be mounted on the shaft. The voltage applied to piezo-patches can be very considerable, of order hundreds of volts (dc) albeit at negligible current levels. One might envisage that the signal would be fed to the rotor via a slip-ring arrangement: the question as to whether the signal would be at full voltage or amplified on the rotor is a question for future development.

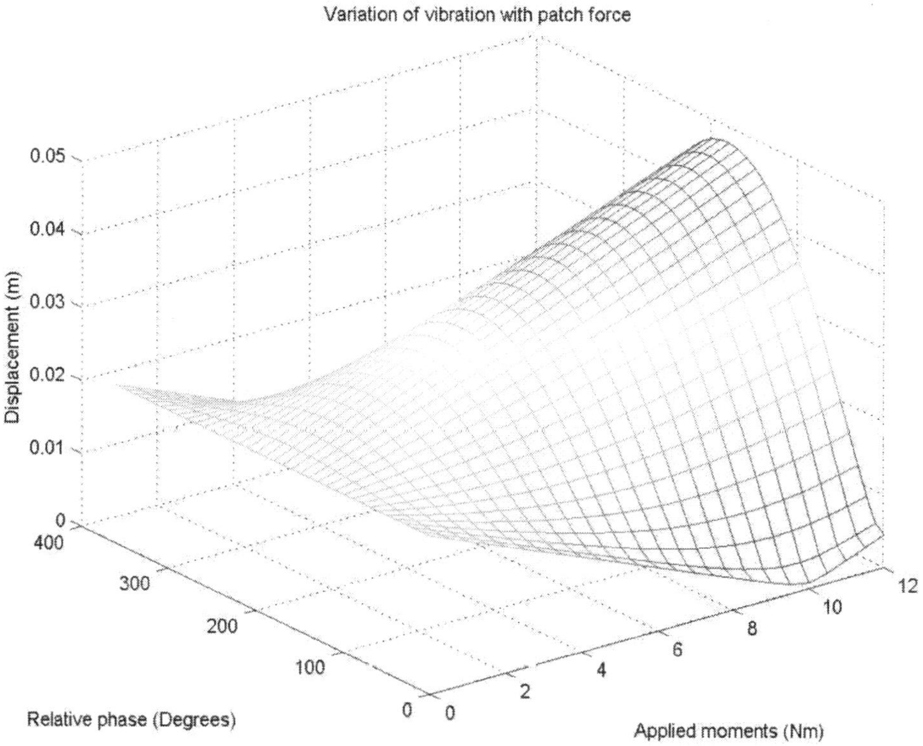

Figure 4

The work presented in this paper is very exploratory. Nevertheless a potential means of controlling flexible rotors has been outlined. In essence the control is effected by imposing a bend to compensate for the imposed imbalance. It is well known that a bend can only be balanced out at a single rotational speed, but in the present case the extent of the bend can be varied by the voltage applied to the piezo-electric patch.

The sample machine taken here is by no means typical: it is extremely flexible but if a machine can be controlled in such a manner, it becomes realistic to operate these devices. This paper has outlined some basic considerations and formulated a method of analysis. Furthermore the possibility of actively controlling a flexible rotor in this way has been established. To develop this idea further the following steps need to be undertaken

a) A review of the range of machine dimensions which would be feasible for this type of technology.
b) Derivation of an appropriate control algorithm
c) The construction and testing of a device

6. CONCLUSIONS

1. It is more realistic to control the forcing on a rotor rather than the dynamic properties. Calculations show that it is difficult to introduce substantial changes in natural frequency by the use of piezo-electric patches. It has been shown that is feasible to counteract an unbalance with an induced bend.

Journal of Physics: Conference Series **181** (2009) 012019 doi:10.1088/1742-6596/181/1/012019

2. Some simple cases have been examined and shown to be feasible. Whilst this is primarily of interest in small machines, the full range of applicability has yet to be established. Note, however, that if the constraint of avoiding critical speeds can be removed then machines can be made much smaller and lighter.
3. An analysis approach has been presented.

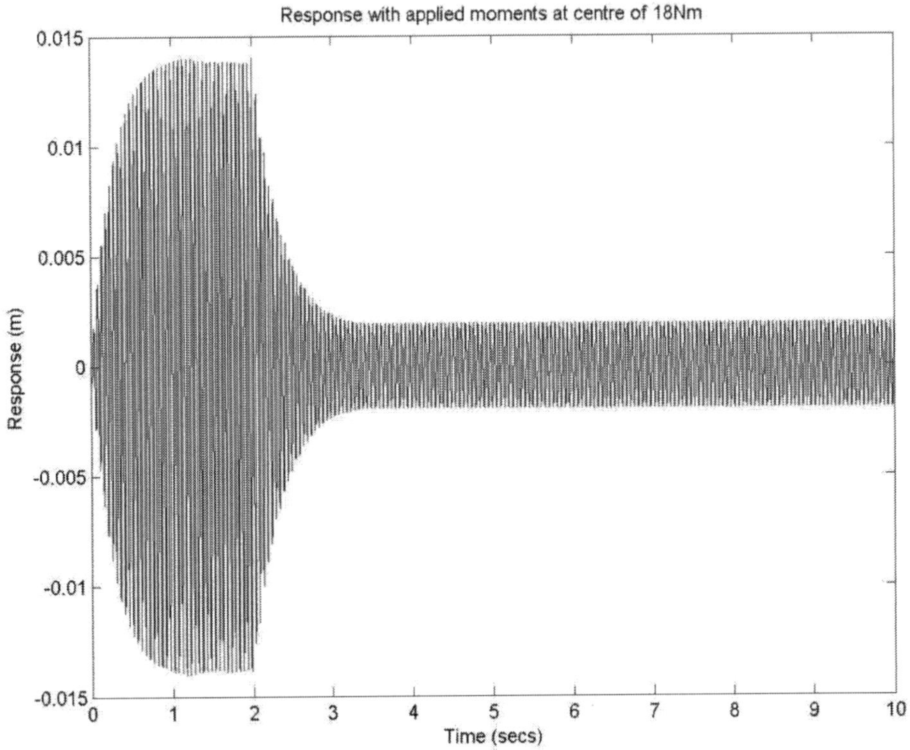

Figure 5

7. REFERENCES

1. A W Lees, S Jana, D J Inman and M P Cartmell, The Control of Bearing Stiffness Using Shape Memory, ISCORMA, Calgary, Alberta, August 2007
2. AJ Zak, MP Cartmell. and WM Ostachowicz, 2003, "Dynamics and Control of a Rotor Using and Integrated SMA/Composite Active Bearing Actuator, *5th International Conference on Damage Assessment of Structures*, University of Southampton, UK, pp 233-240.
3. M.P.Cartmell, Th.Leize, L.Atepor, D.J.Inman, and A.W.Lees, Controlling Flexible Rotor Vibrations by means of an Antagonistic SMA/Composite Smart Bearing, IoP Conference, Modern Practice in Stress and Vibration Analysis, Bath, UK, July 2006.
4. E H Maslan, Smart machine advances in rotating machinery, 9th International Conference on Vibrations in Rotating Machinery, I.Mech.E., Exeter, UK, 8-10 September 2008.
5. M Brennan, 1998, "Experimental Investigation of Different Actuator Technologies for Active Vibration Control", Smart Material Structures, Institute of Physics Publishing Ltd.
6. MI Friswell, JET Penny, SD Garvey and AW Lees , "Introduction to Rotor Dynamics: Modeling and Computation", to be published, Cambridge University Press, 2009.

Journal of Physics: Conference Series **181** (2009) 012020 doi:10.1088/1742-6596/181/1/012020

Finite element analysis of a soft mounted 2-pole asynchronous machine concerning vibration stability

U Werner

Siemens AG, Industry, Drive Technologies, Large Drives, Industry Development, Vogelweiherstr. 1-15, 90441 Nuremberg, Germany

E-mail: werner.ulrich@siemens.com

Abstract. The paper describes a finite element calculation concerning vibration stability of a 2-pole asynchronous machine with sleeve bearings and flexible shaft, mounted on soft foundation elements. The finite element model considers the rotor, stator, end shields and bearing housings, oil film stiffness and damping, the electromagnetic interaction between rotor and stator – deduced from an electromagnetic field calculation –, the gyroscopic effect of the rotor and the influence of the foundation stiffness and damping. The critical mode shapes and natural frequencies are calculated for different rotor speeds and the eigenvalues are analyzed to receive the limit of stability. Therefore the paper shows a calculation method to optimize the vibration stability by considering foundation stiffness and damping.

1. Introduction

In the past foundations of rotating machinery for industrial applications were often designed stiff. But nowadays they become softer, because of plant specific requirements. Therefore the foundation influences the vibration system [1, 3] and therefore has to be taken into consideration. Figure 1 shows exemplary a 2-pole asynchronous machine, mounted on soft foundation elements. The 2-pole asynchronous machine shown in figure 1, with a power output of 2 MW, is designed with a flexible shaft and hydrodynamic sleeve bearings, with cylindrical bearing shells. By analyzing the vibration behavior, not only the critical speeds have to be considered, but also the limit of vibration stability, mainly caused by the oil film characteristics.

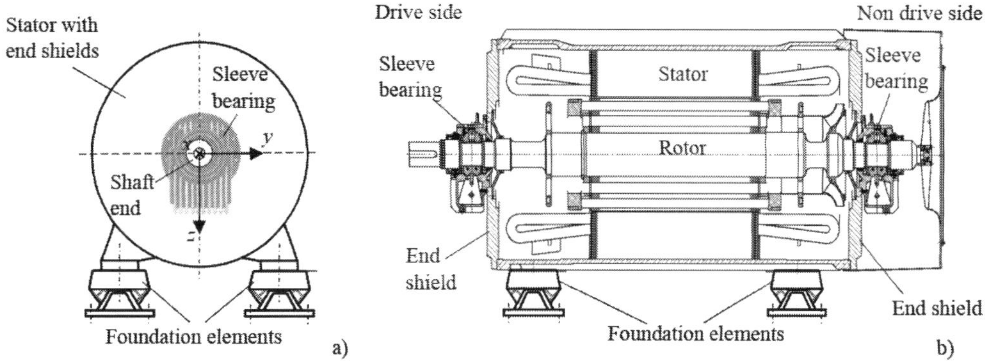

Figure 1: Soft mounted asynchronous machine: a) view of the shaft end; b) longitudinal section

Raising the rotor speed n above the limit of stability, self excited vibrations will occur, which disturb the operation [1, 2]. The specialty of asynchronous machines is the additional electromagnetic coupling between rotor and stator by an electromagnetic field [6-9], which also influences the vibration system and therefore the limit of stability. So the whole vibration system, consisting of the rotor, stator, sleeve bearing, electromagnetic field and the foundation has to be taken into consideration.

2. Data of the 2-pole asynchronous machine and the foundation elements

The data of the asynchronous machine and the foundation elements are described in table 1. In this case the 2-pole asynchronous machine is mounted on soft foundation elements. The longitudinal section of the machine and the soft foundation elements are shown in figure 1.

Table 1. Data of the 2-pole asynchronous machine and the foundation elements

Notation	Symbol	Data
Rated power	P_N	2000 kW
Number of pole pairs	p	1
Rated voltage	U_N	6000 V
Rated frequency	f_N	50 Hz
Rated torque	M_N	6.4 kNm
Rated speed	n_N	2990 r/min
Mass of the stator	m_1	7200 kg
Mass of the rotor	m_2	1900 kg
Type of sleeve bearing	-	Side flange bearing
Shape of the bearing shell	-	Cylindrical
Lubricant viscosity grade of the oil	-	ISO VG 32
Nominal bore diameter of the bearing shell	d_b	110 mm
Bearing width	b_b	81.4 mm
Ambient temperature	T_{amb}	25°C
Lubricant supply temperature of the oil	T_{in}	40°C
Mean relative bearing clearance (DIN 31698)	Ψ_m	1.6 ‰
Vertical stiffness of one foundation element	c_{fz}	10 kN/mm
Horizontal stiffness of one foundation element	$c_{fy} = c_{fx}$	13 kN/mm
Damping ratio of the foundation element	D_f	0.07

3. Finite Element model of the soft mounted asynchronous machine

To calculate the natural vibrations and the limit of vibration stability of the soft mounted asynchronous machine a finite element model is deduced, using the finite element program MADYN. For the modeling of the rotor, which consists of a shaft and a rotor core with squirrel cage, linear beam elements (Timoshenko-beam elements) are used, considering bending and shear. The hydrodynamic oil film is considered by the oil film stiffness matrix $\mathbf{C}_v$ and the oil film damping matrix $\mathbf{D}_v$, which contain the oil film stiffness coefficients (c_{yy}, c_{yz}, c_{zy}, c_{zz}) and the oil film damping coefficients (d_{yy}, d_{yz}, d_{zy}, d_{zz}). These dynamical coefficients are calculated by solving the REYNOLDS Differential equation [4] for each rotor speed, using the calculation program SBCALC from RENK AG. The structural stiffness of the end shield and the sleeve bearing house is considered as a translational spring element (c_{bx}, c_{by}, c_{bz}), which is deduced from a separate finite element calculation. The damping of the end shield and bearing house is neglected, because of the low damping ratio. The stator mass m_1 with its inertias (Θ_{1x}, Θ_{1y}, Θ_{1z}) is concentrated in the centre of gravity. The stator structure is modeled with rigid beam elements, representing a rigid stator structure. This is possible, because the foundation stiffness is assumed to be much lower than the stator structure, which is typical for a soft mounting. The foundation stiffness (c_{fx}, c_{fy}, c_{fz}) and damping (d_{fx}, d_{fy}, d_{fz}) coefficients are considered by a translational spring and damper element under each machine foot. The magnetic spring constant c_m, which describes the magnetic stiffness, – representing the electromagnetic interaction between rotor and stator – is distributed over the core length between the rotor and the stator. The magnetic spring constant c_m is derived by a separate finite element calculation (section 4).

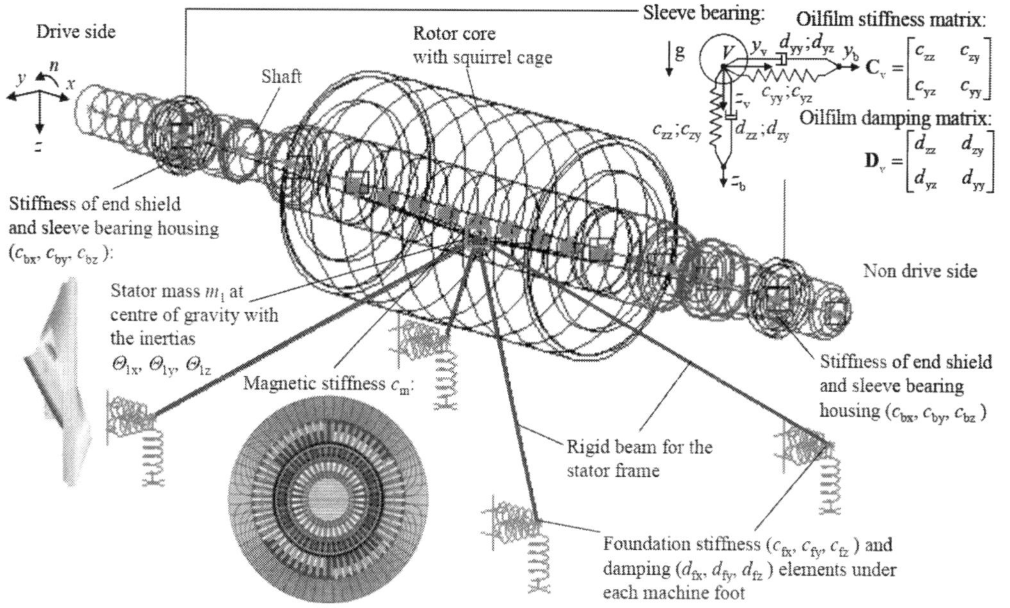

Figure 2: Finite element model of the soft mounted asynchronous machine

4. Finite Element calculation of the magnetic spring constant

To consider the influence of the electromagnetic field, an electromagnetic finite element calculation is deduced, using the two-dimensional electromagnetic field calculation program FEMAG. The electromagnetic model, shown in figure 3a, consists of the stator core containing the three-phase winding, the rotor core with the rotor bars and the cooling slots, the air gap and the rotor shaft. The calculations are done for no-load operation and with rated magnetizing current in the stator winding as static field solution. For the calculation the magnetizing current is impressed as direct current in the stator windings, so no induction in the rotor cage occurs and therefore no field damping by the rotor cage is considered.

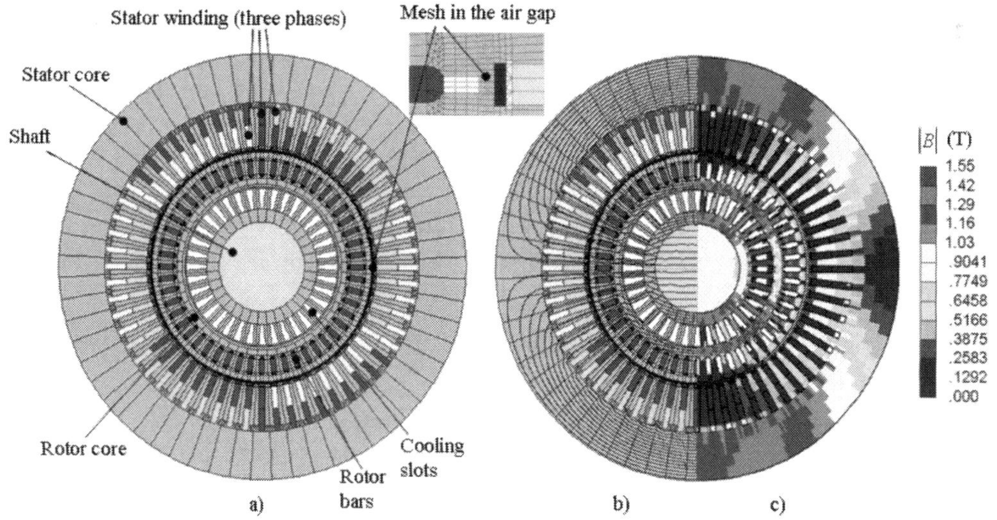

Figure 3: Finite element field calculation of the 2-pole asynchronous machine with a centric rotor; a) Model; b) Flux lines of the field; c) Flux density (absolute value)

Figure 3b shows the field lines and figure 3c the flux density for a certain field position, with a centric rotor. Afterwards the rotor is displaced by different eccentricities e in the two-dimensional model – normal to the field and parallel to the field – and the resulting radial magnetic forces F_r are computed. Without considering field damping effects, these radial magnetic forces are maximal. Field damping leads to lower radial magnetic forces but causes tangential magnetic forces. Mostly these tangential magnetic forces are so low, that they can be neglected for the vibration analysis. The displacement e of the rotor is related to the air gap width δ_0, so a relative displacement $\varepsilon = e/\delta_0$ is derived. The relative eccentricity is analyzed in the range of $\varepsilon = 0 \ldots 0.5$. Figure 4 shows, that for rated magnetization current, there is nearly a linear behavior between the relative eccentricity and the radial magnetic force and there is nearly no difference concerning the direction of the eccentricity, up to an eccentricity of about $\varepsilon = 0.2$ [6, 9].

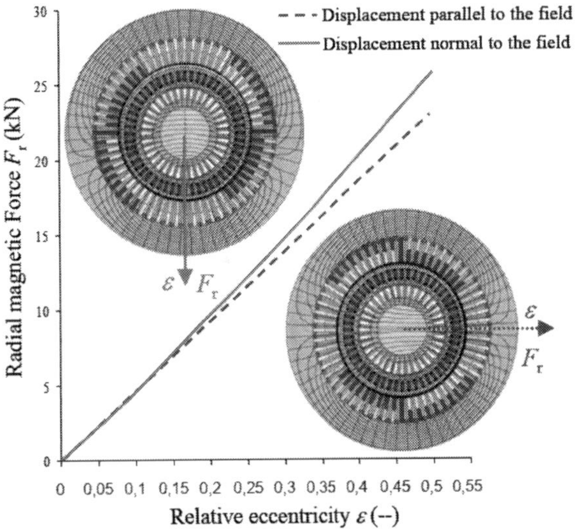

Figure 4: Radial magnetic force F_r for different relative eccentricities ε of the rotor, at rated magnetization current (field lines pictured for $\varepsilon = 0.5$)

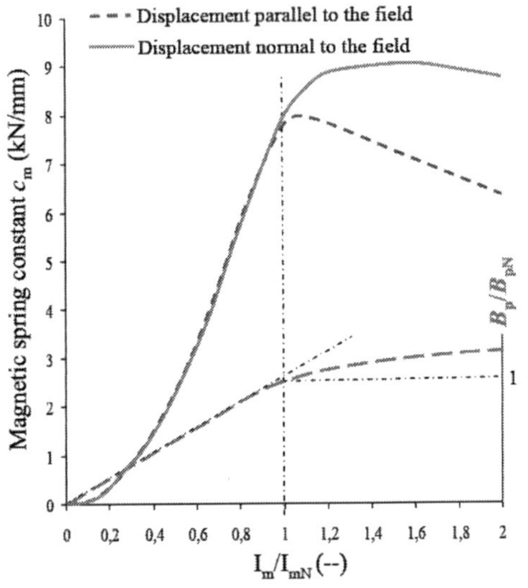

Figure 5: Magnetic spring constant c_m and ration of B_p/B_{pN} for different ratios I_m/I_{mN} (relative eccentricity $\varepsilon = 0.1$)

With raising the relative eccentricity ε far above 0.2 the radial magnetic force is influenced by the direction of the eccentricity. In practice, the relative eccentricity will be less than $\varepsilon < 0.1$ for this kind of asynchronous machine. Therefore a linear behavior between eccentricity and radial magnetic force can be assumed and a magnetic spring constant c_m deduced [6, 9].

$$c_m = F_r / e \tag{1}$$

The magnetic force is acting in the direction of the eccentricity, so the magnetic stiffness is a "negative" stiffness. Figure 5 shows the influence of the magnetic saturation on the magnetic spring constant c_m. The magnetic saturation is described by the ratio of the magnetic induction of the fundamental field B_p to the rated magnetic induction of the fundamental field B_{pN}. Up to the rated magnetization current I_{mN}, there is nearly a linear relation between the stator current and the magnetic induction. Raising the current above the rated value, the saturation of the iron parts – yoke and teeth – increases strongly and leads to a non-linear behavior. Far above the rated magnetization current the magnetic spring value decreases with rising stator current, because the high saturation leads to an increasing "magnetic" air gap. At the rated conditions, the value of the magnetic spring c_m is here 7.6 kN/mm.

5. Oil film stiffness and damping coefficients

The characteristic of a hydrodynamic oil film is well known [1-5] and is shown in Fig. 6. The movement of the journal shaft is described by the vertical displacement z_v and the horizontal displacement

y_v. The movement of the sleeve bearing housing is described by the vertical displacement z_b and the horizontal displacement y_b.

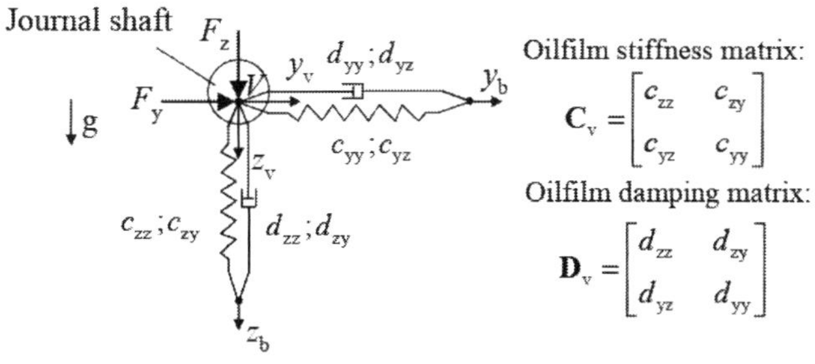

Figure 6: Model for hydrodynamic oil film

The coupling coefficients – stiffness coupling coefficients c_{yz}, c_{zy} and damping coupling coefficients d_{yz}, d_{zy} – cause a coupling between vertical and horizontal movement and the vertical oil film force F_z and the horizontal oil film forces F_y, which is mathematically described in (2).

$$\begin{bmatrix} F_z \\ F_y \end{bmatrix} = \begin{bmatrix} c_{zz} & c_{zy} \\ c_{yz} & c_{yy} \end{bmatrix} \cdot \begin{bmatrix} z_v - z_b \\ y_v - y_b \end{bmatrix} + \begin{bmatrix} d_{zz} & d_{zy} \\ d_{yz} & d_{yy} \end{bmatrix} \cdot \begin{bmatrix} \dot{z}_v - \dot{z}_b \\ \dot{y}_v - \dot{y}_b \end{bmatrix} \tag{2}$$

The stiffness and the damping coefficients of the oil film can be calculated by solving the Reynolds-Differential-Equations [4]. The oil film stiffness coefficients (c_{yy}, c_{yz}, c_{zy}, c_{zz}) and the oil film damping coefficients (d_{yy}, d_{yz}, d_{zy}, d_{zz}) of the sleeve bearings are strongly influenced by the rotor speed n, which is shown in figure 7. For this example they are nearly identical for both bearings. Figure 6 shows clearly, that the coupling damping coefficients are equal ($d_{zy} = d_{yz}$) but the coupling stiffness coefficients are not equal ($c_{zy} \neq c_{yz}$). This leads to an asymmetric oil film stiffness matrix $\mathbf{C}_v$, which is the reason that vibration instability may occur [1, 2].

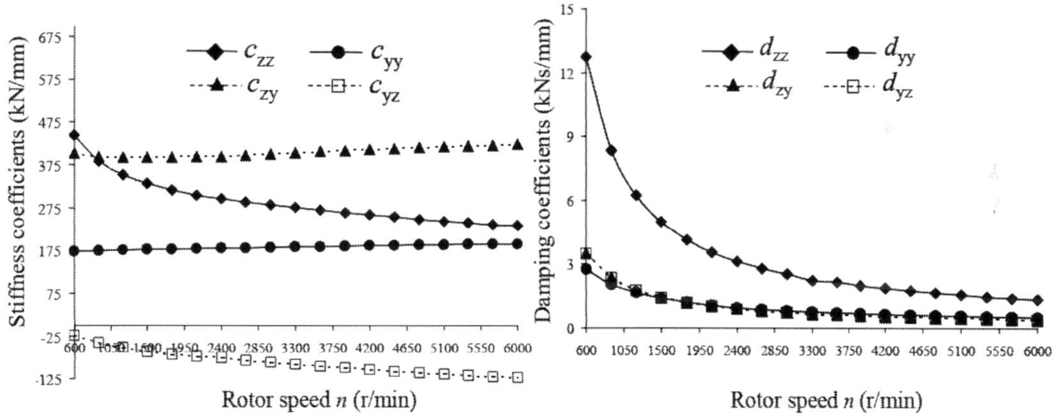

Figure 7: Oil film stiffness and damping coefficients for different rotor speeds

6. Natural vibrations and vibration stability

With the vibration model in figure 2 the natural vibration and the limit of stability can be calculated. Therefore the homogenous differential equation (3), with the mass matrix **M**, the damping matrix **D**, the gyroscopic matrix **G**, the stiffness matrix **C** and the coordinate vector **q**, has to be solved [1].

$$\mathbf{M} \cdot \ddot{\mathbf{q}} + (\mathbf{D} + \mathbf{G}) \cdot \dot{\mathbf{q}} + \mathbf{C} \cdot \mathbf{q} = \mathbf{0} \tag{3}$$

The mathematical ansatz

$$\mathbf{q} = \hat{\mathbf{q}} \cdot e^{\lambda \cdot t} \tag{4}$$

with the eigenvalue λ and the eigenvector $\hat{\mathbf{q}}$, leads to the eigenvalue equation

$$[\mathbf{M}\cdot\lambda^2+(\mathbf{D}+\mathbf{G})\cdot\lambda+\mathbf{C}]\cdot\hat{\mathbf{q}}=\mathbf{0}\ . \tag{5}$$

By solving the determination equation the eigenvalues λ_n can be computed. The eigenvalues occur mostly conjugated complex [1]

$$\lambda_n=\alpha_n\pm i\cdot\omega_n\ , \tag{6}$$

with the real parts $\mathrm{Re}\{\lambda_n\}=\alpha_n$, which describe the decay of each natural vibration and the imaginary parts $\mathrm{Im}\{\lambda_n\}=\omega_n$, which describe the corresponding natural angular frequencies. With this eigenvalues λ_n the corresponding eigenvectors $\hat{\mathbf{q}}_n$ can be calculated. Therefore the solution of the homogenous deferential equation can be described by:

$$\mathbf{q}=\sum_n\hat{\mathbf{q}}_n\cdot k_n\cdot e^{\lambda\cdot t} \tag{7}$$

To investigate the vibration stability the real parts α_n of the complex eigenvalues λ_n have to be analyzed. If one or more real part gets positive, then vibration instability occurs. To get the limit of stability the rotor speed n is raised until at least one real part gets zero. At the limit of stability, that means at the rotor speed n_{limit}, the undamped mode (with $\alpha_{limit}=0$) oscillates with the natural angular frequency of ω_{limit}, as a self exciting vibration [1, 2].

6.1. Limit of vibration stability without foundation damping

By solving the differential equations and analyzing the real part of the complex eigenvalues, the critical mode shape can be found, which gets instable first of all. The other natural vibrations are not getting instable or only at higher rotor speed. Foundation damping is here neglected ($D_F=0$), therefore some natural vibration exist, which have a real part near to zero (e.g. the rigid body modes), but they are here not considered, because a marginal foundation damping would last, that these natural vibrations are not getting instable. The mode shape of the critical natural vibration at the limit of stability is shown in figure 8. Rotor and stator are moving on elliptical orbits and both orbits are run through in the same direction as the rotor rotates. Therefore both orbits are elliptical forward whirls. The mode shape shows additionally, that rotor and stator are vibrating contrary to each other.

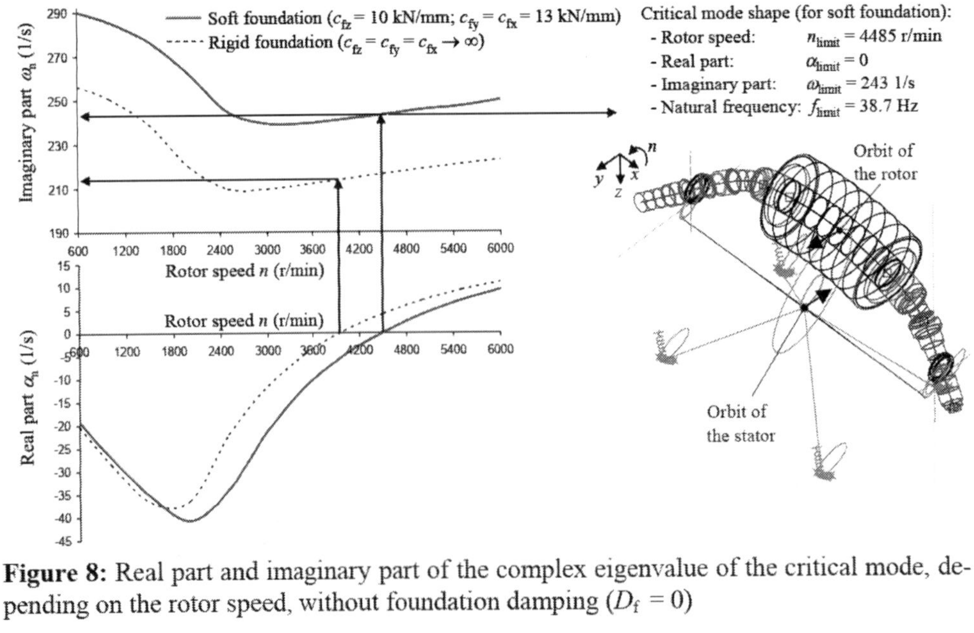

Figure 8: Real part and imaginary part of the complex eigenvalue of the critical mode, depending on the rotor speed, without foundation damping (D_f = 0)

Journal of Physics: Conference Series **181** (2009) 012020 doi:10.1088/1742-6596/181/1/012020

To find the limit of stability n_{limit}, the rotor speed n is raised and the real part of the complex eigenvalue is analyzed. The real part of the complex eigenvalue gets zero ($\alpha_n = \alpha_{\text{limit}} = 0$) at about a rotor speed of $n = n_{\text{limit}} = 4485$ r/min. At this rotor speed, the imaginary part of the complex eigenvalue gets $\omega_n = \omega_{\text{limit}} = 243$ 1/s and the undamped natural frequency becomes $f_n = f_{\text{limit}} = 38.7$ Hz. So the limit of stability occurs at a rotor speed of 4485 r/min. Above this rotor speed, vibration instability occurs. For comparison also a rigid mounting is analyzed. Figure 8 shows clearly, that the limit of stability for the rigid mounted machine ($c_{\text{fx}} = c_{\text{fy}} = c_{\text{fz}} \to \infty$) is reached at a lower rotor speed $n_{\text{limit}} = 3950$ r/min. Here the natural frequency of the critical mode becomes $f_{\text{limit}} = 34.0$ Hz. The vibration shape of the rotor is quite similar to the vibration shape for the soft mounted machine, but without movement of the stator mass, because of the rigid foundation. Therefore the limit of stability can be raised from 3950 r/min to 4485 r/min by putting the asynchronous machine on the soft foundation elements, without considering the damping of the soft foundation elements ($D_{\text{f}} = 0$), which will lead to a still higher limit of stability (section 6.2). The reason for this increase from 3950 r/min to 4485 r/min is the increase of the natural frequency of the critical mode (from 34.0 Hz to 38.7 Hz). The ratio $n_{\text{limit}}/f_{\text{limit}}$ stays nearly the same.

6.2. Influence of foundation damping

In section 6.1 foundation damping was not considered. In this section the influence of the foundation damping on the limit of stability will be shown. In practice, it is often very difficult to derive the real damping of the foundation and often only possible by measurements. But to show the influence of the foundation damping on the limit of stability, different values for the damping coefficients of the foundation are considered here in the example. Referring to [1], the damping coefficients of the foundation under each motor foot d_{fz}, d_{fy}, d_{fx} are here described by the damping ratio D_{f}, the stiffness coefficients c_{fz}, c_{fy}, c_{fx} of the foundation under each motor foot and the stator mass m_1, as a rough simplification.

$$d_{\text{fq}} = D_{\text{f}} \cdot \sqrt{c_{\text{fq}} \cdot m_1} \qquad \text{with: } q = z, y, x \tag{8}$$

Figure 9 shows, that in the range from $D_{\text{f}} = 0...0.15$ the limit of stability n_{limit} can be raised by increasing the foundation damping. The natural frequency f_{limit} of the critical mode shape is only marginal influenced by the foundation damping in the analyzed range. For the rated foundation damping ratio ($D_{\text{f}} = 0.07$), the limit of stability can be shifted to a rotor speed of about 4705 r/min. Compared to a rigid mounting the limit of stability can be increased from 3950 r/min to 4705 r/min by putting the machine on the soft foundation elements. But also the natural frequency of the critical mode shape is shifted to higher values, which is also necessary to consider for the operation speed range.

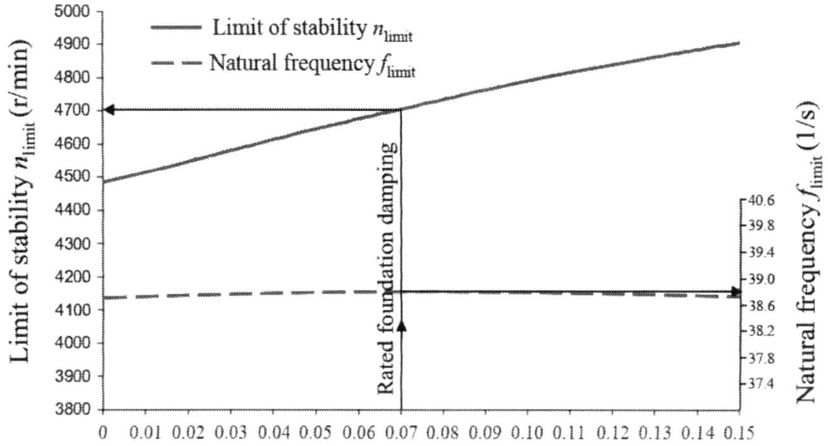

Figure 9: Limit of stability n_{limit} and natural frequency f_{limit} depending on the foundation damping ratio D_{f}

6.3. Influence of magnetic spring constant c_m

Finally the influence of the magnetic spring constant c_m on the limit of stability is analyzed. Figure 10 shows, that with decreasing the magnetic spring constant, the limit of vibration stability is raised. The reason is that a lower magnetic spring constant c_m – e.g. caused by the field damping of the rotor cage – leads to a higher natural frequency of the critical mode shape and therefore to a higher limit of stability. Calculating with the undamped magnetic spring constant is therefore a worst case analysis.

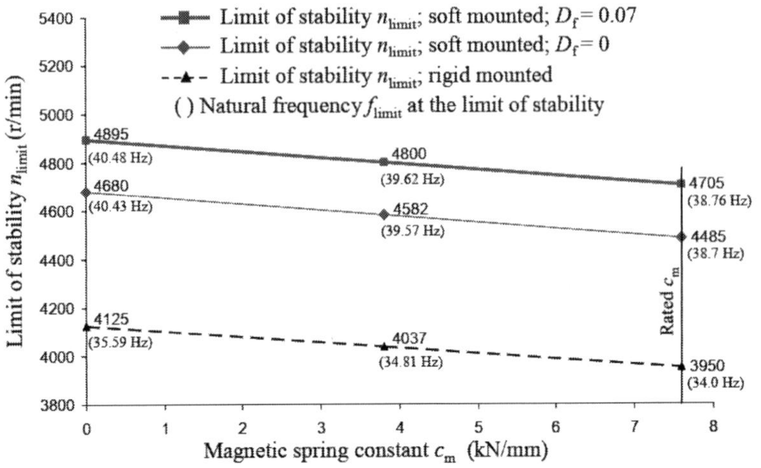

Figure 10: Limit of stability n_{limit} depending on the magnetic spring c_m

7. Conclusion

The paper shows a method for calculating vibration stability of soft mounted asynchronous machines with sleeve bearings and flexible shafts, using the finite element method for the vibration analysis and for the electromagnetic field calculation. It is shown, based on an example, how the limit of vibration stability can be influenced by the foundation stiffness and damping. Additionally the influence of the electromagnetic field on the vibration stability is also derived.

References

[1] Gasch R, Nordmann R and Pfützner H 2002 *Rotordynamik* (Berlin: Springer)
[2] Rao J S 1991 *Rotor Dynamics* (New Delhi: Wiley Eastern Limited)
[3] Kirk R G, DeChowdhury P and Gunter E J 1974 *The effect of support flexibility on the stability of rotors mounted in plain cylindrical bearings* (Proc. IUTAM Symposium Dynamics of Rotors) p 244
[4] Reynolds O 1886 *On the Theory of Lubrication* (Philosophical Transaction of the Royal Society, Vol. 177, London)
[5] Lund J and Thomsen K 1978 *A calculation method and data for the dynamics of oil lubricated journal bearings in fluid film bearings and rotor bearings system design and optimization* (ASME, New York) pp 1-28
[6] Seinsch H O 1992 *Oberfelderscheinungen in Drehfeldmaschinen* (Stuttgart: Teubner)
[7] Belmans R, Vandenput A and Geysen W *Calculation of the flux density and the unbalanced magnetic pull in two pole induction machines* (Archiv für Elektrotechnik, Vol. 70) pp 151-161
[8] Holopainen T P 2004 *Electromechanical interaction in rotor dynamics of cage induction motors* (Ph.D. Thesis, VTT Technical Research Centre of Finland, Helsinki University of Technology)
[9] Werner U 2006 *Rotordynamische Analyse von Asynchronmaschinen* (Dissertation, Technical University of Darmstadt)

Controlling Flexible Rotor Vibrations Using Parametric Excitation

Lawrence Atepor

Department of Mechanical Engineering, University of Glasgow, G12 8QQ, UK

katepor@yahoo.com

Keywords: Piezoelectric actuator, piezoelectric exciter, parametric excitation, flexible rotor, vibration control.

Abstract: This paper presents both theoretical and experimental studies of an active vibration controller for vibration in a flexible rotor system. The paper shows that the vibration amplitude can be modified by introducing an axial parametric excitation. The perturbation method of multiple scales is used to solve the equations of motion. The steady-state responses, with and without the parametric excitation terms, is investigated. An experimental test machine uses a piezoelectric exciter mounted on the end of the shaft. The results show a reduction in the rotor response amplitude under principal parametric resonance, and some good correlation between theory and experiment.

1. Introduction

Unbalance forces are the main source of vibration in rotating machines, but perfect balance is almost impossible to achieve. Moreover the distribution of unbalance can change in time because of wear or depositions all machines are subjected to. Reduction of rotor vibration is very important for safe and efficient functioning of all rotating machines. This paper proposes an active vibration control scheme for controlling transverse vibration of rotor shaft due to mass unbalance and presents both theoretical and experimental studies. The use of piezoelectric actuators in active vibration control has been considered in the past by Palazzolo *et.al.*[1] and Barret *et.al.*[2]. Yabuno *et.al.*[3] used a piezoelectric actuator to stabilize the parametric resonance induced in a cantilever beam and to control bifurcation resulting in the shift of the bifurcation set and the expansion of the stable region. Carmignani *et.al.*[4] developed an adaptive hydrodynamic bearing made up of a mobile housing mounted on piezoelectric actuators. In their work they showed that imposing a harmonic displacement on the mobile bearing, in two orthogonal directions, a rotating force, and a correcting moment can be produced on the shaft of a rotor system to reduce the bending caused by the unbalance. Das *et.al.*[5] proposed an active vibration control scheme for controlling transverse vibration of a rotor shaft due to unbalance. These authors worked on the vibration control of rotors due to unbalance by placing electromagnetic exciters, at convenient location on the span of the rotor away from the bearings. They showed that, locations, distant from discs are in general convenient for exciters as they do not interfering in any way with the rotor operation. Similar electromagnetic exciters were conceptualised by Janik *et.al.*[6] and were used to excite a rotor-shaft system for extracting the modal information by Janik *et.al.*[7] and Irretier *et.al.*[8].

No attempt of vibration control of rotor systems by the use of an axially placed piezoelectric exciter has, however been reported to the authors' knowledge. There are many natural phenomena in which excited parametric and self-excited vibrations interact with one another. Examples are flow-induced vibrations and vibrations in forced rotor systems. The responses of nonlinear excited systems to parametric excitations have been investigated by Nayfeh *et. al.*,[9]. Several authors, including Skalak and Yarymovych [10],

Struble[11], Dugundji *et. al.*,[12], Chester[13] and Cartmell[14] have studied the effects of combined parametric and forced vibrations in dynamic systems. Frolov[15] examined a mechanical system excited simultaneously by parametric and periodic forced excitations. Frolov[15] showed that the resonant amplitude can be reduced by random variation of the system parameters. Mustafa and Ertas[16] theoretically and experimentally examined the effect of a pendulum (attached to the tip of a parametrically excited cantilever beam) whose natural frequency is tuned to be commensurable with a frequency of the beam in order to generate autoparametric resonance. For chosen external and internal resonance combinations, where the excitation frequency is twice the natural frequency of the first beam mode, and the linearised pendulum frequency is one-half that of the first beam mode, the results showed that, in some parametric excitation frequency ranges, the pendulum acts as vibration-absorbing device in the same manner as the pendulum attached to the main system under external excitation. Nguyen and Ginsberg [17] studied vibration control of a simple pendulum using parametric excitation. They showed that with judicious selection of the parametric excitation, a parametric frequency that is very high relative to the highest contemplated excitation frequency can substantially reduce the forced vibration response at any lower excitation frequency. The above ideas have led to the design of the piezoelectric exciter and the deliberate introduction of parametric excitations into a flexible rotor-bearing system axially to moderate the response of the pre-existing mass-unbalance vibration inherent to the rotor. The idea here is to use piezoelectric stack actuator to put axial excitations into the shaft to investigate the interactions between forced vibrations, which emanates from rotor unbalance and parametric excitations which results from the periodic stiffness variation caused by a periodic axial excitations from the actuator. A practically implementable strategy is proposed in which the inherent and predominant instabilities in the flexible rotor-bearing system are manipulated in such a way that their effects on the overall performance of the rotor system can be effectively controlled. In justifying this work, a program of research has been carried out and the results show reductions in the resonant amplitudes for forward whirl in the flexible rotor-bearing system.

2. Theoretical work

2.1 Equations of motion

The rotor is assumed to be simply supported at both ends and the expression for the displacement are expressed in the generalised coordinates using the Rayleigh-Ritz method as $u(y,t) = f(y)q_1(t) = f(y)q_1$ and $w(y,t) = f(y)q_2(t) = f(y)q_2$, where q_1 and q_2 are generalised independent coordinates, u and w are displacements with respect to x and u, $f(y)$ is the displacement function, and it is chosen as the normalised first mode shape of a beam with a constant cross section in bending and simply supported at both ends, and is given as $f(y) = \sin\left(\frac{\pi y}{l}\right)$. The system energies are

$$T = \frac{1}{2}m\left(\dot{q}_1^2 + \dot{q}_2^2\right) + \Omega^2\left(\frac{1}{2}I_{dy} + \rho IL\right) - \Omega a_2\left(\dot{q}_1 \sin\left(\Upsilon q_2\right)\right) + m_u\Omega df\left(l_1\right)\left(\dot{q}_1 \cos\Omega t - \dot{q}_2 \sin\Omega t\right)$$ and

$U = \frac{1}{2}k_s\left(q_1^2 + q_2^2\right)$, where $\Upsilon = g(l_1)$, $m = M_d f^2\left(l_1\right) + \rho s\int_0^l f^2\left(y\right)dy$, $P = I_{dx}g^2\left(l_1\right) + \rho I\int_0^l g^2\left(y\right)dy$,

$a_1 = I_{dy}g^2\left(l_1\right) + \rho I\int_0^l g^2\left(y\right)dy$, $a_2 = I_{dy}g\left(l_1\right) + 2\rho I\int_0^l g\left(y\right)dy$, $f(l_1) = \sin\left(\frac{\pi l_1}{l}\right)$, and $g(l_1) = \frac{\pi}{l}\cos\frac{\pi l_1}{l}$,

$k_s = EI\int_0^l h^2(y)dy + F_0\int_0^l g^2(y)dy$ and is the stiffness of the shaft, $h(y) = \frac{d^2 f(y)}{dy^2}$, l_1 is the position of the disk on the shaft and l is the shaft length and are obtained from the experimental rig. M_d is the mass of disk, s is the cross-sectional area, P, a_1, a_2, a_5 are geometric constants, I is the moment of inertia and ρ is mass per unit volume. The equations of motion are derived by means of Lagrangian analysis as follows,

$$\ddot{q}_1 + \hat{c}\dot{q}_1 - \Omega\,\hat{a}_5\dot{q}_2 + \omega^2 q_1 + \hat{b}q_1^3 - \hat{F}_{act}q_1 = \mu\, d\Omega^2 \sin\Omega\, t \tag{1}$$

$$\ddot{q}_2 + \hat{c}\dot{q}_2 + \Omega\,\hat{a}_5\dot{q}_1 + \omega^2 q_2 + \hat{b}q_2^3 - \hat{F}_{act}q_2 = \mu\, d\Omega^2 \cos\Omega\, t \tag{2}$$

where $\hat{a}_5 = \dfrac{a_5}{m}, \omega^2 = \dfrac{k}{m}, \hat{b} = \dfrac{b}{m}, \hat{c} = \dfrac{c}{m}, \mu = \dfrac{m_u}{m}, \hat{F}_0 = \dfrac{F_{act}}{m}$, $F_{act}q_i$ is the axial excitation force term and the dots denote differentiation with respect to t.

2.2. Solutions to the equations of motion

The approximate solution for the model of the flexible rotor system with and without parametric excitation terms are obtained using the method of multiple scales as

$$\begin{aligned}\bar{q}_1 = {} & 2p\cos\left(\frac{\Omega\, t}{2\omega}\right) - 2q\sin\left(\frac{\Omega\, t}{2\omega}\right) + \frac{\hat{b}}{4\omega^2}p^3\cos\left(\frac{3\Omega\, t}{2\omega}\right) \\ & - \frac{3\hat{b}}{4\omega^2}p^2 q\sin\left(\frac{3\Omega\, t}{2\omega}\right) - \frac{3\hat{b}}{4\omega^2}pq^2\cos\left(\frac{3\Omega\, t}{2\omega}\right) + \frac{\hat{b}}{4\omega^2}q^3\sin\left(\frac{3\Omega\, t}{2\omega}\right)\end{aligned} \tag{3}$$

$$\begin{aligned}\bar{q}_2 = {} & 2r\cos\left(\frac{\Omega\, t}{2\omega}\right) - 2s\sin\left(\frac{\Omega\, t}{2\omega}\right) + \frac{\hat{b}}{4\omega^2}r^3\cos\left(\frac{3\Omega\, t}{2\omega}\right) \\ & - \frac{3\hat{b}}{4\omega^2}r^2 s\sin\left(\frac{3\Omega\, t}{2\omega}\right) - \frac{3\hat{b}}{4\omega^2}rs^2\cos\left(\frac{3\Omega\, t}{2\omega}\right) + \frac{\hat{b}}{4\omega^2}s^3\sin\left(\frac{3\Omega\, t}{2\omega}\right)\end{aligned} \tag{4}$$

$$\begin{aligned}\bar{q}_1 = {} & 2p\cos\left(\frac{\Omega\, t}{2\omega}\right) - 2q\sin\left(\frac{\Omega\, t}{2\omega}\right) + \frac{\hat{b}}{4\omega^2}p^3\cos\left(\frac{3\Omega\, t}{2\omega}\right) - \frac{3\hat{b}}{4\omega^2}p^2 q\sin\left(\frac{3\Omega\, t}{2\omega}\right) \\ & - \frac{3\hat{b}}{4\omega^2}pq^2\cos\left(\frac{3\Omega\, t}{2\omega}\right) + \frac{\hat{b}}{4\omega^2}q^3\sin\left(\frac{3\Omega\, t}{2\omega}\right) - \frac{2\hat{F}_{act}p}{2\Omega_2^2 + 4\Omega_2\omega}\cos\left(\frac{5\Omega_2 t}{4\omega}\right) \\ & + \frac{2\hat{F}_{act}q}{2\Omega_2^2 + 4\Omega_2\omega}\sin\left(\frac{5\Omega_2 t}{4\omega}\right) - \frac{2\hat{F}_{act}\Omega_2\hat{c}kp}{2\Omega_2^2\omega^2 + 4\Omega_2\omega^3}\sin\left(\frac{5\Omega_2 t}{4\omega}\right) \\ & - \frac{2\hat{F}_{act}\Omega_2\hat{c}kq}{2\Omega_2^2\omega^2 + 4\Omega_2\omega^3}\cos\left(\frac{5\Omega_2 t}{4\omega}\right) + \frac{2\hat{F}_{act}\hat{c}kp}{2\Omega_2^2\omega + 4\Omega_2\omega^2}\sin\left(\frac{5\Omega_2 t}{4\omega}\right) \\ & - \frac{2\hat{F}_{act}\hat{c}kq}{2\Omega_2^2\omega + 4\Omega_2\omega^2}\cos\left(\frac{5\Omega_2 t}{4\omega}\right) + \frac{2\hat{F}_{act}\Omega\,\Omega_2\hat{a}_5 kr}{2\Omega_2^2\omega^2 + 4\Omega_2\omega^3}\sin\left(\frac{5\Omega_2 t}{4\omega}\right) \\ & + \frac{2\hat{F}_{act}\Omega\,\Omega_2\hat{a}_5 ks}{2\Omega_2^2\omega^2 + 4\Omega_2\omega^3}\cos\left(\frac{5\Omega_2 t}{4\omega}\right) + \frac{2\hat{F}_{act}\Omega\,\hat{a}_5 kr}{2\Omega_2^2\omega + 4\Omega_2\omega^2}\sin\left(\frac{5\Omega_2 t}{4\omega}\right) \\ & + \frac{2\hat{F}_{act}\Omega\,\hat{a}_5 ks}{2\Omega_2^2\omega + 4\Omega_2\omega^2}\cos\left(\frac{5\Omega_2 t}{4\omega}\right) - \frac{2\hat{F}_{act}Qp}{4\Omega_2^2\omega^2 + 8\Omega_2\omega^3}\cos\left(\frac{9\Omega_2 t}{4\omega}\right) \\ & + \frac{2\hat{F}_{act}Qq}{4\Omega_2^2\omega^2 + 8\Omega_2\omega^3}\sin\left(\frac{9\Omega_2 t}{4\omega}\right)\end{aligned} \tag{5}$$

$$
\begin{aligned}
\bar{q}_2 = {} & 2r\cos\left(\frac{\Omega t}{2\omega}\right) - 2s\sin\left(\frac{\Omega t}{2\omega}\right) + \frac{\hat{b}}{4\omega^2} r^3\cos\left(\frac{3\Omega t}{2\omega}\right) - \frac{3\hat{b}}{4\omega^2} r^2 s\sin\left(\frac{3\Omega t}{2\omega}\right) \\
& - \frac{3\hat{b}}{4\omega^2} rs^2\cos\left(\frac{3\Omega t}{2\omega}\right) + \frac{\hat{b}}{4\omega^2} s^3\sin\left(\frac{3\Omega t}{2\omega}\right) - \frac{2\hat{F}_{act} r}{2\Omega_2^2 + 4\Omega_2\omega}\cos\left(\frac{5\Omega_2 t}{4\omega}\right) \\
& + \frac{2\hat{F}_{act} s}{2\Omega_2^2 + 4\Omega_2\omega}\sin\left(\frac{5\Omega_2 t}{4\omega}\right) - \frac{2\hat{F}_{act}\Omega_2\hat{c}kr}{2\Omega_2^2\omega^2 + 4\Omega_2\omega^3}\sin\left(\frac{5\Omega_2 t}{4\omega}\right) \\
& - \frac{2\hat{F}_{act}\Omega_2\hat{c}ks}{2\Omega_2^2\omega^2 + 4\Omega_2\omega^3}\cos\left(\frac{5\Omega_2 t}{4\omega}\right) - \frac{2\hat{F}_{act}\hat{c}kr}{2\Omega_2^2\omega + 4\Omega_2\omega^2}\sin\left(\frac{5\Omega_2 t}{4\omega}\right) \\
& - \frac{2\hat{F}_{act}\hat{c}ks}{2\Omega_2^2\omega + 4\Omega_2\omega^2}\cos\left(\frac{5\Omega_2 t}{4\omega}\right) - \frac{2\hat{F}_{act}\Omega\Omega_2\hat{a}_5 kp}{2\Omega_2^2\omega^2 + 4\Omega_2\omega^3}\sin\left(\frac{5\Omega_2 t}{4\omega}\right) \\
& - \frac{2\hat{F}_{act}\Omega\Omega_2\hat{a}_5 kq}{2\Omega_2^2\omega^2 + 4\Omega_2\omega^3}\cos\left(\frac{5\Omega_2 t}{4\omega}\right) - \frac{2\hat{F}_{act}\Omega\hat{a}_5 kp}{2\Omega_2^2\omega + 4\Omega_2\omega^2}\sin\left(\frac{5\Omega_2 t}{4\omega}\right) \\
& - \frac{2\hat{F}_{act}\Omega\hat{a}_5 kq}{2\Omega_2^2\omega + 4\Omega_2\omega^2}\cos\left(\frac{5\Omega_2 t}{4\omega}\right) - \frac{2\hat{F}_{act}Qr}{4\Omega_2^2\omega^2 + 8\Omega_2\omega^3}\cos\left(\frac{9\Omega_2 t}{4\omega}\right) \\
& + \frac{2\hat{F}_{act}Qs}{4\Omega_2^2\omega^2 + 8\Omega_2\omega^3}\sin\left(\frac{9\Omega_2 t}{4\omega}\right)
\end{aligned}
\tag{6}
$$

where, $k = \left(\dfrac{1}{-\dfrac{\Omega_2^2}{\omega^2} - \dfrac{2\Omega_2}{\omega}}\right)$ and $Q = \left(\dfrac{1}{-\dfrac{4\Omega_2^2}{\omega^2} - \dfrac{4\Omega_2}{\omega}}\right)$. Equations (3) and (4) are full time-domain solutions of the equations of motion (1) and (2) without parametric excitation terms and equations (5) and (6) are the solutions with parametric excitation terms. p, q, r and s are amplitudes, Ω is the excitation frequency and Ω_2 is the principal parametric resonance frequency and $\Omega_2 = 2\Omega$.

3. Experimental Work

A commercial rotor-kit (Bently Nevada rotor kit RK4) and a piezoelectric exciter specifically developed for this research are used for the experiment. The rotor kit provided a rotor supported by journal bearings, an electrical drive to run the rotor with a separate control box from which the desired rotational speed is selected. The torque is transmitted from the electrical motor to the rotor by means of a solid coupling. Provided are displacement transducers to measure the movements of the rotor, and a rigid V-shaped base, to which any components could be easily attached. The rotor kit is equipped with the piezoelectric exciter designed for active vibration control. The critical parts of the exciter unit are, a piezoelectric actuator supported by a helical compression spring, all placed inside a linear sliding bearing, and an aluminium casing. The piezoelectric actuator is driven by a function generator through a piezoelectric actuator amplifier. To avoid direct contact between the shaft and piezoexciter, and to allow free rotation and movement of the shaft end, a small self-aligned ball bearing is fixed in between the shaft and the piezoexciter. The vibration response of the rotor is then measured by means of a Polytec Laser Vibrometer allowing the displacement responses to be identified and monitored by a multi-channel data acquisition analyser. Figures 1 and 2 show the experimental configuration for activating the flexible rotor system. The leading principle here is to control, axially, the vibrations of the rotor, supported on conventional bearings, by using the piezoelectric actuator.

Figure 1:Close-up of the Piezoelectric Exciter

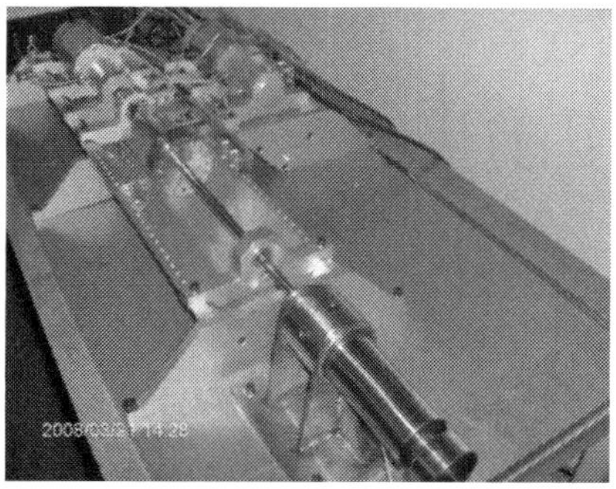

Figure 2:Assembly of the Piezoexciter Test Rig

3.1. Test Setup

The objective of this work has been the design and construction of a test rig to verify the feasibility of active control of vibration in rotor dynamics using a piezoelectric actuator. In particular the possibility of reducing the amplitude of vibrations of a flexible dynamically unbalanced rotor within acceptable levels is investigated. This is carried out by designing a Piezoexciter excited by a high frequency drive. The active Piezoexciter comprises a sliding bearing which houses the piezoelectric stack actuator which is serially attached to a compression spring. Since the actuator operates only in expansion, with small displacement, the reaction spring is set up against it. The spring is adjusted to the required length by the spring compressor and voltage is applied through a piezoelectric voltage amplifier to the actuator which in turn develops the parametric excitation at a frequency of twice the first whirl frequency of the rotor system. The exciter is driven by a function generator through a high voltage amplifier. Activating the piezoelectric actuator at twice the excitation frequency of the rotor system generates the parametric excitation force to be introduced to the shaft, axially. The vibration response of the rotor-bearing system is then measured by means of the laser vibrometer. A multi-channel data acquisition analyser is then used to analyse the response. The compression spring of the exciter unit and the rotor-bearing system are set to their required length and first whirl resonance frequency respectively, and the response of the rotor is measured. The piezoelectric actuator is then activated, first at a frequency twice the first whirl frequency of the rotor system. A series of timed tests are performed and average readings are taken. Sweep tests around the first whirl frequency are then performed, first without activating the piezoexciter, and then with the exciter activated at the parametric excitation frequency.

Journal of Physics: Conference Series **181** (2009) 012021 doi:10.1088/1742-6596/181/1/012021

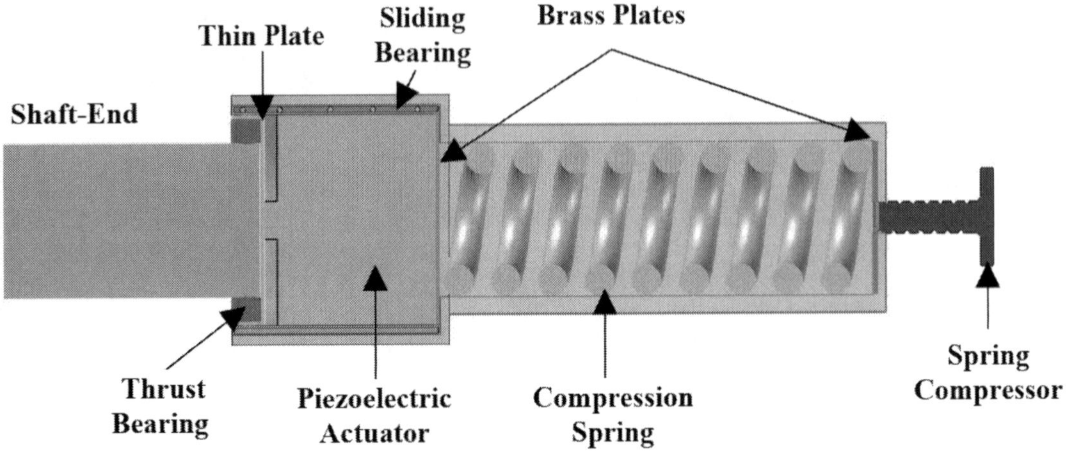

Figure 3: Schematic of the piezoelectric exciter.

3.2. Theoretical Results

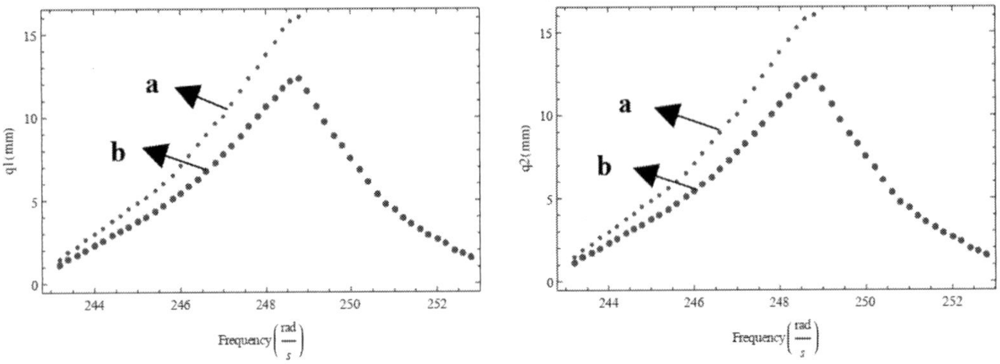

a- without parametric force term, b- with parametric force term

Figure 4: Amplitudes of the response as functions of the frequency

In figure 4, considering plots indicated by **a**, both the responses in the first mode of $\bar{q}_1$ and $\bar{q}_2$ show hardening characteristics, jump phenomena and both stable and unstable solutions when the equations of motion contain no parametric force term. Including the parametric force terms, the results in figure 4, plots indicated by **b** show 23% reduction in amplitude, elimination of the jump phenomena and stable solutions.

3.3. Experimental results

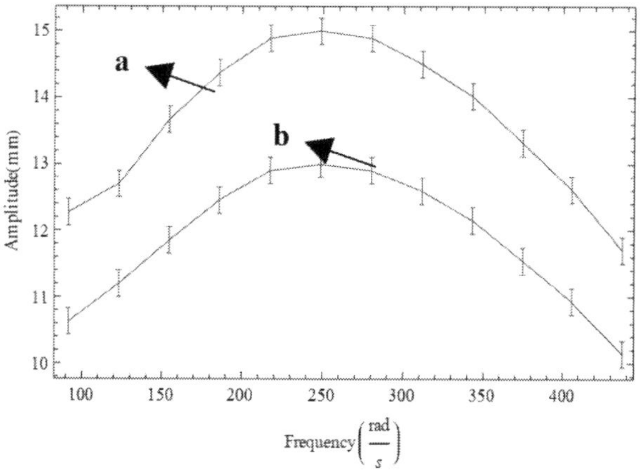

a- piezoexciter unactivated, b- piezoexciter activated at $\Omega_2 = 2\Omega_1$

Figure 5: The amplitudes of disc vibration versus frequency with a spring compression length of 25.2mm.

In figure 5 the combined effects due to the existing forced vibration due to mass unbalance and also an additional parametric excitation in principal parametric resonance provided by the piezoexciter resulted in the moderation of the responses of the pre-existing mass unbalance vibration inherent to the rotor, with an approximately 13% reduction in the critical whirl amplitude.

4. Conclusion

Comparison between the results from the theoretical analysis and the experimental benchmark summarised here shows evidence of a consistent phenomenon. The methods of investigating and identifying the response behaviour of rotor systems have all shown similar trends with regards to the effects of introducing parametric force terms. Prototypical experimental results from tests on rotor systems conclude that the novel piezoelectric exciter concepts could be successfully applied to industrial applications, particularly installations where axial loading on the rotor shaft is also an inherent part of the control actuation.

5. Acknowledgement

The author would like to acknowledge the support of Prof. Matthew P Cartmell of Mechanical Engineering Department, University of Glasgow, GETFund sponsorship and EPSRC grants EP/C531892/1 and EP/C 531906/1.

Journal of Physics: Conference Series **181** (2009) 012021 doi:10.1088/1742-6596/181/1/012021

6. References

[1] Palazzolo A.B., Jogannathan S., Kascak A. F., Montague G.T., and Kiraly L.J., (1993), *Journal of vibration and Acoustics*, **115**, pp. 111-119, Hybrid active vibration control of rotor bearing systems using piezoelectric actuators.

[2] Barret T. S., Palazzolo A.B., Kascak A. F., (1993), *Trans.ASME , J. Eng. Gas Turbine Power* **117** pp.176-187, Active vibration control of rotating machinery using piezoelectric actuators incorporating flexible casing effects.

[3] Yabuno H., Saigusa S., and Aoshima N., (2001), *Nonlinear Dynamics*, **26**, pp. 143-161, Stabilization of the parametric resonance of a cantilever beam by bifurcation control with a piezoelectric actuator.

[4] Carmignani C., Forte P., and Rustighi E., (2001), *ASME Design Engineering Technical Conference and Computers and Information in Engineering Conference, Pittsburgh, September 9-12, 2001*, Active control of rotor vibrations by means of piezoelectric actuators.

[5] Das A.S., Nighil M.C., Dutt J.K., and Irretier H., (2007*), Journal of Mechanism and Machine Theory*, Vibration control and stability analysis of rotor-shaft system with electromagnetic exciters.

[6] Janik T.K., and Irretier H., (1998), *Proceedings of the IFTOMM Conference on Rotordynamics, Darmstadt,* pp. 695-708, New excitation and response measurement techniques for modal testing of flexible rotors.

[7] Janik T.K., and Irretier H., (2000), *International Journal of Rotating Machinery*,**6(1)**,pp.11-18, Experimental modal analysis-a tool for unbalance identification of rotating machines.

[8] Irretier H., and Lindemann S., (2002*), Proceedings of the 6th International Conference on Rotor Dynamics, Sydney, Australia,* pp.388-395, Improvement of damping parameters of flexible rotors by experimental modal analysis and modal updating.

[9] Nayfeh A.H., and Mook D.T., (1995), Nonlinear Oscillations, Wiley, New York.

[10] Skalak R., and Yarymovych M.I., (1960), *Journal of Applied Mechanics*, **27**, pp.159-164, Subharmonic oscillations of pendulum.

[11] Struble R.A., (1963), *ASME Journal of Applied Mechanics*, **30**, pp. 301-303, On the subharmonic oscillations of a pendulum.

[12] Dugundji J., and Chhatpar C.K., (1970), *Rev. Sci. Tech., Ser. Mech. Appl.*, **15(4)**, pp. 741-763, Dynamic stability of a pendulum under parametric excitation.

[13] Chester W., (1975), *Journal of Inst. Math. Applications*, **15**, pp. 289-306, The forced oscillations of a simple pendulum.

[14] Cartmell M.P., (1990), Introduction to linear, parametric and nonlinear vibrations, *Chapman and Hall*, London.

[15] Frolov K.V.,(1967), *Proceedings of the 4th Conference on Nonlinear Oscillations*, pp. 327-336, Parametric and autoparametric oscillations of some nonlinear mechanical systems.

[16] Mustafa G., and Ertas A., (1995), *Journal of Sound and Vibration* , **182(3)**, pp. 393-413, Dynamics and bifurcations of a coupled column-pendulum oscillator.

[17] Nguyen P.H., and Ginsberg J.H., (2001), *Journal of Vibration and Acoustics*, **123**, pp.359-364, Vibration control using parametric excitation.

Controlling rotordynamic response without squeeze-film dampers (SFDs)

Shakir Jiffri, Seamus Garvey
School of Mechanical, Materials and Manufacturing Engineering,
University of Nottingham, UK

eaxmsj@nottingham.ac.uk / seamus.garvey@nottingham.ac.uk

Abstract. SFDs are widely used in rotating machinery to provide damping in order to control rotordynamic response. Although popular, under certain conditions SFDs pose problems such as causing non-synchronous vibration arising from unbalance forces interacting with fluid-film forces affected by cavitation. Furthermore, in the interests of moving towards oil-free rotating machines, the need arises to find alternative means of rotordynamic response control.

In choosing a new vibration control technology, it is first necessary to consider certain general, configuration-independent criteria. For example, does the actuation method provide a limited stroke (eg. piezoelectric or giant magnetostrictive) or is the stroke a "motorised" solution (eg. an ultrasonic motor directly driving the actuator or a pump acting to vary the fill level of closed deformable volumes with incompressible fluid) Is the work per stroke per unit mass of the actuator material sufficient to provide the maximum stroke and force required for the control? What is the bandwidth of the actuator? In the case of electromechanical actuation, what is the coupling factor? Can the elements of the actuator withstand the high temperatures of the operating environment? Is the solution an active or passive one? What are the fatigue properties of the materials used in the actuator? These are some of the questions that need to be considered when evaluating a new control method.

Once the significant properties have been identified, it is necessary to consider each of these in the context of the intended application. If one considers the actuation type, in the limited stroke case it will be required for the actuation to take place at synchronous frequency and the work per stroke per unit mass will determine the quantity of material required. For some applications – particularly aero-engines - one seeks to minimise overall mass and therefore materials with high values of work per stroke per unit mass are attractive. By contrast, in the case of motorised solutions the frequency of operation of the actuator may be orders of magnitude greater than synchronous speed. Since the main source of rotordynamic vibration arises from rotor unbalance – which causes vibration at synchronous speed – the control is also required (at least) at synchronous speed. Therefore, the actuation method should have a bandwidth equal at least to this speed. The coupling factor for piezoelectric and magnetostrictive actuators will determine the amount of energy that should be supplied for a given requirement of work, and therefore high values of this property are desirable. Whether the vibration control method is active or passive will also influence the overall weight and complexity of the machine.

Damping is not the only method by which vibration may be controlled. For example, if it is possible to alter the natural frequencies of the rotor-stator system as the system is approaching a resonance, then the build up of substantial vibration can be avoided as the system effectively never sits on a resonance.

In the present work, various actuation technologies such as piezoelectric, magneto-strictive, electromagnetic, shape memory alloys and ultrasonic motors have been considered. Some of these technologies are considered in more detail as they show more potential of being implemented in a relatively uncomplicated configuration, in the intended application.

Keywords: SFD, damping, response, vibration, control, active control

1. Introduction

The SFD has been a popular means of rotordynamic vibration control since the 1970s, and is widely employed in aircraft gas turbine engines today. The damper manifests as an oil film between the damper journal (a sleeve fitted to the outer race of the bearing) and the stator. Rotation of the journal is prevented by mechanical means, although it is free to undergo whirling motion. The hydrodynamic damping forces generated in the squeeze-film during whirling of the journal serve to oppose the motion of the bearing by dissipating energy from the system.

A paper by Zeidan et. al. [1] presents a detailed account of SFDs, where the authors discuss the historical development of SFDs, theoretical models and experimental results and design procedures. The above paper provides a total of 50 references relating to various aspects of SFDs.

SFDs are well known for their non-linear behaviour. This gives rise to multiple unbalance response solutions for certain speeds. It has been found that the minimum and maximum of these solutions are stable, and thus when operating in this regime the system is in bistable operation. Rabinowitz and Hahn present their work on a theoretical study of the steady-state performance and stability of SFD supported flexible rotors in two separate papers [2, 3]. In [2], the authors conclude that bistable operation can be avoided if the design parameters are chosen appropriately and in [3], the parameters that give rise to instability are discussed. In both the above papers, it is noted that pressurization of the SFD is a prerequisite to stability. The above authors presented a further paper which presents experimental results which validate their previous theoretical work [4]. Zhu et. al. analyse the multiple-solution response of a flexible rotor using the *synchronous circular centred-orbit motion solution, numerical integration method* and *slow acceleration method* in [5]. The authors conclude that there are three basic forms for the multiple-solution response: *resonant, isolated bifurcation* and *swallowtail bifurcation* multiple solutions. References to other analysis methods are given in [5]. An extremely fast method of obtaining the unbalance response of flexible rotors on SFDs is presented in [6], where non-linear forces from the SFD are included in the analysis. The authors report twenty-five orders of magnitude time savings as compared to numerical integration and iterative methods. The stability and bifurcation of unbalance response of flexible rotors on SFDs is analysed in [7], where the cavitation phenomenon of SFDs is included in the model. It is found that unpressurised SFDs promote non-synchronous vibrations.

The multiple response phenomenon under certain conditions means that it is not possible to accurately predict the unbalance response under these conditions, as more than one stable solution can exist. The non-synchronous vibrations that arise under certain conditions due to SFDs give rise to alternating stresses in the rotor that can lead to progressive fatigue, and eventually failure. Both of the above are notable disadvantages of using SFDs to control rotordynamic vibration.

A further disadvantage is that the damper force generated in the SFD will always have a fixed phase relative to the velocity – i.e. it is in phase with the velocity. The authors recognise this as a constraint placed on the control force. The present paper investigates the possibility of achieving better vibration reduction by employing active control forces, where the phase of the forces may not be fixed under all conditions.

2. FE Modelling of a rotor in bearings employing SFDs

In this section, a rotor-stator system is defined, and its behaviour under unbalance excitation explored. The damping of this simple machine is based on SFDs, whose behaviour is modelled here as a linear viscous damper in parallel with the squirrel-cage stiffness (although it is recognised that a more accurate model of an SFD entails a non-linear model) connecting the bearing outer housing to the stator. The bearing outer housing is assumed to be rigidly fixed to the rotor, and it therefore undergoes the same motion as the rotor.

The rotor model is based on the usual linear, second order equations shown below:

Journal of Physics: Conference Series **181** (2009) 012022 doi:10.1088/1742-6596/181/1/012022

$$\mathbf{M}\ddot{\mathbf{q}}(t)+(\mathbf{C}+\Omega\mathbf{G})\dot{\mathbf{q}}(t)+\mathbf{K}\mathbf{q}(t)=\mathbf{f}(t) \tag{1}$$

Where $\mathbf{f}(t)$ is the applied forcing vector (in this case unbalance), $\mathbf{q}(t)$ is the response vector of the system, **M**, **C**, **G** and **K** are the global mass, damping, gyroscopic and stiffness matrices of the system respectively.

The unbalance response of the system in the frequency domain may be calculated as:

$$\mathbf{q}(\Omega)=\left(\mathbf{K}+i\Omega(\mathbf{C}+\Omega\mathbf{G})-\Omega^2\mathbf{M}\right)^{-1}\mathbf{f}(\Omega) \tag{2}$$

The term within brackets is the dynamic stiffness matrix, and will subsequently be referred to as **E**.

2.1 Rotor model

A picture of the FE model that will be used in this paper is shown below:

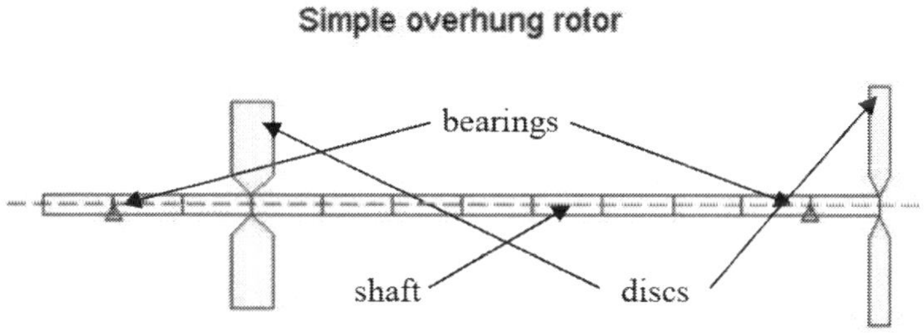

Figure 1 - FE model of a simple overhung rotor in bearings

The model is based on beam elements, where each element has two nodes and each node has 4 Degrees of Freedom (DOFs) – translations in the x and y directions and rotations about these directions. In addition to the stiffness properties of the shaft elements, inertial and gyroscopic contributions from the shaft elements and the discs are incorporated into the model (the discs themselves are assumed rigid). Furthermore, the squirrel cage stiffness and damping from the SFD are applied at the bearings.

The parameters that define the model are stated below:

Table 1 - description of rotor system

Material Properties of the rotor shaft	*Dimensions of the rotor components (mm)*
Young's Modulus – 211 GPa	Shaft (solid) – φ30 throughout
Poisson's Ratio – 0.3	Disc 1 (node 4) – φ300 OD, φ30 ID, 60 thickness
Density (shaft and discs) – 7810 kgm^{-3}	Disc 2 (node 13) – φ350 OD, φ30 ID, 30 thickness
Bearing Properties	
Left bearing stiffness – 4 × 10^5 N/m	Left bearing damping coefficient – 7.202 × 10^3 Ns/m
Right bearing stiffness – 12 × 10^5 N/m	Right bearing damping coefficient – 7.202 × 10^3 Ns/m

For simplicity, it is assumed that the bearing stiffness and damping properties are isotropic. A distribution of unbalance forcing along the length of the rotor is now defined. This will be kept fixed in all of the subsequent modelling. The magnitude of the unbalance forcing is chosen such that the machine will experience deflections at the bearings that are large enough such that the bearing outer housing starts to make contact with the stator. This allows one to model the situation when the hydrodynamic force generated in the SFD is insufficient to drive the rotor back to its axis. It is assumed in this model that the SFD thickness is 0.1 mm. Therefore, when the bearing outer housing makes contact with the stator, the bearing stiffness is iteratively increased to reflect this. The following two figures show the unbalance response at the bearings and the increase in effective lateral stiffness when damper journal-stator contact is made:

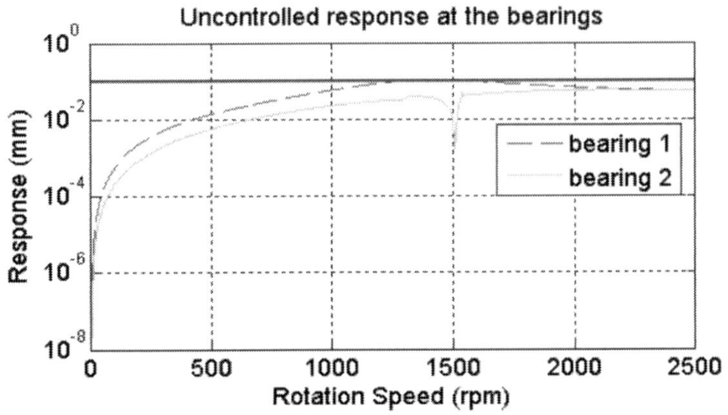

Figure 2 - Rotor response at bearings (SFD case)

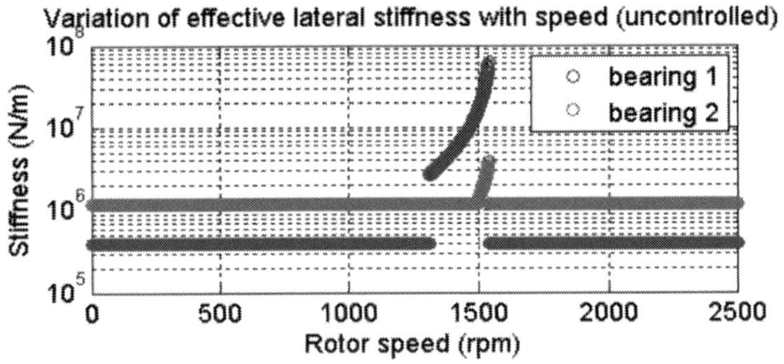

Figure 3 - Increase in effective lateral stiffness to reflect contact of journal with stator

A "*cost function*" can now be defined, which is a measure of the overall rotor vibration. In our case, this may be defined as the sum of squares of translations at all the nodes (in the FE sense) of the rotor. This provides a convenient means of comparing the performance of the SFD with that of an active solution. The following figure shows the cost function as a function of speed for the SFD case:

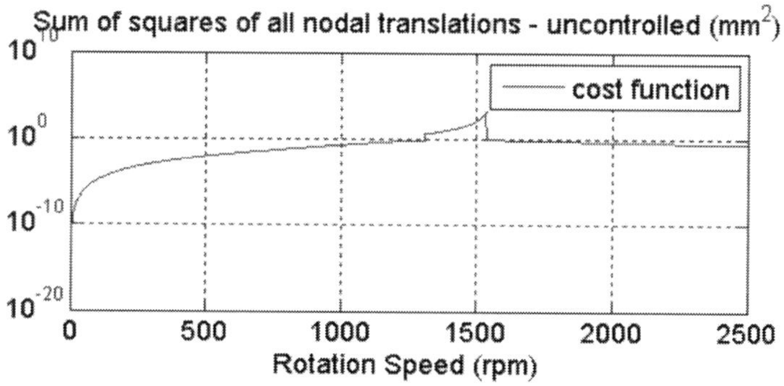

Figure 4 - cost function for SFD case

3. FE Modelling of a rotor in actively controlled bearings

In this simulation, the same model as above is used, with the only change that the SFD is now removed (i.e. the damping value above is now set to zero) and active control forces are applied at the bearings such that the cost function for each individual speed is brought down to its minimum possible value. The optimum active control forces can be computed through the following procedure based on the assumption of a linear model:

From equation (2), we have that

$$\mathbf{q} = \mathbf{E}^{-1}\left[\mathbf{f} + \mathbf{f}_C\right] \tag{3}$$

Where the additional term $\mathbf{f}_C$ contains the control forces. Multiplying both sides by the diagonal weighting matrix $\mathbf{W}$, we have

$$\mathbf{Wq} = \mathbf{WE}^{-1}\left[\mathbf{f} + \mathbf{f}_C\right] \tag{4}$$

The weighting matrix can be used to assign different levels of importance to rotor deflections at individual nodes. In the present analysis, all translations have been assumed equally important. Contributions due to rotations have been ignored by placing zeros in the corresponding entries of the weighting matrix. Now, multiplying equation (4) by its Hermitian transpose, an expression for the weighted sum of squares (i.e. the cost function) is obtained:

$$cost = \left[\mathbf{f} + \mathbf{f}_C\right]^{\mathbf{H}} \mathbf{E}^{-\mathbf{H}}\mathbf{W}^{\mathbf{H}}\mathbf{W}\mathbf{E}^{-1}\left[\mathbf{f} + \mathbf{f}_C\right] \tag{5}$$

Since control forces occur only at the bearings, a constraint may be applied to $\mathbf{f}_C$ to reflect this. Viz.,

$$\mathbf{f}_C = \mathbf{Sp} \tag{6}$$

Substituting (6) into (5), and differentiating with respect to $\mathbf{p}$ and subsequently multiplying by $\mathbf{S}$ gives the optimum control force vector as:

$$\mathbf{f}_C = -\mathbf{S}\left(\mathbf{S}^{\mathbf{H}}\mathbf{E}^{-\mathbf{H}}\mathbf{W}^{\mathbf{H}}\mathbf{W}\mathbf{E}^{-1}\mathbf{S}\right)^{-1}\left(\mathbf{S}^{\mathbf{H}}\mathbf{E}^{-\mathbf{H}}\mathbf{W}^{\mathbf{H}}\mathbf{W}\mathbf{E}^{-1}\right)\mathbf{f} \tag{7}$$

Once the optimal control forces have been found, the new (controlled) response is computed. This is shown in the following figure:

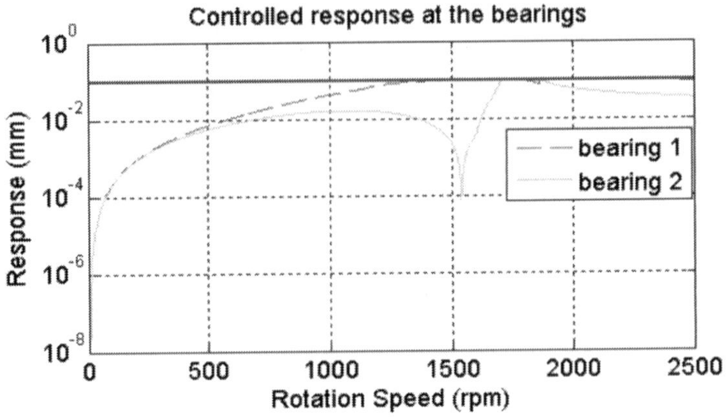

Figure 5 - controlled response at the bearings

It can be seen from the above figure that there still is bearing-stator contact. The cost function in this case is shown below.

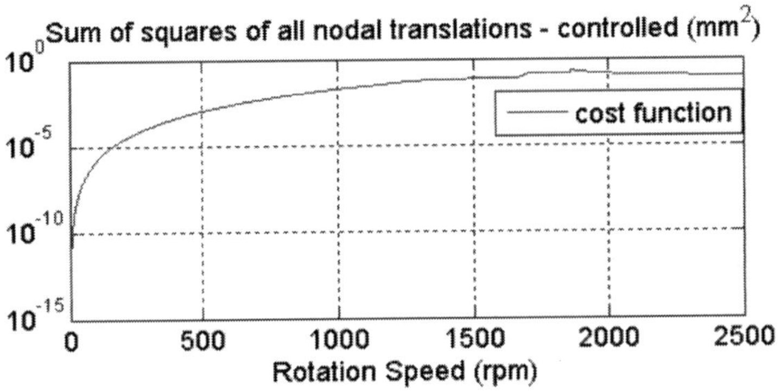

Figure 6 - cost function for controlled case

At this stage, it is appropriate to explain the basis on which the damper values in the SFD case have been chosen. The cost functions for the entire speed range were calculated in the case of many values of damping coefficient, and in each case the maximum cost function from the speed range was taken. A graph of maximum cost function vs. damping coefficient is shown below:

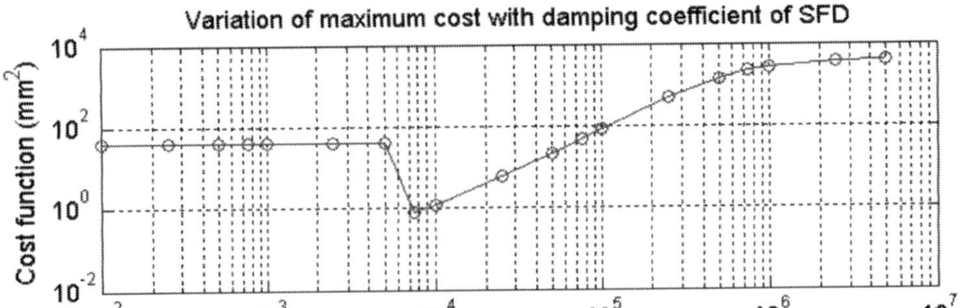

Figure 7 - variation of maximum cost with damping coefficient of SFD

It is found after further simulation that the damping coefficient value giving rise to a minimum value of the maximum cost function is approximately 7202 Ns/m, which is the value that has been used in this work.

It is important to note that the magnitude of control force that the actuator can provide may fall short of the value computed for the optimal control force. This effect has been incorporated into the present model, and it is found that if no such limitation is placed, no bearing-stator contact is observed in the controlled response. The following figure shows the magnitude limitation of the control force, which is chosen to be 50N in the present model.

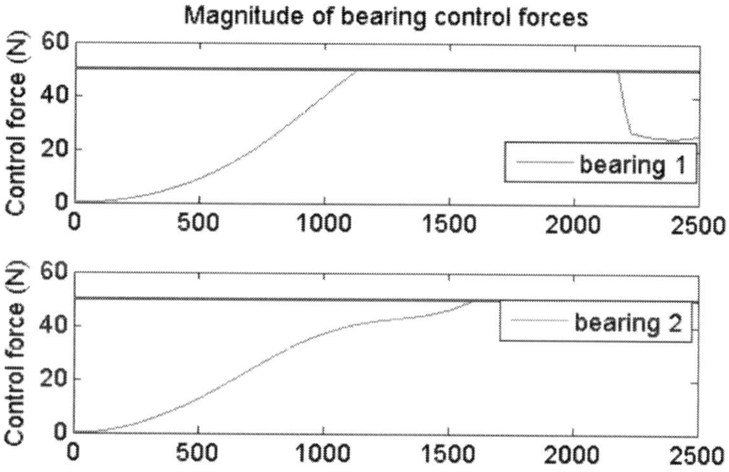

Figure 8 - limitation of control force magnitude

4. Comparing the performance of SFDs with that of an active solution

A comparison between the vibration control performance of the SFD and the actively controlled case can now be made based on the cost function that each case gives rise to. The following figure shows the ratio between the cost function in the controlled case to that in the uncontrolled case:

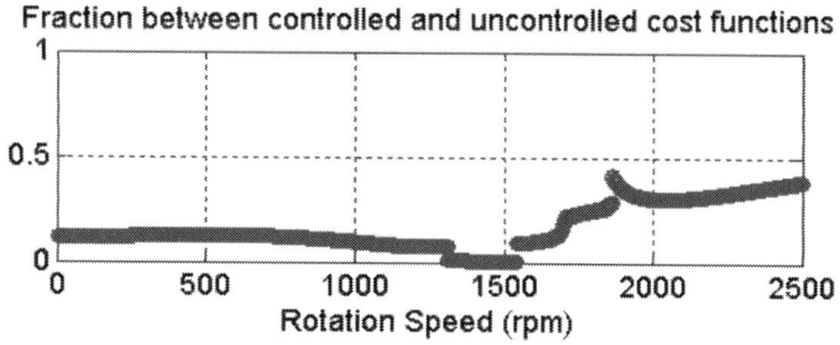

Figure 9 - fraction between controlled and uncontrolled cost functions

It is evident from the above figure that the vibration level with active control forces implemented is much lower than present when using SFDs. One may also compare the magnitude of the damper forces

Journal of Physics: Conference Series **181** (2009) 012022 doi:10.1088/1742-6596/181/1/012022
in the SFD case with the actuator forces in the controlled case. This comparison is shown in the following figure:

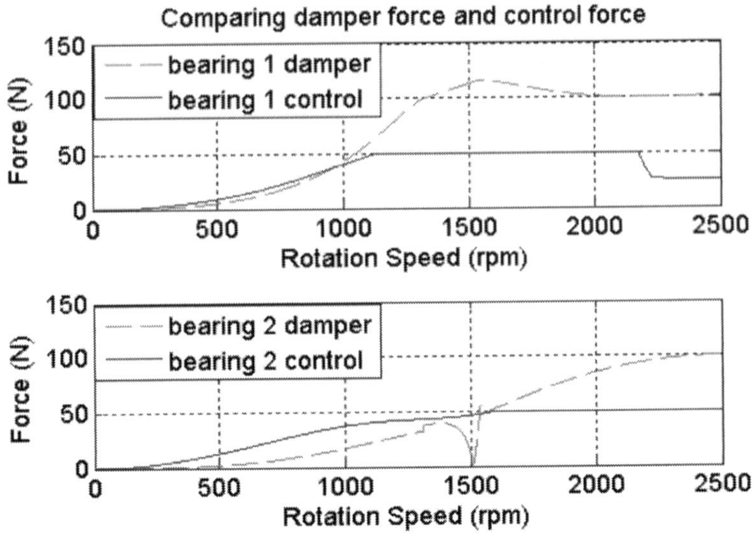

Figure 10 - comparing damper force (SFD case) with actuator force (controlled case)

From the upper half of the above figure, it can be seen that for 0 – 1000 rpm, the damper and active control forces are similar, whereas after this speed the damper force is substantially higher. The lower half of the figure shows that in 0 – 1500 rpm, the control force magnitude is slightly higher and above this speed the damper force is, once again, substantially higher. Therefore one may draw the conclusion that overall, the magnitude of active control forces required for optimum control is lower than that of the damping forces.

The main advantage of using active control forces as compared with SFDs is the freedom to choose the phase of the force with respect to the displacement or velocity. In SFDs, the damping force is always in phase with the velocity. The freedom to change the phase of the control force for various speeds enables one to optimise it for each value of speed in the range, whereas in the case of the SFD, optimisation will only be possible at a single speed.

5. Some candidate technologies and design concepts for active control of vibration

Electro-mechanical devices such as piezoelectrics and magnetostrictives are widely used in actuation, and could provide a viable source of rotordynamic vibration control. It is generally acknowledged in the literature that direct implementation of piezoelectric stacks or magnetostrictives in rotordynamic vibration control is usually infeasible, due to the inherently low stroke produced by these devices. Therefore, some form of mechanical amplification is essential, should actuators of these types be considered for rotordynamic applications.

Vibration control may also be achieved by electromagnetic means. This may be implemented either in the form of an Active Magnetic Bearing (AMB) or an Active Magnetic Damper (AMD). A design for the latter is presented in [8], where the authors implement their design in a scaled test rig producing results that demonstrate a reduction in vibration as a result of applying the AMD. The authors then go on to assess their design on a computational model of a real aero-engine, and have obtained promising results.

6. Conclusions

It has been shown through the simulation results that have been presented in this paper that better rotor-dynamic response control can be achieved by replacing the SFD with an actively controlled solution. This is partly because with the latter, not only is it possible to optimise the control forces for each individual running speed, but it is also possible to vary the phase of the control force with respect to either the displacement or velocity, whereas the damping force generated by an SFD will always be in phase with the velocity.

It is acknowledged that the actuator providing the active control forces may not always be able to generate the required magnitude of control force, simply due to physical/material limitations. This effect has been shown in the relevant simulation results above. However, by choosing the specifications of the actuator such that under most conditions, the demands for optimum control can be met, satisfactory overall control performance can be achieved.

The AMD has been mentioned above as a control method that has been tested and delivered promising results. The authors of the present paper believe that it may be possible to find a cheaper, simpler alternative using piezoelectric technology to achieve the same objectives.

References

[1] Zeidan FY, Andres LS, Vance JM, *Design and application of squeeze film dampers in rotating machinery*. Proceedings of the twenty-fifth Turbomachinery Symposium, 1996: p. 169-188.
[2] Rabinowitz MD, Hahn EJ, *Steady-State Performance of Squeeze Film Damper Supported Flexible Rotors*. Journal of Engineering for Power-Transactions of the Asme, 1977. **99**(4): p. 552-558.
[3] Rabinowitz MD, Hahn EJ, *Stability of Squeeze-Film-Damper Supported Flexible Rotors*. Journal of Engineering for Power-Transactions of the Asme, 1977. **99**(4): p. 545-552.
[4] Rabinowitz MD, Hahn EJ, *Experimental Evaluation of Squeeze Film Supported Flexible Rotors*. Journal of Engineering for Power-Transactions of the Asme, 1983. **105**(3): p. 495-503.
[5] Zhu CS, Robb DA, Ewins DJ, *Analysis of the multiple-solution response of a flexible rotor supported on non-linear squeeze film dampers*. Journal of Sound and Vibration, 2002. **252**(3): p. 389-408.
[6] Yakoub RY, El-Shafei A, *The nonlinear response of multimode rotors supported on squeeze film dampers*. Journal of Engineering for Gas Turbines and Power-Transactions of the Asme, 2001. **123**(4): p. 839-848.
[7] Zhao JY, Linnett IW, McLean LJ, *Stability and Bifurcation of Unbalanced Response of a Squeeze Film Damped Flexible Rotor*. Journal of Tribology-Transactions of the Asme, 1994. **116**(2): p. 361-368.
[8] Tonoli A, Silvagni M, Amati N, Staples B, Karpenko E, *Design of electromagnetic damper for aero-engine applications*, in *Ninth International Conference on Vibrations in Rotating Machinery*. 2008: University of Exeter, UK. p. 761-774.

Untwisting the Campbell diagrams of weakly anisotropic rotor systems

Oleg N. Kirillov

Dynamics and Vibration Group, Technische Universität Darmstadt, Hochschulstr. 1, D-64289 Darmstadt, Germany

E-mail: `kirillov@dyn.tu-darmstadt.de`

Abstract. A brake can be modeled as an axi-symmetric rotor perturbed by dissipative, conservative, and non-conservative positional forces originated at the frictional contact with the anisotropic stator. The Campbell diagram of the unperturbed system is a mesh-like structure in the frequency-speed plane with double eigenfrequencies at the nodes. The diagram is convenient for the analysis of the traveling waves in the rotating elastic continuum. Computing sensitivities of the doublets we find that at every particular node the untwisting of the mesh into the branches of complex eigenvalues is generically determined by only four 2x2 sub-blocks of the perturbing matrix. Selection of the unstable modes that cause self-excited vibrations in the subcritical speed range, is governed by the exceptional points at the corners of the singular eigenvalue surface—'double coffee-filter'—which is typical also in the problems of electromagnetic and acoustic wave propagation in non-rotating anisotropic chiral media. As a mechanical example a model of a rotating shaft is studied in detail.

1. Introduction

It is well known that bending waves can propagate in the circumferential direction of an elastic body of revolution rotating about its axis of symmetry [2, 3, 12, 39]. The frequencies of the waves plotted against the rotational speed are referred to as *the Campbell diagram* [4, 39]. Since the spectrum of a perfect rotationally symmetric rotor at standstill has infinitely many double semi-simple eigenvalues—*the doublet modes*—the Campbell diagram contains the eigenvalue branches originated after the splitting of the doublets by the gyroscopic forces [2]. The branches correspond to simple pure imaginary eigenvalues and intersect each other forming a *spectral mesh* [37] in the frequency-speed plane with the double eigenfrequencies at the nodes [23], Fig. 1(a). Dissipative, conservative, and non-conservative perturbations of the axially symmetric rotor, caused by its contact with the anisotropic stator, generically untwist the spectral mesh of pure imaginary eigenvalues of the Campbell diagram into the separate branches of complex eigenvalues in the $(\Omega, \mathrm{Im}\lambda, \mathrm{Re}\lambda)$-space, see Fig. 1(d). Nevertheless, the eigenvalue branches in the perturbed Campbell diagram can both avoid crossings and cross each other, Fig. 1(e). Moreover, the real parts of the perturbed eigenvalues plotted against the rotational speed—*decay rate plots* [39]—can also intersect each other and inflate into "bubbles", Fig. 1(f). This rather complicated behavior is difficult to predict and even to interpret as it was reported in the studies of numerous mechanical systems, see, e.g, [20, 21, 24, 25, 27, 29, 35, 36, 39, 43, 48]. The present work reveals that the untwisting of the Campbell diagrams is determined by a limited number of singular eigenvalue surfaces.

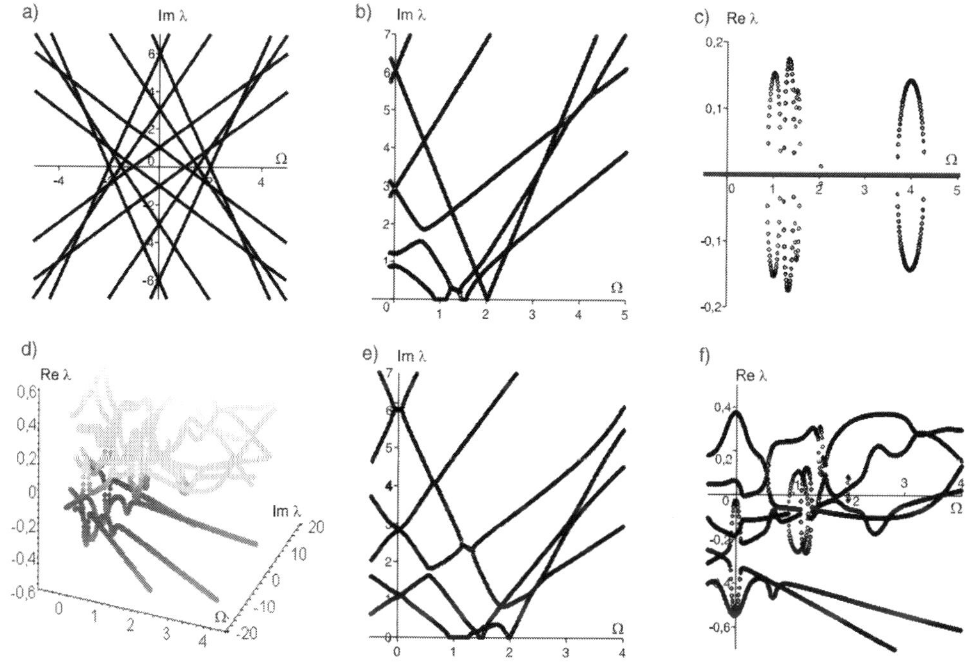

Figure 1. (a) The Campbell diagram of the unperturbed system (2) with 6 d.o.f. in case of $\omega_1 = 1$, $\omega_2 = 3$, and $\omega_3 = 6$; (b) the Campbell diagram and (c) decay rate plots for the stiffness modification $\kappa\mathbf{K}_1$ with $\kappa = 0.2$; (d) untwisting the Campbell diagram in the $(\Omega, \mathrm{Im}\lambda, \mathrm{Re}\lambda)$-space due to perturbation with the matrices $\mathbf{K} = \mathbf{K}_1$, $\mathbf{D} = \mathbf{D}_1$, $\mathbf{N} = \mathbf{N}_1$ and $\kappa = 0.2$, $\delta = 0.1$, and $\nu = 0.2$, (e) the corresponding Campbell diagram and (f) decay rate plots.

2. A model of a weakly anisotropic rotor system

In general, the imperfections in the rotor and stator complicate the linearized equations of motion making them non-self-adjoint with time-dependent coefficients [39]. Nevertheless, an axially symmetric rotor with an anisotropic stator as well as an asymmetric rotor with an isotropic stator are autonomous non-conservative gyroscopic systems [39]. Neglecting the centrifugal stiffness without loss of generality, we consider the finite-dimensional anisotropic rotor system

$$\ddot{\mathbf{x}} + (2\Omega\mathbf{G} + \delta\mathbf{D})\dot{\mathbf{x}} + (\mathbf{P} + \Omega^2\mathbf{G}^2 + \kappa\mathbf{K} + \nu\mathbf{N})\mathbf{x} = 0, \quad (1)$$

which is a perturbation of the isotropic one

$$\ddot{\mathbf{x}} + 2\Omega\mathbf{G}\dot{\mathbf{x}} + (\mathbf{P} + \Omega^2\mathbf{G}^2)\mathbf{x} = 0, \quad (2)$$

where $\mathbf{x} = \mathbb{R}^{2n}$, $\mathbf{P}$=diag$(\omega_1^2, \omega_1^2, \omega_2^2, \omega_2^2, \ldots, \omega_n^2, \omega_n^2)$ is the stiffness matrix, $\mathbf{G} = -\mathbf{G}^T$ is the matrix of gyroscopic forces defined as

$$\mathbf{G} = \mathrm{blockdiag}(\mathbf{J}, 2\mathbf{J}, \ldots, n\mathbf{J}), \quad \mathbf{J} = \begin{pmatrix} 0 & -1 \\ 1 & 0 \end{pmatrix}, \quad (3)$$

and the matrices of damping, $\mathbf{D} = \mathbf{D}^T$, stiffness, $\mathbf{K} = \mathbf{K}^T$, and non-conservative positional forces, $\mathbf{N} = -\mathbf{N}^T$, can depend on the rotational speed Ω. The intensity of the perturbation is controlled by the parameters δ, κ, and ν.

At $\Omega = 0$ the eigenvalues $\pm i\omega_s$, $\omega_s > 0$, of the isotropic rotor (2) are double semi-simple with two linearly independent eigenvectors. The sequence of the frequencies ω_s, where s is an integer index, is usually different for various bodies of revolution. For example, $\omega_s = s$ corresponds to the natural frequency $f_s = \frac{s}{2\pi r}\sqrt{\frac{P}{\rho}}$ of a circular string of radius r, circumferential tension P, and mass density ρ per unit length [25, 27].

Substituting $\mathbf{x} = \mathbf{u}\exp(\lambda t)$ into (2), we arrive at the eigenvalue problem for the operator $\mathbf{L}_0$

$$\mathbf{L}_0(\Omega)\mathbf{u} := (\mathbf{I}\lambda^2 + 2\Omega\mathbf{G}\lambda + \mathbf{P} + \Omega^2\mathbf{G}^2)\mathbf{u} = 0. \tag{4}$$

The block-diagonal structure of the matrices implies eigenvalues of $\mathbf{L}_0$ in the explicit form

$$\lambda_s^+ = i\omega_s + is\Omega, \quad \overline{\lambda_s^-} = -i\omega_s + is\Omega, \quad \lambda_s^- = i\omega_s - is\Omega, \quad \overline{\lambda_s^+} = -i\omega_s - is\Omega, \tag{5}$$

where bar over a symbol denotes complex conjugate. The eigenvectors of λ_s^+ and $\overline{\lambda_s^-}$ are

$$\mathbf{u}_1^+ = (-i, 1, 0, 0, \ldots, 0, 0)^T, \ \mathbf{u}_2^+ = (0, 0, -i, 1, 0, \ldots, 0)^T, \ \ldots \ \mathbf{u}_n^+ = (0, 0, \ldots, 0, 0, -i, 1)^T, \tag{6}$$

where the imaginary unit holds the $(2s-1)$st position in the vector $\mathbf{u}_s^+$. The eigenvectors, corresponding to the eigenvalues λ_s^- and $\overline{\lambda_s^+}$, are simply $\mathbf{u}_s^- = \overline{\mathbf{u}_s^+}$

For $\Omega > 0$, simple eigenvalues λ_s^+ and λ_s^- correspond to the forward and backward traveling waves, respectively, that propagate in the circumferential direction of the rotor. At the angular velocity $\Omega_s^{cr} = \omega_s/s$ the frequency of the sth backward traveling wave vanishes to zero, so that the wave remains stationary in the non-rotating frame. We assume further in the text that the sequence of the doublets $i\omega_s$ has the property $\omega_{s+1} - \omega_s > \Omega_s^{cr}$, which implies the existence of the minimal *critical* speed $\Omega_{cr} = \Omega_1^{cr} = \omega_1$. When the speed of rotation exceeds the critical speed, some backward waves, corresponding to the eigenvalues $\overline{\lambda_s^-}$, travel slower than the disc rotation speed and appear to be traveling forward (reflected waves).

In Fig. 1(a) the mesh of the eigenvalue branches (5) is shown for the 6 d.o.f.-system (2) with the frequencies $\omega_1 = 1$, $\omega_2 = 3$, and $\omega_3 = 6$ that imitate the distribution of the doublets in the spectrum of a circular ring [35]. To illustrate typical untwisting of the Campbell diagram, we plot in Fig. 1(d)-(f) the eigenvalues of the 6 d.o.f.-system (1) with $\kappa = 0.2$, $\delta = 0.1$, $\nu = 0.2$, $\omega_1 = 1$, $\omega_2 = 3$, and $\omega_3 = 6$ for the perturbing matrix $\mathbf{K} = \mathbf{K}_1$, whose non-zero entries are $k_{11} = 1$, $k_{12} = 2$, $k_{13} = 1$, $k_{14} = 2$, $k_{22} = 1$, $k_{23} = 3$, $k_{24} = 4$, $k_{33} = -3$, $k_{44} = -2.5$, $k_{55} = 4$, $k_{66} = 2$, and for the matrices $\mathbf{D} = \mathbf{D}_1$ and $\mathbf{N} = \mathbf{N}_1$, where

$$\mathbf{D}_1 = \begin{pmatrix} -1 & 2 & 1 & 7 & 2 & -2 \\ 2 & 3 & -2 & -4 & 3 & 1 \\ 1 & -2 & 1 & 8 & 2 & 1 \\ 7 & -4 & 8 & 3 & -2 & 3 \\ 2 & 3 & 2 & -2 & 5 & 5 \\ -2 & 1 & 1 & 3 & 5 & 6 \end{pmatrix}, \quad \mathbf{N}_1 = \begin{pmatrix} 0 & -1 & 1 & -1 & -3 & 8 \\ 1 & 0 & 2 & 3 & 2 & 4 \\ -1 & -2 & 0 & 7 & 1 & 3 \\ 1 & -3 & -7 & 0 & 8 & 2 \\ 3 & -2 & -1 & -8 & 0 & 2 \\ -8 & -4 & -3 & -2 & -2 & 0 \end{pmatrix}. \tag{7}$$

In the following we classify and interpret the typical behavior of the eigenvalues of the weakly anisotropic rotor system (1) with the use of the perturbation formula for the double eigenvalues at the nodes of the spectral mesh (5), which we derive in the next section.

3. Perturbation of the doublets

Introducing the indices $\alpha, \beta, \varepsilon, \sigma = \pm 1$ we find that two branches of the spectral mesh $\lambda_s^\varepsilon = i\alpha\omega_s + i\varepsilon s\Omega$ and $\lambda_t^\sigma = i\beta\omega_t + i\sigma t\Omega$ cross each other at $\Omega = \Omega_0$ with the origination of the double eigenvalue $\lambda_0 = i\omega_0$ with two linearly-independent eigenvectors $\mathbf{u}_s^\varepsilon$ and $\mathbf{u}_t^\sigma$, where

$$\Omega_0 = \frac{\alpha\omega_s - \beta\omega_t}{\sigma t - \varepsilon s}, \quad \omega_0 = \frac{\alpha\sigma\omega_s t - \beta\varepsilon\omega_t s}{\sigma t - \varepsilon s}. \tag{8}$$

Let $\mathbf{M}$ be one of the perturbing matrices $\mathbf{D}$, $\mathbf{K}$, or $\mathbf{N}$. In the following, we use the decomposition of the matrix $\mathbf{M} \in \mathbb{R}^{2n\times 2n}$ into n^2 blocks $\mathbf{M}_{st} \in \mathbb{R}^{2\times 2}$, where $s, t = 1, 2, \dots, n$

$$\mathbf{M} = \begin{pmatrix} * & * & * & * & * \\ * & \mathbf{M}_{ss} & \cdots & \mathbf{M}_{st} & * \\ * & \vdots & \ddots & \vdots & * \\ * & \mathbf{M}_{ts} & \cdots & \mathbf{M}_{tt} & * \\ * & * & * & * & * \end{pmatrix}, \quad \mathbf{M}_{st} = \begin{pmatrix} m_{2s-1,2t-1} & m_{2s-1,2t} \\ m_{2s,2t-1} & m_{2s,2t} \end{pmatrix}. \tag{9}$$

Note that $\mathbf{D}_{st} = \mathbf{D}_{ts}^T$, $\mathbf{K}_{st} = \mathbf{K}_{ts}^T$, and $\mathbf{N}_{st} = -\mathbf{N}_{ts}^T$.

We consider a general perturbation of the matrix operator of the isotropic rotor $\mathbf{L}_0(\Omega) + \Delta\mathbf{L}(\Omega)$. The size of the perturbation $\Delta\mathbf{L}(\Omega) = \delta\lambda\mathbf{D} + \kappa\mathbf{K} + \nu\mathbf{N} \sim \varepsilon$ is small, where $\varepsilon = \|\Delta\mathbf{L}(\Omega_0)\|$ is the Frobenius norm of the perturbation at $\Omega = \Omega_0$. For small $\Delta\Omega = |\Omega - \Omega_0|$ and ε the increment to the doublet $\lambda_0 = i\omega_0$ with the eigenvectors $\mathbf{u}_s^\varepsilon$ and $\mathbf{u}_t^\sigma$, is given by the formula $\det(\mathbf{R} + (\lambda - \lambda_0)\mathbf{Q}) = 0$ [33, 34, 44], where the entries of the 2×2 matrices $\mathbf{Q}$ and $\mathbf{R}$ are

$$\begin{aligned} Q_{st}^{\varepsilon\sigma} &= 2i\omega_0(\bar{\mathbf{u}}_s^\varepsilon)^T\mathbf{u}_t^\sigma + 2\Omega_0(\bar{\mathbf{u}}_s^\varepsilon)^T\mathbf{G}\mathbf{u}_t^\sigma, \\ R_{st}^{\varepsilon\sigma} &= (2i\omega_0(\bar{\mathbf{u}}_s^\varepsilon)^T\mathbf{G}\mathbf{u}_t^\sigma + 2\Omega_0(\bar{\mathbf{u}}_s^\varepsilon)^T\mathbf{G}^2\mathbf{u}_t^\sigma)(\Omega - \Omega_0) + i\omega_0(\bar{\mathbf{u}}_s^\varepsilon)^T\mathbf{D}\mathbf{u}_t^\sigma\delta + (\bar{\mathbf{u}}_s^\varepsilon)^T\mathbf{K}\mathbf{u}_t^\sigma\kappa + (\bar{\mathbf{u}}_s^\varepsilon)^T\mathbf{N}\mathbf{u}_t^\sigma\nu. \end{aligned} \tag{10}$$

Calculating the coefficients (10) with the eigenvectors (6) we find the real and imaginary parts of the sensitivity of the doublet $\lambda_0 = i\omega_0$ at the crossing (8) of the branches λ_s^ε and λ_t^σ

$$\begin{aligned} \mathrm{Re}\lambda &= -\frac{1}{8}\left(\frac{\mathrm{Im}A_1}{\alpha\omega_s} + \frac{\mathrm{Im}B_1}{\beta\omega_t}\right) \pm \sqrt{\frac{|c| - \mathrm{Re}c}{2}}, \\ \mathrm{Im}\lambda &= \omega_0 + \frac{\Delta\Omega}{2}(s\varepsilon + t\sigma) + \frac{\kappa}{8}\left(\frac{\mathrm{tr}\mathbf{K}_{ss}}{\alpha\omega_s} + \frac{\mathrm{tr}\mathbf{K}_{tt}}{\beta\omega_t}\right) \pm \sqrt{\frac{|c| + \mathrm{Re}c}{2}}, \end{aligned} \tag{11}$$

where $c = \mathrm{Re}c + i\mathrm{Im}c$ with

$$\begin{aligned} \mathrm{Im}c &= \frac{\alpha\omega_t\mathrm{Im}A_1 - \beta\omega_s\mathrm{Im}B_1}{8\omega_s\omega_t}(s\varepsilon - t\sigma)\Delta\Omega + \kappa\frac{(\alpha\omega_s\mathrm{tr}\mathbf{K}_{tt} - \beta\omega_t\mathrm{tr}\mathbf{K}_{ss})(\alpha\omega_s\mathrm{Im}B_1 - \beta\omega_t\mathrm{Im}A_1)}{32\omega_s^2\omega_t^2} \\ &- \alpha\beta\kappa\frac{\mathrm{Re}A_2\mathrm{tr}\mathbf{K}_{st}\mathbf{J}_{\varepsilon\sigma} - \mathrm{Re}B_2\mathrm{tr}\mathbf{K}_{st}\mathbf{I}_{\varepsilon\sigma}}{8\omega_s\omega_t}, \\ \mathrm{Re}c &= \left(\frac{t\sigma - s\epsilon}{2}\Delta\Omega + \kappa\frac{\beta\omega_s\mathrm{tr}\mathbf{K}_{tt} - \alpha\omega_t\mathrm{tr}\mathbf{K}_{ss}}{8\omega_s\omega_t}\right)^2 + \alpha\beta\frac{(\mathrm{tr}\mathbf{K}_{st}\mathbf{J}_{\varepsilon\sigma})^2 + (\mathrm{tr}\mathbf{K}_{st}\mathbf{I}_{\varepsilon\sigma})^2}{16\omega_s\omega_t}\kappa^2 \\ &- \frac{(\alpha\omega_s\mathrm{Im}B_1 - \beta\omega_t\mathrm{Im}A_1)^2 + 4\alpha\beta\omega_s\omega_t((\mathrm{Re}A_2)^2 + (\mathrm{Re}B_2)^2)}{64\omega_s^2\omega_t^2}. \end{aligned} \tag{12}$$

The complex coefficients A_1, A_2 and B_1, B_2 depend only on those entries of the matrices $\mathbf{D}$, $\mathbf{K}$, and $\mathbf{N}$ that belong to the four 2×2 blocks (9) with the indices s and t

$$\begin{aligned} A_1 &= \delta\lambda_0\mathrm{tr}\mathbf{D}_{ss} + \kappa\mathrm{tr}\mathbf{K}_{ss} + \varepsilon 2i\nu n_{2s-1,2s}, & A_2 &= \sigma\nu\mathrm{tr}\mathbf{N}_{st}\mathbf{I}_{\varepsilon\sigma} + i(\delta\lambda_0\mathrm{tr}\mathbf{D}_{st}\mathbf{J}_{\varepsilon\sigma} + \kappa\mathrm{tr}\mathbf{K}_{st}\mathbf{J}_{\varepsilon\sigma}), \\ B_1 &= \delta\lambda_0\mathrm{tr}\mathbf{D}_{tt} + \kappa\mathrm{tr}\mathbf{K}_{tt} + \sigma 2i\nu n_{2t-1,2t}, & B_2 &= \sigma\nu\mathrm{tr}\mathbf{N}_{st}\mathbf{J}_{\varepsilon\sigma} - i(\delta\lambda_0\mathrm{tr}\mathbf{D}_{st}\mathbf{I}_{\varepsilon\sigma} + \kappa\mathrm{tr}\mathbf{K}_{st}\mathbf{I}_{\varepsilon\sigma}), \end{aligned} \tag{13}$$

where

$$\mathbf{I}_{\varepsilon\sigma} = \begin{pmatrix} \varepsilon & 0 \\ 0 & \sigma \end{pmatrix}, \quad \mathbf{J}_{\varepsilon\sigma} = \begin{pmatrix} 0 & -\sigma \\ \varepsilon & 0 \end{pmatrix}. \tag{14}$$

Therefore, we have identified the elements of the matrices of the perturbation that control the *eigenvalue assignment* [47] near every particular node (Ω_0, ω_0) of the spectral mesh.

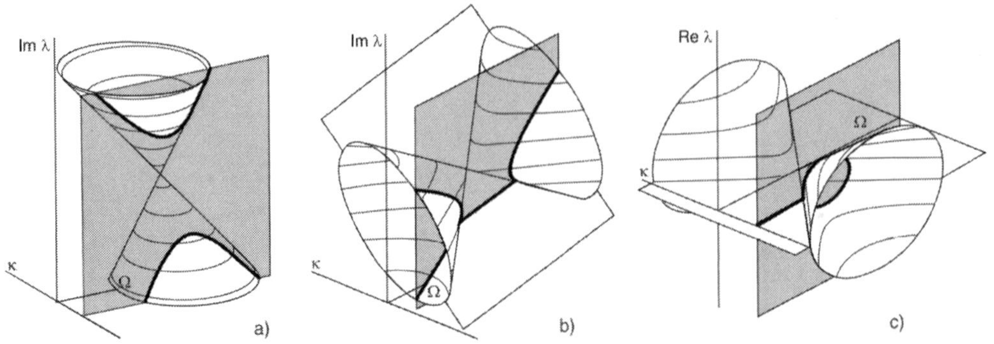

Figure 2. Eigenvalue surfaces (MacKay's cones [15]) and (bold lines) their cross-sections in the plane $\kappa = const$ (grey): (a) a near-vertically oriented cone of imaginary parts in the subcritical range ($\mathrm{Re}\lambda = 0$); (b) imaginary parts forming a near-horizontally oriented cone (15) with the attached membrane (16) and (c) the real parts forming a near-horizontally oriented cone $(\mathrm{Re}\lambda)^2 = -\mathrm{Re}c$ with the attached membrane $\mathrm{Re}\lambda = 0$ in the supercritical speed range.

4. MacKay's eigenvalue cones and instability bubbles due to stiffness modification

Modification of the stiffness matrix induced by the elastic support or by the stationary spring interacting with the rotating continua is typical in the models of rotating shafts [10, 11], computer disc drives [20, 21], circular saws [25, 27, 29], disc brakes [24, 48], and turbine discs [39].

Assuming $\delta = 0$ and $\nu = 0$ in (11) we find that the eigenvalues of the system (5) with the stiffness modification $\kappa\mathbf{K}$ either are pure imaginary ($\mathrm{Re}\lambda = 0$) and form a conical surface in the $(\Omega, \kappa, \mathrm{Im}\lambda)$-space with the apex at the point $(\Omega_0, 0, \omega_0)$, Fig. 2(a),

$$\left(\mathrm{Im}\lambda - \omega_0 - \frac{\kappa}{8}\left(\frac{\mathrm{tr}\mathbf{K}_{ss}}{\alpha\omega_s} + \frac{\mathrm{tr}\mathbf{K}_{tt}}{\beta\omega_t}\right) - \frac{\Omega - \Omega_0}{2}(s\varepsilon + t\sigma)\right)^2 = \mathrm{Re}c \tag{15}$$

or they are complex and in the $(\Omega, \kappa, \mathrm{Re}\lambda)$-space their real parts originate a cone $(\mathrm{Re}\lambda)^2 = -\mathrm{Re}c$ with the apex at the point $(\Omega_0, 0, 0)$, Fig. 2(c). In the $(\Omega, \kappa, \mathrm{Im}\lambda)$-space the corresponding imaginary parts belong to the plane

$$\mathrm{Im}\lambda = \omega_0 + \frac{\kappa}{8}\left(\frac{\mathrm{tr}\mathbf{K}_{ss}}{\alpha\omega_s} + \frac{\mathrm{tr}\mathbf{K}_{tt}}{\beta\omega_t}\right) + \frac{\Omega - \Omega_0}{2}(s\varepsilon + t\sigma), \tag{16}$$

which is attached to the cone (15) as shown in Fig. 2(b).

The existence of eigenvalues with non-zero real part depends on the sign of the product $\alpha\beta$. It is negative only if the crossing in the Campbell diagram is formed by the eigenvalue branch of the reflected wave and by that of either forward- or backward traveling wave. Otherwise, $\alpha\beta > 0$. Due to the property $\omega_{s+1} - \omega_s > \Omega_s^{cr}$ the crossings of the reflected wave with the forward- and backward traveling waves occur only in the *supercritical* speed range $|\Omega| \geq \Omega_{cr}$. The crossings with $\alpha\beta > 0$ are situated in both the super- and *subcritical* ($|\Omega| < \Omega_{cr}$) ranges. Therefore, the eigenvalues with non-zero real part originate only near the supercritical crossings of the eigenvalue branches λ_s^{ε} and λ_t^{σ} with $\alpha\beta < 0$, when the parameters in the (Ω, κ)-plane are in the sector $\mathrm{Re}c < 0$ bounded by the straight lines $\mathrm{Re}c = 0$

$$\kappa = \frac{4(s\varepsilon - t\sigma)(\Omega - \Omega_0)}{\frac{k_{2t-1,2t-1}+k_{2t,2t}}{\beta\omega_t} - \frac{k_{2s-1,2s-1}+k_{2s,2s}}{\alpha\omega_s} \pm 2\sqrt{\frac{(\varepsilon k_{2s-1,2t-1}+\sigma k_{2s,2t})^2+(\varepsilon k_{2s-1,2t}-\sigma k_{2s,2t-1})^2}{-\alpha\beta\omega_s\omega_t}}}. \tag{17}$$

Since for $\alpha\beta < 0$ the cones of the real parts $(\mathrm{Re}\lambda)^2 = -\mathrm{Re}c$ are near-horizontally oriented and extended along the κ-axis in the $(\Omega, \kappa, \mathrm{Re}\lambda)$-space, their cross-sections by the planes $\kappa = const$

are ellipses that are symmetrical with respect to the Ω-axis, as shown in Fig. 1(c) and in Fig. 2(c). Since one-half of the ellipse corresponds to the eigenvalues with positive real parts, it is called the *bubble of instability* [15]. Equation (17) is, therefore, a linear approximation to the boundary of the domain of instability, which is divergence (parametric resonance) for $\Omega_0 = \Omega_s^{cr}$ and flutter (combination resonance) otherwise. The near-horizontal orientation of the corresponding cones of imaginary parts (15) in the $(\Omega, \kappa, \mathrm{Im}\lambda)$-space explains deformation in the presence of the perturbation $\kappa\mathbf{K}$ of the crossings with $\alpha\beta < 0$ into the branches of a hyperbola connected by a straight line in the Campbell diagram, see Fig. 1(b) and Fig. 2(b).

Near the crossings with $\alpha\beta > 0$ the perturbed eigenvalues are pure imaginary (stability). The corresponding cones of imaginary parts (15) are near-vertically oriented in the $(\kappa, \Omega, \mathrm{Im}\lambda)$-space, Fig. 2(a). In the plane $\kappa = const$ this yields the *avoided crossing* [15], which is approximated by a hyperbola shown by the bold lines in Fig. 2(a) (cf. Fig. 1(b)).

The conical singularities of the eigenvalue surfaces in the Hamiltonian systems were known already to Hamilton himself, who predicted the effect of conical refraction of light in birefringent crystals [1, 38]. Later on, the conical singularities of eigenvalue surfaces were found in atomic, nuclear, and molecular physics [6, 7, 22]. Nowadays they bear a name of the Hamilton's *diabolical points* [38]. The existence of the two different orientations of the eigenvalue cones is another fundamental fact of stability theory of Hamiltonian systems established by MacKay in [15] and based on the results of Krein, who introduced the notion of the signature of eigenvalues [14].

To evaluate the Krein signatures, we reduce the system (2) to the form $\dot{\mathbf{y}} = \mathbf{A}\mathbf{y}$, where

$$\mathbf{A} = \begin{pmatrix} -\Omega\mathbf{G} & \mathbf{I}_n \\ -\mathbf{P} & -\Omega\mathbf{G} \end{pmatrix} = \mathbf{J}_{2n}\mathbf{A}^T\mathbf{J}_{2n}, \quad \mathbf{J}_{2n} = \begin{pmatrix} 0 & -\mathbf{I}_n \\ \mathbf{I}_n & 0 \end{pmatrix}, \quad \mathbf{y} = \begin{pmatrix} \mathbf{x} \\ \dot{\mathbf{x}} + \Omega\mathbf{G}\mathbf{x} \end{pmatrix}. \tag{18}$$

The Hamiltonian symmetry of the matrix $\mathbf{A}$ implies its self-adjointness in a Krein space with the indefinite inner product $[\mathbf{a}, \mathbf{b}] = \overline{\mathbf{b}}^T\mathbf{J}_{2n}\mathbf{a}$, $\mathbf{a}, \mathbf{b} \in \mathbb{C}^{2n}$. The matrix $\mathbf{A}$ has the eigenvalues $\lambda_s^{\pm}$ given by the formulas (5) with the eigenvectors

$$\mathbf{a}_s^{++} = \begin{pmatrix} \mathbf{u}_s^+ \\ \lambda_s^+\mathbf{u}_s^+ + \Omega\mathbf{G}\mathbf{u}_s^+ \end{pmatrix}, \quad \mathbf{a}_s^{+-} = \begin{pmatrix} \mathbf{u}_s^- \\ \lambda_s^-\mathbf{u}_s^- + \Omega\mathbf{G}\mathbf{u}_s^- \end{pmatrix}, \tag{19}$$

where the vectors $\mathbf{u}_s^{\pm}$ are determined by expressions (6). Since $i[\mathbf{a}_s^{++}, \mathbf{a}_s^{++}] = i[\mathbf{a}_s^{+-}, \mathbf{a}_s^{+-}] = 4\omega_s > 0$, the eigenvalues λ_s^+ and λ_s^- of the forward and backward traveling waves acquire *positive Krein signature*. The eigenvalues $\overline{\lambda_s^+}$ and $\overline{\lambda_s^-}$ of the reflected waves with $i[\mathbf{a}_s^{-+}, \mathbf{a}_s^{-+}] = i[\mathbf{a}_s^{--}, \mathbf{a}_s^{--}] = -4\omega_s < 0$, have the opposite, *negative Krein signature* [14, 15, 17, 37]. The signature of an eigenvalue in the Campbell diagram coincides with the sign of the doublet at $\Omega = 0$, from which it is branched, and does not change with the variation of Ω. This implies $\alpha\beta > 0$ and near-vertically oriented cones of imaginary parts (15) at the crossings of eigenvalue branches with the *definite* (positive) Krein signature and $\alpha\beta < 0$ and near-horizontally oriented cones of imaginary parts (15) at the crossings with the *mixed* Krein signature [15].

The Krein signature coincides with the sign of the second derivative of the energy, which is a non-degenerate definite quadratic form on the real invariant space associated to a complex conjugate pair of simple pure imaginary non-zero eigenvalues [15]. Interaction of waves with positive and negative energy is a well known mechanism of instability of the moving fluids and plasmas [15, 18]; in rotor dynamics this yields flutter in the supercritical speed range, which is known as the mass and stiffness instabilities [24, 39].

Therefore, in case when anisotropy of the stator is caused by the stiffness modification only, the untwisting of the Campbell diagram is completely described by the one-parameter slices of the two-parameter MacKay's eigenvalue cones. Since there are only two possible spatial orientations of the cones corresponding to either definite or mixed Krein signatures, all one has to do to predict the untwisting of the Campbell diagram into avoided crossings or into bubbles of instability is to calculate the signatures of the appropriate eigenvalues of the isotropic rotor. In the following, we develop the MacKay's theory further and show that even in the presence of non-Hamiltonian perturbations, all the observed peculiarities of the Campbell diagrams and decay rate plots are the one-parameter slices of the eigenvalue surfaces near a limited number of other singularities associated with the definite and mixed Krein signature of eigenvalues.

5. Double coffee-filter singularity near the crossings with definite Krein signature

Understanding the general rules of untwisting the Campbell diagrams of weakly anisotropic rotor systems in the presence of dissipative and non-conservative perturbations is important for the linear stability analysis and for the interpretation of the numerical data in both low- and high-speed applications [39]. In the latter *supercritical flutter and divergence* instabilities are easily excited just by the Hamiltonian perturbations like stiffness modification near the crossings with the mixed Krein signature. Among the low-speed applications the untwisting of the Campbell diagram is directly related to the onset of friction-induced oscillations in brakes, clutches, paper calenders, and even in musical instruments like the glass harmonica [9, 44, 45, 46, 47, 48, 51]. In contrast to the supercritical instabilities, the excitation of the *subcritical flutter* near the crossings with the definite Krein signature by the Hamiltonian perturbations only, is impossible. In this case the non-Hamiltonian dissipative and circulatory forces are required for destabilization.

In general, dissipative, $\delta\mathbf{D}$, and non-conservative, $\nu\mathbf{N}$, perturbations unfold the MacKay's eigenvalue cones (15) and $(\mathrm{Re}\lambda)^2 = -\mathrm{Re}c$ into the surfaces $\mathrm{Im}\lambda(\Omega,\kappa)$ and $\mathrm{Re}\lambda(\Omega,\kappa)$, described by the formulas (11). The new eigenvalue surfaces have singularities at the *exceptional points* [30, 31, 32] that correspond to the double eigenvalues with the Jordan chain that born after the splitting of the double semi-simple eigenvalue $i\omega_0$ at $\Omega = \Omega_0$. In some works numerical methods were developed to find the coordinates of these singularities [16]. Perturbation of the Hamilton's diabolical points is another efficient way to locate the exceptional points [33, 34, 37, 44]. Indeed, from the condition $c = 0$ we easily find their approximate locations in the (Ω,κ)-plane

$$\Omega_{EP}^{\pm} = \Omega_0 + \kappa_{EP}^{\pm}\frac{4\omega_s\omega_t U - \beta\omega_s\mathrm{tr}\mathbf{K}_{tt} + \alpha\omega_t\mathrm{tr}\mathbf{K}_{ss}}{4\omega_s\omega_t(t\sigma - s\varepsilon)}, \quad \kappa_{EP}^{\pm} = \pm\sqrt{\frac{X^2+\alpha\beta(Y^2+Z^2)}{U^2+\alpha\beta(V^2+W^2)}}. \tag{20}$$

The coefficients U, V, W and X, Y, Z in (20) are

$$\begin{aligned} U &= \frac{\mathrm{Re}A_2\mathrm{tr}\mathbf{K}_{st}\mathbf{J}_{\varepsilon\sigma} - \mathrm{Re}B_2\mathrm{tr}\mathbf{K}_{st}\mathbf{I}_{\varepsilon\sigma}}{\alpha\omega_s\mathrm{Im}B_1 - \beta\omega_t\mathrm{Im}A_1}, \quad V = \frac{\mathrm{tr}\mathbf{K}_{st}\mathbf{J}_{\varepsilon\sigma}}{2\sqrt{\omega_s\omega_t}}, \quad W = \frac{\mathrm{tr}\mathbf{K}_{st}\mathbf{I}_{\varepsilon\sigma}}{2\sqrt{\omega_s\omega_t}}, \\ X &= \frac{\alpha\omega_s\mathrm{Im}B_1 - \beta\omega_t\mathrm{Im}A_1}{4\omega_s\omega_t}, \quad Y = \frac{\mathrm{Re}A_2}{2\sqrt{\omega_s\omega_t}}, \quad Z = \frac{\mathrm{Re}B_2}{2\sqrt{\omega_s\omega_t}}. \end{aligned} \tag{21}$$

According to (20), the crossings with the definite Krein signature ($\alpha\beta > 0$) always produce a pair of the exceptional points. For example, for pure non-conservative ($\delta = 0$) and pure dissipative ($\nu = 0$) perturbation of the doublets at $\Omega_0 = 0$, formulas (20) read

$$\Omega_{EP,n}^{\pm} = 0, \quad \kappa_{EP,n}^{\pm} = \pm\frac{2\nu n_{2s-1,2s}}{\rho_1(\mathbf{K}_{ss}) - \rho_2(\mathbf{K}_{ss})}; \quad \Omega_{EP,d}^{\pm} = \pm\delta\frac{\mu_1(\mathbf{D}_{ss}) - \mu_2(\mathbf{D}_{ss})}{4s}, \quad \kappa_{EP,d}^{\pm} = 0, \tag{22}$$

where $\rho_{1,2}(\mathbf{K}_{ss})$ are the eigenvalues of the block $\mathbf{K}_{ss}$ of the matrix $\mathbf{K}$ and $\mu_{1,2}(\mathbf{D}_{ss})$ are the eigenvalues of the block $\mathbf{D}_{ss}$ of the matrix $\mathbf{D}$ [51]. In case of the mixed Krein signature ($\alpha\beta < 0$) the exceptional points exist when the radicand in (20) is positive and does not exist otherwise.

Strong influence of the exceptional points on the stability and their relation to the Ziegler's destabilization paradox due to small damping is well recognized [8, 26, 40, 41, 42, 46]. In

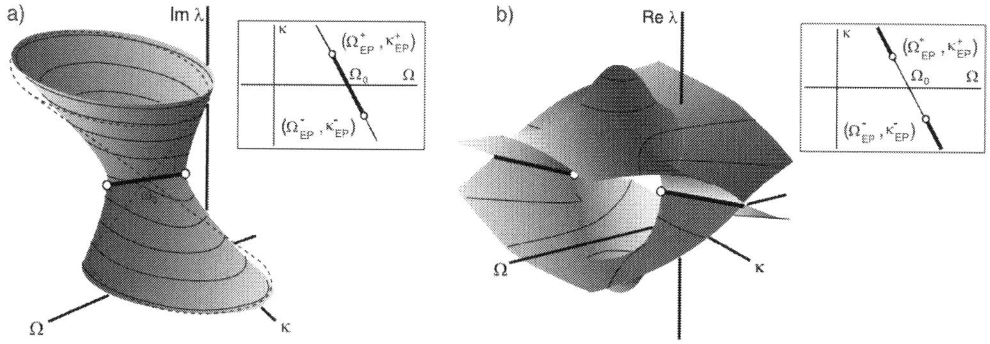

Figure 3. (a) The 'double coffee filter' singular surface $\mathrm{Im}\lambda(\Omega,\kappa)$ with the exceptional points (open circles) and branch cut (bold lines) as a result of the deformation of the MacKay's cone (dashed lines) by the mixed dissipative and circulatory perturbation at any crossing with the definite Krein signature; (b) the corresponding 'viaduct' singular surface $\mathrm{Re}\lambda(\Omega,\kappa)$.

numerous applications in rotor dynamics [20, 21, 24, 25, 27, 29, 39] as well as in hydrodynamics [19], crystal optics [31], acoustics [28], and microwave billiards [30], the generalized crossing scenario in the vicinity of the exceptional points has been observed (visible also in Fig. 1(e,f)) when at the same values of the parameters the imaginary parts of the eigenvalues cross, whereas the real parts don't and vice versa. In our setting, the conditions for coincidence of imaginary parts of the eigenvalues (11) are $\mathrm{Im}c = 0$ and $\mathrm{Re}c \leq 0$ and that for coincidence of the real parts are $\mathrm{Im}c = 0$ and $\mathrm{Re}c \geq 0$. Both real and imaginary parts of the eigenvalues coincide only at the two exceptional points $(\Omega^+_{EP},\kappa^+_{EP})$ and $(\Omega^-_{EP},\kappa^-_{EP})$. The segment of the line $\mathrm{Im}c = 0$ connecting the exceptional points is the projection of the branch cut of a singular eigenvalue surface $\mathrm{Im}\lambda(\Omega,\kappa)$. The adjacent parts of the line correspond to the branch cuts of the singular eigenvalue surface $\mathrm{Re}\lambda(\Omega,\kappa)$. Since simultaneous intersection of the different segments of the line $\mathrm{Im}c = 0$ in the (Ω,κ)-plane is not possible one observes the generalized crossing scenario [30, 33, 34] in the planes $(\Omega,\mathrm{Im}\lambda)$ and $(\Omega,\mathrm{Re}\lambda)$ or $(\kappa,\mathrm{Im}\lambda)$ and $(\kappa,\mathrm{Re}\lambda)$.

For example, in case of pure non-conservative perturbation the real parts of the eigenvalues developing near the doublets with the definite Krein signature at $\Omega_0 = 0$ cross each other in the $(\Omega,\mathrm{Re}\lambda)$-plane at the points of the branch cuts $\kappa^2 > (\kappa^{\pm}_{EP,n})^2$

$$\mathrm{Re}\lambda = \pm\frac{2\nu s n_{2s-1,2s}}{(\rho_1(\mathbf{K}_{ss})-\rho_2(\mathbf{K}_{ss}))\sqrt{\kappa^2-(\kappa^{\pm}_{EP,n})^2}}\Omega + O(\Omega^3), \tag{23}$$

whereas for $\kappa^2 < (\kappa^{\pm}_{EP,n})^2$ they avoid crossing

$$\mathrm{Re}\lambda = \pm\frac{\rho_1(\mathbf{K}_{ss})-\rho_2(\mathbf{K}_{ss})}{4\omega_s}\sqrt{(\kappa^{\pm}_{EP,n})^2-\kappa^2} + O(\Omega^2). \tag{24}$$

At the exceptional points $\kappa = \kappa^{\pm}_{EP,n}$ the eigenvalue branches touch each other at the origin

$$\mathrm{Re}\lambda = \pm\frac{1}{2}\sqrt{\frac{2\nu s n_{2s-1,2s}}{\omega_s}\Omega} + O(\Omega^{3/2}). \tag{25}$$

The degenerate crossing (25) of the real parts has been observed in the model of a rotating circular string passing through the eyelet with friction [25, 44].

Similarly, pure dissipative perturbation of the doublets at $\Omega_0 = 0$ yields crossings of the real parts at the branch cut $\Omega^2 > (\Omega_{EP,d}^{\pm})^2$ in the $(\mathrm{Re}\lambda, \kappa)$-plane and veering of the imaginary parts

$$\mathrm{Im}\lambda = \omega_s \pm s\sqrt{\Omega^2 - (\Omega_{EP,d}^{\pm})^2} + O(\kappa), \quad \mathrm{Re}\lambda = -\frac{\delta \mathrm{tr}\mathbf{D}_{ss}}{4} \pm \frac{\gamma}{16 s\omega_s\sqrt{\Omega^2 - (\Omega_{EP,d}^{\pm})^2}}\delta\kappa + O(\kappa^3), \quad (26)$$

where $\gamma = 2\mathrm{tr}\mathbf{K}_{ss}\mathbf{D}_{ss} - \mathrm{tr}\mathbf{K}_{ss}\mathrm{tr}\mathbf{D}_{ss}$. At the branch cut $\Omega^2 < (\Omega_{EP,d}^{\pm})^2$ the imaginary parts cross and the real parts avoid crossing

$$\mathrm{Im}\lambda = \omega_s + \frac{\mathrm{tr}\mathbf{K}_{ss}}{4\omega_s}\kappa \pm \frac{\gamma}{16 s\omega_s\sqrt{(\Omega_{EP,d}^{\pm})^2 - \Omega^2}}\delta\kappa + O(\kappa^2), \quad \mathrm{Re}\lambda = -\frac{\delta \mathrm{tr}\mathbf{D}_{ss}}{4} \pm s\sqrt{(\Omega_{EP,d}^{\pm})^2 - \Omega^2} + O(\kappa^2). \quad (27)$$

At $\Omega = \Omega_{EP,d}^{\pm}$ the crossings of both the real and imaginary parts are degenerate

$$\mathrm{Re}\lambda = -\frac{\delta \mathrm{tr}\mathbf{D}_{ss}}{4} \pm \frac{1}{4}\sqrt{-\delta\kappa\frac{\gamma}{\omega_s}} + O(\kappa^{3/2}), \quad \mathrm{Im}\lambda = \omega_s \pm \frac{1}{4}\sqrt{-\delta\kappa\frac{\gamma}{\omega_s}} + \frac{\mathrm{tr}\mathbf{K}_{ss}}{4\omega_s}\kappa + O(\kappa^{3/2}). \quad (28)$$

The evolving eigenvalue branches reconstruct the eigenvalue surfaces shown in Fig. 3. The qualitative changes of the eigenvalue branches in the one-parameter slices of the surfaces from the crossing to the avoided crossing due to variation of the parameters Ω and κ are caused by the passage through the exceptional points, where the branches touch each other and the eigenvalue surfaces have Whitney's umbrella singularities. The surface of the imaginary parts shown in Fig. 3(a) is formed by the two Whitney's umbrellas with the handles (branch cuts) glued when they are oriented toward each other. This singular surface is known in the physical literature on wave propagation in anisotropic media as the *double coffee filter* [30, 31]. The *viaduct* singular surface of the real parts results from the gluing of the roofs of two Whitney's umbrellas when their handles are oriented outwards, Fig. 3(b). The double coffee filter singularity is a result of the deformation of the MacKay's eigenvalue cone (shown by the dashed lines in Fig. 3(a)) by the dissipative and non-conservative perturbations. These perturbations foliate the plane $\mathrm{Re}\lambda = 0$ into the viaduct singular surface which has self-intersections along the two branch cuts and an ellipse-shaped arch between the two exceptional points, Fig. 3(b). Both types of singular surfaces appear when non-Hermitian perturbation of Hermitian matrices is considered [13, 33, 34].

Therefore, in a weakly non-Hamiltonian system (1) the fundamental qualitative effect of the splitting of the doublets with the definite Krein signature is the origination of the double coffee filter singular surface of the imaginary parts and the viaduct singular surface of the real parts. Structural modification of the matrices of dissipative and non-conservative forces generically does not change the type of the surfaces, preserving the exceptional points and the branch cuts.

6. Example. Rotating shaft

Simplest mechanical examples described by equations (1) and (2) are some two-degrees-of-freedom models of rotating shafts [5, 10, 11, 39]. In [10] the shaft is modeled as the mass m which is attached by two springs with the stiffness coefficients k_1 and $k_2 = k_1 + \kappa$ and two dampers with the coefficients μ_1 and μ_2 to a coordinate system rotating at constant angular velocity Ω, Fig. 4(a). A non-conservative positional force βr acts on the mass. With u and v representing the displacements in the direction of the two rotating coordinate axes, respectively, the system is governed by the equations [10]

$$\begin{aligned} m\ddot{u} + \mu_1\dot{u} - 2m\Omega\dot{v} + (k_1 - m\Omega^2)u + \beta v &= 0, \\ m\ddot{v} + \mu_2\dot{v} + 2m\Omega\dot{u} + (k_2 - m\Omega^2)v - \beta u &= 0. \end{aligned} \quad (29)$$

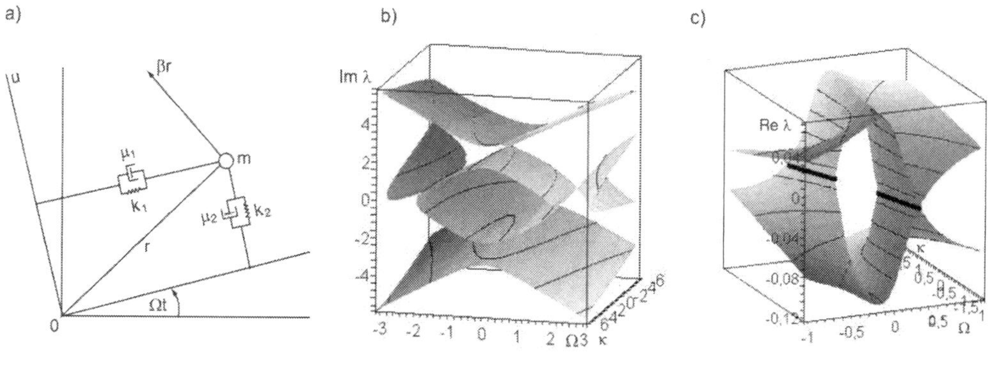

Figure 4. (a) A model of the rotating shaft; (b) four MacKay's cones due to stiffness modification ($\mu_1 = 0$, $\mu_2 = 0$, $\beta = 0$); (c) the viaduct singular surface created by the indefinite damping ($\mu_1 = -0.1$, $\mu_2 = 0.2$) and circulatory force ($\beta = 0.2$).

In Fig. 4(b) we show the numerically found surface of frequencies for the shaft with $m = 1$ and $k_1 = 4$ in the absence of damping and non-conservative forces. The surface has four conical singularities corresponding to the double eigenvalues $\pm 2i$ at $\Omega = 0$ and to the double zero eigenvalues at the critical speeds $\Omega = \pm 2$. The cones in the subcritical speed range are near-vertically oriented while those at the critical speeds are near-horizontal [15]. Consequently, for small stiffness detuning κ the system is stable in the subcritical speed range and unstable by divergence in the vicinity of the critical speeds, where the bubbles of instability in the decay rate plots originate. Addition of the non-conservative forces with $\beta = 0.2$ and indefinite damping with $\mu_1 = -0.1$ and $\mu_2 = 0.2$ yields deformation of the conical surfaces with the apexes at $\Omega = 0$ into the double coffee-filters. The real parts form the viaduct singular surface shown in Fig. 4(c).

7. Conclusion

In the article we aimed at the finding of fundamental properties of a class of non-conservative rotor systems that would unify different linearised models of rotating elastic continua in frictional contact. We have found two types of singular eigenvalue surfaces that exist unavoidably in such models and cannot be destroyed or qualitatively modified by generic changes in the matrices of the system. The double coffee filter and the viaduct singular surface are the imaginary and the real part of the unfolding of any double pure imaginary semi-simple eigenvalue at the crossing of the Campbell diagram with the definite Krein signature. The structure of the perturbing matrices determines only the details of the geometry of the surfaces, such as the coordinates of the exceptional points and the spacial orientation of the branch cuts. It does not yield the qualitative changes irrespective of whether the dissipative and circulatory perturbations are applied separately or in a mixture. For practical applications this means that the behaviour of eigenvalues is described in the same manner for different models of rotor-stator interaction. This will help practitioners and engineers in the development of better models, in their comparison, and in the interpretation of the data both from the numerical and physical experiments. On the other hand, our qualitative results connect the problems of wave propagation in rotating continua with that of electromagnetic and acoustic wave propagation in stationary anisotropic and chiral media [32]. Therefore, the ideas and methods of the latter discipline can be used for more refined investigation of the rotating elastic continua. The qualitative results, the perturbation methodology and the revealed connection to the other physical disciplines can be employed for the improvement of the existing FEM codes aimed at the modeling and optimization of the elements of rotating machinery.

Acknowledgments
The work has been supported by the research grant DFG HA 1060/43-1.

References

[1] Hamilton W.R. 1833 *Dublin Univ. Rev. Q. Mag.* **1** 795.
[2] Bryan G. 1890 *Proc. Cam. Philos. Soc.* **7** 101.
[3] Lamb H., Southwell R.V. 1921 *Proc. R. Soc. London* A **99** 272.
[4] Campbell W. 1924 *Trans. A.S.M.E.* **46** 31.
[5] Kimball A.L. 1924 *Gen. Elec. Rev.* **27** 244.
[6] Von Neumann J., Wigner E.P. 1929 *Z. Phys.* **30** 467.
[7] Teller E. 1937 *J. Phys. Chem.* **41** 109.
[8] Bottema O. 1956 *Indag. Math.* **18** 403.
[9] Spurr R.T. 1961 *Wear* **4** 150.
[10] Shieh R.C., Masur E.F. 1968 *Z. angew. Math. Phys.* **19** 927.
[11] Crandall S.H. 1970 *J. Sound. Vibr.* **11**(1) 3.
[12] Srinivasan A.V., Lauterbach G.F. 1971 *Trans. ASME. Journal of Engineering for Industry* **93** 1229.
[13] Kahan W. 1975 *Proc. Amer. Math. Soc.* **48** 11.
[14] Krein, M.G. 1983 *Topics in Differential and Integral Equations and Operator Theory* (Birkhäuser, Basel).
[15] MacKay, R.S. 1986 *In: Nonlinear phenomena and chaos, ed. S. Sarkar* (Bristol, Adam Hilger).
[16] Jones C.A. 1988 *Quart. J. Mech. Appl. Math.* **41**(3) 363.
[17] Arnold V.I., Avez A. 1989 *Ergodic Problems of Classical Mechanics* (Addison-Wesley)
[18] Stepanyants Yu.A., Fabrikant A.L. 1989 *Sov. Phys. Uspekhi* **32**(9), 783.
[19] Or A.C. 1991 *Quart. J. Mech. Appl. Math.* **44**(4) 559.
[20] Ono K., Chen J.-S., Bogy D.B. 1991 *Trans. ASME, J. Appl. Mech.* **58** 1005.
[21] Chen J.-S., Bogy D.B. 1992 *Trans. ASME, J. Appl. Mech.* **59** 390.
[22] Mondragon A., Hernandez E. 1993 *J. Phys. A: Math. Gen.* **26** 5595.
[23] Dellnitz M., Melbourne I. 1994 *J. of Comput. and Appl. Math.* **55** 249.
[24] Mottershead J.E., Chan S.N. 1995 *Trans. ASME, J. Vibr. Acoust.* **177** 161.
[25] Yang L., Hutton S.G. 1995 *J. Sound Vibr.* **185** 139.
[26] Hoveijn I., Ruijgrok M. 1995 *Z. angew. Math. Phys.* **46** 384.
[27] Tian J., Hutton S.G. 1999 *Trans. ASME. J. Appl. Mech.* **66** 800.
[28] Shuvalov A.L., Scott N.H. 2000 *Acta Mech.* **140** 1.
[29] Xiong L.G., Chen H., Yi J.M. 2002 *J. Mater. Proc. Techn.* **129** 534.
[30] Keck F., Korsch H.J., Mossmann S. 2003 *J. Phys. A: Math. Gen.* **36** 2125.
[31] Berry M.V., Dennis M.R. 2003 *Proc. R. Soc. Lond.* A **459** 1261.
[32] Berry M.V. 2004 *Czech. J. Phys.* **54** 1039.
[33] Kirillov O.N., Mailybaev A.A., Seyranian A.P. 2005 *J. Phys. A: Math. Gen.* **38** 5531.
[34] Kirillov O.N., Mailybaev A.A., Seyranian A.P. 2005 *Dokl. Phys.* **50** 577.
[35] Canchi S.V., Parker R.G. 2006 *J. Sound. Vibr.* **293** 360.
[36] Young T.H., Lin C.Y. 2006 *J. Sound. Vibr.* **298** 307.
[37] Günther U., Kirillov O.N. 2006 *J. Phys. A: Math. Gen.* **39** 10057.
[38] Berry M.V., Jeffrey M.R. 2007 *In: Progress in Optics, 50, E. Wolf, ed.* (Elsevier).
[39] Genta G. 2007 *Dynamics of rotating systems* (Springer)
[40] Kirillov O.N. 2007 *Intern. J. of Non-Lin. Mech.* **42** 71.
[41] Krechetnikov R., Marsden J.E. 2007 *Rev. Mod. Phys.* **79** 519.
[42] Kirillov O.N. 2007 Dokl. Math. **76**(2) 780.
[43] Lesaffre N., Sinou J.-J., Thouverez F. 2007 *Eur. J. Mech. - A/Sol.* **26**(3), 541.
[44] Kirillov O.N. 2008 *Proc. R. Soc.* A **464** 2321.
[45] Kang J., Krousgrill C.M., Sadeghi F. 2008 *J. Sound Vibr.* **316** 164.
[46] Chevillot F., Sinou J.-J., Mazet G.-B., Hardouin N. and Jézéquel L. 2008 *Arch. Appl. Mech.* **78**, 949.
[47] Ouyang H. 2008 *Mechanical Systems and Signal Processing*, doi: 10.1016/j.ymssp.2008.08.001
[48] Spelsberg-Korspeter G., Hochlenert D., Kirillov O.N., Hagedorn P. 2009 *Trans. ASME, J. Appl. Mech.* **76**.
[49] Kirillov O.N. 2009 *Physics Letters A* **373**(10), 940.
[50] Kirillov O.N., Günther U., Stefani F. 2009 *Phys. Rev. E.* **79**(1) 016205.
[51] Kirillov O.N. 2009 *Int. J. of Vehicle Design* **49**(1)

Modal testing and finite element model updating of laser spot welds

N. Abu Husain, H. Haddad Khodaparast, A. Snaylam, S. James, M. Sharp, G. Dearden, H. Ouyang

Department of Engineering, Harrison Hughes Building, University of Liverpool, L69 3GH, United Kingdom

h.ouyang@liverpool.ac.uk

Abstract. Spot welds are used extensively in automotive engineering. One of the latest manufacturing techniques for producing spot welds is Laser Welding. Finite element (FE) modelling of laser welds for dynamic analysis is a research issue because of the complexity and uncertainty of the welds and thus formed structures. In this work, FE model of the welds is developed by employing CWELD element in NASTRAN and its feasibility for representing laser spot welds is investigated. The FE model is updated based on the measured modal data of hat-plate structures and cast as a structural minimisation problem by the application of NASTRAN codes.

1. Introduction

A typical vehicle body-in-white (BIW) is made of metal sheets, which are connected together by thousands of spot welds that are usually manufactured by Resistance Spot Welding (RSW). However, due to strong interest in the use of advanced and lightweight materials (e.g., aluminium alloy, which is very problematic if welded using the conventional RSW) for BIW applications in recent years, the use of Laser Welding for producing spot welds is receiving increased consideration.

The welds in vehicle structures not only provide connections between sheets of metals but also significantly contribute to the vehicle's dynamic characteristics. Due to this fact, precise theoretical tools must be developed to predict their behaviour in order to make the most of the weld joints in automotive applications. While experimental work could offer the required physical understanding concerning the behaviour of the joints, predictive tasks such as design, analysis and evaluation of welded structures are frequently accomplished by computational methods.

Even though computational methods are widely used in predicting the behaviour of welds on a car structure, modelling the weld itself is complicated. This is mainly because of the existence of many local effects (such as geometrical irregularities, material inhomogeneity and defects in the welds) that are not taken into account by FE modelling [1]. Many studies related to FE modelling of laser welding focus mainly on simulating the welding process itself [2, 3]. There are also works being done on fatigue estimation of laser welds [4]. However, to the authors' best knowledge, there is no work on modelling the dynamic behaviour of structures with laser welds.

In this paper, a general connection element called CWELD in MSC.NASTRAN^{TM} [5] is used to represent the weld joints. This element establishes the prospect to generate weld joints that are independent of the mesh, which allows a huge reduction in computational effort and model development time. CWELD is commonly employed to represent the construction of spot welds that is

Journal of Physics: Conference Series **181** (2009) 012024 doi:10.1088/1742-6596/181/1/012024

generally produced by RSW. It is uncertain whether the element can well represent spot welds made by other processes, especially Laser Welding as the construction of the joint is very different from the conventional RSW. It is therefore the aim of this paper to investigate the feasibility of utilising CWELD for representing spot welds made by Laser Welding.

This paper also explains the procedures to correlate FE models regarding the test measurements. The FE models are updated accordingly to minimise the discrepancies between the analytical and experimental results, and the feasibility of CWELD is evaluated.

2. Description of hat-plate structure

The hat-plate structure is designed to represent common structures used in the construction of a car BIW. It consists of a flat plate and a formed hat-like shell (or 'top-hat') joined together by spot welds at the flanges, as shown in Figure 1. The spot welds, which are produced by Laser Welding, are 5 mm in diameter and 60 mm apart. The length of the structure is 564 mm and the width is 110 mm. The plates are made of cold rolled mild steel sheet metal with the thickness of 1.5 mm. A set of nine identical pairs of the structures are built, each having the same nominal dimensions. In order to reduce manufacturing variability, a general procedure as outlined by Mottershead et al. [6] is followed.

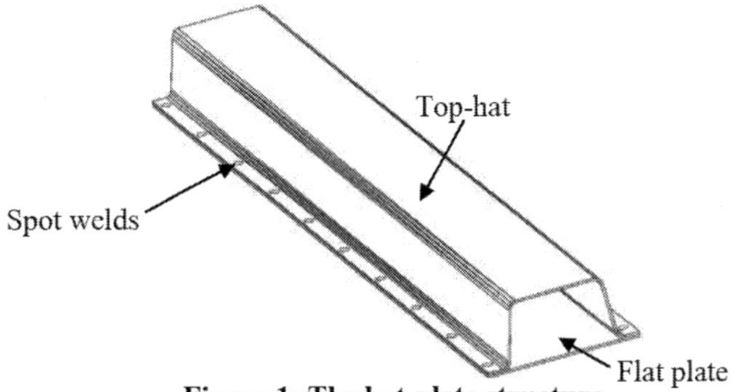

Figure 1: The hat-plate structure

3. Initial FE modelling, modal testing and manual tuning on separate components

Development of the FE models of the components involves four stages [7]: (1) initial FE modelling, (2) modal testing, (3) manual tuning, and (4) model updating. This section describes the development of the initial FE model for the components as well as their modal tests. Then, the manual model tuning is discussed. Model updating is described in Section 4.

3.1. Initial FE modelling

The FE models of the top-hat and flat plate are developed using MSC-NASTRAN shell element (CQUAD4) [8], as depicted in Figure 2. The flat plate's FE model consists of a total of 300 elements and 350 nodes; the top-hat's FE model consists of 8648 elements and 8883 nodes. Nominal values for the material properties (Young's modulus and density) of mild steel are used for both models.

The first five natural frequencies are tabulated in Tables 1 and 2. These FE results are validated in the next section by comparing with the experimental modal data.

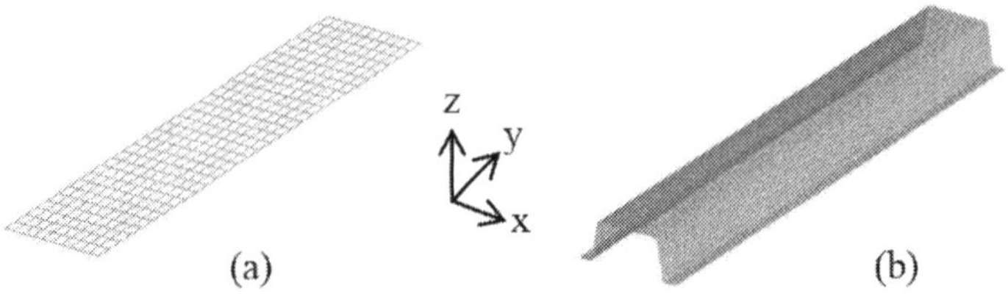

Figure 2: The FE models of (a) the flat plate (b) the top-hat

3.2. Modal testing

Modal testing based on frequency response function (FRF) measurements was carried out to verify the modes of vibration obtained in the initial FE model. Modal testing with free-free boundary condition was conducted on each specimen from both sets of components. The flat plates were tested using one hammer point and two measurement points, while the top-hats were tested using one hammer point and five measurement points. The flat plates were tested in the range of 15 to 300 Hz and the top-hats were tested in the range of 50 to 1000 Hz.

The vibration responses from both tests were measured using 12-channels LMS data acquisition system. Natural frequencies and mode shapes were extracted from the measured FRF by using a PolyMAX [9] curve-fitting procedure from LMS Test.Lab [10] system. The average experimental natural frequencies from the tests are given in Tables 1 and 2. As can be seen from the tables, the computed natural frequencies deviate significantly from the experimental values. This indicates that the initial FE model is not good enough and needs to be tuned and updated.

3.3. FE manual tuning

All modes identified experimentally have their counterparts in the initial FE models. The experimental natural frequencies for the bending modes of the flat plate (i.e., modes 1, 2 and 4) are overestimated, with the biggest frequency error of approximately 4.5%. The frequency errors for the top-hat are also poor, with the highest being about 5.4%.

In order to improve the results, FE manual tuning is carried out. The thicknesses of all nine flat plates are measured at different places and a mean value of 1.4489 mm is used in the manual tuning procedure for both components. Furthermore, a closer inspection on the top-hats reveals that the fold radii are approximately 4 mm, which is 1 mm smaller than the specified radii. The radii corrections are incorporated in the top-hat manual tuning.

When the changes are incorporated in the FE models, the correlations are improved considerably, especially on the flat plate. It is also found that the tuned FE model of the flat plate improves the bending modes but degrades the torsion modes (error of approximately 3.8%). For the top-hat, the error of the three lowest modes is increased (with the minimum value being approximately 6.2%) but the other modes are improved significantly. Due to the big errors, updating has to be carried out in order to improve the results. The frequency errors for the manually-tuned models are presented in Tables 1 and 2.

4. Updating of the models

The updating procedure is performed with the aim to further improve the FE models by minimising the differences between the analytical and experimental models. Therefore, it is cast as a structural optimisation problem [11-13] and the optimisation algorithm of the FE code NASTRAN is used to perform the updating.

Table 1: Correlation of frequencies ω between experimental, initial and manually-tuned FE model for the flat plate

Mode	I Experiment (Hz)	II Initial FE model (Hz)	III Error (%) = |(II – I)/I|	IV Manually-tuned (Hz)	V Error (%) = |(IV – I)/I|
1	24.20	25.14	3.89	24.28	0.34
2	67.10	69.84	4.09	67.46	0.54
3	78.07	77.77	0.39	75.13	3.77
4	132.24	138.26	4.55	133.55	0.99
5	159.54	159.49	0.03	154.07	3.43

Table 2: Correlation between experimental, initial and manually-tuned FE model for the top-hat

Mode	I Experiment (Hz)	II Initial FE model (Hz)	III Error (%) = |(II – I)/I|	IV Manually-tuned (Hz)	V Error (%) = |(IV – I)/I|
1	70.04	70.79	1.07	67.85	3.13
2	272.72	258.12	5.35	255.91	6.16
3	286.67	275.68	3.83	272.86	4.82
4	333.58	348.05	4.34	336.47	0.87
5	394.76	411.91	4.34	388.06	1.70

4.1. Parameter sensitivities

Selecting the updating parameters θ is a vital step in model updating [7, 11, 14]. The number of parameters should be kept to a minimum to avoid problems due to ill-conditioning [7, 11]. Having done a sensitivity analysis, two parameters are selected for the flat plate and four parameters are chosen for the top-hat, as tabulated in Table 3. The parameters are normalised in the form of $\boldsymbol{\delta\theta/\theta}_0$. So, the sensitivities become $(\boldsymbol{\delta\omega/\delta\theta})\boldsymbol{\theta}_0$, where δ denotes a small increment.

Table 3: Normalised sensitivities associated to initial values of updating parameters $\boldsymbol{\theta}_0$

Component	Parameter	Natural frequencies				
		ω_1	ω_2	ω_3	ω_4	ω_5
Flat plate	Young's modulus, E_p	12.08	33.39	37.14	65.62	75.96
	Density, ρ_p	-12.08	-33.38	-37.15	-65.61	-75.98
Top-hat	Fold thickness, t_1	9.63	19.12	25.55	46.76	28.91
	Flange thickness, t_2	1.12	-52.22	-46.24	-30.79	-24.85
	Sidewall thickness, t_3	21.28	19.04	35.35	34.46	57.26
	Top thickness, t_4	34.87	269.77	257.02	63.20	187.46

4.2. Target response selection

Target responses should be selected among those measured. With confidence in the quality of the experimental data, all measured natural frequencies of the flat plate and the top-hat are targeted in the updating process.

4.3. Updating and its results

NASTRAN optimisation algorithm permits the specification of objective function to be minimised in the updating procedure, as follows [15]

$$J = \sum_{i=1}^{n}\left(\frac{\omega_i}{\omega_i^{\exp}} - 1\right)^2 \tag{1}$$

where ω_i is the ith computed natural frequency, $\omega_i^{\exp}$ is the ith experimental natural frequency and n is the number of measured frequencies.

Tables 4 and 5 show the updating results for the flat plate and the top-hat, while Table 6 shows the change in the updating parameters from the initial values. The parameters for the flat plate are allowed to change from 85 to 110% of the initial values, while the change of the top-hat's parameters is from 85 to 120% of their initial values. The updating process for the flat plate converges after two iterations and the top-hat reaches convergence after seven iterations.

The updated results of the flat plate, although reasonable, are not as good as expected. This is probably due to the variations of the material properties along the length of the plate, which is not incorporated in the updating procedure. The top-hat's updating results, however, are improved significantly. The updated FE models of both components are used in modelling the complete welded structure. By ignoring the uncertainties due to manufacturing when welding the complete samples, the modelling uncertainties from the complete model will merely be due to the weld modelling.

Table 4: Flat plate updating - experimental and updated FE model

Mode	I Experiment (Hz)	II Updated FE model (Hz)	VI Error (%) = |(II – I)/I|
1	24.20	24.53	1.37
2	67.10	68.16	1.58
3	78.07	75.91	2.77
4	132.24	134.93	2.03
5	159.54	155.66	2.43

Table 5: Top-hat updating - experimental and updated FE model

Mode	I Experiment (Hz)	II Updated FE model (Hz)	III Error (%) = |(II – I)/I|
1	70.04	70.03	0.01
2	272.72	272.45	0.10
3	286.67	287.09	0.15
4	333.58	333.55	0.01
5	394.76	394.57	0.05

Table 6: Parameter changes due to updating

Component	Parameter	I Initial value	II Updated value	Changes (%) =|(II – I)/I|
Flat plate	Young's modulus, E_p (GPa)	210	212	1.03
	Density, ρ_p (kg/m^3)	7860	7779	1.03
Top-hat	Fold thickness, t_1 (mm)	1.4489	1.4062	2.95
	Flange thickness, t_2 (mm)	1.4489	1.5580	7.53
	Sidewall thickness, t_3 (mm)	1.4489	1.2370	14.63
	Top thickness, t_4 (mm)	1.4489	1.6254	12.18

5. FE modelling of the complete structure

Using the updated FE models of the flat plate and the top-hat, the FE modelling for the complete welded hat-plate is carried out.

5.1. Representation of CWELD

CWELD (Fig. 3) is developed using a 2-noded special shear flexible beam type element [16] with 12 degrees of freedom (six for each node) and each node is connected to its corresponding patch. In modelling the welded hat-plate, 20 CWELD elements are employed to represent the 20 spot welds. The initial values for the material properties and the diameter of the CWELD elements are defined, with the diameter being 5 mm and the material properties are assumed to be the same as the bulk material's nominal values. The potential of CWELD for updating is investigated and briefly described in Section 7.

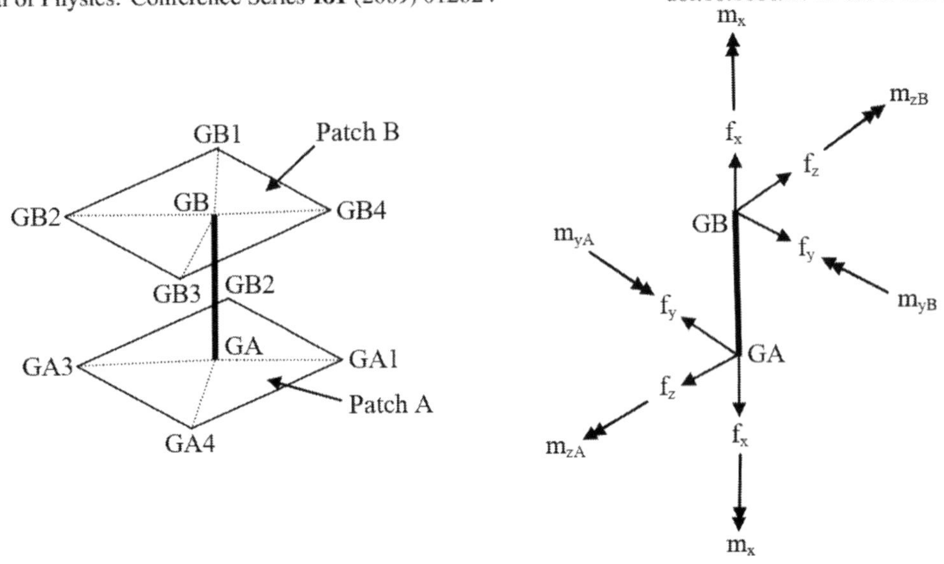

Figure 3: Representation of CWELD element

5.2. Initial FE results

The initial FE results are tabulated in Table 7 (Column II) and the mode shapes are given in Figure 4. These FE results are validated by comparing with the experimental findings, which is explained in the next section.

6. Modal tests on the welded structures

Free-free hammer tests with three hammer points and seven measurement points were performed on the welded structures. As in the previous tests, the vibration response (in the range of 400 to 1000 Hz) was measured using the 12-channels LMS system and extracted using the PolyMAX curve-fitting procedure. The average measured natural frequencies for the samples are shown in Table 7 (Column I).

All the five modes identified experimentally have their counterparts in the initial FE model. Both experimental and numerical findings are compared and the modes are found to be very well correlated. However, the natural frequencies of all the experimental modes are significantly underestimated by the numerical model, with the frequency errors being about 10% and more. The big errors show the critical need for updating to be carried out.

7. Model updating of the complete model

Following the findings explained in the previous section, it is very important for the initial model of the hat-plate to be updated. Several parameters are investigated for updating but only three of them are selected for the procedure, as explained in the following section.

7.1. Parameters sensitivities

Several parameters from the spot weld and the patch of the welds (such as the diameter, thicknesses, Young's moduli, densities and Poisson's ratios) are investigated by carrying out the sensitivity analysis. The patch parameters are included because it is found that the weld parameters alone cannot successfully improve the results of the initial model. Of all the possible parameters, the weld diameter (d_w) and the Young's moduli of the weld and patch (E_w and E_{patch}) are found to be sensitive enough to be selected for the updating process. It is also found that the natural frequencies of the welded structure are all more sensitive to d_w than E_w and E_{patch}, as shown in Table 8.

7.2. Model updating process

Updating is carried out on the basis of the first five measured frequencies from the welded samples, and is completed after three iterations. The natural frequencies of the updated model are presented in Table 7 (Column IV), while the initial and final values of the parameters are shown in Table 9. The values of the weld parameters (i.e., d_w and E_w) are constrained to within a reasonable range. However, the Young's modulus of the patch is allowed to have a very big variation due to uncertainties from the patch properties. The initial value of E_{patch} is made very high to justify for the rigidness of the patch in comparison to the adjacent bulk materials. From the updating procedure, it is found that all the natural frequencies are improved significantly, although the error from the first frequency is not as small as the others.

Table 7: Natural frequencies for the complete structures

Mode (Experiment)	I Experiment (Hz)	II Initial FE model (Hz)	III Error (%) = |(II – I)/I|	IV Updated FE model (Hz)	V Error (%) = |(IV – I)/I|
1	508.12	459.20	9.63	492.39	3.10
2	553.69	472.96	14.58	552.73	0.17
3	575.39	505.31	12.18	577.72	0.40
4	627.45	526.55	16.08	636.18	1.39
5	643.66	532.36	17.29	640.11	0.55

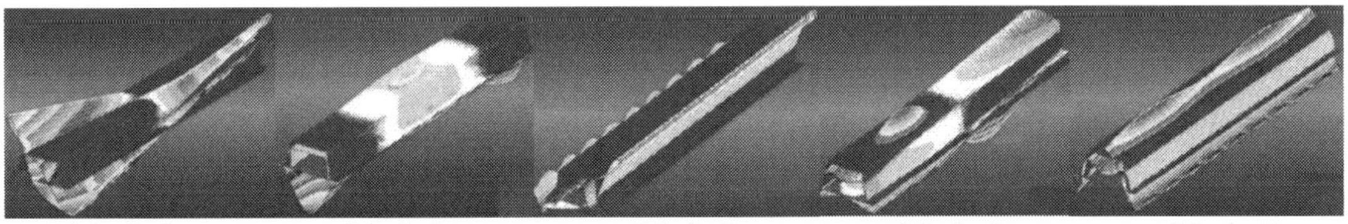

Mode 1 Mode 2 Mode 3 Mode 4 Mode 5

Figure 4: The mode shapes of the hat-plate

Table 8: Initial normalised sensitivities associated with spot weld parameters

Parameter	Natural frequency for corresponding mode				
	ω_1	ω_2	ω_3	ω_4	ω_5
Weld diameter, d_w	4.26	6.35	160.77	148.86	168.09
Weld Young's modulus, E_w	1.19	1.60	4.21	3.73	4.23
Patch Young's modulus, E_{patch}	1.54	2.26	5.33	4.88	5.59

Table 9: Spot weld parameters changes due to updating

Parameter	I Initial value	II Updated value	Changes (%) =|(II – I)/I|
Diameter, d (mm)	5	5.4	8.00
Young's modulus of Weld, E_w (GPa)	210	220	4.50
Young's modulus of Patch, E_{patch} (GPa)	10000	17235	72.35

8. Conclusions

At the beginning of this study, it was uncertain whether CWELD element was capable of representing laser spot welds of the hat-plate structure and leading to reasonably accurate values of natural frequencies. Numerical results show that the initial frequencies were very low in comparison with the experimental ones when CWELD is used and CWELD parameters alone cannot improve the results of the initial model even after updating. Further investigation reveals that when a patch parameter is included in the updating the results improve significantly and combined with this patch CWELD is capable of producing very good results when the right parameters are updated.

The FE model updating procedure is cast as an optimisation problem and handled by the structural optimisation module of NASTRAN. This allows the power of NASTRAN to be exploited and would facilitate application of the method presented in this paper. It is expected that this method can be used for prediction of modal properties and damage assessment of structures with laser spot welds.

Acknowledgment
Discussion with Dr. Zongjun Lu of Jaguar and Land Rover is gratefully acknowledged. Hamed Haddad Khodaparast wishes to acknowledge the support of EC Marie Curie Excellence Project ECERTA under contract number MEXT-CT-2006-042383.

References

[1] Palmonella, M., Friswell, M. I., Mottershead, J. E., Lees, A. W., *Finite element models of spot welds in structural dynamics: review and updating.* Computers and Structures, 2005. **83**: p. 648-661.
[2] De, A., Maiti, S. K., Walsh, C. A., Bhadeshia, H. K. D. H., *Finite element simulation of laser spot welding.* Science and Technology of Welding and Joining, 2003. **8**(5): p. 377-384.
[3] Dong, Z.B., Wei, Y. H., *Three dimensional modelling weld solification cracks in multipass welding.* Theoretical and Applied Fracture Mechanics, 2006. **46**: p. 156-165.
[4] Cho, S.K., Yang, Y. S., Son, K. J., Kim, J. Y., *Fatigue strength in laser welding of the lap joint.* Finite Element in Analysis and Design, 2004. **40**: p. 1059-1070.
[5] MSC.Software Corporation, *MSC.NASTRAN 2005 - Quick Reference Guide.* Vol. II. 2004: MSC.Software.
[6] Mottershead, J.E., Mares, C., James, S., Friswell, M.I., *Stochastic model updating: part 2 - application to a set of physical structures.* Mechanical Systems and Signal Processing, 2006. **20**: p. 2171-2185.
[7] Zivanovic, S., Pavic, A., Reynolds, P., *Finite element modelling and updating of a lively footbridge: The complete process.* Sound and Vibration, 2007. **301**: p. 126 - 145.
[8] The MacNeal-Schwendler (MSC) Corporation, *MSC/NASTRAN Quick Reference Guide V68.* MSC/NASTRAN Quick Reference Guide, ed. M. Reymond and M. Miller. 1994: The MacNeal-Schwendler (MSC) Corporation.
[9] LMS International. *LMS introduces LMS PolyMAX.* 2004 25 May 2008; Available from: http://www.lmsintl.com/LMS-PolyMAX.
[10] LMS International. *LMS Test.Lab.* 2004 15 July 2008; Available from: http://www.lmsintl.com/testlab/lmstestlab.
[11] Kim, G.H., Park, Y.S., *An improved updating parameter selection method and finite element model update using multiobjective optimisation technique.* Mechanical Systems and Signal Processing 2004. **18**(1): p. 59-78.
[12] Jaishi, B., Ren, W., *Finite element model updating based on eigenvalue and strain energy residuals using multiobjective optimisation technique.* Mechanical Systems and Signal Processing, 2007. **21**: p. 2295-2317.
[13] Duan, Z., Liu, Y., Spencer, B. F. *Finite element model updating of structures using a hybrid optimization technique.* in *Proceeding SPIE.* 2005.
[14] Sinha, J.K. (2002) *Reliable finite element modelling design modification and diagnosis using model updating method.* BARC Newletter.
[15] Friswell, M.I., Mottershead, J.E., *Finite element model updating in structural dynamics.* 1995: Kluwer Academic Publishers.
[16] Fang, J., Hoff, C., Holman, B., Mueller, F., Wallerstein, D., *Weld Modelling with MSC.Nastran,* in *2nd MSC Worldwide Automotive Conference.* 2000, MSC.Software Corporation.

Journal of Physics: Conference Series **181** (2009) 012025 doi:10.1088/1742-6596/181/1/012025

Dynamic characterization of the cutting conditions in dry turning

Roger Serra[1], Hicham Chibane[2] and René Leroy[3]

[1] ENI Val de Loire, Université François Rabelais de Tours, Laboratoire de Mécanique et Rhéologie, E.A. 2640, B.P. 3410, 41034 Blois Cedex, France

[2] Université François Rabelais, Laboratoire de Mécanique et Rhéologie, E.A. 2640, B.P. 3410, 41034 Blois Cedex, France

[3] Université François Rabelais, Polytech'Tours, Laboratoire de Mécanique et Rhéologie, E.A. 2640, 7 Avenue Marcel Dassault, 37200 Tours, France

E-mail: roger.serra@univ-tours.fr

Abstract. Machining instability in the form of violent vibrations or chatter is a physical process characterized by extreme cutting force at the cutting point. The process has very negative impact on machine integrity, tool life, surface quality and dimensional accuracy. Thus it could significantly compromise productivity and manufacturing quality. In the present paper, the importance of characterization and identification of dynamic instability in dry turning operation are shown. The stability behaviour of machine vibration or chatter has been examined and the various relevant parameters are studied and discuted. For chatter detection and identification of the transition between stable and unstable states, different methods are used. Results obtained proof the accuracy of these methods.

1. Introduction

Turning is one of the most commonly used machining processes in manufacturing industry. In an attempt to achieve high material removal rates in production, aggressive cutting conditions are often employed. Use of such aggressive cutting conditions leads to chatter, wherein intense vibrations and excessive cutting forces occur at the cutting point. Vibrations and chatter is highly undesirable because it adversely affects the surface quality, limits the dimensional accuracy of a workpiece, leads to higher tool wear rates, creates high levels of noise and could result in damage to the machine tool itself or to the spindle [1].

In order to achieve chatter free machining, conservative cutting conditions are usually chosen resulting in lower metal removal rates and a loss of productivity. For this reason, it is essential to identify chatter free cutting conditions. Knowledge of the stability limits would be necessary to maximize the metal removal rate while maintaining the product quality. The motivation for this work is to identify bounds of dynamic stability and instability in turning process and finally to predict the stability lobes which in turn can be used to determine chatter limits for different cutting conditions [2].

In most cases of pratical interest, the chatter observed in a turning operation is due to the regenerative effect : as the tool cuts a surface, the undulations generated in the previous revolution sustain the tool-workpiece vibrations which are closely coupled to the cutting forces. Chatter prediction models have a long history that began with work by Tobias [3] and

Tlusty [4, 5]. These early efforts concluded that the regenerative effect is the main cause of instability, which leads to the development of chatter. Tlusty and Polacek [6] developed a stability condition in which stability limits can be calculated based upon the system dynamics for orthogonal machining. In the work by Tobias and Fishwick [7] the dynamics of the cutting process were modeled and effects such as process damping were included into the stability model. Early stability lobe diagrams were created by Merrit [8] based upon feedback control theory to model regenerative chatter.

This paper presents a comprehensive dynamic instability model for chatter prediction in a dry turning operation. The first section presents an analysis of chatter during turning operation. The next two sections present the experimental set-up and the procedure used in chatter verification experiments. They are followed by a presentation and discussion of experimental results. The last section outlines the conclusions and recommendations for future work.

2. Analysis of chatter in turning operation

Figure 1 shows the control block diagram of chatter during turning lathe work. In this figure, the uncut chip thickness $h(s)$ is composed of the mean uncut chip thickness $h_0(s)$, the inner modulated cut surface $y(s)$, and the outer modulated workpiece surface $y(s)e^{-\tau s}$, where τ is the time delay between the inner and outer modulated surface (spindle period). The uncut chip thickness $h(s)$ is fed into the cutting process to produce the cutting force $F(s)$ acting on the cutting tool. The transfer function of the cutting process can be expressed as : $K_c b$ where K_c is the specific cutting force and b the cut width.

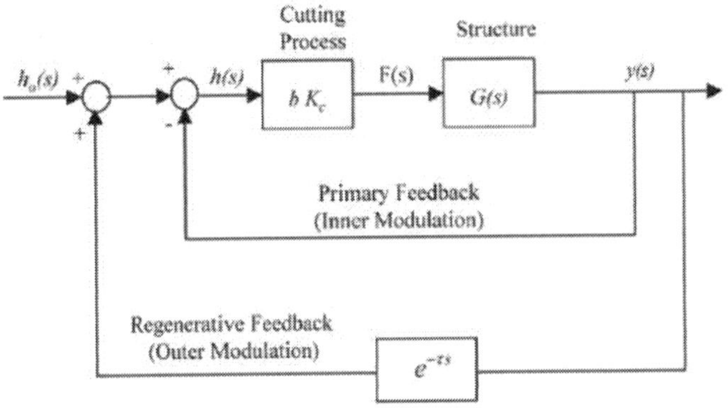

Figure 1. Control block diagram for chatter in turning.

The cutting force $F(s)$ then excites the cutting tool to generate the vibration $y(s)$ between the cutting tool and workpiece. The oriented transfer function of the cutting tool $G(s)$ can be expressed as :

$$G(s) = \sum_{i=1}^{n} u_i \, G_i(s) \tag{1}$$

where n is the number of modes, u_i the directional factor for mode i, and $G_i(s)$ the direct transfer function for mode i, which can also be expressed as :

$$G_i(s) = \frac{1}{m_i \, s^2 + c_i \, s + k_i} \tag{2}$$

where m_i, c_i, and k_i are the mass, damping, and stiffness parameters respectively.

According to the Nyquist stability criterion [9], the control system for turning lathe work is at the stability limit when the gain of the open loop transfer function has the critical value of

-1. The cut width b at the stability limit is called "critical cut width" b_{lim} [10] and can be expressed as :

$$b_{lim} = \frac{-1}{2\,K_c\,\Re\,(G(j\,\omega))_{neg}} \tag{3}$$

where $\Re\,(G(j\,\omega))_{neg}$ is the negative real part of the frequency response function of the cutting tool and ω the chatter frequency in rad/s.

For the sake of simplicity, the cutting tool is taken as a single-degree-of-freedom vibration system. The parameters of the vibration system for the cutting tool are selected as mass, $m = 1kg$, damping ratio, $\zeta = 0.03$ and natural frequency, $f_n = 100\ Hz$. The specific cutting force is equal to $K_c = 1000\ N/mm^2$. Figure 2 shows the real part and magnitude of the frequency response function of the cutting tool. Based on (3), the chatter frequency and the critical cut width b_{lim} can be calculated [10].

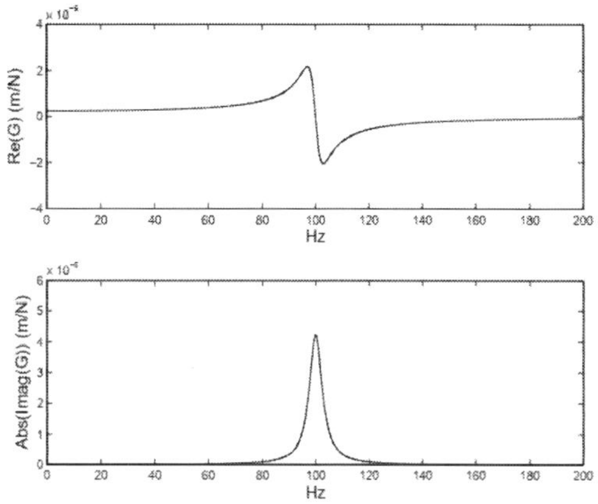

Figure 2. Frequency response function of the simplified cutting tool.

Figure 3 shows the chatter frequency and the critical cut width b_{lim} as a function of spindle speed. Chatter will occur when the cut width b is greater than b_{lim}. However, the critical cut width b_{lim} can be increased if the magnitude of the negative real part of the frequency response function of the cutting tool is reduced (3).

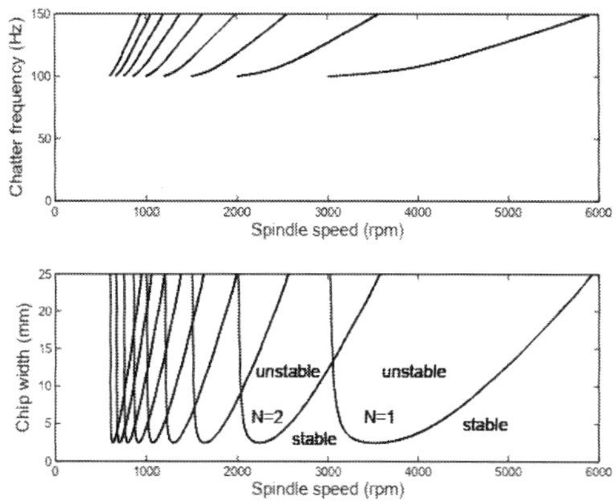

Figure 3. Simulated stability chart.

3. Experimental setup and procedure

Cutting experiments were performed on a CNC lathe Realmeca Conventional plus T400 (Figure 4). The speed of spindle rotation is from 100 to 4000 rpm and the feed rate is from 0.1 to 8 m/min.

Machining tests were carried out by cutting a 42CD4 steel workpieces with a cutting tool SNMG TP200 using diamond shaped CVD coated carbide inserts (120408-MF2) which a nose radius of 0.8 mm, side and end cutting angles of $-6°$ and a rake angle of $93°$. The unmachined round bar of diameter 40 mm and length 500 mm, used in all experiments for dynamic charactrization, was rigidly mounted in the chuck of the lathe with tailstock.

Figure 4. CNC Realmeca T400 turning lathe.

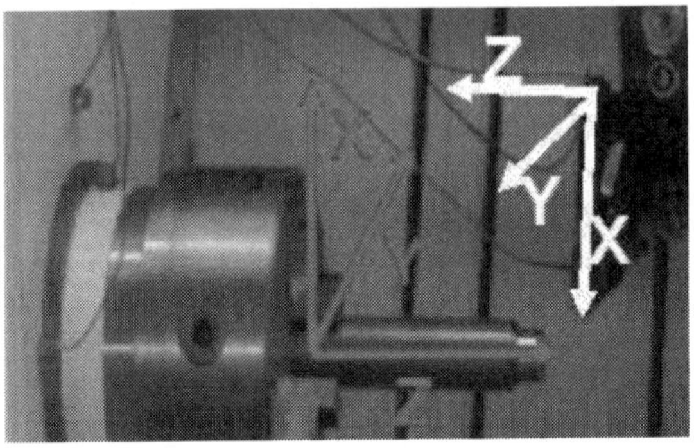

Figure 5. Directions of accelerometers on the workpiece and cutting tool.

The first step in evaluating vibration was to determine the appropriate mounting location for the sensor. Previous work has shown that mounting the sensor on a stationary tool is best. During the validation of the dynamic instability identification, the direct response of the spindle and the cutting tool were instrumented with six B&K unidirectional piezoelectric accelerometers type 4507 with the following characteristics : weight 4.8 g and size 10 mm^3 as shown in Figure 5 with sensitivities given in Table 1 to obtain acceleration signals during machining. The sensors had an upper frequency range of 6 kHz.

Using six accelerometers has allowed us to record as much as possible vibration data from the cutting process. This will generate a large number of features which may be helpful to acquire maximum information about the dynamic of the process and the state of the tool.

	Spindle			Cutting tool		
Direction	X	Y	Z	X	Y	Z
Serial number	2195615	2110703	2195617	2195616	2110695	2195618
Sensitivities (mV/ms^{-2})	10.10	9.86	9.90	10.06	10.14	10.05

Table 1. Sensitivities of accelerometers.

The impulse and displacement responses were recorded on a hard disk PC in a digital form using a B&K Multi-analyser system type 3560 and a B&K Pulse LabShop software. Signal processing was performed with Matlab environment.

The sensor signals were acquired for more than 12 s duration of observation and transmitted through six channels, sampled at 16384 Hz.

For the frequency response of the cutting tool, the impulse was given with a Brüel&Kjaer (B&K) impact hammer (model 8202) equiped with a plastic tip coupled with a piezo-electric force sensor type 2646 of sensitivity 1 mV/g.

The frequency response functions in the axial, radial and feed directions of the cutting tool which were used during validation are given on Figure 6. A first mode at 1900 Hz in the X and Z direction and at 3100 Hz in the Y direction can be clearly seen. A curve fitting technique was used to extract the modal parameters from the transfer function plot.

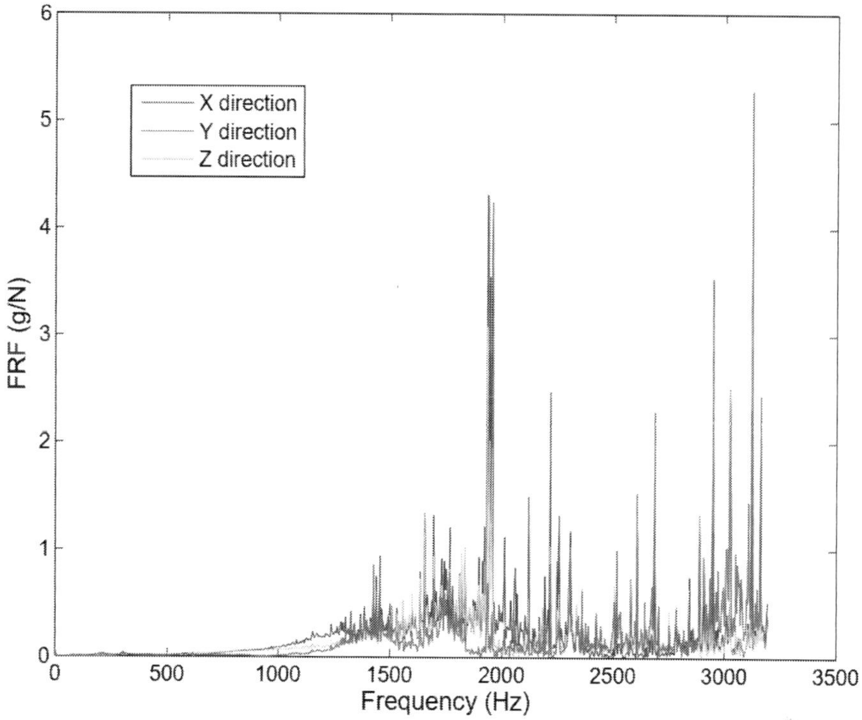

Figure 6. Frequency response function of the cutting tool in the X, Y and Z directions.

This section describes the procedure adopted in conducting experiments for verifying the limits for the unstable-stable turning lathe work. For the sake of clarity, the cutting conditions presented are : a variable depth of cut, a variable spindle speed, a constant feed speed of 0.2 mm/rev and under dry conditions. Once the conditions parameters of turning were determined, several experiments were performed to determine the depth of cut at which machining becomes unstable. Table 2 shows the list of all tests performed.

	Spindle speed (rpm)				
	1500	2000	2500	3000	3500
Depth of cut (mm)	1.4	0.8	0.8	0.8	0.8
	1.8	1.0	1.0	0.9	1.0
	2.3	1.2	1.2	1.0	1.1
	2.5	1.5	1.4	1.2	1.2
	2.9			1.3	

Table 2. List of tests performed.

4. Results and discussion

Figure 7 a) shows the accelerometer signal of the cutting tool in these conditions and without chatter. The signal becomes unstable and Figure 7 b) shows the results with a depth of cut of 1.5 mm.

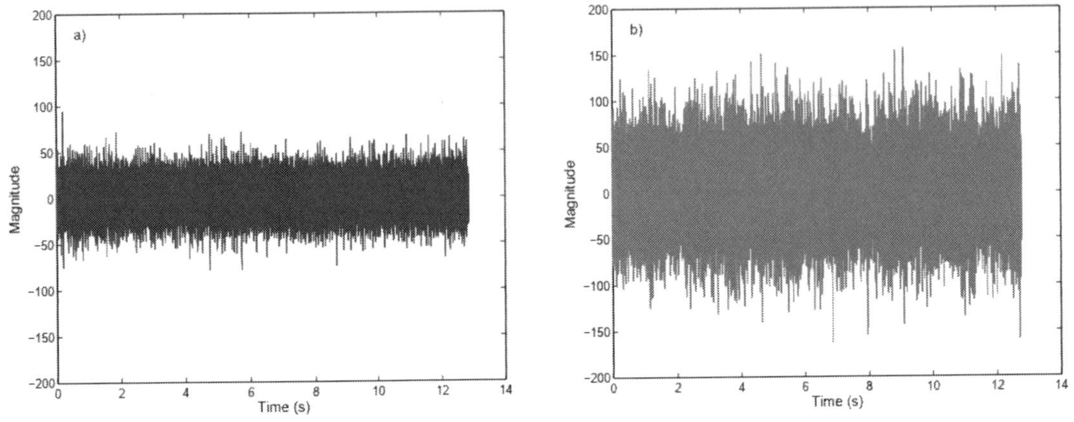

Figure 7. Representative spectra of acceleration signals in Z direction without chatter (a) and with chatter (b).

A representative frequency spectrum is illustred Figure 8 where the evolution of the first mode at 1900 Hz is clearly significant. The time and frequency parameters are very influent regarding the dynamic stability of the process. For clarity, only the feed (or Z) direction of accelerometer mounted on the cutting tool is shown here.

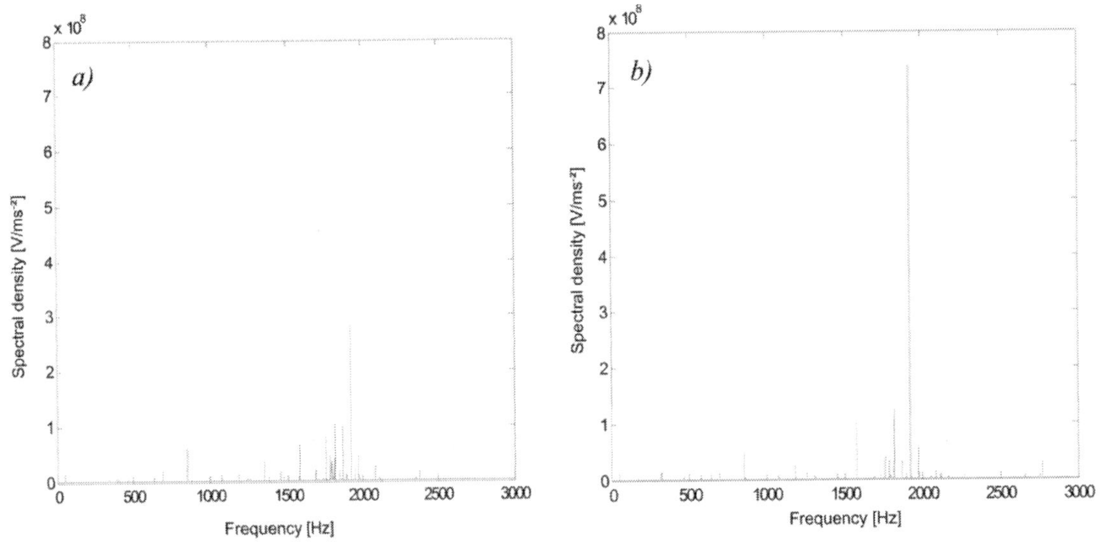

Figure 8. Representative spectrum density of acceleration signals without chatter (a) and with chatter (b).

In Figure 9, it can be seen that the magnitude of the frequency response function is greatly reduced whithout chatter when compared to those with chatter. It is consequently a good indicator of chatter.

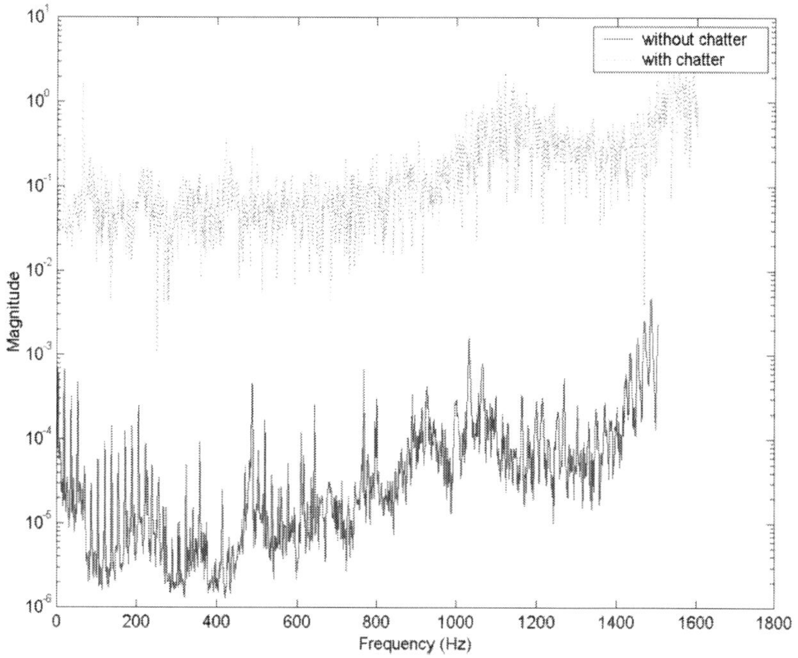

Figure 9. Representative magnitude of acceleration signals with chatter (top) and without chatter (bottom).

The experimental stability chart was created with all machining (see list in Table 2). The resulting diagram is given in Figure 10. In this diagram, stable cutting conditions are represented by a $\circ$ while unstable cutting conditions are shown by an $*$.

5. Conclusions

In this paper, the most relevant parameters of dynamic instability have been explored. Detection of unstable - stable limits from accelerometer signals is improved. Experimental results show that chatter in turning can be almost suppressed and cutting stability increased by an optimization of cutting parameters. Also, as chatter occurs the tool vibrates violently and thus creates an undulated pattern on the workpiece. It creates a rougher surface than in stable cutting, and hence surface roughness is also a good indication of chatter occurence. A future work will try to correlate the limits of instability with the surface roughness data, the tool life, the cutting force and the tool wear.

References

[1] Dimarogonas A., Vibration for Engineers, Prentice-Hall, Englewood Cliffs, NJ, 1996.
[2] Inman D.J., *Vibration with Control, Measurement, and Stability*, Prentice-Hall, London, 1989.
[3] Tobias S.A., *Machine Tool Vibration*, Blackie, London, 1965.
[4] Koenigsberger K., Tlusty J., *Machine Tool Structures*, Pergamon Press, Oxford, 1970.
[5] Tlusty J. *Machine dynamics*, in Handbook of High-Speed Machining Technology, R. I. King, ed. New York: Chapman and Hall, 1985.
[6] Tlusty J., Polacek M., *The stability of machine tools against self excited vibrations in machining*, ASME International Research in Production Engineering, 465474, 1963.
[7] Tobias S.A., Fishwick W., *A Theory of Regenerative Chatter*, The EngineerLondon, 1958.

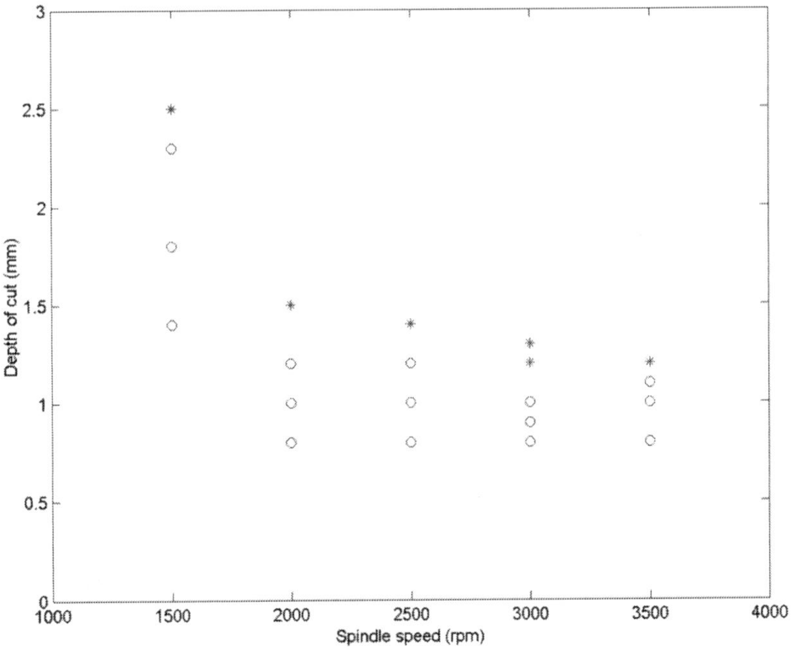

Figure 10. Experimental stability chart (○ - stable, ∗ - unstable).

[8] Merritt H.E., *Theory of self-excited machine tool chatter*, Transactions ASME, Journal of Engineering for Industry 87 (4), 447-454, 1965.
[9] Kuo B.C., *Automatic control systems*, Prentice-Hall, Englewood Cliffs, NJ, 1987.
[10] Tarng Y.S., Lee E.C., *A critical investigation of the phase shift between the inner and outer modulation for the control of machine tool chatter*, Int. Mach. Tools Manuf. 37 (12), pp. 1661-1672, 1997.

Optimisation of variable helix end millings tools by minimising self excited vibration

Ahmad R Yusoff and Neil D Sims[1]

Dept of Mechanical Engineering, University of Sheffield, Mappin Street, S1 3JD, UK

E-mail[1]: n.sims@sheffield.ac.uk

Abstract. During machining, self-excited vibrations known as regenerative chatter can occur. This instability can be avoided by modifying the tool geometry, thereby influencing the time delay terms that arise in the governing equations. The present study uses Differential Evolution (DE) to optimise the tool helix geometry, so as to avoid chatter. The results are compared to those from Sequential Quadratic Programming (SQP). It is shown that the DE approach can significantly increase the chatter stability, and substantially out-performs the SQP algorithm. The performance of the DE approach is due to its ability to perform global optimisation in the presence of significant nonlinearities.

1. Introduction

Aerospace, automotive, mold/die and general manufacturing industries face pressure to ensure lower cost, greater productivity and improved quality in order to encourage the economic growth of the machine tool industry. However, machining productivity using a high material removal rate is inhibited by the dynamic deflection of tool and workpiece. This causes sudden large amplitude vibrations when energy input exceeds the energy dissipated from the system, producing chatter. Chatter is a self excited vibration that occurs in metal cutting if the chip width is too large with respect to the dynamic stiffness of the system, especially when machining with a high material removal rate. It produces a poor surface finish and high tool wear and can even damage machine tools. The boundary of stability (between stable or unstable machining) is a function of depth of cut and spindle speed.

Researchers have created various types of chatter prediction models such as time domain simulation [1], analytical delay differential equations with time periodic coefficients [2], time finite element analysis [3], and semi discretisation methods [4]. To suppress chatter, passive methods such as the use of vibration absorbers or damping treatments have been applied. Active approaches, such as spindle speed modulation or active vibration control have also been considered, but passive methods are often preferred due to their simplicity.

This paper investigates a passive approach to chatter suppression that disrupts the chatter vibration using variable helix tools. Figure 1 illustrates uniform helical end milling tools compared to non-traditional variable helix and variable pitch designs. Variable pitch tools were proposed by Slavicek [5]. Later, Opitz *et al.* [6] studied irregular tooth pitches that produce a higher stable depth of cut. Only Stone [7] applied an irregular *helix* in the early period before 1970s. Variable pitch tools were re-considered by Altintas *et al.* [2] who utilised an invariant time constant and a non uniform multiple regeneration time delay to optimise pitch geometry. Meanwhile, Budak [8] modelled and optimised a non constant pitch cutter using an analytical stability model. Recently, Olgac and Sipahi [9] theoretically maximised the material removal rate by applying an irregular pitch cutter that optimised with the Cluster treatment of Characteristic roots (CTCR) as a mathematical objective function.

This brief overview has demonstrated that current research on passive chatter suppression methods only takes into account variable pitch tools, and has tended to overlook variable helix tools. Besides reducing chatter, the helical angle can also break chip formation and change the line of contact between the tool and workpiece. For the present study, a semi discretisation method (SDM) has been combined with Differential Evolution (DE) to optimise variable helix end milling tools. The target is to reduce chatter, and Sequential Quadratic Programming (SQP) is used to benchmark the optimisation performance.

The paper is presented as follows. First, a summary of SDM applied to the current problem is briefly described. The DE optimisation algorithm and application into SDM are then presented. An optimisation study is then presented for a three flute variable helix milling tool. SQP optimisation is finally carried out to validate the performance of DE.

2. Semi Discretisation Method

The Semi Discretisation Method (SDM) is a well known technique in the Finite Element Method and Computer Fluid Dynamics. SDM analyses stability of linear retarded dynamical systems based up on a discretisation technique. The effects of time delays and time periodicity are considered to produce a high dimensional linear discrete system. SDM is more effective in time and accuracy than fully discrete methods due to the delayed states and time dependent coefficient. Details of SDM are given in ref.[4]. Recently, Sims *et al.* [10] applied SDM to model irregular helix and pitch tools.

The regenerative chatter in milling can be represented as shown in Figure 2. It comprises structural dynamics, regenerative effect and dynamic cutting force coefficients that interact to form a potentially unstable feedback path. The regenerative system includes a time delay term that arises due to the angular spacing between the teeth on the tool. This delay term can be modified by using variable helix or variable pitch tools.

The stability of the feedback path shown in Figure 2 can be determined from the characteristic multipliers (*CM*) or eigenvalues that emerge from the semi-discretisation method. Characteristic Multipliers are obtained for each combination of spindle speed and depth of cut. When the maximum value of *CM* is greater than unity then the system is unstable. The type of instability can characterised by the location where the *CM* crosses the unit circle (cyclic fold bifurcation, secondary hopf bifurcation and period doubling bifurcation). Consequently, the results of the method can be summarised graphically by plotting stability (maximum *CM* greater than unity) on a plot of spindle speed versus depth of cut. Meanwhile, all of the corresponding *CM*'s can be plotted on the complex plane (along with the unit circle) to indicate the relative stability of the system.

3. Differential Evolution

Evolution Algorithms (EA) such as Genetic Algorithm (GA), Evolutionary Programming (EP) and Evolution Strategy (ES) have been researched for several decades. Differential Evolution (DE) was introduced by Price *et al.* [11], and can be considered to be an improved Genetic Algorithm (GA) version with different strategies for faster optimisation. It is similar to other evolution algorithms in which mutation plays the key role, with real valued parameters that directly search for the global optimum. A basic idea in DE is that of adapting the search during the evolution process. Compared to other algorithms, DE has the advantages of simple structure, ease of use, speed and robustness. In machining applications, Saikumar and Shunmugan [12] applied DE to select the best cutting speed, feedrate and depth of cut to achieve optimum surface finish while Krishna [13] applied DE in a grinding process to search for suitable process grinding in minimising surface grinding.

DE can solve objective functions that are non differentiable, non linear, noisy, flat, multi dimension, and with multiple local minima. Such functions are difficult to solve analytically, and the variable helix optimisation problem fits within this scope. DE begins using initial samples at multiple random chosen initial points. With simple algorithms, DE can search for the optimal condition very quickly with minimal control parameters such as mutation, crossover, selection and population. The concept is evolved from GA's with layer population and special evolutionary strategy of self adaptive mutation. However, instead of a binary-coded populations (as in a GA), differential evolution deals with real-coded populations, and uses its own processes of mutation and crossover. The mutation process is created randomly from the selection of three individual vector differences. In the crossover process, any individual population member has equal opportunity to survive in the next generation based on its fitness value. The process of evolving mutation, recombination and selection through generations or new populations is repeated until the optimum solution is achieved. In the present work, the DE source code by Markus Buehren [14] was used. The code is based on the DE algorithm of Price *et al.* [11].

4. Implementation

By combining with the analytical method for chatter stability prediction, the process of DE optimisation can be used to optimise tool/helix geometry. Figure 3 shows the sequence of operations, namely: optimisation setting process, SDM process, objective function evaluation process and DE optimisation process. DE parameters are first set to create an initial population in the optimisation process. To search for the optimum values, DE requires the predicted values from the SDM stability analysis with consideration of the input variables (i.e. helix angles β_i and pitch angles ϕ_i, cutting parameters, etc.). The fitness of each population member is evaluated in terms of the chatter stability: consequently the DE process will strive to obtain tool geometry (helix and pitch angles) that maximize the chatter stability. This process continues until the termination criteria are met.

The values DE parameters used in the present work are given in Table 1. Crossover rate, *CR* is 0.7, scaling factor, *F* is 0.7 and number of population, *NP* is 10 times the dimensions (10*D) and 70 generations are employed in DE optimisation. A 'strategy-7' (DE/rand/1/bin) methodology [13, 12] was implemented owing to its wide application in the literature. This methodology involves random perturbation of a population vector ('/rand'), perturbation of a difference vector for the mutation process ('/1'), and binomial crossover ('/bin').

During the optimisation process the candidate values of tooth helix β_i and pitch ϕ_i, may result in milling cutters whose flutes intersect each other. This is clearly inadmissible from a practical viewpoint. To prevent helical angles intersecting, the helical height different (Δh_β) is introduced as a constraint in the DE optimisation. The helical angle difference calculations for variable helix and variable helix with variable pitch are as follows:

$$\Delta h_\beta = d_c - \frac{d_c - d_c(2N_c)^{-1}}{\tan\beta_i} - \frac{d_c - 2d_c(N_c)^{-1}}{\tan\beta_{i+1}} \quad \text{(variable helix)} \qquad (1)$$

$$\Delta h_\beta = \frac{d_c - d_c(2N_c)^{-1}}{\tan\beta_i} - \frac{\left(d_c - 2d_c(N_c)^{-1}\right)\sin\left(2^{-1}\Delta\phi\right)}{\tan\beta_{i+1}} \quad \text{(variable helix with variable pitch)} \qquad (2)$$

Here, Eqn (1) relates to optimization problems where the free end of the milling cutter has teeth that are uniformly spaced, but the variable helix of each flute means that the teeth are irregularly spaced along the rest of the cutter's axial length. Eqn (2) relates to cutters with both variable helix and a variable pitch at the free end. The Δh_β value is based on the parameters illustrated in Figure 1c, along with the number of cutter teeth, N_c and cutter diameter, d_c, and pitch difference, $\Delta\phi$.

The final optimisation problem can be specified as follows:

Objective function: Minimise the maximum value of $f_{CM}(\beta_i, \varphi_i, n) = CM$ for all values of n and b

Subject to constraints:

Spindle speed	$n_{\min} \le n \le n_{\max}$	(r/min)
Depth of cut	$b_{\min} \le b \le b_{\max}$	(mm)
Helical Angle	$25 \le \beta_i \le 55$ $i = 1,2,3\ldots n$	(°)
Pitch Angle	$\phi + 22.5 \le \phi_i \le \phi + 22.5$ $i = 1,2,3\ldots n$	(°)
Helical height different	$\Delta h_\beta \ge 5$	(mm)

The constraint on β_i (25° to 55°) is to ensure good chip evacuation [8, 9]. The DE needs to search for the suitable value of β_i that produces minimum chatter across the chosen spindle speed and depth of cut range (along with the additional constraints).

In order to benchmark the DE optimisation method, a traditional gradient based opimisation method (Sequential Quadratic Programming (SQP)) was combined with the Semi Discrete Method to minimise chatter and optimise variable helix geometry. The SQP approach is widely used in manufacturing applications: In machining, Kurdi *et al.*[15]compared SQP with Particle Swam Optimisation (PSO) for optimising the surface location error and material removal rate at same time as minimising chatter. The main idea of SQP is to obtain a search direction from a quadratic program solution together with its constraints. The general method to solve constrained optimisation is stated in [16]. In this research, the SQP in Matlab Optimization Toolbox was used by employing the constrained minimisation function [17].

5. Results

The results from a three flute variable helix from Sims *et al.* [10] is initially presented. The parameters from [10] are presented in Table 2. The milling cutter helix geometry is then optimised based on variable helix and variable pitch modifications to reduce chatter. Stability lobes and *CM* results of optimum cutting tools are illustrated and compared with the original design that had been chosen arbitrarily.

The original tool geometry consisted of a three flute variable helix (25,30,35) cutter with variable pitch (120,100,140) at its free end. In Figure 4, a large unstable region can be observed at high depth of cut, giving eigenvalues outside the unit circle. For this low radial immersion scenario three instability conditions are seen: cyclic fold bifurcation, secondary hopf bifurcation and period doubling bifurcation. The optimisation routines were used to adjust the tool helix so as to obtain the most stable chatter performance across the illustrated spindle speed range. Two scenarios were considered: a variable helix with a uniform pitch at the tool's free end, and a variable helix with a variable pitch at the tool's free end.

The performance of the DE and SQP algorithms is summarised in Figure 5. Note that the objective function is a maximum value for a *CM*, so a value less than unity represents complete stability over the chosen spindle speeds and depths of cut.

The DE algorithm achieved complete stability (maximum *CM*= 0.8965) by the second generation. The corresponding variable helix (52,52,41) with variable pitch (107,169,90) stability diagram is shown in Figure 6. Since the maximum *CM* is less than unity, no chatter is observed. In comparison, the SQP algorithm achieved a maximum *CM* of 0.952 after 25 iterations. The corresponding best-case stability diagram is shown in Figure 7. Although this system is also completely stable, the *CM*s are considerably greater than those obtained with the DE algorithm. This indicates a lower margin of stability.

For the scenario where the tool's free-end pitch angle was uniform, Figure 5 shows that a poorer performance was obtained for both optimisation methods. The DE algorithm converged at 50 generations with 1.005 evaluation value, with a variable helix of (53,27,54). The resulting unstable conditions of stability prediction and eigenvalues outside the unit circle are given in Figure 8.

Meanwhile, for the SQP result, the maximum *CM* value higher (1.071) and converges at 5 iterations as shown in Figure 5. The corresponding stability prediction is shown in Figure 9, indicating a very large unstable island (eigenvalues plotted outside the unit circle) in contrast to the DE (Figure 8) and the original (Figure 4) results.

6. Discussions

To recap, the DE algorithm was able to design a three flute variable helix tool (with variable pitch at its free end) that showed a great improvement in chatter stability. However, the importance of also allowing a variable pitch at the free end of the tool has been highlighted. Without this variable pitch at the free end, the variable helix geometry is much less effective because there is less potential to 'disrupt' the time-delay parameters in the governing stability equations.

Both the DE and SQP algorithms were able to improve the chatter stability. However, by referring to the maximum CM value, the optimisation results for the DE algorithm are consistently better than for the SQP algorithm. For the variable helix tool with a uniform helix at the free end, the SQP algorithm is clearly trapped in a local optimum. Furthermore, the performance strongly depends on the chosen initial value because it uses gradient based optimisation [15].

7. Conclusion

This article has illustrated how variable helix milling tools can effectively suppress the chatter vibration of a workpiece during milling processes, using an analytical/numerical study. A differential evolution algorithm was shown to perform much better than a traditional optimisation routine (SQP). The poorer performance of SQP can be attributed to its local nature, in contrast the global optimisation achieved with the DE method.

Although the validation with experimental results has not been presented, the present study suggests that the proposed algorithm can have good potential to optimise variable helix geometries. Further work is needed to validate in the underlying stability algorithms, in order to ensure the capability to increase productivity in practical machining problems.

Acknowledgements

NDS acknowledges to financial support from EPSRC (GRS49841/01). ARY is grateful for PhD studentship sponsored by Ministry of Higher Education of Malaysia and Universiti Malaysia Pahang.

References

[1] Tlusty J and Ismail F 1983 Special Aspects of Chatter in Milling *J. of vib., acoustic, stress and reliability in des., Trans. of the ASME* **157** 24-32

[2] Altintas Y, Engin S and Budak E 1999 Analytical stability prediction and design of variable pitch cutters *J. of Man. Sci. and Eng., Trans. of the ASME* **121** 173-82

[3] Bayly P V, Halley J E, Mann B P and Davies M A 2003 Stability of interrupted cutting by temporal finite element analysis *J. of Man. Sci. and Eng.* **125** 220-5

[4] Insperger T and Stepan G 2002 Semi discretization method for delayed system *Int. J. for Num. Met. in Eng.* **55** 503-18

[5] Slavicek J 1965 The effect of irregular tooth pitch on stability of milling *Adv. in Mac. Tool Des. and Res., Proc. of the 6th MTDR Conference* 15-22

[6] Opitz H, Dregger E U and Rose H 1966 Improvement of the dynamics stability of the milling process by irregular tooth pitch *Adv. in Mac. Tool Des. and Res., Proc. of MTDR Conference* **7** 213-27

[7] Stone B J 1970 The effect on the chatter behaviour of cutters with different helix angles on adjacent teeth *Adv. in Mac. Tool Des. and Res., Proc. of 11th Int. MTDR Con.* **A** 169-80

[8] Budak E 2003 An analytical design method for milling cutters with non-constant pitch to increase stability, Part 1: Theory and Part 2: Application *J. of Man. Sc. and Eng.* **125** 29-38.

[9] Olgac N and Sipahi R 2007 Dynamic and stability of variable pitch Milling *J. of Vib. and Con.* **13** 1031-43
[10] Sims N D, Mann B and Huyanan S 2008 Analytical prediction of chatter stability for variable pitch and variable helix milling tools *J. of Sound and Vib.* **317** 664-86
[11] Price K V, Storn R M and Lampinen J A 2005 *Differential Evolution: A Practical Approach to Global Optimization* (Berlin Heidelberg: Springer)
[12] Saikumar S and Shunmugan M S 2008 Parameter selection based on surface finish in high speed finish in high speed end milling using differential evolution *Materials and Man. Proc.* **21** 341-7
[13] Krishna A 2007 Selection of optimal conditions in the surface grinding process using a differential evolution approach *Proc. of the Inst. of Mech. Engs, Part B: J. of Eng. Man.* **221** 1185-92
[14] Markus B 2008 Differential Evolution.
[15] Kurdi M H 2005 Robust multicriteria optimization of surface location error and material removal rate in high speed milling under uncertainty. In: *Department of Mechanical and aerospace engineering*: University of florida)
[16] Fletcher R 1987 *Practical Methods of Optimization* (West Sussex: John Wilet & Sons)
[17] 2001 *Optimization Toolbox User's Guide: For Use with Matlab* (Natick: The Math Works)

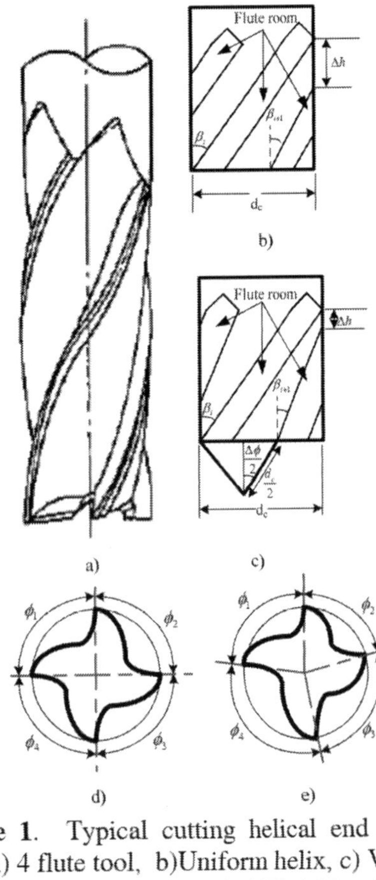

Figure 1. Typical cutting helical end milling tools a) 4 flute tool, b)Uniform helix, c) Variable helix, d) Uniform pitch and e) Variable pitch

Parameter	Value
Number of Generation	70
Population, *NP*	10*D
Crossover rare, *CR*	0.7
Scaling factor, *F*	0.8

Table 1. Typical DE parameters settings

Tool and cutting parameters	
Tool diameter, mm	19.05
Radial immersion, mm	1.00
K_t, MPa	550
K_n, MPa	200
Modal property in x-direction mode	
Natural frequency, *f* (Hz)	169.3
Modal effective mass, *m* (kg)	6.5363
Damping Ratio, ξ	0.0056

Table 2. Cutting, modal and tool parameters for optimization

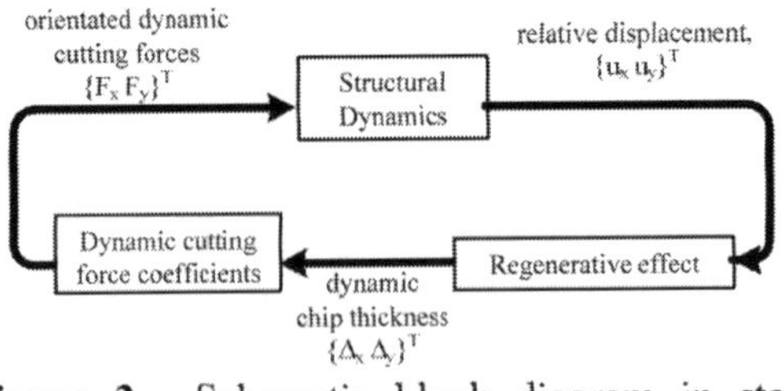

Figure 2. Schematic block diagram in state space form for regenerative chatter

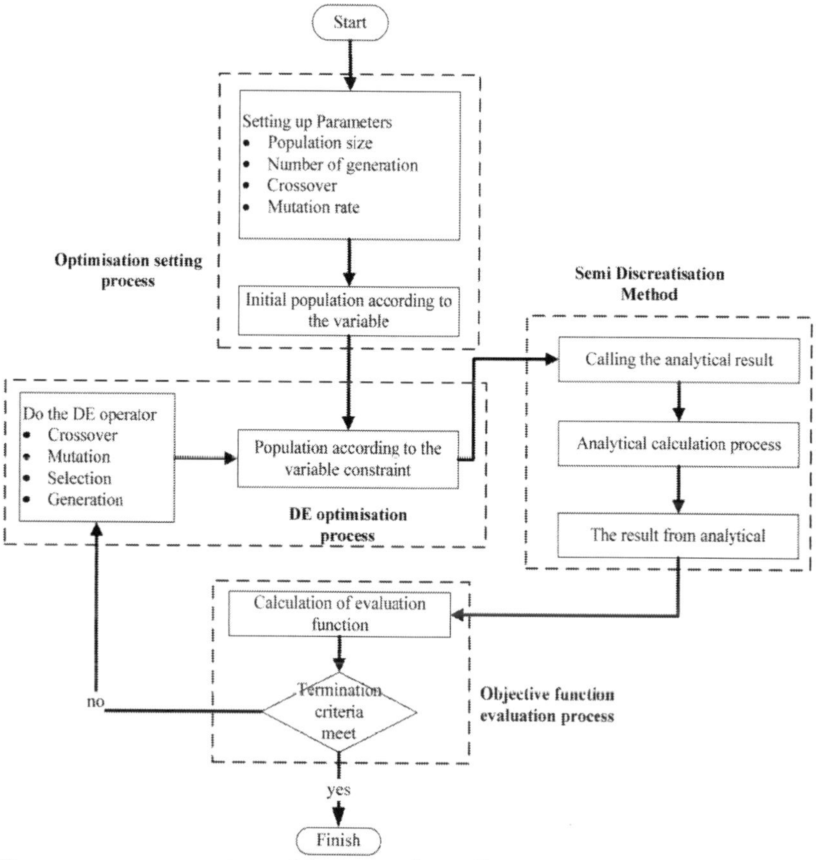

Figure 3. The sequence operation of between Semi Discretisation Method and optimization of Differential Evolution

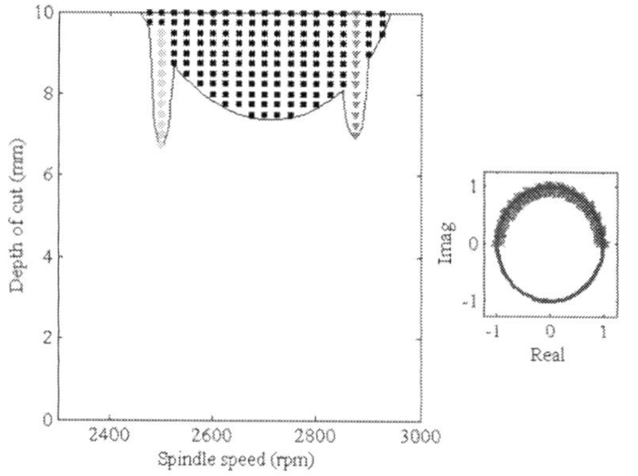

Figure 4. Original stability prediction for variable helix (25,30,35) with variable pitch (120,100,140) [10] (—) stability region, (■) secondary hopf bifurcation, (o) cyclic fold bifurcation and (∇) period doubling bifurcation

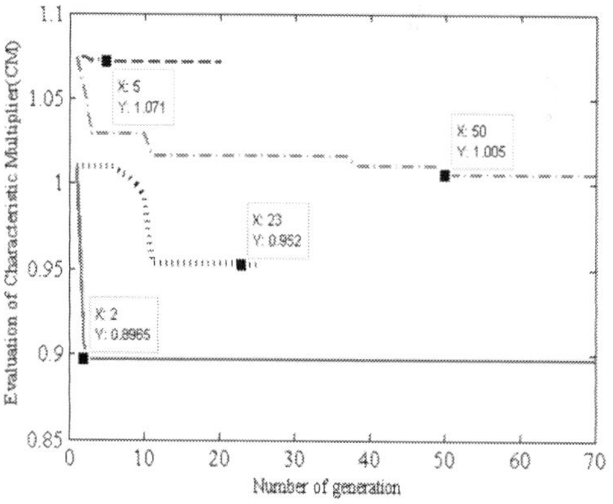

Figure 5. Performance of DE and SQP.
(—) DE, variable helix, variable pitch at end
(....) SQP, variable helix, variable pitch at end
(-.-.) DE, variable helix, uniform pitch at end and (---) SQP, variable helix, uniform pitch at end.

Journal of Physics: Conference Series **181** (2009) 012026 doi:10.1088/1742-6596/181/1/012026

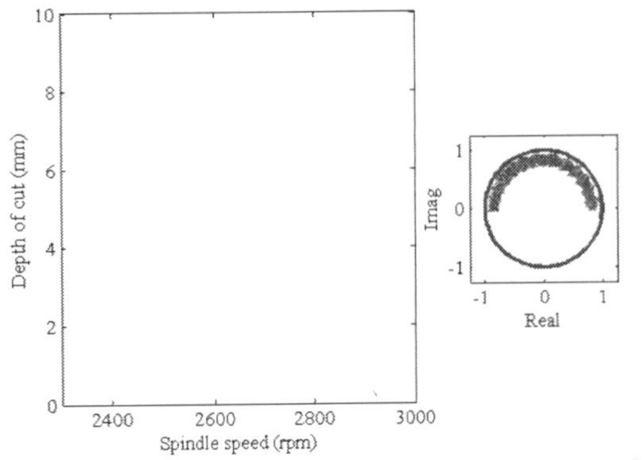

Figure 6. DE Optimised stability prediction for variable helix (52,52,41)with variable pitch (107,163,90)

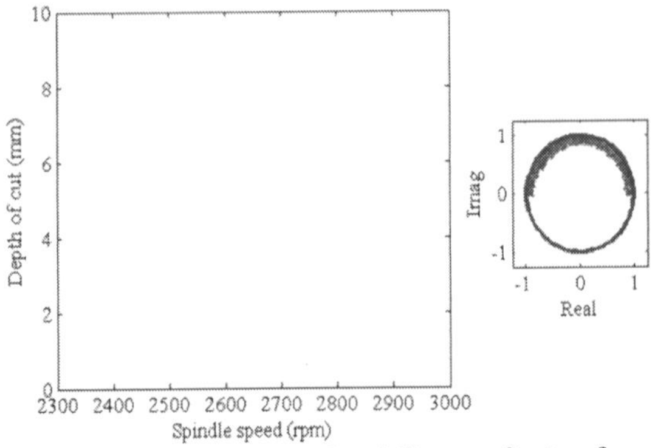

Figure 7. SQP optimised stability prediction for variable helix (25,38,55) with variable pitch (120,105,135)

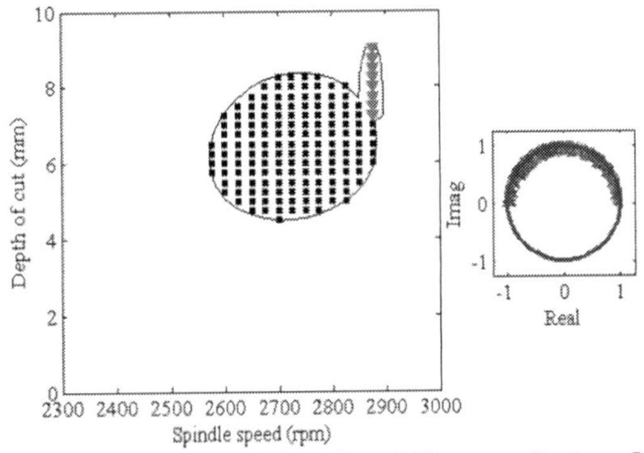

Figure 8. DE optimised stability prediction for variable helix (53,27,54) (—) stability region, (—) stability region, (■) secondary hopf bifurcation and (∇) period doubling bifurcation

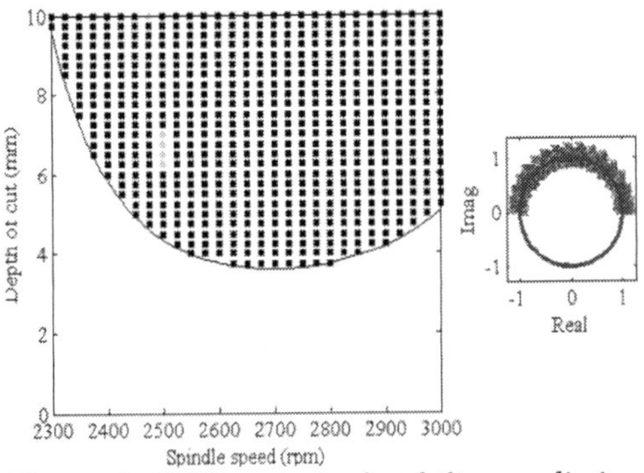

Figure 9. SQP optimised stability prediction for variable helix (28,32,37) (—) stability region, (■) secondary hopf bifurcation and (o) cyclic fold bifurcation

A finite element model of ultrasonic extrusion

M Lucas[1] and Y Daud[2]
[1]Department of Mechanical Engineering, University of Glasgow, G12 8QQ, UK
[2] College of Science and Technology, UTM City Campus, Kuala Lumpur, Malaysia

E-mail: m.lucas@mech.gla.ac.uk

Abstract. Since the 1950's researchers have carried out investigations into the effects of applying ultrasonic excitation to metals undergoing elastic and plastic deformation. Experiments have been conducted where ultrasonic excitation is superimposed in complex metalworking operations such as wire drawing and extrusion, to identify the benefits of ultrasonic vibrations. This study presents a finite element analysis of ultrasonic excitation applied to the extrusion of a cylindrical aluminium bar. The effects of friction on the extrusion load are reported for the two excitation configurations of radially and axially applied ultrasonic vibrations and the results are compared with experimental data reported in the literature.

1. Intruduction

Since the work of Blaha and Langenecker in 1955 [1] researchers have carried out investigations into the effects of applying ultrasonic excitation to metals undergoing elastic and plastic deformations [2], [3]. Experiments have been conducted under various conditions from pure tension and compression, to complex metalworking operations such as wire and strip drawing [2], [4], extrusion [5], and sheet-metal forming. The reported benefits of applying ultrasonic vibrations to tools include a reduction in forming load, an increase in achievable deformation and improvement in the product quality. These benefits have been attributed to volume effects and surface effects. The volume effects deal with the influence of oscillations on the internal stresses during plastic deformation, whereas the surface effects deal with the interfacial contact condition.

Friction at the die-billet interface, particularly for severe metal forming processes such as extrusion and sheet metal forming, significantly influences the forming load and quality of the final product and the use of lubricants is common. There are problems regarding the use of lubricants however, including the difficulty of removal from the finished parts, the possibility for chemical reactions and environmental pollution [6]. Accordingly, by utilising ultrasonic vibration in the forming process, it has been reported that the application of a high viscosity lubricant in wire drawing can be avoided and that the interface friction in wire drawing [4] and extrusion [5] are reduced.

This paper reports a numerical investigation of the effects of the application of ultrasonic excitation in the extrusion of an aluminium cylindrical bar. By utilising finite element (FE) methods, a detailed investigation of the load data for different interface boundary conditions during ultrasonic extrusion can be carried out. In the static and ultrasonic models, the effects of friction on the extrusion load were investigated when radial and axial ultrasonic vibrations were applied to the die. Some previous investigations have reported close agreement between FE results and measured data for conventional extrusion [7] and for ultrasonic drawing [3]. This allows this current study to draw comparisons between the numerical models developed and previously published experimental data, in particular by comparing with the measured data of an experimental study of ultrasonic extrusion [5] in which it was concluded that the extrusion force was significantly reduced, that the effectiveness of ultrasonic excitation in reducing the extrusion force decreased as the extrusion speed increased and that the reduction in extrusion force was related to a reduction in friction. The initial aim of the research presented here is to show that applying ultrasonic excitation to the extrusion process is only effective below a critical extrusion speed. The second aim is to investigate whether the reductions in extrusion force measured in the previously published experimental investigations could be wholly explained by modifications of friction. Recent research which applied ultrasonic excitation in tension and compression tests of aluminium specimens showed that ultrasonic excitation resulted in an acousto-plastic effect which could be modelled as an effective material softening [8]. If this effect also contributed to ultrasonic extrusion experiments reported previously, then adjustments to the friction condition alone in the FE models should not be sufficient to explain the measured reductions in extrusion force.

2. Finite element modelling of ultrasonic extrusion

A series of FE models was developed to simulate the extrusion process under ultrasonic excitation. Based on the previous experimental work by Murakawa [6] and a previous numerical study of radial and axial ultrasonic wire drawing [3], the current FE models could be verified using the formulation for critical drawing speed, V_c. These experimental and numerical studies suggested that applying radial or axial ultrasonic vibration in a drawing process is only effective below a critical speed, V_c. The critical speeds for radial and axial ultrasonic excitation are given by the equations, $V_c = 2\pi af$ for axial ultrasonic excitation and, $V_c = 2\pi af / \tan\alpha$ for radial ultrasonic excitation, where a is the vibration amplitude, f is the ultrasonic frequency and α is the die half angle. Since extrusion is a reverse process of drawing, a similar observation was reported by Petukhow [5] where the effectiveness of applying ultrasonic vibration in an extrusion process was increased at lower speeds.

2.1. Modelling procedure

Finite element simulations were developed using ABAQUS. The extrusion billet was modelled using axi-symmetric 4-node elements, taking advantage of symmetry. A fine uniform mesh was used for the billet to ensure a sufficiently accurate analysis for severe material deformation. The punch and die were assumed to be rigid bodies, modelled by an analytical rigid surface. The deformed and undeformed mesh profiles of the extrusion process are shown in Fig. 1. The material properties used in the model for aluminium were derived from tests in a previous study [9]. The behaviour of aluminium is treated as elastic-plastic isotropic with low strain hardening.

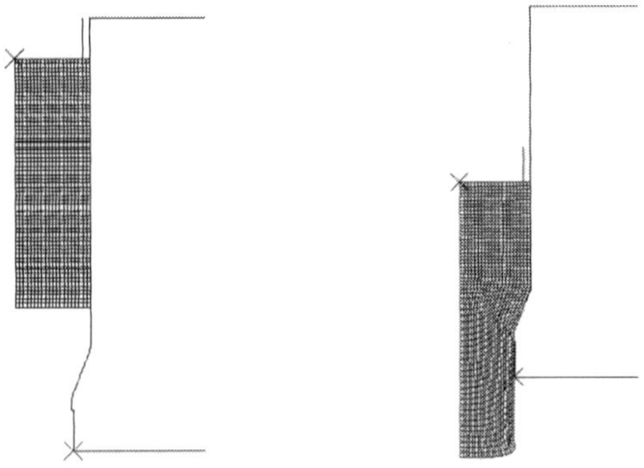

(a) (b)

Figure 1. Finite element model of the billet for extrusion, (a) undeformed and (b) deformed mesh.

The die and billet geometries used were based on a previous numerical study of extrusion by Long et al [7], giving an initial billet diameter of 40 mm, die diameter of 30 mm, providing a extrusion reduction of 43.8 % and a die half angle, $\alpha = 22.5^\circ$. For a smooth transition for the billet to approach the die, the billet was located at a distance of 5 mm from the die opening. An extrusion force was applied by specifying a constant velocity of the plunger.

In the first series of models, radial and axial ultrasonic vibrations were applied for the entire extrusion process. As described earlier, the reduction of extrusion force is significantly affected by the extrusion speed and the ultrasonic vibration mode used. By assuming that extrusion is a reversal of the drawing process, the FE models could be verified by observing the mean of the oscillatory force calculated for various extrusion speeds. This could be achieved when radial and axial ultrasonic excitation was superimposed for speeds ranging from below to above the critical speed.

Radial and axial ultrasonic excitation was superimposed on the die at a frequency of 20 kHz and an amplitude of 3 μm, giving critical speeds of 910 mm/s and 377 mm/s for radial and axial ultrasonic excitation respectively. Therefore, the extrusion speeds set in the FE models were 380 mm/s, 1000 mm/s and 3000 mm/s for radial mode excitation and 300 mm/s, 500 mm/s and 1000 mm/s for axial mode excitation. A constant interface friction coefficient, $\mu = 0.05$, was used in these models.

The effects of interface friction during radial and axial ultrasonic extrusion were numerically investigated in the second series of models. For the first set of simulations, constant coefficients of friction, $\mu = 0.1$, 0.05 and 0 were used throughout the static and ultrasonic intervals. Initially, the material was allowed to deform under static load by defining a constant velocity of the plunger at 10 mm/s. Subsequently, a radial or axial ultrasonic excitation was superimposed on the die when the plunger was displaced a distance of 25 mm. To allow for manageable computational time, the ultrasonic excitation was applied for a short interval at a frequency of 20 kHz and amplitude 3 μm. Models were then created where a coefficient of friction of $\mu = 0.1$ was used during static deformation and was then changed to $\mu = 0.05$ and 0 at the onset of ultrasonic excitation. When the ultrasonic excitation was discontinued, the material was further deformed under static load. The static-ultrasonic extrusion process was completed when the plunger was displaced 30 mm. The numerical force-displacement curves and stress contours were derived from the FE model results.

2.2. Effects of extrusion speed

In an attempt to validate the present FE models, the effect of the extrusion speed during superimposed ultrasonic vibration on the die was investigated and compared with results from a study reported in the literature [3]. The radial and axial ultrasonic excitations were continuously applied in these FE models.

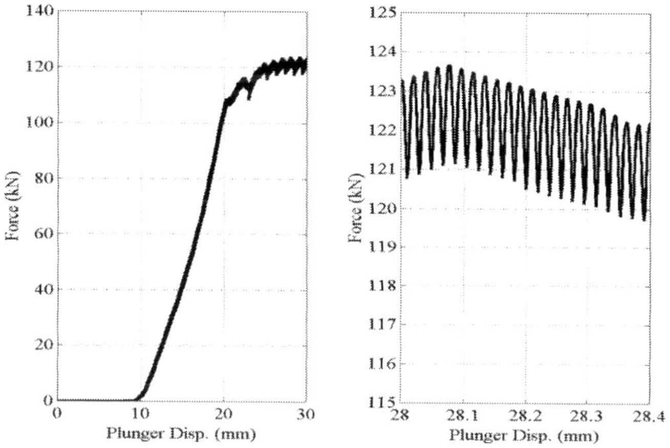

Figure 2. Extrusion speed = 380 mm/s (below critical speed), $\mu = 0.05$, with expanded view of the ultrasonic oscillatory force

Journal of Physics: Conference Series **181** (2009) 012027 doi:10.1088/1742-6596/181/1/012027

Fig. 2 shows the calculated extrusion force for one extrusion speed for radial ultrasonic excitation. Some cycles of the ultrasonic oscillatory force are shown in the figure in an expanded axis for clarity. It can be seen that the oscillatory force is superimposed on a low frequency oscillation caused by the high deformation of some elements during extrusion. This oscillation has been smoothed by ensuring a fine mesh in the vicinity of high deformation zones and could be further smoothed by remeshing in these zones. For the extrusion speed of 380 mm/s, which is below the critical speed for radial ultrasonic excitation, at a plunger displacement of 28.2 mm, the predicted mean extrusion force is 122.0 kN and the magnitude of the peak-peak oscillatory force is 2.3 kN, as illustrated in Fig. 2. For the same plunger displacement the mean forces were higher and the magnitudes of the peak-peak oscillatory forces were reduced for the higher extrusion speeds. The mean force for an extrusion speed of 1000 mm/s was 122.8 kN and increased to 123.0 kN when the extrusion speed was increased to 3000 mm/s, which are higher than critical speed, whilst the magnitude of the peak-peak oscillatory force reduced to 1.5 kN and 0.5 kN for the 1000 mm/s and 3000 mm/s cases respectively. In these simulations, the calculated force-displacement data are consistent with oscillatory stress superposition and in this case the maximum oscillatory force is equal to the static force. In most previous studies only the mean force was measured and it was assumed that the greater the reduction in mean force, the greater the benefit of ultrasonic vibrations. It is difficult to argue that there is any real load benefit from oscillatory stress superposition since the maximum oscillatory load and static load are equal. However, the calculated data agree with previous studies that the achievable mean force reduction is greatest when the extrusion speed is below the critical speed.

The above analysis procedure for radial ultrasonic extrusion was repeated for axial ultrasonic extrusion. In this case, at a plunger displacement of 28.2 mm and an extrusion speed below the critical speed for axial ultrasonic excitation, the mean extrusion force was calculated at 122.6 kN and the peak-peak oscillatory force amplitude was 0.25 kN. By increasing the extrusion speed above the critical speed, 500 mm/s and then 1000 mm/s, the predicted mean force was 123.2 kN for both speeds for the same plunger displacement. The peak-peak oscillatory forces for both extrusion speeds were negligible for axial ultrasonic extrusion.

The above FE results for radial and axial ultrasonic extrusions suggest that the calculated oscillatory forces substantially depend on the extrusion speed. At speeds above the critical speed, the effectiveness of applying the radial and axial ultrasonic excitations is reduced and consequently the drop in the mean oscillatory force calculated due to the ultrasonic load is negligible. In terms of a reduction in the mean oscillatory force compared with the static force, radial ultrasonic vibration is more effective than axial ultrasonic vibration for speeds below the critical speed.

2.3. Effects of adjusting friction

The FE models initially used constant coefficients of friction, $\mu = 0$ (friction free), $\mu = 0.05$, $\mu = 0.1$, during static and ultrasonic intervals. The plunger speed was set at a constant 10 mm/s, well below the critical speed. Fig. 3 shows the predicted force-displacement relationship for static and a short ultrasonic excitation interval applied in the radial mode.

For the ultrasonic interval, the effect is merely the superposition of an oscillatory force on the static force as described by Kirchner [10]. As can be seen in the expanded views of the figure, the FE data agrees with these earlier models and the path of maximum oscillatory force follows the path of the static force, with a reduction in the mean force. Once ultrasonic excitation is discontinued, the force-displacement returns elastically to the static curve.

The magnitude of the predicted peak-peak oscillatory force for radial and axial ultrasonic excitation depends on the interface boundary condition specified and is found to be higher for a high surface friction coefficient compared with a low surface friction coefficient. For all interface conditions a higher peak-peak oscillatory force was calculated for radial ultrasonic excitation than for axial ultrasonic excitation. These peak-peak oscillatory forces for radial and axial ultrasonic load with different surface friction coefficients are summarised in Table 1. Accordingly, the models suggest that the die-billet interface boundary condition effectively changes the magnitude of the peak-peak oscillatory force during ultrasonic extrusion.

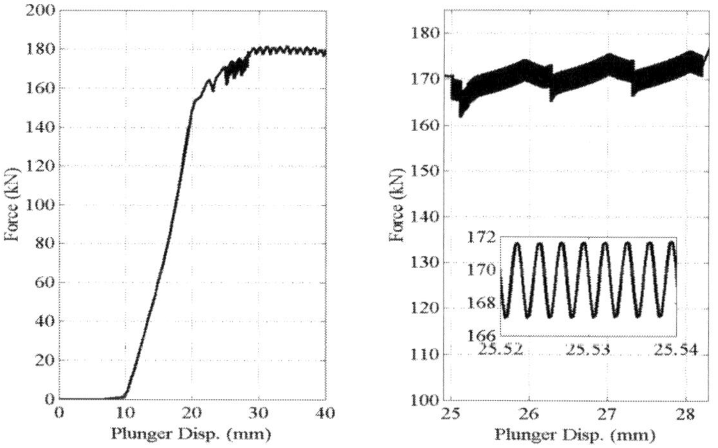

Figure 3. Radial ultrasonic excitation superimposed for a short interval during plastic deformation, plus two expanded views of the oscillatory force, $\mu = 0.1$

Table 1. Peak-peak oscillatory force for radial and axial ultrasonic extrusion

Coefficient of friction	Radial ultrasonic extrusion: peak-peak force (kN)	Axial ultrasonic extrusion: peak-peak force (kN)
$\mu = 0.1$	5.0	3.5
$\mu = 0.05$	3.0	2.5
$\mu = 0$	2.0	0.5

An investigation of the effects of adjusting the coefficient of friction in the FE extrusion model was performed in order to determine if the measured reduction of mean oscillatory force of Petukhow's study [5] could be explained in term of a temporary change in the coefficient of friction during applied ultrasonic excitation. In Petukhow's study, the extrusion force was measured when an axial ultrasonic excitation was superimposed on the die during extrusion of aluminium. The paper reported that by applying an ultrasonic load, the extrusion force was lowered by 35% from the static force. However this result was not accompanied by an explanation due to difficulties of real-time measurement of the oscillatory force response. Also, Petukhow did not clearly define whether the force reduction was attributed to a reduction in friction. Previous investigations of ultrasonic strip drawing [2] and ultrasonic wire drawing [4] suggested that by superimposing either radial or axial ultrasonic excitation on metal strip drawing, and by superimposing axial ultrasonic vibrations on wire drawing, the interface coefficient of friction was significantly reduced in both cases. Rozner [2] reported that a maximum of 40% reduction in the coefficient of friction could be achieved during both radial and axial ultrasonic strip drawing, whilst Susan [4] claimed that the coefficient of friction was lowered by at least 30% during axial ultrasonic wire drawing.

On the basis of Petukhow's experiments, where the force was reduced by 35% during ultrasonic extrusion, FE models were created in an attempt to quantify the friction coefficient which could result in that reduced force. This was achieved by assuming the ultrasonic excitation effectively modifies the interface friction coefficient to a reduction of between 30% and 40% as suggested by Rozner [2] and Susan [4]. For example, if $\mu = 0.1$ during static intervals, then it would be changed to $\mu = 0.07$ (30%

reduction) or to $\mu = 0.05$ (50% reduction) during ultrasonic intervals. Also, the effect of changing the coefficient of friction, $\mu = 0.1$, during static load to a frictionless interface, $\mu = 0$, during ultrasonic load was investigated.

Fig. 4 shows the numerical force-displacement data for the radial ultrasonic extrusion simulation, illustrating how the model effectively changes the boundary condition from the friction condition for static extrusion to the reduced friction condition during the interval of ultrasonic excitation. In the expanded views in the figure, it is clear that the paths of the maximum, mean and minimum oscillatory force are all reduced, departing from the stress superposition model and in common with the reported experimental stress-strain data [2]. The mean force reductions due to modification of the interface friction during ultrasonic excitation of the extrusion models are presented in Table 2.

However, these models only address the issue of interface friction. Therefore some deviation is observed between the measured force data of Petukhow's work and the FE model results. By assuming the ultrasonic excitation significantly reduces the coefficient of friction, say by 50% which is significantly higher than the commonly quoted maximum of 35-40%, the FE model predicts the mean forces were reduced by 25.6% for radial and 25.9% for axial ultrasonic loading. However, Petukhow reported a 35% reduction in the mean force from the static drawing force. This would indicate that friction effects alone may not be sufficient to explain the measured mean force reductions.

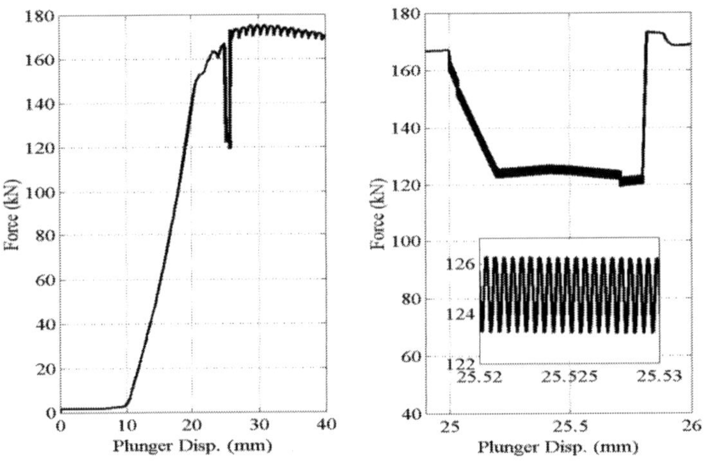

Figure 4. Force-displacement data for static and radial ultrasonic excitation, plus two expanded views of oscillatory force, $\mu = 0.1$ during static intervals changed to $\mu = 0.05$ during ultrasonic interval

Table 2. Mean force reduction due to the reduction of interface friction during ultrasonic extrusion.

Reduction of μ from 0.1 to	Radial mode Force reduction (kN)	Radial mode %	Axial mode Force reduction (kN)	Axial mode %
$\mu = 0.07$ (30%)	27.5	16.7	27.5	16.4
$\mu = 0.05$ (50%)	43.0	25.6	43.5	25.9
$\mu = 0.0$ (100%)	77.0	45.8	76.5	45.5

3. Conclusions

A numerical study of the effects of radial and axial ultrasonic excitation on an extrusion process has been presented. The study supports previous experimental findings that the benefits of applying ultrasonic excitation can only be achieved below a defined critical extrusion speed. By analysis of the force-displacement response from the finite element models has been possible to investigate the effects of adjusting the coefficient of friction at the die-billet interface and hence to confirm previously published experimental studies that have suggested that the measured reduction in the mean forming load is due to an effective reduction in the coefficient of friction during the intervals of ultrasonic excitation. In general it is observed that the measured reductions in forming load reported in the literature are significantly higher than can be achieved by incorporating the commonly quoted friction coefficient reductions into the finite element model and that friction changes alone may not explain the measurement data.

4. References

[1] F. Blaha, B. Langenecker, Tensile deformation of zinc crystal under ultrasonic vibration, Natur., 42, 556, 1955.
[2] A.G. Rozner, Effect of ultrasonic vibration on coefficient of friction during strip drawing, J. of the Acoust. Soc. Am., 49, 1368-1371, 1971.
[3] M. Hayashi, M. Jin, S. Thipprakmas, Simulation of ultrasonic-vibration drawing using the finite element method, J. Mat. Proc. Tech., 140, 30-35, 2003.
[4] M. Susan, and L.G. Bujoreanu, The metal-tool contact friction at the ultrasonic vibration drawing of ball-bearing steels wires, Revesita De Metalurgia, 35, 379-383, 1999.
[5] V.I. Petukhow, O.V. Abramov, A.M Zubko, Yu V. Manegin, Extrusion of aluminium in an ultrasonic filed, Light Metal Age, 31, 6-8, 1973.
[6] M. Murakawa, M. Jin, The utility of radially and ultrasonically vibrated dies in the wire drawing process, J. Mat. Proc. Tech., 113, 81-86, 2001.
[7] H. Long, R. Balendra, FE simulation of the influence of thermal and elastic effects on the accuracy of cold-extruded components, J. Mat. Proc. Tech., 84, 247-260, 1998.
[8] Y. Daud, M. Lucas, Z. Huang, Modelling the effects of superimposed ultrasonic vibrations on compression and tension tests of aluminium, J. Mat. Proc. Tech., 186, 179-190, 2007.
[9] Y. Daud, M. Lucas, Z. Huang, Ultrasonic compression tests on aluminium, Applied Mechanics and Materials, 3-4, 99-104, 2005.
[10] H.O.K. Kirchner, W.K. Kromp, F.B. Prinz, P. Trimmel, Plastic deformation under simultaneous and unidirectional loading at low and ultrasonic frequencies, Mat. Sci. and Eng. 68, 197-206, 1984.

High temperature life prediction of a welded IN718 component

D W J Tanner[1], **A A Becker and T H Hyde**

Department of Mechanical, Materials and Manufacturing Engineering
University of Nottingham, Nottingham NG7 2RD, UK

[1]E-mail: eaxdwjt@nottingham.ac.uk

Abstract. Life predictions from a case study of a welded feature in a generic spoke structure, determined using three-dimensional quasi-static elastic-plastic and creep finite element analyses, are demonstrated. The complete structure consists of multiple Inconel 718 (IN718) TIG-welded features; the welds exhibit noticeably depleted mechanical properties so a multi-material analysis is necessary for accurate predictions. The effect the welds have on the life is investigated for both constant (creep) and cyclic loading (creep-fatigue) conditions at 620°C. Creep damage and Smith, Watson and Topper (SWT) strain parameter lifing methods are used, based upon material properties determined using uniaxial test data. The lower fatigue properties of the welded IN718 material at high temperature had a negative effect on the fatigue life of the structure. The effect of the weld on the life under constant loading at high temperature was found to be more difficult to evaluate due to significant stress relaxation.

1. Introduction

Structures that are subjected to high service temperatures where creep (and possibly creep-fatigue) is significant need well defined material behavior and for all integral structural components, the lifetime requires assessment. For this case study, a thin-section generic spoke structure was analyzed to demonstrate the application of high temperature lifing methodologies to a complicated geometry including a weld with differing material properties. Comparisons are then made to identify how the inclusion of the weld affects the life of the structure.

Life assessment of welded thin-section structures is particularly complex due to residual stresses, distortions and material microstructural changes caused by the welding process. In particular, out-of-plane distortions compromise uniaxial material testing. The weld bead geometry must also be considered (if present) since sharp fillet radii result in high stress concentrations at the weld bead and parent interface. Due to these factors, the weld has a direct effect on the creep and fatigue life [1].

Although it has disadvantages such as those mentioned above, welding (as opposed to casting) reduces the cost of large structural components and can eliminate the need to transport full size cast components from the supplier to the manufacturer, which is also more cost efficient and better for the environment. As such, welding is an integral manufacturing procedure for many components, some of which could not be produced without its application.

The material used throughout this study was sheet Inconel 718 (IN718), a precipitation-hardenable, niobium-modified, nickel-base superalloy that is used extensively in the aerospace, petrochemical, and nuclear industries since it exhibits good strength, excellent resistance to oxidation at high temperatures and favorable weldability. The material properties and failure behavior for both parent and weld material were determined from specimen testing, which is briefly summarized here; more information on the fatigue testing can be found in [2]. The temperature for the testing and the subsequent life assessment analyzes was 620°C, since this is the temperature limit permitted for many critical applications of the material [1].

2. Material property testing

As mentioned above, it is desirable to avoid weld-induced distortions, particularly when material properties need to be determined through conventional uniaxial testing. For test specimen manufacture, the distortions can be effectively mitigated by only welding small plates that are just large enough to form one test specimen. The small size increases the relative stiffness of the plates, and reduces the welding time and therefore the time at which nonuniform thermal expansion exists, which is the driving force of the distortions. Unlike other methods of distortion prevention, such as those involving trailing heat sources or sinks, this solution does not modify the effects of the welding process on the thermal history of the material. If the plates remain flat after welding, the stress-concentrating weld bead excess can be removed by a linishing process.

Both non-welded and TIG butt-welded bow tie specimens were manufactured from a 3.2mm solution heat treated sheet of IN718 provided by Haynes International Ltd. Mechanized, fully penetrating, single pass TIG, square-edge, closed butt welds, were made between the short sides of 100mm x 40mm coupons with matching IN718 filler wire and argon shielding. The small welded plates did not distort and weld bead excess was removed without difficulty. After bow tie profiling and prior to polishing, all specimens were given the same precipitation heat treatment before testing, consisting of: 720°C/8hrs, 50°C/hr furnace cool to 620°C/8hrs, air cool. This heat treatment also effectively relieves welding-induced residual stresses [3].

The following tests were performed on both welded and non-welded specimens at 620°C (all welded specimens were tested in the direction transverse to the weld):

- Uniaxial constant displacement tensile tests
- Constant tensile load uniaxial creep tests at a variety of different nominal stress levels. Notched specimen creep testing was also performed in order to obtain the triaxial stress behavior of the material.
- Uniaxial fatigue tests at a variety of different nominal (maximum) tensile stress levels. The cycles had a trapezoidal (1-1-1-1) form with a frequency of 0.25 Hz and a stress ratio of ≈ 0 (i.e. 1sec ramp up/1sec dwell at max. load/1sec ramp down/1sec dwell at min. load -as can be seen in Figure 6).

The aim of these tests was to determine the required material properties for subsequent finite element (FE) modeling and life prediction.

2.1. Test results

Consistent with tests are different temperatures, the tensile tests resulted in both the 0.2% proof stress and the ultimate strength of the welded specimens exceeding 90% of the values of the non-welded specimens. However, the strain to failure for the welded specimens was significantly less than that of the non-welded specimens (Table 1). The creep tests also showed noticeably lower total strain to failure for the welded specimens than the non-welded specimens, the time to failure was also markedly less for the welded specimens. The fatigue tests resulted in lower failure times for the welded specimens than the corresponding non-welded values for any given nominal stress.

In conclusion, in common with other nickel superalloys, welded IN718 exhibits varied and complex microstructures in the fusion zone and in the heat-affected zone (HAZ). The weld region and the HAZ exhibit slightly lower tensile strength and significantly lower ductility than the parent material. Therefore, although welded IN718 exhibits comparatively little loss of tensile strength, its creep and high temperature fatigue properties are severely decreased [4].

2.2. Weld material property determination

The mechanical properties and stress-strain relationships of the IN718 welds were obtained from the testing mentioned above by calculating the difference in strain of non-welded and cross-welded specimen tests over the gauge length, and determining what the properties of the weld must be in order to give the measured test results. The calculated weld material nominal stress-strain curve at 620°C is shown in Figure 1. The engineering uniaxial tensile properties are given in Table 1. As expected, the 0.2% proof stress calculated for the weld material is lower than that obtained for both the non-welded and welded specimen tests since the weld region is only a small proportion of the welded test specimen gauge length and its properties cause the difference between the non-welded and welded specimen results. FE verification of the material properties was conducted by modeling the uniaxial tensile and creep tests to check the accuracy of the material properties assigned to the weld region. Very good agreement was obtained.

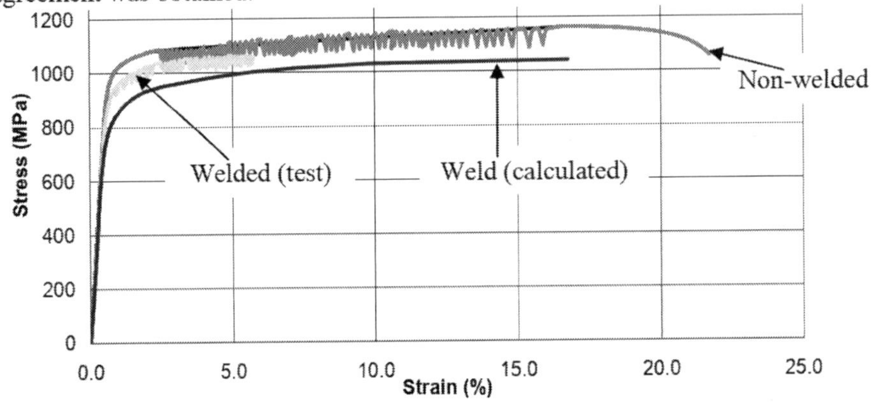

Figure 1. Nominal stress-strain curves for heat-treated IN718 at 620°C.

Table 1. Uniaxial tensile properties of heat-treated IN718 at 620°C.

	0.2% Proof stress (MPa)	Ultimate strength (MPa)	Young's modulus (GPa)	Failure strain (%)
Non-welded	992	1163	166	21.8
Welded (test)	920	1064	165	5.7
Weld (calculated)	808	-----	148	-----

3. Finite element analysis and life assessment of the structure

3.1. Model setup

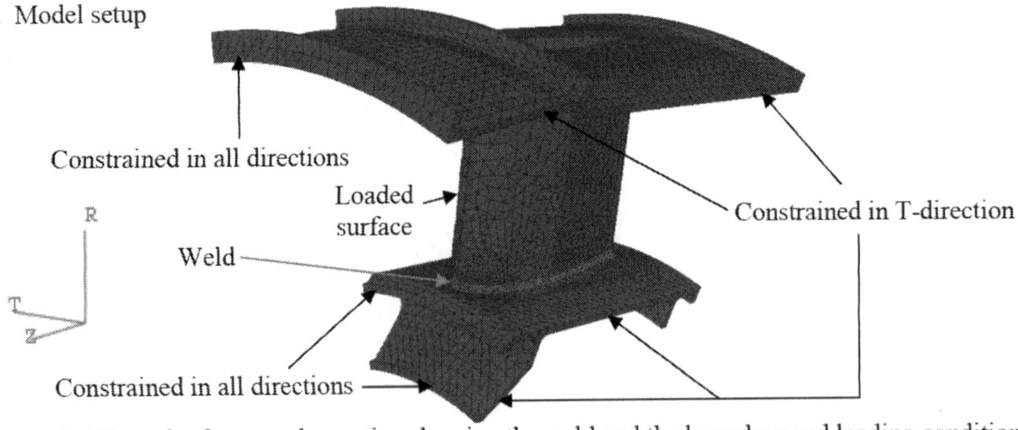

Figure 2. FE mesh of one-spoke section showing the weld and the boundary and loading conditions.

Journal of Physics: Conference Series **181** (2009) 012028 doi:10.1088/1742-6596/181/1/012028

A one-spoke (30°) section of the generic spoke structure was modeled, with a weld positioned in the blend between the spoke and the inner ring, in order to demonstrate the lifing methodology. A 3D solid ABAQUS FE mesh consisting of 20804 quadratic tetrahedron (C3D10) elements and 40587 nodes was generated. The mesh was refined in the regions of high stress that were identified after an initial analysis had been run (Figure 2). A cylindrical coordinate system with the origin at the centre of the complete structure was employed in order to conveniently define the cyclic symmetry along the edge surfaces of the one-spoke section. These surfaces were constrained in the circumferential (T) direction. The front-facing surfaces were constrained in all directions. Load was applied in the negative direction of the axis of rotation of the cylindrical coordinate system (-Z) to the nodes on the front surface of the spoke (Figure 2). The material properties of both parent (non-welded) and weld IN718 metal were obtained from the testing mentioned above. The weld properties were assumed to be uniformly distributed and isotropic, the HAZ was not considered as a separate region due to its small size and the difficulty in obtaining its properties. Two types of FE analysis were performed: a constant load creep analysis at 620°C and a fatigue analysis at 620°C.

3.2. Creep analysis

3.2.1. Procedure and life assessment method. The creep FE analysis was conducted in two steps. Firstly, the load was applied gradually during a nonlinear elastic-plastic static step. It should be noted that global yielding did not occur and plasticity was only included in order to produce a suitably high stress state to cause failure within a reasonable time frame. A nonlinear visco (transient, quasi-static) plastic step followed the loading step, for which the load was held constant. A Kachanov-type continuum damage mechanics approach was used for the creep law and life prediction since it was shown to give more accurate failure times and creep strain curves for the material test data. The following constitutive equations were used [5]:

$$\dot{\varepsilon}_{ij}^{c} = \frac{3}{2} A \left(\frac{\sigma_{eq}}{1-\omega} \right)^{n} \frac{S_{ij}}{\sigma_{eq}} t^{m} \qquad \sigma_{r} = \alpha \sigma_{1} + (1-\alpha)\sigma_{eq} \qquad \dot{\omega} = \frac{B \sigma_{r}^{\chi}}{(1-\omega)^{\phi}} t^{m}$$

where σ_{eq}, σ_1, and σ_r are the equivalent, maximum principal and rupture stresses, respectively. S_{ij} are the deviatoric stress components. ω is a state variable that describes creep damage and varies from 0 (initially no damage) to 1 (failure). At time, $t = 0$, $\omega = 0$ and creep strain, $\varepsilon_{ij}^{c} = 0$. A, n, m, B, χ, ϕ and α are material constants determined from the creep tests, the values are given in Table 2.

Table 2. Creep damage constants of IN718 at 620°C, based on units of stress in MPa and time in hr.

	A	n	m	B	χ	ϕ	α
Parent (non-weld)	2.037E-61	19.300	0.0	4.322E-47	14.728	12.0	0.1
Weld	5.260E-56	17.984	0.0	2.623E-50	16.367	4.0	0.2

3.2.2. *Creep analysis results.* Below are the results from creep FE analyses with and without weld for an applied load of 870kN.

A large amount of stress relaxation occurred during the creep step (Figure 3). The welded material was unable to support as high a stress as the non-welded material and relaxed at a quicker rate initially before steadying at a lower rate. The highest stress occurred in and around the blend between the spoke and the inner ring (Figure 3). This is also the site of the weld. The peak local creep strain was greater in the weld, but not in the region of failure, where the highest stresses caused the most creep damage (Figure 4). Failure of the non-welded case was deemed to have occurred after 137hr in the blend of the spoke-inner ring, where the weld would have been (Figure 4). Failure of the welded case occurred after 365hr on the parent side (inner ring) of the weld-parent interface. The life of the welded case is increased due to the greater stress relaxation of the weld, which causes the failure location to shift into an area that experiences lower stress than the failure location of the non-welded case.

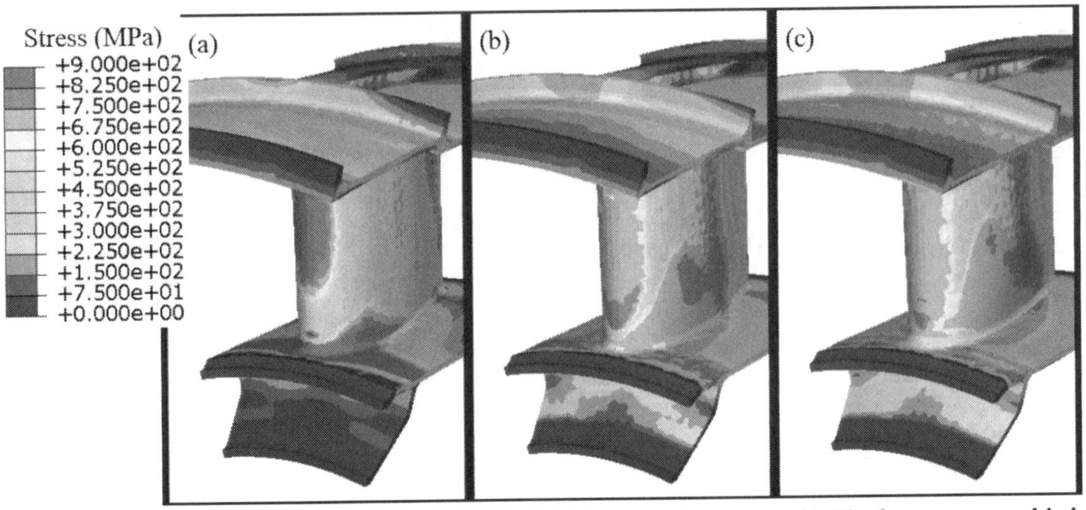

Figure 3. Rupture stress: (a) non-welded case immediately after loading, (b) Final state non-welded, (c) Final state welded case.

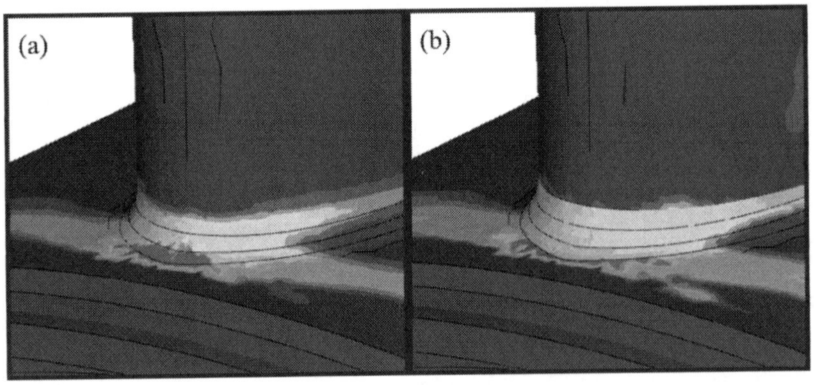

Figure 4. Damage, close-up of failure positions (in red) (a) non-welded case, (b) welded case.

It should be noted that the non-welded case uses properties obtained from wrought material, but this geometry cannot be manufactured without casting or welding. Also, the peak creep strain is very high in the weld and this may cause failure before the creep damage parameter indicates. Failure time determined by the time taken for maximum principal strain to reach 0.1 would give a non-welded life of 53hr and a welded life of 17hr. The failure locations for this criterion are in the lower part of the blend for the non-welded case and in the upper part of the side of the blend (in the weld) for the welded case (Figure 5).

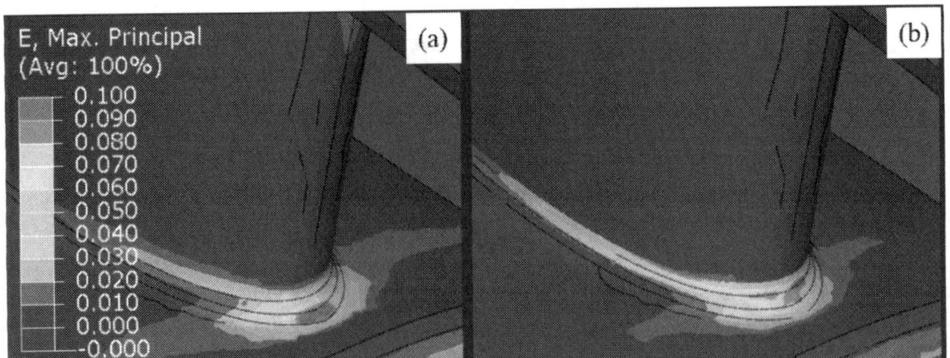

Figure 5. 0.1 maximum principal strain after (a) 53hr non-welded case, (b) 17hr welded.

3.3. Fatigue analysis

3.3.1. *Procedure and life assessment method.* A strain-based parameter lifing method that accounts for mean stress was chosen as in [6]. The Smith, Watson and Topper (SWT) strain range parameter was calculated for each point during the analysis using:

$$\left(\frac{\sigma_{\max}\Delta\varepsilon}{\mathrm{E}}\right)^{\frac{1}{2}}$$

where $\sigma_{\max}$ is the peak stress, $\Delta\varepsilon$ is the total strain range within a cycle and E is Young's modulus. This formula is the strain equivalent to the effective stress presented by Smith, Watson and Topper [7]. This SWT parameter correlates well with cycles to failure using the test data obtained. The fatigue tests were modeled so that peak SWT values could be determined. FE-calculated values averaged over the gauge length gave good agreement with values calculated using test measurements. Therefore, an FE analysis of a structure can be used to calculate the peak SWT value and the corresponding life can be predicted by correlating with the fatigue test data.

The mesh, weld position, boundary conditions, loading position and direction were the same as those used in the creep analysis. The fatigue loading was the same as that used for the material testing (trapezoidal (1-1-1-1) form with a load frequency of 0.25 Hz and a stress ratio of ≈0). The analysis was nonlinear visco (transient, quasi-static) plastic so as to include the effect of creep, the accumulation of which was accounted for during the fatigue cycles using Norton's creep law:

$$\dot{\varepsilon}^{c} = \mathrm{A}\sigma^{n}$$

where A and n are given in Table 2.

Two steps were performed, each equivalent to one cycle; all results were taken from the second step, which is a more stable representation of subsequent cycles than the initial loading cycle. The values of maximum stress, strain range and Young's modulus, taken from the FE models, were converted to SWT strain range and read into the specimen fatigue test lifing curve.

3.3.2. *Fatigue analysis results.* Below are the results from fatigue FE analyses with and without weld for an applied load of 250kN.

Since there is only a short period of hold during each fatigue cycle, little creep and subsequent stress relaxation occurs. Plastic yielding is present during the first cycle and then this stabilizes due to material hardening. The welded material yields more and is unable to support as high a stress as the non-welded material due to its lower yield stress (Figure 6 and Figure 7), therefore it also strains more.

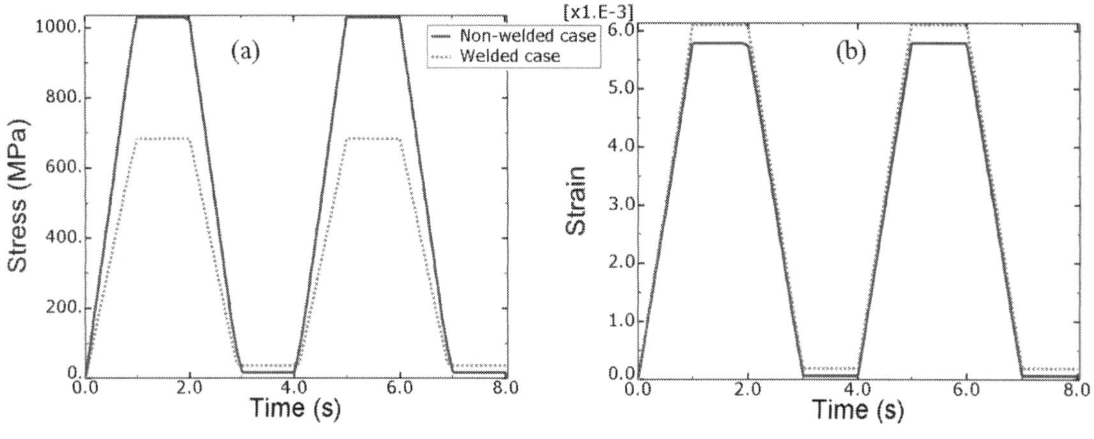

Figure 6. (a) von Mises stress history, (b) maximum principal strain history at peak position.

The predicted location of failure occurs in the blend region between the spoke and the inner ring in both cases, shown by the highest value of SWT in Figure 8. This is also the site of the weld. The value of SWT is slightly higher (<1% increase) for the welded case since the slightly higher strain range is countered by the lower peak stress and Young's modulus. When the predicted values of SWT are compared with those from the test specimens, a life prediction of 108300 cycles is obtained for the non-welded case, while just 1260 cycles for the welded case (Figure 9). This marked reduction of life is expected due to the depleted high temperature fatigue properties of the welded material as discussed previously. To give an expected life of approximately 10000 cycles for the welded case, the applied load would need to be reduced by 8% to 230kN. To give a similar life for the non-welded case (using wrought material properties), a load of 280kN could be applied, which is an increase of 12% from the original load, and over 20% higher than the load for the equivalent welded case.

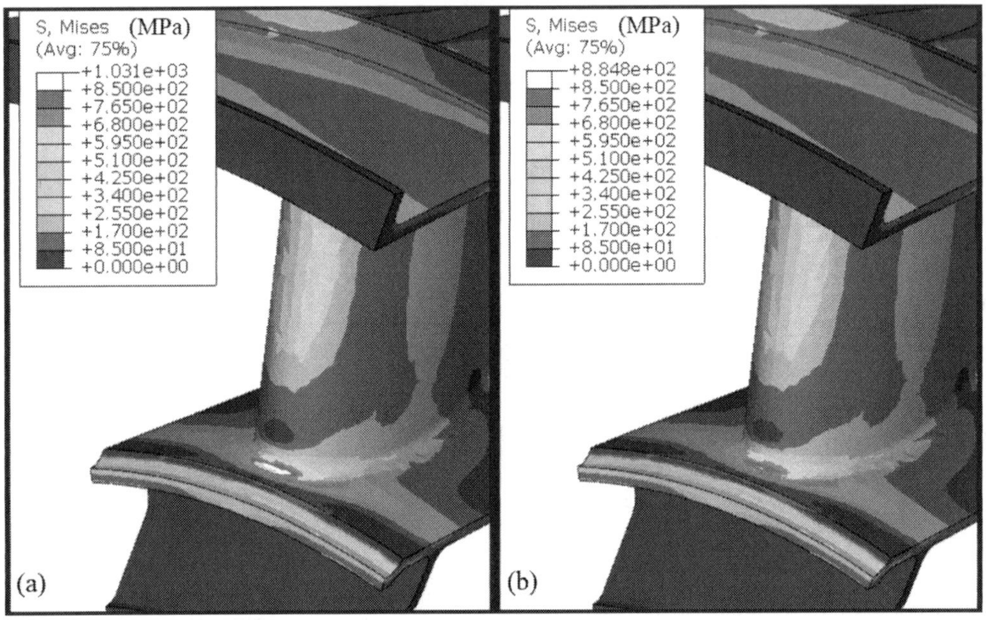

Figure 7. von Mises stress distribution at maximum load (a) non-welded case, (b) welded.

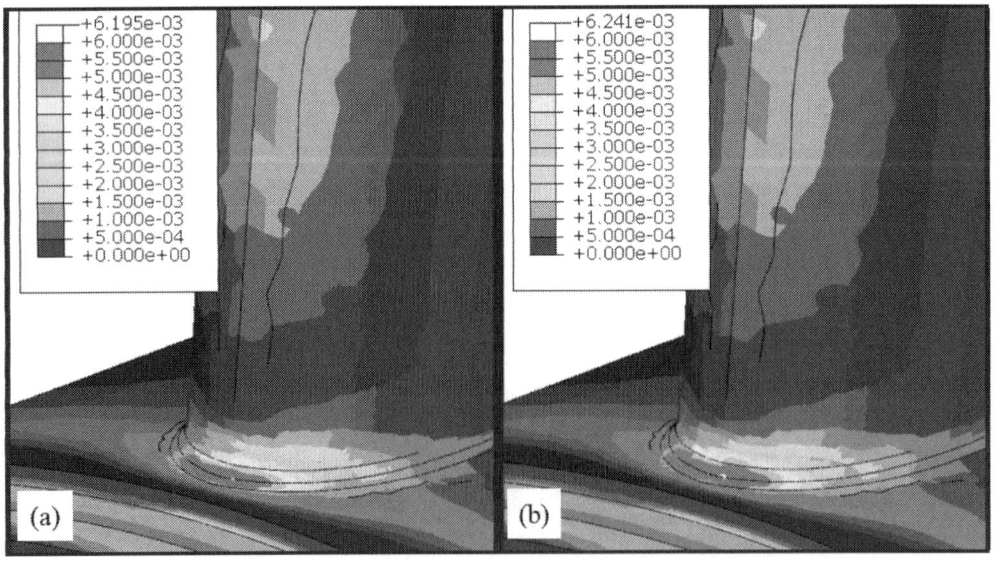

Figure 8. SWT distribution (a) non-welded case, (b) welded.

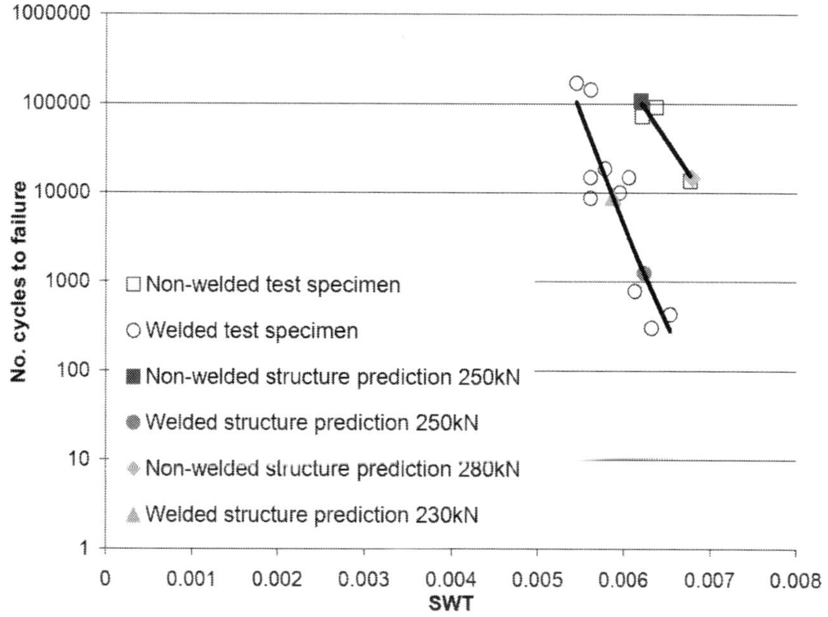

Figure 9. SWT strain parameter against number of cycles to failure.

4. Conclusion

Although welded IN718 exhibits comparatively little loss of tensile strength, its creep and high temperature fatigue properties are severely decreased. The depleted high temperature fatigue properties of the welded IN718 material have a negative effect on the predicted high temperature fatigue life of the structure examined here. The effect of the weld on the life of the structure under constant loading at high temperature was found to be more difficult to quantify due to significant stress relaxation. The predicted failure location was affected by the weld in the constant load case but not in the fatigue load case. Failure would be expected to occur in the region of greatest stress (at the blend where the spoke meets the inner ring) but due to significant stress relaxation in the welded case, the failure location shifts to the parent side (inner ring) of the weld-parent interface.

Acknowledgments

The authors wish to acknowledge the EU VITAL project funding (AIP4-CT-2004-012271).

References

[1] Song Y, Becker A A, Hyde T H and Sun W 2003 *Proc. Int. Conf. on Metal Fabrication and Welding Technology (Nottingham, UK)*, ed A A Becker pp 95-106

[2] Tanner D W J, Becker A A and Hyde T H 2008 *6th Int. Conf. on Low Cycle Fatigue (Berlin, Germany)*, ed P D Portella *et al.* (Berlin, Germany: DVM) pp 461-6

[3] Kyriakoglou Y, Preuss M and Bowen P 2007 Tensile properties and HAZ definition for TIG and EB welds using electronic speckle pattern interferometry (ESPI) *Appl. Mech. Mater.* **7-8** 119-25

[4] Vincent R 1985 Precipitation around welds in the nickel-base superalloy, Inconel 718 *Acta Metall.* **33** 1205-16

[5] Hayhurst D R 1972 Creep rupture under multi-axial states of stress *J Mech Phys Solid* **20** 381-90

[6] Corran R S J and Williams S J 2007 Lifing methods and safety criteria in aero gas turbines *Eng Fail Anal* **14** 518-28

[7] Smith K N, Watson P and Topper T H 1970 A stress-strain function for the fatigue of metals *J Mater* **5** 767-78

Journal of Physics: Conference Series **181** (2009) 012029 doi:10.1088/1742-6596/181/1/012029

A Review of Residual Stress Analysis using Thermoelastic Techniques

A. F. Robinson, J. M. Dulieu-Barton, S. Quinn, R. L. Burguete[*]

University of Southampton, School of Engineering Sciences, Highfield, Southampton, SO17 1BJ, UK

[*]Airbus UK Ltd., New Filton House, Filton, Bristol, BS99 7AR, UK

Abstract

Thermoelastic Stress Analysis (TSA) is a full-field technique for experimental stress analysis that is based on infra-red thermography. The technique has proved to be extremely effective for studying elastic stress fields and is now well established. It is based on the measurement of the temperature change that occurs as a result of a stress change. As residual stress is essentially a mean stress it is accepted that the linear form of the TSA relationship cannot be used to evaluate residual stresses. However, there are situations where this linear relationship is not valid or departures in material properties due to manufacturing procedures have enabled evaluations of residual stresses. The purpose of this paper is to review the current status of using a TSA based approach for the evaluation of residual stresses and to provide some examples of where promising results have been obtained.

1. Introduction

Residual stresses may be introduced into a component during its entire manufacturing process; it is highly unlikely that a component in-service is free of residual stresses. Common routes for introducing residual stress are: (i) during the material production stage via deformation or thermal treatments, (ii) as a result of non-uniform heating or deformation during component manufacture, or (iii) during assembly as a consequence of welding processes or the interference of multiple parts. It is important to understand how residual stresses are distributed in a component to define its performance characteristics. This allows residual stress to be accounted for at the design stage and considered in the component life cycle. If the residual stress distribution is not known, then structural failure may occur due to the combined effect of the residual and applied stresses [1]. A large amount of tensile residual stress would decrease the applied tensile stress that would normally be required to induce plastic deformation, or indeed failure. Conversely a compressive residual stress would be beneficial, enabling the component to withstand a greater tensile stress; this is used in processes such as shot-peening and cold expansion [2]. At present, there are several techniques available for measuring residual stresses. Destructive methods are not always practical for an in-service industrial environment, while the non-destructive methods are typically expensive and time consuming. X-ray and synchrotron diffraction [3] are commonly employed in a research setting, while ultrasonic techniques [4] have proved useful in a manufacturing and industrial environment. Therefore there is a demand for a cheaper and quicker non-destructive, non-contact, full-field residual stress evaluation technique.

Thermoelastic stress analysis (TSA) [5] has been identified as a possible solution for a robust and portable means of non-destructive residual stress evaluation. TSA is now a well established non-contacting analysis technique that provides full-field stress data over the surface of a cyclically loaded component. TSA is based on the small temperature changes that occur when a material is subject to a change in elastic strain, generally referred to as the 'thermoelastic effect'. When a material is subjected to a cyclic load, the strain induced produces a cyclic variation in temperature. The temperature change (ΔT) can be related to the change in the 'first stress invariant', $\Delta\sigma_{kk}$, or the sum of the principal stresses [5]. An infra-red detector is used to measure the small temperature change, which can then be related to the stress using the following equation:

$$\Delta T = -KT_0 \Delta\sigma_{kk} \tag{1}$$

where T_0 is the absolute temperature and K is the thermoelastic constant, $K = \alpha / (\rho C_p)$, where α, ρ, C_p are the material constants of the coefficient of thermal expansion, mass density and the specific heat at constant pressure, of the material respectively.

The analysis that leads to equation (1) is dependent on three important assumptions [5]:

(i) the material behaviour is linear elastic,
(ii) the temperature changes in the material occur adiabatically,
(iii) the relevant material properties are not temperature dependent.

Belgen [6] first observed that this linear equation was not always correct. As a consequence of the temperature dependence of the thermoelastic constant, Belgen proposed that the temperature change was also dependent on the applied mean stress. Experiments by Machin et al [7] confirmed the existence of a mean stress dependence of the thermoelastic response. The assumptions above were reviewed by Wong et al [8], who later proposed a review of the general theory of the thermoelastic effect that did not rely on the previous assumptions. The idea that the material properties are independent of temperature was rejected. As a consequence, a more complicated nonlinear relationship between the temperature difference and the stress components was derived:

$$\dot{T} = \frac{T_0}{\rho C_p}\left[-\left(\alpha + \left(\frac{\nu}{E^2}\frac{\partial E}{\partial T} - \frac{1}{E}\frac{\partial \nu}{\partial T}\right)\sigma_{kk}\right)\dot{\sigma}_{kk} + \left(\frac{(1+\nu)}{E^2}\frac{\partial E}{\partial T} - \frac{1}{E}\frac{\partial \nu}{\partial T}\right)\sigma_{ij}\dot{\sigma}_{ij}\right] \tag{2}$$

where $\dot{T}$ is the rate of change of temperature, E is Young's modulus of the material, ν is Poisson's ratio, σ_{kk} is the first stress invariant and σ_{ij} is the stress tensor.

Equation (2) is known as the 'revised higher order theory' and accounts for the temperature dependence of the material properties that are contained within the thermoelastic constant. It is important to note that assumptions (i) and (ii) are retained in the derivation of equation (2), as it is the temperature dependence of the elastic properties that provide a route for residual stress measurement. In this paper, three approaches for residual stress measurement based on the thermoelastic response are reviewed. The first two are based on the revised higher order theory given by equation (2) and the third is based on changes in material properties contained within K that allows the extent of plastic deformation to be determined and is based on equation (1). Progress to date is detailed in the paper and the limitations and challenges in each approach are identified.

2. Mean Stress Dependence and the Second Harmonic

For the purpose of simplifying the analysis, the simpler case of uniaxial loading (where $\sigma_{11} = \sigma_{kk}$ and $\sigma_{22} = \sigma_{33} = \sigma_{12} = \sigma_{23} = \sigma_{13} = 0$) is examined, so that equation (2) can be reduced to:

$$\dot{T} = -\frac{T_0}{\rho C_p}\left(\alpha - \frac{1}{E^2}\frac{dE}{dT}\sigma_{11}\right)\dot{\sigma}_{11} \tag{3}$$

Equation (3) shows that the rate of temperature change is a function of the applied stress and its rate of change. In TSA, σ_{11} could be regarded as the mean stress, σ_m, and integrating equation (3) over a period from its initial state to final state provides a relationship between the change in temperature and the change in stress:

$$\Delta T = -\frac{T_0}{\rho C_p}\left(\alpha - \frac{1}{E^2}\frac{dE}{dT}\sigma_m\right)\Delta\sigma_{11} \tag{4}$$

The 'revised higher order theory' enables the mean stress dependence of the thermoelastic constant to be accounted for by the temperature dependence of the elastic modulus, and shows that the temperature response is dependent on the mean stress as well as the applied stress as shown in equation (4). Wong et al [9] continued this analysis by considering an experiment whereby a component in a uniaxial system is cyclically loaded about a mean stress, with a sinusoidal stress input such that:

$$\sigma_{11} = \sigma_m + \sigma_a \sin\omega t, \qquad \text{and thus,} \qquad \dot{\sigma}_{11} = \omega\sigma_a \cos\omega t \tag{5}$$

where σ_a is the applied stress amplitude and ω is the frequency of loading.

By substituting equations (5) into equation (3) the following expression was obtained:

$$\dot{T} = -\frac{T_0}{\rho C_p}\left(\alpha - \frac{1}{E^2}\frac{dE}{dT}(\sigma_m + \sigma_a \sin\omega t)\right)\omega\sigma_a \cos\omega t \tag{6}$$

or,

$$\dot{T} = -\frac{T_0}{\rho C_p}\left[\left(\alpha - \frac{1}{E^2}\frac{dE}{dT}\sigma_m\right)\omega\sigma_a \cos\omega t - \frac{1}{2E^2}\frac{dE}{dT}\omega\sigma_a^2 \sin 2\omega t\right] \tag{7}$$

Then, as before, equation (7) can be integrated over a period of time between $t = 0$ and t to give an expression for the temperature at time, t, as follows:

$$T = -\frac{T_0}{\rho C_p}\left[\left(\alpha - \frac{1}{E^2}\frac{dE}{dT}\sigma_m\right)\sigma_a \sin\omega t + \frac{1}{4E^2}\frac{dE}{dT}\sigma_a^2 \cos 2\omega t - \frac{1}{4E^2}\frac{dE}{dT}\sigma_a^2\right] \tag{8}$$

The first term in equation (8), varying with the frequency of the applied load, is dependent on both the applied mean stress and the stress amplitude. The second term in equation (8) varies at twice the frequency and is proportional to the square of the stress amplitude. The third term is not a function of the loading frequency. In conventional TSA it is the practice to reject all data other than that obtained from the fundamental loading frequency. Wong et al [9] observed that by obtaining data at this frequency and at the second harmonic, it would be possible to derive two simultaneous equations where the two unknowns are the cyclic stress amplitude and the mean stress. Subsequently, it was recognised that thermoelastic stress analysis may potentially yield enough information to provide a technique for deriving the residual stress in a component.

In TSA the practice is to sample the minimum and maximum values of T in each cycle such that ΔT is measured, allowing the following expression to be derived from equation (8):

$$\Delta T = a \sin \omega t - b \cos 2\omega t \tag{9}$$

where $$a = -K_0 T_0 \sigma_a \quad , \quad b = -\frac{K_0 T_0}{4\alpha E^2}\frac{\partial E}{\partial T}(\sigma_a)^2$$

and $$K_0 = \frac{1}{\rho C_p}\left(\alpha - \frac{1}{E^2}\frac{dE}{dT}\sigma_m\right)$$

K_0 is the revised version of the thermoelastic constant that is a function of the mean stress and Young's modulus. Obtaining temperature data at both ω *and* 2ω loading frequencies enables the applied cyclic stress amplitude, σ_a, to be calculated from the 2ω component, and subsequently, the mean stress, σ_m, from the following expressions derived from equation (9):

$$\sigma_a = \sqrt{\frac{-b\rho_0 C_p}{T}\frac{4E^2}{\partial E/\partial T}} \quad \text{and} \quad \sigma_m = \left(\frac{a\rho_0 C_p}{T\sigma_a} + \alpha\right)\frac{E^2}{\partial E/\partial T} \tag{12}$$

Validation of the revised thermoelastic theory was provided by Wong et al [9], by comparison of the thermoelastic signal from two uniaxially loaded aluminium specimens. One specimen was undeformed, the other was manufactured curved and straightened to provide a geometrically similar specimen. Strain gauges were used to identify the areas of tensile and compressive stress in the straightened specimen. Results showed that there were significant differences in the thermoelastic data from each specimen; this was attributed to the residual stress and good agreement was made between the data from the strain gauge readings and the equations. Dunn et al [10] confirmed that the mean stress effect is measurable in titanium (Ti-6Al-4V), aluminium (2024) and steel (4340); they observed a declining mean stress dependence of the thermoelastic constant with values of $\partial K / \partial \sigma_m \, K_0^{-1} = 0.45$, 0.31 and 0.11 GPa^{-1} for each metal respectively. Experimentation on a graphite epoxy composite proved inconclusive in terms of measuring the mean stress effect. The original observations regarding the mean stress effect were conducted on specimens made from a titanium alloy (TIMETAL 21S) [7].

Experiments carried out in [11] did not allow the derivation of the mean stress, however, they did confirm that a significant dependence existed for the two titanium alloys (as above) and also for a nickel alloy (Inconel 718). In further work [12], the thermoelastic signal of TIMETAL was found to vary by 21% over a mean stress range of -300 to +300 MPa. Recent work on Nitinol stents by Eaton-Evans et al [13] has confirmed that for Nickel-Titanium alloy, the thermoelastic constant has a high dependence on the mean stress. It was also shown in [13] that the dependence on mean stress in stainless steel was negligible. While this technique of utilising the mean stress effect does appear to provide a potential route for deriving residual stress, it is not without its limitations. Firstly, inspection of equation (2) highlights the sensitivity of the approach. Equation (2) is dominated by the magnitude of the $\left(1/\alpha E^2 \, \partial E/\partial T \, \sigma_m\right)$ term in comparison with unity. The $\partial E/\partial T$ term must be considerable in magnitude if the mean stress is to have any significant influence on the thermoelastic constant. The influence of the parameter can be estimated by setting $\sigma_m = \sigma_{ult}$, i.e. the maximum residual stress possible. It was found that the effect of this governing parameter was much greater in aluminium than for steel; ignoring this term for steel, gave an error of approximately 1.1% [14]. By comparing this to the natural error in the thermoelastic response due to signal noise (that can be up to 10% depending on surface coating and uniformity), and acknowledging that cyclic stresses used in TSA are not usually in the region of failure, it can be seen that this method could not reliably be used for steel components.

A further limitation of the technique relates to the measurement of the second harmonic component of the thermoelastic signal. Gyekenyesi [12] found that the magnitude of the second harmonic of the temperature variation was approximately 2% of the first harmonic. This was established for titanium, which was one of the most sensitive materials tested. Measurement of this small component is difficult, especially as its relative magnitude is not dissimilar to variations caused by signal noise. Subsequently, confidence in the repeatability of results relating to residual stress is an important consideration. While the mean stress effect has been clearly shown in aluminium, nickel and titanium based alloys, the fact that it cannot be reliably used with steel components severely limits the potential of the mean stress approach to thermoelastic residual stress measurement. This combined with the difficulty in measuring the second order effect, prevents this technique being a practical method of residual stress evaluation in metals thus far. The applicability regarding composite materials is unknown.

3. Mean Stress Effect and the Detector Response

An alternative approach to residual stress analysis which utilises the mean stress effect has been explored by Patterson et al [15]. This technique directly relates the detector response to the principal stresses, and most importantly, does not rely on detection of the small second order component of the thermoelastic response. In the presence of residual stress, the effective mean stress is assumed to be the sum of the applied mean stress, σ_{app}, and the residual stress, σ_{res}, such that the mean stress is $\sigma_m = \sigma_{app} + \sigma_{res}$. For an applied cyclic stress $\sigma = \sigma_m + \sigma_a \sin \omega t$, the thermoelastic signal, S, can be expressed using the analysis in [16] as:

$$\frac{S}{\sigma_a} = b_0 + b_1\sigma_{app} + b_1\sigma_{res} \tag{13}$$

where $b_0 + b_1$ are constants directly calculated from a combination of the material properties and the characteristics of the infra-red detector. It should be noted that $b_0 + b_1$ are only expected to be constant in the elastic region, and thus this approach is limited to evaluating residual stress in this region.

From a plot of $(S/\Delta\sigma)$ as a function of the applied mean stress, σ_{app} (Figure 1), a linear regression yields a graph with gradient b_1, and a y-intercept at $b_0 + b_1 \sigma_{res}$. Thus if b_0 was obtained at the experimental temperature for a material with no residual stress, then σ_{res} could be evaluated from the intercept value.

Initial findings [16] based on the previous studies of Machin et al [7] and Gyekenyesi and Baaklini [11] yielded promising results. Good linearity was observed and the regression lines showed a good fit to the data. However, further work [15] exploring the residual stress around cold expanded holes was inconclusive; good agreement of the TSA data with the known stress distribution was found in some areas, while significant scatter and differences were found in others. These findings bring into question the viability of this approach in its current form. The major disadvantage of this approach is that the constants b_0 and b_1 are functions of the detector response and the component temperature. Thus, if there is to be confidence in the repeatability of tests they must be carried out using the same detector in a temperature controlled environment. A further difficulty is that the thermoelastic response must be recorded over a range of applied mean stresses in order to obtain an accurate linear regression, which may not always be possible. If it is possible to correct for temperature variations as described in [17], more accurate results may be possible. Furthermore, if a radiometrically calibrated infra-red system is available, the values of ΔT can be obtained instead of the signal, S, potentially allowing a higher degree of accuracy. Further work with this approach is required to investigate its viability and accuracy. Since the approach utilises the mean stress effect which is governed by the parameter $\partial E / \partial T$ (see equation (4)), it is unlikely to be practical for steel components as $\partial E / \partial T$ is practically zero for most steels.

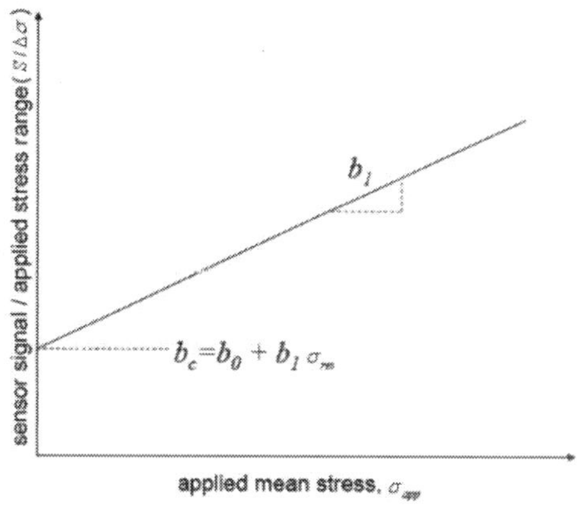

Figure 1: Schematic highlighting the relationships between applied mean stress and the constants b_1 and b_0 using the mean stress method. [15]

4. The Effect of Plastic Deformation on the Thermoelastic Constant

It has been shown [14] that the introduction of plastic deformation modifies the thermoelastic constant in some metals. It has been suggested that this change in thermoelastic constant can be used to estimate the level of plastic strain that a component has induced. Since plastic strain can be directly related to residual stress, there is a clear opportunity to derive a procedure for the assessment of residual stress using TSA, utilising the effect of plastic deformation. Rosenholtz et al [18], and Rosenfield et al [19] have both demonstrated that in steel and aluminium, an application of plastic strain will cause a change in the material property, α, the coefficient of linear thermal expansion. Rosenfield et al [19] also noted that this change in α increases significantly more when subjected to compressive strains, and increases less upon tensile plastic straining.

To indicate if plastic deformation causes a change in thermoelastic constant, a specimen can be loaded in uniaxial tension. If this type of specimen is loaded beyond the material's yield point and then unloaded, it will result in a residual strain; however, there is no residual stress as it can be fully relaxed by the elastic unloading. As a result, σ_m in the modified thermoelastic equation for a uniaxial stress system (equation (4)), becomes zero; therefore any change in the thermoelastic signal would be due to a change in one of the material properties, α, ρ or C_p and not due to a change in the mean stress. Quinn et al [14] conducted tests on steel specimens that had experienced different levels of plastic strain; one specimen was left unstrained, while three specimens were statically strained to give maximum tensile strains of 5%, 6% and 8%, and then unloaded. It was seen that the thermoelastic constant increased from 2.93 x 10^{-6} MPa^{-1} for the unstrained specimen to 3.19 x 10^{-6} MPa^{-1} for the specimen that had experienced 8% plastic strain. The change in the thermoelastic constant is small, but it was repeatable and was seen to increase linearly with the level of plastic strain experienced. Further work in [14] examined a curved beam. One component was machined to shape whilst the other was deformed. The thermoelastic response through the section of both components is shown in Figure 2. It is clear that in the deformed component (Figure 2b) the experimental data shows significant departure from the both the first order and second order theory in the most deformed part of the specimen. For the tensile part of the response this shows good agreement with the modifications observed in K. However, the departures are much greater in the compressive side which are consistent with the observations of [18, 19] where it was shown that exposure to compressive strain changes the coefficient of expansion significantly.

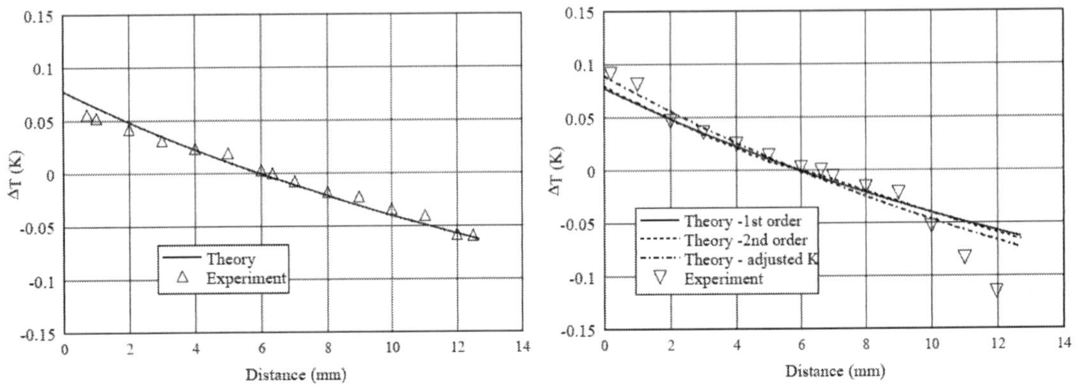

(a) Machined Component (b) Deformed component

Figure 2: Comparison of thermoelastic data to theory for (a) a machined component and (b) a deformed component. [14]

Further work by Quinn et al [20] indicated that the process of strain hardening has a significant effect on the change in thermoelastic constant that can be expected to occur as a result of plastic deformation. It was concluded that the change in thermoelastic constant was dependent on the material dislocation that occurs during strain hardening, and that the change in *K* for a material that does not strain harden would be significantly less than for a material that does. This method shows promise, in that the effect is repeatable and valid for steel components. However, there is a requirement that the residual stress in the component is caused by plastic deformation, and that the material under inspection experiences strain hardening.

5. Conclusions and Future work

Three approaches to residual stress evaluation based on thermoelastic stress analysis have been reviewed. They each show positive results but are not without their limitations, particularly with regards to accuracy and material.

The two approaches utilising the mean stress effect are not appropriate for steel components. Future work will concentrate on quantifying the change in thermoelastic constant resulting from plastic deformation, caused by both mechanical deformation and by heating effects. The variation in thermoelastic response as a consequence of compressive and tensile plastic strain will be explored for steel and aluminium. The possibility of using the revised higher order theory for elastic residual stress will be further explored.

6. Acknowledgement

This work was supported by Airbus UK. The notation adopted in equations (1-8) of section (1) is that of J Eaton-Evans [21].

References

[1] **Withers, P.J.** and **Bhadeshia, H.K.D.H.**, (2001) Overview Residual Stress Part 1 – Measurement Techniques, *Materials Science and Technology*, 17, 355-365.

[2] **Withers, P.J.** and **Bhadeshia, H.K.D.H.**, (2001) Overview Residual Stress Part 2 – Nature and Origins, *Materials Science and Technology*, 17, 366-375.

[3] **Webster, P.J.** and **Withers P.J.**, (2001) Neutron and Synchrotron X-Ray Strain Scanning, *Strain*, 37, 19-33.

[4] **Bray, D.E.**, **Tang, W.** and **Grewal, D.**, (1997) Ultrasonic stress evaluation in a turbine/comprerssor rotor, *J. Testing and Evaluation*, 25, 503-509.

[5] **Dulieu-Barton, J.M.** and **Stanley, P.**, (1998) Development and applications of thermoelastic stress analysis, *J. Strain Analysis*, 33, 93-104.

[6] **Belgen, M.H.** (1968) Infrared radiometric stress instrumentation application range study, *NASA CR-1087.*

[7] **Machin, A.S.**, **Sparrow, J.G.**, and **Stimson, M.G.**, (1987) Mean Stress Dependence of the Thermoelastic Constant, *Strain*, 23, 27-30.

[8] **Wong, A.K.**, **Sparrow, J.G.** and **Dunn, S.**, (1988) On the revised theory of the thermoelastic effect, *J. Physics and Chemistry of Solids*, 49, 395-400.

[9] **Wong, A.K.**, **Dunn, S.A.**, and **Sparrow, J.G.**, (1988) Residual Stress measurement by means of the thermoelastic effect, *Nature*, 332, 613-615.

[10] **Dunn, S.A.**, **Lombardo, D.**, and **Sparrow, J.G.** (1989) The mean stress effect in metallic alloys and composites, *SPIE Stress and Vibration: Recent Developments in Industrial Measurement and Analysis*, 1084, 129-142.

[11] **Gyekenyesi, A.L.**, and **Baaklini, G.Y.**, (1999) Thermoelastic Stress Analysis: The mean stress effect in metallic alloys, *NASA-TM-1999-209376.*

[12] **Gyekenyesi, A.L** (2002) Thermoelastic Stress Analysis: An NDE Tool for Residual Stress Assessment in Metallic Alloys, *J. Engineering for Gas Turbines and Power*, 124, 383-387.

[13] **Eaton-Evans, J.**, **Dulieu-Barton, J.M.**, **Little, E.G.** and **Brown I.A.** (2006) Thermoelastic Studies on Nitinol Stents, *J. Strain Analysis*, 41, 481-495.

[14] **Quinn, S.**, **Dulieu-Barton, J.M.**, and **Langlands, J.M.**, (2004) Progress in thermoelastic residual stress measurement. *Strain*, 40, 127-133.

[15] **Patterson, E. Du, Y.**, and **Backman, D.**, (2008) A new approach to measuring surface residual stress using thermoelasticity, *Proc. SEM XI Int. Congress and Exposition, Orlando, FL.*

[16] **Patterson, E.**, (2007) The potential for quantifying residual stress using thermoelastic stress analysis, *Proc. SEM Conference, Springfield, MA.*

[17] **Dulieu-Barton, J.M.**, **Emery, T.R.**, **Quinn, S.**, and **Cunningham, P.R.** (2006) A temperature correction methodology for quantatative thermoelastic stress analysis and damage assessment. *Materials Science and Technology*, 17, 1627-1637.

[18] **Rosenholtz, J.**, and **Smith, D.**, (1950) The effect of compressive stresses on the linear thermal expansion of magnesium and steel, *J. App. Physics*, 21, 396-399.

[19] **Rosenfield, A. R.** and **Averbach, B. L.**, (1956) Effect of stress on the expansion coefficient, *J. App. Physics*, 27, 154-156.

[20] **Quinn, S.**, **Dulieu-Barton, J.M.**, **Eaton-Evans, J.**, **M.**, **Fruehmann, R.K.**, and **Tatum, P.J**, (2008) Thermoelastic assessment of plastic deformation, *J. Strain Analysis*, 43, 451-468.

[21] **Eaton-Evans, J.** (2006) Thermoelastic stress analysis of intravascular devices, *PhD thesis University of Limerick.*

Collapse of Geometrically Imperfect Stainless Steel Tubes under External Hydrostatic Pressure

Carl T.F. Ross, Thomas Bowler and Andrew P.F. Little.

Department of Mechanical and Design Engineering, University of Portsmouth, Portsmouth, PO1 3DJ, UK.
carl.ross@ntlworld.com

Abstract

This paper reports on an investigation into the buckling behaviour of 5 geometrically imperfect duplex stainless steel tube models, subjected to external hydrostatic pressure. The research was partly experimental and partly theoretical, where in the latter case; both analytical & numerical theoretical analyses were carried out.
The experiments were carried out on five stainless steel tube models of different lengths, using two mild steel end bungs to seal the models. The experimental results showed that the Duplex stainless steel specimens behaved similarly to other isotropic materials tested by other researchers.
The theoretical calculations and analyses were made by MisesNP, a DOS-based computer program, together with the ANSYS finite element structural analysis software.
Combining the results of the present series of models, together with the results of other experimenters, a design chart was produced, which can be used for designing full-scale vessels. It should be emphasised that this design chart has been extended to those from previous studies, so that shorter and thicker vessels can now be designed.

Keywords: buckling, shell instability, stainless steel, design chart, ANSYS.

1. Introduction.

According to Dickens et al [1], there are about 10,000 billion tonnes of methane lying in the deep seas in the form of methane hydrates, from some 3000m to 11000m underwater. The value of this methane is about 536 times the GDP of the USA or about $1,250,000 per person on Earth. The main reason why this methane has not been exploited before is because of the very high hydrostatic pressures in the great depths of the oceans, where in the case of the Mariana Trench, it is about 1,100 bar Ross [2].
Today's materials of construction of submersibles consist in the main, of materials similar to those used over a century ago. Over this period, however, technology has advanced and this is why the present study has been made.
The challenge is that larger Human Occupied Vehicles could be designed and built, with the ability to carry a greater complement at greater depths of water. The Royal Naval submarines can have a displacement of about 16,000 tonnes and a length of about 149.5

Journal of Physics: Conference Series **181** (2009) 012030 doi:10.1088/1742-6596/181/1/012030

metres, which are substantial sizes. However, their operational depth is limited to about 400 m, while the maximum ocean depth is some 29 times this value!

There are not many materials that can withstand high external hydrostatic pressure and this is why this research has been conducted. Titanium, for example, has the highest strength: weight ratio of structural metals; additionally, it has a good corrosion resistance, making it a desirable material for building submersibles; however, it is a very expensive material to use.

When a pressure vessel collapses under external hydrostatic pressure, it usually collapses through non-symmetric bifurcation buckling (lobar buckling) or shell instability [2], as it is sometimes called. When this mode of failure occurs, the buckling pressure is often a small fraction of that for the same vessel to fail under uniform internal pressure.

Another mode of failure for pressure vessels is through axisymmetric yield, with the vessel imploding axisymmetrically, keeping its circular form while collapsing.

To improve the buckling resistance of these vessels to the shell instability mode of collapse, it is usual to stiffen these vessels with ring stiffeners. If the ring stiffeners are not strong enough, the entire ring-stiffened combination can fail by buckling bodily in its flank; this failure is called general instability [2].

In this paper, only collapse by shell instability is studied, because this is the most common mode of failure of such vessels. Both the well-known analytical formula by von Mises [3], as shown by equation (1) and the commercial computer package ANSYS [4] are utilised in the analysis. In the case of the von Mises formula, the number of lobes (n) has to be changed, while calculating the buckling pressure corresponding to each value of 'n', until the lowest buckling pressure is found, namely P_{cr}.

$$p=\frac{Eh}{R}\left[\frac{1}{n^2+\frac{1}{2}\left(\frac{\pi R}{L}\right)^2}\right]+\left\{\frac{\left(\frac{\pi R}{L}\right)^4}{\left[n^2+\left(\frac{\pi R}{L}\right)^2\right]^2}+\frac{\left(\frac{h}{R}\right)^2}{12\left(1-\mu^2\right)}\left[n^2+\left(\frac{\pi R}{L}\right)^2\right]^2\right\} \qquad (1)$$

where,

p = external pressure
n = the number of circumferential waves or lobes that the vessel buckles into.
R = mean radius
H = wall thickness
L = unsupported length of the circular cylinder
E = Young's modulus
μ = Poisson's ratio

Additionally, the thinness ratio, produced by Windenburg and Trilling [5] as shown by equation (2), has to be determined to allow for the effects of initial out-of-circularity, which further decreases the buckling pressure if the vessel fails through inelastic shell instability, where

$$\lambda = \text{Windenburg Thinness ratio} = \{(L/d)^2/(t/d)^3\}^{0.25}\times\sqrt{(\sigma_{YP}/E} \qquad (2)$$

t = Shell thickness

L = Unsupported length of the cylinder
d = Mean shell diameter
E = Young's modulus of elasticity
σ_{YP} = Yield Stress or 0.2% proof Stress

Using the above, it is possible to create a design chart by plotting the reciprocal of 'λ' against the experimentally determined Plastic Knockdown Factors (PKD). Once the design chart is obtained it is possible to predict the theoretical inelastic buckling pressure, by calculating the theoretical elastic buckling pressure (P_{cr}) for a perfect vessel, together with its 'λ', and from the Design Chart to determine PKD for that particular vessel. Once PKD is determined, the predicted collapse pressure (P_{EXP}) for that vessel can be obtained by dividing P_{cr} by PKD, where:

$$PKD = \frac{P_{cr}}{P_{EXP}}$$

P_{cr} = theoretical elastic instability buckling pressure
P_{EXP} = experimental buckling pressure

2. Test Models and Material Properties

2.1 Test Models

The models used for the tests were made from a Duplex stainless steel tube, which was cut into the required five lengths, as shown in Fig. 1.

Figure 1: Five Test Models and End Bungs

The end bungs are shown in Fig. 1. The major dimensions for specimens can be seen in Table 1. The dimensions shown in Table 1 are later optimised for use with ANSYS and MisesNP. The unsupported length was taken as the overall length L_0 of the model, minus the depth of the mid-location of each O-ring from the 'edge'. For both MisesNp and ANSYS, the boundary conditions were assumed to be simply-supported for all the models. The initial out-of-roundness was measured on a Mitutoya Co-ordinate Measuring by an experienced senior technician.

Table 1: Specimens' Dimensions

Model	verall Lengtl L_0 [mm]	Unsupported ,ength L [mm]	Mean Radius [mm]	Thickness [mm]	Out-of-Roundness [mm]
TB1	50	40	34.04	1.62	0.0808
TB2	60	50	34.04	1.62	0.0700
TB3	70	60	34.04	1.62	0.1433
TB4	80	70	34.04	1.62	0.0868
TB5	100	90	34.04	1.62	0.0440

2.2 Material properties of the 'Duplex' stainless steel.

The Young's Modulus was determined by testing a thin-walled circular ring under diametric compression, as shown in Fig. 2.

Figure 2: Method of determining the experimental value for Young's Modulus

The Load-Deflection relationship for the diametrically loaded circular ring of Fig. 2 is shown in Fig. 3.
The calculation for the Young's Modulus was made by using Roark's formulae [6], where flexural, axial and shear deflections were accounted for, but not anticlastic curvature. However, the error in not accounting for anticlastic curvature would have been very small as the decrease in curvature for a diametric deflection of about 1 mm was only about 1.5%.
The measured value of 'E' was as follows:
E = Young's modulus = 146.2 GPa
The yield stress and Ultimate Tensile Strength (UTS) were obtained from two 'flat' tensile test specimens, which were cut along the meridian of an unused length of the tube. The following values of UTS and yield point were found :

UTS1 = 527.9 MPa
Yield Stress1 = 243.7 MPa
UTS2 = 544.1 MPa
Yield Stress2 = 251.0 MPa

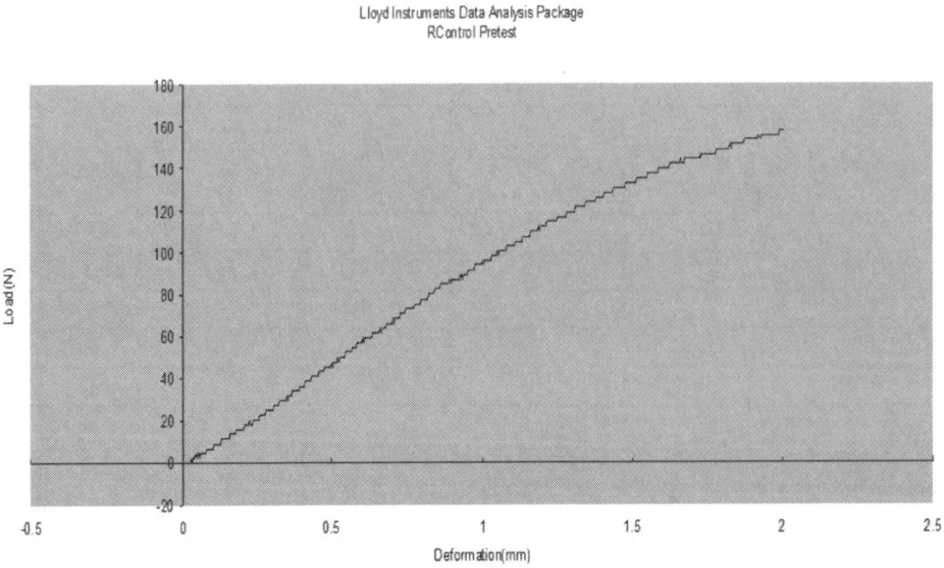

Figure 3: Load-Deflection Relationship for the Diametrically Loaded Circular Ring

The experimentally determined values for UTS and Yield stress agreed with 'book' values and the measured value for 'E' agreed with values obtained by other experiments on different sized rings made from Duplex stainless steel.

3. Experimental Testing

This section records the various steps in the experimental testing of the buckling behaviour of the five Duplex stainless steel models when the external hydrostatic pressure was gradually increased until failure of each specimen took place.

3.1 Equipment

- Test tank; capable of withstanding pressures up to 210 bar
- Hydraulic pump; capable of exerting a pressure of up to 410 bar
- Bourdon tube pressure gauge
- Connecting hose
- Water

3.2 Methodology

The hose of the hand-operated pressure pump was connected to a two-part heavy stable cylinder container ("the tank") in which the models were placed, one at a time, and surrounded with water and covered by the tank top or closure plate. The closure plate was placed on the top of the tank and was sealed by tightening the retaining bolts through the 16 holes distributed equally around the circumference of the closure plate and the top of the tank. The pressure pump was able to exert a maximum pressure of up to 410 bar.
It was important to be able to identify each model and its buckling effects individually throughout the experimental testing and thus, the five models were named TB1 to TB5 respectively, in increasing values of length, as shown in Fig. 1. The reservoir of the pressure

pump was filled with water. The first specimen to be tested, namely, TB5, was a cylindrical shape of stainless steel 100mm length, with an outer diameter of 69.82mm and thickness of 1.62mm. A mild steel bung was inserted firmly into each end of the cylindrical model, to ensure that it remained watertight throughout the experimental testing.

3.2.1 *Preparation.* Model TB5 was placed gently and vertically into the tank. The tank was then filled with water to about 2cm below its rim. The 'O' ring was then fitted into the retaining groove around the circumference of the inside surface of the closure plate. The closure plate was then placed on top of the tank and firmly screwed down by 16 bolts, to seal the system, apart from the bleed screw, which was not in its normal place. With the bleed screw removed, water was pumped into the tank until all the trapped air had escaped through the bleed hole. Each model was then tested to destruction. Each model collapsed with a distinct 'bang' and/or an associated pressure drop. Each model was inspected after collapse and the collapse pressure noted from the reading of a Bourdon tube pressure gauge. Each model was photographed after its collapse, as shown in Fig. 4.

Figure 4: TB5 to TB1 Models (left to right), vertical view after failure

3.3 Experimental Results

The experimental results are given in Table 2, where it can be seen that the shorter models had higher buckling pressures than the longer ones. This is especially true in the case for Specimen TB1, which had a much higher buckling pressure than TB2; this was because its unsupported length was relatively much shorter than that of TB2.

4. Theoretical Investigation

Two theoretical investigations were carried out; one was using MisesNP, a DOS computer program, based on a closed form trivial solution of von Mises. The other theoretical investigation was via the commercial computer package, namely ANSYS; this was based on the Finite Element Method. The circular cylinders were modelled with about 1000 elements for each circular cylinder, where each cylinder was assumed to be perfect. The elements used were the well-known 8-node thin-walled Shell 93 ANSYS quadrilateral elements, which had 6 degrees of freedom per node. The ANSYS analysis was based on a linear Eigen-buckling elastic analysis.

4.1 MisesNP

For this program the input data consisted of the unsupported length of the cylindrical shell, the mean radius of the model, the material thickness, the Young's Modulus, the Poisson's Ratio and the yield stress; it was simple to use The program outputted the buckling pressures, together with the associated number of circumferential waves or lobes, namely 'n'. The program also outputted the Windenburg thinness ratio of Equation (2).

Table 2: Experimental results.

Model	L_o (mm)	L(mm)	P_{EXP} (bar)	Comments
TB5	100	90	144.79	Biggest Noise
TB4	80	70	150.31	Big Noise
TB3	70	60	151.69	Big Noise
TB2	60	50	159.96	Low Noise
TB1	50	40	191.67	No Noise

4.2 ANSYS

ANSYS has many finite element analysis capabilities, enabling it to solve many complex problems and structures in various fields. In this case, the 8-node quadrilateral Shell 93 finite element was used in a linear 'Eigen buckling' analysis. The buckled form of a model is shown in Fig. 5.

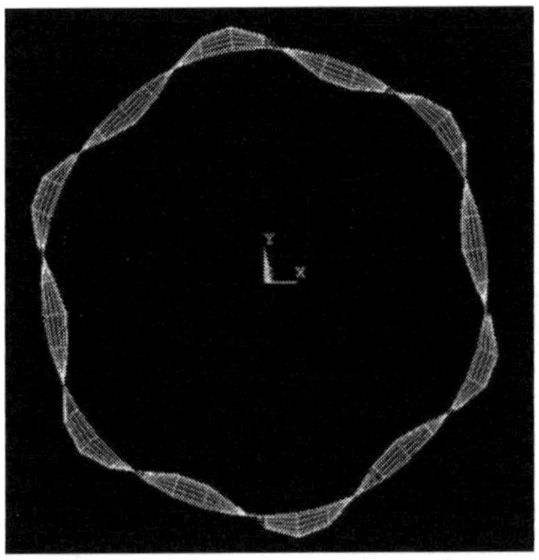

Figure 5: Buckled form of the cross-section of a model, showing its lobar pattern.

5. Theoretical and Experimental Results.

The results from the experimental and theoretical investigations are shown in Table 3, where the symbols 'DTMB' represent the David Taylor Model Basin formula; this is a simplified version of that of the von Mises formula of Equation (1).

From Table 3, it must be noted that the theoretical buckling pressures (P_{cr}) were all based on elastic theory for perfect vessels and this is why the plastic knockdown factor (PKD), from the Design Chart, must be divided into these theoretical values to give the predicted or experimental buckling pressures (P_{EXP}). From Table 3, it should also be noted that ANSYS predicted slightly higher buckling pressures than von Mises and the DTMB formulae; this was because in the ANSYS analysis, the eight node quadrilateral element was used and these could not truly represent the circumferences of the vessels as perfect circles. The problem was worsened because the vessels collapsed around their circumferences. Once the Design Chart is available, it is used as follows:

Calculate P_{cr} and λ from MisesNp and using the value of λ, determine PKD from the design Chart. The predicted buckling pressure, namely P_{EXP}, can then be obtained by dividing P_{cr} by PKD.

Table 3: Theoretical and Experimental buckling pressures.

Model	Length L [mm]	P_{cr}				DTMB [MPa]	λ	P_{EXP} [MPa]
		Von Mises[MPa]	N	ANSYS [MPa]	n			
TB1	40	63.163	5	92.299	6	63.992	0.516	19.167
TB2	50	51.087	5	69.169	5	49.858	0.577	15.996
TB3	60	40.198	4	56.609	5	40.838	0.632	15.168
TB4	70	33.748	4	49.772	5	34.581	0.683	15.031
TB5	90	27.624	4	37.089	4	26.471	0.774	14.479

5.1 Final Design Chart

From Table 3, the Plastic Knockdown Factors (PKD) can now be calculated with the values of the experimentally (P_{EXP}) and theoretically (P_{cr}) obtained buckling pressures and the Design Chart can now be produced, where:

$$PKD = \frac{P_{cr}}{P_{EXP}}$$

Table 4 gives the PKD's and reciprocal thinness ratios for the present series of specimens, which together with the results from other experimenters [2], was used to produce the Design Chart of Fig.6. This Design Chart allows shorter and thicker vessels to be analysed, which was not possible with previous Design Charts [2].

Table 4: Results of plastic knockdown factor, thinness ratio and out-of-roundness

Model	Length L [mm]	P_{cr} [MPa]	λ	P_{EXP} [MPa]	PKD	1/λ	Out of Roundness [mm]
		Von Mises					
TB1	40	63.163	0.516	19.167	3.295	1.937	0.0808
TB2	50	51.087	0.577	15.996	3.194	1.733	0.0700
TB3	60	40.198	0.636	15.168	2.650	1.572	0.1433
TB4	70	33.748	0.683	15.031	2.245	1.465	0.0868
TB5	90	27.624	0.774	14.479	1.908	1.292	0.0440

6. Conclusions

Both the experimental and theoretical investigations of the five Duplex stainless steel tube specimens were performed successfully. The main findings are listed as follows;

- The experimentally obtained Ultimate Tensile Strength and Yield Stress agreed with 'book' values for Duplex stainless steel.

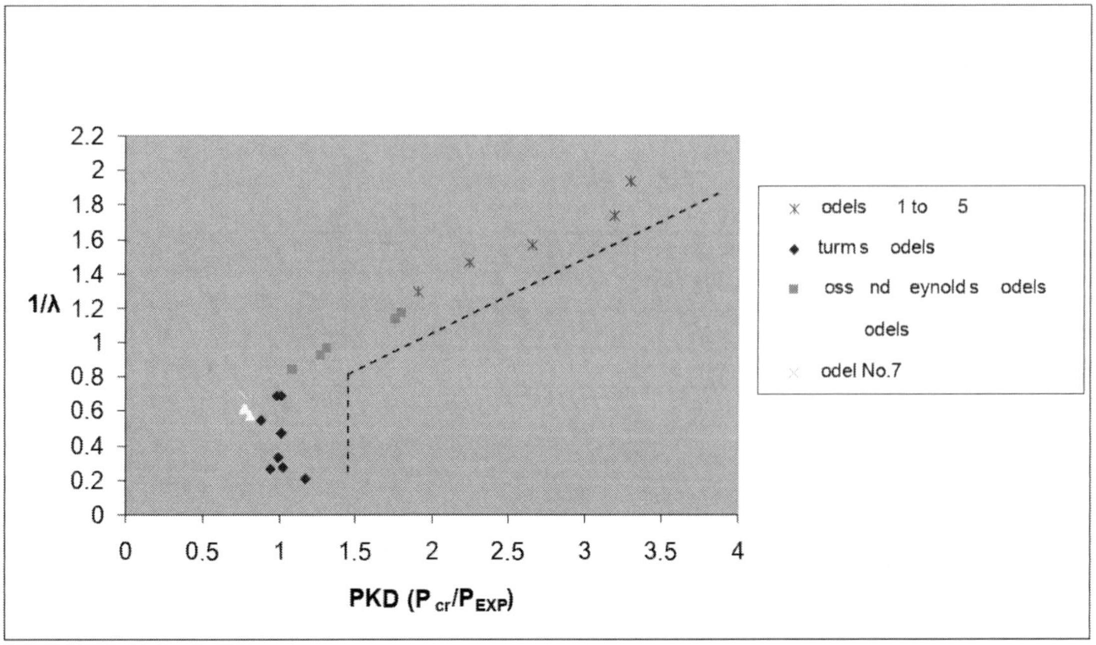

Figure 6: Design Chart using the von Mises results.

- The experimentally obtained Young's Modulus agreed with values obtained by other experimenters, who used circular rings with different dimensions to those used in the present study.
- The design chart was created successfully for use with MisesNP; it appears to be of similar form to other design charts determined at the University of Portsmouth.
- TB1 buckled symmetrically about its cross-section, leaving a near-perfect lobar buckled model, where n = 6.
- The results of MisesNP have been favoured over those of ANSYS as they were generally more accurate when compared with the experimental results.
- ANSYS predicted slightly higher buckling pressures than von Mises; this was because ANSYS used eight node quadrilateral elements, which did not model the circumferences of the specimens as being perfectly circular and it must be emphasised that the vessels collapsed around their circumferences.
- The time and effort needed to complete an ANSYS analysis was significantly greater than when MisesNP was used.
- The design chart created here should prove useful to designers, if the results for new structures have the similar characteristics as those investigated here.
- The design chart also allows shorter and thicker vessel to be analysed, which was not possible before.
- Designers in this field have found BS 5500 to be too conservative in the design of circular cylinders suffering shell instability under external hydrostatic pressure; moreover they find BS 5500 more difficult to use than the method presented here.
- The initial out-of-circularity of the vessels should not exceed 0.16t, where t = the wall thickness of the vessel under consideration.

References.

[1] **Dickens, Gerald R., Paull, Charles, K., Wallace, Paul** and **the ODP Leg 164 Scientific Party.** (1997) Direct measurement of in situ quantities in a large gas-hydrate reservoir, *Nature,* Vol. 385, 30th January.

[2] **Ross, C. T. F.** (2001) Pressure Vessels: External Pressure Technology, *Horwood Publishing Ltd.* Chichester, UK.

[3] **von Mises, R.** (1929) Der Kritische Aussendruck für Allseits Belastete Zylindrische Rohre, Fest zum 70, Geburtstag von Prof. Dr. A. Stodola, Zurich, pp.418-30, 1929.

[4] **ANSYS Incorporations**, Documentation for ANSYS 10.0, Canonsburg; USA, 2005.

[5] **Windenburg, D.F.** and **Trilling, C.,** (1934) Collapse by Instability of Thin Cylindrical Shells Under External Pressure, *Trans.,ASME,* 11, .pp819-825.

[6] **Young, W.C.** (1989) Roark's Formulas for Stress and Strain, McGraw-Hill.

The Influence of Post-Local Buckling Mechanics on the Stress Variations, Axial Stiffness and Ultimate Failure Strength of Uniformly Compressed Thin-Walled I-Section Struts.

J Loughlan N Yidris and P R Cunningham
Department of Aeronautical and Automotive Engineering,
Loughborough University, Loughborough, Leicestershire, LE11 3TU.

Corresponding Author, E-mail: j.loughlan@lboro.ac.uk

Abstract. It is well known that thin-walled compression members are subject to the effects of local buckling and that due to these local effects the compressive carrying capability of short strut members can be significantly reduced. Finite element simulation is employed in this paper to examine the post-buckled response of thin-walled sections giving due consideration to the influence of geometric imperfections and to elasto-plastic material behaviour. The findings from this work highlight the complete loading history of the compression struts from the onset of elastic local buckling through the nonlinear elastic and elasto-plastic post-buckling phases of behaviour to final collapse and unloading. A detailed account of the growth and redistribution of stresses as well as the influence of yielding and yield propagation throughout loading is given in the paper. The results from the finite element simulations are shown to compare well with independent simulations using the finite strip method of analysis.

1. Introduction

The structural performance of thin-walled compression members is governed, essentially, by the effects of buckling and post-buckling behaviour. There are a number of possibilities with regard to the mode of buckling that can occur and this would depend, of course, on the geometrical details of the member cross-section, the member length, end conditions with regard to support and loading and also the member structural material. The buckling mode is characterised, in essence, by half-wavelength such that local buckling of the section walls is associated with a short half-wavelength, distortional buckling by an intermediate half-wavelength and torsional, flexural and torsional-flexural buckling are the overall modes at the longer half-wavelengths. Clearly, interaction is possible and coupled instabilities can be encountered in design whereby local-distortional buckling can occur or, indeed, the interaction between local buckling and overall flexural behaviour may be of influence. The buckling and post-buckling mechanics of thin-walled compression members has been studied by a great many investigators over the years and as a result, our knowledge and understanding of buckling phenomena is now at a fairly detailed level.

Column interaction buckling and post-buckling behaviour has been studied by numerous researchers over the years. The interaction of local buckling with overall column bending has been studied by Loughlan and Howe [1] and by Loughlan [2] for the case of plain channel, lipped channel and I-Sections using a semi-energy post-local buckling analysis procedure. In these works the influence of geometrical imperfections and of local form change during loading are shown to have special significance for columns designed with close local and overall buckling modes. Lau and Hancock [3] and Kwon and Hancock [4] have studied the distortional buckling behaviour of thin-walled channel section columns. Explicit analytical expressions were derived [3] to predict the distortional buckling stress of thin-walled channel section columns and consideration has been given to the post-buckling response [4] of such sections when undergoing local and distortional deformations. The work of Rasmussen [5] and Young and Rasmussen [6] considered the overall bifurcation of locally buckled thin-walled members. The appropriate variational equations developed in these works are applied to doubly symmetric columns [5] and to singly symmetric cross-sections [6] that locally buckle in the fundamental state. For the case of plain channel sections it is shown, using the fundamental state equations, that fixed-ended columns exhibit overall bifurcation behaviour while pin-ended columns do not.

Schafer [7] considered the local, distortional, and Euler buckling of thin-walled columns with regard to their treatment in design. At the time of this work the North American design specifications for cold-formed steel columns ignored local buckling interaction and did not provide an explicit check for distortional buckling. As a result of this, Schafer proposed a new method for design that incorporated, explicitly, the local, distortional and Euler buckling aspects of thin-walled columns. The distortional buckling of cold-formed steel columns is described in comprehensive detail in the AISI research report by Schafer [8]. The report details design methods which account for local and distortional interaction and which permit distortional and Euler interaction. Dinis et al [9] have examined the local-distortional mode interaction behaviour of cold-formed steel lipped channel columns. This work uses the ABAQUS finite element code and 4-node shell elements are employed to simulate behaviour. The numerical post-buckling results reported clearly indicate the evolution of the elastic-plastic column deformations during loading.

Ovesy et al [10-12] have studied the geometric nonlinear compressive behaviour of short struts using the finite strip method of analysis. These works cover channel [10], box [11] and I-sections [12] and give accurate representation of the effects of local buckling on the post-buckling compressional stiffness of the sections in the elastic range of behaviour. In this paper, finite element simulation is employed to examine the post-buckled response of thin-walled I-section struts giving due consideration to the influence of geometric imperfections and to elasto-plastic material behaviour. The findings from this work highlight the complete loading history of the compression struts from the onset of initial compressional loading through the nonlinear elastic and elasto-plastic phases of behaviour to final collapse and unloading of the sections.

2. Finite Element Modelling

The finite element method is used to simulate the compressional post-local buckling behaviour of the I-Section struts. The CQUAD4 quadrilateral shell element of the MSC/PATRAN/NASTRAN finite element software is used to discretise the struts and to formulate the finite element models. The CQUAD4 shell element is particularly well suited for dealing with the post-local buckling behaviour of thin-walled sections. It is readily able to account for the interaction of the local bending and membrane stretching of the section walls which takes place during the post-buckling process. The element has four connecting nodes and is elastically connected to only five of the six degrees of freedom at each of its nodes. It is connected to three translational and two rotational degrees of freedom at each node but does not provide direct elastic stiffness to the sixth degree of freedom which corresponds to the rotation about the normal to the surface of the element. The capability of being able to perform different types of analysis is provided for in MSC/NASTRAN through the appropriate selection of the different solution sequences made available in the software and designed for specific applications. The post-local buckling behaviour of the I-Section struts was determined using the nonlinear static solution sequence involving geometric nonlinearity as well as elasto-plastic material nonlinearity using the simplified elastic-perfectly plastic stress-strain model of the I-Section material.

Suitably refined finite element models with appropriate loading and boundary conditions were developed to simulate the complete compressional loading history of the I-section struts. This covers the onset of initial compression through the nonlinear elastic and elasto-plastic post-buckling phases of behaviour to final collapse and then through the plastic unloading phase of behaviour. The refined models utilised some seven to eleven thousand elements on the full section, depending on cross-sectional geometry, to accurately capture the post-buckling performance of the struts. The use of symmetry in the modelling process does, of course, halve the size of the structural models and reduces, considerably, the computation time involved in solving the appropriate post-buckling equilibrium equations. Results have been obtained in this work using symmetry for computational efficiency. In finite element modelling, boundary conditions and loading conditions are of great importance. Measurably different outcomes can be determined in structural analysis as a result of modifications to the compatibility boundary conditions of a structure. Changes in the boundary conditions cause stiffness enhancement or stiffness relaxation of the structure and thus the deformations and stresses may be quite different for a given level of load application.

2.1 Boundary Conditions

The boundary conditions applied to the I-Section strut models are such that the ends of the constituent flat plate components of the sections are simply supported in nature and thus all nodes at the section ends are constrained in the direction which is normal to the mid-surface of the flat plate component to which they belong. In addition, all nodes on one section end are constrained from moving axially whilst those at the other end are compressed uniformly by the same amount and in an incremental manner. The finite element nonlinear static solution procedure, which utilises the modified Newton-Raphson method, is then able to solve the assembled global equilibrium equations of the strut models for all nodal displacements and this is carried out for each increment of uniform end compression (end displacement) applied to the struts. Knowing the nodal displacements then permits the evaluation of the stress variations within the section walls and the magnitude of the constraint forces at the axial constraint nodes corresponding to each increment of applied end compression. Summing the axial nodal constraint forces then permits the evaluation of the axial load on the strut at each compression increment and thus the complete post-local buckling loading history can be detailed in a step by step manner. The application of end compression (end displacement) increments to the struts and then determining the axial load on the struts at each compression increment by summing the axial nodal constraint forces is a process that lends itself most suitable at the ultimate load condition of the struts when dealing with the plastic unloading phase of behaviour.

Of particular interest are the conditions prevailing at the junctions between the constituent flange and web plates of the cross-sections along the lengths of the struts. Using the finite element method, the nodes at the section junctions are permitted to be free from constraints, in which case, the coupling of the in-plane and out-of-plane deflections of the section walls meeting at the junctions is able to satisfy the natural compatibility and equilibrium conditions at the junctions, leading to waviness of the junctions along the length of the struts. In previous analytical studies and numerical simulation procedures using the finite strip method of analysis it has been assumed that the in-plane and out-of-plane deflections at the section junctions are uncoupled. This means that the plate components meeting at the junctions are assumed to be free to wave in their own plane across their width along the junctions whilst the out-of-plane deflection of each of the plates vanishes. These assumptions clearly violate the compatibility and equilibrium conditions along the junctions and impose stress free in-plane conditions at the junctions in the width direction of each plate. The stress free in-plane approach has been implemented by a great many researchers in the past as an aide to simplifying their analysis and solution procedures when dealing with the post-local buckling behaviour of thin-walled sections. It is an approach whose assumptions are considered to represent fairly valid approximations to actual behaviour up to about three times the critical local buckling load of the section wherein the movements at the section junctions are considered to be negligibly small. The effects of the boundary conditions along the junctions on the compressional behaviour of the I-Section struts have been examined in this paper. Comparisons are given corresponding to the three conditions of natural waviness of the junctions, in-plane stress free conditions at the junctions and immovable straight junctions, not allowed to wave, respectively.

2.2 Local Imperfections and Yielding

Geometric imperfections are taken into account in the finite element modelling process by assuming that the section walls are not perfectly flat. This, of course, leads to a nonlinear compressive response from the onset of loading which would not be the case for the initial loading of the geometrically perfect strut. A fairly common approach to dealing with the local out of flatness of the section walls is to assume that such local imperfections are similar in form to the locally buckled mode shape of the perfect cross-section. This approach has been adopted in the present work due to the fact that it has been shown to give realistic predictions of actual behaviour and also for the ease and simplicity of its application when using the finite element method of analysis. The locally buckled mode shapes are readily determined by subjecting the strut finite element models to the linear static buckling solution sequence of the MSC/PATRAN/NASTRAN finite element software. The local mode shapes thus obtained are then mapped on to the nodal grid system of the finite element model to form the imperfect strut which is then subsequently analysed using the nonlinear static solution sequence of the software to determine the effects of geometric nonlinearity. The magnitude of the imperfection applied to the finite element nodal grid system can be readily altered and thus it is a simple matter of setting the maximum amplitude of the buckled mode shapes to a specific value which is normally taken to be a percentage of the section wall thickness, i.e. maximum imperfection magnitude is set to 0.1t, 0.2t etc. This process allows the examination of the effects of local geometric imperfections in a fairly straightforward manner. With regard to the effects of material nonlinearity, the finite element model is further extended to include the elasto-plastic material constitutive equations into the post-local-buckling analysis of the doubly-symmetric I-section struts. The I-section struts are thus modelled using an isotropic, elastic-perfectly plastic material in conjunction with the Von Mises failure criterion.

3. Findings from Simulations and Discussion

Some findings are presented for a typical doubly-symmetric I-section strut with b_f/b_w shape factor equal to 0.5. Figure 1 shows a typical diagram of the geometry for the I-section struts. The struts have a length, L of 100 mm, the width of the web, b_w is 100 mm and the thickness is 1.5 mm for both the flanges and the web.

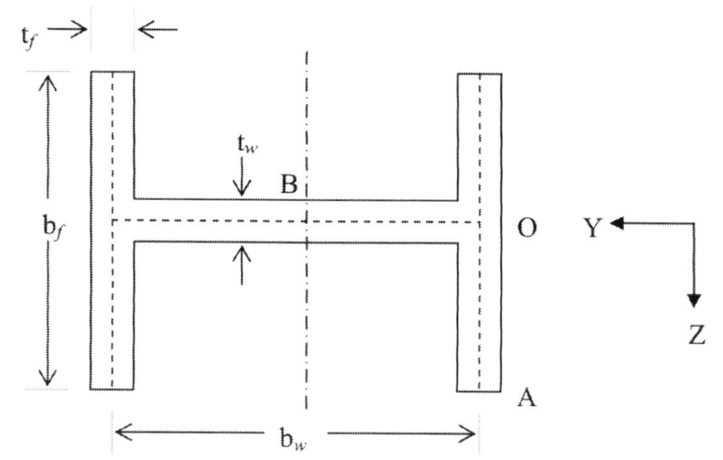

Fig. 1. Geometry of the doubly-symmetric I-section strut.

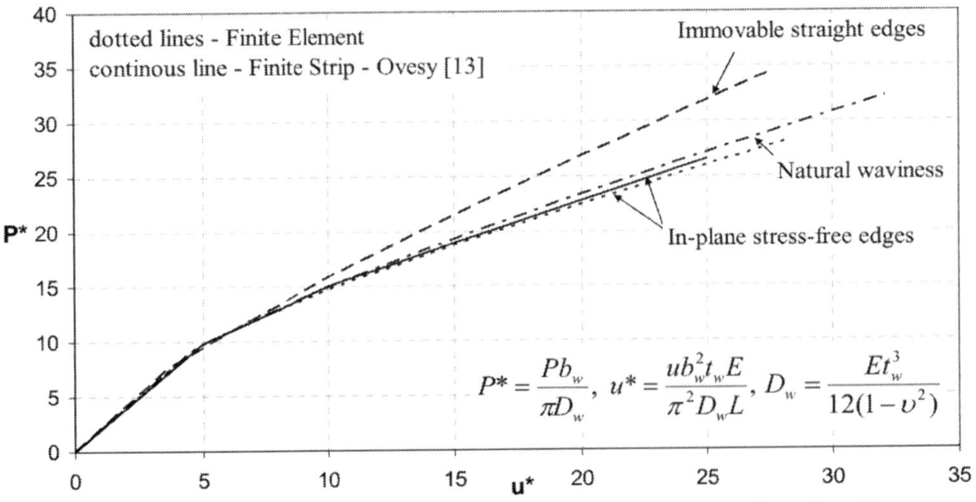

Fig. 2. Non-dimensional Load-End Shortening Variation for a Doubly-Symmetric I-Section Strut

Figure. 2 shows the non-dimensional load, P*, versus the non-dimensional end-shortening, u*, for a doubly-symmetric I-section strut with b_f/b_w shape factor equal to 0.5. It clearly shows that the load-end shortening curve of the in-plane stress-free edge boundary condition at the section junctions bears a close resemblance to the curve of the natural waviness boundary condition for an initial post-local buckling analysis (i.e. up to about 1.5-2.0 times the buckling load). But as the end-compression progresses further the load-end shortening curves move away from each other. Therefore it can be concluded that the assumption of in-plane stress free edges at the junctions is a fairly valid approximation for an initial elastic post-local-buckling analysis. As a comparison with the finite element result, the independent load–end shortening curve [13] of the finite strip method of analysis is plotted. The curve from the finite element solution is seen to be more accurate than that from the finite strip solution. This is due to the fact that the finite element model is a fully converged fine-mesh model which provides a high degree of shape flexibility across the section and along the length of the struts in the post-buckling range. Figure 2 also shows that for the immovable straight edge boundary condition the strut is over constrained and it is clear that this condition does not resemble the actual behaviour of the I-section struts but results in an over-stiff post-local-buckling response. The immovable edge boundary condition was included in this work merely for comparative purposes.

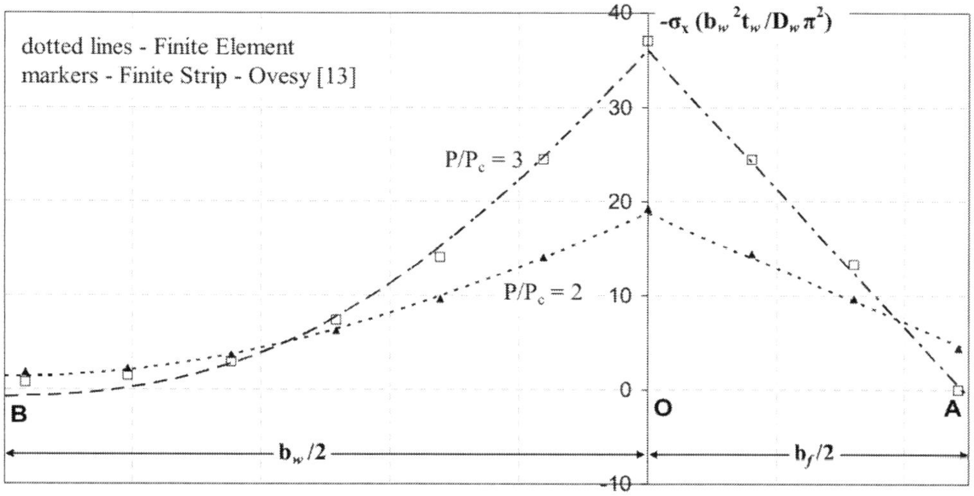

Fig. 3. Longitudinal Membrane Stress Distribution for a Doubly-Symmetric I-Section Strut (at buckle crest)

Figure 3 shows the longitudinal membrane stress distribution across the quarter cross section, i.e. BOA as indicated in Figure 1, for the same I-section strut. The stress distributions are at the half length of the strut at the crest of the buckle. The loads at which the stress distributions are obtained are at twice and three times the buckling load as indicated. As expected, the stress concentrations are towards the junctions of the section in which the total load on the strut is carried more by elements close to the junctions. In addition, the maximum membrane stress occurs at the junctions at the crest of the buckle. The figure also shows good agreement between the finite element method and the finite strip method [13]. The load shedding depicted in Figure 3 is a typical characteristic of post-local buckling mechanics.

With the aim of investigating the effects of material nonlinearity, the work is further extended to include the elasto-plastic material constitutive model into the post-local-buckling analysis of the doubly-symmetric I-section struts. Thus, the I-section strut is modelled as an isotropic, elastic-perfectly plastic material model in conjunction with the Von Mises failure criterion. The yield stress is varied from 250 N/mm^2 to 550 N/mm^2. Figure 4 presents the non-dimensional load-end shortening curves for different levels of yield stress. At all levels of yield stress, the ultimate load of the struts is clearly defined with a subsequent elasto-plastic unloading phase of behaviour. It has been found that the ultimate loads of the struts are closely associated with membrane surface yielding at the section junctions. The influence of local geometrical imperfections is indicated in Figure 4 by the dotted lines for the case of an imperfection maximum amplitude of 0.4t_w and corresponding to the yield stress levels of σ_y = 250 N/mm^2 and 550 N/mm^2 respectively. It is of note that the imperfection curves tend asymptotically to their ideal perfect counterparts far into post-buckling during the plastic unloading phase of behaviour. Of particular interest is the effect of local imperfections on the ultimate carrying capability of the sections. It is clear that the effect is not significant for the higher yield stress section which has a range of stable elastic post-buckling reserve. In this case the loss in carrying capability is less than 3%. For the section designed with near simultaneous buckling and yielding, the effect of imperfections on ultimate load is noted to be significant. For this design the loss in carrying capability is just over 13% and thus the imperfection sensitivity of simultaneous interaction is noted to be an area requiring critical attention.

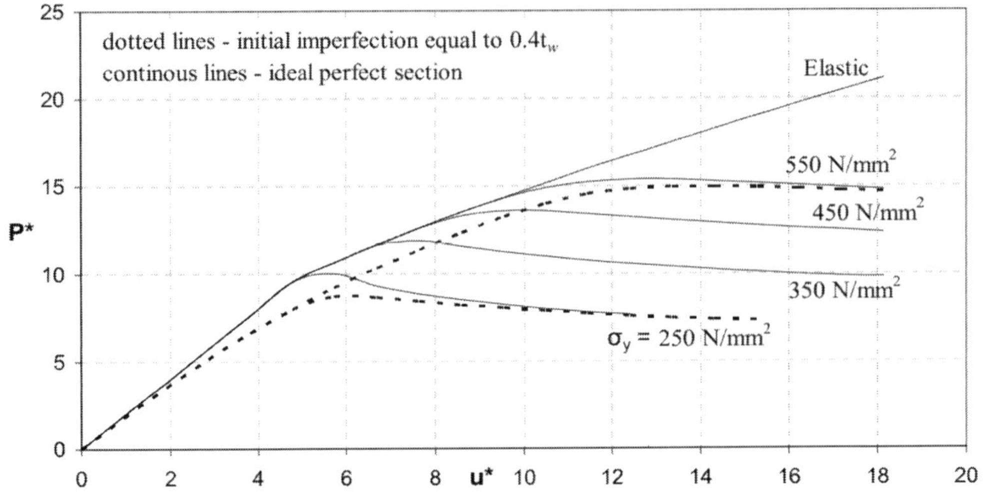

Fig. 4. Non-Dimensional Load-End Shortening Variation (Imperfections and Material Non-Linearity Effects)

4. Concluding Remarks

The compressional stiffness of thin-walled I-section struts is noted to be significantly altered as a result of local buckling of the section walls. The loss in axial stiffness of short strut compression members will depend, to a large extent, on cross-sectional shape and more particularly on section wall thickness. For sections manufactured from high yield materials it is possible to design the struts with a fairly wide range of the ratio of yield stress to local buckling stress σ_y/σ_{cr} of the sections. For the higher values of this ratio considered in this paper it is shown that the post-buckling response involves an initial range of elastic behaviour whereby the nonlinear loss in compressional stiffness is associated with large rotations and local form change during loading.

As loading progresses, surface yielding occurs as a result of the high through-the-thickness bending stresses caused by local buckling. At this point the struts enter into an elasto-plastic phase of behaviour and further loss in compressional stiffness is experienced as a result of the effects of material nonlinearity. The growth of yielding in the elasto-plastic phase of behaviour leads eventually to collapse of the sections at ultimate conditions and subsequently to a post-collapse unloading phase of behaviour. The in-plane stress free boundary condition at the section junctions has been shown to give a response which reflects good agreement with the behaviour of the sections when the junctions are allowed to wave naturally. This is noted to be true in the initial nonlinear elastic stages of behaviour up to about twice the initial buckling load. The load shedding to the section junctions as a result of local buckling is noted to result in junction yielding at the membrane surface of the section walls at, or in the near vicinity, of the ultimate collapse loads of the sections. The influence of local imperfections have been shown to be significant for simultaneous buckling and yielding designs as a result of the unstable equilibrium behaviour associated with simultaneous interaction.

References

[1] Loughlan, J. and Howe, D., The Influence of Local Buckling on the Behaviour of Some Thin-Walled Compression Members., Steel Structures (Advances, Design and Construction), Edited by R. Narayanan, pp. 450-464, Elsevier Applied Science, 1987.

[2] Loughlan, J., Thin-Walled Cold-Formed Sections Subjected to Compressive Loading, Thin-Walled Structures, Vol. 16, No. 1, pp. 65-109, 1993.

[3] Lau, S. C. W. and Hancock, G. J., Distortional Buckling Formulae for Thin Walled Channel Columns, Journal of Structural Engineering, American Society of Civil Engineers, Vol. 113, No 5, May 1987.

[4] Kwon, Y. B. and Hancock, G. J., "Post-Buckling Analysis of Thin-Walled Channel Sections Undergoing Local and Distortional Buckling", Research Report, No. R650, School of Civil and Mining Engineering, The University of Sydney, Australia, 1992.

[5] Rasmussen, K. J. R., Bifurcation of locally buckled members, Thin-Walled Structures, 28 (2) (1997) 117-154.

[6] Young, B. and Rasmussen, K. J. R., Bifurcation of singly symmetric columns, Thin-Walled Structures. 28 (2) (1997) 155-177.

[7] Schafer, B.W., Local, Distortional, and Euler Buckling in Thin-walled Columns. ASCE, Journal of Structural Engineering. 128 (3) 289-299, 2002.

[8] Schafer, B. W., Distortional Buckling of Cold-Formed Steel Columns, Research Report RP00 – 1, Committee on Specifications for the Design of Cold-Formed Steel Structural Members, AISI, Revision 2006, August 2000.

[9] Dinis, P. B., Camotim, D. and Silvestre, N., FEM-based analysis of the local-plate/distortional mode interaction in cold-formed steel lipped channel columns, Computers and Structures, Volume 85 , Issue 19-20, pp. 1461-1474 , 2007.

[10] Ovesy, H. R., and Loughlan J., and GhannadPour, S. A. M., "Geometric Non-Linear Analysis of Channel Sections Under End Shortening, Using Different Versions of the Finite Strip Method", Computers & Structures, 2006, 84, pp 855-872.

[11] Ovesy, H. R., Loughlan J., GhannadPour, S. A. M. and Morada, G., "Geometric Non-Linear Analysis of Box Sections Under End Shortening, Using Three Different Versions of the Finite Strip Method", Thin Walled Structures, 2006, 44, pp. 623-637.

[12] Ovesy, H. R., Loughlan, J. and GhannadPour, S. A. M., 'An Investigation on the Geometric Non-Linear Analysis of I-Section Struts using the Finite Strip Method', 6th International Conference on Steel and Aluminium Structures, Oxford, 24-27 July, 2007, pp. 348-355.

[13] Ovesy, H. R., 'The Development and Application of a Semi-Energy Post-Local Buckling Finite Strip', PhD Thesis, Cranfield University, Bedfordshire, UK, 1998.

Vibration of Workpieces during Aggressive Turning Operations

Xianguo Han[1,3], Minjie Wang[1], Huajiang Ouyang[2]

[1]School of Mechanical Engineering, Dalian University of Technology, Dalian, China
[2] Department of Engineering, University of Liverpool, Liverpool, U.K.
[3] School of Mechanical and Electrical Engineering, Eastern Liaoning University, Dandong, China

h.ouyang@liverpool.ac.uk

Abstract. Turning is a widely used metal cutting operation. Because of the dynamic interaction between the cutter and the workpiece, vibration is excited during turning operations. If the vibration is severe, there will be degradation of surface quality of the workpiece, excessive wear and breakage of the cutter. Chatter can also take place. This paper presents a dynamic model for aggressive turning operations and considers the influence on vibration of the time-dependent reduction of workpiece diameter. The workpiece is modelled as a beam rotating about its longitudinal axis and the cutter provides a moving load. Simulated numerical examples are presented.

1. Introduction

The dynamic model of a rotating beam subjected to different loads has long been studied. Nevertheless, vibration remains a major problem in turning. Lee et al [1] and Katz et al [2] studied vibration of a rotating shaft as a beam based on Euler, Rayleigh, and Timoshenko beam theories under a transverse load moving at constant speed. Huang and his co-workers [3-4] investigated the response of the rotating cylindrical shell and spinning orthotropic beams subjected to harmonic moving loads. Argento and Scott [5] examined a rotating Timoshenko beam excited by a distributed surface force traveling with acceleration. Vibration of a rotating Timoshenko with general boundary conditions subjected to a moving load was investigated by Zu and Han [6]. Sinha [7] studied parametric stability of a rotor-disk system under a nonmoving dynamic axial load. Lee [8] introduced the axial force that was constant and nonmoving and found it had a significant effect on the magnitude of the dynamic response. Zibdeh and Juma [9] considered the moving load as a random force. El-Saeidy [10] analyzed boundary conditions of nonlinear stiffness due to clearance in rolling. He included bending moments in moving load and used the finite element method to study the rotating members subjected to moving load. Sheu and Yang [11] researched the dynamic response of a spinning Rayleigh beam with mass eccentricity but without moving load. Dassanayake and Suh [12] investigated a 3D model as a system of three rotors to simulate workpiece-tool deflections in response to a nonlinear regenerative force with a method of rotor dynamics. He studied different stages of stability for the workpiece and the tool. Ouyang and Wang [13] recently developed a dynamic model for the vibration of a rotating Rayleigh beam subjected to a three-directional load moving in the axial direction. They found that the bending moment induced by the axial force component could have a significant influence on the dynamic response of the beam.

Chatter prediction has a long history that began with works by Tobias and Fishwick [14] and Tlusty and Polacek [15]. Analytical solutions for turning chatter are mainly in regenerative chatter: as the cutter cuts a workpiece surface, the undulations generated in the previous revolution preserve the tool–workpiece vibrations which are closely coupled to the cutting forces. Refs [16] and [17] explained regenerative chatter with a simplified scenario in turning operation. There are several dynamics models of regenerative chatter [18, 19].

Sims [20] developed a new analytical tuning methodology for chatter suppression with vibration absorbers from a regenerative chatter perspective. Chen and Tsao [21] built two models of the workpiece considered as a flexible beam and the tool simplified as a one degree-of-freedom mass-spring-damper system for an orthogonal cutting. They discussed a stability analysis of regenerative chatter for a cantilever beam in turning process.

One disadvantage of the aforementioned moving-load models and regenerative chatter models is that time-dependent reduction of diameter, variable mass and stiffness of the workpiece are neglected, even though chips are removed during a turning operation. Different depths of cut (DOCs) were shown to affect the fundamental natural frequency of a workpiece [22]. Thus, the time-dependent reduction of the shaft diameter would influence the dynamic response and should be studied.

In this paper, a dynamic model is presented for vibration of a rotating shaft as a Rayleigh beam with a time-dependent reducing diameter subjected to a three-directional force that moves in the axial direction. The dynamic response is analysed at different cutting conditions and ratios of radius to length of the shaft. The model represents a simplified tuning operation in metal cutting.

2. Dynamic model

Turning operation has two moving components, a cutter and a workpiece that is fixed to the spindle and pin-mounted at the tailstock. The shaft has a rotary motion while the cutter moves along a straight line. As a result of this operation, the shaft has three obvious sections: un-machined, being-machined and machined, as shown in Figure 1, where l is the full length of the shaft, l_1 is the length of unmachined section of shaft, l_2 is the total length of unmachined and being-machined section of the shaft, and $s(t)$ is the variable length from the spindle end to the location of the cutter.

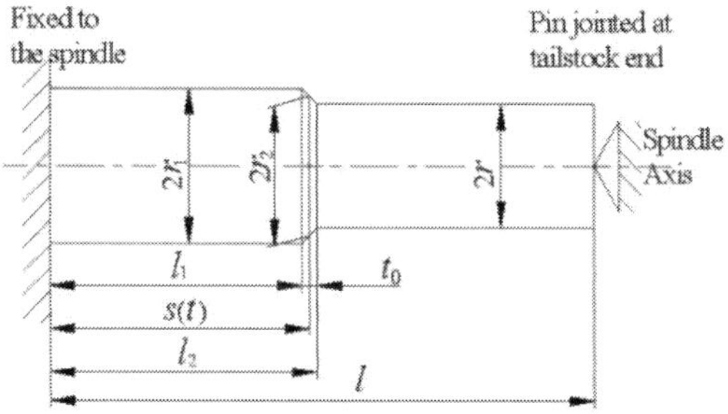

Figure 1. Shaft configuration

Referring to Figure 2, the circular shaft is rotating around its longitudinal axis, and is subjected to a three-directional cutting force along the x axis. A rectangular initial frame is used as the coordinate system. F_x, F_y and F_z are the three perpendicular components of the cutting force acting on the surface of the shaft in the x, y and z directions, respectively, which together play the important role of the shaft material removal. The deflections of the rotating shaft in the y and z directions caused by the cutting force components are represented by v and w.

As a result of the translating axial force F_x, a bending moment M_z is produced as [13]

$$M_z = -F_x\, r_2 \qquad (1)$$

where r_1 is radius of the un-machined section, r is radius of the machined section, and $r_2 = (r_1 + r)/2$ is average radius of the cone of the being-machined section.

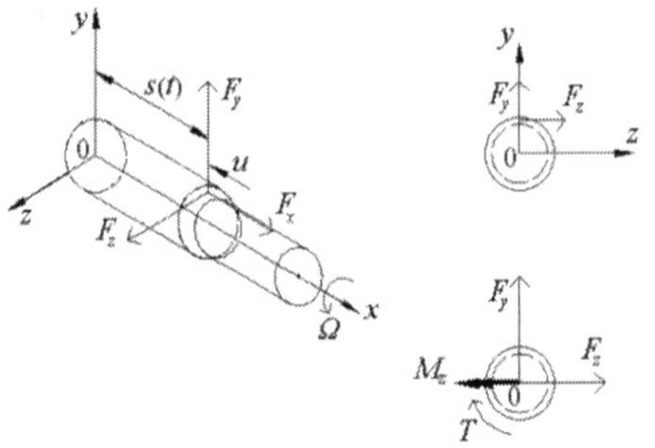

Figure 2. Dynamic shaft model

On the basis of Rayleigh beam theory, the strain energy of the sectioned shaft is [23]:

$$V = \frac{1}{2}\int_0^{l_1} EI_1\left[\left(\frac{\partial^2 v}{\partial x^2}\right)^2 + \left(\frac{\partial^2 w}{\partial x^2}\right)^2\right]dx + \frac{1}{2}\int_{l_1}^{l_2} EI_2\left[\left(\frac{\partial^2 v}{\partial x^2}\right)^2 + \left(\frac{\partial^2 w}{\partial x^2}\right)^2\right]dx + \frac{1}{2}\int_{l_2}^{l} EI_3\left[\left(\frac{\partial^2 v}{\partial x^2}\right)^2 + \left(\frac{\partial^2 w}{\partial x^2}\right)^2\right]\mathrm{d}x - \frac{1}{2}\int_{s(t)}^{l} F_x\left[\left(\frac{\partial v}{\partial x}\right)^2 + \left(\frac{\partial w}{\partial x}\right)^2\right]dx \quad (2)$$

where E is Young's modulus, $I_1 = \pi r_1^4/4$, $I_2 = \pi r_2^4/4$ and $I_3 = \pi r^4/4$ are the second moments of area of the three sections of the shaft, respectively. The unmachined length of the shaft is variable as $l_1 = l - ut$, where u is feed speed of the cutter or the cutting force moving along the longitudinal axis in the opposite direction of the x axis. In Figure 1, variable $l_2 = ut + t_0$ is total length of the unmachined section length plus the being-machined one, where $t_0 = a_p \tan(c_s)$ is the length of the being-machined section, $a_p = r_1 - r$ is DOC, and c_s is the side cutting edge angle of cutter. It should be pointed out that the strain energy of the chips is ignored in this paper.

Excluding the kinetic energy of the chips removed, the kinetic energy of the rotating shaft is [14]

$$T = \frac{1}{2}\int_0^{l_1}\left\{\rho A_1\left[\left(\frac{\partial v}{\partial t}\right)^2 + \left(\frac{\partial w}{\partial t}\right)^2\right] + \rho I_1\left\{\left[\left(\frac{\partial^2 v}{\partial x\,\partial t}\right)^2 + \left(\frac{\partial^2 w}{\partial x\,\partial t}\right)^2\right] + 2\Omega\left(\frac{\partial^2 v}{\partial x\,\partial t}\frac{\partial w}{\partial x} - \frac{\partial^2 w}{\partial x\,\partial t}\frac{\partial v}{\partial x}\right) + 2\Omega^2\right\}\right\}\mathrm{d}x + \frac{1}{2}\int_{l_1}^{l_2}\left\{\rho A_2\left[\left(\frac{\partial v}{\partial t}\right)^2 + \left(\frac{\partial w}{\partial t}\right)^2\right] + \rho I_2\left\{\left[\left(\frac{\partial^2 v}{\partial x\,\partial t}\right)^2 + \left(\frac{\partial^2 w}{\partial x\,\partial t}\right)^2\right] + 2\Omega\left(\frac{\partial^2 v}{\partial x\,\partial t}\frac{\partial w}{\partial x} - \frac{\partial^2 w}{\partial x\,\partial t}\frac{\partial v}{\partial x}\right) + 2\Omega^2\right\}\right\}\mathrm{d}x + \frac{1}{2}\int_{l_2}^{l}\left\{\rho A_3\left[\left(\frac{\partial v}{\partial t}\right)^2 + \left(\frac{\partial w}{\partial t}\right)^2\right] + \rho I_3\left\{\left[\left(\frac{\partial^2 v}{\partial x\,\partial t}\right)^2 + \left(\frac{\partial^2 w}{\partial x\,\partial t}\right)^2\right] + 2\Omega\left(\frac{\partial^2 v}{\partial x\,\partial t}\frac{\partial w}{\partial x} - \frac{\partial^2 w}{\partial x\,\partial t}\frac{\partial v}{\partial x}\right) + 2\Omega^2\right\}\right\}\mathrm{d}x \quad (3)$$

where ρ is the mass density of the shaft, $A_1 = \pi r_1^2$, $A_2 = \pi(r_1 + r)^2/4$, $A_3 = \pi r^2$ are the cross sectional areas of the un-machined, machining, and machined sections, respectively. Ω is rotary speed of the shaft.

The virtual work done by components of cutting forces F_y and F_z and moment M_z is

$$\delta W = F_y\delta v(s,t) + F_z\delta w(s,t) + M_z\left.\frac{\partial\delta v}{\partial x}\right|_{x=s} \quad (4)$$

3. Mathematical formulation and calculation

It is assumed that the deflection of the shaft can be expressed as

$$v(x,t) = \sum_{i=1}^{n}\varphi_i(x)\alpha_i(t) = \boldsymbol{\varphi}^{\mathrm{T}}\boldsymbol{\alpha},\quad w(x,t) = \sum_{i=1}^{n}\varphi_i(x)\beta(t) = \boldsymbol{\varphi}^{\mathrm{T}}\boldsymbol{\beta} \quad (5)$$

where $\boldsymbol{\varphi}^{\mathrm{T}} = \{\varphi_1, \varphi_2, \ldots, \varphi_n\}$, $\boldsymbol{\alpha}^{\mathrm{T}} = \{\alpha_1, \alpha_2, \ldots, \alpha_n\}$, $\boldsymbol{\beta}^{\mathrm{T}} = \{\beta_1, \beta_2, \ldots, \beta_n\}$ and $\varphi_i(x)$ is the spatial function that meets the boundary condition of the shaft and is taken to be the i-th mode shape of the stationary shaft in the analysis, $\alpha_i(t)$ and $\beta_i(t)$ are corresponding modal coordinates.

After Eq. (5) is substituted into Eq. (2) and Eq. (3), one gets

$$V = \frac{1}{2}\int_0^{l_1} EI_1\,[\boldsymbol{\alpha}^{\mathrm{T}}\boldsymbol{\varphi}''\boldsymbol{\varphi}''^{\mathrm{T}}\boldsymbol{\alpha} + \boldsymbol{\beta}^{\mathrm{T}}\boldsymbol{\varphi}''\boldsymbol{\varphi}''^{\mathrm{T}}\boldsymbol{\beta}]\mathrm{d}x + \frac{1}{2}\int_{l_1}^{l_2} EI_2\,[\boldsymbol{\alpha}^{\mathrm{T}}\boldsymbol{\varphi}''\boldsymbol{\varphi}''^{\mathrm{T}}\boldsymbol{\alpha} + \boldsymbol{\beta}^{\mathrm{T}}\boldsymbol{\varphi}''\boldsymbol{\varphi}''^{\mathrm{T}}\boldsymbol{\beta}]\mathrm{d}x + \frac{1}{2}\int_{l_2}^{l} EI_3[\boldsymbol{\alpha}^{\mathrm{T}}\boldsymbol{\varphi}''\boldsymbol{\varphi}''^{\mathrm{T}}\boldsymbol{\alpha} + \boldsymbol{\beta}^{\mathrm{T}}\boldsymbol{\varphi}''\boldsymbol{\varphi}''^{\mathrm{T}}\boldsymbol{\beta}]\,\mathrm{d}x - \frac{1}{2}\int_{s(t)}^{l} F_x\,[\boldsymbol{\alpha}^{\mathrm{T}}\boldsymbol{\varphi}'\boldsymbol{\varphi}'^{\mathrm{T}}\boldsymbol{\alpha} + \boldsymbol{\beta}^{\mathrm{T}}\boldsymbol{\varphi}'\boldsymbol{\varphi}'^{\mathrm{T}}\boldsymbol{\beta}]\mathrm{d}x \tag{6}$$

and

$$T = \frac{1}{2}\int_0^{l_1}\Big\{\rho A_1\big(\dot{\boldsymbol{\alpha}}^{\mathrm{T}}\boldsymbol{\varphi}\boldsymbol{\varphi}^{\mathrm{T}}\dot{\boldsymbol{\alpha}} + \dot{\boldsymbol{\beta}}^{\mathrm{T}}\boldsymbol{\varphi}\boldsymbol{\varphi}^{\mathrm{T}}\dot{\boldsymbol{\beta}}\big) + \rho I_1\big\{\big(\dot{\boldsymbol{\alpha}}^{\mathrm{T}}\boldsymbol{\varphi}'\boldsymbol{\varphi}'^{\mathrm{T}}\dot{\boldsymbol{\alpha}} + \dot{\boldsymbol{\beta}}^{\mathrm{T}}\boldsymbol{\varphi}'\boldsymbol{\varphi}'^{\mathrm{T}}\dot{\boldsymbol{\beta}}\big) + 2\Omega\big(\dot{\boldsymbol{\alpha}}^{\mathrm{T}}\boldsymbol{\varphi}'\boldsymbol{\varphi}'^{\mathrm{T}}\boldsymbol{\beta} - \dot{\boldsymbol{\beta}}^{\mathrm{T}}\boldsymbol{\varphi}'\boldsymbol{\varphi}'^{\mathrm{T}}\boldsymbol{\alpha}\big) + 2\Omega^2\big\}\Big\}\mathrm{d}x + \frac{1}{2}\int_{l_1}^{l_2}\Big\{\rho A_2\big(\dot{\boldsymbol{\alpha}}^{\mathrm{T}}\boldsymbol{\varphi}\boldsymbol{\varphi}^{\mathrm{T}}\dot{\boldsymbol{\alpha}} + \dot{\boldsymbol{\beta}}^{\mathrm{T}}\boldsymbol{\varphi}\boldsymbol{\varphi}^{\mathrm{T}}\dot{\boldsymbol{\beta}}\big) + \rho I_2\big\{\big(\dot{\boldsymbol{\alpha}}^{\mathrm{T}}\boldsymbol{\varphi}'\boldsymbol{\varphi}'^{\mathrm{T}}\dot{\boldsymbol{\alpha}} + \dot{\boldsymbol{\beta}}^{\mathrm{T}}\boldsymbol{\varphi}'\boldsymbol{\varphi}'^{\mathrm{T}}\dot{\boldsymbol{\beta}}\big) + 2\Omega\big(\dot{\boldsymbol{\alpha}}^{\mathrm{T}}\boldsymbol{\varphi}'\boldsymbol{\varphi}'^{\mathrm{T}}\boldsymbol{\beta} - \dot{\boldsymbol{\beta}}^{\mathrm{T}}\boldsymbol{\varphi}'\boldsymbol{\varphi}'^{\mathrm{T}}\boldsymbol{\alpha}\big) + 2\Omega^2\big\}\Big\}\mathrm{d}x + \frac{1}{2}\int_{l_2}^{l}\Big\{\rho A_3\big(\dot{\boldsymbol{\alpha}}^{\mathrm{T}}\boldsymbol{\varphi}\boldsymbol{\varphi}^{\mathrm{T}}\dot{\boldsymbol{\alpha}} + \dot{\boldsymbol{\beta}}^{\mathrm{T}}\boldsymbol{\varphi}\boldsymbol{\varphi}^{\mathrm{T}}\dot{\boldsymbol{\beta}}\big) + \rho I_3\big\{\big(\dot{\boldsymbol{\alpha}}^{\mathrm{T}}\boldsymbol{\varphi}'\boldsymbol{\varphi}'^{\mathrm{T}}\dot{\boldsymbol{\alpha}} + \dot{\boldsymbol{\beta}}^{\mathrm{T}}\boldsymbol{\varphi}'\boldsymbol{\varphi}'^{\mathrm{T}}\dot{\boldsymbol{\beta}}\big) + 2\Omega\big(\dot{\boldsymbol{\alpha}}^{\mathrm{T}}\boldsymbol{\varphi}'\boldsymbol{\varphi}'^{\mathrm{T}}\boldsymbol{\beta} - \dot{\boldsymbol{\beta}}^{\mathrm{T}}\boldsymbol{\varphi}'\boldsymbol{\varphi}'^{\mathrm{T}}\boldsymbol{\alpha}\big) + 2\Omega^2\big\}\Big\}\mathrm{d}x \tag{7}$$

The Lagrange's equations of motion can be derived as

$$\begin{aligned}
&\rho\mathbf{K1}\ddot{\boldsymbol{\alpha}} + \rho\dot{\mathbf{K1}}\dot{\boldsymbol{\alpha}} + [E\mathbf{K3} - F_x\mathbf{B3}]\boldsymbol{\alpha} + \rho\Omega\big(\dot{\mathbf{K2}}\boldsymbol{\beta} + 2\mathbf{K2}\dot{\boldsymbol{\beta}}\big) = F_y\boldsymbol{\varphi}\big(s(t)\big) + M_z\boldsymbol{\varphi}'\big(s(t)\big)\\
&\rho\mathbf{K1}\ddot{\boldsymbol{\beta}} + \rho\dot{\mathbf{K1}}\dot{\boldsymbol{\beta}} + [E\mathbf{K3} - F_x\mathbf{B3}]\boldsymbol{\beta} - \rho\Omega\big(\dot{\mathbf{K2}}\boldsymbol{\alpha} + 2\mathbf{K2}\dot{\boldsymbol{\alpha}}\big) = F_z\boldsymbol{\varphi}(s(t))
\end{aligned} \tag{8}$$

where

$$\mathbf{K1} = A_1\mathbf{A1} + A_2\mathbf{A2} + A_3\mathbf{A3} + I_1\mathbf{B1} + I_2\mathbf{B2} + I_3\mathbf{B3}$$

$$\mathbf{K2} = I_1\mathbf{B1} + I_2\mathbf{B2} + I_3\mathbf{B3}$$

$$\mathbf{K3} = I_1\mathbf{C1} + I_2\mathbf{C2} + I_3\mathbf{C3}$$

and

$$\mathbf{A1} = \int_0^{l_1}\boldsymbol{\varphi}(x)\boldsymbol{\varphi}^{\mathrm{T}}(x)\mathrm{d}x,\ \mathbf{B1} = \int_0^{l_1}\varphi'(x)\varphi^{\mathrm{T}}(x)\mathrm{d}x,\ \mathbf{C1} = \int_0^{l_1}\varphi''(x)\,\varphi''^{\mathrm{T}}(x)\mathrm{d}x$$

$$\mathbf{A2} = \int_{l_1}^{l_2}\varphi(x)\varphi^{\mathrm{T}}(x)\mathrm{d}x,\ \mathbf{B2} = \int_{l_1}^{l_2}\varphi'(x)\varphi'^{\mathrm{T}}(x)\mathrm{d}x,\ \mathbf{C2} = \int_{l_1}^{l_2}\varphi''(x)\varphi''^{\mathrm{T}}(x)\mathrm{d}x$$

$$\mathbf{A3} = \int_{l_2}^{l}\varphi(x)\varphi^{\mathrm{T}}(x)\mathrm{d}x,\ \ \mathbf{B3} = \int_{l_2}^{l}\varphi'(x)\varphi'^{\mathrm{T}}(x)\mathrm{d}x,\ \mathbf{C3} = \int_{l_1}^{l}\varphi''(x)\varphi''^{\mathrm{T}}(x)\mathrm{d}x$$

$$\mathbf{B4} = \int_{s(t)}^{l}\varphi'(x)\varphi'^{\mathrm{T}}(x)\mathrm{d}x \approx \mathbf{B3}$$

Note that **A1**, **A2**, **A3**, **B1**, **B2**, **B3**, **C1**, **C2**, **C3**, **K1**, **K2** and **K3** are all time varying matrices. Eq. (8) will be solved by a fourth-order Runge-Kutta method with a very small time step.

The cutting force is variable with the cutting condition, which can be established from turning experiments [24-25]. To determine the cutting forces under cutting condition, dry turning operations were carried out on a lathe CA6140. The lathe equipped with dynamometer YDX—III9702 (made by a research group at Dalian University of Technology) was used for measuring the cutting force components. The cutter was type PSSNR2525M12 and tool insert was coated cemented carbide. Cutting angles of the tool are: side cutting edge angle $c_s = 45^{\circ}$, inclination angle $i_a = 0^{\circ}$, and normal rake angle $\gamma_n = 10^{\circ}$. The workpiece was 48 mm in diameter and 650 mm in length. The material of the shaft was AISI 1045: its hardness is HB190. The cutting condition and the experimental results are shown in Table 1. The rotational speed, feed rate and depth of cut are carefully chosen in different values as in the subsequent numerical analysis under different cutting condition. In order to simply the analytical model, the influence of vibration on the area of the cutting layer A_c is neglected in this paper.

Table1 Cutting condition and average components of cutting force

Ω (rpm)	DOC (mm)	f (mm/rev)	F_x (N)	F_y (N)	F_z (N)
630	3	0.1	457.29	464.84	973.80
630	2	0.1	349.83	375.55	719.38
630	1	0.1	199.24	224.83	402.49
630	3	0.3	909.64	911.87	2221.98
630	2	0.3	621.10	672.86	1558.09
630	1	0.3	349.19	407.11	863.53
1250	3	0.3	736.65	764.20	1913.61
1250	2	0.3	512.96	586.29	1355.06
1250	1	0.3	296.59	376.21	806.51

4. Numerical results and analysis

The material and geometric properties of the circular shaft are $l = 1.0$ m, $E = 2.07 \times 10^{11}$ Pa and $\rho = 7700$ kg/m^3. Considering the chucked-pinned boundary condition of the shaft in the turning operation, the normalized modes are [26]

$$\varphi_i(x) = \frac{\sin(\beta_i l) - \sinh(\beta_i l)}{\cos(\beta_i l) - \cosh(\beta_i l)}(\cos(\beta_i x) - \cosh(\beta_i x)) - (\sin(\beta_i x) - \sinh(\beta_i x)) \quad (9)$$

where $\beta_i l = [3.9266, 7.0686, 10.2102, 13.3518, 16.4934, \ldots]$ $(i = 1,2,3, \ldots)$. The natural frequency of the shaft (when the shaft is not rotating) is $\omega_i = (\beta_i l)^2 \sqrt{EI_3/(\rho A_3 l^4)}$. Nondimensional parameter $\gamma = a_p/r$ is used, where a_p is depth of cut. Five modes of the shaft are found to produce satisfactory results and hence are used. The numerical results of the dynamic responses v_p and w_p of the shaft at the moving cutter location (in the y and z direction respectively) are depicted in Figures 3-8. Two radii of the circular shaft (r =0.095 m and r =0.019 m) are studied. Please note that the values of r refer to the radius of the shaft after turning and the radius of the original unmachined shaft is r + DOC when different DOCs are studied. So a bigger DOC for a shaft of the same radius actually is thicker before cutting.

Firstly, the effect of decreasing shaft radius on the dynamic response of the shaft is studied. Three cases are considered: (1) the radius of the shaft throughout this particular analysis is taken to be the radius of the shaft after machining ($\gamma 0$) and is a constant, i.e., $r_1 = r_2 = r = 0.019\, m$; (2) the radius of the shaft throughout this particular analysis is taken to be the radius of the original unmachined shaft ($\gamma 1$) and is constant, i.e. $r = r_2 = r_1 = 0.022$ m; (3) the radius of the shaft is time-varying with DOC=3 mm ($\gamma = 0.158$), i.e. $r(t) = r_1 - a_p$. Figure 3 shows the numerical results obtained at $\Omega =$ 1250 rpm, feed speed $f = 0.3$ mm/rev with M_z ignored. The values of the cutting force used in the analysis are from Table 1. Clearly consideration of time-dependent reduction of shaft radius makes a noticeable difference in the dynamic response, as shown in Figure 3.

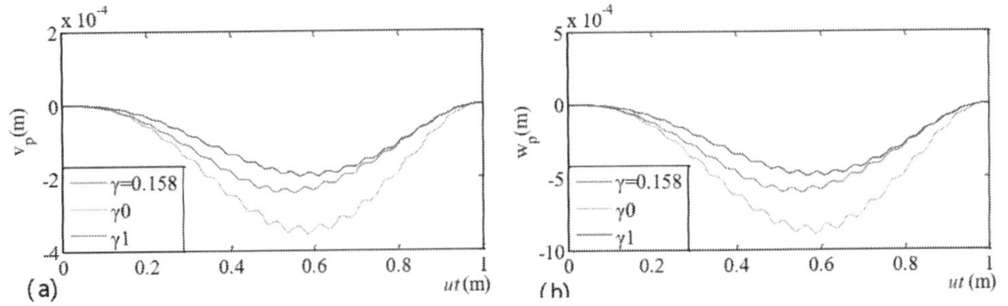

Figure 3. Dynamic response of the shaft (r =0.019 m) subjected to three moving forces at different radii (no M_z, Ω=1250 rpm, f=0.3 mm/rev)

Secondly, different magnitudes of DOC are simulated (DOC=3 mm for $\gamma = 0.032$, DOC=2 mm for $\gamma = 0.021$ and DOC=1 mm for $\gamma = 0.01$) with f=0.3 mm/rev and Ω=1250 rpm, but without M_z. For r=0.095 m, v_p and w_p are shown in Figures 4(a) and (b). For r =0.019 m, they are shown in Figure 5. From Figure 3, reducing shaft diameter has a clear and similar influence on the dynamic response. The bigger γ ($\gamma = a_p/r$) used, which means larger DOC (the shaft radius after cutting r is constant) and thus bigger cutting force, the greater the deflection, as shown in Figures 4 and 5.

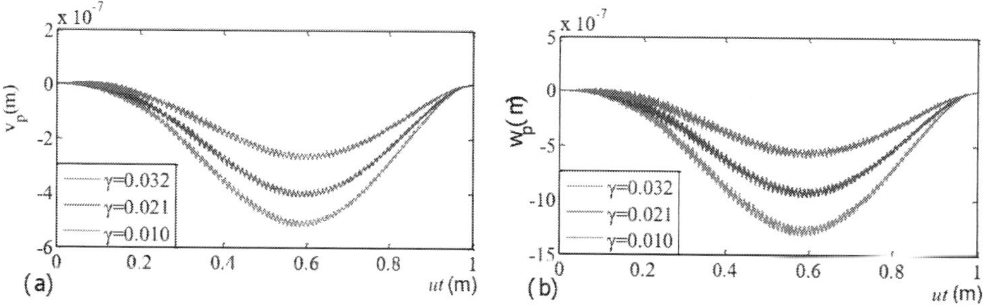

Figure 4. Dynamic response of the shaft(r =0.095 m) subjected to three moving forces at different DOCs (no M_z, Ω=1250 rpm, f=0.3 mm/rev)

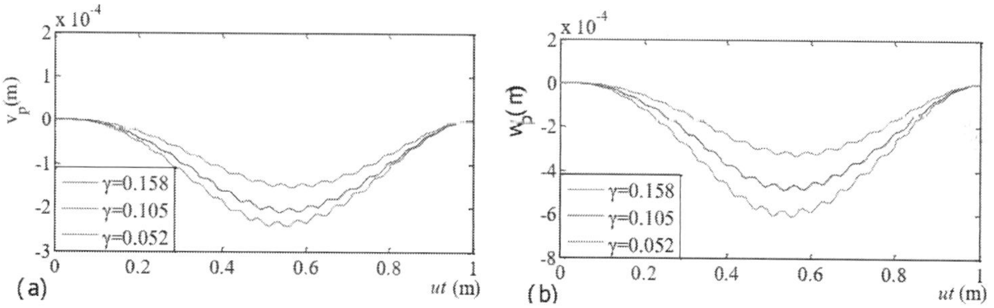

Figure 5. Dynamic response due to three moving forces at different DOCs: r=0.019 m

(no M_z, Ω=1250 rpm, f=0.3 mm/rev)

Next, M_z is considered and the same data are used. At $r = 0.095$ m, numerical results of v_p and w_p are shown in Figure 6(a) and (b), respectively. Again different DOCs have a noticeable influence on dynamic response of the shaft in the y and z axis. Meanwhile, M_z induces a significant fluctuation of the deflection in the y axis of Figure 6(a) than the deflection in z axis of Figure 6(b), this is because M_z acts in the xoy plane (refer to Figure 1) and directly bends the shaft in that plane (causing more deflection in the y direction). On the other hand, using the same cutting condition and $r = 0.019$ m, numerical results also show that DOCs have a significant effect on dynamic response of the shaft shown in Figure 7. The effect of M_z is smaller as M_z itself is smaller for a smaller radius of shaft.

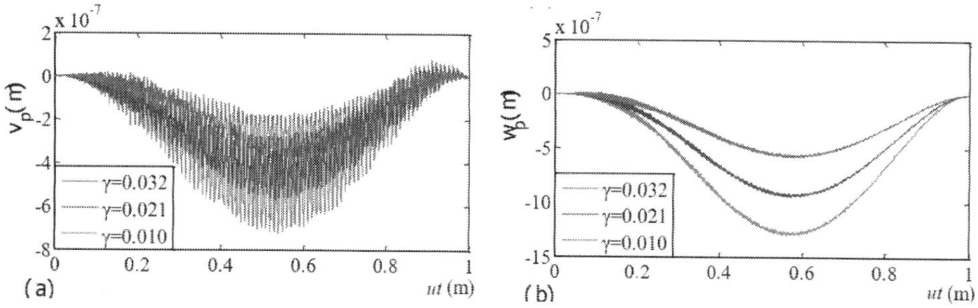

Figure 6. Dynamic response due to three moving forces at different DOCs: r=0.095 m (with M_z, Ω=1250 rpm, f=0.3 mm/rev)

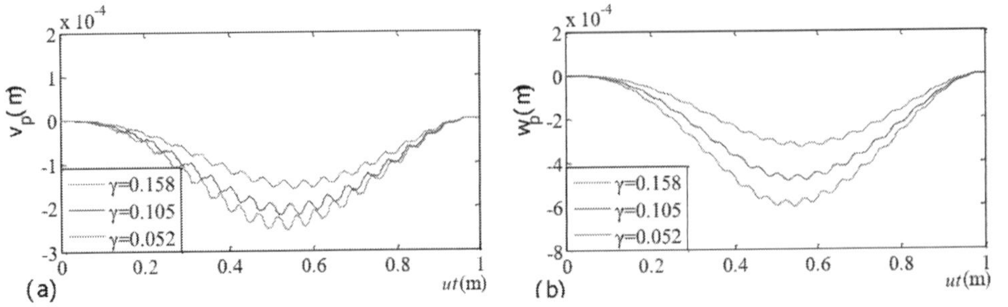

Figure 7. Dynamic response due to three moving forces at different DOCs: r=0.019 m (with M_z, Ω=1250 rpm, f=0.3 mm/rev)

When Ω is decreased to 630 rpm, and the other data remain the same and r=0.095 m is used, numerical results are depicted in Figure 8, from which different DOCs also have a considerable effect on dynamic response v_p and w_p of the large radial shaft in comparison with Figure 6.

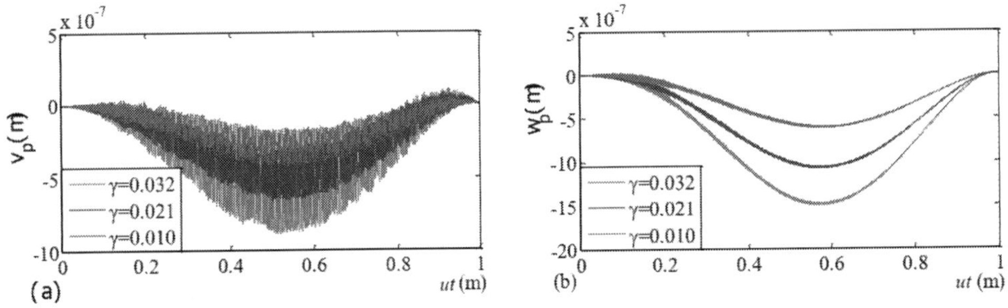

Figure 8. Dynamic response due to three moving forces at different DOCs: r =0.095 m (with M_z, Ω=630 rpm , f=0.3 mm/rev)

M_z is seen to add lower-amplitude higher-frequency oscillation, which is similar to the finding of [13]. The method can be extended to deal with workpieces modelled as Timoshenko beam [27].

5. Conclusions

This paper studies vibration of turning operations. The workpiece is modelled as a clamped-pinned circular Rayleigh beam rotating about its longitudinal axis and the cutter provides a moving load of three perpendicular force components. The dynamic model includes the influence on vibration of the time-dependent reduction of workpiece diameter as a result of material being removed during turning. The numerical results show that the reduction of the workpiece has a significant effect on the dynamic response and the vibration magnitude is greater than that of the unmachined workpiece but smaller than that of the fully machined workpiece, and hence should be considered. It is also found that a bigger depth of cut produces greater vibration magnitude and the moment induced by the axial component of the cutting force results in a big moving-load effect by adding lower-amplitude higher-frequency oscillation to the dynamic response, especially in the range of realistic low feed speeds.

References

[1] Lee, C., Katz, W. R., Ulsoy, A. G., Scott, R. A., Modal analysis of a distributed parameter rotating shaft, *Journal Sound and Vibration*, 1987; 122: 119–130.

[2] Katz, W. R., Lee, C., Ulsoy, A. G., Scott, R. A., The dynamic response of a rotating shaft subject to a moving load, *Journal Sound and Vibration*, 1987; 122:134–148.

[3] Huang, S. C., Chen, J. S., Dynamic response of spinning orthotropic beams subjected to moving harmonic forces, *Journal of Chinese Society of Mechanical Engineers*, 1990; 11: 63–73.

[4] Huang, S. C., Hsu, B. S., Resonant phenomena of a rotating cylindrical shell subjected to a harmonic moving load, *Journal Sound and Vibration*, 1990; 136: 215–228.

[5] Argento, A., Scott, R. A., Dynamic response of a rotating beam subjected to an accelerating distributed surface force, *Journal Sound and Vibration*, 1992; 157(2): 221–231.

[6] Zu, J. W. Z., Han, R. P. S., Dynamic-response of a spinning Timoshenko beam with general boundary-conditions and subjected to a moving load, *ASME Journal of Applied Mechanics*, 1994; 61(1): 52–160.

[7] Sinha, S. K., On general conditions of rotor dynamic stability under combined axial forces and torque, *ASME Journal of Applied Mechanics*, 1992; 59(1): 225–228.

[8] Lee, H. P., Dynamic response of a rotating Timoshenko shaft subject to axial forces and moving loads, *Journal Sound and Vibration*, 1995; 181(1): 169–177.

[9] Zibdeh, H. S., Juma, H. S., Dynamic response of a rotating beam subjected to a random moving load, *Journal Sound and Vibration*, 1999; 223(5): 741–758.

[10] El-Saeidy, F. M. A., Finite element dynamic analysis of a rotating shaft with or without nonlinear boundary conditions subjected to a moving load, *Nonlinear Dynamics*, 2000; 21(4): 377–407.

[11] Sheu, G. J., Yang, S. M., Dynamic analysis of a spinning Rayleigh beam, *International Journal of Mechanical Sciences*, 2005; 47(2): 157-169.

[12] Dassanayake, A. V., Suh, C. S., On nonlinear cutting response and tool chatter in turning operation, *Communications in Nonlinear Science and Numerical Simulation*, 2008; 13(5): 979–1001.

[13] Ouyang, H., Wang, M. J., Dynamics of a rotating shaft subject to a three-directional moving load, *ASME Journal of Vibration and Acoustics*, 2007; 129(7): 386–389.

[14] Tobias, S. A., Fishwick, W., *A Theory of Regenerative Chatter*, The Engineer—London, 1958.

[15] Tlusty, J., Polacek, M., *The Stability of Machine Tools against Self Excited Vibrations in Machining,* International Research in Production Engineering, ASME, 1963, 465–474.

[16] Tarng, Y. S., Young, H. T., Lee, B. Y., An analytical model of chatter vibration in metal cutting, *International Journal of Machine Tools and Manufacturing*, 1994; 34 (2): 183–197.

[17] Olgac, N., Hosek, M., A new perspective and analysis for regenerative machine tool chatter, *International Journal of Machine Tools and Manufacturing*, 1998; 38: 783–798.

[18] Altintas, Y., *Manufacturing Automation: Metal Cutting Mechanics, Machine Tool Vibrations, and CNC Design*, Cambridge University Press, Cambridge, 2000.

[19] Tlusty, J., *Manufacturing Processes and Equipment*, Upper Saddle River, N.J., 2000.

[20] Sims, N. D., Vibration absorbers for chatter suppression: a new analytical tuning methodology, *Journal of Sound and Vibration*, 2007; 301: 592–607.

[21] Chen, C. K., Tsao, Y. M., A stability analysis of regenerative chatter in turning process without using tailstock, *International Journal of Advanced Manufacturing Technology*, 2006; 29: 648–654.

[22] Yu, S. D., Shah, V., Theoretical and experimental studies of chatter in turning for uniform and stepped workpieces, *ASME Journal of Vibration and Acoustics*, 2008; 130(3): 0610051–18.

[23] Chen, L.-W., Ku, D.-M., Dynamic stability of a cantilever shaft- disk system, *ASME Journal of Vibration and Acoustics*, 1992; 181(1): 326–329.

[24] Rao, B., Shin, Y. C., A comprehensive dynamic cutting force model for chatter prediction in turning, *International Journal of Machine Tools and Manufacture*,1999; 39 (10):1631–1654.

[25] Clancy, B. E., Shin, Y. C., A comprehensive chatter prediction model for face turning operation including tool wear effect, *International Journal of Machine Tools and Manufacture*, 2002; 42 :1035–1044.

[26] Goncalves, P. J. P., Brennan, M. J., Elliott, S. J., Numerical evaluation of high-order modes of vibration in uniform Euler–Bernoulli beams, *Journal of Sound and Vibration*, 2007; 301: 1035–1039.

[27] Ouyang, H., Wang, M., A dynamics model for a rotating shaft subjected to axially moving forces, *Journal of Sound and Vibration*, 2007; 308: 674-682.

Modeling and optimization of tensile shear strength of Titanium/ Aluminum dissimilar welded component

E. M. Anawa[1], A. G. Olabi[2] and F. A. Elshukri[3]
1. and 3.Industrial & Manufacturing Eng., Garyounis University, Benghazi, Libya
2. School of Mechanical & Manufacturing Eng., Dublin City University, Dublin 9, Ireland
ezzeddin.hassan2@mail.dcu.ie

Abstract. Titanium was successfully welded to Aluminum using laser welding. Laser welding parameters ranges combinations were experimentally determined using Taguchi approach with the objective of producing welded joint with acceptable welding profile and maximizing the tensile shear strength. Tensile shear strength of dissimilar Al/Ti was evaluated as a response function of the selected laser welding parameters and statistical model was developed to describe it. The result indicates that the developed model can predict the responses satisfactorily. Tensile strength of the joints was found to be same as Al base metal. Furthermore the microstructure and microhardness of the joint have been studied.

1. Introduction

Because of its superior mechanical and metallurgical properties, titanium is used in wide range of engineering and manufacturing applications. On the other hand titanium and its alloys are costly which limits its applications. Moreover, in dissimilar welding titanium forms a brittle intermetalic phase exhibit inferior mechanical properties [1,2]. Many solid-state welding methods for joining these dissimilar materials such as pressure welding [3], diffusion bonding [4] and friction welding [5,6] processes have been studied. Wilden et al. [8] instigated diffusion bonding of pure Al with pure Ti. Selected mechanical testing results of joints showed that diffusion bonding was a suitable process for high strength applications. Laser welding affords a high aspect-ratio bead, low distortion and excellent mechanical properties, which make it possible to be applied in dissimilar welding. The high potential of laser welding technologies has been proven through successful applications in the automotive, aeronautics and aerospace industries [7].

Statistical techniques and Design of Experiment (DOE) are widely used to optimize process parameters. Many researches were applied to optimize process input parameters [8,9,10 and 11]. Taguchi method is DOE technique which used to optimize welding parameters. This method was applied to reduce the number of experiments without disturbing the results [9], due to that the optimization of process parameters can improve quality characteristics and the optimal process parameters; the optimal combination of the process parameters can then be predicted. This work concerns with the effects of welding parameters on the tensile shear strength of titanium/aluminum dissimilar joints and the prediction of the optimal combinations of the selected welding factors. This method was applied to reduce the number of experiments without disturbing the results [9], due to that the optimization of process parameters can improve quality characteristics and the optimal process parameters; the optimal combination of the process parameters can then be predicted.

The objective of this study is to optimize the maximum shear strength of Ti/Al-welded components, through minimizing the laser power and maximizing welding speed in order to optimize the cost and increase the production rate. Additionally the developed models are subjected to ANOVA analysis to test significance on individual model coefficients. The microstructure in the welded zone has been evaluated. Also the microhardness variation in the welded zone and their affect on the tensile strength has been investigated.

2. Experimental design and procedure

Two plates of dissimilar materials mentioned above with their chemical compositions and mechanical properties presented in Tables (1 to 2) are welded together using the CO_2 laser welding process. Lap joint design was applied to join the plates of aluminum and titanium with dimensions of (160 x 80 x 1.5 and 160 x 80 x 1) mm respectively. The titanium plate was positioned above the aluminum plate in direct contact with laser beam during the welding process. A heterogeneous single pass was carried out to join the two plates. Pilot experiments were applied to determine the input welding parameters range. The pilot experiments were carried out by changing one parameter at a time to identify the operating range of the welding parameters under the study.

Table 1: Chemical composition of the Aluminum 5251 H22 and the Titanium G2 (wt %)

Material	Si	Fe	Cu	Mn	Mg	Cr	Zn	H	C	O	N	Al	Ti
Al 5251	**0.407**	**0.509**	**0.154**	**0.154**	**2.152**	**0.154**	**0.15**	-	-	-	**0.03**	**Bal.**	**0.147**
Ti G2	-	**0.3**	-	-	-	-	-	**0.015**	**0.1**	**0.25**	-	**0**	**Bal.**

Table 2: Mechanical and physical properties of the materials

Grade	Tensile Strength, [MPa]	Yield Strength [MPa]	Elongation %	Hardness Brinell [HB], max	Elastic Modulus [GPa]	Thermal Conductivity W/m. °C	Melting Point °C	Thermal Expansion µm/m/°C	Density Kg/m3
Al 5251	**190**	**120**	**6**	**56**	**70**	**134**	**650**	**25**	**2690**
Ti G2	**344**	**275**	**20**	**14.5**	**105**	**16.4**	**1665**	**9.36**	**4510**

The criterion for identification of the operating range of the welding parameters was the approval by visual inspection of the welded joints under the principle of absence of observable welding defects and the presence of a good weld seam. The selected welding parameters for these dissimilar materials are presented in Table 3. The same table shows welding input variables and experiment design levels. The welding experiments were carried out following the Taguchi design matrix generated by Design Expert V7 software, presented in Table 4, in random order. An L16 orthogonal array with three columns and 16 rows was used. Sixteen experiments were required to study the welding parameters using an L16 orthogonal array. Tensile shear strength testing of the jointed specimens was carried out and the results are presented in Table 4. Since the thicknesses of the sheets are not the same, the ultimate tensile shear strength is obtained by dividing the force at fracture of the specimen by the length of the weld line (14 mm) and termed as resistance (N/mm).

Table 3: Process parameters and design levels used

Variables	Code	Unit	Level 1	Level 2	Level 3	Level 4
Laser Power	P	kW	0.9	1.05	1.200	1.35
Welding Speed	S	mm/min	1600	1767	1933	2100
Focus	F	mm	-1.0	-0.67	-0.33	0.00

Table 4: Welding input variables, experiment design levels and tensile shear strength.

Std	Run	P, kW	S, mm/min	F, mm	Shear St, N/mm
1	4	0.9	1600	-1	135
2	5	0.9	1767	-0.67	172
3	12	0.9	1933	-0.33	164
4	1	0.9	2100	0.00	181
5	13	1.05	1600	-0.67	130
6	6	1.05	1767	-1.00	137
7	3	1.05	1933	0.00	164
8	16	1.05	2100	-0.33	147
9	14	1.2	1600	-0.33	120
10	11	1.2	1767	0.00	84
11	7	1.2	1933	-1.00	99
12	9	1.2	2100	-0.67	112
13	2	1.35	1600	0.00	38
14	8	1.35	1767	-0.33	98
15	15	1.35	1933	-0.67	135
16	10	1.35	2100	-1.00	143

Samples for tensile shear testing were cut as shown in Fig 1 (a, b) and prepared from the tensile shear testing. The achieved results, presented in Table 4, were performed in random order to avoid any systematic error and they were tested at room temperature (20 °C). It was noted that during the test the fracture occurred within the welding pool regardless of the input welding parameters. The rupture of the joint was brittle-like fracture, implying that the welding process caused the ductility of the alloy to deteriorate at the WZ.

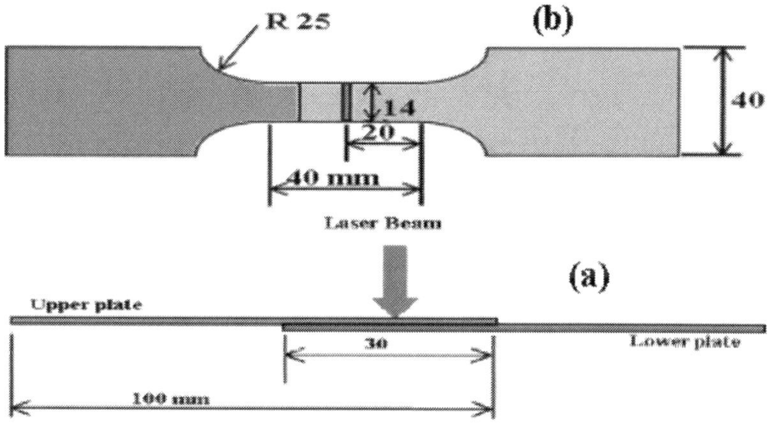

Fig. 1: Schematic diagram illustrating the (a) Center-line welding lap-join, (b) Tensile shear sample [12].

3. Development of a Mathematical Model for Tensile Strength

3.1. Analysis of the result

The raw data, the average tensile shear strength test results are shown in Table 4. To analyze the effects of the welding parameters in detail, ANOVA was conducted; these results are shown in Table 5. The ANOVA results are presented in Table 5; the F_v is used to test the significance of a factor by comparing model variance with residual (error) variance, which is calculated by dividing the model mean square by the residual mean square. As mentioned in the previous chapter the high F_v value for a parameter means that the effect of the parameter on the characteristics is large. The average tensile shear tests appear to be mainly affected by the laser power and focus position as shown in Table 5. The result in Table 5 shows that the highest F_v value in the process was obtained for laser power 'P' equal to 29.20. The F_v value for the focus position 'F' was equal to 1.0, which indicates that the 'F' has a relatively small effect on the process. The F_v value for the welding speed 'S' was not available in the ANOVA analysis, which indicates that the speed has an insignificant effect on the process. Adequate Precision compares the range of the predicted values at the design points to the average prediction error. For this model it was equal to 10.595, as shown in Table 5. Other adequacy measures R^2 and Adjusted R^2 are presented in the same table. All the adequacy measures indicate that an adequate model has been obtained. The final mathematical model for predicting the tensile strength of a dissimilar joint in terms of coded factors and actual factors as determined by Design Expert software are shown below in Eqs. 1 and 2.

Table 5: ANOVA for tensile strength response

Source	Sum of Squares	df	Mean Square	F_v Value	p-value Prob. > F	
Model	16250.2	3	5416.74	16.3743	0.0002	Significant
P	9660.44	1	9660.44	29.2026	0.0002	
F	331.202	1	331.202	1.00119	0.3368	
PF	6258.58	1	6258.58	18.9191	0.0009	
Residual	3969.69	12	330.807			
Cor. Total	20219.9	15				
R-Squared = 0.8037			Adeq. Precision = 15.079			
Adj. R-Squared = 0.7546						

Final Equation in Terms of Coded Factors:

Tensile Strength = 128.65 -32.97*P -6.10*F -35.60*P* F **...(1)**

Final Equation in Terms of Actual Factors:

Tensile Strength = 465.375 -304.741*P +343.792*F -316.445*P*F **...(2)**

3.2. Validation of the model

Fig. 2 shows the actual response versus the predicted response for tensile shear testing result. From this Fig., it can be seen that the model adequately describes the response within the limits of the factors being investigated herein, as the data points are close to the diagonal line. Furthermore, three extra confirmation experiments were carried out using different test conditions, which are presented in Table 5 along with the resulting percentage error. It can be noticed that the average percentage error is almost 11%. The obtained tensile shear stress after laser welding is greater than the base metals value certainly comparing to the aluminum side.

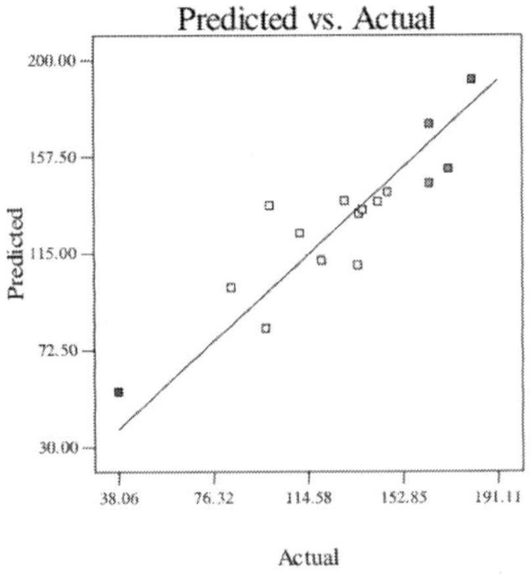

Fig. 2: Predicted Vs Actual for developed tensile shear strength model.

3.3. Effect of process parameters on the response:

1) Laser Power: It is evident from the results that the laser power is the most significant factor associated with the response, as shown in a contour graph in Fig. 3. The highest tensile strength value was 181 MPa, observed to be at a laser power of 0.9 kW as presented in Table 7.12.

2) Focus point position: The results indicate that the focus point position has also has a significant effect on the tensile strength of the laser-welded joint, as shown in Table 4. The model developed indicates that there is an interaction between the two welding parameters (the welding speed and the focus position) as exhibited in Fig. 4.

3) Welding speed: From the ANOVA analysis it can be seen that the welding speed has no obvious effect on the response within the parameter range domain applied. By changing the welding speed the response will not be affected. The relationship between the welding speed and laser power is exhibited in 3D graph Fig. 5 at focus position F = 0.0 mm.

Table 5: Confirmation experiments of the tensile shear strength compared with model results.

Exp. No	P, kW	S, mm/min	F, mm	Tensile strength, MPa		$\lvert E\% \rvert$
				Actual	predicted	
1	1.00	1600	0.00	139	160	13.1
2	1.10	1800	1.00	157	134	17.2
3	0.90	2100	0.0	186	191	2.6

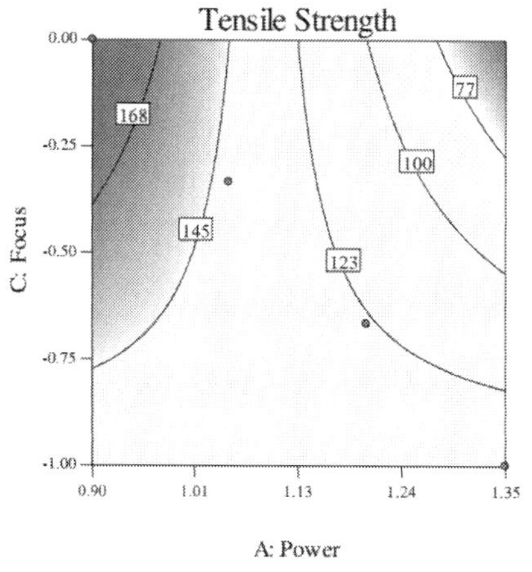

Fig. 3: Contour graph shows the effect Welding parameters At S = 2100 mm/min.

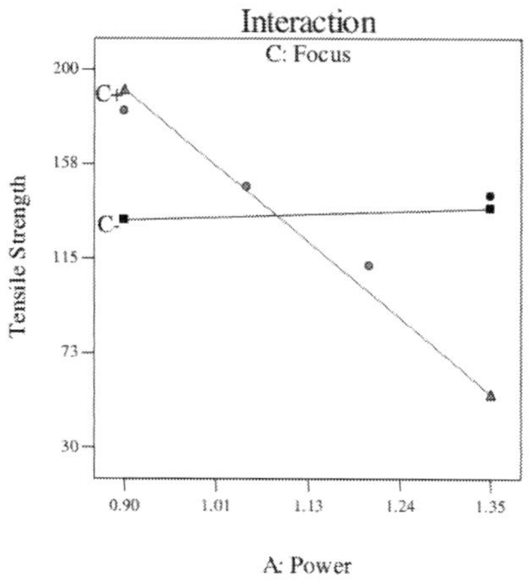

Fig 4: Interaction between the laser power and focus at S = 2100 mm/min.

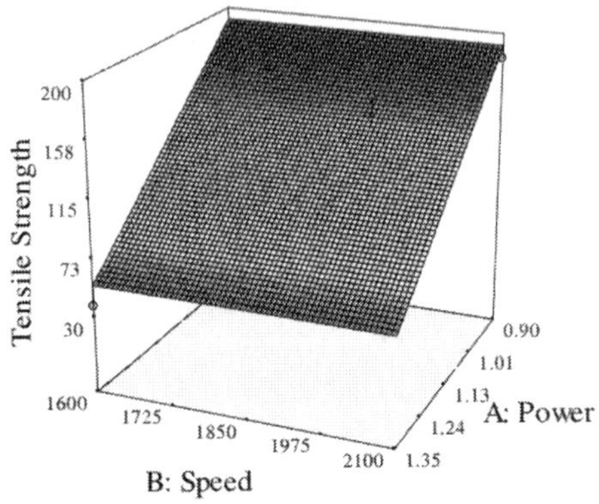

Fig. 5: 3D graph shows the effect of Power and Speed at F = 0.0 mm.

4. Microharness and Microstructure Studies

4.1. Microstructure of dissimilar jointed materials

The welded specimens were prepared for microhardness and microstructure studies by polishing successively in 120, 240, 600, 800 and 1200 SiC paper polishing, followed by a final disc polishing using 3 μm, 1 μm diamond suspension and finished with a SiO_2 suspension to a mirror-like surface aspect. The titanium side of the weldment was etched in Reagent consisting of (10 ml HF and 5 ml HNO_3 in 85 ml of water), and the rest of the regions of the weldment were etched with Keller's reagent (1% HF, 1.5% HCl, 2.5% HNO_3 and H_2O solution).

Different grain textures can be clearly observed in Figs. 6 (a-d) between the upper weld and the lower weld due to the diversity of heat transfer direction. When welding started but before penetration, heat transfer occurred along cross-direction and depth direction, consequently generated the columnar grains whose orientation was perpendicular to the boundary between the fusion zone and the HAZ in titanium plate as it exhibited in Fig. 6 (a). No obvious second phase was observed in the WZ at upper part of welding pool and just the solidification crystals were apparent. Optical microscopy micrograph shows that the HAZ in the vicinity of molten boundary of titanium consists of mainly ά martensite exhibited in Fig. 6 (a). The circular ά in the HAZ was attributed to rapid cooling of the weld metal. The base metal of aluminum 5251 H22 is exhibited in Fig. 6(b). The laser titanium–aluminium dissimilar joining resulted in a complex and heterogeneous microstructures composed of columnar grains and "white solute bands" in the base of welding pool. The welding zones in aluminium are partially melted zones due to the presence of low fusion point elements (magnesium) at the grain boundaries. The Al5251 WZ, HAZ and BM are exhibited in Fig. 6(c, d). It is well established that the microstructure of the joints affect the tensile strength critically. The influence of the heat inputs on the microstructure can be observed on the grain size variation with different heat inputs in Figs. 6(a-d). The higher the heat inputs and greater dwelling time at liquid temperature could accelerate the growing of the grain size and deteriorate the tensile strength of the weldment.

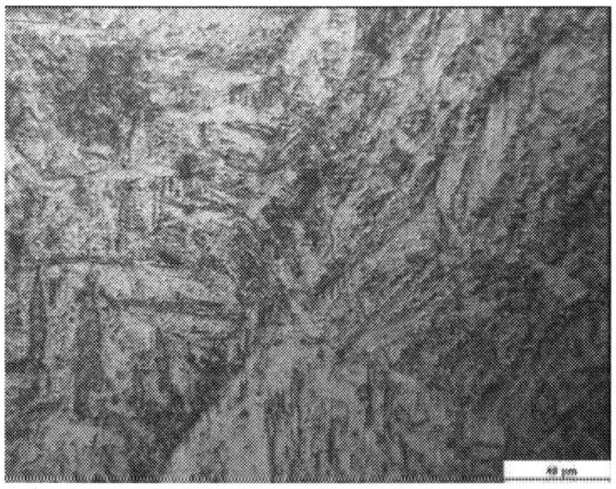
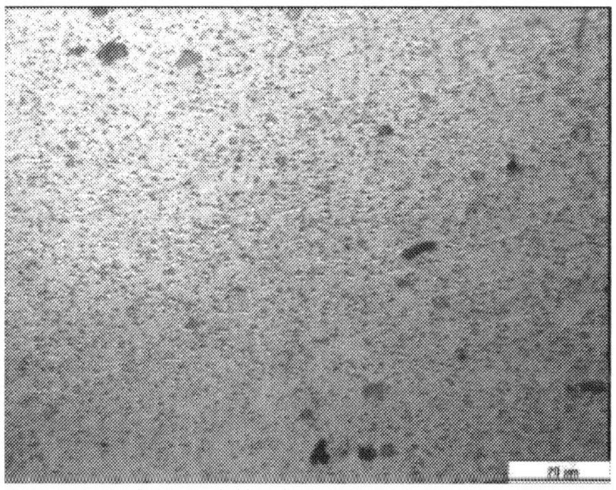

Fig. 6 (a) TiG2 WZ and HAZ Fig. 6(b) Al5251 BM

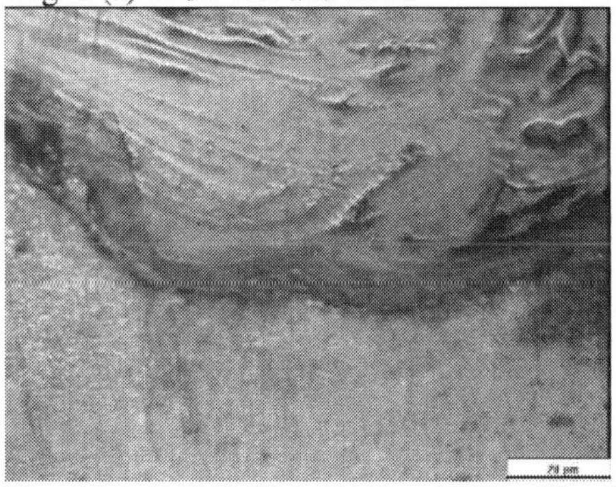
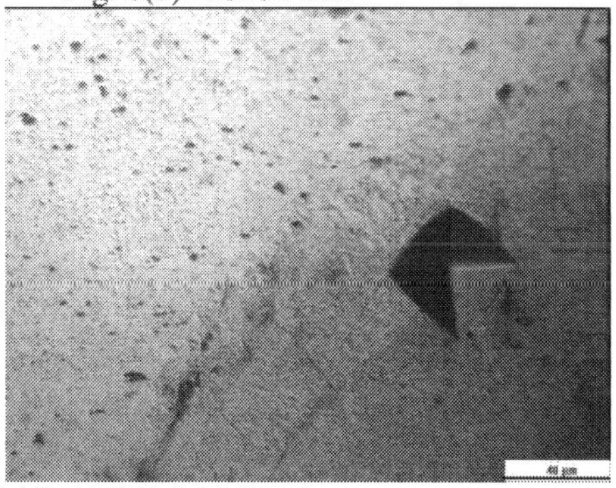

Fig. 6 (c) Al5251 BM, HAZ and WZ Fig. 6 (d) Al5251 BM, HAZ

Figs. 6 (a, b, c and d): The microstructure of lap weld Al5251 to Ti G2 joint.

4.2. Microhardness of dissimilar jointed materials

The specimens selected for microhardness studies were based on heat input calculation. Vickers microhardness measurements with a 50 g loading force test were applied to the selected specimens shown in Fig 7. For each specimen three different positions were subjected to the study (BM, HAZ, and WZ) for each plate of the dissimilar joint as presented in Table 6 and exhibited in microhardness profile of the dissimilar joint Fig. 7. The data on the liquid titanium to liquid aluminium interaction during high solidification rate key-hole laser welding is very limited. In the HAZ of the titanium side, the microhardness increase (232 – 287 HV) and further increase in WZ up to (280- 349 HV), as presented in Table 6, this could be related to the to the quenching effect resulting in a martensitic ά microstructure.

Table 6: Microhardness test result of dissimilar nonferrous to nonferrous materials

Sp No.	**Al 5251 H22**			**Ti G2**		
	BM	**HAZ**	**WZ**	**WZ**	**HAZ**	**BM**
3	95.45	104.2	379.61	347.98	232.7	226.69
8	96.34	103.1	210.2	348.65	263.45	242.75
15	99.7	94.72	149.11	280.71	261.58	245.9
16	98.81	83.85	194.29	327.44	286.99	262.41

Compared to the hardening effect evident in the weld–aluminium interfaces by intermetallic compound generation, the titanium–weld interfaces were not expected to be the weakest point of the assemblies. The fracture was mostly occurring at the aluminum WZ and referring to the microstructure transformation during the welding process which is evident from the microhardness result measured and presented in Table 6 and Fig 7 microhardness profile. The fracture could be interpreted by losing the ductility of aluminium and due to the brittle components formed during the solidification stage in WZ at aluminium side.

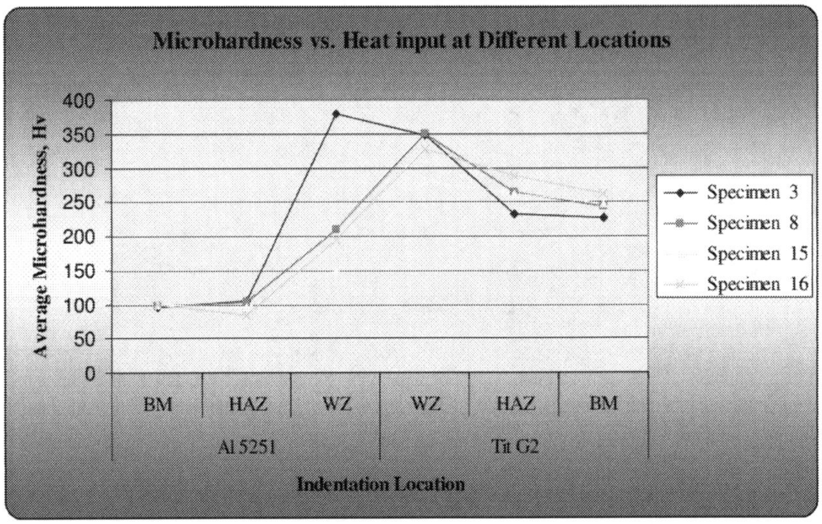

Fig. 7: The Microhardness profile of the dissimilar joint (Al 5251 / Ti G2).

5. Conclusion

- The dissimilar joint between aluminum alloys (Aluminum 5251H22 and titanium G2 alloys were successfully welded by CO_2 laser welding with a single pass and without filler material using the overlap joint design.
- A high-quality joint with high tensile strength was achieved by applying the Taguchi optimization technique. The obtained tensile shear strength values in the optimization step reached up to 186 N/mm when joining Al–Ti. The achieved tensile shear strength was almost same as the Al base metal values. The tensile shear tests indicate that fractures are mostly occurs at the aluminum WZ or HAZ which could be referring to the microstructure transformation accorded during the welding process that was evident from the microhardness result presented in Table 6 and microhardness profile shown in Fig 7.
- The model developed for the considered Al–Ti joints can adequately predict the response within the factors domain.

Acknowledgement

Garyounis University is gratefully acknowledged for the financial support of this research. Technical support from Mr. Martin Johnson, Mr. Michael May and Dublin City University are also gratefully acknowledged.

References

[1]Majumdar B, Galun R, Weisheit A, Mordike BL. Formation of a crack-free joint between Ti alloy and Al alloy by using a high-power CO2 laser. J Mater Sci 1997;32:6 pp 191–200.
[2]Osokin AA. Technological characteristics of the fusion welding of aluminium alloys to titanium alloys. Weld Prod 1976;23: pp18–20.
[3] Saprygin VD. Pressure welding of aluminium–steel and titanium–aluminum transition pieces for low-temperature service. Weld Prod 1975; 22: pp29–31.
[4] Wilden J, Bergmann JP. Manufacturing of titanium/aluminium and titanium/ steel joints by means of diffusion welding. Weld Cut 2004;3:2 pp85–90.
[5]Kim YC, Fuji A. Factors dominating joint characteristics in Ti–Al friction welds. Sci Technol Weld Join. 2002;7:149–54.
[6] Fuji A, Ikeuchi K, Sato YS, Kokawa H. Interlayer growth at interfaces of Ti/Al- 1%Mn, Ti/Al-46%Mg and Ti/pure Al friction weld joints by post-weld heat treatment. Sci Technol Weld Join. 2004;9:507–12.
[7]Z. Li, S. L. Gobbi, Laser welding for lightweight structures Journal of Materials Processing Technology, Volume 70, Issues 1-3, October 1997, pp 137-144.
[8]E.M. Anawa, A.G. Olabi, Using Taguchi method to optimize welding pool of dissimilar laser-welded components, Optics & Laser Technology, Volume 40, Issue 2, March 2008, Pages 379-388.
[9]E.M. Anawa, A.G. Olabi, Optimization of tensile strength of ferritic/austenitic laser-welded components, Optics and Lasers in Engineering, Volume 46, Issue 8, August 2008, Pages 571-577.
[10] Lung Kwang Pan, Che Chung Wang, Ying Ching Hsiao and Kye Chyn Ho, Optimization of Nd:YAG laser welding onto magnesium alloy via Taguchi analysis, Optics & Laser Technology, Volume 37, Issue 1, February 2005, Pages 33-42.
[11]Hyoung-Keun Lee, Hyon-Soo Han, Kwang-Jae Son and Soon-Bog Hong, Optimization of Nd: YAG laser welding parameters for sealing small titanium tube ends, Materials Science and Eng: A, Volume 415, Issues 1-2, 15 January 2006, Pages 149-155.
[12]R.a Borrisutthekul, Y. Miyashita and Y. Mutoh, Dissimilar material laser welding between magnesium alloy AZ31B and aluminum alloy A5052-O, journal of Science and Technology of Advanced Materials, Volume 6, Issue 2, March 2005, pp. 199-204.

Journal of Physics: Conference Series **181** (2009) 012034 doi:10.1088/1742-6596/181/1/012034

Modeling the effect of multiple pulse treatment on deformation processing

G A Tskhondiya[1]

[1]Physics Department, Moscow Institute of Aviation Technologies, 121552, 3 Orshanskaya, Moscow, Russia

E-mail: gigaa@mail.ru

Abstract. The present study reports the application of multiple pulse treatment (MPT) in order to improve plasticity of metal materials. The mechanism of electropulse influence on metal properties is explained by using the permanent strain and the specific energy as the main parameters in the defining relationship. On the one hand, a high current alters the yield stress of alloys. On the other hand, an assessment of the improved plasticity in MPT case is due to computational solutions of rolling treatment. The results of the article seem to have promising applications in metal treatment.

1. Introduction

In the last years, a great effort has been focused on providing an innovative technique that leads to the improved plasticity and maintained strength of metal alloys. The present study represents the MPT technique causing such alterations. MPT is a process of applying a high current of certain duration to metal specimens being, separately or simultaneously, deformed. Since numerous experiments have been carried out, the electropulse influence is clear. For instance, in MPT case of Titanium alloys, the plasticity is three times greater than usual but there is an 8% decrease in strength.

The MPT-caused increase in plastic properties of metals is named the effect of electroplasticity [1]. Physical hypotheses describing the effect vary from the microscopic "electron wind" concept [2] to the macroscopic high local temperature gradient concept [1]; a high local gradient of temperature field at microstructure-level causes mosaic temperature structure at macro-level and leads to the increased plasticity. That is, the local recrystallization, phase changes, the compressive residual stress due to local heating and defects healing should be the main characteristics causing the increase [1-3].

By means of investigating copper specimens, the present paper firstly reports the effect of MPT technique on the variation of the yield stress. The experiments contain the separate high current pulse and loading actions. Some experimental results are highlighted. Secondly, a basic explanation of the effect of electroplasticity is given. Both, the permanent strain and the specific energy (or the amplitude of current density), are assumed as the main parameters altering the defining relationship. Finally, a simple computational model is provided by use of engineering software DEFORM-2D V9.0. Changing adjusting parameters, the alterations of the plasticity in MPT case of rolling are estimated.

2. Experimental

Similar to some research reports, 99% copper alloys were examined in this paper. The twelve as-received polycrystalline samples were taken. Each had a diameter of 1.64mm and a length of 100mm. They were differently pretreated and then uniaxial tensile tests at room temperature were carried out.

A self-made electropulsing generator discharging positive direction multiple pulses was used to apply a high current to the specimens. The surface maximum temperature of the materials was directly detected by the Raytec® MX2 infrared thermoscope.

2.1. Electropulse treatment of 99% copper alloys

The samples were divided into four groups. Samples I, reference specimens, were stretched to 9% plastic deformation then completely unloaded and finally thermally treated (annealing for 5min, T = 673K). Samples II got MPT only. Samples III were stretched to 11.9% plastic deformation, completely unloaded and MP-treated, accordingly. Samples IV were rolled out to a width of 2.16mm (19.2% plastic deformation) then they got MPT. In these experiments, MPT always contained three short electropulses of 0.12s duration each one. The 50Hz mains frequency was maintained in all the cases (table 1).

After different pretreatments, all the samples were air-cooled and stretched with constant speed of 2mm/min to the beginning of necking.

Table 1. MPT testing conditions of 99% copper alloys.

Sample	Frequency (Hz)	Duration (s)	Pulse Number	j_m [a] (A/mm^2)	Temperature (K)
I	-	-	-	-	673
II	50	0.12	3	663	594
III	50	0.12	3	636	575
IV	50	0.12	3	648	582

[a] j_m means amplitude of current density in MPT case

The waveform of electropulsing was detected in situ by a TDS3012 digital storage oscilloscope and it was a damped oscillation wave (figure 1). Here, the waveforms of different electropulsing were similar, but the values of the amplitude of current density varied for different specimens.

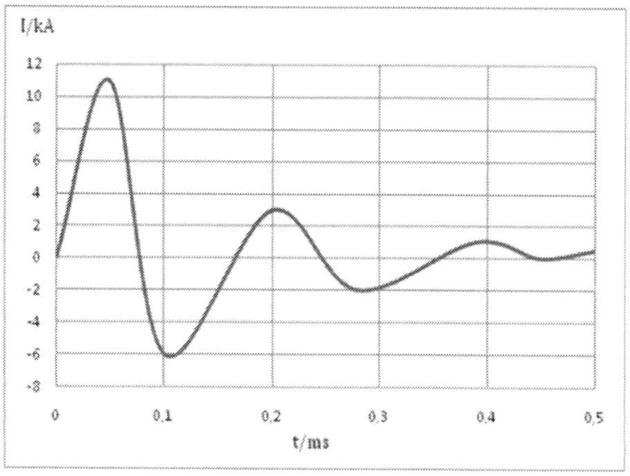

Figure 1. Typical waveform of electropulsing.

2.2. Results

Assuming the sectional area to be unchanged long enough, table 2 was obtained for a single specimen of each group with maximum signs of the effect.

Figure 2 reveals alterations of the yield stress of MP-treated air-cooled samples. Combining with table 1, it can be indicated:

- The yield stress is affected by both electropulsing and permanent strain (plastic strain gained before electropulsing).

Table 2. Yield stress of 99% copper alloys.

Sample №	Sectional area (mm^2)	Load (kg)	Yield stress (kg/ mm^2)
1(I)	2.1	65	30.8
2(II)	2.1	50	23.7
3(III)	1.9	30	15.9
4(IV)	1.8	16	9.0

- The yield stress of MP-treated air-cooled samples decreases. The more permanent strain there is the greater the decrease in the yield stress.
- The temperature during MPT is relatively high (above the recrystallization temperature but less that the melting temperature) similar to a thermal treatment case. However, the comparison of the three MP-treated samples and a purely heated one has shown advantages of MPT. The failure strain of sample №1 was 24% less, on average, than those of in MPT case.
- The decrease in the yield stress for sample №2 can be explained by its hardening features as well as by multiple defects.

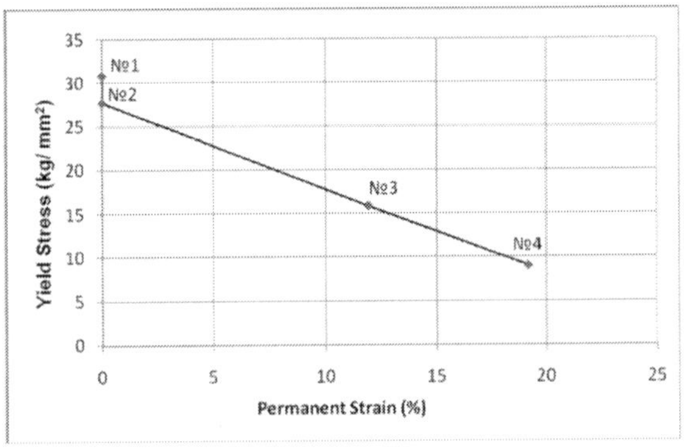

Figure 2. Alterations of the yield stress in MPT case.

3. Discussion

Numerous studies [1, 3-4] have shown that the MPT technique leads to the improved plasticity (the effect of electroplasticity) only since a pulse duration ~ 0.1s and amplitude of current density $\sim 10^9 A/m^2$. Besides, the permanent strain or multiple material defects are vital for the increase and there is no effect without either of these conditions. The basic physical explanation of the increase is the assumption concerning the effect of a high-density current on the occurrence of faults (microcracks and micropores) in the crystalline structure during the process of plastic deformation. Thus, this effect can be explained by the redistribution or concentration of the current and the accompanying temperature field, caused by differences between the physical parameters (conductivity) of the microdefects and those of the bulk of the material, as well as by the boundary geometry of the microdefects. Such a locally nonuniform distribution of physical fields can actually cause macroscopic increases in plasticity, which is observed experimentally [1].

Studies [1, 5-9] have shown that MPT, stimulating the mobility of dislocations, consists of thermal and athermal effects during the applied electropulsing. The thermal effect could be attributed to Joule heating produced by pulsed current. The explanation of the athermal effects claims that the improved plasticity results from mosaic temperature structure at macro-level caused by a high local gradient of temperature field at micro-level [1-3, 10]. In addition, according to [4], a remarkably increased activity of the dislocations implies the electropulsing since dislocation motions are involved in recrystallization.

At the end, a relatively small decrease in the tensile strength of MP-treated metal alloys is remarkable. For instance, in the above-mentioned case of Titanium alloys the decrease is only 8% while the majority of the examined metal materials [2-5] have shown 10% decrease, on average. Thus, for metal materials MPT provides an increase in plasticity as well as rough strength preservation.

3.1. Defining relationship in the case of MPT

Summarizing the experimental data of [1-3] and that of this paper, the defining relationship is obtained since some physical hypotheses are proposed. First, the principle of equidistance: the plastic part of the stress-strain curve moves down along the elastic part without change of shape at all (figure 3). Second, a kinematic parameter of permanent strain and a certain value of specific energy are the principal features of the effect of electroplasticity. Third, the data of tensile tests of the separate electropulsing and deformation cases is generalized in order to derive the defining relationship of the simultaneous electropulsing and deformation actions. In addition, the work of plastic deformation is used as a kinematic parameter as well as the permanent strain.

Since the separate uniaxial tensile tests are of the pattern of loading until the observation of plastic deformation - full unloading - electropulsing - repeated loading - etc, the defining relationship of one-dimensional case is derived as:

$$\bar{\sigma}^* = \bar{\sigma} - \psi(\varepsilon_p) \cdot \chi(q) \tag{1}$$

where $\bar{\sigma}^*$ - a flow stress parameter in MPT case, ε_p - permanent strain, q - specific energy, $\bar{\sigma}$ - a pure flow stress without MPT, $\psi(\varepsilon_p)$ and $\chi(q)$ - experimentally obtained functions. For many-dimensional case of the simultaneous electropulsing-deformation actions, since (1) is used as "a single-step equation" and $\Delta q \to 0$, the defining relationship is in the form of

$$\sigma_i^* = \sigma_i - \int_0^q \psi(\varepsilon_i^{(p)}) \cdot \chi'(q) dq \tag{2}$$

where σ_i^* - effective stress in MPT case, $\varepsilon_i^{(p)}$ - effective plastic strain, q - specific energy ($q \sim \tau \cdot j_m^2$), τ - an electropulse duration, σ_i - effective stress without MPT; the form of functions ψ, χ - might differ from those of above.

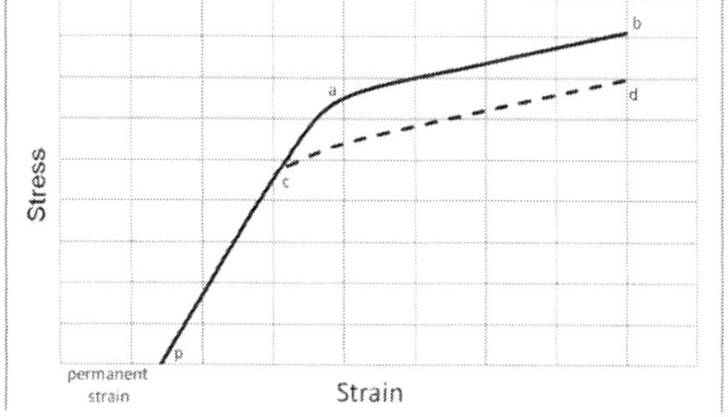

Figure 3. Change of the stress-strain curve in MPT case.

Figure 3: since a pulse applied at point 'p', repeated loading occurs along 'pcd' curve instead of typical 'pab' curve.

An alternative is of a similar origin of (2) but so that the defining relationship for many-dimensional case can be presented in terms of the effective strain

$$\varepsilon_i^* = \varepsilon_i + \int_0^q \frac{\partial f\left(\sigma_i, \psi(\varepsilon_i^{(p)}) \cdot \chi(q)\right)}{\partial(\psi\chi)} \cdot \psi(\varepsilon_i^{(p)}) \cdot \chi'(q) dq \tag{3}$$

It should be mentioned that the equations (2) and (3) are not necessarily equivalent so when it comes to a practical use only one of them is to be adopted as the defining relationship. The difference is mainly because (3) is supposed to depend not only on an effective strain parameter as in the above case but also on a current effective stress.

In addition, using the work of plastic deformation as a kinematic parameter [1], the defining relationship with the same meaning as (1) but in a more complicated form can be derived. In the case of a linear stress-strain curve ($\bar{\sigma} = \sigma_0 + m \cdot \bar{\varepsilon}$)

$$\bar{\sigma}^* = \sigma_0 \cdot \left(1 + \frac{2m}{\sigma_0^2} \cdot \kappa\right)^{1/2} \tag{4}$$

where $\bar{\sigma}^*$ - flow stress in MPT case, σ_0 - the yield stress, m - material constant, κ - a state parameter of structure calculated as

$$d\kappa = \varphi_1(W_p, q) dW_p + \varphi_2(W_p, q) dq \tag{5}$$

where W_p - the work of plastic deformation ($dW_p = \bar{\sigma} \cdot d\varepsilon_p$); the functions $\varphi_1(W_p, q)$, $\varphi_2(W_p, q)$ can be found directly from the experiments since $q = const$, $W_p = const$.

3.2. *Damage factor in the case of MPT*

Since normalized Cockroft-Latham fracture criterion (damage factor) is used to estimate the plasticity in general, it seems pertinent to use it in the case of MPT as well.

The usual damage factor increases while a material is deformed. The fracture occurs when the damage factor has reached its critical value. The critical value of the damage factor D_{cr} must be determined through physical experimentation. The damage factor D_f is defined by

$$D_f = \int_0^{\varepsilon_i} \frac{\sigma_1}{\sigma_i} d\varepsilon_i \tag{6}$$

where σ_1 - maximum principal stress, σ_i - effective stress, $d\varepsilon_i$ - effective strain increment. The plasticity R_p is calculated through

$$R_p = (1 - \frac{D_f}{D_{cr}}) \cdot 100\% \tag{7}$$

In the case of MPT some transformation of the damage factor concept is done. Namely, assuming that the MPT technique increases D_{cr}, the improved plasticity R_p is assessed by means of equations (2) or (3) since directly calculated via (6) and (7). The increment Δ of D_{cr} in the MPT case can be found as

$$\Delta = \begin{cases} D_{0cr} \cdot \ln\left(\dfrac{a \cdot D_{0cr}}{D_f^*}\right), & if \quad D_f^* > 0 \\ 0, \quad if \quad D_f^* = 0 \end{cases} \tag{8}$$

Journal of Physics: Conference Series **181** (2009) 012034 doi:10.1088/1742-6596/181/1/012034

where D_{0cr} - the critical value of the damage factor of pure materials; D_f^* - the MPT damage factor calculated by (6) since σ_i replaced with σ_i^* or ε_i with ε_i^*: if $\sigma_i = \sigma_1$, the usage of (2) is irrelevant but (3) is pertinent; a - experimentally obtained parameter, since the plasticity in MPT case three times greater than usual, $a = 3$.

At the end, it is remarkable that $\Delta = 0$ at once $D_f^* = 3 \cdot D_{0cr}$ and MPT cannot lead to an increase in plasticity any more. Thus, the application of the Cockroft-Latham fracture criterion in order to assess the increased plasticity in the case of MPT seems to be useful.

4. Computational

Nowadays, there is a rapid development of computer technologies used for analyzing different kinds of metal treatments. Among them is DEFORM, an engineering software that enables designers to analyze metal forming, heat treatment, machining and mechanical joining processes on the computer rather than the shop floor using trial and error. Since DEFORM has proven itself to be extremely effective in a wide range of research and industrial applications, the software is used here to model the MPT rolling and assess the plasticity computationally. Thereby, the simplest 2d-model has been created to analyze the effect of electroplasticity and provide an approach for establishing a suitable MPT technology of rolling.

This rolling model is a 2d plane strain case involving two working rolls (cathode), 50mm diameter and two contact rolls (anode), 30mm diameter. The voltage between cathode and anode is 50V. The test sample ("workpiece") is rigid-plastic and has a length of 100mm, a width of 6mm and electric resistance of 0.0018 Ohm-mm. The distance between the centers of the working and contact rolls is 60 mm (figure 4). The angular velocity of the working rolls of 15rad/s is relatively high so it imitates a pulse treatment since constant voltage. In other words, the valid cathode-anode contact, i.e. the direct current duration, of 0.3s is short enough to be considered a pulse treatment for steady flow of duration 0.12s.

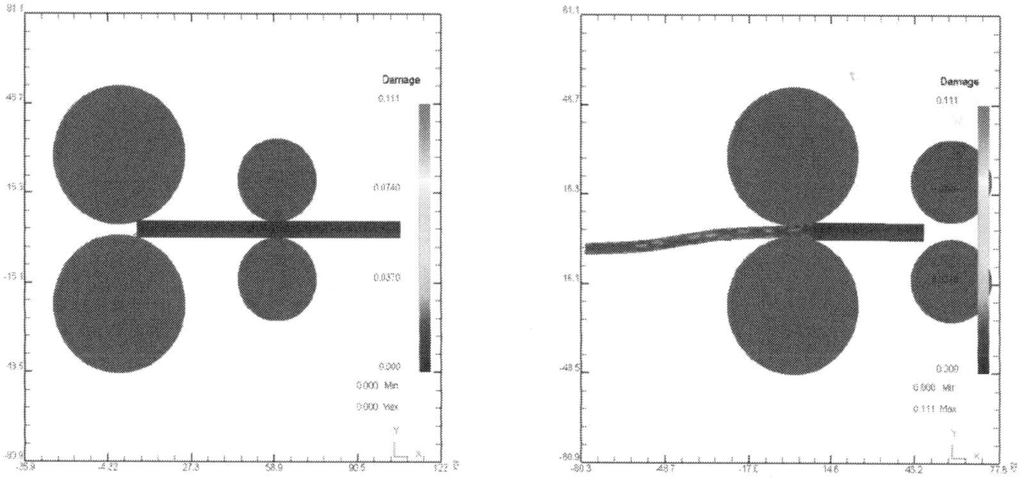

Figure 4. Image of MPT in rolling case.

Some nodes in the mesh have permanent strain before the rolling starts in order to imitate structural nonuniformity. Elastic deformations are neglected. Moreover, focusing on athermal meaning of the effect and taking into account that the duration of energy dissipation [1] in MPT case is considerably less than that of heat transfer, the thermal influence on plasticity is neglected; in a real process the influence of thermal strain can be neglected since intensive cooling of samples during MPT.

Using appropriate equations set [1], since (2) is the defining relationship, the programming algorithm can be presented as

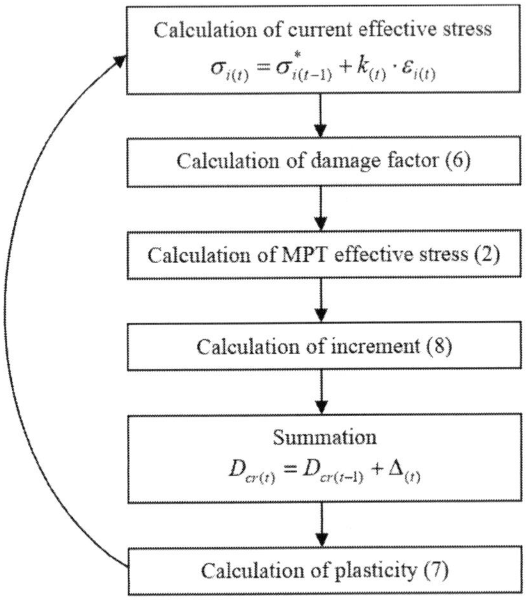

Figure 4. Principle scheme of numerical calculation.

Here, t=1, 2, 3,...- iteration steps, k - approximating parameter, $\sigma^*_{i(0)} = 0$. Since $\Delta_n = 0$, where n is the last iteration step, the calculation ends because of "the failure of material". The damage factor and rest parameters are specified at each element. In addition, such mechanical parameters as stress, strain, strain rate, etc can be easily obtained from solution data if necessary.

Assuming critical value of the damage factor $D_{0cr} = 0.2$, the single passing plasticity R_p (from 6 to 4mm) in the MPT case of rolling was 82.5%. In comparison, using a typical defining relationship $\bar{\sigma} = \bar{\sigma}(\bar{\varepsilon}; \bar{\varepsilon}'; T)$ for Joule (resistance) heating case of rolling, where T - is joule temperature, the single passing plasticity was 44.5%.

5. Conclusion

The effect of electroplasticity, resulting from multiple pulse treatment, is a process of applying a high current of certain duration to metal specimens that leads to an increase in plasticity.

From the experimental data of the MP-treated copper specimens, it should be clear that: the reduction in the yield stress is affected by both electropulses and permanent strain; even though the temperature during the tests is above that of recrystallization, the MPT technique has its advantages against a thermal treatment case in terms of a better-improved plasticity. The failure strain of sample №1(I) was 24% less, on average, than those of in MPT cases.

The effect of electroplasticity can be explained by the redistribution or concentration of the current and the accompanying temperature field, caused by differences between the physical parameters of the microdefects and those of the bulk of the material, as well as by the boundary geometry of the microdefects. The permanent strain or multiple material defects as well as a certain value of specific energy are vital for the increase and there is no effect without either of these conditions. Since some physical hypotheses explaining the effect are proposed, using permanent strain as a kinematic parameter, the defining relationship can be adopted in a form of (2). The increase in plasticity in the case of MPT should be calculated by combination of equations (6-8).

According to the data of numerical solution of a single passing rolling, the plasticity (7) of 82.5% in the case of MPT considerably exceeds that of 44.5% in a resistance heating case of rolling. Thus, MPT should have its advantages against not only thermal treatment but also a resistance heating case of rolling what is probably very useful, especially for hardly-deformed metal materials.

In the end, the MPT technique seems to have promising applications in particular case of rolling as well as in general case of metal treatment.

ACKNOWLEDGEMENTS The author is grateful to Prof. Beklemishev N N for useful discussions, Ph.D. at Institute for Problems in Mechanics of RAS Porohov V A for establishing experiments, Kharlamov A A for practical counseling through the software.

References

[1] Beklemishev N N, Vedenyapin E N and Shapiro G S 1984 Strain of conducting materials under the action of high-density electric currents *Strength of Materials* **16** 128-132
[2] Okazaki K, Kagawa M and Conrad H 1979 *Scripta Metall* **13**, 473
[3] Klushnikov V D and Ovchinnikov I V 1988 Ploskaiya zadacha o vozdeiystvii mgnovennogo tocheclmogo istoclmika tepla *MTT* **4** 118-122
[4] Xu Z, Tang G, Ding F, Tian S and Tian H 2007 The effect of multiple pulse treatment on the recrystallization behavior of Mg-3Al-lZn alloy strip. *Appl. Phys.* A **88** 429-433
[5] Kravchenko V Y 1966 *Fizika Tverdogo Tela* **8**, 927
[6] Okazaki K, Kagawa M and Conard H 1980 *Mater. Sci. Eng.* **45**, 109
[7] Sprecher A F, Mannan S L and Conrad H 1986 *Acta Metall.* **34**, 1145
[8] Zuev L V, Gromov V E and Gurevich L I 1990 *Phys. Stat. Solidi* A **121**, 437
[9] Gromov V E, Zuev L B, Bataronov I L and Gurevich L I 1991 *Fizika Tverdogo Tela* **33**, 3027

Journal of Physics: Conference Series **181** (2009) 012035 doi:10.1088/1742-6596/181/1/012035

Application of spectroscopic ellipsometry for investigations of compaction and decompaction state in Si-SiO_2 systems

W Rzodkiewicz and A Panas

Institute of Electron Technology, Al. Lotników 32/46, 02-668 Warsaw, Poland

E-mail: rzodki@ite.waw.pl

Abstract. In the modern microelectronics technologies, stresses and strains are known to affect process yield and microelectronic device reliability. Therefore, better understanding of generation mechanisms of elastic and non-elastic strains (compaction and decompaction) in the Si-SiO_2 system is needed. The main goal of the work was the elaboration of the analytical functional relationship between refractive index n and density ρ of SiO_2 layers on silicon substrates. Such ρ(n) relationship will allow determination of elastic and non-elastic strains in SiO_2 layers on silicon substrates. For the sake of the quality and shape of silicon substrates surface, before thermal oxidation, all substrate wafers were subjected to interferometric measurements by means of Fizeau interferometer. But, ellipsometric measurements by using Variable Angle Spectroscopic Ellipsometer of J.A. Woollam Company allowed determination of thicknesses and refractive indexes of silica layers. Measured SiO_2 masses and calculated volumes of the layers gave possibility to determine the degree of densification of the oxide layers on silicon substrates. The Hill approximation function curve turned out to be the best fitting curve for the experimental data. The obtained Hill curve indicates saturation for the density of the oxide equalled c.a. 4.53 g/cm^3. This value corresponds to the value near by the one of the crystalline polymorph of silica (stishovite). It seems to be physically established that degree of densification tends to the limiting value.

1. Introduction

Managing internal mechanical stress is an essential issue to provide very good performance and high reliability in advanced CMOS manufacturing. Strain can be generated in MOS devices at many technological steps. Due to scaling down of integrated circuits and rising up functionality in the modern electronics technology, good understanding of the compaction (connected with the elastic strains) and decompaction (connected with permanently irreversible strains) mechanisms in the Si-SiO_2 system is necessary. Additionally, when the structure dimension is going down, their mechanical and optical properties may differ from the corresponding values of the bulk medium. Thus, investigations of both mechanical and optical properties are very crucial.

In connection with this, the main goal of our work was to elaborate the functional relationship between the refractive index n, and the density ρ (as the degree of densification) of silicon dioxide layers on silicon substrates.

2. Experimental details

Czochralski-grown, n-type both <100> and <111> oriented silicon substrates have been applied in our investigations. We started with double-side polished 4-inch wafers. Subsequently, the wafer were subjected to thermal oxidation process at 1000°C in either dry oxygen (in order to grow silicon dioxide layers on <100> silicon substrates with the thickness of approximately 10, 15, 20, 25, 30, 35, 40, 45, 50nm) and water vapour (in order to grow oxide layers on <111> silicon substrates with the thickness of approximately 50, 100, 150, 200 nm). The thickness of SiO_2 layers and their refractive indexes were determined using a variable angle spectroscopic ellipsometer (VASE) with rotated analyser of J.A. Woollam Inc. Co.

Ellipsometric measurements were carried out at 5 points for each wafer in a wide spectral range (250-1000nm by 10nm step) at two angles of incidence (65°, 75°). But, before thermal oxidation, Fizeau interferometer was applied for the investigation of the quality and shape of silicon substrates surface.

2.1. Experiment methodology

The sequence of processing and measurement steps carried out during our research was as follows:

1. Interferometric measurements of silicon substrates.
2. Oxidation of 4-inch, n-type both <100> and <111> oriented Si wafers polished on both sides.
3. Measurements of the mass of oxidized wafers.
4. The ellipsometric measurements (to determine the thickness and refractive index of oxide).
5. Only the removal of back-side SiO_2 layer of each wafer by etching in HF solution.
6. Measurements of the mass of as-etched wafers.
7. The ellipsometric measurements (to determine the changes in refractive index of oxide after removal of back-side SiO_2 layer).
8. The removal of front-side SiO_2 layer of each wafer by etching in HF solution.
9. Measurements of the mass of as-etched wafers.

2.2. Optical model for ellipsometric data analysis

By spectroscopic ellipsometry we get spectral characteristics of Ψ (ratio of amplitudes after reflection for polarizations p and s) and Δ (phase change after reflection for polarizations p and s). To extract quantitative information about a sample, in most cases, it is necessary to construct an investigated structure model consisting of suitable number of layers of specified thicknesses. Then we can fit calculated and experimental Ψ and Δ data to compare generated and measured data. It is known that for the sake of simplicity and convenience, the Cauchy dispersive model is usually being used to fit ellipsometric data for dielectrics and semiconductors in spectral regions where they are transparent (i.e. where the extinction coefficient is zero). The Cauchy model allows determination of optical coefficients (complex refractive index) and thickness of analyzed layers of given structure. The relationship, which allows determination of the SiO_2 refractive index by Cauchy model, is given by the Cauchy dispersion formula [1]:

$$n(\lambda) = A + \frac{B}{\lambda^2} + \frac{C}{\lambda^4} + \ldots \quad (1)$$

where A, B, C are calculated fitting coefficients, and λ stands for wavelength in um.

The spectroscopic ellipsometry analysis for all the investigated spectral range was performed. Sellmeier model which is also being applied to fit experimental and generated data for dielectrics in spectral regions where they are not absorbing has not been considered. However, when the spectral range extends significantly beyond the visible into the mid-infrared, the refractive index of many transparent materials (such as among other things the silicon dioxide) trend downward. Then, the Sellmeier model is strongly recommended for the above mentioned spectral range (mid-infrared range) [2].

The optical model which was used for ellipsometric data analyses consisted of a silicon dioxide layer as a Cauchy layer and a silicon substrate (fixed data taken from [3]). In the applied model, the two parameters of Cauchy (A, B)(1) and SiO_2 layer thickness were fitted (the parameter C was fixed as 0). In opposition to what has been presented in papers [4-6], the interface layer between Silicon substrate and silicon dioxide was not considered. Because, the determined thickness value of the interface layer as a result of the fitting was 0. But, if the thickness was fixed, we obtained much worse value of MSE (Mean Squared Error as a measure of fitting) in comparison to the optical model ultimately assumed in this paper.

3. Results and discussion

For the sake of the quality and shape of silicon substrates surface, before thermal oxidation, all substrate wafers were undergone interferometric measurements by means of Fizeau interferometer (the light source was the laser diode operating at 635nm).
In manufacturing of flat optical elements such as focusing plates, compensators and mirrors one can obtain the very good flatness not going beyond 0.05 of visible wavelength (it can be usually 632.8 nm red line of He-Ne laser) at ratio of the element diameter to its thickness not less than 1:12. In the case of silicon substrates, the ratio is equal to the order of 1:200. Thus, we can observe deformations on the order of several tens of wavelengths on the substrates surfaces after their chemo-mechanical processing. These surfaces are characterised by not axial symmetry (it is clearly illustrated in Fig.1).

Front-side of the silicon wafer **Back-side of the silicon wafer**

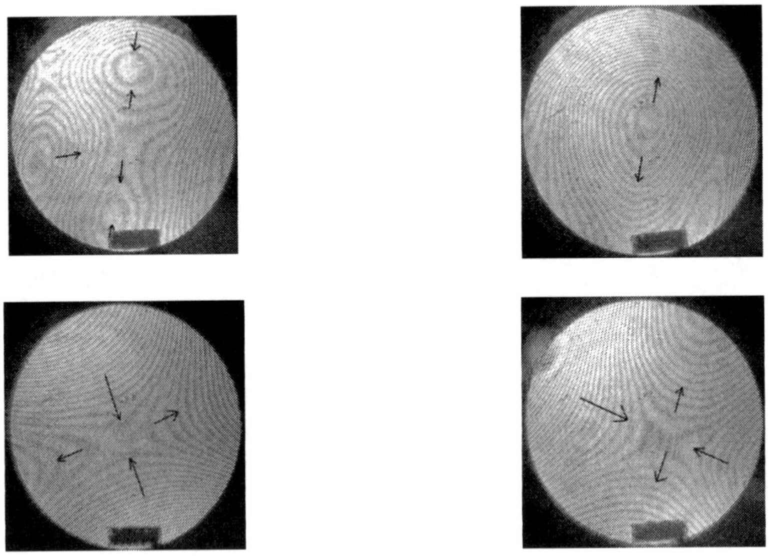

Fig. 1. Interferograms of 4-inch diameter double-side polished silicon substrates (both oriented <100> and <111>) before thermal oxidation process. In these pictures we can evidently see distribution asymmetry of normal deviation of reflected wave beam surface from substrate surface.

Calculations of curvature radiuses determined from the interferograms of silicon substrate surface before oxidation as well as stresses determined on the basis of these calculations by Stoney formula do not yield so such a big stress of substrates. Therefore, we assumed the optical coefficients of silicon substrate in the optical model are constant.
In connection with typical wavelength used to capture single wavelength ellipsometry measurements is 632.8nm (HeNe laser) and the light source of interferometer applied in this studies is 635 nm the changes in refractive index at 630nm were presented.
Analysis of the mass, the volume and the density errors have been performed by the complete differential method [7]. In this analysis, the significant changes in density of SiO_2 layer (of the order of 0.1-0.3 g/cm^3) at least 0.2mg error of the mass were detected. Thus, the accuracy of the oxide mass

measurement played the most important role at the determination of SiO_2 layer mass. On the basis of this analysis we affirmed that the balance with accuracy at least 0.01mg should be used. Therefore, in our investigations such balance was applied.
Once the mass of an SiO_2 layer is measured and its volume is known (calculated as the product of the oxide thickness and surface area of the layer), one may determine directly the density of the layer.
Next, the density of SiO_2 layers obtained in this way were compared with that determined from the value of the refractive index (established in the course of ellipsometric measurements). The Lorentz-Lorenz formula [8-10] (L-L) is one of the common used equation relating oxide density to its refractive index:

$$\Pi = \frac{1}{3} N_A \cdot \alpha = \frac{n^2 - 1}{n^2 + 2} \cdot \frac{M}{\rho} \tag{2}$$

where Π - molar polarization, N_A - Avogadro's number, α - mean polarizibility, M – molecular mass, ρ - density. Substituting the following values: M=60.08 g/mol (molecular mass of SiO_2), ρ=2.2 g/cm3 and n=1.46 (density and refractive index of the relaxed SiO_2, respectively), we obtain molar polarization for SiO_2 layer Π=7.4797 cm^3/mol and (1) becomes:

$$\rho = 8.0324 \cdot \frac{n^2 - 1}{n^2 + 2} \tag{3}$$

Other formulas relating oxide density and its refractive index are: Gladstone-Dale equation [10,11] (G-D):

$$\rho = 4.785 \cdot n - 4.784 \tag{4}$$

the empirical $\rho(n)$ relationship for SiO_2 layer derived by Taniguchi [12] (Ta):

$$\rho = 7.81 \cdot (n-1)^{1.63} \tag{5}$$

and Drude equation [10,13]:

$$\rho = 1.944 \cdot (n^2 - 1) \tag{6}$$

The last formula has usually been applied for study of the optical properties of various liquids (such as chlorobenzene, methanol, water, etc.).
Dependence of silicon dioxide density, ρ_{ox} on the oxide refractive index n for back-side oxide layer against a background of data given in the literature is shown in Fig. 2.

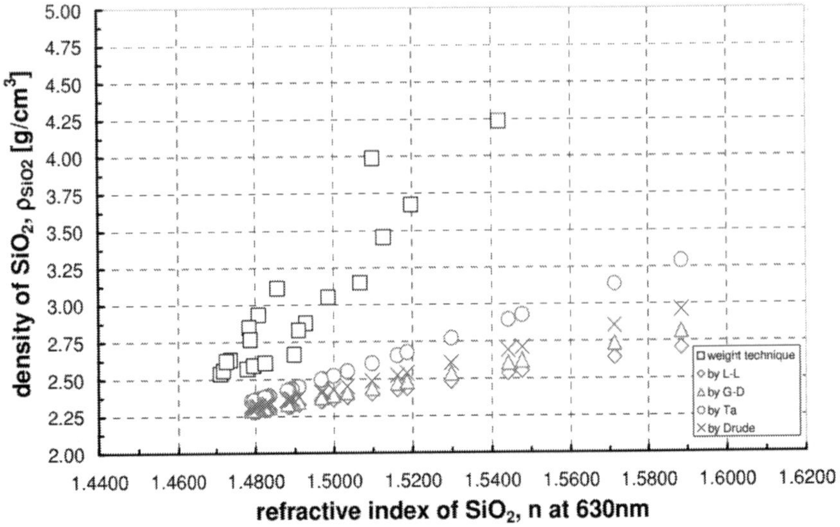

Fig. 2. Dependence of silicon dioxide density, ρ_{SiO_2} (experimentally determined by weight technique) on the oxide refractive index, n (determined by spectroscopic ellipsometry).

The very good evidence for the correctness of the obtained experimental data points is dependence of silicon dioxide density, ρ_{SiO_2} on the oxide thickness, t_{ox} (determined on the assumed optical model) illustrated in Fig. 3. Such regularity results from the observation of two trends of the presented curve. The first, linear dependence is detected for the oxide thicknesses grater than 50nm. But, as the oxides thinner than 50 nm are concerned (closer to the Si-SiO_2 interface), the power dependence $\rho(t_{ox})$ is pointed out.

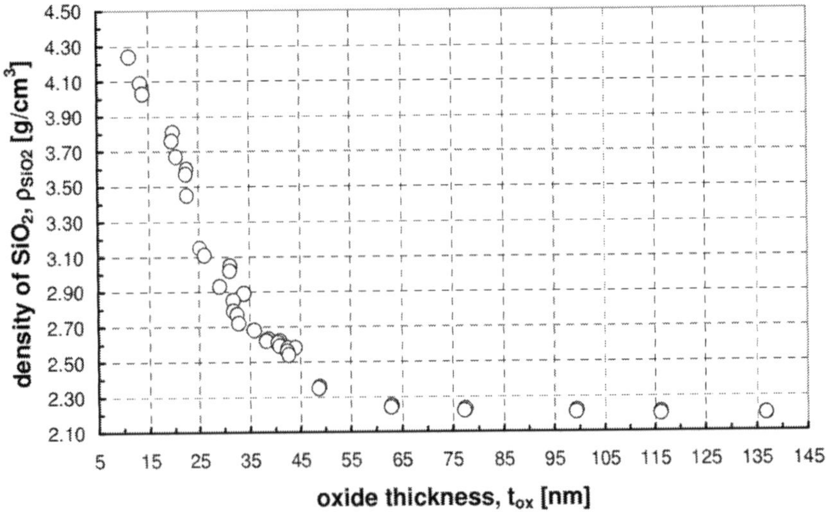

Fig. 3. Dependence of silicon dioxide density, ρ_{SiO_2} (experimentally determined by weight technique) on the oxide thickness, t_{ox} (determined by spectroscopic ellipsometry) on n <100> Si substrate.

On the basis of the density values obtained for the layers during all so far done weight experiments [14] curve fitting analysis was performed. It was necessary for finding of the corresponding curve giving proper relationship between the oxide density and its refractive index. The best curve of several investigated curves (inter alia the logarithmic function with three parameters and Langmuir approximation) turned out to be the Hill function (Fig. 4). The prove for this that correlation coefficient R^2 is greater than 0.9.

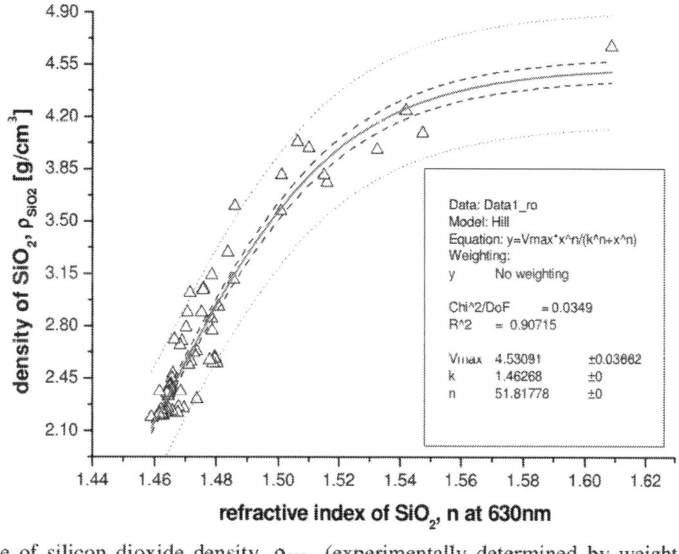

Fig. 4. Dependence of silicon dioxide density, ρ_{SiO_2} (experimentally determined by weight technique) on the oxide thickness, t_{ox} (determined by spectroscopic ellipsometry) on n <100> Si substrate for both front-side and back-side layers as well as the best fitting curve (the Hill approximation function) with 95% confidence and prediction bands.

Dashed and dotted lines in Fig. 4 are confidence and prediction bands (for the 95% confidence level), respectively.

The confidence level is a reliability measure of the fitted regression curve. The 95% confidence level expresses 95% probability of the result being correct. It means that the regression curve determined on the basis of the obtained experimental data will be located between the confidence bands. Whereas,. prediction level shows the measurement of some spread around the determined regression curve (in our case the Hill approximation function represents the regression curve). Therefore, 95% prediction bands determine area in which at least 95% measured experimental data there will be.

The ρ(n) relationship using the Hill approximation function is given by:

$$\rho = 4.531 \cdot \frac{n^{51.818}}{n^{51.818} + 1.4627^{51.818}} \qquad (7)$$

The above ρ(n) relationship indicates on the saturation for the density of the oxide at the value c.a. 4.53 g/cm^3. Such obtained value of density resembles the value of density characteristic for crystalline polymorph of silica (stishovite) given in the literature as c.a. 4.28 [15], 4.29 [16], 4.35 [17] g/cm^3. It seems to be physically explained that degree of densification tends to the limiting value.

4. Conclusions

In this work, we have studied the effect of the strain (degree of the densification) on the optical properties of Si-SiO_2 system by spectroscopic ellipsometry.

On the basis of the carried out results, the following conclusions have been drawn:

- measurement of oxide mass and calculated volume of the oxide layer enabled the densification degree of the oxide layers on silicon substrates to be determined;
- not for each wafer, dependence of the oxide density on its thickness has been obtained properly;
- for the oxide thickness less than 50nm, the significant divergence between the density calculated from the Taniguchi formula and that calculated from Lorentz-Lorenz, Gladstone-Dale, and Drude equations has been confirmed; Because, the three ρ(n) relationships (equations 3, 4 and 6) was derived for only liquids and gases. As Taniguchi formula was concerned [12], the oxide layers with thickness of about 100nm were only considered.
- the Hill function curve with three parameters turned out to be the best fitting curve corresponding to the experimental data determined by weight technique (Fig. 4.). Thus, the three-parameter logarithmic curve giving good results of fitting for previous weight experiments after adding of successive measurement points was getting worse;
- the Hill approximation function curve shows saturation for the density of the oxide to the value c.a. 4.53 g/cm^3. Definite maximum value of density seems to be well-founded physically. Because, the maximum value of stress in the $Si\text{-}SiO_2$ systems occurs at $Si\text{-}SiO_2$ interface. The thinner oxides (closer to the interface) are approaching to this limit.

References

[1] Software for VASE and M-44 ellipsometers – "Guide to using WVAS 32", JA Woollam Co. Inc.
[2] Tiwald T 2009 http://www.jawoollam.com/toolbox.html Newsletters Articles 10 Sellmeier Dispersion.
[3] Woollam JA Co Inc. WVASE 32 program v. 3.441, tabulated at UNL (Lincoln University, Nebrasca, USA).
[4] Aspnes DE, Theetn JB 1979 Phys. Rev. Lett. 43 1046.
[5] Herzinger CM, Johs B, McGahan WA, Woollam JA, Paulson W 1998 J. Appl. Phys. 83 3323.
[6] Zhao C, Lefebvre PR, Irene EA 1998 Thin Solid Films 313-314 286
[7] Thomas JM 1936 Mathematics 22 109.
[8] Banyai L, Gartner P 1984 Phys. Rev. B 29 728.
[9] Oughstun K E, Cartwright N A 2003 11 1541.
[10] Vedam K, Limsuwan P 1978 J. Chem. Phys. 69 4772.
[11] Pliskin W A 1977 J. Vac. Sci. Technol. 14 1064.
[12] Taniguchi K, Tanaka M, Hamaguchi C 1990 J. Appl. Phys. 67 2195.
[13] Drude P 1900 Lehrbuch der Optik. Lepzig.
[14] Rzodkiewicz W, Panas A 2007 Journal of Telecommunications and Information Technology 3/2007 44.
[15] Brazhkin V V, et. al. 2005 J. Phys.:Condens. Matter 17 1869.
[16] Gritsenko V A 2008 Physics-Uspekhi 51 699.
[17] Sheng-Nian L et. al. 2006 Phys.:Condens. Matter 18 659.

Vibration Analysis of Atomising Discs

Huaxia Deng and Huajiang Ouyang

Department of Engineering, University of Liverpool, Liverpool L69 3GH, UK

H.Ouyang@liverpool.ac.uk

Abstract. The centrifugal atomisation of metallic melts using a spinning disc is an important process for powder production and spray deposition. In the manufacturing process the high-temperature melt flows down to the surface of the atomising disc spinning at very high speed. It is observed that there is a hydraulic jump of the melt flow prior to atomisation. In this paper, the dynamic model of the atomising disc as a spinning Kirchhoff plate with this hydraulic jump is established. The flowing melt is modelled as moving mass and weight force in the radial direction. Using a Galerkin method, it is found that the vibration properties of the atomising disc vary with the disc clamping ratio. The amplitude of the vibration is largely raised when the clamping ratio is smaller than the critical jump radius ratio. It is also found that the disc vibration is non-stationary before becoming steady and the amplitude decreases with increasing disc speed.

1. Introduction

Centrifugal atomisation of metallic melts using a spinning disc is an efficient process for powder production and spray deposition. It was originally developed by the Pratt & Whitney Aircraft Group to produce Ni-based superalloy powders [1] and is widely used for manufacturing high quality powders and near-net-shape preforms of a range of advanced metallic materials [2-5]. In the manufacturing process, the metallic melt flows from a nozzle under gravity onto the centre of the atomising disc spinning at high speed and gives the disc an impact. It then spreads quickly to the disc edge due to the large centrifugal force and induces moving weight, friction, thermal stress and pressure to the disc. At the edge of the disc, the melt is thrown off the disc and disintegrated into a spray of droplets with vast heat transfer in subsequence, which solidify in room temperature to form spherical particles with a narrow size distribution. The interaction between the metallic melt and the atomising disc is very complex.

The flow behavior of melts on the surface of the spinning disc has attracted a lot of investigations [6-10]. Previous research has shown that there are three modes of flow behaviour [11, 12]. At low volumetric flow rate, the direct droplet formation is observed at the edge of the disc. With increasing volumetric flow rate, there is a transition from the direct droplet formation to ligament disintegration. Film disintegration emerges at the higher flow rate and the surface of the atomising disc is covered by a continuous film moving from the centre to the circumference of the disc. It is observed that there is a hydraulic jump phenomenon encountered in the moving film prior to the disintegration. Zhao and his coworkers simulated the height distribution of the continuous film which is axisymmetric based on Newtonian fluid theory [13]. The aforementioned studies focused on the flow behaviour, while investigations of the vibration of the atomising disc are rare in the open literatures although it will definitely influence the flow behaviour and subsequently affect the quality of the powder production.

The vibration of atomising discs was first studied by Ouyang [14, 15]. The atomising disc was modeled as a spinning Kirchhoff plate and the continuous film as a moving distributed mass. A numerical example was given by assuming the mode shape of the disc in the radial direction as polynomial functions. The hydraulic jump of the continuous film was not considered in that paper. In this paper, the vibration of spinning atomising disc subjected to the moving film is analysed. The dynamic model is established based on Kirchhoff plate theory and the metallic melt is treated as moving load considering the hydraulic jump. The mode shape of the disc in the radial direction is assumed as a combination of Bessel functions which is more accurate than polynomials and a Galerkin method is employed to solve this problem.

2. Dynamic Model

In centrifugal atomisation, the metallic melt descends to the center of atomising disc spinning at a high speed and is accelerated toward the edge of the disc prior to disintegration. The load on the atomising disc is complicated, involving the gravity of the moving film, fluid inertial force, friction, thermal stress and so on. To establish the dynamic model, several assumptions are made in this paper: (I) the distribution of the melt film is axisymmetric and does not vary in the circumferential direction; (II) the friction between the film and the disc can be ignored; (III) the temperature of the disc is constant and the thermal stress is negligible; (IV) the influence of the vibration of the disc on the film height is assumed to be negligible.

The equation of the motion of the disc subjected to an external distributed load p^* based on Kirchhoff plate [16] is

$$\rho h \frac{\partial^2 w^*}{\partial t^{*2}} + D\nabla^4 w^* - h\frac{\partial}{r^* \partial r^*}(r^* \sigma_r^* \frac{\partial w^*}{\partial r^*}) - h\frac{\sigma_\theta^*}{r^{*2}}\frac{\partial^2 w^*}{\partial \theta^2} = p^*(r,\theta,t) \quad r^* \in [a,b] \tag{1}$$

where the biharmonic differential operator when written in the cylindrical coordinate system is

$$\nabla^4 = \left(\frac{\partial^2}{\partial r^2} + \frac{1}{r}\frac{\partial}{\partial r} + \frac{1}{r^2}\frac{\partial^2}{\partial \theta^2}\right)^2 \tag{2}$$

where ρ, h, D, v, a, b are the mass density, the thickness, the flexural rigidity of the plate, Poisson's ratio, inner radius and outer radius. It must be noticed that this equation is established in the rotating frame fixed to the spinning disc at speed Ω. Dimensionless variables are defined by

$$\begin{gathered} r = r^*/b\ , w = w^*/h\ , t = t^*\sqrt{D/\rho h b^4}\ , \Omega = \Omega^*\sqrt{\rho h b^4/D}, \\ \sigma_\theta = \sigma_\theta^* h b^2/D, \sigma_r = \sigma_r^* h b^2/D, p = p^* b^4/Dh \end{gathered} \tag{3}$$

Under definition (3), Eq. (1) becomes

$$\frac{\partial^2 w}{\partial t^2} + \nabla^4 w - \frac{\partial}{r\partial r}(r\sigma_r \frac{\partial w}{\partial r}) - \frac{\sigma_\theta}{r^2}\frac{\partial^2 w}{\partial \theta^2} = p(r,\theta,t) \tag{4}$$

The in-plane stresses σ_r and σ_θ in Eq. (4) are [17]:

$$\sigma_r = d_1 + \frac{d_2}{r^2} + d_3 r^2, \sigma_\theta = d_1 - \frac{d_2}{r^2} + d_4 r^2 \tag{5}$$

where

$$\begin{gathered} d_1 = \frac{1+v}{8}\frac{(v-1)a_c^4 - (3+v)}{(v-1)a_c^2 - (1+v)}\Omega^2, d_2 = \frac{1-v}{8}\frac{(v+1)a_c^4 - (3+v)}{(v-1)a_c^2 - (1+v)^2}\Omega^2 a_c^2, \\ d_3 = -\frac{3+v}{8}\Omega^2, d_4 = -\frac{1+3v}{8}\Omega^2 \end{gathered} \tag{6}$$

where $a_c = a/b$ is clamping ratio of the atomising disc.

In the region where the melt is present on the disc surface, the external distributed load has two sources: one is due to the fluid inertia force and the other due to the weight of the film

$$p^* = \begin{cases} -\rho_m^* h_m^*(r^*,t^*)(\dfrac{\partial^2 w^*}{\partial t^{*2}} + \mathrm{g}) & a < r^* \le r_m^*(t^*) \\ 0 & r_m^*(t^*) < r^* \le b \end{cases} \tag{7}$$

where ρ_m^*, h_m^* are the mass density and the height distribution of the film. If the film acts as a Newtonian fluid with a constant viscosity which is continuous and axisymmetric, its height distribution can be described as follows [13]

$$h_m^*(r^*,t^*) = \begin{cases} \dfrac{v_k}{n_k w_0 \sqrt{3(2k-1)}} \dfrac{r_0}{r^*} & a < r^* \le r_c^* \\ -\dfrac{1}{n_c}\sqrt{\dfrac{v_k}{\Omega^*}} \ln\left(1 - \sqrt{\dfrac{n_c Q}{\pi n_b \sqrt{v_k \Omega^*}}} \dfrac{1}{r^*}\right) & r_c^* < r^* \le b \end{cases} \tag{8}$$

where v_k and Q are kinematic viscosity and volume flow rate of the melt, constants n_c, n_k and n_b are 0.702, 0.587 and 0.739 respectively. w_0 is the initial velocity of the melt just prior to the impingement, r_0 is radius of the melt stream just before it impinges on the disc, r_c^* is critical jump radius.

The radius of the melt at time t^* is

$$r_m^*(r^*,t^*) = \begin{cases} \displaystyle\int_0^{t^*} u_m(r^*)\mathrm{d}t^* & a < r_m^*(t^*) < b \\ b & b \le r_m^*(t^*) \end{cases} \tag{9}$$

where $u_m(r^*)$ is the radial velocity of the moving film.

$$u_m^*(r^*) = \begin{cases} \dfrac{3 n_k w_0^2 r^*}{v_k} \mathrm{e}^{-\frac{n_k w_0}{v_k} z} & a < r^* \le r_c^* \\ n_b \Omega^* r^* \mathrm{e}^{-n_b\sqrt{\frac{\Omega^*}{v_k}} z}(1 - \mathrm{e}^{-n_b\sqrt{\frac{\Omega^*}{v_k}} z}) & r_c^* < r^* \le b \end{cases} \tag{10}$$

Under the definition of Eq. (3), the dimensionless form of Eq. (7) become

$$p = \begin{cases} -\rho_m h_m(r,t)\dfrac{\partial^2 w}{\partial t^2} - \rho_m^* h_m(r,t)\mathrm{g}\dfrac{b^4}{D} & a_c < r \le r_m(t) \\ 0 & r_m(t) < r \le 1 \end{cases} \tag{11}$$

where ρ_m, h_m and r_m are dimensionless parameters defined as $\dfrac{\rho_m^*}{\rho^*}$, $\dfrac{h_m^*}{h^*}$ and $\dfrac{r_m^*}{b}$.

3. Solutions

The solution of Eq. (1) can be written as

$$w(r,\theta,t) = \sum_{j=-\infty}^{\infty} W_j(r)\mathrm{e}^{\mathrm{i}j\theta} q_j(t) \tag{12}$$

where $\mathrm{i}=\sqrt{-1}$. The modes of the free disc are:

$$W_j(r)=\mathbf{a}^{\mathrm{T}}\boldsymbol{\psi}(r) \tag{13}$$

where

$$\mathbf{a}=\{a_1 \quad a_2 \quad a_3 \quad . \ . \ .\}^{\mathrm{T}}, \qquad \boldsymbol{\psi}(r)=\{\Psi_1(r) \quad \Psi_2(r) \quad \Psi_3(r) \quad . \ . \ .\}^{\mathrm{T}} \tag{14}$$

In Eq. (14) $\Psi_i(r)=R_{mn}(r)$ and $R_{mn}(r)$ is the modes of the static disc, which can be represented by

$$R_{mn}(r)=[\alpha, \ \beta, \ \gamma, \ \delta][J_n(\kappa r), \ Y_n(\kappa r), \ I_n(\kappa r), \ K_n(\kappa r)]^{\mathrm{T}} \tag{15}$$

where $J_n(\kappa r)$ and $Y_n(\kappa r)$ are Bessel functions of the first and second kind of order n, and $I_n(\kappa r)$ and $K_n(\kappa r)$ are modified Bessel functions of the first and second kind of order n. The value of m represents the number of nodal circles while, n represents the number of nodal diameters. The ratios of the constants α, β, γ and δ and the natural frequency parameter κ are determined by the boundary conditions of the disc [18]. The natural frequency parameter κ is given by $\kappa^4=\lambda^2$ where

$$\lambda=\omega\sqrt{\frac{\rho h b^4}{D}} \tag{16}$$

in which ω is the real natural frequencies of the static disc.

By using a Galerkin method, Eq. (4) becomes

$$\begin{aligned}&\sum_{j=-\infty}^{\infty}\int[W_j\mathrm{e}^{\mathrm{i}j\theta}\ddot{q}_j+\nabla^4W_j\mathrm{e}^{\mathrm{i}j\theta}q_j+\mathrm{L}(W_j\mathrm{e}^{\mathrm{i}j\theta})q_j]\delta(W_k\mathrm{e}^{\mathrm{i}k\theta})r\,\mathrm{d}\,r\mathrm{d}\theta\\&=\int p(r,\theta,t)\,\delta(W_k\mathrm{e}^{\mathrm{i}k\theta})r\,\mathrm{d}\,r\mathrm{d}\theta\end{aligned} \tag{17}$$

where

$$\mathrm{L}(w,r,\theta)=-\frac{\partial}{r\partial r}(r\sigma_r\frac{\partial w}{\partial r})-h\frac{\sigma_\theta}{r^2}\frac{\partial^2 w}{\partial\theta^2} \tag{18}$$

Substituting Eq. (13) into Eq. (17) yields

$$\begin{aligned}&\sum_{j=0}^{\infty}\int_{a_c}^{1}[W_j\ddot{q}_j+\nabla^4W_jq_j+\mathrm{L}(W_j)q_j]\delta(W_k)r\,\mathrm{d}\,r\int_0^{2\pi}\mathrm{e}^{\mathrm{i}(k+j)\theta})\mathrm{d}\theta\\&=\int_{ac}^{1}p(r,\theta,t)\,\delta(W_k)r\,\mathrm{d}\,r\int_0^{2\pi}\mathrm{e}^{\mathrm{i}k\theta}\,\mathrm{d}\theta\end{aligned} \tag{19}$$

Because of orthogonality of modes

$$\int_0^{2\pi}\mathrm{e}^{\mathrm{i}(k+j)\theta})\mathrm{d}\theta=\begin{cases}2\pi & k=-j,k=j=0\\0 & \text{else}\end{cases} \tag{20}$$

one gets

$$\sum_{j=-\infty}^{\infty}\mathbf{a}^{\mathrm{T}}\int_{a_c}^{1}[(\boldsymbol{\psi}\boldsymbol{\psi}^{\mathrm{T}})\ddot{q}_j+\omega^2\boldsymbol{\psi}\boldsymbol{\psi}^{\mathrm{T}}q_j+\mathrm{L}(\boldsymbol{\psi})\boldsymbol{\psi}^{\mathrm{T}}q_j]r\,\mathrm{d}\,r=\int_{a_c}^{1}p(r,\theta,t)\boldsymbol{\psi}^{\mathrm{T}}r\,\mathrm{d}\,r \tag{21}$$

Substitution of Eq. (11) into the above equation leads to

$$\sum_{j=-\infty}^{\infty} \mathbf{a}^{\mathrm{T}} \int_{a_c}^{1} [(\mathbf{I} + \rho_m h_m(r,t))(\boldsymbol{\psi}\boldsymbol{\psi}^{\mathrm{T}})\ddot{q}_j + \omega^2 \boldsymbol{\psi}\boldsymbol{\psi}^{\mathrm{T}} q_j + \mathrm{L}(\boldsymbol{\psi})\boldsymbol{\psi}^{\mathrm{T}} q_j] r\,\mathrm{d}\,r$$
$$= \int_{a_c}^{1} -\rho_m^{\ *} h_m(r,t)\, g \frac{b^4}{D} \boldsymbol{\psi}^{\mathrm{T}} r\,\mathrm{d}\,r \tag{22}$$

Since Eq. (22) contains time-dependent parameters, a fourth order Runge-Kutta algorithm is developed to obtain numerical solutions. The initial vertical displacement and velocity of the atomising disc are assumed to be zero.

4. Results and Discussion

The parameters of the atomising disc under study are: $\rho_m = 4100/7850$, $h = 0.005\mathrm{m}$, b=0.2m, $v = 0.3$, $E = 2.1\mathrm{e}11\mathrm{Pa}$, $v_k = 1.263\times10^{-6}\mathrm{m}^2/\mathrm{s}$, $Q = 6.096\times10^{-5}\mathrm{m}^3/\mathrm{s}$.

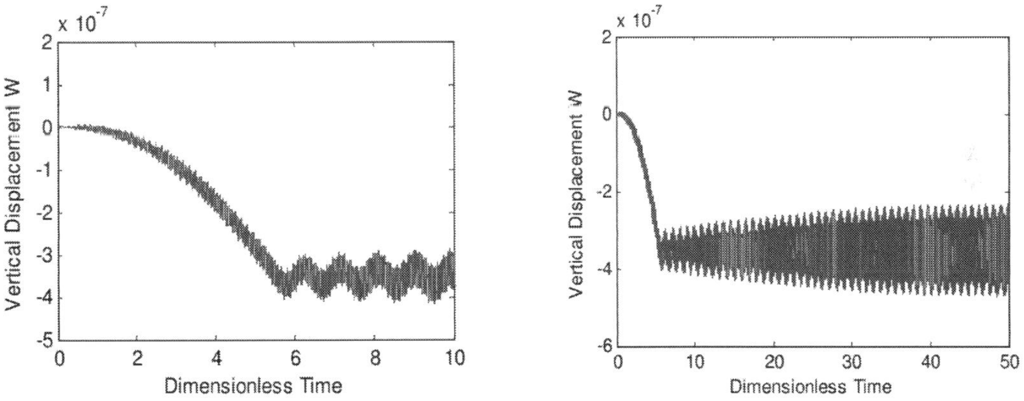

(a) (b)

Figure 1. The disc vibration at $a_c = 0.3$

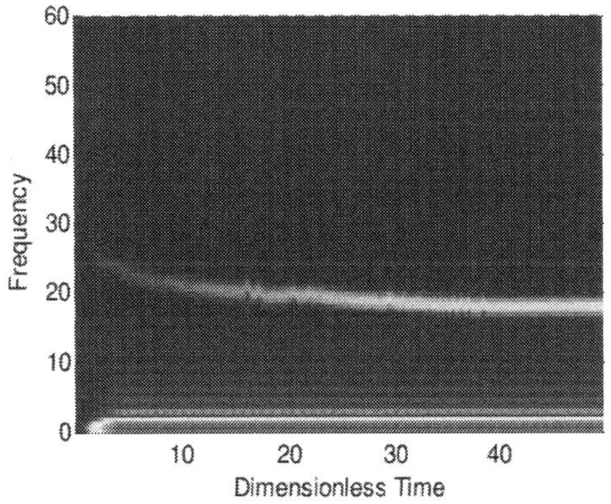

Figure 2. Time-dependent Fourier

As shown in Fig. 1, the vibration of the atomising disc consists of three stages: (I) The downward deflection with flutter-like vibration due to the initial descending weight of the melt; (II) The growing vibration due to the growing mass and weight as the melt film spreads towards the circumference of the disc; (III) The steady vibration after the melt flies off the edge of the disc. The time-dependent

Fourier transformation in Fig. 2 indicates that the frequency decreases initially due to the growing mass when the melt spreads out before reaching the edge of the disc. At the third stage, the melt flow becomes steady and the frequency then also becomes constant as expected. These results are consistent with those reported in [15]. The results shown in Fig.1 are obtained by setting $a_c = 0.3$ and $\Omega = 2$.

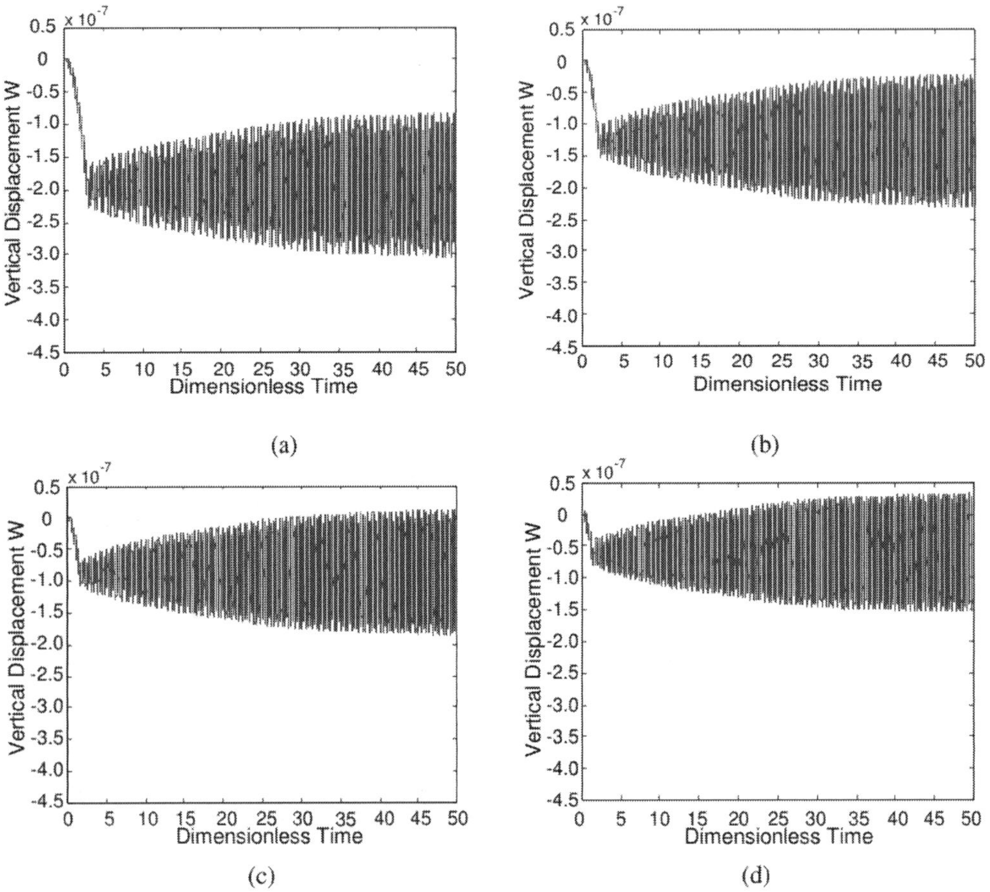

Figure 3. The disc vibration at: (a) Ω=4; (b) Ω=6; (c) Ω=8; (d) Ω=10

The spinning speed of the atomising disc is an important parameter which induces centrifugal force causing the melt film to spread and subsequently determines the size of produced particles. Figure 3 shows the influence of the spinning speed of the atomising disc on the vibration where a_c is equal to 0.3. As the spinning speed increases, the time length of the first vibration stage and the deflection of the disc due to the moving film weight decreases. The amplitude of the steady vibration is also reduced slightly with the increase of the spinning speed. These results, the evidence of the stiffening effect of centrifugal force, agree well with the findings of [14, 15].

Figure 4 shows the vibration of the atomising disc when the clamping ratio a_c changes to 0.1, while Ω remains 2 and other parameters stay the same. It clearly indicates that the vibration properties become quite different compared with Fig. 1. The downward deflection of the disc is very small. Thus, it is difficult to separate the first vibration stage from the second vibration stage. After the growing vibration stage, the second stage, there is a sharply increasing vibration stage prior to the steady stage which does not exist in Fig. 1. In the steady stage, the vibration amplitude is much larger than the

Journal of Physics: Conference Series **181** (2009) 012036 doi:10.1088/1742-6596/181/1/012036

previous one, nearly ten times greater, as shown in Fig. 4. The reasons for this situation are analysed as the following. When the clamping ratio equals 0.3, it is larger than the critical jump radius ratio which is 0.14 specifically. No hydraulic jump emerges in the area of the disc where the deflection and the vibration are induced by the moving film weight. As shown in Fig. 5, the film prior to hydraulic jump is much thinner than the film after hydraulic jump. Thus, the weight of the melt film spreading prior to hydraulic jump is very small. As a consequence, the downward deflection of the disc due to the initial descending weight of the melt is also very small. This leads to the result that the first vibration stage is difficult to tell. After this the melt film keeps spreading out and the vibration grows with increasing weight. Then hydraulic jump emerges. The height of the film suddenly rises up and makes the deflection of the disc sharply increase. After hydraulic jump, the film spreads out smoothly and flies off the atomising disc. The vibration of the atomising disc becomes steady subsequently.

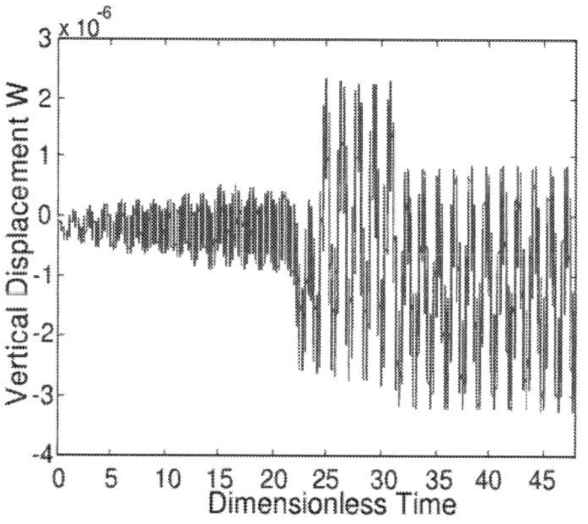

Figure 4. Disc vibration at $a_c = 0.1$

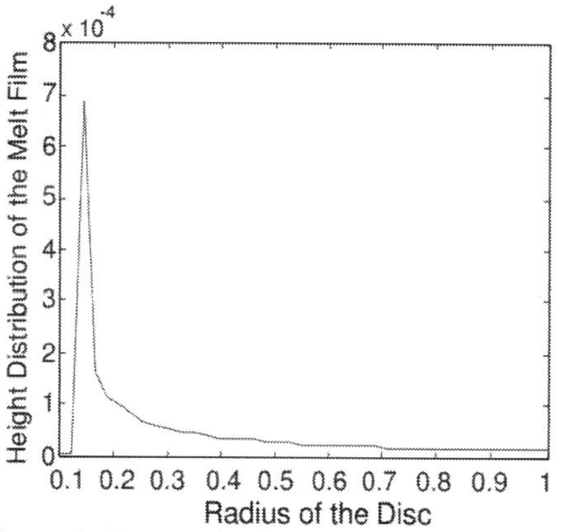

Figure 5. Height distribution of the film at $\Omega = 2$

5. Conclusion

The vibration of the atomising disc spinning at high speed is investigated in this paper. The atomising disc is modelled as a spinning annular Kirchhoff plate clamped at the inner radius and free at the outer radius. The melt flowing on the surface of the disc is treated as moving mass and growing weight. The hydraulic jump phenomenon of the melt film is considered in this paper. Using a Galerkin method to solve the equation of motion, it is found from the simulated example that the vibration properties of the atomising disc vary with the disc clamping ratio. If the clamping ratio is greater than the critical jump radius ratio, the disc motion consists of three vibration stages: a downward oscillatory deflection, growing vibration and finally steady vibration. While the clamping ratio is smaller than the critical jump radius ratio, the amplitude of the vibration is largely raised, nearly ten times greater than the previous one. Moreover, there is a sharply increasing vibration stage between the growing vibration stage and the steady vibration stage due to the hydraulic jump. It is also found that the disc vibration decreases with increasing disc speed.

Acknowledgement

The authors acknowledge the support of their colleagues in the dynamic group at the University of Liverpool. Financial support to the first author by the ORSAS award, China Scholarship Council (CSC) and Duncan Norman Fellowship is gratefully acknowledged.

References

[1] A. Lawley: *Atomization : the production of metal powders*. Princeton, N.J.: Metal Powder Industries Federation, 1992.

[2] S. Ozturk and F. Arslan: Production of rapidly solidified metal powders by water cooled rotating disc atomisation, *Powder Metallurgy*, vol. 44, pp. 171-176, 2001.

[3] R. Angers, R. Tremblay, and D. Dube: Formation of irregular particles during centrifugal atomization of AZ91 alloy, *Materials Letters*, vol. 33, pp. 13-18, 1997.

[4] C. Dogan and S. Saritas: Metal powder production by centrifugal atomization, *International Journal of Powder Metallurgy (Princeton, New Jersey)*, vol. 30, pp. 419-427, 1994.

[5] M. Zdujić and D. Usković: Production of atomized metal and alloy powders by the rotating electrode process, *Powder Metallurgy and Metal Ceramics*, vol. 29, pp. 673-683, 1990.

[6] L. M. Peurrung and D. B. Graves: Film thickness profiles over topography in spin coating, *Journal of the Electrochemical Society*, vol. 138, pp. 2115-2124, 1991.

[7] Y. Y. Zhao: Analysis of flow development in centrifugal atomization: Part I. Film thickness of a fully spreading melt, *Modelling and Simulation in Materials Science and Engineering*, vol. 12, pp. 959-971, 2004.

[8] Y. Y. Zhao: Analysis of flow development in centrifugal atomization: Part II. Disintegration of a non-fully spreading melt, *Modelling and Simulation in Materials Science and Engineering*, vol. 12, pp. 973-983, 2004.

[9] A. G. Emslie, F. T. Bonner, and L. G. Peck: Flow of a Viscous Liquid on a Rotating Disk, *Journal of Applied Physics*, vol. 29, pp. 858-862, 1958.

[10] T. G. Myers and J. P. F. Charpin: The effect of the Coriolis force on axisymmetric rotating thin film flows, *International Journal of Non-Linear Mechanics*, vol. 36, pp. 629-635, 2001.

[11] R. P. Fraser, N. Dombrowski, and J. H. Routley: The filming of liquids by spinning cups, *Chemical Engineering Science*, vol. 18, pp. 323-337, 1963.

[12] J. O. Hinze and H. Milborn: Atomization of Liquids by Means of a Rotating Cup, *Journal of Applied Mechanics*, pp. 145–153, 1950.

[13] Y. Y. Zhao, M. H. Jacobs, and A. L. Dowson: Liquid flow on a rotating disk prior to centrifugal atomization and spray deposition, *Metallurgical and Materials Transactions B: Process Metallurgy and Materials Processing Science*, vol. 29, pp. 1357-1369, 1998.

[14] H. Ouyang: Vibration of an atomising disc subjected to a growing distributed mass, *Journal of the Mechanics and Physics of Solids*, vol. 53, pp. 1000-1014, 2005.

[15] H. Ouyang: Stationary and non-stationary vibration of atomising discs, *Journal of Sound and Vibration*, vol. 308, pp. 699-708, 2007.

[16] H. R. Kim and A. A. Renshaw: Asymmetric, speed dependent tensioning of circular rotating disks, *Journal of Sound and Vibration*, vol. 218, pp. 65-80, 1998.

[17] K. Ono, J.-S. Chen, and D. B. Bogy: Stability Analysis for the Head-Disk Interface in a Flexible Disk Drive, *Journal of Applied Mechanics*, vol. 58, pp. 1005-1014, 1991.

[18] H. Lamb and R. V. Southwell: The Vibrations of a Spinning Disk, *Proceedings of the Royal Society of London. Series A, Containing Papers of a Mathematical and Physical Character*, vol. 99, pp. 272-280, 1921.

A methodology to predict residual life in LCF regime

C Delprete, A Vercelli[1]

Politecnico di Torino, corso Duca degli Abruzzi 24, 10129 Torino

andrea.vercelli@polito.it

Abstract. In this paper a methodology to calibrate a nonlinear constitutive model, implemented in a finite element commercial code and able to correctly predict the cyclic behaviour of material, is proposed. This model allows to take into account all the main phenomena that occur when a metallic material is subjected to cyclic loadings, in particular cyclic hardening or softening, Bauschinger effect, ratcheting and shakedown. High Cycle Fatigue and Low Cycle Fatigue tests under different strain levels have been performed on an aluminum alloy for high temperature applications, and the obtained experimental data have been used to determine the four constitutive parameters of the isotropic and kinematic hardening parts of the model. The comparison between the number of stabilized cycles predicted by numerical simulations and experimental data has been very satisfying. Using a self-made post-processing software, the Basquin-Manson-Coffin, Neu-Sehitoglu, Chaboche and Skelton damage models have been applied to determine the residual life of the specimen and to compare the results.

1. Introduction

Today the continuous improvement in performances demanded from machine components obliges engineers to adopt more and more accurate design methodologies in order to predict, as realistically as possible, the components response under working conditions. This necessity is fundamental especially when the stress state to which the material is subjected, is such as to cause the overcoming of yield stress and, consequently, the plastic strain is not negligible.

In this paper a nonlinear constitutive model implemented in a finite element (FE) commercial code is proposed. The model allows to predict the response of metallic materials under cyclic loadings in plastic field, the typical working conditions of Low Cycle Fatigue (LCF) regime [1, 2]. Using the data obtained from experimental High Cycle Fatigue (HCF) and LCF tests campaigns under different strain levels on a commercial aluminum alloy (2024-T351), the constitutive model has been calibrated and used to reply the fatigue tests on specimens. The results in terms of number of cycles to reach the stabilized conditions and the shape of stabilized cycles, compared with experimentation, have been very satisfactory.

The stress and the strain states calculated by FE simulations have been elaborated in order to predict the residual life of the specimen. By means of a self-made post-processing numerical code, four literature damage models have been applied and, as it has been done for the calibration of the constitutive model, the results have been compared with the actual number of cycles to failure registered on specimens during experimental tests. This activity allows to understand which between

[1] To whom any correspondence should be addressed.

Basquin-Manson-Coffin, Neu-Sehitoglu, Chaboche and Skelton model is the most appropriate and to evaluate the differences in terms of number of cycles predicted by them.

2. Constitutive material model

The constitutive model here presented predicts the nonlinear cyclic response of the material combining two parts. The first one, which takes into account the isotropic hardening, allows to predict the cyclic hardening or softening, while the second one, which considers the nonlinear kinematic hardening, allows to model the Bauschinger effect, the ratcheting and the shakedown [3-6].

The analytical formulation adopted to implement this model in the FE commercial code is derived from the combined hardening model of Chaboche [7], where the final relation related to the isotropic hardening is:

$$dR = b\left(Q - R\right)dp \, , \tag{1}$$

where R is a generic isotropic variable that describes the size of elastic domain, b and Q are material constants that represent respectively the speed of the stabilization and the asymptotic value of R corresponding to hysteresis loop in stabilized conditions, and dp is the increment of the accumulative plastic strain.

The nonlinear kinematic hardening formulation is represented instead by:

$$d\mathbf{X} = \frac{2}{3} C d\boldsymbol{\varepsilon}_{\mathrm{p}} - \gamma \mathbf{X} dp \, , \tag{2}$$

where $\mathbf{X}$ is the back stress tensor that permits to take into account the translation of the elastic field in the stress-strain plane, C and γ are material constants representing respectively the kinematic hardening modulus and the speed of the decrease of C values with the increase of plastic strain, and $\boldsymbol{\varepsilon}_{\mathrm{p}}$ is the plastic strain tensor. The derivative form of equations (1) and (2) is justified because they represent a time evolution of the respective variables.

The formulation used to implement the constitutive model in the FE commercial code derives from equations (1) and (2). In details, the isotropic part computes the evolution of the yield stress by:

$$\sigma_{\mathrm{Y}} = \sigma\big|_0 + Q_\infty \left(1 - e^{-b\varepsilon^{\mathrm{pl}}} \right). \tag{3}$$

Equation (3) is obtained integrating the equation (1) with respect to time [8]: σ_{Y} is the yield stress, $\sigma|_0$ is the value of yield stress resulting from the first application of the load, Q_∞ is equal to Q, b is the same of equation (1), and $\varepsilon^{\mathrm{pl}}$ is the plastic strain.

The nonlinear kinematic hardening part presents instead the following form:

$$\dot{\boldsymbol{\alpha}} = C\left(\sigma_{\mathrm{Y}}\right)^{-1}\left(\boldsymbol{\sigma} - \boldsymbol{\alpha}\right)\dot{\varepsilon}^{\mathrm{pl}} - \gamma \boldsymbol{\alpha} \dot{\varepsilon}^{\mathrm{pl}} . \tag{4}$$

Equation (4) is obtained elaborating the equation (2) and the analogies can be appreciated considering that $\boldsymbol{\alpha}$ is the back stress tensor named $\mathbf{X}$ in equation (2) and $\dot{\varepsilon}^{\mathrm{pl}} = d\varepsilon_{\mathrm{p}}$.

3. Experimental tests campaign

In order to characterize the fatigue behaviour of the 2024-T351 aluminum alloy, a complete experimental tests campaign has been carried out. A first sequence of tensile tests and HCF tests, both at five temperature levels, followed by LCF tests at two temperature levels and three imposed strain levels for each temperature have been performed. In this way, the Young Modulus, the yield stress, the elongation to fracture, the fatigue limit at 10^7 cycles, the parameters of Basquin curve and the parameters of LCF strain-endurance curve, according to the Manson-Coffin relation, have been determined, as it will be explained in paragraph 5.1. In the following, the y axis scale of graphs has been normalized because the data are protected by industrial confidentiality restrictions.

3.1. High Cycle Fatigue tests

The tests have been performed using the 100 kN Amsler Vibrofore of the Department of Mechanics of Politecnico di Torino and the staircase methodology has been followed. The cylindrical geometry of the specimens adopted has been imposed by the user manual of the testing machine and it is reported in figure 1, while in figure 2 the FE model of the specimen is represented.

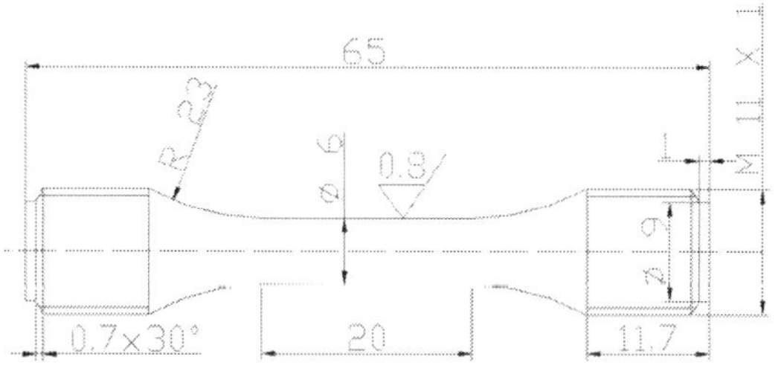

Figure 1. Geometry of the specimen adopted in High Cycle Fatigue tests (dimensions in mm).

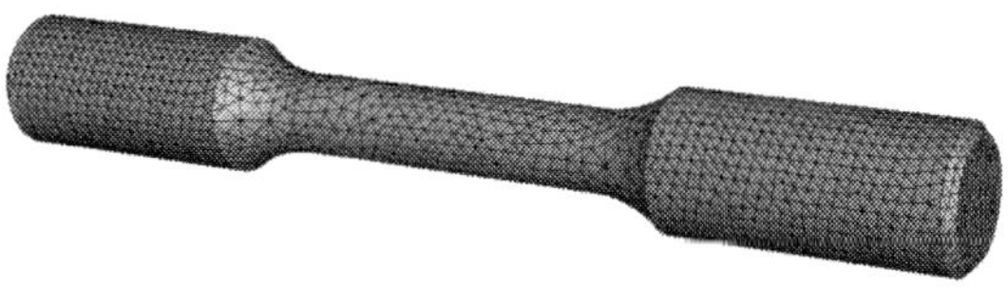

Figure 2. FE model of the specimen adopted in the numerical simulations.

For each temperature level a minimum of 15 meaningful specimens has been tested with amplitude ratio $R = 0.1$ to avoid problems in the phase of load reversal. The results have been then converted in $R = -1$ condition using the Goodman relation. The tests have been carried out with a frequency of about 160 Hz and they have been automatically interrupted when 10^7 cycles have been reached. Figure 3 reports an example of results: the line is the best fitting obtained using the Basquin curve reported in equation (5) where σ_a is the alternate stress, σ'_f and b are material constants, and N_f is the number of cycles. The best fitting has been obtained using the least squares method: on a log-log plot the equation (5) is a straight line, so it is easy to determine the parameters σ'_f and b. This law models the material behaviour in the HCF regime and its parameters are used in the Basquin-Manson-Coffin damage model as it is explained in paragraph 5.1.

$$\sigma_a = \sigma'_f \cdot (2 \cdot N_f)^b . \tag{5}$$

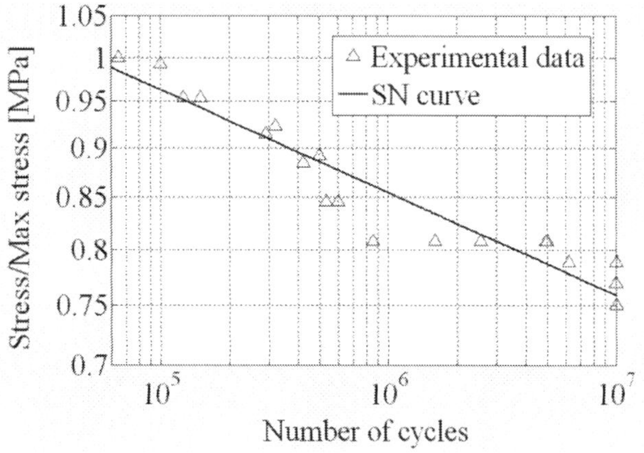

Figure 3. Example of HCF experimental results.

3.2. Low Cycle Fatigue tests

The tests have been performed using, for each temperature and imposed strain level, three specimens of the same geometry of HCF tests but without threads in the grip sections, with a narrow section of 30 mm in length and a total length of 95 mm. This characterization has been done in a laboratory outside the Politecnico di Torino using a 250 kN servohydraulic testing machine which works with a frequency of about 2 Hz. All tests have been conducted with an amplitude ratio $R = -1$ and the data have been acquired in terms of hysteresis loops and number of cycles to failure. An example is reported in figure 4, where are plotted the first ten cycles and then one cycle every ten. As in paragraph 3.1, figure 5 represents the results gathered in terms of plastic strain and number of cycles to failure testing two specimens for each of the three different strain ranges. Now the line is the best fitting obtained using the Manson-Coffin relation reported in equation (6), where ε_{pl} is the plastic strain, ε'_{f} and c are material constants, and N_{f} is the number of cycles. This law models the material behaviour in the field of LCF and its parameters are used by the Basquin-Manson-Coffin damage model, as it is explained in paragraph 5.1.

$$\varepsilon_{\text{pl}} = \varepsilon'_{\text{f}} \cdot \left(2 \cdot N_{\text{f}}\right)^{c}. \tag{6}$$

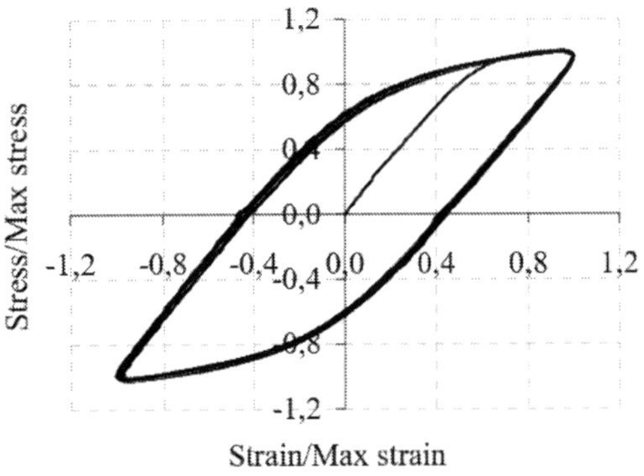

Figure 4. Example of experimental hysteresis loops.

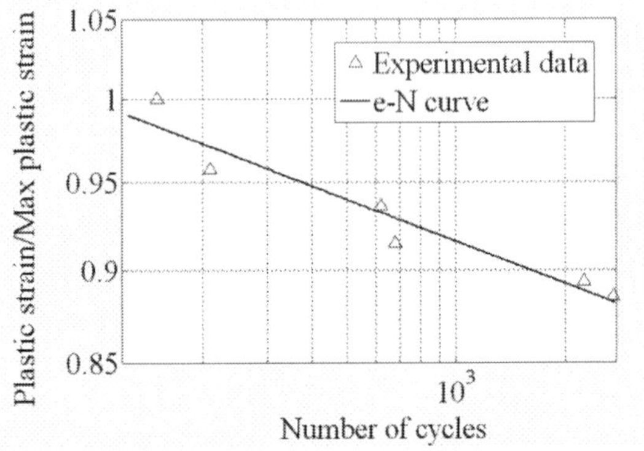

Figure 5. Experimental data and best fitting with the Manson-Coffin relation of equation (6).

4. Constitutive model calibration and comparison between numerical and experimental results

4.1. Model calibration

The nonlinear constitutive model has been calibrated using the LCF experimental data [9] shown, as an example, in figure 4. For each temperature level the experimental stabilized cycle related to the widest strain range applied has been isolated: the couples stress-plastic strain extrapolated from the half stabilized cycle in tension have been used for the calibration of the isotropic part. From the first one to the stabilized one, the hysteresis loops have been used to calibrate the kinematic part of the model. For each cycle have been determined both the amplitude of the elastic field as half the difference between the yield stress in tension and compression, and an equivalent plastic strain range. All these data have been specified in the commercial FE code in tabular form.

4.2. Numerical simulations

The numerical simulations have been realized with the aim to reply the LCF test conditions on the specimen. The FE model used is reported in figure 2 and its geometry is explained in paragraph 3.2.

One of the grip sections has been constrained with a joint on all the nodes, while on the other grip section a symmetric cyclic displacement has been imposed, whose magnitude produces the same strain of the tests. The simulations cover the whole time spent by the material to reach the stabilized conditions, that is when the experimental hysteresis loops do not change anymore their shape and their position on the stress-strain plane. The simulations have been performed on a Linux pc with a Pentium IV processor with 2 GB of RAM, and they take about 45 minutes to be completed.

4.3. Comparison between numerical results and experimental data

In figure 6 a comparison between experimental data and numerical results for one of the imposed strain level investigated is reported. For a better clearance in the plot, only the stabilized cycles are presented, but the simulations cover the whole time in which the cycles come to stabilization and the transient behaviour of the material is correctly predicted. It is possible to appreciate the goodness of numerical results obtained: the two cycles are very close in terms of shape and dimensions, so the material response under cyclic plastic loadings is correctly predicted by the constitutive model.

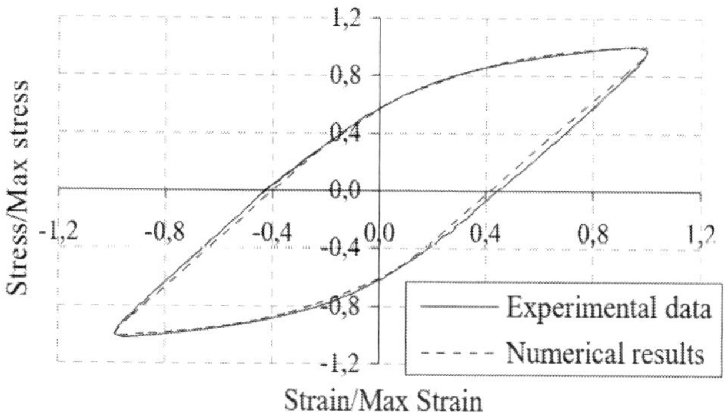

Figure 6. Comparison between the numerical stabilized cycle and the experimental one.

5. Residual life estimation

The results in terms of cyclic stress and strain, calculated through the FE simulation illustrated, have been elaborated to obtain the life prediction under LCF working conditions. In this study four literature damage models have been used: Basquin-Manson-Coffin, Neu-Sehitoglu, Chaboche and Skelton model.

For using these models in a fast and simply way, a numerical code has been developed: this software allows to choose the damage model to use and to import the stress-strain histories of the material, giving back the component life in tabular or graphic form [10]. The number of cycles predicted has been then compared with the actual life registered during the tests.

5.1. Basquin-Manson-Coffin damage model

This is a strain-based model [11] that links the number of cycles N_f with the total strain range $\Delta\varepsilon_{tot}$ which is the sum of an elastic component $\Delta\varepsilon_e$ described by the Basquin law presented in equation (5), and a plastic component $\Delta\varepsilon_p$ governed by the Manson-Coffin relation of equation (6). The whole model is:

$$\frac{\Delta\varepsilon_{tot}}{2} = \frac{\Delta\varepsilon_e}{2} + \frac{\Delta\varepsilon_p}{2} = \frac{\sigma'_f}{E} \cdot (2 \cdot N_f)^b + \varepsilon'_f \cdot (2 \cdot N_f)^c, \quad (7)$$

where E is the Young Modulus, σ'_f and b are material constants related to the elastic part of the total strain, ε'_f and c are material constants related to the plastic part of the total strain.

The determination of the four constants can be done, as in figures 3 and 5, interpolating separately the experimental data related to HCF and LCF tests. If the data are reported in log-log plots, the Basquin and the Manson-Coffin interpolating relations become linear, so it is easy to determine the σ'_f, b, ε'_f and c parameters. The four constants can be also determined analytically [11] from tests at only two fixed strain range levels, for which the measured quantities are stress range and life.

5.2. Neu-Sehitoglu damage model

This model predicts the total damage D_{tot} summing the damage caused by mechanical fatigue D_{mech}, the part caused by creep D_{creep} and the one caused by oxidation D_{ox} [12], as reported:

$$D_{\text{tot}} = D_{\text{mech}} + D_{\text{creep}} + D_{\text{ox}}. \tag{8}$$

The term D_{mech} is calculated by the Basquin-Manson-Coffin relation from which it is possible to determine the number of cycles N_{mech}. Assuming that the fracture happens when the total damage is equal to one, the damage D_{mech} can be calculated as:

$$\frac{\Delta\varepsilon_{\text{mech}}}{2} = \frac{\sigma'_{\text{f}}}{E} \cdot (2N_{\text{mech}})^b + \varepsilon'_{\text{f}} \cdot (2N_{\text{mech}})^c \quad \rightarrow \quad D_{\text{mech}} = (N_{\text{mech}})^{-1} \tag{9}$$

The four parameters σ'_{f}, b, ε'_{f} and c are the same determined for Basquin-Manson-Coffin model presented in paragraph 5.1. The creep damage D_{creep} can be calculated directly:

$$D_{\text{creep}} = \Phi_{\text{creep}} \cdot \int_0^{t_c} A \cdot e^{(-\Delta H/RT)} \cdot \left(\frac{\alpha_1 \bar{\sigma} + \alpha_2 \sigma_{\text{H}}}{K} \right)^m dt \quad \rightarrow \quad D_{\text{creep}} = 1/N_{\text{creep}} \tag{10}$$

where t_c is the whole working time, $\bar{\sigma}$ the equivalent Von Mises stress, σ_{H} the hydrostatic stress, K the maximum stress in tension that acts on the direction of maximum stress, α_1 and α_2 two parameters that take into account the amount of damage in tension or compression conditions, A and m material parameters depending on temperature, ΔH the activation energy, R the universal gas constant and T the working temperature. The material parameters have been found in literature [13, 14] for a similar aluminum alloy and calibrated for a better description of the material of this study. Φ_{creep} is a phase factor between mechanical and thermal strain and it is represented as:

$$\Phi_{\text{creep}} = \exp\left[-0.5\left(\left[\left(\dot{\varepsilon}_{\text{th}}/\dot{\varepsilon}_{\text{mecc}}\right)-1\right]\cdot\left(\xi_{\text{creep}}\right)^{-1}\right)^2\right]. \tag{11}$$

The phase factor Φ_{creep} is an exponential function depending on the ratio $\dot{\varepsilon}_{\text{th}}/\dot{\varepsilon}_{\text{mech}}$ between the thermal and the mechanical strain rate and on the factor of creep sensitivity ξ_{creep}. This function has a bell shape distribution which assumes different values in isothermal fatigue ($\Phi_{\text{creep}} = 4.3937 \cdot 10^{-2}$), in in-phase thermo-mechanical fatigue ($\Phi_{\text{creep}} = 1$) and in out of phase thermo-mechanical fatigue ($\Phi_{\text{creep}} = 3.7267 \cdot 10^{-6}$). The oxidation damage term has not been considered because the constitutive parameters required could not been determined with the available experimental data.

5.3. Chaboche damage model

This model calculates the total damage taking into account stresses [15, 16]. The general formulation is similar to Neu-Sehitoglu, but the contributions to the total damage dD_{tot} are only due to a mechanical damage dD_{mech} and to a creep damage dD_{creep}, without a term linked to oxidation. In equation (12) is presented the general form of the model:

$$dD_{\text{tot}} = dD_{\text{mech}} + dD_{\text{creep}}. \tag{12}$$

Equation (12) is written in an incremental form and this allows to avoid the superposition of effects, formally incorrect in nonlinear problems, and to consider the interactions between mechanical and creep damage, which can accumulate independently one from the other. The components dD_{mech} and dD_{creep} are modelled as:

$$dD_{\text{fat}} = \left[1-(1-D)^{-b+1}\right]^{\alpha} \cdot \left[\frac{\sigma_{\max} - \sigma_m}{A(1-D)}\right]^{-b} dN \qquad dD_{\text{creep}} = \left(\frac{\sigma}{M}\right)^r \cdot (1-D)^{-h} dt \tag{13}$$

where $\sigma_{\max}$ and σ_m are respectively the maximum and the mean stress, D the instantaneous damage, M, α, and b material coefficients, A, r and h parameters depending on temperature. All these six parameters have been deducted from literature [7] for a similar aluminum alloy.

5.4. Skelton damage model

This model calculates the total damage through an energetic approach [17, 18]. The energy density that the material dissipates during hysteresis cycles can be linked to the fatigue crack propagation. Skelton and al. [17] have demonstrated that the fatigue life of materials can be also related to the total dissipated energy density ΔU_{tot}, corresponding to the area circumscribed by hysteresis loops in stress-strain plane, as:

$$\Delta U_{\text{tot}} = \Delta U_{\text{mech}} + \Delta U_{\text{th}} = \oint \sigma \cdot d\varepsilon_{\text{mech}} + \oint \sigma \cdot d\varepsilon_{\text{th}} . \tag{15}$$

Moreover it has been verified that the amount of energy that a material can dissipate before fracture is almost independent from the loading conditions and it is influenced by temperature only. The sum of all the energies U_i calculated for each stabilized cycle [19] is therefore equal to a constant K_s related to the material, as:

$$\sum_{i=1}^{N_f} U_i = \text{constant} = K_s . \tag{16}$$

The value of K_s is computable summing the areas of all the stabilized hysteresis cycles obtained from an LCF test performed until fracture (figure 2). The number of cycles N_f can be then predicted by equation (17), where U_{stab} is the energy density of the generic stabilized cycle:

$$N_f = K_s / U_{\text{stab}} . \tag{17}$$

5.5. Analysis of results

In table 1 the results in terms of number of cycles before fracture are reported and compared with the actual life of specimens tested; with B-M-C it has been indicated the Basquin-Manson-Coffin model.

Table 1. Comparison between actual fatigue life and predictions in LCF regime (in terms of cycles).

Strain range	1st specimen	2nd specimen	Mean	B-M-C	Neu-Sehitoglu	Chaboche	Skelton
$\Delta\varepsilon_{max}/2$	210	290	250	278	269	265	263
$0.8\cdot(\Delta\varepsilon_{max}/2)$	2050	2340	2195	2420	2315	2296	2230
$0.6\cdot(\Delta\varepsilon_{max}/2)$	6890	7510	7200	7615	7480	7390	7110

The results show that the Basquin-Manson-Coffin model is the one that gives the greatest overestimation of the number of cycles. This fact can be justified because its formulation takes into account only the damage related to mechanical fatigue, omitting creep and oxidation. Moreover, the calibration of the four parameters could have been better if more LCF experimental data would have been available. The predictions are however close to the actual life of the first specimen. The Neu-Sehitoglu model is more severe than the Basquin-Manson-Coffin model because also the creep damage is counted for and the predictions are more reliable. Chaboche model is quite similar to Neu-Sehitoglu model in the results and, in general, it is more severe. Unfortunately, the determination of its constitutive parameters has not been rigorous because of the lack of appropriate experimental data; the parameters used have been taken from literature and they refer to an aluminum alloy similar to the one investigated in this work. Finally, the Skelton model permits to obtain the best predictions because it works directly on the actual hysteresis loops of the material.

6. Conclusions

In this paper a methodology to predict life in LCF regime, consisting in the calibration of a combined hardening model to simulate the material cyclic behaviour and in the estimation of the number of cycles to failure, has been proposed. The numerical results of FE simulations are satisfactory in terms of shape and dimensions of the stabilized hysteresis cycle: the areas of the experimental and the numerical cycles are almost the same. The life estimations show that, starting from the most overestimating one, the investigated models can be ordered as it follows: Basquin-Manson-Coffin, Neu-Sehitoglu, Chaboche and Skelton. This is justified considering that the level of detail in the analytical formulation increases following this sequence and all the main phenomena which contribute to the damage are modelled. Consequently, Skelton model seems to be the most appropriate to investigate LCF conditions: working directly on the actual hysteresis loops of the material, all the contributions to the damage are taken into account. The determination of material parameters (omitted as they are protected by industrial confidentiality restrictions), based on the available experimental data, has been rigorous for Basquin-Manson-Coffin, Neu-Sehitoglu and Skelton models, while for Chaboche they have been deducted from literature, so the confidence on the results obtained from this model are a little bit lower with respect to the others. The parameters estimation is the most critical point when damage models have to be used. When appropriate experimental data are available it is possible to perform a rigorous model calibration and the possible mismatch between experiments and predictions can be ascribe to the model formulation. Consequently, it is not obvious that a damage model calibrated through experimental data gives the exact number of cycles to failure.

References

[1] Voyiadjis G Z, Park T 1997 Kinematics of large elastoplastic damage deformation *ASME Mech. Conf.* Evanston IL USA

[2] Habraken A M, Charles J F, Cescotto S 1997 Calibration and validation of an anisotropic elastoplastic damage model for sheet metal forming *ASME Mech. Conf.* Evanston IL USA

[3] Voyiadjis G Z, Woody J W, Chaboche J L 1998 *Damage Mechanics in Engineering Materials* Elsevier Amsterdam NL

[4] Hubel H 1996 Basic conditions for material and structural ratcheting *Nuclear Engineering and Design* **162** 55-65

[5] Guillemer-Neel C, Bobet V, Clavel M 1999 Cyclic deformation behaviour and Bauschinger effect in ductile cast iron *Materials Science and Engineering* **A272** 431-442

[6] Yoshida F 2000 A constitutive model of cyclic plasticity *International Journal of Plasticity* **16** (12) 359-380

[7] Lemaitre J, Chaboche J L 2000 *Mechanics of solid materials* Cambridge University Press New York NY USA

[8] Chun B K, Jinn J T, Lee J K 2002 Modeling the Bauschinger effect for sheet metals, part I: theory *International Journal of Plasticity* **18** (5-6) 571-595

[9] Simple proportional and non-proportional tests *ABAQUS Example Problem Manual version 6.6*

[10] Delprete C, Rosso C, Sesana R, Vercelli A 2008 Sviluppo di un codice di calcolo per la stima della durata a fatica termomeccanica *XXXVII AIAS National Meeting* Roma Italy

[11] Manson S S 1966 *Thermal stress and low-cycle fatigue* Mc Graw-Hill, New York, USA

[12] Neu R, Sehitoglu H 1989 Thermo-Mechanical Fatigue, Oxidation and Creep: Part 2 - Life Prediction *Met. Trans. A* **20A** 1769-1783

[13] Neu R, Sehitoglu H 1989 Thermo-Mechanical Fatigue, Oxidation and Creep: Part 1 - Experiments *Met. Trans. A* **20A** 1755-1767

[14] Takahashi T, Nagayoshi T, Kumano M, Sasaki K 2002 Thermal Plastic-Elastic Creep Analysis of Engine Cylinder Head *SAE Tech. Paper* **2002-01-0585** SAE world congress Detroit USA

[15] Chaboche J L, Policella H, Savalle S 1978 Application of the continuous damage approach to the prediction of high temperature low cycle fatigue *ONERA TP* **1978-70**

[16] Lemaitre J, Chaboche J L 1974 A non-linear model of creep-fatigue damage cumulation and interaction *ONERA TP* **1394**

[17] Skelton R P 1987 *High temperature fatigue: properties and prediction* Elsevier London UK

[18] Skelton R P, Vilhelmsen T, Webster G A 1998 Energy criteria and cumulative damage during fatigue crack growth *International Journal of Fatigue* **9** 641-649

[19] Skelton R P 2004 Hysteresis, yield, and energy dissipation during thermo-mechanical fatigue of a ferritic steel *International Journal of Fatigue* **26** 253-264

7th International Conference on Modern Practice in Stress and Vibration Analysis IOP Publishing
Journal of Physics: Conference Series **181** (2009) 012038 doi:10.1088/1742-6596/181/1/012038

Fatigue crack initiation life prediction of railroad

P Hosseini Tehrani [1], M Saket [2]

1.Department of Railway Engineering, Iran University of Science and Technology, Narmak,16846, Tehran, Iran

E-mail: *hosseini_t@iust.ac.ir*

2. Department of Railway Engineering, Iran University of Science and Technology, Narmak,16846, Tehran, Iran

E-mail: *M_Saket@rail.iust.ac.ir*

Abstract. Study of multiaxial high-cycle fatigue initiation life prediction for railroad is done in this paper. Using ANSYS 11.0 software three dimensional elasto-plastic finite element model of rail/wheel contact is constructed and fine mesh technique in contact region is used to achieve both computational efficiency and accuracy. Stress analysis is performed and fatigue damage in railroad is evaluated numerically using multiaxial fatigue crack initiation model. Using the stress history during one loading cycle and fatigue damage model, the effects of vertical loading, material hardness material fatigue properties and wheel/rail contact situation on fatigue crack initiation life are investigated.

1. Introduction

The recent evolution of railway systems in the world has evidenced the necessity of more efficient management and, at the same time, of developing new, more accurate, design approaches to reducing costs and increasing safety and reliability of railway systems. Among all the sub-systems and the components that are a part of a railway system, the wheel/rail interface is one of the most delicate, both as regards the performances of the train and as regards its safety. Through the wheel/rail interface, in fact, the dynamic and static loads pass from the rail to the wheel through a really small contact area, whose extension and geometry can vary during the in-service period. The behaviour of the wheel/rail interface is fundamental for being certain that adequate comfort, stability and safety is guarantied during the train trip.

This consideration justifies the attention given to these components and the necessity to check the rail surface condition with periodic non-destructive controls [1–4]. However these inspections are expensive and should be, if possible, reduced. This is possible only if new design approaches are developed, able to predict the accurate evolution of the damage mechanisms and, consequently, to indicate the required frequency of non-destructive inspections.

Indeed the problem is complex, due to the fact that damage of the wheel/rail interface depends on many factors and different mechanisms contribute to the deterioration of the contact surfaces.
Wear, rolling contact fatigue and thermal fatigue are the most common types of damage due to the wheel/rail contact. All them are complex phenomena and only recently, thanks to the increased computational potentiality offered by modern computers and software, some successfully tentative of defining methods to understand how damage occurs and to predict its evolution have been developed.

Nowadays, among all damage mechanisms, fatigue is one of the most frequent ones. Unlike the slow deterioration process of wear, fatigue causes abrupt fractures in wheel or the tread surface material loss. These failures may cause damage to rails, damage to train suspensions and, in rare cases, serious derailment of train. In order to accurately describe the stress state under contact condition, analytical solutions, such as the Hertz contact theory [5], are easy to use if the problems satisfy the assumptions in Hertz contact theory. However, some practical problems cannot meet the required assumptions in the Hertz contact theory. For example, this theory assumes that the contact area is small compared to body dimension and surface curvature. For the wheel/rail contact problem, when the contact area is near the wheel flange, the surface curvature is comparable with the contact area and thus the Hertz contact theory is not applicable.

In order to overcome the limitations inherent in the analytical solution, numerical methods for contact analysis, such as finite element method and boundary element method, are widely used.
[6] Use a 2D finite element model and a multiaxial fatigue model developed by [7, 8 and 9] for bearing rolling contact fatigue analysis. [10, 11] Use the Hertz contact theory to calculate the stress response and treat the multiaxial fatigue problem as a uniaxial fatigue problem. The principle stress/strain component in one direction is used for fatigue analysis. [12, 13] developed a semi-analytical approach for stress calculation, which used 3D finite element analysis but applied the contact pressure based on Hertz theory.

In this paper, a 3D finite element model for wheel/rail rolling contact analysis is developed. Fine meshing technique in contact region is used to achieve both computation efficiency and accuracy. A stress response of the numerical simulation of the wheel rolling motion is used for fatigue life prediction and the Von Mises criterion which is the most widely used criterion for multiaxial fatigue of material having ductile behaviour is considered. In order to consider the mean stress effects modified Goodman approach is used. The influences of several parameters are evaluated and several conclusions are drawn based on the present results. The method developed in the current study may be used for fatigue resistance design and inspection planning of railroad.

2. Loading of rails

In order to construct an accurate numerical model of rail fatigue damage a brief understanding of rail loading is needed. Rails are subjected to primary and secondary loading components. The loading by the wheel is applied to the rail as bending stresses, axial stresses, and Hertzian pressure, from rolling contact.

- The bending stresses arise from the static axle load which is usually between 8 and 22.5 tonnes (for heavy haul in Australia up to 40 tonnes) and its dynamic magnification by a moving train. The overall magnitude also depends on the up and down motion of the sleepers, i.e., on the quality of the grounding of the track. Usually, quite different types of vehicles with different axle loads are operated on one track. A contribution to the bending stresses comes also from the weight of the rails itself. Defects in the running surface of the rails such as joints, dips and twists and irregularities in the wheel such as flats and out-of-roundness may play a role as well.

- Axial stresses arise from structural irregularities of the track and from the acceleration and deceleration of the train during train start and stop.
- The loading due to rolling plays a major role in the early crack extension phase. Note, that there is additional loading in lateral direction especially in curved track sections and at switches and crossovers. These forces are also dynamically magnified with increasing speed. For trains equipped with tilting technology local track irregularities in particular in small-radius curves seem to play a major role with respect of lateral load increase in track and bogie as well [14]. Note that the main load case for rails in switches is lateral bending. The primary loads are superimposed by secondary loads, which, by their nature, are thermal and residual stresses. A rail track is installed at a certain ambient temperature, e.g. in springtime or autumn.
- Because built-in rails cannot elongate and shrink with increasing and decreasing temperatures, seasonal differences in the temperature will produce axial thermal stresses. These are tensile stresses at lower and compressive stresses at higher temperatures. Due to Boudnitski and Edel [15] most European rail failures occur at temperatures at or slightly above0°C. At this temperature high tensile thermal stresses are combined with relatively low toughness values of the rail materials.
- Residual stresses in rails are introduced by different mechanisms. Primarily they stem from the manufacturing process, namely from heat treatment and roller straightening [16, 17]. Maximum axial tensile residual stresses of about 200 MPa have been measured in the vertical centre line of the rail below the running surface and in the rail foot whereas the other regions of the rail section are characterized by compressive residual stresses. The residual stresses due to the manufacturing process can be minimized within certain limits by optimizing manufacturing technology. There are differences in the residual stress distributions of naturally hard and head-hardened rails (see, e.g., [18]). Note, that the residual stress state at the ends of the rail bar stocks deviates from that at the central section which is due to differences in the straightening process which has to be realized by other methods than by roller straightening [19].

3. Crack types in rails

Cracks may be induced at or below the surface. Surface cracks are initiated due to high traction forces at high speed rails and they will propagate under the influence of a lubricant in an inclined angle in the direction of the motion of the applied load for rails operated in one direction. Transverse branching may then lead to the complete fracture of the rail. Sub-surface cracks are reported to initiate beneath the rail. The location of crack initiation is about 10–15 mm far from the side surface of the rail which is in contact with the wheel and 6–10 mm below the rail head [20, 21]. These cracks seem to propagate towards the rail surface and to behave like original surface cracks after penetration [22].

Subsurface-initiated failures are the types of failures which are focused on this paper. The final objective of this research is to find an optimized inspection and maintenance plan for the rails in order to reduce fatigue failures.

4. Finite Element Modelling of Wheel / Rail Contact

In order to build a realistic model of wheel/rail contact problem a 3D elasto-plastic finite element model is needed. This model should be able to accurately calculate the 3D stress response in the contact region as well as includes both material and geometric nonlinearity. It may be used to simulate large and complex wheel motions, such as rotation, sliding, hunting movement and even dynamic impact response. Finally, through Fine meshing techniques in contact region, the proposed model is made very efficient in computing and hardware requirements. All the finite element models in this task are built using the commercial software ANSYS 11. The general procedure is described as below.

In order to have more accurate contact analysis, geometry profile of the rail head section and the wheel tread are very important. For this reason UIC60[1] section profile (for the rail) and passenger coach wheel profile is used [23, 24]. The rail's length is a little more than the distance of two sleepers. Fixed boundary conditions (All DOF) are applied to the two area of cross section at the ends of the rail, used 2D Solid42, 3D Solid45 Element for modelling the wheel/rail volume and Target 170, Contact 175 contact element for modelling the Surface to Surface contact.

Due to nonlinearity of contact analysis, the region need to be fine meshed. In this research, a fine mesh that average length is less than 2mm is used in near contact area. Using pilot node at the wheel centre, in this important place a pilot point is connected to the wheel using rigid link element. All the external loading and boundary conditions of the wheel load are applied on the pilot point. These loading and boundary condition can be obtained through field measurement or from numerical simulation of the track system motion analysis.

On the possible contact area of the rail head and the wheel tread, contact elements (Target 170 for rail and Contact 175 for wheel) are used corresponding to the geometry mesh of the wheel. The contact algorithm is augmented lagrangian method. Friction effect is included into material property of the contact element and a Coulomb friction model is used. The coulomb friction coefficient is assumed to be 0.3. The material properties of the wheel and rail are considered to be bilinear kinematic hardening in ANSYS [25]. After that quasi- static analysis is performed and the results for each step are stored. Very fine mesh is applied to the contact area and some depth under contact surface to obtained accurate results. Results accuracy may be shown by comparison with experimental results.

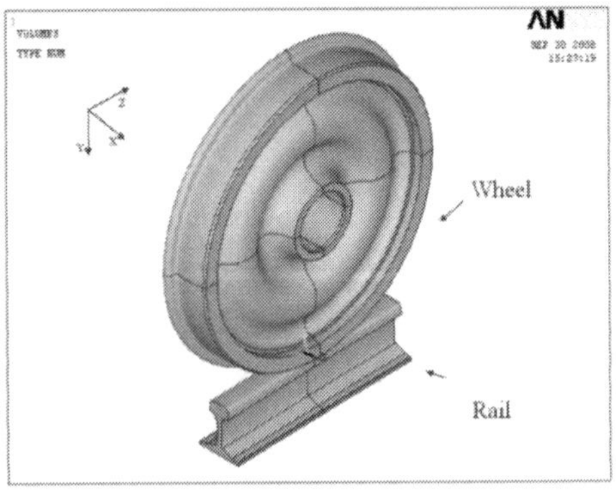

Figure 1. Finite element modelling of wheel/rail contact.

[1] A prevalent rail section that recommended by UIC (International Union of Railways or Union Internationale des Chemins de fer)

5. Numerical Example

The rail profile is chosen according to UIC regulation, the considered rail is the most prevalent UIC60 profile that is shown in the UIC leaflet. The wheel diameter is about 0.89 m and, the wheel profile is chosen according to the AAR standard wide flange contour [26]. The vertical load is assumed to be the maximum design load, which is 146.2 kN. The material properties of the rail and wheel are assumed to be same and are as follow; Young's modulus 205MPa, yielding strength 500MPa, friction coefficient 0.3 tangent modulus 4000MPa, and linear kinematics hardening model is used. In this task the rail length is considered to be 700 mm [25]. The initial contact point is assumed to occur at the railhead centre and wheel tread centre. The results of the static load analysis of the wheel and rail contact are shown in Figs. 2 and 3. Fig. 2 shows the Von Mises stresses from different section views. Fig. 3 shows in-plane shear stress. From the Fig. 2, it is seen that the maximum Von Mises occurs at some depth below the head surface The stress decreases quickly as the depth increases. The high stress occurs within a small region of the contact location. It is seen that the maximum stress occurs about 5-10 mm below the head surface of the rail. This reign may be the sight of underneath crack nucleation. The numerical results predict closely the formation of crack sight similar to experimental results. The stress in the other parts of rail is nearly zero.This indicates that only a small portion of the motion simulation is needed because the stress far away from the contact location is zero. From Fig. 3, a butterfly pattern of the shear stress τ_{xy} is observed. The maximum value also occurs at a depth of about 5-10 mm below the head surface, the maximum τ_{xy} occur at the location some distance away from the contact location.

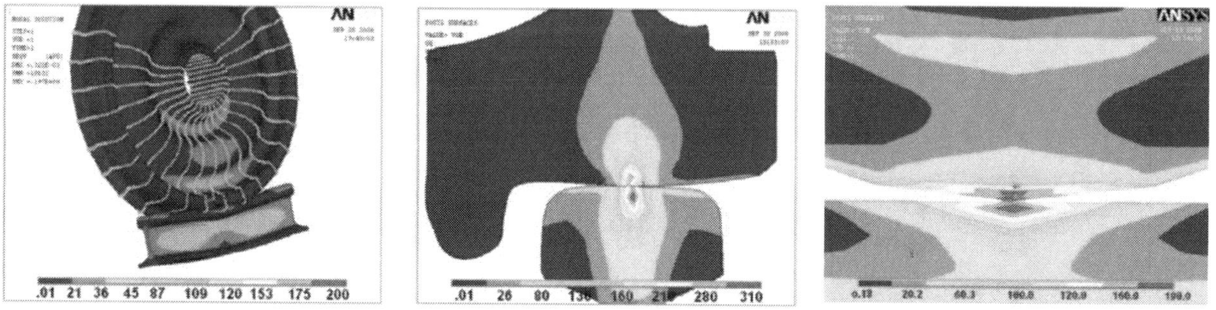

(a) wheel/rail contact (b) Front section view (C) Left section view

Figure 2. Von-Mises stress distribution of wheel/rail contact.(MPa)

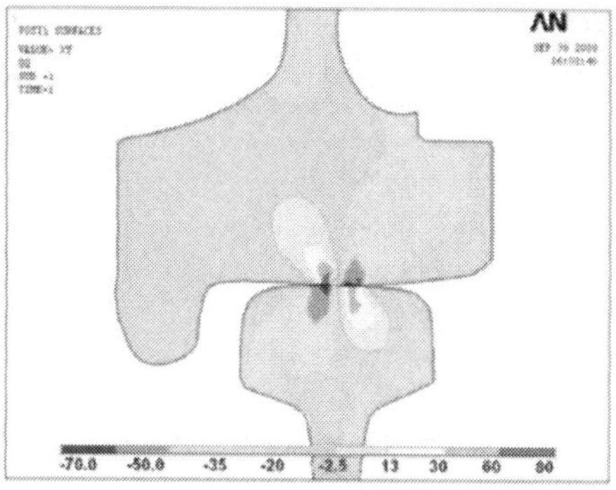

Distribution of τ_{xy}

Figure 3. In-plane shear stress distribution of wheel/rail contact.(MPa)

From Figs. 2 and 3, the stress pattern indicates multiple contact points between the wheel/rail interfaces. After obtaining the stress history of the wheel, the Von Mises method is used as the multiaxial fatigue criterion to calculate the fatigue initiation life. In order to consider the mean stress effects modified Goodman approach is used. For each step vertical load amplitude on rail assumed to be constant and varies between a maximum value and zero. As it is seen in fig. 5(a) in order to examine the effects of load amplitude for different steps the load amplitude varies among 50 kN and 150 kN.

The point of maximum Von-Mises stress is considered as the critical location for fatigue crack initiation. The fatigue S–N curve for uniaxial and torsional loading are plotted in Fig. 4. The fatigue damage distribution versus vertical load and hardness is plotted in Fig. 5. The damage accumulation rate on the wheel section under different vertical loads are calculated and plotted in Fig. 5(a). As expected, the damage accumulation rate increases as the vertical load increases. In this case, the equivalent stress amplitude is lower than the endurance limit for the vertical load below 105 kN. Fig. 5(b) shows that the damage accumulation rate increases as the hardness increases. This may be occurred as a result of decreasing the contact area between wheel and rail which causes higher stresses in contact area.

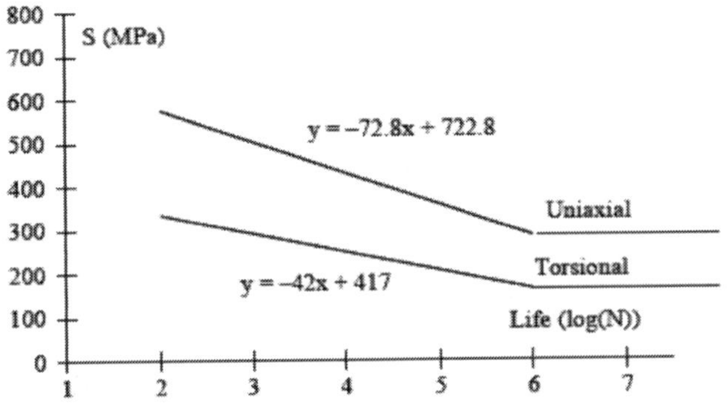

Figure 4. S-N curve for uniaxial and torsional loading.

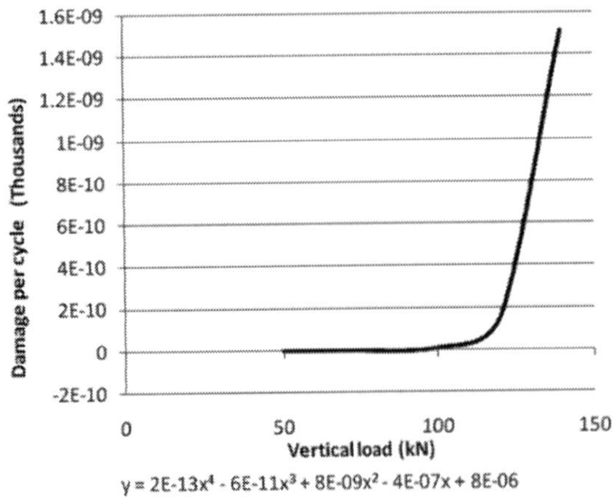

(a)

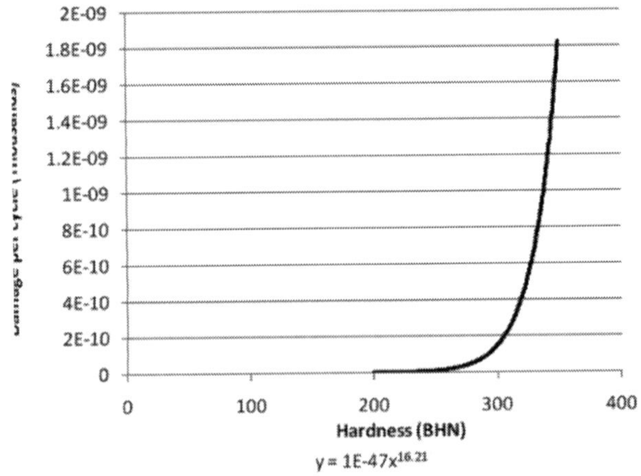

(b)

Figure 5. (a) Relationship between vertical load and Damage per cycle, (b) Relationship between Hardness and Damage per cycle

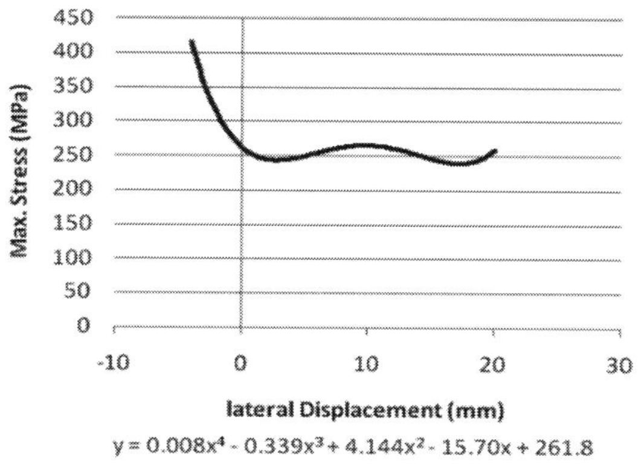

Figure 6. Effects of lateral displacement of wheel on maximum stress occurrence in rail.

Fig. 6 shows effects of lateral displacement of the wheel on maximum stress occurrence in rail.
As it is seen when the contact area is shifted to the left (near the flange of the wheel) stresses in this area will rises dramatically. Besides it is seen that when the contact area is located at the slope change location of the wheel an increasing in stresses occurs. Therefore it is concluded that the worst area for wheel and rail contact is near to the flange of rail also it is better to translate the slope change area of the wheel near to the inside face of the wheel.

6. Conclusion

A multiaxial fatigue life prediction investigation is developed in this paper. Unlike most of the previous studies, the current task is used a three dimensional model and a nonlinear finite element analysis is used for stress computation and the stress history is then used to calculate the fatigue life. The effect of several parameters, namely vertical loads, material hardness, fatigue strength and wheel/rail contact situation, on the fatigue damage in railroad is studied using the Von mises model.
In the current study, the effects of different parameters have been studied individually. Future research needs to consider interactive effects of those parameters because the wheel/rail contact problem is highly nonlinear. Also, other effects, such as residual stress from manufacturing, brake loading, thermal loading, dynamic and impact loadings, material defects, etc. need to be included in the proposed methodology.

7. References

[1] H. Sakamoto, K. Hirakawa, Bull. JSME Ser. A 46 (2003) 613–619.
[2] R. Lunden, Proceedings of the International Wheelset Congress, Sydney, 1992,pp. 163–167.
[3] D.F. Cannon, K.-O. Edel, S.L. Grassie, K. Sawley, Fatigue Fract. Eng. Mater. Struct.26 (2003) 865–887.
[4] O. Orringer, Crack propagation and fracture in contacting bodies, Fatigue Fract.Eng. Mater. Struct. 19 (1996) 1329–1338.
[5] M.R. Ayatollahi, M. Zakeri, M.N. Hassani, Proceeding of the 11 International Conference Fracture (ICF11), Turin, Italy, 2005
[6] Liu Y, Mahadevan S. Multiaxial high-cycle fatigue criterion and life prediction for metals. Int J Fatigue 2005;7(7):790–800.

[7] Howell M, Hahn GT, Rubin CA, McDowell DL. Finite element analysis of rolling contact for nonlinear kinematic hardening bearing steel. ASME J Tribol 1995;117:729–36

[8] Guo YB, Barkey ME. Modeling of rolling contact fatigue for hard machined components with process-induced residual stress. Int J Fatigue2004;26(6):605–13

[9] Fatemi A, Socie DF. multiaxial fatigue damage including out-of-phase loading. Fatigue Fract Eng Mater Struct1988;11:149–65

[10] S ˇraml M, Flasˇker J, Potrcˇ I. Numerical procedure for predicting the rolling contact fatigue crack initiation. Int J Fatigue 2003;25(7):585–95.

[11] ASTM E 647-99. Standard test method for measurement of fatigue crack growth rates. Philadelphia: American Society for Testing and Materials; 1999.

[12] Ringsberg JW. Life prediction of rolling contact fatigue crack initiation. Int J Fatigue 2001;23(7):575–86.

[13] Liu Y, Stratman B, Mahadevan S. Fatigue crack initiation life prediction of railroad wheels. Int J Fatigue 2006;28(7):747–56

[14] Gasemyr H, Normann J. Beurteilung des Einflusses der Gleisqualit€at in Gleisb€ogen mit kleinen Radien auf die Beanspruchung an Radsatzwellen der Drehgestelle f€ur Neigez€uge am Beispiel eines durchgef€uhrten Messprogramms in Norwegen. ZEVrail performance under high axle load conditions. In: Proceedings of the 6th International Conference on Contact Mechanics and Wear of Rail/Wheel systems (CM2003), G€oteburg, Sweden; 2003.

[15] Boudnitski G, Edel K-O. Spannungsintensit€atsfactoren in Schienen. In: Proceedings of Internationales Symposium Schienenfehler, Brandenburg, Germany; 2000 [chapter 11].

[16] Schleinzer G, Fischer FD. Residual stresses in new rails. Mater Sci Engng 2000;A288:280–3.

[17] Schleinzer G, Fischer FD. Residual stress formation during the roller straightening of railway rails. Int J Mech Sci 2001;43:2281–95.

[18] Cannon DF, Pradier H. Rail rolling contact fatigue. Research by the European Rail Research Institute. Wear 1996;191:1–13.

[19] Edel K-O, Boudnitzki G, Schur EA. Literaturanalyse zum Thema ‘‘Bruchmechanik und Eisenbahnschienen’’. Report FH, Brandenburg; 1997.

[20] Clayton P. Tribological aspects of wheel/rail contact: Areview of recent experimental research. In: Proceedings of the 4th International Conference of Contact Mechanics and Wear of Rail/Wheel Systems, Vancouver; 1994.

[21] prEN 13647-1. Railway applications—track. Standard draft submitted to CEN, 2001.

[22] Orringer O, Morris JM, Jeong DY. Detail fracture growth in rails: Test results. Theor Appl Fract Mech 1986;5:63–95

[23] AAR (Association of American Railroads). Manual of standards and recommended practices: section G-wheels and axles. Issue of 1998.

[24] UIC (International Union of Railways) Leaflet 512-3. Technical specification for the supply of solid wheels in rolled non-alloy steel for tractive and trailing stock;also prEN 13262, Draft 1998.& other correlate leaflet(500 series leaflet).

[25] Gupta V, Bastias P, Hahn GT, Rubin CA. Elasto-plastic finite elementanalysis of 2-D rolling plus sliding contact with temperature dependent bearing steel material properties. Wear 1993;169:251–6.

[26] Telliskivi T, Olofsson U, Sellgren U, Kruse P. A tool and a method for FE analysis of wheel and rail interaction. Ansys conference; 2000.

Journal of Physics: Conference Series **181** (2009) 012039 doi:10.1088/1742-6596/181/1/012039

Characterisation of the fatigue properties of cast irons used in the water industry and the effect on pipe strength and performance

H Mohebbi, D A Jesson, M J Mulheron, P A Smith

Faculty of Engineering and Physical Sciences, University of Surrey, UK

E-mail address: H.Mohebbi@surrey.ac.uk

Abstract. As part of an on going programme to characterise the residual properties and understand the failure mechanisms of in-service grey cast iron water pipes, the fatigue crack propagation behaviour of grey cast iron samples has been studied. Specimens were sourced from three ex-service pipes. For each pipe the microstructure and composition were characterised and the fracture toughness was determined. The fatigue behaviour was investigated in terms of the crack growth rate (da/dN) as a function of the applied stress intensity factor range. Clear differences in the fatigue behaviour of the samples from different pipes were observed. The result from these investigations, which indicate that microstructural differences play a role in mechanical behaviour, will support the development of asset management tools for use in the water industry.

1. Introduction

Cast iron is a material that still has significant usage in the water industry. In the context of the United Kingdom, although cast iron pipes are being phased out of distribution and mains systems, a significant portion of current pipe networks are still comprised of the aging cast iron infrastructure that can be up to 150 years old. This cast iron infrastructure is deteriorating at different rates [1-2], which are linked to the geology and other localised conditions. The problem is further complicated by the diversity in production quality as well as the range of diameters and wall thicknesses used. In order to develop a rehabilitation or replacement strategy for the industry, it is important not only to understand the mechanisms by which a pipe can fail, but also to use this understanding to develop tools for asset management. In order to better model these mechanisms and their contribution to service failures an understanding of the initiation and propagation of cracks is necessary. For example, sub-critical cracks could originate from inherent defects present *ab initio* from casting; be generated during installation or arise in service. If this hypothesis is true, any information on the rate of and the factors influencing fatigue behaviour is of relevance. The literature regarding the fatigue of gray cast iron is limited and inconsistent. These inconsistencies are likely in part to be a consequence of the microstructural differences between the gray cast irons used in these studies [3-5]. This is particularly relevant as little has been done by way of comparing the behaviour of different types of cast irons under the same condition.

In the present work the propagation of fatigue cracks in samples of different classes of cast iron were studied in order to investigate the possible role of the microstructure (in particular graphite morphology) on the fatigue behaviour of cast iron pipes. The structure of the paper is as follows. In the next section the test materials and test methods are described. Following this results are presented relating to the microstructure and fracture and fatigue behaviour. Finally the implications for service life are discussed.

2. Experimental Methods

2.1 Material characterisation

Samples were sourced from three water mains which had failed in-service. Three plates were taken from each pipe and test specimens were machined from these plates. Metallurgical examination was carried out to verify that the pipes were a grey cast iron material. Chemical compositions of the samples are given in Table 1.

Table 1 Chemical composition of grey cast iron mains (Wt %)

	C	Si	P	S	Mn	V	Mo
Pipe1	3.23	2.7	0.6	0.12	0.6	0.08	
Pipe2	2.75	2.35	1.2	0.08	0.5	0.1	
Pipe3	3.2	2.36	0.3	0.2	0.7	0.05	0.01

2.2 Microstructure characterisation

Samples of the full pipe wall thickness were cut from each pipe, mounted in conducting Bakelite and polished following a typical schedule to give a 0.25 μm surface finish. The specimens' microstructures were then studied using a Zeiss Axiophot optical microscope. Initially, the samples were studied in the as polished condition; the samples were then etched with 2% Nital (a solution of nitric acid in methanol) and re-examined.

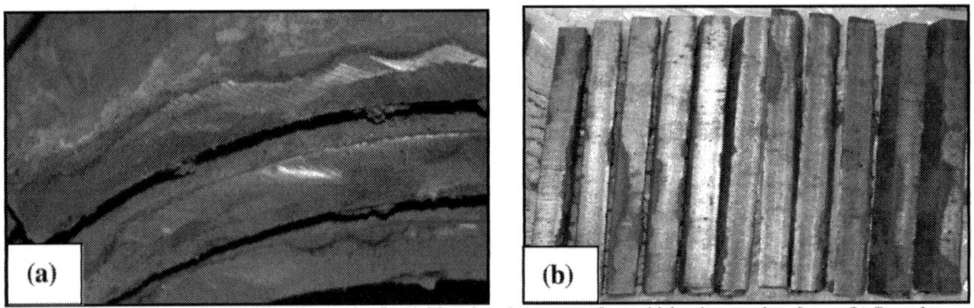

Figure 1 Photographs of the plates from pipe3, (a) Showing the extent of graphitisation on the plates, (b) Samples cut from the plates

2.3 Fatigue and Fracture toughness testing

A single edge notch beam (SENB) configuration was chosen for fatigue testing. To ensure plane strain conditions the ratio of W/B (where B and W are breadth and width) has been chosen according to $1 < W/B < 2$, whilst the minimum breadth of the specimens was chosen to satisfy the requirement of $B > 2.5(K_{IC}/\sigma_{YS})^2$. Specimens were obtained in the form of rectangular cross section bars cut from the plates of ex-service failed mains. To exclude the effects of corrosion, the specimens were machined to remove the surfaces. A notch, nominally 1/3 the width of the specimen was machined into each specimen. Fatigue testing was carried out using an Instron 8511 (servo-hydraulic machine) with an automatic data logger. A three point bending method was employed with a cycling stress ratio of 0.1 and the loading frequency of 10 Hz. At all stages, the requirements of ASTM E647-86a-00 protocol were followed. Specimens were pre-cracked prior to testing. A load shedding method was adopted to obtain a crack growth rate of the order of 10^{-3} and 10^{-5} mm/cycle for pre-cracking and fatigue respectively. Initiation and growth of cracks was observed with the aid of a travelling optical microscope. Crack growth was also monitored with crack propagation sensors (CPA01, VISHAY Measurements Group). A pre-crack was grown, with the maximum cycling load corresponding to ΔK values of 10 $MN/m^{1.5}$, to an approximate length of 2 mm, at which time the test was stopped and micro-photographs of the crack were taken in order to determine the exact length of the generated crack. The test was continued with the maximum cycling load corresponding to ΔK values of 8.5 $MN/m^{1.5}$. A parallel series of experiments were carried out to determine the fracture toughness values of the pipes using an Instron 6025 (5500R) with a loading rate of 1KN/min and 40% sensitivity. At all stages the requirements of ASTM E1820-01 was followed.

3. Results and discussion

3.1 Microstructure analysis

Optical microscopy reveals differences in the microstructures of the specimens taken from different pipes. The microstructures of all pipes show a multiphase matrix of pearlite, ferrite, and phosphide eutectic with different graphite morphology. Microphotographs representing the microstructures seen in samples from pipes 1-3 are presented in figure 2. A summary regarding the graphite morphology (type, size, and form, with reference to ASTM A247-98) is presented in Table 2.

Table 2 Graphite flake identification (with reference to ASTM A247-98) observed in pipes1-3

	The edges				The mid section			
	type	distribution	size class	Maximum flake size/ mm	type	distribution	size class	Maximum flake size/ mm
Pipe1	IIV	Between A (Random orientation) and B (Rosette)	3	0.3	IIV	Between A (Random orientation) and B (Rosette)	2	0.45
Pipe2	IIV	A (Random orientation)	2	0.4	IIV	A (Random orientation)	1	0.9
Pipe3	IIV	B (Rosette)	2	0.4	IIV	B (Rosette)	2	0.4

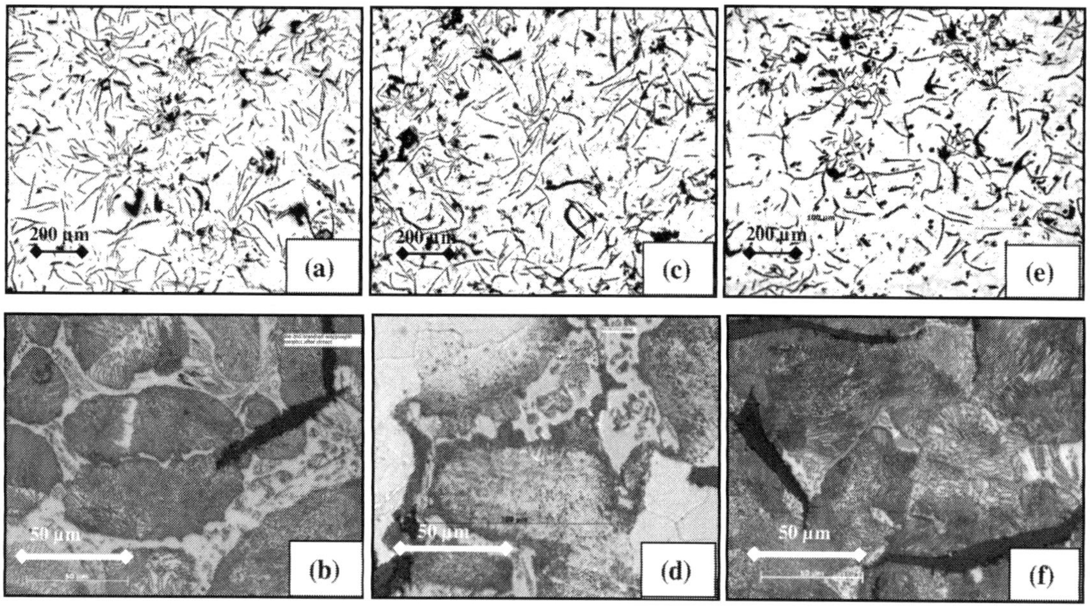

Figure 2 Digital photo-Microphotographs Representing the Morphology of Microstructure seen in Pipes 1-3
a: an example of type B 'rosette' graphite flake morphology observed at Pipe3, b: detail of a
c: an example of type A 'scythe/sickle' graphite morphology observed at Pipe2, d: detail of c
e: an example of mixture of type A and type B graphite morphology observed at Pipe1, f: detail of e

3.2 Fatigue results

The result of crack growth rate versus applied stress intensity factor of all samples is presented in Figure 3. From Figure 3b it is seen that the propagation of cracks in samples from Pipe 3 corresponds to relatively higher ΔK values than for the samples from the other two pipes. This is in agreement with the applied ΔK values for the crack initiation, which were 10.7, 10.2, and 11.4 $MN/m^{1.5}$ for Pipes 1 to 3 respectively. These differences are likely to be due to the microstructural variation of the specimens, which in turn gives rise to differences in the crack growth behaviour of the samples. Experimentally obtained values for fracture toughness also show a variation. The mean fracture toughness values of pipes 1-3 were 15.4, 15.1, and 16.1 $MPa/m^{3/2}$ respectively.

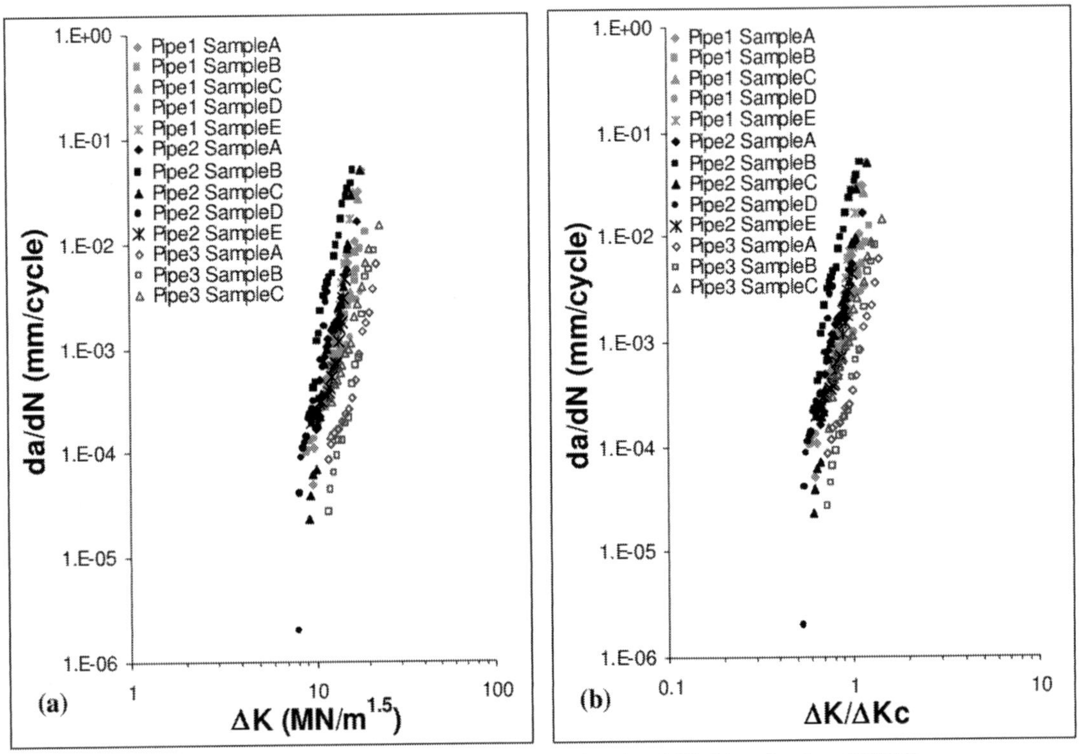

Figure 3 Fatigue crack growth rate of samples from pipes1-3. (a): as a function of ΔK, (b): as function of $\Delta K/\Delta K_c$

Figure 4 presents micro-photographs taken during fatigue testing. From Figure 4a-b, a tendency for interlamellar propagation of the crack as well as crack branching phenomena at the low stress intensity factor regime is observed. The crack growth mechanism seems to vary throughout the fatigue life. At the early stage of the crack development, where ΔK value is near the threshold, the crack follows a preferential path induced by coalescence of the graphite flakes, which act as foci for stress concentration. As the crack grows, crystallographic slip directions seem to be favoured. The mechanism of the crack growth can still be described as the coalescence of the graphite flakes but this time via an intergranular induced path. At this stage, as ΔK increases - and so the crack tip plastic zone becomes sufficiently large - the formation of cracks parallel to the principal crack on planes near the crack tip is observed (Figure 4b-c). Figure 3a-b reveals a fluctuation on the slope of the lines particularly in the intermediate regime of ΔK. This might be due to the structure-sensitive nature of crack growth, which depends on the grain size of the controlling phase. The fracture surface of the intermediate ΔK regime is seen to be smoother in this regime than in the higher ΔK regime (Figure 5). When ΔK increases to a value such that the fracture toughness is approached, the crack starts to grow unstably. The crack at this stage is a high energy one and it does not follow a preferential path. As a result the crack growth occurs via the coalescence of graphite flakes by either a transgranular or an intergranular induced path or some combination of the two.

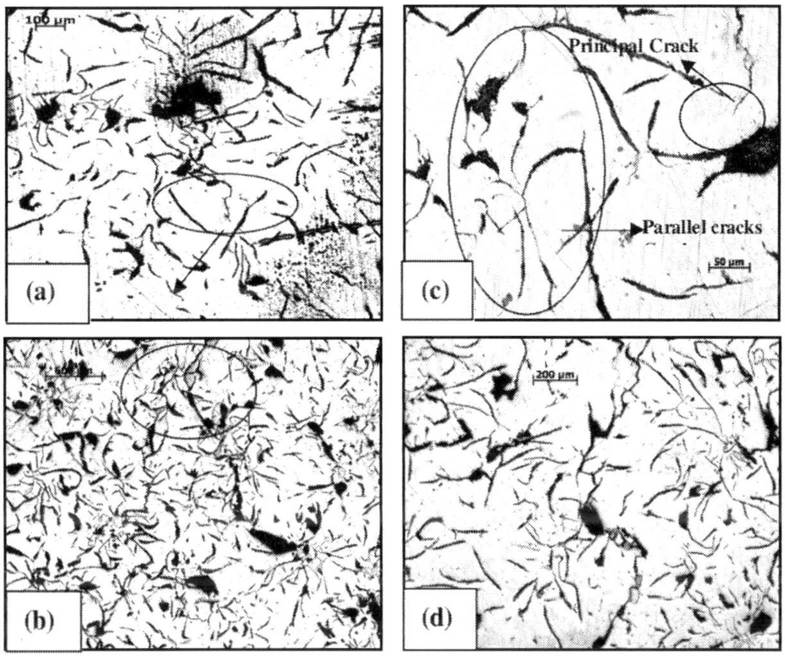

Figure 4 –Digital-microphotographs taken during the fatigue of Sample 17 pipe1 after, (a): 20,000 cycles; crack at an early stage, low ΔK regime, (b): 25,000 cycles an example of the main stream crack branching, (c): 37,500 cycles, example of the initiation of parallel cracks as ΔK increases, (d): 52,500 cycle, an example of graphite flake induced preferential path

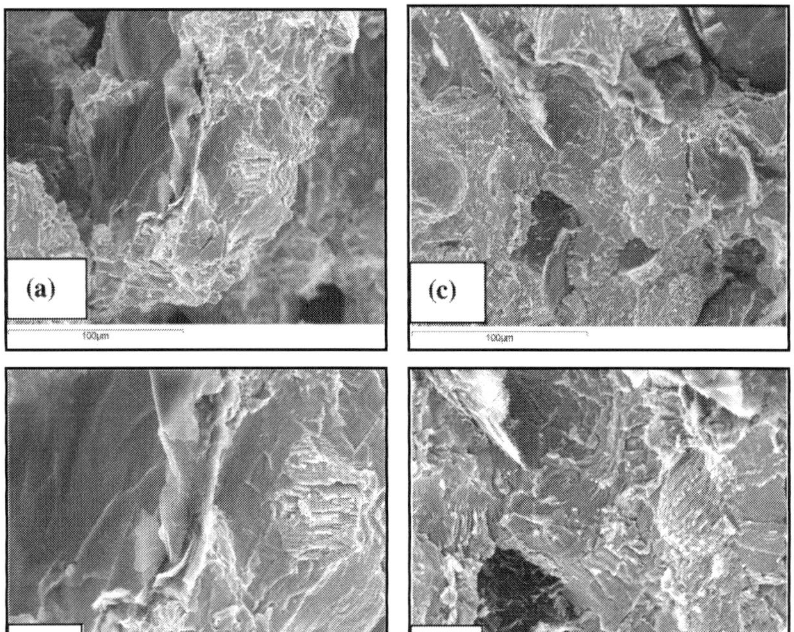

Figure 5 SEM Fractographs of sample4 pipe3 (a): Fatigue fractured surface of stable crack growth regime, (b): Detailed of a (c): Fatigue fractured surface of near failure (d): Details of c

4. Application of Paris relation to model S-N behaviour of the pipes

An S-N model was developed from the fatigue experiment data using the Paris relationship. The fatigue crack growth behaviour of the material is given by the Paris relation as:

$$\frac{da}{dN} = A\Delta K^{m} \qquad (1)$$

Where $\Delta K = K_{MAX} - K_{MIN}$ (2), and A and m are Paris components. The stress intensity factor ,K, for a SENB bend specimen geometry is given by [6] as:

$$K = \frac{PS}{BW^{3/2}} f(a/w) \qquad (3)$$

Where P is the max cycling load, S is the span of the sample, B and W are the breadth and the width of the specimen respectively, and f(a/w) is a function of the ratio of the crack length (a) and W and is calculated according to:

$$f(a/W) = \frac{3(a/W)^{1/2}}{2(1+2a/W)(1-a/W)^{3/2}} \times [1.99 - (a/W)(1-a/W)(2.15 - 3.93a/W + 2.7a^2/W^2)]$$

(4)

The Paris constants for the samples were obtained from log(da/dN)-log(ΔK) plot and averaged for each pipe. The number of cycles to failure for each pipe was obtained by using MATLAB to integrate numerically the Paris relation for various crack configurations of a/W.

$$N_f = \int_{a_i}^{a_f} \frac{da}{A(\Delta K)^m} \qquad (5)$$

It is worth emphasising that S-N plots obtained from the integration of Paris equation are based on the assumption that pipes already contain initial flaws which are either inherent or arise during service Cast iron pipes most commonly suffer from localised corrosion such as corrosion pits. These corrosion pits can be regarded as flaws and hence estimating the remaining in-service fatigue life of the pipe becomes of great importance. Figure 6 presents two different crack length configurations under which the fatigue life modelling was developed. Figure 6a compares a situation where the flaw size is of the same order of magnitude as a graphite flake, thus modelling the case of an inherent flaw. It is seen that for a relatively low applied stress – such as normal service loading conditions - all three pipes are expected to show a fatigue life of approximately 10^7-10^{-8} cycles, with Pipe 3 showing slightly better performance (this difference is more apparent at higher applied stresses). Figure 6b models the situation where the flaw in the pipe is 0.2W (width of the pipe), and in this case the larger remaining fatigue life of Pipe 3 is clearer. The model seems to be valid for intermediate and low applied stress levels (evident from experimental data) but for higher stress levels there are some doubts as to whether the model remains valid.

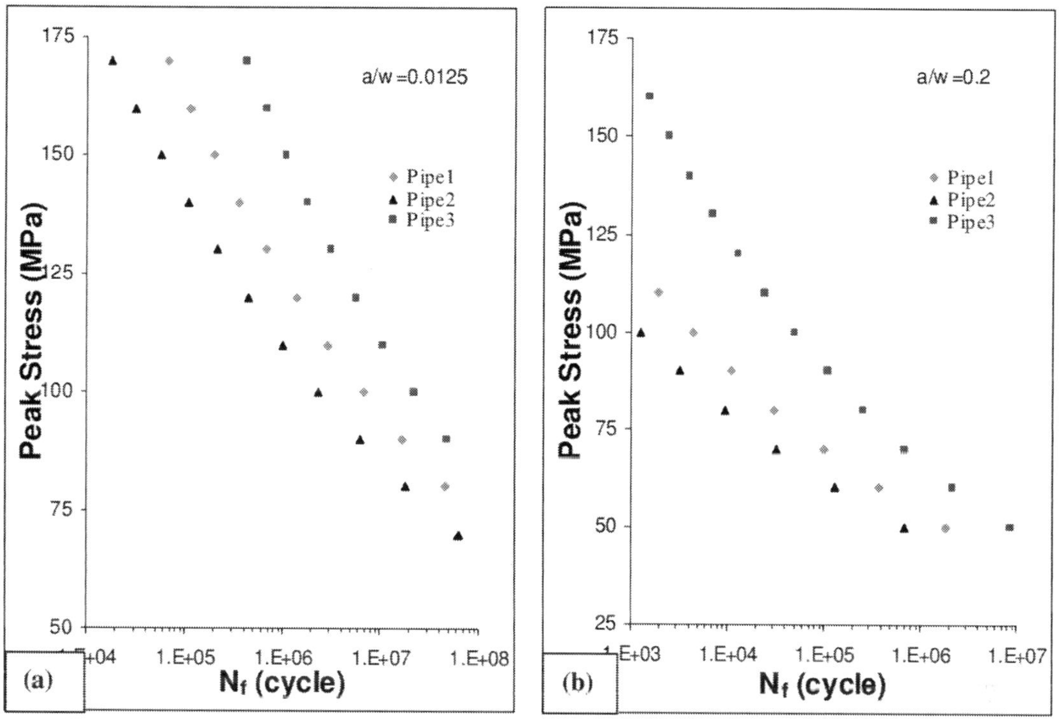

Figure 6, S-N model developed for pipe1-3, a: the flaw size is small (the order of microstructure features) b: the flaw size is large (fifth of the width of the pipe)

5. Concluding remarks

Fatigue of grey cast iron pipes shows microstructural dependent behaviour. In the case of grey cast iron pipe with graphite flake, morphology of the graphite seems to be a key parameter in characterising this behaviour. Further investigation will consider the environmental effects on crack growth mechanisms. Future work will also focus on crack opening displacement and whether this will have pronounced effect on the obtained S-N curve.

Acknowledgment

The authors would like to thank Mr Antony Scratt (Struers Ltd), Mr Anton Chitty (VISHAY measurement groups Ltd UK), Dr. B. Le Page, Mr. C. Burt, Mr. M. Parker, and Mr. P. Haynes for their contribution to the work. The authors would like to acknowledge the support of Thames Water during the course of this work and, in particular, many helpful discussions with Dr Hal Belmonte and Mr Jeff Farrow. The opinions expressed in this paper are those of the authors and are not necessarily endorsed by the University of Surrey or Thames Water.

References

[1] Atkinson, K., Whiter, J. T., Smith, P. A, and Mulheron M. (2002) "Failure of small diameter of cast iron pipes" Urban water Vol **4**(2003), 263-271

[2] Rajani, B. and Makar, J., "A methodology to estimate remaining service life of grey cast iron water mains", Canadian Journal of Civil Engineering, Vol. **27** (2000), 1259-1272.

[3] H M S Belmonte, Smith, P. A, and Mulheron M "Weibull-based methodology for condition assessment of cast iron water mains and its application" Fatigue and fracture of engineering materials and structure, Vol **31**(2008) Issue 5, 370 – 385

[4] M N James, Li Wenfong "fatigue crack growth in austempered ductile and grey cast iron" material science and engineering A265[1999], 129-139

[5] B. M. Kapadia, E. J. Imhof, Jr. "Fatigue crack growth in cast iron and cast steel" , Cast Metals for Structure and Pressure Containment Application, ASME-MPC-11, December, 1979

[6] Richard W. Hertzberg "Deformation and Fracture Mechanics of Engineering Materials" (Hardcover), 4th addition

Journal of Physics: Conference Series **181** (2009) 012040 doi:10.1088/1742-6596/181/1/012040

Numerical analysis of the effect of torispherical head on the buckling of pressure vessels

M.R.Khoshravan, A.Rahmani

University of Tabriz, Faculty of Mechanical engineering, 51666 Tabriz, Iran

rkhosh@tabrizu.ac.ir

Abstract. Effect of torispherical head on the buckling of pressure vessels was investigated by finite element (FE) method. The FE method with use of nonlinear buckling analysis was applied to predict the critical buckling Load. The influences of geometrical parameter such as thickness, knuckle radius and diameter of cylindrical part, on the buckling of heads have been studied. The Arc Length method which can control the load level, the length of the displacement increment and the maximum displacement was been used. By verification performed with the European Convention for Constructional Steelwork (ECCS) code, it was confirmed that the nonlinear buckling analysis could assure accurate results for buckling strength. It was shown that geometrical imperfections had little effect on buckling strength.

1. Introduction

Theories of thin-walled structures applied on pressure vessels were reviewed by Teng et al [1]. Results of numerical evaluation of buckling by using linear and non-linear theories of thin-walled shells of revolution have been presented.

Regarding to existence of non-continue stress in cylinder-head intersection, the choice of head considering the geometrical limitation and production facilities is the most important point in designing of pressure vessels. Torispherical heads are used commonly in pressure vessels because of their simple manufacturing and good strength in high pressure condition (Fig. 1). The buckling strength is one of the most important points in design of pressure vessels [2]. Internal pressurization is often an important loading condition for pressure vessels. Finite element method is often used in the buckling analysis of pressure vessels due to its capability.

The hydrostatic buckling of shells under different boundary conditions (B.C.) has been investigated by using energy method [3]. Results showed that in shells with medium height, under different B.C., buckling load is obtained by applying a scalar coefficient to the buckling load of the pin ended case, but this method is not applicable for the long shells for which the occurred circumferential waves from buckling, are higher than 3.

Teng et al [4] have introduced a numerical model, aided by the method of Eigenmode-affine, in the non-linear analysis of elastic shells. As the shells are sensitive to initial geometrical imperfections, predicting of their buckling resistance would be precise if those imperfections are taken into account.

In torispherical heads by increasing the ratio of knuckle radius per vessel diameter (r/D), dimension of spherical part decreases. Thus, the spherical part as part of the head becomes weaker and in a defined r/D a notable fall in buckling resistance is occurred [5].

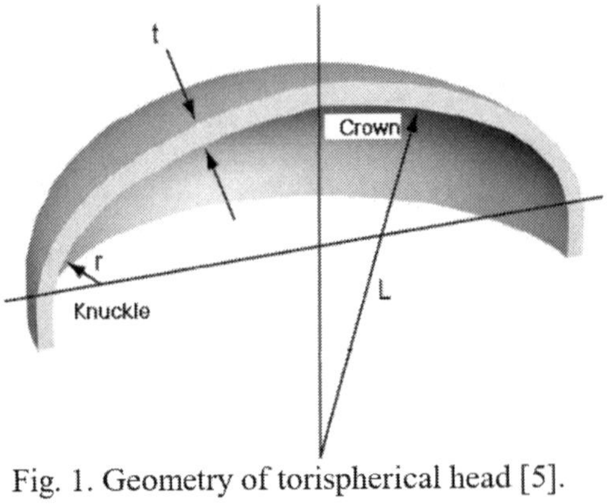

Fig. 1. Geometry of torispherical head [5].

European recommendation ECCS [6] had introduced several experimental relations for design of spherical shells. In following we will discuss about the buckling load and influence of different parameters on it. We will try to bring some propositions for limitation of buckling.

2. Numerical simulation method

Eigenvalue buckling analysis predicts the theoretical buckling strength of an ideal linear elastic structure. Bifurcation buckling refers to the unbounded growth of a new deformation pattern. Imperfection and material non-linearity can not be included in this analysis. Thus, the buckling strength obtained by Eigenvalue buckling analysis may differ from that of a real structure and often yield non conservative results. Therefore, care is needed when using this method in actual evaluation of buckling strength.

Non-linear buckling analysis including geometric and material non-linearity is usually the more accurate approach and is therefore recommended for design or evaluation of actual structures. There are two methods for obtaining buckling strength by non-linear buckling analysis. One basic approach is to constantly increment the applied loads until the solution begins to diverge, which can be obtained from load-controlled buckling analysis. In this approach, a simple static analysis will be done with large deflection extended to a point where the structure reaches its limit load.

3. Modelling

Non-linear finite element method with large deflection analysis was performed using commercial Ansys software. A three dimensional finite element model was generated using Ansys 9.0. For studying of buckling of pressure vessel with torispherical head; we modeled intersection of cylinder-head. The influence of welding and forming on material property were neglected while the effect of welding can be accounted for by modifying the yield stresses. The length of the cylinder was kept at 4λ (λ is the linear elastic meridian bending of half-wave length given by $2.44\sqrt{Rt}$) to ensure that the boundary effects at the far end of the cylinder do not interfere with the behavior of the intersection [7]. The model was meshed with shell 93 element. SHELL93 is particularly well suited to model curved shells. The element has six degrees of freedom at each node: translations in the nodal x, y, and z directions and rotations about the nodal x, y and z-axes. The element has plasticity, stress stiffening, large deflection, and large strain capabilities. The material of the intersections was assumed to have typical properties of steel: an elastic modulus of 1.9×10^5 *MPa*; a Poisson's ratio of 0.26, and yield stress of 206 *MPa*, and exhibit an elastic-perfectly plastic behavior.

For modeling of the geometrical imperfections in ANSYS package, we applied them in the form of initial deformations on the model [8]. For this reason, first we analyzed the model by using the linear method of buckling and then by using the "Update Geom" order, we give values to the magnificent factor. In fact by the resulted displacement of different buckling resolution, a new model with geometrical imperfection was obtained. This factor of geometrical imperfection, which is in fact deviation from the perfect model or the initial deformation, was presented by W_0. The buckled model is illustrated in the Fig. 2.

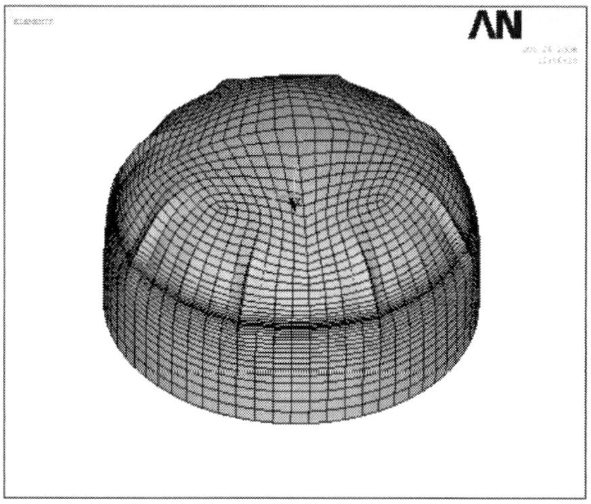

Fig. 2. Finite element model after buckling.

4. Choice of resolving method

For the solution of a nonlinear problem, the choice of solution method and load step (referred to as the time step in the ANSYS software), is very important. It should be made taking into account the anticipated structural behavior and the characteristics of the specific solution method.

Prior to carrying out a nonlinear buckling analysis, it is often beneficial to undertake a linear buckling analysis in order to obtain some appreciation of the buckling behavior. It may also help to identify those regions in the model that will first exhibit nonlinear response and at what load levels these nonlinearities develop.

There are several methods available in ANSYS for the solution of the nonlinear buckling equations. They include the Newton–Raphson and Arc Length methods. For geometrically nonlinear analysis, the Newton–Raphson method has been shown to be one of the best methods available. The most important characteristic of this method is its ability to converge even when the behavior is highly nonlinear. The method is also extremely accurate and generally converges quite rapidly, provided a realistic initial estimate of the displacement vector is used. With this method too, it is possible to control the solution error and estimate the rate of convergence, since for any particular load step the iterations continue until the specified solution error is achieved. Preliminary ANSYS FE analyses of the columns, in which compression loads were applied, showed that the Newton–Raphson method converged quite rapidly. For nonlinear buckling analysis, coarse time steps may be used in the pre-buckling regime, but fine steps are required close to the buckling load and in the post buckling regime. Different time steps may be used in the pre- and post buckling regimes through the multiple 'load steps' option within ANSYS. However, it is not easy to choose an appropriate maximum load level in load controlled analysis using the Newton–Raphson method.

Moreover, the Newton–Raphson method fails when snap-though occurs. The Arc Length method does not have this drawback and allows one to control the load level, the length of the displacement increment and the maximum displacement. Therefore, in the nonlinear buckling analyses, the Arc Length method was used [9].

5. Determination of buckling pressure

In our study we used a similar method as Theng`s study on cone-cylinder intersection [7]. In this approach the curves of load-displacement for nodes in one circumferential path near cylinder-head intersection were plotted. In the initial stage of loading, the curves for all nodes were similar, indicating dominantly axis-symmetric behavior. As the pressure reached a certain value, the curves of nodes at different locations started to diverge from each other. The divergence of these curves is an indication of the growth of non-symmetric buckling deformations. The load corresponded to the divergence point is critical buckling load (Fig. 3).

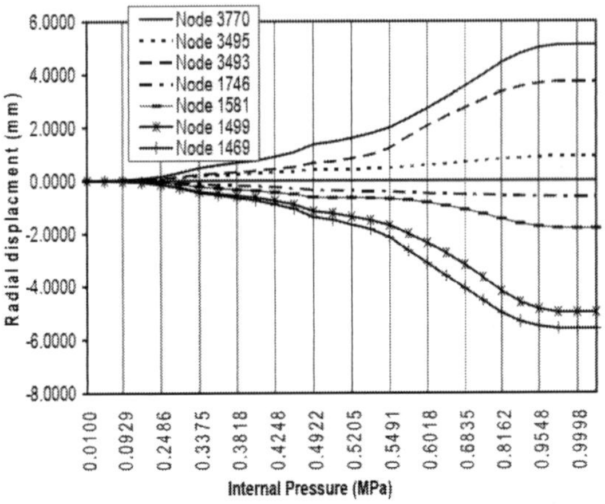

Fig. 3. Determination of buckling pressure.
(Curve of load-radial displacement for torispherical head (t/L=0.002, r/L=0.06)

6. Parametrical study

The Geometry of torispherical head was introduced with the t/L, L/D and r/L parameters in which t is the thickness of vessel which has identical values in heads and cylindrical part, L is the radius of spherical section, r is the Knuckle radius and D is the diameter of cylindrical section. The common heads are used for pressure vessels have radius of sphere equal to diameter of cylinder (L/D=1). Our study was limited to heads with t/L≤500 and r/L≥0.06..

Table 1- Buckling pressure of vessel with torispherical head (L/D=1).

t/L	r/L	P_{Ansys} (MPa)	P_{ECCS} (MPa)	Error %
0.002	0.06	0.20872	0.179	16
	0.08	0.22807	0.276	9
	0.1	0.34149	0.332	2.8
	0.14	0.43883	0.438	0.09
	0.17	0.48337	0.514	5.9
	0.2	0.542663	0.588	7
0.003	0.06	0.397296	0.400	0.67
	0.08	0.46551	0.507	8
	0.1	0.53235	0.610	12.7
	0.14	0.8196	0.805	1.7
	0.17	0.95637	0.944	1.2
	0.2	1.058	1.08	1.9

The data was compared with the design rules given in European Convention for Constructional Steelwork (ECCS) [6]. The ECCS rules are based on buckling of the knuckle and the limit pressure is as given in Eq. (1):

$$\frac{P_b}{F_y} = \frac{120c(\frac{r}{D})^{0.825}}{(D/t)^{1.5}(\frac{L}{D})^{1.15}} \tag{1}$$

Where c=1.0 for crown and segment steel heads and c=1.6 for cold-spun steel heads.

By verification performed with ECCS code, as Table 1, Fig.4 and Fig. 5 illustrate, it was confirmed that the nonlinear buckling analysis could assure accurate results for buckling strength. The discrepancy between numerical analysis (FE) and ECCS [6] corresponded to the geometrical imperfection and residual stress which were taken into account by ECCS code and not by the FE.

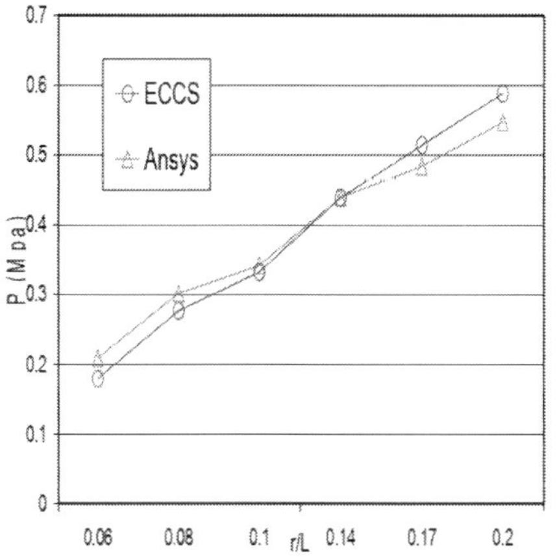

Fig. 4. Curve of critical pressure versus r/L (t/L=0.002)

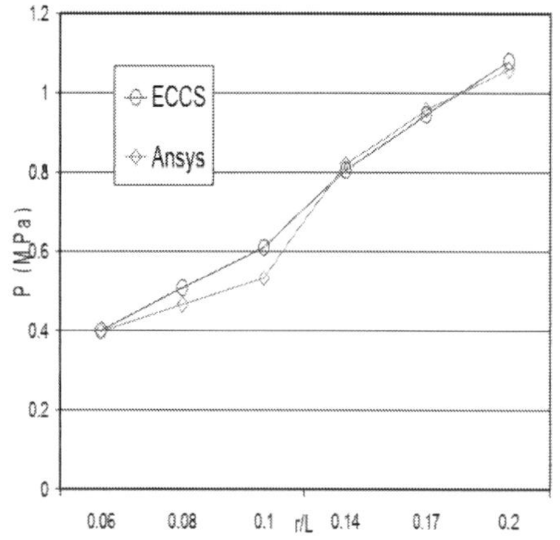

Fig. 5. Curve of critical pressure versus r/L (t/L=0.003)

6.1. Influence of knuckle radius on buckling pressure

In the internal pressure vessels, due to the existence of circumferential tensile stresses in both cylindrical and spherical parts, the intersection is deformed to internal side. Thus, both of the spherical and cylindrical part near the intersection was subjected to circumferential compressive stresses, and so buckling deflection occurred in both of them [10].

The growths of buckles can now be clearly seen in Fig. 6. The number of periodical waves on the ring can be counted from this plot to be 39. It should be noted that this counted number is a rough indication, as the buckling waves are not so uniform.

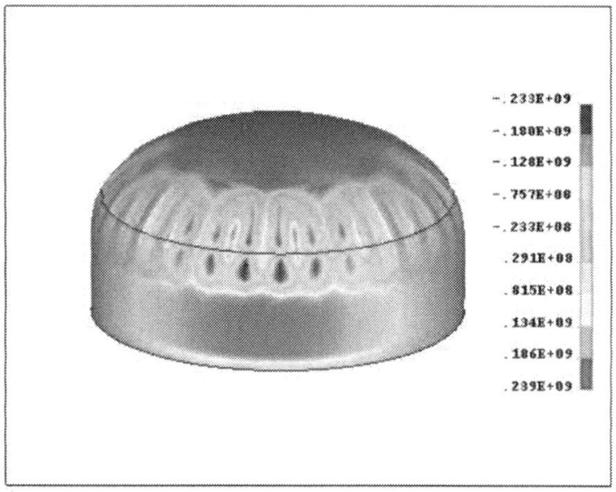

Fig. 6. Influence of knuckle radius t/L=0.002 and r/L=0.06.

Numerical results for the all points of intersection between the head and cylinder have shown clearly that the post-buckling behavior of internally pressurized sphere–cylinder intersections is stable (Fig. 6).

The curves of Fig. 7 show the influence of knuckle radius on the pressure buckling with different thicknesses. In the analysis the value of the radius of spherical part (L) was kept constant (L=0.5 m). For varying the ratio r/L, only the knuckle radius (r) was varied. We observe that for the whole thicknesses, increasing of radius leads to increasing of buckling pressure. So the knuckle radius is an influent parameter for increasing of buckling resistance.

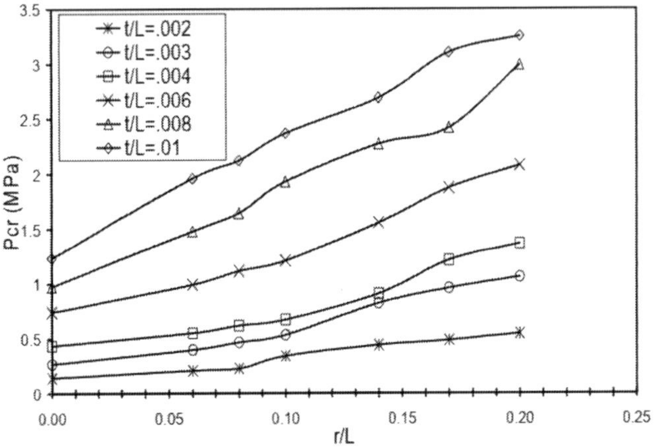

Fig. 7. Influence of knuckle radius on the buckling behavior of the torispherical head.

6.2. Influence of thickness on buckling pressure

Fig. 8 illustrates the influence of t/L on the buckling pressure for different ratios of r/L. Increasing of t/L leads to increasing of buckling pressure. Also the rate of increase, comparing to increasing resulted from the ratio r/L, is higher. The slope of the curves of Fig. 8 comparing to those of the Fig. 7 shows it. Thus the buckling pressure is more sensitive to the thickness.

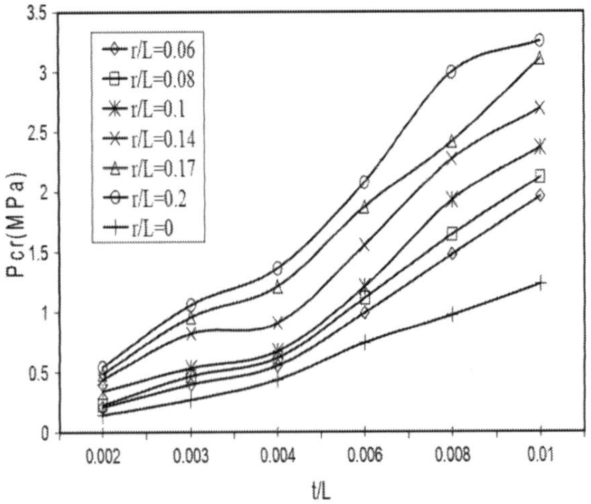

Fig. 8. Influence of t/L on the buckling pressure of torrispherical head.

7. Result and discussion

7.1. Analysis of development of circumferential wrinkles

Due to the buckling, circumferential wrinkles are developed. The number and amplitude of theses wrinkles increase by increasing of internal pressure. The number and amplitude of developed wrinkles are a criterion to evaluate the buckling resistance [11]. Creation and development of wrinkles could be followed by the curve of radial displacement versus distance in a circumferential path in the vicinity of intersection of cylinder and head of the vessel (Fig. 9). Although by accounting of the number of wrinkles in this curve, the number of circumferential wrinkles in the head could be obtained.

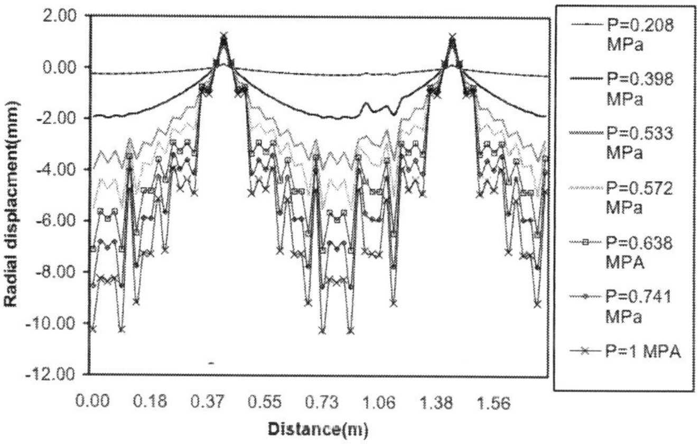

Fig. 9. Development of buckling wrinkle
r/L=0.06 and t/L=0.002.

7.2. Comparing of buckling pressure with limit pressure

The limit pressure of vessel (P_L) with torispherical head is computed with Eq. (2) [12] derived by Shield and Drucker [13].

$$\frac{P_{SD}}{F_y} = (0.33+5.5\frac{r}{D})(\frac{t}{L})+28(1-2.2\frac{r}{D})(\frac{t}{L})^2-0.0006 \quad (2)$$

In which P_{SD} is the limit pressure (psi), $P_{SD} = P_L$ and Fy is the yield stress (ksi).

Fig. 10 shows the critical buckling pressure (P_{cr}) versus the pressure obtained from relation (2). The curve shows that the buckling pressure is most often higher than the limit pressure. It means that the limit pressure is more critical than the buckling pressure. Thus for vessels with torispherical head the buckling pressure as a design criterion will not be sufficient.

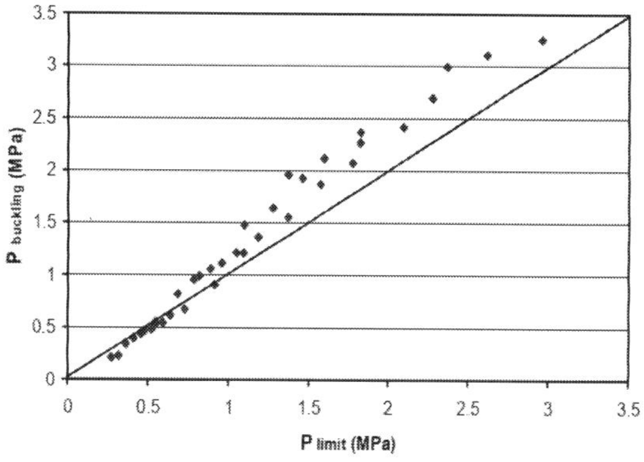

Fig. 10. Curve of buckling pressure versus limit pressure.

Conclusion

In this research, we modeled intersection of cylinder-head of a pressure vessel to **analyze buckling** using the finite element approach. A three dimensional finite element model was generated. Non-linear finite element method with large deflection analysis including geometric and material non-linearity was performed. By resulted displacement of different buckling resolution, a new model with geometrical imperfection was obtained. In result following points were detected:

• The non-linear FE analysis brings the numerical results in the vicinity of experimental. Scatters are generally due the geometrical imperfections and residual stresses in vessel.

• Buckling pressure is influenced by the thickness and height of the vessel. Higher thickness and height lead to a better buckling resistance.

• The influence of knuckle radius in decreasing of compression stresses and also increasing of the buckling pressure is underlined.

• Using of the buckling pressure criterion in design of torispherical heads would be conservative.

References

[1] Teng, J.G. and Hong, T., 1998, "Non-linear thin shell theories for numerical buckling predictions", Journal of Thin-Walled Structures, 31, pp. 89-115.

[2] Wang, X., Xiao, J., and Zhang, Y.C., 2004. "A method for solving the buckling problem of a thiheln walled sl", International journal of pressure vessels and piping, 81, pp. 907-912.

[3] Pinna, R., and Ronald, B.F., 2000, "Hydrostatic buckling of shells with various boundary conditions", Journal of Constructional Steel Research, **56**, pp. 1-16.

[4] Teng, J.G., and Song, C.Y., 2001, "Numerical models for nonlinear analysis of elastic shells with eigenmode-affine imperfections", International journal of Solids and Structures, **38**, pp. 3263-3280.

[5] Blachut, J., 1998, "Buckling of sharp knuckle torispheres under external pressure", Thin walled structure, **30**, No. 1-4, pp 55-77.

[6] European Convention for Constructional Steelwork ECCS, 1998, Technical Working Group 8.4 (Ed.), Buckling of steel shells European Recommendation, 4th Edition, ECCS, Brussels.

[7] Zhao, Y., and Teng, J.G., 2003, "A stability design proposal for cone-cylinder intersections under internal pressure", Int. J. of Pressure and Vessel and Piping; **80**, No.5, 297-309.

[8] Kenny, S., Pegg, N., and Taheri F., 2000, "Dynamic elastic buckling of a slender beam with geometric imperfections subject to an axial impulse, Finite element in analysis and design", **35**, pp. 227-246.

[9] Turvey, G.J., and Zhang, Y, 2006, "A computational and experimental analysis of the buckling, post buckling and initial failure of pultruded GRP columns", Journal of Computers and Structures, **84**, pp. 1527-1537.

[10] Teng, J.G., 1996, "Elastic buckling of cone-cylinder intersection under localized circumferential buckling", Journal of Engineering Structure, **18**, pp. 41.

[11] Teng, J.G., and Hong-Wei, M., 1999, "Elastic buckling of ring-stiffened cone-cylinder intersections under internal pressure", ASME International journal of Mechanical Science, **41**, pp. 357-383.

[12] Miller, C.D., Aug. 2001, "Buckling criteria for torispherical heads under internal pressure", International Journal of Pressure Vessel Technology, **123**, pp. 318-323.

[13] Shield, R.T., and Drucker, D.C., 1961, "Design of thin walled torispherical and toriconical pressure vessel heads", ASME Int. Journal of Applied Mechanics, **28**, pp. 292- 297.

Journal of Physics: Conference Series **181** (2009) 012041 doi:10.1088/1742-6596/181/1/012041

Experimental study of thermal stresses in a bonded joint

A. Deheeger[a,*], C. Badulescu[a], J.D. Mathias[b], M. Grédiac[a]

[a]Laboratoire de Mécanique et Ingénieries, Institut Français de Mécanique Avancée-Université Blaise Pascal Clermont II, Campus des Cézeaux, BP 265, 63175 Aubière Cedex, France

[b]Laboratoire d'Ingénierie pour les Systèmes Complexes, CEMAGREF, Campus des Cézeaux, 24 avenue des Landais - BP 50085, 63172 Aubière Cedex, France

antoine.deheeger@ifma.fr

Abstract. This paper deals with the assessment of the thermal stress distribution in a composite/aluminium bonded joint. An aluminum specimen symmetrically reinforced with two composite patches is heated while both thermal and displacement fields are measured with an infrared camera and a CCD camera, respectively. The grid method is used to obtain the displacement fields in the composite patch. This displacement field is in good agreement with theoretical expectations. This enables us to deduce the longitudinal strain in the composite patch by fitting the displacement field with a suitable model and differentiating it. The shear stress peak which takes place in the adhesive near the free edge of the patch is finally estimated.

1. Introduction

Composite patches are often used to reinforce damaged aeronautical structures. Such structures are subjected to temperature changes which may cause a thermal stress distribution to appear in the bonded joint between composite and metallic substrate. This phenomenon is due to the difference of Coefficient of Thermal Expansion (CTE) between the two adherends.

The aim of this work is to perform full-field measurements in order to characterize the effect of a temperature change in a metal/composite specimen and to deduce the resulting shear stress distribution in the bonded joint. The patched specimen under study is heated. The temperature change is measured on one side of the specimen with an infrared camera and the displacement distribution is obtained on the other side using a CCD camera and a relevant image processing [1] [2]. It is possible to deduce the thermal shear stress distribution in the adhesive using these two types of measurement if they are processed with a suitable model [3] and if the specimen is assumed to be symmetric with respect to its mid-plane.

The experimental set-up is presented below. The first results obtained are discussed and the effect of temperature changes on the displacement/strain fields is experimentally evidenced.

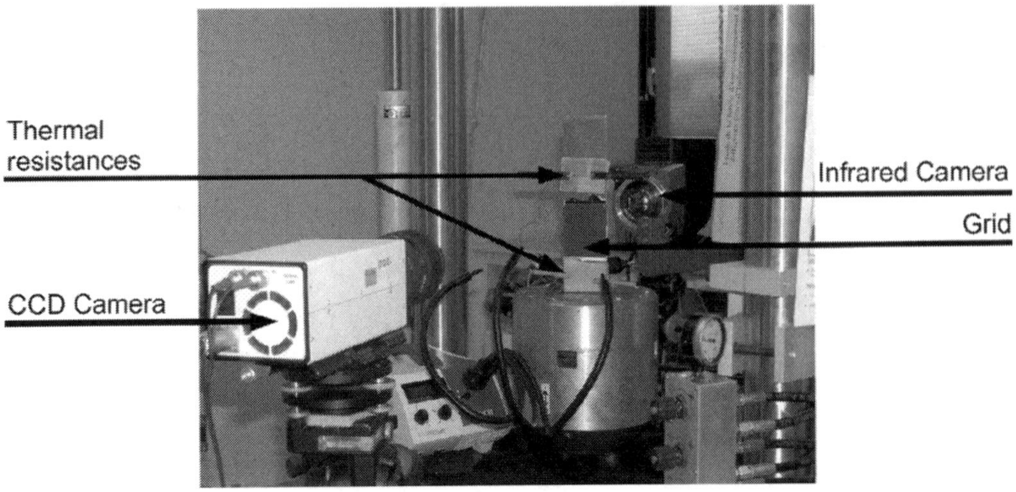

Figure 1 - Experimental set-up.

2. Experimental set-up

2.1. Introduction

The aim of this test is to measure both the thermal and the displacement fields during the same experiment. For this purpose, a CCD camera and an infrared camera are placed in front of each side of the specimen to measure the displacement and temperature fields, respectively (see Figure 1)

2.2. Specimen

The specimen is made of 2024T3 aluminum. Its length, width and thickness are 340 mm, 70 mm and 3 mm, respectively. Aluminum is supposed to be linear isotropic elastic (E_s = 73.8 GPa, v_s = 0.33, α_s=23.6e^{-6} K^{-1}). Composite patches are bonded on each side of the specimen. Their length, width and thickness are 70 mm, 70 mm and 0.5 mm, respectively. The composite is assumed to be orthotropic and elastic (E_x = 181 GPa, E_y = 10.3 GPa, v_{xy} = 0.28, G_{xy} = 7 GPa, α_x=0.02e^{-6} K^{-1}, α_y=22.5e^{-6} K^{-1}). The adhesive used is the Redux 312 supplied by Hexcel. It is assumed to be thermoelastic. Its shear modulus depends on the temperature, as shown in Figure 2 where the real part of the shear modulus G' measured with a suitable rheometer is shown. The thickness of each of the two adhesive layers is 0.15 mm. The composite patches are not tapered near the free edges. This enables us to compare the results observed with some simple analytical models [3] [4]

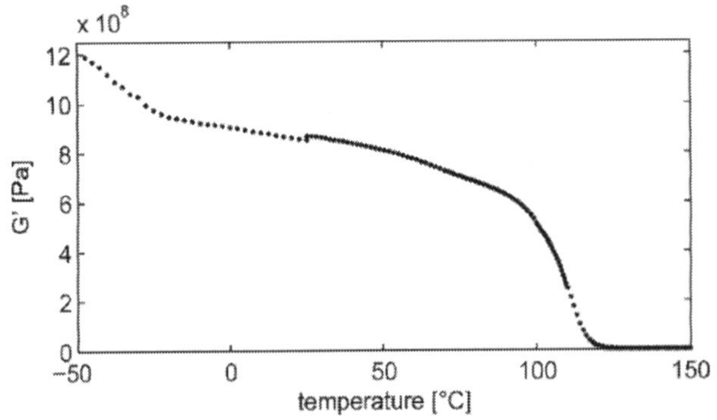

Figure 2 - Evolution of the shear modulus G' of the Redux 312 versus temperature.

2.3. Measurement of the displacement/strain field

The grid method is used to obtain the displacement field on the surface of the patch. This method has been developed by Surrel [2]. It is based on the analysis of the deformation of a grid bonded on the surface of the specimen prior to testing [5]. A camera captures the light intensities on the surface before and after loading. Images are processed by means of an appropriate algorithm [1] [2], which detects very slight variations of the grid pitch caused by the surface deformation. This algorithm calculates the displacement field throughout the surface under investigation. The grid used exhibits a period of 5 lines / mm. The 12 bit camera used exhibits 1376 x 1040 pixels. The distance between the camera and the specimen is adjusted in such a way that 5 pixels are used to discretize one period of the grid. Thus, only the upper-left corner of the patch, corresponding to a 55 mm x 42 mm rectangle, is observed in practice. 9 pixels are used to measure the displacement at a given point. Since one pixel corresponds to 40 μm, spatial resolution is equal to 360 μm.

2.4. Measurement of the temperature field

The second side of the specimen is painted with a black spray to obtain the greatest emissivity as possible and the camera is set to have a global overview of the temperature field on the upper surface of the composite patch. The infrared camera used is a CEDIP Jade. Its sensor exhibits 320 x 240 pixels. The thermal resolution of the camera is 0.02 K. The thermal loading is applied with four Minco 9.8 Ω thermal resistances deposited on each side of the specimen, at each end of the composite patches to obtain a symmetric thermal field.

3. Results

3.1. Temperature field

A typical temperature field on the upper side of the composite is presented in Figure 3(a) and the distribution of the minimum, maximum and average temperatures versus time is presented in Figure 3(b). This is the absolute temperature in °C. The temperature variation is deduced by subtracting the initial temperature which is equal to the room temperature: 20°C.

The temperature field is not strictly homogeneous because the heat flux is transmitted by the four resistances located at the ends of the patch. Hence, there is a variation of about 10°C between the ends and the center of the patch because of the heat exchange with ambient air. For this first approach however, the experimental results will be compared to the analytical results obtained with a constant temperature variation equal to the average temperature variation which is measured.

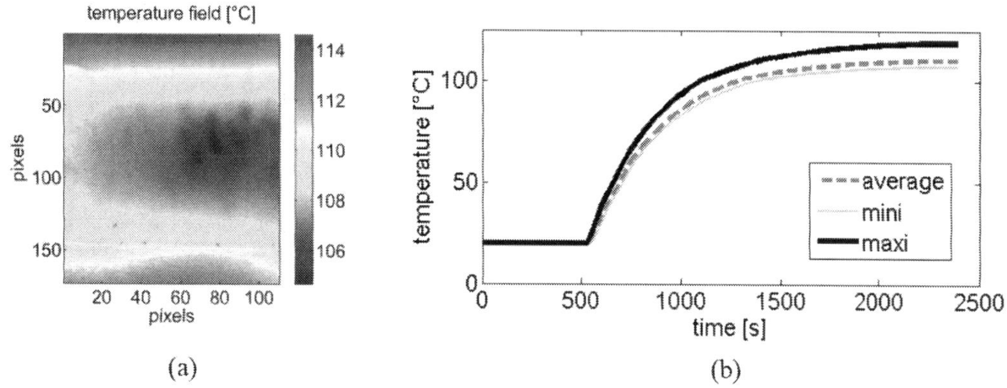

Figure 3 - (a) Final temperature field measured on the patch. (b) Evolutions of the minimum, maximum and average temperatures on the selected area.

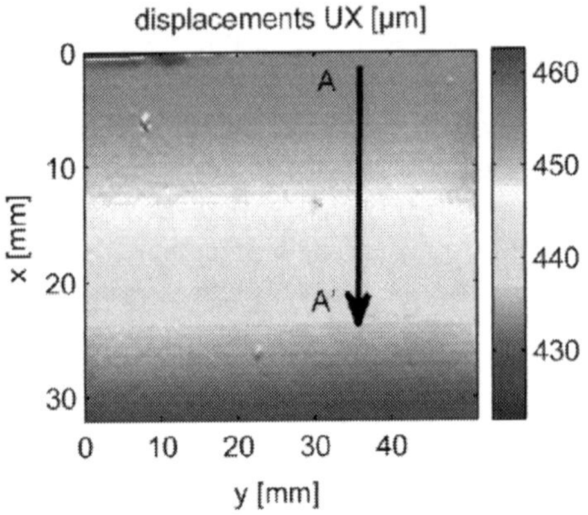

Figure 4 - Displacement field measured in x-direction for ΔT=90°C.

3.2. Displacement and strain fields

A typical displacement field along the x-direction is presented in Figure 4. The results along [AA'] are fitted with a classical shear-lag model suitable for bonded joints [4]. This displacement is described by the following equation

$$u_x^p(x) = \frac{A}{C}e^{Cx} + \frac{A+B}{C}e^{-Cx} + Bx + D \quad \textbf{(1)}$$

where A, B, C and D are optimized to minimize the distance with the experimental displacements.

Three displacement distributions obtained for ΔT = 10°C, 60°C and 90°C are presented in Figure 5(a). The corresponding shear stresses in the adhesive, presented in Figure 5(b) are deduced with the following equation

$$\tau_{xz}^a = e_p \frac{d\sigma_{xx}^p}{dx} = e_p E_p \frac{d\varepsilon_{xx}^p}{dx} = e_p E_p \frac{d^2 u_x^p}{dx^2} = e_p E_p C\left(Ae^{Cx} + (A+B)e^{-Cx}\right) \quad \textbf{(2)}$$

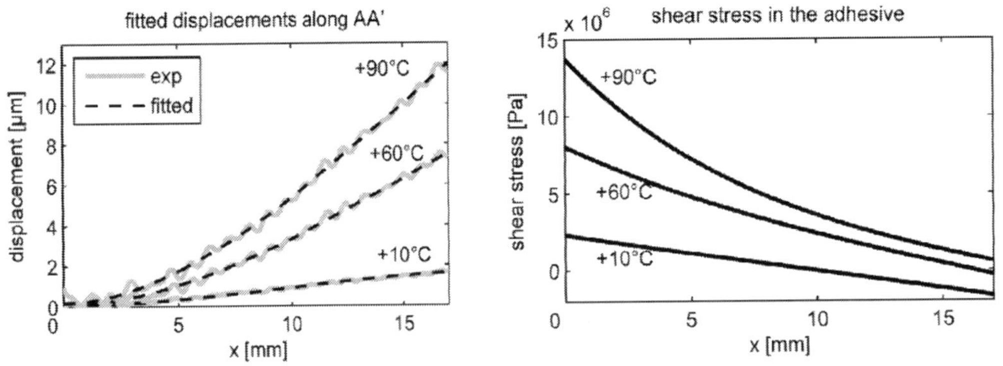

Figure 5 – (a) Fitted displacements along x-direction for three temperature variations (ΔT=10°C, 60°C and 90°C). (b) Resulting shear stress in the adhesive near the free edge for these three temperature variations.

As can be seen in Figure 5(a-b), there is a transfer zone at the free edge since the displacement in the composite patch progressively increases as x increases, whereas the shear stress is maximum in the adhesive near the free edge. This corresponds to a thermal load transfer between the substrate and the composite. The CTE of the substrate is higher than the CTE of the composite which brings about a tensile state in the composite patch. Interestingly, the amplitude of the shear stress peak (at the end of the patch, see Figure 5(b)) is not proportional to the temperature variation. This is due to the fact that the shear modulus of the adhesive is not a constant over the whole temperature variation imposed during the test (20°C to 110°C), as can be seen in Figure 2.

The displacement away from the free edge is linear. This corresponds to a constant value for the strain. This strain obtained at different temperatures at the center of the patch is compared with the strain given by the classical theory for bonded joints [6].

$$\varepsilon_{xx}^{p}(x)=\frac{\chi}{\xi\,E_p}\left[\cosh(\sqrt{\xi}x)+\frac{1-\cosh(\sqrt{\xi}l_x)}{\sinh(\sqrt{\xi}l_x)}\sinh(\sqrt{\xi}x)-1\right] \tag{3}$$

with

$$\begin{cases}\xi=\dfrac{G_a}{e_a}\left[\dfrac{1}{e_pE_e}+\dfrac{1}{e_sE_s}\right]\\ \chi=\dfrac{G_a(\alpha_p-\alpha_s)\Delta T}{e_ae_p}\end{cases} \tag{4}$$

where α_p and α_s are the CTE of the patch and the substrate, respectively, e_p, e_a and e_s, the thicknesses of the patch, the adhesive and the substrate, respectively, l_x the length of the patch, G_a the shear modulus of the adhesive, E_p and E_s the Young's modulus of the patch and the substrate, respectively, and ΔT the temperature variation.

A typical evolution of the composite strain at the center of the patch is shown in Figure 6 when the temperature increases from 20°C to 110°C. The theoretical strain is obtained using Equation 3 for $x=l_x/2$. The shear modulus G_a varies with the temperature as shown in Figure 2. As long as the adhesive is sufficiently rigid or the length of the patch sufficiently long, the transfer length is shorter than the half-length of the patch and the strain obtained at the center of the patch reaches its maximum value, given by

$$\varepsilon_{xx,\max}^{p}=\lim_{l_x\to\infty}\left[\varepsilon_{xx}^{p}(x=\frac{l_x}{2})\right]=\frac{(\alpha_p-\alpha_s)\Delta T}{1+\dfrac{e_pE_p}{e_sE_s}} \tag{5}$$

The evolution of this strain versus temperature in this case is linear and does not depend on the value of G_a. This corresponds to the linear part of the curves in Figure 6, where both experimental and theoretical results are in good agreement up to 80°C.

The second part of the curves, beyond 80°C, includes a peak corresponding to a critical temperature at which even if the temperature variation ΔT increases, the adhesive becomes too flexible to completely transfer the loading and thus the maximum value given by Equation 3 is not reached.

This phenomenon may only appear in the case of an increasing temperature, since the evolution of parameters ΔT and G_a have opposite effects. This would not happen for a decreasing temperature, where both abs(ΔT) and G_a would increase and thus lead to an increasing strain. The theoretical strain peak is in good correlation with the experimental results, even if there is a shift in terms of temperature between the two curves. This shift may be explained by the non-homogeneity of the temperature field or by the uncertainties on the different parameters of the model.

4. Conclusion

An experimental study has been performed in order to highlight the influence of a thermal loading on the strain fields which take place in a bonded joint. Full-field measurement techniques have been used for this purpose. The temperature field is captured by an infrared camera. The displacement field is obtained with the grid method. Results obtained are in good agreement with theoretical expectations, even though some difference can be observed.

Further work is underway to take into account thermo-viscoelasticity properties of the adhesive and temperature gradient in the patch.

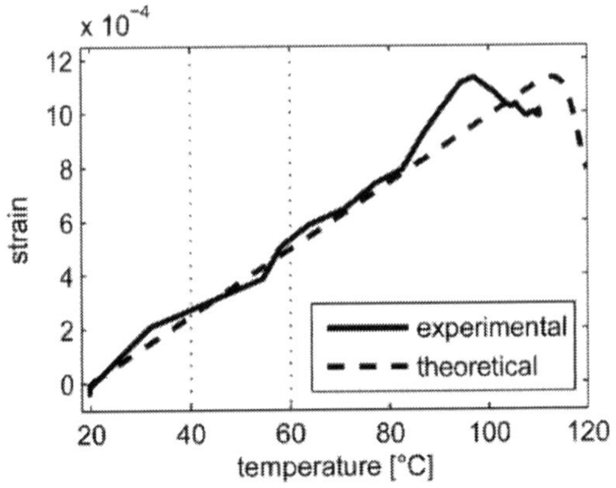

Figure 6 – Measured strain in the middle of the patch compared with its theoretical counterpart based on the average temperature measured.

Acknowledgements
The DGA/French Ministry of Defense is gratefully acknowledged for its support during this study.

References

[1] C. Badulescu, J.D. Mathias, M. Grédiac, and D. Roux. A procedure for accurate one-dimensional strain measurement with the grid method. *Experimental Mechanics*, 2008. Accepted for publication.

[2] Y. Surrel. Moiré and grid methods in optics : a signal-processing approach. In *Interferometry '94 : Photomechanics*, volume 2342, 1994. The International Society for Optical Engineering, SPIE.

[3] A. Deheeger, J.D. Mathias, and M. Grédiac. A closed-form solution for the thermal stress distribution in rectangular metal/composite bonded joints. *International Journal of Adhesion and Adhesives*, 2008. Accepted for publication.

[4] O. Volkersen. Die niektraft in zugbeanspruchten mit konstanten laschenquerschritten. *Luftfahrtforschung*, 15:41-47, 1938.

[5] J.L. Piro and M. Grédiac. Producing and tranferring low-spatial-frequency grids for measuring displacement fields with moiré and grid methods. *Experimental technics*, 28(4), 2004.

[6] L.J. Hart-Smith. Adhesive-bonded double-lap joints. Technical Report, CR-112235, NASA, 1973.

A novel tetrahedal element for static and dynamic analysis of laminated composites

I. A. Jones, W. Ruijter and A. C. Long

Division of Materials, Mechanics and Structures, Faculty of Engineering, University of Nottingham, University Park, Nottingham NG7 2RD

E-mail: arthur.jones@nottingham.ac.uk, woutruijter@gmail.com, andrew.long@nottingham.ac.uk

Abstract. This paper presents a novel approach to formulating finite elements for modelling laminated composites. Most FE systems provide laminated composite elements based upon layered shell elements and some provide layered brick elements, but these require a hexahedral mesh which is difficult to generate for arbitrary geometries. This paper describes a tetrahedral element based on the concept of an equivalent graded material whose properties vary smoothly (rather than stepwise) and whose global behaviour replicates that of the laminate. This allows the possibility of automatically-generated models involving laminated composite behaviour. The element requires a non-standard integration system, and is presented for the cases of elastic, thermoelastic and dynamic behaviour.

1. Introduction

This work is motivated by the desire to undertake finite element analysis of laminated components of non-trivial geometry, specifically where the geometry is defined within a solid modelling system and is difficult to reduce to an essentially prismatic form. Such geometries are often difficult or impractical to mesh with laminated brick or wedge elements, since arbitrary geometries can only be meshed fully automatically via tetrahedral elements.

No evidence could be found of existing laminated tetrahedral elements, so there appears to be little to start from. There have, of course, been element formulations published relating to functionally graded materials (e.g. Panda *et al.* [1]). Conversely, laminated elements have been used to analyse functionally graded materials (e.g. Lee [2]). However, we are not aware of any existing formulation which approximates a laminate as an equivalent graded material.

The present paper describes a novel approach to this problem, in which a conventional laminated formulation is abandoned in favour of one based on a fictitious graded material, which, when subjected to typical loads (mechanical, thermoelastic and dynamic), exhibit equivalent behaviour to the genuine laminated composite material. Integrations within the element can then be carried out over the continuum of the element without any need to discretise the element into awkward sub-regions and (unlike laminated bricks and wedges) with no requirement for element to be aligned with the planes of the true laminate.

2. Elastically-equivalent graded material

The formulation of the element follows a number of discrete stages.

(i) In order to formulate the element, an assumption must be made on the manner in which displacements (and other quantities) vary within the element. Here, a quadratic variation is assumed within a given element and hence within a given region of the laminated material.

(ii) The concept is then established of an "representative sub-laminate" of the true (laminated) material, spanning the extremities of the through-thickness nodal positions of the element whose stiffness is to be evaluated. A unit projected area of this sub-laminate is assumed to be subjected to a strain field varying in the manner defined in item i, analogous to (but with a higher order than) the linear variation of in-plane displacement assumed within classical laminate theory (CLT) [3]:

$$\boldsymbol{\epsilon}(z) = \begin{bmatrix} \epsilon_x & \epsilon_y & \epsilon_z & \gamma_{yz} & \gamma_{zx} & \gamma_{xy} \end{bmatrix}^{\mathrm{T}} = \boldsymbol{\epsilon}^0 + \boldsymbol{\kappa} z + \boldsymbol{\xi} z^2 \tag{1}$$

where $\boldsymbol{\epsilon}^0$, $\boldsymbol{\kappa}z$ and $\boldsymbol{\xi}z^2$ are vectors of coefficients, of which the first two are respectively analogous to the well-known mid-plane strain and curvature terms in CLT.

(iii) The strain energy in the true sub-laminate is evaluated in terms of the stiffness coefficients within each of its N laminae or plies, and in terms of the assumed displacement field. It will be assumed that the elastic behaviour at a given through-thickness position z is defined by the 6×6 stiffness matrix $\mathbf{C}(z)$, which for a real laminate will take vary in a piecewise-constant manner, having the value $\mathbf{C}_k$ within the kth ply. In general, each ply will be orientated arbitrarily, with its off-axis behaviour implicit in the definition of $\mathbf{C}$, without the use of an overbar as is sometimes used in CLT [3]. This leads to an expression of the form:

$$\begin{aligned} \text{S.E.} &= \frac{1}{2}\int_{-h/2}^{h/2} \left(\boldsymbol{\epsilon}^0 + \boldsymbol{\kappa}z + \boldsymbol{\xi}z^2\right) \mathbf{C}(z) \left(\boldsymbol{\epsilon}^0 + \boldsymbol{\kappa}z + \boldsymbol{\xi}z^2\right) dz \\ &= \frac{1}{2}\begin{bmatrix} \boldsymbol{\epsilon}^0 & \boldsymbol{\kappa} & \boldsymbol{\xi} \end{bmatrix} \begin{bmatrix} \mathbf{A}' & \mathbf{B}' & \mathbf{D}' \\ \mathbf{B}' & \mathbf{D}' & \mathbf{F}' \\ \mathbf{D}' & \mathbf{F}' & \mathbf{G}' \end{bmatrix} \begin{Bmatrix} \boldsymbol{\epsilon}^0 \\ \boldsymbol{\kappa} \\ \boldsymbol{\xi} \end{Bmatrix} \end{aligned} \tag{2}$$

where h is the thickness of the sub-laminate (numerically equal to the difference in normal through-thickness position of the uppermost and lowermost nodes of the tetrahedron) and where:

$$\mathbf{A}' = \sum_{k=1}^{N} \mathbf{C}_k \left(z_k - z_{k-1}\right), \quad \mathbf{B}' = \frac{1}{2}\sum_{k=1}^{N} \mathbf{C}_k \left(z_k^2 - z_{k-1}^2\right), \quad \mathbf{D}' = \frac{1}{3}\sum_{k=1}^{N} \mathbf{C}_k \left(z_k^3 - z_{k-1}^3\right),$$
$$\mathbf{F}' = \frac{1}{4}\sum_{k=1}^{N} \mathbf{C}_k \left(z_k^4 - z_{k-1}^4\right), \quad \mathbf{G}' = \frac{1}{5}\sum_{k=1}^{N} \mathbf{C}_k \left(z_k^5 - z_{k-1}^5\right) \tag{3a-e}$$

(iv) The properties (e.g. the coefficients of the elastic stiffness matrix) of the equivalent material are now assumed to be constant within a plane parallel to the planes of the laminate, but to vary with through-thickness position according to some order of polynomial. This polynomial order is chosen so that the number of coefficients (five in this case, defining a quartic) equals the number of independent constants $\mathbf{A}' - \mathbf{G}'$ appearing in equation (2):

$$\mathbf{C}(z) = \mathbf{C}^{(0)} + \mathbf{C}^{(1)}z + \mathbf{C}^{(2)}z^2 + \mathbf{C}^{(3)}z^3 + \mathbf{C}^{(4)}z^4 \tag{4}$$

where $\mathbf{C}^{(0)}$ etc. are 6×6 matrices of coefficients.

(v) The concept is then established of an "equivalent block" of the fictitious graded material, behaving elastically in the same manner as the representative sub-laminate of the material spanning the through-thickness extents of the element. This is then subjected to the various kinematic deformation modes described under section i, and the strain energy determined in terms of the unknown matrix coefficients in equation (4).

$$\text{S.E.} = \frac{1}{2}\int_{-h/2}^{h/2} \begin{bmatrix} \boldsymbol{\epsilon}^0 & \boldsymbol{\kappa} & \boldsymbol{\xi} \end{bmatrix} \begin{bmatrix} \mathbf{C}(z) & \mathbf{C}(z)z & \mathbf{C}(z)z^2 \\ \mathbf{C}(z)z & \mathbf{C}(z)z^2 & \mathbf{C}(z)z^3 \\ \mathbf{C}(z)z^2 & \mathbf{C}(z)z^3 & \mathbf{C}(z)z^4 \end{bmatrix} \begin{Bmatrix} \boldsymbol{\epsilon}^0 \\ \boldsymbol{\kappa} \\ \boldsymbol{\xi} \end{Bmatrix} dz \tag{5}$$

which can be represented in the form:

$$\text{S.E.} = \frac{1}{2} \begin{bmatrix} \boldsymbol{\epsilon}^0 & \boldsymbol{\kappa} & \boldsymbol{\xi} \end{bmatrix} \begin{bmatrix} \mathbf{A}'' & \mathbf{B}'' & \mathbf{D}'' \\ \mathbf{B}'' & \mathbf{D}'' & \mathbf{F}'' \\ \mathbf{D}'' & \mathbf{F}'' & \mathbf{G}'' \end{bmatrix} \begin{Bmatrix} \boldsymbol{\epsilon}^0 \\ \boldsymbol{\kappa} \\ \boldsymbol{\xi} \end{Bmatrix} \tag{6}$$

After evaluation of the integrals, and representing the summations in matrix form, the unknown matrix coefficients can be expressed via the following equation system:

$$\begin{Bmatrix} \mathbf{A}'' \\ \mathbf{B}'' \\ \mathbf{D}'' \\ \mathbf{F}'' \\ \mathbf{G}'' \end{Bmatrix} = \begin{bmatrix} h\mathbf{I} & \mathbf{0} & \frac{h^3}{12}\mathbf{I} & \mathbf{0} & \frac{h^5}{80}\mathbf{I} \\ \mathbf{0} & \frac{h^3}{12}\mathbf{I} & \mathbf{0} & \frac{h^5}{80}\mathbf{I} & \mathbf{0} \\ \frac{h^3}{12}\mathbf{I} & \mathbf{0} & \frac{h^5}{80}\mathbf{I} & \mathbf{0} & \frac{h^7}{448}\mathbf{I} \\ \mathbf{0} & \frac{h^5}{80}\mathbf{I} & \mathbf{0} & \frac{h^7}{448}\mathbf{I} & \mathbf{0} \\ \frac{h^5}{80}\mathbf{I} & \mathbf{0} & \frac{h^7}{448}\mathbf{I} & \mathbf{0} & \frac{h^9}{2304}\mathbf{I} \end{bmatrix} \begin{Bmatrix} \mathbf{C}^{(0)} \\ \mathbf{C}^{(1)} \\ \mathbf{C}^{(2)} \\ \mathbf{C}^{(3)} \\ \mathbf{C}^{(4)} \end{Bmatrix} \tag{7}$$

where the identity and zero matrices are all 6×6. The value of strain energy from equation (6) is then equated to that obtained in equation (2), resulting in the observation that $\mathbf{A}'' = \mathbf{A}'$ etc. Inserting this condition into equation (7) and performing the inversion analytically yields the following expression defining the matrix coefficients in the polynomial expression for elastic stiffness:

$$\begin{Bmatrix} \mathbf{C}^{(0)} \\ \mathbf{C}^{(1)} \\ \mathbf{C}^{(2)} \\ \mathbf{C}^{(3)} \\ \mathbf{C}^{(4)} \end{Bmatrix} = \begin{bmatrix} \frac{3.515625}{h}\mathbf{I} & \mathbf{0} & \frac{-65.625}{h^3}\mathbf{I} & \mathbf{0} & \frac{236.25}{h^5}\mathbf{I} \\ \mathbf{0} & \frac{75}{h^3}\mathbf{I} & \mathbf{0} & \frac{-420}{h^5}\mathbf{I} & \mathbf{0} \\ \frac{-65.625}{h^3}\mathbf{I} & \mathbf{0} & \frac{2205}{h^5}\mathbf{I} & \mathbf{0} & \frac{-9450}{h^7}\mathbf{I} \\ \mathbf{0} & \frac{-420}{h^5}\mathbf{I} & \mathbf{0} & \frac{2800}{h^7}\mathbf{I} & \mathbf{0} \\ \frac{236.25}{h^5}\mathbf{I} & \mathbf{0} & \frac{-9450}{h^7}\mathbf{I} & \mathbf{0} & \frac{44100}{h^9}\mathbf{I} \end{bmatrix} \begin{Bmatrix} \mathbf{A}' \\ \mathbf{B}' \\ \mathbf{D}' \\ \mathbf{F}' \\ \mathbf{G}' \end{Bmatrix} \tag{8}$$

(vi) The modulus matrix for the material can now be reconstructed at any chosen point by evaluating the polynomial expression for $\mathbf{C}(z)$ at the chosen value of z.

3. Thermoelastic behaviour

A broadly similar approach is taken for defining equivalent graded thermoelastic behaviour, though the approach here is to obtain graded thermoelastic properties which result in equivalent distributions of thermal stress for a given temperature distribution.

(i) An analogy is made with the use of "unit force" **N** and "unit moment" **M** vectors in classical laminate theory [3] (corresponding to values of direct and shear force and bending and twisting moment per unit plate width), where the constraint $\sigma_z = \tau_{xz} = \tau_{yz}$ is implied. These quantities are simply the integrals of the stress state $\boldsymbol{\sigma}$ and its product with z through the thickness of the laminate (where, for these equations (9a) and (9b) only, h refers to the thickness of the whole laminate, in accordance with conventional CLT notation):

$$\mathbf{N} = \int_{-h/2}^{h/2} \boldsymbol{\sigma}(z)dz = \int_{-h/2}^{h/2} \bar{\mathbf{Q}}(z)\boldsymbol{\epsilon}(z)dz \tag{9a}$$

$$\mathbf{M} = \int_{-h/2}^{h/2} z\boldsymbol{\sigma}(z)dz = \int_{-h/2}^{h/2} z\bar{\mathbf{Q}}(z)\boldsymbol{\epsilon}(z)dz \tag{9b}$$

(ii) In the present context we define broadly similar vectors of thermal "forces" etc. which will be denoted $\mathbf{N}'$, $\mathbf{M}'$ etc. which are now integrals of the (3D) thermally-induced stress state $\boldsymbol{\sigma}^{(T)}$ and its product with z, and additionally z^2. Some of the terms which result have the physical meaning of the forces and moments which arise when strains $\boldsymbol{\epsilon}^{(T)}$ due to thermal expansion are prohibited, and correspond to the resultant forces and moments required to force the expanded element of material back to its original shape. The terms involving integrals of $z^2\sigma_z$ etc. should be regarded as the summary of the effect of stress variations in a constrained, thermally-loaded material. Specifically, three sets of integrals are defined:

$$\mathbf{N}' = \int_{-h/2}^{h/2} \boldsymbol{\sigma}(z)dz = \int_{-h/2}^{h/2} \mathbf{C}(z)\boldsymbol{\epsilon}^(T)(z)dz = \int_{-h/2}^{h/2} \mathbf{C}(z)\boldsymbol{\alpha}(z)\Delta T(z)dz \tag{10a}$$

$$\mathbf{M}' = \int_{-h/2}^{h/2} z\boldsymbol{\sigma}(z)dz = \int_{-h/2}^{h/2} z\mathbf{C}(z)\boldsymbol{\epsilon}^(T)(z)dz = \int_{-h/2}^{h/2} z\mathbf{C}(z)\boldsymbol{\alpha}(z)\Delta T(z)dz \tag{10b}$$

$$\mathbf{P}' = \int_{-h/2}^{h/2} z^2\boldsymbol{\sigma}(z)dz = \int_{-h/2}^{h/2} z^2\mathbf{C}(z)\boldsymbol{\epsilon}^(T)(z)dz = \int_{-h/2}^{h/2} z^2\mathbf{C}(z)\boldsymbol{\alpha}(z)\Delta T(z)dz \tag{10c}$$

(iii) A temperature distribution is assumed (in the present case, a linear distribution), under which the real and equivalent materials are to be forced to behave in an equivalent manner:

$$\Delta T(z) = \Delta T^{(0)} + z\Delta T^{(1)} \tag{11}$$

(iv) For the true laminate, inserting the assumed temperature distribution into equations (10a)–(10c) results in the following expressions:

$$\mathbf{N}' = \sum_{N}^{k=1} \mathbf{C}_k\boldsymbol{\alpha}_k \int_{z_{k-1}}^{z_k} \left(\Delta T^{(0)} + z\Delta T^{(1)}\right) dz = \begin{bmatrix} \Delta T^{(0)} & \Delta T^{(1)} \end{bmatrix} \begin{Bmatrix} \mathbf{L}_1 \\ \mathbf{L}_2 \end{Bmatrix} \tag{12a}$$

$$\mathbf{M}' = \sum_{N}^{k=1} \mathbf{C}_k\boldsymbol{\alpha}_k \int_{z_{k-1}}^{z_k} z\left(\Delta T^{(0)} + z\Delta T^{(1)}\right) dz = \begin{bmatrix} \Delta T^{(0)} & \Delta T^{(1)} \end{bmatrix} \begin{Bmatrix} \mathbf{L}_2 \\ \mathbf{L}_3 \end{Bmatrix} \tag{12b}$$

$$\mathbf{P}' = \sum_{N}^{k=1} \mathbf{C}_k\boldsymbol{\alpha}_k \int_{z_{k-1}}^{z_k} z^2\left(\Delta T^{(0)} + z\Delta T^{(1)}\right) dz = \begin{bmatrix} \Delta T^{(0)} & \Delta T^{(1)} \end{bmatrix} \begin{Bmatrix} \mathbf{L}_3 \\ \mathbf{L}_4 \end{Bmatrix} \tag{12c}$$

where:

$$\mathbf{L}_1 = \sum_{N}^{k=1} \mathbf{C}_k\boldsymbol{\alpha}_k (z_k - z_{k-1}) \qquad \mathbf{L}_2 = \frac{1}{2}\sum_{N}^{k=1} \mathbf{C}_k\boldsymbol{\alpha}_k \left(z_k^2 - z_{k-1}^2\right)$$

$$\mathbf{L}_3 = \frac{1}{3}\sum_{N}^{k=1} \mathbf{C}_k\boldsymbol{\alpha}_k \left(z_k^3 - z_{k-1}^3\right) \qquad \mathbf{L}_4 = \frac{1}{4}\sum_{N}^{k=1} \mathbf{C}_k\boldsymbol{\alpha}_k \left(z_k^4 - z_{k-1}^4\right) \tag{13a-d}$$

(v) Considering now the fictitious functionally graded material, the coefficient of thermal expansion is henceforward assumed to vary as a polynomial in z, involving as many coefficients as there are independent coefficients $\mathbf{L}_1...\mathbf{L}_4$ in equation (13a-d):

$$\boldsymbol{\alpha}(z) = \boldsymbol{\alpha}^{(0)} + \boldsymbol{\alpha}^{(1)}z + \boldsymbol{\alpha}^{(2)}z^2 + \boldsymbol{\alpha}^{(3)}z^3 \tag{14}$$

It should also be recalled that $\mathbf{C}(z)$ is expressed as a polynomial in z via equation (4). These various expressions for $\Delta T(z)$, $\mathbf{C}(z)$ and $\boldsymbol{\alpha}(z)$ are into the following expressions for thermal forces, moments etc. $\mathbf{N}''$, $\mathbf{M}''$ and $\mathbf{P}''$ in the graded material:

$$\mathbf{N}'' = \int_{-h/2}^{h/2} \mathbf{C}(z)\boldsymbol{\alpha}(z)\Delta T(z)dz, \qquad \mathbf{M}'' = \int_{-h/2}^{h/2} z\mathbf{C}(z)\boldsymbol{\alpha}(z)\Delta T(z)dz,$$
$$\mathbf{P}'' = \int_{-h/2}^{h/2} z^2\mathbf{C}(z)\boldsymbol{\alpha}(z)\Delta T(z)dz \tag{15a-c}$$

After some manipulation including rearrangement into matrix form, the following expressions are obtained:

$$\mathbf{N}'' = \left[\begin{array}{cc} \Delta T^{(0)} & \Delta T^{(1)} \end{array}\right] \left[\begin{array}{cccc} \mathbf{K}_1 & \mathbf{K}_2 & \mathbf{K}_3 & \mathbf{K}_4 \\ \mathbf{K}_2 & \mathbf{K}_3 & \mathbf{K}_4 & \mathbf{K}_5 \end{array}\right] \left[\begin{array}{cccc} \boldsymbol{\alpha}^{(0)} & \boldsymbol{\alpha}^{(1)} & \boldsymbol{\alpha}^{(2)} & \boldsymbol{\alpha}^{(3)} \end{array}\right]^{\mathrm{T}} \tag{16a}$$
$$\mathbf{M}'' = \left[\begin{array}{cc} \Delta T^{(0)} & \Delta T^{(1)} \end{array}\right] \left[\begin{array}{cccc} \mathbf{K}_2 & \mathbf{K}_3 & \mathbf{K}_4 & \mathbf{K}_5 \\ \mathbf{K}_3 & \mathbf{K}_4 & \mathbf{K}_5 & \mathbf{K}_6 \end{array}\right] \left[\begin{array}{cccc} \boldsymbol{\alpha}^{(0)} & \boldsymbol{\alpha}^{(1)} & \boldsymbol{\alpha}^{(2)} & \boldsymbol{\alpha}^{(3)} \end{array}\right]^{\mathrm{T}} \tag{16b}$$
$$\mathbf{P}'' = \left[\begin{array}{cc} \Delta T^{(0)} & \Delta T^{(1)} \end{array}\right]^{\mathrm{T}} \left[\begin{array}{cccc} \mathbf{K}_3 & \mathbf{K}_4 & \mathbf{K}_5 & \mathbf{K}_6 \\ \mathbf{K}_4 & \mathbf{K}_5 & \mathbf{K}_6 & \mathbf{K}_7 \end{array}\right] \left[\begin{array}{cccc} \boldsymbol{\alpha}^{(0)} & \boldsymbol{\alpha}^{(1)} & \boldsymbol{\alpha}^{(2)} & \boldsymbol{\alpha}^{(3)} \end{array}\right]^{\mathrm{T}} \tag{16c}$$

where the 6×6 submatrices $\mathbf{K}_1 ... \mathbf{K}_7$ are as follows:

$$\mathbf{K}_1 = \mathbf{C}^{(0)} + \mathbf{C}^{(2)}\frac{h^3}{12} + \mathbf{C}^{(4)}\frac{h^5}{80}, \qquad \mathbf{K}_2 = \mathbf{C}^{(1)}\frac{h^3}{12} + \mathbf{C}^{(3)}\frac{h^5}{80},$$
$$\mathbf{K}_3 = \mathbf{C}^{(0)}\frac{h^3}{12} + \mathbf{C}^{(2)}\frac{h^5}{80} + \mathbf{C}^{(4)}\frac{h^7}{448}, \qquad \mathbf{K}_4 = \mathbf{C}^{(1)}\frac{h^5}{80} + \mathbf{C}^{(3)}\frac{h^7}{448},$$
$$\mathbf{K}_5 = \mathbf{C}^{(0)}\frac{h^5}{80} + \mathbf{C}^{(2)}\frac{h^7}{448} + \mathbf{C}^{(4)}\frac{h^9}{2304}, \qquad \mathbf{K}_6 = \mathbf{C}^{(1)}\frac{h^7}{448} + \mathbf{C}^{(3)}\frac{h^9}{2304}, \tag{17a-g}$$
$$\mathbf{K}_7 = \mathbf{C}^{(0)}\frac{h^7}{448} + \mathbf{C}^{(2)}\frac{h^9}{2304} + \mathbf{C}^{(4)}\frac{h^{11}}{11264}$$

(vi) Setting $\mathbf{N}' = \mathbf{N}''$ etc., it is apparent from equations (12) and (16a) that the following set of equations can be constructed:

$$\left\{\begin{array}{c} \mathbf{L}_1 \\ \mathbf{L}_2 \\ \mathbf{L}_3 \\ \mathbf{L}_4 \end{array}\right\} = \left[\begin{array}{cccc} \mathbf{K}_1 & \mathbf{K}_2 & \mathbf{K}_3 & \mathbf{K}_4 \\ \mathbf{K}_2 & \mathbf{K}_3 & \mathbf{K}_4 & \mathbf{K}_5 \\ \mathbf{K}_3 & \mathbf{K}_4 & \mathbf{K}_5 & \mathbf{K}_6 \\ \mathbf{K}_4 & \mathbf{K}_5 & \mathbf{K}_6 & \mathbf{K}_7 \end{array}\right] \left\{\begin{array}{c} \boldsymbol{\alpha}^{(0)} \\ \boldsymbol{\alpha}^{(1)} \\ \boldsymbol{\alpha}^{(2)} \\ \boldsymbol{\alpha}^{(3)} \end{array}\right\} \tag{18}$$

This equation system is straightforward to solve in order to find $\boldsymbol{\alpha}^{(0)} ... \boldsymbol{\alpha}^{(3)}$.

4. Equivalent dynamic behaviour

A similar approach is can again be taken for the formulation of the equivalent density of the graded material for use in dynamic calculations, though it is emphasised that this remains untested at the time of writing. The velocity $\dot{u}$ at a point may be expressed in terms of through-thickness position z:

$$\dot{\mathbf{u}}(z) = \dot{\mathbf{u}}^{(0)} + \dot{\mathbf{u}}^{(1)}z + \dot{\mathbf{u}}^{(2)}z^2 \tag{19}$$

The kinetic energy per unit area is as follows:

$$\begin{aligned} \text{K.E.} &= \frac{1}{2}\int_{-h/2}^{h/2} \rho(z)(\dot{\mathbf{u}}(z))^{\mathrm{T}}\dot{\mathbf{u}}(z)dz \\ &= \frac{1}{2}\int_{-h/2}^{h/2} \rho(z) \left[\begin{array}{ccc} \dot{\mathbf{u}}^{(0)} & \dot{\mathbf{u}}^{(1)} & \dot{\mathbf{u}}^{(2)} \end{array}\right] \left[\begin{array}{ccc} \mathbf{I} & z\mathbf{I} & z^2\mathbf{I} \\ z\mathbf{I} & z^2\mathbf{I} & z^3\mathbf{I} \\ z^2\mathbf{I} & z^3\mathbf{I} & z^4\mathbf{I} \end{array}\right] \left\{\begin{array}{c} \dot{\mathbf{u}}^{(0)} \\ \dot{\mathbf{u}}^{(1)} \\ \dot{\mathbf{u}}^{(2)} \end{array}\right\} \end{aligned} \tag{20}$$

For the true laminate, this evaluates to:

$$\text{K.E.} = \frac{1}{2}\left[\begin{array}{ccc} \dot{\mathbf{u}}^{(0)} & \dot{\mathbf{u}}^{(1)} & \dot{\mathbf{u}}^{(2)} \end{array}\right]\left[\begin{array}{ccc} M_1\mathbf{I} & M_2\mathbf{I} & M_3\mathbf{I} \\ M_2\mathbf{I} & M_3\mathbf{I} & M_4\mathbf{I} \\ M_3\mathbf{I} & M_4\mathbf{I} & M_5\mathbf{I} \end{array}\right]\left\{\begin{array}{c} \dot{\mathbf{u}}^{(0)} \\ \dot{\mathbf{u}}^{(1)} \\ \dot{\mathbf{u}}^{(2)} \end{array}\right\} \tag{21}$$

where:

$$M_1 = \sum_{k=1}^{N} \rho_k \left(z_k - z_{k-1}\right) \qquad \ldots \qquad M_5 = \frac{1}{5}\sum_{k=1}^{N} \rho_k \left(z_k^5 - z_{k-1}^5\right) \tag{22a-e}$$

For the graded material, the following variation of density ρ with position z is assumed in order to give an equal number of unknown coefficients as independent terms $M_1 ... M_5$:

$$\rho(z) = \rho^{(0)} + \rho^{(1)}z + \rho^{(2)}z^2 + \rho^{(3)}z^3 + \rho^{(4)}z^4 \tag{23}$$

The expression for kinetic energy can now be evaluated as

$$\text{K.E.} = \frac{1}{2}\left[\begin{array}{ccc} \dot{\mathbf{u}}^{(0)} & \dot{\mathbf{u}}^{(1)} & \dot{\mathbf{u}}^{(2)} \end{array}\right]\left[\begin{array}{ccc} M_1'\mathbf{I} & M_2'\mathbf{I} & M_3'\mathbf{I} \\ M_2'\mathbf{I} & M_3'\mathbf{I} & M_4'\mathbf{I} \\ M_3'\mathbf{I} & M_4'\mathbf{I} & M_5'\mathbf{I} \end{array}\right]\left\{\begin{array}{c} \dot{\mathbf{u}}^{(0)} \\ \dot{\mathbf{u}}^{(1)} \\ \dot{\mathbf{u}}^{(2)} \end{array}\right\} \tag{24}$$

where, in a similar manner to equations (6) and (7), $M_1' = \rho^{(0)}h + \rho^{(2)}\frac{h^3}{12} + \rho^{(4)}\frac{h^5}{80}$ etc. By equating $M_1 = M_1'$ etc., expressing the equations in matrix form and inverting analytically following a very similar procedure to that outlined in section 2, the coefficients $\rho^{(0)} ... \rho^{(4)}$ can be found.

5. Non-standard integration scheme

It is apparent from the above process that, as the material properties vary within the element, the order of variation of strain energy density will be higher than that encountered within conventionally-formulated elements. This implies that, in order to capture the element stiffness accurately, a high integration order must be used within the element. However, this high-order variation occurs only in the through-thickness direction, so it would be wasteful to employ high-order integration in the in-surface directions. A novel integration scheme was therefore developed involving different integration orders in different directions. It was extended from a concept by Rathod *et al.*, [4] who developed integration schemes of arbitrary order by degenerating straightforward hexahedral Gaussian integration schemes. The approach taken here is to degenerate a pentahedral (wedge) integration scheme so that triangular integration is used within planes parallel to the base of the tetrahedron, but an arbitrary order of conventional one-dimensional Gaussian integration is performed in the "height" direction. Full integration can therefore be retained for the direction closest to through-thickness while reduced integration similar to that used in reduced-integration wedge elements is used for integration within planes approximately parallel to the planes of the plies. In practice, the orientation of the element is unknown *a priori* so the integration scheme was further modified by permuting the three local axes. A fourth case, relating to the situation where the "base" of the element maps to the oblique face of the local element, is also obtained. The four possible integration schemes are illustrated in Figure 1. If none of the faces is close to being parallel to the plane of the plies, full integration can of course be used.

6. Trial FE implementation

A full trial implementation of the concept for static mechanical loads has taken place within the LibMesh open-source programming framework [5]. The program makes it possible to compare the results of the present element with those of an existing (hexahedral) element.

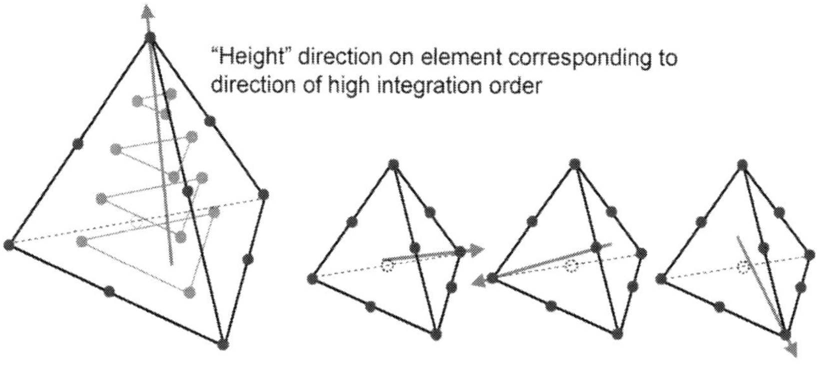

Figure 1. Four variants on the non-standard integration scheme.

Formal verification of the element has not yet taken place, but preliminary verification of the static elastic and thermoelastic behaviour demonstrates the potential usefulness of the element The test case consists of the unit cube, restrained against normal movement on the surfaces $x = 0$, $y = 0$ and $y = 1$, and with the node at $(0, 0, 0)$ restrained against movement in the z-direction. Three sets of properties (given in consistent units) have been chosen:

(i) Isotropic: $E = 100000$, $\nu = 0.3$, $G = 38461$, $\alpha = 13 \times 10^{-6}\,°C$

(ii) Cross-ply asymmetric: $E_1 = 50000$, $E_2 = E_3 = 10000$, $\nu_{12} = 0.3$, $\nu_{23} = 0.35$, $G_{12} = G_{13} = 5000$, $G_{23} = 4000$, $\alpha_1 = 5 \times 10^{-6}\,°C$, $\alpha_2 = \alpha_3 = 15 \times 10^{-6}\,°C$ 4 plies of 0.25 thickness each, with the ply sequence $0°/90°/0°/90°$

(iii) Angle-ply asymmetric: as for cross-ply, but with ply sequence $0°/45°/-45°/90°$

In each case the problem is modelled in two ways: (a) via meshes of hexahedral (brick) elements with $2 \times 2 \times 2$ reduced integration, with separate elements assigned to each layer (b) via a meshes of graded tetrahedral elements, assuming nine layers of 3 Gauss points.

In each case two meshes have been used: one coarse (Hex1, 32 elements, 261 nodes, 783 dof; and Tet1, 6 elements, 27 nodes, 81 dof), the other somewhat finer (Hex2, 128 elements, 785 nodes, 2355 dof; and Tet2, 227 elements, 462 nodes, 1386 dof). The solution noted as "CLT" relates (in the case of the elastic stiffness) simply to the corresponding direct (**A**) and bending (**D**) stiffnesses from classical laminate theory, and (in the case of thermal forces and moments) the numerical integration of thermoelastic stresses following a CLT approach. The true angle-ply laminate situation involves significant interlaminar shearing so the CLT solution was of limited applicability. For reasons of space, only the results relating to the cross-ply laminate are tabulated in detail.

For the mechanical test cases, the surface $x = 1$ is subjected to a prescribed tensile displacement of 1 unit or a bending displacement of 1 radian as appropriate. The reaction forces (in the tension cases) and moments (in the bending cases) are presented in Table 1. Results for the isotropic and cross-ply test cases showed excellent agreement with the reference solutions; for the angle-ply laminate (not shown here) results were still good, with errors of around 1% compared with the Hex2 solution.

Informal verification of the thermoelastic analysis took place by again considering a unit cube, but this time with in-plane displacements constrained to zero, and with a thermal gradient of $\Delta T = 100 + 200z\,°C$ imposed. The reaction forces and moments are again presented in Table 1.

The agreement with the CLT test case is good (typical error up to 1-2%, not quite as good as in the mechanical load examples) except for the angle ply example. Comparison in that case is again more meaningful against the Hex2 example, showing agreement to within a few percent.

Table 1. Static stiffness results: reactions for tension are units of force, reactions for bending are units of moment.

Load case	Solution/mesh Integration scheme	CLT	Hex1 $2 \times 2 \times 2$	Hex2 $2 \times 2 \times 2$	Tet1 9×3	Tet2 9×3
Mechanical	Tension	30549.89817	30549.8	30549.8	30383.2	30552
	diff. from CLT		0.000%	0.000%	-0.546%	0.007%
	diff. from Hex2		0.000%		-0.545%	0.007%
	Bending	2545.824847	2545.81	2545.82	2522.9	2544.97
	diff. from CLT		-0.001%	0.000%	-0.900%	-0.034%
	diff. from Hex2		0.000%		-0.900%	-0.033%
Thermal	Thermal force	-21.7668	-21.7667	-21.7668	-22.2214	-22.0095
	diff. from CLT		0.000%	0.000%	2.089%	1.115%
	diff. from Hex2		0.000%		2.089%	1.115%
	Thermal moment	-3.076205	-3.0762	-3.0762	-3.13285	-3.10479
	diff. from CLT		0.000%	0.000%	1.841%	0.929%
	diff from Hex2		0.000%		1.842%	0.929%

7. Conclusions

(i) A novel approach for formulating a tetrahedral element for the analysis of laminates for mechanical and thermoelastic loading is presented. The approach taken is to create a fictitious functionally-graded material whose elastic and thermoelastic behaviour at structural level is the same as that of the true laminate.

(ii) The element has been implemented in trial form using the LibMesh finite element library. The underlying concept has also been implemented in non-FE form to permit closer examination of the algorithms.

(iii) Some informal verification results have been presented; these are not intended to be complete. In particular, the dynamic aspects still require implementation and testing.

(iv) Results are very encouraging, with errors ranging from negligible to the order of a few percent. No shear locking seems to be apparent.

Acknowledgments

The authors gratefully acknowledge the financial support of the Technology Strategy Board via grant no. TP/5/MAT/6/I/H0473F.

References

[1] Panda S and Ray M C, 2008. Nonlinear finite element analysis of functionally graded plates integrated with patches of piezoelectric fiber reinforced composite. *Finite Elements in Analysis and Design* **44** 493.

[2] Lee H-J, 2005. Layerwise Laminate Analysis of Functionally Graded Piezoelectric Bimorph Beams *Journal of Intelligent Material Systems and Structures*, **16** 365.

[3] Jones R M, 1999. *Mechanics of composite materials* (Philadelphia, PA: Taylor & Francis Inc.)

[4] Rathod H T, Venkatesudu B and Nagaraja K V, 2005. Gauss Legendre Quadrature Formulas over a Tetrahedron. *Numerical Methods for Partial Differential Equations*, **22** 197 (published online by Wiley Interscience, 10 Jun 2005).

[5] http://libmesh.sourceforge.net/

Collapse of Composite Tubes under Uniform External Hydrostatic Pressure

Paul T. Smith, Carl T.F. Ross and Andrew P.F.Little
University of Portsmouth, Portsmouth, PO1 3DJ, United Kingdom
Carl.ross@ntlworld.com

Abstract

This paper describes an experimental and a theoretical investigation into the collapse of 22 circular cylindrical composite tubes under external hydrostatic pressure. The investigations were on the collapse of fibre reinforced plastic tube specimens made from a mixture of three carbon and two E-glass fibre layers. The theoretical investigations were carried out using an in-house finite element computer program called BCLAM, together with the commercial computer package, namely ANSYS. It must be emphasised here that BS 5500 does not appear to exclusively cater for the buckling of composite shells under external hydrostatic pressure, so the work presented here is novel and should be useful to industry.
The experimental investigations showed that the composite specimens behaved similarly to isotropic materials previously tested, in that the short vessels collapsed through axisymmetric deformation while the longer tubes collapsed through non-symmetric bifurcation buckling. Furthermore it was discovered that the models failed at changes of the composite lay-up due to the manufacturing process of these models. These changes seemed to be the weak points of the specimens.

Keywords: circular cylinder, buckling, axisymmetric yield, composite, external pressure, finite elements, BCLAM, ANSYS.

1. Introduction

The surface area of the Earth is about 196.9 million sq.mi (510 million km^2), and about 75% of it is covered by water, but despite this, only about 0.1% of the oceans' bottoms have been explored [1]. Such research has found large quantities of precious metals and minerals, together with large quantities of methane hydrates. In the case of methane, Dickens et al [1] have found deep-sea frozen methane hydrates lying under the ocean bottoms at depths of 2 miles (3.22km) or more. According to Dickens et al, the quantity of this methane is about 10,000 billion tonnes and if it is divided amongst all of mankind, it will amount to about 1,670 tonnes per person on Earth, or in monetary terms about $1,250,000 per person on Earth. Thus, it is necessary to exploit the oceans' bottoms, which are as deep as 7.16 miles (11.52 km).
Now a large submarine can only dive to a depth of about 400m (1312 ft), but the deepest part of the oceans is 29 times deeper than this. As the submarine dives deeper and deeper

into the oceans, the external hydrostatic pressure increases, so that the wall thickness has to be increased. Eventually, the wall thickness becomes so large that the vessel has no reserve buoyancy and will sink like a stone to the very bottom of the ocean [2]. Ross [2] has found that for a submarine of internal diameter 10m (32.81ft) and constructed in high-tensile (HY80) steel, the thickness of its hull will be about 2.3 m (7.58ft), if it is to be designed to dive to the bottom of the Mariana's Trench! The only way to overcome this problem is to use a material with a higher strength: weight ratio than high-strength metals. Such materials are Glass Fibre Reinforced Plastic (GFRP) and Carbon Fibre Reinforced Plastic (CFRP). There are many other composites, which are suitable, but these are cheaper than many of the other composites and have been used successfully for other structures.

2. Background

2.1 Buckling

Under uniform external pressure a long thin-walled circular cylinder can fail by non-symmetric bifurcation buckling or shell instability, at a fraction of that to cause axisymmetric yield [2]. To increase the buckling resistance of such vessels, they are usually stiffened by ring stiffeners, spaced at suitable distances apart. If the ring stiffeners are not strong enough, the entire vessel can collapse bodily by a mode called general instability. Another mode of failure is called axisymmetric deformation, where the vessel keeps its circular form while imploding.

For long thin vessels, the shell instability failure mode usually occurs at a much smaller pressure than that required to cause axisymmetric yield. The extremely long thin circular cylindrical vessel will fail in a flattening or ovalling mode, but the shorter circular cylinder will buckle in a lobar mode. In this paper, we will concern ourselves only with shell instability and axisymmetric deformation.

2.1.1 *Thinness Ratio and Plastic Knockdown Factor.* Windenburg and Trilling [3] introduced a thinness ratio (λ) to consider the fall in theoretical elastic buckling pressures due to the fact that shorter vessels failed by inelastic instability. The effects of initial out-of-roundness of these cylinders can cause a further drop in their buckling pressures than that predicted by elastic theory for perfect vessels. This factor is needed because even slight deviations from the perfect geometry can reduce the buckling pressure of a pressure vessel considerably. With the thinness ratio, it is also required to determine the so-called plastic knockdown factor (PKD) from the experimental results, so that the Design Chart can be made.

2.2 Finite Element Method (FEM)

In the theoretical analyses of these vessels, we will consider shell instability using an in-house finite element computer program called BCLAM, together with the well-known computer package ANSYS.

For both programs, the possibility of using different material models, including orthotropic material properties is allowed for.

2.3 The composite tubes

The models used for the tests were manufactured from fibre reinforced plastic tubes, see Figure 1; they consisted of a mixture of carbon and glass fibre layers. The manufacturer was

Journal of Physics: Conference Series **181** (2009) 012043 doi:10.1088/1742-6596/181/1/012043

Carbon Fibre Tubes Ltd, Hayling Island, UK. The lay-up was 0°/90°/0°/90°/0; the carbon fibres were laid lengthwise (0°) and the E-glass fibres circumferentially (90°). The manufacturer describes the manufacturing process to be rather similar to that of making a Swiss roll. That is the layers were of cloth form and were wrapped around each other, with impregnated resin, in a circumferential manner.

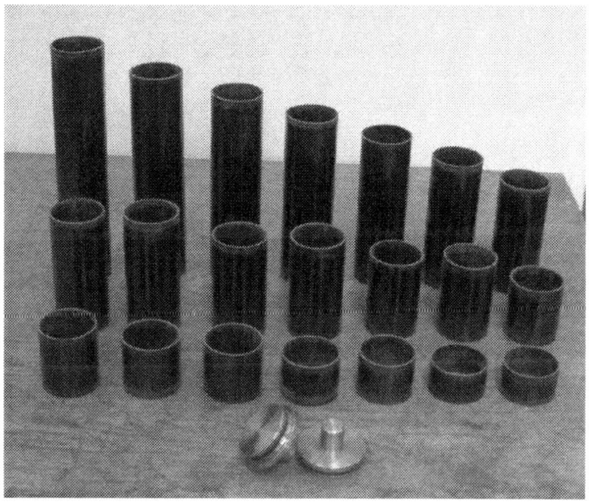

Figure 1: The models and their end bungs

For sealing the two ends of each specimen under external water pressure, two end bungs with 'O' rings had to be manufactured. These end bungs were manufactured out of mild steel and are shown in Figure 1.

The actual number of layers were 3 Carbon layers and 2 E-Glass layers and the layer thicknesses were calculated from the overall measured wall thickness of 1.8 mm,

where

E_1 = Young's modulus of the layer in direction of the fibre

E_2 = Young's modulus of the layer perpendicular to the fibre direction

G_{12} = in-plane shear modulus

2.3.1 *Material properties of the single layers*. The properties of the single layers are shown in Table 1.

σ_{yp} = yield stress in the direction of the fibre

υ_{12} = major Poisson's ratio

2.4 Experimental Analysis and the Design Charts

After calculating a series of theoretical buckling pressures and thinness ratios, with the aid of an in-house computer program, namely MisesNp, for the models made from the present composite, the respective buckling pressures of the models were determined experimentally. MisesNp adopts the von Mises analytical solution for simply-supported ends.

Table 1. Material Properties of the Composite (MPa)

	E_1	E_2	G_{12}	σ_{yp}	ν_{12}	No. of Layers	Thickness (mm)
Carbon	135000	10000	5000	1500	0.3	3	0.3857
E-glass	40000	8000	4000	1000	0.25	2	0.3214

This testing was done in a high-pressure tank, where each model was tested to failure and the experimental buckling pressure was noted. With the experimentally obtained values for the buckling pressure (P_{exp}) of the series it was possible to calculate the PKD:

$$PKD = P_{cr}/P_{exp},$$

where P_{cr}= theoretical buckling pressure.

With the values of PKD and the previously calculated thinness ratios (λ) it was possible to generate a Design Chart, where $1/\lambda$ was plotted against PKD.
This Design Chart could now be used to calculate the predicted (experimental) buckling pressure P_{pred} for a pressure vessel out of the same material for untested vessels.
During the design process the factors PKD and λ have to be calculated. Once the design chart has been produced, it is possible to obtain the PKD from this chart, which then can be used to calculate the predicted buckling pressure, namely P_{pred},, where:

$$P_{pred} = P_{cr}/PKD$$

The new value gives an approximation of the real buckling pressure that can now be used to predict the possible diving depth of the submersible.

3. Experimental Investigation

3.1 Equipment
The experiments were carried out in the laboratories of the University of Portsmouth. The following equipment was used:

- High-pressure test tank (maximum pressure 210 bar).
- Pressure pump, with pressure gauge (maximum pressure 410 bar).

Before the manufactured specimens were tested, they were checked for any defects due to the manufacturing process or any matters that could affect their performance during the testing. It was observed that all the test specimens had minor defects at the ends, when they were cut during manufacture. These defects had the appearance of small delaminations and cracks on the surface. The affected zones could be roughly sized between 0.5cm and 1cm lengths from the ends, for all specimens. Due to the supporting influence of the end bungs, that had shoulders of about 1cm penetrating into the specimens' ends, it was decided that these defects had almost no influence on the results. Some of the specimens, especially the shorter ones, had bigger defects. These could spread almost over the whole length of the specimens. The defects could have had a considerable influence on the strength of the models, particularly those with the larger defects and because of this; it was decided to carefully observe the behaviour of the specimens with the larger defects during the tests.

3.1.1 *Model Failure.* All models were observed before and after failure and pictures were taken for documentation. The most important information has already been given, but in addition to this, three other issues were found, as follows:

(1) Three major failure modes could be indicated during the testing. These failure modes could be related to the length of the models. The three failure modes were:

1. Elastic buckling.
2. Inelastic buckling (see Figure 2).
3. Axisymmetric 'yield' failure (see Figure 3).

The longest models failed due to elastic buckling. In this failure mode the test models buckled elastically, but with a subsequent reduction of pressure the model returned to its original shape without any visible signs of damage.
The medium sized models failed due to inelastic buckling. This meant that the deformation just prior to buckling was large, so that the material had 'yielded' and buckled inelastically. These models did not return to their original shapes after the pressure was decreased to zero.

Figure 2: Inelastic buckling of a model.

The shortest models failed due to axisymmetric 'yield' failure. In this case, the models deformed on the whole circumference; no lobes were formed in this mode, and the vessels collapsed axisymmetrically. This failure mode can be seen in Figure 3.

(2) It was found that the defects that were seen during the pre-test observation had almost no influence on the collapse pressures of the models. All the cracks, fractures or delaminations that could be seen on the models after collapse, were found at different locations to the deformations that were found just after manufacture. A 'micograph picture' of a section, of part of an undamaged tube is shown in Figure 4.

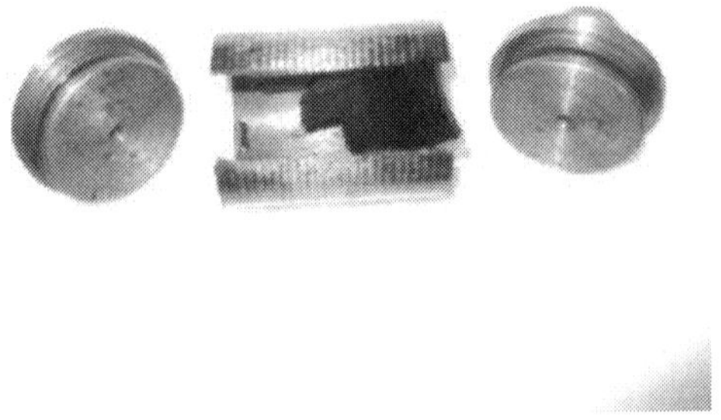

Figure 3: Axisymmetric deformation of a short model.

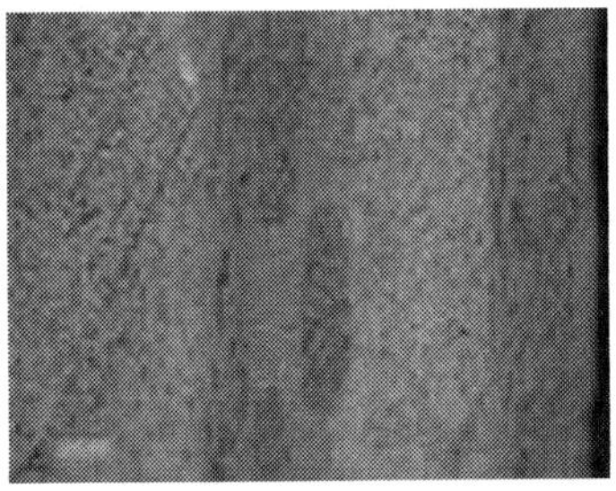

Figure 4: A 'micrograph picture' of the Carbon/Glass material.

(3) After a careful investigation of the failure modes of the models, some of the vessels were subjected to microscopic analysis, via a microscope (See Figure 4), where the layer composition was photographed with magnified pictures of the polished surface perpendicular to the rotational axis of each tube. From these pictures, it could be seen that the lay-up of the different layers varied considerably over the whole circumference.

After a check of the position of these lay-up modifications and the position of post-test fractures, it showed that both were in correlation. This could be seen on all specimens that fractured along the longitudinal axis.

4. Theoretical Investigations

As already stated in the previous section, the theoretical investigation was first done using BCLAM an in-house DOS-based computer program. In this case, a truncated conical shell element was used with two ring nodes at its ends. Each ring node had four degrees of freedom, three translational and one rotational. The assumed displacement form in the azimuthal direction was sinusoidal and this enabled explicit integration in the azimuthal direction, thereby considerably reducing computational time. In contrast to this the ANSYS solution [4], used the Shell99 finite element, which had 8 nodes, with 6 degree of freedoms per node. Three of the degrees of freedom were translational and the other three were rotational degrees of freedom.

The boundary conditions for both analyses were assumed to be simply-supported at the 'ends'.

The results of the BCLAM & Shell99 analyses with the theoretically calculated properties are listed in Table 2, where the figures in parentheses represent the number of circumferential waves or lobes (n) that the vessels buckle into.

<u>Note:</u>

The thinness ratios had been calculated using the equivalent Young's modulus in y-direction and the other parameters listed in Table 2, except for the wall thickness, which was set to 1.8mm for the ME specimens.

5. Theoretical and Experimental Results

From Table 2, it can be seen that the ANSYS results are more stable than those yielded by BCLAM, especially for the very long and the very short vessels. It appears that for the vessels of extreme length, BCLAM was numerically unstable and because of this, no design chart will be produced for BCLAM; only a design chart for ANSYS will be produced, as shown by Figure 5.

For example, with reference to numerical instability, in the case of the longer vessels, 10 conical shell elements were initially used for the BCLAM analysis and when this was increased to 12 conical shell elements, the calculated buckling pressure appeared to converge. However, when the number of elements was increased to 20, the value of the calculated buckling pressure appeared to wildly diverge; similar experiences were observed for the very short vessels. Thus, it was believed that BCLAM was unstable, probably due to a numerically unstable feature commonly called 'locking' [5].

With the experimentally obtained buckling pressures (P_{exp}) and the theoretically calculated buckling pressures (P_{cr}) of the previous sections, the plastic knockdown factors (PKD) for both approaches were calculated by the formula:

$$PKD = \frac{P_{cr}}{P_{pred}}$$

6. Conclusions

- The experimental and theoretical investigations were performed successfully on the described fibre reinforced plastic tubes. The main findings are listed as follows:

- Design Charts were created successfully for use with ANSYS, but not with BCLAM, because of the numerical instability of the latter. It should be emphasised here, that it appears that BS 5500 does not exclusively cater for the buckling of thin-walled shells made from composite materials and this makes the present study a novel one.

- The tested models failed at locations were the lay-up of the single layers changed. These changes seem to reduce the strength of the composite in this location.

Table 2. Experimental Results & the Theoretical Results.

Cylinder No.	Overall Length L_0 (mm)	Mean Diameter (mm)	Layer Thickness (mm)	Out of Circularity (mm)	1/λ	P_{exp} (MPa)	P_{cr} BCLAM (MPa)	P_{cr} ANSYS (MPa)
ME1A	242.755	52.54	1.8	0.4398	.251	1.93	4.025(2)	5.872(6)
ME2	220.295	52.51	1.78	0.3985	.265	2.07	4.19(3)	5.479(2)
ME3	200.27	52.53	1.79	0.3956	.279	2.28	4.380(3)	6.666(2)
ME4	180.49	52.53	1.79	0.3732	.296	2.38	4.546(3)	6.810(4)
ME5	159.795	52.55	1.79	0.3582	.317	2.55	4.808(3)	6.955(4)
ME6	139.625	52.54	1.8	0.3799	.343	2.83	5.282(3)	5.606(3)
ME7A	120.325	52.54	1.8	0.3378	.375	3.59	5.909(3)	6.450(4)
ME7B	119.965	52.52	1.76	0.3484	.375	3.38	5.602(3)	6.700(3)
ME7C	120.015	52.55	1.78	0.3568	.375	3.65	5.759(3)	6.700(3)
ME8A	100.275	52.55	1.78	0.3807	.419	4.34	6.861(3)	8.345(4)
ME8B	100.265	52.55	1.79	0.3546	.419	4.34	6.950(3)	8.345(4)
ME9A	80.29	52.54	1.77	0.3121	.483	5.52	5.039(3)	10.278(4)
ME9B	80.19	52.58	1.81	0.3476	.484	5.65	9.259(4)	10.278(4)
ME10A	60.295	52.56	1.78	0.4182	.591	8.27	16.181(5)	16.418(5)
ME10B	60.44	52.56	1.79	0.4278	.59	8.27	12.695(4)	16.418(5)
ME11A	50.305	52.57	1.8	0.378	.682	10.48	11.210(4)	25.757(6)
ME11B	50.485	52.54	1.78	0.3927	.68	11.45	13.858(4)	25.757(6)
ME12A	40.185	52.53	1.76	0.3889	.835	13.93	18.417(5)	62.830(7)
ME12B	40.39	52.51	1.76	0.4167	.831	16.0	18.298(5)	62.830(7)
ME13A	30.405	52.49	1.72	0.4942	1.164	22.06	26.879(6)	732.160(14)
ME13B	30.22	52.54	1.77	0.3847	1.174	19.70	29.492(6)	732.160(14)
ME14A	34.915	52.55	1.8	0.4882	.972	18.34	24.163(5)	157.160(9)
ME14B	35.05	52.54	1.78	0.4296	.967	18.62	23.267(5)	157.060(9)

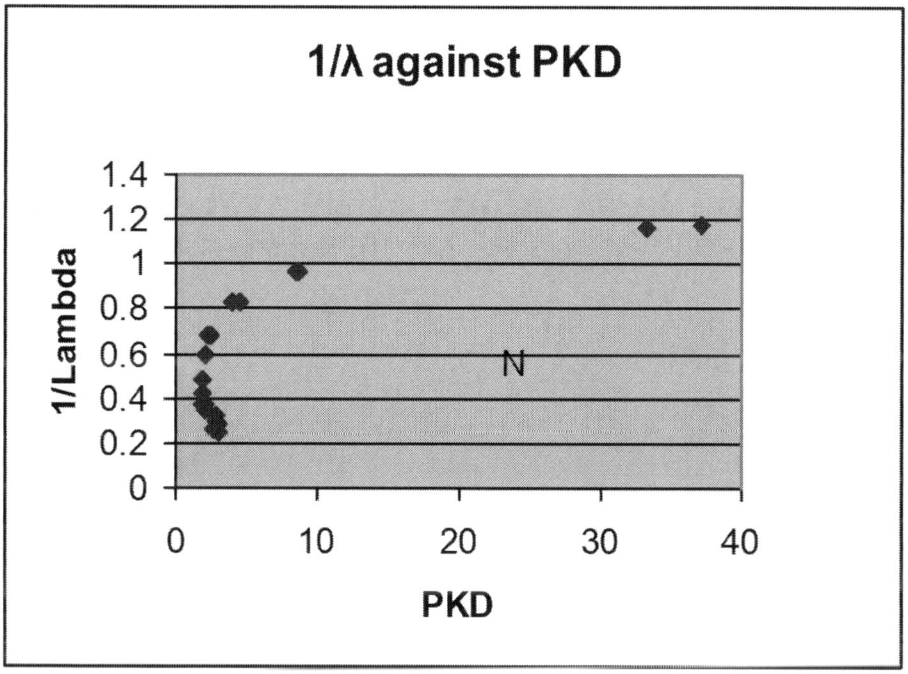

Figure 5: Final Design Chart for ANSYS use

- The Design Chart for the orthotropic composite material has, in general a slightly different shape to design charts obtained for isotropic materials.

- The Design Chart created in this paper should prove useful to designers, providing the structures have the same characteristics as those investigated herein and a bigger safety factor is used.

References

[1] **Dickens, Gerald R., Paull, Charles, K., Wallace, Paul & the ODP Leg 164 Scientific Party,** (1997), Direct measurement of in situ quantities in a large gas-hydrate reservoir, *Nature,* Vol. 385, 30th January.

[2] **Ross, C. T. F.** (2001). Pressure Vessels: External Pressure Technology. *Horwood Publishing Limited,* Chichester, U.K.

[3] **Windenburg, D. F. & Trilling, C.** (1934). Collapse by Instability of Thin Cylindrical Shells Under External Pressure. *Trans., ASME,* 11, pp 819-825.

[4] **ANSYS Incorporated.** (2005). Release 10.0 Documentation for ANSYS. Canonsburg, USA.

[5] **Zienkiewicz, O.C. Taylor, R.L.,** (2000) The Finite Element Method, *Butterworth-Heinemann,* Oxford, U.K.

7th International Conference on Modern Practice in Stress and Vibration Analysis IOP Publishing
Journal of Physics: Conference Series **181** (2009) 012044 doi:10.1088/1742-6596/181/1/012044

Identification of the mechanical behaviour of low density hyperelastic polymeric foams from full-field measurements

F. Pierron, B. Guo

LMPF, Arts et Métiers ParisTech, Rue St Dominique, BP 508, 51000 Châlons en Champagne, France

fabrice.pierron@chalons.ensam.fr;

Abstract. This paper presents a methodology aimed at the identification of the mechanical behaviour of hyperelastic low density polymeric foams using digital image correlation and the Virtual Fields Method. First, a simple bending/shear test enables the identification of the elastic parameters at low strain levels. Then, a uniaxial compression test is performed to identify the parameters of an Ogden-type law (hyperelastic). The strain fields exhibit strong localization caused by the elastic collapse of the cells clearly indicating that an approach based on global force/displacement curves will fail to provide intrinsic material parameters (structural effect). This is still very much a preliminary study but it clearly shows the potential of the approach for this type of materials.

1. Introduction

Low density polymeric foams are widely used in applications that require good low energy absorption capabilities, such as in packaging for instance. However, the determination of their mechanical properties is made difficult because of their hyperelastic behaviour and strong localization effects caused by the local collapse of cells in compression. For the present material, the latter is due to elastic buckling of the cells in compression, with total recovery of the initial cell shape after unloading, resulting in the hyperelastic behaviour. Trying to address such a complex mechanical behaviour with just a global stress/strain curve is extremely difficult. It is therefore essential to bring in more experimental information and optical full-field deformation measurements provide a great opportunity to improve the identification procedures for this type of materials.

The scientific literature provides still only few attempts at using full-field measurements on such foams. Some recent studies can be found but the measurements are often only used as qualitative observations of localization [1,2] or at best to validate finite element simulations [3]. It is also interesting to note that the progress in X-ray tomography has enabled several researchers to digitize the 3D structure of foam specimens. This can be used to build up a fully realistic 3D finite element model [4,5] but more recently, first attempts at producing deformation maps have been released, either using 2D correlation [6] or volume image correlation [7]. However, to the best knowledge of the present authors, full-field measurements on foams have never been used to identify a mechanical behaviour model through an inverse methodology.

The objective of the present paper is to provide some initial ideas and results on how full-field measurements can be coupled to an inverse identification procedure to identify mechanical constitutive parameters of such low density polymeric foams. 2D digital image correlation will enable to measure the in-plane displacement field and the Virtual Fields Method (VFM, [8]) will be used to identify the material parameters.

2. Elastic stiffnesses at low strains

The first case study performed here consists in determining the initial Young's modulus and Poisson's ratio when the foam undergoes very low elastic strains. One of the difficulties in this case is that if this is performed on a usual universal testing machine, the load will be very small and very inaccurate values of the elastic stiffnesses might be obtained (unless a specific machine with adapted load transducer is used). The use of full-field measurements associated to the Virtual Fields Method enables to perform very simple tests without having to worry about the actual stress field in the specimen.

Here, the test is a very simple bending test illustrated on Fig. 1. It consists in clamping a rectangular bar of foam in a manual clamp and loading through small weights positioned at a certain distance from the clamp along the upper surface of the beam. This provides a mixed stress state comprising longitudinal bending stresses and in-plane shear.

(a) Global view of the set up

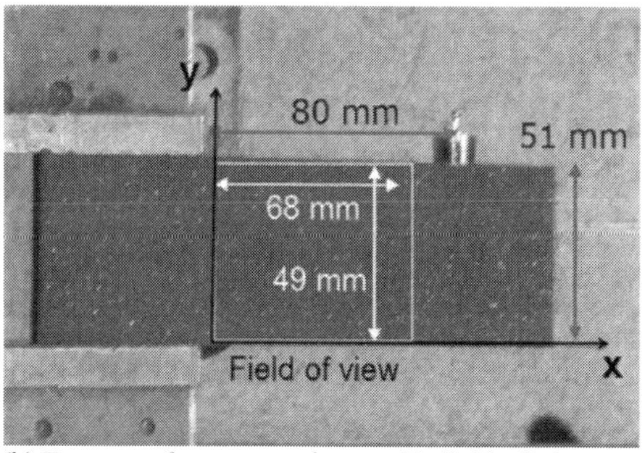

(b) Focus on the test specimen with field of view (thickness: 20 mm)

FIG. 1 – Bending test on foam specimen

The material tested here is low density grey polyurethane packaging foam (27 kg.m^{-3}). The reason for the choice of this foam is that it has been used to produce auxetic (or negative Poisson's ratio) foams [9]. The typical cell size is about 200 µm. A 1.3 Mpixels PCO Sensicam CCD camera was used to record the images of the specimen. A 60 mm focal length lens provided a pixel size of 55 µm. No surface preparation was performed, the natural texture of the foam providing sufficient contrast (though not ideal) for image correlation provided that lateral lighting was used. Images of the field of view represented in Fig. 1 were taken before and after the application of the load and a digital correlation software (Correli, [10]) was used to derive the two components of the in-plane displacement field. A correlation window of 16 by 16 pixels was used together with a shift of 8 pixels, providing and array of 154 x 104 non independent measurements points (because of the 8 pixels shift). In order to calculate the strains, the displacement maps were smoothed using a basis of piecewise functions (triangular 3-noded bilinear finite elements). The nodal displacements were obtained through least-square fitting of the date and the strains derived from the shape functions (for more details, see [11]).

The displacement fields are given on Fig. 2. The noise is rather large because the natural contrast of the foam is far from ideal. Nevertheless, the strain fields (Fig. 3), obtained for an average mesh size of 20 displacement points, clearly show the bending (ε_{xx}) and shear (ε_{xy}) strains coming respectively from the bending moment and the shear force.

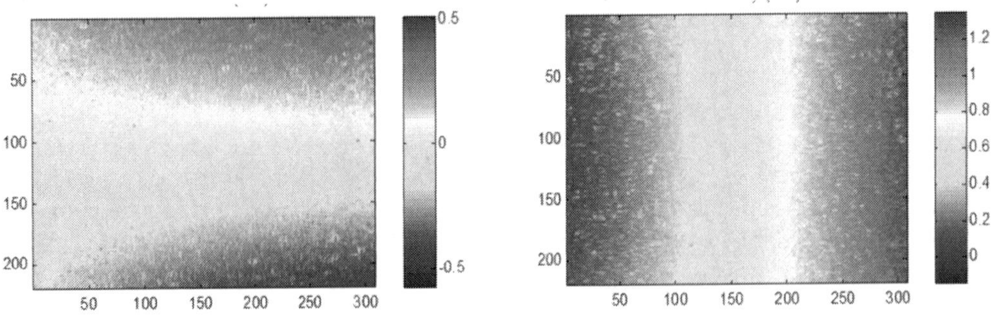

FIG. 2 – Displacement fields (unit: pixels) for a 0.29 N load (30 grams).

The identification was performed using the Virtual Fields Method. This technique enables the identification of the elastic parameters from the strain fields in a direct way. An important issue concerns the choice of the virtual fields. They have a very significant impact on the quality of the identification procedure. Here, optimized piecewise virtual fields have been used. They are defined on the same mesh as that used for the displacement smoothing and an adequate procedure enables to reach the maximum likelihood solution (ie, the solution which is the less sensitive to noise). The details of this procedure can be found in [11]. This procedure is available in the Camfit open source software available on www.camfit.fr.

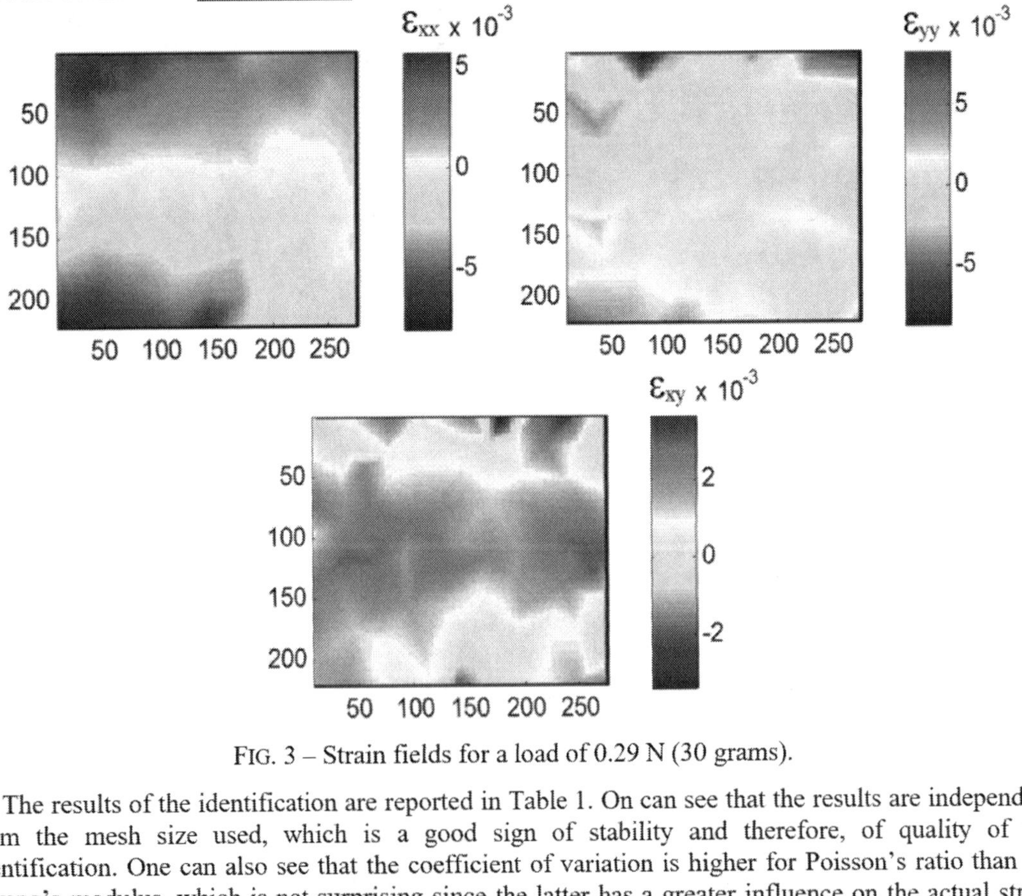

FIG. 3 – Strain fields for a load of 0.29 N (30 grams).

The results of the identification are reported in Table 1. On can see that the results are independent from the mesh size used, which is a good sign of stability and therefore, of quality of the identification. One can also see that the coefficient of variation is higher for Poisson's ratio than for Young's modulus, which is not surprising since the latter has a greater influence on the actual strain field than former. The present authors do not have a reference value for Young's modulus of this foam but Poisson's ratio compares well with values found in [9]. They report a Poisson's ratio of 0.25 for a compressive strain of 20%. They also show that this coefficient decreases sharply with the compressive strain. The present value of 0.33 found here is therefore coherent with their results. The simplicity of the present procedure should be underlined, the use of full-field measurements enabling a less rigorous control of the boundary conditions, the results being insensitive to the actual clamping conditions of the specimen as long as the strain distribution through the thickness remains constant (2D problem). The technique can therefore be used to obtain elastic parameter values for soft materials in a fast and inexpensive way, no tensile testing machine being necessary.

Mesh size	E (kPa)	ν
12	398	0.335
14	413	0.326
16	415	0.346
18	408	0.353
20	417	0.372
Mean	404	0.333
Coef. Var. (%)	3.9	10.7

TAB. 1 – Identified elastic constants for a 0.29 N load (30 grams).

3. Hyperelastic law

One of the main challenges in foam characterization concerns the identification of the hyperelastic behaviour at large strains in the presence of strong strain localization. This paragraph presents a first validation of this approach.

As a first attempt, a simple uniaxial compressive test of a rectangular foam specimen was performed. Fig. 4 shows the experimental setup and the specimen's dimensions.

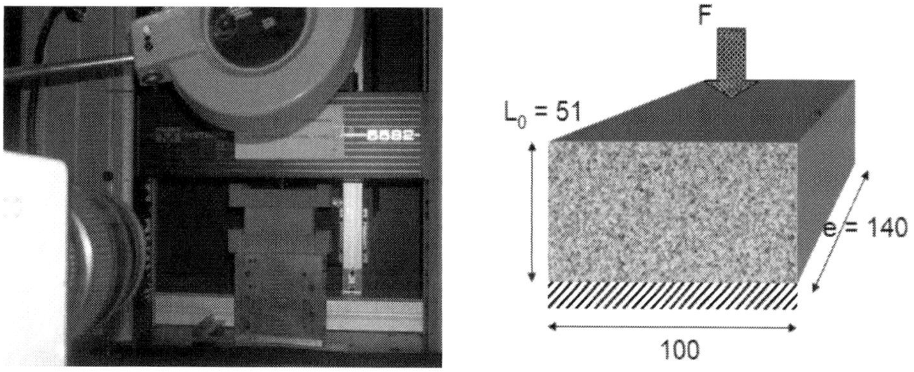

FIG. 4 – Uniaxial compression test (dimensions in mm).

The measurements have been performed with the same parameters as in section 2, except that chalk has been rubbed on the surface to hide the shiny spots and balance out the grey level histogram. Forty images have been recorded throughout the test. The stretches have been obtained through the same finite element smoothing procedure as that described in section 2, with a mesh size of 10 displacement points. It should be noted here that the correlation is performed incrementally, the reference image being updated at each load step. This enables to reach very high deformation levels. The compression strain $\partial u_y / \partial Y$ is plotted on Fig. 5 at different moments of the global load-displacement history (where Y is the vertical coordinate in the undeformed configuration and u_y is the vertical displacement in the deformed configuration). One can see that the global load-displacement curve exhibits a long plateau where the force is constant after a first short linear stage. Then, the load takes up again. The plateau is caused by the elastic buckling of the cells. This happens row by row of cells until the whole of the cells have undergone this phenomenon. Then, the stiffness increases sharply because of the deformation of the cell material (compacted stage). It is interesting to note that during the initial linear part and the final compacted stage, the strain field is reasonably uniform whereas on the plateau, the gradual elastic collapse of horizontal rows of cells produces a very heterogeneous strain field (as also noted in [1]). It is also interesting to note that the collapse of the cells start at the loading surfaces (top and bottom), which is consistent with observations reported in [2].

The idea of this paper is to use the strain fields at each stage of the load history to identify a constitutive behaviour law. For this purpose, a simple two parameters Ogden law has been selected. The main hypothesis here is that the stress is uniaxial (the other stress components have been neglected with respect to σ_{yy}). Also, it was assumed that Poisson's ratio is close to zero. This is consistent with the findings of [9] where it can be seen that Poisson's ration rapidly decreases to zero when the compressive strain increases. It was also shown in [12] that the effect of Poisson's ratio on the identified behaviour of such a foam was not very important. In any case, the assumption of zero Poisson's ratio enables to write that the Cauchy stress component σ_{yy} (deformed configuration) is equal to the Piola-Kirchhoff one π_{yy} (reference configuration). The Ogden law writes [12]:

$$\sigma_{yy} = \frac{1}{\lambda_y}\mu\left[\lambda_y^{\alpha} - 1\right] \tag{1}$$

where λ_y is the stretch in the vertical direction (supposed to be a principal direction of the stretch tensor) and μ and α are the two parameters to identify.

The Virtual Fields Method in the case of non linear constitutive behaviour consists in minimizing the difference between the virtual work of internal and external forces (the latter being obtained from the load cell and the former calculated from the strain fields and constitutive parameters). Details can be found in [13] for the case of elasto-plasticity.

Using a uniform compression virtual field, the cost function can be written as:

$$\Phi(\mu,\alpha) = \sum_{i=1}^{n}\left(F^{(i)}L_0 - e\int \Pi_{yy}^{(i)}(\mu,\alpha)dxdy\right)^2 \tag{2}$$

where L_0 is the initial height of the specimen and e its thickness (or depth, see Fig. 4), $F^{(i)}$ is the load at load step (i) (n steps, here n=40) and $\Pi_{yy}^{(i)}$ the Piola-Kirchhoff stress at load step (i). The integral in the above equation can be approximated by a discrete sum at all the measurement points. Substituting the stress from Eq. 1, the cost function becomes:

$$\Phi(\mu,\alpha) = \sum_{i=1}^{n}\left(F^{(i)} - \frac{eb}{pq}\sum_{p,q}\frac{1}{\lambda_y^{(i)}(p,q)}\mu\left[\lambda_y^{(i)\alpha}(p,q) - 1\right]\right)^2 \tag{3}$$

where p and q are the number of measurement points along the x and y directions and b is the specimen width. The left hand-side term in the squared difference represents the load measured by the load cell and the right hand-side term is the resultant load recalculated from the measured strains and the constitutive parameters.

This cost function was represented graphically in Fig. 6 for a large range of values of the Ogden parameters. A minimum is clearly visible around [μ,α]=[7000 Pa, 300], though above 100 for α, this parameter has little influence on the cost function. On the right hand-side of Fig 6 is a plot of the two terms in the cost function of Eq. 3. It can be seen that the Ogden model fits reasonably well but that the flat plateau is not so well described. This is not surprising since the mechanism of gradual elastic collapse of cell rows will produce a constant force that is not a feature of Ogden's model which is not specific to foams. It would be possible to add up terms in the model but there might be uniqueness problems. It is clear that there is still a lot of work to be done in this area, particularly to identify the parameters on a more complex test with several stress components involved in the specimen's response.

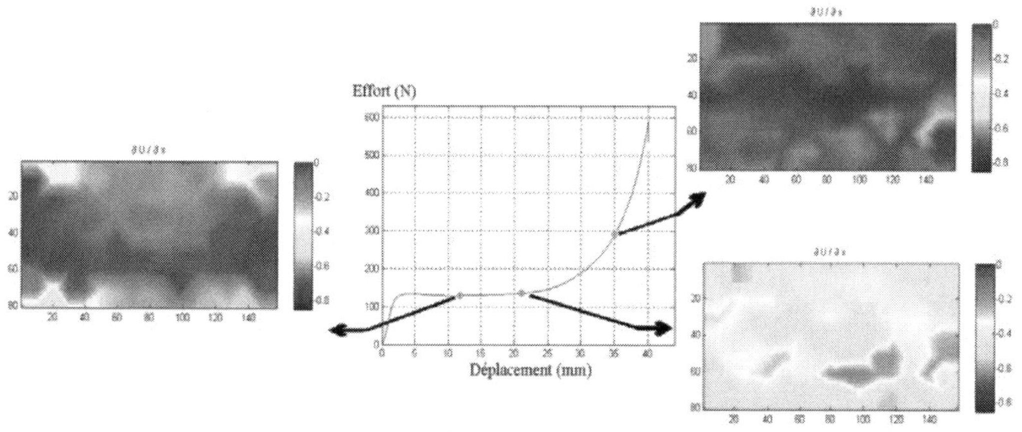

FIG. 5 – Compressive strain fields along the load history.

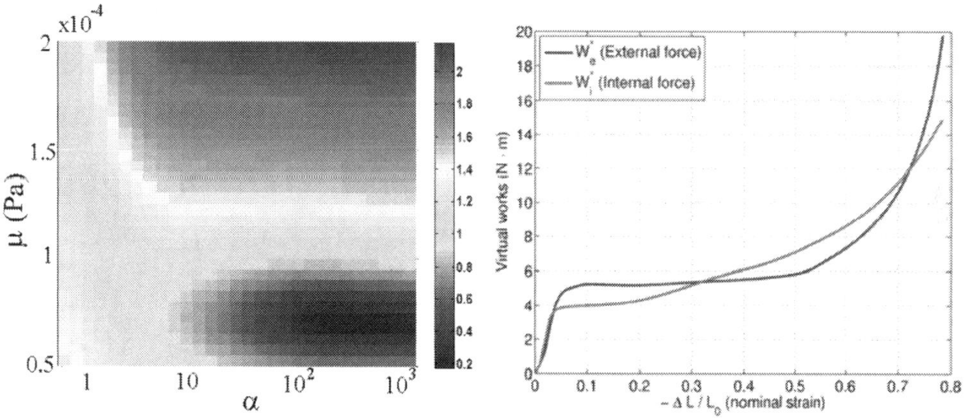

FIG. 6 – Identification results: cost function (left), model fit (right).

4. Evolution of Poisson's ratio with compressive strain of an auxetic foam

The last part of this paper is dedicated to the measurement of Poisson's ratio of an auxetic foam (negative Poisson's ratio). The foam was produced in the group of Dr Fabrizio Scarpa at the University of Bristol from a precursor similar to the foam used in the previous sections. The specimen is an 85 mm long 20 by 20 mm square-base cylinder. It was tested in compression using a manual clamp (see Fig. 7). No force was measured as only Poisson's ratio was sought here. 24 load steps were performed (corresponding to the deformation between the unloaded state in Fig. 7(a) to the maximum compression in Fig. 7(d)). The strain fields were obtained from the displacements by global polynomial fitting. 3rd degree polynomial were selected because there was no evidence of localization here as opposed to what happened in the specimen from section 3. Poisson's ratio was calculated from the use of the virtual fields method with the following virtual field (with axes defined in Fig. 8 and L the total length of the gauge length):

$$\begin{cases} u^* = 0 \\ v^* = -x(x-L)y \end{cases} \quad \begin{cases} \varepsilon^*_{xx} = 0 \\ \varepsilon^*_{yy} = -x(x-L) \\ \varepsilon^*_{xy} = (-x + L/2)y \end{cases} \tag{4}$$

This virtual field is a virtual transverse compression field as represented in Fig. 8. The resulting equation using the virtual fields method [14] is:

$$Q_{xx}\left(-\int_V x(x-L)\varepsilon_{yy}dV+\int_V(-x+L/2)y\varepsilon_{xy}dV\right)+Q_{xy}\left(-\int_V x(x-L)\varepsilon_{xx}dV-\int_V(-x+L/2)y\varepsilon_{xy}dV\right)=0 \quad (5)$$

where Q_{xx} and Q_{xy} are the two isotropic stiffness components. Therefore, Poisson's ratio is directly obtained by:

$$\nu=-\frac{-\int_V x(x-L)\varepsilon_{yy}dV+\int_V(-x+L/2)y\varepsilon_{xy}dV}{-\int_V x(x-L)\varepsilon_{xx}dV-\int_V(-x+L/2)y\varepsilon_{xy}dV} \quad (6)$$

where the integrals can be approximated by discrete sums over the array of measurement points.

FIG. 7 – Compression test on an auxetic foam specimen.

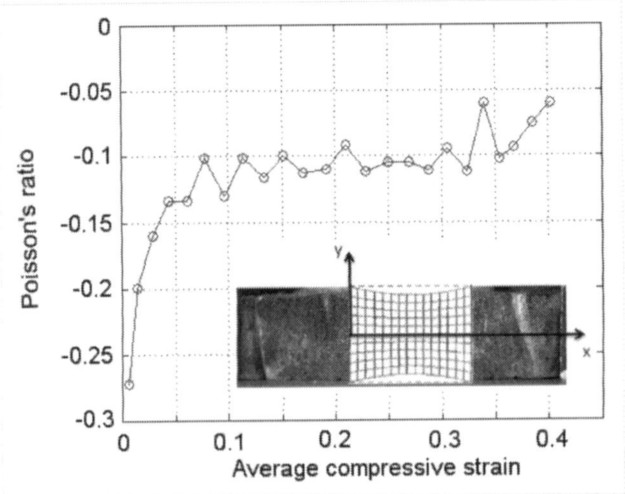

FIG. 8 – Identified Poisson's ratio as a function of compressive strain.

Fig. 8 shows the identified values of Poisson's ratio as a function of the global average compressive strain. The results are consistent with that in [9] where Poisson's ratio decreases sharply from about -0.3 to near zero.

5. Conclusion

This paper has presented an overview of the possibilities of full-field deformation measurements to identify the constitutive behavior of low density polymeric open cell foams. Elastic and hyperelastic behaviour has been considered. An application on an auxetic foam specimen was also shown. There is great potential in the future to enhance the modelling of the mechanical behaviour of such materials thanks to full-field measurements that can tackle both strong localization and very flexible test specimens.

Acknowledgements – The present authors wish to thanks Dr Fabrizio Scarpa for providing the foam specimens, Dr François Hild for the Correli DIC software and Prof. Michel Grédiac for very useful discussions. This study was part of ANR grant « Photofit » (ANR-05-BLAN-0327-01).

References

[1] Y. Wang, A.M. Cuitiño, Full-Field measurements of heterogeneous deformation patterns on polymeric foams using digital image correlation, Int. J. Solids Struct. 39 (2002) 3777-3796.

[2] R. Guastavino, P. Göransson, A 3D displacement measurement methodology for anisotropic porous cellular foam materials, Polym. Test. 26 (2007) 711-719.

[3] H. Jin, W.-Y. Lu, S. Scheffel, T.D. Hinnerichs, M.K. Neilsen, Full-field characterization of mechanical behavior of polyurethane foams, Int. J. Solids Struct. 44 (2007) 6930-6944.

[4] J.H. Kinney, G.W. Marshall, S.J. Marshall, D.L. Haupt, Three-dimensional imaging of large compressive deformations in elastomeric foams, J. Appl. Polym. Sci. 80 (2001) 1746-1755.

[5] J.A. Elliott, A.H. Windle, J.R. Hobdell, G. Eeckhaut, R.J. Oldman, W. Ludwig, E. Boller, P. Cloetens, J. Baruchel, In-situ deformation of an open-cell flexible polyurethane foam characterised by 3D computed microtomography, J. Mater. Sci. 37 (2002) 1547-1555.

[6] S.A. McDonald, N. Ravirala, P.J. Withers, A. Alderson, In situ three-dimensional X-ray microtomography of an auxetic foam under tension, Scripta Mat. 60 (2008) 232-235.

[7] S. Roux, F. Hild, P. Viot, D. Bernard, Three-dimensional image correlation from X-ray computed tomography of solid foam, Comp. Part A 39 (2008) 1253-1265.

[8] M. Grédiac, F. Pierron, S. Avril, E. Toussaint, The virtual fields method for extracting constitutive parameters from full-field measurements : a review, Strain 42 (2006) 233-253.

[9] A. Bezazi, F. Scarpa, Mechanical behaviour of conventional and negative Poisson's ratio thermoplastic polyurethane foams under compressive cyclic loading, Int. J. Fatigue 29 (2007) 922-930.

[10] F. Hild, B. Raka, M. Baudequin, S. Roux, F. Cantelaube, Multiscale displacement field measurements of compressed mineral-wool samples by digital image correlation, Appl. Opt. 41 (2002) 6815-6828.

[11] S. Avril, F. Pierron, General framework for the identification of elastic constitutive parameters from full-field measurements, Int. J. Solids Struct. 44 (2007) 4978-5002.

[12] R.D. Widdle, A.K. Bajaj, P. Davies, Measurement of the Poisson's ratio of flexiblee polyurethane foam and its influence on a uniaxial compression model, Int. J. Eng. Sci., 46 (2008) 31-49.

[13] S. Avril, F. Pierron, Y. Pannier, R. Rotinat, Stress reconstruction and constitutive parameter identification in elastoplasticity using measurements of deformation fields, Exp. Mech. 48 (2008) 403-420.

[14] M. Grédiac, F. Pierron, S. Avril, E. Toussaint, The virtual fields method for extracting constitutive parameters from full-field measurements : a review, Strain 42 (2006) 233-253.

Journal of Physics: Conference Series **181** (2009) 012045 doi:10.1088/1742-6596/181/1/012045

Significance of Low Energy Impact Damage on Modal Parameters of Composite Beams by Design of Experiments

A Shahdin[1], J. Morlier and Y. Gourinat
Département Mécanique des Structures et Matériaux (DMSM), Université de Toulouse, ISAE, 10 av. Edouard Belin BP54032, 31055 Toulouse Cedex 4, France

E-mail: amir.shahdin@isae.fr

Abstract. This paper presents an experimental study on the effects of multi-site damage on the vibration response of composite beams damaged by low energy impacts around the barely visible impact damage limit (BVID). The variation of the modal parameters with different levels of impact energy and density of damage is studied. Vibration tests have been carried out with both burst random and classical sine dwell excitations in order to compare that which of the methods among Polymax and Half Bandwidth Method is more suitable for damping estimation in the presence of damage. Results show that damping ratio is a more sensitive parameter for damage detection than the natural frequency. Design of experiments also highlighted energy of impact as the factor having a more significant effect on the modal parameters. Half Bandwidth Method is found to be unsuitable for damping estimation in the presence of damage.

1. Introduction

The use of fibre-reinforced composite laminates is experiencing an increased growth, namely in the aeronautical, naval and automotive industries, because of their excellent mechanical properties in conjunction with their low weight, and their ability to be tailored for specific applications. Nevertheless, composite materials are very different from metals with respect to their particular failure modes, which may be in the form of matrix cracking, fibre breakage, interlaminar delamination, etc. [1,2]. Delamination, which is a debonding or separation between individual plies of the laminate, frequently occurs in composite laminates. Delamination may arise during manufacturing (e.g., incomplete wetting, air entrapment) or during service (e.g., low velocity impact, bird strikes). In a low energy impact (but high enough to produce damage), only a very small indentation will be seen on the impact surface. This level of damage is often referred to as barely visible impact damage (BVID). Generally, fibre-reinforced composite laminates are very sensitive to medium and low energy impacts. However, the presence of delaminations may significantly reduce the stiffness and strength of the structures and may affect some design parameters such as the vibration characteristics of the structure (e.g., natural frequency and damping ratios). It is therefore important to understand the performance of delaminated composites in a dynamic environment [3,4]. Comprehensive reviews on vibration-based damage detection methods have been presented by Zou et al. [5] on the model-dependent delamination

[1] To whom any correspondence should be addressed.

identification methods for composite structures, and by Doebling et al. [6] and Sohn et al. [7] on general vibration-based damage detection methods.

The delamination-induced decrease of natural frequencies is one of the most commonly practiced approaches in damage detection. A large variety of works can be found in the scientific literature related to the study of structural damage by changes in natural frequencies [8-12]. However, in structures made of composite materials there seems to be a tendency to use damping as a damage indicator tool, as it tends to be more sensitive to damage than the stiffness variations, mainly when delamination is concerned. Therefore damping has been proposed in scientific literature as a more sensitive and attractive damage indicator as compared to natural frequencies [13-16]. Although research works related to damping are fewer in number than those on natural frequency because it is a parameter that is relatively difficult to estimate.

This paper focuses on the correlation of modal parameters and impact damage in composite beams from an experimental point of view based on low energy impacts around the barely visible impact damage limit (BVID). As damping is a phenomenon that is relatively difficult to estimate, so the composite beam specimens have been tested with both burst random (BR) and sine dwell (SD) testing in order to compare the modal damping ratios. Design of experiments is carried out later in order to highlight the factor having the most significant effect on the modal parameters.

2. Material and Specimen

Resin-containing carbon-fibre/epoxy prepregs of T300/914 is used to fabricate the test specimens. The material is supplied by Hexcel composites, the physical properties are set out in (table 1). The specimens are processed in a press. The curing cycle of the laminates is 2 h at 180°C with a warming-up cycle of 0.5 h at 135°C. The laminates are cut into beams using a diamond wheel cutter, following the ASTM D3039/D3470 standards.

Table 1 Physical properties of carbon/epoxy prepreg T300/914

Properties	Symbol	Value
Young's modulus in fibre direction	E_1	144000 MPa
Young's modulus in transverse direction	E_2	10000 MPa
Shear Modulus	G_{12}	4200 MPa
Poisson ratio	$\upsilon_{12}, \upsilon_{23}, \upsilon_{31}$	0.25, 0.3, 0.017
Volume density	ρ	1550 kg/m^3

The composite beam specimens tested in this article have a thickness of 3.12 mm having 24 plies. Their geometric configuration and lay-up is listed in (table 2). The lay-up is chosen as such, in which the delamination is said to have more profound effects on the dynamic characteristics [16].

Table 2 Geometry and lay-up of the composite test specimens

Length	480 mm
Width	50 mm
Thickness	3.12 mm
Number of layers (plies)	24
Thickness of each ply	0.125 mm
Lay-up	$[(0/90/45/-45)_3]s$

The vibration tests are carried out with two steel masses (50 x 30 x 10 mm) attached at the ends [17]. The aim of putting these masses at the ends is to enhance the difference in the modal parameters between the undamaged and the damaged test specimens [18].

3. Experimental Procedure

Two types of experiments are carried out in this work. Vibrations tests for measuring the modal parameters of the composite beams and impact tests in order to induce damage in the specimens. The experimental procedure for these two types of tests is discussed separately.

3.1. High Quality Vibration Tests

The experimental equipment used to obtain the modal parameters discussed in this paper is shown in figure 1. The experimental set-up is based on Oberst method [19]. The Oberst method states that a free-free beam excited at its centre has the same dynamical behaviour as that of a half length cantilever beam. The test specimen is placed at its centre on a B&K force sensor 8200 which is then assembled on a shaker supplied by Prodera having a maximum force of 100N. The fixation system and the end masses are glued to the test specimens with a HBM X60 rapid adhesive. High quality frequency response functions are measured with the help of a Laser Vibrometer OFV-505 provided by Polytec [20]. The shaker, force sensor and the laser vibrometer are manipulated with the help of a data acquisition system supplied by LMS Test Lab. The centre of the test specimens is excited at Point 17 as shown in figure 2 and a high frequency resolution of ($\Delta f = 0.25$Hz) for precise modal parameter estimation is used for both types of excitations i.e., burst random and sine dwell.

Burst random excitation is used which is a broadband type excitation signal [0-1600 Hz]. The signal is averaged 10 times for each measurement point. Hanning windows are used for both the output and the input signals. Response is measured at 33 points that are symmetrically spaced in three rows along the length of the beam (figure 2). The modal parameters are extracted by a frequency domain estimation method (Polymax) based on an automatic extraction using stability diagram.

Whereas, sine dwell excitation is the discrete version of a sine sweep. The frequency is not varied continuously but is incremented by discrete amounts at discrete time points. The modal parameters are extracted by the Half Bandwidth Method (HBM),

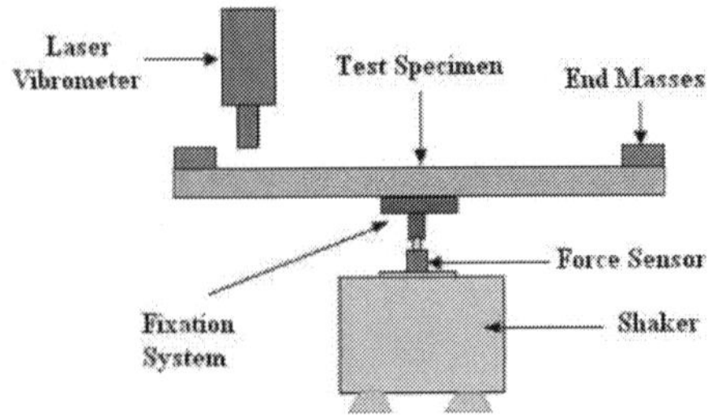

Figure 1. Experimental set-up for vibration testing

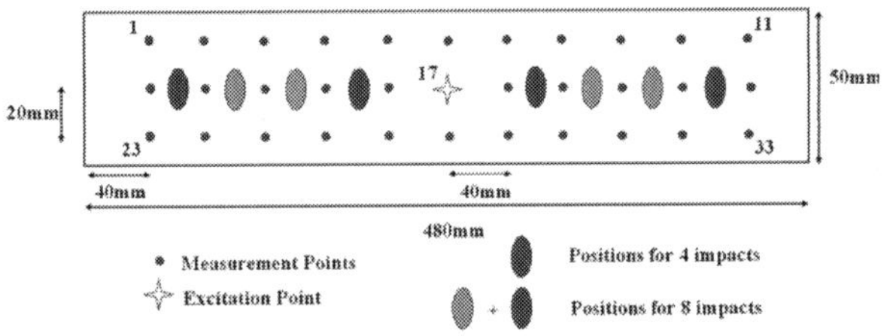

Figure 2. Composite beam specimen with location of damage, excitation and measurement points

3.2. Impact Tests

The impact test system used to damage the composite beams is drop weight system. The impactor tip has a hemispherical head with a diameter of 12.7 mm. The size of the impact window is 80 x 40 mm^2 which allows all the impact points to have the same boundary conditions and all the four ends are clamped. A force sensor (type 9051A) provided by Kistler is placed between the impactor tip and the free falling mass of 2 kg. The impact velocity is measured with the help of an optic sensor. The combined weight of the impact head, freefalling mass, force sensor and the accelerometer is 2.03 kg. In the calculation of impact height, a factor of 1.1 is used to compensate for the losses due to friction between the guidance tube and the drop assembly. Further details on the impact test methodology of this drop tower can be found in the reference [22].

The five composite specimens tested in this article are impacted around the barely visible impact damage limit (BVID). BVID corresponds to the formation of an indentation on the surface of the structure which can be detected by detailed visual inspection and can lead to high damage. In the aeronautical domain, BVID corresponds to an indentation of 0.3 mm after relaxation, aging etc (according to Airbus certifications). In this study, it is decided to take 0.6 mm of penetration depth as detectability criterion just after the impact [22]. Therefore, impact energy of 10 J giving an initial indentation depth of 0.55 mm, shall be considered as the BVID limit. Two of the five specimens are impacted with an impact energy (6 and 8 J) below the BVID limit in order to study the damage that is not visible by naked eye, and two (12 and 14 J) above BVID. The impact parameters for the five composite beam specimens studied in this article are listed in (table 3).

Table 3 Impact test parameters

Beam No	Energy of Impact (J)	Height (mm)	Velocity of impact (m/s) measured
1	6	331.8	2.49
2	8	442.3	2.83
3	10 (BVID)	552.9	3.24
4	12	663.5	3.52
5	14	774.1	3.84

The composite beam specimens have three states. First one is the undamaged state (0), the second is the damage state due to four impacts (4) and the third is the damage state due to eight impacts (8). These impact points are shown in figure 2. Vibration tests are done on the five composite beams after each of these three states.

4. Result and Discussion

4.1. Frequency and Damping change

Frequency and damping changes are studied with the help of the first four bending modes as they have the largest amplitudes for the type of test configuration presented in this article. As discussed previously, delamination induced damage in composites leads to an increase in damping and a decrease in natural frequency. This effect is more significant in the high frequency range [10]. This fact is verified by our experimental results which show that the difference in natural frequencies between the damaged (4 and 8) and the undamaged state (0) for the first mode is very small. But this difference in frequencies increases for the higher modes. For the 2nd and 3rd bending modes, the variation of natural frequency as a function of the undamaged (0) and the two damage states (4 and 8) is presented in figure 3.

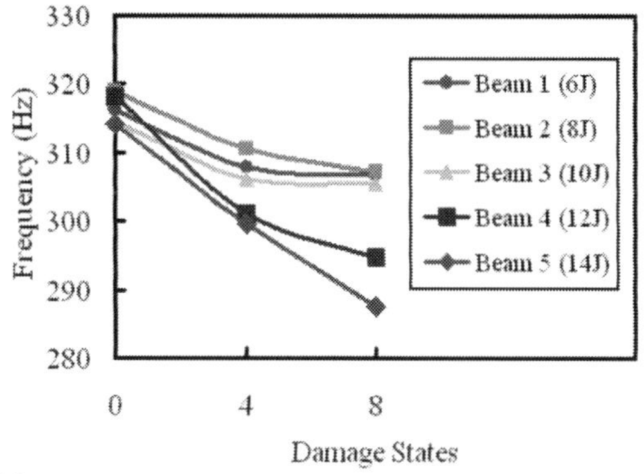

(a)

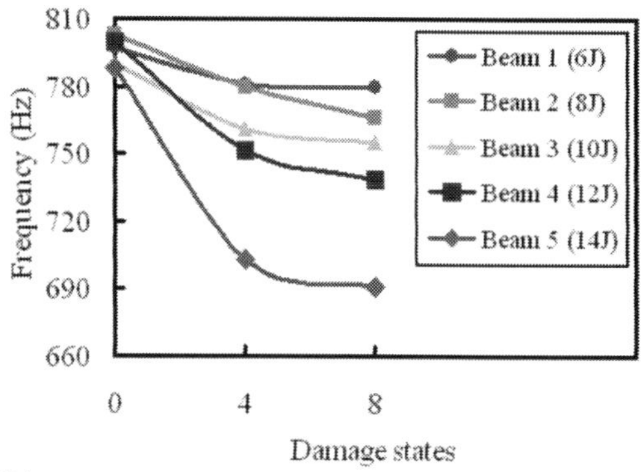

(b)

Figure 3. Variation of damped natural frequencies with damage states for (a) 2nd bending mode and (b) 3rd bending mode: 0 is the undamaged state, 4 is the damaged state at four impact points and 8 is the damaged state at eight impact points

Figure 3 (a) and (b) show that the decrease in natural frequencies with the increase in damage is more significant in case of the higher impact energies. The experimental results also prove that the change in natural frequency between the three states (0,4,8) increases with the increase in impact energy (e.g., increase in damage) for all the five composite beams. Similar results are obtained for the 4th bending mode. The damping ratios estimated by Polymax from burst random testing for the five composite beams are shown in figure 4.

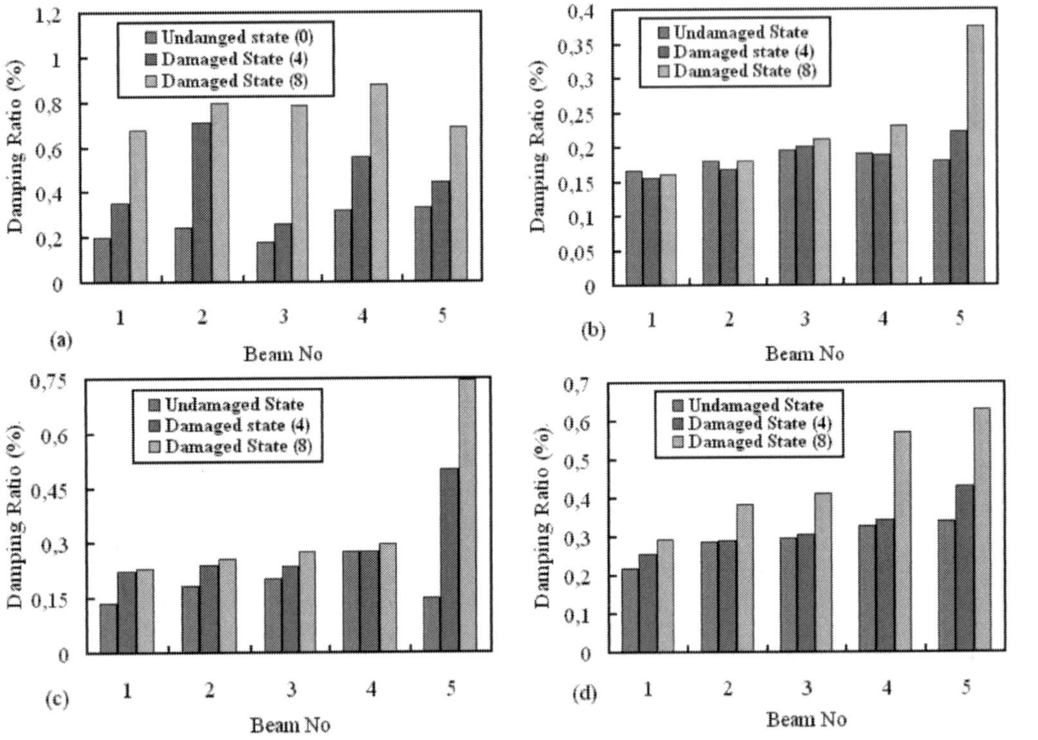

Figure 4. Variation of damping ratios (%) estimated by Polymax for the five composite beams for (a) 1st bending mode (b) 2nd bending mode (c) 3rd bending mode and (d) 4th bending mode, where '0' is the undamaged state, '4' is the damaged state at four points and '8' is the damaged state at eight points

Figures 3 and 4 show a slight discrepancy between the modal parameters at the undamaged state. This anomaly outlines the inherent possibility of false negatives which can arise due to boundary conditions and it gives no indication of damage when it is present, as discussed in the reference [23]. It can be seen from the results in figure 4 that the damping ratio increases with increase in damage in the five beams except for Mode 2 (figure 4b). Furthermore, the change in damping ratios between the three states (0,4,8) for Beams (1-4) is very small for the 2nd and 3rd bending modes (figure 4b and 4c). However unlike natural frequencies, the increase in damping ratio between the damaged and the undamaged states is not always consistent with the impact energy level, due to the complex nature of damping and the difficulties in its estimation. But in case of Mode 4 (figure 4d), damping ratio exhibits quasi linear dependence on the energy of impact. Furthermore, the modal parameter results underline the fact that the damping change ratios are more prominent than the frequency change ratios. In case of Beam 5 impacted at 14 J, the average change in natural frequency (between the damaged and the undamaged cases) for the first four bending modes is 8% whereas in case of damping ratios this average change is as high as 78%. So it is reasonable to assume that damping may be used instead of natural frequency as a damage indicator tool for structural health monitoring purposes [14-16].

For certain measurement points, damage in the composite beams distorts the shape of the resonance peaks and sometimes there is an appearance of twin peaks instead of one in the frequency response functions (FRF) as shown in figure 5. As the Half Bandwidth Method is dependent heavily on the shape of the resonance peaks, so the phenomenon of distorted peak shapes and appearance of twin peaks in case of damage, leads to faulty damping ratios. Due to this reason, the damping ratios estimated by the Half Bandwidth Method in case of sine dwell testing show a significant difference as compared to those estimated by Polymax method. The limitations of the Half Bandwidth Method are explained in reference [21].

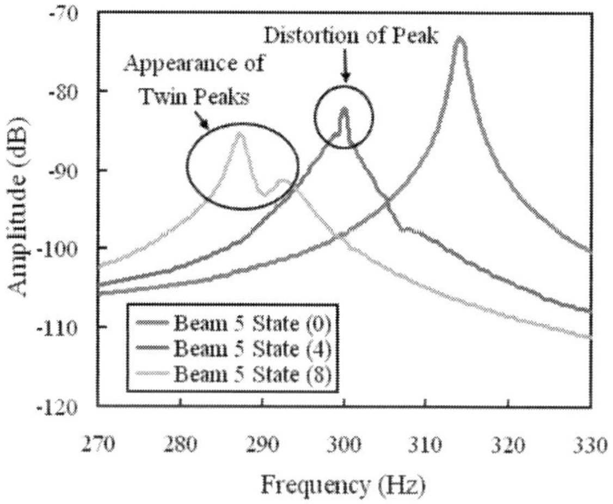

Figure 5. Comparison of the shapes of the FRF at point 2 for Beam 5 for (0) is the undamaged state, (4) is the damaged state at four impact points and (8) is the damaged state at eight impact points

4.2. Design of Experiments (DOE)

Design of experiments (DOE) is a powerful analysis tool for highlighting the influence of key parameters that affect an experimental process and the output of that process [24]. This study is carried out on the modal parameters (natural frequency and damping ratio) of the five composite beam specimens tested by burst random excitation, with an aim to identify the factors which have the most significant effect on the experimentally obtained modal parameters.

The two factors chosen for the design of experiments are the energy of impact (IE) and the density of damage (DD). For the energy of impact there are five levels (6, 8, 10, 12, 14J) and for the density of damage there are two levels (damage at 4 impact points and damage at 8 impact point). By keeping in view the levels of the two factors, a 5 x 2 full factorial design is chosen. The linear regression model associated with a 5 x 2 full factorial design, based on the two variables discussed above is expressed as follows:

$$Y = a_o + a_1.(IE) + a_2.(DD) + a_3.(IE).(DD) + E \quad (1)$$

In equation (1), coefficients represent model constants (a_i) that are the contribution of independent variables on the response. E is the random error term representing the effects of uncontrolled variables, i.e., not included in the model. The model constants (a_i) are determined by multi-linear regression analysis and are assumed to be normally distributed. The error is assumed to be random and normally distributed. These constants (a_i) are obtained with 90% confidence level. The significance of each variable on a given response (modal parameters in our case) is investigated using *t* test values based on Student's distribution. The *t* ratio is the ratio of the parameter estimate (constants) to its standard deviation. A *t* ratio greater than 2 in absolute value is a common rule of thumb for judging significance of the variable. The derived constants (a_i) and *t* ratios for the natural frequencies and the damping ratios are presented in (tables 4 and 5). The *t* ratios greater than 2 are marked in bold in (tables 4 and 5). Negative values of the model constants and *t* ratios indicate that the response decreases with the increase in the value of the parameter. In our case, this is most of the times true for the natural frequencies (table 4) as they decrease with the increase in damage in the specimens.

Table 4 Coefficients and *t* ratios for the natural frequencies (Hz)

	Mode 1		Mode 2		Mode 3		Mode 4	
Term	Constants (a_i)	*t* ratio	Constants (a_i)	*t* ratio	Constants (a_i)	*t* ratio	Constants (a_i)	*t* ratio
IE	0.122	0.66	-7.66	**-5.06**	-38.91	**-6.29**	-53.69	**-10.33**
DD	-0.532	**-4.09**	-2.36	-2.57	-4.52	-1.03	-30.73	**-7.93**
IE x DD	-0.278	-1.51	-2.48	-1.64	-2.17	-0.35	-24.03	**-4.39**

Table 5 Coefficients and *t* ratios for damping ratios (%) estimated by Polymax (burst random testing)

	Mode 1		Mode 2		Mode 3		Mode 4	
Term	Constants (a_i)	*t* ratio	Constants (a_i)	*t* ratio	Constants (a_i)	*t* ratio	Constants (a_i)	*t* ratio
IE	0.014	0.20	0.063	**4.36**	0.186	**2.83**	0.126	**9.80**
DD	0.151	**3.00**	0.026	**2.20**	0.043	0.92	0.066	**7.64**
IE x DD	0.007	0.10	0.008	1.26	0.016	1.03	0.046	1.74

By comparing the *t* ratios for the energy of impact (IE) and the density of damage (DD) in (tables 4 and 5), it can be seen that the energy of impact has a more significant effect on the modal parameters than the density of damage for the 2nd, 3rd and 4th bending modes. However, the density of damage is a more significant factor for the first bending mode. The second order interaction term (IE x DD) is in the majority of the cases insignificant as well.

From the design of experiments, it can be concluded that the energy of impact (IE) is the most significant factor on the natural frequencies and damping ratios (estimated by Polymax method with burst random excitation) for the 2nd, 3rd and 4th bending modes.

5. Conclusion

Vibration tests have been carried out on pristine and damaged composite beam specimens using burst random and sine dwell excitations. The composite beam specimens are impacted with the help of a drop weight system by keeping in view the *barely visible impact damage* limit (BVID). Results show that with the accumulation of damage in the specimens, there is a decrease in natural frequency accompanied by an increase in damping ratio. Results show that damping ratio is a more sensitive parameter for damage detection than the natural frequency. Energy of impact is highlighted as the factor having the most significant effect on the modal parameters by carrying out design of experiments on the experimental data. Furthermore, the Half Bandwidth Method is found unsuitable for damping estimation in the presence of damage.

6. References

[1] Tay T 2001 Characterization and analysis of delamination fracture in composites *Appl. Mech. Rev* **56** 1-31
[2] Matthews F L 1999 Damage in fibre reinforced plastics; its nature, consequences and detection *Proc of DAMAS99* 1-16
[3] Della C N and Shu D 2007 Vibration of delaminated composite laminates: A review *Appl. Mech. Rev* **60** 1-20
[4] Sanders D R, Kim Y I, and Stubbs N 1992 Non-destructive evaluation of damage in composite structures using modal parameters *Exp. Mech.* **32** 240-251
[5] Zou Y, Tong L and Steven G B 2000 Vibration-based model-dependent damage (delamination) identification and health monitoring for composite structures. A review *J. Sound Vib.* **230** 357-378

[6] Doebling S W, Farrar C R and Prime M B 1998 A summary review of vibration-based damage identification methods *Shock Vib. Dig.* **30** 91-105

[7] Sohn H, Farrar C R, Hemez F M, Shunk D, Stinemates D W and Nadler B R 2003 A review of structural health monitoring literature: 1996-2001 *Los Alamos National Laboratory Report* LA-13976-MS

[8] Khoo L M, Mantena P R and Jadhav P 2004 Structural damage assessment using vibration modal analysis *Struc. Health Monitoring* **3(2)** 177-194

[9] Yam L H and Cheng L 2005 Damage detection of composite structures using dynamic analysis. *K. Eng. Mat.* **295-296** 33-38

[10] Kim H Y and Hwang 2002 Effect of debonding on natural frequencies and frequency response functions of honeycomb sandwich beams *Compos. Struct.* **51** 51-62

[11] Adams R D and Cawley P 1979 The localisation of defects in structures from measurements of natural frequencies *J. Strain Anal.* **14** 49–57

[12] Richardson M H and Mannan M A 1993 Correlating minute structural faults with changes in modal parameters *Proc. of SPIE* **923(2)** 893–898

[13] Adams R D 1993 Damping in composites *Mater. Sci. Forum* **119-121** 3-16.

[14] Zhang Z and Hartwig G 2004 Relation of damping and fatigue damage of unidirectional fibre composites *Int. J. Fatig.* **24** 713-738

[15] Gibson R F 2000 Modal vibration response measurements for characterization of composite materials and structures *Compos. Sci. Tech.* **60** 2769-2780

[16] Saravanos D A and Hopkins D A 1995 Effects of delaminations on the damped dynamic characteristics of composites *J. Sound Vib.* **192** 977-993

[17] Vanhoenacker K, Schoukens J, Guillaume P and Vanlanduit S 2004 The use of multisine excitations to characterise damage in structures *Mech. Syst. Signal Process.* **18** 43-57

[18] Shahdin A, Morlier J and Gouriant Y Correlating low energy impact damage with changes in modal parameters: A preliminary study on composite beams *Accepted Jan 09 to SHM*

[19] Wojtowicki J L and Jaouen L 2004 New approach for the measurements of damping properties of materials using Oberst beam *Rev. Sci. Instrum.* **75(8)** 2569-2574

[20] Amraoui M Y and Lieven N A J 2004 Laser vibrometry based detection of delaminations in glass/epoxy composites *J. Sound Vib.* **126** 430-437

[21] Yin H P 2008 A new theoretical basis for the bandwidth method and optimal power ratios for the damping estimation *Mech. Syst. Signal. Process.* **22(8)** 1869-1881

[22] Petit S, Bouvet C, Bergerot A and Barrau J J 2007 Impact and compression after impact experimental study of a composite laminate with a cock thermal shield *Compos. Sci. Tech.* **67** 3286-3299

[23] Mattson S G and Pandit S M 2006 Damage detection and localization based on outlying residuals, *Smart Mater. Strcut* **15** 1801-1810.

[24] Meyers R H and Montgomery D C 1995 *Response surface methodology* (New York: Wiley)

Acknowledgments

The authors gratefully thank research project student Hanno Niemann from TU Braunschweig for his technical support during the experimental work.

Journal of Physics: Conference Series **181** (2009) 012046 doi:10.1088/1742-6596/181/1/012046

Nonlocal conditions for the transition from damage to a localized failure in granular composites under quasistatic loading

Alexey V. Zaitsev

Mechanics of Composite Materials & Structures Dept., Perm State Technical University, Komsomolskiy Ave., 29, 614990, Perm, Russia

e-mail: zav@pstu.ru

Abstract. A two-level structural-phenomenological model was developed for a granular composite with random strength/deformation properties. Materials of this class include, specifically, modern ceramics and composites obtained by power metallurgy methods. The objectives of the model development were: i) to research the major regularities and the mechanisms of strain-induced damages in heterogeneous materials at the strain-softening stage; ii) to formulate nonlocal conditions for transition to a localized failure in anisotropic media. The model allowed describing inelastic quasistatic deformation and failure of heterogeneous solids, accompanied by nucleation and evolution of defects, as a multistage process of damage accumulation, and macrofailure as a result from the loss of stability of the latter process. Changing the mechanisms and scale levels of predominant damage evolution was found and analyzed numerically. New criteria for transition to a localized failure under triaxial quasistatic proportional and nonproportional macrodeformation have been formulated.

1. Introduction

The research of inelastic deformation and failure of composites is associated with the necessity to develop mechanical models for the correct description of the behavior of damaged heterogeneous materials in elements of structures and facilities [1]. When constructing the models of damage accumulation for heterogeneous media, it is necessary to describe not only nucleation and evolution of individual damages, but also the multi-particle interactions in the defect ensemble [2]. These interactions determine the conditions of macrofailure and strength characteristics of the material under loading.

Loading conditions, mechanical properties and responses to changing external mechanical actions on polymers, composites, and rocks differ widely. Notwithstanding this fact, it is possible to set aside general mechanisms of damage accumulation process and unified conditions for transition to macrofailure of the above mentioned materials, when limiting defect-concentration values are reached at a local domain. The results of experiments [3–9] showed that microcrack accumulation appears to differ essentially in those materials, but limiting defect-concentration values are very close to each other (if uniformly distributed failure domains appear). Besides, in the works under review the inference is drawn about the universality of the concentration criterion for transition from equilibrium disperse fracture to nonequilibrium macrofailure.

The results of computational experiments on uniaxial deformation and compression of a model granular composite [10, 11] indicated that the realization of the equilibrium states of the material and sequential changes in stages of damage accumulation process are determined not only by heterogeneity of structure, mutual arrangement of defects, and its concentration. The character of the energy redistribution between a solid and a loading-system is also responsible for that.

As it is shown by experimental researches, macrofailure of a composite is due to an increasing number of cumulative defects if the void fraction of reinforcement aggregates is small or dispersion of their mechanical properties is significant [12]. Otherwise, material tends to brittle failure [13, 14]. Besides, computational experiments [11] showed that the limiting void fraction of damages is found to be significantly dependent on the dispersion of strength/deformation properties of structural elements, presence of initial imperfections, stress-strain state, and the stiffness of the loading-system as well.

It should be noted that damage accumulation process occurs uniformly within the representative volume element of composite at the initial stage of loading only. At the stages before failure, the process is of clearly pronounced localized nature. Therefore, the application of averaged limiting defect-concentration values seems impossible when defining the macrofailure. Considering that these values are functions or functionals that depend on physico-mechanical properties of the material and loading process it testifies the necessity to refine the strength concept for heterogeneous solids based on the assumption that constants of critical defect concentration exist [3–9].

Besides, there is a demand to improve strength analysis procedures that consider the real loading conditions, evolution and character of the collective interaction in the system of defects defining the macrofailure point as a result of loss of stability of damage accumulation. Without understanding major regularities and mechanisms of defect accumulation, without evaluating its stability and determining the conditions of localization beginning, the macrofracture of composites will remain latent and poorly predictable phenomenon of internal structure evolution of the material.

The mechanism of transition to localized failure can be determined from a complex experimental research of the inelastic behavior of heterogeneous materials combined with an analysis of the defect ensemble evolution. The research requires special equipment and is currently performed for the simplest loading modes only (i.e., uniaxial monotonic tension and/or compression). In regard with it, it is profitable to carry out auxiliary computing experiments with representative volume elements of the model granular composite for various triaxial stress-strain states and modes of macrodeformation.

2. Boundary-value problem

Let us consider a heterogeneous body Ω with isotropic scleronomic components geometry, relative position, and the symmetry type of which remain unchanged during deformation and damage accumulation processes; prior to applying the load, the body is in the natural undeformed state. Let us assume that the full adhesion conditions are fulfilled on the contact boundaries of structural elements, i.e.

$$\left[\sigma_{ij}(\mathbf{r})n_j(\mathbf{r})\right]^- = \left[\sigma_{ij}(\mathbf{r})n_j(\mathbf{r})\right]^+ \text{ and } \left[u_i(\mathbf{r})\right]^- = \left[u_i(\mathbf{r})\right]^+ .$$

The quasistatic step-by-step loading of the body Ω is described by the closed system of differential equations in the incremental form, i.e.

$$d\sigma_{ij,j}(\mathbf{r})=0, \quad d\varepsilon_{ij}(\mathbf{r})=\left[du_{i,j}(\mathbf{r})+du_{j,i}(\mathbf{r})\right]/2, \tag{1}$$

$$d\sigma_{ij}(\mathbf{r})=K'(\mathbf{r})d\varepsilon_{kk}(\mathbf{r})\delta_{ij}+2G'(\mathbf{r})d\bar{\varepsilon}_{ij}(\mathbf{r}), \quad \forall\mathbf{r}\in\Omega, \tag{2}$$

$$K'(\mathbf{r})=K(\mathbf{r})\left\{1-\xi\kappa\left[j_\varepsilon^{(1)}(\mathbf{r}), j_\varepsilon^{(2)}(\mathbf{r})\right]\right\}, \quad G'(\mathbf{r})=G(\mathbf{r})\left\{1-\xi g\left[j_\varepsilon^{(1)}(\mathbf{r}), j_\varepsilon^{(2)}(\mathbf{r})\right]\right\}.$$

Herein, $\mathbf{du}(\mathbf{r})$, $\mathbf{d\sigma}(\mathbf{r})$ and $\mathbf{d\varepsilon}(\mathbf{r})$ are the increments of displacements, stresses, and strains which are random functions of the space coordinate $\mathbf{r}$; $K(\mathbf{r})$ and $G(\mathbf{r})$ are the elastic bulk and shear moduli; $\xi=0$ under unloading and after loading until the elastic limit, $\xi=1$ under active loading.

Constitutive equations (2) contain the independent material functions

$$g=\begin{cases}0, & j_\varepsilon^{(2)}(\mathbf{r})<j_{\varepsilon\,\mathrm{cr}}^{(2)},\\ 1, & j_\varepsilon^{(2)}(\mathbf{r})\ge j_{\varepsilon\,\mathrm{cr}}^{(2)},\end{cases} \qquad \kappa=\begin{cases}0, & \left[j_\varepsilon^{(2)}(\mathbf{r})<j_{\varepsilon\,\mathrm{cr}}^{(2)}\right]\vee\left[j_\varepsilon^{(2)}(\mathbf{r})\ge j_{\varepsilon\,\mathrm{c}}^{(2)}\wedge j_\varepsilon^{(1)}(\mathbf{r})\le 0\right],\\ 1, & j_\varepsilon^{(2)}(\mathbf{r})\ge j_{\varepsilon\,\mathrm{cr}}^{(2)}\wedge j_\varepsilon^{(1)}(\mathbf{r})>0,\end{cases} \tag{3}$$

arguments of which are independent invariants of the structural strain tensor,

$$j_\varepsilon^{(1)}(\mathbf{r})=\varepsilon_{kk}(\mathbf{r}) \text{ and } j_\varepsilon^{(2)}(\mathbf{r})=\sqrt{\breve{\varepsilon}_{ij}(\mathbf{r})\breve{\varepsilon}_{ij}(\mathbf{r})}.$$

Herein, $d\breve{\varepsilon}_{ij}=d\varepsilon_{ij}-1/3d\varepsilon_{kk}\delta_{ij}$. If the inequalities (3) are satisfied, the material functions κ and g take its limit values 0 or 1, change the mechanical characteristics step-like, and allow accounting the states of the total or partial loss of strength capacity of a structural element. In regard with the type of the stress-strain state the inequalities describe (the) various mechanisms of the shear damage if the condition $j_\varepsilon^{(2)}(\mathbf{r})\geq j_{\varepsilon\,\mathrm{cr}}^{(2)}(\mathbf{r})$ is satisfied (herein, $j_{\varepsilon\,\mathrm{cr}}^{(2)}(\mathbf{r})$ is the strength constant). The complete failure of the structural element occurs if $j_\varepsilon^{(1)}(\mathbf{r})>0$, while when $j_\varepsilon^{(1)}(\mathbf{r})\leq 0$ the element can still resist hydrostatic compression only.

Let us assume a macrohomogeneous continuum with effective macroproperties corresponding to a real damaged heterogeneous solids. Let us assign a representative volume element Ω_{RVE} in the continuum, state of which will be characterized by the tensors of macrostresses $\boldsymbol{\sigma}^*$ and macrostrains $\boldsymbol{\varepsilon}^*$,

$$\sigma_{ij}^*=\frac{1}{\operatorname{mes}\Omega_{RVE}}\int_{\Omega_{RVE}}\sigma_{ij}(\mathbf{r})\,d\Omega,\qquad \varepsilon_{ij}^*=\frac{1}{\operatorname{mes}\Omega_{RVE}}\int_{\Omega_{RVE}}\varepsilon_{ij}(\mathbf{r})\,d\Omega. \tag{4}$$

The resistance of representative volume element to the external loading determines the relation

$$d\sigma_{ij}^*=\aleph_{ijkl}^*\left(\boldsymbol{\varepsilon}^*,\chi^{(\alpha)}\right)d\varepsilon_{kl}^* \tag{5}$$

between the increments of macrostresses $\mathbf{d\sigma}^*$ and macrostrains $\mathbf{d\varepsilon}^*$, where $\aleph_{ijkl}^*$ are the material functions reflecting the effective properties of the heterogeneous material; $\chi^{(\alpha)}$ are the parameters that define the macrodeformation process. Postulating macroisotropy and absence of deformation anisotropy during the damage accumulation, and assuming that the structure of equations (2) is unchanged at the macrolevel, we will represent the constitutive equations for Ω_{RVE} in an invariant form [10]

$$dJ_\sigma^{(1)}=3K^*\left[1-\xi\kappa^*\left(J_\varepsilon^{(1)},J_\varepsilon^{(2)},\chi^{(\alpha)}\right)\right]dJ_\varepsilon^{(1)},\quad dJ_\sigma^{(2)}=2G^*\left[1-\xi g^*\left(J_\varepsilon^{(1)},J_\varepsilon^{(2)},\chi^{(\alpha)}\right)\right]dJ_\varepsilon^{(2)}. \tag{6}$$

Herein, K^* and G^* are the effective elastic moduli; $J_\sigma^{(\circ)}$ and $J_\varepsilon^{(\circ)}$ are the independent invariants of macrostress $\boldsymbol{\sigma}^*$ and macrostrain $\boldsymbol{\varepsilon}^*$ tensors. Due to the postulated macroscopic heterogeneity of the material the representative volume elements chosen around an arbitrary pair of points have the same effective properties.

The major damage accumulation mechanisms under monotonic proportional and nonproportional loading of Ω_{RVE} will be studied on the basis of the numerical solution of boundary-value problem (1), (2) by the finite element method. The components of the macrostrain $\boldsymbol{\varepsilon}^*$ tensor enclosed in boundary conditions that correspond to displacement-controlled mode

$$\left[d\varepsilon_{ij}^* r_j\right]\Big|_{\partial\Omega_{RVE}}=du_i^0(\mathbf{r}),\qquad \forall\mathbf{r}\in\partial\Omega_{RVE}$$

will be changed; and strength conditions (3) will be verified on each loading step, being determined automatically according to the developed iteration procedure [10]. If the strength conditions are not satisfied then, the strain constants of structural elements change. Herein, $\mathbf{du}^0(\mathbf{r})$ are the displacements specified in every point $\mathbf{r}$ of the surface $\partial\Omega_{RVE}$.

3. Structural failure as a reason to strain-softening

Let us consider some simulation results of damage accumulation process in representative volume elements of model granular composites that occupy cubic domains and contain 6 000 uniform tetrahedral elastic-brittle structural elements with random strength constants (distributed by the Weibull law) and nonrandom elastic moduli $G(\mathbf{r})=4\cdot10^4$ MPa, and $K(\mathbf{r})=6.7\cdot10^4$ MPa.

Figure 1 gives a part of the calculation stress-strain diagram for the model granular composite in an invariant form under pure shape change (i.e., $\varepsilon_{11}^*=\varepsilon_{22}^*=-0.5\varepsilon_{33}^*$, $\varepsilon_{33}^*>0$); the curve does not break at the peak maximum under displacement-controlled mode. The presence of the descending branch in the diagram indicates strain-softening. Multiple jumps and the nonlinear character of the stress-strain curve are induced by structural damage accumulation process that begins on the ascending branch and stops at the stage of residual strength.

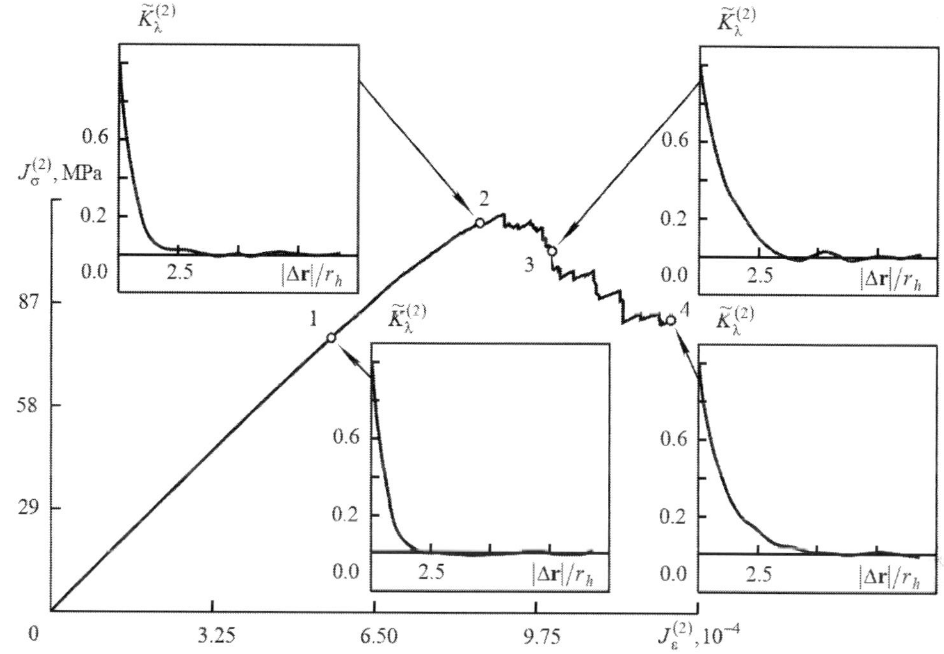

Figure 1. The calculation stress-strain diagram under pure shape change and normalized correlation functions of the damaged structure of the granular composite at different stages

4. Structural failure mechanisms under monotonic proportional loading

Let $\tau(\mathbf{r})$ be the parameter of a random strain field, and $\lambda(\mathbf{r})$ be the random arbitrary unit indicator function that is defined by the determinate radius-vector $\mathbf{r}$ and takes on the value 1 if $\mathbf{r} \in \Omega_D$, and 0 if $\mathbf{r} \in \Omega_{RVE} \setminus \Omega_D$ (Ω_D is a set of damaged grains). Statistical expectation of the multiplication of centered functions $\tau^\circ(\mathbf{r}) = \tau(\mathbf{r}) - \langle \tau(\mathbf{r}) \rangle$ and $\lambda^\circ(\mathbf{r}) = \lambda(\mathbf{r}) - v_D$ (herein, v_D is the void fraction of damages) determined in different points $\mathbf{r}$ and $\mathbf{r}+\Delta\mathbf{r}$ are the correlation functions of the random strain field $K_\tau^{(2)}(|\Delta\mathbf{r}|) \equiv \langle \tau^\circ(\mathbf{r})\tau^\circ(\mathbf{r}+\Delta\mathbf{r}) \rangle$ and of the damaged structure $K_\lambda^{(2)}(|\Delta\mathbf{r}|) \equiv \langle \lambda^\circ(\mathbf{r})\lambda^\circ(\mathbf{r}+\Delta\mathbf{r}) \rangle$.

The authors [16, 17] showed that the developed two-level structural-phenomenological model in combination with the correlation analysis of damage evolution allow describing the mechanisms of inelastic deformation and failure of granular composites under quasistatic loading. They studied the changes in normalized (i.e., related to the dispersions $\langle \tau^\circ(\mathbf{r})\tau^\circ(\mathbf{r}) \rangle$ and $v_D(1-v_D)$) correlation functions $\tilde{K}_\lambda^{(2)}$ and $\tilde{K}_\tau^{(2)}$ during damage accumulation for various modes of proportional macrodeformation in detail.

At the initial stage of pure shape change (i.e., $\varepsilon_{11}^* = \varepsilon_{22}^* = -0.5\varepsilon_{33}^*$, $\varepsilon_{33}^* > 0$) of granular composite damages are accumulated uniformly over the whole representative volume element (state *1*, Fig. 1). The normalized correlation functions $\tilde{K}_\lambda^{(2)}$ corresponding with the deformation stage are local, close to exponential ones and damp completely over the distance $(2 \div 3) r_h$. Hence, r_h is the average heterogeneity size. The subsequent increase in macrostrains gives rise to defect growth (equilibrium states *2* and *3*, Fig. 1). When the strength limit is achieved, localized failure domains (i.e., clusters into which defects begin to coalesce) expand in the composite significantly. The intensification of multy-particle collective interaction in a defect ensemble (the functions $\widetilde{K}_{\lambda}^{(2)}$ in the states *2* and *3* damp over the distances $(4 \div 5) r_h$ and $(5 \div 7) r_h$ correspondingly) leads to the appearance of a periodic term in random fields $\lambda(\mathbf{r})$. A uniform cluster accumulation together with the intensive stress redistribution, and partial or total unloading of undamaged grains surrounding defects, are followed by macrostress decrease that points out the beginning of strain-softening. The transition to the strain-softening associated with the beginning of individual cluster coalescence leads to even greater intensification of interaction in the defect ensemble. At the final deformation stage all damaged domains gradually coalesce into a single branching macrodefect (state *4*, Fig. 1), while the functions $\widetilde{K}_{\lambda}^{(2)}$ damp over the distances $(7 \div 8) r_h$.

Hence, the normalized correlation functions defined at deformation stages make it possible to determine the character of collective multy-particle interactions in the defect ensemble and naturally separate the stages of dispersible and localized failure. Besides, the correlation analysis of the damaged structure allows the authors [16] determining changes of the stages of equilibrium and nonequilibrium failure and proving (in the framework of the model description) the self-similarity of defect evolution at the strain-softening stage for the first time.

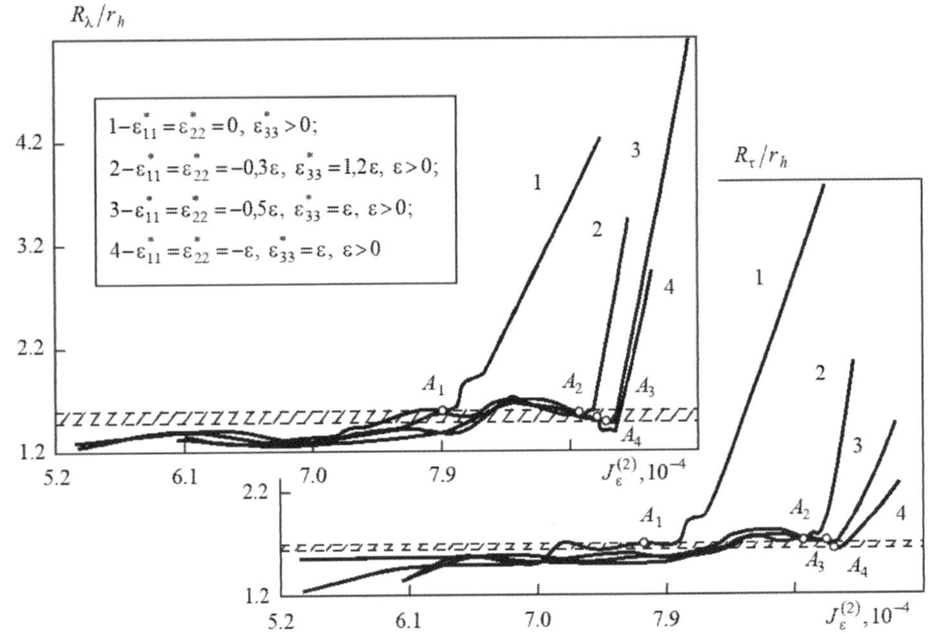

Figure 2. Evolution of correlation integrals for various modes of proportional macrodeformation

5. Nonlocal criterion of transition to localized failure

If the macrofailure is considered to be the critical state, where structural transition under loading results in a loss of continuity in the damaged solids or in reaching of connectivity in the defect ensemble, we can determine parameters widely used in the physics of critical phenomena which characterize collective multi-particle interactions and the mutual arrangement of failed grains. The universal regularities in behavior of stochastic systems that close to critical point (that can be the state of failure localization onset) are the appearance of the long-range interactions in the defect ensemble, which occurs in a sharp increase of the correlation integrals [18]

$$R_\lambda \equiv [\nu_D(1-\nu_D)]^{-1} \int_0^\infty K_\lambda^{(2)}(|\Delta\mathbf{r}|)\, d|\Delta\mathbf{r}| = \int_0^\infty \widetilde{K}_\lambda^{(2)}(|\Delta\mathbf{r}|)\, d|\Delta\mathbf{r}|,$$

$$R_\tau \equiv \left[D_\tau^{(2)}\right]^{-1} \int_0^\infty K_\tau^{(2)}(|\Delta\mathbf{r}|)\, d|\Delta\mathbf{r}| = \int_0^\infty \widetilde{K}_\tau^{(2)}(|\Delta\mathbf{r}|)\, d|\Delta\mathbf{r}|$$

for the random fields $\lambda(\mathbf{r})$ and $\tau(\mathbf{r})$. Herein, $\tau(\mathbf{r})$ is a second invariant of the microstrain tensor.

To define the conditions of start of failure localization in model granular composite, we analyze the changes in the dimensionless values of the correlation integrals R_λ/r_h and R_τ/r_h (Fig. 2 and 3) during the process of deformation under different proportional modes and 6 trajectories as of two-segment polygonal lines. The location of their characteristic points is presented in Table 1.

Table 1

Deformation trajectories for the granular composite as of two-segment polygonal line

Path	Second point location*		
	$\varepsilon_{11}^*, 10^{-3}$	$\varepsilon_{22}^*, 10^{-3}$	$\varepsilon_{33}^*, 10^{-3}$
Pt_1	0,5	0,5	1,5
Pt_2	−1,5	1,5	0,5
Pt_3	−1,5	0,5	1,0
Pt_4	−2,0	0,0	1,0
Pt_5	0,5	−1,5	1,0
Pt_6	0,5	−2,0	1,0

Note: * in the initial state: $\varepsilon_{11}^* = \varepsilon_{22}^* = \varepsilon_{33}^* = 0$, in the final state: $\varepsilon_{11}^* = \varepsilon_{22}^* = \varepsilon_{33}^* = -2.0 \cdot 10^{-3}$

Curve *7* in Fig. 3 corresponds to (with) the mode of proportional macrodeformation $\varepsilon_{11}^* = \varepsilon_{22}^* = -\varepsilon_{33}^*$, $\varepsilon_{33}^* > 0$. As damages accumulate, R_λ/r_h and R_τ/r_h gradually increase. In Fig. 2 and 3 the points A_k denote the dimensionless correlation integrals corresponding with the strength limit and the failure localization onset. Owing to the nonuniform damage resistance of the granular composite [19] and to the dependence of the failure on the deformation process, the values of the second invariant of the macrostrain tensor $J_\varepsilon^{(2)}$, which correspond with the above-mentioned points, are essentially different. The coalescence of individual damages and clusters is accompanied by the enhancement of collective interactions in the defect ensemble (the macrostrains are more and more determined by the damaged grains) and by a sharp increase of R_λ/r_h and R_τ/r_h.

Let us emphasize that the maximum difference in the values R_λ/r_h and R_τ/r_h corresponding with the strength limit (Fig. 2) and with the failure localization onset (Fig. 3) does not exceed 7 % for various modes of triaxial proportional and nonproportional macrodeformation. This behavior confirms the hypothesis (first proposed by the authors of [10, 16, 17]) about the existence of critical value of the correlation integral $R_{\lambda\,\mathrm{cr}}$,

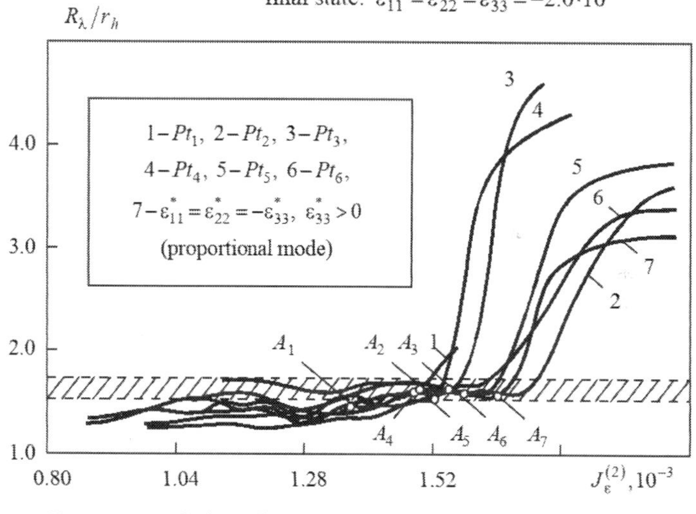

Figure 3. Evolution of correlation integrals for various modes of nonproportional macrodeformation

which is independent of the stress-strain state and quasistatic macrodeformation. The critical value $R_{\lambda\,\mathrm{cr}}$ is a nonlocal constant of the material which accounts the heterogeneity and character of collective multi-particle interaction in the defect ensemble under loading.

Hence, the inequality

$$R_\lambda > R_{\lambda\,\mathrm{cr}} \tag{8}$$

is the nonlocal criterion of transition from the stage of disperse damage accumulation to a localized failure for the damaged model granular composite, independent of the quasistatic deformation modes. The relation found for the nonlocal parameter $R_{\lambda\,\mathrm{cr}}$ and the physico-mechanical constant of material (i.e., the strength limit) allows formulating a new criteria for transition to a localized failure and/or to the strain-softening.

Let us define a critical value of the correlation integral $R_{\lambda\,\mathrm{cr}}^{(\alpha)}$ for each set of structural elements of heterogeneous anisotropic solids, where these elements lost their strength capacity to resist α external action (i.e., transversal or antiplain shear, etc.). Then the fulfillment of the following conditions

$$R_{\lambda}^{(m)} < R_{\lambda\,\mathrm{cr}}^{(m)}$$

corresponds to the strain-hardening and/or with the stage where damages are accumulated uniformly over the whole representative volume element. But should the inequalities

$$R_{\lambda}^{(m)} \geq R_{\lambda\,\mathrm{cr}}^{(k)} \wedge R_{\lambda}^{(m)} < R_{\lambda\,\mathrm{cr}}^{(m)} \tag{9}$$

hold true for each $k \neq m$, then the next equilibrium stress-strain state of the damaged material corresponds to the state when the failure localization starts or when the strain-softening stage according to the k mechanism occurs. Such problems as determining mutual influence of different mechanisms of structural damage accumulation on the transition to failure localization and to strain-softening, and also simultaneity of these transitions according to all mechanisms for anisotropic media require further investigations.

Now we should like to dwell on practical importance of the developed nonlocal criteria (8) and (9). On one hand, the values of the correlation integrals R_{λ} defined at different stages of damage accumulation process exclude defects of some certain kind from consideration. On the other hand, they establish a unique quantitative relation between the regularities of multy-particle interactions in the defect ensemble and mechanical behavior of a model granular composite under quasistatic loading. Therefore, regardless of loading conditions, the study on a certain representative volume element allows determining the deformation stage of a damaged material and predicting the proximity of its macrofailure with the boundary-value problem, being unsolved.

6. Conclusion

The developed two-level structural-phenomenological model of granular composites made it possible to describe a collective multi-particle interaction in a defect ensemble, scale levels of damage accumulation process, major failure mechanisms, and their changes in various modes of triaxial proportional and nonproportional macrodeformation. In the framework of the model description of damage accumulation process in granular composites the nonlocal conditions for transition to a localized failure and/or to strain-softening are formulated independent on the stress-strain state and quasistatic proportional loading modes.

7. Acknowledgements

The author acknowledges the support of the Russian Foundation for Basic Research (Grant RFBR–Urals No 07–01–96056).

References

[1] Krajcinovic, D. (2001) Damage mechanics: accomplishments, trends and needs, Int. J. Mech. Solids Structures, 37, 267–277.
[2] Mishnaevskiy, L.L. (1997) Methods of the theory of complex systems in modelling of fracture: a brief review, *Engin. Fracture Mech.*, 56, 47–56.
[3] Zhurkov, S.N. and Korsukov, V.E. (1974) Mechanism of submicrocracks generation in stress polymers, *J. Polym., A2. Polym. Phys.*, 10, 1509–1520.
[4] Zhurkov, S.N., Kuksenko, V.S., Petrov, V.A. et all. (1980) Concentration criterion of bulk fracture of rocks. In: Physical processes in earthquake sources, Moscow, Nauka, 78–86.
[5] Kuksenko, V.S. and Tamuzs, V.P. (1981) Fracture mechanics of polymer materials, Martinus Nijhoff Publ., Hague.
[6] Kuksenko V.S., Orlov L.G. and Frolov D.I. (1979) Concentration criterion for the crack

coarsening at heterogeneous material. In: Fracture of composite materials: Proc. 1st USA — USSR symposium. (Edited by G.C. Sih and V.P. Tamuzs), Alphenaan den Rijn, Sijthoff and Noordhoff, 25–31.

[7] Kuksenko V.S., Lyashkov A.I., Saveliev V.N. and Frolov D.I. (1982) Physical principles for failure prediction of heterogeneous naterials. In: Fracture of composite materials: Proc. 2nd USA — USSR symposium. (Edited by G.C. Sih and V.P. Tamuzs), Hague, Martinus Nijhoff Publ., 33–38.

[8] Botvina, L.R. (1999) On correlation of various approaches for description of kinetic processes, *Int. J. Fracture*, 99, 131–141.

[9] Botvina, L.R. and Oparina, I. B. (2000) Common features of fracture process on various scale levels, *Int. J. Fracture*, 106, 33–45.

[10] Zaitsev, A.V. (2004) Regularities of damage accumulation processes and conditions of transition to localized failure of grain composites under quasistatic loading. Physical Mesomechanics, 7, 41–50.

[11] Vil'dcman, V.E., Sokolkin, Yu.V. and Zaitsev, A.V. (1997) Damage structure evolution of heterogeneous media at supercritical deformation stage. Mechanics of Composite Materials, 33, 231–238.

[12] Ostoja-Starzewski, M., Sheng, P.Y. and Jasiuk, I. (1997) Damage patterns and constructive response of random matrix-inclusion composites, *Engin. Fracture Mech.*, 58, 581–606.

[13] Barzhenov, S.L., Kuperman, A.M., Puchkov, L.V. et all. (1984) Statistical aspects of the fracture of organic fibre plastics. Mechanics of Composite Materials, 20, 1011–1017.

[14] Bulmanis, V.N., Gusev, Ju.I., Struchkov, A.S. and Antokhonov, V.B. (1985) Experimental investigation of the deformation and fracture features at the tension of the angle-ply winding composites. Mechanics of Composite Materials, 21, 1020–1024.

[15] Ermolenko, A.F. and Protasov, V.D. (1989) The effect of reinforcement on the strength and fracture character of unidirectional fibrous composites. Mechanics of Composite Materials, 25, 276–283.

[16] Wildemann, V.E., Zaitsev, A.V. and Gorbunov, A.N. (1999) Mechanisms and regularities of heterogeneous-solid damage-accumulation at the supercritical stage. Physical Mesomechanics, 2, 37–48.

[17] Tashkinov, A.A., Wildemann, V.E. and Zaitsev, A.V. (1998) Micro-mechanical models of stable damage evolution on post-critical deformation stage and failure of composite materials. In: ECF12 Fracture from Defects. (Edited by M.W. Brown, K.J. Miller and E.R. de los Rios). Warley, EMAS, 3, 1483–1488.

[18] Stanley, H.E. (1971) Introduction to phase transitions and critical phenomena, Oxford University Press, London

[19] Sokolkin, Yu.V., Wildeman, V.E., Zaitsev, A.V. and Rochev, I.N. (1998) Structural damage accumulation and stable post-critical deformation of composite materials. Mechanics of Composite Materials, 34, 171–183.

Journal of Physics: Conference Series **181** (2009) 012047 doi:10.1088/1742-6596/181/1/012047

The dynamic behaviour of a non-stationary elevator compensating rope system under harmonic and stochastic excitations

S. Kaczmarczyk[1], R. Iwankiewicz[2] and Y. Terumichi[3]

[1] School of Applied Sciences, University of Northampton, St. George's Avenue, Northampton NN2 6JD, UK

[2] Institute of Mechanics and Ocean Engineering, Hamburg University of Technology, Eissendorfer Strasse 42 D-21073, Hamburg, Germany

[3] Faculty of Science and Technology, Sophia University, 7-1 KIOI-CHO, CHIYODA-KU, Tokyo, 102-8554 Japan

E-mail: stefan.kaczmarczyk@northampton.ac.uk

Abstract. Moving slender elastic elements such as ropes, cables and belts are pivotal components of vertical transportation systems such as traction elevators. Their lengths vary within the host building structure during the elevator operation which results in the change of the mass and stiffness characteristics of the system. The structure of modern high-rise buildings is flexible and when subjected to loads due to strong winds and earthquakes it vibrates at low frequencies. The inertial load induced by the building motion excites the flexible components of the elevator system. The compensating ropes due to their lower tension are particularly affected and undergo large dynamic deformations. The paper focuses on the presentation of the non-stationary model of a building-compensating rope system and on the analysis to predict its dynamic response. The excitation mechanism is represented by a harmonic process and the results of computer simulations to predict transient resonance response are presented. The analysis of the simulation results leads to recommendations concerning the selection of the weight of the compensation assembly to minimize the effects of an adverse dynamic response of the system. The scenario when the excitation is represented as a narrow-band stochastic process with the state vector governed by stochastic equations is then discussed and the stochastic differential equations governing the second-order statistical moments of the state vector are developed.

1. Introduction

Slender structures such as ropes and cables employed in vertical transport applications are subjected to vibrations caused by various sources of excitation. They include excitations due to the irregularities of the guiding system as well as environmental phenomena such as strong winds and earthquakes [1]. The dynamic characteristics of the system are of a time-variant nature. The natural frequencies of the ropes change with the speed of the transport motion. More importantly, their length variation results in changes of their mass and stiffness characteristics so that *slow* variations of the resonance frequencies occur [2]. An adverse situation arises when the building is excited near its fundamental natural frequency and vibrates periodically [3]. This in turn may result in a passage through resonance in the suspension and compensating rope systems when one of the slowly varying frequencies of the ropes approaches that of the inertial load resulting from the building sway. The study of vibration problems in elevator ropes in high-rise applications due to the low frequency building sway has attracted considerable attention. For example, an analysis of elevator ropes of fixed length in tall buildings subjected to excitation due to wind were investigated in [4] and an experimental study into rope sway control strategy in high-rise elevator installations was conducted in [5]. The main effort was put into the prediction of the overall rope system response and into the development of control strategies based in deterministic models.

In this paper, the results of a study to predict the sway of compensating ropes taking into account the time-variant (non-stationary) nature of the system are presented. The deterministic model is applied to investigate the passage through resonance during the elevator travel. Then, the methodology to account for the stochastic nature of the building motion is discussed and the differential equations governing the second-order statistical moments of the state vector are developed.

2. Dynamic model

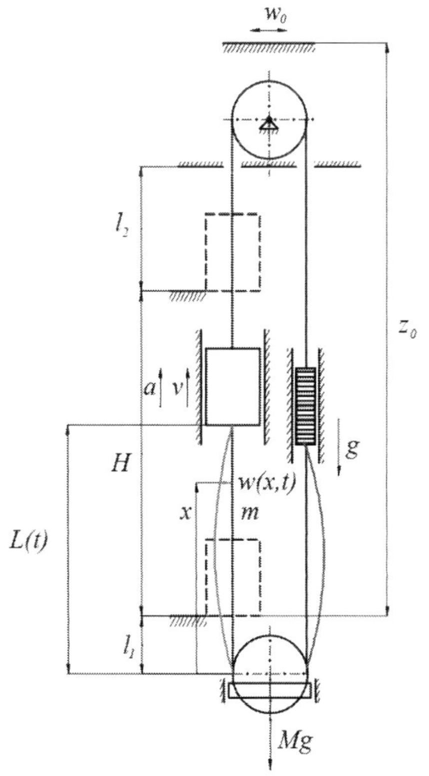

Figure 1

The dynamic response of compensating ropes is studied using the model depicted in Figure 1. This model represents an elevator car with compensating ropes of length $L(t)$ and mass per unit length m, moving at speed v with the acceleration rate a. The system is equipped with a compensator pulley assembly of mass M. The elevator travel height is denoted by H and represents the distance from the bottom (ground floor) landing to the top landing. The distance from the ground floor level to the centre of the compensator pulley is given as l_1 and the overhead height measured from the top landing to the top of the machine room floor is denoted by l_2. The building sway is represented by motion $w_0(t)$ taking place at the height z_0 measured from the ground floor level.

The modern elevator control systems allow an accurately prescribed velocity and acceleration time profiles of the car / counterweight to be realized with their vertical transient oscillations being negligible. Bearing this in mind the tension T of the ropes is expressed as

$$T(x) = m(g+a)x + \frac{1}{2}Mg \tag{1}$$

The lateral displacement $w(x,t)$ of the ropes is governed by the following equation

$$m\left(w_{tt} + 2vw_{xt} + v^2 w_{xx}\right) - T_0 w_{xx} - m(g+a)xw_{xx} - mgw_x = 0, \quad 0 < x < L(t) \tag{2}$$

where $T_0 = \frac{1}{2}Mg$. Assuming that the lateral motions at the car end are not admitted the boundary conditions are given as

$$w(0,t) = 0$$
$$w(L,t) = w_L(t) \tag{3}$$

where $w_L(t)$ represents can be determined if the building sway $w_0(t)$ at z_0 above ground floor level is known (see Figure 1).

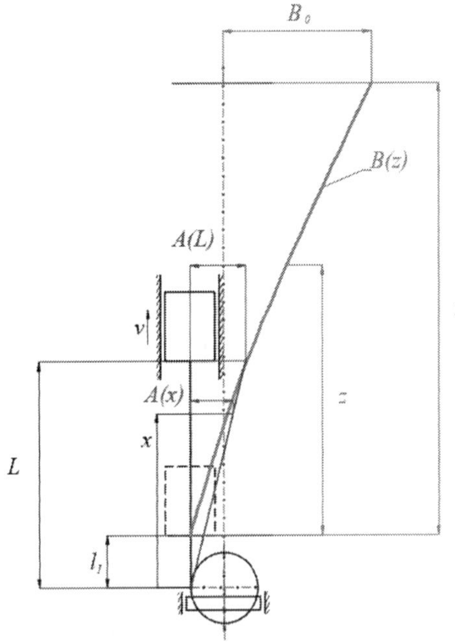

Figure 2.

3. Response to harmonic process

The motion $w_0(t)$ at the height z_0 above the ground floor level caused by the building sway can be treated as a harmonic process defined as

$$w_0(t) = B_0 \sin \Omega t \tag{4}$$

where B_0 is the maximum displacement at z_0 above the ground floor level and Ω is the fundamental natural frequency of the building, respectively.

3.1. Deformations

The building deformation shape $B(z)$ as shown in Figure 2, where z is the coordinate measured from ground floor level, is approximated by a cubic function as

$$B(z)=\frac{B_0}{z_0^3}z^3 \tag{5}$$

Thus, the displacement of the car (taking place together with the building) positioned in the well at $z = L - l_1$ above the ground floor is

$$A(L)=\frac{B_0}{z_0^3}(L-l_1)^3=B_0\left(\frac{L-l_1}{z_0}\right)^3 \tag{6}$$

It is evident that the corresponding shape (displacements) of the compensation rope in terms of the coordinate x (measured from the centre of the compensator pulley assembly, see Figure 2) is then given by a linear function of x as

$$A(x)=\frac{x}{L}A(L)=B_0\left(\frac{L-l_1}{z_0}\right)^3\frac{x}{L} \tag{7}$$

3.2. Discrete equations of motion

The boundary condition (3) are then written as

$$\begin{aligned} &w(0,t)=0\\ &w(L,t)=A(L)\sin\Omega t \end{aligned} \tag{8}$$

where $A(L)$ is given by equation (6). In order to investigate the response due to the building sway the overall lateral displacement of the ropes is expressed as

$$w(x,t)=\bar{w}(x,t)+A(x)\sin\Omega t,\quad 0\le x\le L(t) \tag{9}$$

where $\bar{w}(x,t)$ represents the dynamic deflection relative to the building frame and $A(x)$ is given by equation (7). Thus, the overall displacement is represented as

$$w(x,t)=\bar{w}(x,t)+\left(\frac{L-l_1}{z_0}\right)^3\frac{x}{L}w_0(t) \tag{10}$$

The transformation (10) is substituted into equation (2) and the dynamic deflections are expanded as

$$\bar{w}=\sum_{n=1}^{N}\Phi_n\left(x,L(\tau)\right)q_n(t) \tag{11}$$

where $q_n(t)$ represent the modal coordinates, $\Phi_n(x)=\sin\frac{n\pi}{L(\tau)}x,\ n=1,2,\ldots,N$, are the slowly-varying natural vibration modes of the corresponding taut string. In this formulation the length of the ropes is considered to vary slowly which is observed on the *slow time* scale defined as $\tau=\varepsilon t$, where ε represents a small parameter [2]. After orthogonalising with respect to the natural modes, the following system of ordinary differential equations describing the dynamic behaviour of the compensating ropes is obtained

$$\ddot{q}_r(t)+\sum_{n=1}^{N}C_{rn}(\tau)\dot{q}_n(t)+\sum_{n=1}^{N}K_{rn}(\tau)q_n(t)=Q_r(t;\tau),\quad r=1,2,\ldots,N \tag{12}$$

where the coefficients C_{rn} and K_{rn} are slowly varying. These coefficients and the excitation function Q_r are given as

$$C_{rn}(\tau)=\begin{cases} 2\varsigma_r\omega_r(\tau), & n=r \\ \dfrac{4nrv}{(n^2-r^2)L(\tau)}\left((-1)^{r+n}-1\right), & n\neq r \end{cases},\quad K_{rn}(\tau)=\begin{cases} \omega_r^2(\tau)-\dfrac{r^2\pi^2}{L}\left[\dfrac{v^2}{L}-\dfrac{g+a}{2}\right], & n=r \\ \dfrac{2}{(n^2-r^2)L(\tau)}\left[\dfrac{2rn^3(g+a)}{n^2-r^2}-grn\right]\left[(-1)^{r+n}-1\right], & n\neq r \end{cases}$$

$$Q_r(t;\tau)=\frac{2}{r\pi L(\tau)}\left(\frac{L(\tau)-l_1}{z_0}\right)^3\left\{(-1)^r L\ddot{w}_0(t)-\left[gw_0(t)-2v\dot{w}_0(t)\right]\left[(-1)^r-1\right]\right\}$$

(13)

where $r=1,2,\ldots,N$, $\omega_r(\tau)=\dfrac{r\pi}{L(\tau)}\sqrt{\dfrac{T_0}{m}}$ and ς_r represents the modal damping ratios of the ropes.

3.3. Results

The dynamic response of the compensating system is determined for an installation roped *1:1* of rated speed *8 m/s*, travel height $H = 390\ m$ equipped with *8* compensating ropes of mass per unit length $m = 2.11\ kg/m$ each. The building is subjected to a sway of frequency *0.1 Hz* and amplitude $B_0 = 0.75\ m$ measured at $z_0=387.5\ m$ above the ground floor. The coupled differential equations (12) are solved numerically using $N = 5$ modes in the expansion (11) for masses $M = 500,\ 15{,}000\ kg$ with damping ratios assumed as $\zeta_r = 3\%$ across all modes. The numerical tests are conducted for the lift car ascending with the acceleration/ deceleration rate of $a = 1.2\ m/s^2$ from the lowest landing upwards to the highest level. The MATLAB implementation of an explicit Runge-Kutta (4,5) formula is used to integrate the equations of motion on the fast time scale with the slowly varying coefficients C_{rn} and K_{rn} pre-calculated and evaluated at each integration step. The dynamic deformations relative to the building frame are then determined from the solution using the expansion (11). The fundamental resonance frequencies are determined from the eigenvalues of matrix of the coefficients K_{rn} and accommodate the influence of speed and acceleration of the ropes. They are plotted vs. the length of the ropes in Figures 3(a) and Figure 4(a), respectively, where the frequency of the excitation corresponding to the building sway frequency is also shown. The dynamic deformations of the ropes are then presented in Figures 3(b) and Figure 4(b), respectively. These curves illustrate the deformations and maximum amplitudes of the ropes at various stages of the elevator travel.

It is evident from the plots shown in Figure 3(a) that for the case $M = 15{,}000\ kg$ the frequency of the load coincides with the first (fundamental) frequency of the ropes within the length range of about *382 - 383 m,* when the car approached the top landing. This leads to large motions of the ropes with the amplitudes exceeding *1.5 m* as shown in Figure 3(b). The frequency plots shown in Figure 4(a) for the case $M = 500\ kg$ demonstrate that when a smaller mass of the compensator assembly is used the natural frequencies are lower and in addition to the fundamental resonance (taking place when the rope length is about *105 m*) the possibility of higher (second) resonance exists. The second resonance is predicted when the rope length is about *365 m.* This leads to large displacements of the ropes demonstrating a strong influence of the second mode shape. As shown in Figure 4(b), in this case the amplitudes reached *1.4 m* when the car approached the top landing.

4. Stochastic motion of the building

The motion $w_0(t)$ resulting from the building sway is seldom exactly harmonic. The excitation due to environmental phenomena such as wind action is usually a wide band stochastic process. The response in the fundamental mode of the building is then a narrow-band process with the center frequency equal to the fundamental natural frequency Ω. The stochastic motion of the building can be obtained from the analysis of the building response.

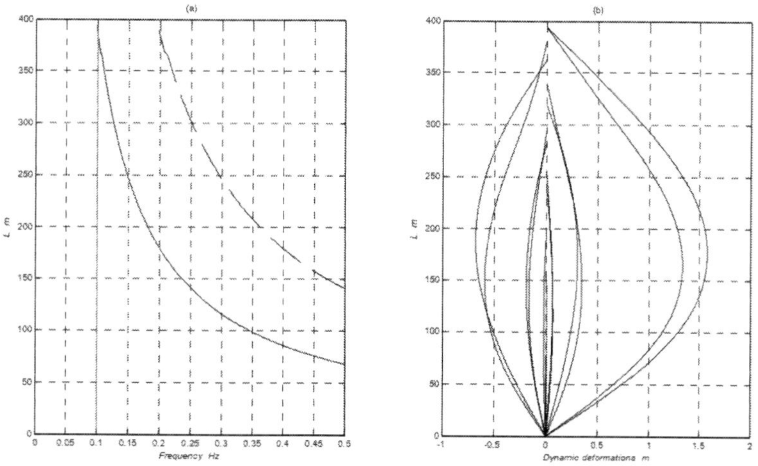

Figure 3. M = *15,000 kg* (a) The variation of the first (solid curve) and second (dashed curve) natural frequencies together with the frequency of the sway (vertical solid line); (b) Deformation shapes of the ropes.

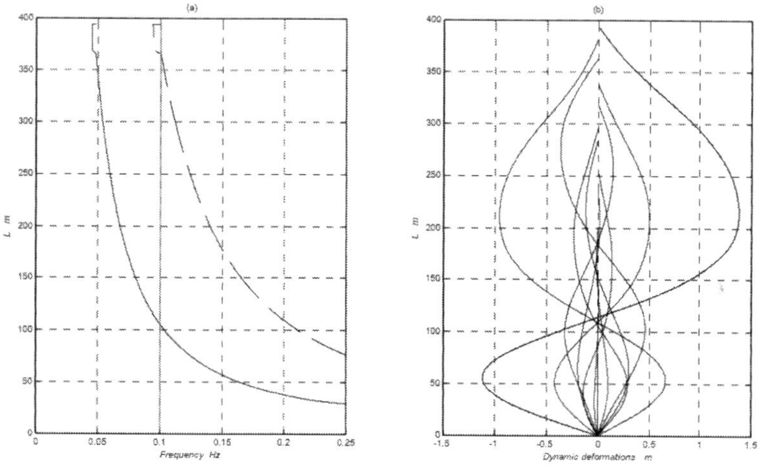

Figure 4. M = *500 kg* (a) The variation of the first (solid curve) and second (dashed curve) natural frequencies together with the frequency of the sway (vertical solid line); (b) Deformation shapes of the ropes.

4.1. Narrow-band process

It may be assumed that $w_0(t)$ is a narrow-band process mean-square equivalent to the harmonic process with the amplitude A_0 and the frequency Ω as given by equation (4). The function $w_0(t)$ together with its first and second derivatives $\dot{w}_0(t)$ and $\ddot{w}_0(t)$ must be continuous [6]. This scenario is adequately idealized by assuming that the motion $w_0(t)$ is the response of the second order auxiliary filter to the process *X(t)*, which in turn is the response of the first-order filter to the Gaussian white noise excitation $\xi(t)$ [6] as illustrated in Figure 5. The governing equations are

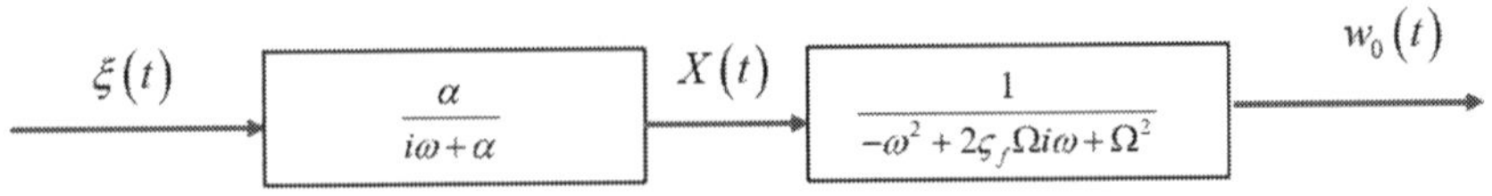

Figure 5. Narrow-band process scenario.

$$\begin{aligned} &\ddot{w}_0(t)+2\zeta_f\Omega\dot{w}_0(t)+\Omega^2 w_0(t)=X(t) \\ &\dot{X}(t)+\alpha X(t)=\alpha\sqrt{2\pi S_0}\xi(t) \end{aligned} \tag{14}$$

where ζ_f denotes the damping ratio of the filter which defines its band width, α is the filter variable, S_0 is the constant level of the power spectrum of white noise $\xi(t)$. Precisely speaking, the response *X(t)* to the Gaussian white noise excitation $\xi(t)$ is a process which is not differentiable in the usual sense, hence the notation $\dot{X}(t)$ is not mathematically meaningful. The second equation in (14) has a clear mathematical sense when it is written in terms of the differential. Accordingly, the governing equations for both filters are written down in a state space form as

$$d\begin{bmatrix} X \\ w_0 \\ \dot{w}_0 \end{bmatrix}=\begin{bmatrix} -\alpha & 0 & 0 \\ 0 & 0 & 1 \\ 1 & -\Omega^2 & -2\zeta_f\Omega \end{bmatrix}\begin{bmatrix} X \\ w_0 \\ \dot{w}_0 \end{bmatrix}dt+\begin{bmatrix} \alpha\sqrt{2\pi S_0} \\ 0 \\ 0 \end{bmatrix}dW(t) \tag{15}$$

where *dW(t)* is the increment of the Wiener process *W(t)* and the Gaussian white noise excitation $\xi(t)$ is the generalized (not in usual sense) derivative of the Wiener process. It is important to note that the choice of parameters in equation (15) should guarantee that the filter variable α is real. The stochastic process $w_0(t)$ is mean-square equivalent to the harmonic process (4) if its standard deviation and variance are given as

$$\sigma_{w_0}=\frac{\sqrt{2}}{2}\mathrm{A}_0, \quad \mathrm{var}(w_0)=\sigma_{w_0}^2=\frac{A_0^2}{2} \tag{16}$$

respectively. It can be shown that this equivalence is achieved if the filter variable α is given as [6]

$$\alpha=\Omega\left(-\zeta_f+\sqrt{\zeta_f^2+\frac{\zeta_f\Omega^3A_0^2}{\pi S_0-\zeta_f\Omega^3A_0^2}}\right) \tag{17}$$

4.2. Stochastic equations

It is evident from the expressions (13) that the excitation function $Q_r(t;\tau)$ can be expressed in terms of the motion $w_0(t)$ and its time derivatives as $Q_r(t;\tau)=\beta_r^{(1)}(\tau)w_0(t)+\beta_r^{(2)}(\tau)\dot{w}_0(t)+\beta_r^{(3)}(\tau)\ddot{w}_0(t)$. Using equation (14) the second derivative of $w_0(t)$ can be expressed as $\ddot{w}_0(t)=X(t)-\Omega^2 w_0(t)-2\zeta_f\Omega\dot{w}_0(t)$ so that the excitation function $Q_r(t)$ can be rewritten as $Q_r(t;\tau)=\gamma_r^{(1)}(\tau)w_0(t)+\gamma_r^{(2)}(\tau)\dot{w}_0(t)+\beta_r^{(3)}(\tau)X(t)$, where $\gamma_r^{(1)}=\beta_r^{(1)}-\Omega^2\beta_r^{(3)}$ and $\gamma_r^{(2)}=\beta_r^{(2)}-2\zeta_f\Omega\beta_r^{(3)}$.

The state vector defined as $\mathbf{Y}(t)=\left[\mathbf{q}(t) \;\; \dot{\mathbf{q}}(t) \;\; X(t) \;\; w_0(t) \;\; \dot{w}_0(t)\right]^T$ is then governed by the following set of stochastic equations

$$d\mathbf{Y}(t)=\mathbf{A}\mathbf{Y}(t)dt+\mathbf{b}dW(t) \tag{18}$$

where

$$\mathbf{A}=\begin{bmatrix}\mathbf{0} & \mathbf{I} & \mathbf{0}\\ -\mathbf{K} & -\mathbf{C} & \mathbf{B}\\ \mathbf{0} & \mathbf{0} & \mathbf{A}_f\end{bmatrix},\quad \mathbf{b}=\begin{bmatrix}\mathbf{0}\\ \mathbf{0}\\ \mathbf{b}_f\end{bmatrix} \tag{19}$$

with **I** denoting the identity matrix and **K** and **C** representing the matrices of coefficients $K_{rn}(\tau)$ and $C_{rn}(\tau)$ appearing in equations (12), respectively. Furthermore, **B** is a matrix with rows comprising coefficients $\beta_r^{(3)},\gamma_r^{(1)},\gamma_r^{(2)}$ and the matrix $\mathbf{A}_f$ and vector $\mathbf{b}_f$ in (19) are defined as follows

$$\mathbf{B}=\begin{bmatrix}\beta_1^{(3)} & \gamma_1^{(1)} & \gamma_1^{(2)}\\ \beta_2^{(3)} & \gamma_2^{(1)} & \gamma_2^{(2)}\\ \cdots & \cdots & \cdots\\ \beta_N^{(3)} & \gamma_N^{(1)} & \gamma_N^{(2)}\end{bmatrix},\quad \mathbf{A}_f=\begin{bmatrix}-\alpha & 0 & 0\\ 0 & 0 & 1\\ 1 & -\Omega^2 & -2\zeta_f\Omega\end{bmatrix},\quad \mathbf{b}_f=\begin{bmatrix}\alpha\sqrt{2\pi S_0}\\ 0\\ 0\end{bmatrix} \tag{20}$$

The differential equations governing the second-order statistical moments of the state vector, i.e. the covariance matrix $\mathbf{R}_{\mathbf{YY}}=E\left[\mathbf{Y}\mathbf{Y}^T\right]$, are then obtained from equation (18) as

$$\frac{d}{dt}\mathbf{R}_{\mathbf{YY}}=\mathbf{A}\mathbf{R}_{\mathbf{YY}}+\mathbf{R}_{\mathbf{YY}}\mathbf{A}^T+\mathbf{b}\mathbf{b}^T \tag{21}$$

Furthermore, higher-order joint statistical moments of the state vector $\mathbf{Y}(t)$ can be determined with the aid of the Itô's differential rule.

5. Conclusions

The analysis of the response to excitation represented as harmonic process showed that when the fundamental mode of the building is tuned to the fundamental mode of the ropes large dynamic deformations of the ropes occur. Large weights of the compensating pulley assembly would be needed to shift the fundamental resonance frequency of the ropes above the frequency of the building excitation far enough to steer away from the fundamental resonance. If a small mass of the compensator assembly is used, a passage through the fundamental resonance will occur at an earlier stage of the elevator up travel and the resulting dynamic displacements are smaller. But, in this case the second mode of the ropes is activated when the car approaches the top landing. The designer of high-rise elevator systems needs both the mean value and the variance to establish bounds for the compensating rope response. Thus, the excitation mechanism can be represented as a narrow-band process mean-square equivalent to the harmonic process and used in the dynamic model to predict the probabilistic characteristics of the rope response.

References

[1] Strakosch G R, 1998 *The Vertical Transportation Handbook* John Wiley, New York.
[2] Terumichi Y, Kaczmarczyk S, Turner S, Yoshizawa M, and Ostachowicz M, 2003 *Mater. Sci. Forum* **440-441**, 497-504.
[3] Sun L, 1995 *Elevator World,* **March 95** pp.133-135.
[4] Blodgett R E and Majumdar A K, 1983 *Trans. ASME J. Vib. Acoustics, Stress and Reliability in Design* **105**, 5-10
[5] Otsuki M, Yoshida K, Nagata K, Fujimoto S and Nakagawa T, *Proc. of the American Control Conf.* (Anchorage, AK, May 8-10, 2002) 238-243
[6] Larsen J W, Iwankiewicz R and Nielsen S R K, 2007 *Probabilistic Eng. Mech.* **22**, 181-193.

An ultrasonic corer for planetary rock sample retrieval

P. Harkness, A. Cardoni and M. Lucas
Dept of Mechanical Engineering, University of Glasgow, Glasgow, G12 8QQ, UK

E-mail: p.harkness@mech.gla.ac.uk

Abstract. Several recent and planned space projects have been focussed on surface rovers for planetary missions, such as the U.S. Mars Exploration Rovers and the European ExoMars. The main functions of similar extraterrestrial vehicles in the future will be moving across planetary surfaces and retrieving rock samples. This paper presents a novel ultrasonic rock sampling tool tuned in a longitudinal-torsional mode along with the conceptual design of a full coring apparatus for preload delivery and core removal. Drilling and coring bits have been designed so that a portion of the longitudinal motion supplied by the ultrasonic transducer is converted into torsional motion. Results of drilling/coring trials are also presented.

1. Ultrasonic Drilling and Coring

Ultrasonic drilling for planetary rock sampling has a number of benefits compared to traditional rotary drilling. These include a reduction in required end-load (an unfeasibly high-mass planetary lander such as the six-ton Luna 24 would be required if large forces were to be generated in low gravity environments), the elimination of the heavy motors and clutches associated with rotating machinery, and mass-reduction of the drillstring itself due to the fact that only a very small mechanical effort must be delivered from the planetary lander on the surface to the bit operating downhole.

A mechanism by which an ultrasonic device can be delivered to the bottom of a shaft by a planetary lander has been set out in a previous work [1]. However, that work focused on a drill system which must, by definition, pulverize all the material which lies before it if the unit is to descend. Such an approach is acceptable for coarse chemical analysis but it is more desirable to recover the strata whole, undamaged, and in their proper lithographic sequence via a coring operation. This is particularly the case where the intention is to return the samples to Earth for analysis, as is envisaged for the future Mars Sample Return mission.

A coring approach is also advantageous when compared to drilling in that not all the rock, soil or ice in the path of the exploratory shaft must be pulverized, as creating the central core requires no power. Despite its huge advantages in terms of preload, reacted torque and system mass, ultrasonic drilling and coring is not expected to be particularly competitive in terms of energy expenditure for shallow-drill operations, requiring 10 to 20 GJ/m^3 in medium-strength rock [1-3], a value which is as much as 50 times higher than the requirements of a conventional drill. However, ultrasonic devices do not require a rotating drillstring and do not experience the friction loads between drillstring and shaft which can absorb up to 80% of the torque applied to deep rotational systems [4], and thus they become progressively more competitive as deeper drilling (> 1m) is considered.

2. System Architecture

The proposed mechanism for sinking and withdrawing the bit, securing the shaft against collapse and shielding against trundling and saltating sand particles has been described in a previous publication [1]. In summary (see figure 1) a coilable tube similar to the Northrop Grumman STEM [5] is proposed to sink the ultrasonic bit complete with transducer, into the surface and secure the resulting shaft. A growing core will pass through the annulus of the coring bit, through the centre of the ring transducers, and up into the bottom section of the STEM.

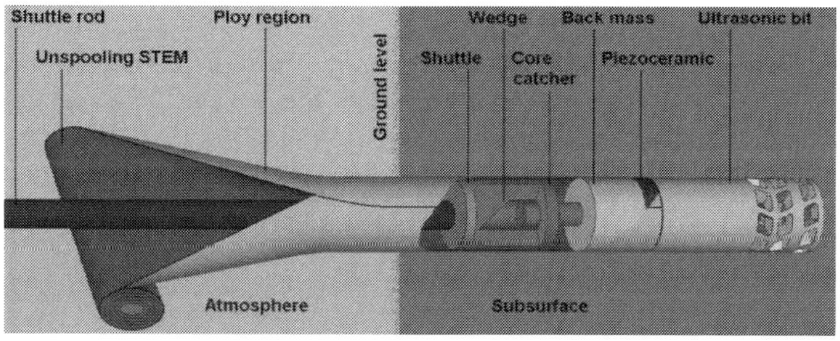

Figure 1. Schematic of system architecture (aligned horizontally)

The core will then be broken and collected by a shuttle with an integral wedge. The load required to break the core will reduce linearly with core length suggesting that the shuttle can be lowered into the hole, touchdown with sufficient force to break off even the strongest rock cores, and be withdrawn to the surface with the core sample secured inside by a core-catcher. Such devices are essentially a ring of triangular doors, arranged so that items can only pass in one direction, rather like a biological valve. It is further proposed that the bottom of the shuttle should be tapered inwards to ensure that fines tend to be moved into the centre of the shuttle rather than building up on top of the ultrasonic bit.

3. Development of the coring bit

In a previous study [1] a longitudinal-torsional motion was developed at the tip of an ultrasonic drilling bit by coupling the first longitudinal (L1) and second torsional (T2) modes of vibration in a cylindrical stepped horn. The step provides the horn with longitudinal amplitude gain whilst some of the longitudinal excitation applied to the base of the horn is transferred to the torsional mode by helical structures, namely diagonal slits, positioned at one of the two torsional nodes. The success of this approach is demonstrated in figure 2.

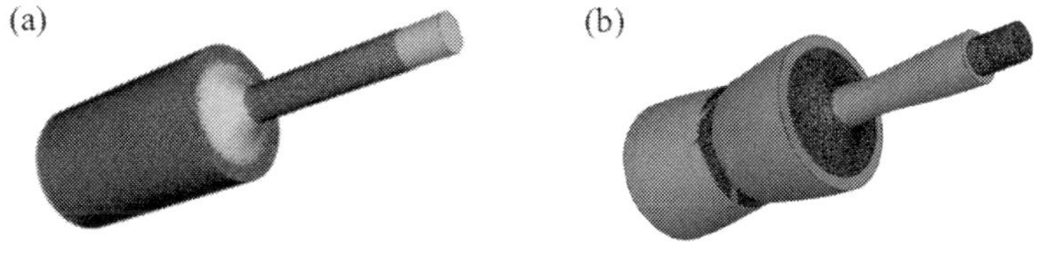

Figure 2. (a) Unslitted, L1 drilling horn, (b) slitted, L1T2 mode-coupled drilling horn.

3.1. Applying the geometric principles of drilling horns to coring horns

The aim in the corer design is to maintain the high degree of L1 amplitude gain obtained by the stepped drillers by similarly reducing the cross-sectional area of the horn from base to tip. It has also been previously shown [1] that coupling the L1 and T2 modes improves the rate of drill penetration in sandstone. The aim is also therefore to preserve a longitudinal-torsional motion in the proposed coring horn by incorporating helical structures to transfer energy from longitudinal to torsional motion.

3.1.1. Reducing cross-section to increase gain. Because both transducer and horn must penetrate the ground, the front surface of the horn must have a working surface area at least equal to any other part of the horn and proposed ring transducer. In seeking to integrate area reduction into the design, this presents some difficulties. However, it can be achieved by creating voids in the front end of the horn and closing them off with a thin disc of material at the tip. This can be done either by a single toroidal void or a number of radial voids, leaving behind the same number of circumferential columns. The gains of these horns are similar but the radial-void geometry is considerably easier to manufacture and so was selected for development.

3.1.2. Inclusion of helical structures. Diagonal slits in the body of the driller allowed torsional motion to be developed in the longitudinally excited horn and so this geometry was again evaluated for the corer. As before, a global T2 mode is developed over the entire horn with base and tip oscillating in phase. However, torsional gain is found to be low and torsion is back-transmitted into the Langevin transducer, with possible negative effects such as heating and fatigue. A better strategy is to develop torsion by controlling the vibration of the columns such that torsional motion is developed towards the tip of the horn and not at the base. This can be achieved by applying a slope to the columns.

3.2. Dealing with flexural column modes.

In corer geometries incorporating radial voids, the interstitial columns tend to exhibit half-wavelength oscillation in the first bending mode which does not assist excitation of the desired torsional vibration. A solution is to create tiers of columns joined by equally-spaced rings. The tiers of columns then oscillate as three discrete units, each exhibiting quarter-wavelength bending. The overall effect is that the corer successfully develops a net change in torsion from the base to the tip of the tiered column structure.

Furthermore, to prevent the constant slope of each column tier transferring energy to the global T2 mode, the direction of slant is reversed in one of the three tiers and thus the desired mode shape can be achieved. The mode shape at eight sequential intervals of one half-cycle is shown in figure 3.

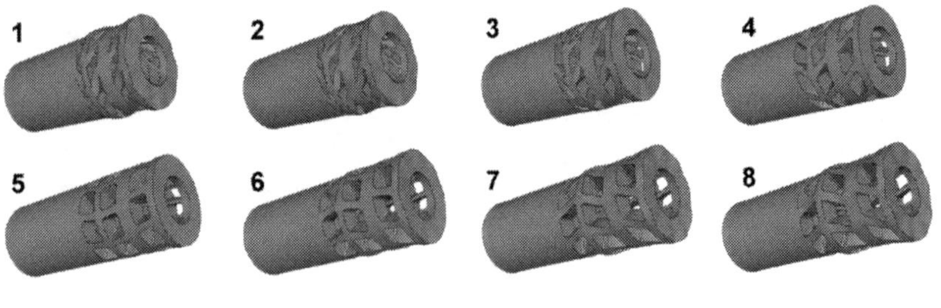

Figure 3. The desired operating mode of the corer over one half-cycle.

4. Validation of the finite-element models and testing of the horn

In order to validate the finite element model, the generic geometry shown in figure 3 was manufactured. The corer has a diameter of 33 mm and a length of 80 mm, which allows it to match the diameter of COTS STEM systems and tune to the 20 kHz frequency of the Langevin transducer.

The modal parameters of the coring horn were ascertained by an experimental modal analysis (EMA), with the vibration response at a grid of points on the corer being measured with a 3D laser Doppler vibrometer. The horn was driven over a frequency range of 0 - 40 kHz and a frequency response function (FRF) measured from the working face of the corer is shown in figure 4 (a). As well as the tuned mode, the modal parameters of several other nearby modes were extracted from the measurement data and these mode shapes, along with the associated FEA predictions, are shown in figure 4 (b). The measured modal frequencies and mode shapes were in close agreement with the finite element model predictions.

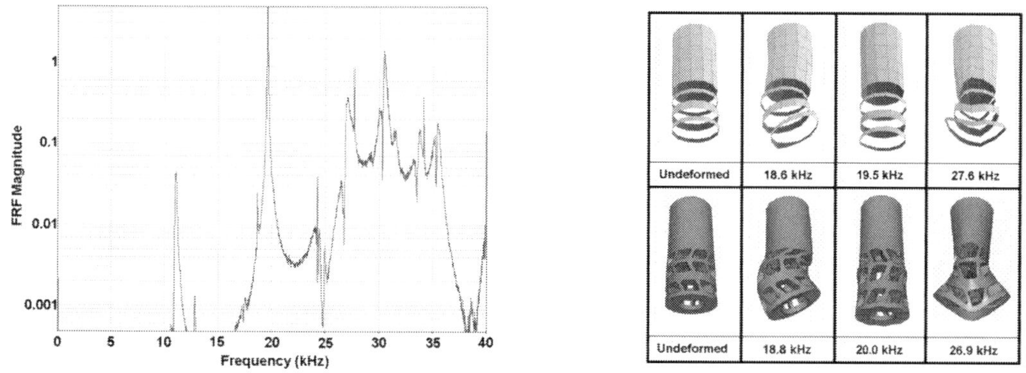

Figure 4. (a) FRF, (b) EMA (top row) and FEA (bottom row) mode shapes.

4.1. Longitudinal and torsional response validation.

The longitudinal and torsional response at 19.5 kHz was measured. The longitudinal excursion at the base and the torsional excursion at the tip are set to 1 and the results compared to FEA in figure 5. Generally agreement was good although there is some discrepancy in the torsional response of the first ring. The results in figure 5(b) also demonstrate that the horn is back-transmitting almost no torsion to the Langevin transducer driving it, a significant advantage over L1T2 mode-coupled designs.

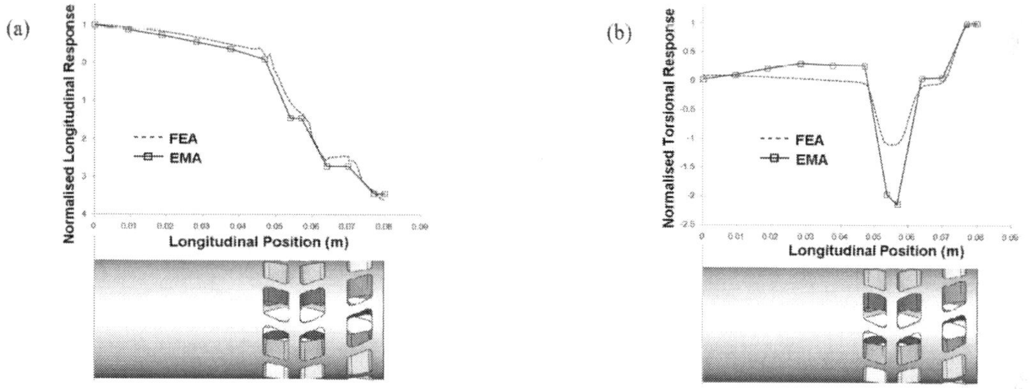

Figure 5. (a) Longitudinal response, (b) torsional response.

4.2. Rock-cutting performance.

The performance of the corer as compared to the ultrasonic drillers was tested and the results are presented in table 1 in terms of the removal rate of sandstone. Not all horns were excited at the same longitudinal amplitude and therefore the removal rate is expressed with respect to the longitudinal amplitude at the tip. Due to a larger tip radius, the scaled tangential velocity of the corer surface is actually higher than that of the driller by a factor of ~1.4.

Table 1. A performance comparison of the drillers and derived coring system in Sherwood sandstone.

	Applied preload (N)	Torsional output (radians T2 output /10 μm L1 input)	Sandstone removal rate (mm^3/μm L1 output/sec)
L1 driller	20	0	0.64
L1T2 driller	20	0.00497	2.57
Corer	20	0.0014	3.00

Journal of Physics: Conference Series **181** (2009) 012048 doi:10.1088/1742-6596/181/1/012048

5. Optimisation of the coring bit geometry

The performance of the ultrasonic corer compared well with the previously described drilling systems. Given a constant inner core diameter (18 mm) and ring thicknesses, the generic geometry has six main variables: (i) the overall length of the horn, in mm; (ii) the height of each tier, in mm; (iii) the slope of the columns, in horn diameters per complete helical revolution; (iv) the degree of cross-sectional area reduction, as a ratio of tip-end area to base-end area; (v) the number of columns in each tier; and (vi) the outer diameter of the horn, in mm. The effects of these parameters on tuned mode frequency, longitudinal gain and torsional response were investigated by FE analysis.

As a starting point, four of the variables were held constant; (iii), (iv), (v) and (vi) at values of 2.5 diameters per revolution, 30% tip-end area with respect to base-end area, 9 circumferential columns and 17 mm outer diameter respectively. The overall length of the horn was varied from 50 mm to 200 mm and the height of each tier from 5 mm to 12 mm, producing a grid of FE data. This process was then repeated with each of the four held variables incremented and decremented in turn, producing nine data grids in total.

Each data grid yielded three results grids, relating to frequency, longitudinal gain and torsional response. To obtain the torsional response data, which is presented in terms of tip torsional amplitude, the FEA models included a longitudinal excitation of amplitude 10 µm at the base of the horn.

The results grids were arrayed such that each horizontal, vertical and diagonal line in the array represents a progression in one variable to the exclusion of all others, with the variables themselves being as laid out in table 2.

Table 2. Parameters of results presented in figure 6, figure 7 and figure 8.

Result plot	Variable (iii) (Column slope, dia/rev)	Variable (iv) (Area ratio)	Variable (v) (Column count)	Variable (vi) (Outer diameter, mm)
1	2.5	30 %	7	17
2	1.5	30 %	9	17
3	2.5	22.5 %	9	17
4	2.5	30 %	9	14
5	2.5	30 %	9	17
6	2.5	30 %	9	20
7	2.5	37.5 %	9	17
8	3.5	30 %	9	17
9	2.5	30 %	11	17

Surface-fitting software was used so that beyond the data points a surface is extrapolated to fill the entire grid space and therefore those regions of the surfaces which are not well-supported by data points should be treated with caution. A data point that is lifted from the surface indicates the degree of discrepancy between the FEA result at that point and the overall fitted surface.

The frequency results are presented in figure 6, longitudinal gain in figure 7 and torsional response in figure 8.

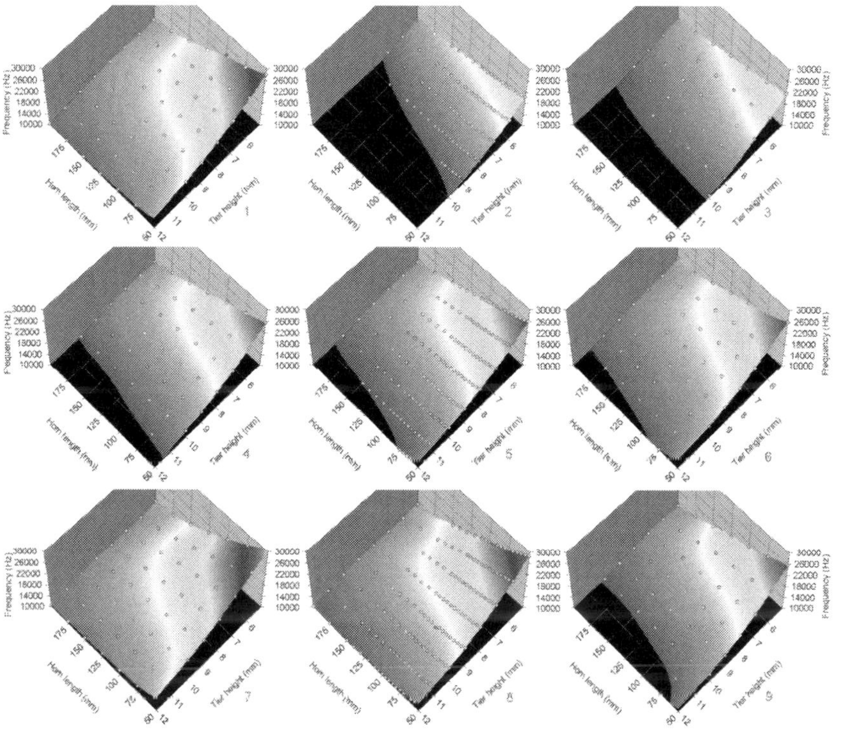

Figure 6. The frequencies predicted for the corer by FEA.

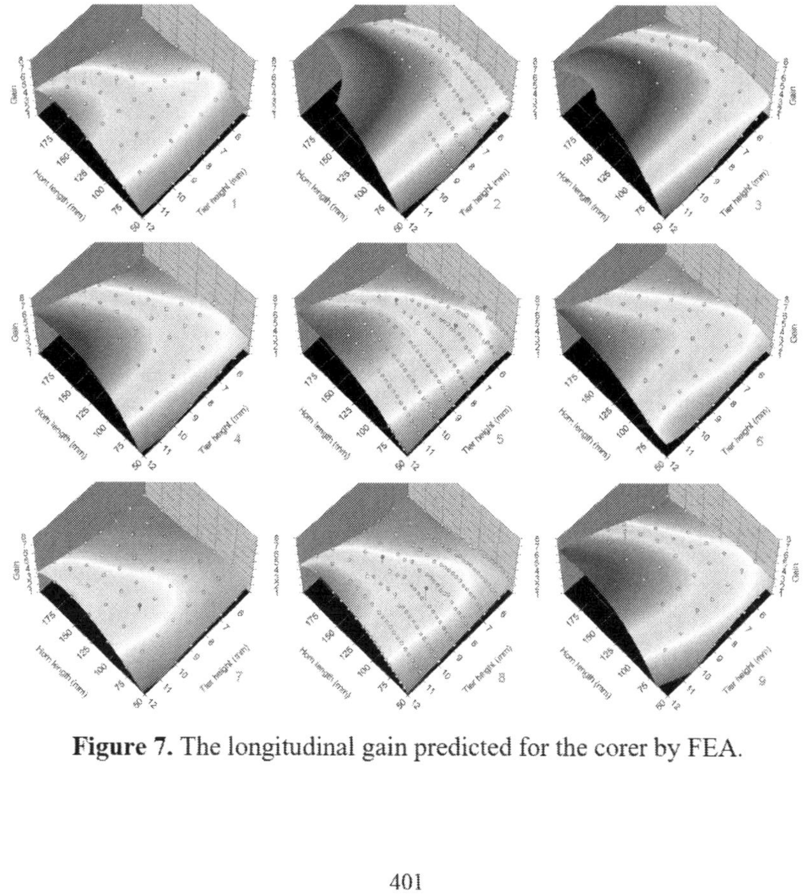

Figure 7. The longitudinal gain predicted for the corer by FEA.

Journal of Physics: Conference Series **181** (2009) 012048 doi:10.1088/1742-6596/181/1/012048

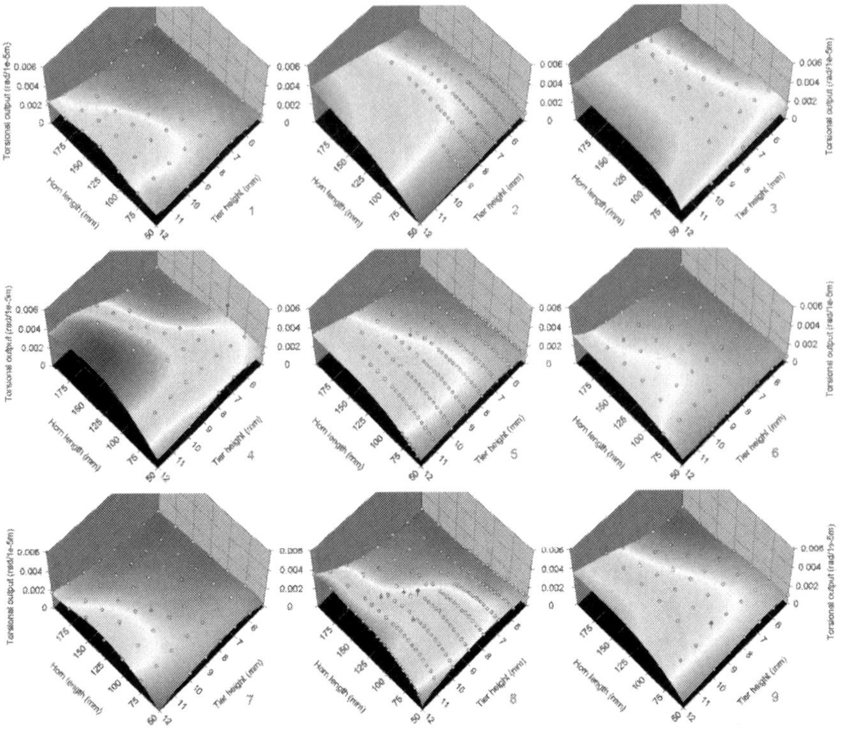

Figure 8. The torsional response predicted for the corer by FEA.

5.1. Evaluation of the FEA results

5.1.1. Geometric influences on frequency. In each subfigure, frequency is increased by reducing the overall length of the horn and the height of each tier. Other factors which tend to increase the frequency of the horn include an increase of area ratio, a decrease of column slope and, to a lesser extent, a reduction in the number of columns. As each of these geometric drivers will result in respectively wider, straighter and deeper columns, we can conclude that the increased longitudinal stiffness in each case serves to increase the frequency of longitudinal vibration in the horn as a whole.

5.1.2. Geometric influences on longitudinal gain. In each subfigure, gain is maximised along the ridge joining small horns with low tiers to long horns with high tiers. We can therefore conclude that there is an optimal ratio of tier height to horn length, which appears to be approximately 7%. A closer inspection of the FEA results than can be presented in this paper indicates that, at this ratio, the longitudinal response plotted along the horn bears a strong similarity to figure 5 (a) in that the L1 node lies very close to the base of the first tier of columns. This arrangement appears to maximise longitudinal gain in the coring horns in much the same manner as it would maximise the performance of a step horn. Further details of this finding are presented in another publication [6].

Gain is also increased in horns with low area ratios and highly sloped columns. It can be concluded that the greater slope of the columns results in a greater proportion of the bending mode of each column being translated into longitudinal amplitude at the front face of the horn.

The maximum longitudinal gain definitively predicted (as opposed to estimated by the interpolated and projected surface) by a corer of figure 7(3) is 6.75. An FEA model of a simple longitudinal stepped horn of the same area ratio (22.5%) was created and the longitudinal gain was calculated to be 4.5, a value which does not vary significantly with changing frequency. The corer therefore significantly outperforms a comparable half-wavelength stepped horn in terms of longitudinal gain.

5.1.3. Geometric influences on torsional response. The torsional response is found to be proportional to the longitudinal response in most cases, but one exceptional result is the torsional response of the smaller-diameter horns in figure 8(4), which are significantly influenced by the reduction in rotary inertia. Generally, corers have achieved torsional responses in the region of 0.001 to 0.002 rad/10μm excitation, with a maximum value of 0.0042 rad/10μm excitation in the aforementioned figure 8(4).

The best mode-coupled drilling horns previously described [1] achieved a value of 0.005 rad/10μm excitation, although this falls to a net torsional differential, from base to tip, of 0.0018 rad/10μm excitation when the torsion at the base due the global T2 mode is subtracted. Therefore, depending on the metric used, corers can approach the levels of torsion generated by mode-coupled drilling horns.

6. Conclusions

The design of mode-coupled drilling horns has been successfully extended to develop a coring horn which delivers a longitudinal performance well in excess of that which can be developed by comparable stepped horns and has a torsional performance to match mode-coupled drillers. The coring horn is also considerably shorter than stepped horns or mode-coupled drillers operating at the same frequency and has the crucial advantage that the external cross-sectional area does not change, despite quite large effective area reductions. This ensures that the large front surface of the corer can operate on the substrate with its full longitudinal and torsional amplitude.

We have further established the effects of design parameters in terms of operating frequency, longitudinal gain and torsional response and shown that, in general, gain can be maximised by increasing the effective area ratio and increasing the slope of the columns. Meanwhile, the torsional response can be increased by reducing the diameter of the horn.

The finite element model of the corer has been validated by experimental modal analysis and the performance of the manufactured ultrasonic corer has been tested on Sherwood sandstone and found to be comparable to mode-coupled ultrasonic drillers.

7. References

[1] Harkness P, Cardoni A and Lucas M, 2008, *Vibration considerations in the design of an ultrasonic driller/corer for planetary rock sampling*, International Conference on Noise and Vibration Engineering, Leuven, Belgium.

[2] Bar-Cohen Y, Sherrit S, Bao X, Badescu M, Aldritch J and Chang Z, 2007, *Subsurface sampler and sensors platform using the ultrasonic/sonic driller/corer (USDC)*, Proc. SPIE Smart Structures and Materials Syposium, San Diego, USA. Paper #6529-18

[3] Liu Y, Weinberg B and Mavroidis C, 2006, *Mechanical design and modelling of a robotic planetary drilling system*, Proc. ASME International Design Engineering Technical Conferences and Computers and Information in Engineering Conference, Philadelphia, USA. DETC 2006-99699

[4] Zacny K et al., *Drilling systems for extraterrestrial subsurface exploration*, 2008, *Astrobiology*, **8**, 665-706

[5] http://www.st.northropgrumman.com/astro-aerospace/SiteFiles/docs/pdf/DS-415-150lbBI-STEM.pdf, 2008

[6] Cardoni A, Harkness P and Lucas M, 2009, *Ultrasonic rock sampling using longitudinal-torsional vibrations*, International Congress on Ultrasonics, Santiago, Chile.

Acknowledgements

This work was funded by Science and Technology Facilities Council (UK), Grant no. ST/F003587/1.

Analysis of flexible fabric structures for large-scale subsea compressed air energy storage

Andrew Pimm[1] **and Seamus Garvey**[2]

[1 & 2] School of Mechanical, Materials & Manuf. Eng., The University of Nottingham, University Park, Nottingham NG7 2RD, UK

[1] eaxajp@nottingham.ac.uk, [2] seamus.garvey@nottingham.ac.uk

Abstract. The idea of storing compressed air in submerged flexible fabric structures anchored to the seabed is being investigated for its potential to be a clean, economically-attractive means of energy storage which could integrate well with offshore renewable energy conversion. In this paper a simple axisymmetric model of an inextensional pressurised bag is presented, along with its implementation in a constrained multidimensional optimization used to minimise the cost of the bag materials per unit of stored energy. Base pressure difference and circumferential stress are included in the optimization, and the effect of hanging ballast masses from the inside of the bag is also considered. Results are given for a zero pressure natural shape bag, a zero pressure bag with circumferential stress and hanging masses, and a nonzero pressure bag with circumferential stress and hanging masses.

1. Introduction

The Integrated Compressed Air Renewable Energy System (ICARES) is a system of harvesting and storing renewable energy in development at the University of Nottingham. ICARES is based on the idea of directly compressing air to around 50bar using excess supply of offshore renewable energy and storing that compressed air in inexpensive flexible fabric structures ("energy bags" or simply "bags") anchored to the seabed at depths of around 500m until demand outweighs supply and it is necessary to retrieve the energy stored in the compressed air. Due to the large hydrostatic pressure of the water column above, high energy densities are possible at such depths (approximately 20MJ/m^3 at 500m) without the need for a strong, expensive container; the water itself acts as the container. Great value is placed upon the importance of energy storage for an efficient and economical power grid and, as penetration of offshore renewable energy grows, a viable means of storing the energy will become increasingly necessary. ICARES is ideally placed to fill this need.

It is anticipated that energy bags will have lobes and/or annuli so that the patches of membrane between the cusps are small enough to react to the pressure difference without rupturing, and to increase the curvature of the membrane and transmit the forces through reinforcements along the cusps. Masses may be suspended between the surface of the bag and the seabed on cables that hang from the inside of the bag so that as it is inflated, progressively larger masses are lifted off the seabed, providing reaction for some of the differential pressure load. In this paper we consider the shape of axisymmetric bags without any annuli, and the lobed bags with annuli will be modelled in future work using a finite element model.

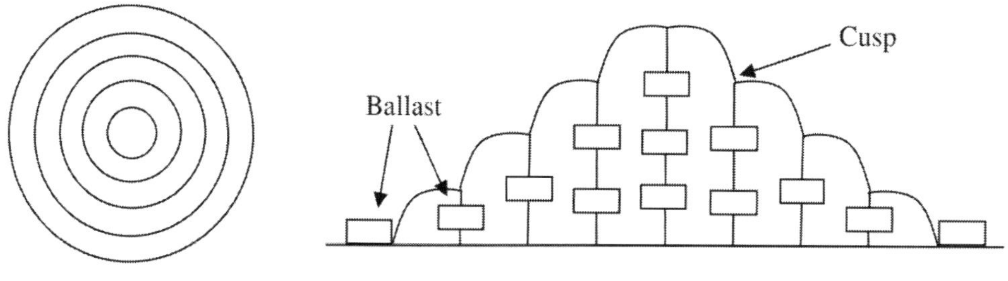

Figure 1. Plan and section views of an energy bag with hanging ballast and annuli.

While storage of thermal energy and compression/expansion machinery will be dealt with elsewhere, in the work presented here we develop and numerically integrate a system of ordinary differential equations to find the shape that a fully-inflated bag will take, and optimise the shape and other parameters to minimise the cost of storing energy. The energy bags are modelled as inextensional axisymmetric membranes in the same way that large scientific balloons have been modelled over the last half century (see [1] and [2]), and a tiered line search optimization routine is used to optimize for pressure difference at the base of the bag, the bag's centre height, the meridional and circumferential stresses in the membrane, and hanging mass.

While the idea of using wind turbines to compress air is not new (see [3] and www.generalcompression.com), no previous research on subsea pressurised membranes has been found and so the field of scientific ballooning is the main area of relevance. No previous work has been found on shape optimisation using axisymmetric balloon models. Much work on compressed air energy storage (CAES) was carried out in the 1970s and 1980s but the only storage vessels considered were caves, disused mines, aquifers, and rigid pressure vessels. CAES is a proven technology and two dedicated plants already exist (in Huntorf, Germany and McIntosh, USA), salt caverns being used as the stores in both plants. Many more CAES plants are currently in the design phase. [3]

2. Derivation of the equilibrium equations

Figure 2 shows a section view of an infinitesimal patch of membrane.

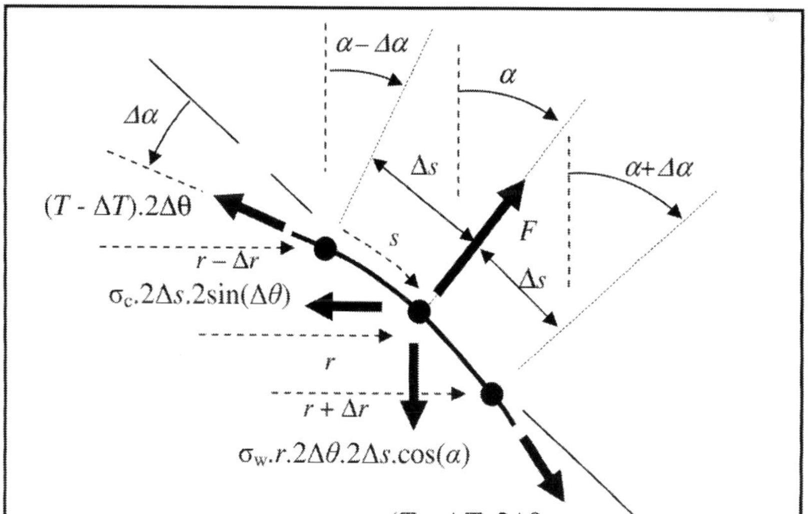

Figure 2. Section view of a patch of membrane subjected to a pressure difference force F.

The patch has meridional length $2\Delta s$, covers a circumferential angle of $2\Delta\theta$, is at radius r and height z, and covers a distance $2\Delta r$ in the radial direction and $2\Delta z$ in the vertical direction. The meridional stress T is in N.rad^{-1}, the circumferential stress σ_c is in N.m^{-1} (force per unit meridional length of membrane), and the hanging load σ_w is in N.m^{-2} (force per unit of projected area, not per unit of membrane area).

The change in radius r with distance s along the membrane is easily derived.

$$\frac{dr}{ds} = \cos\alpha \tag{1}$$

The change in height is found in the same way.

$$\frac{dh}{ds} = -\sin\alpha \tag{2}$$

Subjected to a pressure difference p between inside and outside, the applied force on the patch is

$$F = pr2\Delta\theta 2\Delta s\,. \tag{3}$$

For force equilibrium this must be equal to the sum of reaction forces in the opposite direction:

$$(T + \Delta T + T - \Delta T)2\Delta\theta\sin\Delta\alpha + \sigma_c 2\Delta s 2\sin\Delta\theta\sin\alpha + \sigma_w r2\Delta\theta 2\Delta s\cos\alpha\cos\alpha\,. \tag{4}$$

So the force balance in the direction of the normal, after small angle approximations, is

$$pr2\Delta\theta 2\Delta s = 2T2\Delta\theta\Delta\alpha + \sigma_c 2\Delta s2\Delta\theta\sin\alpha + \sigma_w r2\Delta\theta 2\Delta s\cos^2\alpha\,. \tag{5}$$

Rearranging for $\Delta\alpha$ and dividing through by $4T\Delta\theta$,

$$\Delta\alpha = \frac{pr - \sigma_c\sin\alpha - \sigma_w r\cos^2\alpha}{T}\Delta s\,. \tag{6}$$

Taking the limit as Δs goes to zero gives

$$\frac{d\alpha}{ds} = \frac{pr - \sigma_c\sin\alpha - \sigma_w r\cos^2\alpha}{T}\,. \tag{7}$$

A force balance in the direction of the tangent to the patch moving from top to bottom, yields

$$\begin{aligned}(T + \Delta T)2\Delta\theta\cos\Delta\alpha + \sigma_w r2\Delta\theta 2\Delta s\cos\alpha\sin\alpha \\ = (T - \Delta T)2\Delta\theta\cos\Delta\alpha + \sigma_c 2\Delta s2\sin\Delta\theta\cos\alpha\,.\end{aligned} \tag{8}$$

Taking small angle approximations, rearranging for ΔT, and dividing through by $4\Delta\theta$,

$$\Delta T = \Delta s(\sigma_c - \sigma_w r \sin\alpha)\cos\alpha . \quad (9)$$

Taking the limit as Δs goes to zero,

$$\frac{dT}{ds} = (\sigma_c - \sigma_w r \sin\alpha)\cos\alpha . \quad (10)$$

The shape of the bag is found by solving equations (1), (2), (7), and (10). In carrying out optimization studies it is also necessary to calculate some other quantities that are used in the objective function. The stored energy in the bag depends upon the volume of air contained (V). For simplicity we assume that the cost of carrying a force over a certain distance is proportional to the force-distance product, which we call strength (γ). An explanation of this simplification follows.

Cost of materials equals the product of volume of material required (V_{reqd}) and price of material (per unit volume).

$$\text{Cost} = \text{price}.V_{\text{reqd}} \quad (11)$$

V_{reqd} is the distance over which the force is transmitted (x) multiplied by the required cross-sectional area (A_{reqd}), and A_{reqd} is the force transmitted (F) divided by the material's yield stress (σ_y) upon the factor of safety (FoS).

$$\text{Cost} = \text{price}.xA_{\text{reqd}} = \text{price}\frac{xF.FoS}{\sigma_y} \quad (12)$$

Assuming that price per unit volume scales linearly with yield stress, then

$$\text{Cost} \propto xF. \quad (13)$$

The cost of the bag materials depends upon the surface area (A), and the required strengths (meridional, circumferential, and vertical). These are found by simultaneously solving five more differential equations and taking the final values.

$$\frac{dV}{ds} = 2\pi rh\cos\alpha \quad (14)$$

$$\frac{dA}{ds} = 2\pi r \quad (15)$$

$$\frac{d\gamma_m}{ds} = 2\pi T \quad (16)$$

$$\frac{d\gamma_c}{ds} = 2\pi r\sigma_c \quad (17)$$

$$\frac{d\gamma_w}{ds} = \pi rh\sigma_w \cos\alpha \quad (18)$$

In equation (18), the strength in the vertical direction (γ_w) has been divided by 2: this is done because, in suspending a series of masses along a cable hanging between the membrane and the base of the bag, the average tension in the cable is only half the tension at the very top of the cable (assuming an equal distribution of mass along the cable).

Equations (1), (2), (7), (10), and (14)-(18) are integrated numerically, in this case as an initial value problem. We always set $V_0, A_0, \gamma_{m0}, \gamma_{c0}, \gamma_{w0}$ to zero (subscript 0 indicating an initial value at the top of the bag). An end fitting at the top of the bag requires nonzero r_0 (to represent the radius of the end fitting) and nonzero α_0 (so the weight of the end fitting is reacted by the tension in the membrane), but no end fittings are used here and so r_0 and α_0 are also set to zero. The average density of seawater, ρ_{water} = 1025 kg/m^3. The density of the compressed air at absolute pressure P is calculated using the equation of state for an ideal gas.

$$\rho = \frac{MP}{RT} \tag{19}$$

The molar mass of air, M_{air} = 0.02897 kg.mol^{-1}, and the universal gas constant, R = 8.314472 J.K^{-1}.mol^{-1}. It is assumed that the air is stored at 5°C (so T = 278.15 K).

3. Optimization

3.1. Objective function

We seek to minimise the objective function, $f(x) = \text{cost of a bag/stored energy}$, subject to inequality and equality constraints of the form $g(x) \leq 0$ and $h(x) = 0$. While the total cost of a bag will depend upon the cost of manufacture and deployment, these costs are not taken into account in this study; only the cost of materials is included. This cost breaks down into the sum of the cost of reinforcement, the cost of surface, and the cost of ballast. The cost of reinforcement is the sum of each of the meridional, circumferential, and vertical strengths multiplied by the cost per unit strength of each. The cost of surface is the surface area multiplied by the cost of surface per unit area, and the cost of ballast is the difference between the mass of water and the mass of the air multiplied by the cost of ballast per unit mass.

The energy available in the compressed air store depends upon how the compressed air is expanded. It is conservatively assumed that the air is an ideal gas (the compressibility factor for air at 300K and 60bar is 0.9901 which is very close to 1, the compressibility factor of an ideal gas [4]) and will be expanded isothermally. The work done in the isothermal expansion from stored volume V_A (with absolute pressure P_A) to volume V_B (with absolute pressure P_B – atmospheric pressure) is

$$W_{A \to B} = \int_{V_A}^{V_B} P dV = \int_{V_A}^{V_B} \frac{nRT}{V} dV = nRT \ln \frac{V_B}{V_A}. \tag{20}$$

For an ideal gas, the product PV remains unchanged in an isothermal process, so $\frac{V_B}{V_A} = \frac{P_A}{P_B}$ and

$$W_{A \to B} = P_A V_A \ln \frac{P_A}{P_B}. \tag{21}$$

3.2. Optimization algorithm

Exterior penalty functions [5] are used to transform the constrained problem into a single unconstrained problem. Instead of minimising $f(x)$ subject to the constraints $g_i(x) \leq 0$ for $i = 1,...,m$ and $h_i(x) = 0$ for $i = 1,...,l$, we minimise the auxiliary function $f(x) + \mu\alpha(x)$, where μ is a large positive penalty parameter and α is a penalty function that is zero for feasible points and increasingly positive for increasingly infeasible points. A suitable form for α is

$$\alpha(x) = \sum_{i=1}^{m} [\text{maximum}\{0, g_i(x)\}]^2 + \sum_{i=1}^{l} |h_i(x)|^2 . \tag{22}$$

By squaring both terms, differentiability is ensured at both $g(x) = 0$ and $h(x) = 0$. It is possible to get arbitrarily close to a minimum of $f(x)$ by minimising the auxiliary function for a sufficiently large μ. The following effects are penalised: an upward turn of the meridian and compressive stresses. An upward turn of the meridian leads to undesired looping and very often to a solution which does not reach the seabed (i.e. does not reach $h = 0$).

For a given base pressure difference, energy density increases slightly more than linearly with depth while the forces on the bag (and so material cost) remain almost unchanged. Therefore the objective function (material costs only) will decrease almost linearly with depth. It would be hard to put costs to the extra problems that installing bags at great depths would bring (e.g. installation and piping) without further research, so the effect of depth is not studied here.

The objective function is minimised over $h_0, V, \sigma_c, \sigma_w$, and p_{base} using a tiered line search. This is a multidimensional optimization procedure in which a local stationary point is found by simply tiering line searches, so minimising a series of minima. As an example, in minimising the objective function $f(x_1, x_2)$, we would perform a line search in x_1 for a set value of x_2 to find the minimising x_1 (call this minimising value $\bar{x}_1$) for the certain x_2. $\bar{x}_1, \bar{x}_2$ is found by performing a line search adjusting x_2 (and finding $\bar{x}_1$ for each value of x_2 tried) until the minimum of $f(\bar{x}_1, x_2)$ is found. $\bar{x}_1, \bar{x}_2$ is then a local minimum of $f(x_1, x_2)$. To minimise for more variables, another tier of the same procedure is added. Though V cannot be set directly, the meridional stress at the top of the bag, T_0, which gives a shape that encloses a certain required volume, V_{reqd}, can be found using a root finding algorithm. All of the one-dimensional line searches are carried out using Brent's method: a combination of the golden section search and parabolic interpolation. [6] Other multidimensional search procedures which could be used include Rosenbrock's method and methods that use derivatives in determining the search direction, such as the method of steepest descent and the method of feasible directions [7], as used by Pagitz and Pellegrino in their cutting pattern optimization of lobed superpressure ("pumpkin") balloons. [8]

3.3. Results of the optimization study

In the following analyses, all of the meridional, circumferential, and vertical (hanging cable) stresses are carried through steel at its yield strength (250MPa), with an estimate at the cost of steel of £0.5/kg. The cost of surface is estimated at £2/m^2 and the cost of ballast estimated at £4/tonne. Of course all of the optimum shapes are sensitive to these costs. The optima are all found for a depth of 500m.

3.3.1. *Zero pressure natural shape.* An optimum zero pressure natural shape (ZPNS) bag, which has zero pressure difference at the base and purely meridional stresses, was found and is shown in figure 3. It stores 289m^3 of air compressed to an absolute pressure of 51.28bar, which is 1.62MWh (5.83GJ) of energy. The value of the objective function at this optimum is £1104 per MWh. As mentioned before, this figure only accounts for the costs of reinforcement, surface, and ballast materials.

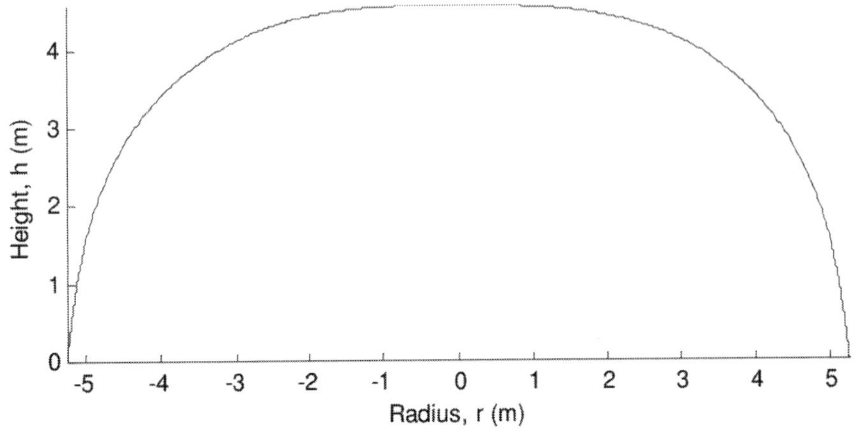

Figure 3. Optimum zero pressure natural shape bag at 500m depth.

It is interesting to note that the optimum ZPNS shape has quite a wide base radius. This is because the pressure difference at the top of a wide, low bag is not as great as it would be at the top of a taller bag with the same base pressure difference. The sides of the bag meet the seabed at angles close to 90° because the smallest possible meridional tension for a given stored volume is that which gives an entry angle of 90°. At angles away from 90°, larger meridional tension is necessary to balance the buoyancy force ($= bV$).

3.3.2. *Zero pressure, nonzero circumferential stress and hanging ballast.* The effects of constant nonzero circumferential stress and constant nonzero hanging ballast on a zero pressure bag are now included in the optimization, and the optimum bag is shown in figure 4. This has zero circumferential stress and hanging ballast of 32kN/m^2, and costs £962 per MWh; a 13% reduction in cost when compared to the ZPNS bag with no hanging masses. It stores 1314m^3 of air compressed to an absolute pressure of 51.28bar, which is 7.36MWh (26.50GJ) of energy. The inclusion of positive circumferential stress increases the value of the objective function and so as compressive circumferential stress is not an option, the optimum value is zero.

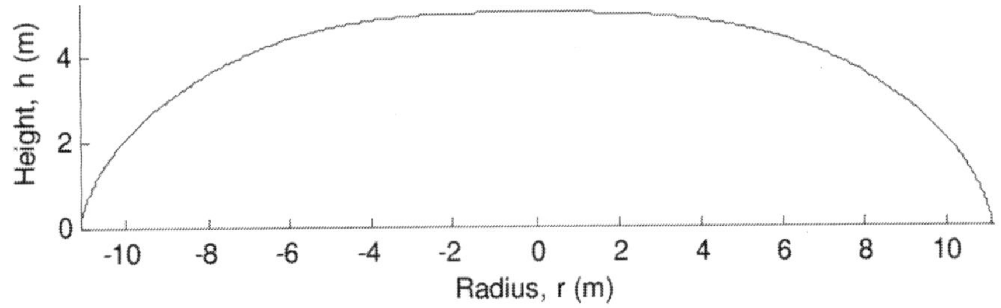

Figure 4. Optimum zero pressure bag with nonzero hanging mass at 500m depth.

It is clear that the inclusion of hanging masses has led to an optimum bag shape which has a greater stored volume and lower profile than the ZPNS bag. Hanging mass counteracts the pressure difference force, particularly in horizontal sections (low α) like at the top of the bag, and so the required net restoring force is reduced. This net restoring force is balanced by curvature of the membrane, and so the required curvature is reduced, and the bag can be wider and store more energy for a given centre height. Once again the sides of the bag meet the seabed at approximately right-angles. It is anticipated that hanging masses will flatten the P-V curve as the bag is inflated and deflated.

3.3.4. *Nonzero pressure, nonzero circumferential stress and hanging ballast.* Now the effect of nonzero base pressure difference is included. Base pressure difference may be positive (like a superpressure balloon) or negative (subpressure). Subpressure balloons have only been considered when studying partial inflation and ascent shapes [9],[10] and underpressurisation of the base of a balloon at float altitude would be hard to achieve, however underpressurizing the base of an energy bag could be achieved by simply pumping in air at pressures lower than the hydrostatic pressure at the base of the bag. The optimum bag is shown in figure 5. This is a subpressure bag with base pressure difference of -32000Pa, zero circumferential stress, and hanging ballast of 10kN/m^2, and costs £907 per MWh. It stores 2249m^3 of air compressed to an absolute pressure of 50.96bar, which is 12.52MWh (45.07GJ) of energy.

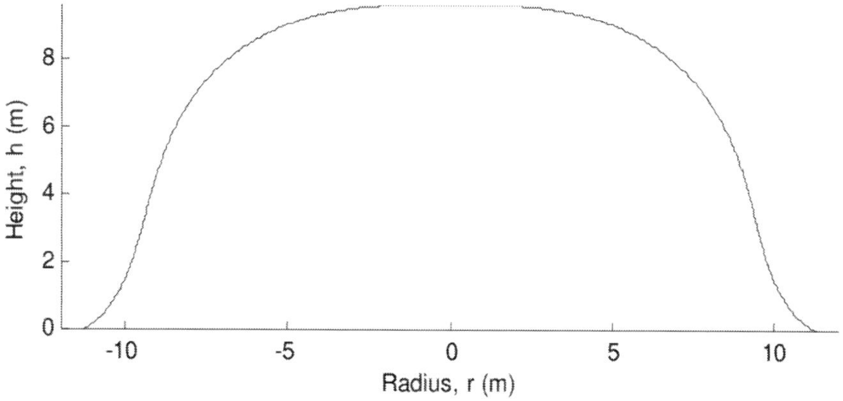

Figure 5. Optimum nonzero pressure bag with nonzero hanging mass at 500m depth.

With the inclusion of nonzero base pressure difference, the base of the optimum bag no longer meets the seabed at right angles because it must take on negative curvature at the base to balance the negative pressure gradient whilst maintaining a fairly low centre height (and so maximum pressure difference).

4. Conclusion

The optimum shape of an inextensible, axisymmetric fabric bag used to store compressed air underwater has been found by modelling the bag with membrane equations and using tiered line searches, Brent's method, and penalty functions to solve the constrained multidimensional optimization problem. The optimum axisymmetric energy bag only has stresses in the meridional direction (so zero circumferential stress), a wide base radius, masses hanging from the inside, and is underpressurised at the base. It costs less than £1000 per MWh though estimated costs have so far only been attributed to materials. Further work with lobed bag models and costs for manufacture and installation will reveal more about the economics of subsea compressed air energy storage.

Acknowledgments

The authors wish to thank the University of Nottingham and E.ON for their financial support and Maxim De Jong of Thin Red Line Aerospace for his useful input and continued enthusiasm for the project.

Journal of Physics: Conference Series **181** (2009) 012049 doi:10.1088/1742-6596/181/1/012049

References

[1] Anon 1951-1956 *Research development in the field of high altitude plastic balloons* NONR-710(01a) reports, Department of Physics, University of Minnesota, MN, USA

[2] Smalley J H 1964 *Determination of the Shape of a Free Balloon – Balloons with Superpressure, Subpressure and Circumferential Stress and Capped Balloons* AFCRL-65-72, Air Force Cambridge Research Labs, Bedford, MA, USA

[3] *Compressed air energy storage power plants* BINE projektinfo 05/07, FIZ Karlsruhe GmbH, Bonn, Germany

[4] Perry R H 1984 *Perry's Chemical engineers' handbook* 6th Edition (New York, USA: McGraw-Hill) table 3-162

[5] Bazaraa M S, Sherali H D and Shetty C M 1993 *Nonlinear Programming: Theory and Algorithms* 2nd Edition (Hoboken, NJ, USA: John Wiley & Sons) p 361

[6] Brent R P 1973 *Algorithms for Minimization without Derivatives* (Englewood Cliffs, NJ, USA: Prentice-Hall) chapter 5

[7] Zoutendijk G 1960 *Methods of Feasible Directions* (Amsterdam, Netherlands: Elsevier)

[8] Pellegrino S and Pagitz M 2007 Shape optimization of "pumpkin" balloons *Proc. AIAA Balloon Systems Conference*, 21-24 May 2007, Williamsburg, VA, USA

[9] Baginski F, Collier W and Williams T 1998 A parallel shooting method for determining the natural shape of a large scientific balloon *SIAM J. Appl. Math.* **58** 961-74

[10] Smalley J H 1966 *Balloon Shapes and Stresses below the Design Altitude* NCAR-TN-25, National Center for Atmospheric Research, Boulder, CO, USA

Journal of Physics: Conference Series **181** (2009) 012050 doi:10.1088/1742-6596/181/1/012050

The effect of inclined soil layers on surface vibration from underground railways using a semi-analytical approach

Simon Jones and Hugh Hunt

University of Cambridge, Department of Engineering
Trumpington Street, Cambridge, United Kingdom CB2 1PZ

E-mail: sj335@cam.ac.uk

Abstract. Ground vibration due to underground railways is a significant source of disturbance for people living or working near the subways. The numerical models used to predict vibration levels have inherent uncertainty which must be understood to give confidence in the predictions. A semi-analytical approach is developed herein to investigate the effect of soil layering on the surface vibration of a halfspace where both soil properties and layer inclination angles are varied. The study suggests that both material properties and inclination angle of the layers have significant effect (± 10dB) on the surface vibration response.

1. Introduction

Underground railways are a proven method for efficiently transporting large numbers of people in densely populated areas: the per-capita power consumption is low and the majority of the infrastructure is located below ground helping to eliminate congestion in urban centers. Unfortunately, ground-borne vibration from these underground railways is a major source of disturbance for individuals either working or living near subway tunnels. Vibration can cause malfunctioning of sensitive equipment [1], damage buildings [2], or impair human comfort and activity [3]. Studies show that inhabitants of urban areas who are subjected to air, road and rail traffic noise report high levels of annoyance and sleep disturbance which negatively impacts quality of life [4–8]. Griefahn et al. [8] reports higher disturbance levels associated with rail noise than from aircraft or ground traffic. This could be attributed to the multiple ways people experience railway induced vibrations: air-borne noise, vibratory motion of the floors, and re-radiated noise both in the room and from household objects such as windows or glassware.

The problem frequency range is between 15Hz to 200Hz [9, 10]; higher frequencies are quickly attenuated by the soil. Low-frequency vibrations arise from the quasi-static load of the train moving along the track [11], general wheel and rail unevenness, and periodic changes in rail-support stiffness when sleepers are present[12].

Acceptable sinusoidal vibration levels for various living and working areas are listed in BS 6472:1992 [13] and depend on many factors such as time of day and building usage. BS ISO 14837 Part 1 [14] provides guidelines on the essential considerations associated with developing prediction models and shows in outline the stages to be observed for new or modified rail systems. Future parts of BS ISO 14837 are meant to quantify acceptable vibration levels from underground railways, but are unavailable at this time.

As a result of more stringent ground-vibration standards, the need for quick and accurate models which can predict the levels of vibration arising from rail traffic is increasing. If predicted levels of vibration are high, changes can be implemented early in the design process rather than using costly retrofits once the project is complete. The models can be used to perform parametric studies to determine which parameters have the greatest effect on ground vibration (e.g. rail pads, continuous slab, tunnel geometry, etc.) so that vibration mitigation measures can be efficiently and economically employed.

Finite element and boundary element methods are common modeling schemes for investigating ground-borne vibration [15–17]. Although FEM/BEM permit accurate modeling of complex geometrical regions (e.g. square tunnels, piled foundations, etc.) these methods suffer from two major drawbacks: input accuracy and speed. The dynamic properties of soils are difficult to measure and can vary significantly over the area of interest [18] thus it is difficult to determine the correct input parameters for the FEM models. It is therefore beneficial to perform parametric studies to cover a range of possible soil parameters; however, common FEM models of underground railways require tens of hours to simulate the response due to a single loading frequency, making in-depth parametric studies intractable.

A more economical approach to simulating ground-vibration due to underground railways involves semi-analytical methods . Researchers at the Universities of Cambridge and Nottingham have developed the PiP software [19] to provide a means by which designers can quickly evaluate design changes and optimize the track, tunnel and foundation to minimize the levels of vibration transmitted to surface. The benefit of the PiP model is its speed, calculating a frequency sweep of 200 Hz in less than a minute, which allows designers to perform hundreds of parametric studies in the time required for a similar FEM/BEM simulation to reach a single solution. The results from PiP have been compared with those of an alternative coupled FEM-BEM method that accounts for a tunnel in a multi-layered half-space; the results were in good agreement for typical soil/tunnel parameters [20].

Although the current PiP model is a useful tool for investigating underground railway vibration problems, the uncertainty associated with some of the modeling assumptions is still not well understood. For instance the semi-analytical approach employed by PiP assumes the soil layering is horizontal. It was previously found by Jones and Hunt[15] that if the interface between soil layers is inclined at small angles ($< 10°$) there is a significant effect on the predicted surface vibration ($\pm$10dB); however, the BEM model used in the previous study required more than 12 hours per computation making it an unacceptable method for implementation into PiP. The work presented herein details a new semi-analytical formulation which can account for inclined soil layers while maintaining fast computational speed.

2. Ground Vibration

Lord Rayleigh[21] found that disturbances within a homogeneous, isotropic halfspace propagate through the medium as compressive waves (P-waves) and shear waves (S-waves) and along the surface as Rayleigh waves. The three waves travel at speeds c_P, c_S and c_R respectively. The compression and shear wave speeds can be written as:

$$c_P = \sqrt{\frac{2G(1-\nu)}{\rho(1-2\nu)}} \qquad \text{and} \qquad c_S = \sqrt{\frac{G}{\rho}} \tag{1}$$

where G, ν and ρ are the shear modulus, Poisson's ratio and density of the medium respectively. The Rayleigh wave speed, c_R, cannot be expressed explicitly in terms of G, ν and ρ but has been shown by Lamb[22] to travel slightly slower than c_S.

Journal of Physics: Conference Series **181** (2009) 012050 doi:10.1088/1742-6596/181/1/012050

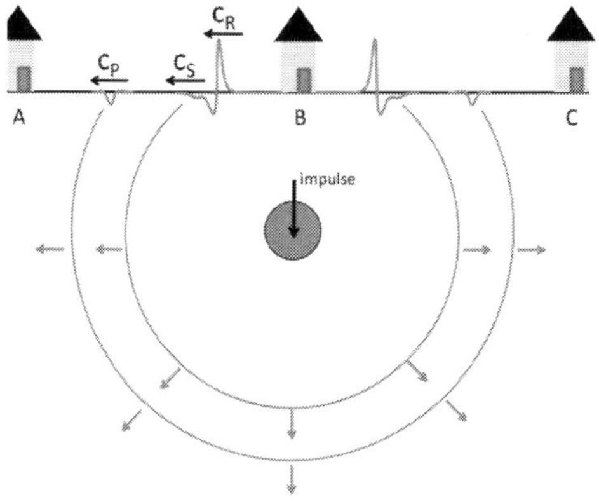

(a) Tunnel in halfspace subjected to impulsive line-load

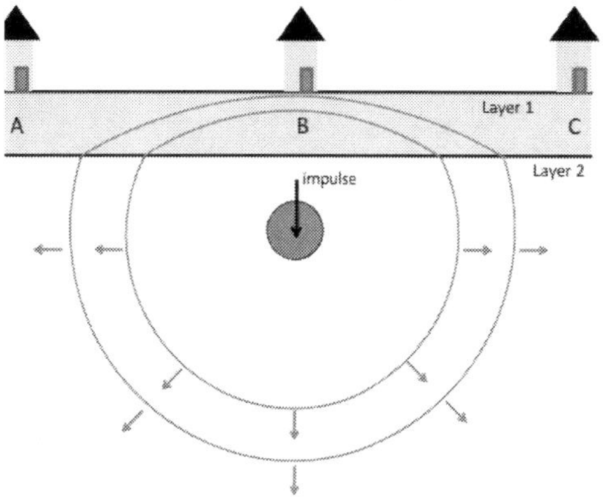

(b) Tunnel in halfspace with horizontal layering subjected to impulsive line-load

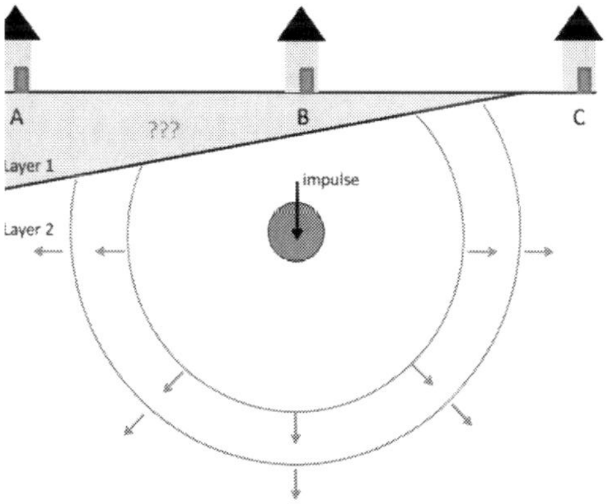

(c) Tunnel in halfspace with inclined layering subjected to impulsive line-load

Figure 1. Wave propagation diagrams for different layering schemes

The effect these waves have on surface vibration is best illustrated through example. Consider the situation depicted in Figure 1(a) of a rigid cylinder subjected to an in-plane, impulsive line load. Surface disturbances develop as the P-wave and S-wave reach the surface, followed by a Rayleigh wave as shown (ignoring reflections). Since wave energy is inversely proportional to the distance traveled[22], the area directly above the tunnel experiences the largest vibration amplitudes (position B). Houses A and C experience the same levels of vibration since they are spaced symmetrically over the tunnel. Peak particle velocity (PPV) shall be used herein to compare vibration levels at various points on the surface.

Consider a layer of softer material added to the surface of the halfspace (Fig 1(b)); Equation 1 states that the wave speeds must reduce as the wavefronts pass through the interface layer, resulting in wave fronts similar to those depicted in Figure 1(b) (ignoring reflections). The house at B still experiences the greatest vibration while levels at A and C are again equivalent due to symmetry.

A final consideration is the inclination of the soil layers. It is common for soil layers to be inclined at angles between 0° and 10° to the surface, as shown in Figure 1(c). Jones and Hunt[15] found that the reflection and refraction of body waves by the inclined interface caused

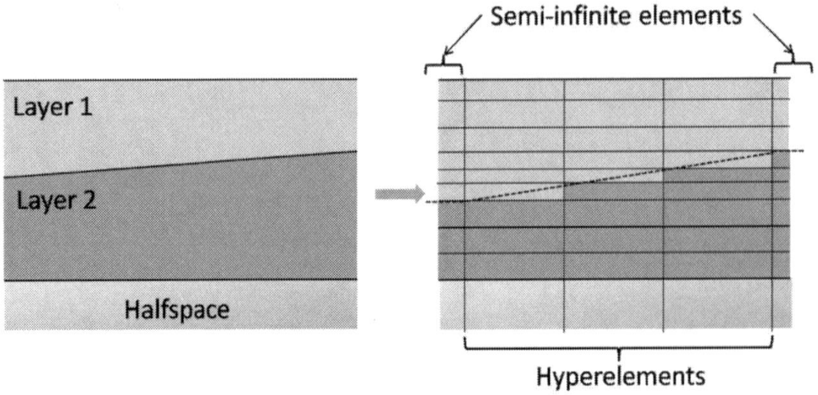

Figure 2. Discretization of inclined soil layer using hyperelements and semi-infinite elements

asymmetric PPV distribution on the surface. Results from the present semi-analytical model are compared to those of the previous BEM model to verify the new model and determine the reduction in processing time.

3. Inclined layers using a semi-analytical approach

The soil layering of interest is presented in Figure 2, containing various soil types and interface inclination angles. To adequately represent this section a "semi-analytical element" approach is employed (programmed in Matlab). The inclined interface is modeled in step-wise increments using hyperelements while semi-infinite elements are included at the extremes. Both element types utilize the analytical solution for wave propagating horizontally while assuming vertical displacements vary linearly through the thickness of the element, as detailed below.

3.1. Semi-infinite elements

The Navier equation governing motion for homogeneous, isotropic, linear elastic bodies, is given by Graff [23] as

$$G\nabla^2\{u\} + (\lambda + G)\nabla\nabla \cdot \{u\} + \rho\{b\} = \rho\{\ddot{u}\} \tag{2}$$

where $\{u\}$ is the displacement vector in the x, y, and z-directions, G and λ are Lame's constants of the solid, ρ is the density of the solid, and $\{b\}$ is the body load vector. Waas[24] shows that solving Equation 2 for plane-strain in the frequency-wavenumber domain results in a transcendental eigenvalue problem which is difficult to solve even for simple geometries. Waas suggested treating the layered region as a continuum in the horizontal direction but to discretize in the vertical direction, resulting in the following equation

$$\left(\mathbf{A}k^2 + i\mathbf{B}k + \mathbf{C}\right)\{v\} = 0 \tag{3a}$$

where

$$\mathbf{C} = \mathbf{G} - \omega^2\mathbf{M} \tag{3b}$$

The vector $\{v\}$ contains the modal coordinates in the x and z directions for the n layers at a given wavenumber, k. The $2n \times 2n$ matrices $\mathbf{A}$, $\mathbf{B}$ and $\mathbf{C}$ consist of the contributions from the n individual layers and are assembled using standard FEA stiffness matrix addition; the submatrices used to construct Equation 3a are given in the Appendix where the subscript "j" refers the the jth layer.

Journal of Physics: Conference Series **181** (2009) 012050 doi:10.1088/1742-6596/181/1/012050

For any given frequency, Equation 3a has a non-trivial solution $\{v\}$ iff

$$\left|\mathbf{A}k^2 + i\mathbf{B}k + \mathbf{C}\right| = 0 \tag{4}$$

This results in a quadratic-eigenvalue problem in k, where the k's are the wavenumbers for the layered region. The solution to Equation 4 consists of $4n$ eigenvalues: a set of $2n$ values with negative imaginary parts ($\mathbf{K}$) and $2n$ values with positive imaginary parts ($\widetilde{\mathbf{K}}$) where it can be shown $\widetilde{\mathbf{K}} = -\mathbf{K}$. The set of eigenvalues with negative imaginary parts correspond to waves traveling to the right while those with positive imaginary parts correspond to waves traveling to the left. The associated eigenvectors (i.e. modeshapes) are designated by ($\mathbf{X}$) and ($\widetilde{\mathbf{X}}$), respectively.

It can be shown that for a semi-infinite region, open to the right

$$\{P\} = \mathbf{R}\{U\} \tag{5}$$

where

$$\mathbf{R} = i\mathbf{AXKX}^{-1} + \mathbf{D} \tag{6}$$

$\mathbf{R}$ is equivalent to the stiffness of the layered region in plane-strain, $\{P\}$ is the external forcing vector, $\{U\}$ is the displacement vector in the frequency domain; the definition of $\mathbf{D}$ is given in the Appendix. For a full derivation refer to Waas[24].

The TLM was developed for a soil layer resting on rigid bedrock, however Andrade[25] later extended this method to include a "halfspace" element allowing for simulation of a soil layer resting on a halfspace. The derivation outlined above remains valid with the addition of a single element in the last rows and columns of the stiffness matrix, given in the Appendix with the subscript "HS".

3.2. Hyperelements

Kausel and Roësset[26] extended the thin-layer method to allow elements of finite length, labeled hyperelements, using the same semi-analytical formulation. This is accomplished by accounting for the waves traveling in both directions through the layer due to nodal loading at both edges of the element.

The method outlined above is first employed to determine the $\mathbf{A}$, $\mathbf{D}$, $\mathbf{X}$ and $\mathbf{K}$ matrices. The stiffness matrix for the hyperelement in plane-strain is then determined as

$$\mathbf{S} = \begin{bmatrix} \mathbf{R} & -\widetilde{\mathbf{R}}\widetilde{\mathbf{J}} \\ -\mathbf{RJ} & \widetilde{\mathbf{R}} \end{bmatrix} \begin{bmatrix} \mathbf{I} & \widetilde{\mathbf{J}} \\ \mathbf{J} & \mathbf{I} \end{bmatrix}^{-1} \tag{7a}$$

where

$$\mathbf{J} = \mathbf{XEX}^{-1} \qquad \widetilde{\mathbf{J}} = \mathbf{TJT} \qquad \widetilde{\mathbf{R}} = \mathbf{TRT} \qquad \mathbf{E} = \mathtt{diag}[e^{-ik_jL}] \tag{7b}$$

where L being the distance between lateral boundaries of the hyperelement and $\mathbf{T}$ is a transform between eigenvectors for left and right traveling waves (see Appendix). For a more detailed discussion see Kausel and Roësset[26].

This procedure is repeated for each column of hyperelements used to mesh the geometry of interest (see Fig. 2); the total stiffness matrix ($\mathbf{S}_{total}$) is assembled from these elementary stiffnesses using standard finite element methods. Thus the nodal displacements due to external loading can be found using

$$\{P\}_{external} = \mathbf{S}_{total}\{U\} \tag{8}$$

where again $\{P\}_{external}$ is the external forcing vector and $\{U\}$ is the nodal displacements in the frequency domain.

Journal of Physics: Conference Series **181** (2009) 012050 doi:10.1088/1742-6596/181/1/012050

Table 1. Boundary element model parameters

(a) Soil Parameters

Property	Layer 1	Layer 2
Poisson's Ratio (ν)	0.44	0.44
Elastic Modulus (G)	$\epsilon_E \times 550(1+0.1i)$ MPa	$550(1+0.1i)$ MPa
Density (ρ)	$\epsilon_\rho \times 2000$ kg/m^3	2000 kg/m^3

(b) Parametric Variables

Variable	Value
ϵ_E	4, 1, $\frac{1}{4}$
ϵ_ρ	4, 1, $\frac{1}{4}$
ϕ	0, 1, 2, 5°

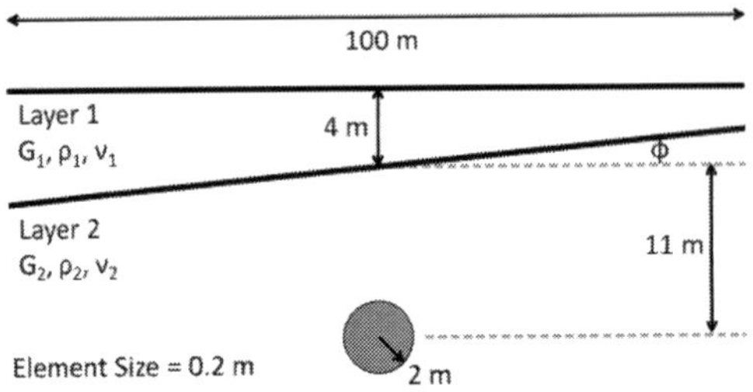

Figure 3. Schematic depiction of boundary element model

4. Modeling

A two-dimensional plane-strain model, almost identical in material and geometry to that used in Jones and Hunt[15], is used to represent the layered halfspace. A schematic of the model is shown in Figure 3 with soil parameters listed in Table 1(a); these parameters are appropriate for soil types found near underground railway tunnels in the London area [27]. Two significant differences exist between the models. First, the TLM soil material includes soil damping, where no damping was included in the BEM model (see Table 1(a) where a damping coefficient of 0.1 is included in the elastic modulus). This slight damping helps stabilize the solution of the eigenvalue problems. Second, the hole in the soil representing the tunnel is not included in the TLM model; instead 16 points spaced evenly at the tunnel radius are used to apply the loading. It is assumed the presence of this extra material will have negligible effect on the results (see Section 5 for a discussion of this assumption).

Table 1(b) lists the scaling factors ϵ_E and ϵ_ρ used for the parametric study. These result in wave speeds in Layer 1 which are factors of ± 2, or equal to that of Layer 2, as given by Equation 1. Layers with wave speeds that differ by a factor of two are common (ie. London Clay overlaid by a gravel layer). Table 1(b) also lists the layer inclination angles investigated. As depicted in Figure 3, the inclination pivot point is located over the center of the tunnel at a depth of 4 meters. The vertical discretization into thin-layers is adapted as loading frequency increases to ensure at least four elements per shear wavelength. The incline is broken into ten steps, which results in a step height of approximately 35cm for a 2° inclination over 100 meters. The disturbance is modeled as an impulsive 1cm vertical displacement of a rigid tunnel.

5. Results

A few representative results from the current TLM simulation are compared to the respective BEM model results[15] in Figure 4. Figure 4(a) presents the surface peak-particle velocity (PPV) for a 4m layer resting on a halfspace with a horizontal interface layer. The results compare well in magnitude and general trend, including the peak PPV directly over the tunnel and trailing off as distance from center increases. Two discrepancies are evident, however. First, the TLM results are slightly higher at the center ($\sim$ 3dB). This is likely due to the lack of "hole" in the soil representing the tunnel (see Section 4). The continuous material transfers the wave energy from the bottom of the tunnel more efficiently to the surface; it can propagate directly upward

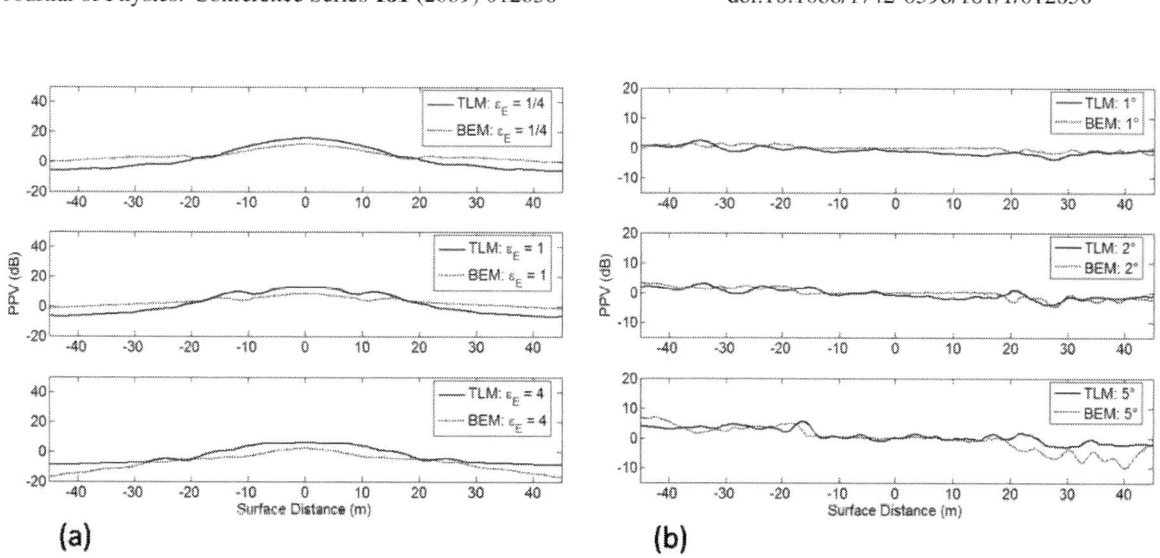

Figure 4. Comparison of thin-layer method (TLM) results and boundary-element method (BEM) results: (a) layer stiffness variation; (b) inclination angle variation

rather than having to travel around the soil void. Therefore, neglecting the tunnel in the model appears to significantly effect the results and must be rectified in future models. Secondly, the TLM results are lower at large distances from center. This is attributed to material damping in the TLM soil which gradually reduces the wave energy reaching the surface.

Figure 4(b) presents the surface PPV insertion gain for a layer with an inclined interface resting on a halfspace; the upper layer has an elastic modulus one quarter that of the halfspace. Insertion gain is calculated by subtracting the surface PPV for the inclined case from that of the horizontal case with equivalent layer properties. Again, the results agree well in magnitude to the BEM results. These results suggests that even small differences in layer properties and inclination angles can have a significant effect on the surface PPV ($\pm$10dB).

The BEM model required approximately 13 hours to compute the response of the region for 100 frequency steps using 0.2 meter elements. This corresponds to approximately 8 minutes per frequency. The TLM model required only 18 minutes to compute the response for 150 frequency steps, corresponding to approximately 7 seconds per frequency. Although this is approximately 150 times faster than the BEM model, it does not meet the required standard for incorporation into the PiP model ($\sim$2 seconds per frequency). The bulk of the computation time is for evaluating Equation 8 using the standard Matlab matrix division operator ($x = A\backslash b$). It is thought a custom solver suited to the banded nature of the TLM could potentially reduce the processing time to meet the PiP specifications.

6. Conclusions and Future Work

The thin-layer method, a semi-analytical formulation, has been used to simulate ground-borne vibration from underground railways using both semi-infinite elements and hyperelements. The model can account for varying material properties, loading parameters and soil-layer inclination angles. The results are comparable to a previous boundary-element model and can be computed approximately 150 times faster. Future work to include the tunnel stiffness and to increase the computational speed of the model are currently in progress.

Appendix

$$[A]_j = \frac{h_j}{6}\begin{bmatrix} 2(2G_j+\lambda_j) & 0 & (2G_j+\lambda_j) & 0 \\ 0 & 2G_j & 0 & G_j \\ (2G_j+\lambda_j) & 0 & 2(2G_j+\lambda_j) & 0 \\ 0 & G_j & 0 & 2G_j \end{bmatrix} \quad [G]_j = \frac{1}{h_j}\begin{bmatrix} G_j & 0 & -G_j & 0 \\ 0 & (2G_j+\lambda_j) & 0 & -(2G_j+\lambda_j) \\ -G_j & 0 & G_j & 0 \\ 0 & -(2G_j+\lambda_j) & 0 & (2G_j+\lambda_j) \end{bmatrix}$$

$$[D]_j = \frac{1}{2}\begin{bmatrix} 0 & \lambda_j & 0 & -\lambda_j \\ G_j & 0 & -G_j & 0 \\ 0 & \lambda_j & 0 & -\lambda_j \\ G_j & 0 & -G_j & 0 \end{bmatrix} \quad [B]_j = \frac{1}{2}\begin{bmatrix} 0 & (G_j-\lambda_j) & 0 & (G_j+\lambda_j) \\ -(G_j-\lambda_j) & 0 & (G_j+\lambda_j) & 0 \\ 0 & -(G_j+\lambda_j) & 0 & -(G_j-\lambda_j) \\ -(G_j+\lambda_j) & 0 & (G_j-\lambda_j) & 0 \end{bmatrix}$$

$$[M]_j = \frac{\rho_j h_j}{6}\begin{bmatrix} 2 & 0 & 1 & 0 \\ 0 & 2 & 0 & 1 \\ 1 & 0 & 2 & 0 \\ 0 & 1 & 0 & 2 \end{bmatrix} \quad \mathbf{T} = \begin{bmatrix} 1 & & & & \\ & -1 & & & \\ & & \ddots & & \\ & & & 1 & \\ & & & & -1 \end{bmatrix}$$

$$[A]_{HS} = \frac{iGC_s}{2\omega\alpha^3}\begin{bmatrix} (\alpha-2)\alpha^2 & 0 \\ 0 & (1-2\alpha) \end{bmatrix} \quad [B]_{HS} = G\frac{(1-2\alpha)}{\alpha}\begin{bmatrix} 0 & -1 \\ 1 & 0 \end{bmatrix} \quad [C]_{HS} = i\omega\rho\begin{bmatrix} C_s & 0 \\ 0 & C_p \end{bmatrix}$$

$$[D]_{HS} = G\frac{(1-2\alpha)}{\alpha}\begin{bmatrix} 0 & 1 \\ -1 & 0 \end{bmatrix} \quad \alpha = \frac{C_s}{C_p}$$

References

[1] Gordon C 1999 *Proc. SPIE* **3786** 22–39
[2] Hildebrand R 2001 *Report at Department of Vehicle Engineering* **Stockholm**
[3] Institution B S **BS 6841:1987**
[4] Schultz T 1978 *J. Acoust. Soc. Am.* **64** 377–405
[5] Kryter K 1982 *J. Acoust. Soc. Am.* **72** 1212–1242
[6] Fidell S, Barber D and Schultz T 1991 *J. Acoust. Soc. Am.* **89** 221–233
[7] Miedema H and Vos H 1998 *J. Acoust. Soc. Am.* **104** 3432–3445
[8] Griefahn B, Marks A and Robens S 2006 *Journal of Sound and Vibration* **295** 129–140
[9] Grootenhuis P 1977 *Journal of Sound and Vibration* **51** 443–448
[10] Greer R and Manning C 1998 *Acoustics Bulletin* **23** 13–17
[11] Sheng X, Jones C and Thompson D 2003 *Journal of Sound and Vibration* **267** 621–635
[12] Forrest J 1999 *Modelling of ground vibration from underground railways* Ph.D. thesis University of Cambridge
[13] Institution B S **BS 6472:1992**
[14] Institution B S **BS ISO 14837-1:2005**
[15] Jones S and Hunt H 2008 *Proceedings of the international conference on noise and vibration engineering, ISMA 2008* ID 456
[16] Müller K, Grundmann H and Müller G 2008 *Proceedings of the 7th European conference on structural dynamics, EURODYN 2008* E257
[17] Degrande G, Gupta S and Fiala P 2007 *Proceedings of the 3rd international symposium on evironment vibrations: prediction, monitoring, mitigation and evaluation, ISEV 2007* pp 53–62
[18] Hunt H 1988 *Measurement and modelling of traffic-induced vibration* Ph.D. thesis University of Cambridge
[19] Hussein M and Hunt H 2007 *Journal of Sound and Vibration* **305** 401–431
[20] Hussein M, Hunt H, Rikse L and *et al* 2007 *9th International Workshop on Railway Noise*
[21] Rayleigh L 1885 *Proc. London Mathematical Society* **17** 4–11
[22] Lamb H 1904 *Phil. Trans. Roy. Soc. (London)* **203** 1–42
[23] Graff K 1991 *Wave motion in elastic solids* (Dover Publications)
[24] Waas G 1972 *Linear two-dimensional analysis of soil dynamics problems in semi-infinite layered media* Ph.D. thesis Engineering, University of California Berkeley
[25] Andrade P 1999 *Implementation of second-order absorbing boundary conditions in frequency-domain computations* Ph.D. thesis University of Texas at Austin
[26] Kausel E and Roësset J 1977 *Journal of the engineering mechanics division, ASCE* **103** 569–588
[27] Hussein M 2004 *Vibration from underground railways* Ph.D. thesis University of Cambridge

Journal of Physics: Conference Series **181** (2009) 012051 doi:10.1088/1742-6596/181/1/012051

Finite-element analysis of thermal-induced stresses around a cased injection well

A Ferla[1], A Lavrov[2] and E Fjær[2]

[1] Department of Civil Engineering, Catholic University of Leuven, Kasteelpark Arenberg 40, 3001 Heverlee, Belgium
[2] Formation Physics Department, SINTEF Petroleum Research, 7465 Trondheim, Norway

Corresponding author's e-mail address: andries_ferla@hotmail.com

Abstract. Injection of surface fluids (sea water, CO_2, steam) into hydrocarbon reservoirs induces thermal stresses in the wellbore structures and in the near-well area. These stresses may endanger the integrity of the casing, the cement and the surrounding formation. Therefore, an accurate assessment of injection-induced thermal stresses and the associated risk of failure are of utmost importance for a safe and environmentally secure oil production. A coupled finite-element model has been developed and tested as a tool for assessing the probability and extent of failure caused by thermal-induced stresses around a cased wellbore. A feature of the model is an option for an automatic mesh refinement using Nikishkov elements and a quad-tree data structure. The refinement is automatically provided in the regions of rapid change of the nodal variables (displacements, temperature).

1. Introduction

Injection of surface fluids (sea water, CO_2) into hydrocarbon reservoirs is a part of the production technology during enhanced oil recovery and underground storage of greenhouse gases. During underground production of heavy oil, steam-assisted gravity drainage and cyclic injection of steam are major technologies used at present. All these types of technology involve thermal and mechanical interaction between the injection fluid and the wellbore structures at depth that were previously in thermal equilibrium with the formation and thus had the temperature equal to the formation temperature at the depth of interest. Subjecting the wellbore structures and the near-well area to the temperature of the injected fluid induces thermal stresses and strains in the wellbore casing, cement and in the surrounding rocks [1, 2]. This may, and often does, result in an integrity loss of the structures. There is no need to say that an integrity loss can have negative economic and environmental consequences. Therefore, a quantitative analysis of the risk and extent of thermally-induced damage is of utmost importance for a prolonged, environmentally secure hydrocarbon production. This is especially true in the new millennium when conventional easy-to-develop reserves are being gradually depleted, and more and more difficult fields are involved in production, e.g. deepwater fields and high temperature high pressure reservoirs.

The most common types of casing failure in wells with cyclic steam injection are caused by heating/cooling-induced tensile stresses or by casing buckling. In this work, only the former mechanism is investigated. This mechanism works as follows. During injection of the hot fluid (steam), compressive axial stress is induced in the casing since the casing is fixed at its ends. The compressive stresses can measure hundreds of MPa and thus can induce yielding in the casing material. During subsequent cooling, and due to the hysteresis of the casing material, tensile axial stresses are induced in the casing, which may and often do result in the failure at the casing coupling, or in tensile fracturing of the casing.

A common cure against casing failure caused by heating/cooling-induced tensile stresses is to set a pre-stressed casing. In that technology, the casing is stressed in tension in the axial direction before it is cemented. After the cementation is completed, the initial tensile stresses will prevent the casing from yielding when heating wants to induce compressive stresses in the casing. As a result, the risk of casing failure in tension will be reduced during a subsequent cooling.

Stress analysis in casing subjected to thermal loading is usually done without taking into account the effect of the surrounding rock formation (e.g. [3-6]). It is conceivable that the heterogeneous reservoir rock, with interbedding softer and harder rocks, might additionally aggravate the stress environment in the wellbore structure and thus further reduce the well integrity.

The main objectives of the work presented in this paper were to develop and test a finite-element model of thermal induced stresses generated in the wellbore casing, cement and near-well area during injection of surface fluids; and to study the usually neglected effect that the presence of hard or soft rock layers might have on stresses and failure of casing and cement. Also, we aimed to investigate the effect of pre-stressed casing on the stability of the cement and the near-well rock. Also, we aimed to test an automatic mesh refinement algorithm that makes use of Nikishkov finite elements, described further in the paper. It should be emphasized that all materials are linear elastic in our model. Thus, we cannot model the heat-induced yielding and the subsequent generation of tensile stresses during cooling. But the model allows us to estimate whether the yielding will occur or not, and if yes, what parts of the wellbore structures and the formation are most prone to thermal-induced failure.

2. Finite-element model

The finite-element method (FEM) is one of the most popular methods for engineering stress analysis, at least of the problems that do not involve stress singularities or material fragmentation. Amongst the reasons for the wide-spread use of FEM in solid and structural mechanics are its flexibility and relative ease of implementation in a computer code, good capabilities for implementing even the most complex material models, and a solid and well-established theoretical background of FEM [7].

Our wellbore model is a two-dimensional model and represents an axisymmetric problem. Both steady-state and transient simulations are possible. In the transient mode, the Crank-Nicolson scheme for time integration is used. Coupling between the thermal and mechanical models is one-way coupling, i.e. the results from the temperature field computation are fed into the mechanical model, but the results of the stress analysis do not in any way affect the thermal model. The mechanical model computes the mechanical stresses induced by both mechanical and thermal loading of the system.

2.1. Geometry

The geometry of the problem is schematically shown in figure 1. As evident in figure 1, the wellbore section being analysed is always straight and parallel to the greatest in-situ far-field principal stress. Due to the nature of FEM, the model has to be finite, even though in reality the rock formation has an infinite extent in horizontal direction. It is common to assume that keeping the ratio between the model radius and the wellbore radius at or greater than 10 provides a good approximation to the infinite case. In our simulations, this ratio was equal to approximately 20.

The dimensions of the model used in our simulations are listed in table 1.

2.2. Elements, mesh and remeshing

Two types of mesh can be used in the model implementation. One makes use of triangular 3-node elements. The same mesh is used for the thermal and mechanical analyses, thus the accuracy is first order in temperature and displacements. The other type of mesh geometry makes use of rectangular

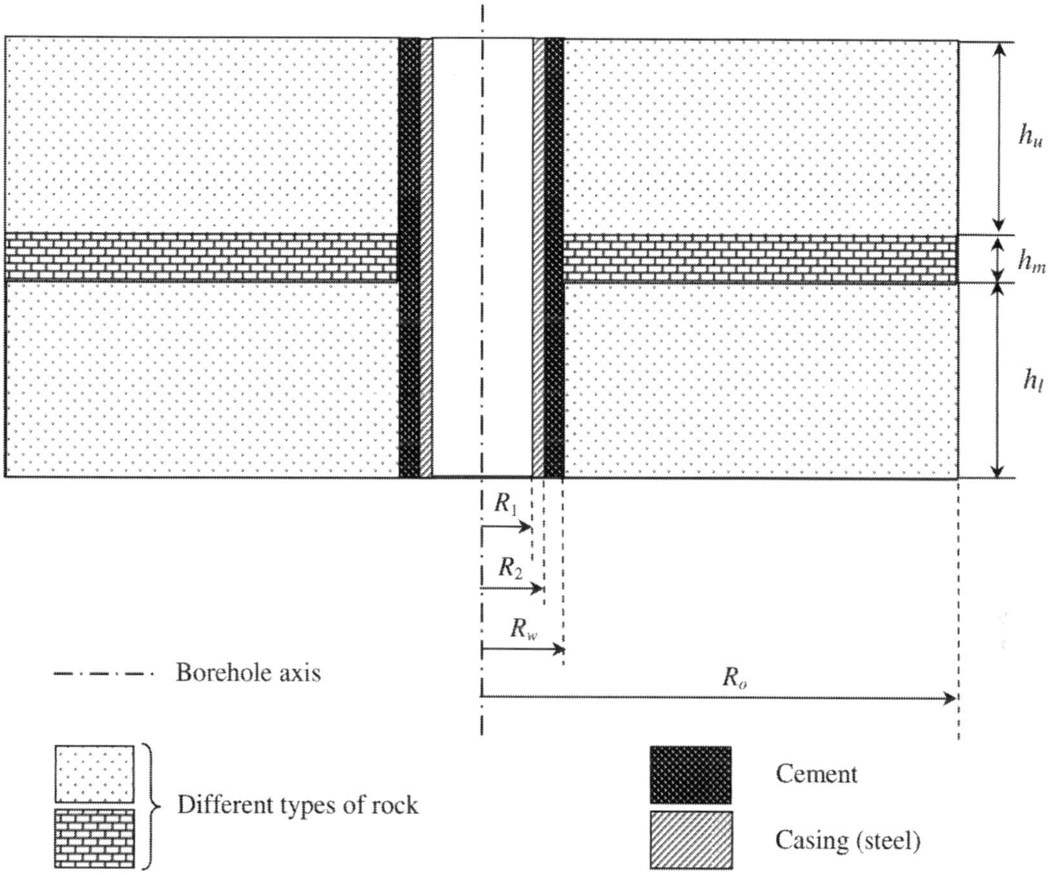

Figure 1. Schematic plot of cross-section along the borehole axis.

Table 1. Model dimensions as indicated in Figure 1

Dimension	Value, m
R_1	0.10
R_2	0.12
R_w	0.14
R_o	2.66
h_u	1.28
h_m	0.08
h_l	1.20

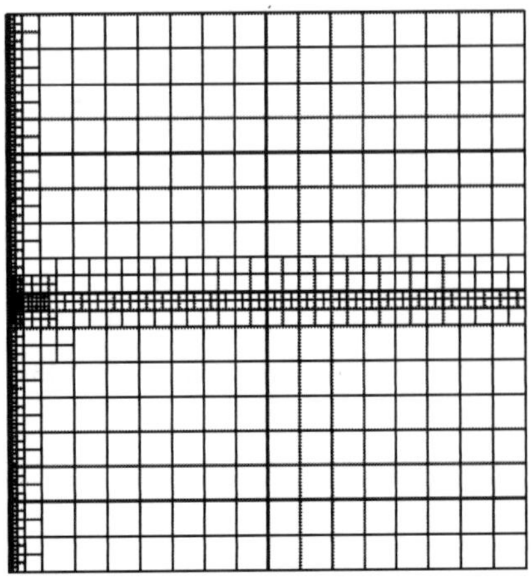

Figure 2. Finite-element mesh with rectangular Nikishkov elements..

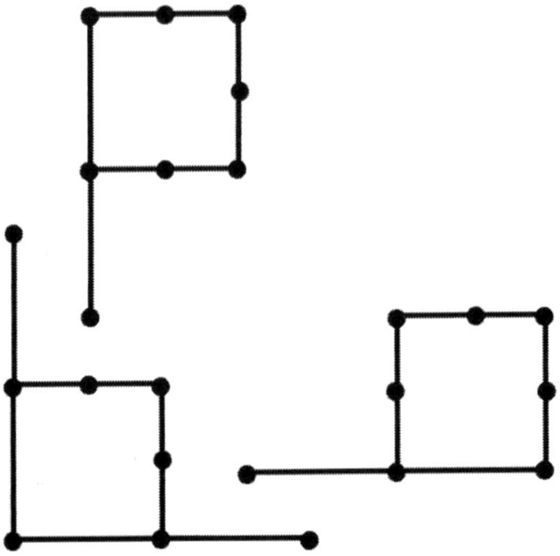

Figure 3. Nikishkov refinement elements.

elements and is used when automatic remeshing is opted for. An example of this type of mesh is shown in figure 2. Only this latter type of mesh was used in the simulations described in this paper. The rectangular mesh may initially include both the conventional 8-node serendipity elements and the Nikishkov elements. Examples of the Nikishkov refinement elements are shown in figure 3. An extensive discussion of these elements and their shape functions can be found in [8]. When automatic refinement is performed, the elements are contained in a quad-tree data structure. After refinement, the so-called mesh balancing is carried out, in order to ensure that no two elements having a common boundary have a generation difference in excess of 1.

2.3. Material properties

All materials are assumed to be linear elastic herein. Poromechanical effects in the rock are neglected. Thermal and elastic properties of four materials that represent casing, cement and two types of rock are listed in table 2. These parameters are used for the finite-element stress computation. During post-processing, appropriate failure criteria are checked for every part of the model. The parameters that are used in this failure analysis are listed in table 3. It should be emphasized that the yield stress of steel and the parameters of the Mohr-Coulomb failure criterion listed for cement and rocks are used for the evaluation of the failure risk only, and do not serve as input in the FEM stress analysis.

2.4. Mechanical initial and boundary conditions

The largest in-situ principal stress is assumed to be parallel to the wellbore axis. For the sake of clarity, this stress will be termed 'vertical' in what follows. Accordingly, the wellbore is vertical, too. The top boundary of the model is assumed to be located at a depth of D = 500 m. The initial vertical stress is assumed to be equal to the weight of the overburden having a density of 2500 kg/m^3. The initial vertical stress is thus equal to 12.5 MPa. Two other initial in-situ stresses, i.e. the horizontal stresses, are assumed to be equal to each other, in order to assure axial symmetry of the problem. Roller boundary conditions are set at the top, bottom and at the far boundary of the model. These are the most common boundary conditions used for this type of problems in rock mechanics. The initial conditions are applied in the sequence that should mimick as accurately as possible what is happening during completion of a cased well.

Table 2. Thermal and mechanical properties of materials used in the FEM simulations.

Material	Density, kg/m^3	Specific heat, J/(kg·K)	Thermal conductivity, W/(m·K)	Thermal expansion coefficient, 10^{-6} K^{-1}	Young's modulus, GPa	Poisson's ratio
Steel	7800.	1000.	45.0	25.	200.0	0.3
Cement	2300.	900.	0.8	12.	25.0	0.2
Hard rock	2500.	900.	2.4	15.	30.0	0.3
Soft rock	2500.	900.	0.6	15.	0.3	0.3

Table 3. Failure criteria and strength properties of materials used in the post-processing of the FEM results.

Material	Failure criterion	Yield stress, MPa	Cohesion, MPa	Internal friction angle, °
Steel	Von Mises	600.	-	-
Cement	Mohr-Coulomb	-	15.	30.
Hard rock	Mohr-Coulomb	-	20.	30.
Soft rock	Mohr-Coulomb	-	5.	30.

Just after the wellbore has been drilled, and before the casing is set, the radial σ_{rr} and circumferential $\sigma_{\theta\theta}$ stresses in the model can be evaluated using a well-known analytical solution from the theory of elasticity as follows [9]:

$$\sigma_{rr} = \frac{\sigma_o R_o^2 - \sigma_w R_w^2}{R_o^2 - R_w^2} - \frac{R_o^2 R_w^2}{R_o^2 - R_w^2} \frac{\sigma_o - \sigma_w}{r^2} \quad (1)$$

$$\sigma_{\theta\theta} = \frac{\sigma_o R_o^2 - \sigma_w R_w^2}{R_o^2 - R_w^2} + \frac{R_o^2 R_w^2}{R_o^2 - R_w^2} \frac{\sigma_o - \sigma_w}{r^2} \quad (2)$$

where r is the radial coordinate, i.e. the distance from the borehole axis; R_w is the radius of the well; R_o is the external radius of the model; σ_w is the wellbore pressure; σ_o is the radial stress acting at the external boundary of the model. We assume σ_o to be equal to the theoretical in-situ horizontal stress given by $\sigma_v \nu/(1-\nu)$, where ν is the Poisson's ratio of the rock, and σ_v is the vertical stress applied to the model and equal to the overburden pressure. We assume σ_w to be equal to the 'completion stress', p_c, which is the wellbore pressure at the time just before the casing is installed. It can be assumed to be equal to e.g. 6 MPa for a 500 m deep well.

The stresses given by (1) and (2) are applied as initial stresses inside the rock. Inside the cement, the initial stress state is assumed to be hydrostatic. In order to ensure the radial stress continuity between cement and rock, this hydrostatic stress has to be equal to p_c. Finally, in the casing, the initial radial stress is equal to p_c, and the initial vertical stress is equal to zero (in a casing that has not been pre-stressed) or to some positive value (in a pre-stressed casing). The initial wellbore pressure at the inner wall of the casing is equal to p_c, and so is the initial radial stress at the outer wall of the casing. This ensures continuity of the initial radial stress across the boundary between the casing and the cement.

During injection, rollers are retained as boundary conditions at the top, bottom and far vertical boundaries of the model. At the wellbore wall, a normal stress equal to the downhole injection pressure is applied, e.g. 7 MPa.

2.5. Thermal initial and boundary conditions

The initial temperature distribution is homogeneous, the temperature being equal to the virgin formation temperature which should eventually be evaluated based on the geothermal gradient and the depth, D. It is assumed to be equal to 30°C in the simulations presented in this paper.

Two thermal modes are available: steady state and transient. In the steady state mode, the stress problem is solved for a homogeneous temperature field, with the temperature equal to the downhole steam temperature in all parts of the model, i.e. casing, cement and rock. In the transient mode, the temperature is rapidly increased at the inner wall of the casing so as to acquire a profile along the wellbore axis. As a special case, for a relatively short section of the well, the temperature along the wellbore wall may well be assumed to be constant. In the simulations of a relatively short wellbore section described in this paper the downhole steam temperature is assumed to be equal to 280°C along the entire wellbore section. In general, the temperature distribution along the wellbore axis should be provided by a wellbore flow simulation as an input to our finite-element model.

In the transient simulations, two types of boundary conditions, i.e. Dirichlet (isothermal) or Neumann (adiabatic), can be applied at the bottom, top and far vertical side of the model during injection. In the simulations presented in this paper, we apply the adiabatic boundary condition at the top and bottom boundaries, and isothermal condition at the far vertical side of the model. On that latter boundary, the temperature is maintained constant and equal to the original formation temperature, i.e. 30°C.

3. Results

Eight test cases have been studied, summarized in Table 4. The test cases differ in lithology, thermal regime and casing pre-tension. In the transient simulations, the stress distributions have been analysed after 10 hours of injection. At that time, the temperature front propagates to between 0.5 and 1.0 m from the borehole axis into the formation. The casing pre-tension, if any, was chosen so as to match the expected compressive stresses caused by the temperature increase.

Table 4. Test cases.

Case ID	Lithology			Casing pre-tension	Thermal regime
	Upper layer	Middle layer	Bottom layer		
A	Hard rock	Soft rock	Hard rock	No	Steady state
B	Soft rock	Hard rock	Soft rock	No	Steady state
C	Hard rock	Soft rock	Hard rock	No	Transient
D	Soft rock	Hard rock	Soft rock	No	Transient
E	Hard rock	Soft rock	Hard rock	Yes	Steady state
F	Soft rock	Hard rock	Soft rock	Yes	Steady state
G	Hard rock	Soft rock	Hard rock	Yes	Transient
H	Soft rock	Hard rock	Soft rock	Yes	Transient

Figure 4 shows the vertical (axial) stress distributions for the steady state thermal regime. All vertical stresses shown here are taken along the borehole and in the middle of the casing. In both cases shown in the figure the casing is in compression due to the thermal stresses but the rock formation has a great influence on the stresses in the casing. In case A where a soft layer of rock is located in between stiffer layers, a very large negative peak in the vertical stress is observed at the location of the layer. In case B were a stiff layer is lying in between softer rocks the casing is less compressed in the

Journal of Physics: Conference Series **181** (2009) 012051 doi:10.1088/1742-6596/181/1/012051

neighbourhood of the layer but in the region of the soft rock the compression is larger. Thus, a thin layer of a soft rock in between stiffer layers might be more dangerous for the borehole integrity than the opposite lithology.

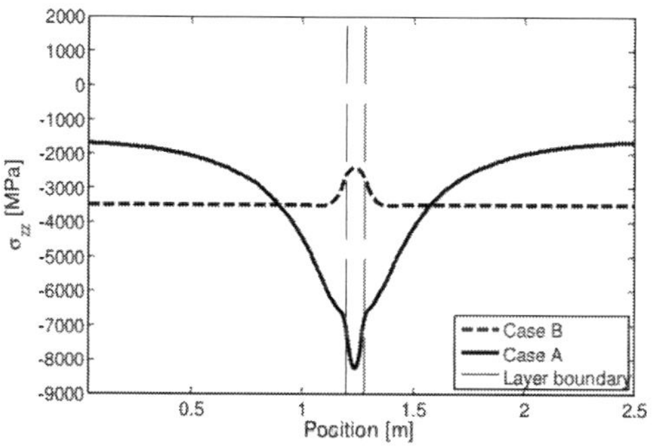

Figure 4. Vertical stresses in the casing under a steady state thermal regime.

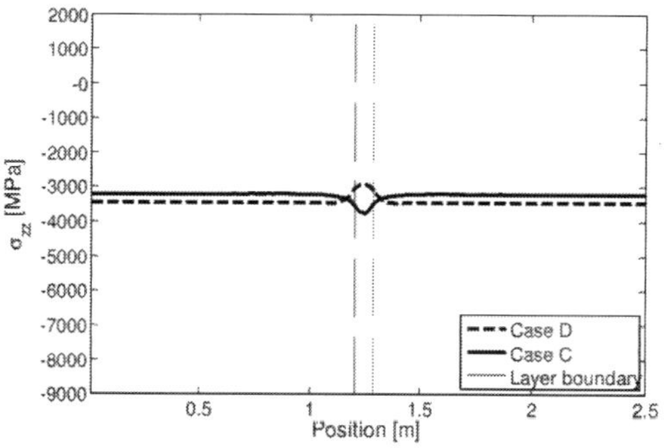

Figure 5. Vertical stresses in the casing after 10 hours of injection.

At the start of injection (figure 5), the stresses are smaller than during the steady state (figure 4), but all the trends seen in figure 5 are present also in figure 4.

Pre-tensioning the casing should prevent compression in the casing. As can be seen in figure 6 the applied pretension that was calculated based on the injection temperature without taking into account the effect of the surrounding formation, was not sufficient to prevent large axial compressive stresses from developing in the casing, especially in Case C were a thin layer of soft rock is formed between stiffer layers. It can be concluded that the amount of pre-tension needed depends on the rock formation the casing is installed in.

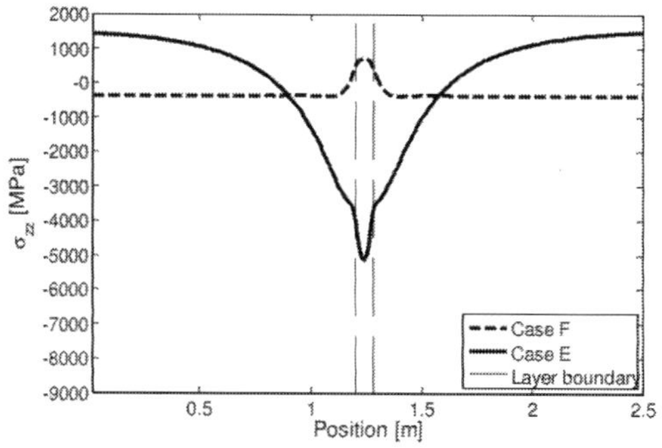

Figure 6. Vertical stresses in a pretensioned casing with a steady state thermal regime.

The distribution of the axial stress in the cement indicates the same trends as those in the casing, although the magnitude of the generated stresses is an order of magnitude lower in the cement. This is due to the fact that the elastic modulus of cement is an order of magnitude smaller in cement. Since the strength of the cement is however much lower than that of steel, the induced thermal stresses might have much more devastating results in the cement than they might have in steel. Cement failure may represent a serious threat for the wellbore integrity since this might create a hydraulic communication up along the well.

Analysis of the radial stresses has shown that in case A tensile radial stresses develop at the interface between the casing and the cement in the vicinity of the boundaries between the rock layers. These radial stresses are found in the simulations with and without casing pre-tensioning. Analysis of the shear stress produced at the interfaces casing/cement and cement/rock indicates that increased shear stresses are produced near the central layer in our model. Both elevated radial tensile stresses and shear stresses at the interfaces may result in various forms of failure in and around the casing. Applying the strength data shown in table 3 it was found that in the case of a soft middle layer the soft rock will fail, as also will the cement in the vicinity of the soft layer. In the case of a stiff middle layer, no failure would occur in the model, but the safety factor is higher in the stiff layers and the casing than they are in the soft layer and the cement.

4. Conclusions

A finite element code has been developed to model the thermal induced stresses in a steel casing, cement and the near-well area caused by fluid injection as part of e.g. enhanced oil recovery or thermal production of heavy oil. The finite element code makes use of a special type of element, the Nikishkov element, that allows an automatic adaptive refinement of 8-node serendipity elements using a quad-tree data structure. Numerical simulations of test cases with the developed code have proven that interbedding layers of relatively soft and relatively stiff rocks have a major effect on the thermal stress distribution along the casing as well as on the effect a casing pre-tension might have on the reduction of thermal compressive stresses. A complex stress environments form along the well that may include elevated axial stresses, shear stresses along the boundaries casing/cement and cement/rock, or even tensile radial stresses between the casing and the cement. These stress conditions may result in the material failure which can jeopardize the well integrity during steam injection. It should be noted that the simulations presented in this paper have been carried out for a very short casing section. Further work is required in order to conclude whether the effect found out herein will still be as well pronounced in more realistic, longer casing sections. Also, poroelastic effects in the rocks should be further included in the model.

References

[1] Dusseault M B, Bruno M S and Barrera J 2001 *SPE Drilling and Completion*, June, 98-107.
[2] Collins P M 2007 *SPE Reservoir Evaluation and Engineering*, August, 367-375.
[3] Rodríguez-Prada H 1990 Improved calculation of casing strains and stresses in a steam injector well. *Society of Petroleum Engineers* paper 21080.
[4] Maharaj G 1996 Thermal well casing failure analysis. *Society of Petroleum Engineers* paper 36143.
[5] Plácido J C R, Ademar P Jr, Paulo L A B F, Pasqualino I P and Estefen S F Stress analysis of casing string submitted to cyclic steam injection. *Society of Petroleum Engineers* paper 38978.
[6] Heathman J and Beck F E 2006 Finite element analysis couples casing and cement designs for HP/HT wells in East Texas. *Society of Petroleum Engineers* paper 98869.
[7] Zienkiewicz O C, Taylor R L and Zhu J Z 2005 *The Finite Element Method: Its Basis and Fundamentals. Sixth Edition.* (Amsterdam: Elsevier).
[8] Nikishkov G P 2005 *ANZIAM J.* **46**(E) C15-C28.
[9] Fjær E, Holt R M, Horsrud P, Raaen A M and Risnes R 2008 *Petroleum Related Rock Mechanics. Second Edition.* (Amsterdam: Elsevier).

Journal of Physics: Conference Series **181** (2009) 012052 doi:10.1088/1742-6596/181/1/012052

Wavelet analysis of beam-soil structure response for fast moving train

Piotr Koziol[1], Zdzislaw Hryniewicz[1], Cristinel Mares[2]

[1] Koszalin University of Technology
Department of Civil and Environmental Engineering
Sniadeckich 2, 75-453 Koszalin, Poland

[2] Brunel University
School of Engineering and Design
Uxbridge, Middlesex, UB8 3PH, UK

Piotr.Koziol@wbiis.tu.koszalin.pl

Abstract. This paper presents a wavelet based approach for the vibratory analysis of beam-soil structure related to a point load moving along a beam resting on the surface. The model is represented by the Euler-Bernoulli equation for the beam, elastodynamic equation of motion for the soil and appropriate boundary conditions. Two cases are analysed: the model with a half space under the beam and the model where the supporting medium has a finite thickness. Analytical solutions for the displacements are obtained and discussed in relation to the used boundary conditions and the type of considered loads: harmonic and constant. The analysis in time-frequency and velocity-frequency domains is carried out for realistic systems of parameters describing physical properties of the model. The approximate displacement values are determined by applying a wavelet method for a derivation of the inverse Fourier transform. A special form of the coiflet filter used in numerical calculations allows to carry out analysis without loss of accuracy related to singularities appearing in wavelet approximation formulas, when dealing with standard filters and complex dynamic systems.

1. Introduction

The problem analysed in this paper is related to the abatement of vibrations generated by high-speed trains. The investigation of the traffic interaction with the environment plays a significant role in the construction of new tracks and trains [1, 2, 3] and is one of most important subjects of EU research in the area of surface transportation.

Modern high-speed trains move with velocities which increase very dangerously the level of vibrations [4, 5]. The Rayleigh velocity is usually treated as a critical velocity and the theoretical investigations and experimental tests show that the ground vibrations grow significantly when the trains' speed reaches this critical value [6, 7]. Although many results concerning railway tracks built on the surface were described in the literature [3, 6], new analytical and numerical approaches to the problem of vibrations abatement are still needed, with special interest in ground vibration analysis [7, 8, 9].

The type of the load and properties of the supporting layer strongly influence the level of accuracy of numerical approximations of displacement. Two different cases of the same model are presented in this paper: the first one with a half space under the beam and the second one when the supporting layer has finite thickness. The complexity of solutions is discussed in relation to the velocity and frequency of the load. The level of vibrations generated by the load is computed in the cases of harmonic and constant point loads moving along the beam resting on the surface of viscoelastic medium. The case of the harmonic load is very important for the analysis of surface vibrations generated by moving load and usually shows a strong response of the system for even its small velocities. It does not show any singularities in mathematical formulas and can be effectively solved by using carefully adopted calculation methods. In order to avoid numerical instabilities appearing for some systems of parameters, a wavelet approximation method [10] and a specially chosen coiflet filter [11] are used for derivation of the inverse Fourier transform. This method is very effective for high velocities and high load frequencies and allows to reduce numerical problems for calculations that prevent effective analysis of the system due to the big amount of time needed for calculations [12].

2. A load moving along a beam

An infinitely long beam is resting on viscoelastic surface and the point load is moving along the beam. Two cases are analyzed: finite thickness h of supporting medium (Fig.1) and a half space under the beam (Fig.2).

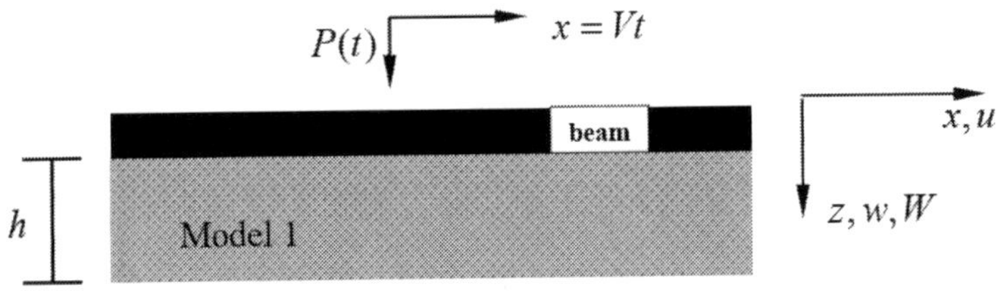

Figure 1: Finite thickness of the medium.

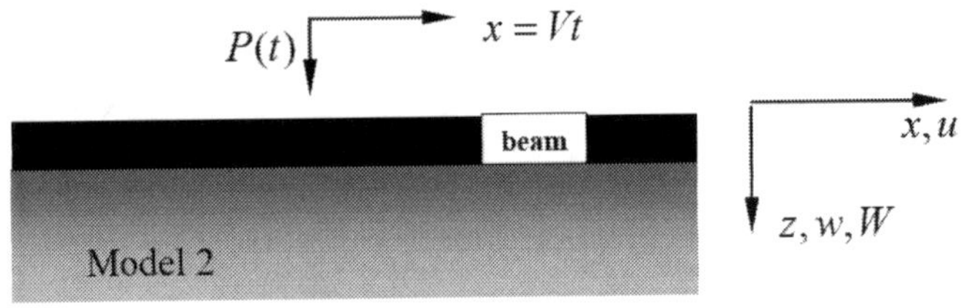

Figure 2: Half space under the beam.

The equation for the vertical motion of the beam can be written as Euler Bernoulli equation:

$$EI\frac{\partial^4 W(x,t)}{\partial x^4} + \rho_B \frac{\partial^2 W(x,t)}{\partial t^2} = P(t)\delta(x - Vt) - a\sigma_{zz}(x,0^+,t) \quad (1)$$

where $W(x,t)$, $\sigma_{zz}(x,z,t)$, $P(t)$, EI, ρ_B, a and $\delta(\cdot)$ are the vertical displacement of the beam, the vertical stress, the vertical point load, the bending stiffness, the mass per unit length of the beam, the thickness of the beam in y direction and the Dirac delta function, respectively.

The motion of the viscoelastic layer can be described by the following elastodynamic Navier's equation:

$$(\lambda+\mu+(\lambda^*+\mu^*)\frac{\partial}{\partial t})\nabla_{xz}(\nabla_{xz}\mathbf{u})+(\mu+\mu^*\frac{\partial}{\partial t})\nabla_{xz}^2\mathbf{u}=\rho\frac{\partial^2\mathbf{u}}{\partial t^2} \tag{2}$$

where $\mathbf{u}(x,z,t)=[u(x,z,t),0,w(x,z,t)]$, ρ and λ, μ are the displacement vector, the mass density of the soil and Lame' constants, respectively. The boundary and continuity conditions are:

$$u(x,0,t)=0\,, \tag{3}$$

$$w(x,0,t)=W(x,t)\,, \tag{4}$$

$$u(x,h,t)=0\,,\ w(x,h,t)=0 \text{ (for model 1)} \tag{5a}$$

$$\lim_{z\to\infty}u(x,z,t)=0\,,\ \lim_{z\to\infty}w(x,z,t)=0 \text{ (for model 2).} \tag{5b}$$

The interpretation of these conditions can be described in the following way: the beam does not move horizontally, the vibrations of the beam and the vertical displacements of the soil at the surface are the same and the radiated waves vanish at $z=h$ (model 1) and at the infinity (model 2). The Fourier transform can be used for obtaining the steady state response of the surface for these systems [4]. The described model presents the system based on the assumptions formulated in [4, 12] where the problem of ground vibrations due to a load moving in a tunnel is investigated for harmonic and constant loads. Accordingly the reformulated equations and conditions lead to the investigation of the soil response for the load moving along the beam resting on the surface similar to the models published in [8, 13, 9] where the main direction of investigation is focused on derivation of critical velocities. One should note that the present paper analyses the two-dimensional model, whereas the analysis in three dimensions would allow to solve the problem in a more general case with y direction and possibly simplify some calculations by introduction of other systems of coordinates.

3. Analytical solution

The formulas for displacements and stress components can be rewritten by using Lame' potentials:

$$u=\frac{\partial\varphi}{\partial x}+\frac{\partial\psi}{\partial z},w=\frac{\partial\varphi}{\partial z}-\frac{\partial\psi}{\partial x},\sigma_{zz}=\left(\lambda+\lambda^*\frac{\partial}{\partial t}\right)\left(\frac{\partial^2\varphi}{\partial x^2}+\frac{\partial^2\varphi}{\partial z^2}\right)+2\left(\mu+\mu^*\frac{\partial}{\partial t}\right)\left(\frac{\partial^2\varphi}{\partial z^2}-\frac{\partial^2\psi}{\partial x\partial z}\right) \tag{6}$$

Instead of equation (2) one can write two scalar equations:

$$\rho\frac{\partial^2\varphi}{\partial t^2}=\left(\lambda+2\mu+(\lambda^*+2\mu^*)\frac{\partial}{\partial t}\right)\left(\frac{\partial^2\varphi}{\partial x^2}+\frac{\partial^2\varphi}{\partial z^2}\right),\ \left(\mu+\mu^*\frac{\partial}{\partial t}\right)\left(\frac{\partial^2\psi}{\partial x^2}+\frac{\partial^2\psi}{\partial z^2}\right)=\rho\frac{\partial^2\psi}{\partial t^2}. \tag{7}$$

Applying the Fourier transforms

$$\tilde{\tilde{f}}(k,\omega)=\int_{-\infty}^{+\infty}\int_{-\infty}^{+\infty}f(x,t)e^{i(\omega t-kx)}dxdt\,,\quad f(x,t)=\frac{1}{4\pi^2}\int_{-\infty}^{+\infty}\int_{-\infty}^{+\infty}\tilde{\tilde{f}}(k,\omega)e^{-i(\omega t-kx)}d\omega dk \tag{8}$$

gives the following representation of the system in the transform domain:

$$(EIk^4-\rho_B\omega^2)\tilde{\tilde{W}}(k,\omega)=\int_{-\infty}^{+\infty}P(t)e^{i(\omega-kV)t}dt-a\tilde{\tilde{\sigma}}_{zz}(k,0^+,\omega) \tag{9a}$$

$$\frac{d^2\tilde{\tilde{\varphi}}}{dz^2}-R_L{}^2\tilde{\tilde{\varphi}}=0\,,\ \frac{d^2\tilde{\tilde{\psi}}}{dz^2}-R_T{}^2\tilde{\tilde{\psi}}=0 \tag{9b}$$

where $R_L{}^2=k^2-\omega^2/(c_L{}^2-i\omega(\lambda^*+2\mu^*)/\rho)$, $R_T{}^2=k^2-\omega^2/(c_T{}^2-i\omega\mu^*/\rho)$ and c_L, c_T are velocities of the longitudinal and the shear waves in the layer, respectively.

The boundary and continuity conditions (3)-(5) can be rewritten accordingly:

$$\tilde{\tilde{u}}(k,0,\omega)=0\,,\ \tilde{\tilde{w}}(k,0,\omega)=W(k,\omega)\,,\ \tilde{\tilde{u}}(k,h,\omega)=\tilde{\tilde{w}}(k,h,\omega)=0\,,$$

$$\lim_{z\to\infty}\tilde{\tilde{u}}(k,z,\omega)=\lim_{z\to\infty}\tilde{\tilde{w}}(k,z,\omega)=0\,.$$

The solutions for equations (9b) are:

$$\tilde{\tilde{\varphi}} = A_1(k,\omega)e^{R_L z} + A_2(k,\omega)e^{-R_L z}, \ \tilde{\tilde{\psi}} = A_3(k,\omega)e^{R_T z} + A_4(k,\omega)e^{-R_T z}, \tag{10}$$

where coefficients A_1 and A_3 become zeros when the supporting layer has infinite thickness (model 2). Then one can derive the solutions for displacements and stresses in the transform domain:

$$\tilde{\tilde{u}} = ik(A_1(k,\omega)e^{R_L z} + A_2(k,\omega)e^{-R_L z}) + R_T(A_3(k,\omega)e^{R_T z} - A_4(k,\omega)e^{-R_T z}) \tag{11a}$$

$$\tilde{\tilde{w}} = R_L(A_1(k,\omega)e^{R_L z} - A_2(k,\omega)e^{-R_L z}) - ik(A_3(k,\omega)e^{R_T z} + A_4(k,\omega)e^{-R_T z}) \tag{11b}$$

$$\begin{aligned}\tilde{\tilde{\sigma}}_{zz} = [(\tilde{\lambda} + 2\tilde{\mu})(R_L)^2 - \tilde{\lambda}k^2][A_1(k,\omega)e^{R_L z} + A_2(k,\omega)e^{-R_L z}] - \\ -2ik\tilde{\mu}R_T[A_3(k,\omega)e^{R_T z} - A_4(k,\omega)e^{-R_T z}]\end{aligned} \tag{11c}$$

and hence, by using boundary conditions, one obtains the system of algebraic equations with respect to A_i that can be solved accordingly to the Cramer's rule:

$$A_j(k,\omega) = (\int_{-\infty}^{+\infty} P(t)e^{i(\omega-kV)t}dt)D_j(k,\omega)/D(k,\omega) \tag{12}$$

with D and D_j the associated determinants. The solution for vertical displacement is represented by the equation:

$$\tilde{\tilde{w}}(k,0,\omega) = \tilde{\tilde{w}}_0(k,\omega)\int_{-\infty}^{+\infty} P(t)e^{i(\omega-kV)t}dt \tag{13}$$

where $\tilde{\tilde{w}}_0(k,\omega) = (R_L(D_1 - D_2) - ik(D_3 + D_4))/D$ for model 1 and $\tilde{\tilde{w}}_0(k,\omega) = (-R_L D_2 - ikD_4)/D$ for model 2. Hence, by the Fourier transform (8), one can obtain the integral form of solution for the displacements at the surface:

$$w(0,0,t) = \frac{1}{4\pi^2}\int_{-\infty}^{+\infty}\int_{-\infty}^{+\infty} \tilde{\tilde{w}}_0(k,\omega)(\int_{-\infty}^{+\infty} P(t_1)e^{i(\omega-kV)t_1}dt_1)e^{-i\omega t}dkd\omega. \tag{14}$$

When the moving load is harmonically varying in time with the load frequency Ω, $P(t) = P_0\cos(\Omega t)$, formula (14) takes the following form:

$$w(0,0,t) = \frac{P_0}{4\pi}\int_{-\infty}^{+\infty}[\tilde{\tilde{w}}_0(k,kV-\Omega)e^{-i(kV-\Omega)t} + \tilde{\tilde{w}}_0(k,kV+\Omega)e^{-i(kV+\Omega)t}]dk. \tag{15}$$

The case of $\Omega = 0$ can be treated as a load constant in time. For this type of load the integrand in (15) has a strong singularity that prevents direct numerical calculation of the vertical displacement for both models. Therefore, in this paper, the displacements for relatively low load frequency are derived in order to approximate the constant point load. Instead of classical numerical integration, the alternative method [10] based on wavelet approximation of the inverse Fourier transform is used for calculation of integral (15). This method was previously successfully applied for the solution of number problems related to ground born vibrations analysis and investigation of interactions of fast railway transportation with the environment [12, 15, 16]. This method is very efficient for the analysis of ground and beam vibrations generated by loads moving with velocities near critical values and relatively high or low load frequencies [12, 15]. The wavelet approach allows to analyse detailed features of dynamic systems which might be lost when using numerical integration.

The method adopted in this paper for calculation of integral (15) is based on wavelet expansion of functions in $L^2(\mathbf{R})$, with multiresolution coefficients $c_{n,k}$ and $d_{j,k}$ obtained by using specially constructed wavelet filters:

$$f(x) = \sum_{k=-\infty}^{+\infty} 2^{n/2}c_{n,k}\Phi(2^n x - k) + \sum_{j=n}^{\infty}\sum_{k=-\infty}^{+\infty} 2^{j/2}d_{j,k}\Psi(2^j x - k). \tag{16}$$

Φ and Ψ are the scaling function and the wavelet function, respectively, derived in this paper by using coiflet filter p_k listed in Appendix. The average value of these coefficients ($M_2 = \sum_{k=0}^{M_1} kp_k$) differs from any integer and therefore allows to take into account all terms of approximating sequence (16), as opposed to coiflet filters used in previous publications [12, 15, 16]. In those cases, another singularity appeared in wavelet formulas and due to that fact, an additional approximating procedure was needed in order to eliminate troublesome terms. This modified system of coiflets improved the approximation accuracy and eliminated doubts related to analytical nature of the used sequences.

For the coiflets, the approximating formula (16) takes the following form:

$$f(x) = \lim_{n\to\infty} f_n(x) = \lim_{n\to\infty} \frac{2^{-n-2}}{\pi} \prod_{j=1}^{j_0} \left(\sum_{k=0}^{M_1} p_k e^{ikx/2^{j+n}}\right) \sum_{k=k_{min}}^{k_{max}} \tilde{f}((k+M_2)2^{-n}) e^{ixk2^{-n}} . \tag{17}$$

M_1, M_2, j_0 are integers and $k_{min}(n) = \omega_{min} 2^n - 16$, $k_{max}(n) = \omega_{max} 2^n - 1$ define the interval $[\omega_{min}, \omega_{max}]$ covering the set of variable ω in transform domain which has important meaning for the nature of original function. Formula (17) can be used for derivation of original function f from its Fourier transform $\tilde{f}$, with significantly large n. Usually, it is sufficient to assume $n = 4$ and f_4 almost coincides with the original function. Nevertheless, the term f_{17} is used in this paper, due to the fact that considered, relatively low, load frequencies shows strong sensitivity of the system for its dynamic changes. Numerical simulations show that for $n > 16$ the sequence (17) stabilises and further terms do not differ significantly. One should note that a similar procedure can be applied for a wide spectrum of daublets and coiflets filters which were specially constructed for numerical applications. These carefully considered and modified, if needed, systems of coefficients give approximation with appropriate accuracy depending on the nature of the analysed systems.

4. Wavelet evaluation of displacements – numerical examples

The following system of parameters [4, 6, 12, 15, 16] is considered for parametric analysis: Young's modulus $E = 3\cdot 10^7$ N/m^2; the mass density $\rho = 1700\,\text{kg/m}^3$; $\mu^* = \lambda^* = 3\cdot 10^4\,\text{kg/ms}$ and Poisson's ratio $\nu = 1/3$ for the soil; the mass density $\rho_B / a = 3\cdot 10^4\,\text{kg/m}^2$; the bending stiffness $EI/a = 10^9\,\text{Nm}$, the width in y direction $a = 4\,\text{m}$ and the vertical point load $P_0 = 5\cdot 10^4\,\text{N}$ for the beam. The harmonic load is moving with three different velocities: $35\,\text{m/s} \approx 0.46\cdot c_R$, $65\,\text{m/s} \approx 0.86\cdot c_R$, $80\,\text{m/s} \approx 1.05\cdot c_R \approx 0.98\cdot c_T$. These velocities allow to highlight some features of the system response for harmonic and constant loads. The values $c_R \approx 76\,\text{m/s}$ and $c_T \approx 82\,\text{m/s}$ are the Rayleigh velocity and the velocity of the shear waves in the soil, respectively. The thickness of the soil in model 1 is assumed to be $10\,\text{m}$ and $10^4\,\text{m}$, and those cases are compared with model 2. The load frequency $f_\Omega = \Omega/2\pi$ is equal to $0.1/2\pi\,\text{Hz}$, $0.01/2\pi\,\text{Hz}$ and $10^{-10}/2\pi\,\text{Hz}$ for the representation of the estimated constant load [4].

Numerical simulations show (Fig. 3) that vertical displacement for the constant load can be approximated by the harmonic one with relatively low load frequency. One can show that the response characteristics stabilises when the frequency load increases below $10^{-6}\,\text{Hz}$ for the considered system. Therefore the value $\Omega = 10^{-10}$ of the frequency for the harmonic load is considered in parametric analysis for the investigation of the load constant in time.

Figures 4 and 5, show plots of the vertical displacements for different sub-critical velocities and both considered models. One can observe that even relatively high thickness of the supporting layer in model 1 does not give a good approximation of the more realistic model when the soil under the beam is a half space (Figs. 1 and 2), especially for high velocities near critical values (Fig. 5). Therefore one should analyse very carefully models with finite supporting layer [4] because complex dynamic variations of the system can give wrong results when applying this assumption. However the waves' reflections might influence obtained results for this model, the propagation of reflected waves can be negligible due to viscoelastic properties of layer and its appropriately chosen thickness for the approximating procedure.

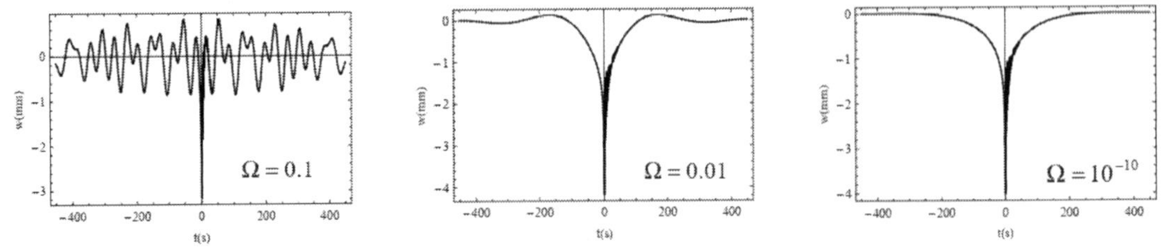

Figure 3: The vertical displacement for model 1 and $h = 10^4\,\mathrm{m}$.

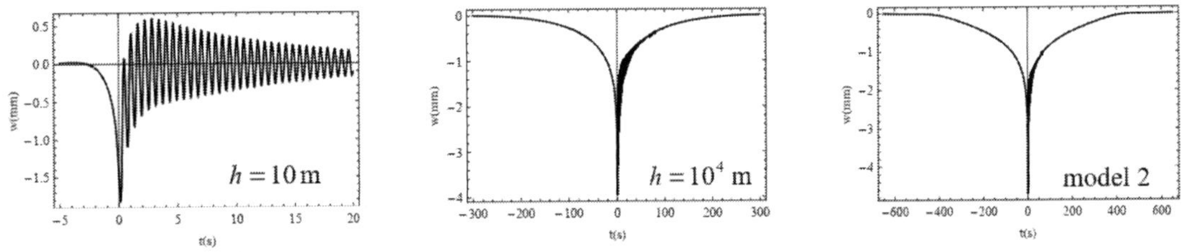

Figure 4: The vertical displacement for the velocity $V = 35\,\mathrm{m/s}$.

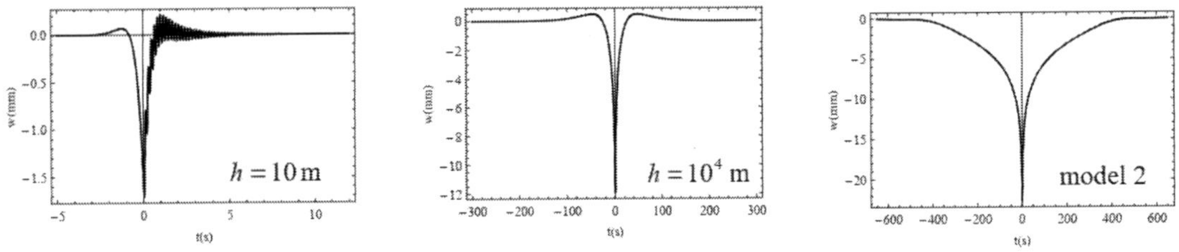

Figure 5: The vertical displacement for the velocity $V = 80\,\mathrm{m/s}$.

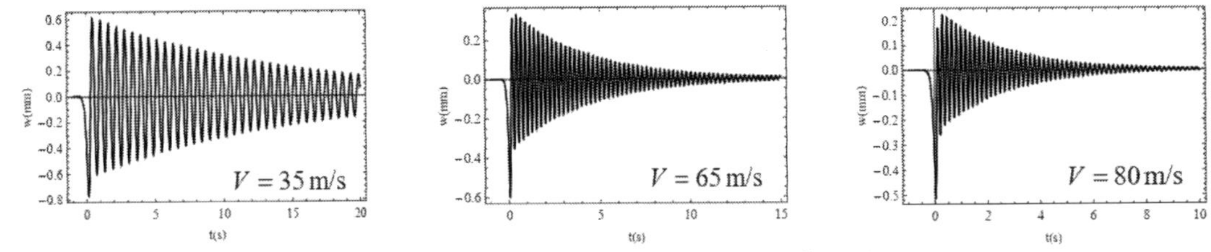

Figure 6: The vertical displacement for $h = 10\,\mathrm{m}$.

Figure 6 shows that the maximum amplitude of vertical vibrations decreases for higher velocities when the frequency load is equal to $\Omega = 10^{-10}$ which means that the assumption of the thickness $h = 10\,\mathrm{m}$ for the supporting layer is not realistic and leads to wrong results. This tendency is reversed when the value h increases (Fig. 7) and these correct observations of the influence of velocity on train induced vibrations are confirmed by the analysis of more realistic model 2 (Fig. 8) and by the theory [3, 4, 6, 7, 12]. The wavelet method adopted in this paper for the derivation of vertical displacements allows further detailed analysis of the system, especially in the area of high velocities and the point load constant in time.

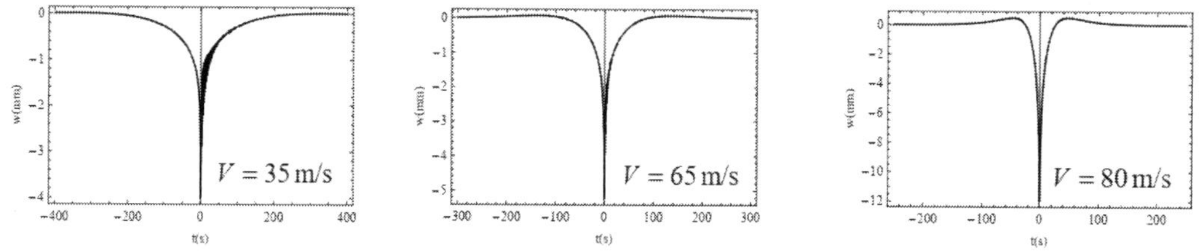

Figure 7: The vertical displacement for $h = 10^4$ m.

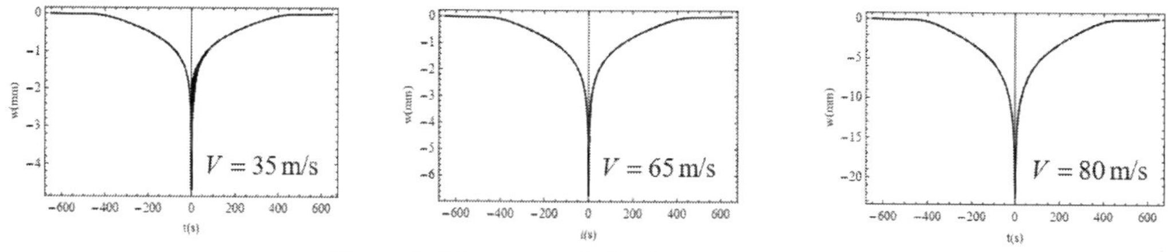

Figure 8: The vertical displacement for model 2.

The maximum amplitude derived by using the coiflet based wavelet approximation for a wide range of velocities shows that critical value of velocity for the constant load can be estimated around $80\,\mathrm{m/s}$ for model 2 (Fig. 9). One can observe that the vibrations can reach the relatively high level of $40\,\mathrm{mm}$ when the constant load moves with velocity near critical value.

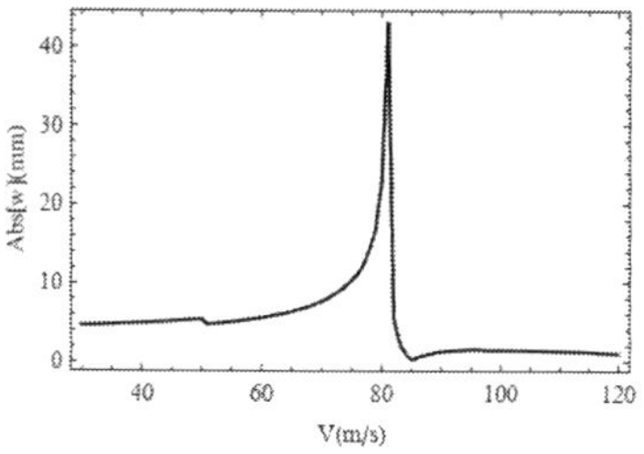

Figure 9: Critical velocity for model 2 and constant load – the maximum of vertical displacement.

5. Conclusions

The problem of viscoelastic soil vibrations generated by a load moving along a beam resting on the surface was analysed in two cases: with finite thickness of supporting layer and with half space under the beam. A special wavelet method was adopted for the derivation of vertical displacements at the surface. This method allowed to carry out the analysis in the area of high velocities and low frequencies leading to the approximation of the point load constant in time. The applied estimation allowed to alleviate the problem of singularities appearing in the integrated formulas when the constant load was considered. The modified coiflets filter improved the accuracy of wavelet approximation compared to previously published results. The parametric analysis of the system was carried out depending on a number of factors, e.g. the thickness of supporting layer and the velocity of moving load and the critical velocity was numerically estimated for the constant load and a half space under the beam. The analysis of this system can be extended by using the wavelet approach for more complex situations.

Appendix

The following set of coiflets is used for numerical calculations performed in this article:

{0.003401479882015607, -0.004130806329954543, -0.03536170269249431, 0.05747767104264993, 0.3843902644404712, 0.5358632409346619, 0.1908760013178301, -0.1321131305836887, -0.05295999083912471, 0.05813917906468963, 0.009758111187504831, -0.01825628044991493, 0.0002608645070967113, 0.00327048515783943, -0.0003823627249285679, -0.0002646325745805278, 0.000017334234085592, 0.000014273738297708887}.

References

[1] Hermsworth B., (2000). Reducing groundborne vibrations: state of the art study. Journal of Sound and Vibration, 231(3), 703-709.

[2] Chen Y.H., Huang Y.H., (2003). Dynamic characteristics of infinite and finite railways to moving load. Journal of Engineering in Mechanics, 129(9), 987 – 995.

[3] Fryba L., (1999). Vibrations of Solids and Structures under Moving Loads. Thomas Telford Ltd., London.

[4] Metrikine A.V., Vrouwenvelder A.C.W.M, (2000). Surface ground vibration due to a moving train in a tunnel: two – dimensional model. Journal of Sound and Vibration, 234(1), 43-66.

[5] Dieterman H.A., Metrikine A., (1996). The equivalent stiffness of a half space interacting with a beam. Critical velocities of a moving load along the beam. European Journal of Mechanics A/Solids, 15, 67-90.

[6] Krylov V.V. (ed.), (2001). Noise and Vibrations from High-Speed Trains. Thomas Telford Ltd, London.

[7] Madshus C., Kaynia A.M., (2000). High speed railway lines on soft ground: dynamic behaviour at critical train speed. Journal of Sound and Vibration 231(3), 689-701.

[8] Dieterman H.A., Metrikine A., (1996). The equivalent stiffness of a half space interacting with a beam. Critical velocities of a moving load along the beam. European Journal of Mechanics A/Solids 15, 67-90.

[9] Kononov A.V., Wolfert R.A.M., (2000). Load motion along a beam on a viscoelastic half-space. European Journal of Mechanics and Solids 19, 361-371.

[10] Wang J., Zhou Y., Gao H., (2003). Computation of the Laplace inverse transform by application of the wavelet theory. Communications in Numerical Methods in Engineering 19, 959-975.

[11] Beylkin G., Coifman R., Rokhlin V., (1991). Fast wavelet transforms and numerical algorithms. Communications on Pure and Applied Mathematics, 44, 141-183.

[12] Koziol P., Mares C., Esat I., (2008). Wavelet approach to vibratory analysis of surface due to a load moving in the layer. International Journal of Solids and Structures 45, 2140-2159.

[13] Dieterman H.A., Metrikine A., (1997). Critical velocities of a harmonic load moving uniformly along an elastic layer. Transaction of the ASME Journal of Applied Mechanics 64, 596-600.

[14] Achenbach J.D., (1984). Wave Propagation in Elastic Solids, Elsevier, Amsterdam.

[15] Koziol P., Mares C., Esat I., (2007). Analysis of surface vibrations generated by underground fast train. International Conference on Engineering Dynamics, Carvoeiro, Algarve, Portugal, 16-18 April 2007.

[16] Koziol P., Mares C., Esat I., (2008). Wavelet analysis of a solid vibration due to a load moving along a beam resting on a surface. International Conference on Noise and Vibration Engineering, Leuven, ISMA2008.

Solving time-dependent problems by an RBF-PS method with an optimal shape parameter

Ana M. A. Neves[1], C. M. C. Roque[1], A. J. M. Ferreira[1], C. M. M. Soares[2], R. M. N. Jorge[1]

[1] Departamento de Engenharia Mecânica e Gestão Industrial, Faculdade de Engenharia da Universidade do Porto,
Rua Dr. Roberto Frias, 4200-465 Porto, Portugal
[2] IDMEC - Instituto de Engenharia Mecânica - Instituto Superior Técnico, Av. Rovisco Pais, 1096 Lisboa Codex, Portugal

E-mail: ana.m.neves@fe.up.pt, croque@fe.up.pt, ferreira@fe.up.pt, cristovao.mota.soares@dem.ist.utl.pt, rnatal@fe.up.pt

Abstract. An hybrid technique is used for the solutions of static and time-dependent problems. The idea is to combine the radial basis function (RBF) collocation method and the pseudospectral (PS) method getting to the RBF-PS method. The approach presented in this paper includes a shape parameter optimization and produces highly accurate results.

Different examples of the procedure are presented and different radial basis functions are used. One and two-dimensional problems are considered with various boundary and initial conditions. We consider generic problems, but also results on beams and plates. The displacement and the stress analysis are conducted for static and transient dynamic situations. Results obtained are in good agreement with exact solutions or references considered.

1. Introduction

Both pseudospectral (PS) method ([1, 2]) and radial basis function (RBF) method ([3] to [7]) are good solvers for PDEs. Combining the two methods we can extend the high accuracy of the results to complex geometries and keep it simple to implement.

Ferreira and colleagues used the multiquadrics RBF ([8]) to solve time-dependent problems including structural problems. Fasshauer, Ferreira, and colleagues ([9] to [14]) have already used with success the RBF-PS method for the solution of some problems. This paper extends the application of the RBF-PS method to the transient analysis of structural problems including an optimization of the shape parameter for the radial basis functions, allowing an user-independent analysis.

2. RBF-PS method for time-dependent problems

Suppose you want to approximate a function that you want to differentiate or to approximate the solution $u(x)$ or $u(x,t)$ of a given a differential equation with boundary conditions. The approximation considered is a finite sum of very smooth and global basis functions,

$$u(x) = \sum_{k=0}^{N} \lambda_k \phi_k(x), \text{ in the case of static problems} \tag{1}$$

$$\text{or} \quad u(x,t) = \sum_{k=0}^{N} \lambda_k(t)\phi_k(x), \quad \text{for time-dependent problems} \tag{2}$$

where the basis functions $\phi_k(x)$ can be for example trigonometric functions or polynomials, such as Chebyshev polynomials. Then, you differentiate these functions exactly ([2]).

When using pseudospectral method, if you are given a set of grid points x_i and corresponding function values $u_i = u(x_i)$, you can use this data to approximate the derivative of u via *differentiation matrices.* Writing u as a column vector, you can find a square matrix D such that, at x_i, you have

$$u' = D.u. \tag{3}$$

Finding the derivative of a vector of data becomes a *matrix* $\times$ *vector* multiplication. We just need some manipulations to get to D.

You must evaluate (1) at the grid points x_i and get

$$u(x_i) = \sum_{k=0}^{N} \lambda_k \phi_k(x_i), \tag{4}$$

or in matrix-vector notation

$$u = A\lambda, \tag{5}$$

where λ is the column vector of the coefficients λ_k, matrix A has entries $A_{ik} = \phi_k(x_i)$, and u is as before. If you ensure that A is invertible, you get

$$\lambda = A^{-1}.u. \tag{6}$$

Recall that the invertibility of matrix A depends both on the basis function chosen and the location of the points x_i. For univariate polynomials with a set of distinct points invertibility is ensured.

On the other hand, differentiating both sides of (1), you get

$$\frac{d}{dx}u(x) = \sum_{k=1}^{N} \lambda_k \frac{d}{dx}\phi_k(x) \tag{7}$$

Evaluating at the grid points x_i we get in matrix-vector notation

$$u' = M\lambda, \tag{8}$$

where λ is the column vector of the coefficients λ_k, matrix M has entries $M_{ik} = \frac{d}{dx}\phi_k(x_i)$, and u is as before.

So, using (6) in (8), we obtain $u' = M.A^{-1}.u$ so that the differentiation matrix D we were looking for in (3) is

$$D = M.A^{-1}. \tag{9}$$

D_{ij} is the derivative of the jth curve at x_i.

In this paper we use (infinitely smooth) radial basis functions in a spectral framework. The basis function expansion $\phi_k(x)$ in (1) will take the form $\phi_k(x) = g(||x - x_k||, \epsilon) = g(r, \epsilon)$, chosen from a list more extense than the following ([14] among others), e.g.:

$g(r,\epsilon) = e^{-(\epsilon r)^2}$;	gaussian	(10)
$g(r,\epsilon) = 1/\sqrt{(1+(\epsilon r)^2)}$;	inverse multiquadric	(11)
$g(r,\epsilon) = e^{-\epsilon r}(15 + 15\epsilon r + 6(\epsilon r)^2 + (\epsilon r)^3)$;	cubic matérn	(12)
$g(r,\epsilon) = \sqrt{1+(\epsilon r)^2}$;	multiquadric	(13)
$g(r,\epsilon) = max(1-\epsilon r, 0)^8(32(\epsilon r)^3 + 25(\epsilon r)^2 + 8\epsilon r + 1)$	Wendland's $\varphi_{3,3}$	(14)

being r the (Euclidean) distance and ϵ a free parameter.

For the RBF-PS technique, matrix A in (5) and (9) has entries

$$A_{ij} = g(r_j, \epsilon)_{|x=x_i} = g(||x_i - x_j||, \epsilon). \tag{15}$$

Furthermore, the entries of M in (8) become $\frac{d}{dx}g(r,\epsilon)_{|x=x_i}$.

In all dynamic problems, depending on the nature of the problem, at each time t we can approximate $u(x,t)$ or $u(x,y,t)$ considering the *forward Euler method* or the *leap frog method*, for example:

$$\frac{\partial u}{\partial t} \approx \frac{u(x,t_{n+1}) - u(x,t_n)}{\Delta t} \tag{16}$$

$$\frac{\partial u}{\partial t} \approx \frac{u(x,t_{n+1}) - u(x,t_{n-1})}{2\Delta t} \tag{17}$$

$$\frac{\partial^2 u}{\partial t^2} = \frac{\partial}{\partial t}\left(\frac{\partial u}{\partial t}\right) \approx \frac{u(x,t_{n+1}) - 2u(x,t_n) + u(x,t_{n-1})}{(\Delta t)^2} \tag{18}$$

being $\Delta t = t_{n+1} - t_n$. This allows us to march in time. The key to the solution of the problems is the approximation of the spatial derivative, both in dynamic or static problems.

To solve a PDE like

$$\frac{\partial}{\partial t}u(x,t) + \frac{\partial}{\partial x}u(x,t) = 0 \tag{19}$$

using the rbf-spectral differentiation matrix D to express the spatial derivative and (16) leads to

$$u(x,t_{n+1}) = u(x,t_n) - \Delta t.D.u(x,t_n) \tag{20}$$

The procedure just described can be generalized to more complex linear differential operators. In this paper we are interested in those such as the ones involved in the following PDE's

$$\frac{\partial}{\partial t}u(x,t) = \frac{\partial^2}{\partial x^2}u(x,t) \tag{21}$$

$$\frac{\partial^2}{\partial t^2}u(x,t) = \frac{\partial^2}{\partial x^2}u(x,t) \tag{22}$$

If we use (16) for the time derivative and the second order rbf-spectral differential matrix D^2 for the spatial derivative, (21) leads to

$$u(x,t_{n+1}) = u(x,t_n) + \Delta t.D^2.u(x,t_n) \tag{23}$$

and if we use (18) for the time derivative and the second order rbf-spectral differential matrix D^2 for the spatial derivative, (22) leads to

$$u(x,t_{n+1}) = 2 * u(x,t_n) - u(x,t_{n-1}) + (\Delta t)^2 * D^2 * u(x,t_n); \tag{24}$$

where D^2 is defined as $D^2 = M^2.A^{-1}$ being $M^2_{ik} = \frac{d^2}{dx^2}g(||x_i - x_k||,\epsilon)$.

In general, D^p_{ij} is the pth derivative of curve number j at x_i:

$$D^p = M^p.A^{-1} \tag{25}$$

being $M^p_{ik} = \frac{d^p}{dx^p}g(||x_i - x_k||,\epsilon)$.

Table 1. Solution errors (%) for problem 3.1.1 with 11 Chebyshev points, $\Delta t = 0.001$, and Wendland C6 RBF

time	$E_{x0=1}$	E_{x1}	E_{x2}	E_{x3}	E_{x4}	E_{x5}
0.1	1.169e-1	1.167e-1	1.125e-1	9.735e-2	6.853e-2	3.477e-2
0.2	8.727e-2	8.716e-2	8.505e-2	7.722e-2	6.102e-2	3.898e-2
0.3	6.484e-2	6.478e-2	6.369e-2	5.959e-2	5.102e-2	3.909e-2
0.4	5.170e-2	5.166e-2	5.113e-2	4.914e-2	4.496e-2	3.913e-2
0.5	4.498e-2	4.496e-2	4.472e-2	4.379e-2	4.185e-2	3.917e-2

For the solution of a two-dimensional problem involving $\frac{\partial}{\partial x}u(x,y,t)$, $\frac{\partial}{\partial y}u(x,y,t)$, $\frac{\partial^2}{\partial x^2}u(x,y,t)$, $\frac{\partial^2}{\partial x \partial y}u(x,y,t)$, or $\frac{\partial^2}{\partial y^2}u(x,y,t)$ we can use the same approximation of the time derivative but we must use different differentiation matrix for the approximation of each spatial derivative. We consider the following approximations, e.g.:

$$\frac{\partial}{\partial x}u(x,y,t) \approx D_x.u(x,y,t) \qquad D_x = \frac{\partial}{\partial x}g(r,\epsilon)_{|(x,y)=(x_i,y_i)}.A^{-1} \tag{26}$$

$$\frac{\partial^2}{\partial x^2}u(x,y,t) \approx D_{xx}.u(x,y,t) \qquad D_{xx} = \frac{\partial^2}{\partial x^2}g(r,\epsilon)_{|(x,y)=(x_i,y_i)}.A^{-1} \tag{27}$$

$$\frac{\partial^2}{\partial x \partial y}u(x,y,t) \approx D_{xy}.u(x,y,t) \qquad D_{xy} = \frac{\partial^2}{\partial x \partial y}g(r,\epsilon)_{|(x,y)=(x_i,y_i)}.A^{-1} \tag{28}$$

(29)

being $g(r,\epsilon)$ the chosen RBF and A as in (15).

The question of the invertibility of the matrix A remains unsolved for some cases. Fasshauer presents detailed information on the subject in his book [14].

The optimization of the RBF shape parameter is the same used in [12, 13, 14] and a fairly detailed exposition is available in these references.

3. Numerical examples

3.1. One-dimensional problems

3.1.1. Initial-boundary-value problem 1

$$\begin{cases} \text{PDE} & \frac{\partial u}{\partial t} = \frac{\partial^2 u}{\partial x^2},\ 0 \le x \le 1,\ t \ge 0 \\ \text{BC} & \frac{\partial u}{\partial x}(0,t) = \frac{\partial u}{\partial x}(1,t) = 0,\ t \ge 0 \\ \text{IC} & u(x,0) = 9 + 3\cos(\pi x) + 5\cos(4\pi x),\ 0 \le x \le 1 \end{cases} \tag{30}$$

The PDE was implemented for the RBF-PS method as $u_{t+1} = u_t + \Delta t.D2.u_t$.

In tables 1-2 we present the results obtained with present method. We compare with the exact solution $u(x,t) = 9 + 3e^{-\pi^2 t}\cos(\pi x) + 5e^{-16\pi^2 t}\cos(4\pi x)$. Results are in good agreement, the biggest error being approximately 0.2%.

Using 11 Chebyshev points, $\Delta t = 0.001$, and Wendland C6 RBF, the optimized shape parameter is $\epsilon = 0.100064$. For the same RBF but using 21 Chebyshev points and $\Delta t = 0.0001$, we get $\epsilon = 0.160453$. If we use 11 Chebyshev points, $\Delta t = 0.001$, and the Matérn Cubic RBF, we obtain $\epsilon = 0.276603$. Using 21 Chebyshev points, the Matérn Cubic RBF, and $\Delta t = 0.0001$, we get $\epsilon = 1.250831$.

Using 21 Chebyshev points the error achieves a value smaller than $5 * 10^{-5}$%.

Table 2. Solution errors (%) for problem 3.1.1 with 11 Chebyshev points, $\Delta t = 0.001$, and Matérn Cubic RBF

time	$E_{x_0=1}$	E_{x_1}	E_{x_2}	E_{x_3}	E_{x_4}	E_{x_5}
0.1	1.175e-1	1.173e-1	1.132e-1	9.823e-2	6.971e-2	3.611e-2
0.2	8.809e-2	8.800e-2	8.594e-2	7.824e-2	6.225e-2	4.038e-2
0.3	6.590e-2	6.584e-2	6.477e-2	6.075e-2	5.229e-2	4.046e-2
0.4	5.286e-2	5.283e-2	5.231e-2	5.036e-2	4.625e-2	4.046e-2
0.5	4.618e-2	4.617e-2	4.593e-2	4.503e-2	4.313e-2	4.046e-2

3.1.2. Initial-boundary-value problem 2

$$\begin{cases} \text{PDE} & \frac{\partial^2 u}{\partial t^2} = \frac{\partial^2 u}{\partial x^2},\ 0 \le x \le \pi,\ t \ge 0 \\ \text{BC} & u(0,t) = u(\pi,t) = 0,\ t \ge 0 \\ \text{IC} & \begin{cases} u(x,0) = \pi x - x^2,\ 0 \le x \le \pi \\ \frac{\partial u}{\partial t}(x,0) = 0,\ 0 \le x \le \pi \end{cases} \end{cases} \tag{31}$$

To solve this problem we considered $\Delta t = 1.5625 * 10^{-5}$ and 81 Chebyshev points for $t \in [0, 4]$. Results obtained with the RBF-PS method, with an optimized shape parameter, both for Matérn Cubic and Wendland C6 RBF are in good agreement with the exact solution, which is $u(x,t) = \frac{8}{\pi}\sum_{odd\ n} n^{-3}\sin(nx)\cos(nt)$. The error is lower than 1% except for the boundary as the exact solution here is equal to zero for every t and around $t = 1.6$ which corresponds to values near zero for the exact solution.

The values for the shape parameter are quite different as we change the RBF: for the Matérn Cubic RBF the optimal shape parameter obtained was $\epsilon = 4.142739$ and for the Wendland C6 was $\epsilon = 0.962844$.

3.1.3. Transient analysis of a beam For the transient dynamic study of a beam in bending we are using the first-order shear deformation theory (FSDT) ([15]), with shear correction factor $K = 5/6$. When applied to beams, the equations of motion are

$$KGbh\left(\frac{\partial^2 w_0}{\partial x^2} + \frac{\partial \theta_x}{\partial x}\right) + bq = bI_0\frac{\partial^2 w_0}{\partial t^2} \tag{32}$$

$$EI\frac{\partial^2 \theta_x}{\partial x^2} - KGbh\left(\frac{\partial w_0}{\partial x} + \theta_x\right) = bI_2\frac{\partial^2 \theta_x}{\partial t^2} \tag{33}$$

Here, $w = w(x,t)$ is the transverse displacement, $\theta_x = \theta_x(x,t)$ is the rotation about the x axis, $K = 5/6$ is the shear correction coefficient, and q is the total transverse load. The remaining terms are obtained from given constants that characterize both the material properties and the structural properties of the beam.

We use the static solution of bending equilibrium as the initial conditions.

For the time-steping procedure we considered equations (32) and (33) divided by bI_0 and bI_2, respectively, and used the *forward Euler method.*

We consider an isotropic beam with both ends simply-supported and material properties $E = 10920$; $\rho = 1$; $\nu = 0.25$. The dimensions of the beam are $a = 1$, $b = 1$, $h = 0.1$, being a the length, h the thickness, and $b * h$ the cross section dimensions.

We used 11 Chebyshev points along the beam, $x \in [0, 1]$, the Cubic Matérn RBF (see (12)), and $\Delta t = 5 * 10^{-5}$. The optimal RBF shape parameter obtained was $\epsilon = 0.276603$.

In figure 1 we present the transverse displacement of the central point of the beam for $t \in [0, 1]$.

3.2. Transient analysis of a plate

Consider now an isotropic square plate in bending, clamped at all edges. Length of each side is $a = 2$ and side-over-thickness ratio is $a/h = 10$. The material properties are $\rho = 1$, $E = 10920$, and $\nu = 0.3$.

In the present study the First-Order Shear Deformation Theory (FSDT) is used ([15]). When applied to plates, the equations of motion are

$$D_{11}\frac{\partial^2\theta_x}{\partial x^2} + D_{16}\frac{\partial^2\theta_y}{\partial x^2} + (D_{12}+D_{66})\frac{\partial^2\theta_y}{\partial x\partial y} + 2D_{16}\frac{\partial^2\theta_x}{\partial x\partial y} + D_{66}\frac{\partial^2\theta_x}{\partial y^2} + D_{26}\frac{\partial^2\theta_y}{\partial y^2} + \\ - kA_{45}\left(\theta_y + \frac{\partial w}{\partial y}\right) - kA_{55}\left(\theta_x + \frac{\partial w}{\partial x}\right) = I_2\frac{\partial^2\theta_x}{\partial t^2} \quad (34)$$

$$D_{16}\frac{\partial^2\theta_x}{\partial x^2} + D_{66}\frac{\partial^2\theta_y}{\partial x^2} + (D_{12}+D_{66})\frac{\partial^2\theta_x}{\partial x\partial y} + 2D_{26}\frac{\partial^2\theta_y}{\partial x\partial y} + D_{26}\frac{\partial^2\theta_x}{\partial y^2} + D_{22}\frac{\partial^2\theta_y}{\partial y^2} + \\ - kA_{44}\left(\theta_y + \frac{\partial w}{\partial y}\right) - kA_{45}\left(\theta_x + \frac{\partial w}{\partial x}\right) = I_2\frac{\partial^2\theta_y}{\partial t^2} \quad (35)$$

$$\frac{\partial}{\partial x}\left[kA_{45}\left(\theta_y + \frac{\partial w}{\partial y}\right) + kA_{55}\left(\theta_x + \frac{\partial w}{\partial x}\right)\right] + \\ \frac{\partial}{\partial y}\left[kA_{44}\left(\theta_y + \frac{\partial w}{\partial y}\right) + kA_{45}\left(\theta_x + \frac{\partial w}{\partial x}\right)\right] + q = I_0\frac{\partial^2 w}{\partial t^2}, \quad (36)$$

where $w = w(x, y, t)$ is the transverse displacement, $\theta_x = \theta_x(x, y, t)$ and $\theta_y = \theta_y(x, y, t)$ are the rotations about the x and y axis, respectively, $K = 5/6$ is the shear correction coefficient, and q is the total transverse load.

The initial conditions are the static solution of bending equilibrium. A mesh of 81 points was used, corresponding to 9 equally spaced points per side, in [-1,1], and $\Delta t = 10^{-5}$. The RBF used was the Matérn Cubic (see (12)) and the optimal RBF shape parameter obtained was $\epsilon = 0.104817$.

In figure 2 we present the transverse displacement of the central point of the plate considered along the time $t \in [0, 1]$.

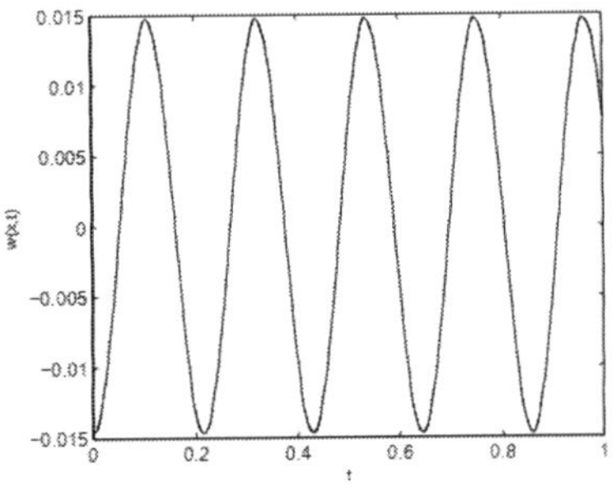

Figure 1. Central displacement of the beam

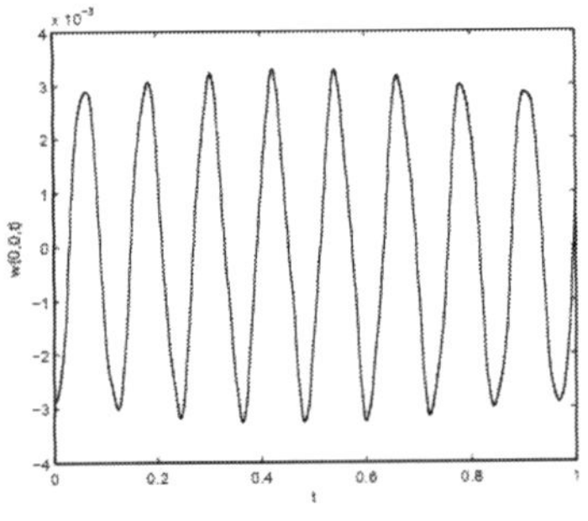

Figure 2. Central displacement of the plate

4. Conclusions

This paper addresses the solution of several PDE problems using a technique that combines the radial basis function (RBF) collocation technique and the pseudospectal (PS) method with the optimization of the RBF shape parameter. This allows the extension of the accurate results to complex geometries, keeping it simple to implement.

Several numerical tests were performed using some radial basis functions, boundary conditions, and initial conditions, for both one and two-dimensional problems. We extended previous work to the transient analysis of a beam and a plate.

Results obtained demonstrate that the method produces good results which are in good agreement with exact solutions or references considered.

Further studies, including the application of the method to composite structures and with more complex geometries are to be made.

References

[1] Trefethen L N 2000 *Spectral Methods in Matlab* (Philadelphia: SIAM)
[2] Fornberg B 1998 *A Practical Guide to Pseudospectral Methods* (Cambridge: Cambridge University Press)
[3] Kansa E J 1990 *Computers & mathematics with applications*, **19(8-9)** 127–145
[4] Kansa E J 1990 *Computers & mathematics with applications*, **19(8-9)** 147–161
[5] Larsson E and Fornberg B 2003 *Computers & Mathematics with Applications*, **46 (5-6)** 891–902
[6] Fasshauer G E 1997 *Proc. of the 3rd Int. Conf. on Curves and Surfaces* **2** 131–138
[7] Sarra S A 2005 *Applied Numerical Mathematics* **54 (11)** 79–94.
[8] Ferreira A J M, Martins P A L S and Roque C M C 2005 *Journal of Sound and Vibration* **280** 595–610
[9] Ferreira A J M and Fasshauer G E 2007 *Composite Structures* **79** 202–210
[10] Ferreira A J M and Fasshauer G E 2006 *Computer Methods in Applied Mechanics and Engineering* **196** 134–146
[11] Fasshauer G E 2005 *Boundary Elements XXVII* 47–56
[12] Ferreira A J M and Fasshauer G E 2007 *Advances in Meshfree Techniques* 283–310
[13] Ferreira A J M, Fasshauer G E, Batra R C and Rodrigues J D 2008 *Composite Structures* **86** 328–343
[14] Fasshauer G E 2007 *Meshfree Approximation Methods with Matlab* (Singapore: World Scientific Publishers)
[15] Reddy J N 1997 *Mechanics of laminated composite plates* (New York: CRC Press)

Volterra Methods for Constructing Structural Dynamic Observables for Nonlinear Systems: an Extended Calculation.

G. Manson, K Worden

Dynamics Research Group, Department of Mechanical Engineering, University of Sheffield, Mappin St, Sheffield S1 3JD

graeme.manson@sheffield.ac.uk, k.worden@sheffield.ac.uk

Abstract. Although a great deal of work has been carried out on structural dynamic systems under random excitation, there has been a comparatively small amount of this work concentrating on the calculation of the quantities commonly measured in structural dynamic tests. Among the existing work, the Volterra series, a means of predicting nonlinear system response for weakly nonlinear systems, has allowed the computation of various measurable quantities of interest for structural dynamics, including: auto- and cross-spectra, FRFs, coherences and higher-order spectra. These calculations are quite intensive and are typically only possible using computer algebra. A previous calculation by the authors for the coherence for a Duffing oscillator yielded results which showed some qualitatitive disagreement with numerical simulation; the object of the current paper is simply to extend the calculation in order to see if better agreement can be achieved.

1. Introduction

The Volterra series is well-known as a means of predicting nonlinear system response for weakly nonlinear systems. An attractive feature of the series is that it allows the calculation of closed-form expressions (albeit approximations based on the truncated series) for various quantities of interest to structural dynamicists; these include a number of objects commonly encountered in dynamic testing. The research of the current authors began with [1], in which the Frequency Response Function (FRF) for a Single-Degree-of-Freedom (SDOF) system – the randomly excited Duffing oscillator – was computed up to a certain order of approximation i.e. to a series order quadratic in P, the white noise intensity. The results were interesting; the computed FRF showed good qualitative agreement with an FRF calculated by numerical simulation in that the 'resonance' frequency increased with increasing levels of excitation. The approximation also shed light on the known observation that the Hilbert transform applied to a nonlinear system FRF under random excitation does not produce distortions. The analysis was extended to a Multi-DOF system in [2]. The calculations were very demanding and could only be carried out with the aid of computer algebra. In [3], the Volterra methodology was applied to the approximate computation of the coherence function, an often-used tool in structural dynamics for diagnosing nonlinearity. In this case, a Padé or rational approximation in P was used rather than the polynomial expansions used in [1] and [2]. Two systems were considered in [3]: a symmetric Duffing oscillator with cubic stiffness only and an asymmetric oscillator with quadratic and cubic terms. In the case of the asymmetric oscillator with quadratic term dominant, a (1,1) approximation (ratio of linear functions) gave good quantitative and qualitative agreement with the results of numerical solution. However, even when a (2,2) approximation (ratio of quadratics) was computed for the symmetric oscillator, the expression under-predicted the results of numerical simulation and also failed to show some of the appropriate qualitative behaviour. The objective of the current paper is simply to extend the analysis of the coherence to a higher order in order to see if closer agreement with numerical simulation can be obtained. Recent developments in the use of the Volterra series in structural dynamics include [4] and [5] where analytical approximations to the bispectrum are obtained and applied.

The layout of the paper is as follows. Section Two reviews the necessary background on the Volterra series in general, and in the particular case of the SDOF Duffing oscillator. Section Three discusses how the Padé approximations are formed from the various cross-spectra involved and also describes the extended approximation for the symmetric case of the Duffing system. The paper concludes with some discussion in Section Four.

2. FRFs and Cross-Spectra

The coherence for a linear system is constructed as the ratio of certain cross-spectra and auto-spectra, or alternatively in terms of the FRF and the auto-spectra. Before proceeding to the construction of the coherence function for a nonlinear system in terms of a Volterra series expansion, it is useful to recall the definitions of the FRFs and cross-spectra for such systems in terms of the Volterra series.

It is well-known that many nonlinear systems or input-output processes $x(t) \longrightarrow y(t)$ can be realised as a mapping [6,7],

$$y(t) = y_1(t) + y_2(t) + y_3(t) + \ldots + y_n(t) + \ldots \tag{1}$$

where

$$y_n(t) = \int_{-\infty}^{+\infty} \ldots \int_{-\infty}^{+\infty} d\tau_1 \ldots d\tau_n \; h_n(\tau_1, \ldots, \tau_n) x(t-\tau_1) \ldots x(t-\tau_n) \tag{2}$$

Among the conditions for such a series to apply are time-invariance of the system of interest and polynomial nature for the nonlinearity.

This is the *Volterra series* and the functions h_n are the *Volterra kernels*. The dual frequency-domain representation is based on the *Higher-order FRFs* (HFRFs) or *Volterra kernel transforms*, $H_n(\omega_1, \ldots, \omega_n)$, $n = 1, \ldots, \infty$, which are defined as the multi-dimensional Fourier transforms of the kernels.

$$H_n(\omega_1, \ldots, \omega_n) = \int_{-\infty}^{+\infty} \ldots \int_{-\infty}^{+\infty} d\tau_1 \ldots d\tau_n \; h_n(\tau_1, \ldots, \tau_n) e^{-i(\omega_1\tau_1 + \ldots + \omega_n\tau_n)} \tag{3}$$

The definition of the FRF of a linear system based on the input/output cross-spectrum, $S_{yx}(\omega)$, and input autospectrum, $S_{xx}(\omega)$, is also well-known,

$$H(\omega) = H_1(\omega) = \frac{S_{yx}(\omega)}{S_{xx}(\omega)} \tag{4}$$

where the spectral quantities are defined as expectations i.e. $S_{yx}(\omega) = E[\overline{Y}(\omega) X(\omega)]$. The expectations are obtained by averaging discrete or fast Fourier transforms from neighbouring time segments.

The *composite FRF* $\Lambda_r(\omega)$, of a nonlinear system under random excitation, is defined similarly,

$$\Lambda_r(\omega) = \frac{S_{yx}(\omega)}{S_{xx}(\omega)} \tag{5}$$

Using the Volterra series representation in (1) results in the expression

$$\Lambda_r(\omega) = \frac{S_{y_1x}(\omega) + S_{y_2x}(\omega) + \ldots + S_{y_nx}(\omega) + \ldots}{S_{xx}(\omega)} \tag{6}$$

$\Lambda_r(\omega)$ was approximated in [1] by obtaining expressions for the various cross-spectra between the input and the individual output components. The general term used was,

$$S_{y_{2n-1}x}(\omega) = \frac{(2n)!S_{xx}(\omega)}{n!\,2^n(2\pi)^{n-1}} \int_{-\infty}^{+\infty} \ldots \int_{-\infty}^{+\infty} d\omega_1 \ldots d\omega_{n-1} \times$$

$$H_{2n-1}(\omega_1, -\omega_1, \ldots, \omega_{n-1}, -\omega_{n-1}, \omega) S_{xx}(\omega_1) \ldots S_{xx}(\omega_{n-1}) \tag{7}$$

Now, given that the input autospectrum is constant over all frequencies for a white noise input (i.e. $S_{xx}(\omega) = P$), the composite FRF for random excitation follows. Substituting the constant S_{xx} into (7) gives

$$\Lambda_r(\omega) = \sum_{n=1}^{n=\infty} \frac{(2n)!P^{n-1}}{n!\,2^n(2\pi)^{n-1}} \int_{-\infty}^{+\infty} \ldots \int_{-\infty}^{+\infty} d\omega_1 \ldots d\omega_{n-1} H_{2n-1}(\omega_1, -\omega_1, \ldots, \omega_{n-1}, -\omega_{n-1}, \omega) \tag{8}$$

The system of interest here is going to be the basic symmetrical Duffing oscillator with equation of motion,

$$m\ddot{y} + c\dot{y} + ky + k_3y^3 = x(t) \tag{9}$$

where m is the mass and c is the damping constant. k and k_3 are the linear and cubic stiffnesses respectively. This system is time-invariant as required for the application of the Volterra series.

The relevant higher-order FRFs of this system for the forthcoming analysis are:

$$H_1(\omega) = \frac{1}{-m\omega^2 + ic\omega + k} \tag{10}$$

$$H_3(\omega_1, \omega_2, \omega_3) = -k_3 H_1(\omega_1) H_1(\omega_2) H_1(\omega_3) H_1(\omega_1 + \omega_2 + \omega_3) \tag{11}$$

$$H_5(\omega_1, \omega_2, \omega_3, \omega_4, \omega_5) =$$

$$\frac{3}{10} k_3^2 H_1(\omega_1 + \omega_2 + \omega_3 + \omega_4 + \omega_5) H_1(\omega_1) H_1(\omega_2) H_1(\omega_3) H_1(\omega_4) H_1(\omega_5) \times$$

$$\{H_1(\omega_1 + \omega_2 + \omega_3) + H_1(\omega_1 + \omega_2 + \omega_4) + H_1(\omega_1 + \omega_2 + \omega_5) + H_1(\omega_1 + \omega_3 + \omega_4) + H_1(\omega_1 + \omega_3 + \omega_5) +$$

$$H_1(\omega_1 + \omega_4 + \omega_5) + H_1(\omega_2 + \omega_3 + \omega_4) + H_1(\omega_2 + \omega_3 + \omega_5) + H_1(\omega_2 + \omega_4 + \omega_5) + H_1(\omega_3 + \omega_4 + \omega_5)\} \tag{12}$$

or, alternatively,

$$H_5(\omega_1,\omega_2,\omega_3,\omega_4,\omega_5) =$$
$$\frac{3}{10}k_3^2 H_1(\omega_1+\omega_2+\omega_3+\omega_4+\omega_5)H_1(\omega_1)H_1(\omega_2)H_1(\omega_3)H_1(\omega_4)H_1(\omega_5)\times$$
$$\{H_1(\omega_1+\omega_2+\omega_3)+\text{ 9 more combinations of three frequencies from five}\} \tag{13}$$

which were all needed for the calculation in [3]. Finally, the extended calculation requires,

$$H_7(\omega_1,\omega_2,\omega_3,\omega_4,\omega_5,\omega_6,\omega_7) =$$
$$-\frac{3}{70}k_3^3 H_1(\omega_1+\omega_2+\omega_3+\omega_4+\omega_5+\omega_6+\omega_7)\times$$
$$H_1(\omega_1)H_1(\omega_2)H_1(\omega_3)H_1(\omega_4)H_1(\omega_5)H_1(\omega_6)H_1(\omega_7)\times$$
$$\{H_1(\omega_1+\omega_2+\omega_3+\omega_4+\omega_5)H_1(\omega_1+\omega_2+\omega_3)+\text{ 219 other } H_1(.)H_1(.)\text{ products}$$
$$+H_1(\omega_1+\omega_2+\omega_3)H_1(\omega_4+\omega_5+\omega_6)+\text{69 other } H_1(.)H_1(.)\text{ products}\} \tag{14}$$

The last term shows that the extended calculation here will be many times more demanding than those previously carried out. For the symmetric Duffing oscillator, all even-order kernels and their transforms are easily shown to vanish. The first stage in the calculation is to form the higher-order Padé approximation for the coherence.

3. Padé Approximations for the Coherence

The well-known expression for the coherence of a linear system is,

$$\gamma^2(\omega) = \frac{S_{yx}(\omega)S_{xy}(\omega)}{S_{xx}(\omega)S_{yy}(\omega)} = \frac{|H(\omega)|^2 S_{xx}(\omega)}{S_{yy}(\omega)} \tag{15}$$

where the interpretation is that the coherence is the ratio of the power in the output *linearly correlated with the measured input* divided by the power in the output. For a linear system, one expects this ratio to be unity. It is a textbook calculation [6], to show that the coherence is always less than unity if measurement noise is present on the input or output. It is also well-known that the coherence falls below unity if the system in question is nonlinear, although this belief is considerably less well-supported theoretically.

Assuming that the coherence is computed from the auto and cross-spectra as in the linear case, the expression in the nonlinear case is simply,

$$\gamma^2(\omega) = \frac{S_{yx}(\omega)S_{xy}(\omega)}{S_{xx}(\omega)S_{yy}(\omega)} = \frac{|\Lambda_r(\omega)|^2 S_{xx}(\omega)}{S_{yy}(\omega)} \tag{16}$$

with reference to equation (15). In terms of the individual Volterra components of the signal y, this becomes,

$$\gamma^2(\omega) = \frac{(S_{y_1x}(\omega)+S_{y_3x}(\omega)+S_{y_5x}(\omega)+\ldots)(S_{xy_1}(\omega)+S_{xy_3}(\omega)+S_{xy_5}(\omega)+\ldots)}{S_{xx}(\omega)(S_{y_1y_1}(\omega)+S_{y_1y_3}(\omega)+S_{y_3y_1}(\omega)+S_{y_2y_2}\ldots)} \tag{17}$$

and this expression takes into account the fact that terms of the form $S_{y_n x}(\omega)$ can be shown to vanish identically when n is even as do terms $S_{y_m y_n}(\omega)$ where $m+n$ is odd.
In terms of the dependence on the input level P, (with a slight abuse of notation) this looks like,

$$\gamma^2(\omega) = \frac{(O(P)+O(P^2)+O(P^3)+\ldots)(O(P)+O(P^2)+O(P^3)+\ldots)}{P(O(P)+O(P^2)+O(P^2)+O(P^3)\ldots)} \tag{18}$$

so both the numerator and denominator are expanded in powers of P. This means that an approximation to the coherence will have the general Padé form,

$$\gamma^2_{(m,n)} = \frac{\sum_{i=0}^{m} a_i P^i}{\sum_{i=0}^{n} b_i P^i} \tag{19}$$

where m and n are the orders of approximation of the numerator and denominator respectively. For the purposes of this paper it will be assumed that $m = n$.

To order (0,0), the expression for the coherence is,

$$\gamma^2_{(0,0)}(\omega) = \frac{S_{y_1 x}(\omega) S_{x y_1}(\omega)}{P S_{y_1 y_1}(\omega)} = \frac{P S_{y_1 y_1}(\omega)}{P S_{y_1 y_1}(\omega)} = 1 \tag{20}$$

on using the first lemma of Appendix A in [3]. This shows that to this order the coherence is unaffected by the nonlinearity. To order (1,1), one obtains from equation (17) (suppressing the argument ω),

$$\gamma^2_{(1,1)} = \frac{S_{y_1 x} S_{x y_1} + S_{y_1 x} S_{x y_3} + S_{y_3 x} S_{x y_1}}{P(S_{y_1 y_1} + S_{y_1 y_3} + S_{y_3 y_1} + S_{y_2 y_2})} = \frac{S_{y_1 x} S_{x y_1} + 2\mathrm{Re} S_{y_1 x} S_{x y_3}}{P(S_{y_1 y_1} + 2\mathrm{Re} S_{y_1 y_3} + S_{y_2 y_2})} \tag{21}$$

and using the first lemma, from Appendix A in [3], this becomes,

$$\gamma^2_{(1,1)} = \frac{S_{y_1 x} S_{x y_1} + 2\mathrm{Re} S_{y_1 x} S_{x y_3}}{S_{y_1 x} S_{x y_1} + 2\mathrm{Re} S_{y_1 x} S_{x y_3} + P S_{y_2 y_2}} = \frac{A_{11}}{A_{11} + B_{11}} \tag{22}$$

where,

$$A_{11} = S_{y_1 x} S_{x y_1} + 2\mathrm{Re} S_{y_1 x} S_{x y_3} \tag{23}$$

and,

$$B_{11} = P S_{y_2 y_2} \tag{24}$$

One can immediately observe that if the Duffing oscillator is symmetric, then $y_2 = 0$, $S_{y_2 y_2} = B_{11} = 0$ and the expression for the coherence to this order collapses to $\gamma^2_{(1,1)} = 1$. So for the symmetric system, the coherence is still insensitive to the nonlinearity to this order. If the oscillator is asymmetric, one would expect some distortion to occur and this is shown in [3]. In the symmetric case, higher-order approximations are required. The (2,2) Padé approximation will be considered next,

$$\gamma^2_{(2,2)} = \frac{S_{y_1x}S_{xy_1} + S_{y_3x}S_{xy_3} + 2\mathrm{Re}S_{y_1x}S_{xy_3} + 2\mathrm{Re}S_{y_1x}S_{xy_5}}{P(S_{y_1y_1} + S_{y_1y_3} + S_{y_3y_1} + S_{y_3y_3} + S_{y_1y_5} + S_{y_5y_1})} \tag{25}$$

Analysis, based on the appendices in [3], eventually yields,

$$\gamma^2_{(2,2)} = \frac{A_{22}}{A_{22} + B_{22}} \tag{26}$$

where,

$$A_{22} = |S_{y_1x}|^2 + |S_{y_3x}|^2 + 2\mathrm{Re}S_{y_1x}S_{xy_3} + 2\mathrm{Re}S_{y_1x}S_{xy_5} \tag{27}$$

and,

$$B_{22} = \frac{6P^4}{(2\pi)^2}\int_{-\infty}^{\infty}\int_{-\infty}^{\infty} d\omega_1 d\omega_2\,|H_3(\omega_1,\omega_2,-\omega-\omega_1-\omega_2|^2 \tag{28}$$

This shows that to (2,2) order, the coherence for the symmetric Duffing oscillator finally shows some sensitivity to the nonlinearity. The relevant integrals for the (2,2) Padé approximation to the coherence were computed in [3]. For coefficient values of $m=1$, $c=20$, $k=10^4$ and $k_3=5\times10^9$, the results for $P=$ 0.005, 0.01 and 0.02 are shown in Figure 1.

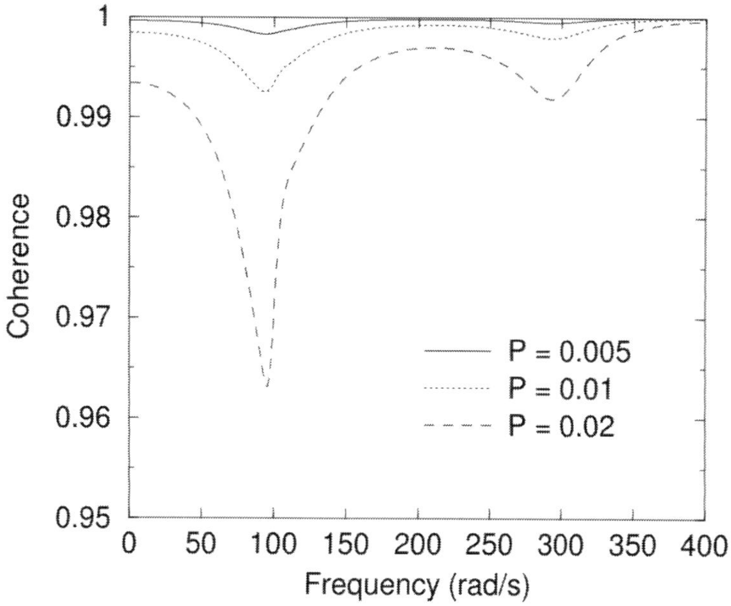

Figure 1.Coherence from (2,2) Padé approximation for symmetric Duffing oscillator.

A comparison with numerical simulation is made in Figure 2; the equation of motion (9) was integrated numerically using a Fourth-order Runga-Kutta scheme. The coefficient values were the same as in the Padé approximation. The excitation was a Gaussian white noise sequence with RMS values chosen to correspond to the P values above. This in itself was not trivial as the Gaussian sequence was filtered into the interval [0,200] Hz in order to ensure correct behaviour of the Runge-Kutta routine. This lowered the RMS from the required value and the final simulation used a corrected RMS to account for this. 190000 points of data for each case were taken and the coherence was estimated from a 512-line FFT; this gave 185 averages. It is seen that the Volterra approximation considerably under-predicts the coherence distortion. The Padé approximation generates some of the correct qualitative features, i.e. is less than unity in the neighbourhood of the natural frequency and its third harmonic. However, it fails to predict the shift upwards in these dips which is shown in the numerical results. This was assumed to be a higher-order effect.

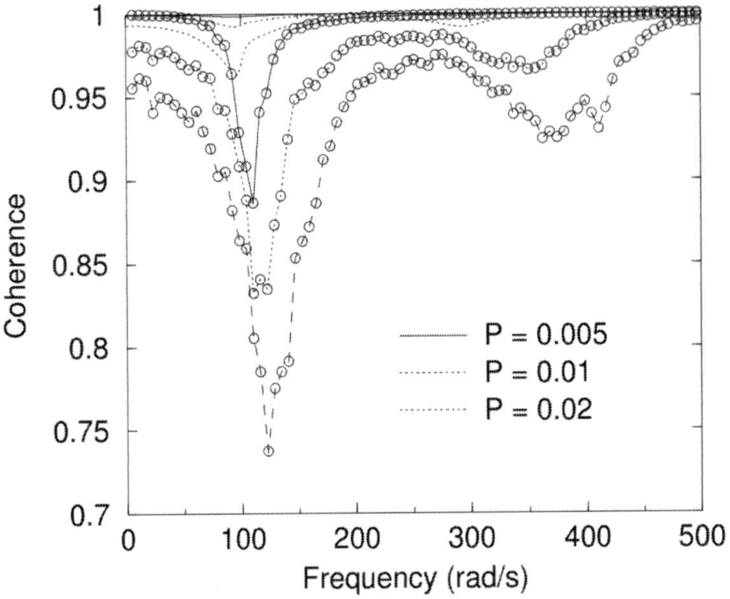

Figure 2.Coherence from (2,2) Padé approximation for symmetric Duffing oscillator. Comparison with numerical simulation (lines with circles).

The (3,3) approximation demands considerably more work. For the sake of brevity, only the final results will be given here. As before, the approximation takes the form,

$$\gamma^2_{(3,3)} = \frac{A_{33}}{A_{33} + B_{33}} \tag{29}$$

where,

$$\begin{aligned} A_{33}(\omega) = 1 &- \frac{3k_3 P}{\pi}\mathrm{Re}\{H_1(\omega) I_1\} + \frac{9k_3^2 P^2}{4\pi^2}\mathrm{Re}\{|H_1(\omega)|^2 I_1^2 + 2H_1(\omega)^2 I_1^2 + 4H_1(\omega)[I_1 I_2 + I_3]\} \\ &- \frac{27k_3^3 P^3}{4\pi^3}\mathrm{Re}\{H_1(\omega)^3 I_1^3 + |H_1(\omega)|^2 [2I_1^2 I_2 + H_1(\omega) I_1^3 + I_1 I_3] + 4H_1(\omega)^2 [I_1^2 I_2 + I_1 I_3] \\ &+ H_1(\omega)[2I_1^2 I_4 + I_1^2 I_5 + 4I_1 I_2^2 + 4I_1 I_6 + 4I_1 I_7 + 4I_8 + 2I_1 I_9 + 8I_{10} + 4I_{11} + 4I_{12} + \frac{2}{3} I_{13}]\} \end{aligned} \tag{30}$$

and,

$$B_{33}(\omega) = \frac{3k_3^2 P^2}{2\pi^2} I_{14} - \frac{9k_3^3 P^3}{2\pi^3}\mathrm{Re}\{|H_1(\omega)|^2 I_1 I_{14} + 2I_1 I_{15} + 4I_{16} + I_1 I_{17} + 2I_{18}\} \tag{31}$$

The integrals I_1 to I_{18} are listed in the Appendix. Each integral was computed using the calculus of residues within the Maple computer algebra system [8]. The closed-form expressions for the integrals are far too long to give here, and in any case, do not appear to yield a great deal of insight into the problem.

The (3,3) approximation for the coherence is shown in Figure 3 for various values of P.

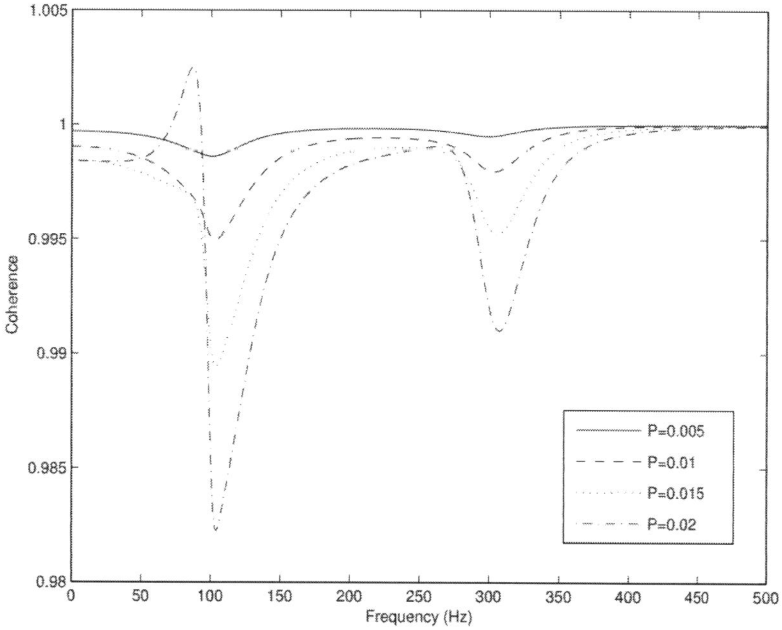

Figure 3. Coherence from (3,3) Padé approximation for symmetric Duffing oscillator.

The results are interesting, if a little disappointing in some respects. The maximum predicted drop in the coherence (at $P = 0.02$) is now of the order of 0.2, a decrease on the (2,2) approximation, which gave about 0.4. The higher-order result is actually moving away from the result of the numerical simulation (Figure 2). A positive result of the calculation is that the higher-order approximation is now predicting an upward frequency shift in the coherence minima with increasing white-noise intensity. However, this shift is markedly lower than that observed in the numerical simulation. One aspect of the approximation which requires comment is the fact that the predicted coherence goes above unity, something which cannot occur; this is presumably due to the fact that the series is truncated, one assumes that still higher-order terms would correct this effect.

4. Discussion and Conclusions

The results of the paper are summarised in the transition between Figures 1 and 3. In Figure 1, the (2,2) Padé approximation to the coherence function for a randomly-excited Duffing oscillator is given, as computed in a previous paper. The calculation has been extended to give the (3,3) Padé approximation here, and the result is given in Figure 3. The (2,2) approximation under-predicted the drops in coherence observed from numerical simulations at the resonance frequency and third-harmonic. In defiance of expectations, the (3,3) approximation under-predicts more severely. One possible explanation for this is that the Volterra series is taking an alternating form, converging on its ultimate limit through a sequence of under and over-estimates. If this is the case, then the limit is presumably somewhere between the envelope provided by the (2,2) and (3,3) approximations and will itself be an under-estimate of the observed (simulated) coherence. While disappointing, this is not entirely unexpected as the calculation for the FRF in [1] showed such an underestimate of frequency shift with increasing excitation that one might conclude that the Volterra series only reproduces qualitative behaviour with confidence. In qualitative terms, the current calculation may be considered a success as the new approximation to the coherence does reproduce the shifts in the minima which appear in the numerical simulation. However, like the frequency shifts in the FRF, the shifts in the coherence are under-estimated by the Volterra approximation.

The current calculation is the most demanding that the authors have so far attempted, and the apparent conclusion seems to be that qualitative prediction accuracy with the Volterra series is rather elusive. However, it should be noted that the approximate coherence for the asymmetric Duffing oscillator in [3] did give good quantitative agreement with numerical simulation. This suggests that any immediate further work in this area should be directed towards trying to systematically determine the circumstances under which the Volterra series might be expected to give quantitative predictions rather than pursuing even higher-order calculations.

References

[1] Worden K and Manson G 1998 Random vibrations of a Duffing oscillator using the Volterra series *Journal of Sound and Vibration* **217** 781-789.

[2] Worden K and Manson G 1999 Random vibrations of a multi degree-of-freedom nonlinear system using the Volterra series. *Journal of Sound and Vibration* **226** pp.397-405.

[3] Worden K and Manson G 2005 A Volterra series approximation to the coherence of the Duffing oscillator. *Journal of Sound and Vibration* **286** pp.529-547.

[4] Marzocca P, Nichols JM, Milanese A, Seaver M and Trickey ST 2008 Second-order spectra for quadratic nonlinear systems by Volterra functional analysis: analytical description and numerical simulation. *Mechanical Systems and Signal Processing* **22** pp.1882-1895.

[5] Nichols JM, Marzocca P and Milanese A 2008 On the use of the auto-bispectral density for detecting quadratic nonlinearity in structural systems. *Journal of Sound and Vibration* **312** pp.726-735.

[6] Volterra V 1959 *Theory of Functionals and Integral Equations*. Dover Publications, New York.

[7] Schetzen M 1980 *The Volterra and Wiener Theories of Nonlinear Systems*. John Wiley Interscience Publication. New York.

[8] *Maple 12 - software for algebraic computing*. Maplesoft Ltd. 2009.

Appendix A.

(The limits of all integrals here are from $-\infty$ to ∞.)

$$I_1 = \int d\omega \, |H_1(\omega)|^2$$
$$I_2 = \int d\omega H_1(\omega)^2 H_1(-\omega)$$
$$I_3(\omega) = \iint d\omega_1 d\omega_2 H_1(\omega+\omega_1+\omega_2) \, |H_1(\omega_1)|^2 |H_1(\omega_2)|^2$$
$$I_4 = \int d\omega H_1(\omega)^3 H_1(-\omega)$$

$$I_5 = \int d\omega | H_1(\omega) |^4$$

$$I_6(\omega) = \iint d\omega_1 d\omega_2 H_1(\omega + \omega_1 + \omega_2) | H_1(\omega_1) |^2 H_1(\omega_1) | H_1(\omega_2) |^2$$

$$I_7(\omega) = \iint d\omega_1 d\omega_2 H_1(\omega + \omega_1 + \omega_2) | H_1(\omega_1) |^2 H_1(-\omega_1) | H_1(\omega_2) |^2$$

$$I_8 = \iiint d\omega_1 d\omega_2 d\omega_3 H_1(\omega_1 + \omega_2 + \omega_3) | H_1(\omega_1) |^2 H_1(\omega_1) | H_1(\omega_2) |^2 | H_1(\omega_3) |^2$$

$$I_9(\omega) = \iint d\omega_1 d\omega_2 H_1(\omega + \omega_1 + \omega_2)^2 | H_1(\omega_1) |^2 | H_1(\omega_2) |^2$$

$$I_{10}(\omega) = \iiint d\omega_1 d\omega_2 d\omega_3 H_1(\omega + \omega_1 + \omega_2) H_1(\omega + \omega_1 + \omega_3) | H_1(\omega_1) |^2 | H_1(\omega_2) |^2 | H_1(\omega_3) |^2$$

$$I_{11}(\omega) = \iiint d\omega_1 d\omega_2 d\omega_3 H_1(\omega + \omega_1 + \omega_2) H_1(\omega_1 + \omega_2 + \omega_3) | H_1(\omega_1) |^2 | H_1(\omega_2) |^2 | H_1(\omega_3) |^2$$

$$I_{12}(\omega) = \iiint d\omega_1 d\omega_2 d\omega_3 H_1(\omega + \omega_1 + \omega_2) H_1(-\omega_1 - \omega_2 + \omega_3) | H_1(\omega_1) |^2 | H_1(\omega_2) |^2 | H_1(\omega_3) |^2$$

$$I_{13} = \iiint d\omega_1 d\omega_2 d\omega_3 | H_1(\omega_1 + \omega_2 + \omega_3) |^2 | H_1(\omega_1) |^2 | H_1(\omega_2) |^2 | H_1(\omega_3) |^2$$

$$I_{14}(\omega) = \iint d\omega_1 d\omega_2 | H_1(\omega - \omega_1 - \omega_2) |^2 | H_1(\omega_1) |^2 | H_1(\omega_2) |^2$$

$$I_{15}(\omega) = \iint d\omega_1 d\omega_2 | H_1(\omega - \omega_1 - \omega_2) |^2 | H_1(\omega_1) |^2 H_1(\omega_1) | H_1(\omega_2) |^2$$

$$I_{16}(\omega) = \iiint d\omega_1 d\omega_2 d\omega_3 H_1(\omega + \omega_1 - \omega_2) | H_1(\omega - \omega_2 - \omega_3) |^2 | H_1(\omega_1) |^2 | H_1(\omega_2) |^2 | H_1(\omega_3) |^2$$

$$I_{17}(\omega) = \iint d\omega_1 d\omega_2 | H_1(\omega - \omega_1 - \omega_2) |^2 H_1(\omega - \omega_1 - \omega_2) | H_1(\omega_1) |^2 | H_1(\omega_2) |^2$$

$$I_{18}(\omega) = \iiint d\omega_1 d\omega_2 d\omega_3 H_1(\omega_1 + \omega_2 + \omega_3) | H_1(\omega - \omega_2 - \omega_3) |^2 | H_1(\omega_1) |^2 | H_1(\omega_2) |^2 | H_1(\omega_3) |^2$$

Journal of Physics: Conference Series **181** (2009) 012055 doi:10.1088/1742-6596/181/1/012055

The new numerically-analytic method for integrating the multiscale thermo elastoviscoplasticity equations with internal variables

Vladimir Kukudzhanov

Ishlinsky Institute for Problems in Mechanics of the Russian Academy of Sciences, prospekt Vernadskogo 101-1, Moscow, Russia

E-mail: kukudz@ipmnet.ru

Abstract. Integration of the constitutive equations of ductile fracture models is analyzed in this paper. The splitting method is applied to the Gurson's and Kukudzhanov's models. The analysis of validity of this method is done. It was shown that Kukudzhanov's model describes a large variety of materials since it involves residual stress and viscosity.

1. Introduction

Finite strain problems, such as metal forming, where plastic strains are by two to three orders of magnitude larger than elastic strains, involve both the problems: choosing of the appropriate material model and integration scheme. In this paper we consider a material model for void-containing ductile solids that has been developed by Gurson [3], and micromechanical model of fracture that was suggested by Kukudzhanov [4, 6]. A brief description of these models is given in Section 2. The distinctive feature of the constitutive equations of the plasticity theory is that in addition to differential relations, these equations involve also a finite relation that constrains the stress tensor invariants. Owing to this, various mathematical methods can be used. Here we use the splitting method, proposed by Kukudzhanov [5]. This method is based on splitting of the constitutive relations with respect to physical processes. In the paper a numerical example of a uni-axial tension test on a square element is considered to illustrate the basic model features.

2. Elastoviscoplastic constitutive relations

2.1. General formulation of the elastoviscoplastic constitutive relations

We assume the strain-rate additivity

$$\dot{\varepsilon} = \dot{\varepsilon}^{el} + \dot{\varepsilon}^{pl}, \tag{1}$$

For the case of linear elasticity

$$\dot{\sigma} = \mathbf{D} : \dot{\varepsilon}^{el}, \tag{2}$$

where $\mathbf{D}$ is the fourth-order elasticity tensor. In the following, we assume linear isotropic elasticity, so that

$$D_{ijkl} = 2G\,\delta_{ik}\delta_{jl} + \left(K - \tfrac{2}{3}G\right)\delta_{ij}\delta_{kl},$$

where G and K are the elastic shear and bulk moduli respectively, and δ_{ij} is the Kronecker delta.

The yield function involves the first and second invariants of the stress tensor and is given by

$$\Phi(p, q, H) = 0, \tag{3}$$

where $p = -\frac{1}{3}\boldsymbol{\sigma} : \mathbf{I}$ is the hydrostatic stress, $q = \left(\frac{3}{2}\mathbf{s} : \mathbf{s}\right)^{1/2}$ is the equivalent stress and H_i, $i = 1, 2, \ldots, n$, is a set of internal state variables, $\mathbf{I}$ is the second order identity tensor and $\mathbf{s}$ is the stress deviator. The function Φ is defined such that when $\Phi < 0$, the response is purely elastic. The flow rule is written as

$$\dot{\varepsilon}^{pl} = \dot{\lambda}\frac{d\Phi}{d\boldsymbol{\sigma}}, \tag{4}$$

where λ is a positive scalar. Using (3), the flow rule becomes

$$\dot{\varepsilon}^{pl} = \dot{\lambda}\left(-\frac{1}{3}\frac{\partial\Phi}{\partial p}\mathbf{I} + \frac{\partial\Phi}{\partial q}\mathbf{n}\right), \tag{5}$$

where $\mathbf{n} = \frac{3}{2q}\mathbf{s}$. According to [1] $\mathbf{n}$ can be determined as

$$\mathbf{n} = \frac{3}{2q^{el}}\mathbf{s}^{el}.$$

Using this notation, the stress tensor can be written as

$$\boldsymbol{\sigma} = -p\mathbf{I} + \frac{2}{3}q\mathbf{n}. \tag{6}$$

The plasticity model is completed by describing the evolution of the state variables

$$\dot{H}_i = h_i(\dot{\varepsilon}, \boldsymbol{\sigma}, H_j). \tag{7}$$

2.2. *Gurson's plasticity model*

We briefly discuss the material model for void-containing ductile solids. Based on a rigid-plastic upper-bound solution for spherically symmetric deformations around a single spherical void, Gurson [3] proposed the yield condition

$$\Phi = \left(\frac{q}{\sigma_y}\right)^2 + 2q_1 f \cosh\left(-\frac{3}{2}\frac{q_2 p}{\sigma_y}\right), - \left(1 + q_3 f^2\right) \tag{8}$$

where σ_y is the equivalent tensile flow stress representing the actual microscopic stress state in the matrix material, and f is the current void volume fraction. The parameters q_1, q_2 and q_3 were introduced by Tvergaard [7] to make the predictions of Gurson's equations agree with numerical studies of the materials containing periodically distributed circular cylindrical voids. The yield surface given by Equation (8) becomes that of von Mises (which is assumed in Gurson's model to hold for the matrix material) when $f = 0$.

In Gurson's model the microscopic equivalent plastic strain $\bar{\varepsilon}_m^{pl}$, to vary is assumed according to the equivalent plastic work expression

$$(1 - f)\sigma_y\, d\bar{\varepsilon}_m^{pl} = \boldsymbol{\sigma} : d\varepsilon^{pl}. \tag{9}$$

The change in void volume fraction during an increment of deformation is partly due to the growth of existing voids and partly due to the nucleation of new voids

$$df = df_{nuc} + df_{gr}. \tag{10}$$

The matrix is assumed to satisfy the plastic incompressibility condition but, because of the existence of voids, the macroscopic response does not. The change of the void volume fraction due to growth is related to the change of the total volume as

$$df_{gr} = (1-f)d\varepsilon^{pl} : \mathbf{I}. \tag{11}$$

In this model, voids nucleation is considered to be controlled by plastic strain

$$df_{nuc} = A\, d\bar{\varepsilon}^{pl}_{m}. \tag{12}$$

As suggested by Chu and Needleman [2] the parameter A is chosen so that the nucleation strain follows a normal distribution with mean value ε_N and standard deviation s_N:

$$A = A(\bar{\varepsilon}^{pl}_{m}) = \frac{f_N}{s_N\sqrt{2\pi}} \exp\left(-\frac{1}{2}\left(\frac{\bar{\varepsilon}^{pl}_{m} - \varepsilon_N}{s_N}\right)^2\right), \tag{13}$$

where f_N is the volume fraction of void nucleating particles.

2.3. *Micromechanical model of damage and fracture*

We briefly discuss the micromechanical model of ductile fracture, as suggested by Kukudzhanov [4, 6]. The model is based on the phenomenological theory of dislocations and micro-defect development. According to this model, the plastic deformation and damage is a single process caused by the motion of dislocations and also by generation and development of microdefects. The deformation process is assumed to consist of two stages.

In the *first stage*, the material is subjected to plastic deformation due to the motion of dislocations in crystal grains of the material. Part of these dislocations that have accumulated at the grain boundaries, leads to residual microstresses (it is assumed that the deviatoric components of the residual stress tensor are proportional to the tensor of accumulated flux of dislocations). On the macrolevel, it leads to an elastoviscoplastic flow with simultaneous hardening of the material. The stress tensor consists of active and residual parts

$$\boldsymbol{\sigma} = \boldsymbol{\sigma}^{a} + \mathbf{s}^{r}. \tag{14}$$

Active stress is determined by using the associated flow rule

$$\dot{\varepsilon}^{pl} = \dot{\lambda}\frac{d\Phi}{d\boldsymbol{\sigma}^{a}} \tag{15}$$

and the von Mises type flow condition for elastoviscoplastic material with *kinematic* and *isotropic hardening*

$$\begin{aligned} \Phi &= q^{a} - \sigma_y = 0, \\ \sigma_y &= \left(1 + 3\mu_1\bar{\varepsilon}^{pl}\right)^{N} + \psi(\tau_1\dot{\bar{\varepsilon}}^{pl}), \end{aligned} \tag{16}$$

where $q^{a} = \sqrt{\frac{3}{2}\mathbf{s}^{a} : \mathbf{s}^{a}}$ is the equivalent active stress; $\bar{\varepsilon}^{pl} = \sqrt{\frac{2}{3}\mathbf{e}^{pl} : \mathbf{e}^{pl}}$ is the equivalent plastic strain; function determines the influence of strain rate on the yield stress, $\psi(0) = 0$. For residual stress we have the well-known law of kinematic hardening

$$\mathbf{s}^{r} = 2\alpha\mathbf{e}^{pl}. \tag{17}$$

The *second stage* begins after the intensity of the dislocation flux accumulated at the boundaries of grains attains a critical level, beyond which the process of annihilation of dislocations begins (it is assumed that the flux of annihilation of dislocations is proportional to the amount of dislocations accumulated at the inclusions). This process is accompanied by disclinations of grains with the formation of voids between them. On the macrolevel, this process leads to relaxation of the internal stresses and softening of the material. The appearance of voids changes the behavior of the material essentially. In this case the material consists of the matrix and the voids, i.e., it is a two-phase material with the flow condition for active stress

$$\Phi = \left(\frac{q^a}{\sigma_y}\right) + 2q_1 f \cosh\left(\frac{3}{2}\frac{q_2 p^a}{\sigma_y}\right) - \left(1 + q_3 f^2\right) = 0. \tag{18}$$

Voids nucleates once the equivalent residual stress q^r achieves the critical value q^r_c. The volume of nucleated voids is considered to be a small disturbance. The relaxation of residual stress is described by equation

$$\dot{\mathbf{s}}^r + 2\mu_2 \frac{Q\left(q^r - q^r_c\right)}{\tau_2 q^r}\mathbf{s}^r = 2\alpha\dot{\mathbf{e}}^{pl}, \tag{19}$$

function Q, that characterizes the relaxation, is supposed to be a linear function of its argument, τ_2 is relaxation time of residual stresses ($\tau_2 \gg \tau_1$).

3. Splitting method

Splitting method for the integration of equations governing the behavior of elastoviscoplastic media was proposed by Kukudzhanov [5]. The idea of this method is based on the fact that an additive operator can be replaced by a multiplicative operator on a small time interval.

It was shown that for the classical elastoplastic medium independent of the time scale, the splitting leads to an algebraic powerlaw equation for the correction coefficient of the elastic solution. For elastoviscoplastic media, the splitting leads to a differential equation for the correction coefficients, which in this case are functions of time. The general scheme of splitting looks as follows. We decompose the equation

$$\dot{\boldsymbol{\sigma}} = \mathbf{D} : (\dot{\varepsilon} - \dot{\varepsilon}^{pl}). \tag{20}$$

The predictor is taken at $\dot{\varepsilon}^{pl} = 0$, which corresponds to the elastic material. In this case, for each step Δt, one should solve the elastic problem

$$\dot{\boldsymbol{\sigma}} = \mathbf{D} : \dot{\varepsilon} \tag{21}$$

subjected to the initial conditions obtained at the previous step for the complete elastoplastic problem.

The corrector is taken at $\dot{\varepsilon} = 0$ in equation (20). Using equation (4), equations (20) and (7) becomes

$$\dot{\boldsymbol{\sigma}} = -\dot{\lambda}\mathbf{D}\frac{\partial\Phi}{\partial\boldsymbol{\sigma}}, \quad \dot{H}_i = h_i(-\dot{\lambda}\frac{\partial\Phi}{\partial\boldsymbol{\sigma}}, \boldsymbol{\sigma}, H_j). \tag{22}$$

The second relation in (22) provides the evolutionary equations for the internal parameters that characterize the structure of the material, hardening, damage, porosity and other properties.

For the classical (equilibrium) elastoplastic medium, the properties of which are independent of the change in the time scale, one can eliminate the time t from equations (22) and proceed to the variable λ, to obtain

$$d\boldsymbol{\sigma} = -d\lambda\mathbf{D}\frac{\partial\Phi}{\partial\boldsymbol{\sigma}}, \quad dH_i = \tilde{h}_i(-d\lambda\frac{\partial\Phi}{\partial\boldsymbol{\sigma}}, \boldsymbol{\sigma}, H_j). \tag{23}$$

Solving equations (23) subjected to the initial conditions resulting from the solution of the elastic problem $\boldsymbol{\sigma}(\lambda_0) = \boldsymbol{\sigma}^{el}$, $H_i(\lambda^n) = H_i^n$, we find the variables $\boldsymbol{\sigma}^{n+1}$ and H_i^{n+1} as functions of λ, $\boldsymbol{\sigma}^{el}$, H_i^{el}

$$\boldsymbol{\sigma} = \boldsymbol{\sigma}(\lambda, \boldsymbol{\sigma}^{el}, H_i^{el}), \quad H_i = H_i(\lambda, \boldsymbol{\sigma}^{el}, H_j) \quad \text{for} \quad \lambda = \lambda^{n+1}. \tag{24}$$

Substituting these relations into the plasticity condition (3), we obtain the equation

$$\Phi(p(\lambda), q(\lambda), H_i(\lambda), \lambda) = 0. \tag{25}$$

Solving this equation and substituting the resulting λ into (24) we obtain the final solution of the problem.

For the elastoviscoplastic medium, the plasticity condition is of differential type. In this case, equation (3) for λ leads to the differential equation

$$\Phi(p(\lambda), q(\lambda), H_i(\lambda), \lambda, \dot{\lambda}) = 0. \tag{26}$$

This splitting scheme is stable, if the predictor scheme for the solution of the elastic problem is stable and there exist solutions of equations (25) and (26).

4. Integration of the constitutive relations of GTN model by the splitting method

For simplicity we assume that there in no nucleation of voids, i.e. $df_{nuc} = 0$, and that the yield stress is constant, $\sigma_y = \sigma_{y0} = const$. During the constitutive calculations, where stresses and state variables are updated, the total strain e is known. The aim of this integration is to find stresses and state variables at the end of the time increment $t + \Delta t$.

At the predictor step one has to solve the elastic problem.

$$\dot{p}^{el} = -K\dot{\varepsilon} : \mathbf{I}, \quad \dot{q}^{el} = 2G\dot{\varepsilon} : \mathbf{n}. \tag{27}$$

We integrate these equations taking the solutions at the previous step t as the initial conditions. The result can be written as

$$p^{el} = p_t + \Delta p^{el}, \quad q^{el} = q_t + \Delta q^{el}, \tag{28}$$

At the correction step, the first equation in can be written as

$$\frac{d\sigma}{d\lambda} = -K\frac{\partial\Phi}{\partial p}\mathbf{I} - 2G\frac{\partial\Phi}{\partial q}\mathbf{n}. \tag{29}$$

Projecting this equation onto $\mathbf{I}$ and $\mathbf{n}$, and using equation (6), we find

$$\frac{dp}{d\lambda} = -K\frac{\partial\Phi}{\partial p}, \qquad \frac{dq}{d\lambda} = -6Gq, \qquad \frac{df}{d\lambda} = -(1-f)\frac{d\Phi}{dp}. \tag{30}$$

Initial conditions for equations (30) are taken from the solution of the elastic problem at the predictor step: $p(\lambda_0) = p_0 = p^{el}$, $q(\lambda_0) = q_0 = q^{el}$, $f(\lambda_0) = f_0 = f_t$.

Integrating this equation with initial conditions, we find

$$dp = \frac{K}{(1-f)}df, \qquad f = 1 - (1 - f_0)\exp\left(-\frac{p - p_0}{K}\right). \tag{31}$$

Eliminating $d\lambda$ from equations (30), we find

$$\frac{dp}{dq} = \frac{K}{6G}\frac{1}{q}\frac{\partial\Phi(p,q)}{\partial p} \quad \text{for} \quad \Phi(p,q) = \varphi(p)\psi(q). \tag{32}$$

The problem of integrating the elastoplastic equations reduces to the solution of one differential equation and one non-linear algebraic equation

$$\frac{dp}{dq} = \alpha\frac{1}{q}\psi(p), \quad p(q_0) = p_0, \tag{33}$$

$$\Phi(p,q) = q^2 + \phi(p) = 0, \tag{34}$$

where $\alpha = \frac{K}{6G}$, $\psi(p) = \frac{\partial\Phi}{\partial p}$, $\phi(p) = 2q_1 f(p)\cosh\left(\frac{3}{2}q_2 p\right) - (1 + q_3 f^2(p))$.

Writing this equation in the incremental form, we obtain a non-linear algebraic equation. Thus, the problem reduces to the solution of the set of two non-linear equations

The computational advantages of the splitting method, as compared with standard methods, are achieved due to at the correction step some equations can be integrated analytically.

5. Integration of the constitutive relations of the micromechanical model

We use the splitting method to integrate the constitutive equations of plasticity by Mises with kinematical hardening.

Relaxation equations (15) and (16) give

$$\dot{\varepsilon}^{pl} = \dot{\lambda}\frac{3}{2q^\alpha}\mathbf{s}^\alpha = \dot{\Lambda}\mathbf{s}^\alpha, \quad \dot{\Lambda} = \frac{3}{2q^\alpha}\dot{\lambda}. \tag{35}$$

Equation (23) for stresses relaxation can be written as

$$d\mathbf{s}^\alpha = -2G\mathbf{s}^\alpha d\Lambda. \tag{36}$$

Integrating this equation with initial condition $\Lambda = \Lambda_0$, $\mathbf{s}^\alpha(\Lambda_0) = \mathbf{s}^{el}$, we find

$$\mathbf{s}^\alpha = \mathbf{s}^{el}X, q^\alpha = q^{el}X, \tag{37}$$

where $X = \exp(-2G(\Lambda - \Lambda_0))$ is the correction coefficient of the elastic solution.

Using equations (35) and (37), and taking into account that $\dot{\varepsilon}^{pl} = \dot{\mathbf{e}}^{pl}$ one can find

$$\dot{\bar{\varepsilon}}^{pl} = \sqrt{\frac{2}{3}\dot{\mathbf{e}}^{pl} : \dot{\mathbf{e}}^{pl}} = \frac{2}{3}\dot{\Lambda}q^{el}l^{-2G(\Lambda-\Lambda_0)} = -\frac{1}{3G}q^{el}\dot{X}. \tag{38}$$

Integrating this equation we obtain

$$\bar{\varepsilon}^{pl} = \bar{\varepsilon}^{pl}_t - \frac{1}{3G}q^{el}(X-1), \tag{39}$$

Substituting the values $\bar{\varepsilon}^{pl}$ and $\dot{\bar{\varepsilon}}^{pl}$ in the matrix flow condition (16), we have

$$\Phi = q^{el}X - \left[1 + 3\mu_1\left(\bar{\varepsilon}^{pl}_t - \frac{1}{3G}q^{el}(X-1)\right)\right]^N - \psi\left(-\frac{1}{3G}q^{el}\tau_1\dot{X}\right) = 0, \tag{40}$$

where $\psi(x) = \alpha\tanh(x)$.

To solve this differential equation we approximate $\dot{X} = \frac{X-1}{\Delta t}$ and obtain a non-linear equation for X which is solved by means of Newton's method. Using equation (17), we find the equivalent residual stress

$$q^r = 3\alpha\bar{\varepsilon}^{pl}. \tag{41}$$

In the *first stage* of deformation, the stress tensor can be written as

$$\boldsymbol{\sigma} = -p^\alpha\mathbf{I} + \frac{2}{3}(q^\alpha + q^r)\mathbf{n}, \tag{42}$$

where $p^\alpha = -K(\varepsilon : \mathbf{I})$ is the elastic pressure.

To integrate the constitutive relations at the *second stage* of deformation, after nucleation of voids, we generalized the numerical method [1] for the rate-dependent plasticity.

Flow rule (15) in the incremental form at the end of the increment is written

$$\Delta\varepsilon^{pl} = \frac{1}{3}\Delta\varepsilon_p\mathbf{I} + \Delta\varepsilon_q\mathbf{n}, \quad \Delta\varepsilon_p = -\Delta\lambda\frac{\partial\Phi}{\partial p^\alpha}, \quad \Delta\varepsilon_q = -\Delta\lambda\frac{\partial\Phi}{\partial q^\alpha}. \tag{43}$$

$$\Delta\varepsilon_p\frac{\partial\Phi}{\partial q^\alpha} + \Delta\varepsilon_q\frac{\partial\Phi}{\partial p^\alpha} = 0. \tag{44}$$

We can rewrite equation (2) in incremental form

$$\boldsymbol{\sigma}^\alpha = \boldsymbol{\sigma}^{el} - K\Delta\varepsilon_p\mathbf{I} - 2G\Delta\varepsilon_q\mathbf{n}. \tag{45}$$

Projecting the elasticity equation (45) onto $\mathbf{I}$ and $\mathbf{n}$, and using equation (6), we find

$$p^\alpha = p^{el} + K\Delta\varepsilon_p, \qquad q^\alpha = q^{el} - 3G\Delta\varepsilon_q, \tag{46}$$

where $p^{el} = p_t + \Delta p^{el}$, $q^{el} = q_t + \Delta q^{el}$.

Taking into account that $f = f_t + \Delta f$, we can write the expression for porosity

$$f = \frac{f_t + \Delta\varepsilon_p}{1 + \Delta\varepsilon_p}. \tag{47}$$

For an equivalent plastic strain in the matrix material, we can write

$$\Delta\bar{\varepsilon}^{pl}_m = \frac{-\Delta\varepsilon_p p^\alpha + \Delta\varepsilon_q q^\alpha}{(1-f)\sigma_y(\bar{\varepsilon}^{pl}_m)}, \tag{48}$$

$$\sigma_y = (1 + 3\mu_1\bar{\varepsilon}^{pl}_m)^N + \alpha\tanh(\Delta\bar{\varepsilon}^{pl}_m\tau_1/\Delta t), \tag{49}$$

where $\bar{\varepsilon}^{pl}_m = (\bar{\varepsilon}^{pl}_m)_t + \Delta\bar{\varepsilon}^{pl}_m$.

Equations (48) and (49) are solved with given values of $\Delta\varepsilon_p$ and $\Delta\varepsilon_q$.

The problem of integrating the elastoviscoplastic equations reduces to the solution of the set of equations: (18), (44), (46) – (49). These equations are solved using Newton's method.

Projecting equation (19) onto $\mathbf{n}$ and taking into account that $\mathbf{n}$ is collinear with s^r/q^r and $\dot{\mathbf{e}}^{pl} = \dot{\varepsilon}_q\mathbf{n}$, we find

$$\dot{q}^r + \frac{2\mu_2}{\tau_2}Q\left(q^r - q^r_f\right) = 3\alpha\dot{\varepsilon_q}. \tag{50}$$

We use the backward Euler scheme to integrate this equation. Taking into account that Q is supposed to be a linear function of its argument, we can write

$$\Delta q^r + 2\mu_2\frac{\Delta t}{\tau_2}\left(q^r - q^r_f\right) = 3\alpha\Delta\varepsilon_q, \tag{51}$$

where we suppose that $2\mu_2 = G$. Solving this equation, we find

$$q^r = q^r_f + \left(q^r_0 - q^r_f\right)/(1+y), \tag{52}$$

where $q^r_0 = q^r_t + 3\alpha\Delta\varepsilon_q$, $= G\frac{\Delta t}{\tau_2}$.

Residual stress tensor is written in the form

$$\mathbf{s}^r = \frac{2}{3}q^r\mathbf{n}. \tag{53}$$

Finally, the solution of the problem is determined by equations (14), (45) and (48).

6. Numerical example

The results of the finite element analysis of a micromechanical model are represented in Figure 1. The elastic-plastic properties of the matrix material common for all calculations are specified by $\overset{\circ}{A} = 500$, $\nu = 0.3$, $\sigma_{y_0} = 1$, $\mu_1 = 5$, $N = 0.4$, $a = 1$, $q_1 = 1.5$, $q_2 = 0.2$, $q_3 = 2.25$, $q_f^r = 0.01$; for (a), (b) and (c): $a = 1$, $q_c^r = 0.05$; for (a) and (b): $r_2 = 30$; for (c): $n = 5$; for (d): $a = 5$, $q_c^r = 0.3$, $\tau_1 = 1$, $\tau_2 = 10$.

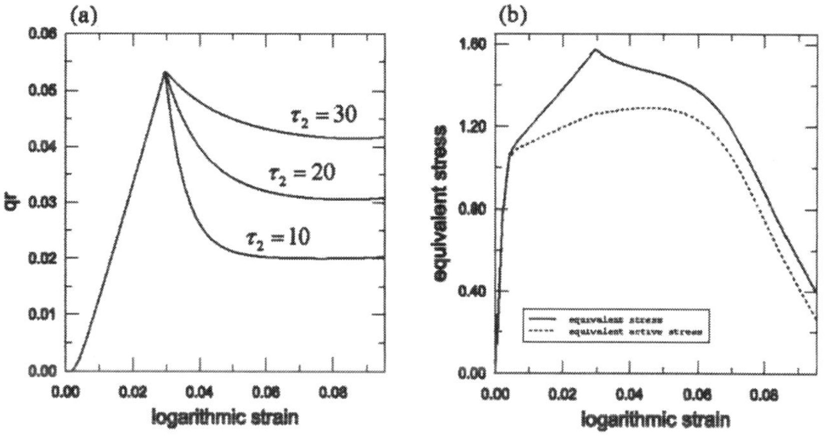

Figure 1.

Figure 1 a shows that smaller stain rate leads to larger voids growth, i.e. material damage occurs sooner in smaller strain rates. Figure 1 b illustrates a correlation between the active and full equivalent stress. The micromechanical model describes a large variety of material properties.

7. Conclusions

Application of the splitting method to the Gurson's and Kukudzhanov's models has been demonstrated in this paper. It was shown that the computational advantages of splitting method (as compared with standard methods) are achieved due to at the correction step, some equations can be integrated analytically. Constitutive relations of the micromechanical model, based on the phenomenological theory of dislocations, were integrated using splitting and backward Euler methods. It was shown that this model describes a large variety of materials as it involves kinematic hardening with residual stress relaxation after voids nucleation and viscosity.

References

[1] Aravas N 1987 On the Numerical Integration of a Class of Pressure-Dependent Plasticity Models *Intern. J. Numer. Methods in Eng-ing* **24** (7) 1395–1416
[2] Chu C C and Needleman A 1980 Void nucleation effects in biaxially stretched sheets *J. Eng-ing Mater. Technol.* **102** 249–256
[3] Gurson A L 1977 Continuum theory of ductile rupture by void nucleation and growth: Part I - Yield criteria and flow rules for porous ductile materials *J. Eng-ing Mater. Technol.* **99** 2–15
[4] Kukudzhanov V N 1999 Micromechanical model of fracture of an inelastic material and its application to the investigation of strain localization *Mech. Solids* **34** (5) 58–69
[5] Kukudzhanov V N 2004 Decomposition method for elastoplastic equations *Mech. Solids* **39** (1) 73–80
[6] Kukudzhanov V N 2006 Coupled models of elastoplasticity and damage and their. integration *Mechanics of Solids* **41** (6) 83–109
[7] Tvergaard V 1981 Influence of voids on shear band instabilities under plane strain condition *Int. J. Fracture* **17** (4) 389–407

Journal of Physics: Conference Series **181** (2009) 012056 doi:10.1088/1742-6596/181/1/012056

Low dimensional models for stick-slip vibration of drill-strings

Marcos Silveira and Marian Wiercigroch

Centre for Applied Dynamics Research, School of Engineering, University of Aberdeen, Aberdeen, AB24 3UE, UK

E-mail: m.silveira@abdn.ac.uk, m.wiercigroch@abdn.ac.uk

Abstract. Effective reduction of drill-string vibration is still a major problem in drilling industry and therefore robust predictive tools need to be developed. In this paper we study two low dimensional nonlinear models. The first is a 1-DOF torsional model of the botom-hole assembly (BHA). The second model is a 3-DOF torsional system having in addition to the BHA a rotary table, which allows simulation of interactions for which there is experimental evidence. Three different friction models with increasing levels of complexity are applied to determine their influence in the dynamical responses. Comparison between the dynamic responses for three friction models shows that the dangerous stick-slip limit-cycles do not change qualitatively. Simulations show that, if appropriately controlled, large amplitude stick-slip limit-cycles can change to small amplitude limit-cycles in Model 2. In Model 1, with constant velocity of the rotary table, it goes from a large amplitude stick-slip limit-cycle to a fixed point. Bifurcation diagrams confirm the existence of a set of parameters in which the system operates without stick-slip vibration.

1. Introduction

Drilling a borehole for oil and gas extraction is done by means of rotary motion of a drill-bit against the rock. The rotary motion is transmitted to the bit from a motor, usually at the surface of the well, by the use of a drill-string. This drill-string is made of tube sections with threaded connections. It can reach lengths of various kilometers, making it a very slender structure. Torsional vibration is one of the most important vibration modes in drill-string, and the main phenomenon related to it is stick-slip vibrations.

The lower part of the drill-string, comprised of drill-collars, stabilizers, vibration absorvers, MWD tools and the drill-bit is called the bottom-hole assembly (BHA). The dynamic behaviour of the BHA is greatly responsible for the most harmful vibrations, represented by the torsional, axial and lateral modes and the coupling mechanisms between them. The drill-string vibrations can either be induced by drill-bit—formation and drill-string—bore-hole interactions, or be self-induced [1]. Although the coupling mechanisms are highly nonlinear, the study of each of these modes sepparately is important to stablish the relevance and influence that they have over the full dynamics of the system. The complexity of the drill-string dynamics comes in part from the coupling mechanisms, which converts energy from the different vibration modes. The drill-bit, which relates the axial force to torque is an important player in these mechanisms.

For this study, low dimensional models are presented that focus on torsional vibrations of the drill-string. Although not being able to reproduce fully the complex dynamics of the drilling system, they permit the study of the main characterisitcs of the stick-slip vibration effect. The assumption that the drill-string behaves as a torsional spring, meaning it has a dominant fundamental mode, is valid for a high ratio of mass moment of inertia of the BHA relative to the drill-string [5, 6].

2. Modelling stick-slip vibrations

2.1. *Friction models for bit-rock interaction*

The torque-on-bit (T_b) is a friction law, a function that defines the resistive torque on the bit and models the interaction between the drill-bit and the rock. It can be defined in various ways, following the form:

$$T_{b_1} = \begin{cases} T_{st} & if \quad \dot{\phi}_1 = 0, \\ T_{sl} & if \quad \dot{\phi}_1 \neq 0. \end{cases} \tag{1}$$

The simplest model for T_b is a piecewise Coulomb-like friction law (Friction Model 1), with a value for static friction (T_{st}) and another, lower, value for sliding (or dynamic) friction (T_{sl}), which in this case is independent of the velocity $\dot{\phi}_1$. More realistic friction laws account for dependency on the velocity and weight-on-bit (WOB), and are usually discontinuous at null velocity. A comprehensive survey of various friction models is given by Wojewoda et al. [2].

A more complex friction model (Friction Model 2) accounts for dependency of the dynamic coefficients, μ_0 and μ_1, on the contact velocity, and is given by [3]:

$$T_{b_2} = sgn\left(\dot{\phi}_1\right) W_b \left(\frac{\mu_o - \mu_1}{1 + \lambda_1 \left|\dot{\phi}_1\right|} + \mu_1 + \lambda_2 \dot{\phi}_1^2 \right), \tag{2}$$

where λ_1 and λ_2 are dry friction constants and W_b is the WOB.

A third friction model (Friction Model 3) incorporates an exponential dependency on the contact velocity, allowing better accuracy near null velocity, and is given by [4]:

$$T_{b_3} = T_{sl} + (T_{st} - T_{sl})\, e^{-\left|\dot{\phi}_l/\omega_{sl}\right|^{\delta_{sl}}} + b_l \left|\omega_l\right|, \tag{3}$$

where T_{st} and T_{sl} are the static and dynamic friction torques and ω_{sl}, δ_{sl} and b_l are dry friction constants.

2.2. *1-DOF torsional pendulum model - Model* 1

As a first approximation, the drill-string can be modelled as a 1-DOF torsional pendulum. In this arrangement, the equivalent inertia of the system is represented by a mass at the bottom, the drill-pipes are represented by a torsional spring, the equivalent damping acting on the system is represented by a torsional damper, corresponding to the lower portion of figure 1. Also as a first approximation, the angular velocity of the rotary table is constant, as the amplitude of its oscillations are considerably smaller than the amplitude of oscillations of the velocity of the bit. The equation of motion for this system is:

$$J_1 \ddot{\phi}_1 + c_1 \dot{\phi}_1 + k_1 (\phi_1 - \phi_2) - T_b = 0, \tag{4}$$

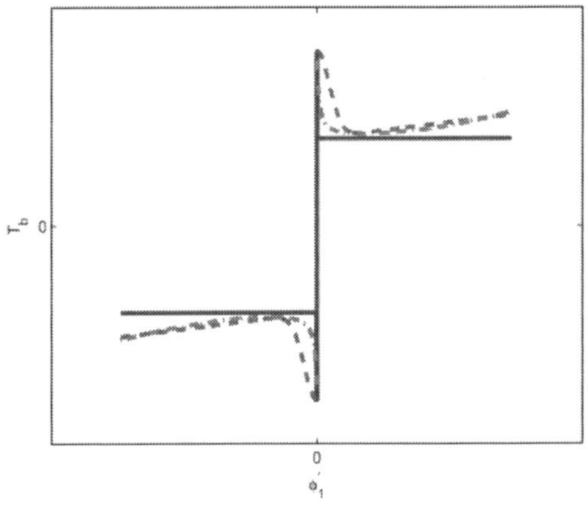

Figure 1. Friction models showing dependency of friction torque on rotary velocity (T_{b_1} -blue, T_{b_2} -.red, T_{b_3} - -green)

where J_1 is the equivalent mass moment of inertia of the drill-string, c_1 is the equivalent damping coefficient along the drill-string, k_1 is the equivalent torsional stiffness of the drill-pipes, ϕ_1 is the angular position of the bit at the bottom of the drill-string, ϕ_2 is the angular position of the rotary table at the top of the drill-string, T_b is the torque-on-bit and a dot denotes differentiation over time.

In this problem, the interest is in the relative displacement between the top and the bottom, as the torque stored in the spring (representing the drill-pipes) depends on this quantity. With this in mind, the relative angular displacement between the rotary table and the bit is defined as:

$$\phi = \phi_2 - \phi_1, \tag{5}$$

then (4) can be written as:

$$J_1\ddot{\phi} + c_1\dot{\phi} + k_1\phi - T_b = 0. \tag{6}$$

2.3. 1-DOF torsional pendulum with parametric excitation of the angular velocity of the rotary table – Model 2

As observed in field data [5–7], the velocity of the rotary table oscillates around the nominal velocity when the bit is experiencing stick-slip oscillations. The oscillations of the rotary table have smaller amplitude and higher frequency than the stick-slip oscillations. One simple manner of introducing this oscillation in the model is to apply a parametric excitation in the velocity of the rotary table, in the form of:

$$\phi_2 = \Omega_2 t + A cos\left(\omega t\right), \tag{7}$$

where A is the amplitude of oscillations and ω is the frequency of oscillations. It is possible then to tune the values of A and ω to emulate the behaviour of the velocity of the rotary table. It is clear that nulling these two parameters makes this model similar to Model 1.

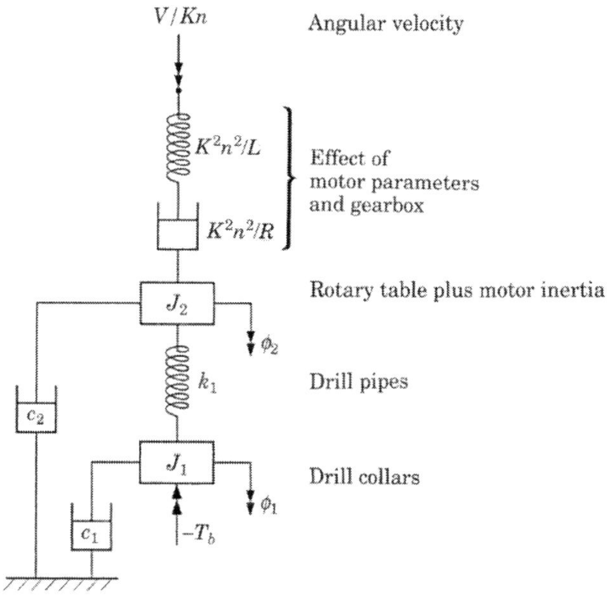

Figure 2. Schematic representation of the drill-string system [5]

2.4. 3-DOF torsional pendulum model – Model 3

In order to include the dynamics of the drive system, a 3-DOF electromechanical model is used. In this model are included the characteristics of the motor and the rotary table. It is assumed that the connection between the rotary table and the motor is stiff. The equations of motion for this system are:

$$\begin{aligned} &J_1\ddot{\phi} + c_1\dot{\phi} + k_1\phi - T_b = 0, \\ &J_2\ddot{\phi} + c_2\dot{\phi} + k_1\phi - KnI = 0, \\ &L\dot{I} + Kn\dot{\phi} + RI - V = 0, \end{aligned} \tag{8}$$

where J_2 is the equivalent mass moment of inertia of the drive system, c_2 is the equivalent damping coefficient of the drive system, I is the armature current, K is the electric motor constant, L is the armature inductance, n is the combined gear ratio of transmission, R is the armature resistance, V is the armature voltage and a dot denotes differentiation over time. Other parameters are as in the 1-DOF model.

3. Numerical results

3.1. Simulating discontinuous systems

The models considered are highly non-linear, exhibiting relaxation vibrations, and are also discontinuous because of the friction models used. Therefore, the investigations were performed with numerical simulations of the differential equations, using a 4th order Runge-Kutta solver in conjunction with the bisection method to improve accuracy.

Extensive work has been done in simulation of discontinuous systems, and specifically in system undergoing stick-slip vibrations [3, 8–11]. These simulations have to tackle the problem by separating the discontinuous phase space into a series of adjacent continous regions. In the present case, two regions are necessary, one for the stick mode and one for the slip mode, with the stick region being a straight line, which is visible in the usual stick-slip limit-cycle. A switch function is used to perform the transitions from stick to slip modes. This is done by monitoring the velocity and the torque built in the torsional spring to determine in which mode the system is, and selecting the equations of motion accordingly.

A total of six points were selected along the stick-slip limit-cycle in order to evaluate the response for the different models, as shown in figure 2. Points 1 and 5 are of great importance for the simulations as they enclose the limits of each mode. Therefore, the evolution of states depends on the accuracy of these two points. Points 2, 3 and 4 represent the region that encloses the states of the system during the cycle. Point 6 gives an indication of long-term convergence, as it is the final position after the whole simulation, which is 100 s in this case.

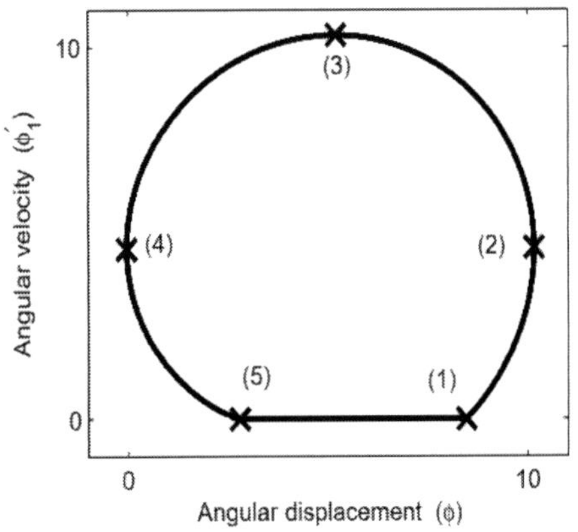

Figure 3. Chosen points on a typical stick-slip limit-cycle

The results of the convergence study shows that all 6 points chosen converge similarly for decreasing step size. The point showing more sensitivity to step size is point 6, as was expected. The point showing less sensitivity is point 5, although this value must be taken carefully as for higher step sizes this point on the cycle degenerates. As a good balance between accuracy and speed, the step size chosen for the simulations was 0.01, as for this value both points 1 and 5, which define the transition from stick to slip modes, are already stabilised and the other points are close to stabilisation.

3.1.1. Bisection method In order to improve the accuracy of the values at the transition from stick mode to slip mode (point 1 in figure 2), the bisection method with 10 iterations was used. This means that the original interval in which the transition values were calculated is divided in 10 parts, and the values are narrowed in one of these parts. One comparison was made with Model 2 taking $\omega = 8\omega_d$ and $A = 0.02$, where ω_d is the natural frequency of the torsional pendulum, and using Friction model 1. The interval containing the transition angular displacement ϕ calculated without the bisection method is $[8.477, 8.533]$. This interval is narrowed to $[8.4566, 8.4568]$ using the bisection method as discussed. This comparison shows that the method gives a considerable gain in accuracy ($1E-4$ compared to $1E-1$), although the increase in computing time is negligible (0.37%).

3.1.2. Comparison between the friction models The comparison between friction models is carried out by selecting five representative points along the stick-slip limit-cycle, these being the first five points chosen on the stick-slip limit-cycle, shown in figure 2. Model 2 was used for this comparison, varying the parameters A, ω and Ω_2.

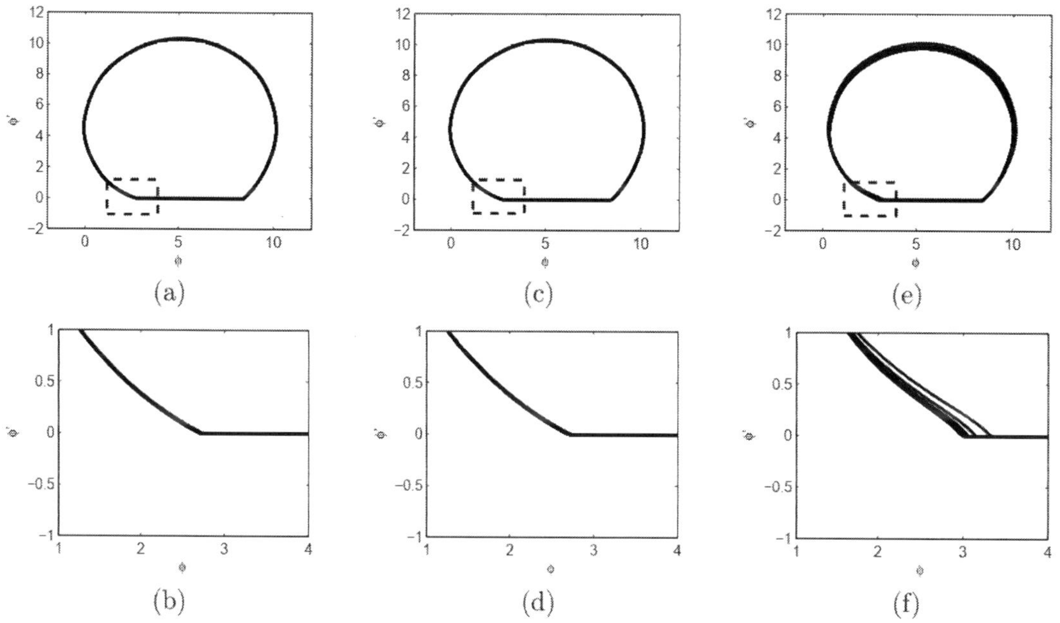

Figure 4. Phase-plane of stick-slip vibrations (a) Friction Model 1, (b) Friction Model 2, (c) Friction Model 3, (d)–(f) are zoom-up of the bottom left regions showing switching from slip to stick of respective figures (a)–(c)

Results of the comparison between the three friction models show small difference between a more simple friction model and a more complex one, confirming that the main aspect of the system in order to exhibit stick-slip vibrations is the difference between the static and dynamic friction characteristics. The small oscillations on the velocity of the top (Model 2) introduces quasi-periodicity to the system. The point on the limit-cycle most affected by this is the transition between the slip mode into stick mode. Bifurcation diagrams confirm that there is a region in parameter space that allows the system to operate without stick-slip vibrations. In figure 5 it can be observed that for values of Ω_2 less than $\Omega_2^* = 5.227$ the system presents large amplitude stick-slip limit-cycle, while for values of Ω_2 greater than Ω_2^* the system presents small amplitude limit-cycle. This region is related to a combination of high angular velocities of the rotary table and high damping coefficient of the drill-string. A control system aiming to suppress such vibrations should enclose the parameters in this region.

3.1.3. Sliding surface The limits of the sliding surface, which determine the conditions for stick mode, are given in this system by the torque in the drill-string, result of the twist. Expressing the torque in terms of ϕ results in the upper and lower limits of this state variable that represent the sliding surface [12, 13], which is shown in figure 5. It is possible to demonstrate that the

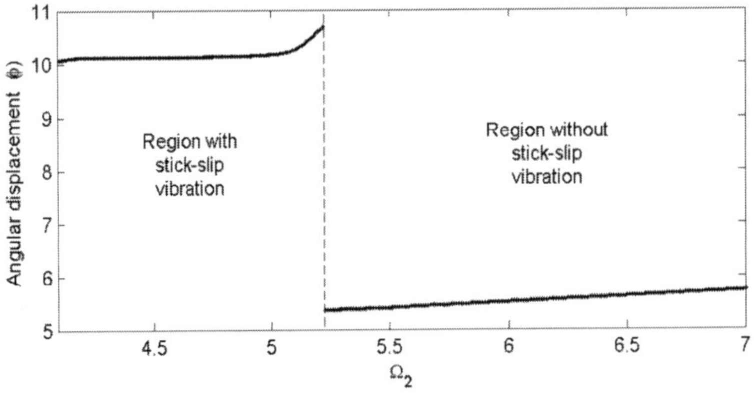

Figure 5. Bifurcation diagram for varying Ω_2 showing the switch from stick-slip to stick-slip free region

sliding surface does not depend on the state of the system while this is at stick mode. The upper and lower limits are calculated by:

$$\begin{aligned} \phi_{upper} &= \frac{-c_1\dot{\phi}_1+T_{f_{max}}}{k_1}, \\ \phi_{lower} &= \frac{-c_1\dot{\phi}_1-T_{f_{max}}}{k_1}, \end{aligned} \tag{9}$$

where $T_{f_{max}}$ is the maximum torque obtained with the friction model being used, usually being equal to T_{st}. While at stick mode, $\dot{\phi}_1 = 0$. Moreover, $T_{f_{max}}$ and k_1 do not depend on the system's state, and ϕ_2 depends only on A and ω. It follows that ϕ_{lower} and ϕ_{upper} will not change for varying Ω_2, while the dependency on A and ω clearly comes through the excitation present in ϕ_2 in Model 2.

4. Conclusions

The comparison between the three friction models used shows small difference between more simple and more complex models, which confirms that the main aspect responsible for a system to exhibit stick-slip vibrations is the difference between the static and dynamic friction characteristics. The point on the limit-cycle most affected by oscillations of the velocity of the rotary table is the transition between the slip mode into stick mode, as can be seen in figure 3. It can also be seen that the system using Friction Model 3 has more pronounced difference. However, this small difference is attenuated during the stick phase and it does not propagate in the long term dynamics. Bifurcation diagrams, such as the one in figure 5, allowed to find Ω_2^* above which the system operates without stick-slip vibrations. More importantly, Ω_2^* does not change substantially with different friction models. Also, the sliding surface does not change for different friction models, as the maximum value for friction does not depend on the contact velocity. A combination of angular velocities and damping on the system can also avoid stick-slip vibrations. A control-system aiming to suppress such vibrations should enclose the parameters in this region. For further analysis, a small scale experimental setup is being constructed, which will enable the investigations of correlating the numerical simulations to experimental data.

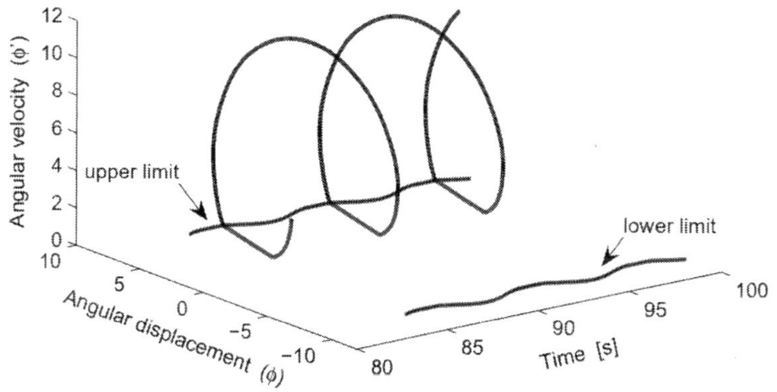

Figure 6. Three-dimensional space of stick-slip vibrations showing the lower and upper limits of the sliding surface

References

[1] Spanos PD, Chevallier AM, Politis NP, Payne ML. Oil well drilling: a vibrations perspective. The Shock and Vibration Digest. 2003;35(2):81–99.

[2] Wojewoda J, Stefanski A, Wiercigroch M, Kapitaniak T. Hysteretiv effects of dry friction: modelling and experimental studies. Phil Trans R Soc A. 2008;366:747–765.

[3] Wiercigroch M. A note on the switch function for the stick-slip phenomenon. Journal of Sound and Vibration. 1994;175(5):700–704.

[4] Wouw N, Nijmeijer H, Mihajlovic N. Friction-induced limit cycling in flexible rotor systems: an experimental drill-string system. In: Proceedings of IDETC/CIE; 2005. p. 200–300.

[5] Jansen JD, Steen L. Active damping of self-excited torsional vibrations in oil well drillstrings. Journal of Sound and Vibration. 1995;179(4):257–279.

[6] Steen L. Suppressing stick-slip induced drill-string oscillations: a hyperstability approach [PhD Thesis]. University of Twente; 1997.

[7] Payne M. Drilling bottom-hole assembly dynamics [PhD Thesis]. Rice University; 1992.

[8] Galvanetto U. Bifurcations and chaos in a four-dimensional mechanical system with dry friction. Journal of Sound and Vibration. 1997;204(4):690–695.

[9] Galvanetto U. Some discontinuous bifurcations in a two-block stick-slip system. Journal of Sound and Vibration. 2001;248(4):653–669.

[10] Galvanetto U. Non-linear dynamics of multiple friction oscillators. Computer Methods in Applied Mechanics and Engineering. 1998;178:291–306.

[11] Galvanetto U, Bishop SR. Computational techniques for nonlinear dynamics in multiple friction oscillators. Computer Methods in Applied Mechanics and Engineering. 1998;163:373–382.

[12] di Bernardo M, Kowalczyk P, Nordmark A. Bifurcations of dynamical systems with sliding: derivation of normal-form mappings. Physica D. 2002;170:175–205.

[13] Kowalczyk P, di Bernardo M. Two-parameter degenerate sliding bifurcations in Filippov systems. Physica D. 2005;204:204–229.

7th International Conference on Modern Practice in Stress and Vibration Analysis IOP Publishing
Journal of Physics: Conference Series **181** (2009) 012057 doi:10.1088/1742-6596/181/1/012057

A computationally efficient software application for calculating vibration from underground railways

M.F.M. Hussein[1] and H.E.M. Hunt[2]
[1] Department of Civil Engineering, University of Nottingham, Coates Building, University Park , Nottingham, NG7 5PP, UK

[2] Department of Engineering, University of Cambridge, Trumpington Street, Cambridge, CB2 1PZ, UK.

E-mail of contact author: mohammed.hussein@nottingham.ac.uk

Abstract. The PiP model is a software application with a user-friendly interface for calculating vibration from underground railways. This paper reports about the software with a focus on its latest version and the plans for future developments. The software calculates the Power Spectral Density of vibration due to a moving train on floating-slab track with track irregularity described by typical values of spectra for tracks with good, average and bad conditions. The latest version accounts for a tunnel embedded in a half space by employing a toolbox developed at K.U. Leuven which calculates Green's functions for a multi-layered half-space.

1. Introduction

Vibration from underground railways has a significant environmental impact on nearby buildings and their occupants. Vibration is generated at the wheel-rail interface and propagates to nearby buildings where it causes vibration of walls and floors. Vibration of buildings can cause annoyance to occupants and malfunctioning of sensitive equipment. Vibration of walls and floors is also reradiated as noise which may result in a serious problem especially in sensitive buildings such as a concert hall or a recording studio.

Modelling vibration from underground railways has recently gained a special attention. There is an increasing demand nowadays for numerical models that can be used in the design stage of underground tunnels to ensure that the levels of noise and vibration in buildings will be acceptable. The models are also needed to investigate the effect of vibration countermeasures in existing tunnels before these are implemented. There are number of models to simulate vibration from underground railways. These are based on numerical methods such as the coupled Finite Element-Boundary Element (FE-BE) method [1-3] or the Finite Difference (FD) method [4]. These models can account for tunnels with circular and non-circular cross-sections and can account for layering and unusual details of the soil. The main disadvantages of these models are that they take long time to complete the computations and they require large computational resources. Therefore these models are useful to understand vibration from underground tunnels but they are difficult to be used by design engineers who normally would like to investigate the results due to several design parameters in a short time using personal computers.

[1] To whom any correspondence should be addressed.

The Pipe-in-Pipe (PiP) model was first introduced to simulate vibration from a circular tunnel embedded in a full space [5-7]. The tunnel wall and its surrounding soil are modelled as two concentric pipes, see Figure 1. The inner pipe accounts for the tunnel and this is modelled using the thin shell theory. The outer pipe accounts for the soil and this is modelled using the elastic continuum theory. The main advantages of the PiP model are its accuracy in accounting for the dynamic tunnel-soil interaction and its computational efficiency; the model completes the calculations in a significantly less time compared with other models, see [8] for example, and with much less demand on computational resources.

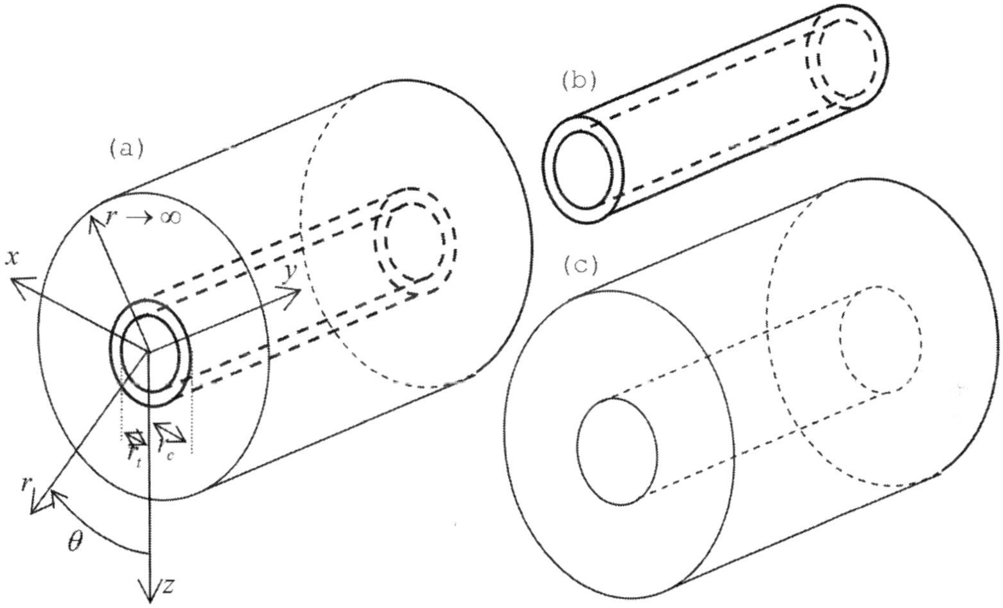

Figure 1: The PiP model in its basic formulation accounts for (a) two concentric pipes. The inner pipe shown in (b) accounts for a tunnel wall and the outer pipe shown in (c) (with external radius tends to infinity) accounts for the surrounding soil [5-7].

The PiP model was developed into a software application with a user friendly interface [9,10]. The software was developed as a numerical tool for engineers responsible for the designing of vibration countermeasures in underground tunnels. The software was used to calculate vibration due to a train of infinite length running along a tunnel at a constant velocity. The output of the software was the Power Spectral Density (PSD) of the vertical displacement at a point in the soil due to a roughness excitation described by a white noise spectrum. The first three versions of the software were based on a model of a tunnel embedded in a full space. Therefore it was only used to assess the performance of vibration countermeasures rather than to obtain absolute predictions. Version 3 of the software had many features such as calculating vibration from a tunnel with rigid bedrock underneath. It was also used to plot a contour map of vibration around the tunnel at a constant excitation frequency.

In this paper, a new version of PiP, version 4, is presented. The new version accounts for an important development of the model as it simulates vibration from a tunnel embedded in a half space. The new version also calculates vibration due to some typical spectra of track irregularity rather than for only white noise spectrum as in the previous versions. The next two sections of this paper present the new version of the software and discuss the developments that will appear in future versions.

2. PiP version 4

Version 4 of the software accounts for an important development of the model. The tunnel wall is embedded in a half space rather than full space as in the previous versions. The new model is reported in [11] and is based on the assumption that the displacements at the tunnel-soil interface are not affected by the existence of a free-surface. The validity of this assumption has been checked in [11] and a good match is found when comparing the results calculated by the new model and those calculated by the coupled FE-BE method. The assumption enables performing calculations of vibration from a tunnel embedded in a half space in 3 steps. First, the displacements at the tunnel-soil interface are calculated by the original PiP model, i.e. using a model of a tunnel embedded in a full space. Second, the forces in a full space that generate the same displacements at the virtual cylindrical surface (corresponding to the tunnel-soil interface) are calculated using a variant to the PiP model which accounts for a full space. Third, the forces are used along with Green's functions for an elastic half space to calculate the displacements in the far field.

To calculate Green's functions for an elastic half space, the ElastoDynamics toolbox [12,13], developed at K. U. Leuven, has been built in as a part of the PiP software. The toolbox performs a fast and efficient calculation of Green's functions by using the direct stiffness method. The method considers the dynamic equilibrium in the frequency-radial wavenumber domain. The toolbox has the capability to calculate Green's functions for a multi-layered half space. However, it is used in this version of PiP to do the calculations for a half space only.

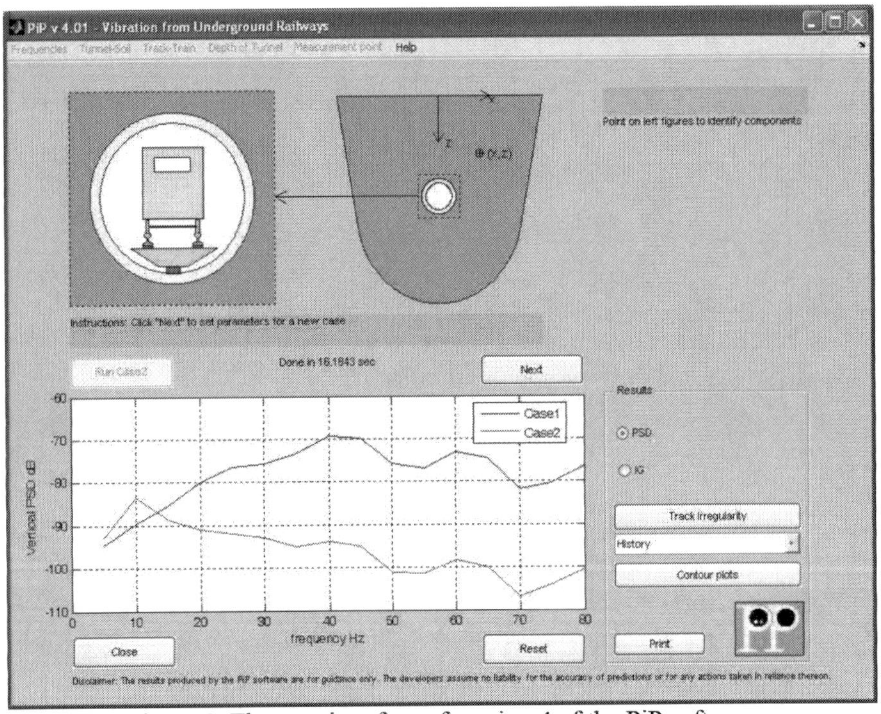

Figure 2: The user interface of version 4 of the PiP software

Figure 2 shows the software user interface. The user can use the default parameters or can enter the appropriate parameters for the soil, tunnel, floating-slab track, the train, and the point in the soil where the PSD is required. The software performs the calculations for a white-noise spectrum of the track at first. To demonstrate the computation efficiency of the software, the running time to produce any of the curves shown in the plot window of the interface in Figure 2, i.e. to calculate vibration at a point in the soil in the frequency range 5-80 with a 5 Hz interval is 16 seconds. This is done on a normal PC with 2GHz processor and 2GB RAM.

One of the improvements in the new version of PiP is in the tunnel model. The tunnel has been modelled using the elastic continuum theory rather than the thin shell theory as in previous versions.
In the same way as the previous versions, the new version calculates vibration due to a train of infinite length running on a floating slab-track supported at the tunnel bottom. To improve the running time further in this version, calculations of forces at the wheel-rail interface are performed assuming a rigid tunnel. This allows calculations of these forces from Green's functions of a double beam on elastic foundation as described in [14]. These forces are then used as an input to the track-tunnel-soil system to calculate vibration in the far field.

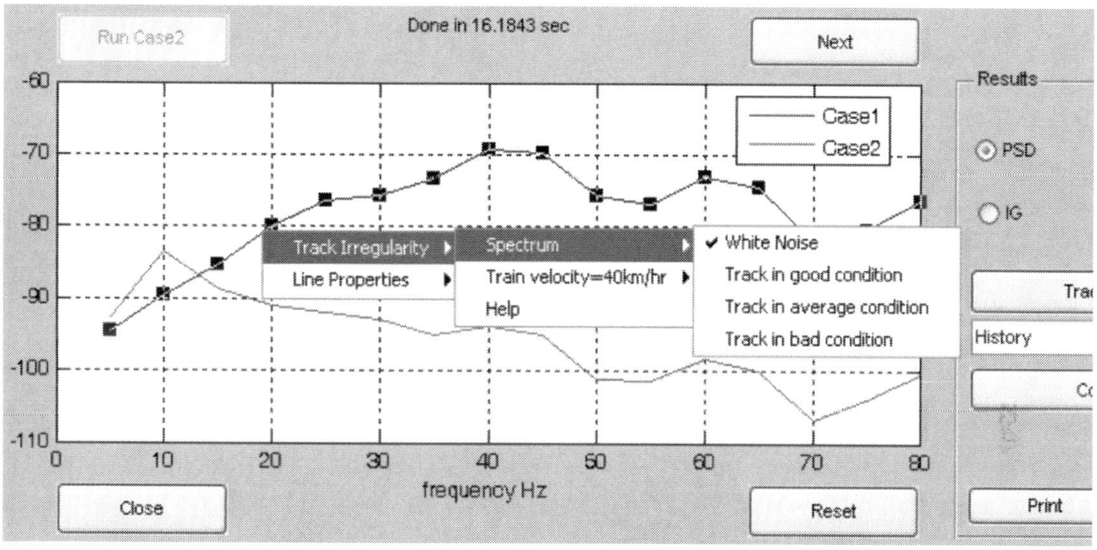

Figure 3: The displacement PSD can be calculated for tracks in good, average and bad conditions

The user can get the PSD for typical track irregularity, see Figure 3, according to the situation of the track; whether it is in its worst, average or best situation. The values of the PSD of track irregularity S_δ are those used by Forrest and Hunt [6] which are derived by Frederich [15]. Those are calculated by the following equation

$$S_\delta(f) = \frac{1}{v}\frac{a}{(b + f/v)^3} \tag{1}$$

where a and b are unevenness and waviness of the track respectively and these are given in Table 1, f is the frequency and v is the velocity of the train.

The software also calculates the PSD of the vertical displacement at a mesh of points around the tunnel. Figure 4 demonstrates this where vibrations are calculated due to 80 Hz excitation at the ground surface down to 40 m depth. In Figure 4, the top left plot shows vibration using the default parameters (with a 40 Hz floating slab) and the top right plot shows the effect of using a 10 Hz floating-slab track. The bottom plot shows the Insertion Gain (IG) which is used as an indicator to illustrate how vibration is changed due to a change in one of the parameters or due to the use of vibration countermeasure.

Table 1. Values of unevenness a, and waviness b [15].

Condition	$a\ [mm^2(1/m)^2]$	$b\ [1/m]$
Worst	9.39×10^{-1}	6.89×10^{-2}
Average	1.31×10^{-2}	2.94×10^{-2}
Best	1.90×10^{-4}	9.71×10^{-3}

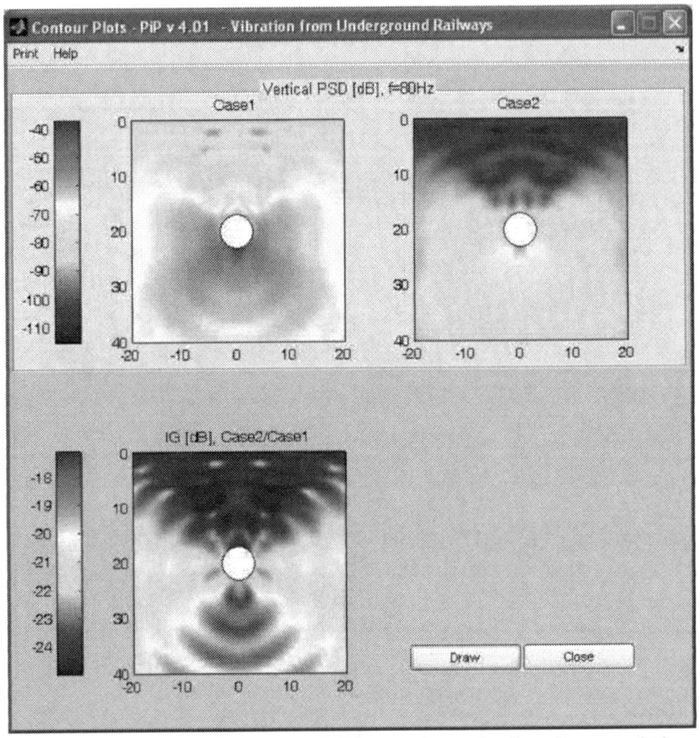

Figure 4: The displacement PSD can be calculated around the tunnel at a fixed frequency.

3. Future developments of PiP

The current version of PiP calculates the vertical displacements in soil due to a train of infinite length running on a floating-slab track in a tunnel embedded in a half-space. Future versions will account for horizontal and longitudinal components of the displacements. The software will also account for a tunnel embedded in a multi-layered half space. This will be done by taking advantage of other capabilities of the ElastoDynamics toolbox. The list of future developments includes considering a train with a finite length rather than infinite length; accounting for other types of tracks such as ballasted tracks and trough tracks; coupling buildings on pile foundations; accounting for inclined layers of soil; and accounting for floating-slab tracks with discontinuous slabs.

4. Summary

This paper has reported on a software application with a user-friendly interface for calculating vibration from underground railways. The software is fast running on personal computers and therefore it can be used as a tool to predict vibration from tunnels and to investigate parameters and measures to reduce vibration from underground tunnels. The paper has highlighted the features of version 4 of the software and has discussed developments which are planned for future versions.

References

[1] Sheng X, Jones C J C and Thompson D J, Prediction of ground vibration from trains using the wavenumber finite and boundary element methods 2006, *Journal of Sound and Vibration* **293, 3-5**, 575-586.

[2] Clouteau D, Arnst M, Al-Hussaini T M, Degrande G, Freefield vibrations due to dynamic loading on a tunnel embedded in a stratified medium 2005, *Journal of Sound and Vibration* **283, 1-2**, 173-199.

[3] Degrande G, Clouteau D, Othman R, Arnst M, Chebli H, Klein R, Chatterjee P, Janssens B, A numerical model for ground-borne vibrations from underground railway traffic based on a periodic finite element–boundary element formulation 2006, *Journal of Sound and Vibration* **293, 3-5**, 645-666.

[4] Thornely-Taylor R M, The prediction of vibration, ground-borne and structure-radiated noise from railways using finite difference method- Part1- theory 2004, *Proceeding of the Institute of Acoustics*, **26, 2**, 69-79.

[5] Forrest J A, Hunt H E M, A three-dimensional model for calculation of train-induced ground vibration 2006, *Journal of Sound and Vibration* **294, 4-5**, 678-705.

[6] Forrest J A, Hunt H E M, Ground vibration generated by trains in underground tunnels. *Journal of Sound and Vibration* 2006, **294, 4-5**, 706-736.

[7] Hussein M F M, Hunt H E M, A numerical model for calculating vibration from a railway tunnel embedded in a full-space 2007, *Journal of Sound and Vibration* **305, 3**, 401-431.

[8] Gupta S, Hussein M F M, Degrande G, Hunt H E M, Clouteau D. A comparison of numerical models for the prediction of vibrations from underground railway traffic 2007, *Soil Dynamics and Earthquake Engineering*, **27, 7**, 608-624.

[9] The PiP model 2009; http://www.pipmodel.com.

[10] Hussein M F M, Hunt H E M. The PiP model, a software for calculating vibration from underground railways 2007, *the International Congress of Sound and Vibration*, Cairns, Australia.

[11] Hussein M F M, Gupta S, Hunt H E M, Degrande G, Talbot J P. An efficient model for calculating vibration from a railway tunnel embedded in a half-space 2006, *the 13th International Congress of Sound and Vibration*, Vienna, Austria.

[12] The ElastoDynamics Toolbox for Matlab, 2009, http://www.kuleuven.be/bwm/edt/index.html

[13] Schevenels M and Degrande G. ElastoDynamics Toolbox (EDT) for Matlab. Version 2.0. User's manual BWM-2007-07, 2007, Department of Civil Engineering, K.U. Leuven.

[14] Hussein M F M, Hunt H E M, Modelling of floating-slab tracks with continuous slabs under oscillating-moving loads 2006, *Journal of Sound and Vibration* **297, 1-2**, 37-54.

[15] Frederich F, *Die Gleislage – aus fahrzeugtechnischer Sicht [Effect of track geometry on vehicle performance]*. Zeitschrift fur Eisenbahnwesen und Vekehrstechnik – Glaser Annalen, 1984, **108, 12**, 355-362.

7th International Conference on Modern Practice in Stress and Vibration Analysis IOP Publishing
Journal of Physics: Conference Series **181** (2009) 012058 doi:10.1088/1742-6596/181/1/012058

Raising of a heavy semi-circular vault on a rigid centering

Dmitry A. Parshin
Ishlinsky Institute for Problems in Mechanics, Russian Academy of Sciences,
Vernadsky Ave. 101 bldg 1, Moscow, 119526 Russia
E-mail: parshin@ipmnet.ru

Abstract. The modelling of a process of a semi-circular vault raising on a rigid cylindrical centering is conducted. Within the bounds of accreted solids mechanics the evolution of the vault stress-strain state under gravity action during the raising and after its completion is investigated. The vault material is described with defining relationships for elastic or aging viscoelastic materials. The formulation of the non-classical quasi-static initial-boundary accretion problem is given. Its analytical solution in series and quadratures is built. An effective method of calculation of residual stresses in the finished vault after supporting centering removing is offered.

1. Introduction

Everybody knows that proper weight of large constructions makes an essential, and more often than not determinative, contribution to their stress-strain state. So when designing such constructions one cannot but take into account gravity action.

It is also known that vast majority of large-size constructions is not installed ready-made on a base prepared but is being installed gradually, element by element. It is easily seen that the weight of every element newly added is sure to cause a supplementary strain of the present construction. That is why the stress-strain state of a construction gradually raised under continuous gravity action must greatly differ from the state of this construction if we would install it on the base and then expose the entire construction to gravity action.

Besides, it is also clear that if the structural material used possesses rheological properties than both any actual and the final state of the raised object should considerably depend on the raising time mode. During a raising process of an object of such kind there are two tendencies that are continuously interacting. The first one is the tendency of permanent loading the construction with weight of additional elements and the second one is the stress rearrangement in elements so far added due to strain procrastination under changing construction geometry. The type and the result of this interaction are determined by various particular factors.

A certain technological process of raising may prove to cause the situation when stresses having initially arisen concentrate in some areas of the construction while progressing and result in inadmissibly high level of technological stresses in some zones of the finished object. In this case it is necessary to take special measures to control the stress-strain state of the being raised object.

From the speculations above it becomes obvious that not only the organization of the control mentioned but even an ordinary strength calculation of gradually raised heavy objects cannot be

Journal of Physics: Conference Series **181** (2009) 012058 doi:10.1088/1742-6596/181/1/012058

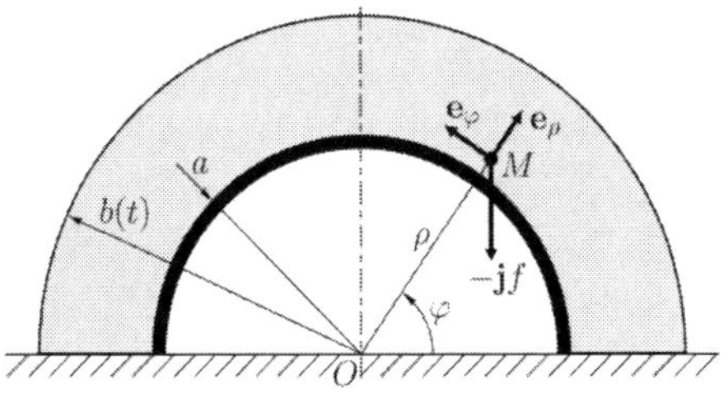

Figure 1. A vault raised on a centering.

implemented in principle by means of a design model dealing only with the final configuration of the object considered. Reasonable results cannot be obtained either if we use classical equations and boundary conditions of mechanics in a time-varying spatial domain. It is caused by the fundamental fact that deformation process of an accreted solid greatly differs from that of a regular solid. Consequently, description of the former process should be performed not by traditional approaches of deformable solids mechanics that do not take into consideration features of accretion processes but in the framework of accreted solids mechanics concepts.

See the information about the current state and some solved problems of accreted solids mechanics, for instance, in [1–9]. And the present paper is intended to one problem of gradual raising of a heavy semi-circular vault that manifests properties of creep and aging. So far this problem has not been solved.

2. Statement of the problem

Let the process of raising of a semi-circular vault supporting on a smooth rigid base start at time $t = t_1$. The raising is implemented by means of vault erection due to continuous attachment of uniformly thick elementary layers of additional material to its external surface. During the process the internal surface of the vault is supported by a rigid circular centering which was previously placed on the base. The very first layer of the material is directly attached to the centering and adhere to it. The process lasts up to time $t = t_2 > t_1$ and after it is completed the centering is removed. During the process and after its completion the vault is attached to the base by a sliding mount that prevents the feet of the vault to be separated from the base but does not prevent their sliding along it. The material that is originally free of stresses is assumed homogeneous isotropic aging linear and viscoelastic.

Let us state a problem to study the process of creating of the stress-strain state of the vault in question under gravity action during its raising procedure considered above and the evolution of this state from the instant of raising completion to the instant of centering removing. Besides, let us study the residual stresses taking place in the completed vault even if the centering were removed long ago.

The process of erecting vault deformation is assumed to be quasistatic. We perform our consideration in small strains in the case of a plane strain state.

Let a be the internal vault radius (the external centering radius) and $b(t)$ its external radius at time t, $b(t_1) = a$ (Fig. 1). Taking into account the fact that the strain is small suppose the strictly monotone increasing function of time $b(t)$ is known. As we consider the process of continuous accretion of the vault, the function $b(t)$, $t \in [t_1, t_2]$ is to be continuous. Moreover, suppose this function is piecewise continuously differentiable. Let $b(t) \equiv b_{\text{fin}} = b(t_2)$ when $t > t_2$.

We associate the vault transverse cross-section plane with the circular cylindrical coordinate system (ρ, φ, z) with the right orthonormal local frame $\{\mathbf{e}_\rho, \mathbf{e}_\varphi, \mathbf{k}\}$ where ρ is the polar radius counted from the arch central axis, φ is the polar angle counted upwards from the base (Fig. 1) and z is the longitudinal coordinate. By $\mathbf{j}$ denote the unit normal to the base. Then the cubic density of the gravity forces acting on the vault equals to $\mathbf{f} = -\mathbf{j}f$, where $f = \text{const}$ is the material specific weight.

We describe the material used to erect the vault in the framework of the linear theory of viscoelasticity of homogeneously aging isotropic media [1,10], i.e. we start from the state equation

$$\mathbf{T}(\mathbf{r},t) = G(t)\big(\mathcal{I} + \mathcal{N}_{\tau_0(\mathbf{r})}\big)\big[2\,\mathbf{E}(\mathbf{r},t) + (\varkappa - 1)\,\mathbf{1}\,\mathrm{tr}\,\mathbf{E}(\mathbf{r},t)\big]. \tag{1}$$

Here $\mathbf{T}(\mathbf{r},t)$ and $\mathbf{E}(\mathbf{r},t)$ are the stress and small strain tensors, respectively, $\mathbf{1}$ is the unit tensor of rank 2; $G(t)$ is the elastic shear modulus and $\varkappa = (1-2\nu)^{-1}$, $\nu = \mathrm{const}$ is Poisson's ratio. The viscoelasticity operator $\mathcal{I} + \mathcal{N}_s$ in (1) is determined by the relations

$$\mathcal{I} + \mathcal{N}_s = (\mathcal{I} - \mathcal{L}_s)^{-1},$$
$$\mathcal{L}_s f(t) = \int_s^t f(\tau)K(t,\tau)\,d\tau, \quad \mathcal{N}_s f(t) = \int_s^t f(\tau)R(t,\tau)\,d\tau,$$
$$K(t,\tau) = G(\tau)\,\partial\Delta(t,\tau)/\partial\tau, \quad \Delta(t,\tau) = G(\tau)^{-1} + \omega(t,\tau),$$

where $\mathcal{I}$ is the identity operator, $\mathcal{L}_s$ and $\mathcal{N}_s$ are the integral Volterra operators with a parameter s; $K(t,\tau)$ and $R(t,\tau)$ are the creep and relaxation kernels, respectively, and $\Delta(t,\tau)$ and $\omega(t,\tau)$ are the specific strain function and the creep measure in the case of pure shear. $\tau_0(\mathbf{r})$ is the time at which stresses appear at the point with the position vector $\mathbf{r}$. Time t is assumed to be counted from the material fabrication instant.

It should be noted that the equation (1) was suggested by the academician of the Armenian Academy of Sciences N. Kh. Arutyunyan especially for description of creeping processes in concrete and is widely used for this purpose. Meanwhile, it also well comes to agreement with experimental data for creep of some kinds of polymers, rocks, soil, and ice.

Let us introduce the integral linear operator $\mathcal{H}_s = (\mathcal{I} - \mathcal{L}_s)G(t)^{-1}$ and its reverse operator $\mathcal{H}_s^{-1} = G(t)(\mathcal{I} + \mathcal{N}_s)$ with the parameter s and for an arbitrary function $g(\mathbf{r},t)$ of time t and point $\mathbf{r}$ of the erected body denote $g^\circ(\mathbf{r},t) = \mathcal{H}_{\tau_0(\mathbf{r})}\,g(\mathbf{r},t)$. Then constitutive relation (1) can be written in another form [2]

$$\mathbf{T}^\circ(\mathbf{r},t) = 2\,\mathbf{E}(\mathbf{r},t) + (\varkappa - 1)\,\mathbf{1}\,\mathrm{tr}\,\mathbf{E}(\mathbf{r},t). \tag{2}$$

Note that the specific strain function for the pure shear $\Delta(t,\tau)$ describes the increase during time t of the angle of shear in the sample when we create at time instant τ and then hold a constant stress state of pure shear of unit intensity. The summand $G(\tau)^{-1}$ determines the sample's instantaneous elastic reaction to the force action indicated and the summand $\omega(t,\tau)$ is the specific creep strain taking place in the sample at considered time. Thus, by definition, we get the identity $\omega(\tau,\tau) \equiv 0$, $\tau \geqslant 0$. Taking into account this identity the specific strain function can be presented as follows:

$$\Delta(t,\tau) = \mathcal{H}_\tau 1. \tag{3}$$

In the problem under consideration stresses appear in the body points at the time we include these points in the body structure. It means that

$$\tau_0(\mathbf{r}) \equiv \tau_*(\rho), \tag{4}$$

where $\tau_*(\rho)$ is the time of attachment of a layer with radius ρ to the vault. It is clear that the equation $\tau_*(\rho) = t$ describes the external surface of the continuously erecting vault $\{\rho = b(t)\}$, i.e. its instantaneous erecting surface. In other words, we get the identity

$$\tau_*\big(b(t)\big) \equiv t. \tag{5}$$

Journal of Physics: Conference Series **181** (2009) 012058 doi:10.1088/1742-6596/181/1/012058

3. Boundary value problem for the raised vault

During the process of accretion the elements of additional material are added to the growing solid in the course of its strain motion in space. It is clear that the entire body formed in such a way cannot in general have the original unstrained configuration. It is this fact that is decisive in the strain process for any accreted solid and essentially distinguishes the mechanical behavior of such solids from the behavior of solids of constant composition (classical bodies in continuum mechanics) and of solids whose boundary is variable because of the removal of the material. Owing to this characteristic property, it is impossible to define the strain measure of an accreted solid by the method usually adopted in continuum mechanics; therefore, the Cauchy formulas do not hold for the total strain tensor components, and hence the Saint-Venant conditions of their compatibility are not satisfied.

Note, however, that the particles of additional material after its adhesion to the surface of growth continue their motion further as part of the solid. This means that, in the space region occupied by the entire accreted solid at a given time, a sufficiently smooth velocity field of its particles is uniquely determined. Therefore, we can expect that the problem of deforming of such a body can be well defined for the displacement velocities.

Starting from this, we write out the velocity analog of the constitutive relation (2) for velocities of changes of tensor quantities this relation ties together:

$$\mathbf{S}(\mathbf{r},t) = 2\mathbf{D}(\mathbf{r},t) + (\varkappa - 1)\,\mathbf{1}\,\mathrm{tr}\,\mathbf{D}(\mathbf{r},t). \tag{6}$$

Here we introduce the tensor $\mathbf{S} = \partial\mathbf{T}^\circ/\partial t$ [2] and the strain rate tensor

$$\mathbf{D} = (\nabla\mathbf{v}^{\mathrm{T}} + \nabla\mathbf{v})/2, \tag{7}$$

where $\mathbf{v}(\mathbf{r},t) = \mathbf{e}_\rho v_\rho + \mathbf{e}_\varphi v_\varphi + \mathbf{k}\,v_z$ the velocity field of the accreted body particles. In the considered case of plane strain, we have

$$\mathbf{k}\cdot\mathbf{v} = 0, \quad \partial\mathbf{v}/\partial z = \mathbf{0}. \tag{8}$$

The equation for the introduced above tensor $\mathbf{S}$ the region occupied at the current time by the accreted body can be obtained by applying the linear operator $\mathcal{H}_{\tau_0(\mathbf{r})}$ to the equilibrium equation $\nabla\cdot\mathbf{T}+\mathbf{f}=\mathbf{0}$ and by differentiating the result with respect to time t. Due to the fact that the lower limit in the integral tensor mentioned depends on a point of the body this integral operator does not commute with the divergence operator. But one can show [2] that in the accretion processes studied in the present paper, for the stress tensor we can write $(\nabla\cdot\mathbf{T})^\circ = \nabla\cdot\mathbf{T}^\circ$. In this case the following analog of the equilibrium equation is satisfied in the entire accreted solid both during and after the process of its erecting:

$$\nabla\cdot\mathbf{T}^\circ + \mathbf{f}^\circ = \mathbf{0}. \tag{9}$$

Differentiate it with respect to t: $\nabla\cdot\mathbf{S}+\partial\mathbf{f}^\circ/\partial t = \mathbf{0}$. By calculating with (3) the vector-function $\partial\mathbf{f}^\circ/\partial t = -\mathbf{j}\,h(\rho,t)$, where $h(\rho,t) = f\,\partial\omega\big(t,\tau_*(\rho)\big)/\partial t$, we obtain the equation

$$\nabla\cdot\mathbf{S} = \mathbf{j}\,h(\rho,t). \tag{10}$$

Note that according to properties the creep measure should have (see, for example, [1]), $h(\rho,t)\geqslant 0$ when $t > \tau_*(\rho)$.

Consider now a condition that is to be given on the erection surface of the body. As we suppose the material to be attached is originally free of stresses the complete stress tensor is required to be equal to zero on the accreted body external surface [11]:

$$\mathbf{T} = \mathbf{0}, \quad \rho = b(t), \quad t\in[t_1,t_2]. \tag{11}$$

A remarkable fact is that (11) can be transformed to the boundary condition for components of the tensor $\mathbf{S}$ which has the form analogous to a classical condition for forces on the solid surface [2]. The set of such conditions (11) for all instants of the continuous growth as an initial condition for the stress tensor in points of having been accreted solid can be written as follows:

$$\mathbf{T}\big(\mathbf{r}, \tau_*(\rho)\big) = \mathbf{0}, \quad \rho \in [a, b_{\text{fin}}]. \tag{12}$$

Noticing that from (4) the equality

$$g^\circ\big(\mathbf{r}, \tau_*(\rho)\big) = g\big(\mathbf{r}, \tau_*(\rho)\big)/G\big(\tau_*(\rho)\big) \tag{13}$$

follows for the stress tensor $\mathbf{T}$, taking (12) into account, we get

$$\mathbf{T}^\circ\big(\mathbf{r}, \tau_*(\rho)\big) = \mathbf{0}, \quad \rho \in [a, b_{\text{fin}}]. \tag{14}$$

Apply the divergence operator to (14):

$$\big[\nabla \cdot \mathbf{T}^\circ(\mathbf{r}, t)\big]\Big|_{t=\tau_*(\rho)} + \nabla\tau_*(\mathbf{r}) \cdot \mathbf{S}\big(\mathbf{r}, \tau_*(\rho)\big) = \mathbf{0}.$$

Then by determining the gradient $\nabla\tau_*(\mathbf{r}) = \mathbf{e}_\rho(\varphi)\,\tau_*'(\rho)$, and using (9), and taking (13) into account we obtain $-\mathbf{f}/G\big(\tau_*(\rho)\big) + \mathbf{e}_\rho(\varphi) \cdot \mathbf{S}\big(\mathbf{r}, \tau_*(\rho)\big)\,\tau_*'(\rho) = \mathbf{0}$. The last thing to do is to single out from this conditions family on surfaces $\{\rho = \text{const}\}$ the one that corresponds to the considered time instant $t \in [t_1, t_2]$, i.e. to suppose $\rho = b(t)$. To do it we use the identity (5) as well as the same identity having been differentiated with respect to t, i.e. the identity $\tau_*'\big(b(t)\big) \equiv b'(t)^{-1}$. After all we have the following condition for the tensor $\mathbf{S}$ on the instantaneous erecting surface at $t \in [t_1, t_2]$:

$$\mathbf{e}_\rho \cdot \mathbf{S} = \mathbf{f}\,b'(t)/G(t), \quad \rho = b(t). \tag{15}$$

After we stop erecting the vault its external surface remains unloaded. Therefore we can use an ordinary condition of stress absence for $t > t_2$. An evident analog of this condition for tensor $\mathbf{S}$ is the boundary condition $\mathbf{e}_\rho \cdot \mathbf{S} = \mathbf{0}$, $\rho = b_{\text{fin}}$. As we suppose that $b(t) \equiv b_{\text{fin}}$ when $t > t_2$ the form (15) of the condition for the external surface of the vault remains formally valid even after erecting completing and consequently can be used for any $t \geqslant t_1$. However, it is necessary to understand that nature of this condition during the vault continuous erecting stage and after its completing is different in essence.

Consider now conditions on other parts of boundary surface of erected vault. At its internal surface $\{\rho = a\}$ we have to require meeting of kinematic condition of tough adhesion of the material and the centering. To set a problem in velocities this condition should be

$$\mathbf{v} = \mathbf{0}, \quad \rho = a. \tag{16}$$

For the feet of the vault $\{\varphi = 0, \pi\}$, as we mentioned when stating the problem, we use a mixed condition for a sliding mount $\mathbf{e}_\varphi \cdot \mathbf{T} \cdot \mathbf{e}_\rho = 0$, $\mathbf{e}_\varphi \cdot \mathbf{v} = 0$. Its analog for the tensor $\mathbf{S}$ is

$$\mathbf{e}_\varphi \cdot \mathbf{S} \cdot \mathbf{e}_\rho = 0, \quad \mathbf{e}_\varphi \cdot \mathbf{v} = 0, \quad \varphi = 0, \pi. \tag{17}$$

Thus combining the relations (10), (6), (7), (8), (16), (15), (17), we get the following boundary value problem which describes a process of erected vault deforming up to the instant of removing the centering some time later after the erecting process has been completed:

$$\begin{gathered} \nabla \cdot \mathbf{S} = \mathbf{j}\,h(\rho, t), \quad \rho \in \big(a, b(t)\big), \quad \varphi \in (0, \pi), \quad t > t_1; \\ \mathbf{S} = 2\mathbf{D} + (\varkappa - 1)\,\mathbf{1}\,\mathrm{tr}\,\mathbf{D}, \quad \mathbf{D} = (\nabla\mathbf{v}^{\mathrm{T}} + \nabla\mathbf{v})/2; \quad \mathbf{k} \cdot \mathbf{v} = 0, \quad \partial\mathbf{v}/\partial z = \mathbf{0}; \\ \mathbf{v}|_{\rho=a} = \mathbf{0}; \quad \mathbf{e}_\rho \cdot \mathbf{S}|_{\rho=b(t)} = -\mathbf{j}\,q(t); \quad \mathbf{e}_\varphi \cdot \mathbf{S}|_{\varphi=0,\pi} \cdot \mathbf{e}_\rho = 0, \quad \mathbf{e}_\varphi \cdot \mathbf{v}|_{\varphi=0,\pi} = 0. \end{gathered} \tag{18}$$

Here we introduce the notation $q(t) = f\,b'(t)/G(t) \geqslant 0$. The form of this task is equivalent to a classical boundary value problem of linear elasticity theory where the vector $\mathbf{v}$ and the tensors $\mathbf{D}$ and $\mathbf{S}$ play parts of the displacement vector, the small strain tensor and the stress tensor divided into shear modulus. Time t is this problem parameter.

4. Solution of the problem. Determining of stresses in the raised vault

Let us look for the radial and circular components of the velocity vector field $\mathbf{v}$, which is the solution of the boundary value problem (18), as the cosine and sine Fourier series expansions of angles that are multiples of $\varphi \in [0, \pi]$ respectively. Because of the structure symmetry we have $v_\rho\big|_\varphi = v_\rho\big|_{\pi-\varphi}$, $v_\varphi\big|_\varphi = -v_\varphi\big|_{\pi-\varphi}$. Therefore

$$v_\rho = c_0(\rho, t) + \sum\nolimits_{n=1}^{\infty} c_n(\rho, t) \cos 2n\varphi, \qquad v_\varphi = \sum\nolimits_{n=1}^{\infty} d_n(\rho, t) \sin 2n\varphi. \tag{19}$$

Note that these expansions satisfies to the homogenous mixed boundary conditions on the arch feet $\{\varphi = 0, \pi\}$ contained in (18).

After we use (19) in (18) that satisfies to the Lame equations we are to solve systems of linear inhomogeneous ordinary differential equations with respect to ρ of the second order with variable coefficients for unknown functions $c_0(\rho, t)$, $c_n(\rho, t)$, $d_n(\rho, t)$. On solving them we get

$$2\,c_0/\rho = -\varkappa^{-1} B_{40} - B_{10},$$
$$2\,c_n/\rho = \big[(n + \varkappa^{-1}) B_{2n} - B_{3n}\big]/(2n-1) + \big[(n - \varkappa^{-1}) B_{4n} - B_{1n}\big]/(2n+1),$$
$$2\,d_n/\rho = \big[(n - 1 - \varkappa^{-1}) B_{2n} + B_{3n}\big]/(2n-1) - \big[(n + 1 + \varkappa^{-1}) B_{4n} + B_{1n}\big]/(2n+1);$$

$$\mathbf{S} = \mathbf{e}_\rho \mathbf{e}_\rho\, S_\rho + \mathbf{e}_\varphi \mathbf{e}_\varphi\, S_\varphi + \mathbf{k}\,\mathbf{k}\,\nu (S_\rho + S_\varphi) + (\mathbf{e}_\rho \mathbf{e}_\varphi + \mathbf{e}_\varphi \mathbf{e}_\rho)\, S_{\rho\varphi},$$
$$S_{\rho/\varphi} = \pm B_{10} - B_{40} + \sum\nolimits_{n=1}^{\infty} \big[\pm B_{1n} \mp (n \pm 1) B_{2n} \mp B_{3n} \pm (n \mp 1) B_{4n}\big] \cos 2n\varphi,$$
$$S_{\rho\varphi} = \sum\nolimits_{n=1}^{\infty} \big[B_{1n} - nB_{2n} + B_{3n} - nB_{4n}\big] \sin 2n\varphi.$$

Here we have introduced the functions

$$B_{1m}(\rho, t) = K_m^+ \Phi_{m+1}(\rho, t) + \big[b(t)/\rho\big]^{2m+2} \alpha_{1m}(t), \quad B_{2m}(\rho, t) = \varkappa \Phi_m(\rho, t) + \big[b(t)/\rho\big]^{2m} \alpha_{2m}(t),$$
$$B_{3m}(\rho, t) = K_m^- \Psi_{m-1}(\rho, t) + (\rho/a)^{2m-2} \alpha_{3m}(t), \qquad B_{4m}(\rho, t) = \varkappa \Psi_m(\rho, t) + (\rho/a)^{2m} \alpha_{4m}(t);$$

$$\Phi_m(\rho, t) = \frac{k}{2m-1} \int_a^\rho (\xi/\rho)^{2m} h(\xi, t)\, d\xi, \qquad \Psi_m(\rho, t) = \frac{k}{2m+1} \int_\rho^{b(t)} (\rho/\xi)^{2m} h(\xi, t)\, d\xi.$$

$K_m^\pm = \big[m(2m \pm 3)/(2m \mp 1)\big] \varkappa \pm 1$ and $k = 2/\big[\pi(\varkappa + 1)\big]$ are numerical multipliers. Functions $\alpha_{jm}(t)$ are to be found from the system of linear equations that we obtain on satisfying the boundary conditions for (18) on the internal and external erected vault surfaces. We do not present the expressions for α_{jm} because they are extremely cumbersome.

After the problem (18) is solved, the tensor $\mathbf{T}^\circ$ evolution in every point of the completed vault can be restored from the found velocity of changes of this tensor by means of the initial condition (14): $\mathbf{T}^\circ(\mathbf{r}, t) = \int_{\tau_*(\rho)}^t \mathbf{S}(\mathbf{r}, \tau)\, d\tau$, $t \geqslant \tau_*(\rho)$. Then using the known evolution of the tensor $\mathbf{T}^\circ$ it is possible according to its definition to determine the evolution of the stress tensor $\mathbf{T}$, by solving the integral Volterra equation of the second kind

$$\mathbf{g}(\mathbf{r}, t) - \int_{\tau_*(\rho)}^t \mathbf{g}(\mathbf{r}, \tau) K(t, \tau)\, d\tau = \mathbf{T}^\circ(\mathbf{r}, t) \tag{20}$$

with the parameter $\mathbf{r}$ for the unknown tensor-function of time $\mathbf{g}(\mathbf{r}, t) = \mathbf{T}(\mathbf{r}, t)/G(t)$. In analytical form will be

$$\mathbf{T}(\mathbf{r}, t) = \mathcal{H}_{\tau_*(\rho)}^{-1} \mathbf{T}^\circ(\mathbf{r}, t) = G(t) \left[\mathbf{T}^\circ(\mathbf{r}, t) + \int_{\tau_*(\rho)}^t \mathbf{T}^\circ(\mathbf{r}, \tau) R(t, \tau)\, d\tau\right], \quad t \geqslant \tau_*(\rho). \tag{21}$$

However, if the expression for the resolvent $R(t, \tau)$ of the kernel $K(t, \tau)$ is too complicated to calculate using (21), then it seems more desirable to solve (20) numerically, for example, by the quadrature method [12].

5. Residual stresses in the finished vault after centering removing

Above we constructed the solution of the problem on the deformation of a viscoelastic aging vault raised under gravity action on a smooth rigid base on a supporting rigid centering which remains adhered to the internal vault surface. This problem makes sense for any arbitrary large time $t > t_1$. Let it be determined by known functions $\mathbf{v}(\mathbf{r},t)$, $\mathbf{D}(\mathbf{r},t)$, $\mathbf{S}(\mathbf{r},t)$, $\mathbf{T}(\mathbf{r},t)$. Knowing the last one we can determine the centering reaction, that is the evolution of distribution of contact stresses on the internal vault surface $\mathbf{t}_a(\varphi,t) = -\mathbf{e}_\rho \cdot \mathbf{T}|_{\rho=a}$. Then the problem (18) for functions $\mathbf{v}(\mathbf{r},t)$, $\mathbf{D}(\mathbf{r},t)$, $\mathbf{S}(\mathbf{r},t)$ can be replaced by equivalent boundary value problem

$$\begin{gathered}\nabla\cdot\mathbf{S} = \mathbf{j}\,h(\rho,t), \quad \rho\in\big(a,b(t)\big), \quad \varphi\in(0,\pi), \quad t>t_1;\\ \mathbf{S} = 2\mathbf{D} + (\varkappa-1)\,\mathbf{1}\,\mathrm{tr}\,\mathbf{D}, \quad \mathbf{D} = (\nabla\mathbf{v}^{\mathrm{T}} + \nabla\mathbf{v})/2; \quad \mathbf{k}\cdot\mathbf{v} = 0, \quad \partial\mathbf{v}/\partial z = \mathbf{0};\\ \mathbf{e}_\rho\cdot\mathbf{S}|_{\rho=a} = -\mathbf{p}(\varphi,t); \quad \mathbf{e}_\rho\cdot\mathbf{S}|_{\rho=b(t)} = -\mathbf{j}\,q(t); \quad \mathbf{e}_\varphi\cdot\mathbf{S}|_{\varphi=0,\pi}\cdot\mathbf{e}_\rho = 0, \quad \mathbf{e}_\varphi\cdot\mathbf{v}|_{\varphi=0,\pi} = 0.\end{gathered} \tag{22}$$

Here we introduce the notation $\mathbf{p} = \partial\mathbf{t}_a^\circ/\partial t$. Adjusted for equality $\tau_*(a) = t_1$ we write down

$$\mathbf{p}(\varphi,t) = \frac{\partial\mathcal{H}_{t_1}\mathbf{t}_a(\varphi,t)}{\partial t} = \frac{\partial}{\partial t}\left[\frac{\mathbf{t}_a(\varphi,t)}{G(t)} - \int_{t_1}^{t}\frac{\mathbf{t}_a(\varphi,\tau)}{G(\tau)}K(t,\tau)\,d\tau\right]. \tag{23}$$

Now we shall assume that, starting from some time $t = t_{\det} > t_2$, when the formation of the vault has been already completed, its internal surface stresses start changing in a way that differs from that prescribed by a presence of a rigid centering in the structure. It changes arbitrary according to a law $\tilde{\mathbf{t}}_a(\varphi,t)$, $t > t_{\det}$. Suppose this loading is a continuous time function that has piecewise continuous derivative at any time period and $\tilde{\mathbf{t}}_a(\varphi,t_{\det}+0) = \mathbf{t}_a(\varphi,t_{\det})$. When $t \leqslant t_{\det}$ we assume $\tilde{\mathbf{t}}_a(\varphi,t) \equiv \mathbf{t}_a(\varphi,t)$.

By $\tilde{\mathbf{v}}(\mathbf{r},t)$, $\tilde{\mathbf{D}}(\mathbf{r},t)$, $\tilde{\mathbf{S}}(\mathbf{r},t)$, $\tilde{\mathbf{T}}(\mathbf{r},t)$ denote functions that describe the stress-strain state changing of the vault which internal surface is loaded from the instant of starting of its erecting to the arbitrary large instant with the distributed loading $\tilde{\mathbf{t}}$. The first three of these functions are evidently solutions of the boundary value problem

$$\begin{gathered}\nabla\cdot\tilde{\mathbf{S}} = \mathbf{j}\,h(\rho,t), \quad \rho\in\big(a,b(t)\big), \quad \varphi\in(0,\pi), \quad t>t_1;\\ \tilde{\mathbf{S}} = 2\tilde{\mathbf{D}} + (\varkappa-1)\,\mathbf{1}\,\mathrm{tr}\,\tilde{\mathbf{D}}, \quad \tilde{\mathbf{D}} = (\nabla\tilde{\mathbf{v}}^{\mathrm{T}} + \nabla\tilde{\mathbf{v}})/2; \quad \mathbf{k}\cdot\tilde{\mathbf{v}} = 0, \quad \partial\tilde{\mathbf{v}}/\partial z = \mathbf{0};\\ \mathbf{e}_\rho\cdot\tilde{\mathbf{S}}|_{\rho=a} = -\tilde{\mathbf{p}}(\varphi,t); \quad \mathbf{e}_\rho\cdot\tilde{\mathbf{S}}|_{\rho=b(t)} = -\mathbf{j}\,q(t); \quad \mathbf{e}_\varphi\cdot\tilde{\mathbf{S}}|_{\varphi=0,\pi}\cdot\mathbf{e}_\rho = 0, \quad \mathbf{e}_\varphi\cdot\tilde{\mathbf{v}}|_{\varphi=0,\pi} = 0.\end{gathered} \tag{24}$$

Here $\tilde{\mathbf{p}} = \partial\tilde{\mathbf{t}}_a^\circ/\partial t$. The extensive variant of this equality is analogous to (23).

Introduce the vector-functions $\mathbf{t}_{a\Delta} = \tilde{\mathbf{t}}_a - \mathbf{t}_a$, $\mathbf{p}_\Delta = \tilde{\mathbf{p}} - \mathbf{p}$, $\mathbf{v}_\Delta = \tilde{\mathbf{v}} - \mathbf{v}$, $\mathbf{D}_\Delta = \tilde{\mathbf{D}} - \mathbf{D}$, $\mathbf{S}_\Delta = \tilde{\mathbf{S}} - \mathbf{S}$, $\mathbf{T}_\Delta = \tilde{\mathbf{T}} - \mathbf{T}$. As $\mathbf{t}_{a\Delta} \equiv \mathbf{0}$ when $t \leqslant t_{\det}$ we get

$$\begin{gathered}\mathbf{p}_\Delta(\varphi,t) = \frac{\partial}{\partial t}\left[\frac{\mathbf{t}_{a\Delta}(\varphi,t)}{G(t)} - \int_{t_1}^{t}\frac{\mathbf{t}_{a\Delta}(\varphi,\tau)}{G(\tau)}K(t,\tau)\,d\tau\right] = \begin{cases}\mathbf{0}, & t < t_{\det},\\ \partial\mathcal{H}_{t_{\det}}\mathbf{t}_{a\Delta}(\varphi,t)/\partial t, & t > t_{\det};\end{cases}\\ \int_{t_{\det}}^{t}\mathbf{p}_\Delta(\varphi,\tau)\,d\tau = \mathcal{H}_{t_{\det}}\mathbf{t}_{a\Delta}(\varphi,\tau)\Big|_{\tau=t_{\det}}^{t} = \mathcal{H}_{t_{\det}}\mathbf{t}_{a\Delta}(\varphi,t) - \frac{\mathbf{t}_{a\Delta}(\varphi,t_{\det})}{G(t_{\det})} =\\ = \mathcal{H}_{t_{\det}}\mathbf{t}_{a\Delta}(\varphi,t), \quad t \geqslant t_{\det}.\end{gathered} \tag{25}$$

Subtracting relations (22) from respective relations (24) we obtain the problem

$$\begin{gathered}\nabla\cdot\mathbf{S}_\Delta = \mathbf{0}, \quad \rho\in\big(a,b(t)\big), \quad \varphi\in(0,\pi), \quad t>t_1;\\ \mathbf{S}_\Delta = 2\mathbf{D}_\Delta + (\varkappa-1)\,\mathbf{1}\,\mathrm{tr}\,\mathbf{D}_\Delta, \quad \mathbf{D}_\Delta = (\nabla\mathbf{v}_\Delta{}^{\mathrm{T}} + \nabla\mathbf{v}_\Delta)/2; \quad \mathbf{k}\cdot\mathbf{v}_\Delta = 0, \quad \partial\mathbf{v}_\Delta/\partial z = \mathbf{0};\\ \mathbf{e}_\rho\cdot\mathbf{S}_\Delta|_{\rho=a} = -\mathbf{p}_\Delta(\varphi,t); \quad \mathbf{e}_\rho\cdot\mathbf{S}_\Delta|_{\rho=b(t)} = \mathbf{0}; \quad \mathbf{e}_\varphi\cdot\mathbf{S}_\Delta|_{\varphi=0,\pi}\cdot\mathbf{e}_\rho = 0, \quad \mathbf{e}_\varphi\cdot\mathbf{v}_\Delta|_{\varphi=0,\pi} = 0.\end{gathered} \tag{26}$$

Under (25) the solution of this problem is trivial when $t < t_{\rm det}$. In particular, $\mathbf{S}_\Delta \equiv \mathbf{0}$ when $t < t_{\rm det}$. Therefore,

$$\begin{aligned} \mathbf{T}^\circ_\Delta(\mathbf{r},t) &= \int_{\tau_*(\rho)}^t \mathbf{S}_\Delta(\mathbf{r},\tau)\,d\tau = \begin{cases} \mathbf{0}, & \tau_*(\rho) < t < t_{\rm det}, \\ \int_{t_{\rm det}}^t \mathbf{S}_\Delta(\mathbf{r},\tau)\,d\tau, & t \geqslant t_{\rm det}; \end{cases} \\ \mathbf{T}_\Delta(\mathbf{r},t) &= \mathcal{H}^{-1}_{\tau_*(\rho)}\mathbf{T}^\circ_\Delta(\mathbf{r},t) = G(t)\left[\mathbf{T}^\circ_\Delta(\mathbf{r},t) + \int_{t_{\rm det}}^t \mathbf{T}^\circ_\Delta(\mathbf{r},\tau)R(t,\tau)\,d\tau\right] = \\ &= \mathcal{H}^{-1}_{t_{\rm det}}\mathbf{T}^\circ_\Delta(\mathbf{r},t), \quad t \geqslant t_{\rm det}. \end{aligned} \tag{27}$$

In the region occupied by the having been erected vault we determine the functions $\mathbf{u}_\Delta(\mathbf{r},t) = \int_{t_{\rm det}}^t \mathbf{v}_\Delta(\mathbf{r},\tau)\,d\tau$, $\mathbf{E}_\Delta(\mathbf{r},t) = \int_{t_{\rm det}}^t \mathbf{D}_\Delta(\mathbf{r},\tau)\,d\tau$, $t \geqslant t_{\rm det}$. By integrating all relations in problem (26) with respect to the parameter from $t_{\rm det}$ to $t \geqslant t_{\rm det}$, and by applying the operator $\mathcal{H}^{-1}_{t_{\rm det}}$ to the integrating result those of them that originally contained the tensor $\mathbf{S}_\Delta$, and taking into account (27), (25) we obtain the following boundary value problem:

$$\begin{gathered} \nabla\cdot\mathbf{T}_\Delta = \mathbf{0}, \quad \rho\in(a,b_{\rm fin}), \quad \varphi\in(0,\pi), \quad t\geqslant t_{\rm det}; \\ \frac{\mathbf{T}_\Delta}{G(t)} = (\mathcal{I}+\mathcal{N}_{t_{\rm det}})\big[2\mathbf{E}_\Delta + (\varkappa-1)\,\mathbf{1}\,\mathrm{tr}\,\mathbf{E}_\Delta\big], \quad \mathbf{E}_\Delta = \frac{\nabla\mathbf{u}_\Delta^{\rm T} + \nabla\mathbf{u}_\Delta}{2}; \quad \mathbf{k}\cdot\mathbf{u}_\Delta = 0, \quad \frac{\partial\mathbf{u}_\Delta}{\partial z} = \mathbf{0}; \\ \mathbf{e}_\rho\cdot\mathbf{T}_\Delta|_{\rho=a} = -\mathbf{t}_{a\Delta}(\varphi,t); \quad \mathbf{e}_\rho\cdot\mathbf{T}_\Delta|_{\rho=b_{\rm fin}} = \mathbf{0}; \quad \mathbf{e}_\varphi\cdot\mathbf{T}_\Delta|_{\varphi=0,\pi}\cdot\mathbf{e}_\rho = 0, \quad \mathbf{e}_\varphi\cdot\mathbf{u}_\Delta|_{\varphi=0,\pi} = 0. \end{gathered} \tag{28}$$

We see that the problem is a classical problem of plain deforming for a weightless vault that satisfies the constitutive relation (1), attached to a smooth rigid base by a sliding mount and loaded by a time-dependent distributed loading at its internal surface at time $t = t_{\rm det}$. In this problem, the displacement vector, the small strain tensor, and the stress tensor are replaced by the variables $\mathbf{u}_\Delta$, $\mathbf{E}_\Delta$, $\mathbf{T}_\Delta$, respectively. According to the known *correspondence principle* of the linear theory of aging solids viscoelasticity [1] the tensor $\mathbf{T}_\Delta$, representing the solution of problem (28), can be found at each time $t \geqslant t_{\rm det}$ from the solution of the corresponding elastic problem which should be got by substituting in (28) the operator $\mathcal{I} + \mathcal{N}_{t_{\rm det}}$ with the identity operator $\mathcal{I}$ and contains t as a parameter. If $t \to +\infty$ we get that the tensor $\mathbf{T}_\Delta(\mathbf{r},+\infty)$ satisfies to the elastic problem with the loading $\mathbf{t}_{a\Delta}(\varphi,+\infty)$ and the shear modulus $G(+\infty)$.

If the new law $\tilde{\mathbf{t}}_a(\varphi,t)$ of changing stresses at the internal surface of the originally considered erected vault leads to the exemption of this surface, i.e. to the centering removing, in other words if $\tilde{\mathbf{t}}_a(\varphi,+\infty) = \mathbf{0}$, then $\mathbf{t}_{a\Delta}(\varphi,+\infty) = -\mathbf{t}_a(\varphi,+\infty)$. In this case the tensor $\tilde{\mathbf{T}}(\mathbf{r},+\infty) = \mathbf{T}(\mathbf{r},+\infty) + \mathbf{T}_\Delta(\mathbf{r},+\infty)$ is a tensor of residual stresses in the completed vault after centering removing. Thus we have proved

PROPOSITION. *Let the stress tensors $\mathbf{T}_{\rm res}(\mathbf{r})$ and $\mathbf{T}_\infty(\mathbf{r})$ define the long-term states that take place in a heavy aging viscoelastic vault under consideration after the process of its raising is completed and the centering is removed or, respectively, the vault is exposured on this centering for a long time. In the first case we have $\mathbf{e}_\rho\cdot\mathbf{T}_{\rm res}|_{\rho=a} \equiv \mathbf{0}$, and in the second case $\mathbf{t}_{a\infty}(\varphi) = -\mathbf{e}_\rho\cdot\mathbf{T}_\infty|_{\rho=a}$ is the steady-state centering reaction. Let the stress tensor $\mathbf{T}_{\rm el}(\mathbf{r})$ define the state of a weightless elastic vault that is built without residual stresses, has the same dimensions as the having been erected viscoelastic vault in question and the same elastic characteristics as this vault has after a long exposure. The elastic vault mentioned is loaded at its internal surface by the distributed loading $\mathbf{t}_{a\infty}$, with the conditions of feet attaching being the same as for the erected vault. Then $\mathbf{T}_{\rm res} = \mathbf{T}_\infty - \mathbf{T}_{\rm el}$.*

6. Conclusions

In the present paper a process of gradual raising of a heavy semi-circular vault on a smooth rigid base is considered. The raising is implemented by means of continuous attachment of additional material layers to the vault external surface while its internal surface is supported by a rigid circular centering. The material used for the constructing possesses properties of creep and aging.

We have formulated a linear quasistatic problem of accreted solids mechanics that describes plane deforming of the vault under gravity action during its raising and after the raising stops but the completed vault is still being supported by the centering. This problem is transformed to a boundary value problem which mathematical form matches with that of the classical problem in elasticity theory. An analytical solution of the latter problem in series is built. By means of this solution the evolution of the stress-strain state in every point of the completed vault is exposed by time integrating procedure and by solving a Volterra integral equation of the second kind.

A proposition about structure of residual stresses in the completed vault after the centering removing is proved. According to the proposition these stresses can be determined as difference between the stresses corresponding to the accretion problem solved in the paper and stresses corresponding to a certain classical problem of elasticity theory with zero mass forces.

Acknowledgments
This work is supported by the Russian Foundation for Basic Research (under grants Nos. 08-01-00553-a, 08-01-91302-IND_a, 09-08-01180-a, and 09-08-01194-a), by the Department of Energetics, Mechanical Engineering, Mechanics and Control Processes of the Russian Academy of Sciences (Program No. 13 OE), and by the Russian Science Support Foundation.

References
[1] Arutyunyan N Kh and Manzhirov A V 1999 *Contact Problems in the Theory of Creep* (Erevan: NAN RA) [in Russian]
[2] Manzhirov A V 1995 The general non-inertial initial-boundaryvalue problem for a viscoelastic ageing solid with piecewise-continuous accretion *J. Appl. Math. Mech.* **59** (5) 805–816
[3] Manzhirov A V and Chernysh V A 1992 Strengthening of a buried arch by the way of accretion *Izv. Akad. Nauk. Mekh. Tverd. Tela* (5) 25–37
[4] Manzhirov A V and Parshin D A 2006 Accretion of a viscoelastic ball in a centrally symmetric force field *Mech. Solids* **41** (1) 51–64
[5] Manzhirov A V and Parshin D A 2006 Modeling the accretion of cylindrical solids on a rotating mandrel with centrifugal forces taken into account *Mech. Solids* **41** (6) 121–134
[6] Manzhirov A V and Parshin D A 2007 Modeling of the deformation process of accreted conic solids *Vestn. Samarsk. Gos. Univ. Estestv.-Nauch. Ser.* **54** (4) 290–303
[7] Manzhirov A V 2007 Mechanics of accreted solids: state-of-the-art, problems, and perspectives *Topical Problems in Continuum Mechanics* (Proc. Int. Conf. 25–28 September 2007 Tsakhkadzor) (Erevan: Erevan State Univ. Architecture and Civil Eng.) pp 243–246 [in Russian]
[8] Parshin D A 2007 Strengthening of a conic column by the way of accretion *Modern Problems in Continuum Mechanics* (Proc. XI Int. Conf. 26–29 November 2007 Rostov-on-Don) vol 1 (Rostov-on-Don: TsVVR) pp 205–209 [in Russian]
[9] Manzhirov A V and Parshin D A 2008 New results in the classical problems of Earth and mining sciences *Modern Problems in Continuum Mechanics* (Proc. XII Int. Conf. 1–5 December 2008 Rostov-on-Don) vol 1 (Rostov-on-Don: TsVVR) pp 145–149 [in Russian]
[10] Arutyunyan N Kh 1966 *Some Problems in the Theory of Creep* (Oxford: Pergamon Press)
[11] Arutyunyan N Kh and Metlov V V 1983 Nonlinear problem of the creep theory of accreted ageing bodies *Izv. Akad. Nauk SSSR. Mekh. Tverd. Tela* (4) 142–152
[12] Polyanin A D and Manzhirov A V 2008 *Handbook of Integral Equations* 2nd ed (Boca Raton, London: Chapman and Hall / CRC Press)

Deformation and fracture of plates of finite thickness

Andrei Kotousov, Steven Harding and John Codrington

School of Mechanical Engineering, University of Adelaide, SA 5005, Australia

e-mail: Andrei.Kotousov@adelaide.edu.au

Abstract. Plane theories of elasticity are the theoretical foundation of many important developments in engineering. However, the understanding of the elastic solutions derived under plane stress or plane strain assumption, is rather far from complete. In particular, it is not clear how adequate the classical two-dimensional solutions of the plane theories of elasticity are when applied to the analysis of actual plate components made of sufficiently brittle material and having a finite thickness. So far there is no generally accepted criterion for identifying what thickness would qualify for plane-stress or plane-strain states and, in general, what effect on the stress distribution the plate thickness has. In this work we make an attempt to review some recent numerical studies and analytical efforts in order to throw light on this problem.

1. Introduction

The solutions to the plane theories of elasticity have served as a basis for many engineering design procedures, standards and failure assessment techniques over a long period of time. The popularity of the plane theories can be attributed to their simplicity when compared to the three-dimensional equations of elasticity, which are not particularly amenable to analytical treatment. The plane theories of elasticity simplify the actual state of stress in a plate subjected to in-plane loading by adopting either plane stress (zero transverse stress) or plane strain (zero transverse strain components) assumptions. When dealing with plates of finite thickness, the dominant state of stress is often selected using a simple rule. For relatively thin plates the stress state is considered to be plane stress, or plain strain otherwise. However, currently there is no generally accepted criterion for determining what thickness corresponds to either stress states. As a consequence there exists a significant level of empiricism when determining whether the particular thickness of a plate for a particular elastic problem qualifies for either plane stress or plain strain state. More importantly, many exact three-dimensional solutions do not obey this simple rule, despite strong correlation between the state of stress and plate thickness. For instance, when considering an infinite plate with a circular hole loaded by internal pressure the stress state remains plane stress regardless of the plate thickness. In problems incorporating singularities (e.g., cracks, angular corners and multi-material joints), the region close to the stress singularity is widely regarded to be under plane strain conditions. However, the true three-dimensional state of strain in this region is considerably more complex and can deviate significantly from the plane strain solution, as has been demonstrated in recent numerical, theoretical and experimental studies (refer to references).

In this paper we attempt to address several previous experimental studies and recent analytical investigations that describe various three-dimensional phenomena for typical problems of elasticity

Journal of Physics: Conference Series **181** (2009) 012059 doi:10.1088/1742-6596/181/1/012059

widely encountered in scientific and engineering practices. Furthermore, recent numerical findings will be compared to the corresponding results obtained from plane theories of elasticity and first order plate theory. From this comparison some general conclusions can be derived in regards to the applicability of these theories to engineering analysis of plate components.

2. Non-singular solutions

To demonstrate the difference between the three-dimensional exact solutions and the two-dimensional counterparts, consider the elementary problem of an infinite plate of finite thickness 2h loaded by uniaxial stresses on infinity containing a circular hole as shown in Figure 1a. This problem has been studied by a number of researchers. The earlier studies were based on the plane theories of elasticity, the first being by Kirsch [1] in 1899, due to the difficulties posed by the three-dimensional equilibrium equations of elasticity. Related problems involving multiple and different shaped holes were solved by Muskhelishvilli [2] using the complex variable method and conformal mapping techniques. Comprehensive reviews on this subject have been given by Neuber [3] and Savin [4].

However, it has been found that in the case where the radius of the hole, R, is of a comparable order of magnitude to the plate thickness, 2h, three-dimensional effects can significantly influence the elastic stress distribution and, consequently, the failure of such components. In this instance the very thick or very thin assumptions of the plane theories of elasticity no longer provide an accurate estimate.

There are many experimental evidences indicating the plate thickness to have an effect in plate problems. One such example is that for relatively thin plates subjected to an in-plane tensile loading, the crack either originates at the corner, i.e. the interface between the hole and free surface of the plate, or at the centre of the plate as shown in Figure 1b. The opposite situation takes place for relatively thick plates where the crack typically originates in the vicinity of the corner [5], as depicted in Figure 1c. Of course, the presence of imperfections or discontinuities within the structure strongly influences the crack origin and so one explanation for the phenomenon is the likely presence of such discontinuities within the vicinity of the free surface. Another possible contributing factor to this phenomenon is that in the corner or mid-plate regions the stress concentration factor may differ from the value (SCF = 3) predicted by the plane theories of elasticity.

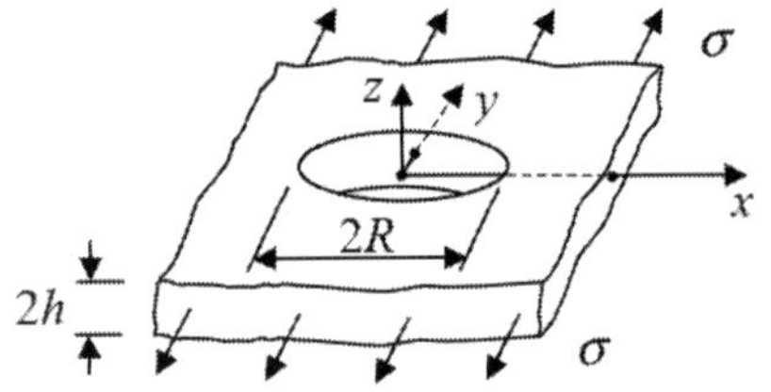

Fig.1a. Plate with a circular hole

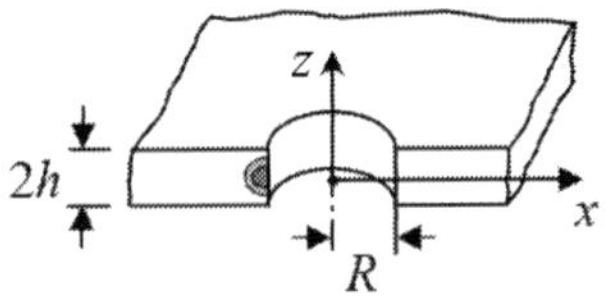

Fig.1b. Thin plate: failure location

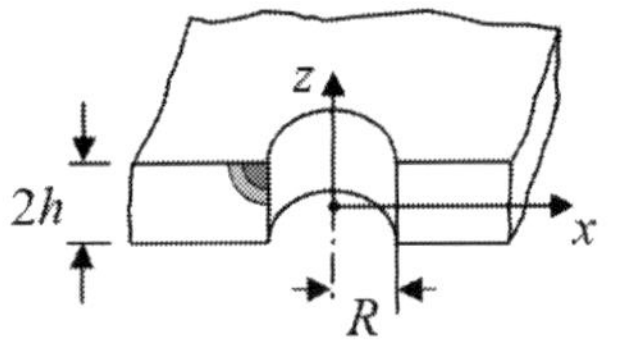

Fig.1b. Thick plate: failure location

In an effort to obtain a more accurate solution, various three-dimensional studies have been conducted including Green's solution in an infinite series form [6], approximate solutions given by Sternberg and Sadowsky [7] and numerical solutions by Alblas [8]. In particular, in 1962 Reiss [9] used a perturbation analysis to obtain a solution, which yielded three-dimensional corrections to that of the plane result, for small values of thickness to radius ratios, which substantiate the findings of Alblas. Later, by utilizing the first order plate theory, Kotousov and Wang [10] obtained an exact analytical solution for the problem within this theory. Finally, an analytical solution utilizing Kantorovish and Krylov [11] and Fourier transform methods was derived by Folias and Wang [12]. From their study, the maximum stress concentration factor distribution across the thickness at the critical locations $x = \pm R$ and $y = 0$, see Figure 1a., are shown in Figure 2, where $SCF_{max} = \sigma_{max}/\sigma$.

These theoretical results are convincing when explaining the preferable crack formation regions for different plate thicknesses as for relatively thin plates, R/h >1, the maximum stress occurs at the mid-plane ($z = 0$) where failure is most likely to originate. Conversely for relatively thick plates, R/h<0.5, the maximum stress is located in the vicinity of the free surfaces and hence is the preferred crack initiation region for thick plates

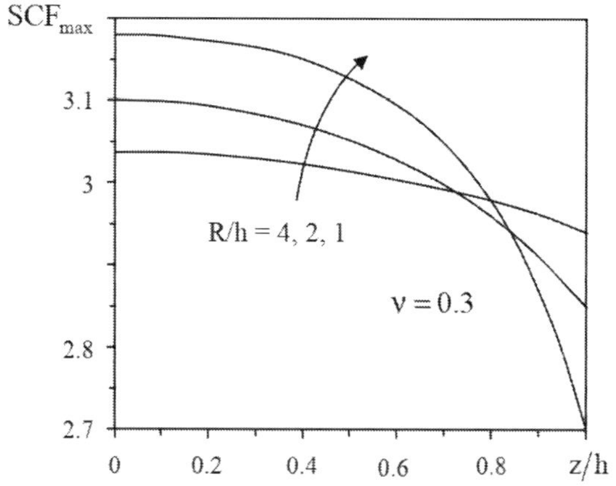

Fig.2a. Thin plate.

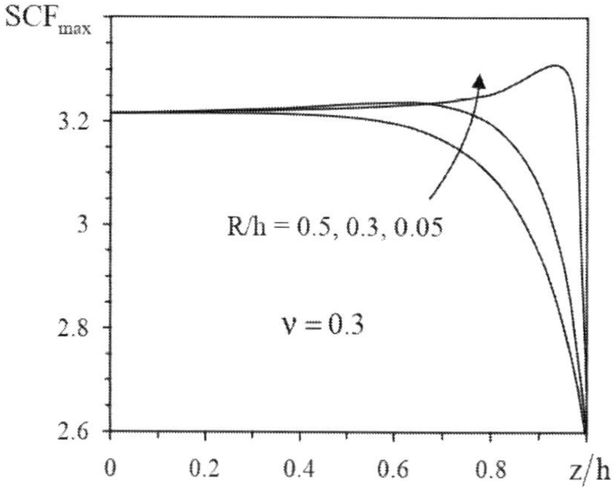

Fig. 2b. Thick plate.

Dependence of the stress concentration factor (SCF) across the plate thickness at critical location ($x = \pm R$ and $y = 0$)

Over the last two decades major progress in the analysis of three-dimensional solutions has been achieved with the application of the finite element method (FEM). Finite element results from various studies normally confirm the main features of the plane solutions and reveal new three-dimensional effects. For instance, Figure 3 shows the minimum and maximum stress concentration factors for $x = \pm R$ and $y = 0$ as a function of Poisson's ratio, ν [13].

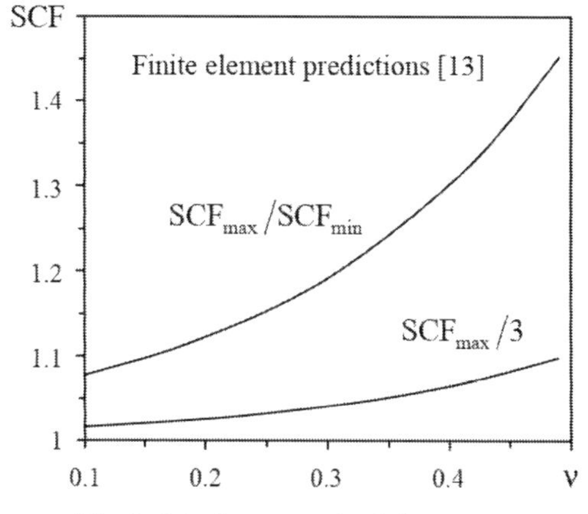

Fig.3. Maximum and minimum stress concentration factor at $x = \pm R$, $y = 0$ as a function of ν

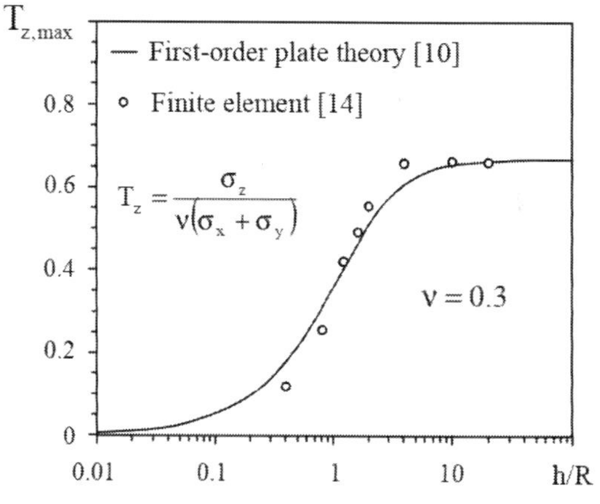

Fig.4. Maximum constraint factor as a function of half-thickness to hole radius ratio

Although there is only small a disparity between the maximum stress concentration factor and its plane stress or plane strain counterpart (SKF = 3), a significant variation of the stresses exists across the plate thickness, as given by the ratio of the maximum and minimum SCF's, which is significantly affected by Poisson's ratio, ν, potentially reaching 50 percent for high Poisson's ratio. In the case of elliptical holes subjected to uniaxial tension, recent three-dimensional numerical investigations have shown the difference between the plane theory of elasticity and the corresponding three-dimensional solutions to reach 30 percent in term of the maximum SCF, and 100 percent in terms of the variation of SCF across the plate thickness [13].

In an effort to characterize the dominant stress state in plate components, a constraint factor defined as the ratio of the out-of-plane stress, σ_z, to the sum of the in-plane stress components, $T_z = \sigma_z / \nu(\sigma_x + \sigma_y)$, is commonly employed to quantify the out-of-plane stress components [10-13]. In particular the constraint factor is especially useful in fracture mechanics and composites where delamination, which is strongly associated to the transverse stress components, is a principal failure mechanism. The constraint factor ranges from zero corresponding to the plane stress assumption, to one for the plane strain state. Numerical and analytical results are shown in Figure 4. for a Poisson's ratio of 0.3. This figure demonstrates that for the practically important situations ($0.1 < h/R < 10$) neither the plane stress or plane strain assumptions provide reasonable estimates of the transverse stress components.

It is worthwhile to briefly outline the primary features of the first order plate theory, also known as the Kane and Mindlin theory or generalized plane strain theory [10]. The theory was first developed by Kane and Mindlin in their work on high frequency extensional vibrations of short cylinders [15] in 1956, where the plane stress and plane strain assumptions under and over estimate the extensional natural frequencies observed experimentally [15]. The governing equations of the theory retain the transverse stress components, while still maintaining the simplicity of a two-dimensional model, to provide a good approximation to the experimentally observed results. The importance of this theory for the analysis of three-dimensional plate problems stems from the very fact that the first-order plane theory is, probably, the only elementary extension of the classical plane theory of elasticity, which allows for analytical three-dimensional solutions to be obtained for non-trivial geometries. Within the literature, there are now many examples of analytical and semi-analytical solutions derived within the theory. For instance, the theory provides an accurate estimate of the effect of the plate thickness on the stress distribution in a plate component [10] as seen from Figure 4.

3. Singular solutions

In this section we consider the classical singular problem of the theory of elasticity, namely an elastic wedge subjected to in-plane loading, which is also known as the Williams' problem. This problem was first systematically studied by Williams in 1952, where the in-plane stress components at the vertex of an isotropic corner were first shown to be singular [16], see Figure 5a. Later, by employing the plane theory of elasticity in conjunction with eigenfunction expansion approaches or Mellin and Fourier transform methods, various researchers investigated this singular mode for multi-material junctions [17], as well as for inelastic, anisotropic and inhomogeneous materials subjected to various boundary conditions. The solution is fundamental to the asymptotic methods of failure assessment that are becoming very popular to predict the integrity and life of various structural components (bi-materials, welded and adhesive joints), machine parts (contact and sliding pairs) as well as in the quantification of failure mechanisms (fracture, fatigue and fretting fatigue) [18-20].

Later a number of researchers, led by Benthem (1979), adopted a finite difference scheme in conjunction with the eigenfunction expansion method to demonstrate that at the intersection of a crack front and the free surface the square root Williams singularity disappears, where a separate and distinct 3D corner singularity develops in its place [21], see Figure 5b. As with the Williams' singularity, this result was generalized for various geometries, material models and multi-material junctions. For the specific case of crack-like profiles, experimental studies have found the corner singularity to strongly influence fatigue crack growth behavior [22].

In 2005 Kotousov first identified and investigated the out-of-plane mode for arbitrary wedges subjected to in-plane loading with reference to the first order plate theory [23], see Figure 5c. This singular mode was found to be associated with the Poisson's effect and corresponds to the out-of-plane shear stress components. The formation mechanism of this singular mode is illustrated in Figure 6 for the particular configuration of a zero vertex angle wedge (crack) subjected to anti-symmetric (in-plane shear) loading. Intuitively this loading will create both compressive and tensile regions along the opposing free edges of the crack faces, where Poisson's effect will encourage a scissor motion of the two crack faces in a similar manner as the tearing mode encountered in fracture mechanics (see Figure 6). This intuitive analysis is further supported by the corresponding out-of-plane displacements given by the classical asymptotic equations of the plane stress solution for a crack subjected to shear loading. However, in the past the out-of-plane singular mode has presumably been overlooked as the displacements at the crack tip are predicted to become infinite within the bounds of the plane stress theory of elasticity and so the corresponding plane solution cannot be utilised for a quantitative analysis of this singular mode. It is clear that the same mechanism will also generate the out-of-plane singularity for non-zero vertex angles.

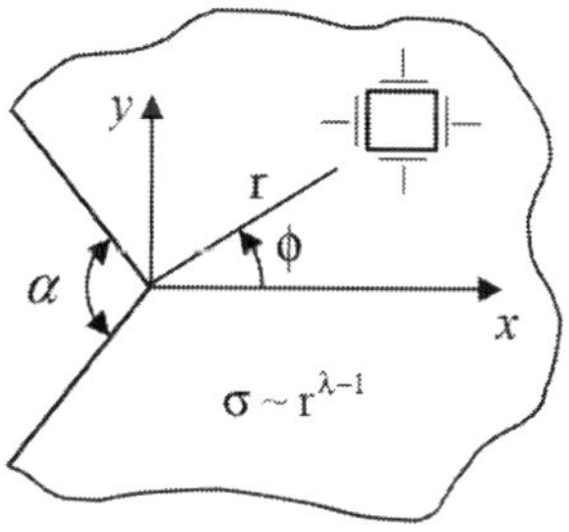

Fig.5a In-plane singularity

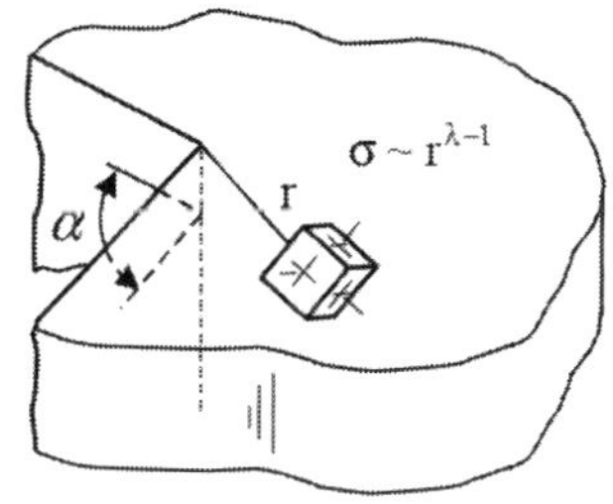

Fig.5b Corner singularity

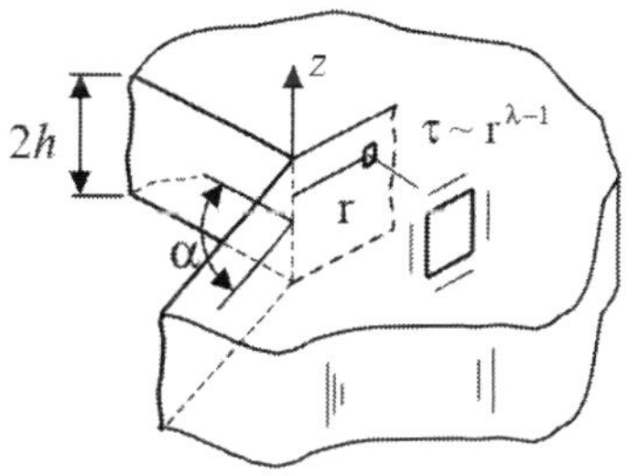

Fig.5b. Out-of-plane singularity

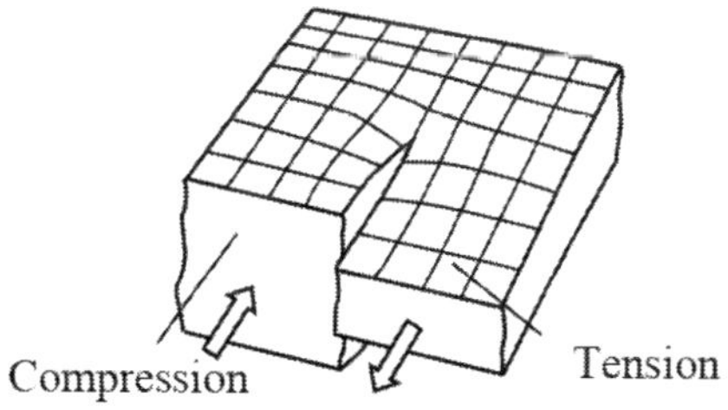

Fig.6 Crack in shear: Poisson's effect

The minimum values of λ $(\lambda > 0)$ which correspond to the strongest power of the singularity are shown in Figure 7a for the in-plane normal (mode I), shear (mode II) and out-of-plane (mode O) modes. It can be readably seen that the out-of-plane singular power matches that for the fracture mode III. However, there exist essential differences between the two. For example, the out-of-plane singular mode is symmetric about the mid-plane and coupled with the applied shear mode (mode II). It has been investigated numerically in the recent work of Harding and Kotousov [24]. Figure 7b depicts the dependence of the notch stress intensity factors as defined by $K = \lim_{r \to 0} \sigma (2\pi r)^{\lambda - 1}$ for a 90° notch.

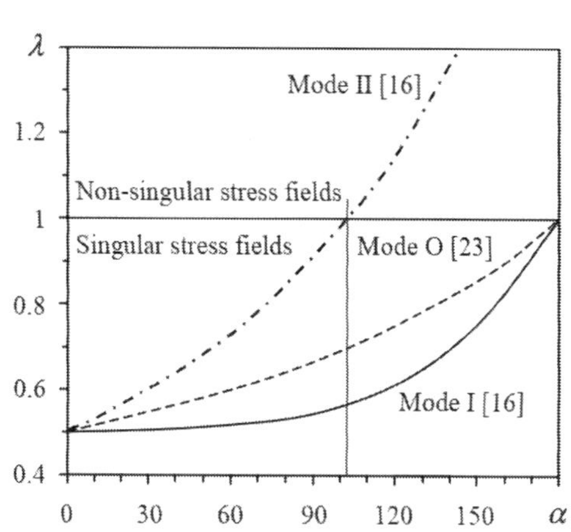

Fig.7a. Strength of singularity for in-plane mode I and II and O-mode

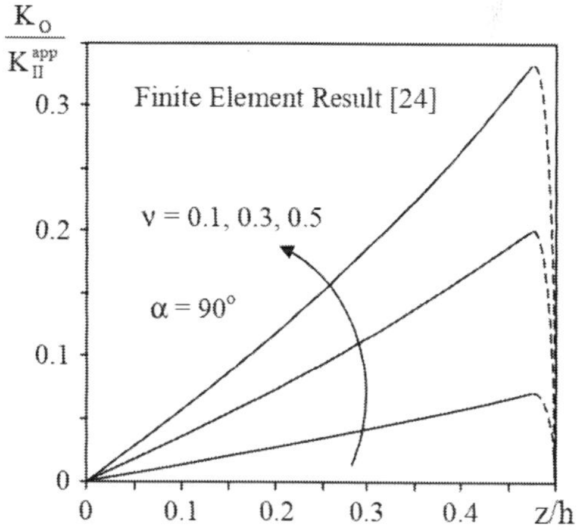

Fig. 7b. The ratio of intensity of O-mode to II-mode across the thickness for α = 90° and various Poisson's ratios in $m^{0.243}$

As shown in Figure 7a, the Poisson's ratio significantly affects the notch stress intensity factor, which can rival the magnitude of the shear mode in the case of a crack (zero notch angle). The most interesting finding of the numerical results is the existence of the out-of-plane singularity for vertex angles exceeding 102.4°, despite the disappearance of the in-plane shear singular mode [24]. This fact was overlooked in many studies that state that for notch angles exceeding this critical value no line singularity exists to contribute to the failure of notched components loaded in mode II.

4. Crack Problems

The crack geometry is important in many practical applications and so is discussed here in some detail. Two cases are considered; a crack stressed in mode I and in mode II. In other words we consider an infinite plate of thickness 2h, containing a through-the-thickness crack, subjected to K_I^{app} or K_{II}^{app} stress intensity factors. Due to space considerations, only the most interesting features of the three-dimensional solutions for these problems are discussed.

4.1 Crack stressed in mode I

The transition from plane strain ($T_z = 1$) to plane stress conditions ($T_z = 0$) for a semi-infinite crack is shown in Figure 8a below. Although the stress field at the crack tip generally follows the plane strain solution, the basic assumption, $\varepsilon_z = 0$, of the theory is not valid within close vicinity of the crack tip. The results of various numerical, analytical and experimental studies are shown in Figure 8b, which indicate the free surface displacements at the crack tip to be finite. Thus, the actual three-dimensional displacement and strain fields cannot be approximated by either plane stress or plane strain assumptions in the vicinity of the crack.

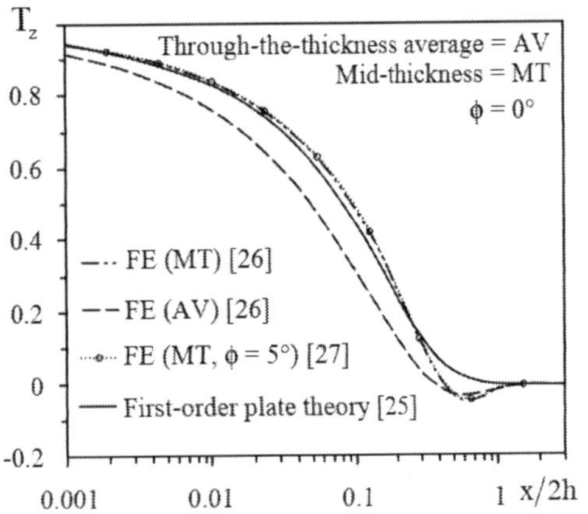

Fig. 8a. Constraint factor in the vicinity of the crack tip (x is the distance from the crack tip)

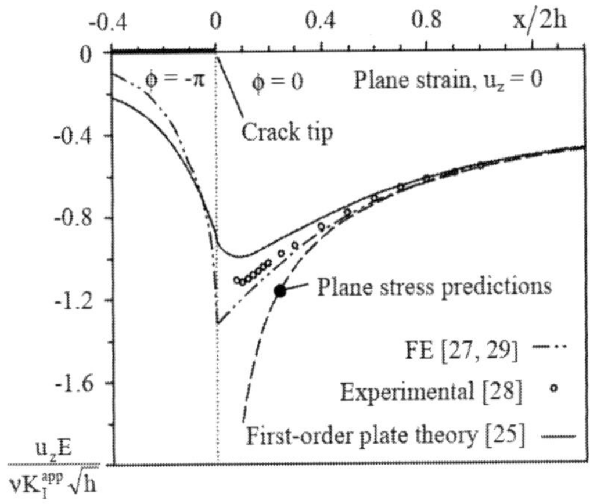

Fig.8b. Out-of-plane displacement, u_z, on the free surface, $z = \pm h$ near the crack tip

4.2 Crack stressed in mode II

From the various results available for cracks stressed in mode II, the dependence of the O-mode as a function of crack length is shown in Figure 9a, while Figure 9b depicts the variation of the stress intensity factor close to the plate surface [24].

Typical experimental studies are limited to h/a < 1. From Figures 9a and 9b there is a strong presence of three-dimensional effects as shown by the increase of the local stress intensity factor K_{II} and a significant magnitude of K_O close to the plate surface. It can be argued that these effects are responsible for the preferred region of failure initiation for cracks loaded in mode II, which is experimentally observed to be close to the plate surface, and the decreased fracture toughness for mode II fractures. The difference between the two can reach up to 50% depending on the specimen geometry and material, according to past experimental studies.

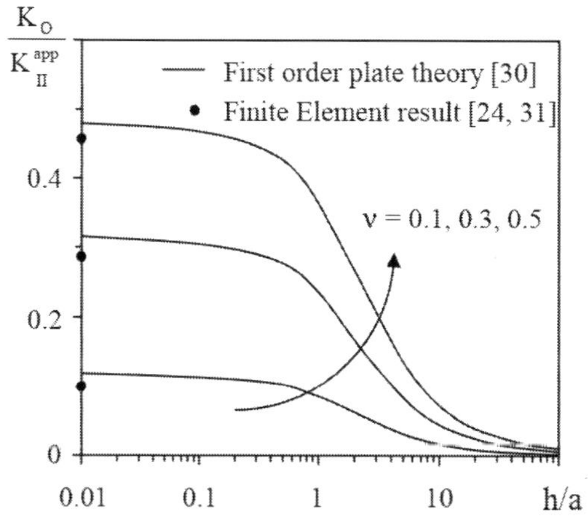

Fig. 9a. Intensity of O-mode as a function of the ratio of the crack length to the plate thickness

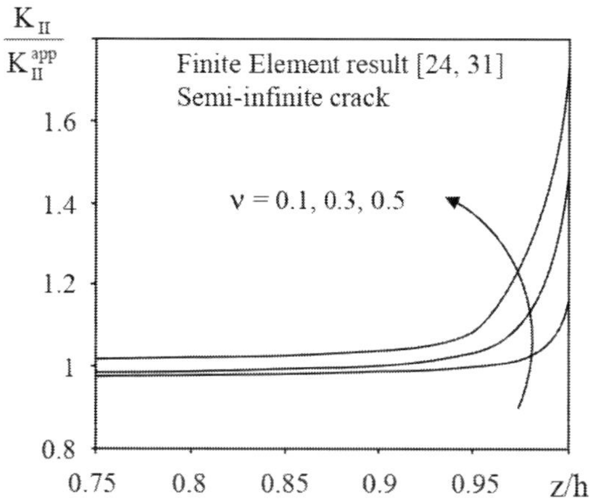

Fig.9b. Rise of the intensity of shear mode II in the vicinity of the plate surface

5. Conclusion

The purpose of this article has been to summarize several important and new features of the elastic three-dimensional solutions of plate problems. Significant differences were identified between the exact three-dimensional solutions and the corresponding solutions of the plane theory of elasticity. The distribution of in-plane stresses were shown to vary significantly through the plate thickness, in particular with problems incorporating singularities. For various important problems the plane stress theory of elasticity is unable to provide even a rough estimate of the transverse stress components. When applied, it is unsurprising that this theory sometimes provides peculiar results as it represents an approximate three-dimensional theory, even when the plane stress equations are solved exactly.

The first order theory was shown to provide a reasonable estimate of the in-plane and transverse stress components averaged across the plate thickness and led to the discovery of a new singular mode for notched components. This out-of-plane mode is coupled with the shear mode (mode II) and is related to the transverse shear stress components. This mode is significantly affected by Poisson's ratio, remains symmetric about the plate mid-plane and has zero magnitude at the plate surface and mid-plane. From recent numerical studies the theoretical findings have been confirmed and indicate that the mode could significantly influence fracture of cracked or notched components, contact or sliding mechanisms and bi-material or adhesive joints.

References

[1] Kirsch, G. (1898): Die Theorie der Elastizitat and die Bediirfnisse der Festigkeitslehre. Z. Vereines Deutscher Ing. 42, 797-807
[2] Muskhelishvili, N. I. (1953): Some basic problems of mathematical theory of elasticity. Groningen: Nordhoff
[3] Neuber, H. (1985): Kerbspannungslehre. Berlin, Heidelberg, New York: Springer
[4] Savin, G. N. (1961): Stress concentration around holes. New York: Pergamon Press
[5] Broek, D. (1974): Elementary engineering fracture mechanics. In: Chapt. 14: Plates with holes, 4th edn., pp. 361-370. Groningen: Nordhoff
[6] Green, A. E. (1948): Three-dimensional stress system in isotropic plates, I. Trans. R. Soc. London, Ser. A. 240, 561
[7] Sternberg, E.; Sadowsky, M. A. (1949): Three dimensional solution of the stress concentration around a circular hole in a plate of arbitrary thickness. J Appl Mech 16, 27-38

[8] Alblas, J.B. (1957): Theorie van de driedimensionale spanningstoestand in een doorboovde plaat. Dissertstion. Technische Hogeschool Delft, H. J. Pairs, Amsterdam.
[9] Reiss, E. L. (1963): Extension of an infinite plate with a circular hole. J Soc Ind Appl Math 11,840.
[10] Kotousov, A., Wang, C.H. (2002): Three-dimensional stress constraint in an elastic plate with a notch. Int J of Solids and Struct 39, 4311-4326.
[11] Kantorovich, L. V. and Krylov, V.L. (1964): Approximate methods of higher analysis, Groningcr: Nordhoff.
[12] Folias, E. S. and Wang, J.-J. (1990): On the three-dimensional stress field around a circular hole in a plate of arbitrary thickness. Comp Mech 6, 379-391.
[13] Yu, P., Guo, W., She, C. and Zhao, J. (2008) The influence of Poisson's ratio on thickness-dependent stress concentration at elliptic holes in elastic plates. Int J Fatigue 30, 165–171.
[14] Li, Z., Guo, W. and Kuang, Z. (2000): Three-dimensional elastic stress fields near notches in finite thickness plates. Int J Solids Struct 37, 7617-7631.
[15] Kane, T.R., Mindlin, R.D. (1956): High frequency extensional vibrations of plates. J Appl Mech 23, 277–283.
[16] Williams, M.L., (1952): Stress singularities resulting from various boundary conditions in angular corners of strips in extension. J Appl Mech 74, 526 – 528.
[17] Bogy, D.B. (1971): Two edge-bonded elastic wedges of different materials and wedge angles under surface tractions. J Appl Mech 38, 377 – 386.
[18] Nowell D, Dini D, Hills D.A. (2006): Recent developments in the understanding of fretting fatigue. Eng Fracture Mech 73, 207-222.
[19] Lazzarin P., Lassen T., Livieri P (2003): A Notch Stress Intensity approach applied to fatigue life predictions of welded joints with different local toe geometry. Fatigue Fracture of Eng Mat Struct 26, 49-48.
[20] Kotousov, A. (2007): Effect of a thin plastic adhesive layer on the stress singularities in a bi-material wedge. Int J Adhesion and Adhesives 27, 647-652.
[21] Benthem J.P. (1963): A Laplace transform method for the solution of semi-infinite and finite streip problems in stress analysis. Quarterly J Mech and Appl Math, XVI, 413-429.
[22] Heyder, M., Kolk, K., Kuhn, G. (2005): Numerical and experimental investigations of the influence of corner singularities on 3D fatigue crack propagation. Eng Fracture Mech 72, 2095-2105.
[23] Kotousov, A. (2005): On stress singularities at angular corners of plates of arbitrary thickness under tension. Int J Fracture 132, L29–L36.
[24] Harding, S. and Kotousov, A. (2009): A three-dimensional study of Williams' problem, Int J Solids Struct (submitted)
[25] Codrington, J, Kotousov, A. and Ho, S.Y. (2008): Out-of-plane stress and displacement for through-the-thickness cracks in plates of finite thickness. J Mechs Mat Struct 3, 261-270
[26] She, C. and Guo, W. (2007): The out-of-plane constraint of mixed-mode cracks in thin elastic plates, Int. J. Solids Struct 44, 3021–3034.
[27] Nakamura, T. and Parks, D. M. (1988): Three-dimensional stress field near the crack front of a thin elastic plate, J. Appl. Mech 55, 805–813.
[28] Humbert, L., Valle, V. and Cottron, M. (2000): Experimental determination and empirical representation of out-of-plane displacements in a cracked elastic plate loaded in mode I. Int J Solids Struct 37, 5493–5504
[29] Pfaff, R. D., Washabaugh, P. D. and Knauss, W. G. (1994): An interpretation of Twyman-Green interferograms from static and dynamic fracture experiments, Int J Solids Struct 32, 939–955.
[30] Kotousov, A. (2007): Fracture in plates of finite thickness. Int J Solids Struct 44, 8259 – 8273.
[31] Nakamura, T., Parks, D.M. (1989): Antisymmetrical 3-D stress field near the crack front of a thin elastic plate. Int J Solids Struct 25, 1411-1426.

Journal of Physics: Conference Series **181** (2009) 012060 doi:10.1088/1742-6596/181/1/012060

Modelling of Crack Deflection at Core Junctions in Sandwich Structures

J. Jakobsen[1], J.H. Andreasen[2] and O.T. Thomsen[2]

[1] Siemens Wind Power A/S, Blade Development, Assensvej 11, DK-9220 Aalborg East, Denmark

[2] Department of Mechanical Engineering, Aalborg University, Pontoppidanstraede 101, DK-9220 Aalborg East, Denmark

E-mail: ott@me.aau.dk

Abstract. The paper treats the problem of crack propagation in sandwich panels with interior core junctions. When a face-core interface crack approaches a trimaterial wedge, as it may occur at a sandwich core junction, two options exist for further crack advance; one is for the interface crack to penetrate the wedge along the face-core interface, and the second is deflection along the core junction interface. Crack deflection is highly relevant and a requirement for the functionality of a newly developed peel stopper for sandwich structures. The physical model presented in this paper enables the quantitative prediction of the ratio of the toughnesses of the two wedge interfaces required to control the crack propagation, and the derived results can be applied directly in future designs of sandwich structures. The solution strategy is based on finite element analysis (FEA), and a realistic engineering practice example of a tri-material composition corresponding to face and core materials is presented.

1. Introduction

A sandwich material is a layered assembly made from two thin, strong, and stiff face sheets bonded to a lightweight compliant core material. This creates a stiff, strong and lightweight structural element. The sandwich concept has been vastly utilized by the aerospace industry for decades due to the requirements for high stiffness and strength and at the same time low weight. However the marine, transport, and wind turbine industries have developed a significant interest in adopting the sandwich concept during the last 20-30 years due to increased mechanical performance requirements in these sectors, the possibility of tailored design as well as cost effectiveness [4, 5]. Failure in sandwich structures has been and still is an important design issue due to their sensitivity towards the application of strongly localised loads and poor damage tolerance [1-3, 5].

Delamination/debonding failure and damage tolerance of sandwich structures have been considered previously by the authors of this paper [1-3], and a new concept to stop face-core debonding has been developed. This concept has been referred to as a peel stopper [1-3], see Fig. 1, which appears as a "wedge" like core constituent made from a specially chosen material. The principal idea behind the proposed peel stopper concept is to ensure that when a core-face interface crack is approaching the wedge of the peel stopper it will deflected away from face-core interface by the wedge. This particular mechanism has proven to be a very effective way of confining propagating debonding cracks [1-3].

This paper is associated with the further development of the proposed peel stopper in general, and with the development of an in-depth physical understanding of the crack deflection process in particular. The paper presents a physical model, based on Linear Elastic Fracture Mechanics (LEFM) that is able to predict crack deflection or wedge penetration. The developed model may in the future be used to improve the peel stopper wedge design.

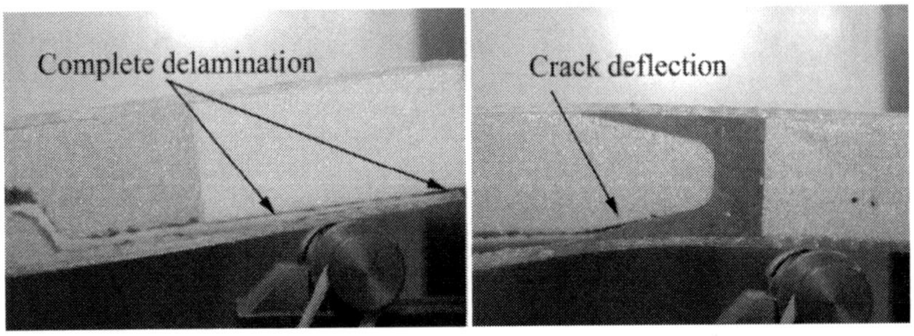

Figure1. Photos taken from a three point bending test with the debonding area in focus. Left: Shear crack initiated in the green polymer foam core (left); the crack kinked into a face-core interface crack underneath the yellow foam core followed by a complete core-face delamination/separation. Right: Peel stopper was embedded between the two foam cores. The wedge of the peel stopper deflected the interface crack into the core, where the crack was finally arrested [3].

The characterisation of failure phenomena in sandwich structures, weather they are considered from a failure initiation point of view or from a damage tolerance point of view, has been addressed by many researchers. Debonding failures due to the layered composition of sandwich structures have been studied intensely by a number of researchers. This includes experimental investigations of sandwich debonding and crack propagation [4-10], fatigue failure of sandwich structures [11,12] and local effects indiuced at core junctions, inserts and load introductions [13-17].

Cracks at bi-material interfaces between isotropic materials (i.e. like face-core interface cracks) were studied in [18-26]. Among the crack configurations relevant for the present study are edge cracks on bi-material interfaces [21-24], where [24] also considered crack kinking out of an interface. [25] considered a crack in a homogenous isotropic media approaching an interface with the aim of predicting whether the crack would penetrate the interface or be deflected by it. Finally [26] treats the stress singularities at a tri-material wedge for both fully bonded but also unbonded interfaces.

Several approaches have been proposed to characterise the tendency of a crack to propagate. The classical approach is to estimate the stress intensity factors of a given crack. Another method is to consider the energy release rate. The energy release rate may for linear elastic material behaviour be expressed in terms of the stress intensity factors. In this work the energy release rate is determined by evaluating the path independent J-integral around the crack tip [27]. The full details of the work presented herein may be found in [29] which again expands on the work in [2,3,28].

2. Modelling

A physical model has been proposed inspired by [21,22,25]. The model describes weather an approaching interface crack will be deflected by a tri-material wedge or if it will penetrate it. The model is based on LEFM, which limits the model to linear elastic material behaviour and small crack process zones. Moreover, the model presented is for homogenous isotropic materials only. However, the model can be expanded to orthotropic materials [29].

The aim is to establish interface toughness requirements for a broad band of mode mixities of an approaching interface crack. Furthermore a larger range of wedge angles will be considered. The work has been limited to only one interface composition, which is an isotropic approximation for a glass fibre (GFRP) face laminate bonded to Divinycell H60 and a H200 cross linked PVC foam core materials. This forms a generic trimaterial wedge type of problem as seen in many foam cored sandwich structures. The problem has been approached by studying the stress intensity factors a short distance (a) before and after the tri-material wedge (see Figure 2). This approach (or model) is here referred to as the "physical model". The physical model consists of three sub-models, which are named "Crack Approach", Crack Deflection", and "Crack Penetration", respectively, see Figure 2.

The thickness of the face sheet is modelled as finite and set to 2mm; whereas the other dimensions are modelled as "semi-infinite". The distance (a) indicated in the physical model (Figure 2) is considered as a characteristic distance of the problem, and it has been set to 1.26mm (see calculation of a in [28]). The choice of this value was based on the plastic yield zone size of the more compliant material, which for this tri-material composition is core-2 (H60). It should be mentioned that the sensitivity with respect to the face sheet thickness and characteristic distance values has not been analysed.

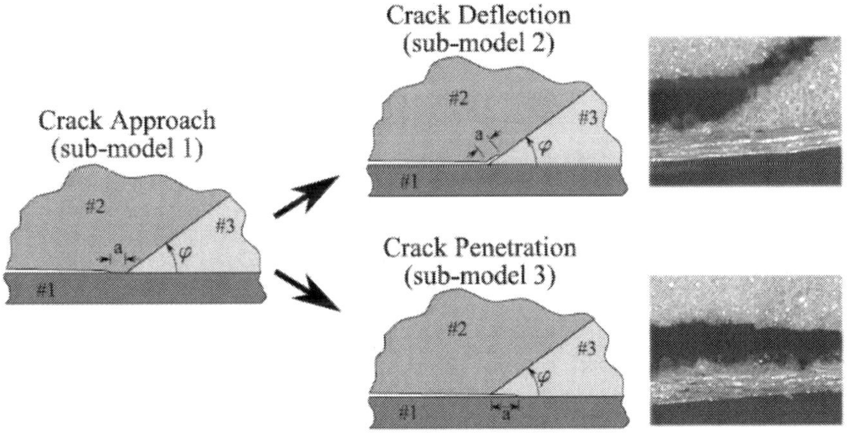

Figure 2. Physical model with two stages: (1) "Crack Approaching" and (2) "Crack Deflection" or "Crack Penetration". The wedge angle is φ.

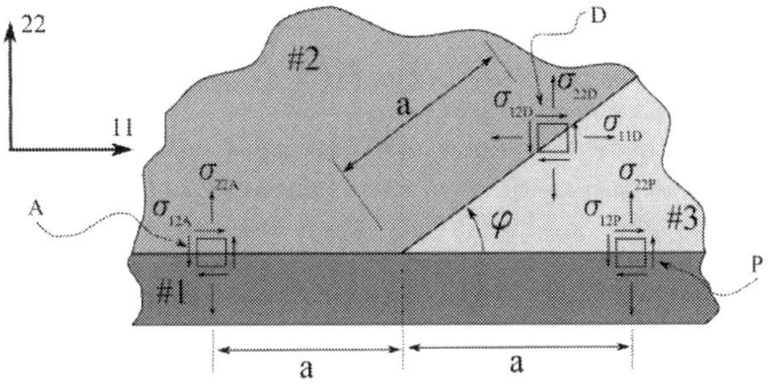

Figure 3. Illustration of method adopted for estimating the magnitude of the far field stresses. The stress level is determined at a distance (a) from the tri-material wedge. No cracks appear in this sub-model, which is referred to as the "Far field stress" model or sub-model 4.

The far field stresses may have a significant role in the interface toughness requirements (to provoke crack deflection), and they have been taken into account through a fourth sub-model (see

Figure 3). This sub-model is characterized by not having any crack included but it is being loaded in a similar way to the three other sub-models. The far field stresses are considered in three locations (A, D and P). These locations have been chosen to be identical to the crack tip positions for the "Approaching", "Deflected", and "Penetrated" crack stages, respectively.

In location A and P the transverse normal stresses (σ_{22}) and shear stresses (σ_{12}) will contribute to the singular stress field, whereas this is not the case for the normal stresses parallel to the interface (σ_{11}). For this particular reason only σ_{22} and σ_{12} are considered at locations A and P. Regarding location D the full 2D stress tensor is considered because all three stress component will contribute to the singular stress field at this location.

The physical model shown in Figure 2 and the far field stresses shown in Figure 3 have been formulated mathematically in terms of the stress intensity factors as shown in Eqs. (1) [29]:

$$\begin{aligned} K_{23} &= cK_{12}a^{i(\varepsilon_{12}-\varepsilon_{23})} + d\overline{K_{12}}a^{i(\varepsilon_{23}-\varepsilon_{12})} + b_D\sigma_{11D}a^{\frac{1}{2}-i\varepsilon_{23}} \\ &\quad + g_D\sigma_{22D}a^{\frac{1}{2}-i\varepsilon_{23}} + g_A\sigma_{22A}a^{\frac{1}{2}-i\varepsilon_{23}} + h_D\sigma_{12D}a^{\frac{1}{2}-i\varepsilon_{23}} + h_A\sigma_{12A}a^{\frac{1}{2}-i\varepsilon_{23}} \\ K_{13} &= eK_{12}a^{i(\varepsilon_{12}-\varepsilon_{13})} + f\overline{K_{12}}a^{i(\varepsilon_{13}-\varepsilon_{12})} \\ &\quad + m_P\sigma_{22P}a^{\frac{1}{2}-i\varepsilon_{13}} + m_A\sigma_{22A}a^{\frac{1}{2}-i\varepsilon_{13}} + n_P\sigma_{12P}a^{\frac{1}{2}-i\varepsilon_{13}} + n_A\sigma_{12A}a^{\frac{1}{2}-i\varepsilon_{13}} \end{aligned} \tag{1}$$

In Eq. (1) the stress intensity factors are named according to the interface combinations such that the stress intensity factor for the interface crack between material #1 and #2 is denoted K_{12}. Furthermore, the complex notation; $K_{12}=K_{1,12}+iK_{2,12}$, for mode-I and –II stress intensity factors has been utilized.

The stress components in Eqs. (1) refer to the far field stresses shown in Figure 3. The parameters c, d, b_D, g_D, g_A, h_D, h_A and e, f, m_P, m_A, n_P, n_A are complex coefficients, which depend on the six elastic constants involved in a tri-material problem and also the wedge angle (φ). ε_{ij} is the oscillatory index characterising each of the three interfaces, and (a) is the characteristic dimension of the problem (i.e. distance from the tri-material wedge). This distance has been given the same value along the three interfaces but it is possible to specify three independent distance values as well.

The results presented herein (see [29] for full details) have been solved in the absence of any far field stresses, but it should be mentioned that the influence from far field stresses may be significant. For that reason future work and development of this model should include determination of the complex functions associated with the far field stresses.

Similar to the crack kink problem summarized [28], the energy release rate for an interface crack can be determined from the stress intensity factors according to [30]. The energy release rates for the deflected and penetrated crack configurations are then formulated as shown in Eqs. (2):

$$G_{23} = \frac{\left(\frac{1-\nu_2}{\mu_2} + \frac{1-\nu_3}{\mu_3}\right) K_{23}\overline{K_{23}}}{\cosh(\pi\varepsilon_{23})}$$

$$G_{13} = \frac{\left(\frac{1-\nu_1}{\mu_1} + \frac{1-\nu_3}{\mu_3}\right) K_{13}\overline{K_{13}}}{\cosh(\pi\varepsilon_{13})} \tag{2}$$

Crack deflection may be predicted from the criterion stated in Eq (3), which was inspired by the work presented in [25]. Accordingly, crack deflection is predicted if the inequality condition in Eqs. (3) is fulfilled.

$$\frac{G_{23}}{G_{13}} \geq \frac{\Gamma_{23}(\psi_{23})}{\Gamma_{13}(\psi_{13})}$$

$$\frac{\left(\frac{1-\nu_2}{\mu_2}+\frac{1-\nu_3}{\mu_3}\right)K_{23}\overline{K_{23}}\cosh(\pi\varepsilon_{13})}{\left(\frac{1-\nu_1}{\mu_1}+\frac{1-\nu_3}{\mu_3}\right)K_{13}\overline{K_{13}}\cosh(\pi\varepsilon_{23})} \geq \frac{\Gamma_{23}(\psi_{23})}{\Gamma_{13}(\psi_{13})} \tag{3}$$

In Eqs. (3) G_{ij} and Γ_{ij} are the energy release rate and the interface toughness for the interface ij, respectively. In the absence of far field stresses the interface toughness ratio can be fully determined from the stress intensity factor of the "Crack Approaching" stage. The toughness ratio requirement may then be used as a design tool for a further development of the peel stopper proposed in [3].

3. Solution and generation of "prediction surface"

The solutions for the complex coefficients *c*, *d*, *e* and *f* can be obtained by analyzing and solving the four sub-models (See Figures 2 and 3) analytically, and using the output/results as input to the formulation of a set of algebraic equations which may subsequently be solved. See [29] for details.

However an approximate solution for the complex functions *c*, *d*, *e* and *f* was determined by solving a linear elastic fracture mechanics problem (LEFM) for the crack models sketched in Figures 2 and 3. Finite Element solutions for the LEFM problem and the different crack models were derived using ABAQUS v6.6®. Thus, FE-models corresponding to the three different crack configurations, See Figure 4 for the converged mesh distributions.

The models were analysed for a range of wedge angles (φ) starting from 10° and ending at 90°. This range was chosen because of its relevance to future peel stopper designs for sandwich structures [2,3]. Only solutions for positive mode mixities of the approaching crack have been performed. The reason for this is related to the materials commonly used in the experimental testing of the sandwich peel stopper [3]. These types of core materials will not allow an interface crack to propagate toward the wedge with negative mode mixities, because a negative mode mixity will make an approaching crack kink out of the interface before reaching the wedge. For full details see [29].

The solutions of the complex functions and the criterion in Eqs. (3) can be used to create a "prediction surface" as shown in Figure 5. Crack deflection is then predicted if the interface toughness ratio is below the red surface shown in Figure 5.

The prediction surface has not been drawn for the full range of mode mixities (ψ_{12}) because parts of the range yield negative values, with the physical implication that the crack faces overlap. This is (of course) not physically reasonable, and therefore the surface in Figure 5 has been cut of ("truncated"). For this particular tri-material composition only values (ψ_{12}, φ) within the green area represent a physically meaningful design space. By studying the prediction surface it is seen that for low core junction angles (φ) the physical model predicts the highest interface toughness ratio, which further leads to a lower manufacturing quality requirements (easier to fulfill the interface toughness ratio requirement). Moreover it may be seen that the physical model predicts physical results for the entire range of studied mode mixities (ψ_{12}). Based on these results, it is recommended that small wedge angles should be recommended for future development of the peel stopper, and this is in close agreement with the combined experimental findings and modeling results presented in [2,3,28].

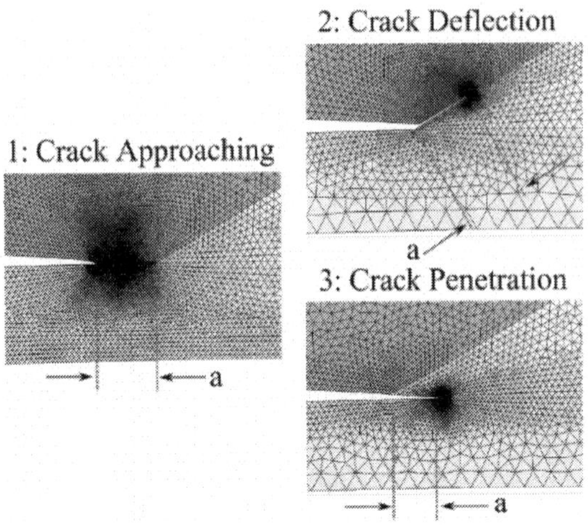

Figure 4. Mesh distributions in the vicinity of the tri-material wedge and the crack tip for the three crack stages [29].

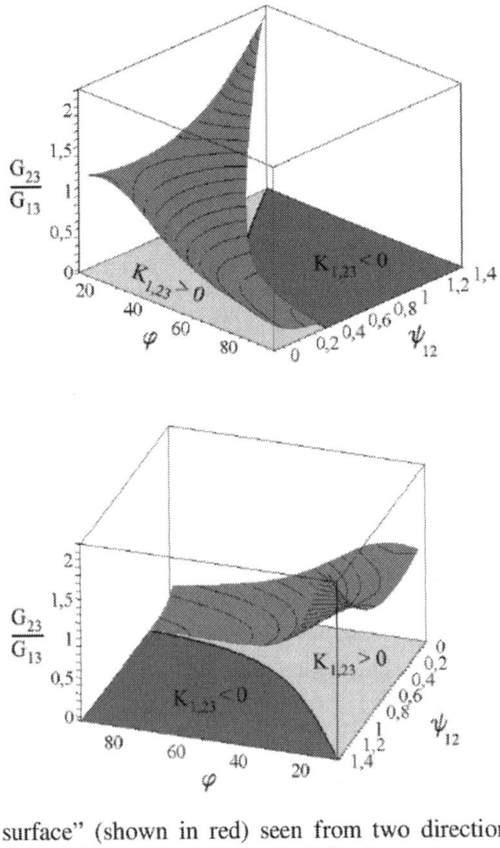

Figure 5. "Prediction surface" (shown in red) seen from two directions. Interface toughness ratios (G_{23}/G_{13}) below the surface will yield crack deflection, whereas crack penetration will occur for G_{23}/G_{13} ratios above the surface. The surface is cut off (truncated) as it only is physically meaningful for $K_{1,23}>0$ [29].

4. Conclusions

A physical model, based on LEFM and the assumption of isotropic material constituents, that is able to predict crack deflection or penetration for tri-material wedge problems has been presented. In principle the model relates the stress intensity factors of an approaching interface crack to the stress intensity factor of a deflected or penetrated crack through the determination of transition functions. These functions are formulated in terms of complex coefficients that depend on the actual composition of the tri-material wedge and the wedge angle. A solution to the complex coefficients and a FE-based solution approach has been demonstrated for a specifically chosen tri-material wedge composition. The selected composition is a close approximation to a tri-material wedge consisting of a glass fibre composite laminate, and two polymer foam core materials of different densities (PVC, Divinycell H200 and H60) as discussed in [2,3,28]. The practical background for this particular choice of material combination and geometry is a newly proposed device for crack deflection and crack arrest in sandwich structures, a so-called peel stopper, where the chosen tri-material configuration represents a sandwich face sheet and two adjoining core materials.

The purpose of the FE based approach is to create input to the physical model in such a way that the complex coefficients can be determined. The results presented only cover solutions to the singular part of the model/problem.

The complex coefficients determined from the FE analyses were used to construct a "prediction surface" for the analysed composition of the tri-material wedge through the proposed physical model. The "prediction surface" may be used as a design tool for identifying and choosing proper wedge angles, which further may be used to improve future peel stopper geometries as discussed in [2,3]. It was found that small wedge angles allow the highest interface toughness ratio, which means lower fabrication demands if an effective "peel stopper" is to be designed and manufactured.

5. Acknowledgements

The work presented was supported by:

- The Danish Research Council for Technology and Production Sciences (FTP), Grant/Award No. 26-04-0160 "Structural grading - a novel concept for the design of sandwich structures".
- The Innovation Consortium "Integrated Design and Processing of Lightweight Composite and Sandwich Structures" (abbreviated "KOMPOSAND") funded by the Danish Ministry of Science, Technology and Innovation and the industrial partners Composhield A/S, DIAB ApS (DIAB Group), Fiberline Composites A/S, LM Glasfiber A/S and Vestas Wind Systems A/S.
- US Navy, Office of Naval Research (ONR), Grant/Award No. N000140710227: "Influence of Local Effects in Sandwich Structures under General Loading Conditions & Ballistic Impact on Advanced Composite and Sandwich Structures". The ONR program manager was Dr. Yapa Rajapakse.

The support received is gratefully acknowledged.

6. References

[1] Jakobsen J., Bozhevolnaya E., Thomsen OT and Jakobsen, L. 2006 Peel Stopper for Sandwich Components *Patent Application PA 2006 01619, Aalborg University, Denmark..*

[2] Jakobsen J, Bozhevolnaya E, Thomsen OT. 2007 *Composites Science and Technology*, Vol. **67**, pp. 3378-3385.

[3] Bozhevolnaya E., Jakobsen J, Thomsen O.T. 2007 *Strain - An International Journal for Experimental Mechanics*, In press (doi:10.1111/j.1475-1305.2007.00413.x).

[4] Carlsson L, Matteson R, Aviles F, Loup D 2005 *Composites Science and Technology*, Vol. **65** (15-16) pp. 2612-2621.

[5] Avilés F and Carlsson L 2006 *Journal of Sandwich Structures and Materials*, Vol. **8**(1), pp.7-31.

[6] Berggreen C and Simonsen BC 2005 *Journal of Sandwich Structures and Materials*, Vol. **7**(6), pp. 483-517.

[7] Berggreen C, Simonsen BC and Borum KK 2007 *Journal of Composite Materials*, Vol. **41**(4), pp. 493-520.

[8] Nøkkentved A, Lundsgaard-Larsen C and Berggreen C 2005 *Journal of Sandwich Structures and Materials*, (2005), Vol. **7**(6), pp. 461-482.

[9] Lundsgaard-Larsen C, Sørensen BF, Berggreen C and Østergaard RC 2008 *Engineering Fracture Mechanics*, Vol. **75**(8), pp. 2514-2530.

[10] Østergaard, R.C., Sørensen B.F. and Brøndsted P., "Measurement of interface fracture toughness of sandwich structures under mixed mode loadings", *Journal of Sandwich Structures and Materials*. (2007), Vol. 9(5), pp. 445-466.

[11] Burman M and Zenkert D *International Journal of Fatigue*, Vol. **19**(7), pp. 551-561.

[12] Burman M and Zenkert D 1997 *International Journal of Fatigue*, Vol. **19**(7), pp. 563-578.

[13] Thomsen OT, Bozhevolnaya, E and Lyckegaard A 2005 *Composites Part A: Applied Science and Manufacturing*, Vol. **36**, pp. 1397-1411.

[14] Bozhevolnaya E and Thomsen OT 2005 *Composite Structures*, Vol. **70**(1), pp. 12-23.

[15] Bozhevolnaya E and Thomsen OT 2005 *Composite Structures*, Vol. **70**(1), pp. 1-11.

[16] Bozhevolnaya E and Lyckegaard A 2006 *Composite Structures*, Vol. **73**(1), pp. 24-32.

[17] Lyckegaard A, Bozhevolnaya E and Thomsen OT 2006 *Journal of Sandwich Structures and Materials*, Vol. **8**(5), pp. 423-435.

[18] Sih G and Rice J 1964 *Journal of Applied Mechanics*, Vol. **31**, pp. 477-482.

[19] Rice J and Sih G 1965 *Journal of Applied Mechanics*, Vol. **32**, pp. 418-423.

[20] Rice J 1988 *Journal of Applied Mechanics*, Vol. **55**, pp. 98-103.

[21] He MY and Hutchinson JW 1989 *Journal of Applied Mechanics*, Vol. **56**(2), pp. 270-278.

[22] He MY, Bartlett A, Evans AG and Hutchinson JW 1991 *Journal of the American Ceramic Society*, Vol. **74**, pp. 767-771.

[23] He MY and Hutchinson JW 1989 Kinking of a Crack out of an Interface – Tabulated Solution Coefficients, *Harvard University Report MECH-113*.

[24] Suo Z and Hutchinson JW 1990 *International Journal of Fracture*, Vol. **43**(1), pp. 1-18.

[25] He MY and Hutchinson JW, 1989 *International Journal of Solids and Structures*, Vol. **25**(9), pp. 1053-1067.

[26] Pageau SS, Joseph PF and Biggers SB 1994 *International Journal of Solids and Structures*, Vol.**31**(21), pp. 2979-2997.

[27] Rice J 1968 *Journal of Applied Mechanics*, Vol. **35**, pp. 379-386.

[28] Jakobsen J, Andreasen JH and Bozhevolnaya E 2009 *Engineering Fracture Mechanics*. In press (doi:10.1016/j.engfracmech.2008.06.03)

[29] Jakobsen J, Andreasen JH and Thomsen OT 2009 *Engineering Fracture Mechanics*. In press.

[30] Malyshev B and Salganik R 1965 *International Journal of Fracture Mechanics*, Vol. **1**(2), pp. 114-128.

Journal of Physics: Conference Series **181** (2009) 012061 doi:10.1088/1742-6596/181/1/012061

Numerical and experimental analysis of the directional stability on crack propagation under biaxial stresses

Rodríguez-Martínez, R.[1], **Urriolagoitia-Calderón, G.** [1], **Urriolagoitia-Sosa, G.**[1], **Hernández-Gómez, L. H.**[1], **Merchán-Cruz, E. A.**[2], **Rodríguez-Cañizo, R. G.**[2], **Sandoval-Pineda, J. M.**[2]

[1]**INSTITUTO POLITÉCNICO NACIONAL** Sección de Estudios de Posgrado e Investigación (SEPI), Escuela Superior de Ingeniería Mecánica y Eléctrica (ESIME), Edificio 5. 2do Piso, Unidad Profesional Adolfo López Mateos "Zacatenco" Col. Lindavista, C.P. 07738, México, D.F. México.

[2]**INSTITUTO POLITÉCNICO NACIONAL** Sección de Estudios de Posgrado e Investigación (SEPI), Escuela Superior de Ingeniería Mecánica y Eléctrica (ESIME). Unidad profesional, AZCAPOTZALCO, Av. de las Granjas No. 682, Col. Sta. Catarina Azcapotzalco, C.P. 02550, México D.F. México.

rrodriguezm@ipn.mx, urrio332@hotmail.com, guiurri@hotmail.com., luishector56@hotmail.com, eamerchan@gmail.com, ricname@hotmail.com, jsandovalp@ipn.mx

Abstract. In this paper, the case of Single Edge Notch *(SEN)* specimens subject to opening/compressive loading was analyzed; The loads are applied in several ratios to evaluate the influence of the specimen geometry, and the Stress Intensity Factor *(SIF)* K_I values on the directional stability of crack propagation. The main purpose of this work is to evaluate the behaviour of the fracture propagation, when modifying the geometry of the *SEN* specimen and different relationships of load tensión/compressión are applied. Additionally, the precision of the numerical and experimental analysis is evaluated to determine its reliability when solving this type of problems. The specimens are subjected to biaxial opening/compression loading; both results (numerical and experimental) are compared in order to evaluate the condition of directional stability on crack propagation. Finally, an apparent transition point related to the length of specimens was identified, in which the behaviour of values of *SIF* changes for different loading ratios.

1. Introduction

Many studies of fracture mechanics [1-4] have shown how the form of the stress field at the crack tip in Single Edge Notch Specimens (*SEN*), obey global conditions such as the specimen and applied loads. Initially this statement was based on single loading, however the problem is more complicated when mixed loading is involved.

When considering mixed loading, the crack propagation stability is an engineering problem that has attracted great attention. Since the pioneer work of Erdogan and Sih [5], where was considered the crack initiation direction for an angled crack which predicted to be driven under a maximum circumferential stress criterion, since then several additional factors that are involved in the fracture process have been found.

Several methodologies have been developed for the understanding of crack propagation under mixed loading. For example crack extension stability was studied by Cotterell [6-7] applying eigenfunction series expansion. Regarding crack path extension, Kipp and Sih [8] investigated the effects caused by the applied load direction, the curvature crack radius and the complete stress (energy field). Additionally, applications of numerical and experimental techniques have been proposed to evaluate crack growth instability [1-3], all these work is based on *SEN* specimens under biaxial loading. An approach using an energetic criterion based on S-Theory trying to understand crack initiation and propagation in beams with edge crack under mixed loading was analyzed by Nobile [9] and Sih [10].

From the investigations presented above, it has been proven that there are many factors involved. To be able to evaluate these factors, it is very important then to be separated in order to establish their individual contribution to the fracture process. There are two sets of factors that influence the fracture process (global and local). The global conditions involve geometry, loading conditions, environment, etc. while the local conditions involve flaws and material imperfections that developed in the neighborhood of the crack path.

Directional stability of crack extension is highly dependent on global conditions, making the fracture initiation to be highly stable, stable or unstable. As the loading rises, the local conditions may show its influence [3]. So, if the fracture process is dominated by the global conditions, the path crack growth could be predicted. However, if the local conditions dominate, the crack path could become indeterminable. For the *SEN* specimens, on square geometry show a better directional stability than a narrow one. Nevertheless, compressive and transversely applied stresses could serve as stabilizing factors.

This work is focused on a *SEN* specimen subject to a mixed opening and compressive loading. Where the loads are applied at several different ratios, from which it could be possible to evaluate the influence of the specimen geometry on the propagation of the crack. Additionally, the precision of the numerical and experimental analysis is evaluated and compared to determine its reliability when solving this type of problems. Finally, an apparent transition length of specimens was also identified, where the behaviour of *SIF* values changes for different loading ratios. Cases presented in this work have practical application, like in the analysis of cracked structural components, which are subjected to combined loads. Such are the cases presented in biomechanical analysis, for bones, arteries and tissues that become affected by cracks.

2. Study cases

To carry out numerical-experimental analysis, seventeen photoelastic models were built with the dimensions and the applied loads shown in table 1. The parameters L, w, a, Popen y Pcomp are defined in figures 1 and 2.

TABLE 1.
CASES OF STUDY

Geometry of plates			Loading condition (N)							
			Case 1a		Case 1b		Case 1c		Case 1d	
L (m)	*w* (m)	*a* (m)	P_{open}	P_{comp}	P_{open}	P_{comp}	P_{open}	P_{comp}	P_{open}	P_{comp}
0.1016	0.1016	0.0635	446	0	446	223	446	2230	446	4460
0.1143										
0.1270										
0.1397										
0.1524										
0.1651										
0.1778										
0.1905										
0.2032										
0.2159										
0.2286										
0.2413										
0.2540										
0.2667										
0.2794										
0.2921										
0.3048										

3. Numerical simulation and results

Numerical simulations of sixty eight cases (see table 1) were carried out. Different geometric conditions with different loading ratios were considered. The compressive and opening load values and loading ratios are presented in table 2. In table 3 are shown the mechanical properties for policarbonate PSM-1 used in the numerical evaluation, the analysis was performed in a 2D manner and under elastic conditions.

TABLE 2.
LOADING RELATION

Compressive Load (N)	Opening Load (N)	Ratio P_{comp}/P_{open}
0	446	0
223	446	0.5
2230	446	5
4460	446	10

TABLE 3.
MECHANICAL PROPERTIES OF SPECIMENS

Material	Propertie	Value
Policarbonate PSM-1	Elastic modulus E	2.50 GPa
	Poisson's ratio ν	0.38

3.1 Finite element model

For this work, *FEM* was used to determine the *SIF* in the components under loading conditions that are stated as follows. Each specimen was simulated considering conditions as follows; constant width w = 101.6 mm and a variable length L from 101.6 up to 304.8 mm (figure 1). The crack size a is considered with a constant value of 63.5 mm for all cases. The loads were applied with a constant opening value of 446 N and four different transverse compressive load valuated at 0, 223, 2230 and 4460 N respectively for the different study cases (see table 1). It is important to highlight that although a biaxial load mode was applied, shear load doesn't take place at the crack tip due to the symmetry of the specimen in regard to compressive load [11]. Finite element analysis was carried out using the computational program *ANSYS V 7.0* described in refs. [11-12] and the finite element model is shown in the figure 2, where the boundary and loading conditions can be observed.

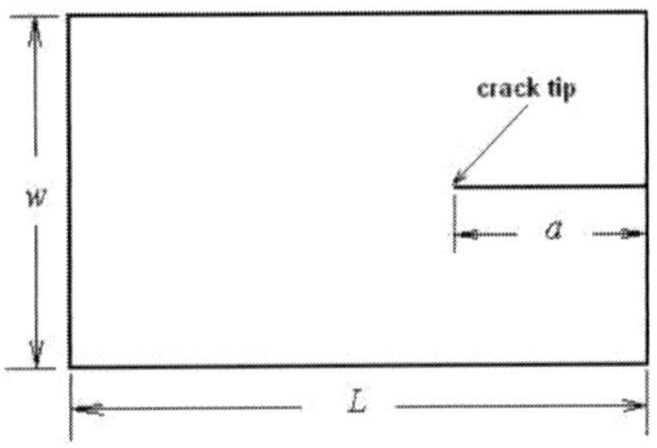

Figure 1 Model of cracked plate (*SEN* specimen).

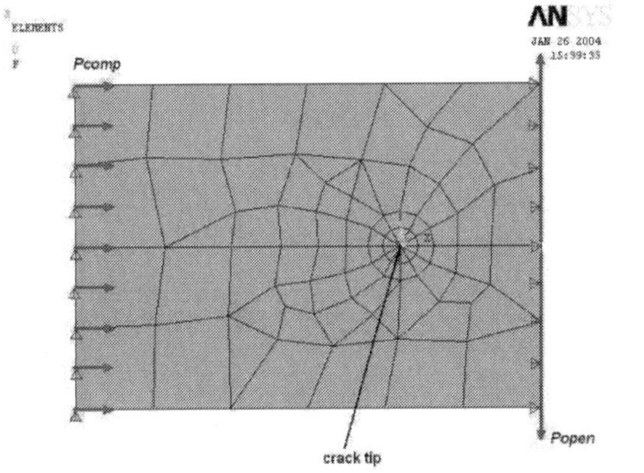

Figure 2 Finite element model.

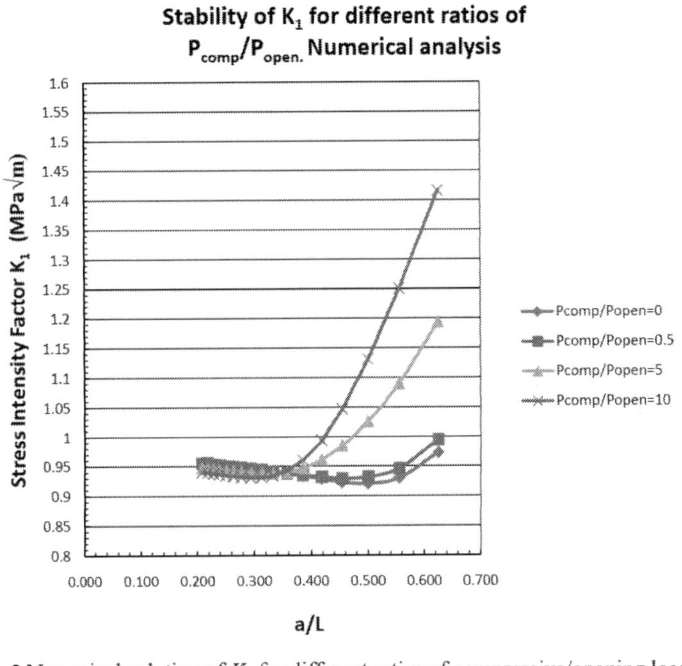

Figure 3 Numerical solution of K_I for different ratios of compressive/opening loads.

Figure 3 shows the *SIF* values evaluated for a loading conditions given in table 2, against the relation of crack length/specimen length *a/L* of the specimen. Figure 4 shows the results obtained at the zone in which the behaviour of K_I is unstable and it is possible to observe how fast the values change. It seems to be a transition length *L* of the specimen that coincides for all study cases, which has a value of 0.18161m and a geometric ratio of *a/L*=0.3497. It can be observed as well in figure 4, that the transition zone seems to be common for all loading condition cases. Results shown in figure 4 seem to be are significant only in a qualitative way, since it is in this zone where seemingly occurs a change in the tendency of the values of K_I.

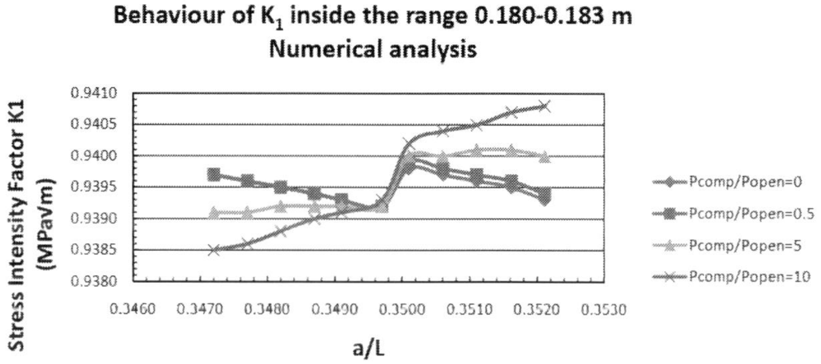

Figure 4 Numerical solution of K_1 in the transition zone. Length range from 0.180 to 0.183 m.

4. Experimental procedure and results

To carry out this work an experimental loading rig was used, it was designed and built by Urriolagoitia et. al. [2] (figure 5). The rig is a metallic structure designed to apply uniform loads on a rectangular SEN specimen made with a photo-elastic material. The specimen was prepared with a crack along the longitudinal axis. The applied loads are of two kinds; an opening load and a longitudinal compressive pressure. A Schematic drawing of the rig is shown in figure 6. The opening load it is applied to the specimen through a group of eigth fastener blocks. These fastener blocks have were instrumented with strain gauges. These blocks are forced to separate by means of a screw, which produces the opening load in the specimen. This opening load can be measured using strain gauges that have been placed on the blocks (figure 6). The compressive force required for this experimental work is applied by means of eight individual fingers placed at the right end of the plate (figure 6). These fingers are operated by horizontal screws. Each finger is provided with a couple of strain gauges placed in such a way that any kind of flexural stress is suppressed. This characteristic allows a good control of the complete stress field. It is very important to state, that this study was carried out based on the results reported in other similar works and on the numerical analysis shown previously [1-4, 13-14]. Actually, other methods have been used for determining fracture parameters, which are very interesting and provide important information on the stress field at the crack tip [15].

4.1 Experimental tests and computing of K_I

Procedure for experimental tests was to mount the plate in the loading rig, which was held by means of screws. The next steps was placing the loading rig in the polariscope and proceed to connect strain gages to strain indicator, to be able to measure the loads applied to the specimen through the deformation of the load cells. The study cases were the same ones that were analyzed by means of numerical simulation (see tables 1 and 2) where dimensions and loads applied to the specimens are shown. The crack was introduced making a cut by means of a very thin hacksaw blade, and the crack tip was made by means of a sharp blade. For experimental determination of K_I the following model was used, which is described in ref.[16].

$$K_I = \sqrt{2\pi r_i}\left[\frac{(2\tau_{\max})_i - (2\tau_{\max})_j}{1-(r_i/r_j)^{1/2}}\right] \quad (1)$$

To apply this equation it is necessary to measure two radios that correspond to two consecutive isochromatic loops r_i and r_j, (fig. 7) as well as their respective maximum shear stresses $(\tau_{máx})_i$ y $(\tau_{máx})_j$, which are determined by means of the optical stress law [16] given as

$$\tau_{\max} = \frac{Nf_\sigma}{2t} \quad (2)$$

Now, different values of stress intensity factors are calculated establishing all possible permutations for any couple of isochromatic fringes. From these results it is obtained the stocking $\overline{K_I}$ and the standard deviation S_K. Finally, values that are outside of $K_I \pm S_K$ are eliminated and another value $\overline{K_I}$ is computed. This procedure is carried out for all study cases.

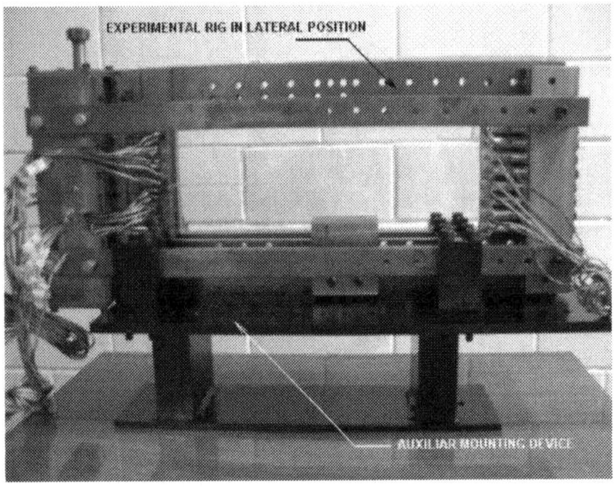

Figure 5 Experimental loading rig.

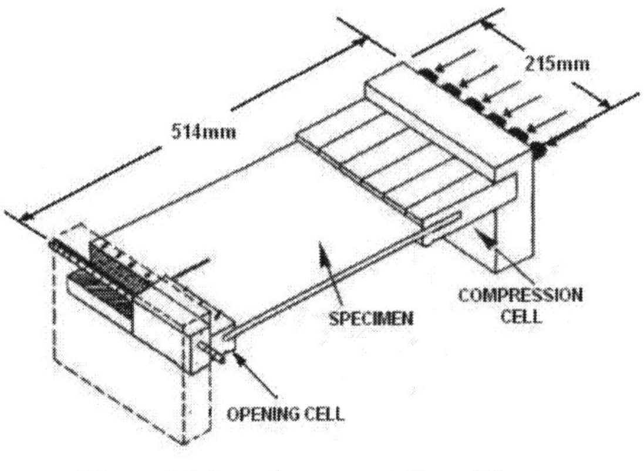

Figure 6 Schematic representation of the rig.

Experimental results are shown in figure 8. It is important to point out that it was not possible to carried out experimental tests in the transition zone of L, due to the small variation in the magnitude of this parameter (smaller than 1mm), which didn't allow to build these specimens. At the end of each experiment, the opening load was increased to the point of producing fracture so as to observe the direction of propagation which coincides with the stability approach presented for KI values. The values of the failure loads for some specimens are shown in figures 9 and 10.

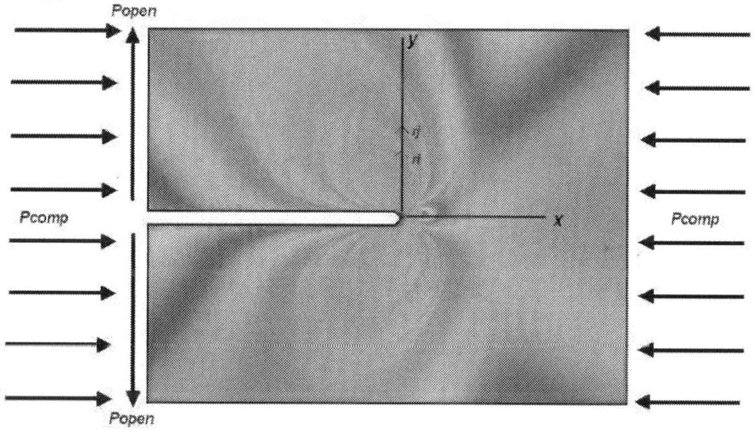

Figure 7 Isochromatic fringe pattern in a photoelastic specimen that defines shear stress state.

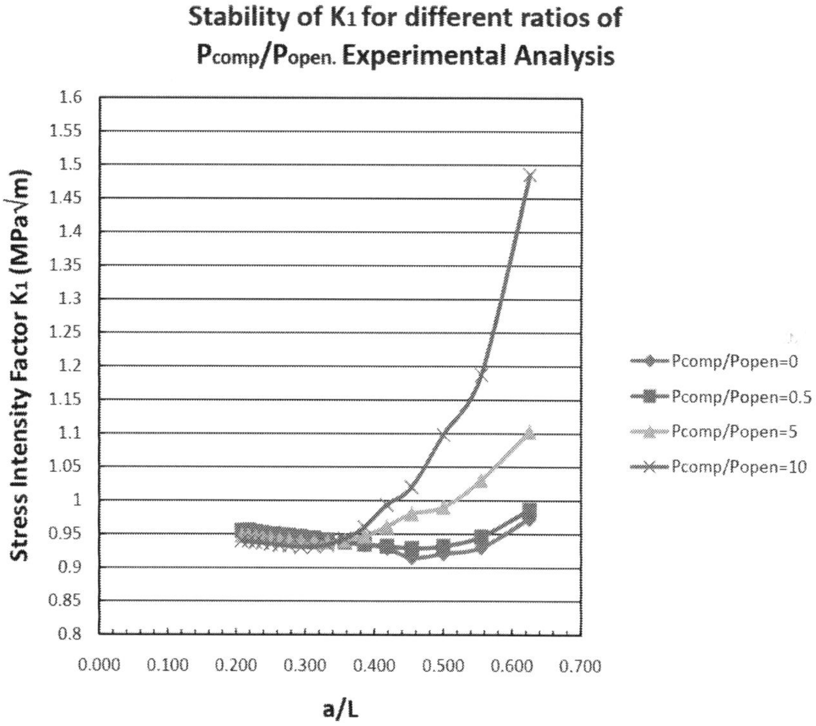

Figure 8 Experimental solution of K_I for different ratios of compressive/opening loads.

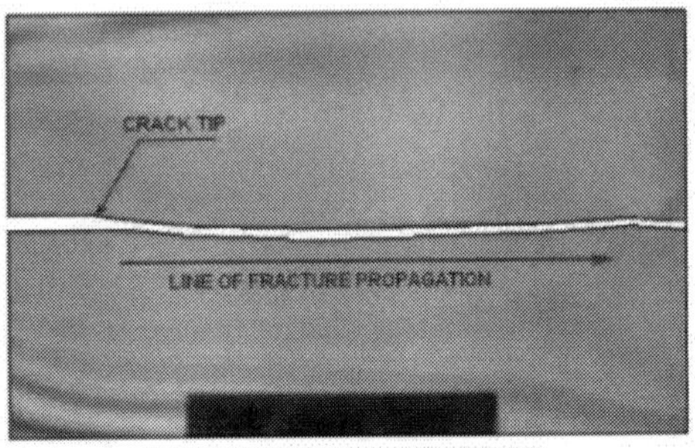

STABLE DIRECTION OF PROPAGATION
SPECIMEN OF 0.3048x0.1016 m
(a)

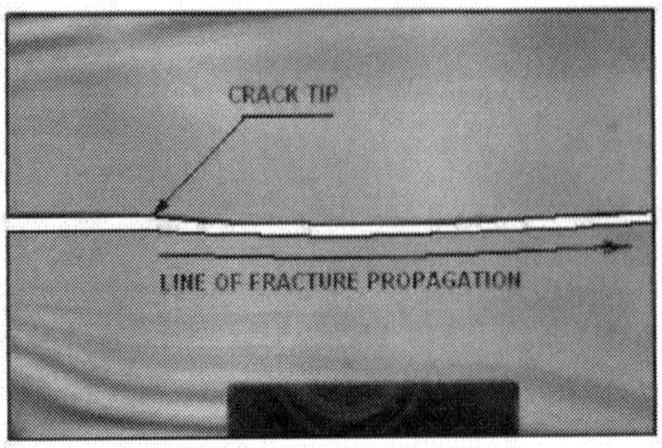

STABLE DIRECTION OF PROPAGATION
SPECIMEN OF .2032x0.1016 m
(b)

Figure 9 Stable fracture propagation in specimens with relationship $a/L < 0.3497$. Figure (a) shows the stress state at the crack tip when failure occurs, that is 1670 N opening load and 4460 N compressive load. Figure (b) shows the stress state at the crack tip when failure occurs, that is 1480 N opening load and 4460 N compressive load.

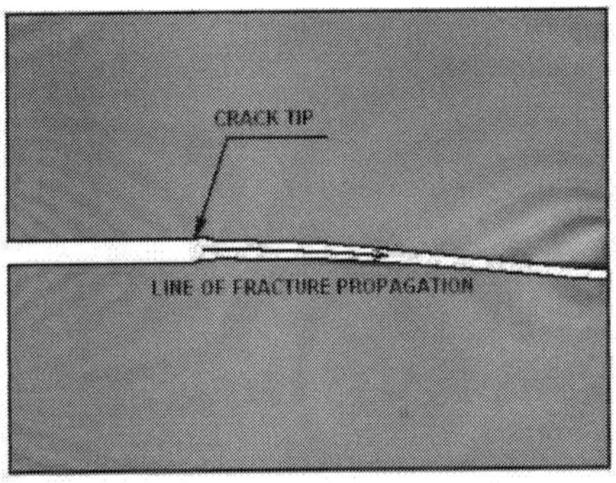

UNSTABLE DIRECTION OF PROPAGATION
SPECIMEN OF 0.1524x0.1016 m
(a)

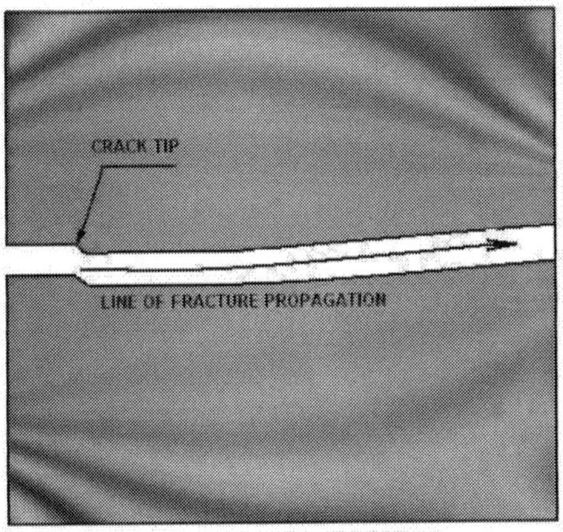

UNSTABLE DIRECTION OF PROPAGATION
SPECIMEN OF 0.1016x0.1016 m
(b)

Figure 10 Unstable fracture propagation in specimens with relationships $a/L > 0.3497$. Figure (a) shows the stress state at the crack tip when failure occurs, that is 1290 N opening load and 4460 N compressive load. Figure (b) shows the stress state at the crack tip when failure occurs, that is 1115 N opening load and 4460 N compressive load.

5. Discussion

The results that were obtained for the values of the *SIF* have shown clearly the influence of fundamental parameters such as the specimen geometry and the relation between the opening and transverse loads. Zanganeh et. al. [17] mentions that this effect is due to the stress term known as T-stress which is defined as the second non-singular term in Williams's crack tip stress field solution. Also refers to the Cotterell studies [18] where he says that the T-stress controls the stability of the crack direction as well as assumed that the ideal direction for crack growth is along the line of symmetry of the stress distribution ahead of the crack tip and concluded that if the sign of the T-stress is negative then the crack path has a tendency to return to its original ideal path (referred to as 'directionally stable'). However, if the T-stress is positive the crack path does not return to this original ideal path. Additionally, Pacey [19] says that a fringe pattern in the crack tip may be produced in specimens by subjected them to compressive loads in line with de crack growth direction and can also be attributed to the specimen geometry and the mode of loading. He also affirms that must be some point in specimen where a transition occurs from tensile to compressive loading and where the fringe order is zero.

In figures 3, 8 and 9 it can be observed, as if the length of the specimen is increased, more stable is the determined values of K_I, and the directional stability of crack propagation is also stable. Numerical and experimental results were compared and it can be observed like experimental results are very similar to numerical results, although some values differ due to unavoidable errors in the experimental process.The findings obtained were consistent in a general manner for all rates of loading conditions P_{comp}/P_{open}, which also show that the conditions of fracture propagation are stable. On the other hand, the shorter specimens are more sensitive to changes in the compressive loading. Characteristic can be observed clearly for the relationship of loads P_{comp}/P_{open}= 10. Seems to exist a transition length of the specimen that coincides for all cases of study, which has a value of 0.18161 m and a geometric ratio of a/L = 0.3497. For the geometric relationships of a/L greater than this value, and loading conditions of P_{comp}/P_{open}= 5 and P_{comp}/P_{open}= 10, the values of K_I increase drastically, as well as the direction of fracture propagation which seem to become more unstable. Analyses using the finite element method was carried out on transition length, for this studies the results are shown in figure 4 and it can be corroborated the influence that of the load and geometry ratios in each case of study. By *FEM* the exact size of transition length is determined as well which is of 0.18161m and a geometric relationship of a/L=0.3497.

Conclusions

When concluding the numeric and experimental work, as well as when analyzing the results, quite interesting conclusions were reached regarding the tendency of the K_I values for specimens subjected to biaxial tensile-compressive stresses as well as their influence on the fracture propagation stability. The conclusions of the work are:

1. The *SEN* specimens with length $L > 0.18161$ m and with geometric relationships $a / L < 0.3497$ are stable in the direction of fracture propagation.
2. The *SEN* specimens with length $L < 0.18161\ m$ with geometric relationships $a / L > 0.3497$ are unstable in the direction of fracture propagation.
3. The *Y* factor in *SEN* specimens subjected to compressive and opening load, is an a / L function as well.
4. The geometry of the rectangular specimens influences directly the K_I values and the directional stability of fracture propagation.
5. The compressive load increases the stress intensity factor K_I considerably in short specimens up to *45%* for P_{comp}/P_{open} = 10, while in longer specimens K_I is increased by *1.6%* for the same ratio of loads.
6. The uncertainty in the K_I values for short specimens and big rates of load P_{comp}/P_{open}, produce uncertainty in the direction of fracture propagation, this seems to indicate that both parameters are related in a direct way.
7. Without compressive load, a long specimen is more stable for the direction of fracture propagation.

8.- Acknowledgement

The authors gratefully acknowledge the financial support from the Mexican government by de Consejo Nacional de Ciencia y Tecnología and the Instituto Politécnico Nacional, Escuela Superior de Ingeniería Mecánica y Eléctrica.

9.- References

[1] Urriolagoitia-Calderón G and Hernández-Gómez L H, 1996, Evaluation of crack propagation stability with the Williams stress function-I. Stress field analysis, *International Journal Computers and Structures*, Vol. 61, No. 4, pp 775-780.

[2] Urriolagoitia-Calderón G and Hernández-Gómez L H, 1997, Evaluation of crack propagation stability-II. Numerical Analysis, *International Journal Computers and Structures*, Vol. 63, No. 5, pp 1007-1014.

[3] Urriolagoitia-Calderón G and Hernández-Gómez L H, 1997, Experimental analysis of crack propagation stability in single edge notch specimens, *Theoretical and Applied Fracture Mechanics*, Vol. 28, pp 57-68.

[4] Sauceda-Meza I, 2000, Análisis Numérico-Experimental de la Estabilidad Direccional de Propagación de Fractura bajo Campos Biaxiales de Esfuerzos, Ph D Thesis, Instituto Politécnico Nacional SEPI-ESIME.

[5] F. Erdogan, G.C. Sih, 1963, On crack extension in plates under plane loading and tranasverse shear, *J. Basic Eng. Trans. ASME* 85(D), pp. 519-527

[6] B. Cotterell, 1965, On brittle fracture paths, *Int. J. Fract. Mech.* I, pp. 96-103

[7] B. Cotterell. 1966, Notes on the paths and stability of cracks, *Int. J. Fract. Technol.* 2, pp. 526-533

[8] M. E. Kipp. G. C. Sih, 1975, The strain energy density failure criterion applied to notched elastic solids, *Int. J. Solids Struct.*, 11, pp. 153-173

[9] L. Nobile, 2000, Mixed mode crack initiation and direction in beams with edge crack, *Theor. Appl. Fract. Mech.*, 33, pp. 107-116

[10] G. C. Sih, 1974, Strain density factor applied to mixed mode crack problems, *Int. J. Fract.*, 10, pp. 305-321

[11] ANSYS Structural Analysis Guide, 2004, Chapter 10, Fracture Mechanics, SASP IP Inc, Ed.

[12] ANSYS Element Reference, 2004, Chapter 4, Element Library, SAS IP Inc.

[13] Rodríguez-Martínez R, 2004, *Análisis* Numérico-Experimental de Estabilidad Direccional de Fractura bajo Esfuerzos Biaxiales, MSc Thesis, Instituto Tecnológico de Tlalnepantla, México.

[14] Urriolagoitia-Calderón G, 1976, Directional Stability of Cracks Under Biaxial Stresses, Ph D Thesis, Department of Mechanical Engineering, Imperial College of Science and Technology, University of London.

[15] Wang C H and Luo P F, 2008, An Experimental Study on the Elastic-Plastic Fracture in a Ductile Material Under Mixed-Mode loading, *Strain*, Vol. 44, No. 3, pp 223-230, June.

[16] Dally J. W. and Riley W. F., 1991, Experimental Stress Analysis, McGraw-Hill, Inc., p.428 and p.511

[17] Zanganeh, M, Tomlinson, R A, and Yates J. R., 2008, T-stress determination using thermoelastic stress analysis, *Journal of Strain Analysis for Egineering Design*, 43, pp. 529-537

[18] Cotterell, B, 1966, Notes on the paths and stability of cracks, *International Journal Fracture*, 2(3), pp. 526-533

[19] Pacey, M N, James, M N, and Patterson, E A, 2005, On the use of photoelasticity to study crack growth mechanisms, *Experimental Mechanics*, 45

Journal of Physics: Conference Series **181** (2009) 012062 doi:10.1088/1742-6596/181/1/012062

Curve fitting of mixed-mode isopachics

R.I. Hebb, J. M. Dulieu-Barton, K. Worden[+], P. Tatum*

University of Southampton, School of Engineering Sciences, Highfield, Southampton, SO17 1BJ, UK

[+] Department of Mechanical Engineering, University of Sheffield, Mappin Street, Sheffield, S1 3JD, UK

[*]AWE, Aldermaston, Reading, Berkshire, RG7 4PR, UK

Abstract

Recent work has focused on exploiting the observation that the stress-sum contours (isopachics), obtained from TSA, in the vicinity of the tip take the form of a simple curve – the cardioid. The analysis made use of the cardioid nature of the isopachics by deriving expressions for the SIFs in terms of the cardioid area and the positions of certain tangents to the curve. Both Genetic Algorithms (GAs) and Differential Evolution (DE) have also proved successful for parameter estimation, but some of the curve-fits indicated that the cardioid form was inappropriate for the base model, particularly for mixed-mode cracks. The effect of crack-tip interaction has been explored and shows this has a small effect on the cardioid form. New, higher resolution infra-red detectors have become available since the original data was collected, so the object of the current paper is to use new techniques to extract the cardioid form and use a GA to perform the curve fitting.

1. Introduction

Thermoelastic Stress Analysis (TSA) is one of a group of full-field techniques for experimental stress analysis that has proved to be extremely effective for studying stress fields in the vicinity of crack-tips. An understanding of such fields is vital to the development of effective diagnosis and prognosis algorithms for Non-Destructive Testing (NDT) and Structural Health Monitoring (SHM). A means of undertaking crack-tip studies using TSA is based on the observation that the stress-sum contours (isopachics) in the vicinity of the tip take the form of a simple curve – the cardioid. This treatment is based on the first order Westergaard equations [1], or equivalently, the singular terms in the eigenfunction expansion of Williams [2], and was demonstrated in [3] where the Stress Intensity Factors (SIFs) were obtained for simulated cracks (in the form of spark eroded slots) in mode I and mixed mode opening. The analysis in [3] made use of the cardioid geometry of the isopachics by deriving expressions for the SIFs in terms of the cardioid area and the positions of certain tangents to the curve. This treatment was fully exploited in [4] by developing algorithms and software that automatically obtained the SIFs from the thermoelastic data and demonstrated that the technique could be applied to actual cracks. Other researchers have developed successful techniques for crack-tip stress studies, e.g. [5, 6] based on the Mushkelishvili stress function.

Journal of Physics: Conference Series **181** (2009) 012062 doi:10.1088/1742-6596/181/1/012062

Recent work [7] has allowed the estimations of crack-tip SIFs by direct curve-fitting of the cardioid form to measured isopachics from TSA. One advantage of this approach is that the curve-fit also provides an accurate estimate of the crack-tip location. Both Genetic Algorithms (GAs) [7] and Differential Evolution (DE) [8] proved successful for the actual parameter estimation, but some of the curve-fits indicated that the cardioid form was inappropriate for the base model. One of the possible explanations for this is that the cardioid form is theoretically only suitable for an *isolated* crack-tip stress field, as derived from the Westergaard equations. In the experimental data for the previous studies, a crack was placed centrally in a plate and there will therefore have been two (potentially) interacting crack-tips involved (note that the term *interaction* is used rather loosely here, it simply means that the observed field is a linear superposition of fields from two crack-tips).

The object of this current paper is to demonstrate that the techniques developed for extracting the isopachics from the thermoelastic data in [7] can be applied to data obtained from high resolution infra-red systems. Since the data used in [7] was collected, higher resolution systems have become available which present new challenges in data extraction. Furthermore, the algorithm developed in [7] was targeted at using a selected number of isopachics. In this work the possibility of using the entire captured data set is explored. To do this, data has been collected using a high resolution device and a new approach to data extraction developed, which is described in the paper. The paper describes the progress made in applying the new approach by obtaining the curve-fitted isopachics.

2. Theory

Thermoelastic stress analysis is based on the measurement of small temperature changes that occur in solids on the application of a cyclic stress. It can be shown that the temperature changes are proportional to the changes in the sum of principal stresses [9]. The temperature changes are measured using a highly sensitive infra-red detector. By applying an appropriate calibration factor A, the detector output, S, known as the thermoelastic signal, provides a measurement of the stress-sum via:

$$AS = \Delta(\sigma_1 + \sigma_2) \tag{1}$$

There are various assumptions underlying this assertion: these are outlined in [9], where a review of the applications of TSA is presented. In order to extract information relating to SIFs, appeal is made to the Westergaard equations [1], and it can be shown that in the vicinity of the crack-tip (but not the immediate vicinity where plasticity is a factor and Linear Elastic Fracture Mechanics (LEFM) does not apply) the stress sum is related to the mode 1 and mode 2 SIFs K_I and K_{II} by:

$$\Delta(\sigma_1 + \sigma_2) = \Delta(\sigma_x + \sigma_y) = \frac{2K_I}{\sqrt{2\pi r}}\cos\left(\frac{\theta}{2}\right) - \frac{2K_{II}}{\sqrt{2\pi r}}\sin\left(\frac{\theta}{2}\right) \tag{2}$$

where r and θ are polar coordinates centred at the crack-tip with θ measured anti-clockwise from the crack-line. An elementary rearrangement of this formula yields:

$$r = \frac{K_I^2 + K_{II}^2}{\pi A^2 S^2}\left[1 + \cos(\theta + 2\phi)\right] = \frac{r_0}{2}\left[1 + \cos(\theta + 2\varphi)\right] \tag{3}$$

where $\varphi = \tan^{-1}(K_{II} / K_I)$ is the rotation of the cardioid and r_0 is the maximum radial coordinate of the cardioid curve. Equation (3) shows that a curve of constant signal, S, is a cardioid. The method used in [4] to determine the SIFs was to carry out a nonlinear least-squares curve-fit of equation (3) to constant S contours extracted from the thermoelastic images, and thus determine r_0 and φ for given S. Given this data, the SIFs are computed by solving the following simultaneous equations:

$$K_I^2 + K_{II}^2 = \pi r_c A^2 S^2 \quad \text{and} \quad \tan\phi = \frac{K_{II}}{K_I} \tag{4}$$

In the current work it is proposed to used the GA approach described in [7].

3. TSA System

Infra-red imaging equipment is constantly being improved as the underlying technology quickly advances. The system used to collect the data described in this paper is the Silver 480M system manufactured by Cedip Infrared Systems. The detector elements in the camera are made from indium antimonide with an array size of 320 x 256. Two arrays are present in the detector, with one array over-layed on top of the other. Each array has its own internal buffer used for data storage. This set-up allows the arrays to be used sequentially, with one array collecting the incoming photons, and the other transferring data to its internal buffer. Consequently, a faster frame rate and a more continuous data capture process than using a single array is possible (e.g. the Deltatherm system [9]). The maximum frame rate of the entire system (with processing) is 383Hz, although a faster frame rate can be achieved using windowing. Windowing operates by only sending data from a certain region on the array, thus processing time is reduced as less data is supplied. For the processing, a dedicated processor housed within the camera itself digitizes the voltage from each detector element, before sending it to the computer as a digital signal with units of 'digital level' (DL) or S as given in equation (1). The data are output in a Comité Consultatif International des Radiocommunications (CCIR) phase-alternate line (PAL) video format, and sent to a Cedip FG9800 frame grabber card in the computer.

The Silver system has been radiometrically calibrated, so the temperature in an image can be automatically obtained from the DL data output by the detector. To do this four calibration files have been supplied by the manufacturer for four separate temperature ranges within an overall temperature range of 5 to 200°C. To calibrate the ΔDL data to ΔT, the mean temperature of the specimen is measured at each pixel, and the gradient of the calibration curve for that temperature is used to calculate the ΔT value [10].

4. Test Specimens and set-up

The dimensions of the plate specimen considered here is shown in Figure 1. The plate was manufactured from Duralumin® and has dimensions 300 x 300 mm with a thickness of 1.7 mm. At the centre of the plate a small hole has been drilled and a spark eroded slot of total length 50 mm has been produced from the central hole. The 'crack angle', β, (see Figure 1) is 45°. To spread the load as evenly as possible across the width of the plate when mounted, 15 holes (at both the top and bottom of the plate) of 6 mm diameter so that the load is imparted into the plate through pins fixed in the holes. Bonded to the front and back at the top and bottom are 4 Duralumin® tabs 6 mm thick and 30 mm wide. The purpose of these tabs is to locally strengthen the area around the holes. The specimen was mounted in an Instron 8800 servo-hydraulic machine using jigs that comprised 4 sheets of steel 12 mm thick. Two sheets are bolted either side both top and bottom of the specimen. Each sheet has a 16 mm hole through which connecting pins are slotted, so that another 12 mm steel tab can be connected to the plates. This tab is then gripped by the jaws of the machine. Figure 2 shows the specimen with the top mounts connected (only the top mounts have been shown for clarity).

Before experiments can be performed, the surface of the specimen being viewed by the infra-red detector needs to be treated. Previous work has shown that it is extremely difficult to obtain workable results from a plain untreated metallic surface [11], because of reflected radiation and variations in the surface. To prepare the surface, it is firstly cleaned with acetone and dried. If required, the surface can also be polished to obtain the smoothest surface possible. Once clean and smooth, two thin layers of matt black paint are applied (with the first layer being allowed to dry before application of the second). The specimen was mounted centrally in the servo-hydraulic test machine, and the infra-red detector was positioned to examine the crack-tip. The frequency and amplitude of the cyclic loading was controlled by the test machine.

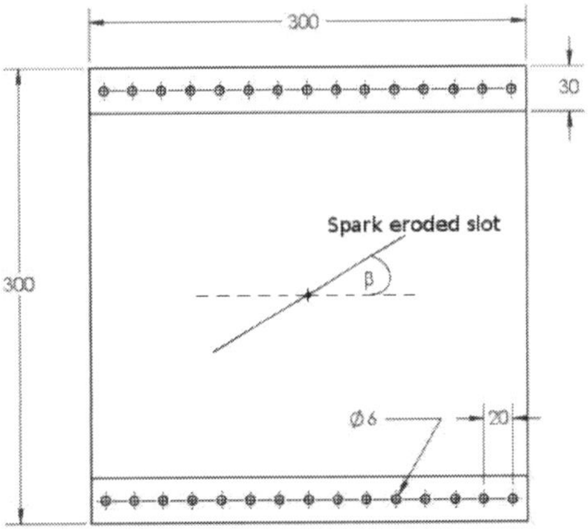

Figure 1: Specimens tested

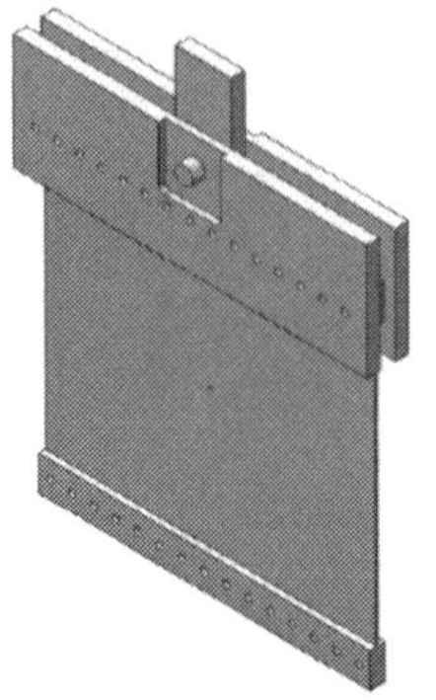

Figure 2: Loading jig

5. Extracting Isopachics from the Thermoelastic Data

The GA fits a curve to the coordinates of the isopachics, which are input as a list of (x, y) values. To extract these values the thermoelastic data is first loaded into Altair (software supplied with the Cedip system). Altair has an 'area' function, where a rectangular area can be selected in the data, and various forms of information obtained from that area. One function is to copy the the thermoelastic data from each pixel to an external text file. Therefore, by choosing an area around the region of interest, the values of relevant pixels can be obtained. The number of desired results (N) is input to the program at this stage (this decides how many 'sets' of values to create). The header of the copied text file contains the minimum and maximum values present in the area selected. The program uses the minimum and maximum values to calculate the range of values present in the list and then divides this range into N equally spaced bins, creating a blank text file for each bin.

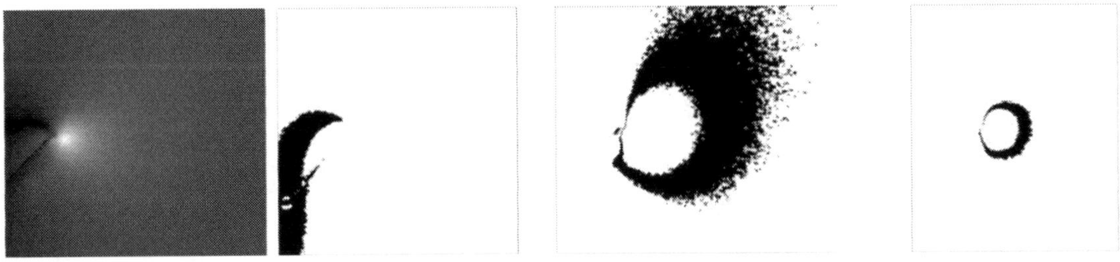

a) Full field data b) Example of an incomplete curve c) Complete curve with data extending beyond the array d) Complete data set contained within the array

Figure 3: Processing the thermoelastic data to input into GA

Moving through the list, the program compares the value of each (x, y) coordinate to the bins, and appends the x and y values to the relevant bin's text file. An example of the process can be seen in Figure 3 where (a) is the thermoelastic data used and the other three images are three of the bins produced. Image (b) is an example of an incomplete curve in the data sets. The data points for this set continue through the bottom left corner and to the other tip of the slot. Similarly, (c) is an example of a data set where the range continues outside of the image boundary. Cases like (b) and (c) are not used for curve-fitting process. Image (d) is an example of a complete closed curve which would be suitable for analysis. However, the GA requires coordinates of a single curve, as opposed to a range of curves. Therefore, the coordinates of a line that would bisect each area of pixels is required. If the data points lay on a horizontal line as opposed to a closed curve, then all that would be required is a simple averaging procedure over the length of the line.

When the polar coordinates of a circle (with its centre on the origin) are plotted as r against θ the end result is a constant. The same process is performed on the data sets by transforming the coordinates of the pixel locations from Cartesian to the polar coordinate system. In Altair, a point near to the crack tip is chosen (as this will be near the centre point of each closed curve) and this point is defined in the program to be the origin of the polar coordinate system. The program then loads each data set in turn. Running through the currently loaded data set, the (x, y) positions of the data are transformed to polar coordinates and put into a new array which is ranked according to increasing θ. The number of points (P) required in the final curve is decided (e.g. 100). Using the newly created array, the full range of θ is divided into P bins, with the mean r of the points contained being calculated for each bin. Each bin is therefore assigned a θ and r coordinate. Once all P bins have been averaged, the polar coordinates of the bins are transformed back into Cartesian coordinates. Here the origin is defined as the bottom left corner of the original image, as previously the origin was the top left which is the standard method in images. Figure 4 illustrates the process. Firstly a dataset is loaded as shown in Figure 4a. All of the points are transformed to polar coordinates and the mean r is then calculated for each bin (the over layed blue points)- see Figure 4b. Both sets of mean r and the original data are then converted back to the Cartesian coordinate system (Figure 4c), with the image in Figure 4d showing the resulting isopachic. It is these coordinates that are used in the GA for the curve-fitting process.

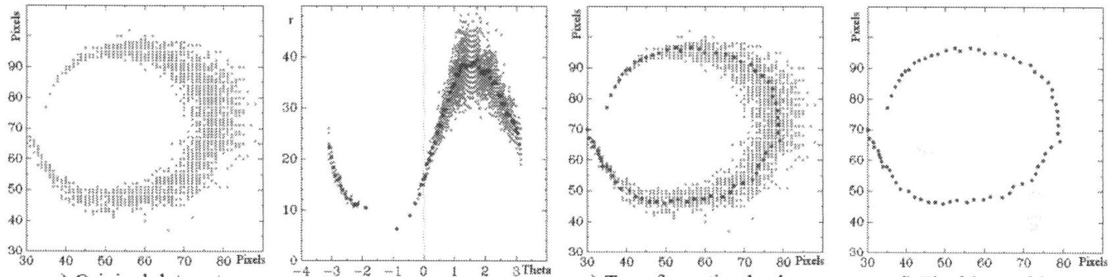

a) Original data set
b) Finding the mean of the polar coordinates
c) Transformation back to Cartesian
d) Final isopachic

Figure 4: Demonstration of the averaging technique to reduce a data set to a single isopachic

6. Results

Figure 3(a) is from the plate with $\beta = 45°$. The applied stress range is 9.81 MPa with a loading frequency of 10 Hz. This view differs to the actual data obtained only in the fact that it is a crop of the full field. The data was analysed to produce 20 isopachics, and Figure 5 shows the odd numbered isopachics that result from the averaging process (only the odd numbered curves have been shown for clarity). One immediately striking observation is that not all of the isopachics take the cardioid form. Curves 1 and 2 are explained by the fact that the regions of constant S that they represent do not make a complete curve, with an example being Figure 3(b). Curve 3 has squared edges because the region of constant S is larger than the selected area in the original image. Data sets of this kind (either incomplete curves or having values outside of the boundary) are not used in the GA. It can also be noticed that the curves also provide negative y values. The original data doesn't have any negative y-values for positions, as image coordinates start at (0, 0) and are always positive. This is an issue that needs to be rectified.

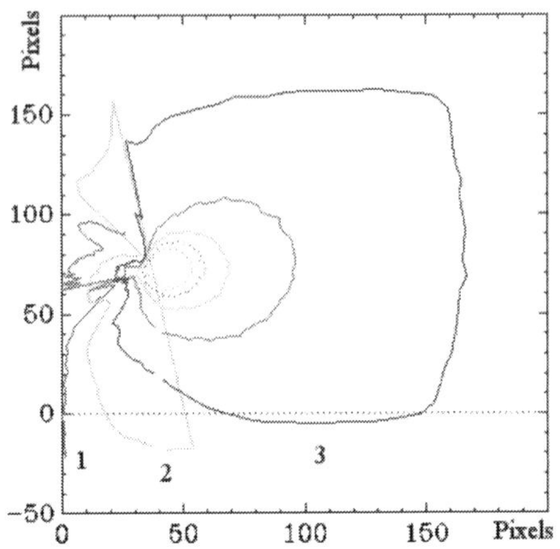

Figure 5: Odd numbered isopachics for the data shown in 4a

Results of running the GA using the isopachics from Figure 5 are provided in Figure 6. Here the isopachics have been extracted (circular data points) and the GA performed curve fitting to them. Table 1 shows the results of the GA for each data set (position of the cusp in terms of pixels, r_0, rotation angle and fitness) with the cardioid numbers set for 1 being the largest cardioid and 5 the smallest. The fitness parameter provides a means of comparing the suitability of each individual to the rest of the population by assigning a single number to each individual. Individuals are selected for mating based on their fitness so the fitter an individual is then the more likely it is to be selected for the mating process to provide better solutions in subsequent generations. The fitness function used is based on a mean-square error (MSE) for the curve fit. If the measured points on the cardioid are denoted by (r_i, θ_i), $i = 1, \ldots, N$, the MSE is defined by:

$$M = \frac{100}{N\sigma_r^2}\sum_{i=1}^{N}\left(r_i - r_{ie}\right)^2 \qquad (5)$$

where r_{ie} is the estimate from the curve-fit and σ_r^2 is the variance of the array of measured radii. The fitness is the reciprocal of the normalised MSE, with the normalization of the MSE chosen so that a value of less than 5 for the MSE indicates a good fit to the data, and less than 1 being an excellent fit.

The actual location of the slot tip can be obtained examining the thermal image of the specimen, and it occurs at (34, 77) pixels; the estimated error on this value is around ± 2 pixels. For the (x, y) position of the cusp, all of the generated cardioids have their x-position shifted to the right of the slot-tip by as much as 10 pixels which equates to 3 mm. Figure 7 shows a zoomed region around the slot tip. Arrow A points to the actual location of the slot-tip, and arrow B points to (41, 80), which is the mean of the generated cusps' coordinates. An inspection of Figure 6 shows that the GA fits much better curves to the smaller isopachics than the larger curves. This may be explained by the fact that the Westergaard equation is only fully valid for situations where $r << a$ [1], so as the size of the isopachic increases, so the accuracy in the generated solution decreases. Cardioid 1 had a maximum radial distance from the cusp of 26.7 mm i.e. > than the 25 mm slot length and its cusp is located furthest away from actual slot-tip location and from the other slot-tips. Neglecting this data improves matters giving the average location of the slot-tip at (40, 78) shown by arrow C in Figure 7. Using the three smallest cardioids provides a relative position of the slot tip to within less than 1 mm, which is comparable to that achieved in previous work [7, 8]. The actual average position of the slot is (10.2, 23.1) mm and it can be seen that the y location has been located precisely although the x location disagrees by about 2 mm. This warrants further investigation but is encouraging.

Table 1: Results of the GA fitting curves to isopachics

Cardioid Number	x (pixels (mm))	y (pixels (mm))	r_0 (pixels (mm))	2φ (radians)	Fitness
1	45.75 (13.7)	89.30 (26.8)	89.02 (26.7)	0.289	0.35978
2	41.99 (12.6)	80.79 (24.2)	53.86 (16.2)	0.393	0.49101
3	40.23 (12.1)	76.82 (23.0)	37.98 (11.4)	0.338	0.61657
4	39.41 (11.8)	77.39 (23.2)	29.48 (8.8)	0.442	0.58084
5	39.30 (11.8)	77.11(23.1)	23.82 (7.1)	0.497	0.55064

It can also be noted that the cardioids tend to rotate more as they approach the slot tip. This rotation is a current area of study, as the GA works on an idealised cardioid as the base model, which may not be technically correct. The cause of the rotations is currently not fully understood so at this stage the SIFs for these results have not been calculated.

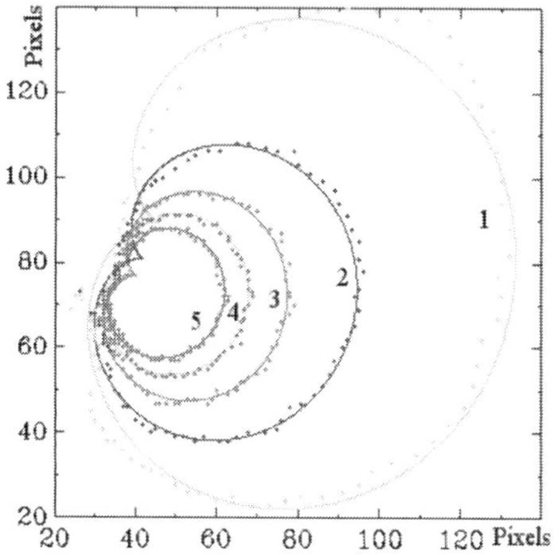

Figure 6: Curve fitting to the isopachics (units in pixels)

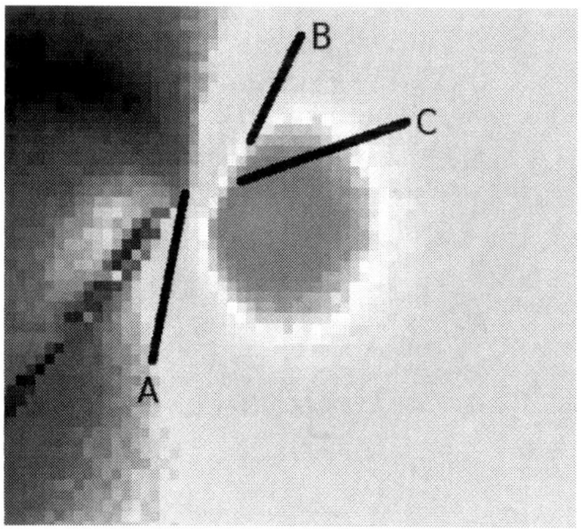

Figure 7: Actual slot tip location (A) compared to the location calculated by the GA (B). Arrow C shows the location when cardioid 1 is neglected.

7. Conclusions and Future Work

The work in this paper has demonstrated that it is possible to use the existing GA routine for curve-fitting isopachics using data from the whole field. An approach has been developed for extracting data from the supplier's software that can be input into the GA algorithm. The approach has been demonstrated on a 45° slot. The results are encouraging in that the estimation of the position of the slot-tip is good using cardioids close to the slot tip. However, the cardioids are not 'nested' but are rotated. It is considered that this inaccuracy centres on the use of the first order Williams stress function and that an improved fit can be obtained using the full Williams stress function:

Journal of Physics: Conference Series **181** (2009) 012062 doi:10.1088/1742-6596/181/1/012062

$$AS = \frac{2\sqrt{|K_I^2 + K_{II}^2}}{\sqrt{2\pi r}}\cos\left(\frac{\theta}{2} + \phi\right) + T_S + 2\sqrt{A_I^{\,2} + A_{II}^{\,2}}\sqrt{r}\cos\left(\frac{\theta}{2} - \tan^{-1}\frac{A_{II}}{A_I}\right) + O\left(\sqrt{r}\right) \qquad (5)$$

where A_I and A_{II} are geometry-dependent constants.

Using the GA on thermoelastic data obtained for various r and β will enable an understanding of the cause of the cardioid rotations observed in the data.

It is also considered that the use of a slot instead of an actual crack is not desirable therefore in future work new specimens will be manufactured that contain mixed mode cracks. The method to obtain plates will involve spark eroding a horizontal slot in larger plate and then subjecting it to a fatigue load until the desired crack length is grown. Then a smaller plate will be cut out to give the necessary crack angle.

References

1 Ewalds, H L and Wanhill, R J. *Fracture Mechanics*. Arnold, London, 1991.

2 Williams, M L. On the stress distribution at the base of a stationary crack. *Journal Applied Mechanics*, 24 (1957).

3 Dulieu-Smith, J M and Stanley, P. Progress in the thermoelastic evaluation of mixed-mode stress intensity factors. In *Proc. SEM Spring Conf. on Expt. Mech.* (Dearborn 1993), 617-629.

4 Dulieu-Barton, J M, Fulton, M C, and Stanley, P. The analysis of thermoelastic isopachic data from crack tip stress fields. *Fatigue Fract. Engng. Mater. Struct.*, 23 (1999), 301-313.

5 Tomlinson, R A, Nurse, A D, and Patterson, A. On determining the stress intensity factors for mixed mode cracks from thermoelastic data. *Fatigue Fract. Engng. Mater. Struct.*, 20, 2 (1997), 217-226.

6 Tomlinson, R A and Marsavina, L. Thermoelastic investigations for fatigue life assessment. *Exp. Mech.*, 44 (2004), 487-494.

7 Dulieu-Barton, J M and Worden, K. Genetic identification of crack-tip parameters using thermoelastic isopachics. *Meas. Sci. Technol.*, 14 (2002), 176-183.

8 Dulieu-Barton, J M and Worden, K. Identification of crack-tip parameters using thermoelastic isopachics and differential evolution. *Key Engineering Materials*, 245-246 (2003), 77-86.

9 Dulieu-Barton, J M and Stanley, P. Development and applications of thermoelastic stress analysis. *Journal of Strain Analysis for Engineering Design*, 33 (1998), 93-104.

10 Fruehmann, R K, Dulieu-Barton, J M, and Quinn, S. On the thermoelastic response of woven composite materials. *Journal of Strain Analysis for Engineering Design*, 43 (2007), 435-450.

11 Dulieu-Barton, J M and Quinn, S. Identification of the sources of non-adiabatic behaviour for practical thermoelastic stress analysis. *Journal of Strain Analysis for Engineering Design*, 37, 1 (2002), 59-71.

Journal of Physics: Conference Series **181** (2009) 012063 doi:10.1088/1742-6596/181/1/012063

Identification of material stiffness and damping in vibrating plates using full-field measurements

A. Giraudeau, F. Pierron

LMPF, Arts et Métiers ParisTech, Rue St Dominique, BP 508, 51000 Châlons en Champagne, France

alain.giraudeau@chalons.ensam.fr;fabrice.pierron@chalons.ensam.fr

Abstract. This study concerns the identification of complex stiffness components (ie, stiffness and damping) on isotropic vibrating plates. One of the main difficulties in damping measurements comes from the connection of the specimen to its environment (joints, clamps etc...) causing extra energy dissipation. The present procedure based on full field slope measurements from a deflectometry set up together with an inverse identification technique called the Virtual Fields Method is capable of identifying the material damping regardless of the dissipation coming from the specimen boundary conditions. This paper will present some experimental results confirming this statement.

1. Introduction

The measurement of material elastic stiffness and damping parameters, essential for the prediction of the vibrating or vibro-acoustic behaviour of a large range of structures, is common in material testing laboratories. The identification of the stiffness parameters is usually performed using tension, bending or torsion tests on rectangular coupons leading to simple stress states that can be expressed as functions of the specimen geometry and the applied load through a closed-form solution of the mechanical problem. Nevertheless, these procedures exhibit certain drawbacks. First, experimental boundary conditions must comply with that of the mechanical model, which is not always easy to achieve. Then, only a small number of parameters can be retrieved from a specific test because of the very simple stress state. As a result, several tests usually have to be performed to identify the full set of material parameters, increasing the cost of the procedure. As an alternative, several authors have tried to use the resonance frequencies of bending plates to identify the full stiffness tensor. More recently, the above approach was refined by using not only modal frequencies but also mode shapes.

An alternative to these methods was suggested by Grédiac et al. [1] making use of the measurement of slope fields at the surface of bent plates and performing the identification through a particular application of the principle of virtual work, the so-called Virtual Fields Method (VFM). This procedure uses the global equilibrium of the observed part of the coupon. One advantage of this technique is that stiffnesses are obtained directly (no iterations, no optimization scheme) and that restrictions on specimen geometry and boundary conditions are less critical than with other methods.

The measurement of damping parameters is a more complex problem than stiffness because of all the parasitic dissipation that is usually added in a classical mechanical test. One another main advantage of the present extension of this method is to greatly minimize the parasitic effects of the boundary conditions.

2. Theory – Virtual Fields Method (VFM.)

2.1. Considered case

The considered coupon of the isotropic test material is a rectangular thin plate (thickness h), free on its external boundary and clamped approximately in its centre to an excitation device which imposes a sinusoidal driving movement in the out-of-plane direction (inertial excitation). The external boundary of the plate is free. Assuming linear viscoelastic behaviour of the material and thin plate theory, the vibrating response of the plate is pure harmonic bending at the same frequency [2]. This response can be described by the actual deflection field $w(x,y,t)$ such that: $w(x,y,t)=\overline{w}(x,y)\cdot cos(\omega t+\phi(x,y))$ or using complex notation: $w(x,y,t)=\Re e[(w_r+jw_i)\cdot \exp(j\omega t)]$.

2.2. Global equilibrium

The plate can be virtually divided in two parts: a small zone called Π surrounding the clamping area and the second part Ω which is free on its external boundary. The virtual border between these two parts Π and Ω shall be called Γ. In the case of small perturbations, the local equation of equilibrium can be written at any point M of Ω (using the convention for summation on repeated indices):

$$\sigma_{ij,j}+f_i=\rho\cdot\gamma_i \tag{1}$$

where σ is the stress tensor, f the vector of volume forces, ρ the density and γ the acceleration vector (with i and j belong to {1,2,3}).

Each term can be multiplied by a u^* function which is user selected and must be continuous over Ω and differentiable. This function can be seen as a displacement field which is virtually imposed all over Ω:

$$\sigma_{ij,j}u_i^*+f_iu_i^*=\rho\gamma_iu_i^* \tag{2}$$

Equation (2) can then be integrated over Ω and leads to:

$$\int_{\Omega}\sigma_{ij,j}u_i^*dV+\int_{\Omega}f_iu_i^*dV=\int_{\Omega}\rho\gamma_iu_i^*dV \tag{3}$$

Using the divergence theorem for the calculation of the first term and assuming $\varepsilon*$ is the virtual strain field related to the virtual displacement field $u*$, Eq. 3 can be rearranged in a more useful expression :

$$-\int_{\Omega}\sigma_{ij}\varepsilon_{ij}^*dV+\int_{\partial\Omega}T_iu_i^*dS+\int_{\Omega}f_iu_i^*dV=\int_{\Omega}\rho\gamma_iu_i^*dV \tag{4}$$

where T is the vector of boundary tractions over $\partial\Omega$. It can be seen each term of Eq. 4 could be viewed as a virtual work under the action of the virtual displacement $u*$ and the related $\varepsilon*$ virtual strain. Respectively from the left to the right one can find, the virtual work of the internal forces, of the junction forces along the Γ border, of the volume forces and finally of the acceleration forces. Due to the assumption of the linear behaviour of the plate material, the acceleration of the harmonic response field of the plate can be simply expressed as $\gamma_i=\omega^2u_i$. Finally assuming that the volume forces can be neglected, the global equilibrium of the part Ω, free along its external boundary, can be written as:

$$-\int_{\Omega}\sigma_{ij}\varepsilon_{ij}^*dV+\int_{\Gamma}T_iu_i^*dS=-\rho\omega^2\int_{\Omega}w.w^*dV \tag{5}$$

where w is the out-of-plane deflection (it is assumed that the inertial forces generated by the in-plane deformations can be neglected since they are at least one order of magnitude lower than w) and w* the virtual out-of-plane deflection.

2.3. Virtual Fields Method (V.F.M.)
The principle of the Virtual Fields Method is to replace the stress components in the above equation by the actual elastic strains through the constitutive equations which parameters are to be identified. Then, by selecting 'appropriate' virtual fields, $u^*(x,y)$, it is possible to derive equations relating the materials constitutive parameters to integral functions of the actual strains.

In the present case, the only external forces acting on Ω are the connection forces on the Γ border. If a virtual displacement field u^* is selected such that it cancels out the virtual work of these connection forces then the second term of Eq. 5 is null and only the first and the last integrals of the latter remain. It can be shown that the first term can be expressed according to the Love-Kirchhoff theory, and Eq. 5 can then be rewritten as:

$$-\int_S (D_{ij} + j\omega B_{ij}) k_j k_i^* dS = -\rho h \omega^2 \int_S w.w^* \, dS \tag{6}$$

where w and k are the actual deflection and curvatures fields to be measured on the surface S of Ω whereas w^* and the corresponding k^* are the virtual deflection and curvature fields to be selected for the identification (here, i and j belong to {1,2,6} with the usual rule of contracted indices). $[D]$ and $[B]$ are respectively the elastic bending stiffness and viscous damping matrices, whose D_{11}, D_{12}, B_{11} and B_{12} components are the unknown material parameters. It must be pointed out that this relationship is valid whatever the excitation frequency, at resonance or out of resonance.

By separating the real and imaginary parts of Eq. 6 it can be shown that the latter can be split into two independent equations, presented in Eq. 7, where G_p and H_p denotes combinations of products between virtual and actual curvatures. These quantities are computed using the real or imaginary parts of the actual curvature fields according to whether the p index is re or im.

$$\begin{aligned} D_{11}G^{re} + D_{12}H^{re} - \omega B_{11}G^{im} - \omega B_{12}H^{im} &= \rho h \omega^2 \int_S w^{re} w^* dS \\ D_{11}G^{im} + D_{12}H^{im} + \omega B_{11}G^{re} + \omega B_{12}H^{re} &= \rho h \omega^2 \int_S w^{im} w^* dS \end{aligned} \tag{7}$$

with

$$G^p = \int_S \left(k_1^p k_1^* + k_2^p k_2^* + \tfrac{1}{2} k_6^p k_6^* \right) dS; \; H^p = \int_S \left(k_2^p k_1^* + k_1^p k_2^* - \tfrac{1}{2} k_6^p k_6^* \right) dS \tag{8}$$

The introduction of two independent deflection virtual fields w_1^*, w_2^* and related virtual curvatures fields k_1^*, k_2^* and k_6^* leads to a linear system of four equations whose unknowns are the unknown parameters: D_{11}, D_{12}, B_{11} and B_{12}.

2.4. Virtual fields selection
The selection of the virtual fields is a major issue for the identification method. It is carried out using three successive criteria.

- First, using piecewise virtual fields [3], virtual deflection and curvature fields are generated such that they are null along the Γ border leading to the cancellation of the virtual work of the connection forces.
- Then, to ensure independence between the four equations issued from Eq. 3, a particular application of a work by Grédiac *et al* [4] is used to select the 'special' virtual deflection fields $w_{(1)}^*$, $w_{(2)}^*$ which verify particular values of G_p and H_p and lead to very simple calculations of the parameters.
- The final selection is achieved using an adaptation of the work by Avril *et al* [5] providing 'optimized' virtual fields which minimize the effects of the measurement noise on the identification results.

3. Experimental results

3.1. Set-up

As presented in the previous section, the application of the Virtual Fields Method to the present case requires an excitation arrangement providing an inertial out-of-plane excitation of the plate which is free at its boundaries and a measurement set-up based on an optical method providing the curvature fields and the out-of-plane displacement fields on the whole surface of the plate. These measurements must be taken at two particular times: in-phase and at $\pi/2$ lag with the driving movement of the plate. These two particular positions correspond respectively to the real and imaginary parts of the displacement field of the coupon.

Tests have been carried out using an acrylic 200 x 160 x 3 mm^3 specimen. The sinusoidal out of plane inertial excitation is provided by a dedicated device where the coupon is clamped in its centre between the end of the driving rod of the device and a rigid steel washer of the same diameter, tightened by an axial screw. An accelerometer mounted on the rod is used to control the imposed driving movement (see Fig. 1).

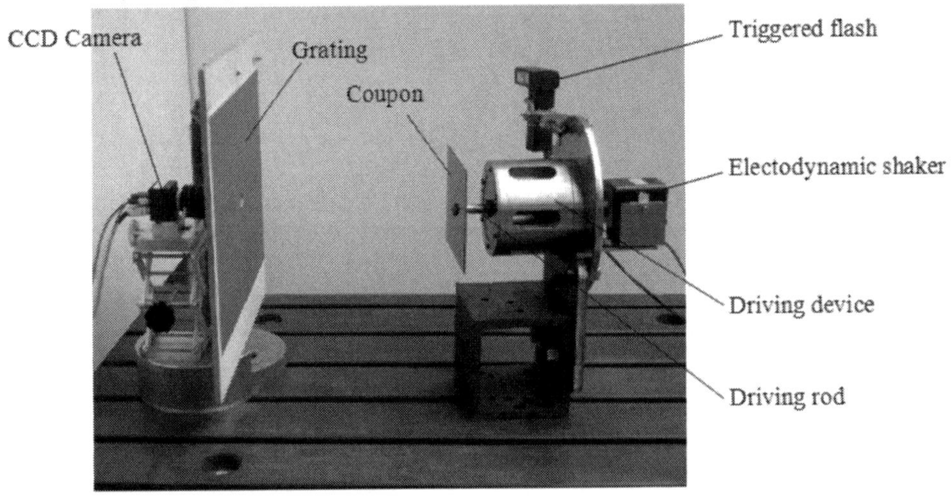

Fig. 1 – Experimental set up

3.2. Slope measurements by deflectometry

To avoid noise problems on curvatures arising from a double spatial differentiation of measured deflection fields, measurement of the slope field is preferred through deflectometry [6,7]. Images are frozen using a flash triggered by the driving movement with a *0* or *π/2* lag to get the images respectively related to the real or imaginary parts of the response.

Deflection and curvature fields are respectively obtained from spatial numerical integration and differentiation of polynomial fits to the slope fields (see Fig. 2).

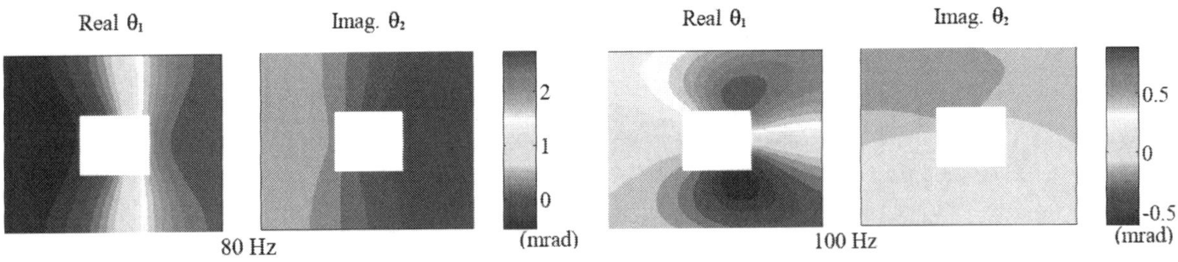

Fig. 2 – Examples of measured slope fields at 80 Hz and 100 Hz

Journal of Physics: Conference Series **181** (2009) 012063 doi:10.1088/1742-6596/181/1/012063

3.3. Identification results

Identification has been carried out using Eq. 7 with a set of two displacement virtual fields selected such that they are null along the rectangular border Γ of the Π zone surrounding the clamping area (central area in white on Fig. 2). The two stiffness parameters D_{11}, D_{12} and the two viscous damping parameters B_{11}, B_{12} of the tested material are first extracted. According to [8], material elastic properties, Young modulus E and Poisson ratio ν and their related loss factors $\eta = \tan(\delta)$ and η_v are presented in Table 1 below for five excitation frequencies. It must be noticed the 100 Hz excitation is close to a resonance of the tested plate and the four other frequencies are out of resonance. The identified values are compared with the Young modulus and its loss factor obtained from, respectively, vibrating tests performed by the authors in the same frequency range on a beam and classical DMTA tests. Coefficients of variation resulting from eight measurements at each tested frequency are noted in italic blue.

Table 1 – First identification results

	70 Hz		80 Hz		90 Hz		100 Hz		110 Hz		**Beam/DMTA**	
E (GPa)	4.62	*1.8%*	4.63	*1.3%*	4.71	*1.2%*	4.63	*2.4%*	4.69	*0.6%*	4.9	*2.4%*
υ	0.33	*1.3%*	0.33	*0.5%*	0.33	*0.8%*	0.34	*1.7%*	0.34	*0.6%*		
η_E=tan(δ)	0.065	*2.9%*	0.052	*1.8%*	0.050	*1.8%*	0.064	*7.1%*	0.068	*0.3%*	0.054	*3.5%*
η_υ	-0.024	*10%*	0.019	*4.7%*	-0.014	*6%*	-0.018	*9.1%*	0.021	*0.8%*		

The results for the elastic constants seem to be consistent and no significant differences are noticeable between the results issued from measurements at resonance or out of resonance. Some improvements are needed to increase the accuracy of image freezing in the time domain. The present trigger process is probably responsible for the higher scatter in the two loss factors compared to the elastic constants.

4. Insensitivity to boundary conditions

As mentioned in section 2, the selection of appropriate virtual fields leads to the cancellation of the virtual works of the junction forces along the Γ border. Therefore if dissipation phenomena occur in the clamping area, this does not affect the global equilibrium of the Ω observed part of the coupon as mentioned in Eq. 6. As a consequence, the identification process should not be disturbed.

To prove this very important feature of the method, a second experimentation was carried out where the same coupon was alternatively mounted at the end of the driving rod, first rigidly clamped as previously described in section 3. Secondly the same fixture is used with two rubber washers (1.5 mm thick) inserted between each face of the plate and the rigid parts of the driving device. In the first case, due to the rigid clamping, the displacement of the rod and the driving movement of the plate are the same. In the second case, a miniature accelerometer was mounted directly on the rear face of the plate, as closely as possible to the clamping area (see Fig. 3). This transducer provides a good evaluation of the out of plane solid movement of the plate. The comparison between the two acceleration signals provided by this accelerometer and the one mounted on the driving rod shows the amplitude of the solid motion of the plate is 14% lower than the amplitude of the movement of the rod and the phase lag is 34°. These results are obviously due to the dissipative behaviour of the rubber washers.

Fig. 3 – Details of second tested attachment of the plate

Identifications were carried out using images taken at 100 Hz with the two configurations. The results are presented in Table 2 in the same manner as above.

Table 2 – Identification results at 100 Hz with the two different attachments

100 Hz	Initial clamping		Rubber washers	
E (GPa)	4.63	*2.4%*	4.73	*1.5%*
υ	0.34	*1.7%*	0.33	*0.7%*
η_E=tan(δ)	0.064	*7.1%*	0.055	*5.2%*
η_υ	-0.017	*9.1%*	-0.018	*6.4%*

These results confirm that only the material properties of the Ω area of interest are identified without effects of the external conditions, which is one of the strengths of the method.

5. Conclusion

Previous studies have shown the principle of the Virtual Fields Method applied to vibrating plates and demonstrated the experimental feasibility. The last tests have proved that the method greatly minimizes the parasitic effects due to experimental conditions. This property obviously deserves major attention since it is of great interest for the identification of material damping. Very important potential applications of the method concern the identification of local material damping in a structure (plate or shell with stiffeners, for instance) and the possibility of obtaining local damping values at different locations of a plate or structure to be used as damage indicators. This is presently underway on composite plates.

References

[1] Grédiac M., (1996) Direct identification of elastic constants of anisotropic plates by modal analysis: theoretical and numerical aspects. *J. of Sound and Vib.*, 193(3):401-415.
[2] Giraudeau A. and Pierron F. (2005) Identification of stiffness and damping properties of thin isotropic plates using the virtual fields method. *J. of Sound and Vib.*, 284:757-781.
[3] Toussaint E., Grédiac M., and Pierron F. (2006) The virtual fields method with piecewise virtual fields. *Int. J. of Mech. Sci.*, 48:256-264.
[4] Grédiac M., Toussaint E. and Pierron F. (2002) Special virtual fields for the direct

determination of material parameters with the virtual fields method *Int. Journal of Solids and Structures,* 39(10):2691-2705.

[5] Avril S., Grédiac M., and Pierron F. (2004) Sensitivity of the virtual fields method to noisy data. *Comp. Mech.*, 34(6):439-452.

[6] Giraudeau A., Guo B. and Pierron F. (2006) Stiffness and damping identification from full field measurements on vibrating plates. *Exp. Mech.* 46(6):777-787.

[7] Surrel Y., (2004) Deflectometry: a simple an efficient non interferometry method for slope measurement. *In SEM Annual Congress on Experimental Mechanics*, 7-10 June in Costa-Mesa, California, USA.

[8] Pritz T. (2007) The Poisson's loss factor of solid viscoelastic materials. *J. of Sound and Vib.*, (306):790-802.

Journal of Physics: Conference Series **181** (2009) 012064 doi:10.1088/1742-6596/181/1/012064

High Speed Image Correlation for Vibration Analysis

Thorsten Siebert*[1], Rob Wood[2], Karsten Splitthof[1],

[1]Dantec Dynamics GmbH
Kässbohrerstr. 18, D-89077 Ulm, Germany

[2]Dantec Dynamics Ltd.
Garonor Way, Royal Portbury, Bristol BS20 7XE, United Kingdom

Thorsten.Siebert@Dantecdynamics.com

Abstract. Digital speckle correlation techniques have already been successfully proven to be an accurate displacement analysis tool for a wide range of applications. With the use of two cameras, three dimensional measurements of contours and displacements can be carried out. With a simple setup it opens a wide range of applications.

Rapid new developments in the field of digital imaging and computer technology opens further applications for these measurement methods to high speed deformation and strain analysis, e.g. in the fields of material testing, fracture mechanics, advanced materials and component testing. The high resolution of the deformation measurements in space and time opens a wide range of applications for vibration analysis of objects. Since the system determines the absolute position and displacements of the object in space, it is capable of measuring high amplitudes and even objects with rigid body movements. The absolute resolution depends on the field of view and is scalable. Calibration of the optical setup is a crucial point which will be discussed in detail. Examples of the analysis of harmonic vibration and transient events from material research and industrial applications are presented. The results show typical features of the system.

1. Introduction

In recent years optical full-field measuring techniques are increasingly being used in research and industry as development and design tools for improved characterization of materials and components. Besides moiré and interferometer based techniques the digital image correlation method is widely used for full-field and three dimensional displacement measurements [1-4].

Digital Image Correlation (DIC) is a full-field image analysis method, based on grey value digital images, that allows determination of the contour and surface displacements of an object under load in three dimensions. Due to rapid new developments of high resolution digital cameras and computer technology, the applications for these measurement methods have broadened. Since the system determines the absolute position and displacement of the object in space, deformation measurements with very high resolution are possible even under the presence of large deformation amplitudes and macroscopic rigid body movements.

The displacement resolution depends on the field of view (typically down to 10^{-5} of the field of view) and therefore is scalable, e.g. a few µm for an A4 paper size field of view depending on the number of pixels used. The measurable displacement range for this configuration is from μm up to several cm.

Out of the contour and displacement measurement, the image correlation technique provides material parameters far into the range of plastic deformation. Further data analysis tools allow the determination of the location and the amplitude of the maximum strain, the global strain distribution and behavior of crack growth, which are important parameters for material testing and fracture mechanics. Analyzing the temporal displacement by using e.g. Fourier analysis can give information about modal shape and vibration amplitudes.

2. Principles of Digital Image Correlation

Two imaging sensors, looking from different positions at an object, similar to human vision, provide enough information to perceive the object as three dimensional. Using a stereoscopic camera setup (figure 1, left), each object point is focused on a specific pixel in the image plane of the respective camera.

With the knowledge of the imaging parameters for each camera (intrinsic parameters: focal length, principle point and distortion parameters) and the orientations of the two cameras with respect to each other (extrinsic parameters: rotation matrix and translation vector), the position of each object point in three dimensions can be calculated.

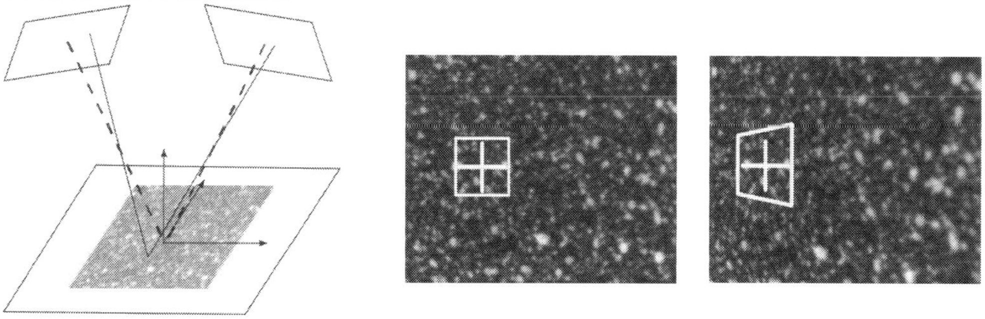

Fig. 1. Principle of stereoscopic setup (left) and grey value pattern in different images on the camera.

Using a stochastic intensity pattern on the object's surface, the position of each surface point in the two images can be identified by applying a correlation algorithm. By using this correlation algorithm, a matching accuracy of the original and the transformed facet of better than 0.01 pixel can be achieved [5].

The stochastic pattern can be applied easily with a colored spray paint, printing or other transfer methods. The size and distribution does influence the achievable resolution. An investigation about an optimum pattern and evaluation parameters can be found e.g. at Lecompte et al [6].

2.1. Calibration

The quality of the measurement is ensured by the exact knowledge of the intrinsic and extrinsic imaging parameters of the system. The imaging model is based on a pinhole model (figure 2). Thus the projection of the object point on the CCD is defined by the intersection of the line from the object point through the principle point of the imaging system and the CCD. The distance of the principle point to the image plane is the focal length f, the projection on the image plane the position of the optical axis on the CCD.

In addition to these pinhole model parameters, the distortion is also taken into account (radial as well as tangential distortion parameters are calculated). The calibration of the imaging parameters is easily done by taking a series of images of a calibration target under different perspective views.

The calibration process is integrated in the measurement software. During the calibration process the calibration points, which might be dots or intersection of lines, are detected automatically and displayed online on screen. Both cameras capture the images of the calibration target at the same time, the calibration parameters from both cameras will be calculated.

Typically eight images are sufficient to calculate all calibration parameters accurately. An online procedure and the direct user feedback allow an easy, reliable and fast calibration of the system.

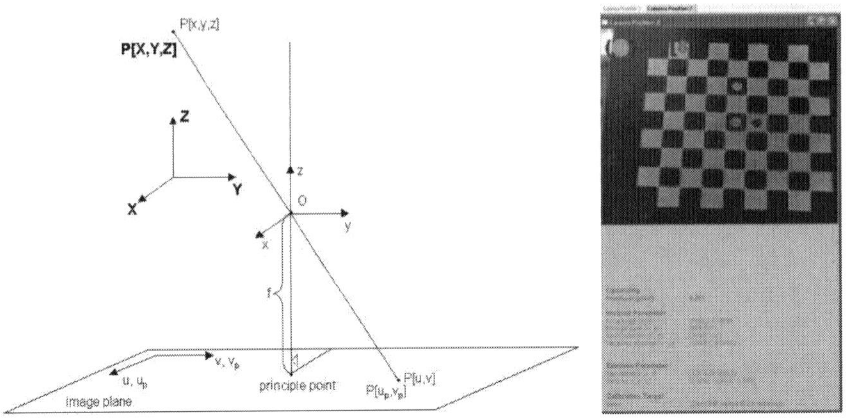

Fig. 2. Pin-hole camera model (left) and
calibration target and calculated calibration parameters (right).

2.2. Acquisition

Depending on the type of camera which is used for the image acquisition, several parameters and adjustments of the cameras can be made, e.g. exposure time, frame rate, brightness, contrast and area of interest. Typically a series of a few up to several hundreds of images are acquired and saved during the experiment. The acquisition can be started manually or fully automatically using various types of triggering. These images, combined with the calibration parameters, are the base for the evaluation.

2.3. Evaluation

By applying the correlation algorithm, the position of all object points can be identified in the image from both cameras. Using the intrinsic and extrinsic parameters of the system, the 3-dimensional coordinates for each object point can be calculated leading to a 3-dimensional contour of the object.
Following the changes of the grey value pattern for each camera along the loading steps, the surface displacements of the object are calculated. The matching accuracy of the correlation algorithm resulting in a displacement resolution down to 1/100 000 of the field of view for a MegaPixel CCD camera. For an A4 paper size this gives a resolution in displacement of down to 3-4 μm. [5-8]

3. Applications

Digital image correlation opens a wide range of potential applications. Since this technique is based on the analysis of a series of grey value images, deformation and strain analysis of static as well as of dynamic events can be performed. In the dynamic applications different types of excitation are used: a single frequency or a frequency sweep excitation, a shock excitation using a shock event and the use of noise excitation. A comparison between measurements using DIC, accelerometers, strain gauges and simulation have been successfully carried out on measurements of a bending beam using different excitation types [8, 9].
In the following we present three dynamic applications: a vibration analysis of a membrane, a shock excitation of an aluminium plate and an excitation of a circuit board using white noise.

3.1. Vibration of a membrane

During this experiment the vibration of a rubber membrane in front of a loudspeaker, which is excited close to a resonance frequency, was investigated. The size of the evaluated area is about 300 x 260 mm^2. The used rubber material has a porous surface, which gives a statistical grey value distribution, thus no extra preparation of the surface was necessary. A white colour was used to increase the amount of diffusely reflected light. The loudspeaker was excited at 59.9 Hz where a membrane was vibrating in resonance.

For image acquisition, high-speed cameras (Nanosense MK III) with 1280 x 1024 pixels resolution where used. The maximum frame rate at full resolution of these cameras is 1000Hz, for this measurement the frame rate was reduced to 625Hz. In figure 3 the setup including the high-speed cameras and the loudspeaker, is shown.

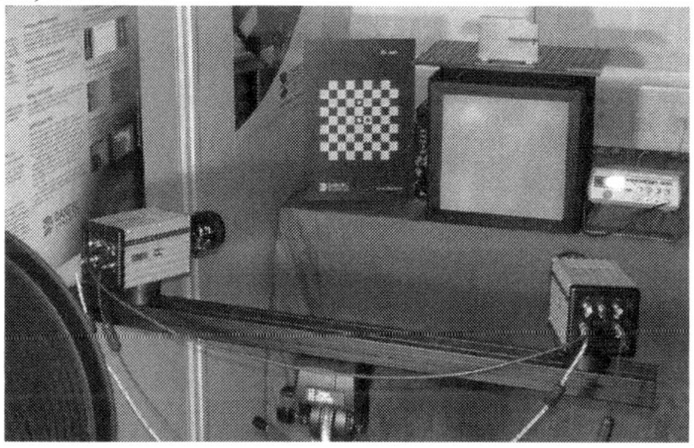

Fig. 3. Experimental setup with high-speed cameras, calibration target, loudspeaker and frequency generator for vibration measurement.

The image acquisition was started previous to the excitation of the loudspeaker. For the evaluation 150 images corresponding to 0.25 seconds of testing were processed.

The results of the measurement are displayed as a 3-dimensional model. On the surface of the object's model different data can be mapped in order to display the full field information. The different data are contour information (e.g. distances to Best-Fit plane), displacements (e.g. absolute X-, Y- or Z-deformation) or strains (e.g. principal strains). Figure 5 shows the full field display of the out of plane deformation of the membrane at the maximum elongation during the vibration. The colours represent the amplitudes of displacements. In the lower part of the graphic a spatial profile along the horizontal direction is displayed.

The maximum amplitude in this measurement is about 4 mm. The centre point and the circular area around the centre are vibrating out of phase. In order to get more detailed information the displacements of a marked centre point versus time is shown (figure 4). The evaluated series of images covers about 14 cycles of vibration. As it can be seen from the curve, it takes only 1 to 2 excitation cycles until the system is vibrating in a stable and stationary mode.

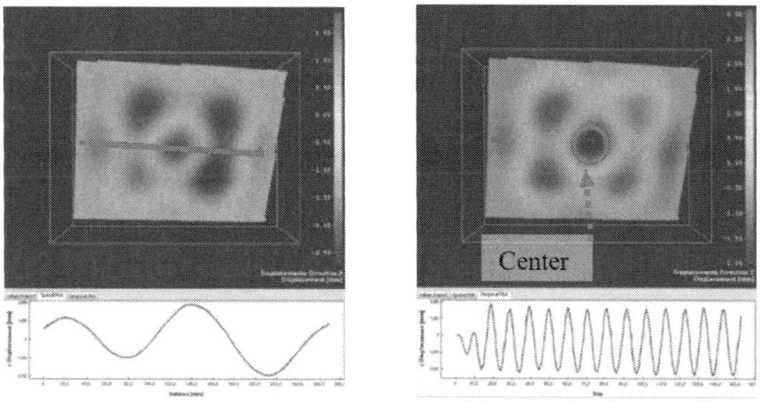

Fig. 4. Full field out-of-plane deformation of the membrane at maximum deflection of the centre (top left and top right). Spatial profile along a horizontal line (bottom left) and temporal plot of the displacement of the centre point (bottom right).

For the next measurement the frequency of excitation was set to another resonance at 36.5 Hz. Figure 5 shows the amplitude of vibration of the centre point and of a point in the lower left corner. The point in the centre vibrates with amplitude of about 5 mm symmetrically in the positive and negative direction. At the corner point the membrane is fixed on the housing of the loudspeaker, thus this point moves only in the positive direction. The amplitude is just about 0.05 mm. This example demonstrates the dynamic range, which can be covered with a single full field measurement.

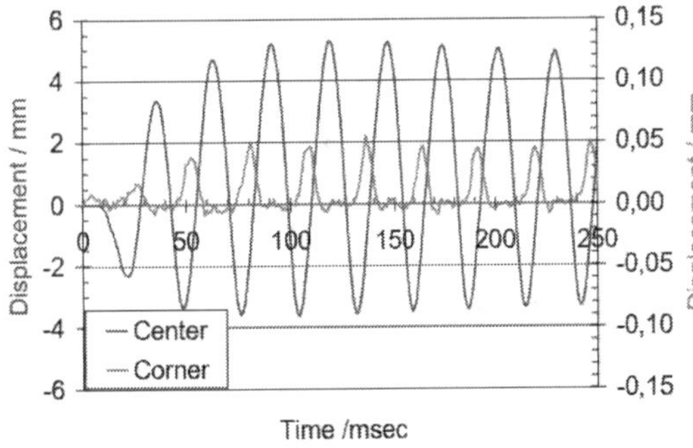

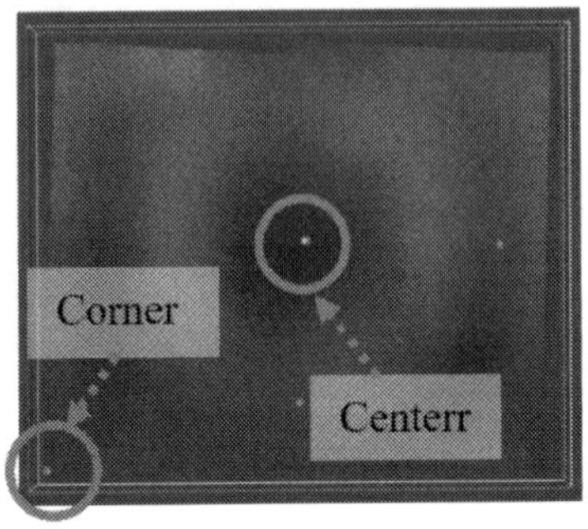

Fig. 5. : Out-of plane displacement of points in the centre and in the corner over time.

3.2. Shock excitation

Besides harmonic events the technique of image correlation is suitable to analyze transient events. For the measurement which is presented here, an explosive charge was used for excitation, which was positioned behind the backside of a square shaped aluminium plate with a size of 70x70 cm^2. The first measurement was performed with an image acquisition rate of 1 kHz, for a second measurement the frame rate was increased to 5 kHz. With both measurements the full field behaviour of the plate could be analyzed.

The series of images from this event can be processed like the images from the harmonic vibration before. As an example figure 6 shows the evolution of the displacement of the plate up to 5 msec after firing of the explosive charge. For a better understanding the displacements are scaled by a factor of 300.

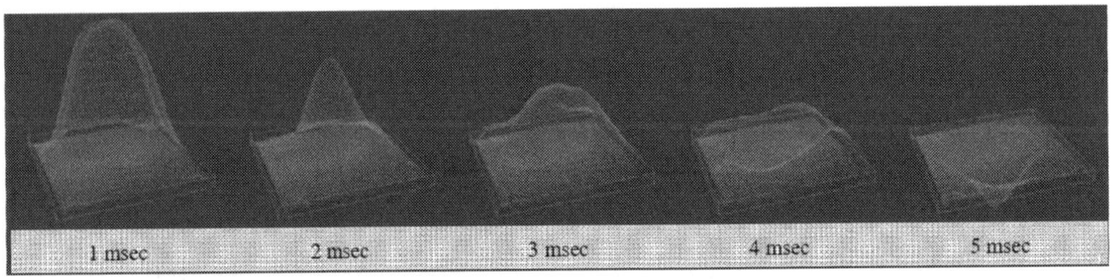

Fig. 6. : Full field displacement (scaled by a factor of 300) of the plate.

From this full field information single points can be selected, analyzing their temporal evolution. The displacements of the centre point and a point close to the border over a period of 125 msec are plotted in figure 7. The maximum amplitude of the centre is about 1.5 mm; it is about 0.5 mm of the point near the border. Due to the strong damping the amplitude decreases quickly down to 0.3 mm after 100 msec. Because of the presence of higher frequencies of vibration, which were stimulated by the shock excitation as well, the temporal displacement curves are not completely smooth.

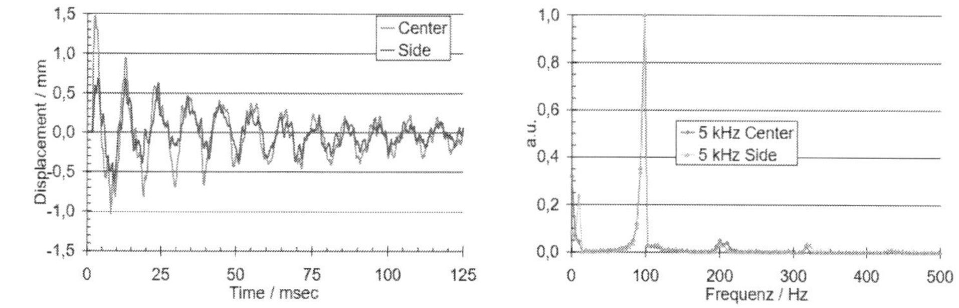

Fig. 7. : Out-of plane displacement of two points up to 125 msec after firing the explosive charge (left) and the spectrum (right).

A Fourier Transformation from these temporal displacements clearly reveals these vibration frequencies. Figure 8 shows the main frequencies for the point in the centre and a point close to the border. The main frequency for both points is not very sharp and is between 92 and 97 Hz. The amplitudes of the higher frequencies are much smaller also with a broad bandwidth.

Using the time evolution of the displacement of individual points out of the full field data, the different frequencies of vibration, which were excited with the shock event, were clearly resolved. Since the data of all object points are recorded simultaneously (no scanning is required), this method could be advantageously used for the study of transient events. The mode shape of the vibration could easily be analyzed.

3.3. Noise excitation

A circuit board was mounted on a shaker in a horizontal position. In an area of about 65 x 30 mm the movement of the circuit board was recorded at 2 kHz frame rate. The shaker was driven using white noise in a frequency range between 200 and 1 kHz. In Fig. 8 the image of one camera is showing the area under investigation and the full-field result as well as the displacement of a single point over time. The displacement of the point is in the range of ca. 40 µm.

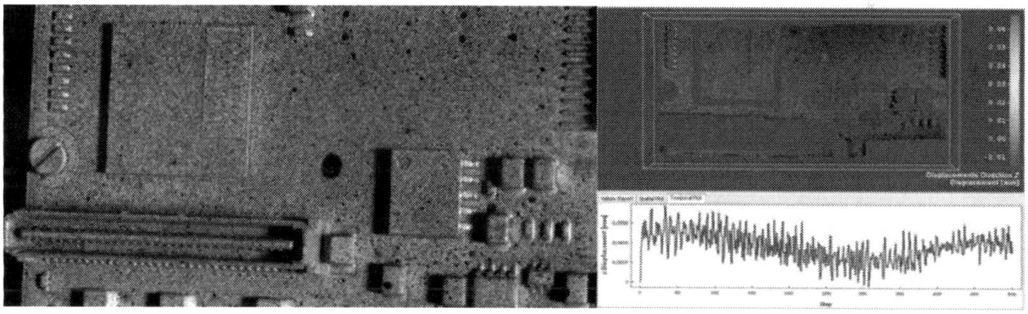

Fig. 8: Field of view (left), full-field displacement at one measurement step (right top) and displacement of marked point (cross) over measurement steps (right bottom).

By applying a frequency analysis at different points the main frequencies of the object can be determined. In this case an accelerometer indicates a main frequency at about 770 Hz. This frequency was found mainly in the centre of the board where the accelerometer was placed and no elements are located. The reconstruction of the mode shape at this frequency is shown in Fig. 9. The amplitude is in the range of 10 µm with a smooth shape which does not show areas of critical stressing of the board.

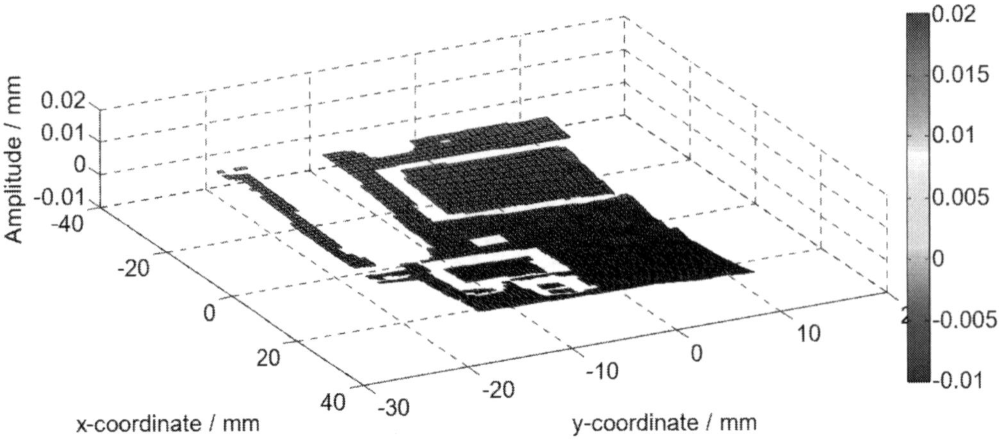

Fig. 9: Reconstructed mode shape at 770 Hz

A frequency analysis at a point close to the longitudinal connector showed another peak at about 535 Hz. The reconstructed mode shape is shown in Fig. 10. The amplitude of about 20 μm doubles the amplitude at 770 Hz. The shape of this mode generates also gradients of displacement and strain in an area of the connector and small electronic elements on the board. This frequency at 535 Hz will lead to higher chance of failure than other frequencies.

The application of this full-field method in combination with a frequency analysis allows the identification and reconstruction of the different mode shapes. This gives quantitative information about the behavior and damage of the sample at different frequencies.

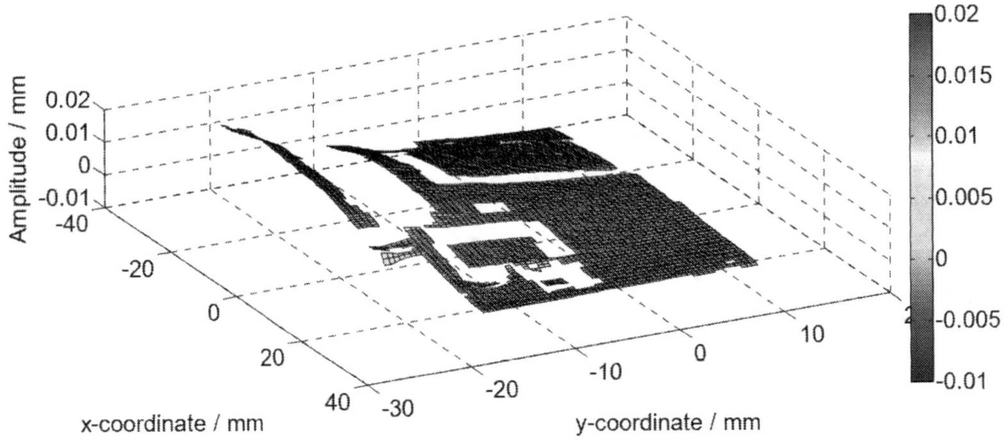

Fig. 10: Reconstructed mode shape at 535 Hz

4. Conclusion

This paper shows recent developments and applications of digital image correlation. Improvements in the calibration procedure significantly ease the handling of the system, while decreasing calibration time and enhancing the reliability of the results. Due to the modular concept of the hardware and the software this system can be used for a wide range of applications. Based on a full field method, this technique allows the determination of displacement amplitudes and vibration phases with high spatial and temporal resolution and within a large dynamic range. Amplitudes from m to μm and frequencies up to several tens of kHz could be covered, giving rise to applications of dynamic displacement and dynamic strain measurements. The use of high speed cameras offers the chance of measuring full field deformation and strain fields for harmonic as well as transient events. Applications of harmonic vibration, a shock phenomenon and noise excitation were shown, where accurate full field measurements of displacements have been demonstrated. From these measurements different mode shapes can be extracted and displayed.

References

[1] Sutton M.A., McNeil S.R., Helm J.D., Chao Y.J, "Advances in 2-D and 3-D computer vision for shape and deformation measurements", in Photomechanics, P.K. Rastogi, Ed. Topics in Applied Physics, 77, 323-372, Springer Verlag, New York (2000)

[2] Mguil-Touchal S., Morestin F., Brunet M., "Various Experimental Applications of Digital Image Correlation Method", CMEM 97, Rhodes, May, 45-58

[3] Winter D., "Optische Verschiebungsmessungen nach dem Objektrasterprinzip mit Hilfe eines flächenorientierten Ansatzes", Dissertation in der Fakultät Maschinenbau und Elektronik an der Universität Braunschweig, 1993

[4] Sutton M.A., Orteu J-J., Schreier H.W., „Image Correlation for Shape, Motion and Deformation Measurements", Springer Science+Business Media, LLC, ISBN 978-0-387-78746-6 (2009)

[5] Becker T, Splitthof K., Siebert T., Kletting P., "Error estimations of 3D digital image correlation measurements", Speckle06, SPIE Vol. 6341, 63410F, (2006)

[6] Lecompte D., Sol H, Vantomme J, Habraken A., "Analysis of speckle patterns for deformation measurements by digital image correlation", Speckle06, SPIE Vol. 6341, 63410E, (2006)

[7] Schmidt T., Tyson J., Galanulis K., Revilock D., Melis M., "Full-field dynamic deformation and strain measurements using high-speed digital cameras", 26th International Congress on High-Speed Photography and Photonics, SPIE Vol. 5580, 174-185, (2005)

[8] Siebert T., Becker T., Splitthof K., Neumann I., Krupka R., „High-speed digital image correlation: Error estimations and applications", Optical Engineering, 46(5):051004 (2007)

[9] Neumann, I., „Schwingungsmessungen in der Umweltsimulation mit einem Bildkorrelationsmesssystem", Diploma Thesis, Hochschule Ravensburg-Weingarten, Germany, 2006

Journal of Physics: Conference Series **181** (2009) 012065 doi:10.1088/1742-6596/181/1/012065

Characterization of the deformation behaviour of PCBs under dynamic loading conditions

P. F. Fuchs[1], **Z. Major**[1, 2], **R.W. Lang**[1, 3]

1 Polymer Competence Center Leoben GmbH, Leoben, A

2 Institute of Polymer Product Engineering, University of Linz, Linz, A

3 Institute of Materials Science and Testing of Plastics, University of Leoben, Leoben, A

E-mail: fuchs@pccl.at

Abstract. Printed circuit boards (PCBs) are frequently exposed to a complex combination of external and internal thermo-mechanical loads (static, cyclic and impact loads superimposed by local and global temperature effects). In this study, instrumented impact tests of the PCBs were performed and characterized. In addition to the acceleration measurement of the impact, the deformation behaviour was analyzed using a high speed camera and strain gauges. To reduce the testing effort for future PCB design the oscillation behavior of the PCB after impact was simulated using a finite element (FE)-software. Dynamic mechanical analysis experiments were performed and a linear viscoelastic material model was defined. Furthermore, the results of the simulation were compared to the measured values. In spite of the difference in the measured and simulated frequency values over the time, the viscoelastic effects including the damping behaviour were reflected accurately in the simulation.

1. Introduction and Objectives

Printed circuit boards (PCBs) are the heart of all common electronic devices. The area of application of these electronic devices is wide spread. Consequently, the PCBs are exposed to various application dependent complex loads. To meet the increasing demands on the PCBs the build ups and the materials used have to be optimized [1].

In this study the impact loading situation was selected for detailed analysis. Impact is a frequent failure initiator in PCBs and therefore of high practical interest [2] [3]. Due to the load the PCBs are exited to a damped oscillation which was examined and characterized.

To accelerate the optimization process, a finite element (FE) - software was used to simulate the impact. In order to describe the deformation behaviour in the simulation, the material properties had to be determined. Unlike previous studies on the drop test simulation, e.g. [4], which used either elastic or elastic-plastic material models, in this case a viscoelastic material model was chosen. Thus the time dependent damping of the oscillation amplitude and the frequency development could be simulated. Finally the results from the simulation were compared to the experimental results.

The objectives of this paper are to:

- conduct instrumented impact tests on PCBs identifying the loading situation.
- determine the material properties (time dependent modulus values).
- perform a data reduction identifying the parameters for the material model.
- simulate the instrumented impact test.
- compare the simulation with the experiment.

2. Experimental

The specimen characterized was a simplified PCB plate consisting of glass fiber reinforced epoxy layers without intermediate copper layers.

2.1. Instrumented Impact test

Instrumented impact tests were performed using a drop tower and subsequently analyzed. The loading was recorded by an acceleration sensor and the deformation behaviour was characterized by strain gauges applied on the surface of the plates and by a high speed camera (Photron Ultima 512, San Diego, CA, USA). Figure 1 shows the tested plate mounted on the impact fixture with the strain gauges applied. To measure the impact loading the acceleration sensor was fixed at the base plate.

Figure 1. Instrumented impact test with specimen and applied strain gauges and acceleration sensor.

The PCB deformed and the post-impact oscillations measured by the high speed camera are shown in figure 2a. The deformation amplitude is determined at various deformation stages and plotted over the time in figure 2b.

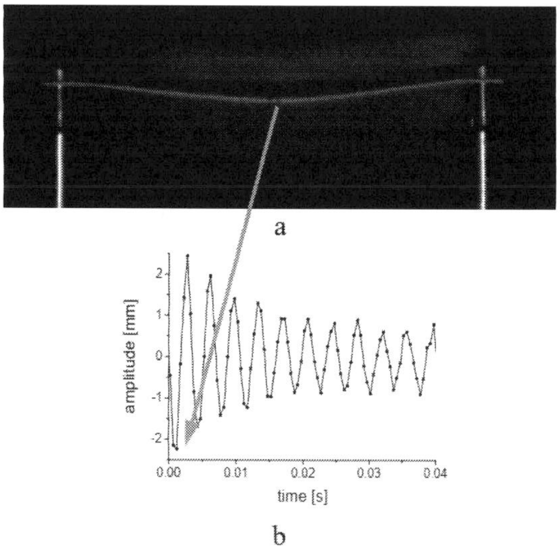

Figure 2. High speed camera image at a deformation stage selected (a) and the oscillogram related to the post-impact oscillations (b).

Due to the limited optical resolution of the high-speed camera, the amplitude decrease could be monitored only during the first 50 ms after the impact. Data from strain gauges was used to cover larger times.

Furthermore, the strain gauges provided information on the local strains. The strain gauges were applied both at the center of the plate and at the half distance between the edge and the mid of the plate. The strain gauge at the center was a biaxial type and additionally to the longitudinal strain the transversal strain was measured.

2.2. Mechanical Analysis

The plate specimen consists of fiber reinforced epoxy resin layers and was therefore supposed to show viscoelastic deformation behavior. To determine an adequate viscoelastic material model for the FE simulation, dynamic mechanical analysis (DMA) experiments were performed under bending load over a wide frequency and temperature range (from 0.01 up to 10 Hz and at the temperatures of -40,-20, 0, 23, 60, 125 °C). The plate was mounted on a 3 point bending fixture and exposed to sinusoidal oscillations by the actuator of the dynamic testing machine (figure 3). An electrodynamic testing machine (BOSE 3450, BOSE Co, MN, USA) was used for these tests.

Figure 3. Bending fixture for the dynamic mechanical analysis with clamped plate specimen.

2.3. Data Reduction

The data obtained from the DMA is the dynamic bending modulus over the frequency for the different temperatures. To get the parameters for the linear viscoleastic material model used in the FE simulation, the following steps were performed:

- Creation of a master-curve using the time (frequency)-temperature shift principle. The master-curve approach is well defined and described in the polymer science literature e.g. [5], [6].
- Calculation of the dynamic tensile modulus from the dynamic bending modulus considering the plate stiffness.
- Usage of the linear elastic equations to determine the dynamic shear modulus from the dynamic tensile modulus assuming a constant bulk modulus and a Poisson's ratio value from the literature [7].
- Transformation of the frequency basis to a time basis [8].
- Prony series fit for the dynamic shear modulus master-curve. A detailed description on the physical and mathematical background of the Prony series can be found in literature [8].

2.4. Simulation

The deformation behavior during the impact process of the test plate was simulated using commercial FE software Abaqus 6.7.1 (Simulia, Providence, RI; USA). The plate was modelled as a homogenous solid and the boundary conditions were set corresponding to the mounting situation on the impact fixture (see figure 1). The specimen was loaded directly by the acceleration values measured using the acceleration sensor. The material model was assumed as linear viscoelastic and defined by the Prony series parameters calculated from the DMA data. The deformation amplitude as a function of time was determined in the simulation and compared to the experimental results.

3. Results and Discussion

First, the data measured in the DMA experiments for the generation of the material model is presented. The frequency dependence of the dynamic bending modulus values for three different temperatures over the frequency range investigated (from 0.01 up to 10Hz) are shown in figure 4.

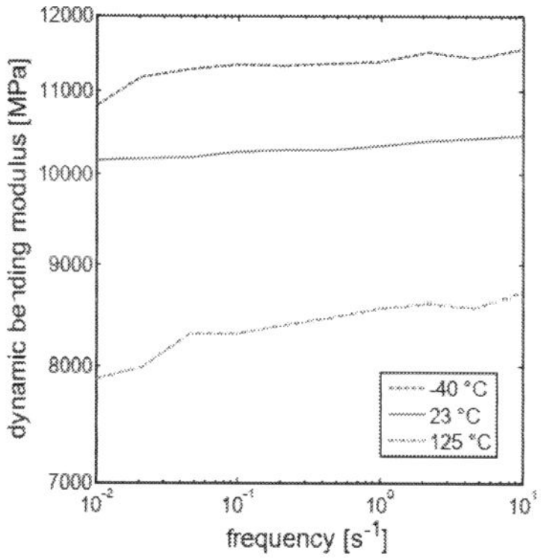

Figure 4. The frequency dependence of the dynamic bending modulus for three different temperatures (-40°C, 23°C and 125°C).

As the experimentally determinable frequency range is limited, to obtain dynamic modulus values for higher frequency ranges, a reference temperature was selected (in this case it was the room temperature, 23 °C) and a master-curve was created.

The master-curve was generated by shifting the dynamic modulus measured at higher temperatures to lower frequencies and the dynamic modulus measured at lower temperatures to higher frequencies. The concept is based on the physical effect, that the influence on the material behavior is similar regarding a change of the loading time and a change of the temperature.

The time-dependent dynamic shear modulus was calculated from the frequency-dependent dynamic bending modulus. The resulting master-curve was shown in figure 5. A Prony series measured was fitted to the data and the relaxation time values and the modulus values were calculated.

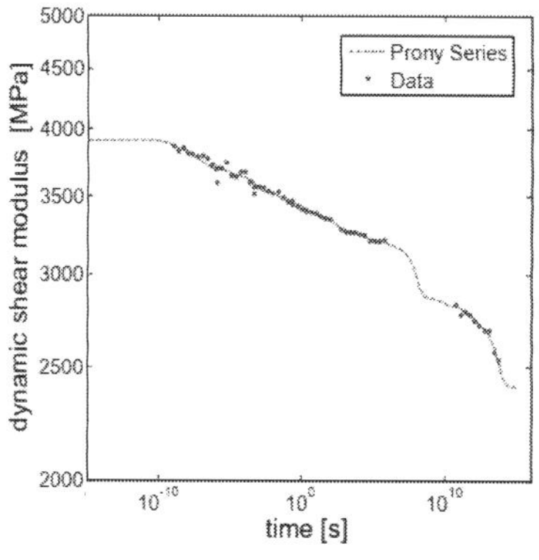

Figure 5. Master-curve of the dynamic shear modulus with the Prony series fit.

The envelop curves of both the measured deformation amplitudes and the simulated deformation amplitudes at the center of the plate over the time were shown in figure 6.

The deformation amplitude was calculated from the longitudinal strain measured by the strain gauge applied at the center of the specimen. The maximum value of the deformation measured by the high speed camera to the maximum value of the longitudinal strain was taken as proportionality factor. The second curve shows the results of the FE simulation. Over the measured time range of 0.25s, the experimental and the simulated envelop curves reveal only small, nearly negligible differences. It is concluded that the damping behaviour of the material in terms of amplitude decay is accurately described by the Prony series and by the master curves generated using the DMA experiments.

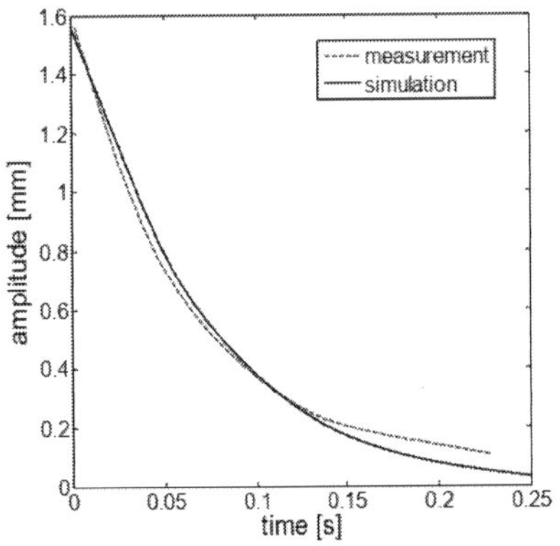

Figure 6. Deformation amplitude development of the post impact oscillation.

The frequency of the damped oscillation was found to decrease with increasing time both in the experiments and in the simulation. For comparison the results are shown in figure 7. Along with the amplitude, the loading rate is also decreasing over the time. Due to the viscoelastic properties of the epoxy matrix, the decline of the loading rate leads to a lower stiffness. As the frequency of the oscillation depends on the specimen stiffness [9], this effect could possibly explain the sinking frequency.

However, the experimental results reveal significantly smaller values over the entire time range than the simulation (the initial values are at about 300 Hz and 500 Hz respectively). No plausible reason for this discrepancy was found up to now. One reason of the difference between the experiments and the simulation could be measurement errors due to the very small forces at the dynamic bending test which were difficult to record. To adress this problem, further experiments will be performed and theoretical considerations regarding the viscoelastic model used will be done.

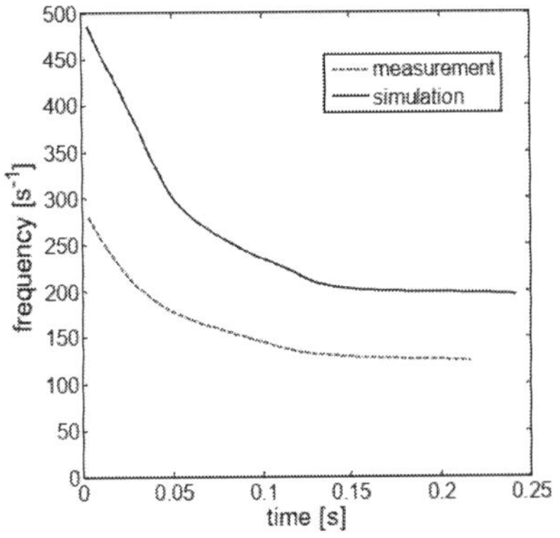

Figure 7. Frequency development over the time of the post impact oscillation.

4. Summary and Conclusion

Instrumented impact tests were performed on PCB plates and the subsequent oscillations were recorded by optical devices and by strain gauges over a wide time range.

To characterize the material behavior and to determine material parameters for FE simulation dynamic mechanical analysis (DMA) tests were carried out. A linear viscoelastic material model was defined by the parameters of a Prony series and applied.

FE simulations were performed on PCB plates using the viscoelastic material models and the results of the simulation in terms of amplitude and frequency were compared.

While the results of the simulation reveal a good correlation to the experimental results regarding the amplitude of the deformation behaviour, significant difference was found between the measured and the calculated frequency range for the damped oscillations.

To gain more insight into the frequency change during the damped oscillations, both further experiments will be performed and theoretical considerations regarding the viscoelastic material model and the FE model used will be done.

References

[1] Ehrler S 2002 Properties of new printed circuit board base materials *Circuit World* **28**/4 38

[2] Jeng S T, Sheu H S, Yeh C L, Lai Y S, Wu J D 2007 High G drop impact response and failure analysis of a chip packaged printed circuit board *International journal of impact engineering* **34** 1655

[3] Liu F, Meng G, Zhao M 2007 Viscoelastic influence on dynamic properties of PCBs under drop impact *Journal of electronic packaging* **129** 266

[4] Wang Y, Low K H, Pang H L J, Hoon K H, Che F X, Yong Y S 2006 *Modeling and simulation for a drop-impact analysis of multi-layered printed circuit boards* Microelectronics Reliability **46** 558

[5] Wineman A S, Rajagopal K R 2000 *Mechanical Response of Polymers An Introduction* (Cambridge:University Press)

[6] Schwarzl F R 1990 *Polymermechanik* (Berlin: Springer)

[7] Walter H 2003 *Morphologie-Zähigkeitskorrelation von modifizierten Epoxidharzsystemen mittels bruchmechanischer Prüfmethoden an Miniaturprüfkörpern* (Halle-Wittenberg: Martin-Luther-Universität)

[8] Tschoegel N W 1989 *The Phenomenological Theory of Linear Viscoelastic Behavior* (Berlin: Springer)

[9] Pilkey W K 1994 *Formulas for Stress, Strain, and structural Matrices* (New York: Wiley–Interscience)

Effect of Impact Damper on SDOF System Vibrations under Harmonic and Impulsive Excitations

S. Mehdi Zahrai[a], Amir F. Rod[b]

a Associate Professor-School of Civil Engineering

The University of Tehran, Tehran-Iran

Email: mzahrai@ut.ac.ir

b Dean of Research Unit, Asian Seismic Risk Reduction Center (ASRC)

Tehran-Iran

Email: amirfrod@yahoo.com

Abstract. Impact damper is a highly sensitive instrument, which belongs to the category of passive vibration devices used to attenuate the vibration of discrete and continuous systems.
In this paper, the effect of single unit impact damper on the single degree of freedom system under two different excitations is investigated. A study is made of the general behavior of a single particle impact damper; with the main emphasis on sinusoidal and impulse excitations. The effects of mass ratio, coefficient of restitution, and gap size are determined by using a MATLAB program.
It is found that the impact damper is an efficient device for reducing the vibration of structures subjected to impulsive and harmonic excitations. Numerical investigations are compared with each other. Effect of this damper shows that the system under consideration can serve as an efficient damper.
It is observed that the response of the system varies with small changes in the particle's mass, the coefficient of restitution and particle's clearance. There are critical values of the particle's parameters which an increase or a decrease does not contribute significantly toward amplitude reduction of the system.

1. Introduction

Structural control technology offers many new ways to protect structures from natural and other types of hazards. Semiactive structural control technology, appears to combine the best features of both passive and active control systems and to offer a viable means of protecting civil engineering structural systems against earthquake and wind load.[1]

An impact damper generally consists of a mass which is allowed to travel freely between two defined stops. Under the right conditions, the vibration of the system to which the impact damper is attached will cause the mass of the impact damper to strike the structure. Therefore, there is a need to understand the particle's motion, at varying parameters, as it travels between the container's boundaries to explore the potential of the impact damper.[2] The mass of the damper, the distance of travel, the excitation frequency, the peak values of modal amplitude at the location of the impact damper should be further investigated.

For more than three decades, researchers have studied the possibility of using impact damper to improve approaches to reduce structural responses (In 1973 S. F. Masri studied that chain-type impact dampers offer a simple and reliable method for attenuating wind-induced vibrations of tall flexible structures and also he performed experiments on a class of nonlinear dissipative cantilever and simply supported beams subjected to external sinusoidal excitation. In addition the damper was more effective when located away from the node of the mode shape[3]; Ranjit K. Roy and Richard D. Rocke and J. Eari Foster referred when the impact damper attached to a vibrating mechanical system the collision of the particle with the container boundaries results in a reduction of the vibration amplitude of the primary system through momentum transfer in 1975[4]; Akl and Butt (1994) performed effect of impact damper in vibration control of flexible structures; [5]; R. Chalmers and S. E. Semercigl were mentioned that a properly designed impact damper, is more effective and less sensitive to the changes in the external load and the system's parameters in 1990[6]; Aamir S. Butt and Fred A. Akl studied that impact damper can reduce the response of a vibrating continuous system if used effectively, and that its effectiveness depends on the behavior of the particle in 1997[7]; R.D. Nayeri and S.F. Masri and J.P. Caffrey investigated a comprehensive study concerning the development and evaluation of practical design strategies for maximizing the damping efficiency of multiple-unit particle dampers under random excitation in 2007[8];). It is found that the impact damper is an efficient device for reducing the vibrations of multidegree-of-freedom systems, particularly in structures such as tall buildings.[9]

While numerous analytical and experimental studies have been conducted over many years to develop strategies for modeling and controlling the behavior of impact dampers, no guidelines currently exist for determining optimum strategies for maximizing the performance of particle dampers, whether in a single unit or in arrays of dampers, under different excitation.[10] Numerical simulation studies for SDOF demonstrate the feasibility, reliability, and robustness of the control method. We studied an extensive set of numerical simulation on SDOF under harmonic and impulsive excitations. The mass of damper, the distance of travel, and the coefficient of restitution were monitored by different values. The Figures Indicate that SDOF system is sensitive for a range of gaps, masses, and coefficients of restitution.

2. Triangular Impulsive Load

For the first study, the SDOF system is affected by triangular impulsive load, with the initial force F0, which after td it will be changed to zero in a linear situation.
Mass of the oscillator is m and the impulsive force is:

$$P(t) = F_0\left(1 - t/t_d\right) \quad \textbf{(1)}$$

Initial position of free mass is in the middle of container which is better than resting to one of its two walls. Because when the force is applied to the same direction of mass damper, it will stick for a while to the wall and when force is changed, it starts moving; by passing d the first impact will be occurred; so the initial position of freemass is located in the middle of container.

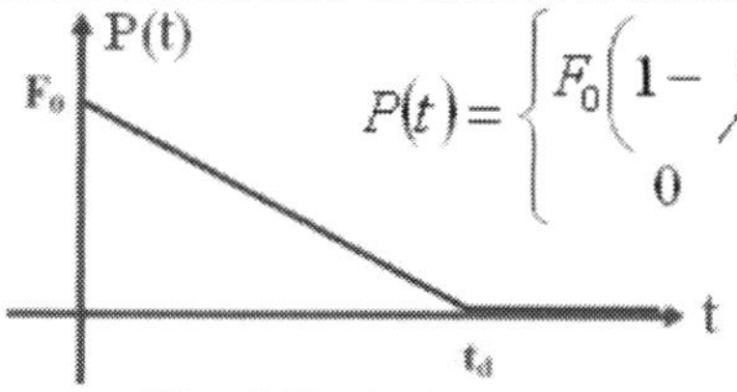

$$P(t) = \begin{cases} F_0\left(1 - t/t_d\right) & t < t_d \\ 0 & t > t_d \end{cases}$$

Fig. 1 Variation of impulsive load

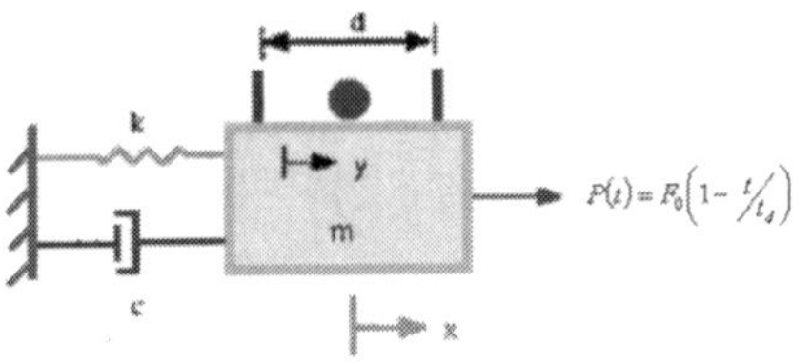

Fig. 2 Initial position of free mass is in the middle of container subjected to impulsive excitation

In numerical simulation the effect of mass damper was seen just at impacts, so during the moving of mass damper its velocity is constant and does not have any changes, thus it is assumed that oscillator's surface is frictionless; but in experimental model it is impossible to omit the friction, so there should be some differences between mechanical model and numerical simulation results. In this study, the momentum which causes at impacts is considered by:

$$\dot{y}^{+} = -e\dot{y}^{-} \qquad 0\langle e\langle 1 \tag{2}$$

The value of e is between zero and one, when it is closer to zero, the amplitude of system will be damped sooner. By following the equation:

$$(m+M)\dot{x}^{-} + m(1+e)\dot{y}^{-} = (M+m)\dot{x}^{+} \tag{3}$$

Which M is mass damper and an impact helps to reduce of excitation when at the time of impact $\dot{y}^{-}$ and $\dot{x}^{-}$ become the opposite of each other, so for this purpose if d is chosen too small, the undesired impacts will be increased. In this situation total number of impacts will be decreased and desired impacts would be increased, so that by decreasing e, damping of the system will be increased for two reasons, first if the value of e is closer to zero and the direction of impact is undesired (it means $\dot{y}^{-}$ and $\dot{x}^{-}$ are not opposite of each other), the raising of momentum will be decreased (term of $m(1+e)\dot{y}^{-}$ will be smaller) and second, if the direction of impact is desired for the system, the decreasing velocity of mass damper with the change of direction will cause the next desired impact, so in this condition the best value for e with considering the results will be 0.5.

3. Harmonic Excitation

The SDOF system is utilized by mass damper $M = \mu m$ which is a coefficient of oscillator mass m, so μ is mass ratio and system is excited by sinusoidal load.

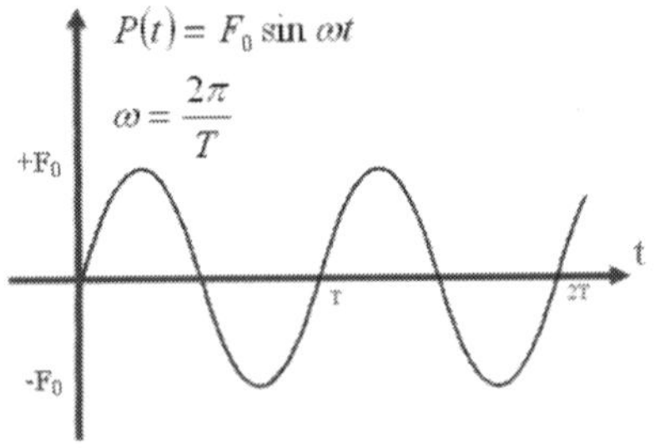

Fig. 3 Harmonic excitation applied to SDOF

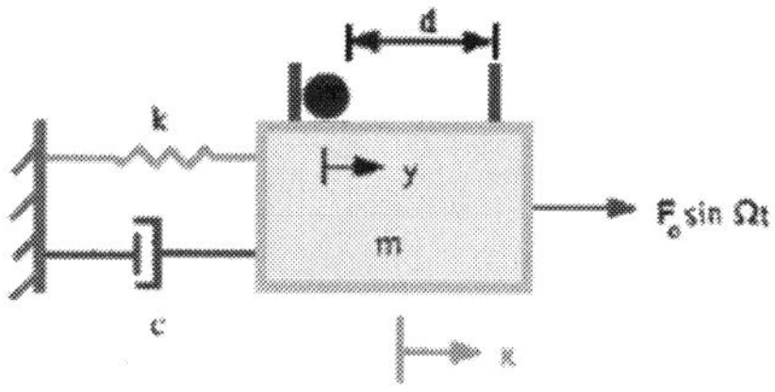

Fig. 4 Model of single unit impact damper Subjected to harmonic excitation

The initial displacement and velocity values for the primary and secondary masses are considered to be zero. The free mass particle is considered to be resting at the left wall of the container. When excitation is applied to the oscillator, the free particle starts moving relative to the primary mass. It is assumed that the oscillator's surface which comes in contact with the particle is frictionless. Therefore, the velocity of the particle stays constant between impacts. The motion of an oscillator to a sinusoidal excitation, between impacts, is given by the differential equation:

$$\ddot{X} + 2\xi\omega_n\dot{X} + \omega_n^2 X = \frac{F_0}{m}\sin\Omega t \tag{4}$$

Which $\dot{X}$ is oscillator velocity and ω_n is natural frequency and Ω is frequency of the forced harmonic excitation. If $\dot{Y}$ is the absolute velocity of the particle just after the previous impact and $\dot{X}(t)$ is the velocity of the oscillator at time t, then the relative velocity of the particle is:

$$\dot{y}(t) = \dot{Y} - \dot{X}(t) \tag{5}$$

The acceleration of the particle during the time interval Δt is calculated from:

$$\ddot{y}(t) = \frac{\dot{y}(t) - \dot{y}(t-\Delta t)}{\Delta t} \tag{6}$$

The total displacement $s(t)$ traveled by the particle since the previous impact is calculated:

$$s(t) = \sum \underbrace{s(t-\Delta t)} + \underbrace{s(\Delta t)} \tag{7}$$

Which $s(t)$ is total displacement and $s(t-\Delta t)$ is summation of displacement by the particle during each time interval except the last and $s(\Delta t)$ is displacement by the particle during the last time interval Δt.

From the restitution model the relationship between the particle's relative velocity before and after an impact was mentioned in equation (2), and from the conservation of momentum:

$$\dot{X}_+ + \mu(\dot{X}_+ + \dot{y}_+) = \dot{X}_- + \mu(\dot{X}_- + \dot{y}_-) \tag{8}$$

By substituting the two above equations, the velocity of the oscillator just after the impact is obtained as:

$$\dot{X}_+ = \dot{X}_- + \frac{\mu(1+e)\dot{y}}{(1+\mu)} \tag{9}$$

And the absolute velocity of the particle is stored as:

$$\dot{Y} = \dot{X}_+ + \dot{y}_+ \tag{10}$$

The positions of the two masses do not change at impact because it is assumed to occur during an infinitesimal increment of time. The velocity and displacement of the oscillator just after the previous impact are used as initial conditions to obtain its time history solution until the next impact.[11]

4. Analytical Procedures

The efficiency of impact damper has led to a number of investigations to study its behavior. Theoretical analyses in this paper have been confined to the case of an impact damper attached to a single degree of freedom main system.

It should be again noted that the absolute velocity of the particle is constant between impacts since the container's surface which comes in contact with the particle is assumed to be frictionless. The position of mass does not change at impact because it is assumed to occur during an infinitesimal increment of time. It had been stated that an impact damper effectively reduces the amplitude of the system only during impacts when the direction of the particle is against of the system. The momentum transfer between the particle and the system becomes maximum at a specific gap. That value produces the maximum reduction in the system's velocity at each impact, thus lowering its overall amplitude. It must be noted that the system exhibits a different value of the optimum gap at each mode of vibration. That value of the gap is smaller for modes occurring at higher frequencies. It is observed that the tendency of the particle to be suspended between the container boundaries in a larger gap. This is due to the particle having to travel further to reach the other wall, which allows the system to go through several cycles of motion before contact.

5. Numerical Observations

If the gap between the auxiliary mass and the end stops is properly adjusted, the mass collides with the stops causing the attenuation of the vibration amplitude of the main system. It is also observed that, too small or too large gap does not cause a reduction in the magnitude of the system's velocity; while a specific value produces maximum damping in the structure. If the gap is too small, it does not allow the system to reverse direction after a previous impact. On the other hand if it is too large, the system reverses its direction twice before coming in contact with the particle, as shown in Figs. 5 to 9 respectively for harmonic and impulsive excitations.

Note: The solid blue line (x) is oscillator displacement without impact damper, the dash green line (X) shows oscillator displacement which utilizes by impact damper, the red solid line (y) indicates relative particle displacement and the dash red line (Y) is absolute particle displacement. μ is mass ratio (particle's mass divided by the oscillator's mass, d is allowable particle travel and e is coefficient of restitution which exists between particle and container walls.

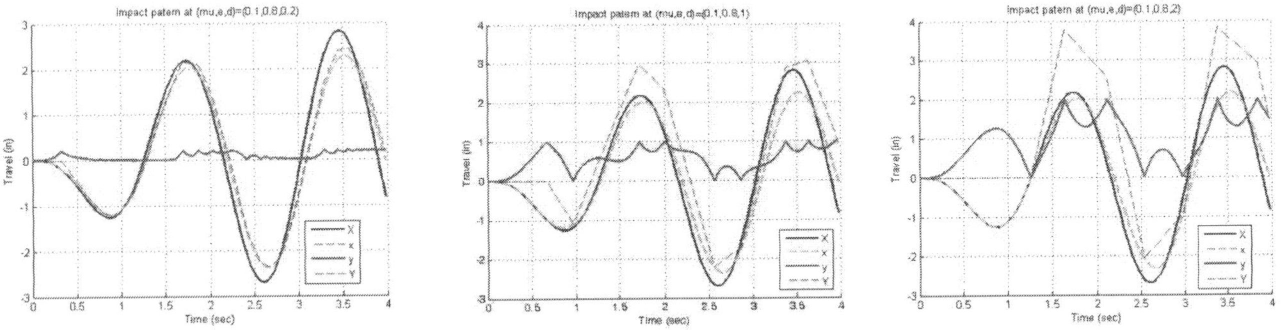

Fig. 5 Displacement response function
with $\mu = 0.1$ $e = 0.8,$ $d = 0.2$ 1.0 2.0
for harmonic excitation

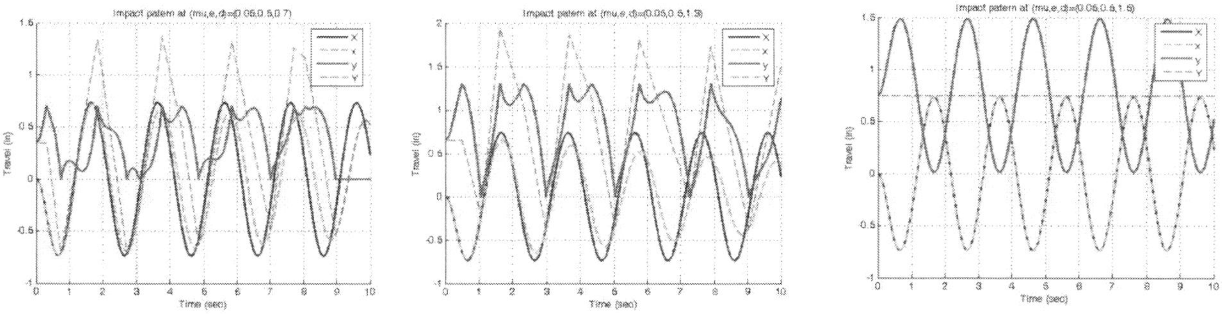

Fig. 6 Displacement response function

with $\mu = 0.05$ $e = 0.5,$ $d = 0.7$ 1.3 1,5

for impulsive excitation

This is achieved through energy dissipation in impacts and through the counteraction of the external excitation by the result of returning forces.

Figures 7 and 8 show that there is a critical value of the particle's mass beyond which an increase in the mass does not improve damping. This phenomenon can be explained by considering the conservation of momentum between the particle and the system. As the particles mass is increased, its absolute velocity immediately after the impact decreases, which in turn reduces its relative velocity and it takes longer to travel toward the other wall of the container.

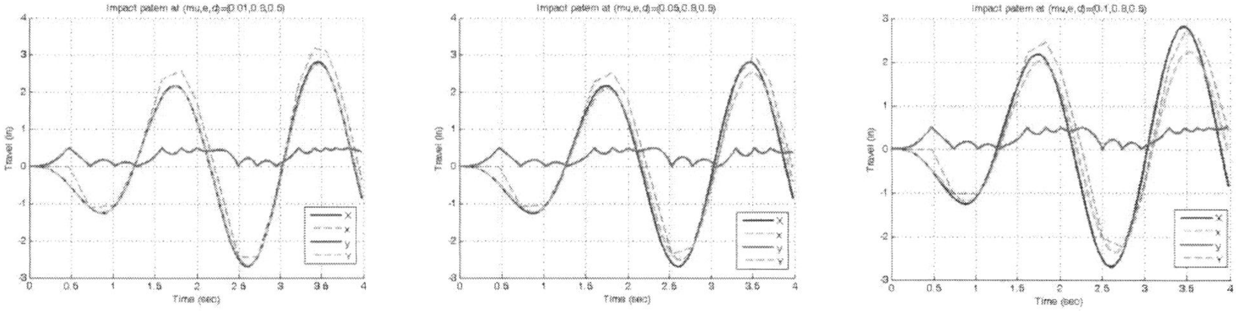

Fig. 7 Displacement response function

with $\mu = 0.01$ 0.05 0.1 $e = 0.8,$ $d = 0.5$

for harmonic excitation

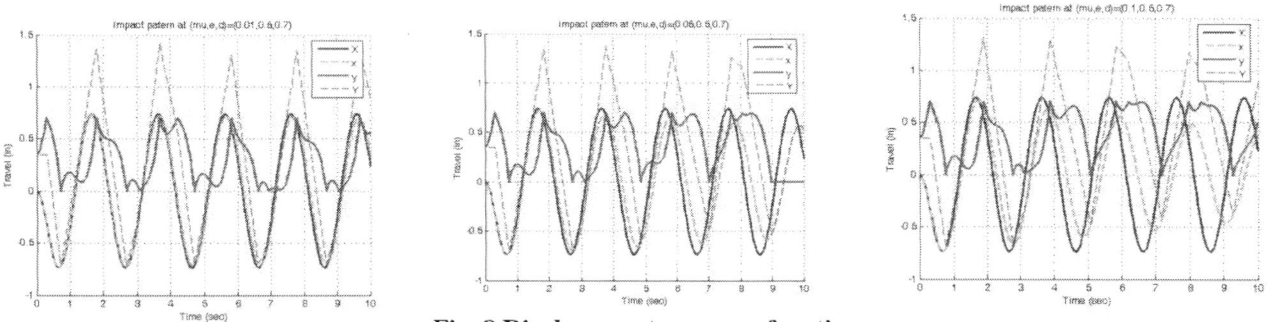

Fig. 8 Displacement response function

with $\mu = 0.01$ 0.05 0.1 $e = 0.5,$ $d = 0.7$

for impulsive excitation

If the force of friction exists, it also contributes to the reduction in the velocity while the particle is in motion. As the mass is increased beyond a certain value, its relative velocity immediately following the impact does not allow it to overcome the friction force while in motion. It then comes to rest relative to the system before approaching the other wall. At that point, if the system resumes its motion in the same direction and its acceleration is large enough to overcome the force of friction, the particle starts traveling in the opposite direction relative to the system. If a similar situation arises before approaching the container boundary, the particle reverses its direction again. Therefore, it is possible for the particle to be suspended between the container's boundaries while the system goes through several cycles of motion before making the next impact. Since the number of impacts decreases with an increase in mass, the overall effect of the damper stabilizes. Here it must be noted that the relative acceleration of the particle is independent of its mass and depends only upon the coefficient of kinetic friction; therefore, a heavier particle does not slow anymore rapidly than a lighter one does.

The coefficient of restitution "e" which is the ratio between the relative velocity immediately following the impact to the relative just before impact determines the rebound velocity of the particle and. It depends on the type, shape, and surface area of the materials coming in contact. At low values of the coefficient of restitution, the particle's rebound velocity does not allow it to escape from the wall on which it makes an impact. In the absence of any friction, the particle breaks free at the instant the velocity of the system decreases. (Figs. 9 and 10)

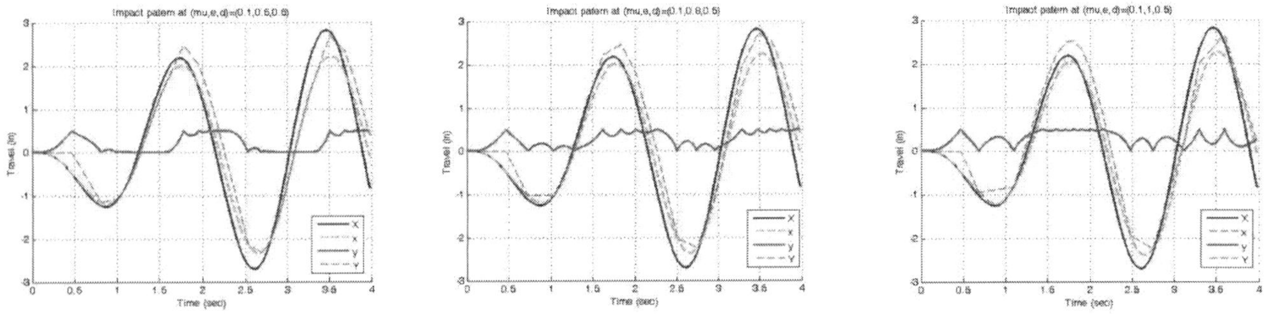

Fig. 9 Displacement response function
with $\mu = 0.1 \quad e = 0.5, \quad 0.8 \quad 1.0 \quad d = 0.5$
for harmonic excitation

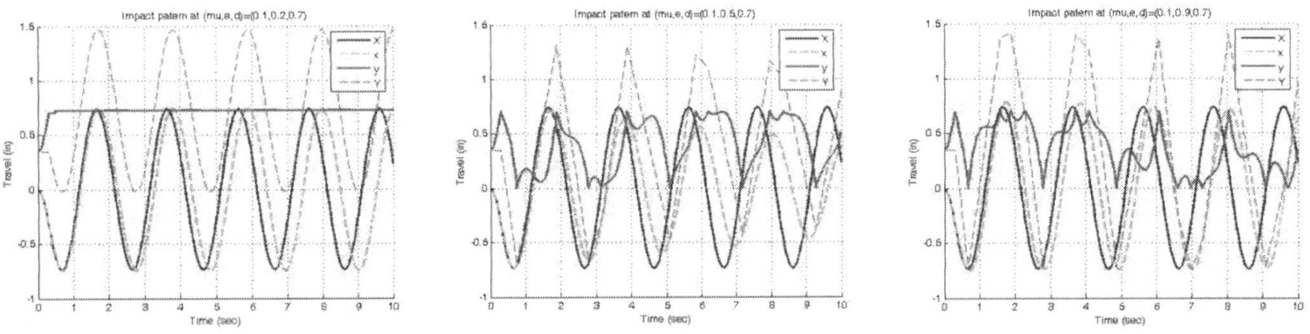

Fig. 10 Displacement response function
with $\mu = 0.1 \quad e = 0.2, \quad 0.5 \quad 0.9 \quad d = 0.7$
for impulsive excitation

If an impact pattern exists such that each impact decreases the velocity of the system, then an increase in the coefficient of restitution will further decrease the system's amplitude. This is due to an increase in the velocity of the particle at impact. At high values of the coefficient of restitution, the particle may reach the other wall of the container too quickly because of a high relative velocity. Then, the impact pattern of two impacts per cycle will not exist. However, the situation can be remedied by increasing the gap, which will allow the particle to travel further before making the impact. At a certain value of the gap, the impact pattern will return to two impacts per cycle.

6. Conclusions

The following conclusions are drawn from the study reported here in.

(i) In this paper was showed that impact damper can serve as an efficient damper to investigate SDOF system vibration under harmonic and impulsive excitations.

(ii) It was investigated that the system's parameters such as mass, gap and coefficient of restitution are the main factors to reduce the vibration amplitude.

(iii) When an impact damper is attached to a damped driven system, it is observed that the response of the system varies with small changes in the particle's mass, the coefficient of restitution and particle's clearance. There are specific values of mass, gap and coefficient of restitution which improve damping.

(iv) For harmonic and impulsive excitations, damp of the system will be improved by increasing of mass ratio, (regarding to Figures, range of 5% to 10% is more reasonable).

(v) In impulsive excitation there is a specific value of d which is efficient for the system, and any decreasing or increasing will not help to the system, (based on the results from the Figs.: $0.7<d<1.3$).

(vi) Coefficient of restitution in impulsive excitation has also a specific value to reduce vibration amplitude (Figs. show the best value is $0.2<e<0.9$).

References

[1] B. F. Spencer Jr., S. Nagarajaiah, (2003), "State of the Art of Structural Control", Journal of Structural Engineering, July 2003, pp.845-856.

[2] Aamir S. Butt and Fred A. Akl, (1997), "Experimental Analysis of Impact-Damped Flexible Beams", Journal of Engineering Mechanics, April 1997, pp 376-383.

[3] S. F. Masri, (1973), "Steady-State Response of a Multidegree System with an Impact Damper", Journal of Applied Mechanics, March 1973, pp 127-132.

[4] Ranjit K. Roy and Richard D. Rocke and J. Eari Foster, (1975), "The Application of Impact Dampers to Continuous Systems", Journal of Engineering for Industry, Nov. 1975, pp1317-1324.

[5] Akl and Butt (1994), Application of Impact Dampers in Vibration Control of Flexible Structures, NASA report, Structures and Mechanics Division, August 1994.

[6] R. Chalmers and S. E. Semercigl, (1991), Impact Damping the Second Mode of a Cantilevered Beam, Journal of Sound and Vibrations, (1991), Vol. 146, pp157-161.

[7] Aamir S. Butt and Fred A. Akl, (1997), "Experimental Analysis of Impact-Damped Flexible Beams", Journal of Engineering Mechanics, April 1997, pp 376-383.

[8] R.D. Nayeri and S.F. Masri and J.P. Caffrey, (2007), Studies of the Performance of Multi-Unit Impact Dampers Under Stochastic Excitation", Journal of Vibration and Acoustics, April 2007, Vol. 129, pp239-251.

[9] S.F. Masri, (1973), "Steady-State Response of a Multidegree System with an Impact Damper", Journal of Applied Mechanics, March 1973, pp 127-132.

[10] R.D. Nayeri, S.F. Masri, J.P. Caffrey (2007), "Studies of the Performance of Multi-Unit Impact Dampers", Journal of Vibration and Acoustics, Vol.129, pp 239- 251.

[11] Aamir S. Butt, (1994), "Dynamics of an Impact-Damped Harmonic Oscillator", Developments in Theoretical and Applied Mechanics, Vol17, pp 298-308.

Experimental verification and comparison of mode shape-based damage detection methods

M Radzieński[1], M Krawczuk[1,2]

[1] Technical University of Gdansk, Faculty of Electrical and Control Engineering, Narutowicza 11/12, 80-952 Gdansk, Poland

[2] Institute of Fluid Flow Machinery, Polish Academy of Sciences, ul. Fiszera 14, 80-952 Gdansk, Poland

Email: Maciej.Radzienski@gmail.com

Abstract. This paper presents experimental verification and comparison of damage detection methods based on changes in mode shapes such as: mode shape curvature (MSC), modal assurance criterion (MAC), strain energy (SE), modified Laplacian operator (MLO), generalized fractal dimension (GFD) and Wavelet Transform (WT). The object of the investigation is to determine benefits and drawbacks of the aforementioned methods and to develop data preprocessing algorithms for increasing damage assessment effectiveness by using signal processing techniques such as interpolation and extrapolation of measured points. Noise reduction algorithms based on moving average, median filter, and wavelet decomposition are also tested. The experiments were performed on an aluminium plate with riveted stiffeners. Damage was introduced in a form of damaged rivets and a saw cut in the angle bar. Measurements were made using a non-contact Scanning Laser Doppler Vibrometer (SLDV) at 101 points in two rows, distributed over the structure height and positioned along two reinforcing ribs.

1. Introduction

In the last two decades non-destructive examination (NDE) and structural health monitoring (SHM) has received a considerable amount of interest in the literature. Among them vibration analysis for damage detection was the most propagate due to its implementation simplicity [1, 2]. Development of Laser Scanning Doppler Vibrometer (SLDV) devices makes those techniques are even more attractive because of non-contact, high resolution vibration measurements in short time periods.

Vibration-based damage detection methods make use of the dynamic responses of a structure, such as natural frequencies and mode shapes. Occurrence of damage leads to changes in structural dynamics. These changes may be used for damage localization and its severity assessment.

Recently, many studies were made proposing different vibration-based damage detection methods. However, in the majority of cases those methods were tested on numerical results. Additionally, it is hard to evaluate the utility of a specific method without comparison of results obtained from the same set of data.

Thus, the purpose of the present study is to determine benefits and drawbacks of the most popular and promising methods, relying on direct comparison of them. Furthermore, it is shown that digital signal processing of the measured data may enhance results in damage localization.

2. Damage identification methods

In this section, six different damage detection methods are presented. The first three of them (2.1-2.3) are based on comparison between two sets of measurements. The first is a reference state and the second is a current state of the structure. Another group of damage detection methods, described in sections 2.4-2.6, make use of only current state data for damage detection and localization. However, using additional reference data may improve effectiveness.

2.1. Mode shape curvature (MSC) and damage index (DI)

Following [3, 4, 5] Pandey in 1991 proposed the use of the second derivative of the mode shape as a tool for damage detection. It has been reported that damage of the structure causes changes in mode shape curvature and that it can be a good indicator for detecting and localizing the damage.

This method is founded on determining the absolute difference of mode shape curvature (MSC) or mode shape curvature squares (MSCS) for the undamaged and damaged object as follows:

$$MSC_i = \sum_{j=1}^{N_m} \left| \phi''^{d}_{i,j} - \phi''_{i,j} \right|, \quad MSCS_i = \sum_{j=1}^{N_m} \left| \left(\phi''^{d}_{i,j}\right)^2 - \left(\phi''_{i,j}\right)^2 \right|, \tag{1}$$

where ϕ'' and ϕ''^{d} are the mode shape second derivatives of undamaged and damage states respectively and N_m is the number of measured modes and i is measurement point index.

Stubbs and Kim proposed damage index (DI), which is also based on changes in mode shapes curvatures and it is defined as [4, 5]:

$$DI_i = \sum_j \frac{\left(\left(\phi''^{d}_{i,j}\right)^2 + \sum_{i=1}^{N}\left(\phi''^{d}_{i,j}\right)^2\right) \cdot \sum_{i=1}^{N}\left(\phi''_{i,j}\right)^2}{\left(\left(\phi''_{i,j}\right)^2 + \sum_{i=1}^{N}\left(\phi''_{i,j}\right)^2\right) \cdot \sum_{i=1}^{N}\left(\phi''^{d}_{i,j}\right)^2}. \tag{2}$$

2.2. Modal assurance criterion (MAC)

One of the methods for localizing damage in the structure consists in using the modal assurance criterion (MAC) and co-ordinate modal assurance criterion (COMAC), which were presented by Allemang and Brown (1982) and Lieven and Ewins (1988), respectively. These indexes indicate the correlation between two sets of mode shapes. The MAC is a global index of damage detection and gives information about which modes have been changed the most. The COMAC compares mode shapes in a point-wise manner giving the possibility of locating the damage [5, 6]:

$$MAC_{j,k} = \frac{\left| \sum_{i=1}^{N_o} \left(\phi_{i,j}\phi^{d}_{i,k}\right) \right|^2}{\left(\sum_{i=1}^{N_o} \left|\phi_{i,j}\right|^2\right)\left(\sum_{i=1}^{N_o} \left|\phi^{d}_{i,k}\right|^2\right)}, \quad COMAC_i = \frac{\left(\sum_{j=1}^{N_m} \left|\phi_{i,j}\phi^{d}_{i,j}\right|\right)^2}{\left(\sum_{j=1}^{N_m} \left|\phi_{i,j}\right|^2\right)\left(\sum_{j=1}^{N_m} \left|\phi^{d}_{i,j}\right|^2\right)} \tag{3}$$

where: $\phi_{i,j}$ and $\phi^{d}_{i,k}$ stands for a value of mode shape j and k at point i, in the undamaged and damage object respectively. N_o is a number of points, N_m is a number of modes.

The diagonal value of the MAC matrix varies from 0 to 1 and indicates correlation between the *j'th* mode shape of undamaged structure and *k'th* damage structure, where 0 is for absence of correlation and 1 for complete correlation.

2.3. Strain energy damage index (SEDI)

Cornwell [6] suggested that a structure can be divided into small areas. If there is a damage in one of them, changes in flexural rigidity for a specific area may be used to localize damage.

For a beam-like structure the strain energy damage index is given by:

$$\beta_i = \sum_{j=1}^{N_m} \frac{\int_{a_i}^{a_{i+1}} \left(\frac{\partial^2 \phi^d{}_j}{\partial x^2} \right)^2 dx}{\int_0^l \left(\frac{\partial^2 \phi^d{}_j}{\partial x^2} \right)^2 dx} \Big/ \sum_{j=1}^{N_m} \frac{\int_{a_i}^{a_{i+1}} \left(\frac{\partial^2 \phi_j}{\partial x^2} \right)^2 dx}{\int_0^l \left(\frac{\partial^2 \phi_j}{\partial x^2} \right)^2 dx}. \tag{4}$$

2.4. Modified Laplacian operator (MLO)

This method is based on localizing changes in slope of a mode shape caused by the damage. Because experimental mode shapes are a series of measured points distributed in space, therefore the second difference of this signal can be estimated by Laplacian difference equation. The one dimensional Laplacian Δ_i of discrete mode shape ϕ_i is given by:

$$\Delta_i = \phi_{i-1} - 2\phi_i + \phi_{i+1}. \tag{5}$$

Ratcliffe [4] introduced a modified Laplacian operator to enhance the results for one dimensional structures and Qiao et al.[7] extended this for two dimensional structures. This method is based on calculating the difference between Laplacian and a cubic fit polynomial to the Laplacian. A cubic function is determined for every point of the Laplacian, using only two points on either side of the considered element for coefficients calculations.

2.5. Generalized fractal dimension (GFD)

The fractal dimension (FD) was originally proposed by Mandelbrot [8]. In 1988 Katz's defined estimation of FD for curvature using a sequence of points. However, this method of estimating FD could give false peaks in higher mode shapes in maximum and minimum points of its first derivative. To overcome this problem Wang and Qiao [2, 9] proposed using scale factor in the FD algorithm. They call this method generalized fractal dimension (GFD) and it is defined as follows:

$$GFD_M(x) = \frac{\log(n)}{\log(n) + \log(ds(x_i, M) / Ls(x_i, M))} , \tag{6}$$

$$Ls(x_i, M) = \sum_{j=1}^{M} \sqrt{(y(x_{i+j}) - y(x_{i+j-1}))^2 + s^2(x_{i+j} - x_{i+j-1})^2} , \tag{7}$$

$$ds(x_i, M) = \max_{1 \le j \le M} \sqrt{(y(x_{i+j}) - y(x_i))^2 + s^2(x_{i+j} - x_i)^2} . \tag{8}$$

2.6. Wavelet transform (WT)

A wavelet is a function used to decompose a signal $f(x)$ into series components $\psi_{u,s}(x)$ derived from a mother wavelet $\psi(x)$ by scaling and translating, as given by:

$$\psi_{u,s}(x) = \frac{1}{\sqrt{s}} \psi(\frac{x-u}{s}) \tag{9}$$

Considering the mode shape ϕ as a spatial one-dimensional signal, the continuous wavelet transform (CWT) can be obtained as [10, 11]:

$$W\phi(u,s) = \frac{1}{\sqrt{s}} \int_{-\infty}^{+\infty} \phi(x)(\frac{x-u}{s})\, dx . \tag{10}$$

The wavelet transform has the ability to find singularities in considered functions. Therefore, an abrupt change or peak in wavelet coefficient can be used as an indicator of the damage location.

3. Experimental setup and results

Figure 1 shows the experimental set-up used in measuring mode shapes and frequency response functions (FRFs) of the aluminium plate by a Polytec PSV-400 Scanning Laser Doppler Vibrometer (SLDV). An electromechanical shaker (GWV B100) was used for vibration excitation.

In the experiment an "FFT" acquisition was firstly performed to obtain FRFs using a periodic chirp excitation. This provides natural frequency determination by finding local maxima in an averaged FRF graph. Those frequencies were used successively as a single-frequency excitation, namely "FAST SCAN", for measuring corresponding mode shapes. The noise level of measured mode shapes is mostly determined by utilizing frequency bandwidth B_w and the signal level. The noise level is proportional to $\sqrt{B_w}$ but the time of acquisition is inversely proportional to B_w, so it should be considered to find a middle ground between acquisition time and precision [12]. The signal can be maximized by coating the measured surface with light back-scattering paint, such as bright acrylic paint used in this research. In addition a scanning head should be positioned away from the examined structure in laser visibility maximum if possible.

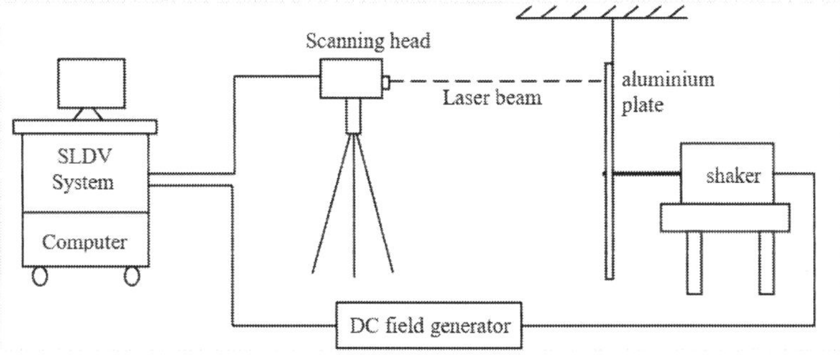

Figure 1. Experimental set-up for measuring aluminium vibrations using a PSV-400 scanning laser vibrometer.

In order to compare presented damage detection methods experimentally, a 70cm x 100cm x 0.4cm aluminum plate with two riveted stiffeners was tested. The test structure was fixed to the ceiling by two sets of 1mm steel cables connected to the bracing, which is an approximate of four free boundary conditions. Stiffeners were equal-sided 70cm x 4cm x 0.4cm aluminum angles and they were jointed to the plate by two rows of 27 aluminum rivets in each of them (totally 108 rivets in the whole structure). Two damage scenarios were tested. At first, 6 and 12 rivets were cut in one or both stiffeners. Next, the damage was made by a saw cutting the orthogonal leg of the right angle in the middle of its length by 25% and 50% depth. Measurements were made in two rows between rivets along the plate height in 101 points each side. The specimen schematic drawing and a photograph are shown in figure 2.

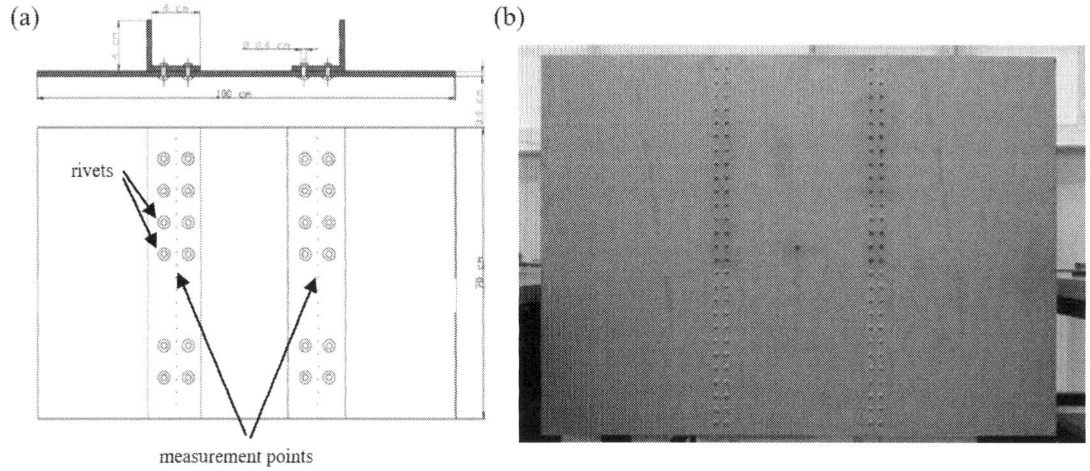

Figure 2. Aluminum plate: (a) schematic drawing, (b) real object.

Journal of Physics: Conference Series **181** (2009) 012067 doi:10.1088/1742-6596/181/1/012067

3.1. Signal processing

Despite conditioning measurements for minimal noise level, there could still be some disturbances that may bring inaccurate or even false damage localizations. In order to overcome this problem, different noise reduction methods were tested based on moving average, median filter and wavelet decomposition.

Without de-noising the signal, only two methods could localize the damage, namely GFD and WT. This proves that those methods are less noise sensitive than others. Median and moving average filtering are able to reduce noise levels. However, it does not improve the effectiveness of the damage detection methods. Moreover, after filtering only WT was able to indicate damage position. De-noising process based on wavelet decomposition was the only one which benefits in more accurate damage detection in all described methods. For this purpose *wden* MATLAB® function was used with level of decomposition $L=2$ for detail coefficients obtaining by wavelet *'sym8'* and the threshold selection rule set for heuristic without rescaling. After de-noising, mode shapes were interpolated using a cubic spline algorithm to increase number of measured points three times and that gives more accurate damage localization. The last operation in data preprocessing was extrapolating the signal by a few points on each side. This operation helps to minimize the negative boundary effect in some of the damage localization methods. After calculating specific damage index, additional points on edges were cut off.

An example of the mode shape before and after data preprocessing is shown in figure 3. In spite of some subtle changes, the results of damage localization are completely different, which is presented in figure 4, exemplified by damage index calculated for 50% saw cut case, using 8 mode shapes. Using raw data leads to very jagged output function with the maximum value in 0.4 of structure length. The same index computed for preprocessed data gives clearly information about damage localized in the middle of the plate stiffener, which corresponds to the real state.

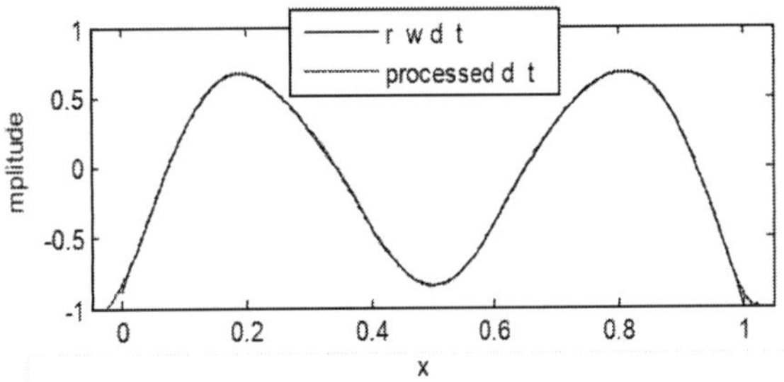

Figure 3. Mode shape comparison for raw and preprocessed data.

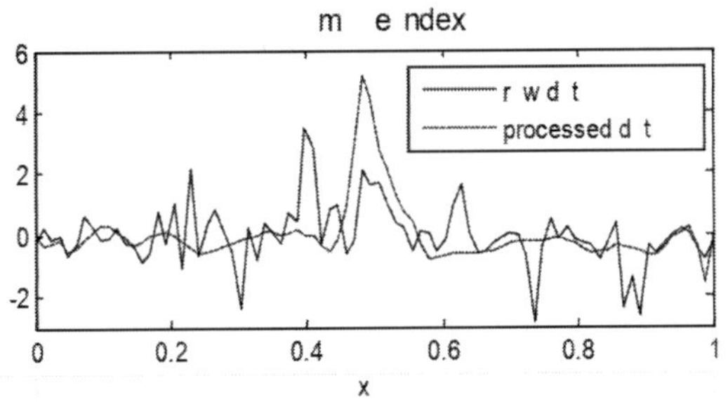

Figure 4. Damage index of preprocessed and raw data comparison.

3.2. Results

Because of limited space, results from only one damage scenario will be presented. Figure 5 presents FRFs of examined plate and its changes caused by the damage occurrence. Natural frequencies read out from FRFs are shown in table 1. All results from relative damage detection methods, for the case of 50% deep saw cut in the middle of the right stiffener are shown in figure 6.

Table 1. Natural frequencies change caused by saw cut in the right stiffener.

$f_{reference}$ [Hz]	$f_{25\%cut}$ [Hz]	$\Delta f_{25\%cut}$ [%]	$f_{50\%cut}$ [Hz]	$\Delta f_{50\%cut}$ [%]
247,66	247,50	0,06	247,19	0,19
318,75	318,59	0,05	316,72	0,64
566,09	565,47	0,11	564,22	0,33
657,34	656,41	0,14	654,22	0,48
994,69	994,06	0,06	992,81	0,19
1177,81	1176,25	0,13	1173,75	0,34
1406,56	1405,00	0,11	1404,69	0,13
1581,88	1581,88	0,00	1581,88	0,00

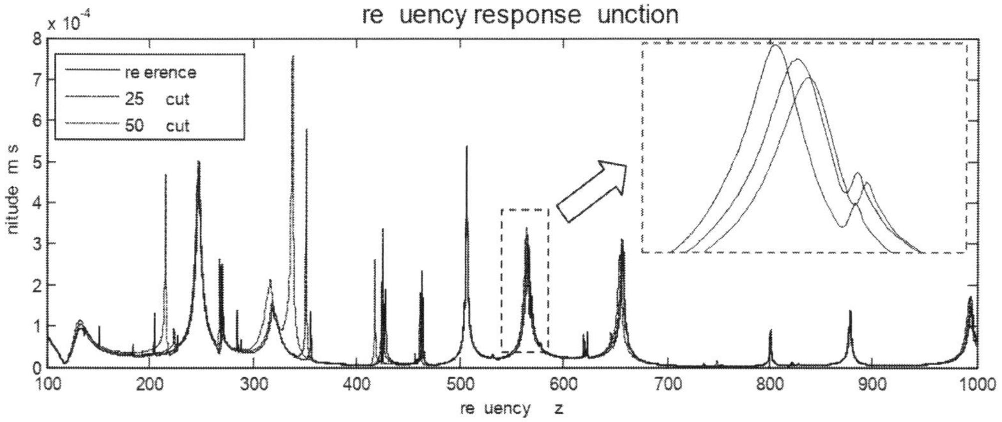

Figure 5. Frequency response function.

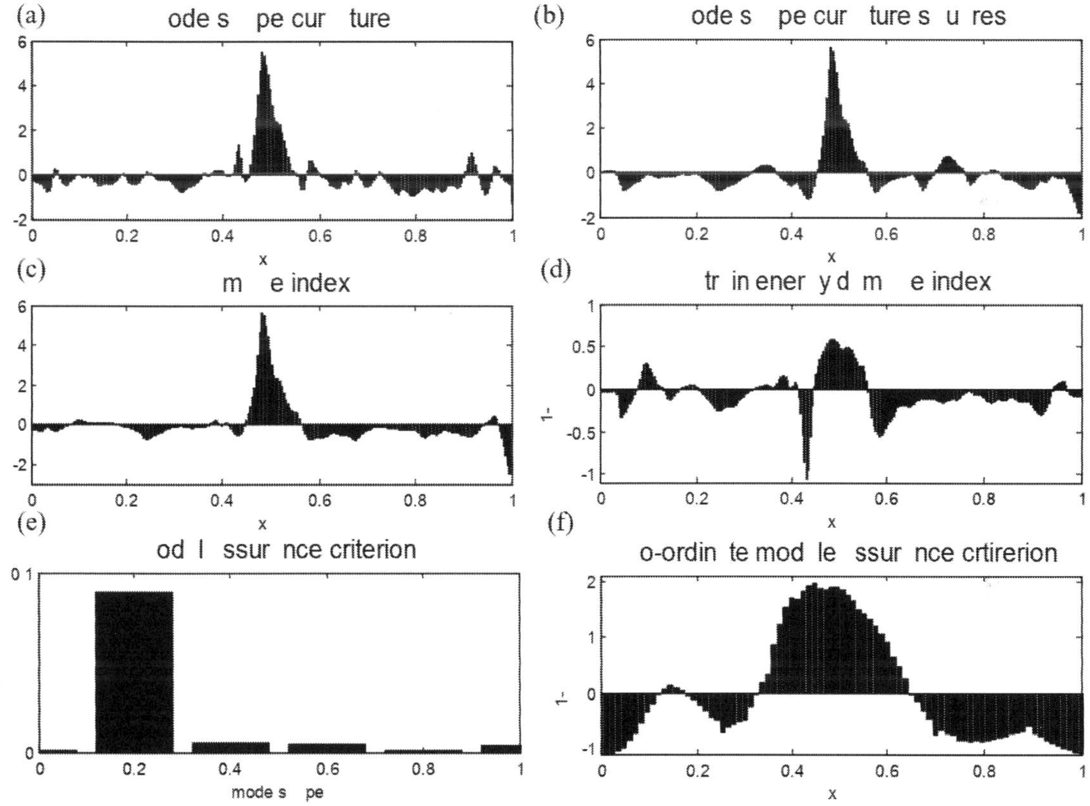

Figure 6. Damage detection relative indexes: (a) mode shape curvature, (b) mode shape curvature squares, (c) damage index, (d) strain energy damage index, (e) modal assurance criterion, (f) co-ordinate modal assurance criterion.

All indexes besides MAC and SEDI were statistically normalized using z-score transformation. This provides easier result analysis and reliable comparison.

Figure 7 shows results obtained from three absolute damage detection methods described in section 2. Additionally, the same methods were used as relative indexes through calculation difference of results for damaged and undamaged plates mode shapes. They are presented in figure 8. All indicators from figure 7 and 8 were obtained by summing indexes estimated for 8 mode shapes separately.

GFD were calculated with scale s = 100 and sliding window size of M = 6.

For WT 'gaus4' mother wavelet was used with scales s from 1 to 50. For this method interpolation was changed from 3 for each point to 10, and extrapolation point number was established as a half of maximum scale parameter on both sides in order to minimize the boundary effect.

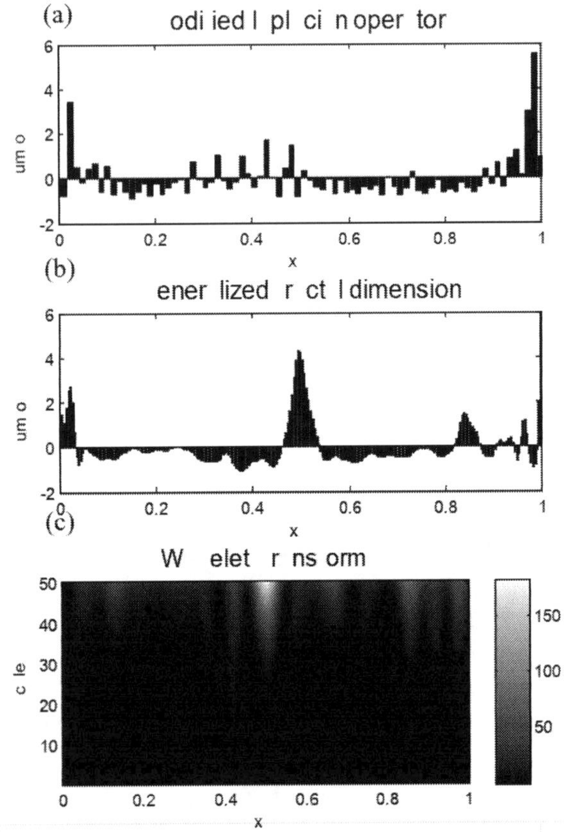

Figure 7. Damage detection absolute indexes: (a) modified Laplacian operator, (b) generalized fractal dimension, (c) wavelet transform.

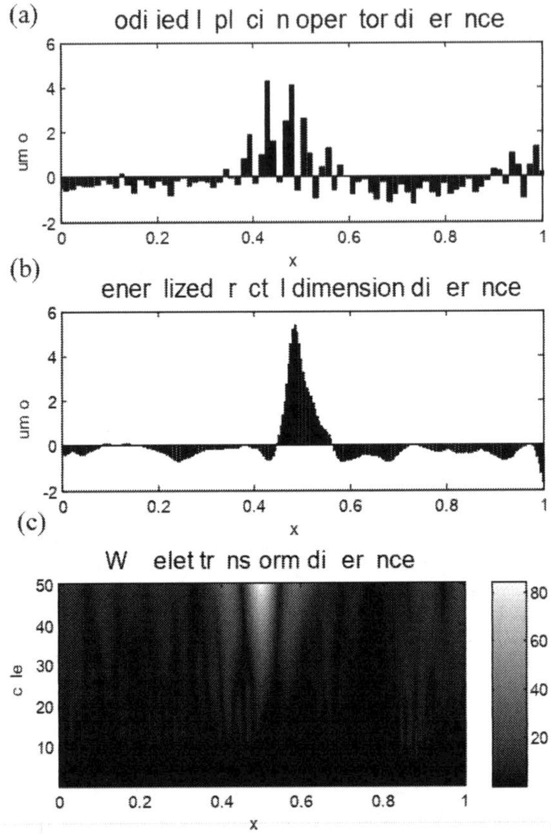

Figure 8. Damage detection absolute indexes used as relative indexes: (a) modified Laplacian operator, (b) generalized fractal dimension, (c) wavelet transform.

3.3. Discussion

All relative damage detection methods were able to correctly pinpoint the damage in the middle of the measured distance. However, COMAC and SEDI are not giving as sharp peaks in the crack position as other methods.

It is worth noticing that an increasing number of measured points does not necessarily provide better results, due to noise propagation in mode shapes second derivatives. This effect was described in detail by Sazonov and Klinkhachorn [13]. However, when de-noise techniques are used, they may take advantage of high spatial measurements.

As far as the absolute damage detection methods are concerned, both GFD and WT clearly identify the right position of damage. Nevertheless, MLO was not producing correct results. Calculation difference between indices for reference and current state leads to the increased effectiveness of those methods. Thus GFD difference computing can remove three peaks near the edges of the structure leaving only one correctly localized damage position. In case of using WT as a relative index, it also brings in more explicit information about damage location.

Similar results of presented damage detection methods for a steal cantilever beam with a notch were obtain by Radzieński *et al.* [14].

4. Conclusions

The presented work is devoted to comparing damage detection techniques based on vibration analysis. Six various methods are verified and compared on real measured data which lead to the following conclusions and suggestions:

- both measured structure surface preparation and proper positioning of the scanning head are important for providing high quality measurements,
- digital signal processing may hardly enhance the effectiveness of most damage detection methods,
- in relative damage detection methods the bests results are received from MSC and DI computing,
- WT seems to be the most effective, noise independent and versatile damage detection method,
- In case of WT and GFD based methods, primary state of structure data may be used for increasing damage assessment effectiveness if possible,
- Further studies on experimental verification of damage detection methods for multiple damages and for 2D signal measurement should be conducted.

References

[1] Doebling S W, Farrar C R, Prime M B 1998 A summary review of vibration-based damage identification methods *Shock and Vibration Digest* **20** 91-105

[2] Wang J, Qiao P Z 2007 Improved Damage Detection for Beam-type Structures using a Uniform Load Surface *Structural Health Monitoring* **6(2)** 99-112

[3] Pandey A K, Biswas M, Samman M M 1991 Damage detection from changes in curvature mode shapes *Journal of Sound and Vibration* **145** 321-332

[4] Ratcliffe C P 1997 Damage detection using a modified Laplacian operator on mode shape data *Journal of Sound and Vibration* **204** 505-517

[5] Allemang R J, Brown D L 1982 A correlation coefficient for modal vector analysis *Proc. 1st Int. Modal Anal. Conf.* **1**, 110–116.

[6] Lieven N A J, Ewins D J 1988 Spatial correlation of mode shapes the coordinate modal assurance criterion (COMAC*) Proc. 6th Int. Modal Anal. Conf.* **1** 690–695

[7] Qiao P Z, Lu K, Lestari W 2008 A Combined Static/Dynamic Technique for Damage Detection of Laminated Composite Plates *Experimental Mechanics* **48** 17–35

[8] Mandelbrot B B 1967 How long is the coast of Britain? Statistical self-similarity and fractional dimension *Science* **156** 636-638

[9] Wang J, Qiao P Z 2008 On irregularity-based damage detection method for cracked beams *International Journal of Solids and Structures* **45** 688–704

[10] Rucka M, Wilde K 2006 Application of continuous wavelet transform in vibration based damage detection method for beams and plates *Journal of Sound and Vibration* **297** 536-550

[11] Morlier J, Bos F,Castera P 2006 Diagnosis of a portal frame using advanced signal processing of laser vibrometer data, *Journal of Sound and Vibration* **297** 420–431

[12] Jin S, Pai P F 2000 Locating Structural Defects Using Operational Deflection Shapes, *Journal of Intelligent Material Systems and Structures* **11** 613-632

[13] Sazonov E, Klinkhachorn P 2005 Optimal spatial sampling interval for damage detection by curvature or strain energy mode shapes, *Journal of Sound and Vibration* **285** 783–801

[14] Radzieński M, Krawczuk M, Ostachowich W 2009 Experimental verification and comparison of mode shape-based damage detection methods, *Proceedings of the 8th International Conference on Damage Assessment of Structures (DAMAS2009)*

A strain-damage coupled model and its application to near crack tip localization analysis

V A Kovalev[1], **Y N Radayev**[2]

[1] Department of Applied Mathematics, Moscow City Government University of Management, 28 Sretenka Str., Moscow, 107045, Russia

[2] Department of Continuum Mechanics, Samara State University, 1 Akad. Pavlov Str., Samara, 443011, Russia

E-mail: kovalev@migm.ru

E-mail: radayev@ssu.samara.ru

Abstract. Three-dimensional equations of a strain-damage coupled model for ductile deformation processes in solids are presented in an attempt to find approaches for study of the anisotropic damage effect on the perfect plastic flow. Damage state is represented by a symmetric damage tensor. A modified Tresca yielding condition, associated flow and damage rules are used to formulate the strain-damage coupled constitutive equations, thus allowing to take account of that effect. The principal axes of stresses are chosen as a local frame for representing the static, kinematic and constitutive equations of the coupled model. The closed system of equations represented in the local principal frame needed for formulation of the strain-damage coupled model is obtained. This system is shown belong to the hyperbolic type. Theory of characteristic lines of the strain-damage coupled equations is developed. A numerical solution of the plane strain problem of the strain-damage localization near a mode I crack tip is obtained. It shows the marked effect of damage induced anisotropy on the localization of plastic strains near a crack tip and allows to estimate an extension of the damaged zone ahead of the crack tip. Gradual distortion of the slip line geometry is also determined thus demonstrating how the anisotropic damage affects plastic flow.

1. Introduction

A formulation of continuum damage model requires before all a description and approximation of discontinuous and often nondeterministic distributions of microcracks, generated in the course of the brittle deformation process, internal structure of the metals, strained at high temrerature, which are greatly affected by the nucleation of holes and their growth within the grains or on the grain boundaries (all these phenomena are usually referred to as the creep damage), and slip lines systems (for a ductile damaging process in a perfectly plastic solid).

A relevant aspect of damage state analysis is the damage induced anisotropy. Anisotropic damage state is to be primarily described by a directional distribution of damage measured in some way. Damage tensors can be then extracted in a regular way from the directional damage distribution.

An approach to representation of three-dimensional anisotropic damage state has been developed in [1], [2]. A general thermodynamical model of three-dimensional anisotropic damage state based on the canonical formalism has been discussed in [3], [4]. The canonical damage state variables, defined therein, provide a very simple representation of the thermodynamic damage state potentials. Their derivation needs averaging the local directional damage, so that the importance of averaged directional damage for the formulation of continuum damage model can be seen.

The aim of this paper is to give a reminder of the mathematical description of an anisotropic damage state, discuss requisite mathematical relations for the averaged directional damage and estimate its variation limits for an arbitrary type of damage induced anisotropy, obtain modified yielding criteria, associated flow and damage rules in order to formulate strain-damage coupled constitutive equations which are necessary for study of the anisotropic damage effect on the near crack tip perfectly plastic flow. kinematic and constitutive equations of the coupled model.

After the Introduction, the paper includes (Section 2) a short review of damage state description by the symmetric second rank damage tensor, as proposed in [2] with the objective of establishing a reference framework of notations and terminology. The definition of the second rank damage tensor is reestablished in order to elucidate its origin from the classical Finger strain tensor.

In Section 3 average values of the directional damage are considered. We define and then compute them (those considerations will be restricted by the second order approximations of directional damage distribution).

In Section 4 we give kinematic and constitutive equations of the coupled model. The closed system of equations represented in the local principal frame needed for formulation of the strain-damage coupled model is obtained assuming a nontrivial modification of incremental equations of the perfect plasticity. This system is shown belong to the hyperbolic type. The hyperbolicity enables us to develop theory of characteristic slip lines of the strain-damage coupled equations. Particularly this holds in the case of plane strain thus allowing to generalize slip lines theory known from plane strain perfect plasticity to coupled states.

In Section 5 a numerical solution of the plane strain problem of the strain-damage localization near a mode I crack tip is obtained showing the marked effect of damage induced anisotropy on the localization of plastic strains near a crack tip and allows to estimate an extension of the damaged zone ahead of the crack tip. Gradual distortion of the slip line geometry is also determined thus demonstrating how the anisotropic damage affects plastic flow.

2. The symmetric second rank damage tensor of continuum damage mechanics

An adequate description of anisotropic damage state requires an appropriate directional damage variable $\varsigma(\mathbf{n})$, where $\mathbf{n}$ is a unit three-dimensional vector often referred to as director. In the actual damaged state the value of ς associated with the direction pointed by $\mathbf{n}$ is the damage measured in appropriate terms. Since the early classical work of L M Kachanov [5], the variables representing the damage are traditionally interpreted in terms of effective load-carrying area reduction. In a perfectly plastic solid the effective area reduction is caused by micronecking. Corresponding slip line mechanism and kinematic are shown by Fig. 1.

In the current damaged state, due to distributed microcracks, cavities or slip lines mechanism, the load carrying area $dA^*(\mathbf{n})$ of a plane element normal to $\mathbf{n}$ is less than its observed geometrical area $dA(\mathbf{n})$. Thus, the definition of damage variable $\varsigma(\mathbf{n})$ appears as

$$\varsigma(\mathbf{n}) = \frac{dA^*(\mathbf{n})}{dA(\mathbf{n})}. \tag{1}$$

The directional damage variable $\varsigma(\mathbf{n})$ is of primary importance for the all following consideration. As ς is decreasing when the damage is growing, this variable may be more correctly called as the directional integrity.

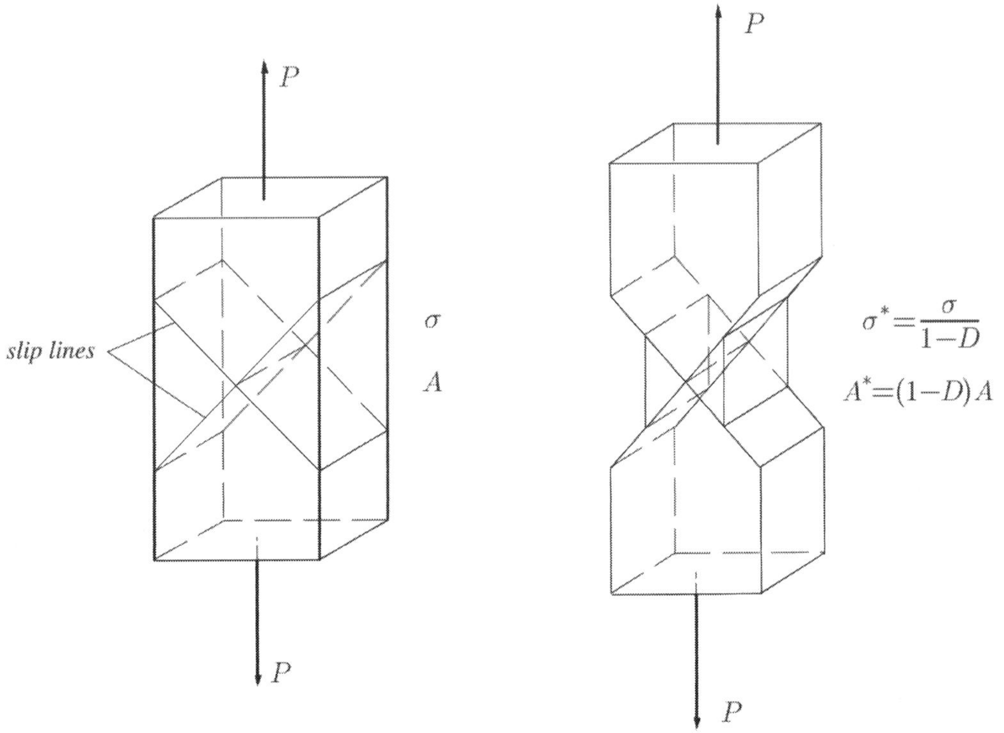

Figure 1. Effective net area reduction in a perfectly plastic solid due to slip line mechanism (plane strain state, the Onat–Prager (1955) scheme).

As it was elucidated in our previous discussion [2], the symmetric second rank damage tensor **D** determines the reduction of the effective area of the plane element normal to the director **n** according to the relation

$$\varsigma = \sqrt{\mathrm{tr}\,[(\mathbf{I}-\mathbf{D})^2\mathbf{n}\otimes\mathbf{n}]}\,. \tag{2}$$

Symmetry of the damage tensor provides a clear mechanical interpretation for the damage principal directions and values (the latter is called as principal damages).

In view of symmetry, the damage tensor can be represented in the spectral form

$$\mathbf{D} = \sum_{\alpha=1}^{3} D_{(\alpha)}\mathbf{d}_{(\alpha)}\otimes\mathbf{d}_{(\alpha)}, \tag{3}$$

wherein $\mathbf{d}_{(1)}$, $\mathbf{d}_{(2)}$, $\mathbf{d}_{(3)}$ are vectors of the orthonormal eigenbasis, and $D_{(1)}$, $D_{(2)}$, $D_{(3)}$ — the damage tensor eigenvalues (principal damages).

Substitution of the spectral decomposition (3) into equation (2) gives

$$\varsigma = \sqrt{(1-D_{(1)})^2 n_{(1)}^2 + (1-D_{(2)})^2 n_{(2)}^2 + (1-D_{(3)})^2 n_{(3)}^2}, \tag{4}$$

where $n_{(i)}$ are the components of the unit vector **n** with respect to the damage eigenbasis $\mathbf{d}_{(1)}$, $\mathbf{d}_{(2)}$, $\mathbf{d}_{(3)}$.

For a plane element, orthogonal to the principal axis of damage labelled by γ, we obtain from the latter equation the following formula

$$D_{(\gamma)} = \frac{dA_{(\gamma)} - dA^*_{(\gamma)}}{dA_{(\gamma)}} \qquad (\text{no sum on } \gamma;\ \gamma = 1, 2, 3), \tag{5}$$

that accords to the classical Kachanov notion [5] of a damage variable.

If one renumber the principal damages in the following order

$$D_{(3)} \leq D_{(2)} \leq D_{(1)}, \tag{6}$$

then for an arbitrary orientation $\mathbf{n}$ the following double sided estimation is derived:

$$D_{(3)} \leq 1 - \varsigma \leq D_{(1)}. \tag{7}$$

The principal damages can be represented in terms of the principal damage stretches $L^D_{(\alpha)}$ (see discussion in [3]) as follows

$$1 - D_{(\gamma)} = \frac{L^D_{(1)} L^D_{(2)} L^D_{(3)}}{L^D_{(\gamma)}}, \tag{8}$$

or vice versa

$$\begin{aligned} L^D_{(1)} &= \sqrt{\frac{(1-D_{(2)})(1-D_{(3)})}{(1-D_{(1)})}}, \quad L^D_{(2)} = \sqrt{\frac{(1-D_{(1)})(1-D_{(3)})}{(1-D_{(2)})}}, \\ L^D_{(3)} &= \sqrt{\frac{(1-D_{(1)})(1-D_{(2)})}{(1-D_{(3)})}}. \end{aligned} \tag{9}$$

3. Directional average of the second order damage state approximation

Consider the mean of the directional damage variable ς as represented by the second order approximation (4). We will follow [6] and shall use the notations θ and φ for the spherical angles. Let $C_{(j)} = (1 - D_{(j)})^2$ $(j = 1, 2, 3)$ and a pair of triangle brackets $<>$ denotes the averaging over the sphere of unit directions.

The following iterated integral

$$<\varsigma>= \frac{1}{4\pi} \int_0^{2\pi} \int_0^{\pi} \sqrt{C_{(1)} \sin^2\theta \cos^2\varphi + C_{(2)} \sin^2\theta \sin^2\varphi + C_{(3)} \cos^2\theta} \sin\theta d\theta d\varphi \tag{10}$$

by making use of the result [7] (formula 2.597.2) becomes

$$<\varsigma>= \frac{2}{\pi} \int_0^1 \mathbf{E}(k) \sqrt{(C_{(3)} - C_{(2)})\tau^2 + C_{(2)}}\, d\tau, \quad k = \sqrt{\frac{(1-\tau^2)(C_{(2)} - C_{(1)})}{(C_{(3)} - C_{(2)})\tau^2 + C_{(2)}}}, \tag{11}$$

where $\mathbf{E}(k)$ is the complete elliptic integral of the second kind, k is the modulus.

The number of the independent parameters in (11) can be reduced by introducing ratios of the eigenvalues $C_{(1)}$, $C_{(2)}$, $C_{(3)}$:

$$\frac{\pi <\varsigma>}{2\sqrt{C}} = I(p_1, p_2) \qquad (p_1 = C_{(1)}/C_{(2)},\ p_2 = C_{(2)}/C_{(3)},\ C = C_{(3)}), \tag{12}$$

where

$$I(p_1, p_2) = \int_0^1 \sqrt{(1-p_2)\tau^2 + p_2}\, \mathbf{E}(\sqrt{p_2(1-\tau^2)(1-p_1)((1-p_2)\tau^2 + p_2)^{-1}})\, d\tau.$$

The following expressions for the variables p_1, p_2, C in terms of the principal damage stretches (see equations (9)) are valid:

$$\sqrt{p_1} = \frac{L^D_{(2)}}{L^D_{(1)}}, \quad \sqrt{p_2} = \frac{L^D_{(3)}}{L^D_{(2)}}, \quad \sqrt{C} = L^D_{(1)} L^D_{(2)}. \tag{13}$$

The mean of the integral $I = I(p_1, p_2)$ is found as (numerical result)

$$<I> = \int_0^1 \int_0^1 I(p_1, p_2) dp_1 dp_2 = 1.157790\ldots.$$

The average value (over the unit square) of the ratio $<\varsigma>/\sqrt{C}$ can be numerically obtained as

$$\left\langle \frac{<\varsigma>}{\sqrt{C}} \right\rangle = \frac{2}{\pi} <I> = 0.737072\ldots.$$

One more useful representation for $<\varsigma>/\sqrt{C}$ is

$$\frac{\pi <\varsigma>}{2\sqrt{C}} = q^{-1}(1 + p^2q^2) J(p, q),$$

where the improper integral

$$J(p, q) = \int_0^q \frac{k\mathbf{E}(k)dk}{\sqrt{q^2 - k^2}\,(1 + p^2k^2)^2}$$

depends on the parameters

$$p = \sqrt{\frac{L^{D\,2}_{(1)} L^{D\,2}_{(2)} - L^{D\,2}_{(1)} L^{D\,2}_{(3)}}{L^{D\,2}_{(1)} L^{D\,2}_{(3)} - L^{D\,2}_{(2)} L^{D\,2}_{(3)}}}, \quad q = \sqrt{1 - \frac{L^{D\,2}_{(2)}}{L^{D\,2}_{(1)}}}.$$

A more symmetric formula for the ratio $<\varsigma>/\sqrt{C}$ must be remarked

$$\frac{\pi <\varsigma>}{2\sqrt{C}} = (1 + p^{*2}) J^*(p^*, q^*), \tag{14}$$

wherein

$$J^*(p^*, q^*) = \int_0^1 \frac{k^*\mathbf{E}(q^*k^*)dk^*}{\sqrt{1 - k^{*2}}\left(1 + p^{*2}k^{*2}\right)^2}, \quad p^* = \sqrt{\frac{L^{D\,2}_{(2)}}{L^{D\,2}_{(3)}} - 1}, \quad q^* = \sqrt{1 - \frac{L^{D\,2}_{(2)}}{L^{D\,2}_{(1)}}}.$$

This latter integral J^* for a comparatively weak damage anisotropy (*i.e.* $q^* \to 0$) after asymptotic evaluations is obtained as

$$J^*(p^*, q^*) = \frac{\pi}{2p^{*4}}(1 + \frac{1}{4}q^{*2}p^{*-2}) \left\{ \frac{1}{4}(1 + p^{*-2})^{-3/2} \ln\frac{\sqrt{1 + p^{*-2}} + 1}{\sqrt{1 + p^{*-2}} - 1} + \frac{1}{2p^{*-2}(1 + p^{*-2})} \right\} - $$
$$- \frac{1}{8p^{*4}} \frac{q^{*2}}{\sqrt{1 + p^{*-2}}} \ln\frac{\sqrt{1 + p^{*-2}} + 1}{\sqrt{1 + p^{*-2}} - 1} + O(q^{*4}).$$

If a damage state is axially symmetric (when for the case $q^* = 0$), the above equation provides the following exact formula for integral J^* in equation (14):

$$J^*(p^*,\ 0) = \frac{\pi}{2p^{*4}} \left\{ \frac{1}{4}(1 + p^{*-2})^{-3/2} \ln\frac{\sqrt{1 + p^{*-2}} + 1}{\sqrt{1 + p^{*-2}} - 1} + \frac{1}{2p^{*-2}(1 + p^{*-2})} \right\}. \tag{15}$$

Journal of Physics: Conference Series **181** (2009) 012068 doi:10.1088/1742-6596/181/1/012068

4. A strain-damage coupled model

The equations required for mathematical modeling damaging processes in a perfectly plastic solid we formulate in isostatic co-ordinate net.

We start from the spectral decomposition of the stress tensor $\boldsymbol{\sigma}$

$$\boldsymbol{\sigma} = \sigma_1 \mathbf{l} \otimes \mathbf{l} + \sigma_2 \mathbf{m} \otimes \mathbf{m} + \sigma_3 \mathbf{n} \otimes \mathbf{n}, \tag{16}$$

where σ_1, σ_2, σ_3 — principal stresses, $\mathbf{l}$, $\mathbf{m}$, $\mathbf{n}$ — orthonormal system of stress eigenvectors. Then the static equilibrium equation

$$\boldsymbol{\nabla} \cdot \boldsymbol{\sigma} = \mathbf{0} \tag{17}$$

by introducing

$$d_1 = \mathbf{l} \cdot \boldsymbol{\nabla}, \quad d_2 = \mathbf{m} \cdot \boldsymbol{\nabla}, \quad d_3 = \mathbf{n} \cdot \boldsymbol{\nabla}$$

can be represented in the isostatic co-ordinate net as follows

$$\begin{cases} d_1\sigma_1 + \kappa_{23}(\sigma_1 - \sigma_2) + \kappa_{32}(\sigma_1 - \sigma_3) = 0, \\ d_2\sigma_2 + \kappa_{31}(\sigma_2 - \sigma_3) + \kappa_{13}(\sigma_2 - \sigma_1) = 0, \\ d_3\sigma_3 + \kappa_{12}(\sigma_3 - \sigma_1) + \kappa_{21}(\sigma_3 - \sigma_2) = 0, \end{cases} \tag{18}$$

where κ_{ij} are the principal curvatures of the co-ordinate surfaces.

The compatibility equation for strain increments is

$$d\mathbf{S} = \boldsymbol{\nabla} \times d\boldsymbol{\varepsilon} \times \boldsymbol{\nabla} = \mathbf{0}.$$

The incompatibility tensor $d\mathbf{S}$ can be given by its physical components in the isostatic co-ordinate net. Here we present the minimally necessary set of them (others can be obtained by the usual cyclic permutations)

$$\begin{aligned} dS_{<11>} = {} & -d_2d_2d\varepsilon_3 - d_3d_3d\varepsilon_{<22>} + \left(\kappa_{21}^2 - \kappa_{31}^2\right)(d\varepsilon_3 - d\varepsilon_{<22>}) + \\ & + d_3\left(\kappa_{21}\left(d\varepsilon_3 - d\varepsilon_{<22>}\right)\right) - d_2\left(\kappa_{31}\left(d\varepsilon_3 - d\varepsilon_{<22>}\right)\right) - \\ & - \kappa_{23}\kappa_{32}\left(d\varepsilon_{<22>} + d\varepsilon_3 - 2d\varepsilon_{<11>}\right) - \kappa_{31}d_2d\varepsilon_3 - \\ & - \kappa_{21}d_3d\varepsilon_{<22>} - \kappa_{32}d_1d\varepsilon_{<22>} - \kappa_{23}d_1d\varepsilon_3 + \\ & + 2\kappa_{32}d_2d\varepsilon_{<12>} + (2\kappa_{32}\kappa_{13} + \kappa_{23}\kappa_{31} + \kappa_{32}\kappa_{31} + (d_2\kappa_{32}))d\varepsilon_{<12>}, \end{aligned} \tag{19}$$

$$\begin{aligned} dS_{<12>} = {} & d_2d_1d\varepsilon_3 + d_2\left[\kappa_{32}\left(d\varepsilon_3 - d\varepsilon_{<11>}\right)\right] + \kappa_{31}d_1\left(d\varepsilon_3 - d\varepsilon_{<22>}\right) - \\ & - \kappa_{23}d_2d\varepsilon_3 + \kappa_{31}\left(d\varepsilon_3 - d\varepsilon_{<11>}\right)\left(\kappa_{32} - \kappa_{23}\right) + \\ & + d_3d_3d\varepsilon_{<12>} + (\kappa_{21} + \kappa_{12})d_3d\varepsilon_{<12>} + \\ & + (-\kappa_{31}^2 + \kappa_{32}\kappa_{23} + 2\kappa_{13}\kappa_{31} + 2\kappa_{21}\kappa_{12} - \kappa_{21}^2 - d_2\kappa_{31} + d_3\kappa_{12})d\varepsilon_{<12>}. \end{aligned} \tag{20}$$

We proceed our considerations to the notion of effective stress. For a simple case of uniaxial tension of damaging specimen the effective stress σ^*, the actual stress σ and the damage D are involved in the equation

$$\sigma^* = \frac{\sigma}{1 - D}. \tag{21}$$

As $\sigma^* > \sigma$ in view of $0 < D < 1$, the latter equation in the simplest form takes account of the stress magnifying effect in a progressively damaging solid.

The effective stress concept is easily generalized to the three-dimensional case by using as previously the isostatic co-ordinate representation

$$\sigma_j^* = \frac{\sigma_j}{1 - D_j} \quad (\text{no sum on } j). \tag{22}$$

The generalized effective stress notion is then applied to formulation of the yielding condition that describes those zones in a solid which are progressively damaging in the course of plastic flow. A general form of the yielding condition for an isotropic perfectly plastic solid

$$f(\sigma_1, \sigma_2, \sigma_3) = 0$$

changes into a new form of the principal effective stresses σ_j^*

$$f(\sigma_1^*, \sigma_2^*, \sigma_3^*) = 0, \tag{23}$$

which enters the framework of our mathematical model of a damage state in perfectly plastic body. Thus we obtain, for example, a formulation of the Tresca yielding condition taking account of anisotropic damage:

$$\frac{\sigma_1}{1 - D_1} - \frac{\sigma_2}{1 - D_2} = 2k \tag{24}$$

for the facet of the Tresca prism, and

$$\frac{\sigma_1}{1 - D_1} - \frac{\sigma_3}{1 - D_3} = 2k, \quad \frac{\sigma_2}{1 - D_2} - \frac{\sigma_3}{1 - D_3} = 2k \tag{25}$$

for the edge.

The damage accumulation law is given by linear equations for principal damage and strains increments as follows

$$dD_j = \mathrm{K}_j \mathrm{sgn}\left(d\varepsilon_j^P\right) d\varepsilon_j^P \quad (\text{no sum on } j), \tag{26}$$

where K_j are material constants.

For given yielding condition (23) the associated flow rule in the principal axes of stress (and also damage) tensor is

$$d\varepsilon_j^P = (d\Lambda)\frac{\partial f}{\partial \sigma_j} \tag{27}$$

with an undetermined increment $d\Lambda$.

The associated flow rule can be formulated in terms of the principal effective stresses, which leads to

$$d\varepsilon_j^P = (d\Lambda)\frac{1}{1 - D_j}\frac{\partial f}{\partial \sigma_j^*} \quad (\text{no sum on } j).$$

In the following discussion we shall consider plane strain coupled states. In this case the flow rule is represented by

$$d\varepsilon_1^P = \frac{d\Lambda}{1 - D_1}, \quad d\varepsilon_2^P = -\frac{d\Lambda}{1 - D_2}, \quad d\varepsilon_3^P = 0.$$

In view of (26) assuming $dD_3 = 0$ we have

$$d\Lambda = \frac{(1 - D_1)^2 \left[(\beta - 1)\, d\sigma_2 - d\sigma_1\right]}{\mathrm{K}_1 \mathrm{sgn}\left(d\varepsilon_1^P\right)\sigma_1 + (\beta - 1)^3\, \mathrm{K}_2 \mathrm{sgn}\left(d\varepsilon_2^P\right)\sigma_2}, \tag{28}$$

where

$$\beta - 1 = \frac{1 - D_1}{1 - D_2}.$$

Hence, for plane strain states the flow rule is obtained as

$$\begin{cases} \dfrac{d\varepsilon_1^P}{F} = (\beta - 1)\, d\sigma_2 - d\sigma_1, \\ \dfrac{d\varepsilon_2^P}{F} = -(\beta - 1)^2\, d\sigma_2 + (\beta - 1)\, d\sigma_1, \end{cases} \tag{29}$$

whith the following notation

$$F = \frac{1 - D_1}{\mathrm{K}_1 \mathrm{sgn}\left(d\varepsilon_1^P\right)\sigma_1 + (\beta - 1)^3 \mathrm{K}_2 \mathrm{sgn}\left(d\varepsilon_2^P\right)\sigma_2}.$$

Strain incompatibility equation reads (κ_1, κ_2 — the stress principal lines curvatures)

$$\begin{aligned} d_2 d_2 d\varepsilon_1 - (\beta - 1) d_1 d_1 d\varepsilon_1 - ((1 + \beta)\kappa_2 + 2 d_1 \beta) d_1 d\varepsilon_1 + \\ + (1 + \beta)\kappa_1 d_2 d\varepsilon_1 + (d_1 d_1 \beta + \kappa_2 d_1 \beta - \\ - \kappa_1 d_2 \beta - \beta d_1 \kappa_2 + \beta d_2 \kappa_1 - \beta \kappa_2^2 + \beta \kappa_1^2) d\varepsilon_1 = 0. \end{aligned} \tag{30}$$

The Cauchy formulae for displacements increments $du_{<j>}$ are

$$\begin{aligned} d\varepsilon_1^P &= \kappa_1 du_{<2>} + d_1 du_{<1>}, \\ d\varepsilon_2^P &= \kappa_2 du_{<1>} + d_2 du_{<2>}. \end{aligned} \tag{31}$$

Thus we obtain the hyperbolic system of incremental equations. The characteristic lines of the system are determined by the following ordinary differential equation

$$\frac{ds_1}{ds_2} = \pm\sqrt{\frac{1 - D_1}{1 - D_2}}, \tag{32}$$

where ds_j is an elementary lenght of the characteristic line. The latter equation is of crucial importance for numerical analysis of plane strain problems formulated within the framework of the developed strain-damage coupled model.

5. Numerical analysis of strain-damage localization near a mode I crack tip

Localization of plastic strains and anisotropic damage near a linear mode I semi-infinite crack tip in a perfectly plastic solid under plane strains conditions is of particular interest as a sample application of the developed strain-damage coupled model to study of these localization phenomena. Perfectly plastic analysis of this problem (see [8] for details) gives the near crak tip stress distribution in the form

$$\left\{ \begin{array}{c} 0 \le \varphi \le \pi/4 \\ \dfrac{\sigma_{rr}}{2k} = \dfrac{1}{2} + \dfrac{\pi}{2} - \dfrac{1}{2}\cos 2\varphi, \quad \dfrac{\sigma_{\varphi\varphi}}{2k} = \dfrac{1}{2} + \dfrac{\pi}{2} + \dfrac{1}{2}\cos 2\varphi, \quad \dfrac{\sigma_{r\varphi}}{2k} = \dfrac{1}{2}\sin 2\varphi; \end{array} \right.$$

$$\left\{ \begin{array}{c} \pi/4 \le \varphi \le 3\pi/4 \\ \dfrac{\sigma_{rr}}{2k} = \dfrac{\sigma_{\varphi\varphi}}{2k} = -\varphi + \dfrac{1}{2} + \dfrac{3\pi}{4}, \quad \dfrac{\sigma_{r\varphi}}{2k} = \dfrac{1}{2}; \end{array} \right.$$

$$\left\{ \begin{array}{c} 3\pi/4 \le \varphi \le \pi \\ \dfrac{\sigma_{rr}}{2k} = \dfrac{1}{2} + \dfrac{1}{2}\cos 2\varphi, \quad \dfrac{\sigma_{\varphi\varphi}}{2k} = \dfrac{1}{2} - \dfrac{1}{2}\cos 2\varphi, \quad \dfrac{\sigma_{r\varphi}}{2k} = -\dfrac{1}{2}\sin 2\varphi. \end{array} \right.$$

Here k is the shear yield stress. This stress distribution is employed as initial.

Slip lines geometry and perfectly plastic strains localization zone are shown by Fig. 2. Very small localization zone of this kind is used as an intial approximation.

Numerical analysis of the strain-damage localization is based on the plane strain coupled equations given in the previous section of the paper taking account of the strong hyperbolicity

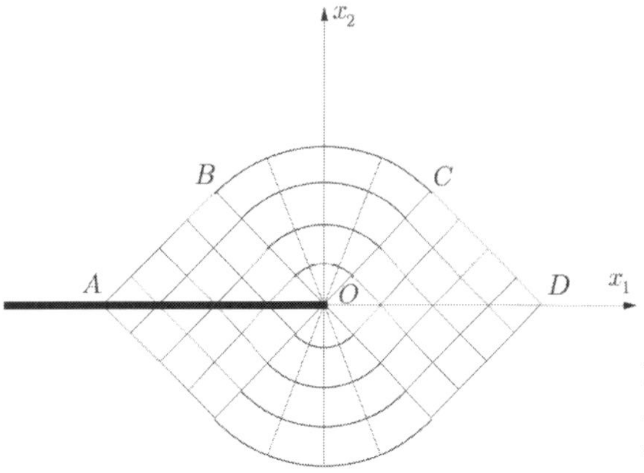

Figure 2. Slip lines and plastic flow zone near a linear mode I crack tip under plane strains conditions (perfectly plastic analysis).

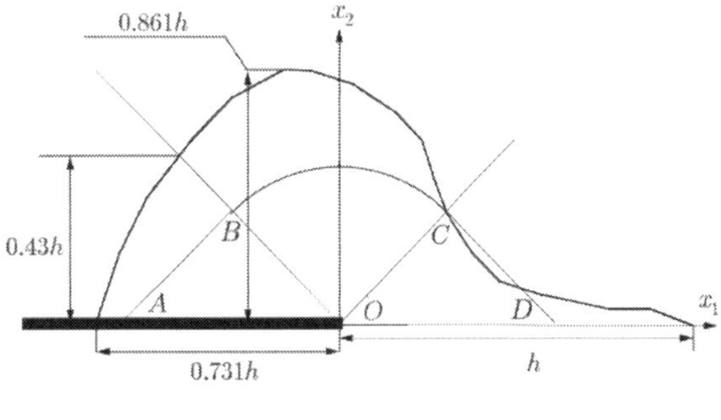

Figure 3. Plastic flow zones near a linear mode I crack tip under plane strains conditions (strain-damage coupled model, $K_1 = 0.05$, $K_2 = 1.5$).

of equations of the coupled model. The values of the dimensionless constants are: $K_2 = 1.5$ and comparatively small value of the constant $K_1 = 0.05$. The standard finite difference technique has been applied to the coupled differential equations. The Courant necessary condition is verified by the aid of equation (32). Results of the numerical analysis show a marked effect of anisotropic damage on linear dimensions of near crack tip localization zone. If h denotes the linear horisontal dimension of the near crack localization zone before the crack tip (see Fig. 3) then the following data are obtained: $0.731h$ for linear dimension of the localization zone on the crack surface; $0.43h$ for vertical dimension of the "triangle" localization zone ahead of the crack tip; $0.861h$ maximal vertical dimension of the strain-damage localization zone. An important characteristics of the strain-damage localization is the plastic strains concentration coefficient. Numerical estimation gives the value 1.214 for this coefficient.

References

[1] Murakami S 1988 Mechanical modeling of material damage *J. Appl. Mech.* **55** No. 2 280

[2] Murakami S and Radayev Y N 1996 Mathematical model of three-dimensional anisotropic damage state *Izvestiya Rossijskoi Akademii Nauk. Mech. of Solids* **4** 93 (in Russian)

[3] Radayev Y N 1996 Thermodynamical model of anisotropic damage growth. Part I. Canonical damage state variables of continuum damage mechanics and thermodynamical functions of three-dimensional anisotropic damage state *J. Non-Equilib. Thermodyn.* **21** 129

[4] Radayev Y N 1996 Thermodynamical model of anisotropic damage growth. Part II. Canonical damage growth rate equations and theory of damage invariants *J. Non-Equilib. Thermodyn.* **21** 197

[5] Kachanov L M 1958 On the creep rupture time *Izvestiya Akademii Nauk SSSR. Otd. Tekh. Nauk* 8 26 (in Russian)

[6] Radayev Y N 2004 On directional average of the local anisotropic damage *Int. J. Fracture* **128** 293

[7] Gradshteyn I S and Ryzhik I M 1965 *Table of Integrals, Series, and Products* translation from the Russian ed. by A Jeffrey (New York, London: Academic Press)

[8] Kachanov L M 1969 *Foundations of the Theory of Plasticity* (Moscow: Nauka) (in Russian)

Journal of Physics: Conference Series **181** (2009) 012069 doi:10.1088/1742-6596/181/1/012069

Numerical and experimental evaluation of the residual stress relaxation and the influence zone due to application of the crack compliance method

Sandoval-Pineda, J. M.[1♣], Urriolagoitia-Sosa, G.[2♠], Urriolagoitia-Calderón, G.[2], Hernández-Gómez, L. H.[2], García-Lira, J.[1], Beltrán-Fernández, J. A.[2] and Rodríguez-Martínez, R.[2]

[1]INSTITUTO POLITÉCNICO NACIONAL

Sección de Estudios de Posgrado e Investigación, Escuela Superior de Ingeniería Mecánica y Eléctrica (ESIME). Unidad profesional, Azcapotzalco, Av. de las Granjas No. 682, Col. Sta. Catarina Azcapotzalco, C.P. 02550, México D.F. México.

[2]INSTITUTO POLITÉCNICO NACIONAL

Sección de Estudios de Posgrado e Investigación (SEPI), Escuela Superior de Ingeniería Mecánica y Eléctrica (ESIME). Edificio 5. 2do Piso, Unidad Profesional Adolfo López Mateos "Zacatenco" Col. Lindavista, C.P. 07738, México, D.F. México

♣ jsandovalp@ipn.mx, ♠ guiurri@hotmail.com

Abstract. This paper presents the results concerning an evaluation of the crack compliance method. The research was focused on the relaxation caused by a cut induced to obtain the data required to calculate the residual stress field. The main objective in this research is to establish the optimum place to cut in a specimen that has suffered a failure and how extended is the zone of relaxed stresses. It has been recognized that a crack vanishes the beneficial or detrimental effects of the residual stress fields. This research has been performed in a numerical and experimental way, so results can be compared and *FEM* on this topic can be assessed.

1.- Introduction

One of the most complicated problems in structural analysis is the measurement of residual stresses and their effect in the structural integrity. The relevance to determine, predict and control the residual stress field introduced in a component and their effect is extremely important. Tensile residual stresses induce premature failures by activating cracks; they also induce stress by corrosion and are a cause of distortion in the elements. Thus, the subsequent understanding of residual stresses on a macro and micro scale is vital to determine the quality and reliability of mechanical components [1-4]. Residual stresses are always present in manufactured pieces, but are never evaluated directly, which is done is a measure of the strain caused on a macro and micro level and this is transformed to stress. Therefore diverse techniques for their evaluation have been developed. The techniques are divided in destructive, non-destructive and semi-destructive. One of these methods is the Crack Compliance Method (*CCM*) which is based on Fracture Mechanics theory [5-7]. The *CCM* considers the destruction of the equilibrium state of the residual stress field [2 and 8]. The CCM has been widely used to quantify residual stress fields; nevertheless there are no records on a deep and exhaustive evaluation in its application and on the possible expected error under real conditions [5]. Therefore in this work, it is presented the numerical and experimental evaluation of the *CCM*.

The main reason for the development of a numerical evaluation of the *CCM*, resides in obtaining data by inducing a crack or cut of progressive growth into a mechanical component. This crack tends to break the existing equilibrium between the induced residual stresses, causing relaxation that will result in a structural change of the component and by measuring this change in the geometry it can be obtained the original residual stress state. A numerical simulation will prevent the action of external factors, and allow the evaluation of the cut effect alone as well as their consequences in the neighborhood [9]. It is also extremely important to corroborate the numerical simulation by an experimental evaluation, as there is not enough information on the optimal aspects to apply the *CCM*. Moreover, when the cut is performed the procedure employed must be selected carefully, since this can alter the original state of the residual stresses [10]. It is important to select an experimental procedure which could be carried out and reproduce the original residual stress in the component.

Furthermore, this paper will use the numerical evaluation of the *CCM* to facilitate its correct experimental application and assure that the obtained results are within the expected parameters. The numerical evaluation of the CCM will help in the correct selection of the parameters needed to measure the relaxed strain, the optimum location of the strain gauges and the sizes of the relaxed area in the specimen. Consequently, these studies will make it possible to reduce costs for the experimental evaluation and accelerate the development of the test.

2.- Background for the Crack Compliance Method [1, 5, 11 and 12]

In this section a brief summary of the theory relative to the CCM is presented. Let the unknown residual stress distribution in the beam be represented by the summation of an nth order polynomial series as [13]:

$$\sigma_y(x) = \sum_{i=0}^{n} A_i P_i(x) \tag{1}$$

where A_i are constant coefficients and P_i are a power series, x^0, x^1, x^2, ... x^n etc. The strain $\varepsilon(a,s)$ (where a = crack length and s = is the distance between the location of the strain gauge and the crack plane) due to the stress fields $P_i(x)$ is known as the compliance function $C_i(a_j,s)$ and is given by:

$$C_i(a_j,s) = \frac{1}{E'}\int_0^{a_j} K_I(a)Z(a)da \tag{2}$$

$K_I(a)$ is the stress intensity factor due to the residual stress field when the crack depth in the beam is equal to a and $K_{IF}(a)$ is the stress intensity factor corresponding to the same depth due to a pair of virtual forces F applied tangentially at a position on the beam where strain measurements will be taken during the *CCM* cutting of the slot (where $Z(a)$ is a geometry dependant function (equation (3))):

$$Z(a) = \frac{\partial^2 K_{IF}(a,s)}{\partial F \partial s} \tag{3}$$

By following the weight function approach [7-9], $K_I(a)$ and $K_{IF}(a)$ can be expressed as:

$$K_I(a) = \int_0^a h(x,a)\sigma_y(x)dx \tag{4}$$

$$K_{IF}(a) = \int_0^a h(x,a)\sigma_{yF}(x)dx \tag{5}$$

where $\sigma_{yF}(x)$ is the stress field due to the virtual force F. Wu and Carlson [14], provide equations for the evaluation of $h(x,a)$ for given values of a and x. Equation (5) was integrated numerically by using Simpson's rule [12] in order to obtain $C_i(a_j,s)$. Once the $C_i(a,s)$ solutions are determined the expected strain due to the stress components in equation (1) can be obtained as [1 and 5]:

$$\varepsilon(a_j,s)=\sum_{i=0}^{n} A_i C_i(a_j,s) \tag{6}$$

The unknown terms A_i are determined so that the strains given by equation (6) match those strains measured in the experiment during cutting i.e., $\varepsilon(a_j,s)_{actual}$. In order to minimise the average error over all data points for an nth order approximation, the method of least squares is used to obtain the values of A_i. The number of cutting increments m is therefore chosen to be greater than the order of the polynomial, i.e. $m > n$. This work used $n = 7$ with 8 constants A_i and $m = 9$, this being the number of experimental slot cutting depths at which strain readings were collected. The least square solution is obtained by minimising the square of the error relative to the unknown constant A_i, This gives a linear set of simultaneous solutions from which A_i values are determined and equation (1) is then used to determine the residual stress distribution.

3.- CCM evaluations

3.1. Numerical simulation of the residual stress field by FEM

In this section, it is presented a numerical simulation of the crack compliance method and some consequences that might appear in its experimental application. The research was focused on the strain relaxation that occurs in a mechanical component in which it is induced a residual stress field when applying the *CCM* and its auto equilibrium after performing a cut. Evaluating as well if the component can be reused as an assessment specimen and if is possible to relate it to cases where failure has already occurred. The primary target is to determine the influence zone of the cut in the contiguous sections parallel to the notch. The research was performed by simulating a beam under four point bending in a 2D analysis.

Bilinear mechanical properties and a kinematic hardening rule were used (figure 1). Yield stress and yield strain were set at *420 MPa* and *0.002* respectively, additionally maximum stress and maximum strain at *540 MPa and 0.01* respectively. The Young's modulus was *210 GPa* and a Poisson ratio of *0.29*. Element Plane 183 was used and a multistep analysis was performed. Two concerted loads of *180 N* were applied at superficial nodes on the beam, which are high enough to induce a residual stress state into the beam by bending (figure 2a). Because, four point bending procedure was applied, a constant profile in the induction of the residual stress at the center of the beam along the x axis can be observed and can be used to evaluate the *CCM*.

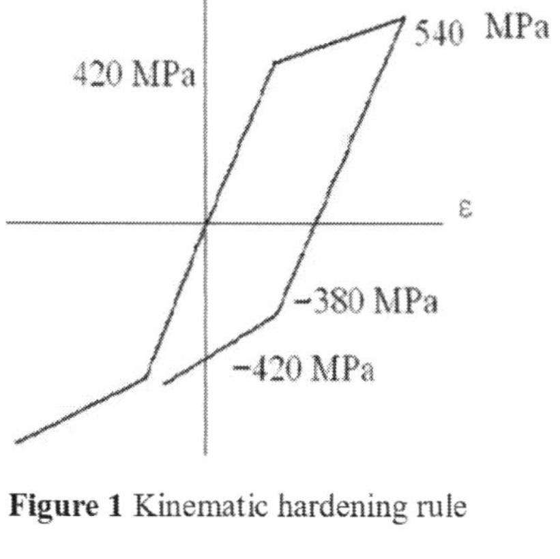

Figure 1 Kinematic hardening rule

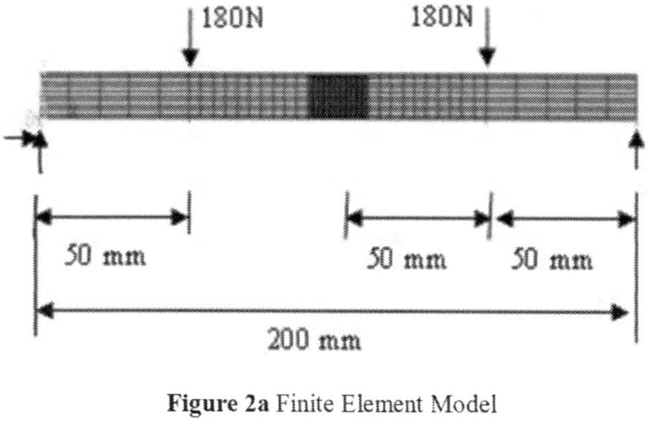

Figure 2a Finite Element Model

As reference the dimension of the whole mesh for the center of beam is shown in figure 2a. Each element of the discretized mesh are *0.5 mm* wide by *1 mm* high, this elements are later used to simulate the cut (figure 2b). Under these conditions, a maximum stress value of *84.1 MPa* in tension and compression respectively, at the surfaces of the beam were found. In figure 3 the total profile of

Journal of Physics: Conference Series **181** (2009) 012069 doi:10.1088/1742-6596/181/1/012069

the residual stress field, acting in the central zone of the beam, obtained by *FEM* is shown. In figure 4 the value of the residual stress field obtained by *FEM* and induced at the central section of the beam is shown as well. All results have been parameterized with respect of the thickness of the beam (x/t).

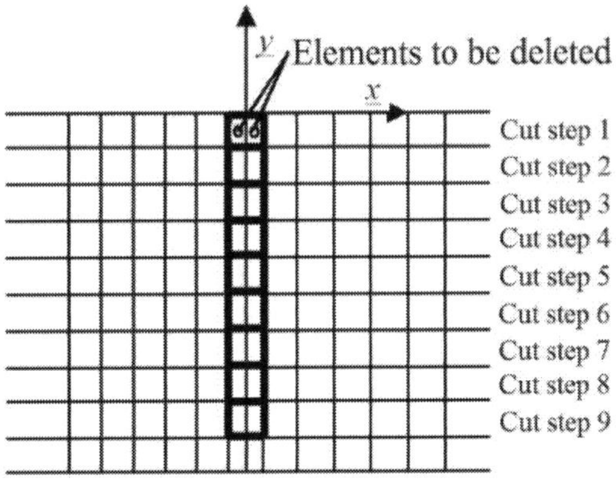

Figure 2b Incremental slot

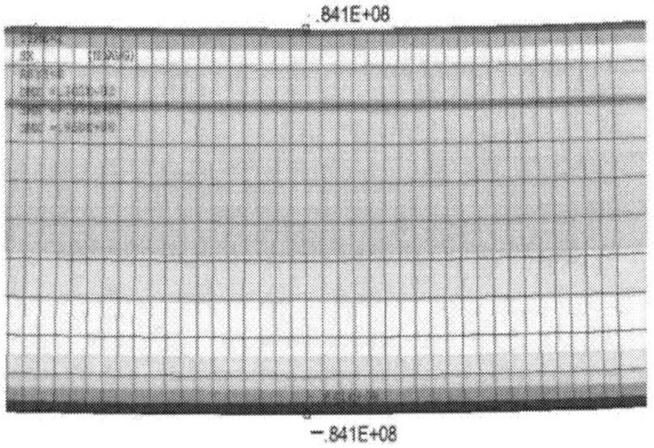

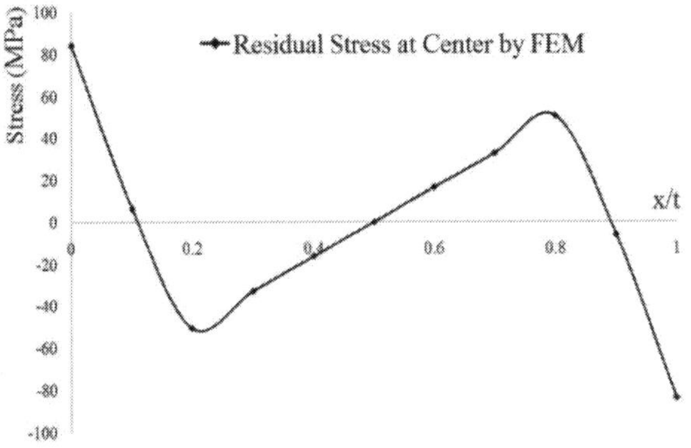

Figure 3 Residual stress field by *FEM*

Figure 4 Residual stress field by *FEM*

3.2. Numerical evaluation of the CCM by determining the original residual stress field

In figure 5 it is shown the general stress relaxation obtained by numerical simulation of the total cut, from which it is possible to determine by *CCM* the original residual stress field that is acting in the beam [15]. The cut was performed by deleting a pair of elements (figure 2b) in a single step, until 9 steps were reached. Strain data was collected after pair of elements was deleted at the bottom.

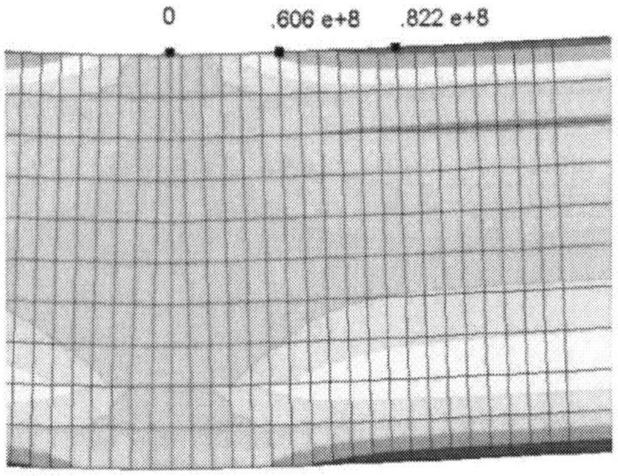

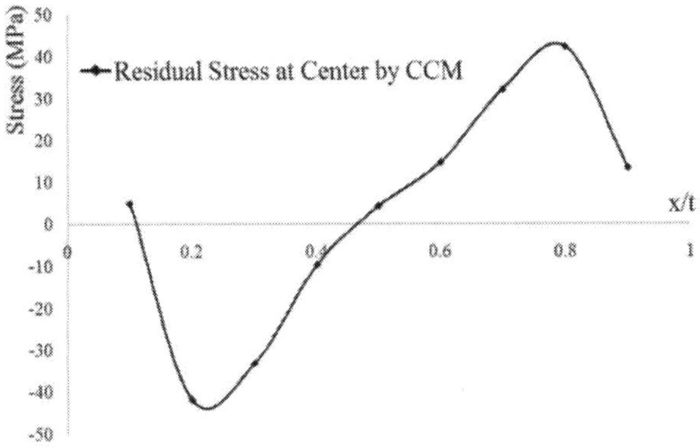

Figure 5 General stress relaxation

Figure 6 Residual stress field obtained by *CCM*

The strain data obtained after completing the 9 steps was used by the *CCM* to obtain the original residual stress profile in the beam. In figure 5 it can be observed that at the center of the beam the stress value is *0 MPa*, this is due to the slot induction. It must be emphasized that in figure 5, additionally it can be observed that at a distance of *3 mm* along the *x* axis from the cut, the surface value of the stress field decreases to *60.6 MPa*. On the other hand, for a *6 mm* length the superficial residual stress value tends to reach its original value (*82.2 MPa*). Nevertheless, by inducing the cut it is possible to determine the original value of the residual stress field using the *CCM* (figure 6). In figure 7 it can be observed a comparison between the data for both, the *FEM* and the *CCM*, those at the center of the beam.

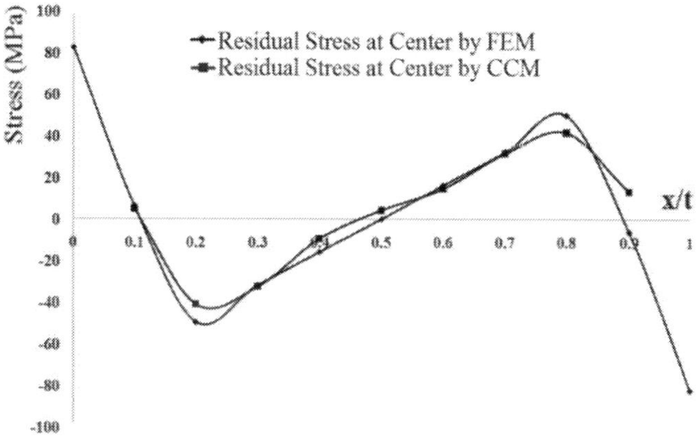

Figure 7 Comparison between residual stresses at the center of the beam

3.3. Relaxation zone 3 mm away from the center

A new cut was simulated 3 mm away along the *x* axis from the initial deletion of elements, as show in figure 8. This operation was performed to determine the precise effect that the first cut had to the original residual stress field. This second cut is used to once again determine the residual stress field existing at this zone. In figure 9 are presented the values of the residual stress fields obtained for both the *FEM* and the *CCM* when the second cut was perform.

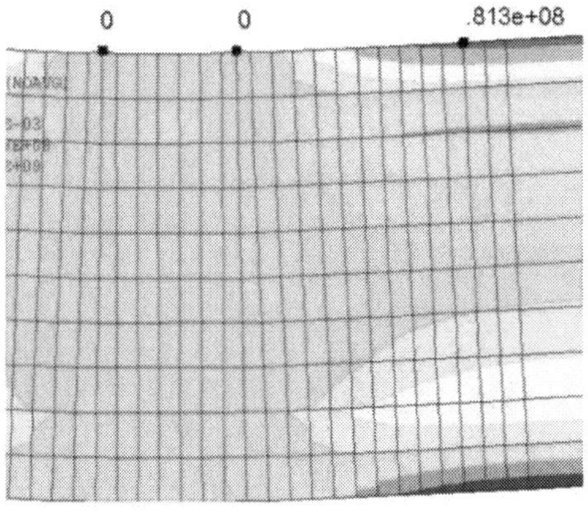

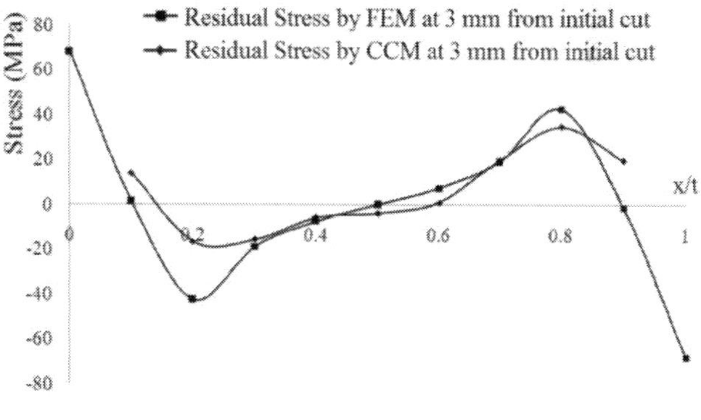

Figure 8 Influence zone at 3 mm from center **Figure 9** Comparison of residual stress at 3 mm from initial cut

3.4. Relaxation zone 8 mm away from the center

A third case is now considered, in which the residual stress field obtained when a second cut is induced 8 mm away along the *x* axis from the first cut (figure 10). In figure 11 it is analyzed the case in which the residual stress field is maintained and then it is compared against the original residual stress state for this center zone. In these figures it can be observed the accuracy of the *CCM* to determine the desired residual stress field for each case and the zones of possible application. The main purpose of this numerical evaluation is based on the necessity to determine the effect of the residual stress fields in those engineering components that have suffered a crack failure. In real life, all engineering mechanical elements are built up through a manufacturing process and residual stresses are induced. So, when a failure occurs it is important to establish, if the stress state in the piece has something to do with it. Additionally, the numerical analysis could provide information about where and how to perform the experimental analysis. For example in this research, it can be seen in figure 11 that the residual stress field obtained at a distance 8 mm away along the x axis did not suffer from the relaxed effect of the cut, maintaining its original magnitude and could provide information in where to performed the cut in mechanical elements that have fail from a crack.

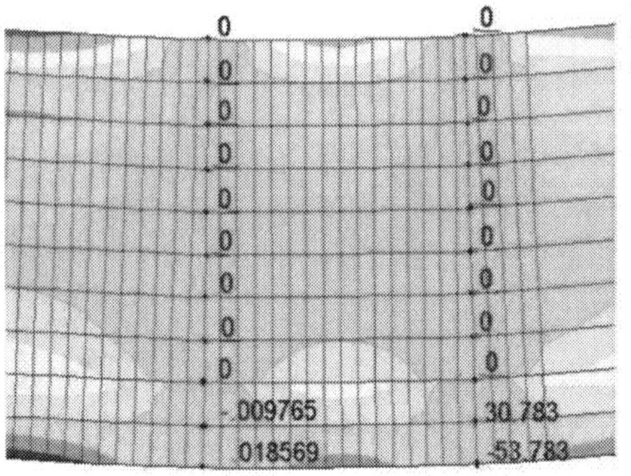

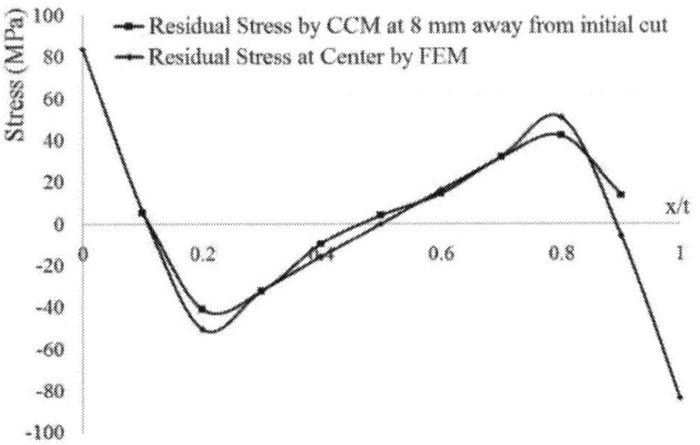

Figure 10 Influence zone at 8 mm from center **Figure 11** Comparison of residual stress at 3 mm from initial cut

3.5. *Experimental evaluation of the CCM*

All residual stress calculations were carried out using the *CCM* as described above. The materials used for this purpose were 18 specimens of *EN 8* steel with the shape of a beam (dept 38.1, mm high 12.7 mm and length 250 mm). The materials were annealed and strain gauges were attached to each specimen (6 specimens with only one gauge at the center and 12 specimens as highlighted in figure 12). The specimens were then mounted on a four point bending rig and plastically bent (bending moment of 975 Nm). The bending operation (figure 13) produced residual stress profiles that can be estimated by using the loading and unloading bending stress approach [16]. This allows the evaluation of residual stresses using the *CCM*. Once the bending procedure was completed, the specimen was cut on the *EDM*.

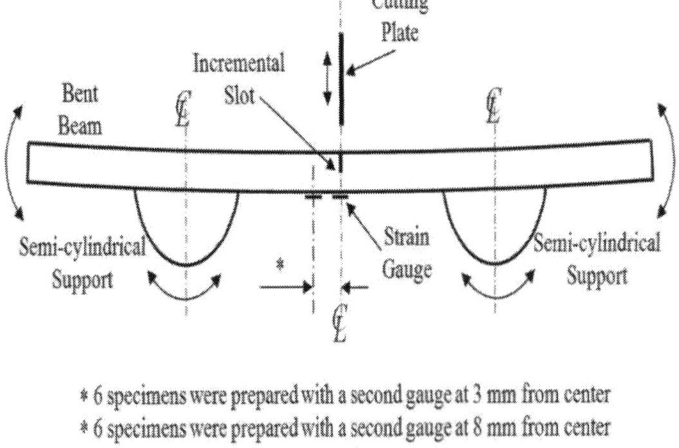

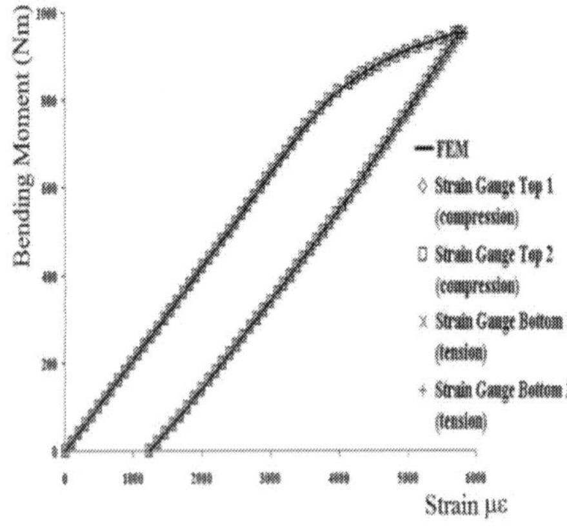

Figure 12 Experimental preparation of the specimen **Figure 13** Bending results

The results of the residual stress field, determined using the *CCM*, are presented in figure 14. The residual stress results obtained by *FEM* and *superposition* are also indicated in this figure. It can be seen that all results agree reasonably well. In figure 15 it is presented the determination of the residual stress field when a cut is placed 3 mm away along the *x* axis from the initial cut. It can be seen in figure 15, that a relaxation of the residual stress field has occurred (as it was predicted by FEM), but it is not possible to determine the profile near to the cut.

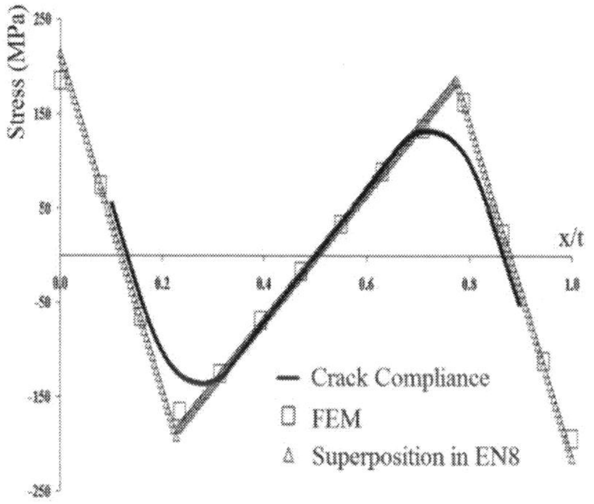

Figure 14 Residual stress evaluations at center

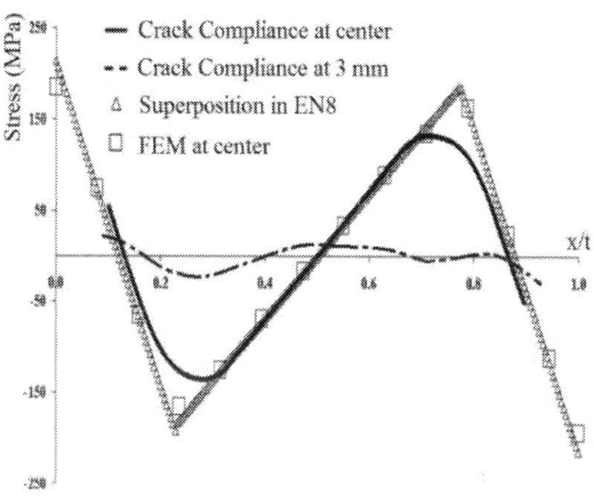

Figure 15 Residual stress evaluations at 3 mm

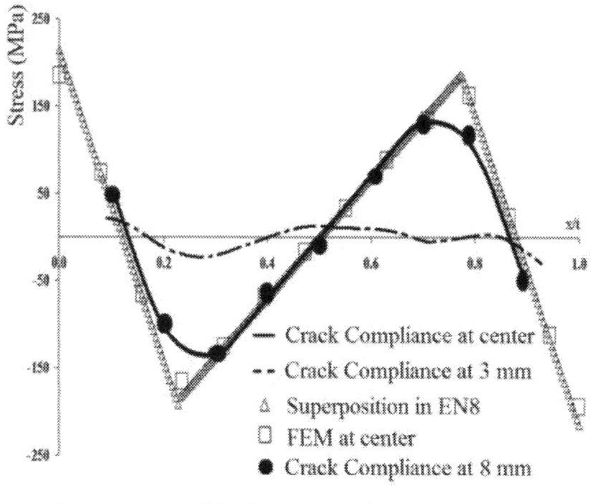

Figure 16 Residual stress evaluations at 8 mm

In figure 16 it is presented the residual stress profile obtained at 8 mm away along *x* axis from the initial cut performed at the center of the beam. Here it can be observed, that the residual stress stands when the cut is 8 mm away from the initial cut. This is a significant result, as it provides information on where to apply the *CCM* when specimens have failed as a consequence of a crack.

4.- Conclusions

The obtained results applying *CCM*, fully validate the value for the residual stress field obtained by the *FEM*. It has been established as well that the residual stress field maintains its value at a zone 8 mm away along *x* axis from the original cut. The experimental results validate the *FEM* procedure and could open new opportunities for the application of numerical simulations. *FEM* has proved to be an important tool for the validation and optimization of the *CCM*. Also, this work is focus in bring to the scientist community some of the advantages and beneficial aspects that the *CCM* could provide on the evaluations of residual stress.

5.- Acknowledgement

The authors gratefully acknowledge the financial support from the Mexican government by the Consejo Nacional de Ciencia y Tecnología and the Instituto Politécnico Nacional, Escuela Superior de Ingeniería Mecánica y Eléctrica. In particular to Mr. Michael Hartman from Oxford Brookes University.

6.- References

[1] Urriolagoitia-Sosa G, 2005, *Analysis of prior strain history effect on mechanical properties and residual stress in beams*, Ph D Thesis, Oxford Brooks University, pp 129-142.
[2] Kandil F A, Lord J D, Fry T A and Grant P V, 2001, A review of residual stress measurement methods – A guide to technique selection, *Project CPM 4.5, Measurement of residual stress in components NPL Report MAT(A) 04.*
[3] Martin P, 1998, A review of mechanical reliability, *Proceedings of the Institution of Mechanical Engineers, Part E: Journal of Process Mechanical Engineering*, Vol. 212, No. 4, pp 281-287.
[4] Semenski D, Bozic Z and Wolf H, 2006, A crack growth analysis in critical structural components, *Materiali in Tehnologije,* Vol. 40, No. 4, pp 123-128.
[5] Cheng W, Finnie I, Gremaud M and Prime M, 1994, Measurement of near surface residual stresses using electric discharge wire machining, *Journal of Engineering Materials and Technology,* Vol. 116, No. 1, pp 1-7.
[6] Prime M B, 1999, Residual stress measurement by successive extension of a slot: The crack compliance method, *Applied Mechanics Reviews*, Vol. 52, No. 2, pp 75-96.
[7] Nowell D, Hills D A and Tochilin S, 2000, Use of the crack compliance method for the measurement of residual stress, *Proceedings of the Sixth International Conference on Residual Stresses*, pp 845–852.
[8] Wyatt J E and Berry T J, 2005, A new technique for the determination of superficial residual stresses associated with machining and other manufacturing processes, *Journal of Materials Processing Technology*, Vol. 1, pp132-140.
[9] Urriolagoitia-Sosa G, 2008, Numerical evaluation of the crack compliance method (CCM) in beams with and without prior history, *Applied Mechanics and Materials,* Vols. 13-14, pp 173-182.
[10] Milan M T, Filho W W, Malafaia A M S, Pelizer B C and Silva C P, 2006, Slot machining effects on residual stress measurements using the crack compliance method, *Journal of Testing and Evaluation,* Vol. 34, No. 2, pp 93-97.
[11] Urriolagoitia-Sosa G, Durodola J and Fellows N A, 2007, Effect of strain hardening on residual stress distribution in beams determined using the crack compliance method, *Journal of Strain Analysis for Engineering Design,* Vol. 42, No. 2, pp 115-121.
[12] Urriolagoitia-Sosa G, Durodola J F and Fellows N A, 2003, Determination of residual stress in beams under Bauschinger effect using surface strain measurements, *Strain*, Vol.39, No. 4, pp 177-185.
[13] Chen P C T, 1986, Bauschinger and hardening effects on residual stress in autofrettaged thick-walled cylinders, *Journal of Pressure Vessels Technology*, Vol. 108, February, pp 108–112.
[14] Wu X R and Carlson A J, 1991, *Weight function and stress intensity factor solutions,* Pergamon Press, Oxford, England.
[15] Sandoval-Pineda J M, 2008, *Análisis del efecto de predeformado en el comportamiento de vigas sometido a cargas cíclicas*, Ph D Thesis, Instituto Politécnico Nacional, SEPI, ESIME, pp 96-106.
[16] Timoshenko S P and Gere J M, 1991, *Mechanics of materials,* Third SI Edition, pp 250–340.

Dynamic Response Analysis of a Heavy Commercial Vehicle Subjected to Extreme Road Operating Conditions

Chinnaraj K.[1], Dr. Sathya Prasad Mangalaramanan[2] and Dr. Lakshmana Rao C.[3]

[1] Research Scholar, Department of Applied Mechanics, IIT Madras, India
Senior Manager, Computer Aided Engineering, Ashok Leyland Ltd, Chennai, India
[2] Section Head, Advanced Engineering, Ashok Leyland Ltd, Chennai, India
[3] Professor, Department of Applied Mechanics, IIT Madras, India

chinnarajk.alvvc@ashokleyland.com, prasad.alc@ashokleyland.com, lakshman@iitm.ac.in

Abstract: Wheel excitations measured on a heavy commercial vehicle by driving it through extreme road operating conditions, are considered as inputs to perform dynamic response analysis in a simulated laboratory and computational environment. From initial modal analysis results using finite elements, critical vehicle frame rail locations are identified for dynamic laboratory strain measurements on a six poster road load simulator that employs dynamic wheel excitations as input. Dynamic stresses calculated from measured strain values are then compared with computationally obtained stress results on each of these locations. This study also points out all geometric locations and vibration modes that may affect the design behavior of the frame members under extreme road operating conditions. The results obtained from this work can be considered for further fatigue life prediction and design optimization of chassis frame rail assembly.

1. Introduction

In the design of automotive structures, if a structural member is to withstand millions of cycles of dynamic load applications, it is important to understand the dynamic stress-time histories due to repeated dynamic loads caused by uneven, extreme road operating conditions. Traditionally in automotive industry, road load specifications for design, verification and validation tests are directly given by Original Equipment Manufacturers (OEMs), which are generated based on an envelope of generic customer usage profiles [1]. These generic road load specifications in many cases are over specified, often leading to failure modes different from those experienced in field. Therefore, in recent times, the road load specification for a particular vehicle is generated based on real-time road load measurements.

The direct strain measurement at various frame rail locations by driving the vehicle on an accelerated road-test may be considered as one of the options for evaluating dynamic response of frame rail assembly. However, carrying out physical testing in accelerated test environment may not be feasible in all cases due to reasons such as complexity in instrumentation, driver fatigue and chances for unnoticed failures during testing, etc; Hence a new approach combining experimental and computational methods, is proposed as an efficient and reliable means for the dynamic response analysis of a heavy commercial vehicle in its early development stage. From the dynamic analysis results, durability of automotive structure can also be performed thereof, for any given road operating conditions.

In the first step of this approach, a prototype vehicle is used once to capture the road load excitations on extreme road operating conditions. In the second step, road load excitations are filtered and processed for giving it as an input signal for six poster road load simulator and finite element model. In the third step, critical frame rail locations are identified for strain-history measurement based on analytical modal analysis results. In the fourth step, transient dynamic analysis is carried out using modal super position technique to predict dynamic stress histories on the identified frame rail locations. This technique is chosen in the analysis due to its applicability to automotive structures especially when the structural mass is large, the stiffness is low and the frequencies of the operating loads pass through number of natural frequencies of the structure [2]. Finally from the strain history measured from six poster road load simulator, dynamic stress histories are calculated and compared with analytically predicted dynamic stress histories on those critical locations.

2. Structural Response in Time Domain and Vehicle Load Specification

It has been established that most vehicle loads can be generally described by a stationary random process [3]. Vehicle surface road loads to automotive structures are typically random in nature, as evidenced from Figures 1 to 6. The stationary random process of a vehicle can be easily described in terms of time or frequency domain format. In this work, data obtained in time-domain format is directly considered for carrying out vehicle structural response analysis.

2.1. Structural Response in Time Domain

Equation of motion of a linear automotive structural system, in general, is expressed in matrix format using equation (1). The system of differential equations can be solved directly in the physical coordinate system, corresponding to each load-time step:

$$[M]\{\ddot{x}(t)\}+[C]\{\dot{x}(t)\}+[K]\{x(t)\}=\{p(t)\}, \qquad (1)$$

where $[M]$, $[C]$, $[K]$ are mass, damping and stiffness matrices, respectively and $\{p(t)\}$is an applied load vector. From the displacement vector $\{x(t)\}$ obtained by solving the system of differential equations (1), the other transient response variables such as acceleration, strain and stress can be obtained for each time step [4,5].

2.2. Road Load Specification for Six Poster Simulator and Computational Model

The dynamic road load excitations at all wheels are measured by driving the prototype vehicle at a constant speed of 30 km/h, on various test road conditions featuring Belgian Pave, Corrugation, Herring Bone, Setts and Pothole roads. The measured road load excitations in the form of acceleration is then filtered and processed using SoMat Ease3 data processing software [6], before taking it as acceleration input signal to six poster road load simulator for experimental strain measurement, and FE model for numerical simulation.

In the current work, only a stretch of 10 sec road acceleration data on Belgian pave road representing extreme road operating conditions is considered as input for the ease of handling of huge computational data. Figures 1 to 6 show time history of acceleration profiles measured using accelerometers at all axle ends of prototype vehicle on Belgian pave test track, available in Ashok Leyland Technical Center.

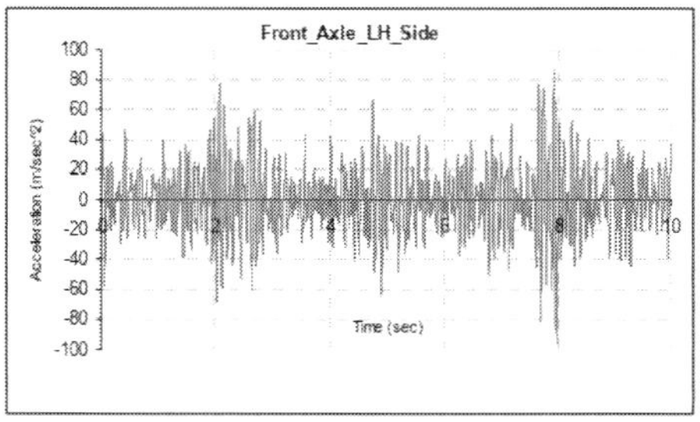

Figure 1. Pave Signal on Front Axle LH Side

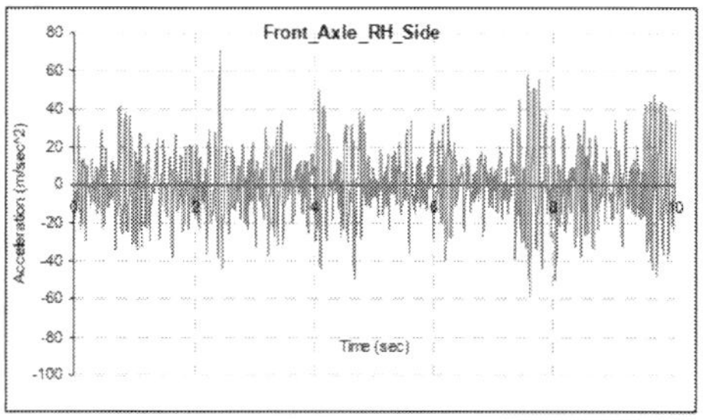

Figure 2. Pave Signal on Front Axle RH Side

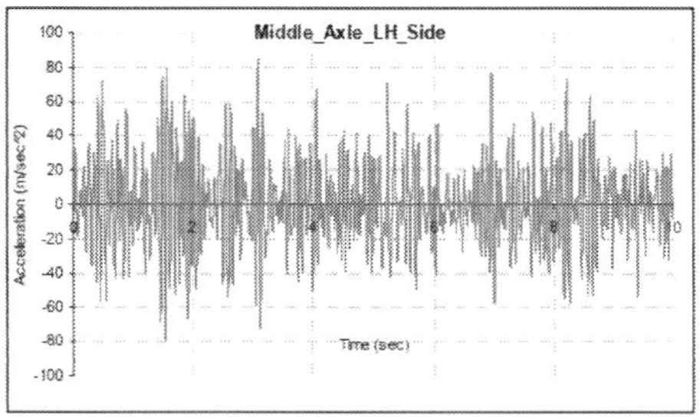

Figure 3. Pave Signal on Middle Axle LH Side

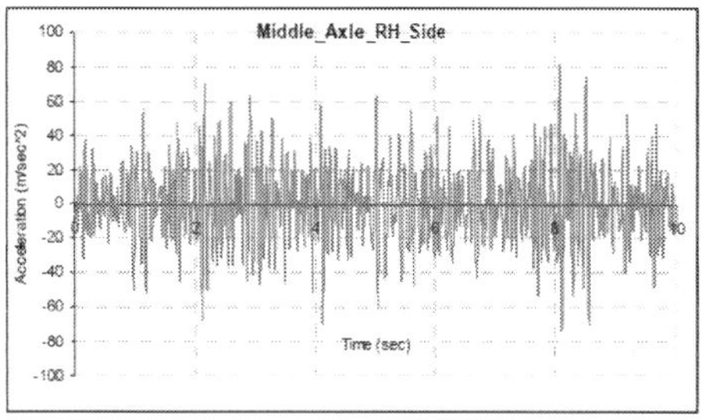

Figure 4. Pave Signal on Middle Axle RH Side

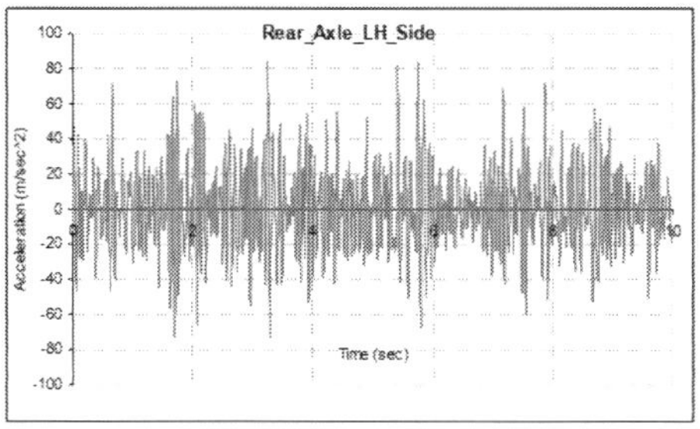

Figure 5. Pave Signal on Rear Axle LH Side

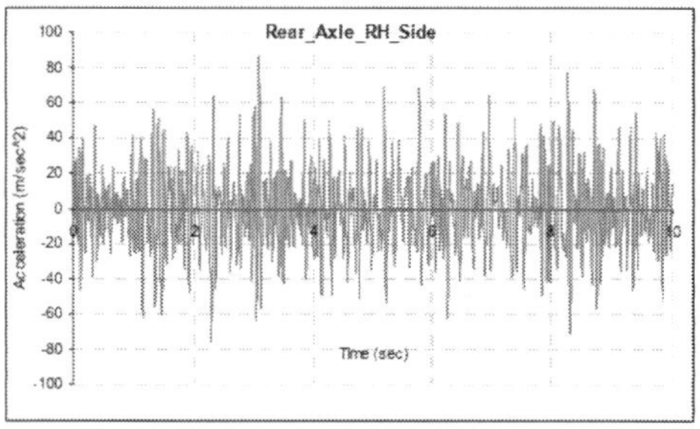

Figure 6. Pave Signal on Rear Axle RH Side

3. Computational Analysis

3.1. Fundamental Vibration Modes and Failure Locations

In order to identify the critical failure locations in load bearing structural members, the truck systems are modeled using finite elements as shown in figure 7. The frame side and cross members are modeled by maintaining smooth flow of shell elements throughout the geometry. Bolt connections in the assembly are simulated by using beam and spider elements. The section and material properties are specified for beam elements to get correct stiffness and realistic mass of bolt joints. While commercial CAD software CATIA [7] is used to model the geometry, Hyperworks [8] preprocessor is employed to mesh the model before taking into MSC Nastran [9] finite element solver.

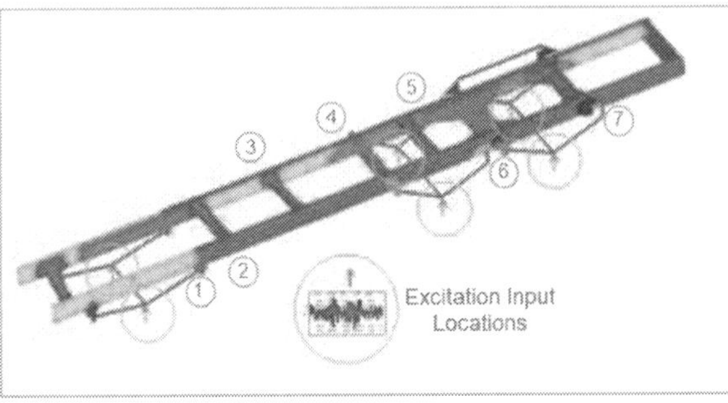

Vehicle Specification

Vehicle Model	Multi Axle Truck
Wheel base	3850mm
GVW	25 Tons
Overall Length	8300mm
Front Overhang	1200mm
Rear Overhang	1800mm
Vehicle Width	2432mm

Figure 7. FE Model with Seven Strain Locations and Six Input Locations

The effect of tyres, leaf springs and bushes are considered in the analysis by measuring the stiffness properties from component tests and idealizing them as linear spring elements in the FE Model. Other non-load bearing structures and major vehicle assemblies such as engine, cab, fuel tank and payload are represented by lumping their respective masses at appropriate center of gravity locations and connecting to relevant bolt holes of load bearing structural members.

The predicted modal frequencies and mode shapes for the first six low vibration modes by employing MSC Nastran code is shown in figures 8 and 9. From the modal analysis results, seven critical locations are identified, where the presence of residual stresses and high loading stresses are expected.

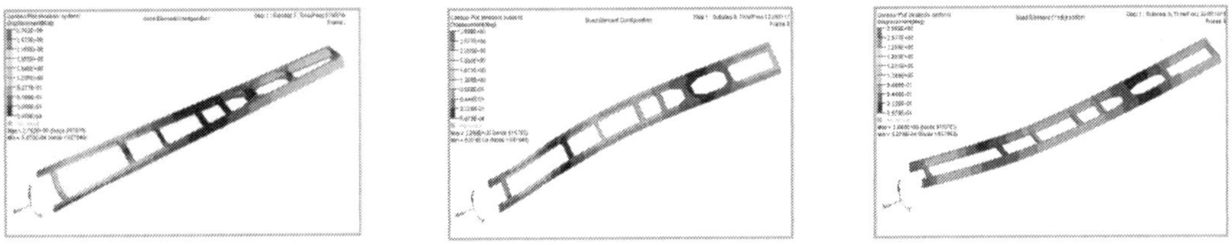

Figure 8. Fundamental Torsion, Lateral Bending and Vertical Bending Modes
(1st, 2nd and 3rd natural Frequencies - 5.35 Hz, 12.20 Hz and 20 Hz, respectively)

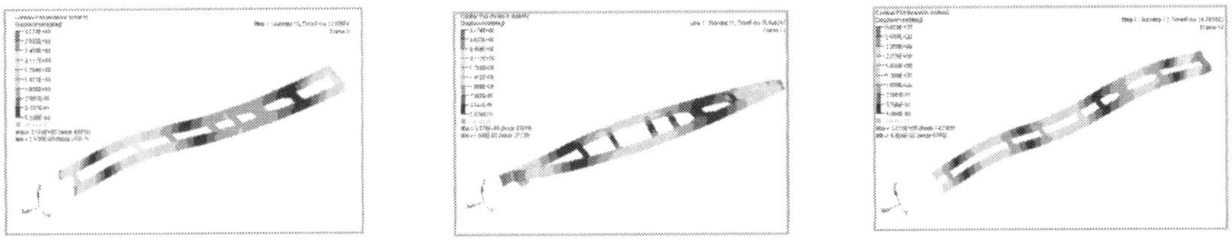

Figure 9. Full Cycle Lateral Bending, Combined Torsion & Bending and 1.5 Cycle Lateral Bending Modes (4th, 5th and 6th natural Frequencies - 22.84 Hz, 25.41 Hz and 39.76 Hz, respectively)

These identified frame rail locations are strain gauged using 45^0 strain rosettes to experimentally measure the strain-history on six poster road load simulator.

3.2. Computational Transient Response Analysis

The finite element model used in modal analysis is supplemented with appropriate constraints and degree of freedoms for giving excitation load in vertical direction on all six wheel locations. Apart from vertical road load excitations, inertia loads of major assemblies are taken into consideration by applying mass and mass moment of inertia of all assemblies at its corresponding center of gravity locations. While the direct transient analysis is not feasible for automotive applications as it uses prohibitive computation time and needs considerable amount of memory to store output results, the modal superposition approach, is used for carrying out dynamic analysis.

Modal superposition approach is an effective technique to obtain the transient response of any structure. This approach evaluates stress histories by linearly combining the modal participation histories of each mode of interest, with the corresponding modal influence coefficient as obtained from the modal analysis. While using this technique, two important parameters that need to be considered are the number of modes to be extracted and accounted for, and the size of the time step to be used when obtaining the participation histories.

4. Experimental Analysis

Figure 10 shows the six poster road load simulator available in Ashok Leyland Technical Center. It consists of 6 identical vertical load modules using servo hydraulic actuators fitted with precision hydrostatic bearings, providing the support to reproduce any vertical road surface events accurately in laboratory environment. Using this simulator, any road simulations such as durability tests could be easily performed on 6 wheel vehicles up to 600 KN GVW (Gross Vehicle Weight). Besides durability tests, it is also used to test vehicles for squeaks and rattles; evaluate the ride behavior of different vehicles with a set of standard road profiles and test vehicles as they roll off the production line.

For vehicle road load simulation using this system, the actual road surface events on vehicle wheels are measured using accelerometers or string potentiometers for displacements. The measured signals are then analyzed and digitized in terms of actuator displacements using remote parameter control software before feeding into servo-controller of the simulator to execute vehicle simulation test.

Figure 10. Six Poster Road Load Simulator

In the laboratory experiments, the identified critical frame rail locations from analytical modal analysis are strain gauged using 45^0 strain rosettes to measure dynamic strain histories on six poster road load simulator. Experimentally measured and processed 10 sec Belgian pave road load data is given as input to six poster simulator and corresponding strain time histories with sampling rate of 100 Hz are recorded on those identified frame rail locations.

From the measured dynamic strain histories, the dynamic stress histories are calculated and compared with the computationally obtained dynamic stress histories at all critical frame rail locations.

5. Comparison of Experimental and Computational Results

The comparison of dynamic stress values obtained from road load simulator measurement and computationally predicted stress values on all identified locations are shown in Figures 11 to 17.

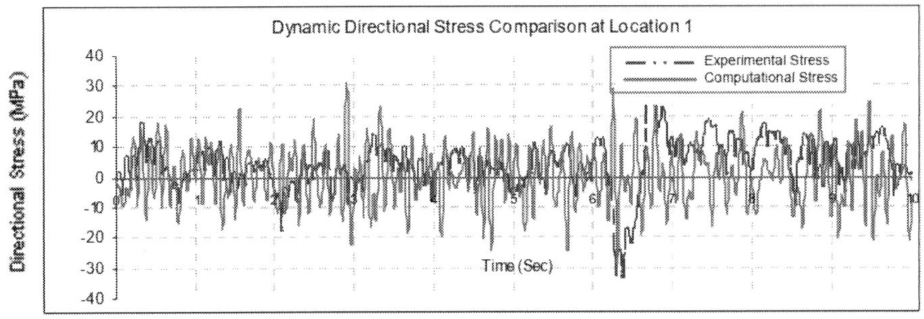

Figure 11. Experimental vs. Computational Dynamic Directional Stress at Location 1

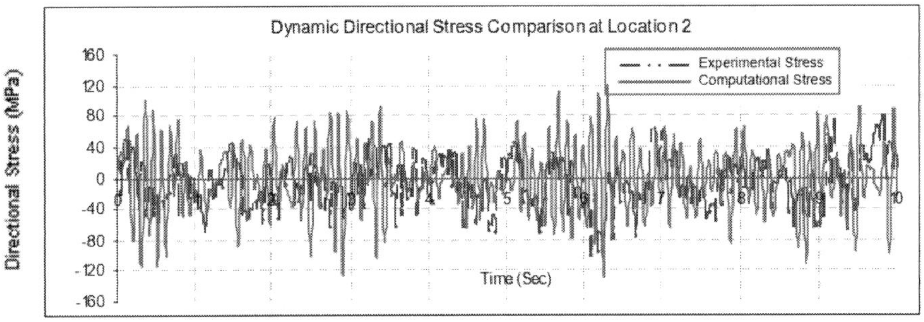

Figure 12. Experimental vs. Computational Dynamic Directional Stress at Location 2

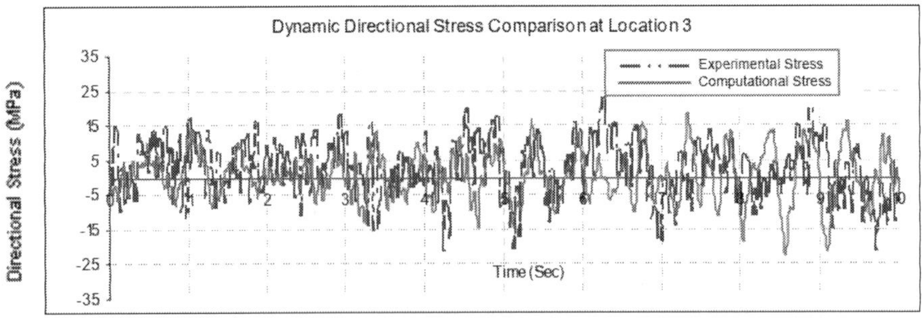

Figure 13. Experimental vs. Computational Dynamic Directional Stress at Location 3

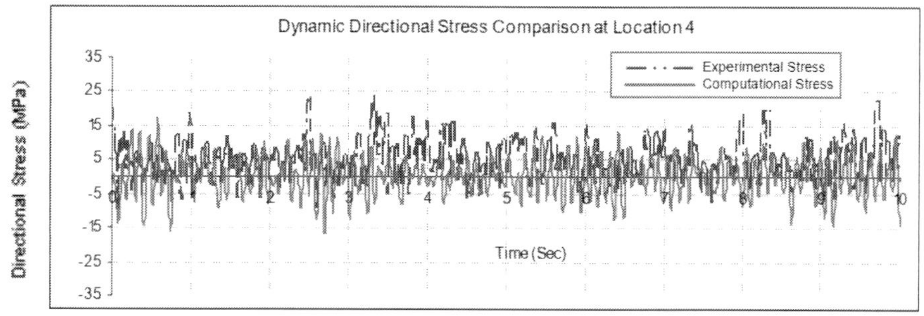

Figure 14. Experimental vs. Computational Dynamic Directional Stress at Location 4

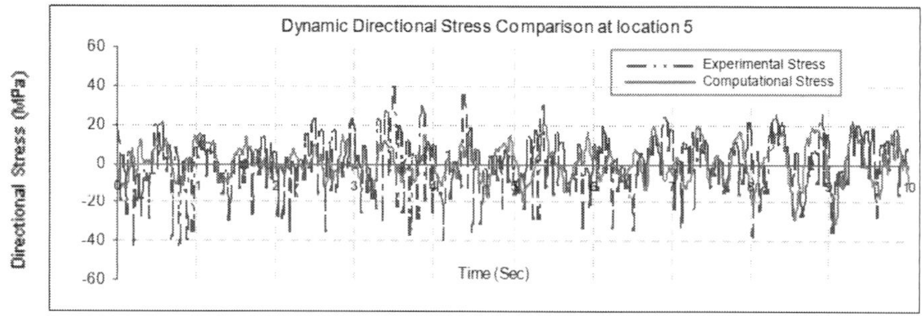

Figure 15. Experimental vs. Computational Dynamic Directional Stress at Location 5

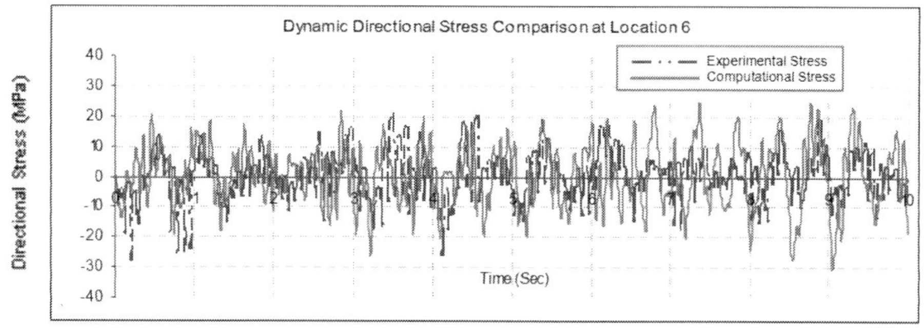

Figure 16. Experimental vs. Computational Dynamic Directional Stress at Location 6

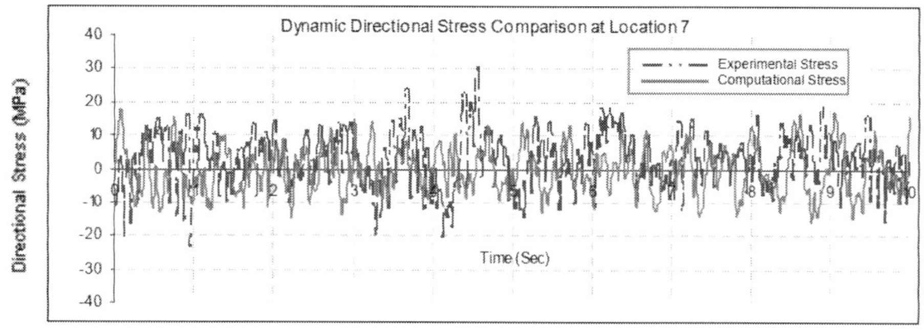

Figure 17. Experimental vs. Computational Dynamic Directional Stress at Location 7

From the comparative dynamic stress plots, it could be observed that the computational stress amplitudes are higher than the experimental stress amplitudes at locations 1 and 2 and lower at locations 5 and 7.

Among other locations, we could observe close correlation between computational and experimental stresses in locations 3 and 6, although there are slight deviations at the end.

The likely causes of the difference between simulated and measured stresses in locations 1, 2, 5 and 7 could be the residual stresses resulting from the fabrication and assembling of frame structure. These residual stress effects could not be accounted in the computational method employed in this work and currently research is under progress to validate the hypothesis.

6. Summary and Conclusions

This work presents an efficient methodology for the integrated dynamic response analysis of a heavy commercial vehicle consisting of accelerated road load measurement, road load simulation on a six poster, dynamic stress analysis on experimental as well as computational environment.

With the demonstrated methodology, dynamic response analysis of any prototype vehicle can be performed efficiently in simulated environment for any road load or extreme road operating conditions. As this method is faster and at the same time tests are performed with real time road inputs, it would be a preferred method for OEMs especially during initial stage of vehicle development.

This study also opens up an interesting hypothesis that the residual stresses resulting from the fabrication and assembling of frame structure may have significant effect on experimental strain measurement. These effects are being investigated separately as a future work.

Acknowledgments

The authors would like to thank M/s Ashok Leyland Ltd, India, especially Mr.R.R.G.Menon, Executive Director, Product Development, for his valuable support to carry out this exercise at their Technical Centre. The authors also thank Mr. P.Dayalan and Mr. S.Nandakumar of M/s Ashok Leyland Ltd for their help during experimental measurement.

References

[1] MIL-STD-810 Committee, Environmental Test Methods and Engineering Guides, MIL-STD-810F, 2000

[2] H.S.Kim, H.J.Yim and C.B.Kim, "Computational Durability Prediction of Body Structures in Prototype Vehicles" Intl Journal of Automotive Technology, Vol. 3, No. 4, 2002, pp. 129-135

[3] Su, H., Rakheja, S. and Sankar, T.S., "Stochastic Analysis of Nonlinear Vehicle Systems Using a Generalized Discrete Harmonic Linearization technique" Journal of Probabilistic Engineering Mechanics, Vol 6, No.4, 1991, pp 175-183

[4] Hong Su, "Vibration Test Specification for Automotive Products based on Measured Vehicle Load Data" SAE Paper No 2006-01-0729

[5] Su, H., "Automotive CAE Durability Analysis Using Random Vibration Approach," MSC 2nd Worldwide Automotive Conference, Dearborn, MI, Oct. 2000.

[6] SoMat Ease Version 3 - User's Guide, 1999.

[7] Catia V5R17 - User's Design Manual, 2008.

[8] Hyperworks 9.0 - User's Guide, 2008.

[9] MSC-Nastran 2007r1 - User's Guide, 2007.

Strain limit dependence on stress triaxiality for pressure vessel steel

Yang-chun DENG[1], Gang CHEN[2], Xiao-feng YANG[1],Tong XU[2]

1 Hubei Special Equipment Safety Inspection and Research Institute, No.352 Xudong St. Wuhan,China,430077

2 China Special Equipment Inspection and Research Institute, Beijing, China

The corresponding author's e-mail:deng_ych@126.com

Abstract. In this paper, the failure characteristics of pressure vessel materials were investigated, and measurement and analysis approaches for ductile fracture strains were studied. Based on uniaxial tensile tests of notched round bar specimens, combined with finite element analyses and microscopic observations of fracture surface, the relationships between the stress triaxiality factor and the ductile fracture strain are proposed for three typical Chinese pressure vessel steels, 16MnR, Q235 and 0Cr18Ni9. The comparison of experimental fracture strains with the multiaxial strain limit specified in ASME Ⅷ-2 2007 shows that the strain limit criterion of ASME is suitable for carbon steels but not suitable for austenitic stainless steels for Chinese pressure vessel steels. To improve the calculation accuracy for fracture strain of materials and to develop the strain limit criterion for Chinese pressure vessel materials, more experimental studies and numerical analyses on fracture strain are necessary.

1. Introduction

Based on many experimental results of material properties, Bridgman[1] concluded in 1952 that the ductility of materials is related to the stress state and can be significantly improved by applying a higher pressure. In 1959, Davis and Connelly[2] investigated the effect of the stress state on the fracture strain of cylinders subject to internal pressure and tensile loads, and proposed the concept of stress triaxiality factor. In damage mechanics the stress triaxiality factor is a very important concept, defined as

$$TF = \frac{\sigma_m}{\bar{\sigma}} = \frac{(\sigma_1 + \sigma_2 + \sigma_3)/3}{\sqrt{\left[(\sigma_1 - \sigma_2)^2 + (\sigma_2 - \sigma_3)^2 + (\sigma_1 - \sigma_3)^2\right]/2}} \tag{1}$$

where TF is the stress triaxiality factor, σ_m is the average stress, $\bar{\sigma}$ is the Mises equivalent stress, $\sigma_1 \sim \sigma_3$ are the principal stresses. McClintock[3] in 1968, and Rice and Tracey[4] in 1969, investigated the mechanical mechanisms of the growth of a cylindrical hole and spherical hole respectively. In these two papers, they gave the quantitative relationship between the growth of holes and stress triaxiality factor, which are recognized as a milestone for the development of damage mechanics of materials. Void initiation, growth, and coalescence are three typical processes in the ductile fracture of metal materials[5]. The fracture strain of materials directly correlates with the critical size of voids.

Hancock and Mackenzie[6], and Mackenzie[7] performed uniaxial tension tests on the same batch of notched round bar specimens and carried out the microscopic and macromechanical analyses of the specimens with an assumption of uniform stress and strain distributions at the notch root of the specimens. Experimental results showed that the equivalent stress dropped rapidly after reaching a maximum value on the equivalent stress vs. equivalent strain curve. If unloading started immediately after the maximum equivalent stress, cracks were observed from microscopic analyses; if unloading started before the maximum equivalent stress, discrete voids was observed. Therefore, the equivalent strain corresponding to the maximum equivalent stress was considered as the fracture strain of materials. Macromechanical behaviour analyses showed that there existed a nonlinear relationship between the stress triaxiality factor and the fracture strain, the bigger the stress triaxiality factor the smaller the fracture strain was.

Whether it is global plastic instability or local ductile fracture, the failure of materials is determined by the stress state of structures. The possibility for both failure modes should be investigated in the prediction of the failure mode of structures.

In this paper, we present a pilot study on the ductile fracture of pressure vessel materials by using three typical Chinese pressure vessel steels.

2. Specimens for ductile fracture measurement of pressure vessel materials

2.1 Experimental specimens

The ductile fracture of materials often occurs locally. Normally, the equivalent strain is used as a measurement of the ductility of materials, it is a material parameter related to the stress triaxiality factor but independent of specimens. Since the ductile fracture under triaxial stress states is very complicated, there is no unified conclusion made and no standard available for measuring the ductile fracture strain of materials under triaxial stress states. The notched round bar tension specimen is simple, easy to test and can be used to obtain different stress triaxiality factors by changing the notch radius. Therefore, the tension test on the notched round bar has been widely employed by researchers as a standard method to develop the relationship of the stress triaxiality factor and the strain.

Bridgman[1] made a milestone contribution to the stress analysis of notched round bar tension tests. He investigated the stress distribution of a round bar tension specimen near the neck before fracture. Bridgman's method has been widely used for the stress analysis of notched round bar tension specimens, often referred to as Bridgman notch specimens or the Bridgman model. In the Bridgman model, it was assumed that near the neck the specimen profile along the longitudinal direction was a circle-like shape with a diameter of R, the deformation of its cross section was uniform with a radius of a, as shown in Fig.1.

The centre of the round bar has the maximum stress triaxiality factor written as[1]

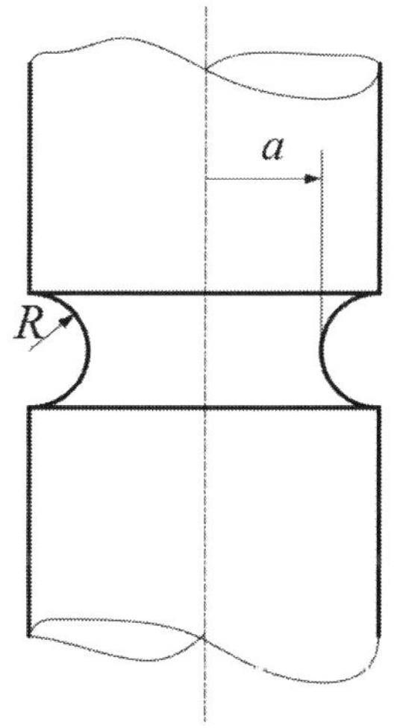

Fig.1 Diagram of the main dimensions at the neck of a Bridgman tension specimen

$$TF_{r=0} = \left[\frac{\sigma_m}{\bar{\sigma}}\right]_{r=0} = \frac{1}{3} + \ln\left(\frac{a}{2R} + 1\right) \tag{2}$$

where a is the root radius of the notch, and R is the notch radius. With an assumption of uniform stress and strain distributions at the cross section of the notch root, the effective strain and stress are written as

$$\bar{\varepsilon} = 2\ln\frac{a_0}{a} \tag{3}$$

$$\bar{\sigma} = \frac{P}{\pi a^2} \tag{4}$$

where $\bar{\varepsilon}$ is the efficient strain at the cross section of the notch root, a_0 is the initial radius of the notch root before loading, and a is the notch root radius changing with the tension load. In this paper, we employed notched round bar tension specimens to measure the equivalent strain of materials at ductile fracture under triaxial stress states.

2.2 Materials of specimens

Three typical pressure vessel materials were tested, 16MnR, Q235, and 0Cr18Ni9. The technical specifications and chemical compositions of the materials are listed in Table 1.

Table 1 Specification and chemical composition of specimens

Standard	Material	No.	Element proportions by weight (%)						
			C	Mn	Si	S	P	Ni	Cr
GB6654	16MnR	1	0.18	1.51	0.353	0.002	0.011		0.006
GB6654	Q235	2	0.151	0.45	0.201	0.003	0.013		0.150
GB/T 4237	0Cr18Ni9	3	0.046	1.198	0.477	0.003	0.022	8.13	18.35

Six different notch radii were chosen: 0.25mm, 0.5mm, 1.0mm, 1.5mm, 2.0mm, 2.5mm and 3.0mm. The root radius of the notched round bars was 4 mm. 2 to 3 specimens were fabricated for each size.

3. Experimental result analysis for ductile fracture of pressure vessel materials

The tension test was performed with an MTS-880 tensile test machine under displacement controlled loading. The loading speed was 0.5mm/min, which was considered to be a quasic-static load, that is, the loading speed effect could be neglected[8,9]. In the test, the load was recorded, the change of the notch radius at the neck was measured by a horizontal extensometer and the length change of specimens was measured by a vertical extensometer.

Since pressure vessel materials exhibit good ductility, plastic instability may still occur in a notched round bar tension test. Combined with experimental results, finite element analyses (FEAs) were used to investigate whether the failure mode of the specimen was ductile fracture. ANSYS 9.0, a commercial FE program, was used for all FEAs in this study. Due to the symmetry of the geometry, axisymmetric models with Plane82, an 8-node quadrilateral element type, were employed. The isotropic material with an isotropic hardening law was assumed. The arc-length method, large deformation and the Von Mises yield criterion are employed for stress and strain analyses.

From notched round bar tension tests, we can directly obtain $P-\Delta$ curves, where P is the tensile load and Δ is the radius reduction at the neck of specimens. Using Eqs (3) and (4), we can convert $P-\Delta$ curves to equivalent stress and strain curves, $\bar{\sigma}-\bar{\varepsilon}$. Based on FEA results from the node of the outer surface at the neck, we can plot $P-\Delta$ curves. Fig.2(a) and Fig.2(b) show the comparison of experimental and FEA $P-\Delta$ curves for specimens with notch radius of 0.5mm and 2.0mm respectively. No-notched round bar tensile tests were performed on 3 specimens for each kind of material, and 3 true stress-strain curves were obtained for the same material. Therefore, three sets of material parameters corresponding to 3 true stress-strain curves of tensile test data were used in FEM analysis. In another words, 1-05-11, 1-05-12, and 1-05-13 shown in Fig.2(a) are FEM analysis results for notch radius 0.5mm corresponding to 3 sets of material parameters respectively, whereas 1-20-11, 1-20-12, and 1-20-13 in Fig.2(b) are results for notch radius 2.0mm..

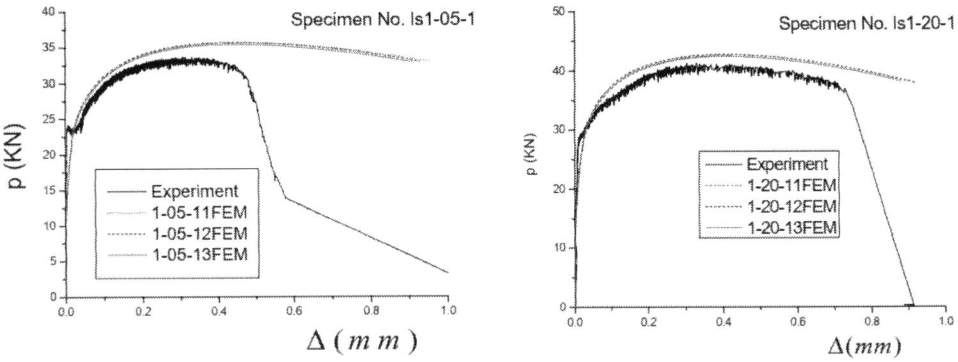

(a) notch radius 0.5mm (b) notch radius 2.0mm

Fig.2 Curves of $P-\Delta$

As shown in Fig.2(a), the experimental curve at the earlier loading stage matches well with that from FEAs, while the experimental curve at the late stage of loading drops much quicker than that from FEAs. Due to the assumption of a continuous body in FEAs, the maximum value of P on FEA curves corresponds to the load at plastic instability. For experimental specimens, the void initiation after plastic deformation will reduce the real net cross-section area at the neck, resulting in a decrease of the load-carrying capacity of specimens. The maximum value of P on the experimental curve in Fig.2(a) is less than that on the FEAs curve, that is, less than the load at plastic instability from FEAs. In addition, Δ value corresponding to maximum P from test is less than that obtained from FEAs. The reason that Δ from test is smaller is: during the test, before sample met the plastic instability, ductile fracture damage arising while lack of material ductility. Therefore, the failure mechanism for the specimens with 0.5mm notch radius is ductile fracture.

Fig.2(b) shows that the $P-\Delta$ curve from FEAs has a reasonable match with the experimental curve. The difference between the experimental curve and the FEAs curve may come from the real net cross-section area reduction at the neck due to the void initiation in the test, which cannot be considered via FEA unless damage mechanics is included in the model. Δ value corresponding to maximum P from test is close to that obtained from FEAs. The two Δ values went through a large plastic deformation process after the maximum load, and the $P-\Delta$ curve from the test and that from finite element analyses are still very close, which indicates that plastic instability is reached first before the damage, that is the plastic instability failure. Therefore, the failure of the specimen with 2mm notch radius is plastic instability, and the experimental results from this specimen cannot be used for ductile fracture criterion studies.

The optical microscopic images of the fracture surfaces for the two specimens are illustrated in Fig.3. Dimples are observed on the fracture surfaces of both specimens; these are evidence of voids before fracture. The dimple distribution of 1-05-1 specimen with a notch radius of 0.5mm is very dense and linked together to form cracks causing fracture in materials; the dimple distribution of 1-20-1 specimen with a notch radius of 2.0mm is relatively loose, their existence may only reduce the load-carrying capacity of materials, and the failure of the specimen is plastic instability due to its insufficient strength.

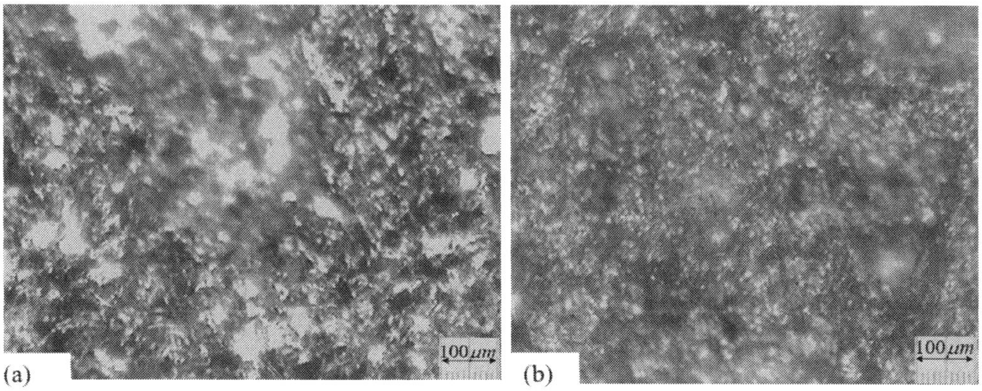

Fig.3 Detailed inspection of fracture surfaces of tensile specimens
(a) notch radius 0.5mm (b) notch radius 2.0mm

4. The relationship of stress triaxiality factor and strain under ductile fracture

Until now, although more studies are still required the Bridgman's approach is still a widely used method for calculating the equivalent strain (Eqs.(3)) and the stress triaxiality factor (Eq.(2)) in engineering applications. Using the approach described in Section 3 we investigated the relationship of the stress triaxiality factor and the equivalent strain of materials under ductile fracture by analysing more than 60 notched round bar tension tests and comparing them with FEA results. It was found that, for 16MnR and Q235 specimens, the failure mode was ductile fracture when the notch radii of round bar specimens were less than or equal to 1.5mm, and the failure mode of specimens was plastic instability when the notch radii of specimens were greater than 1.5mm; for 0Cr18Ni9 specimens, the failure mode was ductile fracture when their notch radii were less than or equal to 2.0mm, and the failure model was plastic instability when their notch radii were greater than 2.0mm. By using the relationship of $\varepsilon_f = A \cdot \exp(-B \cdot TF)$, we obtained the relationships of the stress triaxiality factor and the equivalent strain under ductile fracture for the three materials, as follows:

16MnR: $\varepsilon_f = 0.633 \cdot \exp(-0.510TF)$, $1.18 \le TF \le 2.53$

Q235: $\varepsilon_f = 1.133 \cdot \exp(-0.813TF)$, $1.18 \le TF \le 2.53$

0Cr18Ni9: $\varepsilon_f = 1.398 \cdot \exp(-0.882TF)$, $1.01 \le TF \le 2.53$

In ASME Ⅷ-2 2007[10] pressure vessel and boiler standard, the strain limit to protect against local failure is introduced and written as :

$$\varepsilon_f = \varepsilon_{fu} \cdot \exp\left[-\left(\frac{\alpha_{sl}}{1+m_2}\right)\left(TF - \frac{1}{3}\right)\right] \tag{5}$$

where ε_L is the limiting triaxiality strain, i.e. the fracture strain ε_f; ε_{Lu} is the uniaxial strain limit, i.e. fracture strain ε_{fu} which is the maximum value among m_2, specified elongation and specified reduction area; m_2 is related with σ_s/σ_b, for ferritic steel, $m_2 = 0.6 \cdot (1-\sigma_s/\sigma_b)$ and $\alpha_{sl} = 2.2$, for austenitic stainless steel, $m_2 = 0.75 \cdot (1-\sigma_s/\sigma_b)$ and $\alpha_{sl} = 0.6$. Using Eq. (5), the fracture strain under triaxiality stress states, ε_f, can be easily obtained from a round bar uniaxial tension test by measuring σ_s、σ_b, specified elongation and specified reduction area.

To verify the reliability of the strain limit, three materials, 16MnR, Q235 and 0Cr18Ni9, were investigated. For facilitating the comparison with experimental results, we rewrote the strain limit, Eq. 5, as

$$\ln\left(\frac{\varepsilon_L}{\varepsilon_{Lu}}\right) = -\left(\frac{\alpha_{sl}}{1+m_2}\right)\left(TF - \frac{1}{3}\right) \tag{6}$$

If the rewritten form of the strain limit, Eq. (6), is plotted in the coordinates, $\ln\left(\frac{\varepsilon_L}{\varepsilon_{Lu}}\right)$ vs. $TF - 1/3$, then it is a straight line with a slope of $-\frac{\alpha_{sl}}{1+m_2}$. Fig.4(a-c) show the comparison of experimental results and Eq (6) for the three materials. As shown in Fig.4(a & b), the experimental curves of two ferritic steels, 16MnR and Q235, are above the strain limit line defined by Eq.(6). That is, it is safe and conventional to use the strain limited specified in ASME Ⅷ-2 2007 for 16MnR and Q235. Fig.4(c) shows that the experimental curve of 0Cr18Ni9, an austenitic material, is below strain limit line defined by Eq.(6). That is, it is not safe to use the strain limited specified in ASME Ⅷ-2 2007 for 0Cr18Ni9.

5. Conclusion

The notched round bar tension specimen can be used for the fracture strain measurement of materials, but the failure mode of specimens, such as plastic instability or ductile fracture, should also be determined.

Based on a large amount of experimental data from notched round bar tension tests, combined with FEA results and microscopic observations of fracture surface, the

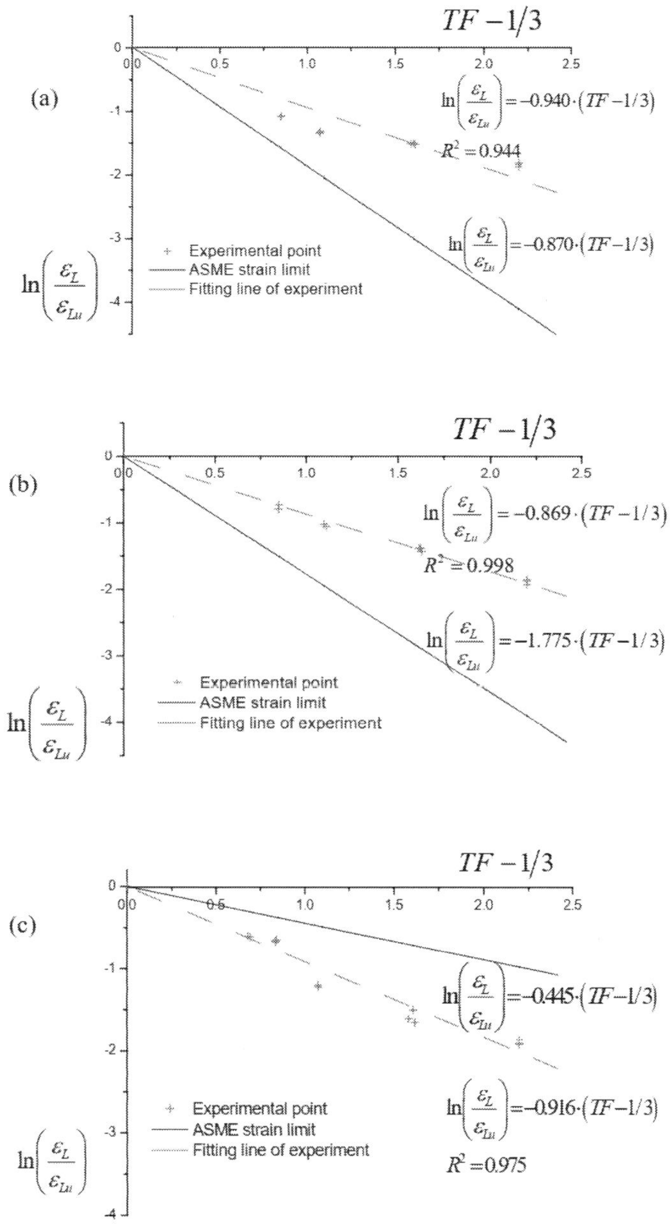

Fig.4 Strain limit from experimental data and ASME strain criterion

(a) $16MnR$ (b) $Q235$ (c) $SUS304$

relationships between the stress trixaixlity factor and the strain under ductile fracture for three steels commonly used in Chinese pressure vessel are proposed.

The fracture strains of 16MnR, Q235 and 0Cr18Ni9 were calculated from the strain limit specified in ASME Ⅷ-2 2007, and they were compared with experimentally measured failure strains. It was found that the strain limit criterion of ASME Ⅷ-2 2007 was safe and quite conservative for 16MnR and Q235 steel but not safe for 0Cr18Ni9 steel. Therefore, the strain limit criterion of ASME Ⅷ-2 2007 can be used directly in carbon steels for Chinese pressure vessels, and more research will be needed before the strain limit criterion is used for austenitic steels for Chinese pressure vessels.

To improve the calculation accuracy of fracture strains for different materials and to develop the strain limit criterion for Chinese pressure vessel materials, more experimental studies and numerical analyses on fracture strain are necessary. The experimental principle, measurement method and analysis methods used in this study provided a foundation for further studies on ductile fracture of pressure vessel materials.

Acknowledgments

This paper is sponsored by the 11th Five-year China National Key Technology R&D Programme, No.2006BAK02B02.

References

[1] P.W.Bridgman, Studies in Large Plastic Flow and Fracture, First Edition,1952 the McGraw-Hill Company,Inc.

[2] E.A.Davis and F.M.Connelly. Stress distribution and plastic deformation in rotating cylinders of strain-hardening material. Journal of Applied Mechanics, 1959,V26,No.1:25-30

[3] F.McClintockhe.A Criterion for Ductile Fracture by the Growth of Holes,Journal of Applied Mechanics,1968,35(2):363-371

[4] J.R.Rice and D.M.Tracey.On the Ductile Enlargement of Voids in Triaxial Stress Fileds.Journal of the Mechanics and Physics of Solids,1969,17:201-217

[5] G.Mirone. Role of stress triaxiality in elastoplastic characterization and ductile failure prediction. Engineering Fracture Mechanics,2007,Vol.74:1203-1221.

[6] J.Hancock and A.Mackenzie.On the mechanisms of ductile failure in high-strength steels subjected to multi-axial stress-states. Journal of the Mechanics and Physics of Solids,1976,24:147-169.

[7] A. Mackenzie J.W.Hancock D.K.Brown, On the Influence of State of Stress on Ductile Failure Initiation in High Strength Steels, Engineering Fracture Mechanics, 1977,Vol.9:167-188.

[8] O.S.Hopperstad T.Borvik et al. On the influence of stress triaxiality and strain rate on the behaviour of a structural steel.Part Ⅰ.Experiments. European Journal of Mechanics A/Solids, 2003,22:1-13

[9] T.Borvik O.S.Hopperstad T.Berstad. On the influence of stress triaxiality and strain rate on the behaviour of a structural steel.Part Ⅱ. Numerical stuy. European Journal of Mechanics A/Solids, 2003,22:15-32

[10] 2007 ASME Boiler & Pressure Vessel Code,Ⅷ-Division 2,Alternative Rules, Rules for Construction of Pressure Vessels.

Journal of Physics: Conference Series **181** (2009) 012072 doi:10.1088/1742-6596/181/1/012072

Finite element prediction of the ultimate axial load capacity of V-section band clamps

S M Barrans[1] and M Muller[1]

[1] University of Huddersfield, School of Computing and Engineering, Queensgate, Huddersfield, HD1 3DH, UK

E-Mail: s.m.barrans@hud.ac.uk

Abstract. Band clamps with a flat bottomed V-section are used to connect a pair of circular flanges to provide a joint with significant axial strength. Despite the wide application of V-band clamps, their behaviour is not fully understood and the ultimate axial strength is currently only available from physical testing. This physical testing has indicated that the ultimate strength is determined by two different types of structural deformation, an elastic deformation mode and a plastic deformation mode. Initial finite element analysis work has demonstrated that analysis of this class of problem is not straightforward. This paper discusses the difficulties encountered when simulating this type of component interaction where contact is highly localised and contact pressures are high.

1. Introduction

V-section band clamps are used in a wide range of applications in the aerospace and automotive industries. As stated by Shoghi [1] they were invented during the Second World War by the Marmon Corporation and are nowadays used to connect together the housings of diesel engine turbochargers and also to assemble satellites to their launching device [2]. The process of assembling V-band clamps to a pair of circular flanges has recently been investigated using classical analysis [3], finite element analysis [4] and experimental validations [5]. Figure 1 shows a V-section band clamp assembled to a pair of flanges in an experimental test rig.

Figure 1: V-section band clamp assembled to a pair of circular flanges for initial experimental testing

Despite their wide application, once assembled to a pair of flanges little is known about the interaction between flange and band. Moreover the failure mode of V-band clamps when undergoing an axial load is not fully understood. Work has therefore been initiated to generate a finite element model able to predict the ultimate axial load capacity and deformation of V-band Clamps.

2. Initial Finite Element Analyses

To get a first understanding of the structural behaviour of V-band clamps initial finite element analyses were generated using the package ABAQUS. The bands were approximated as being axisymmetric and the plane of symmetry between the two flanges was used to reduce the model size. Whilst Shoghi [1] has demonstrated that the contact load between the bands and flanges is not uniform around the band, the load variation was judged to be small for this first approximation. A sliding contact interaction between the flange and the band clamp was defined and to simulate tightening of the band, it was shrunk on to the flanges. The band clamp was meshed with 2D axisymmetric linear plain strain elements with reduced integration as suggested by Dassault Systems [6] because as observed by Konter [7], second order elements do not distribute the contact pressure uniformly between the mid-side and corner nodes. No meshing was needed for the flange because it was set up as an analytical rigid body. For these initial analyses an implicit solver and a penalty contact algorithm were used with the knowledge that penalty contacts tend to converge more easily and to be more stable as stated by Konter [8].

The material for the band clamp was AISI 304 stainless steel, with a Young's Modulus of 227 GPa, and a Poisson's Ratio of 0.29. The material was modelled as being elastic-plastic with the hardening approximated as linear as explained by Dixit and Dixit [9]. Material data was taken from experimental tests carried out by Shoghi [1]. Using Ludwik's theorem as presented by Meyer [10] these were transformed from engineering to true stress and strain values.

These finite element analyses were set up for a range of different flange diameters, D_f, from 55 mm to 2000 mm with a flange contact edge radius R of 0.1 mm. In practice flange contact radii vary from 0.1 to 0.7 mm. Dassault Systems [11] state that the contact algorithm in ABAQUS requires that the master surface should not contain sharp edges. In the first step of each analysis a boundary condition (BC) was applied to the band at its symmetrical edge to prevent it from moving in the axial direction as shown in Figure 2.

The flange was restrained at its reference point (RP) to prevent it from moving in the radial and axial directions and to prevent it from rotating. Initially there was a gap of about 0.1 mm between the band and the flange. The band was then shrunk onto the flange using an applied reduction in temperature to generate a thermal contraction. This was done to simulate the assembly procedure of tightening the T-bolt nut, which generates an axial load. Using several steps the flange was then moved in the axial direction to simulate a failure of the V-band clamp joint. For these contact interactions the V-band clamp was set as the deformed slave-surface and the flange was set to be the deforming rigid, master-surface. Throughout this paper only the flange diameter D_f rather than the band clamp diameter D_b is used because D_b highly depends on the T-bolt load. After performing these analyses the maximum value of the reaction force in the axial direction at the flange's reference point (RP) could be recorded. This represents the ultimate axial load capacity of the joint. These results are summarized in Figure 3 which indicates that the highest load capacity exists with a flange diameter, D_f of 250 mm.

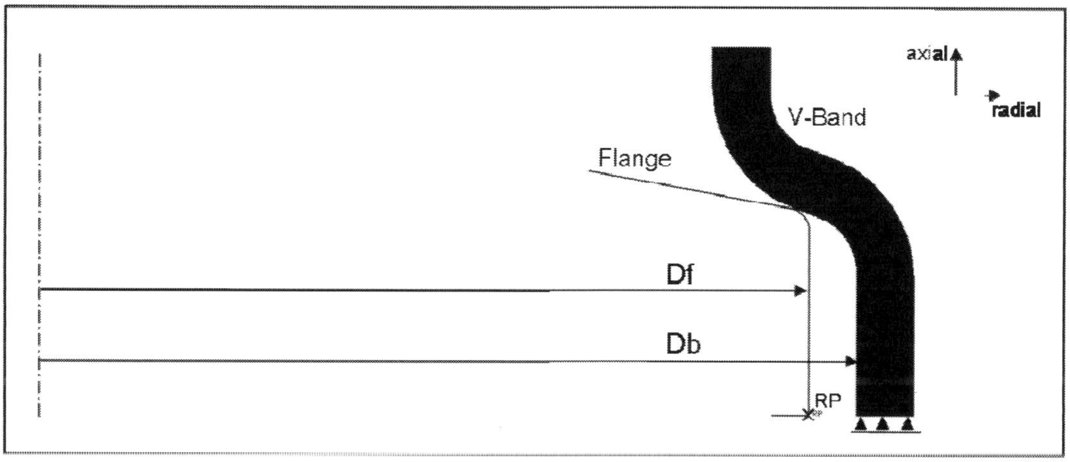

Figure 2: Axisymmetric model arrangement.

It is worth mentioning at this point that these initial analyses were carried out with no friction, a single edge contact radius, R, and a relatively low mesh density using a free, unstructured mesh.

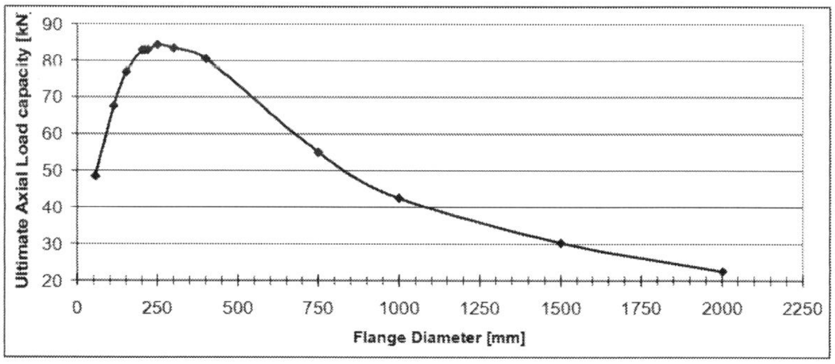

Figure 3: Predicted Axial Load Capacity of V-Section Band Clamps over varying Flange Diameters for Initial FE-Analyses.

3. Implicit Analysis

Using a fixed flange diameter D_f = 250 mm, analyses were carried out for coefficients of friction of 0 and 0.3, for contact edge radii R of 0.3 and 0.5 mm, and for several element thicknesses along the sliding contact surface. Although a contact edge radius of 0.1 mm was used in the initial analyses, these analyses failed with a coefficient of friction of 0.3 and a change in mesh density applied.

3.1. Axial Load Capacity

The analyses generated appeared to be highly dependent on the mesh type and mesh element thickness especially along the contact sliding surface. Figure 4 shows the results for the ultimate axial load capacity for a model with R = 0.5 mm and a coefficient of friction μ of 0.3. As can be seen in this figure the graph for the structured mesh seems to converge to a certain load value for decreasing element size. The graph for the free mesh oscillates with decreasing element size.

Journal of Physics: Conference Series **181** (2009) 012072 doi:10.1088/1742-6596/181/1/012072

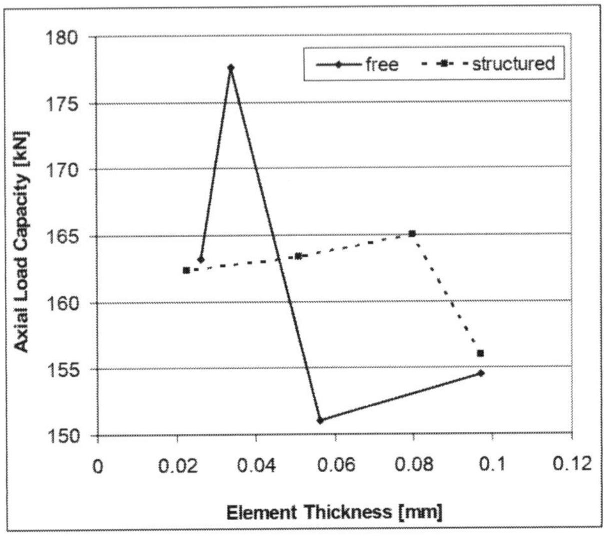

Figure 4: Axial Load Capacity depending on Element Thickness for free and structured mesh, μ = 0.3 and R = 0.5 mm.

Figure 5 indicates the problems that occurred during this analysis work and therefore needed further investigation. Figure 5a shows the axial reaction force at the flange reference point. The very high peak at 255 kN is clearly erroneous. The relative positions of flange and band at this peak are shown in Figure 5b. It can be seen that once the flange starts to slide the elements along the band clamp surface become heavily deformed, especially as the flange slides over the surface. In further analyses these erroneous peaks were noted but were not reported as ultimate axial load capacity. Ignoring this peak, the ultimate axial load capacity of this model is 162 kN.

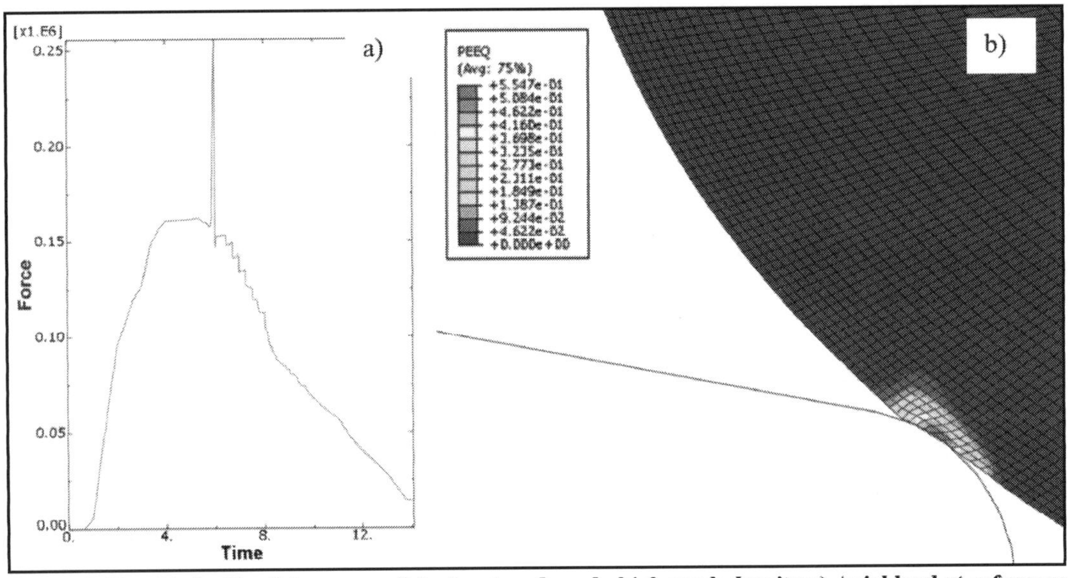

Figure 5: Results for R = 0.5 mm, μ = 0.3, structured mesh, high mesh density a) Axial load at reference point during failure b) position of flange at peak axial load.

Figure 6 shows the results for the same model but with a free, unstructured mesh with nearly the same element size. Again, there is an erroneous peak at the same position in the failure process but it is just at 175kN (Figure 6a), whereas the more reliable axial load capacity seems to be at 163kN almost the same as for the structured mesh. Figure 6b also indicates the same phenomenon of elements starting to deform as the flange starts to slide.

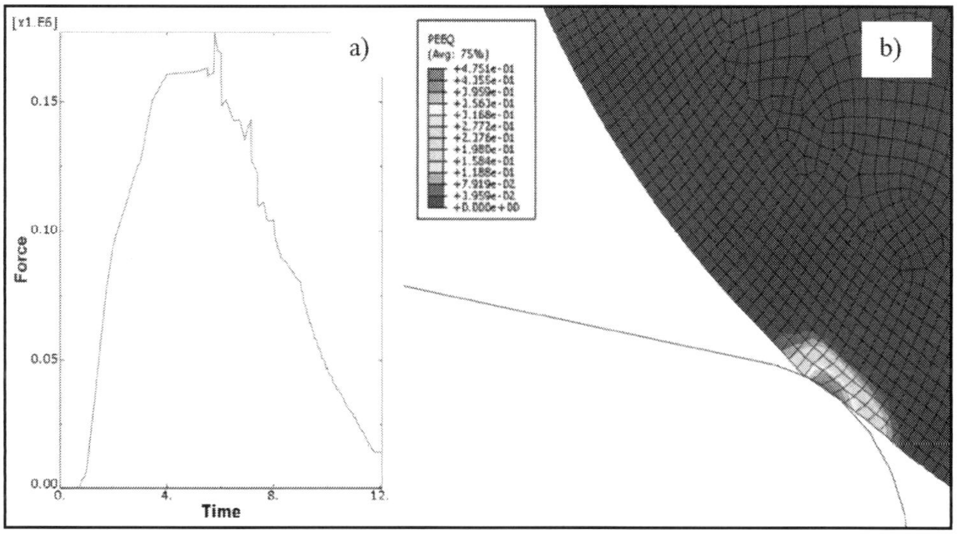

Figure 6: Results for R = 0.5 mm, μ = 0.3, free mesh, high mesh density a) graph of axial load during failure, b) detailed view of position of flange at axial load.

In both cases mentioned above the peak axial load capacity occurs before the flange starts to slide. This shows that as soon as plastic deformation starts in the V-band material, as shown in Figure 5b the ultimate axial load capacity has been reached. This phenomenon has also been seen for models with R = 0.5 mm and μ = 0.

3.2. Plastic Strain

In this section the effect of element thickness along the contact sliding surface of the V-section band clamp on the development of plastic strain in the band clamp will be analysed, for a model with R = 0.3 mm and μ = 0.3. Unlike the models for R = 0.5 mm these models generated many more problems. Figure 7 identifies the position on the contacting surface of the maximum plastic strain ε_{ps}. These values are shown in Figure 8. The two graphs show that the maximum plastic strain increases as the element size decreases. This is due to the one integration point of the reduced integrated linear elements getting closer to the contacting surface, giving much more accurate results. The two meshing methodologies can be seen to give very similar results.

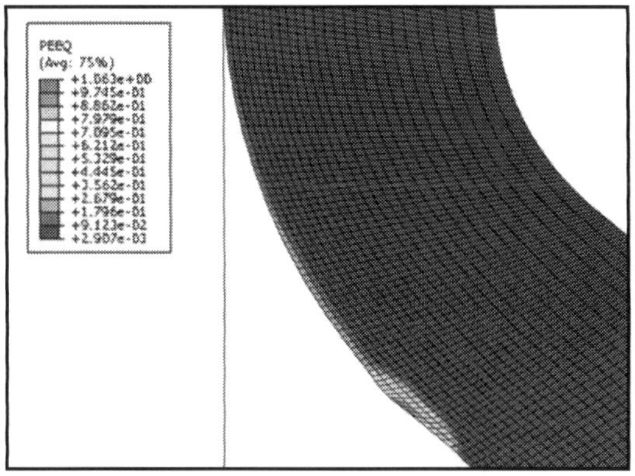

Figure 7: Plastic strain development at contact surface.

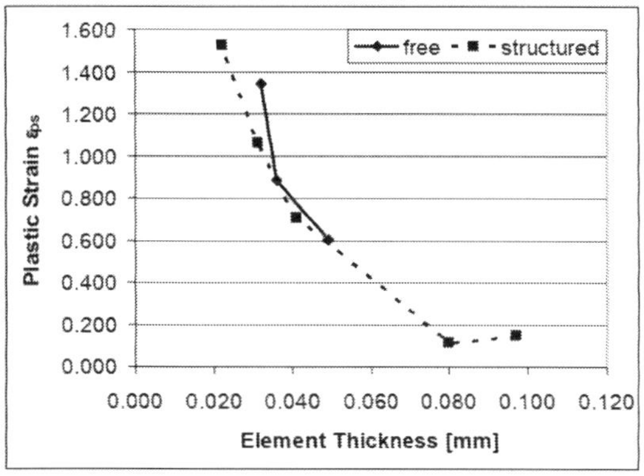

Figure 8: Maximum plastic strain along contact sliding surface for free and structured mesh for R = 0.3 mm and μ = 0.3.

The peak plastic strain described above is only of academic interest as it is not responsible for ultimate failure of the band. Figure 9 identifies the point on the band where large plastic strains ε_{pr} will lead to substantial band deformations and hence joint failure. This point stays the same for all models used in this analysis work and as shown in industrial applications of band clamps [12], is very likely to be in the region where a crack through the whole thickness of the band clamps starts to develop. Although the two graphs in Figure 10 seem to differ slightly, they still indicate the same tendency of plastic strain development. Again the plastic strain increases as the element size in this region has been reduced. The results for the structured mesh seems to be slightly more reliable because the element thickness along the whole inside of the band clamp was reduced whereas for the free mesh only the element size along the contact surface was decreased.

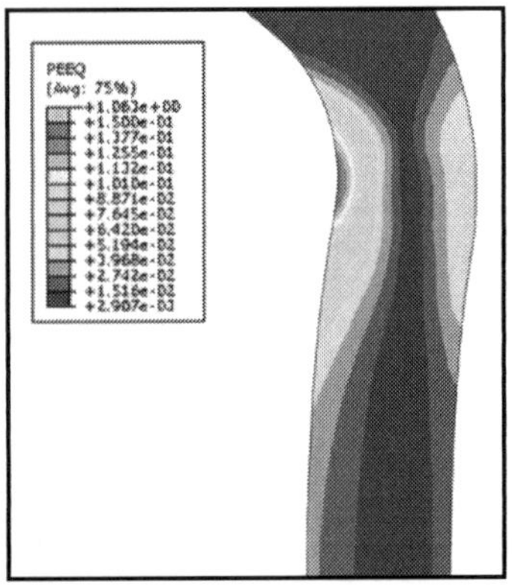

Figure 9: Position of maximum plastic strain sampling point for Figure 10

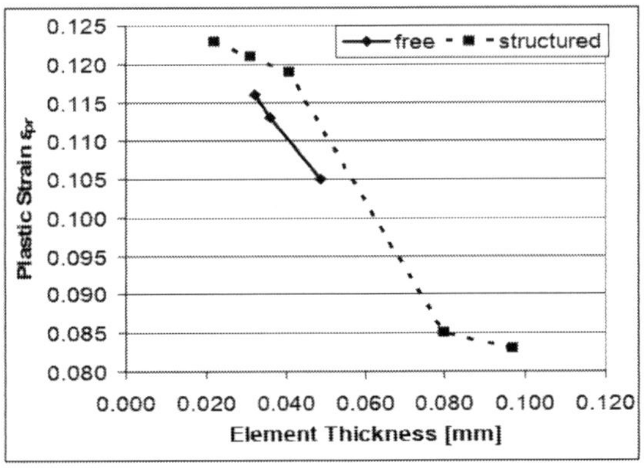

Figure 10 Maximum plastic strain taken from end point of radius for R = 0.3 mm and μ = 0.3.

4. Comparison of Explicit and Implicit Analysis

In this section the analyses for contact edge radii R = 0.1, 0.3 and 0.5 mm along with coefficients of friction μ = 0, 0.15 and 0.3, with a flange diameter D_f = 111 mm will be discussed and compared. The analyses were each solved using explicit and implicit solvers. The time periods for applying load were taken from an initial frequency analysis of the band where the first natural frequency f_n was predicted to be about 6850 cyc/time. The corresponding time period was t_n = 0.000146 sec, which in this paper is referred to as normal time. Time periods ten and fifty times slower (t_{10} = 0.00146 sec and t_{50} = 0.0073 sec) were also used.

4.1. Explicit Analysis Mesh Structure

Figures 11a and b show models that have reached the full failure state having the two extreme cases of very small radius R = 0.1mm combined with a very high coefficient of friction μ = 0.3, and a model with a relatively large radius R = 0.5mm and μ = 0.3, using an explicit solver. As can be seen in a) the small radius and high friction create a large distortion of the elements along the contact surface of the band clamp generating large plastic strain ε_{ps} = 180 whereas in b) the contact surface seems to be deformed as expected with a maximum plastic strain of ε_{ps} = 0.934, which seems very much more realistic. This demonstrates that the reliability and accuracy of the explicit analyses is highly dependent on the contact edge radius. During this analysis work it has been experienced that a higher edge radius R results in a more stable solution giving more realistic results.

As can be seen in Figure 12a and b also the coefficient of friction has a significant influence to the accuracy of the results. Figure 12b indicates that for the same contact edge radius R = 0.3 the higher coefficient of friction μ = 0.3 results in significantly deforming the elements along the contact surface and giving too large plastic strain results.

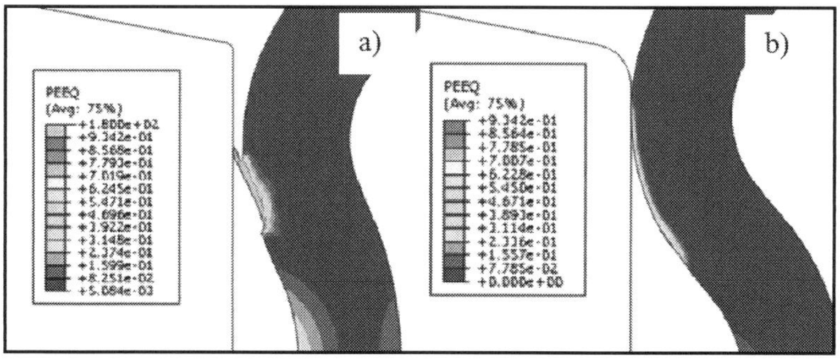

Figure 11: Full failure state using explicit solver and free mesh, 10 times slower a) R=0.1 mm $\mu = 0.3$ and b) R = 0.5 mm $\mu = 0.3$

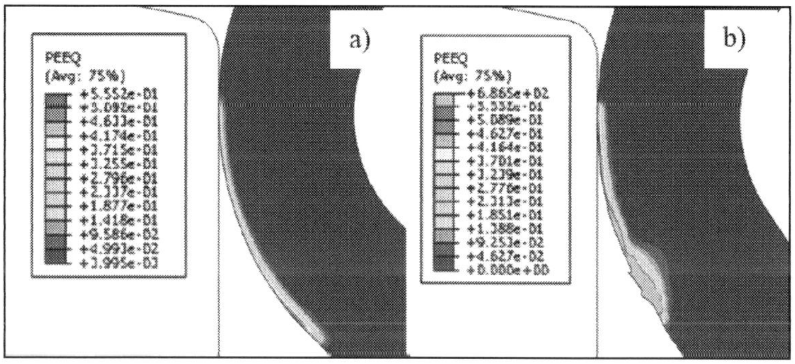

Figure 12: Model of V-section band clamp reached full failure mode using explicit solver for free mesh, 50 times slower a) R = 0.3 mm $\mu = 0$ and b) R = 0.3 mm $\mu = 0.3$

Several explicit analyses have shown that the type of mesh had only a slight effect on the accuracy of the results. Comparing the maximum plastic strain values along the contact surface of two explicit models, the strain values are nearly the same for both. This proofs that the mesh type does have almost no effect for a time t_{10} on the accuracy of the results generated.

As obtained in this subsection the results for contact in an explicit analysis highly depend on the radius of the rigid master surface (flange), the coefficient of friction, whereas the mesh type has no significant influence.

4.2. CPU (Run) Time

All analyses using an implicit solver and a free mesh were always running faster than their explicit counterpart which was set to run 10 times slower. Moreover the run time for the explicit solver increases as the mesh was changed from free to structured. This increase was due to an increase of elements as for the free mesh only the elements along the sliding surface were kept small and getting bigger the more they were away from this surface and for the structured mesh all elements in the model were kept the same size as at the contacting surface.

4.3. Reliability of Results

In Table 1 all explicit and implicit results for the models with a flange diameter $D_f = 111$ mm are compared, for contact edge radius R = 0.5 mm, all possible coefficients of friction and 10 times slower for the explicit.

As stated in the subsection before the CPU time increases a lot as the mesh density gets finer but comparing the results for ε_{pr} to the implicit results there is no significant difference. The results obtained here along with the results from both subsections before support the knowledge of the accuracy and reliability increasing as the coefficient of friction decreases and the contact edge radius R increases. The mesh type for explicit analyses has only a slight influence on accuracy but significantly increases the CPU time.

Table 1: Comparison of implicit and explicit solver for flange diameter D_f = 111 mm.

Contact Edge Radius R [mm]	Coefficient of Friction μ	Max. Plastic Strain at Corner End ε_{pr}	Mesh Type	Load Time (only for Explicit)	Solver Type	CPU (Run) Time [min]
0.5	0.3	0.1381	free	---	Implicit	9.0
0.5	0.3	0.1512	free	10	Explicit	19.0
0.5	0.3	0.1519	structured	10	Explicit	32.0
0.5	0.15	0.1021	free	---	Implicit	9.5
0.5	0.15	0.1182	free	10	Explicit	14.0
0.5	0.15	0.1183	structured	10	Explicit	31.0
0.5	0	0.0773	free	---	Implicit	9.2
0.5	0	0.0804	free	10	Explicit	14.0
0.5	0	0.0812	structured	10	Explicit	30.0

5. Conclusion

In this paper a model capable of predicting the ultimate axial load capacity and structural deformation of V-band Clamps using finite element analysis has been generated. It has been shown that the accuracy of analyses using implicit and explicit solver were highly depending on the contact edge radius R. Moreover the coefficient of friction μ had a significant influence on the stability of implicit analyses. Using explicit or implicit solver for contact analysis the contacting radius should be smooth enough. The accuracy of the results also depended on the element size and amount of elements along the sliding contact surface. Moreover the results were found to be depending on the type of mesh.

6. Further Work

The work presented in this paper has provided an understanding of how a model should be set up to investigate the ultimate axial load capacity of V-section band clamps. In further work the model should be improved to predict the interplay of elastic and plastic deformation mode, taking into account the influence of several V-band diameters.

References

[1] Shoghi K 2003 Stress and Strain Analysis of flat and V-section band clamps *PhD Thesis* Huddersfield
[2] NASA 2000 *Marman clamp system design guidelines*. Guideline GD-ED-2214, Goddard Space Flight Centre
[3] Shoghi K, Barrans S M, Rao H V 2004 Stress in V-section band clamps *Proceedings of the Institution of Mechanical Engineers* **218** 251-261
[4] Shoghi K, Barrans S M, Rao, H. V. 2003 Classical and finite element analysis of V-band retainers *NAFEMS World Congress* (Orlando Florida)
[5] Shoghi K, Barrans S M, Ramasamy P 2006 Axial load capacity of V-section band clamp joints *8th International Conference on Turbochargers and Turbocharging* (London) pp273-285
[6] Dassault Systemes 2007 Getting Started with Abaqus, v6.7 Section 12.4
[7] Konter A 2000 *How to Undertake a Contact and Friction Analysis* (Glasgow: NAFEMS) p 12
[8] Konter A 2005 *Advanced Finite Element Contact Benchmarks* (Glasgow: NAFEMS) p 19
[9] Dixit P M and Dixit U S 2008 *Modeling of Metal Forming and Machining Processes by Finite Element and Soft Computing Methods* (London: Springer) p 101
[10] Meyers M A and Chawla K K 1999 *Mechanical Behavior of Materials* (London: Prentice-Hall International UK Ltd) p 116
[11] Getting Started with Abaqus 2007 v6.7 Section 12.4 Dassault Systemes
[12] Brown I 10-12-2008 Teconnex Ltd. Keighley UK personal conversation

Journal of Physics: Conference Series **181** (2009) 012073 doi:10.1088/1742-6596/181/1/012073

Magnetic Shape Memory (MSM) Actuators in Practical Use

Katarzyna Majewska[1], Arkadiusz Zak[1], Wieslaw Ostachowicz[1,2]

[1]Institute of Fluid-Flow Machinery, Polish Academy of Sciences, 14 Fiszera St., Gdansk, Poland

[2]Gdynia Maritime University, Faculty of Navigation, 3 Jana Pawla II Ave., Gdynia, Poland

E-mail: k.majewska@imp.gda.pl

Abstract. Limited practical applications of MSM actuators motivates the authors to start experimental and theoretical research in the field of multifunctional materials. The authors present a concept of using MSM actuators for control, altering and tuning of forced vibrations of a rotor. The main goal of their experimental research is to show how the activation of MSM actuators can influence forced vibration responses of a rotor system in terms of altering and tuning selected rotor resonant frequencies and vibration amplitudes. Experimental results show that MSM actuators can be successfully applied for vibration reduction and vibration control in the case of rotor systems.

1. Introduction

In recent years a considerable increase has been observed in scientific interest in the properties, development and applications of new structural materials of controllable mechanical properties. Because of the exceptional and controllable properties such materials like: piezoelectrics, electro and magnetorheological fluids, and most of all shape memory alloys (SMAs) are of the highest interest. Within the shape memory alloys (SMAs) two special types of alloys can be distinguished:

- Shape Memory Alloys (SMAs) are the alloys that 'remember' their shape and after deformation can return to the low temperature shape by heating and subsequent cooling.
- Magnetic Shape Memory Alloys (MSMAs) are the ferromagnetic alloy exhibiting large changes in their shape and size under applied magnetic field.

Exceptional physical properties of the shape memory alloys (SMAs) allow them to be integrated within other materials in order to achieve properties that cannot be produced by any conventional technological processes. In such a way new multifunctional materials of fully controllable properties can be made, which next integrated with elements of structures enable one to control their static or dynamic characteristics. By the use of the shape memory alloys (SMAs), as such multifunctional materials, static and dynamic characteristics like: static deflection and shape, critical loading and stability, natural frequencies and modes of vibrations, resonant frequencies and amplitudes of forced vibrations, or structural damping can be controlled, altered or tuned. Investigations on rotor systems equipped with smart bearings utilising shape memory alloys (SMAs) were reported in the literature [1-3], however the use of magnetic shape memory alloys (MSMAs) for that purpose has not been investigated.

2. Magnetic shape memory effect

For the first time the magnetic shape memory effect (MSME) was observed and reported by Ullakko in 1996 [4]. MSMAs as materials that undergo large and completely reversible non-linear deformation influenced by an external magnetic field have been known for a relatively short time. However, since that time many theoretical models have been developed and used for modelling their quasi-static behaviour, magnetic-field-induced strains, all related with the MSME [5, 6].

The MSME can be considered as a significant and completely reversible non-linear deformation in the martensitic phase driven by an external magnetic field. Full understanding of the MSME behaviour requires investigation at micro and macro scale levels. Both scales should be examined because the macroscopic deformation observed in this kind of alloys is strictly connected with certain changes at the micro scale level.

Until now the most known MSMA is the alloy of NiMnGa (nickel, manganese and gallium). In MSMAs the crystallographic structure of the martensitic phase is tetragonal (unit cell length a and c) and the austenitic phase is regular (unit cell length a_0) – see Fig. 1. In the martensitic phase three martensitic variants can be marked out. Each of them is magnetised along a preferred crystallographic direction named the magnetic easy axis. This direction is aligned with the short edge c of the tetragonal unit cell. Magnetisation can be oriented in either positive or negative direction of the easy axis and aligned with the direction of the external magnetic field vector.

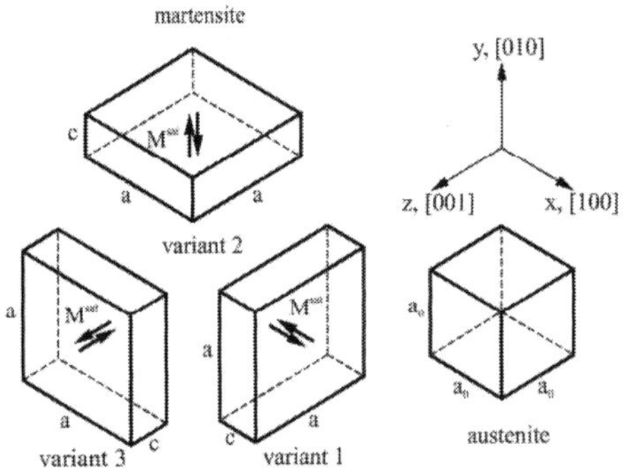

Figure 1. Crystallographic structure of NiMnGa alloy variants [6]

Let us consider a sample made out of a MSMA that initially in the austenitic phase and is cooled down under constant and compressive stress σ_{xx} such that $\sigma_{sv} < \sigma_{xx} < \sigma_b$. The symbol σ_{sv} is the minimum stress value necessary to obtain a single variant martensitic configuration and σ_b is the blocking stress above which the MSME is not observed. As a result of the acting compressive stress σ_{xx} and the martensitic transformation a single martensitic variant is obtained in the alloy – *variant 1* – and the sample is slightly shortened. At the micro scale, due to cooling, several magnetic domains can be distinguished in the single variant state as it can be seen in Fig. 2a. The direction of the magnetic polarisation M is connected with the direction of the acting stress σ_{xx}. Subsequent application of an external magnetic field H_y orthogonal to the applied stress σ_{xx} starts the nucleation and growth of *variant 2* by the motion of magnetic domain walls, reorientation, and migration of twins. The growth of *variant 2* occurs at the cost of *variant 1* and causes strain reorientation and elongation. This process is named as the magnetic field-induced martensitic variant reorientation in MSMAs.

2.1. One-dimensional model of the magnetic shape memory effect

Experimental observations reported in the literature allows one to make the following simplifications in the phenomenological modelling of MSMAs:

- value of magnetic anisotropy can be assumed as infinite [6],
- strains connected to magnetostriction can be neglected,
- MSME depends only on the process of martensitic variant reorientation.

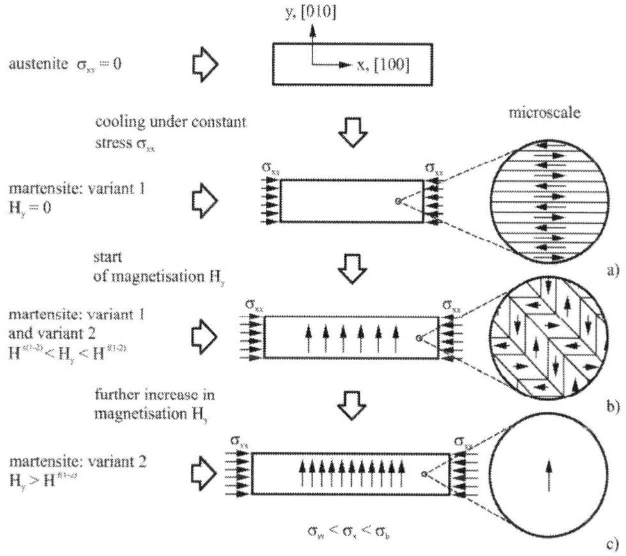

Figure 2. Schematic diagram of the magnetic shape memory effect in a NiMnGa alloy [6]

The total macroscopic strain ε related to the MSME can be expressed as a sum of the thermoelastic strain ε^{te}, the reorientation strain ε^{r}, and the transformation strain ε^{tr}:

$$\varepsilon = \varepsilon^{te} + \varepsilon^{r} + \varepsilon^{tr} \tag{1}$$

The thermoelastic strain ε^{te} is a sum of the thermal strain ε^{t} and the elastic strain ε^{e}:

$$\dot{\varepsilon}^{te} = \dot{\varepsilon}^{e} + \dot{\varepsilon}^{t} = \dot{\varepsilon}^{e} - \frac{\Theta(\xi)}{E(\xi)}\dot{T} = \dot{\varepsilon}^{e} + \alpha(\xi)\dot{T} \tag{2}$$

where T is temperature, Θ is a thermal coefficient, E is Young's modulus and α is a linear expansion coefficient. In MSMAs the reorientation strain ε^{r} associated with the MSME is proposed as proportional to the rate of changes in the martensitic variant volume fraction [6, 7]:

$$\dot{\varepsilon}^{r} = -\frac{\Lambda(\xi)}{E(\xi)}\dot{\xi}_s^{(2)} = \varepsilon_r\,\dot{\xi}_s^{(2)} \tag{3}$$

where Λ is a reorientation coefficient, ε_r is the maximum elongation connected with martensitic variants reorientation and $\xi_s^{(2)}$ is the stress-induced *variant 2* martensitic phase. Similar to thermal SMAs the transformation strain ε^{tr} is proposed as proportional to the rate of changes in the strain-induced martensitic phase ξ_s [7]:

$$\dot{\varepsilon}^{tr} = -\frac{\Omega(\xi)}{E(\xi)}\dot{\xi}_s = \varepsilon_l\,\dot{\xi}_s \tag{4}$$

where Ω is a transformation coefficient and ε_l is the maximum elongation connected with martensitic transformation driven by temperature. In the phenomenological model proposed here the constitutive equation describing changes in the stress σ is postulated as analogous to thermal SMAs [7-9]:

$$(\sigma-\sigma_0)=E(\varepsilon-\varepsilon_0)+\Theta(T-T_0)+\Omega(\xi_s-\xi_{s0})+\Lambda\left(\xi_s^{(2)}-\xi_{s0}^{(2)}\right) \tag{5}$$

where σ, ε and ξ_s denote the stress, strain and relative volume fraction of the martensitic phase, E_m is Young's modulus of the martensitic phase and '0' denotes a certain starting level. Assuming that the martensitic reorientation process takes place under constant temperature and that a state of single martensitic variant orientation has been established the thermal effects, as well as the effects due to the martensitic transformation, can be neglected. Because of that Eq. (5) can be rewritten in a simpler form:

$$(\sigma-\sigma_0)=E_m(\varepsilon-\varepsilon_0)+\Lambda\left(\xi_s^{(2)}-\xi_{s0}^{(2)}\right) \tag{6}$$

This form of equation can be used a base form of the constitutive relation for changes in the stress σ connected with the MSME.

3. Magnetic Shape Memory (MSM) Actuators in Practical Use

3.1. Concept of a special rotor rig with a MSM actuator
A Bently Nevada Rotor Kit 4RK is a high quality machine unit enabling one observation and investigation of transverse vibrations of rotating machinery. The main parts of the Bently Nevada Rotor Kit 4RK are: base on which the driving motor (1) is mounted, helical coupling (3), 560 mm long and 10mm of the diameter shaft (4) with standard (5) and 'smart' (6) bearings – for more details refer to Fig. 3 and Fig. 4.a.

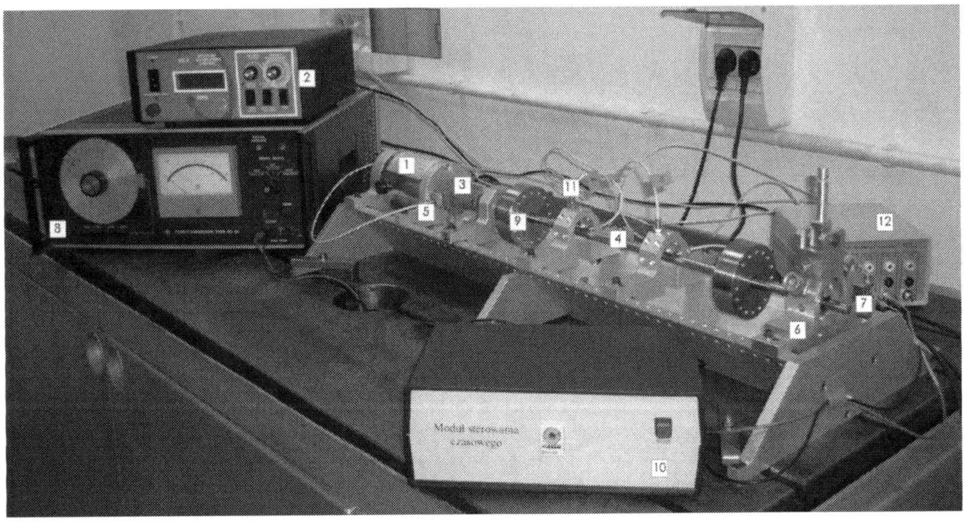

Figure 3. A special rotor rig with a smart bearing support

The motor run is controlled by a driving unit (2) that allows to achieve the rotation speed of 10000 rpm for at a given speed ratio. The shaft is attached to the motor by the helical coupling that compensates all misalignments resulting from the rotor assembly. Additionally it is possible to attach two mass wheels (9) – 75mm external diameter, 10mm internal diameter, 25mm width, 800g weight. The position and mass unbalance, in the range from 0.1 g to 2.0 g, can also be fully controlled and modified. By standard the shaft is supported by two self-lubricating sliding bearings. For the purpose of the experimental research one shaft support with the sliding bearing has been replaced by 'smart' bearing support (6) – Fig. 3 and Fig. 4.a – equipped with a MSM actuator (7) presented in Fig. 4.b. The MSM actuator is driven and controlled by a generator unit (8) and a time module controller (10). The smart bearing support controls the vibrations of the rotor in the horizontal plane, while the vibrations in the vertical plane are uncontrolled. The position of each bearing can be altered and modified.

Figure 4. Smart bearing assembly (a) with the AdaptaMat(MSM) actuator (b)

In order to measure the transverse vibrations of the shaft a set of 2 proximity probes (11) were used. The signals from the probes were observed via oscilloscope or computer thanks to the use of a proximitor assembly unit (12). During operation the rotation speed of the rotor can be changed and this may result in multiple passes through the resonance regions of high vibration amplitudes. In these cases the use of the smart bearing support helps to control the level of the vibration amplitudes or ti reduce them to a safe level.

3.2. Experiment

The custom modified rotor rig described above allowed the authors to carry experimental measurements. The transverse vibrations of the rotating shaft were observed and tuned by the use of the MSM actuator.

The experiment was conducted for the shaft with one additional mass wheel placed at the mid-span of the shaft for the following three different settings: (a) shaft with one mass wheel rotating at the speed of n =1000 rpm, which is equal to frequency f = 16.67 Hz, (b) shaft with one mass wheel rotating at the speed of n = 1750 rpm, which is equal to frequency f = 29.17 Hz, (c) shaft with one mass wheel rotating at the speed of n = 4600 rpm, which is equal to frequency f = 76.67 Hz.

All measurements were performed in the following manner. By the use of the motor driving unit the required shaft rotating speed was established. Next the working frequency of the MSM actuator work was selected by the use of the generator and the time module controller. The frequency of the MSM actuator could vary from 10 Hz to 350 Hz with a step of 10 Hz. The activation time was chosen as t = 8 s. Changes in rotor vibration characteristics due to activation of the MSM actuator were observed on the oscilloscope.

Figure 5 shows the obtained amplitude-frequency characteristics of the shaft with one mass wheel. It indicates the regions of the resonant frequencies in u and v direction. The dashed line denotes the frequencies for which the experimental measurements were carried out.

In order to show the effectiveness of the MSM activation an indicator parameter has been defined and used. In the case of all results of numerical calculations presented below the indicator I is used defined based on the following formula:

$$I = \sqrt{\frac{A_1}{A_2}}, \quad A_i = \int_0^{f_N} P_i(f)\,df, \quad i = 0,1 \tag{7}$$

where f_N denoting the frequency, P_i is the signal power, 0 and 1 denote the case of inactive and active MSM actuator, respectively.

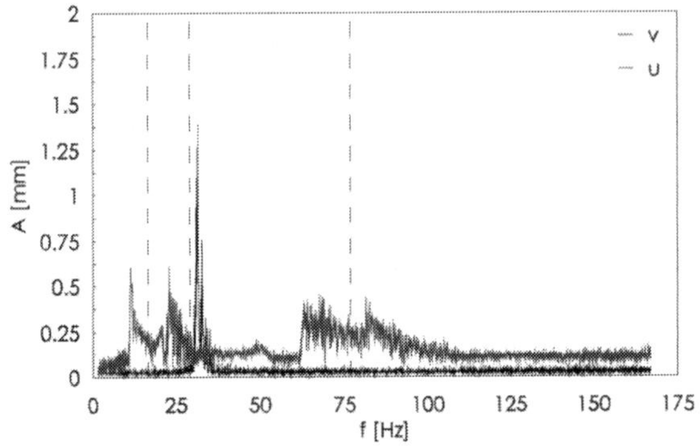

Figure 5. Amplitude-frequency characteristic for the shaft with one mass wheel

Results for the shaft with additional mass wheel rotating at the speed of n = 1000 rpm
Figure 6 shows the effectiveness of the MSM activation as a function of the activation frequency f and changes in vibration reduction.
Minimal vibration reduction was achieved for the MSM actuator activation frequency equal to $f = 30$ Hz and maximal for $f = 120$ Hz.
Results for the shaft with additional mass wheel rotating at the speed n = 1750 rpm
Figure 7 shows the effectiveness of the MSM actuator activation as a function of the activation frequency f and changes in vibration reduction. Minimal vibration reduction was achieved for the MSM actuator activation frequency equal to $f = 40$ Hz, and maximal for $f = 260$ Hz.
Results for the shaft with additional mass wheel rotating at the speed n = 4600 rpm
Figure 8 shows the effectiveness of the MSM actuator activation as a function of the activation frequency f and changes in vibration reduction. Minimal vibration reduction achieved for the MSM actuator activation frequency equal to $f = 10$ Hz, and maximal for $f = 190$ Hz.

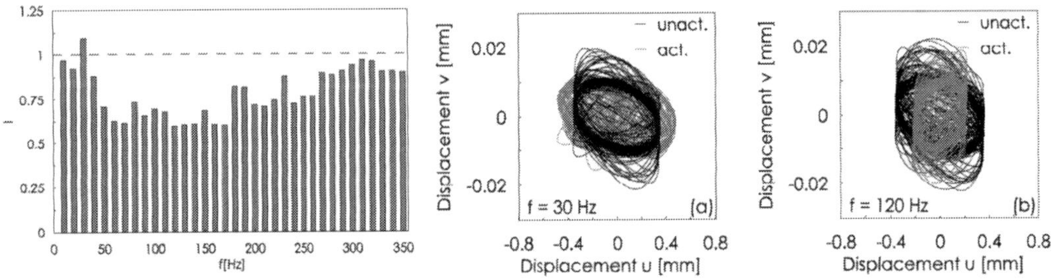

Figure 6. Indicator I changes as a function of the MSM actuator activation frequency f and minimal (a) and maximal (b) vibration reduction

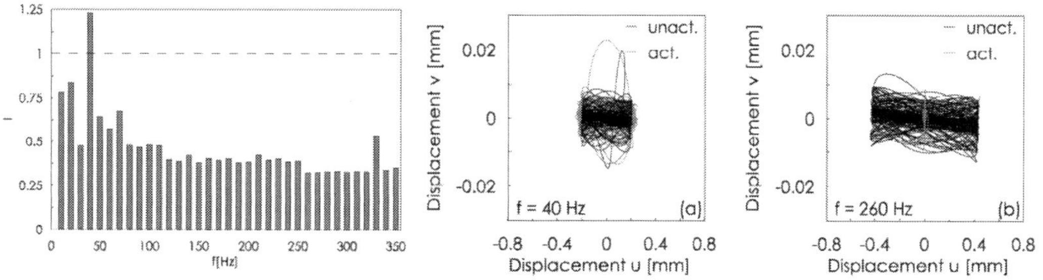

Figure 7. Indicator I changes as a function of the MSM actuator activation frequency f and minimal (a) and maximal (b) vibration reduction

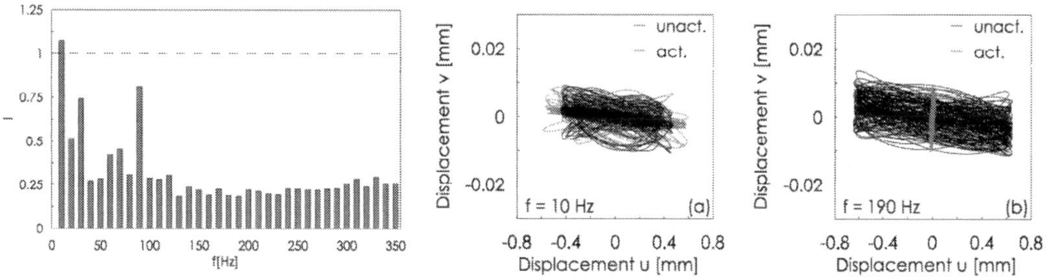

Figure 8. Indicator I changes as a function of the MSM actuator activation frequency f and minimal (a) and maximal (b) vibration reduction

3.3. Comparison of experiment and numerical simulation

Numerical simulations were made for the shaft with one mass wheel and for 3 different rotating speeds. The FEM has been applied for the analysis of the effectiveness of the MSM actuator activation. For shaft modelling 10 beam finite elements were used with 2 nodes and 4 degrees of freedom per each node. The mass wheel was modelled a rigid body with 6 degrees of freedom.

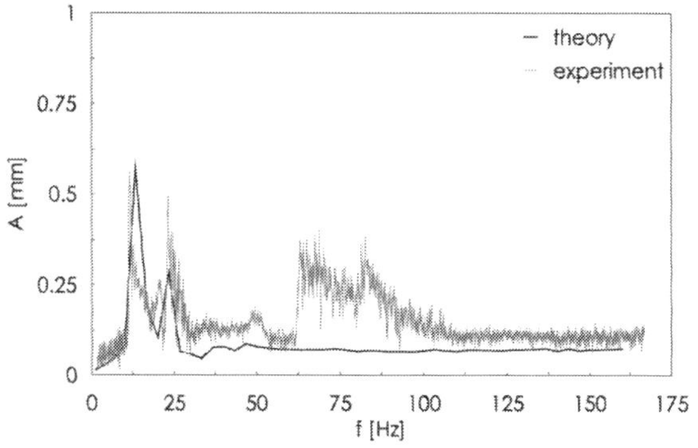

Figure 9. Amplitude-frequency characteristic obtained from experimental measurements and from numerical simulations

First comparison is connected with the amplitude-frequency characteristic. Figure 9 shows the amplitude-frequency characteristic obtained from the experimental measurements (grey) and from the numerical simulations (black). Only the horizontal direction was considered because this direction was the direction of work of the 'smart' bearing.

Second comparison is connected with the effectiveness of the MSM actuator activation. Figure 10 shows the effectiveness of the MSM actuator activation obtained from experimental measurements (grey) as well as for numerical simulations (black). It is shown in terms of the indicator *I* defined by equation (7) in the following way. Starting form left results for the shaft with one mass wheel rotating at the speed of n = 1000 rpm are presented, next for the shaft with one mass wheel rotating at the speed of n = 1750 rpm, and finally for the shaft with one mass wheel rotating at the speed of n = 4600 rpm.

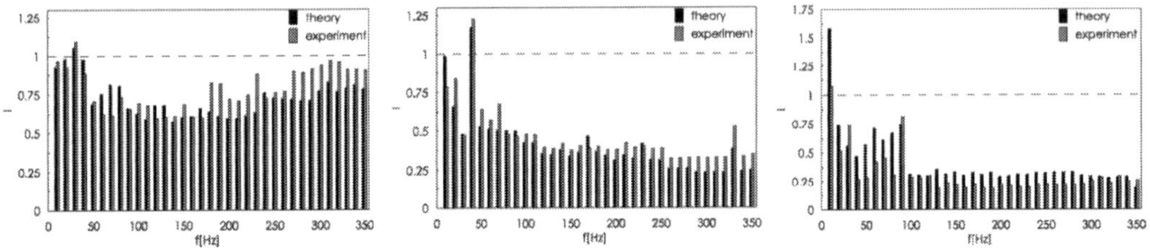

Figure 10. Changes in indicator *I* as a function of the MSM actuator activation frequency *f* obtained from experimental measurements and from numerical simulations

4. Results

The obtained result from experimental measurements and numerical simulations allow the authors to formulate the following general conclusions:

- MSM actuators can be successfully used for reduction of forced vibrations of rotor systems.
- Both experimental and numerical results show that activation of the MSM actuator significantly influences forced rotor vibration responses.
- In certain cases the observed effectiveness of the MSM actuator activation is lower. This is when the MSM actuator activation frequency is equal to a multiple of the natural frequency of the rotor.
- In certain cases the MSM actuator activation cause an increase in the forced vibration amplitudes of the rotor. This happens when the MSM actuator works as an additional source of excitation.

References

[1] Zak A, Cartmell M and Ostachowicz W 2003 *Key Engineering Materials* **245–246** 233–240
[2] Inman DJ, Cartmell MP, Lees AW, Leize Th and Ateport L 2006 *Applied Mechanics and Materials* **5–6** 29–36
[3] Lees AW, Jana S, Inman DJ and Cartmell MP 2007 Proceedings of ISCORMA-4, Calgary, Alberta, Canda
[4] Ullakko K 1996 *International Journal of Materials Engeenerig and Performance* **5** 405–409
[5] Kiang J and Tong L 2005 *Journal of Magnetism and Magnetic Materials* **292** 394412
[6] Kiefer B and Lagoudas D C 2005 *Philosophical Magazine* **Special Issue** 1–34
[7] Brinson L 1993 *Journal of Intelligent Material Systems and Structures* **4** 229–242
[8] Tanaka K 1986 *Res. Mechanica* 18 251–263
[9] Liang C and Rogers C 1990 *Journal of Intelligent Material Systems and Structures* **1** 207–234

Load estimation from photoelastic fringe patterns under combined normal and shear forces

Venketesh N Dubey and Gurtej S Grewal

School of Design, Engineering and Computing, Bournemouth University
Talbot Campus, Fern Barrow, Poole, BH12 5BB, UK

E-mail: vdubey@bmth.ac.uk

Abstract Recently there has been some spurt of interests to use photoelastic materials for sensing applications. This has been successfully applied for designing a number of signal-based sensors, however, there have been limited efforts to design image-based sensors on photoelasticity which can have wider applications in term of actual loading and visualisation. The main difficulty in achieving this is the infinite loading conditions that may generate same image on the material surface. This, however, can be useful for known loading situations as this can provide dynamic and actual conditions of loading in real time. This is particularly useful for separating components of forces in and out of the loading plane. One such application is the separation of normal and shear forces acting on the plantar surface of foot of diabetic patients for predicting ulceration. In our earlier work we have used neural networks to extract normal force information from the fringe patterns using image intensity. This paper considers geometric and various other statistical parameters in addition to the image intensity to extract normal as well as shear force information from the fringe pattern in a controlled experimental environment. The results of neural network output with the above parameters and their combinations are compared and discussed. The aim is to generalise the technique for a range of loading conditions that can be exploited for whole-field load visualisation and sensing applications in biomedical field.

1. Introduction

Photoelasticity has been conventionally used for experimental stress analysis, however, use of this technique is not very well explored in sensing applications, especially by analysing the fringe patterns. The main challenge involved is to extract load information from the fringe patterns, as there may be infinite number of load settings for the same resulting patterns. This may be further compounded due to generalised loading conditions and out of plane deformation of the photoelastic material. The technique, however, looks ever more promising with the advent of high computing power and low straining photoelastic material that can lend to dynamic and wave propagation studies. Advantage of using this technique is to get whole-field visualization of the stress field, which may provide load information of the entire field as opposed to employing strain gauges, or load cells that only offer discrete load information. Photoelasticity has been used in developing a number of discrete signal-based sensors [1-3], however, as the material develops fringe patterns that contain details of the loading conditions this can be exploited for developing a whole-field sensor. If this technique is fully developed for a sensing device, it could find application in many biomedical sensing areas such as early detection of diabetic foot ulceration or assessment of pressure sores in disabled subjects, thereby enacting prevention strategies by suitable footwear or bed designs.

One of the main impediments in assessment of diabetic foot ulceration is the difficulty of resolving the forces acting on the sole of the foot during standing and walking. The shear force acting during walking is considered to be more harmful than the vertical force alone [4] as it occludes the blood flow more severely, thus resulting in tissue dystrophy and rapid ulceration. However, the mechanism of ulceration due to biomechanical forces is not very well understood [5, 6], partly due to unavailability of appropriate sensors that can resolve these forces and partly because the forces cannot be quantified. Some interesting works have been reported on the risk analysis of walking foot pressure in diabetic patients using pedobarograph [7, 8], however, they have looked at foot pressure not the components of forces under foot. This research aims to tackle these problems by analysing normal and shear force data from the image under actual condition of loading. This requires inverse problem of force extraction to be solved i.e. finding the applied force from the generated fringe patterns. In our earlier work [9] we proposed use of neural networks for extraction of vertical load using intensity information from photoelastic fringes. It was shown that in absence of analytical solutions, neural network is a better approach for such inverse problems. In this paper we consider various geometric and statistical parameters of the image to train the neural network for extracting vertical as well as shear forces under controlled experimental conditions. The results from various combinations of these parameters have been compared and discussed.

2. Experimental Setup and Loading Conditions

The setup used for this experiment is shown in Fig. 1. Vertical load was applied by directly placing the weights over the indenter disc set over the photoelastic sensor plate while the shear force was applied by pulling the disc over a pulley using weights connected to a nylon cord as shown in the figure. The optical setup used in this experiment included a digital camera (Olympus, SP500) with high image resolution of 6M, a cold cathode fluorescent tube (CCFL) as a light source and a plane polariscope constructed using linear polariser [10]. The photoelastic material used was PS-4 of 3mm thickness with suitable coatings and adhesives [11, 12].

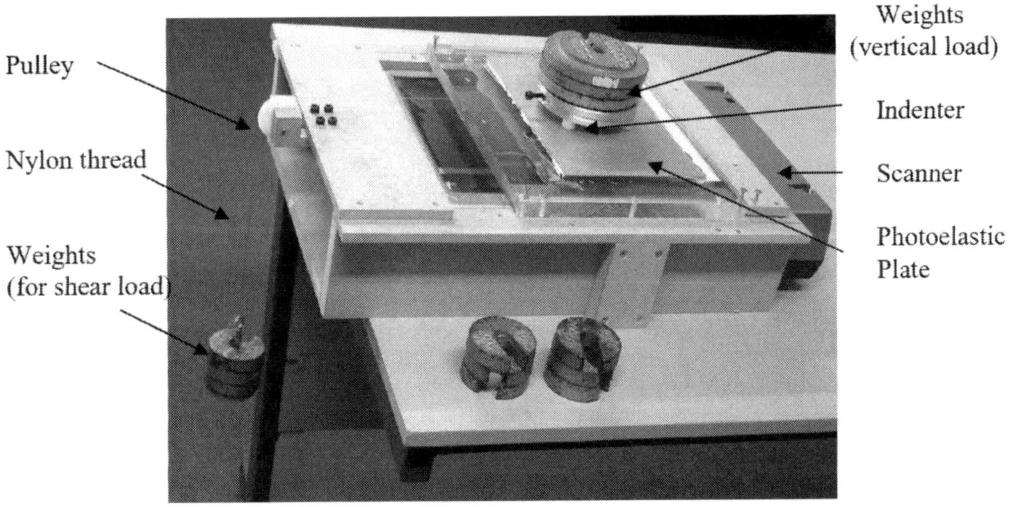

Figure 1: Experimental Setup for vertical and shear loading

Figure 2 shows fringe patterns developed on the surface of the photoelastic plate due to vertical and shear forces applied by the specifically designed indenter. The vertical force deforms the material and fringes propagate radially from point of application of load in a symmetrical way. The fringe patterns appear concentric circles of different colours under white light. When shear force is introduced in conjunction with the vertical force the fringe patterns smear in direction of application of force. This appreciable shift in the fringe pattern can be advantageously exploited for shear force sensing.

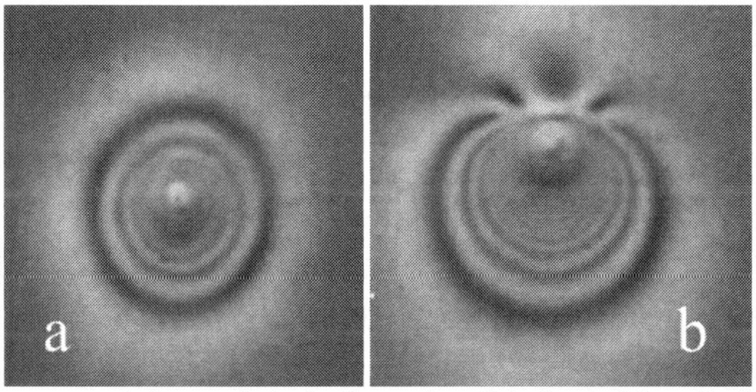

Figure 2: Fringe pattern generated on a photoelastic sensing plate due to (a) vertical force (b) due to combined vertical and shear forces. The effect of shear force results in appreciable shift in the image that can be exploited for developing a shear force sensor.

3. Fringe Characteristics

As stated earlier that it is not straightforward to extract load information from the fringe pattern hence a neural network based approach was used for normal force information [9]. In the current work to extract vertical as well as shear forces images were recorded at various combination of these forces to train the neural network. Table 1 shows a part of the range of vertical and shear forces applied on the material for this experiment. The vertical load was incremented in steps of 100g starting from 500g (excluding the weight of indenter) and the shear load was incremented in steps of 200g. The limiting value of shear at a particular vertical load was experimentally determined by estimating the co-efficient of friction between the two contacting surfaces.

Table 1: Applied Vertical loads with a range of Shear loads

Vertical force (N)	**8.68**	**9.66**	**10.64**	**11.62**	**.......**	**21.43**	**22.42**	**23.4**
Shear force (N)	1.47	1.47	1.47	1.47		1.47	1.47	1.47
	3.43	3.43	3.43	3.43		3.43	3.43	3.43
	5.39	5.39	5.39	5.39		5.39	5.39	5.39
	7.35	7.35	7.35	7.35		7.35	7.35	7.35
			9.32	9.32		9.32	9.32	9.32
						11.28	11.28	11.28
						13.24	13.24	13.24
						15.20	15.20	15.20
						17.16	17.16	17.16
								19.13

The error in loading was minimised by measuring the diameter of the circular fringe by repeating the experiment several times. For ease of implementation this was achieved by converting the images to HSV plane (this ensures that the original colour information does not change with change in fringe gradient or slight variation in light intensity) and segmenting it to measure the diameter of the outermost fringe by the developed algorithm [13]. The measured diameter for 9 sets of experiments was averaged to get as an ideal diameter (the fringe diameters are represented in pixels and for each experiment). The percentage error for all the measured diameters was evaluated against the known diameter. The average percentage error for all the vertical loads for 9 experiments was found to be 2.557%. Thus the calculated error in input data would affect the accuracy of results by a minimum of 2.557% as this much error has been added in the input at the outset inevitably. The various factors that may have contributed to this error include unsymmetrical fringe patterns due to uneven load transfer, unsymmetrical design of weights and disorientation of the indenter while shear force was applied.

3.1 Response to Vertical Force

As discussed earlier the fringe patterns pan out due to increase in the vertical load. Figure 3 shows the influence of increasing vertical load on the diameter of fringe patterns, this was obtained from 9 repeated experiments. Every point plotted in each data line is average measured diameter from 9 experiments. It is evident that the plot of averaged diameter has significant differences. This means that using just one geometric parameter in determination of load from the fringe patterns may not be a reliable method.

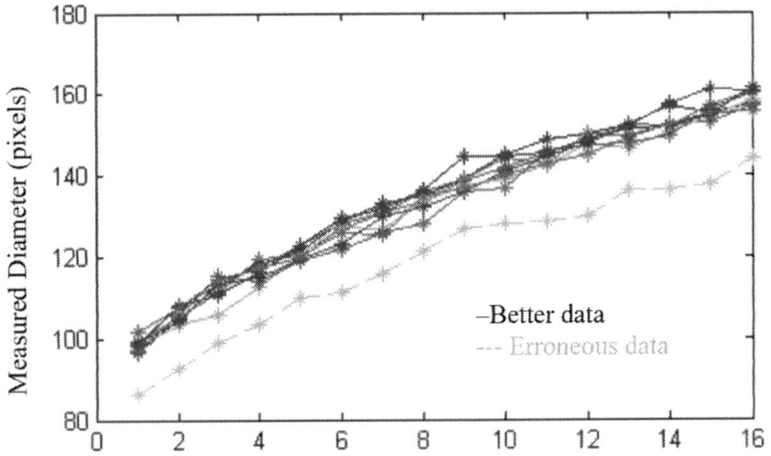

Figure 3: Influence of Vertical load on measured diameter of fringe patterns

4. Data Conditioning and Analysis

The fringe data acquired in the form of images was used for training and testing of the neural networks. Since an entire image cannot be used as input due to high input dimensionality, it was necessary to condition the data and optimise the dimensionality of input being fed to the network [14]. A large number of parameters were extracted from the fringe images that were used as input to the network. A region of interest (ROI) was selected encompassing fringe patterns and the following parameters were extracted (i) Mean pixel intensity (ii) Median pixel value (iii) Standard deviation (iv) Kurtosis (v) Skewness of data (vi) Horizontal radius- horizontal stretch in fringe from point of load (vii) Vertical radius- vertical stretch in fringe from point of load (viii) Pixel area of segmented ROI

(ix) Intensity information from region of interest. These parameters were fed in combination and individually to reach to an optimal input for network training. Principal component analysis (PCA) was used as data reduction technique [15]. Since training data is limited by the number of experiments, it was important to narrow down to an optimal dimensionality through PCA whilst maintaining subtle differences in fringe patterns under consecutive loads.

4.1 Data Analyses using Statistical Parameters

The input training data used for the network was derived from statistical analysis performed on the acquired images. Each image was stacked down to individual planes and intensity data from only one plane was used for statistical analysis. Since there is no considerable difference between fringe patterns in different planes (R, G & B), the plane with highest contrast was considered for analysis. Different statistical parameters, kurtosis, skewness, mean pixel intensity, standard deviation, median were extracted and fed as input to neural networks to map the shear force. Figure 4 shows the optimal results that were achieved using statistical parameters as input to the network with 23 random test images.

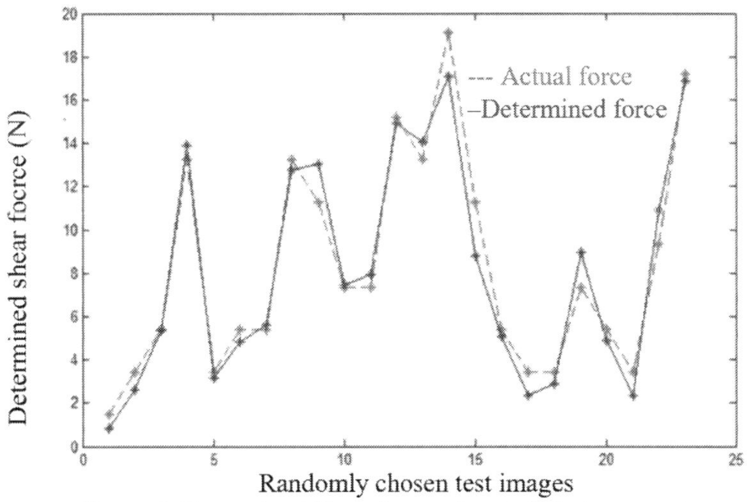

Figure 4: Determined Shear force using statistical parameters

The figure shows the actual shear force compared to the determined shear force at different vertical load values. The average percentage error found for 23 test images was 13.15% and for the entire load range (112 test images) was 13.45%. The results obtained were reasonable to continue using neural networks for force determination but were not accurate enough. As can be seen from the figure, for some test images the error is too high due to insufficient input data and training patterns. Many of the statistical parameters like mean pixel intensity and standard deviation did not show much variation with change in load values.

4.2 Data Analyses using Geometrical Parameters

As shear force smear the image in the direction of force thus measuring geometrical parameters could be a better approach that could lead to meaningful results. The measured parameters included segmented area from hue plane and radius measured in both horizontal and vertical axis since they are controlled by different forces. Horizontal radius being influenced mostly by the vertical load, whereas vertical radius is influenced by the shear load which stretches the fringe patterns in direction of application of load starting from point of loading. The results showed that the percentage error was 8.15 % for 23 test images and 11.4 % for entire range of loading (112 images). The results obtained were better than what were achieved with statistical parameters, since the geometrical parameters were influenced more significantly with the change in load. However, the accuracy desired was not achieved and thus combined statistical and geometric data was tried for training the network.

4.3 Combined Statistical and Geometric Parameters

Since the encouraging results were obtained using statistical and geometric data, it was envisaged that error would reduce if the training data was combined from the two. Thus a network was constructed and trained with combined statistical and geometric data. The results improved dramatically by almost 4% as the total averaged percentage error for the entire range was found to be 8.43% and 7.229% for 23 test images. To move a step further in an attempt to achieve higher accuracy the network was trained with intensity data extracted from load images. However, the results gave an average error of 19% for the entire range and 13.3% for 23 test images thus a lower accuracy compared to the previously fed input of statistical or geometrical values. This obviously is not a better combination of fringe parameters for load determination. The reason why intensity data came out with such a low accuracy was that the input fed as intensity from a small region of interest would not be as relevant and crisp as features extracted from geometrical and statistical parameters. To improve results the entire image was considered for extracting input rather than just a region of interest or line of interest, this however marginally improved the error by 5.99% at the cost of high dimensionality of the network [14].

5. Modified Strategy

The determination of shear force through different approaches like statistical parameters, geometric parameters and intensity data did provide clear indication for implementation of neural networks but the results desired were still not very accurate. A modified approach was adopted in order to improve the accuracy and efficiency of the system. A network was used to determine the vertical force from image data using intensity information and for the determination of shear force the data was fed to a second network as a combination of previously tested parameters and the vertical force was determined by the first network.

5.1 Determination of Vertical Force

In order to analyse the fringe patterns as accurately as possible both vertical and shear forces were determined from the same fringe patterns. This will also be the actual situation when using this technique for sensing applications. A network trained from the intensity data extracted from image was tested for optimal results for vertical force determination. Since the desired numbers of targets were considerably fewer compared to the shear force, the network was trained efficiently and it required less computational time. Figure 5 shows the results obtained for the determination of vertical load from fringe patterns.

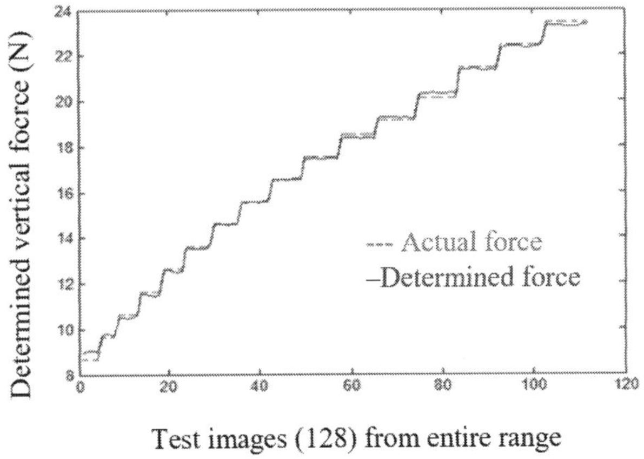

Figure 5: Determination vertical force for 128 test images

The averaged percentage error for 128 test images was found to be 0.67%, thus giving an accurate value of the applied vertical load. The results appeared stepped up in the plot because at each vertical load different shear loads were applied (Table 1). In absence of shear load the graph would be a straight line for each vertical load. The vertical load determined from the network can now be combined with other input parameters (statistical, geometric, intensity) and fed to a new network for determination of shear load.

5.2 Statistical and Geometric Parameters with Vertical Load Input

The network was trained with combined statistical and geometric parameters as input but an additional input parameter, the determined vertical load, was added as another relevant input. Figure 6 shows the results obtained from the trained network tested over 23 test images. The error was reduced drastically from averaged 8% to 5.37% for the entire range of load and 3.86% for the 23 test images as shown in the plot.

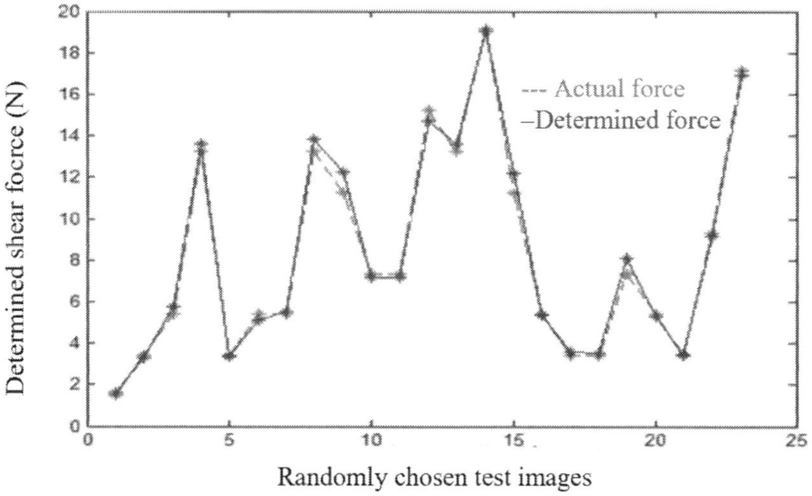

Figure 6: Determined Shear force using statistical, geometric and vertical input

5.3 Intensity Data with Vertical Input

The determined vertical force input was added to the conditioned intensity data. The results obtained improved slightly from 5.1% error to 5.08% but the network dimension increased. This option was not considered viable. The addition of determined vertical force as input improved the results significantly for the network trained with statistical and geometrical parameters and the error reduced dramatically (8% to 5%). However, for the network trained with intensity data there was no significant improvement in results (5.1% to 5%) with additional input of vertical force. This was because the intensity input was extracted by considering the entire image, thus it represented the whole image and addition of similar image data (vertical force) was unlikely to change the results. On the other hand input of the vertical force to statistical and geometric parameters was found to improve the error as it complemented the whole image. However, the network fed with intensity data from entire image also gave comparable results to this, thus the network trained with entire intensity data was considered to be optimal and efficient.

6. Conclusions

This paper focused on analysing the fringe patterns obtained under vertical and shear loads in order to determine the load characteristics. Various strategies were adopted to refine the results by taking various image parameters for implementation. Since an image has enormous data in itself it cannot be used as a whole for training neural networks of moderate size. Thus a range of data were extracted from the images including statistical, geometric, intensity and fed as training data. A mix of input parameters were also tried to achieve the best possible network with desired accuracy. The accurate results in determination of shear force from fringe patterns were found in two networks, one trained with the combination of statistical and geometric parameters with determined vertical force and second trained with intensity data from entire image. Both networks gave comparable results of 5.37% and 5.1% error respectively, however, the later was found to be the right network as it represented the image better. The training of network in this case was easy and required no feature extraction or specific processing of data. Therefore the later network was chosen to be a better choice due to most relevant input to the network with high tolerance.

Acknowledgments
This research was supported by the bursary awarded by the School of Design, Engineering & Computing, Bournemouth University.

References
[1] Dubey, V. N. and Crowder, R. M., 2006 "A dynamic tactile sensor on photoelastic effect" Sensors and Actuators, 128: 217-224.
[2] Dario, P., Fermi, D. and Vivaldi, F., 1987 "Fiber-optic catheter-tip sensor based on photoelastic effect" Sensors and Actuators 12: 35-47.
[3] Jacobsen, S. C., Wood, J. E., Knutti, D. F. and Biggers, K. B., 1986 "The Utah/MIT dextrous hand: work in progress" in D.T. Pham and W.B. Heginbotham edited Robot Grippers, pp 341-389.
[4] B. Davis, "Foot ulceration: hypothesis concerning shear and vertical forces acting on adjacent regions of the skin", Medical Hypothesis (1993), 40:44-7.
[5] P. Cavanagh, J. S. Ulbrecht, and G. M. Caputo, "New developments in the biomechanics of the diabetic foot", Diabetes Metab. Res. Rev (2000), 16(Sl)1:S6-S10.
[6] M. Curryer, and. E. Lemaire, "Effectiveness of various materials in reducing plantar shear forces: a pilot study", J. Am Podiatr Med. Assoc. (2000) 90(7): 346-53.
[7] V.J. Thomas, K.M. Patil, S. Radhakrishnan, V.B.N. Murthy, and R. Parivalavan [2003] "Diabetic feet at risk of plantar ulcers: a new method of analysis of walking foot pressure images with changes in foot sole properties" IEEE TENCON Conference on Convergent Technologies for the Asia-Pacific Region, Vol.4, 1320-4.
[8] G. Charanya, K.M. Patil, V.B. Narayanamurthy, R. Parivalavan, and K. Visvanathan [2004] "Effect of foot sole hardness, thickness and footwear on foot pressure distribution parameters in diabetic neuropathy" Proceedings of the Institution of Mechanical Engineers, Part H (Journal of Engineering in Medicine), v 218, n H6, 431-43.
[9] G. S. Grewal, V. N. Dubey, [2007] "Inverse Problem of Photoelastic Fringe Mapping using Neural Networks", Measurement Science & Technology, Vol. 18, No. 5, pp. 1361-1366.
[10] Edmund, O (2004). Determining fast axis of quarter wave plate, Edmund Optics. 2004.
[11] Vishay Measurement Groups [1992], Photoelastic Materials, Bulletin S-116-G.
[12] V. N. Dubey, G. S. Grewal [2007] "Tactile Whole-field Imaging Sensor on photoelasticity", ASME International Design Engineering Technical Conference, DETC2007-35264, Las Vegas, Nevada (USA), Sept. 4-7, 2007.
[13] G. S. Grewal, [2007] "Digital Photoelasticity in Biomedical Sensing" PhD Thesis, Bournemouth University.
[14] V. N. Dubey, G. S. Grewal and D. J. Claremont [2007] "Load Extraction from Photoelastic Images Using Neural Networks" Experimental Mechanics, Vol. 47, N0. 2, pp.263-270.
[15] C. M. Bishop, [1995], Neural networks for pattern recognition, Oxford University Press.

Journal of Physics: Conference Series **181** (2009) 012075 doi:10.1088/1742-6596/181/1/012075

Thermoelastic stress analysis of a 2D stress field using a single detector infrared scanner and lock-in filtering

G Pitarresi[1], A Normanno and L D'Acquisto

Dipartimento di Meccanica, Università degli Studi di Palermo, Viale delle Scienze, 90128 Palermo, Italy

E-mail: pitarresi@dima.unipa.it

Abstract. A low resolution, low cost, fast infrared scanner is used to acquire the temperature change along a line on the surface of cyclically loaded samples. The temperature signal is sampled versus time by exploiting the raster scanning movement of the thermocamera single detector. The temperature data is then post-processed by a Fast Fourier Transform based lock-in algorithm implemented in MATLAB®, in order to filter out the thermoelastic effect induced temperature change. A procedure is also implemented in order to extend the data sampling time by opportunely synchronising successively grabbed data frames. The effectiveness of such synchronisation procedure is first demonstrated by performing the signal processing analysis on unidirectional tensile samples made of polycarbonate. The ability of the implemented one-dimensional FFT lock-in algorithm to analyse also two-dimensional stress fields is then assessed by processing the temperature data acquired on a polycarbonate disc under cyclic compression loads.

1. Introduction

The Thermoelastic Effect describes the temperature changes in loaded matter as induced by the volume changes under linear elastic behavior [1]. When such transformation is adiabatic and the material is homogenous and isotropic, a simple linear relationship can be derived where the temperature change, measured on the free surface of a body, is proportional to the change of the first stress invariant as follows [2,3]:

$$\Delta T = -T_o \cdot K \cdot \Delta\left(\sigma_x + \sigma_y\right) \quad \text{with } K = \alpha / \rho C_p \tag{1}$$

where T_o is the absolute temperature and K the thermoelastic constant function of the coefficient of thermal linear expansion α, the density ρ and the specific heat C_p of the bulk material. Thermoelastic Stress Analysis techniques (TSA) are aimed at acquiring stress information by measuring temperature changes on the surface of bodies when simple thermoelastic laws of the like of eq. 1 apply [4]. Three most common features in many TSA set-ups include: the application of cyclic loads for the material to achieve adiabatic conditions, the use of high thermal resolution IR sensors for non-contact full field temperature measurements, the use of lock-in signal analysis techniques in order to filter out the thermoelastic effect induced temperature change from other noisy components [5-8]. The special needs associated with the last two features have advised TSA practitioners to using only on purpose

[1] To whom any correspondence should be addressed.

developed commercial hardware, so far available at elevated costs and with a limited choice of suppliers and configurations.

Aim of this work is to present an experimental methodology using fast linear IR scanners to measure the thermoelastic signal on the free surface of a body with a generic stress field. These systems, non conventional in TSA applications, use a single detector to measure the temperature versus time along a chosen line. A fast raster scanning mode is realized by means of an oscillating mirror which changes the optical path between the sensor and the measured spot. The use of a single detector and of a simpler optical hardware, as compared for instance to early single detector TSA systems such as SPATE, proposes such IR systems as a potential tool for TSA, with costs that can be estimated as three to four times lower than actual high thermal resolution focal plane array systems.

The ability of fast single detector IR cameras to sample temperature versus time with a resolution adequate for deriving the thermoelastic signal was first demonstrated in [9]. The methodology proposed was though limited to one-dimensional stress field analyses. Moreover the high scanning speed of the IR camera used in [9] requires the sample to be stressed at very high frequency loads (over 100 Hz). The possibility to overcome these limitations was recently discussed in [10], by means of the use of slower linear IR scanners whose detector movement is kept sweeping along the same line in space. One further feature of the approach presented in [10] was the adoption of a lock-in correlation algorithm more effective in filtering out the thermoelastic signal component.

This work implements two substantial improvements on the methodology presented in [10]: the synchronization of different acquired frames to extend the time window of sampled data, and the adoption of a lock-in filter implementing a one-dimensional Fast Fourier Transform (FFT) analysis to measure the thermoelastic signal on a surface with a generic two-dimensional stress field. Both these features will be further discussed in the following sections and supported with experimental analyses performed on two polycarbonate samples: a tensile specimen loaded in traction and a circular compression loaded disc developing a 2D stress field. The experimental results obtained clearly demonstrate the validity of the proposed methodology for measuring the thermoelastic signal.

2. Description of the IR equipment

The IR thermocamera employed in this study is a Varioscan 3022 by JENOPTIK GmbH. This system employs a single IR detector, a thermoelectrically cooled MCT sensor with a thermal resolution of 0.12 K [11]. Two rotating mirrors allow the signal to be collected from different points, horizontal-wise and vertical-wise, collecting the signal on a full field frame array of 360×240 pixels. The system acquires and save each frame in about 0.9 seconds, and a small temporal gap is spent between the acquisition of two succeeding frames during which the signal is not acquired. One feature of the employed system is the possibility to fix the mirror which enables the vertical scanning so that each row in the data frame is acquired on the same horizontal line in the field of view. The rotational movement of the mirror sweeping the horizontal line consists of an oscillation of the mirror around its axis. In the first half period of this alternating rotation the line is swept from left to right, during which the signal from 360 points is acquired in 1/270 sec, and the same number of points is acquired in the second half period moving now from right to left (figure 1a). With this raster scanning mechanism 240 rows from the same line are collected and piled up to build one data frame array. All acquired frames can then be automatically captured, stored and displayed on a PC by means of a proprietary software (Irbis© V.2.0) allowing a complete remote control of the system through an Ethernet connection.

It can be noted that the time employed to acquire one single measure is very fast. It is also known from IR thermography theory that a good signal to noise ratio, end hence thermal resolution, depends on the time the sensor remains focused on the measuring spot. Despite the low thermal resolution of the system available in this work, poor thermal resolution should not be an issue in this approach since higher resolution fast IR scanners are available. For instance thermal resolution values of 0.03 K are achieved by the Varioscan 3021 models [11] or by the AGEMA system employed in [9]. This resolutions, achieved with the fast scanning mechanism described above, are close to the values reached by commercial focal plane array systems routinely employed in TSA [7,8].

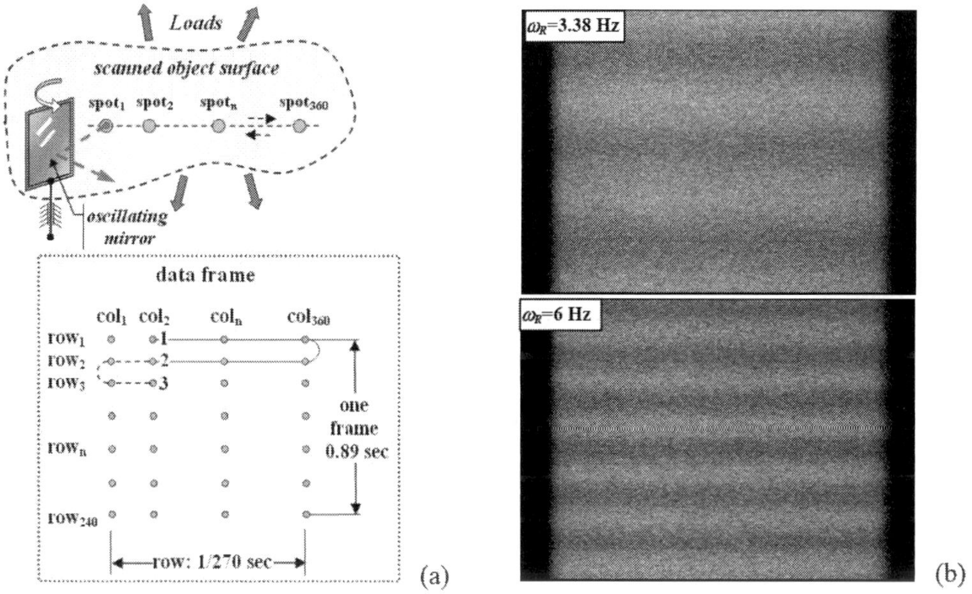

Figure 1. (a) Schematic representation of the single line raster scanning mechanism and data frame formation; (b) gray scale images of thermogram frames from tensile samples.

2.1. Time-temperature sampling.

In this work the fixed-line scan mode is used. The temperature from a pixel on the scanned line is acquired periodically whenever the optical path of the sensor focuses on it. The information from one pixel location is than stored in a column of the data frame array. The column-wise information is then a measure of temperature versus time. This enables a procedure by which the temperature-time variation is sampled over 360 aligned points for a total time of 0.9 sec (i.e. the frame rate). If the body is loaded at sufficiently slow frequencies, the thermoelastic effect induced temperature change can be sampled over a finite number of periods during a single frame acquisition. Figure 1b shows two thermograms from a tensile sample loaded at 3.38 Hz and 6 Hz. It is possible to distinguish alternating darker and lighter horizontal fringes caused by the thermoelastic effect induced temperature change. The horizontal orientation of such thermal fringes is caused by the high speed of the sensor line swap, such that the time shift on the thermoelastic signal from different columns is non distinguishable. In the case of a cyclic loading frequency f_L, a simple relationship can be established between the number of horizontal thermal fringes shown in one frame, n_S, and the camera frame rate f_R:

$$n_S = f_L / f_R = f_L \cdot (0.888 \text{ sec}) \qquad (2)$$

where 0.888 s is the exact time spent by the camera employed in this work to acquire a frame [11].

3. Lock-In algorithm

The thermoelastic signal from each of the 360 points on the scanned line is derived by performing a lock-in filtering operation of the temperature versus time data. A scheme of this lock-in procedure is shown in figure 2. In order to make the filtering mechanism clear, some equations are reported on the scheme, which describe the trigonometric operations of signal processing performed if the signal is represented by continuous harmonics. In reality the signal is digitized by the sampling mechanism described above and the filtering operations are performed by applying FFT analyses. As shown in figure 2 the core of the lock-in technique consists in mixing the measured signal, named S, with a reference signal which is a carrier of the loading frequency, i.e. the useful frequency. This frequency named ω_R can be assigned as an input data or can be derived from S if the corresponding harmonic is easily identified on the frequency domain with the FFT analysis. Once ω_R is obtained, two pure cyclic reference signals named F and G, with F in quadrature with G, are built and multiplied to S. This operation results in two signals which contain the information of interest as DC components, and hence a low-pass filtering operation performed with a new FFT analysis is able to filter out such components named X and Y. These represent the thermoelastic signal components in phase and in quadrature with the reference signal and by combining them the thermoelastic signal amplitude A is readily obtained.

The above procedure is implemented in this work with an algorithm using the Fast Fourier Transform functions available in Matlab®. The approach here proposed implements a 1D FFT analysis of each column of the data frame array, differing from the global 2D FFT approach proposed in [10]. This 1D lock-in treatment allows the thermoelastic signal to be derived on each scanned point independently, enabling the immediate application of this analysis to the case of a 2D stress field. In fact although the 2D FFT based treatment from [10] is believed to be more effective in noise rejection, it is much more elaborate for analyzing 2D stress fields.

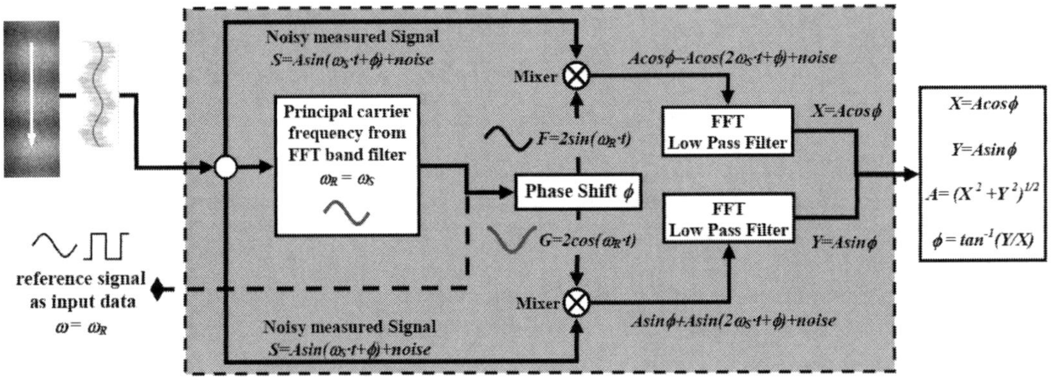

Figure 2. Scheme of the implemented lock-in analysis.

3.1. Synchronisation of data frames.

The FFT analysis requires that the sampled data are equally spaced in time. This is not the case for the data along the columns of the acquired frame arrays. In fact from figure 1a it is possible to follow the movement of the mirror acquiring the signal. Lets consider the first three measures of column 2 in this scheme: the time spent by the mirror to go from point 1 (row1,col2) to point 2 (row2,col2) is longer than from point 2 to point 3 (row3,col2). This is due to the longer distance swept by the mirror from point 1 to 2. By further proceeding along column 2, it is observed that the time interval is the same every two measures, i.e. considering alternate points. In light of this the sampled data useful for the FFT analysis need to be taken by considering only odd rows or, alternatively, even rows. The total number of samples is then reduced from 240 to 120. The corresponding time interval Δt becomes equal to one oscillation period of the mirror, i.e. Δt=2·(1/270) sec, and the sampling frequency is then f_S =1/Δt= 135 Hz. The maximum loading frequency applicable for the FFT analysis to work, i.e. the Nyquest frequency, will be half the sampling frequency, f_{Ny}=f_S/2=67.5 Hz. The total number of points sampled in one frame, N=120, and the time interval between two succeeding acquisitions on the same point, Δt, affects the value of the fundamental frequency f_F. This is the first order harmonic frequency determining the discrete resolution of the frequency domain from the FFT analysis, and is given by:

$$f_F = \frac{1}{N \cdot \Delta t} = \frac{135}{120} = 1.125 \text{ Hz} \tag{3}$$

The thermoelastic signal in TSA is usually heavily corrupted by noise. The small number of points from a single frame, useful for the FFT analysis to extract the thermoelastic signal, is likely not enough to perform the filtering operation very effectively. It is observed that if the total sampling period were extended, a higher N value and a smaller f_R would be obtained (see eq. 3), with further benefits being a smaller influence of border effects, and a more effective noise to signal reduction.

With the IR scanner in use in this work, in order to sample the signal over a longer time window, the information from more than one acquired frame has to be opportunely collected. As noticed before, this is not straight forward because there is a small time gap between two succeeding frames during which the camera is not acquiring the signal. So if these are joined by simply attaching one frame after the other, a new array is formed with its columns doubled in length, but a discontinuity is then inserted in the temperature versus time data at the point of attachment of the frames. This discontinuity introduces further noise which can affect the analysis. An algorithm is proposed in this work to solve the above problem and to be able to attach more frames in a sequence, in order to build a longer array. The steps of this so called synchronization algorithm is here described and graphically supported by considering a data frame consisting of a pure sinusoidal signal with a frequency of 3 Hz (see figure 3):

i) Frame 2 is attached with frame 1 starting from row 1 of frame 2;

ii) The FFT analysis of frame 1+2 is performed and the amplitude of the harmonics of the resulting signal plotted in the frequency domain. The amplitude of the harmonic at the reference frequency f_L is determined in particular, and named $A(1)$;

iii) Steps *i* and *ii* are repeated attaching frame 2 to frame 1 starting from row2 of frame 2;

iv) Steps from *i* to *iii* are repeated for *n*-times, and at each iteration frame 2 is attached to frame 1 starting from the *n-th* line of frame 2;

v) For each combination of attached frames an FFT analysis is performed as done in step *ii*, and the power distribution in the frequency domain plotted. The amplitude of the harmonic at the f_L frequency from this plot is $A(n)$;

vi) The optimum number of lines to remove from frame 2 to get the best *synchronisation* with from 1 is n^* such that $A(n^*)=\max(A(n))$.

vii) Steps from *i* to *vi* are repeated to synchronise frame 2 with frame 3, frame 3 with frame 4 and so on. For each couple of frames the optimum number n^* of lines to remove is obtained.

viii) When the total number of couples of frames have been synchronised, the final array is built up by simply joining together all the frames removing from each frame its first n^* rows as determined in step *vii*.

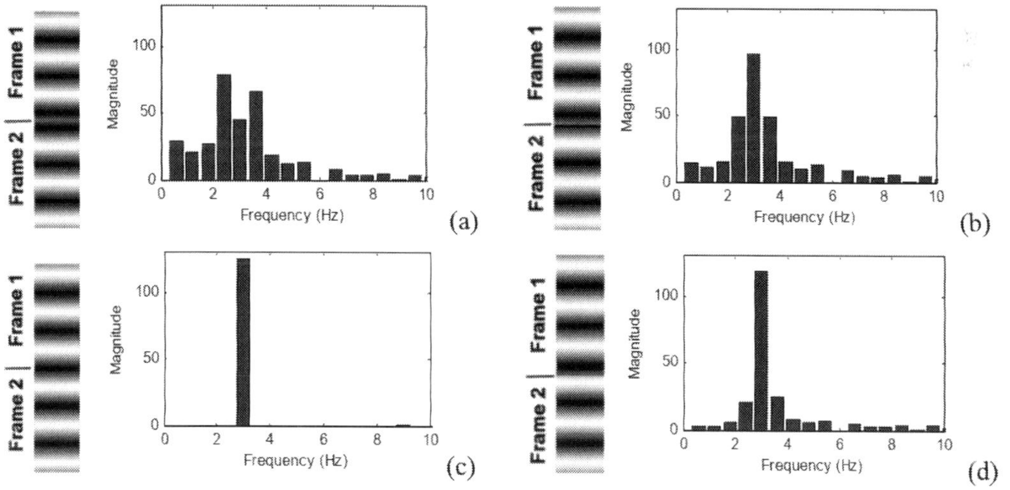

Figure 3. Main steps of the synchronization algorithm applied to join two pure sinusoidal signal frames. (a,b) n<n*; (c) n=n*; (d) n>n*.

Journal of Physics: Conference Series **181** (2009) 012075 doi:10.1088/1742-6596/181/1/012075

The logic behind this procedure is that the best synchronisation is achieved when the harmonic of the thermoelastic signal has a maximum amplitude, because this means that a smooth attachment is obtained and the power is not dispersed over other harmonics as would be the case if a steep jump is present at the attached location. The effectiveness of this procedure is demonstrated in figure 3 for a pure sinusoidal signal, and in figure 4 for synchronised experimental data frames from tensile samples.

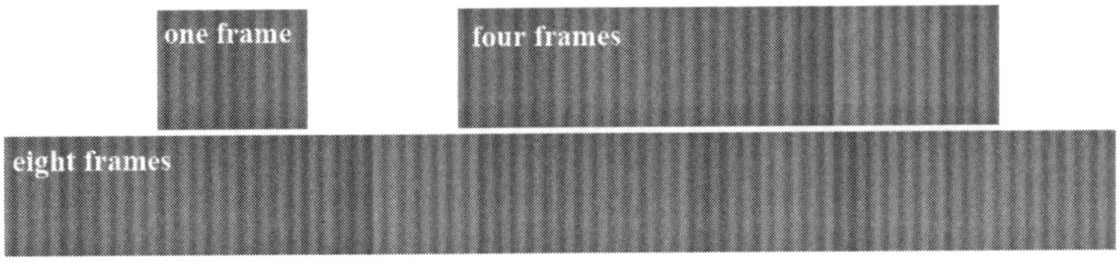

Figure 4. Example of synchronised frames from experimental data.

4. Experimental results and discussion

The temperature-time sampling mechanism based on the fast linear IR scanner, the 1D FFT based lock-in filtering algorithm and the synchronization procedure of acquired frames are the main features of the TSA technique presented in this work. These have been implemented for the experimental evaluation of the thermoelastic signal on cyclically loaded polycarbonate samples. This material has a high thermoelastic constant (e.g. $K \approx 46 \times 10^{-6}$ MPa^{-1}) and a relatively wide elastic range enabling the possibility to generate high values of the thermoelastic signal. This is evaluated and plotted in figure 5a in uncalibrated units corresponding to grey scale levels. In fact the acquired frames are 8 bit gray scale images. Since the IR system in use gave no other possibility to access the measured raw data, the bitmap images were imported in Matlab after an optimization of the brightness level and contrast range parameters performed with the Irbis© software [11].

Two types of samples have been tested by applying cyclic sinusoidal loads, generating one and two dimensional stress fields as described below.

4.1. One-dimensional stress field analysis.

This stress case was investigated by determining the thermoelastic signal on tensile samples at different traction-traction load amplitudes. A plot of the stress amplitude versus the thermoelastic signal amplitude is shown in figure 5a. In this plot the thermoelastic signal is reported for each stress amplitude and data are then interpolated with a linear regression curve. The correlation coefficient of the linear regression line is used as a parameter to quantify the linear behavior as foreseen by the theory (eq. 1), and used as an index of the quality of the analysis results. Data plotted in figure 5a are in particular derived applying the lock-in analysis to a single column on an extended array of data obtained by joining 24 acquired frames with the synchronization procedure of section 3.1. Figure 5b,c shows the correlation coefficient of the linear regression performed on the thermoelastic signal versus stress amplitude results from all the points scanned. In particular the results in figure 5b are referred to an extend sampling time of 24 synchronized frames, while figure 5c consider the analysis on a single grabbed frame. It is seen that performing the analysis on a higher number of synchronized frames dramatically improves the linear trend between the thermoelastic signal and stress amplitude. This is confirmed in figure 5d which plots the average correlation coefficient and its standard deviation variation range at increasing number of synchronized frames. It must be noted that for the present case of a uniform stress field each scanned point should deliver the same thermoelastic signal. The analysis can then be improved by averaging the thermoelastic signal obtained from each point. Indeed it has been found that this further averaging operation improves results, but these are not shown or further considered since the final scope of the work is to analyse two-dimensional stress fields for which each scanned point deliver different thermoelastic signals and so the quality of the analysis should rely only on the filtering of the data measured on that point. Tests at frequencies of 3 Hz and 9 Hz, have also been performed showing very similar results to those reported in figure 5.

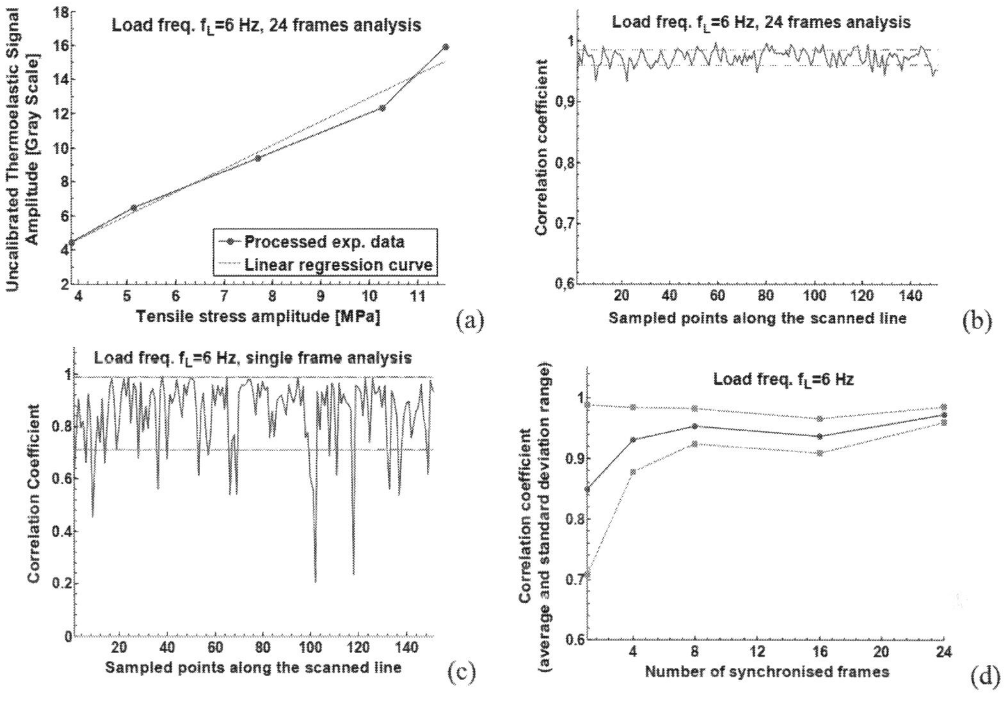

Figure 5. (a) Thermoelastic signal measured versus stress amplitude; (b,c) Correlation coefficient of the linear regression as shown in figure 5a, for all sampled points; (d) Correlation coefficient at varying number of synchronised frames (average value and standard deviation calculated from all sampled points on the scanned line).

4.2. Two-dimensional stress field analysis.

For this stress case a polycarbonate disc (100 mm diameter per 7.5 mm thickness) was used. Concentrated compression loads were applied varying sinusoidally (average load of -2 kN and load amplitude of 1.8 kN). The elastic theoretical solution of the in-plane stress field for this shape and load case is readily available in the literature. Figure 6a shows a contour map of the isopachics normalized by dividing for the absolute value of the stress invariant at the center of the disc (about 6 MPa in this analysis). The experimental analysis consisted in deriving the thermoelastic signal from the points along the diametral line aligned with the load (marked in figure 6a). The disc was loaded at a frequency of 3 Hz and the IR camera was tilted on its side by a 90° angle to allow the scanned line to coincide with the diametral loading direction of the sample. The analysis considered only half diameter of the disc, the one resting against the fixed plate, to minimize the influence of the deformation movement of the disc, bigger on the side of the disc resting against the moving actuator. One further reason was an evident disturbance of the heat coming from the moving actuator affecting the signal on the area of high stress concentration near the loading point.

The thermoelastic signal was determined on a built up array of 40 synchronized grabbed frames. The infrared image of the disc was used to position the line scan and to identify the pixels corresponding to the center of the disc and the point of contact with the fixed plate. This segment comprised about 110 pixels. The thermoelastic signal derived with the lock-in algorithm is plotted in figure 6b, after normalization performed dividing by an average value of the signal from the points near the center of the disc. This signal is plotted together with the theoretical trend. It can be seen that the experimental curve agrees well with the theoretical stress gradient, demonstrating the ability of the implemented technique to analyse this 2D stress field. The thermoelastic signal evaluation is plotted up to about 2.5 mm away from the border of the disc. In fact nearer points did not follow correctly the theoretical trend. This was likely due to some unknown influences such as local plasticity and flattening deformation on the contact point changing the local theoretical stress distribution.

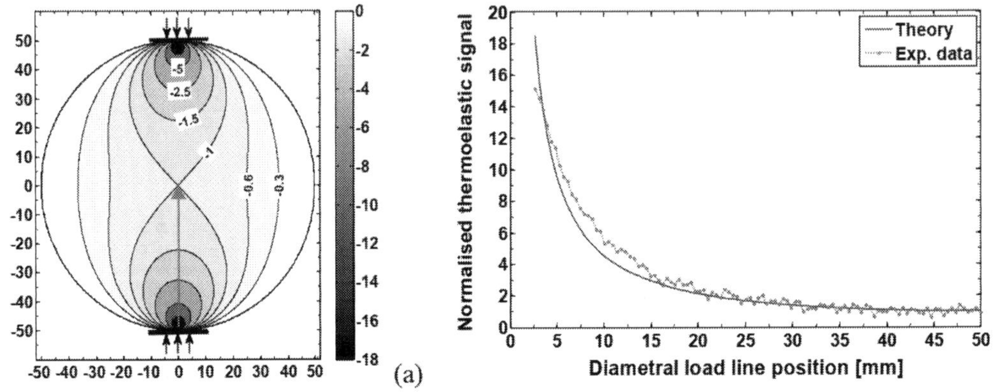

Figure 6. (a) Contour map of isopachichs in a compression disc (scale normalised by the absolute value of the isopachich at the center of the disc); (b) comparison of theoretical and experimental normalised isopachics along the half diametral load line marked in figure 6a.

5. Conclusions

A new thermoelastic stress analysis set-up has been implemented based on three main features: a temperature-time sampling technique exploiting the raster scanning mechanism of fast linear IR scanners, a 1D-FFT based lock-in filtering algorithm, and a procedure for synchronizing several successively acquired frames to extend the sampling time. Based on these features a thermoelastic stress experimental analysis of 1D and 2D stress fields has been performed. In both cases the measured thermoelastic signal agreed well with the theoretical predictions when the data sampling and analysis was performed on more grabbed frames. The intrinsic quality of the results obtained is to a certain extent affected by the low thermal resolution of the IR scanner available for this work. This should though not be a concern for the methodology presented since IR scanners with higher thermal resolution are available in the market, using the same scanning principle. It is also believed that a customization of such IR scanners to better suit the scopes and the features of the present thermoelastic stress analysis could easily determine further improvements in both performances and hardware costs.

References

[1] Thomson W (Lord Kelvin) 1878 *Philosophical Magazine* 5 4-27.
[2] Biot M A 1956 *J. Applied Physics*, 27 (3) 240-53.
[3] Pitarresi G and Patterson E A 2003 *J Strain Anal Eng* **38(5)** 405-17.
[4] Dulieu-Barton J M and Stanley P 1998 *J Strain Anal Eng* **33(2)** 93-104.
[5] Belgen M H 1968 *NASA Contractor Report: CR-1067.*
[6] Harwood N and Cummings W M 1991 *Thermoelastic Stress Analysis* (Bristol: Adam Hilger IOP Publishing).
[7] Lesniak J R, Boyce, B R and Sandor B I 1991 *NASA Contract Report CR-NASA-19262.*
[8] Brémon P and Potet P 2001 Lock-In Thermography: a tool to analyze and locate thermomechanical mechanisms in materilas and structurres. Proc. SPIE 2001, 4360(560).
[9] Audenino A L, Crupi V and Zanetti E M 2004 *Exp Tech* **28** 23-8.
[10] Pitarresi G, D'Acquisto L and Siddiolo A M 2008 *J. Strain Analysis* **43** 493.
[11] Varioscan high resolution Model 3021, 3021-st and 3022 manual (2000), edited by JENOPTIK Laser, Optik, Syteme GMBH.

Combining digital image correlation and projected fringe techniques on a multi-camera multi-projector platform

T Nam Nguyen[1], Jonathan M Huntley[1], Richard L Burguete[2] and C Russell Coggrave[3]

[1] Loughborough University, Wolfson School of Mechanical and Manufacturing Engineering, Loughborough, LE11 3TU, UK

[2] Airbus UK, Filton, Bristol, BS99 7AR, UK

[3] Phase Vision Ltd, Innovation Centre, Epinal Way, Loughborough, LE11 3EH, UK

E-mail: T.N.Nguyen@lboro.ac.uk

Abstract. This paper presents how a shape measurement system (SMS) based on projected fringes can be combined with a 2-D digital image correlation (DIC) technique to accurately measure surface profile and 3-D displacement fields at the same time. Unlike traditional 3-D DIC techniques, the proposed method can measure discontinuous surfaces as easily as smooth ones. The method can also be extended to a multi-camera multi-projector system, and thus complete 360° 3-D displacement fields can be obtained within a single global coordinate system. Details of the algorithm are presented together with experimental results.

1. Introduction

Among optical techniques available to accurately measure shape and deformation of 3-D surfaces, projected fringe and digital image correlation are two well-established techniques that have been widely used in industry. The projected fringe technique is recognised for its high accuracy and spatial resolution in measuring shape and out-of-plane displacement fields [1], but it cannot identify in-plane displacements reliably since information of surface features is not used. On the other hand, 2-D DIC is a proven technique for accurately measuring in-plane deformation, which has later been extended to 3-D DIC for 3-D deformation measurements [2]. However, the performance of 3-D DIC in measuring surface profile is inferior to that of the projected fringe technique. Many applications require fast, accurate and concurrent measurements of both surface profile and deformation for manufacturing quality control and structural testing purposes. Therefore, it is desirable to combine the strengths of the projected-fringe and DIC techniques on one measurement system.

The literature contains a small number of works that describe combinations of projected fringe and DIC techniques. Reference [3] describes a method using two projectors and two cameras equipped with colour filters to separately capture fringe patterns and speckle patterns that are simultaneously present on the object surface. The projected fringe technique is then applied to estimate surface profile and out-of-plane displacements, while the 2-D DIC technique is used to determine in-plane displacements. References [4] and [5] describe a simpler system using one camera and one projector, in which the fringe pattern is isolated from the speckle pattern during the computation stage using a

Fourier transform. In reference [6], sinusoidal fringes and speckle pattern are captured in two temporally discrete steps and processed in a similar way to the above-mentioned methods. In each of these works, the authors have described the system performance on continuous surfaces.

In this work, a one-camera one-projector system based on fringe projection and temporal phase unwrapping techniques is used to measure shapes and 3-D displacement fields of objects with complex surface discontinuities. In order to achieve this, the conventional image correlation technique is modified to cope with discontinuities. The method can be extended to a multi-camera multi-projector system so that complete 360° surface profile and 3-D displacement fields can be obtained within a single global coordinate system. The accuracy of the current system has been validated with an experiment on a specimen undergoing rigid-body deformation.

2. Surface profile estimation and segmentation

The shape measurement system can be configured to include multiple camera-projector sensors. The simplest arrangement consists of a sensor containing one camera and one projector. A sequence of patterns consisting of sinusoidal intensity fringes is projected onto the specimen surface during the measurement process. The position of a scattering point on the surface is encoded as spatial distortions of the fringe patterns in the intensity image. Phase estimation and temporal phase unwrapping [7] are applied at each camera pixel independently of its neighbours to estimate 3-D coordinates of scattering points on the specimen surface. In this way, surfaces with complex geometries and discontinuities can be measured as easily as simple ones. The authors have given details of the measurement algorithm in reference [8] with further information about the calibration method in reference [1].

To demonstrate the ability of the system to measure discontinuous surfaces, shape measurement results for a planar test specimen containing a round hole and a rectangular step are given in Figure 1. The texture image is shown in Figure 1-a, and the results of *X*-, *Y*- and *Z*-coordinates of nearly one million scattering points are presented in Figures 1-b, 1-c and 1-d, respectively. The specimen was located near the origin of the calibrated measurement volume and was positioned with the planar surface at approximately 10° with respect to the *XY*-plane. The Cartesian coordinate system was orientated with the *Z*-axis pointing toward the sensor (see Figure 5). Within the hole region, measured data correspond to the background instead of the specimen of interest, thus resulting 3-D coordinates have been automatically masked out using a depth threshold. The shadow region to the left of the step is automatically detected and omitted from the generated surface model. In addition, the dark regions that can be seen around the edges of the texture image have a very low fringe visibility and are also automatically removed from the final result. A mask of valid pixels lying within a region of interest is shown in Figure 2-a. Thus, the resulting 3-D model of the specimen is reconstructed with high confidence despite the presence of the surface discontinuities.

Once the 3-D shape is obtained, it is straightforward to detect discontinuities and segment discrete continuous regions of the surface. A pixel can be identified as corresponding to a discontinuity if it satisfies one of the following conditions,

$$\sqrt{(X_x - X_{x-1})^2 + (Y_x - Y_{x-1})^2 + (Z_x - Z_{x-1})^2} \geq \rho_x , \tag{1.1}$$

$$\sqrt{(X_y - X_{y-1})^2 + (Y_y - Y_{y-1})^2 + (Z_y - Z_{y-1})^2} \geq \rho_y , \tag{1.2}$$

$$mask(x, y) = 0 , \tag{1.3}$$

where x and y are respectively the horizontal and vertical image coordinates of the pixel (as shown in Figure 1-a); the constants ρ_x and ρ_y are the thresholds of 3-D distance between two neighbouring pixels in horizontal and vertical directions, respectively; and *mask* is the logical mask specifying whether a pixel contains a valid 3-D coordinate (see Figure 2-a). Discontinuities identified in this way may define closed boundaries between continuous surface regions that can then be segmented and labelled using standard image processing techniques [9]. Figure 2-b shows the resulting label map consisting of two continuous regions labelled with positive integers (corresponding to the plane and the step) and a background labelled as zero.

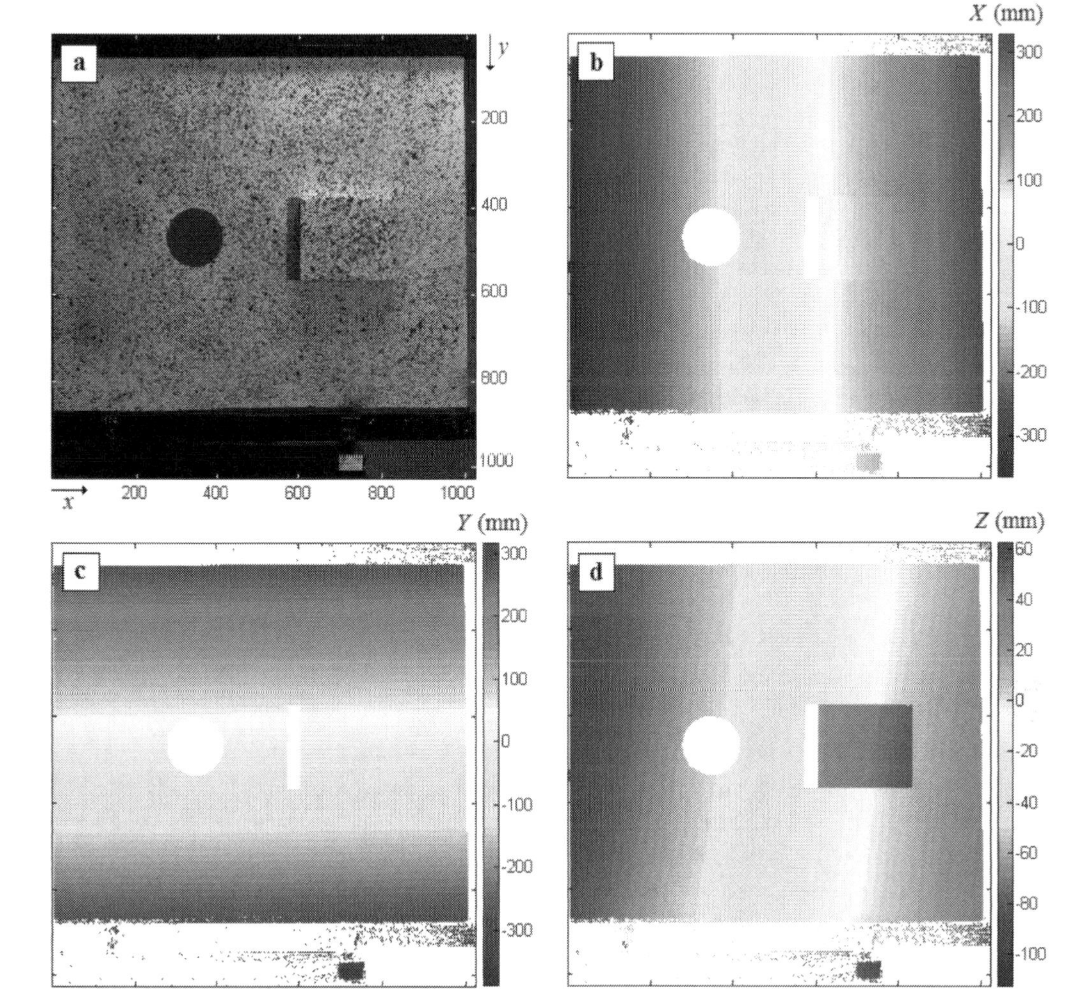

Figure 1: Surface profile estimation. (a) 1024×1024 texture image of the planar specimen with a hole and a step. (b, c and d) Colour range maps of *X*-, *Y*- and *Z*-coordinates of scattering points corresponding to all pixels, where invalid pixels are shown in white colour.

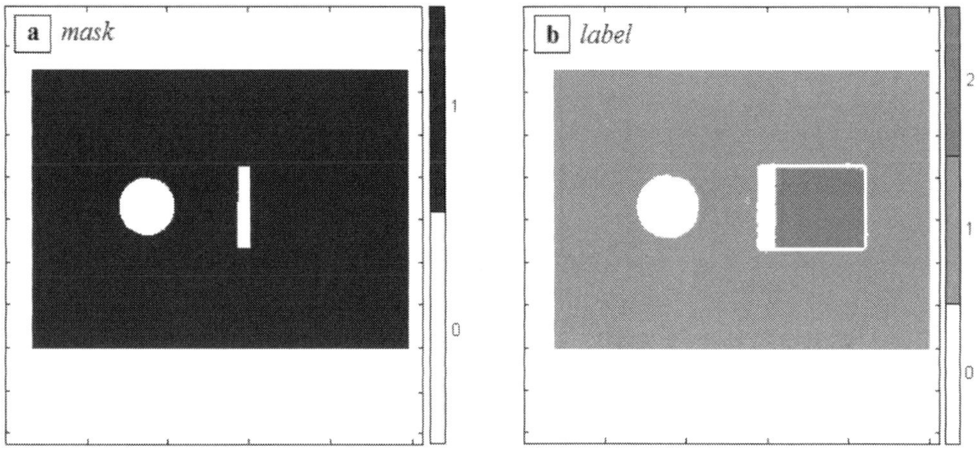

Figure 2: Surface segmentation. (a) Mask of valid pixels within region of interest. (b) Label map of connected regions (b).

3. Digital image correlation of discontinuous sub-images

The 2-D DIC technique is used to estimate the displacement vectors of sub-images within the region of interest during the deformation process. A small region of a reference image (called reference sub-image) is matched against a region of a deformed image (called deformed sub-image). In practice, the specimen surface is prepared with a high-contrast random speckle pattern (as can be seen in Figure 1-a) to aid the matching algorithm. A successful match of two sub-images is obtained when their correlation score (e.g. cross-correlation or squared distance) meets some predefined criteria. In conventional DIC techniques [2], the average of relative displacement of the matched sub-images is assumed to be the displacement of the pixel at the centre of the reference sub-image. As a consequence, all pixels within the sub-images are assumed to lie on the same continuous region, which is the case of sub-image 1 illustrated in Figure 3. However, this assumption may be invalid for a sub-image that intersects multiple continuous regions (e.g. sub-image 2 and 3) as each region may undergo a different deformation that is generally not the same as that of the sub-image.

In this work, the image correlation algorithm is modified so that the correlation score is computed for the labelled regions independently. The cross-correlation score of a displacement candidate (u,v) for the kth region of a reference and deformed sub-image can be computed by

$$c(u,v,k) = \sum_{x,y} f(x,y,k)g(x+u,y+v,k), \tag{2}$$

where f and g are respectively the reference and deformed sub-image whose DC term has been removed to exclude the effects of changes in ambient lighting, and (x,y) specifies a pixel within the region under investigation. Equation 2 can be calculated more efficiently in the frequency domain by

$$c(u,v,k) = F^{-1}\{F(f')^{*}.F(g')\}, \tag{3}$$

where F is the Fourier transform, the asterisk indicates the complex conjugate, and f' and g' are respectively the sub-images f and g padded with zeros around their edges to allow sub-pixel displacement estimation. In addition, pixels that belong to other regions and the background are set to zeros in f' and g' so that they do not contribute to the correlation score. An optimised correlation procedure based on that described in reference [10] is used to compute the image displacement (u,v) with a sub-pixel accuracy of up to one-hundredth of a pixel.

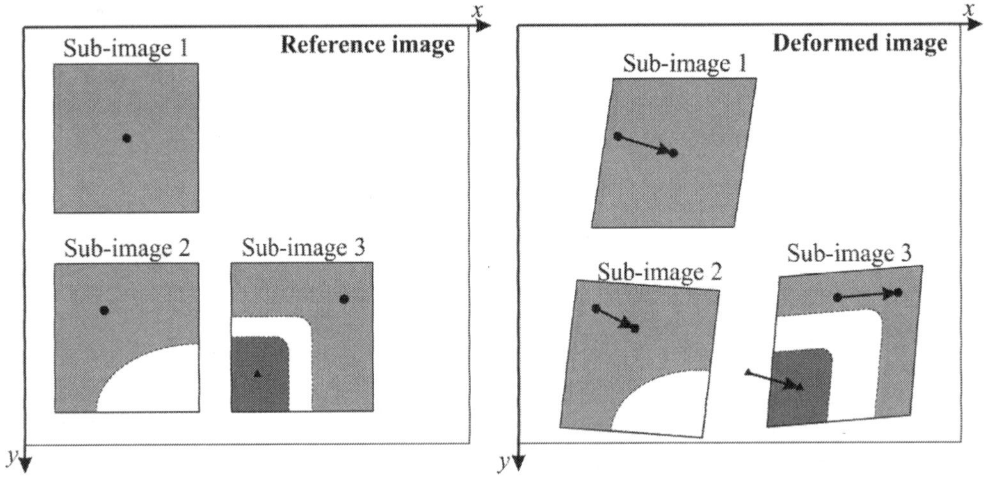

Figure 3: Image correlation of various sub-image scenarios. Sub-image 1 contains a continuous region of the plane, sub-image 2 contains a top-left portion of the hole, and sub-image 3 corresponds to the top-right corner of the step. The "dot" and "triangle" represent the centroid of region 1 and 2, respectively. The arrow represents the image displacement vector of the region centriod between the reference and deformed states. Conventional image correlation algorithms only deal robustly with sub-image 1, while the proposed algorithm is more general and should be robust with all three scenarios.

4. Measurement of 3-D displacements

4.1. 3-D displacements by a one-camera one-projector sensor

Surface profile and 3-D displacement fields can be estimated using a sensor consisting of one camera and one projector. The analysis process is illustrated in Figure 4.

At a reference state, the object surface is measured with the shape measurement system using projected fringes as introduced in Section 2, generating a dense cloud of 3-D points corresponding to the scattering points imaged at camera pixels. A white-light texture image of the speckle pattern on the object surface is also captured by the camera. From the measured point cloud, discontinuities are detected and continuous regions are segmented. At each subsequent loading state, the deformed 3-D point cloud and texture image are also obtained in the same way.

A region of interest is selected in the reference image, in which a grid of reference sample points is populated. Using the correlation technique described in Section 3, those 2-D sample points are matched with corresponding 2-D points in the deformed texture image.

The 3-D coordinates of the reference sample points can be extracted easily from the reference 3-D point cloud as they correspond to integer pixels in the reference image. Image coordinates of a deformed sample point are generally non-integers, and hence it is necessary to interpolate 3-D coordinates from the neighbouring pixel sites using a bicubic interpolation. The systematic error induced by the interpolation process is not usually significant, since the geometric distances between the interpolation points are small (due to the high density of the point cloud),

Finally, the 3-D displacements are computed by direct subtraction of the 3-D coordinates of the deformed and reference sample points. Alternatively, the singular value decomposition (SVD) technique may be used to match subsets of neighbouring 3-D points [11], thus generating both displacement and rotation components.

4.2. Extension to multi-camera multi-projector system

The shape measurement system used in this work can easily be extended to a multi-camera multi-projector system due to the modular design. More cameras and/or projectors can be added in order to inspect different parts of the object, since the present calibration technique [1] is able to automatically bring 3-D point clouds measured by different camera-projector pairs together into a single global coordinate system,

Combining 3-D displacement fields requires two steps. (i) Displacement fields measured by the same camera (but different projectors) are merged together. Since 2-D reference sample points of those pairs are identical, most of their 3-D displacements have approximately the same values and thus can be statistically combined to improve the estimated displacement field. (ii) Displacement fields generated by different cameras are automatically transformed into the common global coordinate system.

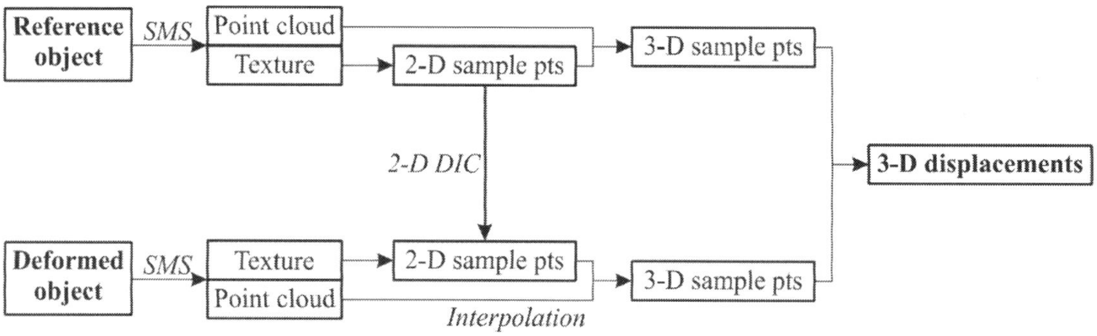

Figure 4: 3-D displacement estimation procedure using a one-camera one-projector sensor.

5. Experimental results

Figure 5 depicts the experiment arrangement for measuring surface profile and 3-D displacement fields. The planar specimen containing a hole and a step, which has been described in Section 2, was mounted on a test rig consisting of a translation stage that moved the specimen in its plane and mostly in the *X*-direction. The specimen was translated by a series of 18 displacements varying from 0 mm to 1.016 mm with an increment of 0.127 mm and from 1.016 mm to 10.160 mm with an increment of 1.016 mm. The translation stage has a resolution of 2.54 μm. A one-camera one-projector sensor was used in this experiment. The camera has an image resolution of 1024×1024 pixels which corresponds to a field of view of approximately 700×700 mm^2.

The surface profile and 3-D displacement fields were estimated for each true displacement prescribed by the translation stage. Figure 1 shows the results of the surface profile corresponding to the reference state (with the true displacement of 0 mm). A grid of nearly 1,000 sample points was selected within a region of interest in the reference image and was automatically redistributed to adapt with discontinuities present on the measured surface (see Figure 6-a). Figure 6-b shows an estimated 3-D displacement magnitude field that corresponds to the true displacement of 10.160 mm. The 3-D displacement magnitude d of a sample point was computed by

$$d = \sqrt{d_X^2 + d_Y^2 + d_Z^2} , \tag{4}$$

where d_X, d_Y and d_Z are respectively the *X*-, *Y*- and *Z*-components of the 3-D displacement estimated at the sample point. For visualisation purpose, the 3-D displacement magnitude field has been interpolated for each pixel from the 3-D displacement magnitudes of the sample points. The image displacement fields u and v are given in Figure 6-c and 6-d, respectively. It can be seen that all regions of the specimen have almost the same displacement magnitude (with the maximum difference of about 0.15 mm) as the specimen undergoes rigid-body displacement. Meanwhile, the image displacement field u has two distinct regions due to different magnifications caused by the presence of the step.

To quantify the accuracy of each 3-D displacement field measurement, the mean value of the estimated displacement magnitudes is compared against the true displacement. The mean 3-D displacement magnitude is computed by

$$\bar{d} = \frac{1}{n}\sum_{i=1}^{n} d_i , \tag{5}$$

where i represents a sample point and n is the number of sample points. The mean image displacements $\bar{u}$ and $\bar{v}$ are computed in the same way. Figure 7-a shows the variations of these mean displacements with the true displacement. The "ideal $\bar{d}$" and "estimated $\bar{d}$" curves are effectively superimposed on this graph. Figure 7-b therefore presents the error of the mean 3-D displacement magnitude which is computed by

$$\bar{e} = \bar{d} - d_{true} , \tag{6}$$

where d_{true} is the true displacement. The standard deviation of 3-D displacement magnitudes of the sample points is represented as an error bar. It is shown that the estimated 3-D displacements have the maximum error of about 0.1 mm (corresponding to about 1 part of 7,000 of the measurement volume side) and the maximum standard deviation of about 0.15 mm, for rigid-body displacements of up to 10.160 mm.

6. Concluding remarks

The projected fringe and digital image correlation techniques can be combined on a one-camera one-projector system to accurately measure both profile and 3-D deformation of surfaces with complex geometrical discontinuities. Experimental results show that this method currently achieves the accuracy of 1/7,000 of the measurement volume side for rigid-body 3-D displacements of up to about 10 mm. The method can easily be extended to a multi-camera multi-projector system to obtain complete 360° measurement. However, the method may be restricted to specimens undergoing relatively static deformation due to the use of the temporal phase unwrapping technique.

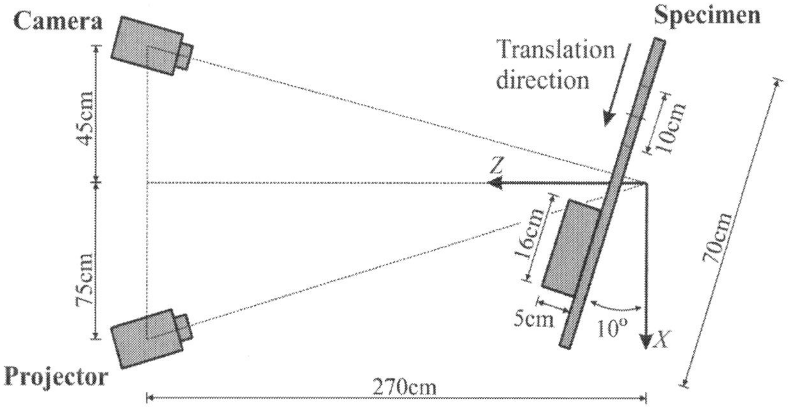

Figure 5: Experiment arrangement. The dimensions are approximate and not up to scale.

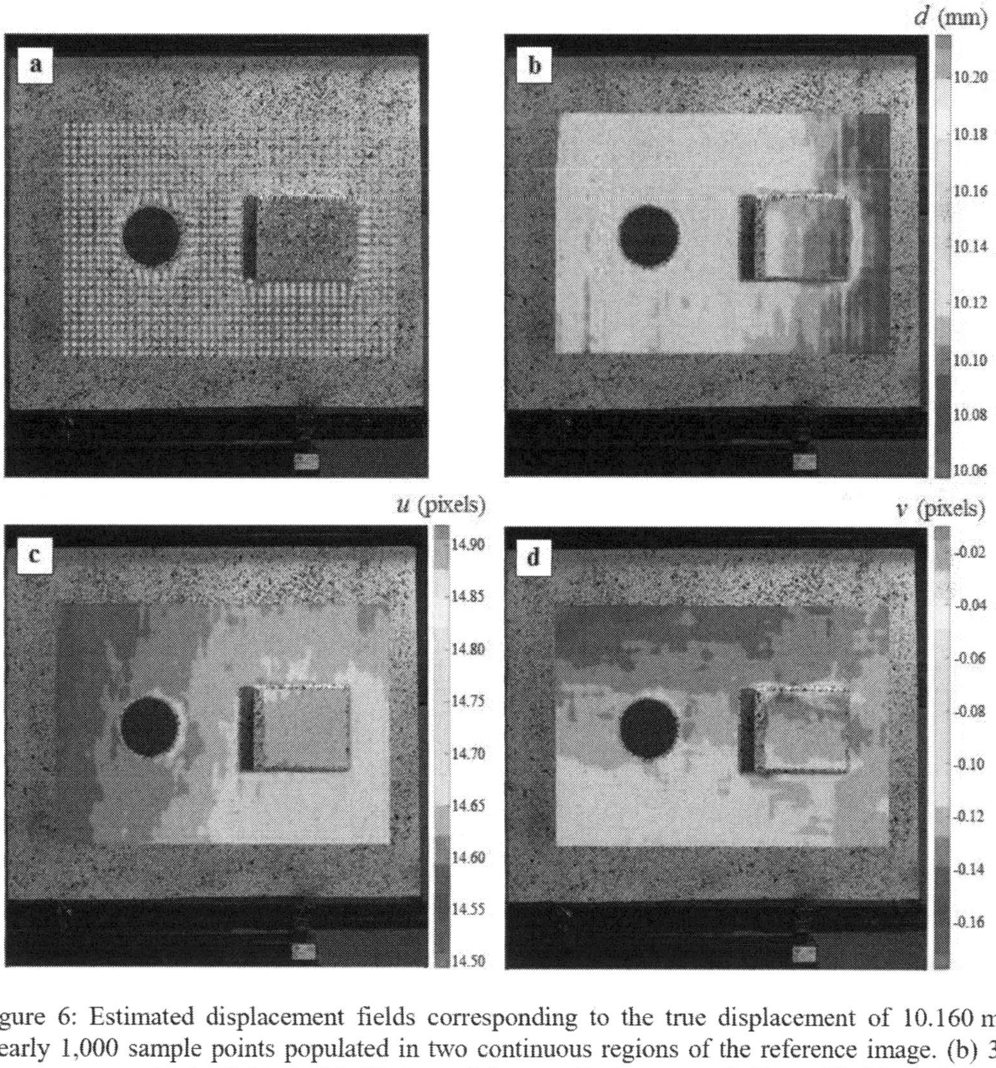

Figure 6: Estimated displacement fields corresponding to the true displacement of 10.160 mm. (a) Nearly 1,000 sample points populated in two continuous regions of the reference image. (b) 3-D displacement magnitude field d. (c) Horizontal image displacement field u. (d) Vertical image displacement field v.

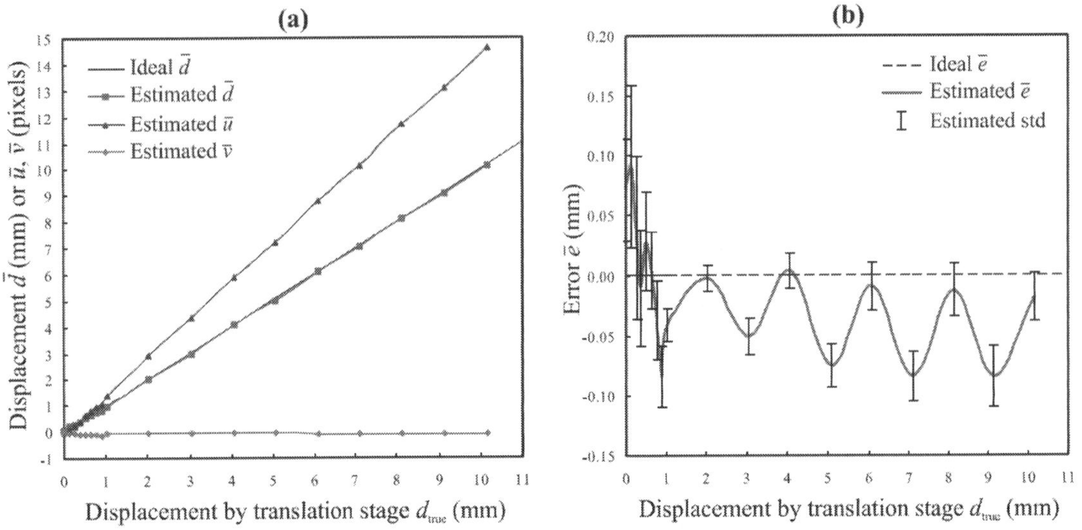

Figure 7: Comparison of measured displacements and true displacement prescribed by translation stage. (a) Mean 3-D displacement magnitude $\bar{d}$ and mean image displacements $\bar{u}$ and $\bar{v}$. (b) Error $\bar{e}$ of mean 3-D displacement magnitude $\bar{d}$ from true displacement. Geometric terms $\bar{d}$ and $\bar{e}$ are expressed in mm, while image-based terms $\bar{u}$ and $\bar{v}$ are expressed in pixels.

References

[1] JM Huntley, T Ogundana, RL Burguete and CR Coggrave, Large-scale full-field metrology using projected fringes: some challenges and solutions, *Proc. SPIE* 6616, 66162C, 2007.

[2] MA Sutton, SR McNeill, JD Helm and YJ Chao, Advances in two-dimensional and three-dimensional computer vision, *Photomechanics – Topics in Applied Physics*, PK Rastogi (Ed.), Springer, 323-372, 2000.

[3] H Weber, R Lichtenberger and T Wolf, The combination of speckle correlation and fringe projection for the measurement of dynamic 3-D deformations of airbags caps, *IUTAM Symposium on Advanced Optical Methods and Applications in Solid Mechanics*, Poitiers (France), 31 Aug – 4 Sep, 1998.

[4] CJ Tay, C Quan, T Wu and YH Huang, Integrated method for 3-D rigid-body displacement measurement using fringe projection, *Opt. Eng.* 43 (5), 1152-1159, 2004.

[5] C Quan, CJ Tay and YH Huang, 3-D deformation measurement using fringe projection and digital image correlation, *Optik* 115 (4), 164-168, 2004.

[6] J Klein, P Lointier, V Chalvidan and F Hild, Mesure de forme et de champ de déplacement par association de la projection de franges et de la corrélation d'images, *5ème Colloque Méthodes et Techniques Optiques pour l'Industrie*, Saint-Etienne (France), 15-19 Nov, 2004.

[7] HO Salder and JM Huntley, Temporal phase unwrapping: application to surface profiling of discontinuous objects, *Applied Optics* 36 (13), 2770-2775, 1997.

[8] CR Coggrave and JM Huntley, High-speed surface profilometer based on a spatial light modulator and pipeline image processor, *Opt. Eng.* 38 (9), 1573-1581, 1999.

[9] RC Gonzalez and RE Woods, *Digital Image Processing*, 2nd edition, Pearson Education, 2001.

[10] JM Huntley, An image processing system for the analysis of speckle photographs, *J. Phys. E: Sci. Instrum.* 19, 43-49, 1986.

[11] M Sjödahl and BF Oreb, Stitching interferometric measurement data for inspection of large optical components, *Opt. Eng.* 41 (2), 403-408, 2002.

Measurement of the shear wave speed in a submerged plate

Andrew J Hull and Benjamin A Cray
Autonomous and Defensive Systems Department
Naval Undersea Warfare Center, Newport, Rhode Island 02841 USA

E-mail: andrew.hull@navy.mil, benjamin.cray@navy.mil

Abstract. This paper develops an inverse method to estimate, in water, the shear wave speed in an isotropic, thick, elastomeric plate. The submerged plate is mechanically shaken and a scanning laser vibrometer is used to measure normal velocity on one surface. The temporal domain measurements are transformed into the frequency domain using a Fourier transform, then, the spatial domain measurements are transformed into the k_x,k_y wavevector domain using two Fourier transforms. Once the data is in the wavevector-frequency domain, the propagation wavenumber of each specific wave type can be estimated by fitting a circle to each collection of spectral peaks. Using this measured estimate of the wavenumber corresponding to the propagating wave, the Newton-Raphson gradient method is applied (inserting the estimated wavenumber into to the theoretical dispersion curve equation for wave propagation in a fluid-loaded plate,) hence resulting in an estimate of the shear wave speed. An experiment is included to illustrate the method and statistical properties of the measurement are discussed.

1. Introduction

Inverse methods involving wave propagation are typically broken into two methods. The first is a natural frequency or resonant method [1-3], where natural frequencies of a structure are measured and equated to the corresponding analytical natural frequencies. The analytical natural frequencies are usually functions of material modulus and this resultant expression can be solved to produce a value for the modulus at each natural frequency. The second method is a transfer function method [2,4-6], which typically involve equating measured data with an analytical model of the system. This mathematical expression is rewritten so that the estimated parameters become functions of the data. Historically, most inverse methods have been developed in air (or in vacuo). The loading of underwater structures is more complicated than that of air loaded testing. Fluid forces apply a compressional load whose effects usually need to be incorporated into the model. Failure to properly include fluid loading will often produce inaccuracies.

Inverse methods do exist to measure the material properties of fluid-loaded plates. In one type of submerged inverse problem [7,8], the dilatational and shear wave speed and attenuation was determined by exploiting the phase angle measurement via an insertion loss test. In another, the complex dilatational wave speed was found by applying a system model to an echo reduction experiment [9]. These methods involve actively insonifying the plate, however, all acoustic test tanks have lower frequency limitations. Measurements below 15 kHz are typically not possible.

Elastic plate theory has been extensively developed [10-12]. Thick plates have not been used to measure material properties because they support multiple wave types propagating in the medium, and any measurement technique has to have the ability to discern between each wave type and its contribution to the measurement. Transfer function methods that measure one output (at a single location) versus a fixed input do not have the capability to separate various wave types and their associated response levels. It is precisely this property of multiple waves that this paper exploits to measure the shear wave speed in a submerged thick plate. This is accomplished by shaking the plate at a fixed frequency while simultaneously measuring the normal velocity of the plate across its entire surface. These spatial domain measurements are transferred into a wavevector (two-wavenumber) domain by means of Fourier transforms. Individual waves are identified in this domain and the resulting free wave propagation wavenumbers are accurately estimated. Once they are known, they are inserted into a Newton-Raphson iterative solver applied to the theoretical dispersion equations for the propagation of plate waves. This results in estimates of the shear wave speed. An experiment is included to illustrate the technique when applied to an underwater elastomeric plate.

2. System equations

The theory of wave motion in a submerged, isotropic, elastic thick plate of infinite extent has been previously derived [12,13] and the transfer functions developed. The objective of this work is to estimate the shear wave speed in an underwater elastic plate using the underlying theoretical dispersion curve developed for the plate with fluid-loaded boundary conditions on the front and back. This dispersion curve equation is

$$\begin{aligned} g(k_d, k_s, k_f) = {} & p_1 \cos\left[h\left(k_d^2 - k^2\right)^{1/2}\right]\cos\left[h\left(k_s^2 - k^2\right)^{1/2}\right] + \\ & 2f_1 \cos\left[h\left(k_d^2 - k^2\right)^{1/2}\right]\sin\left[h\left(k_s^2 - k^2\right)^{1/2}\right] + 2f_2 \sin\left[h\left(k_d^2 - k^2\right)^{1/2}\right]\cos\left[h\left(k_s^2 - k^2\right)^{1/2}\right] + \\ & (p_2 + f_3)\sin\left[h\left(k_d^2 - k^2\right)^{1/2}\right]\sin\left[h\left(k_s^2 - k^2\right)^{1/2}\right] - p_1 = 0, \end{aligned} \tag{1}$$

where h is the thickness of the plate (m), k_s is the shear wavenumber ($\mathrm{rad\,m^{-1}}$), k_d is the dilatational wavenumber ($\mathrm{rad\,m^{-1}}$), k_f is the fluid wavenumber ($\mathrm{rad\,m^{-1}}$), and k is the measured free propagation wavenumber ($\mathrm{rad\,m^{-1}}$) corresponding to a specific wave traveling in the plate. (For the experiments in this paper, the measurements of k will either correspond to a F(0) flexural wave, L(0) longitudinal wave, or a FE(1) flexural wave.) The relationship between the shear wavenumber and the shear wave speed is $k_s = \omega / c_s$ where ω is the angular frequency ($\mathrm{rad\,s^{-1}}$) and c_s is the shear wave speed ($\mathrm{m\,s^{-1}}$); the relationship between the dilatational wavenumber and the dilatational wave speed is $k_d = \omega / c_d$ where c_d is the dilatational wave speed ($\mathrm{m\,s^{-1}}$); and the relationship between the fluid wavenumber and fluid wave speed is $k_f = \omega / c_f$ where c_f is the fluid compressional wave speed. In equation (1), the constants are given by

$$p_1 = -8\left(k_d^2 - k^2\right)^{1/2}\left(k_s^2 - k^2\right)^{1/2}\left(k_s^2 - k^2\right)^2 k^2 , \tag{2}$$

$$f_1 = \mathrm{i}\rho_f \rho^{-1}\left(k_f^2 - k^2\right)^{-1/2}\left(k_d^2 - k^2\right)^{1/2}\left(k_s^2 - 2k^2\right)^2 k_s^2 , \tag{3}$$

$$f_2 = 4\mathrm{i}\rho_f \rho^{-1}\left(k_f^2 - k^2\right)^{-1/2}\left(k_s^2 - k^2\right)^{1/2}\left(k_d^2 - k^2\right)k^2 k_s^4 , \tag{4}$$

$$p_2 = \left(k_s^2 - 2k^2\right)^4 + 16\left(k_d^2 - k^2\right)\left(k_s^2 - k^2\right)k^4 , \tag{5}$$

and

$$f_3 = \rho_f^2 \rho^{-2}\left(k_f^2 - k^2\right)^{-1}\left(k_d^2 - k^2\right)k_s^8 , \tag{6}$$

where ρ is the density of the plate ($\mathrm{kg\,m^{-3}}$) and ρ_f is the density of the fluid ($\mathrm{kg\,m^{-3}}$). Equations (1) thru (6) define the wavenumber-frequency dispersion curve of the plate with a fluid load on both sides. Using known values of h, k_d, k_f, k, ρ, and ρ_f, the value of k_s (and c_s) will be estimated. It is noted that in the absence of a fluid load, $f_1 = f_2 = f_3 = 0$ and the zeros of equation (1) are at the exact location of the zeros of the Rayleigh-Lamb equation for wave propagation in a plate.

To understand the estimation process, it is useful to display equation (1) as a surface with respect to the dilatational and shear wavenumber and examine its functional behavior. Figure 1 is a plot of the frequency equation for the propagation of waves in a fluid loaded elastic plate versus shear and dilatational wavenumber using a free propagation wavenumber of k = 115.6 $\mathrm{rad\,m^{-1}}$. In this figure, the right plot is a plot of the function with respect to the dilatational wavenumber with the shear wavenumber fixed at 100 $\mathrm{rad\,m^{-1}}$ and the bottom plot is a plot of the function with respect to the shear wavenumber with the dilatational wavenumber fixed at 15 $\mathrm{rad\,m^{-1}}$. The plate thickness h was 0.0254 m, the density of the plate ρ was 1100 $\mathrm{kg\,m^{-3}}$, the density of the fluid ρ_f was 1000 $\mathrm{kg\,m^{-3}}$, the fluid wavenumber k_f was 12.84 $\mathrm{rad\,m^{-1}}$, and the frequency ω was $(2\pi)3000$ $\mathrm{rad\,s^{-1}}$.

Note that the function is varying with respect to the shear wavenumber and essentially flat with respect to the dilatational wavenumber. This important feature reveals three pertinent facts with respect to the estimation process: (1) the shear wavenumber (and hence the shear wave speed) can be accurately estimated, (2) this estimation process may be multi-valued due to more than one relative minimum of the function, and (3) the dilatational wavenumber cannot be accurately estimated using this method. Furthermore, the estimation of the shear wavenumber is relatively invariant with respect to the dilatational wavenumber. A previous measure of the dilatational wave speed of this material will be used in the estimation process.

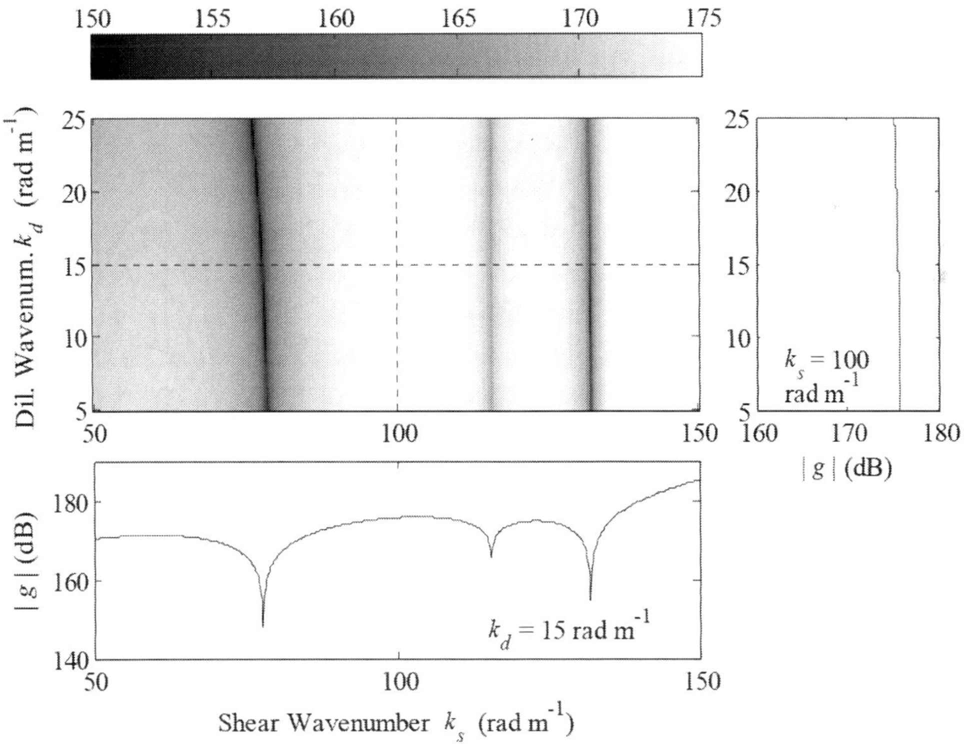

Figure 1. A plot of the dispersion equation for waves in a fluid-loaded elastic plate at 3 kHz.

A Newton-Raphson method can be applied to equation (1) for the estimation of the shear wavenumber. To eliminate the ambiguity of both positive and negative wavenumbers, the estimation process is applied to the square of the wavenumber, rather than the wavenumber itself. This produces

$$\left[k_s^2\right]_{j+1} = \left[k_s^2\right]_j - \left[\frac{\partial g(k_d, k_s, k_f)}{\partial (k_s^2)}\right]_j^{-1} g(k_d, k_s, k_f)_j , \tag{7}$$

where j is the iteration number at every fixed measurement location. Equation (7) is evaluated using the estimated free propagation wavenumber k, and the result is an estimate of the square of the shear wavenumber. Finally, the relationship $k_s = \omega / c_s$ is used to find the shear wave speed.

3. Experiment

An experiment was undertaken to validate the measurement technique. This estimation process uses the following assumption: (1) The return energy from the reflections at the edge of the plate does not interfer with the measurement process, and (2) the particle motion of the plate and fluid are linear. A plate was molded using Cytech Industries EN-6, a two-part urethane that consists of a mixture of a prepolymer and a curing agent. The plate was 0.780 m by 0.755 m by 0.0254 m thick and had a mass of 16.6 kg. The dilatational wave speed was previously measured as 1421 $\mathrm{m\,s^{-1}}$. The plate was mounted on four corners with bungee cords and a Wilcoxon Model F3/Z602WA electromagnetic shaker was attached to the backside near the middle. When the shaker was turned on, the front side was interrogated with a scanning Polytec LDV PSV-200 Doppler laser vibrometer that measured the normal velocity of the plate. The entire experiment was submerged in the Acoustic Test Facility at the Naval Undersea Warfare Center in Newport, Rhode Island, USA. This tank measures approximately 18.3 m by 12.2 m by 10.2 m deep and holds 2.46 million liters of fresh water. The experiment was conducted at a water temperature of 16.1 degrees Celsius. This experimental setup is shown in figure 2.

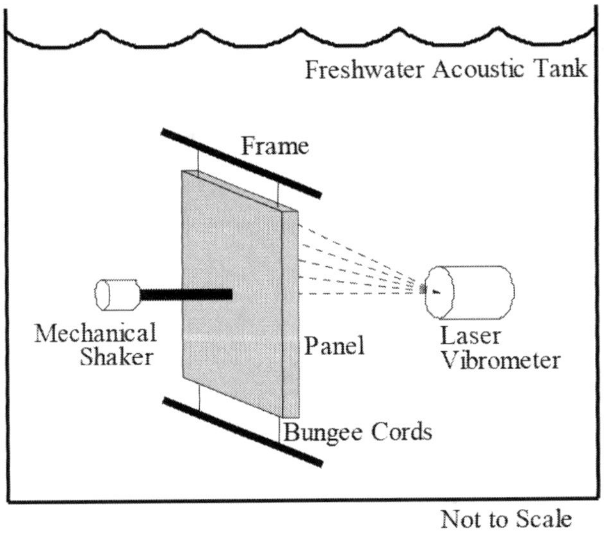

Figure 2. Experimental Setup.

A square grid of 90 by 90 points with a point-to-point spacing of 0.0082 m was used to collect 8100 spatial domain data points per frequency bin. After the data set was collected it was transformed into the frequency domain using a fast Fourier transform. Next, it was zero padded and transformed into the k_x, k_y wavevector domain using a two-dimensional 512 by 512 point fast Fourier transform.

Once this was accomplished, three waves were identified based on their relative spectra maxima. Isotropic elastic plate theory predicts that every wave will be circular in the k_x, k_y wavevector domain and centered at $k_x = k_y = 0$, thus a circle was fit to each set of relative maxima points using an ordinary least square estimate to the wavevector domain data to estimate the propagation wavenumber. This was done at 1 to 6 kHz in increments of 1 kHz. This circular data set allows statistical properties of the estimation process to be computed.

Figure 3 is a plot of the measured wave propagation locations in the k_x, k_y wavevector domain at (a) 4 kHz (left) and (b) 6 kHz (right). The F(0) flexural wave data is denoted with an x marker, the L(0) longitudinal wave data is denoted with a + marker, the FE(1) flexural wave is denoted with a o marker, and the circles fit to the markers are denoted with solid lines. For clarity, the markers have been decimated by eighty nine percent. Once the propagation wavenumbers are known, the shear wave speed can be estimated using equation (1). At the frequency of 5 kHz, the flexural wave is beginning to become incoherent across the major dimensions of the plate. The results of this estimation procedure are shown in table 1 for all of the waves measured during the test at all experimental frequencies. Additionally, the in air test data are included in this table for comparison. (The measured propagation wavenumber of the flexural wave at 3 kHz was used to construct figure 1.) The average value for the shear wave speed estimate for the flexural wave was 246.9 $\mathrm{m\,s^{-1}}$ and the average shear wave speed estimate for the longitudinal wave was 211.6 $\mathrm{m\,s^{-1}}$. This indicates a dispersion of the shear wave speed with respect to wave type. This is shown in figure 4, where all measurements are compared and displayed with their first standard deviation using an error bar plot. The average shear wave speed for all underwater measurements was 229.2 $\mathrm{m\,s^{-1}}$. Using this average value and the value of the dilatational wave speed, the dispersion curve in the wavenumber-frequency plane can be calculated. This is displayed as figure 5 along with the each data point. The slight mismatch between theory and experiment that is observed is due to the variation of the shear wave speed.

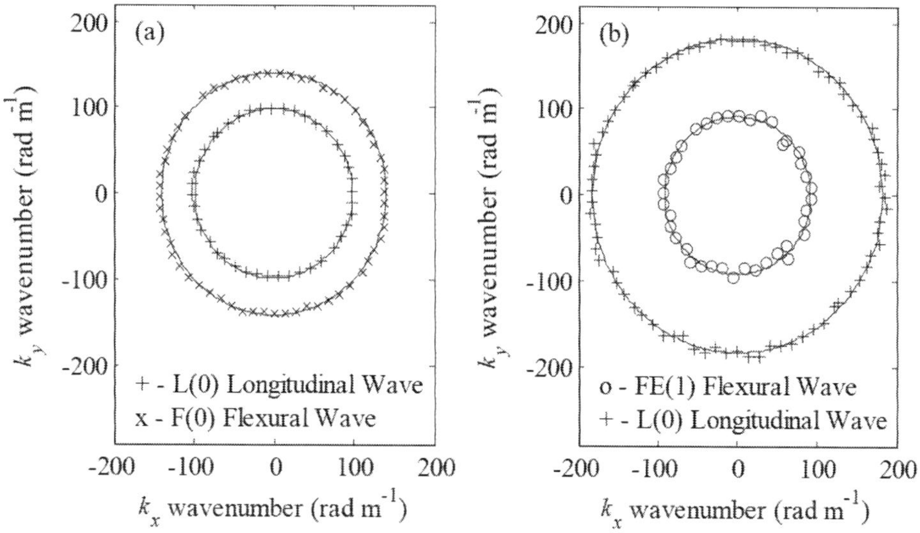

Figure 3. Locations of Lamb waves in the k_x, k_y wavevector plate for (a) 4 kHz and (b) 6 kHz.

Each of the measurements was statistically analyzed by calculating the standard deviation of the radius of the data points for each wave at every frequency. Once this was known, the shear wave speeds were estimated at plus and minus one standard deviation away from the mean. These results are shown for each individual wave in table 2. For the analysis using minus one standard deviation, the average shear wave speed was estimated to be 237.4 $\mathrm{m s^{-1}}$, and for the analysis using plus one standard deviation, the average shear wave speed was estimated to be 221.8 $\mathrm{m s^{-1}}$. These estimates are off by the original estimate of 229.2 $\mathrm{m s^{-1}}$ by 3.6 and 3.2 percent respectively, which indicates that the estimation process is extremely stable and the propagation of errors thru the analysis are small.

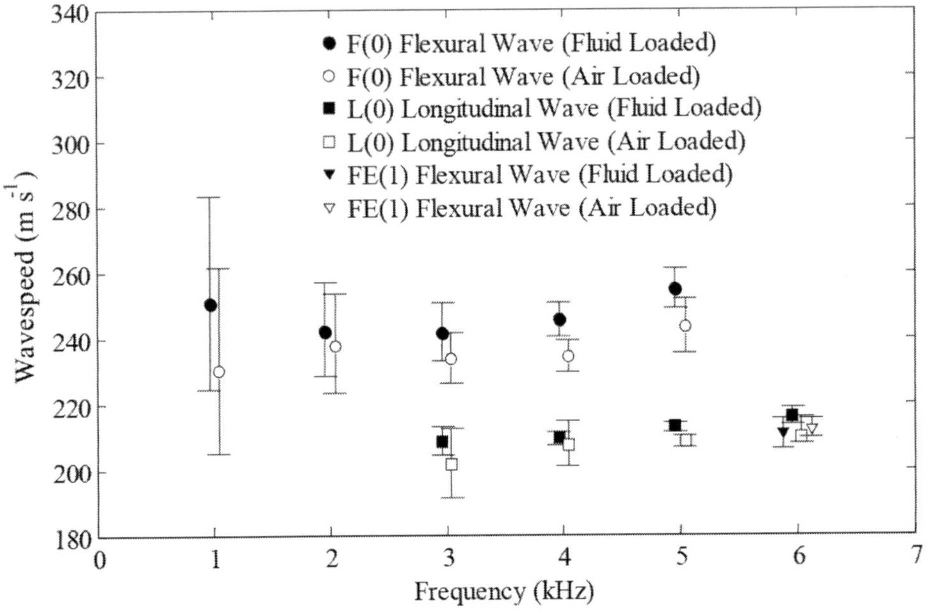

Figure 4. Shear wave speed measurements showing one standard deviation.

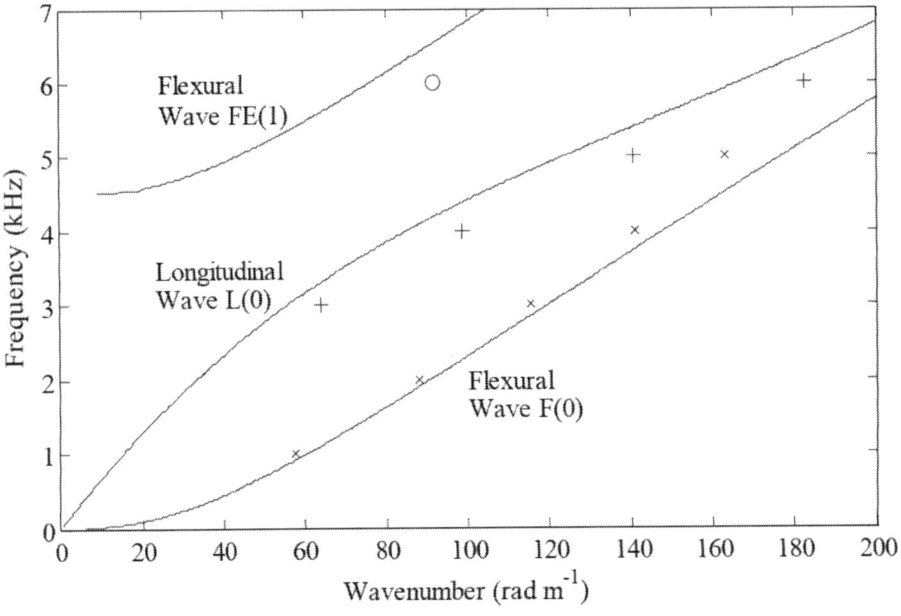

Figure 5. Dispersion curve of fluid-loaded plate showing data (markers) and theory (solid lines).

Table 1. Estimated shear wave speeds for all measured waves.

Wave Name	Frequency kHz	Measured k rad m^{-1}	Estimated k_S rad m^{-1}	Estimated c_S (water) m s^{-1}	Estimated c_S (air) m s^{-1}
F(0)	1	57.7	25.1	250.8	229.9
F(0)	2	88.1	52.0	241.9	237.3
F(0)	3	115.6	78.0	241.6	233.6
F(0)	4	141.1	102.4	245.4	234.2
F(0)	5	163.1	123.3	254.8	243.2
L(0)	3	63.6	90.5	208.4	201.3
L(0)	4	98.7	120.0	209.4	207.1
L(0)	5	140.6	147.8	212.6	208.2
L(0)	6	182.6	174.5	216.0	209.5
FE(1)	6	91.4	178.9	210.7	212.0

Table 2. Statistical analysis of estimated shear wave speeds.

Wave Name	Frequency kHz	Standard Deviation rad m^{-1}	Estimated c_S (−1 Std. Dev.) m s^{-1}	Estimated c_S (+1 Std. Dev.) m s^{-1}
F(0)	1	3.5	283.2	224.4
F(0)	2	3.2	257.0	228.5
F(0)	3	2.9	250.6	233.1
F(0)	4	2.2	250.6	240.4
F(0)	5	3.0	260.9	249.0
L(0)	3	1.9	212.9	204.2
L(0)	4	1.6	211.4	207.3
L(0)	5	1.6	214.0	211.1
L(0)	6	2.8	218.3	213.7
FE(1)	6	4.1	215.4	206.2

4. Conclusions

The shear wave speed of a submerged isotropic plate can be accurately estimated by the inverse method developed in this paper. This approach consists of shaking an underwater plate, measuring the normal velocity of one surface, transforming these spatial measurements into the k_x, k_y wavevector domain, identifying the major propagation wave types, and then using these wavenumbers to estimate the shear wave speed via an inverse method applied to the theoretical dispersion curve equations for propagation of waves in a fluid-loaded plate. Experimental measurements show that the use of this method provides accurate estimates of the shear wave speed. Additionally, it was shown that the dilatational wave speed cannot be accurately estimated using this technique, as the dispersion curve equations of motion do not have well defined relative minima with respect to the dilatational wavenumber.

5. Acknowledgements
This work was funded by the Office of Naval Research. The authors wish to thank Program Manager Dr. David M. Drumheller for sponsoring this effort. Additionally, the authors wish to thank Walter Boober, Rene LaFleur, and Gorham Lau of the Naval Undersea Warfare Center Acoustic Test Facility for collecting the spatial domain data.

References
[1] Madigosky W M and Lee G F 1983 *J. Acoust. Soc. Am.* **73** 1374-1377.
[2] Oyaniji S O and Tomlinson G R 1985 *J. Sound Vib.* **101** 277-298.
[3] Garrett S L 1990 *J. Acoust. Soc. Am.* **88** 210-220.
[4] Buchanan J L 1987 *J. Acoust. Soc. Am.* **81** 1775-1786.
[5] Hull A J 1996 *J. Sound Vib.* **195** 545-551.
[6] Larsson P O 1991 *J. Sound Vib.* **146** 111-123.
[7] Piquette J C and Paolero A E 2003 *J. Acoust. Soc. Am.* **113** 1518-1525.
[8] Piquette J C 2004 *J. Acoust. Soc. Am.* **115** 2110-2121.
[9] Gartland G J 2007, MS Thesis, Michigan State University, East Lansing, Michigan.
[10] Graff K F 1975 *Wave Motion in Elastic Solids* (New York: Dover Publications).
[11] Brekhovskikh L M 1980 *Waves in Layered Media* (San Diego, California: Academic Press).
[12] Hull A J 2005 *J. Sound Vib.* **279** 497-507.
[13] Osborne M F M and Hart S D 1945 *J. Acoust. Soc. Am.* **17** 1-18.

Residual stress relaxation measurements across interfaces at macro- and micro-scales using slitting and DIC

A Blair[1], N Daynes[1], D Hamilton[1], G Horne[1], PJ Heard[2], DZL Hodgson[1], TB Scott[2], A Shterenlikht[1*]

[1] Department of Mechanical Engineering, University of Bristol, Bristol BS8 1TR, UK
[2] Interface Analysis Centre, University of Bristol, Bristol BS2 8BS, UK
[*] corresponding author, email mexas@bristol.ac.uk

Abstract. In this paper digital image correlation is used to measure relaxation of residual stresses across an interface. On the macro scale the method is applied to a tri-layer bonded aluminium sample, where the middle layer is in tension and the top and the bottom layers are in compression. High contrast speckle pattern was sprayed onto the surface. The relaxation was done with the slitting saw. Three dimensional image correlation was used. On the micro scale the technique was applied to a heat treated large grain brass loaded in tension. Mechanical and electro polishing was used for surface preparation. A focused ion beam was used for slitting across a grain boundary and for imaging. Grain orientation was measured using electron back-scattering diffraction. Two dimensional image correlation was employed. In all macro- and micro-scale experiments the range of measured relaxation was sub-pixel, almost at the limit of the resolution of the image correlation algorithms. In the macro-scale experiments, the limiting factor was low residual stress, due to low shear strength of the Araldite glue used for bonding. Finite element simulation of the relaxation agreed only qualitatively with the experimental results at both size scales. The methodology is intended for use with inverse methods, i.e. the measured relaxation is applied as the boundary conditions to an appropriate FE model which produces stresses equal to the relaxed residual stresses, but with opposite sign. The main conclusion is that the digital image correlation method could be used to measure relaxation caused by slitting in heterogeneous materials and structures at both macro- and micro-scales. However, the repeatability of the techniques needs to be improved before residual stresses can be determined confidently.

1 Introduction

Measurement of residual stress in areas of high stress gradient is an active research area. The most successful technique to date is the crack compliance, or slitting, method, which involves introduction of a slot to a material under stress and measuring the relaxation caused [1, 2]. In this work an attempt is made to use the digital image correlation (DIC) technique for relaxation measurement at macro- and micro-scales. The Istra 4D DIC system by Dantec Dynamics [3] has been used in this work.

2 Macro-scale experiments

The specimens were machined from aluminium alloy 7050-T7351 provided by Abaqus UK. A Young's modulus and a yield stress of 73.3 GPa and 440 MPa respectively were measured for this alloy, and a Poisson's ratio of 0.33 was assumed. The specimen was prepared as follows: (1) three strips of

Journal of Physics: Conference Series **181** (2009) 012078 doi:10.1088/1742-6596/181/1/012078

aluminium, one long and two short, cross section 6 mm × 12 mm, were prepared, (2) the long strip was loaded to 10 kN in tension, producing a nominal stress of 139 MPa, (3) with the long strip still under load, the two short strips, side plates, were glued to the central strip on both front and back sides using Huntsman's Araldite 2010-1 adhesive, with a nominal shear strength of 23 MPa, (4) after the glue had cured the load was removed, thus creating a specimen with highly non-uniform residual stress – tensile in the central strip, and compressive in the side plates. Fig. 1 schematically shows the completed specimen. Note the outer edges of the side plates were chamfered along the width to reduce the mode II stress intensity factor at the interface corners.

The speckle pattern was applied to the front of the try-layer beam, Fig. 2(c)-(d), and two 6mm long strain gauges were applied to the two side faces. It is generally accepted that the front face gauge is only sensitive to the first 5-10% of the slot depth and the back face gauge gives useful data from 5-10% to 95% of the specimen thickness [2]. The specimen temperature was monitored with a thermocouple. The temperature drift never exceeded 0.5°C. In addition, the gauges were temperature compensated for aluminium.

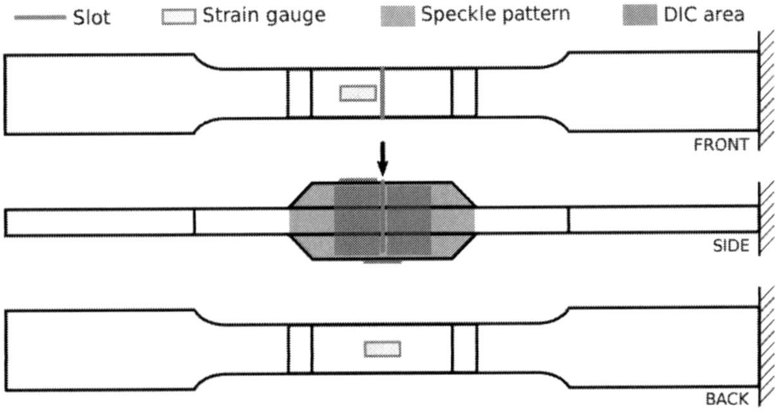

Figure 1: Schematic of the macro-scale experiment. Note the black arrow indicating the direction of slitting, and the chamfered edges of the side plates.

Typically electro discharge machining (EDM) is used to introduce a slot at the macro-scale. However, the EDM process requires the area near the cut to be submerged in oil, which would destroy or at least invalidate the DIC pattern. Therefore a 0.4 mm thick slitting saw was used in this work. The cutting was done in 0.25 mm increments. Strain gauge readings and DIC images were taken at each increment and before and after the entire slot was created, see Fig. 2(a)-(b). The position of the specimen relative to the cutting blade was controlled by a milling machine stage. Digital readouts from a pair of LVDTs allowed for positional accuracy of 10 μm. This allowed the specimen to be returned to a reference position for DIC image acquisition after slotting, which significantly reduced rigid body motion.

The imaging was done with the resolution of approximately 20 μm/px. 3D DIC measurement was used with a facet size of 60 px and a facet overlap of 1/3. The DIC relaxation displacement field normal to the slot is shown in Fig. 3, and in Fig. 4(a) it is plotted on the lines parallel to the slot. Even though the DIC displacement field has the expected shape and magnitude, and the profiles on either side of the slot match within the error bars, it was not possible to extract a meaningful strain field from it. This is probably due to the fact that displacement measured were very small, the whole range of relaxation being sub-pixel.

The relaxation strain gauge data is shown in Fig. 2(e) and the FE simulated relaxation strain is shown in Fig. 2(f). The FE predicted residual stress is shown in Fig. 4(b). The FE strain was averaged over the size of the strain gauge, 6 mm. Apart from showing three distinct regions in the back face data, the experimental and the simulated curves do not agree. The shapes of the curves differ, and the FE strain is roughly 500% higher. One possible explanation is that the glue was not fully cured when the loading was removed from the specimen. This would have have reduced the locked-in stresses substantially.

Journal of Physics: Conference Series **181** (2009) 012078 doi:10.1088/1742-6596/181/1/012078

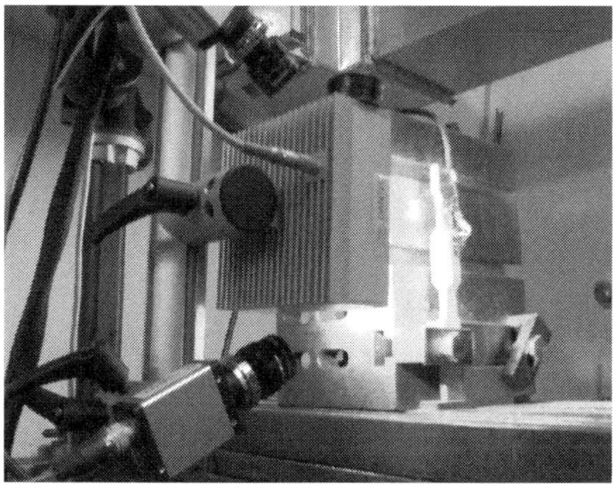

(a) DIC setup. Note that specimen was returned to the initial position for each image.

(b) Slitting of specimen. The front face strain gauges are visible above the blade. The saw is traversing from the rear of the specimen towards the camera, to protect the speckle pattern.

(c) Left hand CCD image. Note the large areas out of focus away from the slot due to the limited focal range of the cameras.

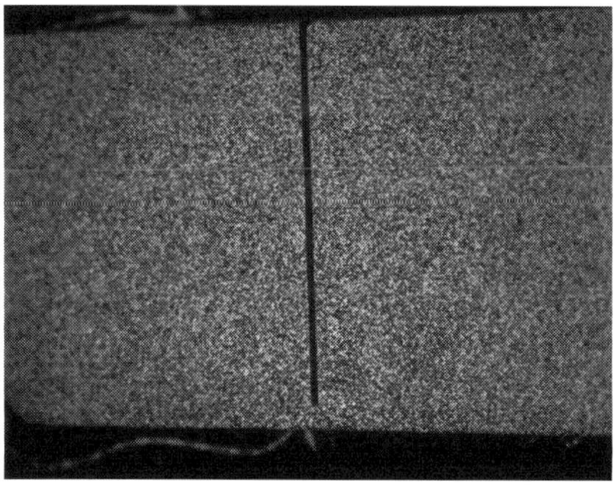

(d) Right hand CCD image. Note the connective ligament at the bottom of the slot—thought to be the reason why the displacement profiles are not precisely symmetrical at each end.

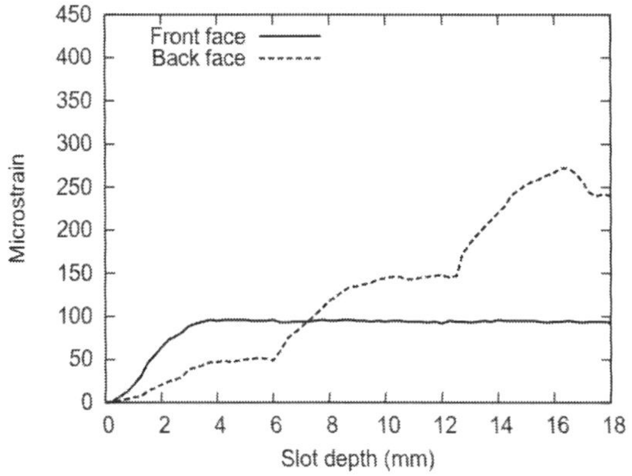

(e) Strain gauge relaxation results. Note the front strain gauge is sensitive to only first 4 mm. Passing the interfaces, at approx. 6 mm and 12 mm, causes sharp peaks in the back face strain.

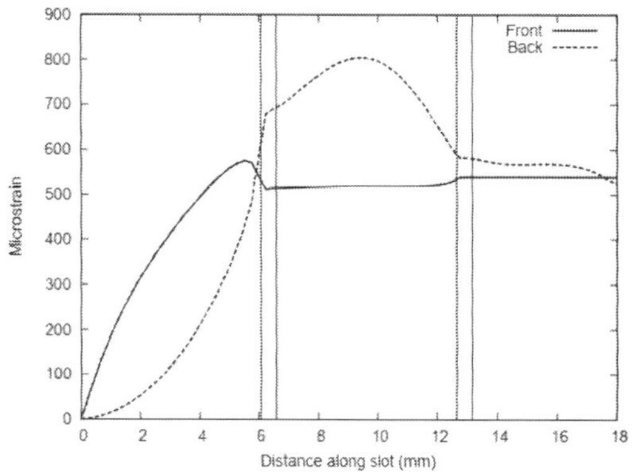

(f) FE simulated strain gauge relaxation results. Values were averaged over the area where the strain gauges were placed in the experiment. The vertical lines indicate the layers of Araldite at slot depths of 6 mm and 13 mm.

Figure 2: Macro-scale slitting experiment and FE results.

Journal of Physics: Conference Series **181** (2009) 012078 doi:10.1088/1742-6596/181/1/012078

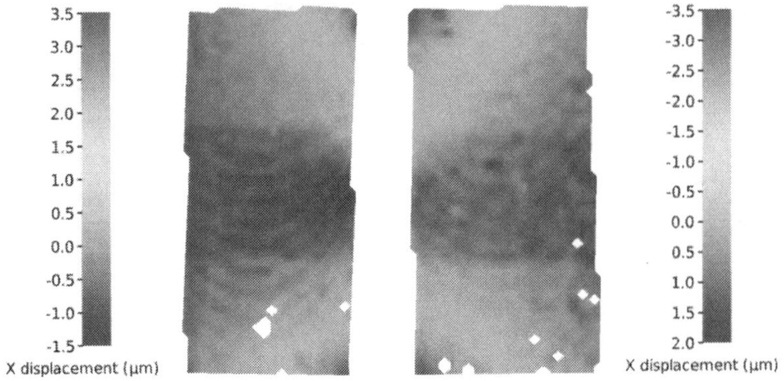

(b) Left side (c) Right side

Figure 3: DIC relaxation displacements normal to the slot. Note the concentric circles on the plot, which indicate the pre-experiment milling of the surface has some effect on the displacements. Also clearly visible are the interface lines, one third and two thirds of the way down the image. The maximum side plate relaxation is at the outer edges of the specimen. The maximum relaxation of the central strip is at the centre of the specimen.

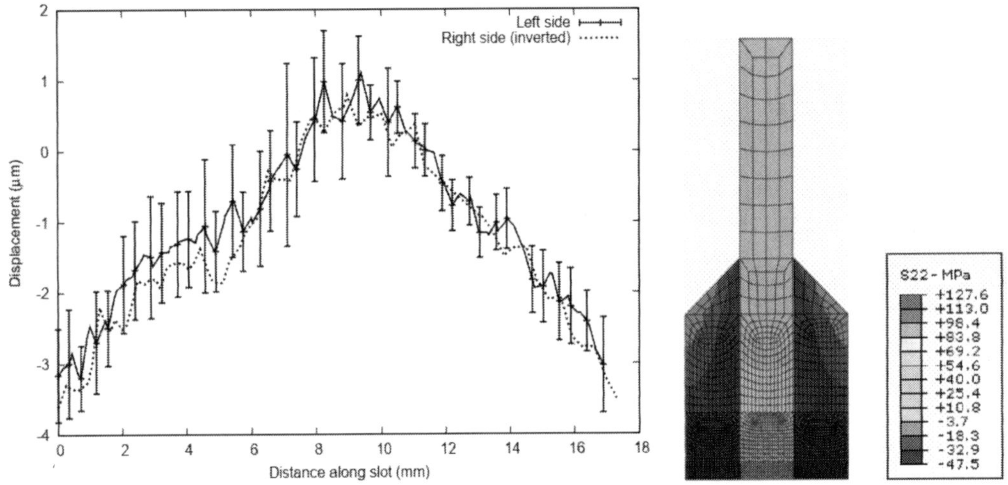

(a) DIC relaxation displacement normal to the slot.

(b) FE prediction of residual stress.

Figure 4: (a) DIC relaxation; and (b) FE residual stress. The maximum tensile stress in the central strip is 74 MPa. The maximum compressive stress in the side bars is 23 MPa. The maximum shear stress is 22 MPa, just below the shear strength of the Araldite, 23 MPa.

3 Micro-scale experiments

At the micro- and the nano-scales the focused ion beam (FIB) was used for slotting and imaging. For the basics of FIB refer to [4]. In this work FEI Strata FIB201 system was used [5].

A combination of FIB and DIC has been used successfully to measure residual stresses at micro-scale. However, most results to date have been obtained on special materials, e.g. non-metallic thin films [6, 7, 8]. Conducting measurements in metals could be significantly more complex due to an effect called 're-deposition', a condition where material removed from the slot is being re-deposited on the surface near the slot, thus preventing successful application of DIC. This effect was observed in mild steels in previous work [9].

In this work 40% zinc brass was used. It was heat treated to 875°C and quenched, resulting in purely β phase large grain brass. A 5 mm × 4 mm specimen was machined from heat treated brass.

Given the complexity of creating a controlled residual stress sample at the micro-scale, tensile stress was locked in the specimen using a specially machined loading device, shown in Fig. 5(a).

3.1 Mechanically polished sample

The sample was mechanically polished to 1 μm, and stressed to 200 MPa. In contrast to mild steel, there was no re-deposition on brass. In fact, the gas assisted etching (GAE), used in FIB milling of steel to alleviate the re-deposition problem, made FIB imaging of brass worse, see Fig. 5(b)-(c); note these are two different experiments on the same sample.

Because FIB imaging uses a single camera, FIB imaging is inherently two-dimensional. Therefore a 2D version of the Istra DIC software was used. The DIC results, Fig. 5(d)-(f), show the maximum relaxation at the centre of the slot, dropping to zero towards the ends of the slot and decreasing to the noise, 'reference', level away from the slot. Once again, these results are at the very limit of the DIC resolution.

3.2 Electro-polished sample

It is frequently claimed that mechanical polishing can itself introduce residual stress in the surface layers. To investigate this effect, the specimen was electro-polished (details are given in [10]).

Results of slitting the electro-polished sample are shown in Fig. 6(a)-(e). The surface now had much less detail, which led to poorer DIC results. The best DIC relaxation displacement field is shown in Fig. 6(g). However, only a very general relaxation pattern can be seen.

FE modelling of the grain boundary interface required grain orientations, which were measured with electron back scattering diffraction (EBSD) method, Fig. 6(f). The outline of the FE model is shown in Fig. 7(a), and the FE predicted relaxation displacements are shown in 7(b). Orthotropic copper material properties were assumed with E_{1111}=169 GPa, E_{1122}=122 GPa and E_{1212}=75.3 GPa.

In the absence of reliable DIC results, FE relaxation displacements could not be verified. One of the assumptions in the FE model was that the grain boundary is normal to the surface. This is based on Fig. 6(d). However, in other experiments oblique grain boundaries have been observed, Fig. 6(e). In such cases new FE models, with correct grain boundary angles, must be created.

Conclusions

The proposed tri-layer residual stress specimen is useful as a macro-scale highly non-uniform residual stress test case. The DIC was successful in producing qualitative measurement of relaxation displacement caused by the removal of residual stress. However, the DIC resolution remains one of the major limiting factors. As the DIC resolution is linked to to the quality of imaging, significant efforts were directed towards obtaining images rich in detail and of high contrast. At the macro-scale the best pattern was produced using spray paint. At the micro-scale mechanical polishing left substantially more surface detail than electro polishing. The macro-scale results were slightly better than the micro-scale, probably due to a superior surface pattern. However, at both scales the relaxation displacement ranges were sub-pixel, at the limit of the DIC resolution.

An obvious route for improvement is to use a bigger CCD, which would give a better image and DIC resolution. Another interesting direction is an investigation of alternative cut geometries, e.g. ring core, see Fig. 8, has an advantage that the whole of the material inside the ring is in the fully relaxed state, provided the slot is sufficiently deep.

Acknowledgments

The authors gratefully acknowledge Airbus UK for provision of materials. They thank Dr Richard Burguete, Airbus UK, and Prof Peter Flewitt, Department of Physics, University of Bristol, for advice in all aspects of experimental work.

Journal of Physics: Conference Series **181** (2009) 012078 doi:10.1088/1742-6596/181/1/012078

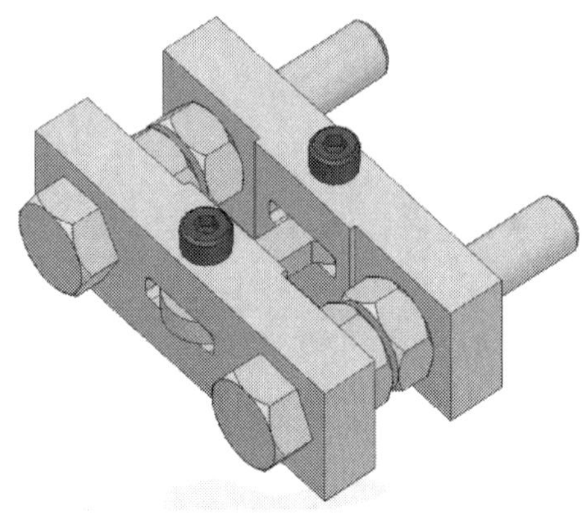

(a) Micro-scale tensile specimen assembly

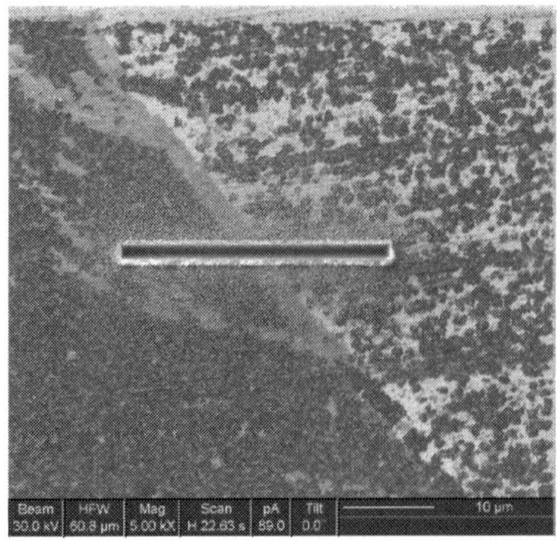

(b) FIB image showing surface deposition and damage after ion-milling slot using iodine GAE.

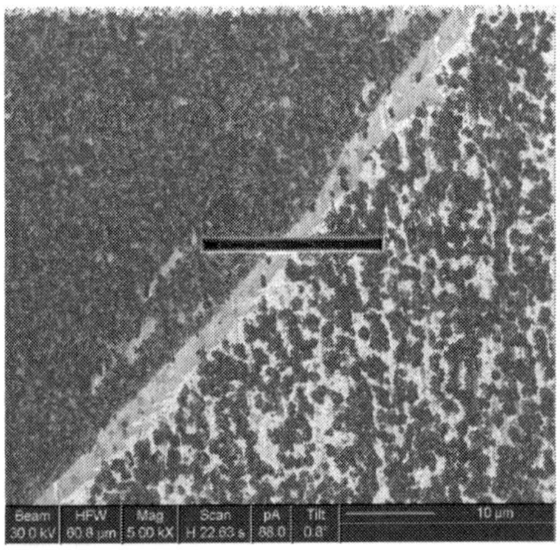

(c) FIB image showing pristine surface after ion-milling slot, no GAE. The two grains can be seen clearly. Note the grain boundary is 2-3 μm thick.

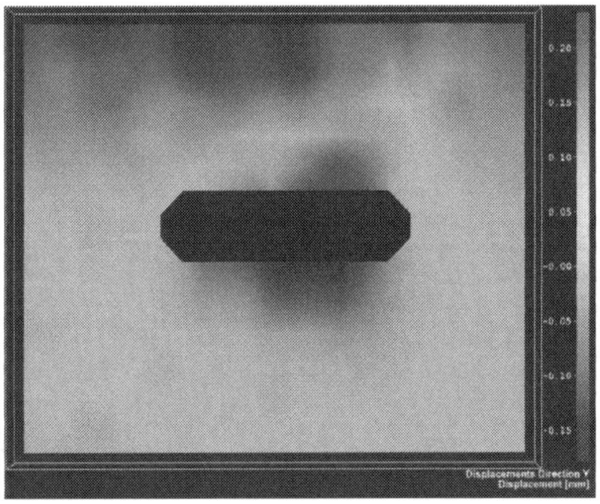

(d) Vertical displacement field, relaxation caused by the FIB slot. Note the units are in pixel.

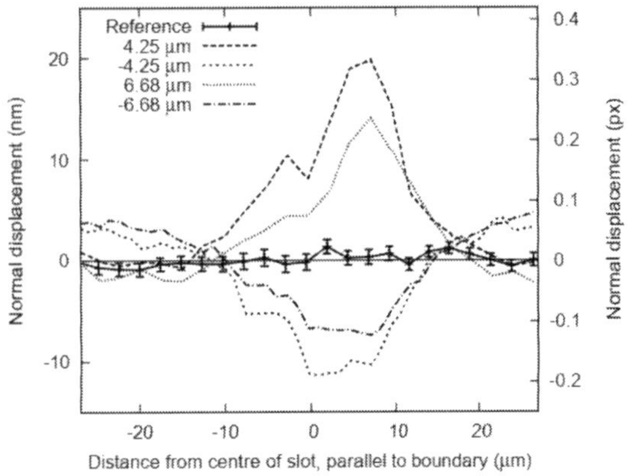

(e) Relaxation extracted from DIC on lines parallel to the slot, distance from the slot centre is shown in the legend. Slot length is 20 μm.

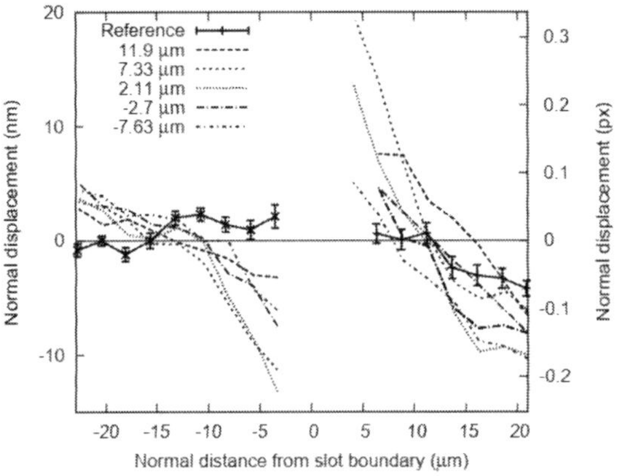

(f) Relaxation extracted from DIC on lines normal to the slot, distance from the slot centre is shown in the legend. Note the maximum displacement is only about 0.3 pixels.

Figure 5: Mechanically polished micro-scale slitting experiment and FE results.

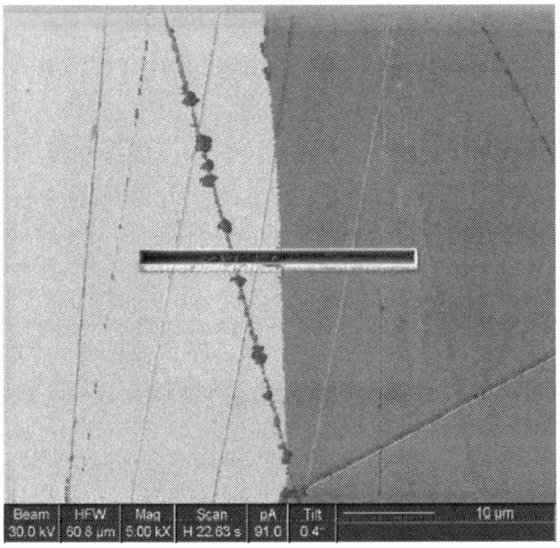

(a) ×5k image, a faint outline of the ×10k image location can be seen, due to exposure to the damaging ion beam.

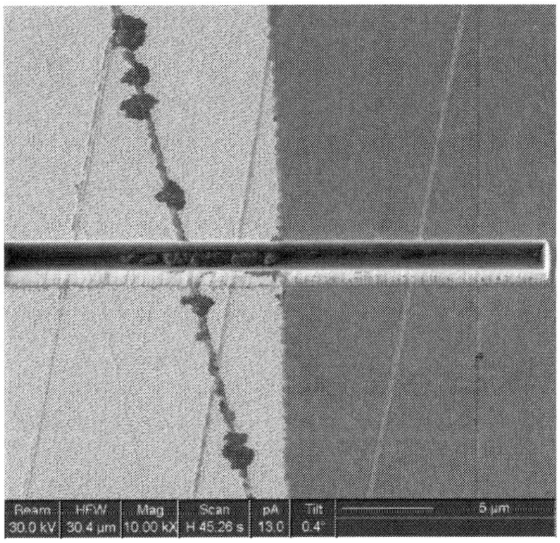

(b) ×10k image of the same slot.

(c) Side slot profile. Note a typical 'V' shape.

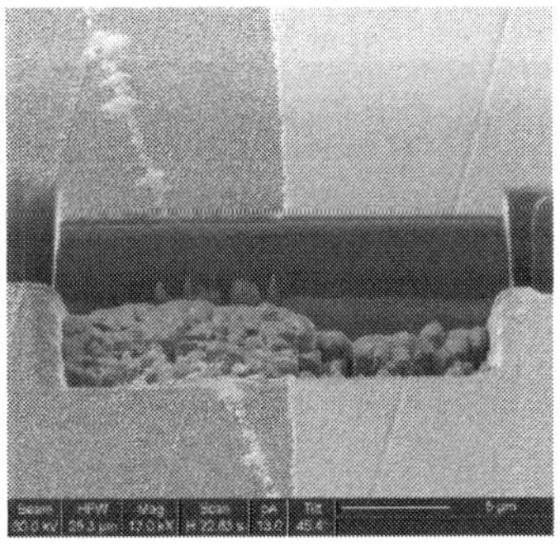

(d) Slot depth profile. Note irregular depth in the left grain.

(e) Detail of unrelated FIB slot, note grain boundary at oblique angle to the surface.

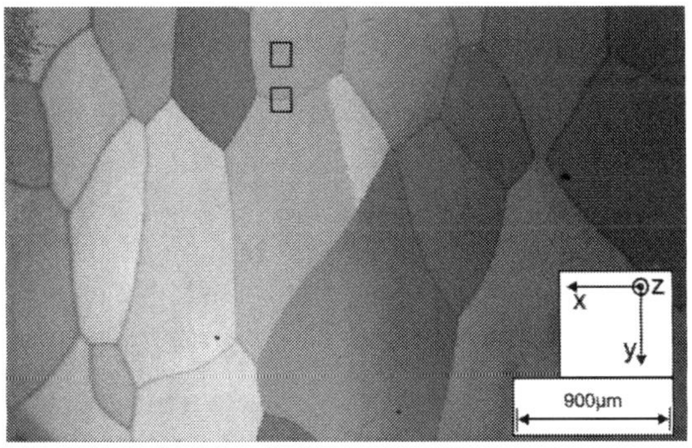

(f) EBSD inverse pole plot. Colours represent crystal orientations. Locations of the FIB images for slotting are marked.

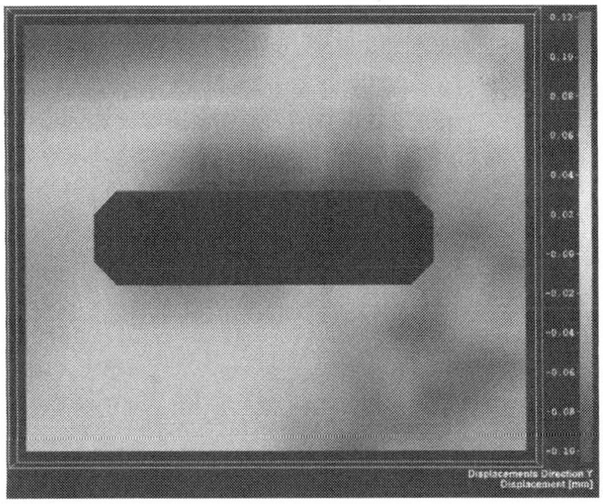

(g) DIC relaxation displacement in pixels. Note the range is only 0.2 pixels.

Figure 6: Electro polishing micro-scale slitting experiment.

References

[1] W. Cheng and I. Finnie. *Residual Stress Measurement and the Slitting Method.* Springer, New York, 2007.

Journal of Physics: Conference Series **181** (2009) 012078 doi:10.1088/1742-6596/181/1/012078

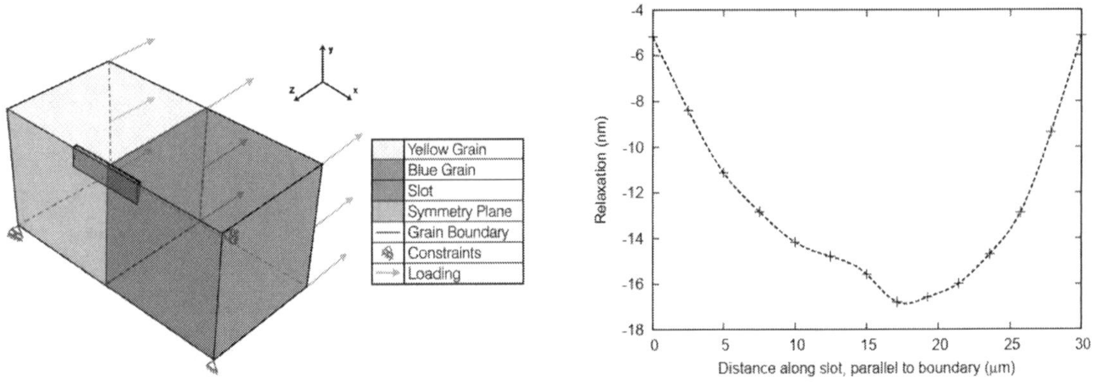

(a) Schematic of micro-scale FE model

(b) Predicted relaxation.

Figure 7: Micro-scale FE model and predicted relaxation.

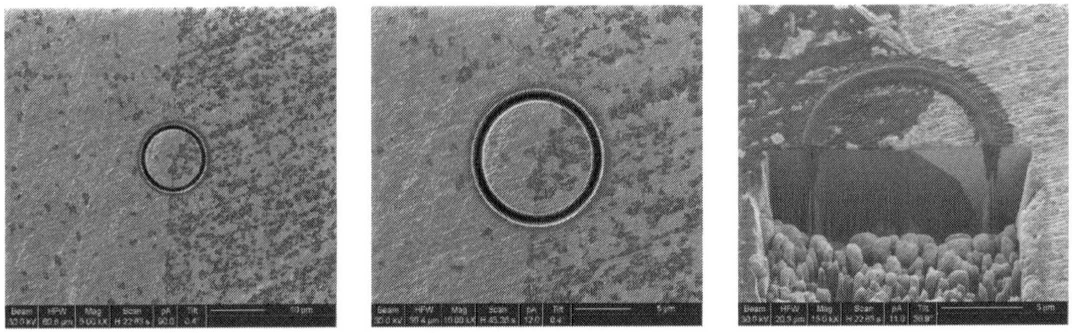

(a) ×5k

(b) ×10k

(c) Profile

Figure 8: FIB images of ring core. Note the grain boundary crosses the sample surface at an oblique angle.

[2] M B Prime. Comparison with other residual stress measurement methods, 2007. Available from: http://www.lanl.gov/residual/compare.shtml (last accessed 15/03/09).

[3] Dantec Dynamics. Digital image correlation for deformation measurement, 2009. http://www.dantecdynamics.com/Default.aspx?ID=1030 (last accessed 22/02/09).

[4] C A Volkert and A M Minor. Focused ion beam microscopy and micromachining. *MRS Bulletin*, 32:389–399, 2007. Available from: http://www.mrs.org/bulletin (last accessed 08/04/08].

[5] FEI Company. Focused ion beam technology, capabilities and applications, 2004. www.fei.com/uploadedFiles/Documents/Content/2006_06_FIBCapabilities_td.pdf.

[6] K.J. Kang, N. Yao, M.Y. He, and A.G. Evans. A method for in situ measurement of residual stress in thin films by using the focused ion beam. *Thin Solid Films*, 443(1-2):71–77, 2003.

[7] N. Sabaté, D. Vogel, A. Gollhardt, J. Marcos, I. Gràcia, C. Cané, and B. Michel. Digital image correlation of nanoscale deformation fields for local stress measurements in thin films. *Nanotechnology*, 17(20):5264–5270, 2006.

[8] Yo sep Yang, Jong Gu Bae, and Chan Gyung Park. Measurement of residual stress in thin-sized steel wires by using focused ion beam and digital image correlation method. *Advanced Material Research*, 26–28:1187–1190, 2007.

[9] N. Daynes, G. Horne, P. J. Heard, D. Z. L. Hodgson, and A. Shterenlikht. Microscale residual stress measurement in steel using focused ion beam slotting and digital image correlation. In *Proceedings of the 2008 International Conference on Residual Stresses, Denver, Colorado, USA*, 2008.

[10] A. Blair, N. Daynes, D. Hamilton, and G. Horne. Measurement of residual stress in heterogeneous materials. Masters Project Report M2009/12, Department of Mechanical Engineering, University of Bristol, 2009.

Journal of Physics: Conference Series **181** (2009) 012079 doi:10.1088/1742-6596/181/1/012079

Capturing higher modes of vibration of micromachined resonators

Venkatesh K P and Rudra Pratap

CranesSci MEMS Lab, Department of Mechanical Engineering,Indian Institute of Science, Bangalore, India.

E-mail: pratap@mecheng.iisc.ernet.in

Abstract. MEMS resonators have potential applications in the areas of RF-MEMS, clock oscillators, ultrasound transducers, etc. The important characteristics of a resonator are its resonant frequency and Q-factor (a measure of damping). Usually large damping in macro structures makes it difficult to excite and measure their higher modes. In contrast, MEMS resonators seem amenable to excitation in higher modes. In this paper, 28 modes of vibration of an electrothermal actuator are experimentally captured— perhaps the highest number of modes experimentally captured so far. We verify these modes with FEM simulations and report that all the measured frequencies are within 5% of theoretically predicted values.

1. Introduction

Micromachined resonators have potential applications in RF-MEMS, clock oscillators, resonant sensors such as biological mass detector, optical communications, displays, barcode readers and biomedical imaging [1, 2]. To use these resonators effectively, one needs to find a suitable actuation mechanism that is compatible with MEMS technology. Several actuation schemes are explored by different research groups across the world. They can be classified into electrostatic, piezoelectric, electromagnetic and electrothermal types [3, 4]. The response of the resonator to any actuation mechanism depends on two important characterstics of the resonator — its resonant frequency and quality factor (Q) [5]. Its an easy task to estimate the resonant frequencies using numerical tools but it is challenging to capture the higher modes of vibration of the resonator experimentally [6] as the amplitude of vibration reduces enormously at higher modes.

A resonator can be used in variety of applications depending on the mode shape and resonant frequency. For example an electrothermal actuator can be used as an accelerometer in the first mode and as a micro mirror in second and third modes [3]. So it is essential to determine the resonant frequencies and mode shapes which dictate the suitability of their application. Many researchers are exploring ways to capture the higher modes of vibration, Polytec researchers [4] have reported successful extraction of data for the first four modes of a MEMS device, while the same group in collaboration with MEMUNITY also reported data till seventh mode for a compressor [7]. In 2006, Liang-chia chen [8] reported data till seventh mode of vibration of an AFM cantilever. Mitchell [9] in early 1998 has reported the data for a macro scale plate (with dimensions 18 in $\times$ 18 in $\times$ $\frac{1}{8}$ in) upto the ninth normal mode. Martarelli [10] was successful in capturing the frequencies till 23rd mode of vibration of turbine blade using Laser Surface

Velocimeter. To the best of our knowledge this is the maximum number of modes captured and reported in the literature till now in the history of structural vibrations.

The current study is partly motivated by a need to explore higher modes of vibration in MEMS devices for sensing purposes. For high Q materials like silicon, most of the damping in oscillations of these devices comes from squeeze film effects [11, 12, 13]. It has been shown that it is possible to reduce this kind of damping and increase the Q of the MEMS resonators by exciting them in higher modes [1]. Of course, the study of higher modes of oscillation is an important area in its own right. What is interesting here is that because of the small scale of the vibrating structure, we are able to capture fairly high modes of oscillation relatively easily. In our experiments we have been able to capture up to 28 modes of out-of-plane vibrations of a resonator from the response to a single excitation. Subsequent numerical simulations have confirmed that the captured modes are indeed what is expected and the measured frequencies are very close to the theoretically predicted values.

In this study we use a MEMS resonator fabricated by SOI MUMPs process [14]. This resonator is actually an electrothermal actuator. Structure of the fabricated device is shown in Figure 1. The structure consists of a proof mass suspended by a set of meandering beams. The beam segments that span the meander width are called span beams, while those connecting the span beams are called connector beams [15]. The meanders are connected to the substrate by means of a set of straight beams. Table 1 provides the details of geometric and material properties of the structure used in simulations. Details of experiments and numerical simulations carried out are provided in the second section.

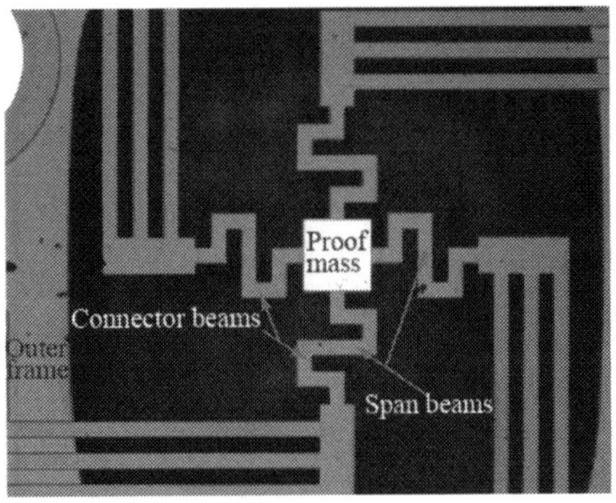

Figure 1. SEM image of the micromachined electrothermal actuator.

Table 1. Mechanical and physical parameters of the electrothermal actuator.

Parameter	Value
Length of straight beams connecting to the substrate, L	1000 μm
Width of straight beams connecting to the substrate, w	45 μm
Thickness of structure, t	25 μm
Length of span beams, l_s	200 μm
Length of connector beams, l_c	100 μm
Proof mass size	$200 \times 200\ \mu m^2$
Young's modulus, E	130 GPa
Poisson ratio, ν	0.22
Density, ρ	2300 kg/m^3
Co-efficient of thermal expansion, α	2.33×10^{-6} /°C

2. Theory

Here, we discuss the governing equations of elastic structural vibration. We also outline the numerical and experimental procedure to calculate the natural frequencies of the resonator.

2.1. Numerical procedure

In this work, the numerical results are obtained by carrying out modal analysis of the structure in finite-element software, ANSYS [16]. The numerical computation of natural frequencies and the mode shapes involves three main steps. In the first step, the geometric modelling is done in MEMS specific design tool, CoventorWare, by specifying the mask data and fabrication steps involved in realizing the structure. The 3D model is imported into ANSYS and the modal analysis is carried out to estimate the eigenvalues and eigenvectors. For this analysis, the structure is meshed with 25000 eight-noded brick elements. The zero-displacement and zero rotation boundary conditions are applied on the substrate. Then, the theory of elastic dynamics is used for determining the natural frequencies and the corresponding mode shapes of an undamped system.

The general governing equation of motion for free vibration can be written as [5]:

$$[M]\{\ddot{w}\} + [K]\{w\} = 0 \tag{1}$$

where $[M]$ is the structural mass matrix and $[K]$ is the stiffness matrix. If the displacement $w(x, y, z)$ is written in terms of n eigenvectors $\phi_i(x, y)$ and the time dependent modal coordinate $q_i(t)$ [17] as

$$w(x, y, t) = \sum_{i=1}^{n} \phi_i(x, y) q_i(t) \tag{2}$$

then the eigenvalues and the corresponding eigenvectors can be calculated from the equilibrium equations:

$$[K]\{\phi_i\} = \omega_i^2 [M]\{\phi_i\} \tag{3}$$

Based on the orthogonal properties of the eigenvectors, we obtain the expressions for resonant frequencies and corresponding mode shapes:

$$\{\phi_i\}^T [K]\{\phi_i\} = \omega_i^2 \{\phi_i\}^T [M]\{\phi_i\}. \tag{4}$$

Let $\{\phi_i\}^T[K]\{\phi_i\} = K_i$ and $\{\phi_i\}^T[M]\{\phi_i\} = M_i$. Then we get

$$\omega_i^2 = \frac{K_i}{M_i} \tag{5}$$

where M_i and K_i are the mass and stiffness of the structure in i^{th} mode corresponding to the eigenvector ϕ_i and the resonant frequency ω_i.

Finally, the natural frequencies and the corresponding mode shapes (shown in Fig. 4) for all out-of-plane modes are extracted in the postprocessing step. The numerical values of frequencies for the first 28 modes are given in Table 2.

2.2. Experimental procedure

Use of a *Laser Doppler Vibrometer* (LDV) [4] is an attractive method for noncontact dynamic displacement measurements with high positional accuracy. LDVs measure out-of-plane velocities by interferometrically measuring the change in frequency and phase of the back scattered laser light reflected from the surface. In the microsystems analyzer (MSA), the laser light is fed through a microscope to reduce the focused spot size of the laser light [4, 18]. This is important since the velocity is averaged over the spot area and the motion of the small features can only be measured with a small spot size [4]. Single point LDVs require multiple measurements for establishing the mode shapes. Since it is not possible to measure input forces directly, one common technique is to use the excitation voltage as a reference [4]. Response measurements can be sequentially taken at several points, and then combined to provide mode shapes [18]. We have used a Scanning Laser Vibrometer for the experiments conducted in this study.

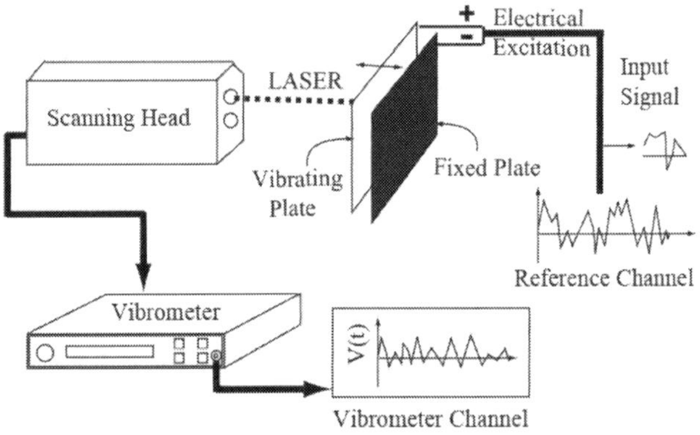

Figure 2. Working principle of the laser doppler vibrometer.

The working principle of the set-up is schematically shown in Fig. 2. The experimentation involves electrical excitation (say $V_{\text{input}} = V_{\text{dc}} + V_{\text{ac}}$, where $V_{\text{ac}} \ll V_{\text{dc}}$) that causes the suspended structure to vibrate. The laser spot from the interferometer in the scanning head is positioned on a scan point on the object by means of mirrors and is scattered back. The back scattered laser light interferes with the reference beam in the scanning head. A photo detector records the interference. A decoder in the vibrometer provides a voltage which is proportional to the velocity of the scanned point parallel to the measurement beam. The voltage is digitized and processed as the vibrometer signal. The output signal can be obtained as velocity or displacement signal using the velocity or the displacement decoder.

The MEMS device under investigation here is an electrothermal actuator. In such devices, hot and cold junctions are generated on application of voltage due to the Peltier effect. Due to thermal expansion, the structure tends to bend. Upon application of a sinusoidal signal, structure starts to vibrate about its equilibrium position. To capture the modal parameters (resonant frequency and damping) of the structure, we apply a pseudorandom signal of voltage 8 ± 2 V during the experimentation. After averaging the FRF of the output signal over 5 times, we determine the resonant frequency of the structure using the frequency response curve, and the corresponding mode shapes are obtained from presentation mode of the vibrometer. Figure 3 shows the frequency response of the structure, natural frequencies and the corresponding mode shapes for the first 28 out-of-plane modes shown in Figure 4.

Journal of Physics: Conference Series **181** (2009) 012079 doi:10.1088/1742-6596/181/1/012079

3. Results and discussion

We now present the results of the numerical and experimental studies. In contrast to such studies at macroscales, we use the experimental results here to validate the numerical results since the finite element model of the structure is likely to have more errors due to uncertainties in material properties and geometric parameters. Figure 3 shows the frequency response of the resonator. From the FRF the modal parameters are extracted and the natural frequencies corresponding to the first 28 modes are as listed in Table 2, and the corresponding mode shapes are shown in Fig. 4. Since the structure is symmetric we can notice many degenerate modes (pseudorepeated modes) [19] such as second and third modes of vibration. Similarly 9th and 10th, 16, 17, 18 and 19th modes are again degenerate. The difference between eigenvalues of degenerate modes in

Table 2. Comparison between experimental and numerical results.

Mode No.	f_{exp} (kHz)	f_{num} (kHz)	% error $\lvert\frac{f_{\text{exp}}-f_{\text{num}}}{f_{\text{exp}}}\rvert \times 100$	Mode No.	f_{exp} (kHz)	f_{num} (kHz)	% error $\lvert\frac{f_{\text{exp}}-f_{\text{num}}}{f_{\text{exp}}}\rvert \times 100$
1	32.94	33.97	3.13	15	374.80	377.77	0.79
2	53.75	54.08	0.62	16	395.00	409.64	3.71
3	54.42	54.85	0.79	17	403.30	413.23	2.46
4	73.98	74.14	0.21	18	410.50	413.82	0.81
5	101.80	104.77	2.92	19	424.20	418.26	1.40
6	173.30	174.31	0.58	20	430.20	434.55	1.01
7	183.30	184.44	0.62	21	448.40	445.68	0.61
8	263.90	257.90	2.27	22	482.20	472.75	1.96
9	288.40	296.52	2.81	23	581.60	557.66	4.11
10	295.60	306.24	3.59	24	730.00	731.57	0.21
11	299.50	310.08	3.53	25	740.50	744.76	0.57
12	303.10	314.22	3.67	26	769.40	783.96	1.89
13	324.70	326.12	0.44	27	800.90	791.71	1.14
14	360.20	358.23	0.55	28	878.60	920.52	4.77

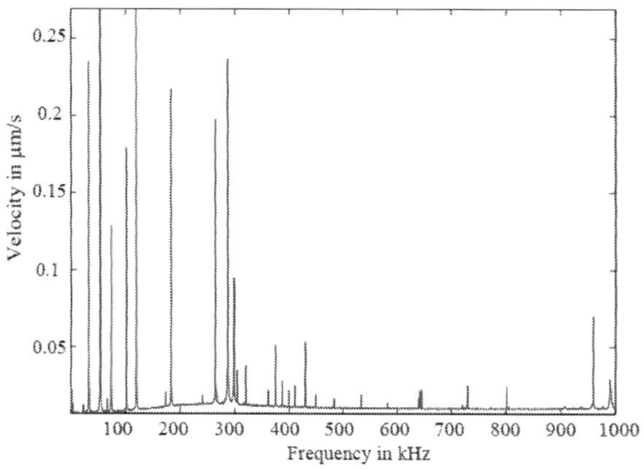

Figure 3. Frequency response function from experimental studies.

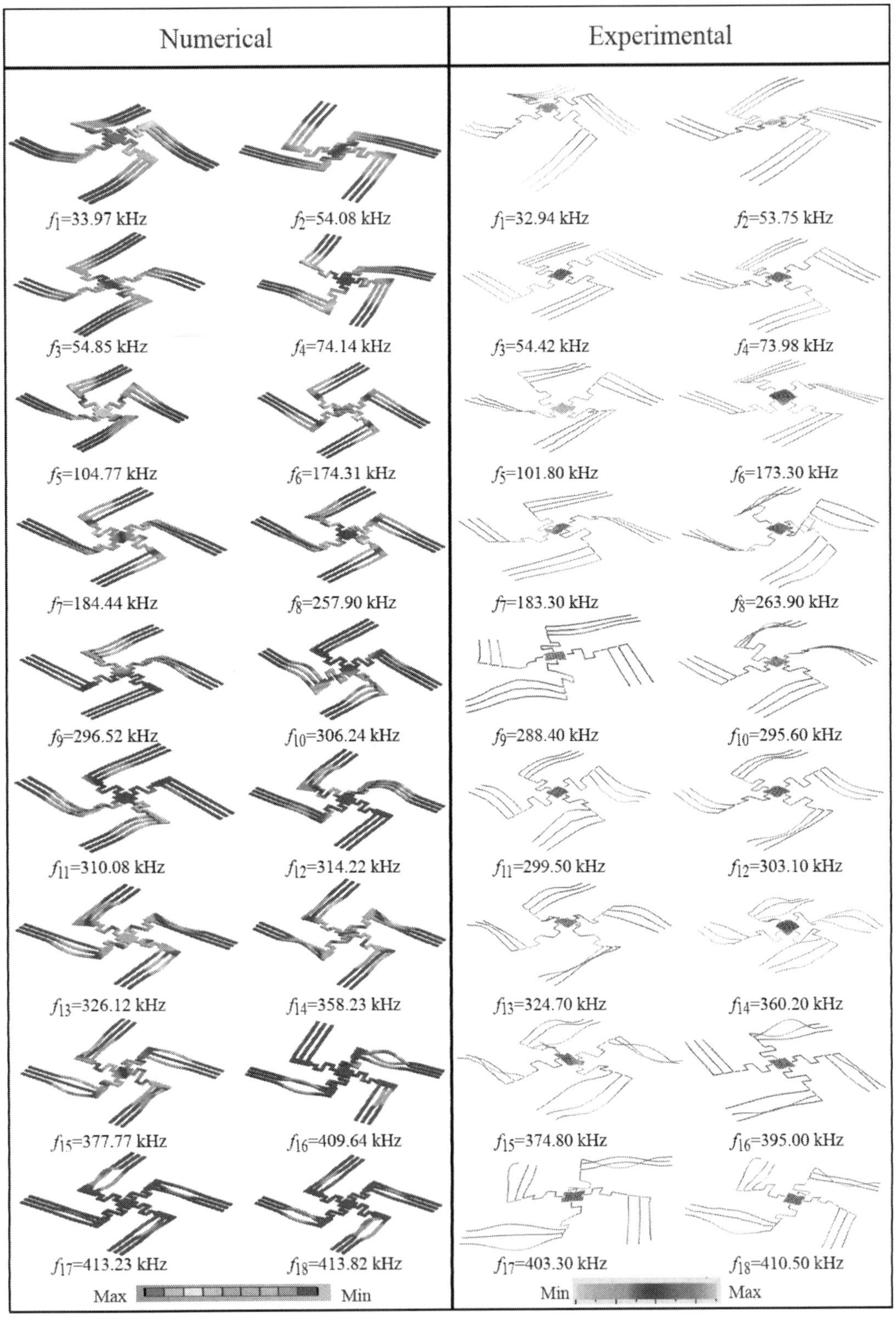
Numerical
Experimental
f1=33.97 kHz
f2=54.08 kHz
f1=32.94 kHz
f2=53.75 kHz
f3=54.85 kHz
f4=74.14 kHz
f3=54.42 kHz
f4=73.98 kHz
f5=104.77 kHz
f6=174.31 kHz
f5=101.80 kHz
f6=173.30 kHz
f7=184.44 kHz
f8=257.90 kHz
f7=183.30 kHz
f8=263.90 kHz
f9=296.52 kHz
f10=306.24 kHz
f9=288.40 kHz
f10=295.60 kHz
f11=310.08 kHz
f12=314.22 kHz
f11=299.50 kHz
f12=303.10 kHz
f13=326.12 kHz
f14=358.23 kHz
f13=324.70 kHz
f14=360.20 kHz
f15=377.77 kHz
f16=409.64 kHz
f15=374.80 kHz
f16=395.00 kHz
f17=413.23 kHz
f18=413.82 kHz
f17=403.30 kHz
f18=410.50 kHz
Max
Min
Min
Max

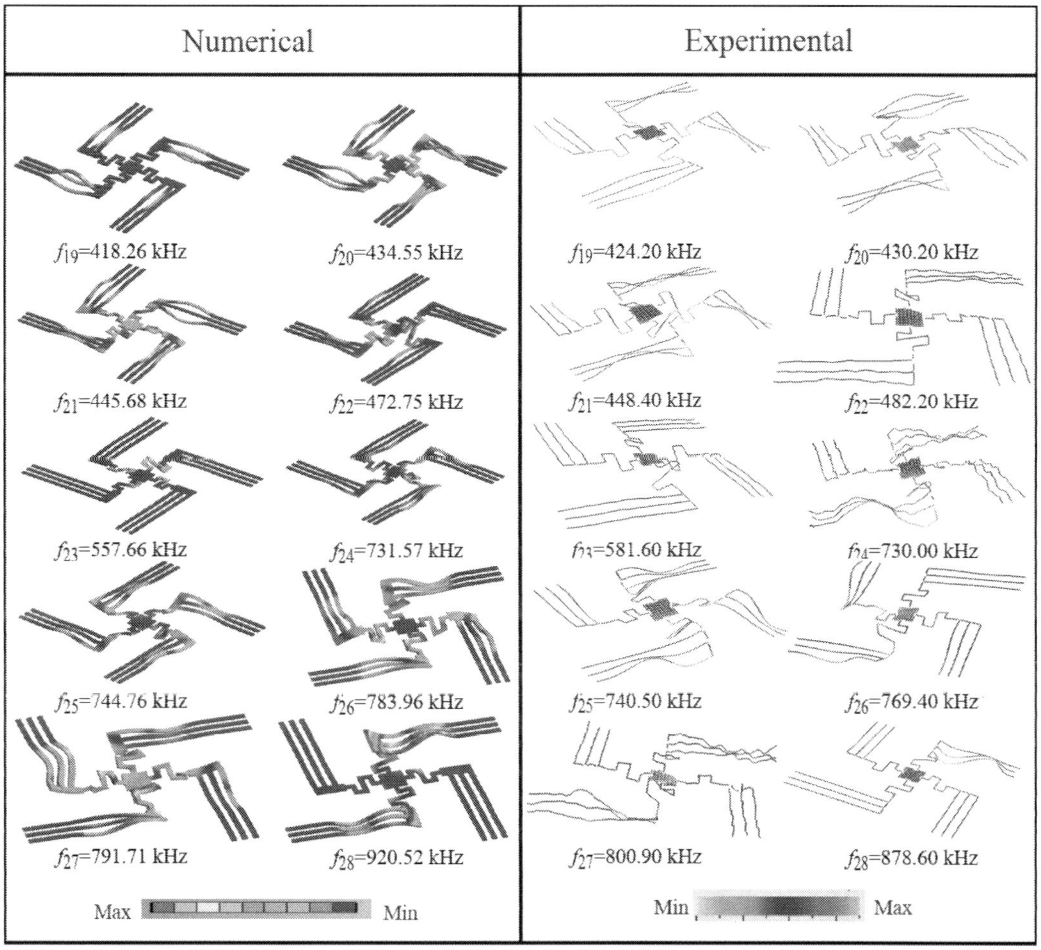

Figure 4. Comparison between numerical and experimental mode shapes.

the symmetric structure indicates of nonuniformities in the modal parameters such as mass and stiffness of the structure [19]. These nonuniformities will further change the dynamic performance characteristics of the system. For example, nonuniformities in mass may result in changes in inertial forces, whereas nonuniformities in stiffness may result in unexpectedly large deflections at certain frequencies in the harmonic response of the system [19]. It can be noticed from Table 2 that numerical and experimental values are very close, with an error less than 5% for all the modes, and the degenerate modes are very close to each other hence reducing the nonuniformities present in the system.

4. Conclusions

We have experimentally determined 28 modes of out-of-plane vibrations of a MEMS resonator. The experimentally measured frequencies are compared with numerically computed values from the modal analysis of the resonator carried out using FEM analysis. The numerically found frequencies match the experimental values within 5% error — can be incredible result especially for higher modes. Our work shows that micro resonators or micro mechanical structures can be excited in fairly higher modes, in contrast to their macro counterparts, rather easily. The accuracy of results shows that one can confidently employ the higher modes for sensing in MEMS devices. This is particularly significant in the light of the reported results in the literature that Q-factor of a resonator goes up in higher modes if the Q is dominated by squeeze film damping effects.

Acknowledgments

We thank Prof. Navakanta Bhat and Mr. Prasad from Electronics and Communication Engineering Department, IISc., for providing the fabricated structure. This work is supported by a grant from University Grants Commission.

References

[1] A K Pandey and Rudra Pratap 2007 Effect of flexural modes on squeeze film damping in MEMS cantilever resonators *J. Micromech. Microengg.* **17** 2475-2484.
[2] M. Madou 1997 *Fundamentals of Microfabrication* (Florida: CRC Press).
[3] Lijie Li, et.al 2007 Design, simulation and characterization of a MEMS optical scanner *J. Micromech. Microengg.* **17** 1781-1787.
[4] Polytec GmbH- Advanced measurements using light., `http://www.polytec.com`.
[5] S. S. Rao 1995 *Mechanical Vibration* (New York: Wesley Publishing Company).
[6] Chi-Hung Huang and Chien-Ching Ma 2001 Experimental measurement of mode shapes and frequencies for vibration of plates by optical interferometry method *J. Vibrations and acoustics* **276** 276-280.
[7] www.polytec.com/automotive
[8] Liang-chia chen, et.al 2006 High bandwidth dynamic full-field profilometry for nano-scale charecterization of MEMS *J. Physics.* **48** 1058-1062.
[9] L D Mitchell, et.al 1998 An emerging trend in experimental dynamics: merging of laser based three dimensional structural imaging and modal analysis *J. Sound and vibration* **211(3)** 323-333.
[10] M Martarelli, et.al 2001 Automated modal analysis by scanning laser vibrometry: problems and uncertainities associated with the scanning system calibration *J. Mechanical systems and signal processing* **15** 498-508.
[11] R. Pratap, S. S. Mohite and A. K. Pandey 2007 Squeeze-Film Effects in MEMS Devices *J. Indian Institute of Science* **87:1**.
[12] Kwok P. Y., Weinberg M. S. and Breuer K. S 2005 Fluid effects in vibrating micromachined structures. *J. Micromechanical Systems* **14(4)**: 770-781.
[13] Minhang Bao and Heng Yang 2007 Squeeze film air damping in MEMS. *Sensors and Actuators A: Physical* **136**: 327.
[14] SOIMUMPs multi-user MEMS process, MEMSCAP Inc. `http://www.memsrus.com/nc-mumps.soi.html`.
[15] G K. Fedder 1994 *Simulation of micromechanical systems* Doctoral thesis University of California.
[16] ANSYS 10 – *Finite element solver for multiphysics problems.*, `http://www.ansys.com`.
[17] D J Ewins 2000 *Modal Testing: Theory and Practice* (Taunton: Research Studies Press).
[18] Ozdoganlar O Burak, et.al 2005 Experimental modal analysis for microelectromechanical systems *J. Experimental mechanics* **45** 498-506.
[19] Brandon J. Jellison, et.al 2004 Nondestructive evalustion of parts with denerate modes using pseudorepeated roots *J. Dynamic systems, Measurement and control* **126** 498-508.

A study on the precision of a confocal probe from its fork frequency and amplitude

Z R Qiu, K N Yao, B X Chen, X H Li and Y Zhong,

State key laboratory of precision measuring technology and instruments,

Tianjin University, Tianjin, 300072, P. R. China

E-mail: qzr@tju.edu.cn

Abstract. The confocal probe is a kind of non-contact measurement probes. Through an optical system containing an oscillating tuning fork, it converts displacement measurement to time difference measurement. For the measuring precision, the oscillation stability and swing of the tuning fork are important. Based on analysis, the relationship between the measuring precision and the frequency and amplitude of the tuning fork is described, and the means to obtain stable frequency and amplitude is introduced.
Key words. Resonance Frequency and Amplitude, Confocal probe, Measuring Precision Non-contact measurement.

1. INTRODUCTION

Optical non-contact probes have the advantages of no measuring force, high speed, etc [1]. However, surface conditions, including color, surface, roughness, and inclination greatly affect the accuracy of the probes [2]. As the traditional optical probes, out-of-focus probes have very high sensitivity near the focus, but they have tiny measuring ranges [3-5]. Therefore, the dynamic active confocal measuring instruments have been developed. They have high accuracy, a wide measurement range, and good dynamic performance. More important, they are suitable of measuring highly reflective surfaces, such as high-accuracy metallic and optical pieces [6].

The confocal probe we study works by time difference measuring. The measuring signal is obtained through the light focusing point modulated by the oscillating fork scanning on the surface twice periodically. Both the signal period and the laser focus' syntonic movement are determined by mechanical resonance of the fork. To get high measuring precision, we study the stability of the tuning fork, and the means to improve the stability.

2. PRINCIPLE OF THE CONFOCAL PROBE

Figure 1~3 shows the principle of confocal probe. It converts displacement measurement to time difference measurement [7]. In figure 1, a laser beam emits from light source 1. It is focused on measured surface 8 by lens 5 attached to a tuning fork 6, which is used to modulate the focal position of the optical system. The beam return from the surface 8, and then is focused on the pinhole 3. Detector 4 receives the beam behind the pinhole 3 and transforms the beam to an electrical signal. A peak signal is formed on receiving element 4, when the focal plane coincides with surface 8. Since the fork is oscillating, the light focusing point moves accordingly and scans on the surface twice periodically. Therefore, in one period, two peaks (hump signal) are formed as figure 2 shows. In figure 2, Δt is the time difference between the time when the focal plane is in the neutral position and the time when the focal plane coincides with the surface measured. The displacement of surface measured leads to the difference of Δt. The relationship between the both is described as follows.

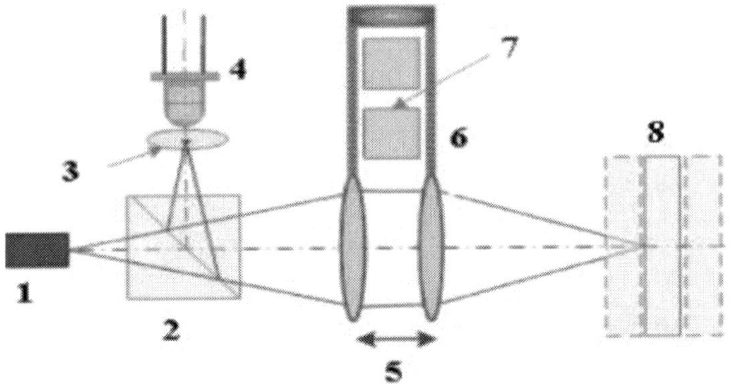

Figure 1. Working principle of confocal probe.

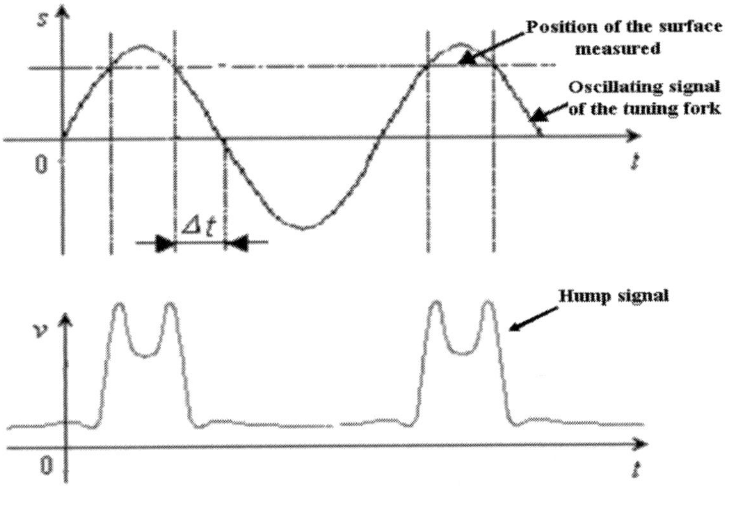

Figure 2. The method of time difference measurement.

The oscillation equation of lens is

$$\Delta w = A\sin\omega_0(t_0 + \Delta t) \tag{2.1}$$

where Δw is the displacement of lens, A is amplitude, ω_0 is angular frequency, Δt is difference of time, t_0 is original time when the fork is in the neutral position, therefore, $t_0 = 0$. When the focal plane coincides with the measured surface,

$$\Delta w = A\sin\omega_0\Delta t \tag{2.2}$$

Then the time difference of lens to reach this position is

$$\Delta t = \frac{1}{\omega_0}\arcsin\left(\frac{\Delta w}{A}\right) \tag{2.3}$$

When the lens move, the focal plane moves accordingly. The relationship between the displacement of the focal plane and the displacement of lens can be described by equation (2.4) as figure 3 shows [6].

$$\Delta' = \frac{f'^2 \Delta w}{2\Delta w^2 - (f' + f' - b)\Delta w + f'^2} + \Delta w \tag{2.4}$$

where f' is foci of L_1, L_2, b is distance between the two lenses when they are static, Δw is displacement of the two lenses.

Therefore, through measuring Δt from the hump signal [8], we could obtain the displacement of lens, and then obtain the displacement of surface measured by equation (2.4). The displacement measurement converts to time difference measurement.

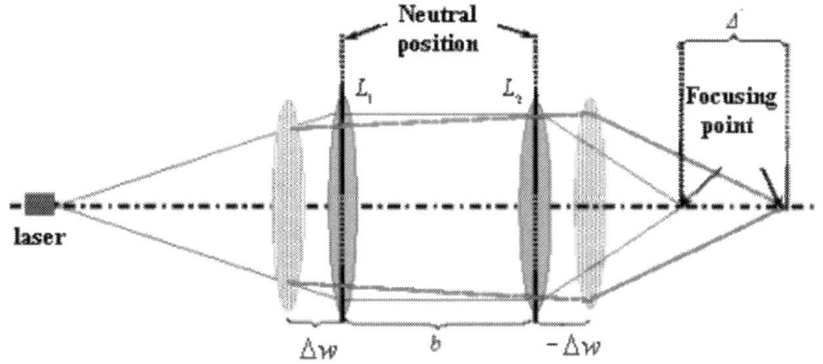

Figure 3. Relationship between the displacement of the focal plane and the displacement of lens.

3. VIBRATION OF FORK AND DESIGN

3.1. Principle of the Fork Self-Oscillation

Figure 4 shows the fork self-oscillation system, which consists of two parts. One is fork and lens, and the other is positive feedback circuit composed by coils and amplifiers. The sensing coil senses the amplitude of the fork and this signal is amplified and fed back to excite the fork. When the energy consumed by fork equals the energy supplied by the circuit, the system achieves stable state. Since the amplitude and frequency of the fork are relatively constant, they can be used for measuring.

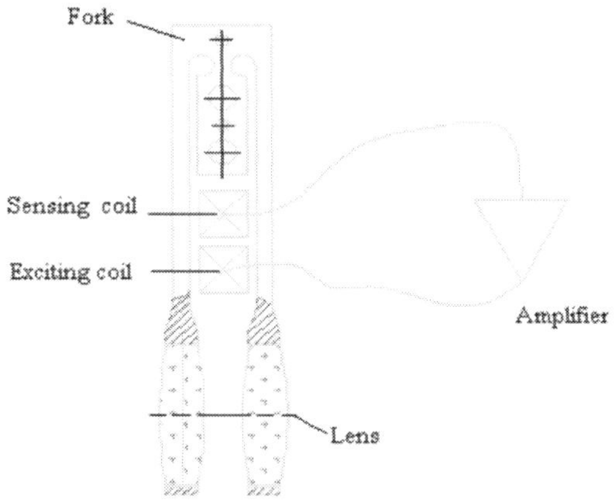

Figure 4. Principle of the fork self-oscillation.

3.2. Analysis of the Measuring Precision from the Fork

As equation (2.3) shows, there will be error, if the amplitude and frequency vary. Both the factors are analyzed as follows.

When the frequency of fork oscillating changes $d\omega_0$, the time error accordingly is

$$d\Delta t = -\frac{1}{\omega_0^2}\arcsin\left(\frac{\Delta w}{A}\right)d\omega_0 \tag{3.1}$$

Since the nonlinearity of sinusoid, the precision decreases as Δw increases. The ultimate value of Δw used in our system is $0.8A$. So that, $\arcsin\left(\frac{\Delta w_{\max}}{A}\right) = 0.927$ and equation (3.1) is changed to

$$d\Delta t_{\max} = -\frac{0.927 d\omega_0}{\omega_0^2} \tag{3.2}$$

When the amplitude of fork oscillating changes dA, the time error accordingly is

$$d\Delta t = -\frac{\Delta w}{\omega_0 A\sqrt{A^2 - \Delta w^2}}dA \tag{3.3}$$

Since $\Delta w_{\max} = 0.8A$, equation (3.3) is changed to

$$d\Delta t_{\max} = -\frac{4dA}{3\omega_0 A} \tag{3.4}$$

As it is analyzed above, equation (3.2) and equation (3.4) indicate that the amplitude of fork has more obvious effect on the precision of measuring system than the frequency. The following section will show the details to obtain the stability of fork oscillating.

3.3. Design of the Fork

To obtain stable frequency, we choose 3J53 [$Ni_{42}CrTiAl$] to make the fork. 3J53 is one kind of constant elastic alloys. These alloys have stable natural frequency, low temperature coefficient of elastic modulus and good time stability. They are widely used as exactitude elasticity component or standard frequency component in precision instrument, communication technology, and computer technology [9].

When the material is selected, the capability of fork is determined by its figuration. Figure 5 shows the theoretical oscillating model of cantilever beam, where l is length of the beam, a is height of the section of the beam, b is width of the section of the beam, m is mass of the beam, and M is mass of the end of the beam. The natural angular frequency of cantilever beam is described by equation (3.5).

$$\omega^2 = \frac{3EI}{l^3(M + (33/140)m)} \tag{3.5}$$

where E is elastic modulus of cantilever beam, and I is inertia moment of cantilever beam. Since the section of this cantilever beam is rectangular, equation of inertia moment is

$$I = \frac{ba^3}{12} \tag{3.6}$$

For convenience to design the size of fork, the angular frequency ω is replaced by frequency f. We change equation (3.5) to equation (3.7).

$$f = \frac{1}{4\pi}\left\{\frac{Eba^3}{l^3[M + (33/140)balp]}\right\}^{\frac{1}{2}} \tag{3.7}$$

where ρ is density of material.

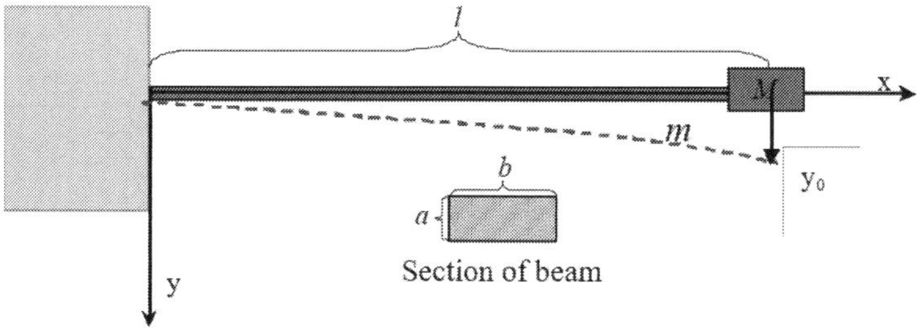

Figure 5. Theoretical model of cantilever beam.

Higher natural frequency is beneficial to oscillation stability and measuring speed of the system, but the amplitude will be small, which limits the measuring range of the system. Hence we choose the natural frequency as high as possible under the premise that the amplitude of fork is large enough for measurement.

We select the amplitude of fork to be 0.1mm to 0.2mm, while its natural frequency is about 300Hz to 500Hz. The fork designed is shown in figure 6. Two cantilever beams of fork are connected by the pedestal, and there are screwed holes to fix the fork. Two beams are designed in the same shape to eliminate effect on oscillation of the pedestal [10]. To get more stable frequency, we conducted heat treatment to the fork. The parameters of the fork are listed below: l is 49.5mm, b is 5.5mm, a is 2mm, M is 0.00194kg. Elastic modulus of 3J53 is 1.85×10^{-5}Mpa, and density is 7900kg/m^3. The theoretical frequency of the designed fork is about 380Hz by equation (3.7). Due to the complexity of oscillating system, practical frequency was tested in experiment. This will be shown in the section 3.5.

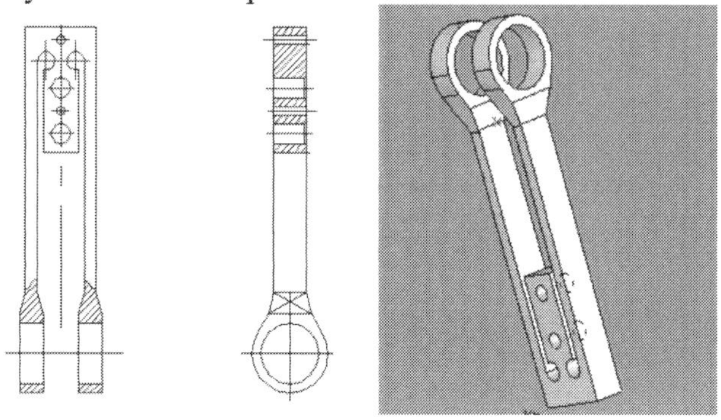

Figure 6. Figuration of the fork.

3.4. Design of the Fork Self-Oscillation

The base circuit of the fork self-oscillation is shown in figure 7. It is an amplifier circuit with the core OP07. The exciting coil is joined between pin 1 and pin 2, while the sensing coil is joined between pin 3 and pin 4. This circuit is simply designed according to the section 3.1, but the amplitude of fork presents inferior stability in practice. Consequently, we improved the circuit as figure 8 shows.

The major improvement is that an AD5280 is added to the circuit to work together with Rw1. AD5280 is one kind of DCP (Digitally Controlled Potentiometers). It can offer resistance value varied from 1kΩ to 200kΩ.

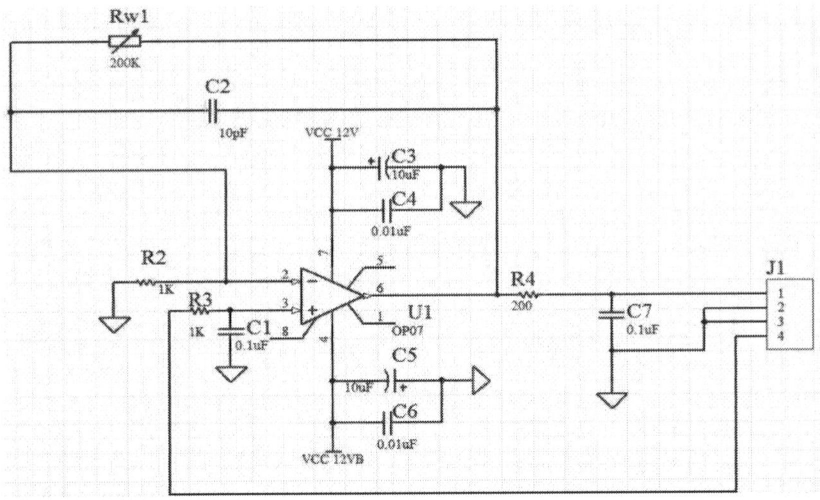

Figure 7. Base circuit of the fork self-oscillation.

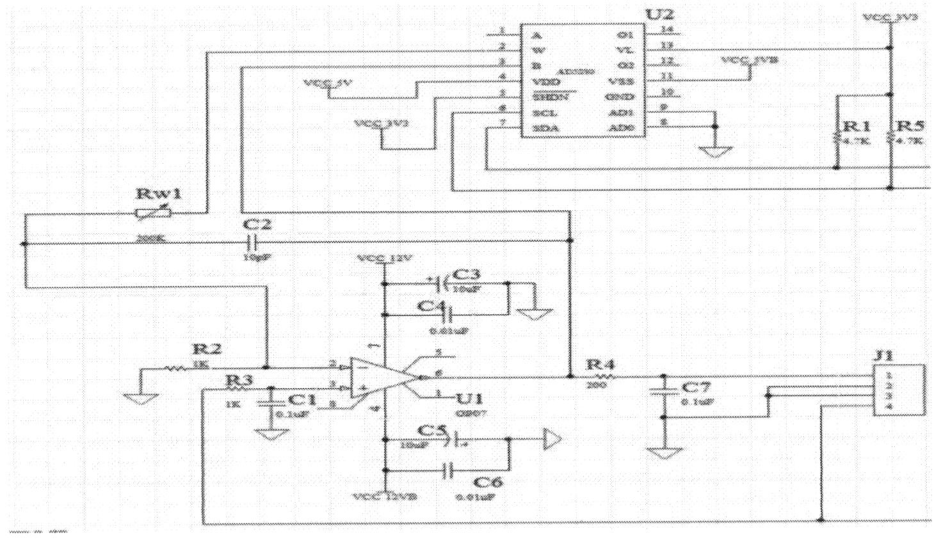

Figure 8. Improved circuit of the fork self-oscillation.

To make the resistance value of AD5280 vary automatically, we design a control system as figure 9 shows. There is a threshold value set in FPGA (Field-Programmable Gate Array). When the sensing coil signal is greater than threshold value, the value of AD5280 will be adjusted lower; when the signal is less than the threshold value, the value of AD5280 will be adjusted higher. The gain of amplifier can be adjusted in real time through the control system to keep the amplitude of fork stable.

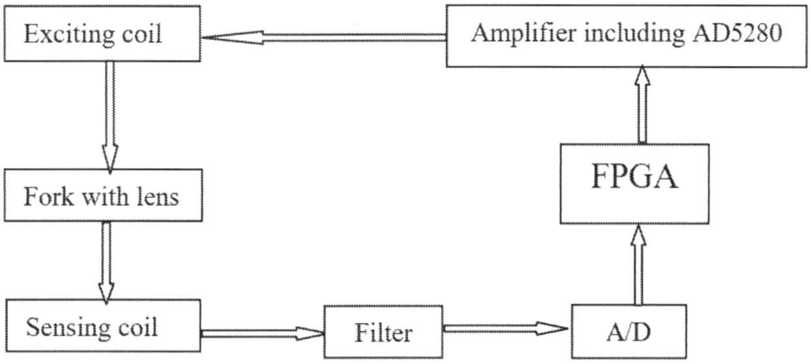

Figure 9. Control system for fork oscillation.

3.5. Experiment

Since the sensing coil senses the oscillation of the fork, the sensing coil signal was detected to reflect the oscillation of the fork. Figure 10 shows the frequency of the fork.

The frequency of the fork is about 343.785Hz, and it approaches the theoretical frequency. The maximum offset value of frequency is 0.045Hz in 6 hours' running time. The relative error is 1.31×10^{-4}. From equation (3.2), we calculated the maximum time error $d\Delta t_{max}$, 3.53×10^{-7}s. Therefore, the error of frequency can be ignored in measuring system.

Figure 11 shows the amplitude of the fork. The value is about 0.107mm. The maximum offset value of amplitude is $1.3\mu m$ in 4 hours' running time. The relative error is 1.2×10^{-2}. From equation (3.8), we calculated the time error $d\Delta t_{max}$, 4.7×10^{-5}s. Therefore, the error of amplitude can also be ignored in measuring system. As a comparison, we do the experiment according to the circuit in figure 7. The maximum offset value of amplitude is about $20\mu m$ in a shorter running time, and it affects the precision greatly. Therefore, the improved circuit is much better for the confocal probe.

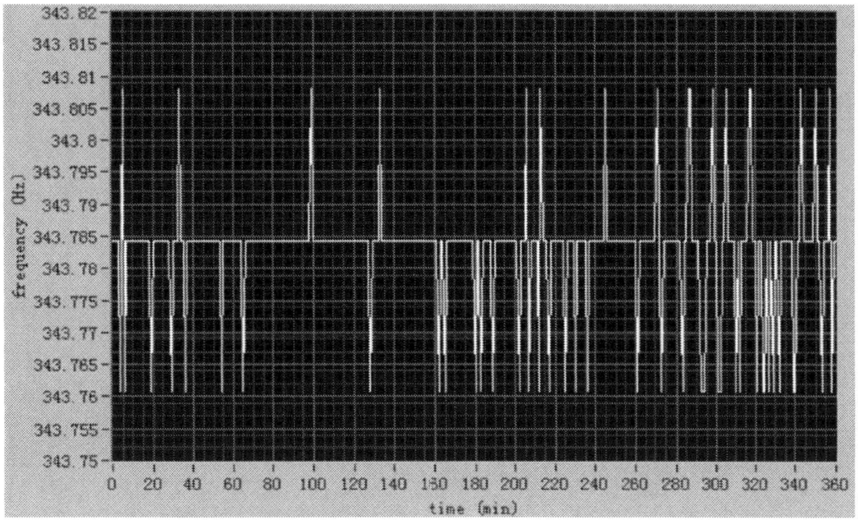

Figure 10. The frequency of oscillating fork

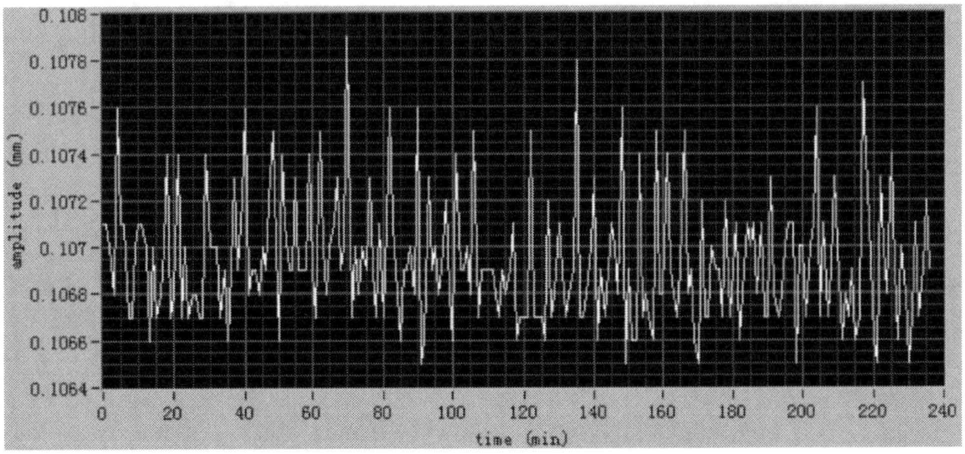

Figure 11. The amplitude of oscillating fork

4. CONCLUSION

(1) The precision of confocal probe has more effect from the amplitude of its fork rather than the frequency of its fork.
(2) The effect of frequency of the fork on measuring precision can be diminished negligible through proper design of fork.
(3) Since the amplitude of the fork can be maintained stable through controlling the amplification gain of amplifier in circuit continuously, the effect of amplitude of the fork on measuring precision can also be ignored in the system.

References

[1] Yamazaki A 1993 Non-contact probe for continuous measurement of surface inclination and position using dynamic irradiation of the light beam *Annals of the CIRP*. 42 (1):586~588
[2] Zhang G X. 2001 A system for measuring high-reflective sculptured surfaces using optical non-contact probe *Annals of the CIRP*. 50 (1):369~372
[3] Jordan H J, Tiziani H 1998 High accurate characterization of engineering surfaces using confocal microscopy *Meas.Sci.Technol.* 9:1142~1151
[4] Tiziani H J 1994 Three-dimensional analysis by a microlens array confocal arrangement *Applied Optics*. 33(4):567~572
[5] Wilson T 1990 *Confocal Microscopy*. London, UK: Academic Press
[6] Gao X 2002 A study on the optical non-contact probe for surface measurement *Ph.D. Dissertation* Tianjin, P.R.China: 33~40
[7] Zhang G X, Fan Y M, Gao X, and et al 2004 A confocal probe based on time difference measurement *CIRP ANN-MANUF TECHN.* Vol.53:417~420
[8] Qiu Z R, Qin J, Zhang H W and G X Zhang 2006 Structure Design of A Laser Confocal Probe For Measuring the Micro-Sphere *Journal of Physics: Conference Series.* Vol.48:641~646
[9] Wang S L 1990 Development and application of constant elastic alloys *Journal of Shanghai Iron and Steel Research. 2:43~49*
[10] Liu Y M, Tan Y C, and Zhao Z Y 2003 Study on time stability of properties of ferromagnetic alloys with invariant elasticity *Metallic Functional Materials.* Vol.10, No.6: 22~25

The Assessment of Vibration Absorption Capacity of Elevator's Passengers

I Herrera[1] and S Kaczmarczyk[2]

[1] Grupo Ingelev, Escuela de Ingenierías Industriales, Universidad de Extremadura, Avda. De Elvas s/n, 06071 Badajoz, Spain, ingelev@unex.es
[2] Division of Engineering, The University of Northampton, St George's Avenue, Northampton, United Kingdom, stefan.kaczmarczyk@northampton.ac.uk

Abstract. The transfer function expressions for damped two degrees of freedom systems with damped vibration absorbers were used to model and simulate the dynamic response of an elevator's car. The numerical simulation is implemented in MATLAB and experimental tests are conducted. The elevator's car under various load conditions with passengers wearing different type of shoes was excited over a range of frequencies and amplitudes. The agreement of the results from experimental tests with the model predictions is analysed. The simulation model is validated and the passenger's role to act as a dynamic absorber is assessed.

1. Introduction

The analysis of the dynamic behaviour of the elevator car system plays an important role in elevator engineering. Nowadays, passengers demand high ride quality of elevators. In particular low vibration and noise levels must be ensured in the modern design of elevators. International standards encourage the development of uniform, reliable and precise measurement and processing techniques to be applied within the elevator industry [1].

Possibly, the main difficulty in developing a tool, such as software for elevator dynamic analysis, is to model the dynamic characteristics and behaviour of passengers who are the most complex and important elevator mechanical "component", in sufficient detail. The main goal of this paper is to propose a simplified model to predict the dynamics of an elevator car with passengers load.

An elevator is a complex mechanical system [2]. Great errors results occur when trying to isolate the problem to the one directly involved component without prior justification and modeling. So we must refer to the Elevator System 'ES' for mechanical behavior of the elevator purposes [3]. On the opposite extreme, detailed and rigorous description on the dynamics of the human body parts requires complex and sophisticated models like those applied to dummies for crash tests on elevators [4] or on vehicle cars. This may be too much difficult in our first attempt in understanding the ES, that is, the elevator car mechanics.

The ISO standard for lift ride quality measurements [1] states the aspects that influence the dynamic measurements results for comfort. First, coupling of the instrumentation to the car floor must be similar to that for a person by a minimum feet pressure of 60kPa. Secondly, on personel traveled within the instrumentation, is stated that not more than 2 persons should be present. Additionally, the instrumentation shall be place on "any car floor coverings which are normally present". Consequently, the ISO states the components that can affect the acceleration measurements in the car: the instrument, the persons and the floor coverings.

Experimental measurements on real elevators with standard equipment for lift ride quality and subsidiary analysis based on more than 15 geared elevators under several running conditions (travel direction, speed, number of persons, starting floor, ...[5] have revealed that the vertical car vibrations are mainly based on a single frequency that slips slightly with the traveled direction, independently of the number of passenger's traveled.

First we tried with the simplest undamped one degree of freedom expression, considering the suspension ropes and terminal springs as the elastic constant, and the suspended mass (car plus frame plus passenger traveled) as the equivalent mass; but is not capable to explain the constancy of the vibration frequency when varying the number of passengers. So, it is clear at this point, that we need a damped model as simple as possible and some laboratory experiments to understand the results on real elevators.

2. Simplified mechanical model

In a typical elevator system the car frame is suspended by sling and the car is mounted within the car frame on isolation blocks (see Figure 1). By considering vertical displacements only the elevator system comprising the car and the passengers/weights is represented by a two degrees of freedom model with y_{PA} and y_{CA} defining the passengers and car displacements, respectively (see Figure 2).

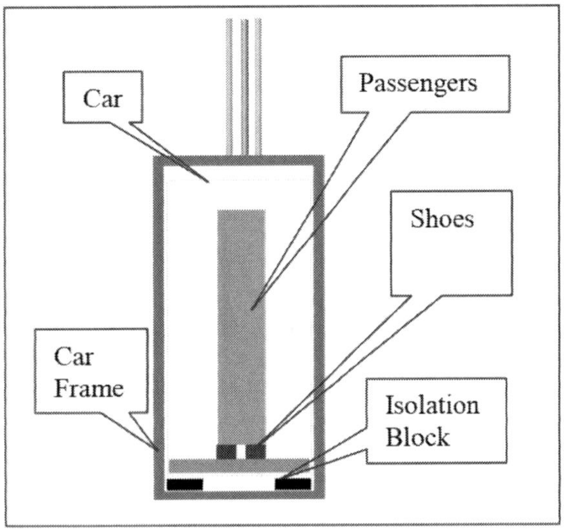

Figure 1. The car-passengers subsystem.

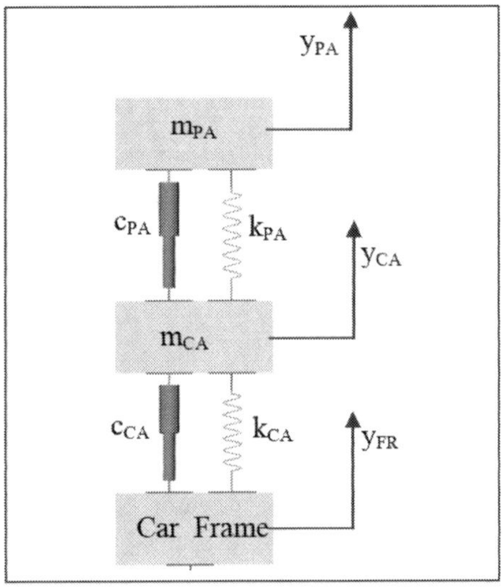

Figure 2. Discrete model.

The model consists of mass m_{PA} representing the passengers and of mass m_{CA} representing the car. The passenger mass is supported at the car floor via a spring-damper element of stiffness coefficient k_{PA} and viscous damping coefficient c_{PA}, respectively. The car is isolated from the frame by isolation blocks represented by a spring-damper element of stiffness coefficient k_{CA} and viscous damping coefficient c_{CA}, respectively.

The stiffness coefficient k_{PA} and viscous damping coefficient c_{PA} are defined as

$$\frac{1}{k_{PA}}=\frac{1}{k_{pa}}+\frac{1}{k_{sh}}+\frac{1}{k_{fl}};\frac{1}{c_{PA}}=\frac{1}{c_{pa}}+\frac{1}{c_{sh}}+\frac{1}{c_{fl}} \tag{1}$$

where k_{pa} denotes the average stiffness of the human body in the vertical direction, k_{sh} denotes the shoes vertical stiffness, k_{fl}, denotes the vertical stiffness of the covering floor of the car, c_{pa} denotes the average viscous damping coefficient of the human body in the vertical direction, c_{sh} denotes the shoes viscous damping coefficient in the vertical direction and c_{fl} denotes the covering floor viscous damping coefficient of the car in the vertical direction.

The elevator car is attached to the car frame by a number of isolation blocks and their combined stiffness and damping coefficients are determined as

$$k_{\mathrm{PA}} = nk_{\mathrm{sb}}; c_{\mathrm{PA}} = nc_{\mathrm{sb}} \tag{2}$$

where k_{sb} and c_{sb} are the spring/damper constants for a single isolation block, respectively and n the number of isolation blocks.

As shown in Figure 2, the car frame is subjected to a harmonic base motion excitation y_{FR}. A frequency sweep is applied to investigate the forced response of the system and to identify the resonance frequencies.

Then, the differential equations of motion of the forced-movable base system in matrix form are given as:

$$\begin{bmatrix} m_{\mathrm{CA}} & 0 \\ 0 & m_{\mathrm{PA}} \end{bmatrix} \begin{bmatrix} \ddot{y}_{\mathrm{CA}} \\ \ddot{y}_{\mathrm{PA}} \end{bmatrix} + \begin{bmatrix} c_{\mathrm{CA}} + c_{\mathrm{PA}} & -c_{\mathrm{PA}} \\ -c_{\mathrm{PA}} & c_{\mathrm{PA}} \end{bmatrix} \begin{bmatrix} \dot{y}_{\mathrm{CA}} \\ \dot{y}_{\mathrm{PA}} \end{bmatrix} + \begin{bmatrix} k_{\mathrm{CA}} + k_{\mathrm{PA}} & -k_{\mathrm{PA}} \\ -k_{\mathrm{PA}} & k_{\mathrm{PA}} \end{bmatrix} \begin{bmatrix} y_{\mathrm{CA}} \\ y_{\mathrm{PA}} \end{bmatrix} = \begin{bmatrix} c_{\mathrm{CA}} \dot{y}_{\mathrm{FR}} + k_{\mathrm{CA}} y_{\mathrm{FR}} \\ 0 \end{bmatrix} \tag{3}$$

Now we introduce the dimensionless constants

$$\omega_{\mathrm{r}} = \frac{\omega_{\mathrm{nPA}}}{\omega_{\mathrm{nCA}}} = \frac{1}{\sqrt{m_{\mathrm{r}}}} \sqrt{\frac{k_{\mathrm{PA}}}{k_{\mathrm{CA}}}}; m_{\mathrm{r}} = \frac{m_{\mathrm{PA}}}{m_{\mathrm{CA}}}; \Omega = \frac{\omega}{\omega_{\mathrm{nCA}}}; \omega_{\mathrm{nCA}}^2 = \frac{k_{\mathrm{CA}}}{m_{\mathrm{CA}}}; \omega_{\mathrm{nPA}}^2 = \frac{k_{\mathrm{PA}}}{m_{\mathrm{PA}}}$$

and the additional constants for the dimensionless time, and damping factors, respectively,

$$\tau = \omega_{\mathrm{nCA}} t; 2\zeta_{\mathrm{CA}} = \frac{c_{\mathrm{CA}}}{m_{\mathrm{CA}} \omega_{\mathrm{nCA}}}; 2\zeta_{\mathrm{PA}} = \frac{c_{\mathrm{PA}}}{m_{\mathrm{PA}} \omega_{\mathrm{nPA}}}$$

Then the expression by (3) transforms into:

$$\begin{aligned} &\frac{\mathrm{d}^2 y_{\mathrm{CA}}}{\mathrm{d}\tau^2} + (2\zeta_{\mathrm{CA}} + 2\zeta_{\mathrm{PA}} m_{\mathrm{r}} \omega_{\mathrm{r}}) \frac{\mathrm{d}y_{\mathrm{CA}}}{\mathrm{d}\tau} + (1 + m_{\mathrm{r}} \omega_{\mathrm{r}}) y_{\mathrm{CA}} - 2\zeta_{\mathrm{PA}} m_{\mathrm{r}} \omega_{\mathrm{r}} \frac{\mathrm{d}y_{\mathrm{PA}}}{\mathrm{d}\tau} - m_{\mathrm{r}} \omega_{\mathrm{r}}^2 y_{\mathrm{PA}} - 2\zeta_{\mathrm{CA}} \frac{\mathrm{d}y_{\mathrm{FR}}}{\mathrm{d}\tau} - y_{\mathrm{FR}} = 0 \\ &\frac{\mathrm{d}^2 y_{\mathrm{PA}}}{\mathrm{d}\tau^2} + 2\zeta_{\mathrm{PA}} \omega_{\mathrm{r}} \frac{\mathrm{d}y_{\mathrm{PA}}}{\mathrm{d}\tau} + \omega_{\mathrm{r}}^2 y_{\mathrm{PA}} - 2\zeta_{\mathrm{PA}} \omega_{\mathrm{r}} \frac{\mathrm{d}y_{\mathrm{CA}}}{\mathrm{d}\tau} - \omega_{\mathrm{r}}^2 y_{\mathrm{CA}} = 0 \end{aligned} \tag{4}$$

After computing the Laplace transform of equations (4) and solve for $Y_{\mathrm{CA}}(s)$ and $Y_{\mathrm{PA}}(s)$, which are the transformed functions of $y_{\mathrm{CA}}(\tau)$ and $y_{\mathrm{PA}}(\tau)$ respectively, we get after following similar steps as those from Balachandran[8]:

$$\begin{aligned} Y_{\mathrm{CA}}(s) &= \frac{K_{\mathrm{FR}}(s) E_{\mathrm{PA}}(s)}{D_{\mathrm{PA}}(s)} \\ Y_{\mathrm{PA}}(s) &= \frac{K_{\mathrm{FR}}(s) C(s)}{D_{\mathrm{PA}}(s)} \end{aligned} \tag{5}$$

where

$$K_{FR}(s) = (2\zeta_{CA} + 1)Y_{FR}(s) + 2\zeta_{CA} y_{FR}(0)$$
$$E_{PA}(s) = s^2 + 2\zeta_{PA}\omega_r s + \omega_r^2$$
$$D_{PA}(s) = s^4 + [2\zeta_{CA} + 2\zeta_{PA}\omega_r m_r + 2\zeta_{CA}\omega_r]s^3 + [1 + m_r\omega_r^2 + \omega_r^2 + 4\zeta_{PA}\zeta_{CA}\omega_r]s^2 + [2\zeta_{PA}\omega_r + 2\zeta_{CA}\omega_r^2]s + \omega_r^2$$
$$C(s) = 2\zeta_{PA}\omega_r s + \omega_r^2$$

where $Y_{FR}(s)$ is the Laplace transform of $y_{FR}(\tau)$, the frame displacement. In our case, for the sinusoidal frequency sweep, we state:

$$y_{FR}(\tau) = y_0 \sin(\Omega_0 \tau) \tag{6}$$

where $\Omega_0 = \omega_0/\omega_{nCA}$, ω_0 is the angular frequency of the shaker, and y_0 is the displacement amplitude of the shaker.

The time domain solution of the system can be obtained through the inverse Laplace transforms of $Y_{CA}(s)$ and $Y_{PA}(s)$ by using MATLAB.

3. Test Setup

The laboratory car-frame was excited by a shaker actuator with closed loop displacement control and varying amplitude, phase and frequency.

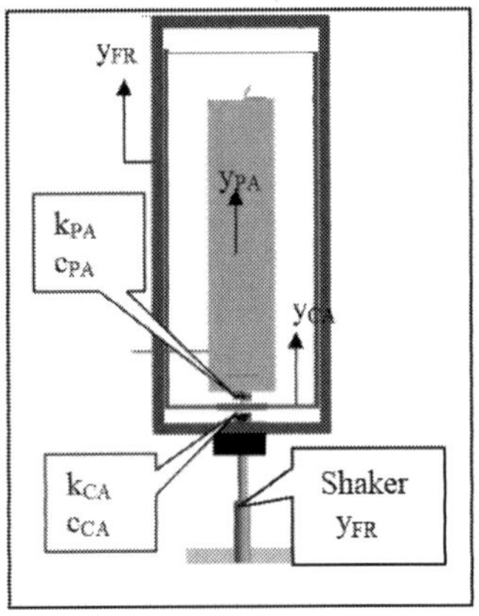

Figure 3. Test Setup.

The test setup is shown in Figure 3. A closed loop controlled actuator was fixed to a rigid frame and to the car by a rigid base. The car consists of a floor and a ceiling joined by four rigid angular bars. A special fixture has been developed to distribute the load due to weights on four pairs of shoes standing on the car floor. Two arrangements were done: one for traveling persons and the other for rigid masses. In the arrangement for rigid masses the shoes are fitted with wood-foot fixtures which are supporting a rigid base where the rigid masses or weights are held. In this way, the shoes are subjected to similar boundary conditions as the human feet.

Every test must specify the span or displacement amplitude, the frequency bounds of the sweep, the car and passenger/weight masses, the type of shoes and covering floor and the spring/damper constants of the isolation blocks. The basic tests outputs consist of a plot of the shaker/frame displacement transmissibility in µm/µm versus frequency in Hz, and other plot of the car to shaker/frame displacement phase versus frequency in Hz.

For transmissibility measurements a CSI 2130 portable analyzer with two IMI 603C1 piezoelectric accelerometers was fix to the actuator and to the rigid base of the car.

4. Results, Theoretical Comparison and Discussion

A MATLAB routine was developed to determine the theoretical car displacement transmissibility, which is defined as the ratio of the amplitude of the car response to that of the frame/excitation motion:

$$|G(\Omega)| = \left|\frac{Y_{CA}(\Omega j)}{Y_{FR}(\Omega j)}\right| \tag{7}$$

and derived from (5) taking $y_{FR}(0)=0$ and substituting s by Ωj:

$$|G(\Omega)| = \left|\frac{(1+2j\zeta_{CA}\Omega)(\omega_r^2 - \Omega^2 + 2j\zeta_{PA}\omega_r\Omega)}{\Omega^4 - \left[\left(\omega_r^2 + 2j\zeta_{PA}\omega_r\Omega\right)(1+m_r) + (1+2j\zeta_{CA}\Omega)\right]\Omega^2 + (1+2j\zeta_{CA}\Omega)\left(\omega_r^2 + 2j\zeta_{PA}\omega_r\Omega\right)}\right| \tag{8}$$

If the passengers/weights mass is very small compared to the car mass then $m_r \to 0; \omega_r \to \infty$, the expression (8) becomes:

$$|G^1(\Omega)| = \left|\frac{1+2j\zeta_{CA}\Omega}{-\Omega^2 + 1 + 2j\zeta_{CA}\Omega}\right| \tag{9}$$

which is the amplification factor for a single degree of freedom system with the car itself.

The frequencies where the magnitude of G(Ω) will peak is of much interest in lift ride quality measurements. These correspond to the resonant frequencies that, generally, are not the same as the undamped or damped natural frequencies [7].

The transmissibility given by expression (8) is analyzed using a typical 312 kg car with 6 persons capacity mounted on 4 rubber isolation blocks (m_{CA}=312kg; k_{sb}=0.267MN/m; ζ_{sb}=0.032; $k_{CA}=4k_{sb}$ =1.07MN/m; $\zeta_{CA}=0.032=.5\cdot(4c_{sb})/(k_{CA}m_{CA})^{1/2}$) in the frequency range of interest between 0 and 25Hz.

For one passenger of mass m_{1p}=82kg and stiffness k_{1p}=76kN/m [6]: $m_{PA}=m_{1p}$, $k_{PA}\approx k_{1p}$ and the dimensionless constants is found to be m_r=82/312=0.26 and ω_r=0.52. The corresponding transmissibility plots are shown in Figure 4 for different ζ_{PA} ranging from 0.001 to 0.6.

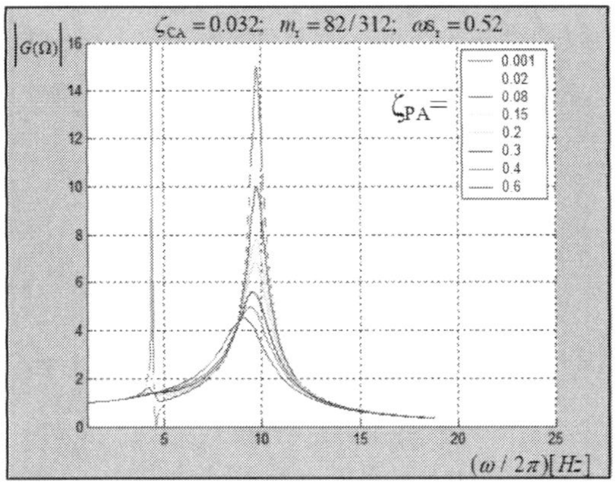

Figure 4. Transmissibility for 1 passenger

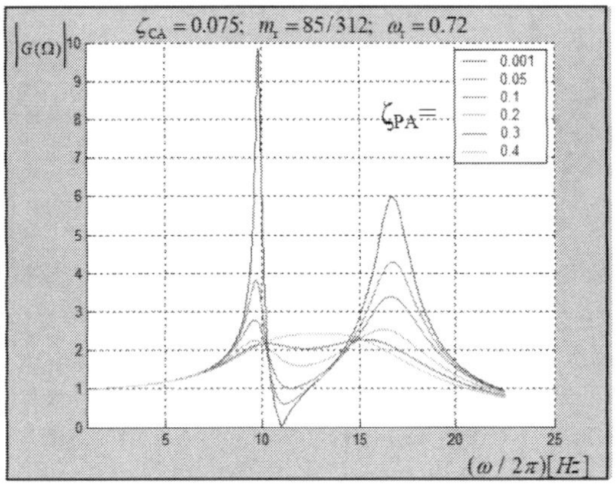

Figure 5. Transmissibility for 2 peaks.

Only if ζ_{PA} is less than 0.2 the first expected peak for two degrees of freedom system appears (see the zone for $\omega/2\pi \approx 5$Hz). Similar intensity of the two frequency peaks is only possible if ζ_{PA} is very small ($\zeta_{PA} \leq 0.02$) or if the dimensionless frequency ratio ω_r be larger than 0.52 (see Figure 5 for ω_r=0.72, 1.4 times the actual value of 0.52, using more robust isolation blocks: k_{IB} =0.7MN/m; ζ_{IB}=0.075; k_{CA}=2.8MN/m; ζ_{CA}= 0.075 and 1 passenger of 85kg mass). Thus, it can be concluded that the first resonance is damped out for the high damping ratio. The two masses tend to become locked together and appear to behave like a single mass. Consequently, only one resonant frequency is produced and its value coincides with that of the original primary system that the car is.

The effect of increasing the number of the passengers n, can be demonstrated with the assumption that all of them move in phase with each other and at the same amplitude for equal mass m_{1p}, stiffness k_{1p} and damping factor ζ_{1p}.

$$\left.\begin{aligned} \frac{1}{k_{PA}} &= \frac{1}{nk_{1p}} \\ \frac{1}{c_{PA}} &= \frac{1}{nc_{1p}} \end{aligned}\right\} \zeta_{PA} = \frac{c_{PA}}{2\sqrt{m_{PA}k_{PA}}} = \frac{nc_{1p}}{2\sqrt{n \cdot m_{1p} \cdot n \cdot k_{1p}}} = \zeta_{1p} \tag{10}$$

The above expression (10) means that the damping ratio ζ_{PA} is independent of the number of ideal passengers. The plots presented in Figure 6 shows the effect of increasing number of passengers together with the experimental results and the type of shoes worn. It is found that maximum amplitudes in the empty car resonance frequency are increasingly attenuated with the number of passengers and that sport shoes are preferred to moccasins type. For all Ω, ζ_{PA} and number of ideal passengers:

$$\left|G^0(\Omega)\right| \geq \left|G(\Omega, \zeta_{PA})\right| \tag{11}$$

that is: the new curves, basically, are "covered" by that for the empty car G^0 and the amplification factor is lowered for all r by increasing passengers.

Experimental results in Figure 6 are denoted by 'x' dot in the same colour as the corresponding theoretical curve. Good agreement is found in the predicted resonance frequencies and amplitudes for the empty car and 1 passenger cases.

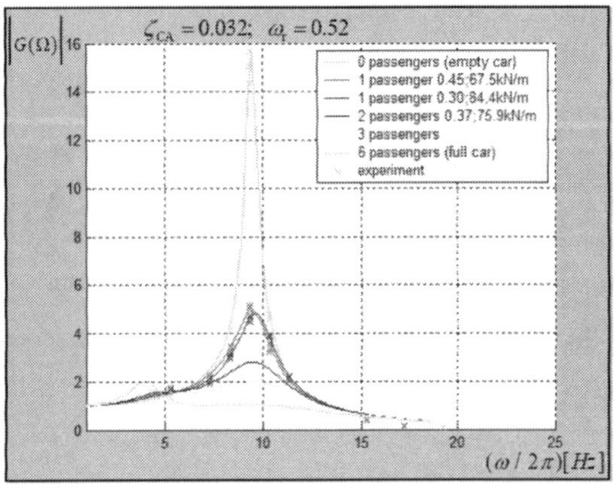

Figure 6. Influence of the passengers and type of shoes and experimental data.

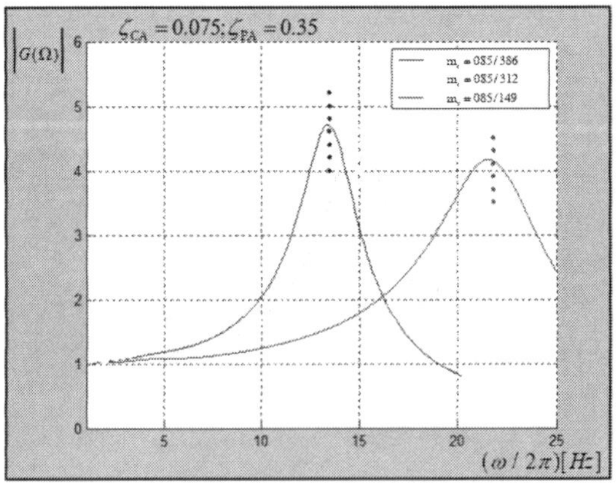

Figure 7. Varying the car mass with a single passenger.

Journal of Physics: Conference Series **181** (2009) 012081 doi:10.1088/1742-6596/181/1/012081

The results for different car masses, m_{CA}, with a single passenger are shown in Figure 7. It can be observed the coincidence of the peaks frequency with those of the empty car (denoted by vertical lines), and a small decrease of the peak amplitudes with the decreasing car mass. This confirms the experimental data on real lifts [5] that the attenuation of the car vibration acceleration rate when traveling passengers is effective with different car masses and even more efficient with increasing ratio of passengers to car mass m_r.

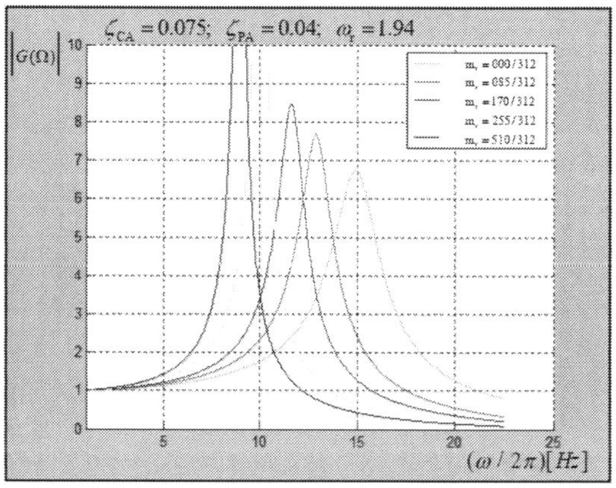

Figure 8. Substitution of persons by weights.

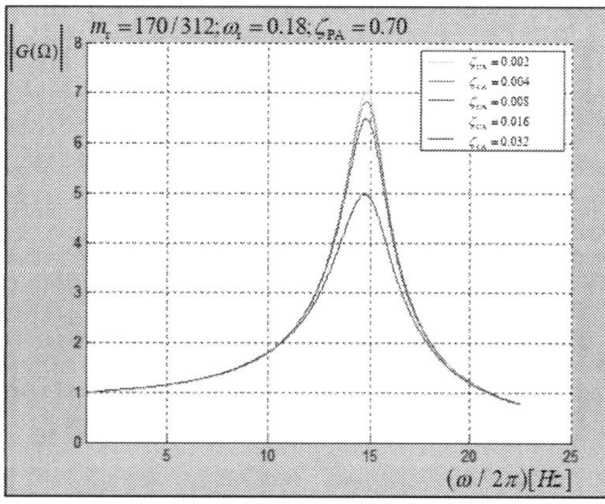

Figure 9. Varying the car damping for two idealized passengers.

The effect of rigid masses instead of persons is predicted by (8) if $\omega_r \to \infty; \zeta_{PA} \to 0$:

$$\left|G^*(\Omega)\right| = \left|\frac{1+2j\zeta_{CA}\Omega}{-(1+m_r)\Omega^2+1+2j\zeta_{CA}\Omega}\right| \tag{12}$$

which is the amplification factor for the car with its mass increased accordingly. In this case, the special fixture described in the test setup was used with a total weight mass of 200kg (m_{PA}=200kg) standing on four pairs of moccasins type shoes (k_{PA}=2.5MN/m; ω_r=3.79; $\zeta_{PA}\to 0$). Applying (12) we get the resonance frequency of the system f, given as:

$$f(m_{PA}=200kg;\zeta_{PA}=0)=\frac{1}{2\pi}\sqrt{\frac{nk_{sb}}{m_{CA}+m_{PA}}}\simeq 11.7Hz$$

Clearly different from that for the car itself:

$$f(m_{PA}=0)=\frac{1}{2\pi}\sqrt{\frac{nk_{sb}}{m_{CA}}}\simeq 15Hz$$

Using the MATLAB routine for equation (8) and the experimental configuration (m_r=200/312; ω_r=1.94) in a wide range of ζ_{PA}, was obtained a resonance frequency of 11.56Hz. The effect of traveling weights 'over' shoes instead of passengers is illustrated in Figure 8. Both, the resonance peaks are slightly amplified and the resonance frequency is lowered with increasing weight mass.

Finally, the effect of the increasing car damping is illustrated in Figure 9 for two idealized passengers wearing moccasins. The expected benefit for better attenuation is shown.

5. Conclusions

The amplification factor for a damped two degrees of freedom model of an elevator car was analyzed to predict the effect of passenger vibrational behaviour and characteristics on the dynamic response of an elevator's car. It is evident that rigid weights, even when supported on moccasins or sports shoes, don't diminish the resonance response of the car. The effect of lowering the resonance frequency by coupling the weight mass to the car mass is predicted. However, in the case of the car with the passengers load it is observed that the resonance response is reduced. This effect becomes more prominent when sports shoes are used instead of moccasins-type shoes as predicted by the amplification factors.

It is concluded that the dynamic response of the elevator car can be substantially influenced by the passenger's characteristics and behaviour during the elevator travel. Thus, more consideration should be given in BS ISO 18738 standard to the requirements and methodology for the measurement of elevator ride quality. In particular, a more in-depth analysis is needed to specify the type of shoes the person can wear when measuring elevator ride quality during elevator travel. In addition, the issues of the number of persons present in the car and the type of the car floor covering should be assessed.

References

[1] ISO 2003 Lifts-Measurement of lift ride quality. ISO 18738:2003(E)
Gómez A, Galán A, Lizaranzu M and Pelegay JL 2003 Confort en Instalaciones de Ascensor. Evaluación y Análisis de Factores de Influencia *Tecnología del Ascensor* I Herrera (Badajoz IAEE) 53-64.

[2] Fukui D, Kenichi O, Hideaki K and Takeshi N 2006 Analysis of Elevator Dynamics with Building Block Model Method *Moss 2006 Conference* S Kackmarczyk (Northampton The University of Northampton).
Herrera I and Gragera C Effective Safety Of The Electromechanical Elevator Safety Gear 2002 *Elevator Technology* **12** A Lustig (Milán IAEE) 110-19.

[3] Herrera I. 2008 Structure Of Vibrelev 1G: Simulation Software for the Elevator Dynamic Response *Elevator Technology* **17** A Lustig (Thessaloniki IAEE) 160-8.

[4] Kiyoshi F et als 2008 Influence of Elevator Acceleration Induced Loading on Injury Levels *Elevator Technology* **17** A Lustig (Thessaloniki IAEE) 87-98.

[5] Herrera I 2004 On the experimentally determined mechanical characteristics of Passengers Wearing Different Types of Shoes and Car Floor Coverings for the Elevator *Report for Mac-Puarsa I+D Full Elevator Department* (Badajoz Ingelev Group Extremadura University) 33.

[6] Herrera I 2008 Influence Of The Passengers In The Elevator's Dynamic Response *Elevator Technology* **17** A Lustig (Thessaloniki IAEE)152-9.

[7] Silva C 2007 *Vibration, Damping, Control and Design* C Silva (Lonodon CRC Press) 7.51-4.

[8] Balachandran B and Magrab E 2005 *Vibraciones* Balachandrán B and Magrab E (Mexico Thomson) 464-5.

Journal of Physics: Conference Series **181** (2009) 012082 doi:10.1088/1742-6596/181/1/012082

Analytical modelling of the transient dynamics of pipes with flowing medium

Andrei Kotousov and Roslina Mohammad

School of Mechanical Engineering, University of Adelaide, SA 5005, Australia

e-mail: Andrei.Kotousov@adelaide.edu.au

Abstract. The primary purpose of this work is to present an innovative approach to investigate the effect of a flowing medium on the transient response of a simply supported pipe due to impulsively applied loading. The importance of this problem is manifested in numerous applications in the field of transportation and power energy. In this paper we describe a simplified mathematical model of the pipe-filled-medium interactions adopting the classical Bernoulli-Euler beam theory. Further we develop an analytical procedure based on the perturbation method to analyse the dynamic response of the system due to an impulse loading and consider particular cases of the applied loading.

1. Introduction

Pipes are extensively used across many industries including power, nuclear, gas and petroleum to transport flammable or non-flammable liquids or gas. In the case of an accident the pipe structures are often subjected to impulse loading. Thus, the transient response of pipes subjected to impulse loading is important for better understanding of the potentially dangerous consequences associated with accidents so that they could be avoided or minimized [1,2]. In a number of experimental and theoretical papers it was demonstrated that the structure-filled-medium interactions significantly influence the dynamic behavior [3-6]. For example, the flowing medium can lead to principally new phenomena such as the pipe instability in the absence of the applied axial loading or pipe whip behavior [3]. The latter has the same origin as the chaos of an unrestrained garden hose with running through water. In the case of a full-bore failure of a high-pressure above ground pipeline the pipe whip phenomenon has potential for considerable damage to any property in the vicinity and could also potentially destroy a substantial length of pipeline [4].

Most of the previously published works dealt with the situation when the filled medium is stationary and not many studies have addressed to the pipe-flowing-medium interactions [5,6]. The main reason for that is, probably, a very complex nature of such interactions, which is not very amenable to numerical and analytical modeling techniques. On the other hand the experimental studies with flowing medium are quite difficult to conduct and in many previous studies the flowing media effects were ignored. Therefore, the primary purpose of this work is to develop analytical tools needed to facilitate the study of such systems. This paper is focused on the investigation of the transient response of a simply supported pipe with flowing medium to an impulse loading. To model the transient response, we adopt the classical Bernoulli-Euler beam theory and extend this theory to incorporate the flowing medium effects: inertia of the flow, centrifugal force due to the pipe curvature as well as the Coriolis force due to the flow rotation. The derived governing equation represents a fourth order nonhomogeneous partial differential equation (PDF). A dimensionless form of the governing equation contains only two dimensionless parameters controlling the system mechanical behavior. In the case of a small influence of the flowing medium on the mechanical behavior, the governing equation can be solved using an approach based on the standard perturbation method. The equations and approach can be extended to other types of pipe constraints as well as dynamic loading conditions.

This paper is organized as follows. The actual beam model incorporating the flowing media is specified in Section 2. The general approach to the solution of the initial-value problem is presented in Sections 3. Results, including the application of the approach for few special cases are given in Section 4. This is followed by brief conclusions.

2. Simplified Governing Equation

In this section we formulate the simplified governing dynamic equation of a simply supported pipe with flowing gas or liquid [7]. Consider a pipe element of length $d\bar{x}$ as shown in the Figure 1.

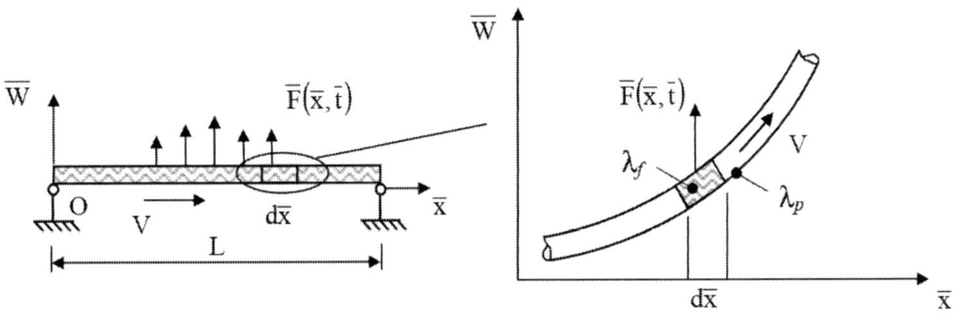

Fig.1 Dynamic equation of the system pipe-flowing liquid

Let λ_f and λ_p be the medium and pipe densities per unit length. In the case of the pipeline's transverse motion the element will be subjected to the inertia force of intensity

$$-\frac{\partial^2 \overline{W}}{\partial \bar{t}^2}(\lambda_f + \lambda_p) \quad (1)$$

Because the flow rotates with angular speed $\frac{\partial^2 \overline{W}}{\partial \bar{x} \partial \bar{t}}$, the Coriolis force acting on the element $d\bar{x}$ will be given by

$$-2\frac{\partial^2 \overline{W}}{\partial \bar{x} \partial \bar{t}} \lambda_f V d\bar{x} \quad (2)$$

where V is the velocity of the medium. With the same sign we can write the centrifugal force acting on the element due to the curvature of the gas flow

$$-\frac{\partial^2 \overline{W}}{\partial \bar{x}^2} \lambda_f V^2 d\bar{x} \quad (3)$$

It is assumed for simplicity that the dynamics of the pipe do not affect significantly the flow characteristics such as velocity and density. Thus, we decouple the structural and fluid mechanics components of the problem.

Journal of Physics: Conference Series **181** (2009) 012082 doi:10.1088/1742-6596/181/1/012082

Taking into account the elastic response of the pipe element $d\bar{x}$, the governing dynamic equation can be written in the following form

$$EI\frac{\partial^4 \bar{W}}{\partial \bar{x}^4}+(\lambda_p+\lambda_f)\frac{\partial^2 \bar{W}}{\partial \bar{t}^2}+2V\lambda_f\frac{\partial^2 \bar{W}}{\partial \bar{x}\partial \bar{t}}+V^2\lambda_f\frac{\partial^2 \bar{W}}{\partial \bar{x}^2}=\bar{F}(\bar{x},\bar{t}) \quad , \tag{4}$$

E represents Young's modulus, I is the second moment of inertia,. The right-hand side term $\bar{F}(x,t)$ represents a driving force per unit length. We assume that $0 \le \bar{x} \le L$ and $\bar{t} \ge 0$.

It is always advantageous (and often necessary) to rewrite a governing equation in a dimensionless form. To do so, we introduce the following scaling transformations

$$x=\frac{\bar{x}}{L}, \quad t=\frac{\bar{t}}{\tau}, \quad W(x,t)=\frac{\bar{W}(x,t)}{L} \tag{5}$$

where T is a parameter which will be defined later in this paper. Then,

$$\frac{EI}{L^3}\frac{\partial^4 W}{\partial x^4}+\frac{(\lambda_p+\lambda_f)L}{\tau^2}\frac{\partial^2 W}{\partial t^2}+\frac{2V\lambda_f}{\tau}\frac{\partial^2 W}{\partial x\partial t}+\frac{V^2\lambda_f}{L}\frac{\partial^2 W}{\partial x^2}=\bar{F}(x,t) \tag{6}$$

It is straightforward to show the dimensional consistency of the equation. To proceed with the solution of the governing equation, it is convenient to recast this equation into a more compact form. Dividing both sides by EI/L^3, we obtain

$$\frac{\partial^4 W}{\partial x^4}+\frac{(\lambda_p+\lambda_f)L^4}{EI\tau^2}\frac{\partial^2 W}{\partial t^2}+\frac{2V\lambda_f L^3}{EI\tau}\frac{\partial^2 W}{\partial x\partial t}+\frac{V^2\lambda_f L^2}{EI}\frac{\partial^2 W}{\partial x^2}=F(x,t) \tag{7}$$

now we define

$$\tau=L^2\sqrt{\frac{\lambda_p+\lambda_f}{EI}}, \quad \varepsilon=\frac{VL\lambda_f}{\sqrt{EI(\lambda_p+\lambda_f)}}, \tag{8}$$

and

$$\beta=VL\sqrt{\frac{(\lambda_p+\lambda_f)}{EI}}, \qquad F(x,t)=\frac{L^3}{EI}\bar{F}(x,t), \tag{9}$$

then Eq (6) becomes

$$W_{,xxxx}+W_{,tt}+\varepsilon(2W_{,xt}+\beta W_{,xx})=F(x,t), \tag{10}$$

In Eq (10) we introduced the following notation for derivative $\partial W/\partial x \equiv W_{,x}$. The similar rule is applied to the variable t as well as throughout higher-order derivatives. Observe also that following the scaling transformation (8) and (9) the space variable x varies between 0 and 1. Finally, note that for our analysis, we will impose the initial and boundary conditions corresponding to a simply supported pipe can be written

$$W(0,t)=W_{,xx}(0,t)=W(1,t)=W_{,xx}(1,t)=0. \tag{11}$$

We now proceed to describe a solution of (10) for an external force $F(x,t)$, which will subsequently be specialised to a short-duration temporal impulse.

3. Approach

3.1 Non-homogeneous classical beam equation

We start the transient analysis with the classical non-homogeneous beam equation subjected to the boundary conditions (11) at $x = 0$ and $x = 1$:

$$W_{,xxxx} + W_{,tt} = F(x,t). \tag{12}$$

We also assume that the driving load distribution $F(x,t) \equiv 0$ at $t < 0$. Several solution techniques can be applied to analyse the beam's transient response to an impulse perturbation including Laplace-Laplace transforms, the method of undetermined coefficients, variation of parameters, and the Joint transform scheme [8]. An analytical solution of Eq (12) with boundary conditions (11) obtained by the separation of variables technique and Laplace transform [9] can be written as

$$W(x,t) = \sum_{k=1}^{\infty} Y_k(x) \int_0^1 du Y_k(u) \int_0^t d\tau F(u,\tau) \frac{\sin(\omega_k (t-\tau))}{\omega_k} \tag{13}$$

where Y_k are the set of normalized eigenfunctions corresponding to the solution of the homogeneous equation $W_{,xxxx} + W_{,tt} = 0$, subjected to the specified boundary conditions:

$$Y_k(x) = \sqrt{2}\sin(\lambda_k x), \qquad \lambda_k = k\pi \tag{14}$$

and

$$\omega_k = \lambda_k^2 = (k\pi)^2. \tag{15}$$

Coefficient $\sqrt{2}$ in eigenfunctions, Eq (14), is the normalization constant that ensures that the eigenfunctions are not only orthogonal but also orthonormal. Eq (13) provides the general analytical expression for the transient spatio-temporal dynamics of a simply supported beam driven by arbitrary load distribution $F(x,t)$.

As example, consider a spatio-temporal impulse driving force at time $t = 0$ with a magnitude f_a (in scaled variables). We use Dirac's time-and-space delta function to model this impulse. Then, $F(x,t)$ takes the form

$$F(x,t) = f_a \cdot \delta(x^*)\delta(t), \tag{16}$$

where x^* represents the position along the pipe at which the driving impulse is applied.

Carrying out the space and time integrations and invoking the basic properties of Dirac's delta function, we obtain:

$$W(x,t) = 2f_a \sum_{k=1}^{\infty} \sin(\lambda_k x)\sin(\lambda_k x^*) \frac{\sin(\omega_k (t-\tau))}{\omega_k}. \tag{17}$$

The above equation is well known and describes the mechanical response of a simply supported beam subjected to a spatio-temporal impulse and this equation represents the exact analytical solution of the problem [10].

3.2 Perturbation method

We now proceed with the case of a simply supported pipe with flowing medium. In this case the simplified governing equation was derived in the previous section and can be written as

$$W_{,xxxx} + W_{,tt} + \varepsilon(2W_{,xt} + \beta W_{,xx}) = F(x,t) \tag{18}$$

where two parameters, ε and β fully control the transient response of the system. As a result the derived governing equation (18) can be used to investigate various dynamic phenomena using a reduced size physical model. Such a scale modelling is appropriate if the values of ε and β are kept the same for the scale model and reference conditions. This can potentially result into substantial benefits if experimental approach is adopted for the investigation.

At fixed β Eq (18) reduces to the classical Bernoulli-Euler equation when both products ε is limiting to zero. Further, the smallness of this parameter will be exploited in the following perturbation analysis.

In contrast to the previously considered case, an attempt at a standard separation of variables approach to determine the dynamic response fails because it is impossible to satisfy the boundary conditions with a nontrivial solution. However, the problem can be solved as a regular perturbation problem with ε as a small parameter, that is, the solution has the form

$$W = W_0 + \varepsilon W_1 + \varepsilon^2 W_2 + \ldots \varepsilon^m W_m + \ldots \qquad (19)$$

If the expansions (19) are substituted into Eq (18) and like powers of ε are equated, then the following expressions for the deflection W, correct up to order ε^m, are obtained,

$$\left.\begin{array}{ll} W_{0,xxxx} + W_{0,tt} = F(x,t) & \\ W_{1,xxxx} + W_{1,tt} = F_0(x,t) & F_0(x,t) = -2W_{0,xt} - \beta W_{0,xx} \\ W_{2,xxxx} + W_{2,tt} = F_1(x,t) & F_1(x,t) = -2W_{1,xt} - \beta W_{1,xx} \\ \ldots & \\ W_{m,xxxx} + W_{m,tt} = F_{m-1}(x,t) & F_{m-1}(x,t) = -2W_{m-1,xt} - \beta W_{m-1,xx} \end{array}\right\} \qquad (20)$$

These equations are subjected to the specified boundary conditions (11). The above equations can be considered as a recurrent system that can be solved through a step-by-step integration using the general solution for the simply supported beam subjected to arbitrary load distribution (13), which can be written as

$$W_m(x,t) = 2\sum_{k=1}^{\infty} \sin(\lambda_k x) \int_0^1 du \sin(\lambda_n u) \int_0^t d\tau F_{m-1}(u,\tau) \frac{\sin(\omega_k (t-\tau))}{\omega_k} \qquad (21)$$

The above recurrent equations (20) and general solution for non-homogeneous beam equation (21) provide an analytical tool for the analysis of the effect of the flowing media on the transient response of pipes. Further, we consider several examples to demonstrate the developed approach.

4. Particular cases

4.1 Harmonic impulse driving force

First consider a harmonical impulse driving force at time $t = 0$ with a magnitude f_n (in scaled variables), i.e.

$$F(x,t) = f_n \sin(\lambda_n x)\delta(t), \qquad (22)$$

where n is a positive integer number and λ_n is defined by Eq (15).

Then, in accordance with Eqs (20)

$$W_{n0}(x,t) = f_n \sin(\lambda_n x) \frac{\sin(\omega_n t)}{\omega_n}, \qquad (23)$$

which is also describes the transient response of pipe with a stationary filled media. In accordance with Eqs (20):

$$F_{n0}(x,t) = -2f_n\lambda_n \cos(\lambda_n x)\cos(\omega_n t) + \beta \sin(\lambda_n x)\sin(\omega_n x) \quad (24)$$

Then, caring out all integrations and after simplifications, the first order correction of the deflection function can be written as a sum of two terms

$$W_{n1}(x,t) = W_{n1}^{Cor}(x,t) + W_{n1}^{Cen}(x,t) \quad (25)$$

where

$$W_{n1}^{Cor} = -4\lambda_n f_n \sum_{k=1}^{\infty} \sin(\lambda_k x) \frac{k(1-(-1)^{n+k})}{\pi(k+n)(k-n)} \frac{\cos(\omega_k t) - \cos(\omega_n t)}{(\omega_n + \omega_k)(\omega_n - \omega_k)}$$

$$W_{n1}^{Cen}(x,t) = \frac{\beta f_n}{2\omega_n^2} \sin(\lambda_n x)(\sin(\omega_n t) - \cos(\omega_n t)\omega_n t).$$

The first term, $W_{n1}^{Cor}(x,t)$, represents the contribution of Coriolis force and the second term, $W_{n1}^{Cen}(x,t)$ is due to the centrifugal acceleration of the flowing media through the curved pipe. It is interesting to note that the first term, $W_{n1}^{Cor}(x,t)$, is limited in time; however the second term, $W_{n1}^{Cen}(x,t)$, in the first order correction is growing with time demonstrating a synergetic effect of the flowing media in the structural response of the system. Both terms are continuous functions with zero values at $t = 0$. This means that the effect of the flowing medium on the transient response is small within relatively small time intervals from the start of the impulse loading characterised by the dimensionless parameter τ, see Eq (8).

4.2 Arbitrary driving impulse force

In the case of arbitrary driving impulse force, the spatial component can be represented in Fourier series as

$$F(x,t) = \delta(t)\sum_{n=1}^{\infty} f_n \sin(\lambda_n x) \quad (26)$$

Using the principle of superposition, the general solution can be found by the simple summation of the particular solutions for each harmonic. For example, if the system is subjected a spatio-temporal impulse driving force, i.e.

$$F(x,t) = f_a \cdot \delta(x^*)\delta(t), \quad (27)$$

where f_a is the amplitude of the driving impulse. Then, the ε^0 - order solution for the pipe deflection is

$$W_0(x,t) = 2f_a \sum_{n=1}^{\infty} \sin(\lambda_n x)\sin(\lambda_n x^*)\frac{\sin(\omega_n t)}{\omega_n} \quad . \quad (28)$$

The first order correction of this solution, which incorporates the flowing medium effects, can be found from Eq (25) as

$$W_1(x,t) = \sum_{n=1}^{\infty} f_n W_{n1}(x,t)$$

where W_{n1} is given by Eq (25) and $f_n = 2f_a \sin(\lambda_n x^*)$.

5. Conclusion

This study has developed a robust analytical tool to enable efficient modelling of the transient response of a simply supported pipe with a flowing media to an impulse loading. The analytical approach was developed from a paradigm, which combines the general analytical solution of non-homogeneous Bernoulli-Euler beam equation and an application of the perturbation method. We have provided a self-contained, detailed derivation of this analytical approach. This approach may contribute to shedding light on the effect of the transported media on the dynamic response of pipelines subjected to impulse loading. In particular, the preliminary results obtained in this work have demonstrated that the effect of the flowing media is negligible within small time intervals from the start of the load impulse and can have a synergistic effect for large time intervals.

References

[1] Brown, S., Forensic engineering: Reduction of risk and improving technology (for all things great and small), Eng Failure Analysis, 2007,14: 1019–1037

[2] Milne, I., The importance of the management of structural integrity, Eng Failure Analysis, 1994, 1:171-181,

[3] Dundulis G., Uspuras, E., Kulak R. F., and Marchertas, A. Evaluation of pipe whip impacts on neighboring piping and walls of the Ignalina, Nuclear Power Plant Nuclear Eng and Design, 2007, 237: 848–857.

[4] Kotousov', A. Tuft, P. and Schultz, R. Analysis of pipe whip phenomenon associated with rupture of high pressure aboveground pipeline, Int J Pressure Vessels and Piping, 2009, submitted.

[5] Lu, G.Y., Zang, S.Y., Lei, J.P and Yang, J.L. Dynamic responses and damages of water-filled pre-pressurized metal tube impacted by mass, Int J Impact Engineering 34 (2007) 1594–1601.

[6] Jones, N. and Birch, R.S. Influence of internal pressure on the impact behaviour of steel pipelines, In: Nishioka T. editor. The 1995 joint ASME/JSME pressure vessels and piping conference, 1995. p.99-199, Trans. ASME, J Pressure Vessel Technol 1996; 118(4), 464

[7] Feodosiev, V.I., Advanced Stress and Stability Analysis, Springer, Berlin, 2005. 471.

[8] Kreyszig, E. Advanced Engineering Mathematics, John Wiley, 1999.

[9] Barhen, S. Analytical and numerical modelling of the transient dynamics of a microcantilever sensor, Physics Letters A 372 (2008), 947-957.

[10] Timoshenko, S., Young, D., and Weaver, W. Vibration problems in engineering, John Wiley, 1974.

Journal of Physics: Conference Series **181** (2009) 012083 doi:10.1088/1742-6596/181/1/012083

Numerical analysis of the effect of isotropic and kinematic hardening of anisotropic targets in impact loading

M N Krivosheina, S V Kobenko, M A Kozlova

Institute of Strength Physics and Materials Science, SB RAS, Tomsk, 634021, Russia
E-mail: marina_nkr@mail.ru, sergeyvk@inbox.ru, kozlova_ma@mail.ru

Abstract. The paper presents numerical simulation of the dynamic loading of a metal transtropic target with steel projectiles of various shapes. Equations of the continuum mechanics are used. The target loading velocities are 300m/sec and 600m/sec. The change of the target material's yield stress in various directions in the process of elastoplastic straining was analysed using the isotropic hardening model. The influence of the effect of an isotropic kinematic hardening on the target's deflected mode when the target is under the impact loading was studied. The behaviour and features of the material's limits of plasticity zones changes distribution when it is tensioned and compressed and when the target is subject to impact loading are shown.

1. Introduction
The paper discussed the yield stress zones formation in the initially anisotropic material of the target when the target is under impact loading. The framework of isotropic and kinematic hardening models is employed. Various material hardening models are used to clarify the defining relations describing the behaviour of anisotropic mediums. The hypotheses of isotropic, kinematic and combined (isotropic-kinematic) hardening have proven to be the most useful.

In the isotropic hardening law formulated by Hill and Hodge, one assumes that when subject to plastic flow, the yield surface is expanding, keeping its shape and its position with respect to the origin of the stress space unchanged. Isotropic hardening law does not account for the Bauschinger effect that was observed in the materials studied.

A hardening law is called kinematic if under this law the initial yield surface undergoes transfer, keeping its sizes and shape constant. Experimental studies [1, 2] showed that initially anisotropic aluminium alloys satisfy kinematic hardening based on Ziegler law.

The goal of our work is to study, using Ziegler hardening law, the influence of the effects of isotropic and kinematic hardening on the behaviour and features of the distribution of changes in the target of the plasticity limits changes zones on tension and compression in the target when the target is under impact loading.

2. The basic equations of the model
The axes of Cartesian coordinate system in which the stress and strain tensors are described are chosen to coincide with the main axes of the material anisotropy. The system of equations describing non-stationary adiabatic motions of compressible medium includes the continuity equation, continuous medium motion equations and energy equation [3] both for isotropic and anisotropic mediums:

-continuity equation

$$\frac{\partial \rho}{\partial t} + div\rho\bar{\upsilon} = 0, \tag{1}$$

-equation of motion

$$\rho\frac{d\upsilon^k}{dt} = \frac{\partial \sigma^{ki}}{\partial x_i} + F^k, \tag{2}$$

-energy equation

$$\frac{dE}{dt} = \frac{1}{\rho}\sigma^{ij} e_{ij}. \tag{3}$$

Here ρ – is the medium density, $\bar{\upsilon}$ – is the velocity vector, F^k – are components of the vector of mass force, σ_{ij} – are contravariant components of the symmetric stress tensor, E – is the specific integral energy.

$$e_{ij} = \frac{1}{2}(\nabla_i \upsilon_j + \nabla_j \upsilon_i), \tag{4}$$

e_{ij} – are the components of the symmetric tensor of strain rates, υ_i – are the components of the velocity vector, $i, j = \overline{1,3}$. In addition it is necessary to write down the equations that characterize physical properties of examined medium. We shall assume that the total strain can be represented as the sum of elastic and plastic strains, the elastic strain is given by generalized Hooke's law:

$$\frac{d\sigma_{ij}}{dt} = C_{ijkl} e_{kl}, \tag{5}$$

C_{ijkl} - are the elastic constants.

In order to construct the defining relationship of the plasticity for anisotropic medium, it is necessary to determine the initial yield condition, hardening law, describing the change of the loading surface, and flow rule, that describes the position of the plastic strain incremental vector. As for the loading surface, the gradient law is valid, i.e. vector increments of the plastic strain are normal at the regular points on the yield surface:

$$d\varepsilon_{ij}^p = d\lambda \frac{\partial F}{\partial \sigma_{ij}},$$

$d\lambda$ - positive increment of scalar function which depends on $d\sigma_{ij}$, σ_{ij}, ε_{ij}, F - plasticity function, $i, j = \overline{1,6}$.

We shall assume that the following conditions are satisfied in the course of elastoplastic strain of anisotropic materials modelling [4]: plastic material flow does not depend on hydrostatic stress (for materials that have a low anisotropic degree), elastic behaviour does not depend on the plastic strain, and the material is incompressible in the plastic domain. In this case when stress and strain tensors are decomposed into spherical and deviatoric parts, there are 5 independent components of deviator stress tensor and 5 independent components of deviator strain tensor. Therefore, the loading process can be considered in the 5-dimensional space of stress and strain deviators, taking the hydrostatic stress into account. Since the loading velocity does not exceed 600m/sec., barotropic medium model is used in the course of hydrostatic stress calculation.

The loading and strain processes of orthotropic material are presented in the 5-dimensional vector spaces of the stress S_i and strain $Э_i$ in agreement with the theory of A.A. Ilyushin. Instead of 6 interdependent functions S_{ij} A.A. Ilyushin introduces 5 independent functions S_i so that the transformations are one-to-one and linear [5]. The transformations of the components of stress deviator from 6-dimensional space to 5-dimension space can be written in following way:

$$S_1 = \sqrt{\frac{3}{2}} S_{11}, \quad S_2 = \sqrt{2}(\frac{1}{2} S_{11} + S_{22}), \tag{8}$$

$$S_3 = \sqrt{2} S_{12}, \quad S_4 = \sqrt{2} S_{23}, \quad S_5 = \sqrt{2} S_{31}.$$

In equation (8) S_i and S_{ij} are the components of stress deviators in 5- dimensional and 6-dimensional Euclidean real spaces respectively. Components of strain deviators, written in 5-dimensional and 6- dimensional spaces, are transformed according to the same rule.

Since the considered aluminium alloys demonstrate properties of both isotropic and kinematic hardening [1] in the process of plastic strain we can write the law of the flow surface variation that takes into account isotropic and kinematic hardening in the 5-dimensional space of stress deviators. Kovalchuk-Kosarchuk plasticity condition [2], taking into account isotropic and kinematic material hardening, when parameter η varies, is transformed into Mises-Hill plasticity condition (when $\eta = 1$), or into Tresca plasticity condition, modified for anisotropic materials (when $\eta = 0$):

$$F(S_i,\alpha_i,R) = [\eta\sqrt{\left(\frac{(S_1-\alpha_1)^2}{r_1^2}+\frac{(S_2-\alpha_2)^2}{r_2^2}\right)}+(1-\eta)(C_1(S_1-\alpha_1)+C_2(S_2-\alpha_2))]^2+$$

$$+\frac{(S_3-\alpha_3)^2}{r_3^2}+\frac{(S_4-\alpha_4)^2}{r_4^2}+\frac{(S_5-\alpha_5)^2}{r_5^2}-R^2=0,$$

$$R(\psi)=1+\xi\psi,$$

where $\psi = \int (dЭ_i^P dЭ_i^P)^{\frac{1}{2}}$, ψ - is the accumulated plastic strain, R- is the function of isotropic hardening, ξ – is the constant of isotropic material hardening, $i = \overline{1,5}$.

In this paper $\eta = 1$ and this corresponds to the plasticity condition by Mises-Hill.

Hooke's law for anisotropic mediums in the 5-dimentional space is:

$$dS_j = D_{kj}\left(dЭ_k - dЭ_k^p\right) = D_{kj}\left(dЭ_k - d\lambda\frac{\partial F}{\partial S_k}\right),$$

where $j,k=\overline{1,5}; D_{kj}$ are the components of the elastic constants matrix for 5-dimentional space; $Э, Э^p$ - total and plasticity strain respectively.

For materials whose properties do not depend on the hydrostatic stress value, the Ziegler law of kinematic hardening can be written [6]:

$$d\alpha_i = d\zeta(S_i-\alpha_i),$$

where $d\zeta \geq 0$ is the functional of the loading history, α_i are the components of displacement vector of the loading surface centre in the space of stress in the direction of i.

$\alpha = \sqrt{\alpha_i\alpha_i}$ is the magnitude of the displacement of the loading surface centre in the stress space. If $d\varepsilon_{ij}^p = 0$, then $d\zeta = 0$.

Using the experimental data obtained for aluminium alloys the papers presents:

$$\alpha_i = A_0\frac{S_T(k_v)}{S_{T_0}}\psi^p,$$

$S_T(k_v)$ is the yield stress of the initial material when loaded along the ray trajectory, which coincides with the direction of instantaneous stress vector; k_v are the components of the unit vector, which coincides with the direction of instantaneous stress vector; S_{T_0} is the yield stress when it is tensioned in the direction of axis S_1; A_0 and p which are the constants of the material.

$$d\alpha = 0, \text{ if } d\psi = 0.$$

The functional of the loading depends on the accumulated plastic strain, accumulated plastic strain increment, stress deviator components, the components of displacement vector of the loading surface centre and material constants:

$$d\zeta = pA_0\frac{S_T(k_v)}{S_{T_0}}\psi^{p-1}d\psi\frac{1}{\sqrt{(S_i-\alpha_i)(S_i-\alpha_i)}}.$$

Thus, the displacement law of the loading surface of stress deviators in the 5-dimentional space can be written:

$$d\alpha_j = pA_0 \frac{S_T(k_v)}{S_{T_0}} \psi^{p-1} d\psi (S_j - \alpha_j) \frac{1}{\sqrt{(S_i - \alpha_i)(S_i - \alpha_i)}}.$$

Stresses, identified in the element, rigidly rotated in space, are recalculated using the Jaumann derivative and are written in coordinates:

$$\frac{D\sigma^{ij}}{Dt} = \frac{d\sigma^{ij}}{dt} - \sigma^{ik}\omega_{jk} - \sigma^{jk}\omega_{ik},$$

where $\omega_{ij} = \frac{1}{2}\left(\nabla_j \upsilon_i - \nabla_i \upsilon_j\right)$.

3. Problem statement

The numerical modelling of impact was carried out in a three-dimensional formulation by the finite element method [7]. The projectile material is isotropic steel. The contact surface between the projectile and the target has the frictionless sliding.

The cylindrical target thickness is 30mm. The material of the target is 2024 transtropic aluminium alloy with the following elasticity and plasticity specifications:

$\rho = 2700\, kg/m^3$, $E_1 = 92.1$GPa, $E_2 = E_3 = 86.7$GPa, $\nu_{12} = 0.34$, $\nu_{31} = 0.32$, $\nu_{23} = 0.33$, $G_{12} = G_{13} = 31$GPa, $G_{23} = 33$GPa, $\sigma_{1T} = 350$MPa, $\sigma_{2T} = \sigma_{3T} = 290$MPa, $\tau_{12T} = \tau_{13T} = 300$MPa, $\tau_{23T} = 240$MPa, $\eta = 1$.

Here E_i - Young modulus, G_{ij} - shear modulus, ν_{ij} - Poisson ratios.

The target loading is modelled by spherical and cylindrical shape projectiles. In the course of study of the zones of isotropic hardening into the target material the diameters of cylindrical projectile are 7mm and 4.94mm, the length of cylinders are 7mm and 14mm respectively and the spherical projectile diameter is 8mm. The cylindrical shape projectiles are used to study the kinematic hardening zones formation. In the first case, the target diameter and length are 15mm, in the second case, the diameter is 30mm if the masses are identical. The elastoplastic flow of the projectile material is described by Prandtl-Reuss model. The steel yield stress is 1GPa. As a numerical method the method of final elements modified by G.R.Johnson for impact problems is used [8].

4. Formation of isotropic hardening zones

The formation of isotropic hardening zones inside the target material that occurs as a result of the impact loading, depends on physical-mechanical properties of the target material, as well as kinematic and geometric parameters of the target loading. When projectiles of various configurations, but of identical mass and velocity are used, the hardening zones inside the target material differ in the degree of the hardening and in 3-dimensional configuration of these zones. The paper does not consider the destruction of the target material in order to exclude the influence of the used anisotropic destruction criterion type on the results. As a consequence, elastoplastic material deformation is also considered in the domain where the material can be destroyed.

For the cases of the target loading with projectiles of various configurations at a 300m/sec., projectile decay curves are demonstrated in fig. 2a. The differences in projectile shapes result in significant differences in the projectile rebound times and in the values of the accumulated plastic strain in the target and projectile. At the beginning of the process, out of three cases presented, the minimal contact area between the projectile and target occurs in the case of spherical projectile and it causes a noticeable deformation of the projectile and the lowest projectile deceleration at the first 5 μs of the process. The maximal hardening of the target material occurs when it is loaded with spherical projectile. Fig. 3 presents the hardening zones of the material, where the maximal accumulated plastic strain (in the target zone, adjacent to the contact surface) is 0.2. The target material hardening in the

Journal of Physics: Conference Series **181** (2009) 012083 doi:10.1088/1742-6596/181/1/012083

case of the loading with spherical projectiles is very deep. It is explained by the small contact surface and by the extension of the zone of the hardening deep into the target at the beginning of the process.

Compact projectile has the maximal contact surface at the beginning of the process. Therefore, the projectile deceleration is faster – over 8 μs. At 20 μs of the process there are the rebounds of the projectiles of all forms from the target. Fig. 3a, 3b, 3c demonstrate 3-dimensional configurations of spherical, compact and extended projectiles and the distribution of the zones of hardening inside the target sections. For the cases of target loading with cylindrical projectile, the values of the maximal accumulated plastic strain under the projectile inside the target coincide – 0.1, while the zone of hardened target material when loaded with compact projectile increases 3-4 times in depth.

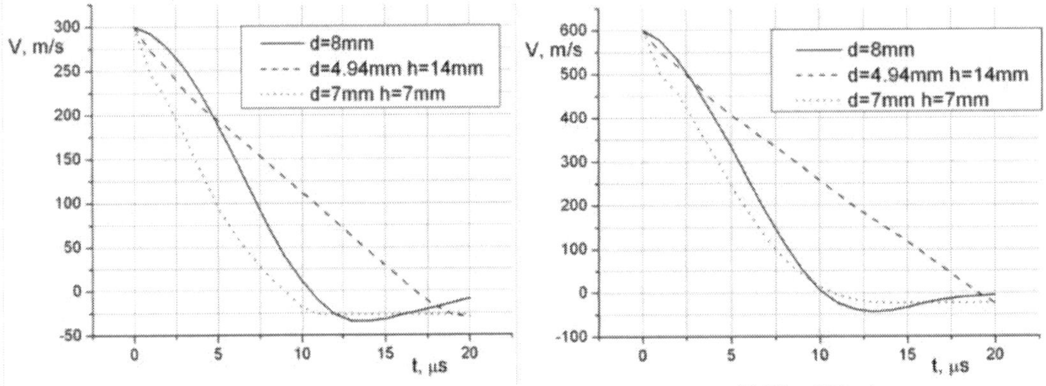

a) V_0=300m/sec b) V_0=600m/sec

Fig. 2 Projectiles velocity decay curves in the interaction with the target.

a) b) c)

Fig. 3 Zones of the hardened material inside the target at V_0=300m/sec.

Braking deceleration curves for the cases of loading of the target by the identical configuration projectiles at V=300m/s are shown in fig. 2b. In this case the large strain of projectiles of all shapes is observed and the first occurred rebound from the target was of the spherical projectile. The contact area of both spherical and compact projectiles and target eventually become approximate, and if the projectiles masses are identical then projectile velocity decay curves also become close.

At time 2μs the maximal accumulated strain inside the target is observed when it is loaded by spherical projectile and is equal to 0.22 (fig. 4a). The volume of the hardened material zone of the target for the cases of loading by three projectile shapes at time 2μs is maximal for compact projectile (fig. 5a). At 5μs into the process, the maximal accumulated plastic strain inside the target is observed when it is loaded by spherical projectile and is equal to 0.27. Eventually the hardened material zone inside the target extends greater when it is loaded by spherical projectile (fig. 4b). It is explained by greater deformation of the spherical projectile at the beginning of the process and therefore by extending of the contact area of the projectile and target. Within the framework of the model of isotropic hardening of a material, the change of the yield stress occurs when tensioned and compressed. Because of the decrease of the plastic strain values inside the target and as the distance from the contact area increases the degree of the target material hardening decreases, therefore in this part of the target the yield stress of the material remains almost constant. For extended shape projectiles the rebound from a target occurs at time 18μs, thus the volume of the hardened material of the target is minimal. In this case in contrast to spherical projectile the contact area of extended projectile and target increases slower (fig. 6a). Thus the maximal accumulated plastic strain at time 5μs of the process is equal 0.2.

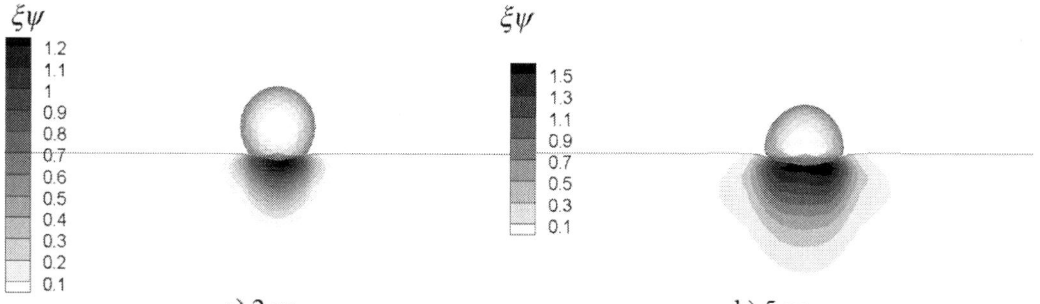

a) 2 μs b) 5 μs

Fig. 4. Zones of the hardened material inside the target at V_0=600m/sec.

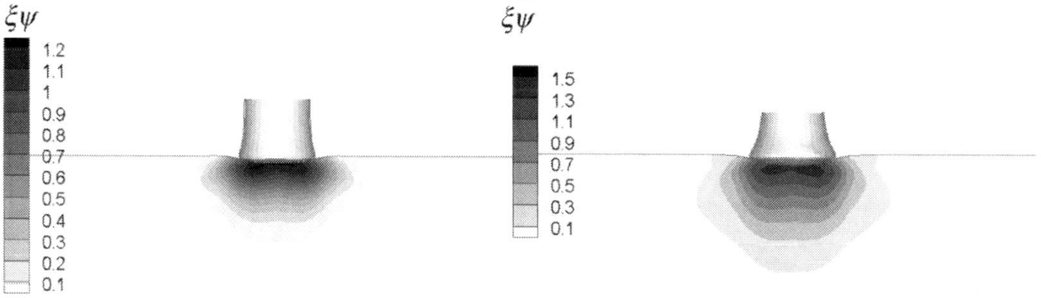

a) 2 μs b) 5 μs

Fig. 5. Zones of the hardened material inside the target at V_0=600m/sec.

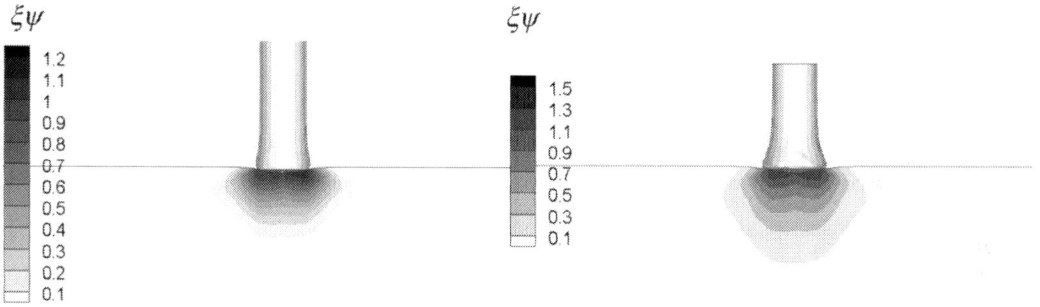

a) 2 μs b) 5 μs

Fig. 6. Zones of the hardened material inside the target at V_0=600m/sec.

5. The formation of zones of kinematic hardening

The formation of zones of kinematic hardening inside the material of the target is investigated assuming the absence of the isotropic hardenings. Fig. 7 demonstrates the solid line of the projectiles velocity decay curve (d=30mm and d=15mm) during the interaction with the target for the case when kinematic hardening of the material of the target is not taken into account, i.e. α=0. The dashed line is for projectiles velocity decay curve when kinematic hardening of the material of the target is taken into account, i.e. α >0.

Figures 7a and 7b shows that the accounting of kinematic hardening does not influence integral characteristics of the process of the impact loading of the target.

The distribution of the fields of total strains at 10μs and 20μs of the process of the loading by disk-shaped projectile is shown in fig. 8a and 8b. In contrast to the fields of stresses which distribution occurs in the wave-like fashion, the change of the fields of strains occurs with increasing of their values during the entire process in the directions which form a zone under projectile and in a direction of conical surface.

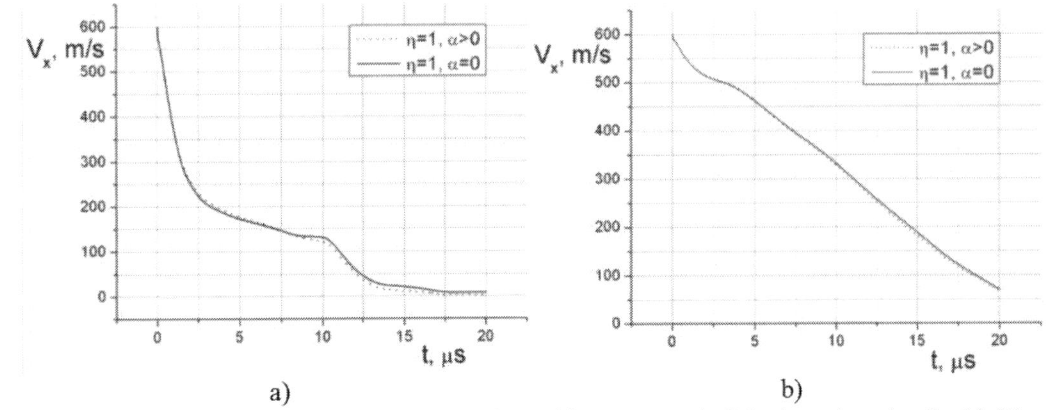

Fig. 7 Change of projectile velocity in interaction with a target: a) disk-shaped projectile (d=30mm), b) compact projectile (d=15mm).

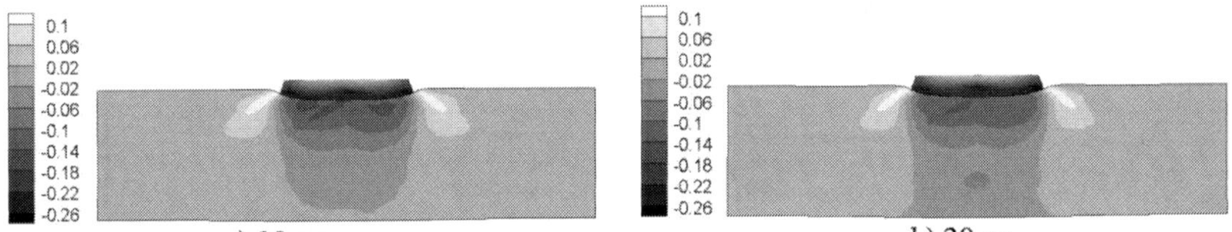

a) 10 μs b) 20 μs

Fig. 8. Distribution of ε_x isolines in cross section ZX.

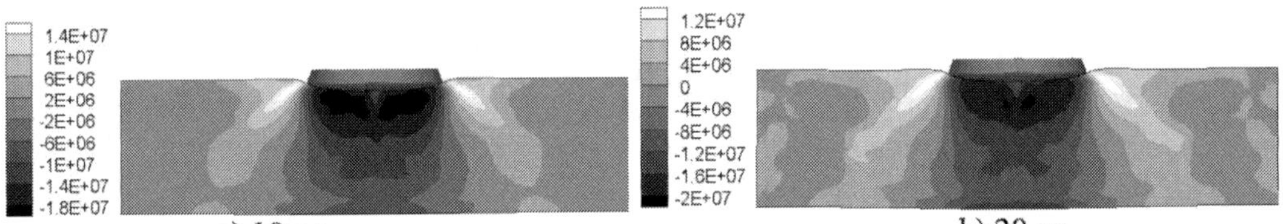

a) 10 μs b) 20 μs

Fig. 9. Distribution of α_1 isolines in cross section ZX

Fig. 9 presents the distributions of additional stresses α_1 in at 10μs and 20μs after the impact. The distributions of additional stresses, as well as total strains are not related to the wave-like distribution of the stresses inside the target. The distribution α_2 in the directions which are perpendicular load axis at 10μs and 20μs of the process are shown in fig. 10. Distribution α_2 has the character which is identical to α_1, but has an opposite sign. It is explained by the method of the impact loading of the target.

During the entire process of the impact loading under the projectile the zone of compression strain in the direction of the axis of the impact loading and the zone of tension strain - in the perpendicular direction are formed.

Figures 9a, 9b and 10a and 10b demonstrate the hardening when compressed in the direction of one axis and the hardening when it is tensioned in the perpendicular direction in the identical points of the target. Therefore the accounting of kinematic hardening irrespective of the level of additional stresses does not influence integral characteristics of the process of the loading, i.e. the projectile deceleration (fig. 7).

The zone of the hardening of the material when it is tensioned is formed and extended in the direction of the back surface of the target from the moment of the plastic strain onset. Eventually the zone of the hardening of the material when it is compressed is formed inside the target at the angle of 45°. Thus, there are additional stresses of the opposite signs inside the target in the orthogonally related directions (fig. 9 and 10). It causes the change of the values of the yield stress of the material. Such character of the change of the target material's yield stress is explained by the method of loading.

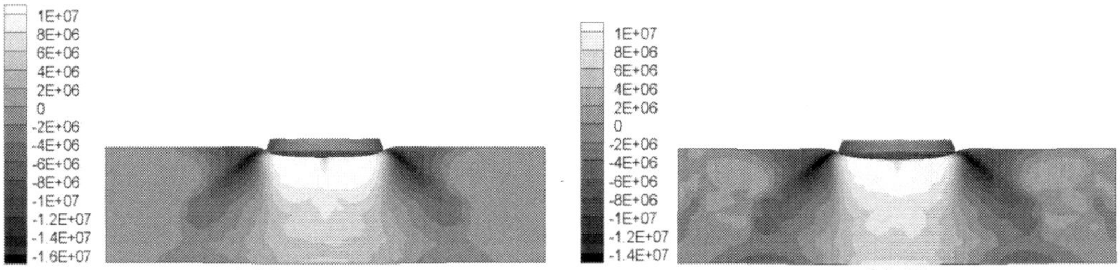

a) 10 μs b) 20 μs

Fig. 10. Distribution of additional stresses inside the target in the direction of the axis OY at 10μs and 20μs respectively.

In this case the material of the target has the plastic compression strain in the impact direction, and it has the plastic tension strain in the perpendicular directions.

6. Conclusion

1. In the considered loading velocity range (300m/sec – 600m/sec) the increase of loading velocity leads to the extending of the contact area of projectile and target and therefore to the increase of the volume of the hardened material inside the target.
2. The numerical simulation of the impact loading of the target with regard to kinematic hardening of the transtropic material by the example of 2024 aluminum alloy shows, that:
   a) effects of kinematic hardening of the transtropic material of the target does not influence integral characteristics of the process of the loading because at points of the target in the orthogonally related directions there are additional stresses of the opposite signs;
   b) the values of the tensor components of additional stresses (α_{ij}) are not related to the wave-like pattern of the stressed state of the target, but they are related only to the distribution of the values of the components of the total strain tensors in the corresponding direction.

References

[1] Kosartshuuk V V, Kovaltshuk B I, Lebedev A A 1982 *Prob. Prochnosti* № 9 3-9.
[2] Kosartshuuk V V, Kovaltshuk B I, Melnikov S A 1991 *Prob. prochnosti.* №11 19 - 25.
[3] Sedov L I 1976 *Mexanika sploshnix sred.* M. Nauka Vol. 2. 574.
[4] Krivosheina M N 2006 *Fiz. Mezomech.* Vol. 2 № 2 37–42.
[5] Ilyushin A A 1963 *Plastichnost. Osnovi obshei matematicheskoi teorii* M. P. AS USSR 271.
[6] Ziegler H 1959 *Quart. Appl. Math.* 17 № 1 55 – 65
[7] Radchenko A V, Krivosheina M N, Kobenko S V 2004 *Space Debris*, Kluwer Academic Publishers Vol. 2 № 4 319-330.
[8] Johnson G.R. 1977 *Journal of Applied Mechanics* №3 95-100.

Journal of Physics: Conference Series **181** (2009) 012084 doi:10.1088/1742-6596/181/1/012084

Vibration of a continuous beam excited by a moving mass and experimental validation

Danut Stăncioiu, Simon James, Huajiang Ouyang, John E Mottershead

Department of Engineering, University of Liverpool, The Quadrangle, Liverpool L69 3GH, U.K.

h.ouyang@liverpool.ac.uk

Abstract. The work presented in this paper deals with the vibration of a continuous slender beam excited by a mass moving at various speeds along it. An experimental model is designed and set up to study this problem. This model, which consists of a four-span continuous beam traversed by a moving mass at a constant speed, is used to build a theoretical model for the supporting structure. A series of tests designed to assess the accuracy of the model are carried out. The final section of the paper is dedicated to the numerical and experimental results and discussion.

1. Introduction

The problem of an MDOF system travelling along a flexible structure has long been studied in relation to the vehicle-bridge interaction dynamics. The supporting structure has commonly been modelled as a beam whereas the moving structure is modelled with different degrees of complexity in order to take into account various geometric and dynamic characteristics. Although this type of interaction is a complex phenomenon, which is influenced by many parameters as mentioned in [1], there might be some limiting cases where the interest rests mainly in the dynamics of the supporting structure and the flexibility of the moving structure can be neglected. In many of these cases, the moving structure can be assimilated with a moving mass with an important reduction in the number of parameters that may influence the model.

Although a lot of work has been dedicated to the theoretical study of this problem, yet few studies were concerned with any experimental investigation. Many of the experimental results previously mentioned in the literature focused on the direct validation of the model by comparing it with the experimental results obtained when a real vehicle moves along a bridge [2, 3]. To the authors' best knowledge, few references in literature were dedicated to studying small-scale models. In [4], an experimental model consisting of a single-span aluminium beam subjected to a moving mass train was used to asses the capabilities of a FEM formulation to predict the dynamic response of a beam under a moving continuous load. Bilello [5] used a scaled model to asses the dynamic response of a mass moving along an aluminium beam and came to the conclusion that the measured response is basically due to the first beam mode. A great body of work was carried out in studying the inverse problem of contact force identification using the experimental data, for example, in a series of papers by Zhu and Law [6-8].

The object of the following study is the experimental investigation of the vibration of a multiple-span continuous beam excited by a rigid body modelled as a mass moving at constant speed over it.

2. Experimental set up

The moving load experiment uses a 90 mm diameter steel ball rolling along a multi-span steel strip 3.64 m long, 100 mm wide and 3.12 mm thick. Each of the four spans is 910 mm long. A schematic of the test setup is shown in Fig. 1. The first roadway pivot is fixed in the longitudinal, transverse and vertical directions. The remaining roadway pivots allow free movement in the longitudinal direction but are fixed in the transverse and vertical directions. All pivots are free to rotate, their pivot centres being at the roadway thickness centre.

The ball is released from some point on the acceleration ramp and enters the first span of the roadway. As the ball proceeds along the roadway the vertical flexing of the roadway is measured by the eight laser displacement transducers.

The ball is kept on the roadway centre by a 'U' shaped plastic track glued to the roadway surface. At the end of its run, the ball is caught in a sand trap.

Laser beams across the roadway at ball centre height are positioned above each roadway pivot. As the ball breaks each laser beam a timing signal is sent to the data logger, with the first one being used as a trigger to start data logging. Time domain data is saved from the laser beam switches and from the eight laser displacement transducers for the period that the ball is on the roadway and for some seconds following to record the roadway vibration decay.

The data logger is an LMS Scadas III system using LMS TestLab software. All nine channels used are recorded simultaneously.

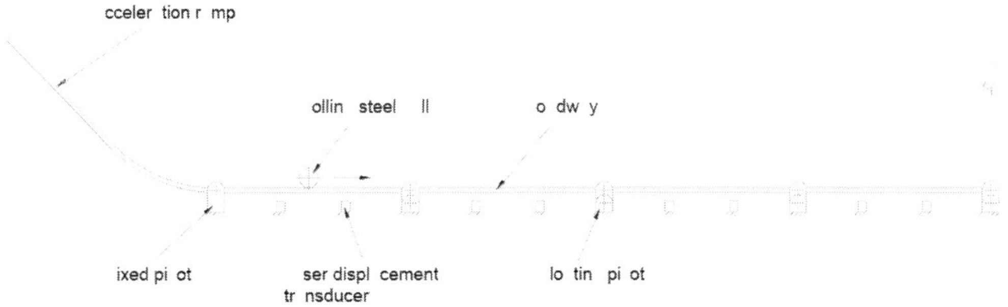

Figure 1. Schematic arrangement of moving load experiment

3. Theoretical Model

3.1. Assumptions and free vibration response

The supporting structure presented in Fig. 1 is modelled under the assumptions: the moving mass travels on the central axis of the structure only without exciting any torsional vibration, the structure has a linear elastic behaviour with small deflections, the effects of shear deformation and rotary inertia are neglected as well as the effect of the initial sag, the contact between the two structures is assumed smooth without friction. The assumption made about the supports is that they restrain all degrees of freedom of the beam except the rotation in the vertical plane. The model proposed is a four-span continuous beam. The flexural rigidity EI and mass per unit length ρA are assumed constant on the whole length of the beam L. An experimentally determined frequency-dependent damping denoted by the letter c will be considered.

The dynamic behaviour of an Euler-Bernoulli beam oriented along the x-axis when crossed by a mass m moving from left to right at constant speed v is governed by equation (1):

$$EI\frac{\partial^4 w}{\partial x^4}(x,t)+\rho A\frac{\partial^2 w}{\partial t^2}(x,t)+\rho Ac\frac{\partial w}{\partial t}(x,t)=-m(\ddot{w}(vt,t)+g)\delta(x-vt) \tag{1}$$

where δ is Dirac delta function and the over-dots represents the total derivative with respect to time t.

The transverse deflection $w(x, t)$ of the vibrating continuous beam will be approximated as a finite series expansion of the form:

$$w(x,t) = \sum_{k=1}^{N} \psi_k(x) q_k(t) \tag{2}$$

The modal shape function corresponding to mode k = 1 to N restrained to each span for a multi-span beam with transverse displacement constrained at the supports can be written as:

$$\psi_k(x) = A_j \sin(\beta_k x) + C_j \sinh(\beta_k x) + B_j(\cos(\beta_k x) - \cosh(\beta_k x)) \tag{3}$$

where coordinate x varies from zero to the span length L_s. The coefficients A_j, B_j and C_j for each span j = 1 to 4 are to be determined by imposing continuity conditions for displacements, slopes and moments at the supports [9]. For different continuous beam configurations, the modal shape functions are tabulated or presented in a detailed form in many references [9, 10]. A slight inconvenience worth mentioning when using modal beam functions is that they contain hyperbolic functions whose accuracy is poor when the terms $\beta_k x$ or $\beta_k L_s$ increases and special attention should be paid to this [11, 12]. Different approaches to deal with a continuous beam were presented in [13, 14].

Generally, any commercial software with symbolic capabilities is good enough to solve the four equations with all nine remaining constraints both for frequencies and for coefficients. For this study, some capabilities of Mathematica to simplify or reduce parts of a symbolic expression were used. The frequency equation determined for the case of a four span beam contains three factors:

$$\begin{gathered} \sin(\beta L_S) \\ \cos(\beta L_S)^2 + 2\frac{\sin(\beta L_S)}{\sinh(\beta L_S)} + \coth(\beta L_S)^2 \sin(\beta L_S)^2 - 2\coth(\beta L_S)\sin(2\beta L_S) \\ \coth(\beta L_S)\tan(\beta L_S) - 1 \end{gathered} \tag{4}$$

which when equated with zero give three sets of solutions:

- $\beta_k L_S = k\pi$ (exact solution) which corresponds to the modes 1+4k which are asymmetric about all three support points (first row of Figure 2);
- $\beta_k L_S \approx k\pi + \frac{\pi}{12}$ and $\beta_k L_S \approx k\pi + \frac{5\pi}{12}$ which correspond to modes 2+4k and 4+4k respectively with symmetry about the midpoint (second and forth rows of Figure 2) ;
- $\beta_k L_S \approx k\pi + \frac{\pi}{4}$ that correspond to modes 3+4k which are anti-symmetric about the midpoint and symmetric about the first and third supports (third row of Figure 2).

The natural frequencies can be further calculated from:

$$f_k = \frac{1}{2\pi}\beta_k^2\sqrt{\frac{EI}{\rho A}}. \tag{5}$$

The experimentally and theoretically determined natural frequencies of the four-span beam are shown in

Table 1 and the first four mode shapes in Fig. 2. In order to obtain a better accuracy of the theoretical frequency values, the beam's thickness, which, for a rectangular cross section beam, controls the flexural stiffness to unit mass ratio, is modified to 3.045 mm.

A brief analysis of this result shows that the continuous beam model considered can predict the free vibration response of the experimental structure with a high degree of accuracy.

Table 1. **Natural frequencies of the four-span continuous beam**

	Experimental frequency (Hz)	Theoretical frequency (Hz)	Relative Error [%]	Modal damping (%)
1	8.37	8.41	0.44	2.29
2	9.80	9.81	0.10	3.07
3	13.25	13.13	0.91	1.58
4	17.21	16.97	1.39	1.16
5	33.38	33.63	0.75	0.77
6	36.55	36.50	0.14	0.89
7	42.78	42.56	0.51	0.56
8	49.57	49.10	0.95	0.51

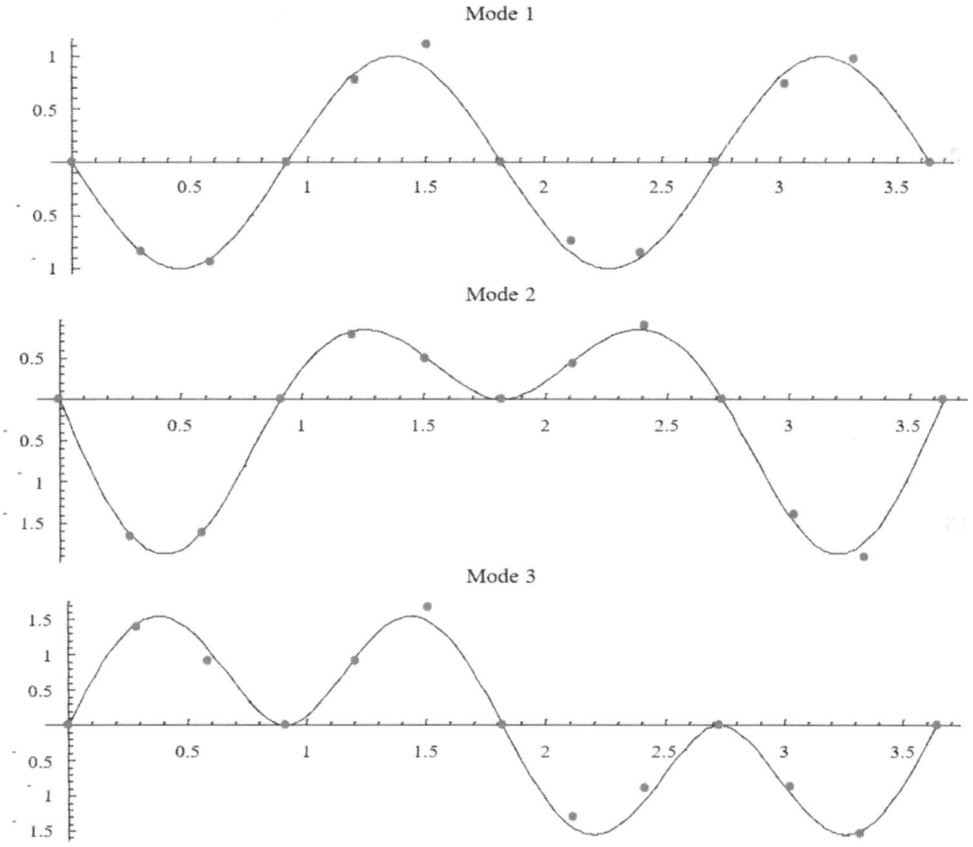

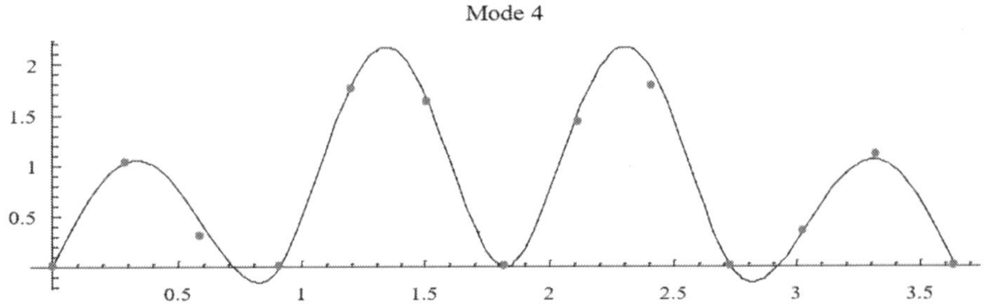

Figure 2. Mode shapes. continuous line: – theoretically determined, dots : experimental results

3.2. Theoretical Model for the interaction of two structures

The beam's differential equation (1) can be solved by replacing the deflection function $w(x, t)$ with the series expansion shown in (2). This results in a linear system of ordinary differential equations with variable coefficients whose matrix form can be written as:

$$(\mathbf{M}+\mathbf{\Delta M}(vt))\ddot{\mathbf{q}}(t)+(\mathbf{D}+\mathbf{\Delta D}(vt))\dot{\mathbf{q}}(t)+(\mathbf{K}+\mathbf{\Delta K}(vt))\mathbf{q}(t)=-mg\boldsymbol{\psi}(vt) \tag{6}$$

where the constant matrices **M**, **D**, and **K** elements are defined as:

$$\begin{aligned} &M_{kk}=\int_0^L \psi_k^2(x)\mathrm{d}x, \qquad D_{kk}=c(f_k)M_{kk}, \qquad K_{kk}=\beta_k^4 M_{kk}; \\ &M_{ki}=0, \qquad D_{ki}=0 \qquad K_{ki}=0\,(k\neq i) \end{aligned} \tag{7}$$

with $c(f_k)$ the frequency-dependent damping mentioned in section 3.1, and $i, k = 1,2, \ldots N$; and

$$\begin{aligned} &\Delta M_{ki}(vt)=m\psi_k(vt)\psi_i(vt), \qquad \Delta D_{ki}(vt)=2mv\psi_k(vt)\psi_i'(vt), \\ &\Delta K_{ki}(vt)=mv^2\psi_k(vt)\psi_i''(vt). \end{aligned} \tag{8}$$

In equation (6), the mode shape functions and the modal coordinates are assembled into two vectors:

$$\boldsymbol{\psi}(x)=\begin{pmatrix}\psi_1(x) & \psi_2(x) & \ldots & \psi_N(x)\end{pmatrix}^T, \quad \mathbf{q}(t)=\begin{pmatrix}q_1(t) & q_2(t) & \ldots & q_N(t)\end{pmatrix}^T. \tag{9}$$

4. Numerical Results

A series of tests are run for different speeds (ranged from 1.8 m/s to 3.7 m/s) of the moving mass. The speed shows very small variations along the beam's length. Therefore it is assumed constant all over the beam.

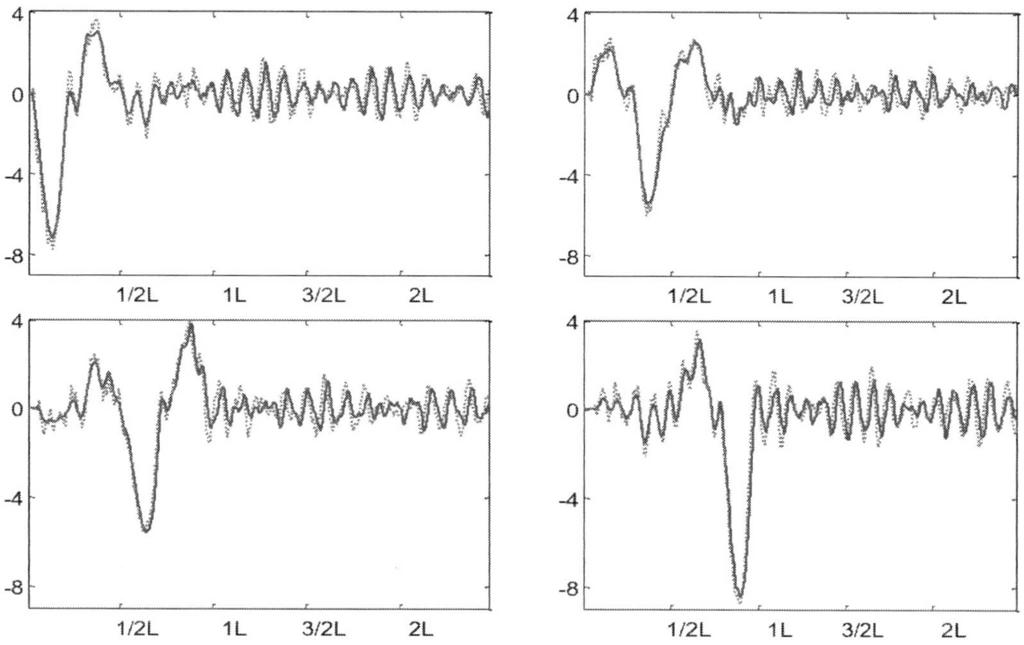

Figure 3. Time history of beam deflection (mm) versus vt at four different points along the beam (- - experiment, - theory)

These tests are used to compare the theoretical results obtained using the method presented in section 3 with the experimental results. The time history of the beam deflection (mm) at v=3.32 m/s at four points (at approximately $2L_s/3$ $5L_s/3$ $7L_s/3$ $11L_s/3$) is shown in Figure 3. For this model, the weight of the beam and the initial sag together with their influence on the results have been neglected in relation with the dynamic effects induced by the moving mass. The mass exits the beam at $1L$ and the beam's oscillations continue freely, from $1L$ to above $2L$.

The graphical comparison of the figures shows that the model slightly underestimates the experimental results. There are also noticeable oscillations of the experimentally obtained signals that are "cut" by the theoretical model (the early history in the bottom left graph of Fig. 3), and this is not due to the number of modes considered for approximation. In fact, the theoretically obtained deflections show a fast convergence when the number of modes $N > 8$. In this case, 16 modes are considered.

It is also visible that the experimentally measured signal shown in Figure 3, presents a beat-like phenomenon whose nodes, around $1L$, $3L/2$ and $2L$ are poorly captured by the theoretical model.

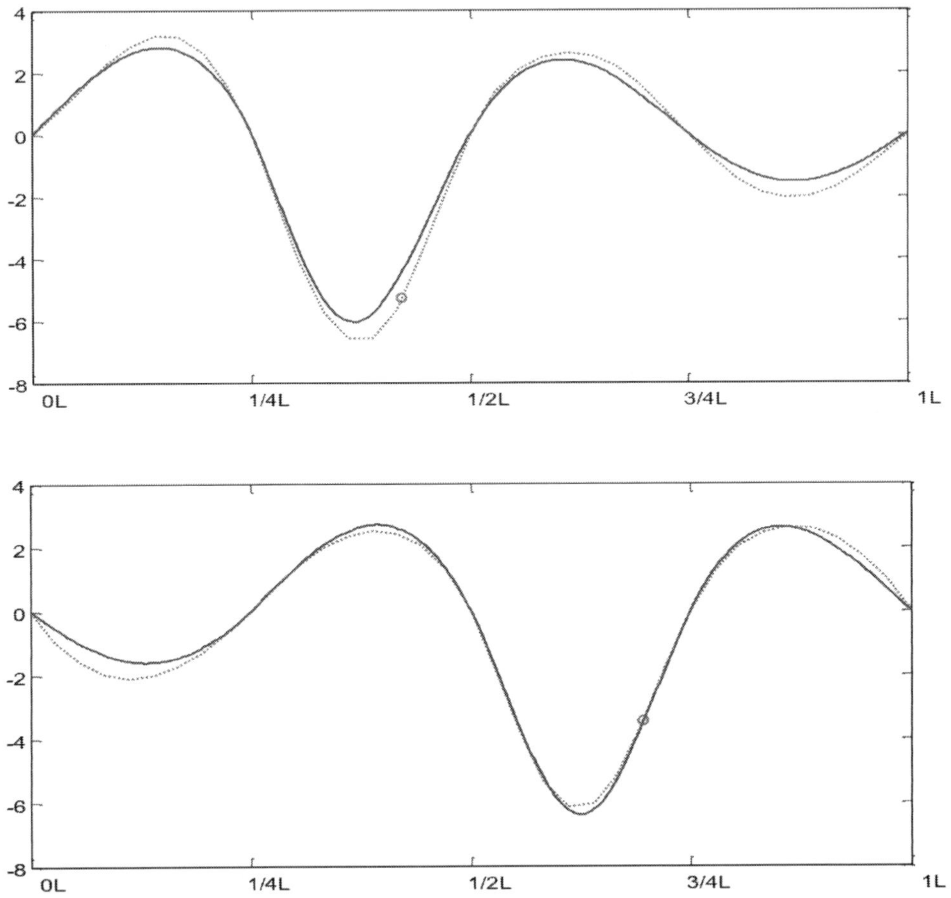

Figure 4. Beam's deflection at $t = 0.38$ and 0.68s

The accuracy of the results decreases with the increase in the moving speed and the results shown in Figure 3 presents one of the worst matches obtained.

The beam deflections (in mm) at time 0.38 and 0.68 are shown in Figure 4, along with the mass position at that time (marked by a small red circle). The experimental results are taken at eight equally spaced points along the beam. In this figure, the experimental results are replaced by a spline-interpolated curve shown in dashed line. The curve fits exactly through the discrete experimental data points, but it may show a slightly different slope at the entry point (0L) and exit point (1L). The support positions are marked by ticks on the horizontal axis.

There are some discrepancies between the experimental data and the numerical results although these differences are less than 1 mm. The contact force theoretically estimated (solution approximated with 20 modes) shows only positive values, which suggests that there is permanent contact between the supporting structure and the moving mass.

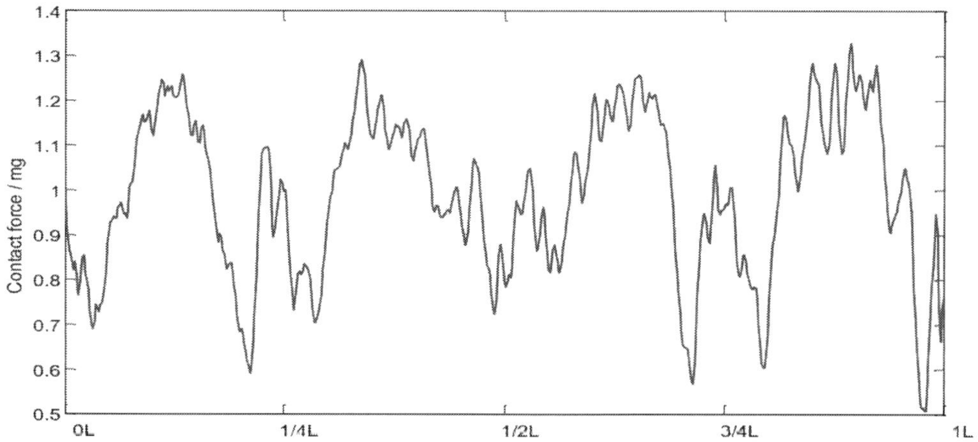

Figure 5. The contact force to beam weight ratio.

A method that might be thought to be simpler in some respects to deal with the problem of a continuous beams subjected to a moving structure is presented in [15] where the beam is considered to have a single span and the supports are introduced as elastic constraints whose elasticity is increased to a very high value. That method presents the advantage of dealing with the modes corresponding to only one-span beam. The results in terms of displacements obtained with both methods are similar.

Within the speed range tested, no separation between the moving ball and the beam structure is found. It is believed that higher moving speeds should cause separation with a subsequent change of the dynamic response [15-19]. The limit speed that allows separation, based on the theoretical model is determined to be 3.9 m/s. Further tests at higher moving speeds are now underway. When the stationary structure has many spans or it is complicated structure, the numerical (the finite element method) and analytical combined method [20] should be used.

5. Conclusions

The main object of this paper is to study the possibility of this model to represent the dynamic effects that occur when a four-span continuous beam is excited by a moving mass. Both experimentally obtained time history of the beam deflection and beam deflection shape at different fixed time instants are compared with the theoretical results. Although there are some discrepancies, their correlation is reasonably good. Free vibration of the continuous beam is also studied. The method proposed shows a good capability in representing the mode shapes and the natural frequencies experimentally obtained. It can be concluded that the theoretical model is more accurate at low speeds and the displacement response converges when the number of modes is greater than 8. The contact force variation determined based on the theoretical model shows that no separation occurs in spite of the high flexibility of the beam and the high travelling speed.

Acknowledgement

This work is sponsored by the Engineering and Physical Sciences Research Council (grant number EP/D057671/01: *Moving Load Distributions in Structural Dynamics*).

References

[1] J.L. Humar, A.H. Kashif, Dynamic analysis of slab-type bridges, *Journal of Structural Engineering* 121 (1) (1995) 48-62.

[2] L. Fryba, *Vibration of Solids and Structures under Moving Loads*, Thomas Telford, London (1972)

[3] M.F. Green, D. Cebon, Dynamic response of highway bridges to heavy vehicle loads: theory

and experimental validation, *Journal of Sound and Vibration* 170 (1) (1994) 51-78.

[4] J.R. Rieker, M.W. Trethewey, Experimental evaluation of beam structures subjected to moving mass trains, in: Proceedings of the IMAC-XII International Modal Analysis Conference, Honolulu, Hawaii, USA (1994) Vol. II 968-974.

[5] C. Bilello, L.A. Bergman, D. Kuchma, Experimental investigation of a small-scale bridge model under a moving mass, *Journal of Structural Engineering* 130 (5) (2004) 799-804.

[6] X.Q. Zhu, S.S. Law, Moving Forces identification on a multi-span continuous bridge, *Journal of Sound and Vibration* 228 (2) (1999) 377-396.

[7] X.Q. Zhu, S.S. Law, Moving forces identification on a multi-span continuous bridge, *Journal of Sound and Vibration* 228 (2) (1999) 377–396.

[8] X.Q. Zhu, S.S. Law, Orthogonal function in moving loads identification on a multi-span bridge, *Journal of Sound and Vibration* 245 (2) (2001) 329–345.

[9] D.J. Gorman, *Free Vibration Analysis of Beams and Shafts*, John Wiley & Sons Inc., New York, 1975.

[10] T.H. Chan, B.D. Ashebo, Theoretical study of moving force identification on continuous bridges, *Journal of sound and Vibration* 295 (3-5) (2006) 870-883.

[11] Y. Tang, Numerical evaluation of uniform beam modes, *ASCE Journal of Engineering Mechanics* 129 (12) (2003) 1475-1477.

[12] P.J.P. Gonçalves, M.J. Brennan, S.J. Elliott, Numerical evaluation of high-order modes of vibration in uniform euler-bernoulli beams, *Journal of Sound and Vibration* 301 (3-5) (2007) 1035-1039.

[13] K. Henchi, M. Fafard, Dynamic behaviour of multi-span beams under moving loads, *Journal of Sound and Vibration* 199 (1) (1997) 33-50.

[14] T. Hayashikawa, N. Watanabe, Dynamic behaviour of continuous beams with moving loads, *ASCE Journal of the Engineering Mechanics Division* 70 (EM1) (1981) 229-246

[15] D. Stancioiu, H. Ouyang, J.E. Mottershead, Separation between a beam with multiple elastic supports and a moving two-axle system, Proceedings of the 7th European Conference on Structural Dynamics, 7-9 July 2008, Southampton.

[16] U. Lee, Separation between the flexible structure and the moving mass sliding on it, *Journal of Sound and Vibration* 209 (1998) 867-877.

[17] D. Stancioiu, H. Ouyang, J.E. Mottershead, Vibration of a beam excited by a moving oscillator considering separation and reattachment, *Journal of Sound and Vibration* 310 (4-5) (2008) 1128-1140.

[18] D. Stancioiu, H. Ouyang, J.E. Mottershead, Vibration of a continuous beam with multiple elastic supports excited by a moving two-axle system with separation, *Meccanica* 43 (2008) 1-11.

[19] L. Baeza, H. Ouyang, Dynamics of a truss structure and its moving-oscillator exciter with separation and impact-reattachment, *Proceedings of the Royal Society of London, Series A* 464 (2008) 2517-2533.

[20] L. Baeza, H. Ouyang, Vibration of a truss structure excited by a moving oscillator, *Journal of Sound and Vibration* 321 (2009) 721–734

Journal of Physics: Conference Series **181** (2009) 012085 doi:10.1088/1742-6596/181/1/012085

Approximation of the Duffing oscillator frequency response function using the FPK equation.

E J Cross, K Worden

Department of Mechanical Engineering, University of Sheffield, Mappin St
Sheffield S1 3JD

k.worden@sheffield.ac.uk

Abstract. Although a great deal of work has been carried out on structural dynamic systems under random excitation, there has been a comparatively small amount of this work concentrating on the calculation of the quantities commonly measured in structural dynamic tests. Perhaps the most fundamental of these quantities is the Frequency Response Function (FRF). A number of years ago, Yar and Hammond took an interesting approach to estimating the FRF of a Duffing oscillator system which was based on an approximate solution of the Fokker-Planck-Kolmogorow equation. Despite reproducing the general features of the statistical linearization estimate, the approximation failed to show the presence of the poles at odd multiples of the primary resonance which are known to occur experimentally. The current paper simply extends the work of Yar and Hammond to a higher order of approximation and is thus able to show the existence of a third 'harmonic' in the FRF.

1. Introduction

Among the many techniques developed for the analysis of nonlinear systems under random excitation, the Fokker-Planck-Kolmogorov (FPK) equation approach is attractive for a number of reasons; among them is the property recognised by Yar and Hammond [1,2], of having considerable potential with regards to obtaining significant response statistics.

When the excitation to a system can be idealised as Gaussian white noise, the solution of the corresponding FPK equations is the transition probability density function. This is considered the most complete statistical description of the process, and when expressed as an expansion of the eigenfunctions of the corresponding FPK operators, can be easily manipulated to give the autocovariance and spectral density functions of the system. Unfortunately, exact solutions to the FPK equations can rarely be found for nonlinear systems, instead methods to approximate the solution must be used.

This paper follows the work of Yar and Hammond [1,2] closely, in using a variational approach to approximate the eigenfunctions of the FPK operators of the Duffing oscillator under Gaussian white noise excitation $n(t)$,

$$\ddot{x} + 2c\dot{x} + x + \varepsilon x^3 = n(t) \tag{1.1}$$

where, $E[n(t)] = 0\,, \qquad E[n(t)n(t')] = 2D\delta(t-t')$

Here, c represents viscous damping, ε a cubic spring stiffness, D is a constant which measures the white noise intensity, and δ is the Dirac delta function; *E[]* is the expectation operator.

In this paper, just as in Yar and Hammond's paper, a first-order approximation is initially employed, which is sufficient to reproduce the general features of the statistical linearisation estimate, but not accurate enough to indicate any presence of the poles at odd multiples of the primary resonance which are known to occur from simulation and experiment [3]. The work is then extended to a third-order approximation which successfully shows the existence of a secondary resonance at three times the natural frequency. The paper follows Yar and Hammond [1] very closely in its description of the basic theory; this is because the paper [1] contains a number of typographical errors which are corrected here. This theory will then form the basis of the extended calculations.

2. The Fokker Planck Kolmogorov (FPK) Equations

The transition probability density $p(x, y, t \mid x_0, y_0, 0)$ for the Duffing oscillator system of equation (1.1) satisfies the backwards and forwards FPK equations,

$$\frac{\partial p}{\partial t} = L(p) \qquad \text{(forward)}$$

$$\frac{\partial p}{\partial t} = L^*(p) \qquad \text{(backward)} \tag{2.1}$$

with the initial condition: $\underset{t \to \tau}{Lt}\, p(x, y, t \mid x_\tau, y_\tau, \tau) = \delta(x - x_\tau)\delta(y - y_\tau)$.

In the above, L and L^* are adjoint differential operators defined as follows,

$$L(p) = -y\frac{\partial p}{\partial x} + \frac{\partial}{\partial y}(2cy + x + \varepsilon x^3)p + D\frac{\partial^2 p}{\partial y^2}$$

$$L^*(p) = y_\tau\frac{\partial p}{\partial x_\tau} - \frac{\partial}{\partial y_\tau}(2cy_\tau + x_\tau + \varepsilon x_\tau{}^3)p + D\frac{\partial^2 p}{\partial y_\tau{}^2} \tag{2.2}$$

2.1. Eigenfunction Expansion of the FPK

Following the work of Johnson and Scott [4], the transition probability density function can be expressed as an expansion of the eigenfunctions of the FPK operators L and L^* as follows

$$p(x, y, t \mid x_0, y_0, 0) = \sum_{i=1}^{\infty} e^{-\lambda_i t} u_i(x, y) u_i^*(x_0, y_0) \tag{2.3}$$

where,

$$Lu_i + \lambda_i u_i = 0 \tag{2.4a}$$

$$L^* u_i^* + \lambda_i u_i^* = 0 \tag{2.4b}$$

Since L and L^* are adjoint, their corresponding eigenvalues λ_i are the same. This leads to the following biorthogonality relation which will prove useful later,

$$\int\!\!\int u_i u_j^* dxdy = \delta_{ij} \tag{2.5}$$

Where δ_{ij} is the Kronecker delta which is unity when $i = j$, and zero else.

2.2. The stationary probability density

An important solution to (2.4a) occurs when $\lambda_0 = 0$, i.e. $L(u_0) = 0$. u_0 or p_s is called the stationary probability density and can be used to find the lower-order statistical moments of the process, such as the displacement variance σ_x^2. For the Duffing oscillator

$$u_0 = p_s(x, y) = Ae^{-\frac{c}{D}(y^2+x^2+\frac{\varepsilon}{2}x^4)} \tag{2.6}$$

where the normalisation constant A is given by,

$$A^{-1} = \sqrt{\frac{\pi D}{c}} \frac{\Gamma(\frac{1}{2})}{\left(\frac{c\varepsilon}{D}\right)^{\frac{1}{4}} 4e^{\frac{1}{4}\frac{c}{\varepsilon D}}} U\left(0, \sqrt{\frac{c}{\varepsilon D}}\right)$$

and $U(a,b)$ is the parabolic cylindrical function.

2.3 The Operator G

Once the stationary probability density function has been found, as in the work of Caughey [5], Yar and Hammond [1,2] and Johnson and Scott [4], it is found to be advantageous to consider the operator G which is related to L by the following relation,

$$G(w_i) = \frac{L(p_s(x, y)w_i)}{p_s} \tag{2.7}$$

From which G is found to be,

$$G(w) = -y\frac{\partial w}{\partial x} + (-2cy + x + \varepsilon x^3)\frac{\partial w}{\partial y} + D\frac{\partial^2 w}{\partial y^2} \tag{2.8a}$$

where,

$$G(w) + \lambda w = 0 \tag{2.8b}$$

Now, since,

$$G(w) + \lambda w = 0 \ \Rightarrow \frac{L(p_s w)}{p_s} + \lambda w = 0 \ \Rightarrow L(p_s w) + \lambda p_s w = 0$$

it follows that if $w_i(x, y)$ is an eigenfunction of G then $w_i(x, y)p_s(x, y)$ is an eigenfunction of L. It is also clear here that G may be obtained from L^* on simply replacing y by $-y$. Therefore, if $w_i(x, y)$ is an eigenfunction of G, $w_i(x, -y)$ is an eigenfunction of L^*. Using these two observations equation (2.3) may now be re-expressed as,

$$p(x, y, t \mid x_0, y_0, 0) = p_s(x, y)\sum_{i=1}^{\infty} e^{-\lambda_i t} w_i(x, y)w_i(x_0, -y_0) \tag{2.9}$$

and the relation (2.5) can be re-expressed as,

$$\int_{-\infty}^{\infty}\int_{-\infty}^{\infty} p_s(x,y)w_i(x,y)w_j(x,-y)dxdy = E[w_i(x,y)w_j(x,-y)] = \delta_{ij} \qquad (2.10)$$

2.4 Autocovariance and Spectral Density Functions.

Expressions for the autocovariance function and spectral density function in terms of the eigenfunctions $w_i(x,y)$ may now be easily derived.

The autocovariance function, $\phi_{xx}(\tau)$ is defined as,

$$\phi_{xx}(t_1 - t_0) = E[x(t_0)x(t_1)] = E[x_0 x_1]$$

In an obvious notation. In full,

$$\phi_{xx}(t_1 - t_0) = \int_{-\infty}^{\infty}\int_{-\infty}^{\infty} p_j(x_0, y_0, t_0; x_1, y_1, t_1)x_0 x_1 dx_0 dx_1 dy_0 dy_1 . \qquad (2.11)$$

where $p_j(x_0, y_0, t_0; x_1, y_1, t_1)$ is the complete joint probability density function of (x_0, y_0) and (x_1, y_1). When a stationary solution of the FPK exists, the joint probability density function can be related to the transition probability density by,

$$p_j(x, y, t; x', y', t') = p(x', y', t' - t \mid x, y, 0) p_s(x, y) \qquad (2.12)$$

(which is essentially the definition of a conditional probability [6]).

Using this relation, equation (2.11) becomes,

$$\phi_{xx}(\tau) = \int_{-\infty}^{\infty}\int_{-\infty}^{\infty}\int_{-\infty}^{\infty}\int_{-\infty}^{\infty} p(x_1, y_1, \tau \mid x_0, y_0, 0) p_s(x_0, y_0) x_0 x_1 dx_0 dx_1 dy_0 dy_1$$

which, on substituting equation (2.3), eventually yields,

$$\phi_{xx}(\tau) = \sum_{i=1}^{\infty} e^{-\lambda_i \tau} \gamma_i \eta_i \qquad (2.13)$$

where,

$$\gamma_i = \int_{-\infty}^{\infty}\int_{-\infty}^{\infty} x w_i(x,y) p_s(x,y) dxdy = E[x w_i(x,y)]$$

$$\eta_i = \int_{-\infty}^{\infty}\int_{-\infty}^{\infty} x w_i(x,-y) p_s(x,y) dxdy = E[x w_i(x,-y)]$$

The spectral density function is obtained by taking the Fourier transform of the autocovarience function and is straightforwardly found to be,

$$S_{xx}(\omega) = 2\sum_{i=1}^{\infty} \gamma_i \eta_i \frac{\lambda_i}{(\lambda_i^2 + \omega^2)} \qquad (2.14)$$

where γ_i and η_i are as defined above. One can then trivially express the FRF (magnitude) as,

$$|H(\omega)| = \sqrt{\frac{S_{xx}(\omega)}{2D}} \tag{2.15}$$

3. Approximating the eigenfunctions of the FPK operators

Since the eigenfunctions of the FPK equations for the Duffing system can not be found exactly, an approximation technique must be used. A variational method based on the Rayleigh Ritz method is used here as employed by Yar and Hammond [2] and Atkinson [7].

As shown in references [2,5,7], approximate eigenfunctions $w^*(x,y)$ of the operator G can be expressed in the form,

$$w^*(x,y) = \sum_{i=1}^{N} C_i z_i(x,y) \tag{3.1}$$

where $\{z_i(x,y), i = 1,\dots,N\}$ is a set of independent trial functions orthonormal with respect to the stationary density $p_s(x,y)$, and C_i are constants to be found. The constant N is fixed by the degree of approximation required. Now, upon noting that,

$$p_s(x,y) = ae^{-\frac{c}{D}y^2}be^{-\frac{c}{D}(x^2+\frac{\varepsilon}{2}x^4)} = p_x(x)p_y(y) \tag{3.2}$$

where a and b are appropriate normalisations, one can see immediately that the trial functions will factorise as,

$$w^*(x,y) = \phi_i(x)H_j(y) \tag{3.3}$$

where the $\{\phi_i(x)\}$ will be a set of functions orthonormal with respect to $p_x(x)$, and the $\{H_j(y)\}$ will be a set of functions orthonormal with respect to $p_y(y)$. The approximation is further simplified by the observation that only eigenfunctions that are odd polynomials in x and y will contribute to the autocovariance and spectral density functions [2]. A set of basis functions $\{z_k\}$ consisting of products of the odd polynomials of $\{\phi_i(x), H_j(y)\}$ can therefore be used in the place of the full set in the approximation (3.1).

Once the basis functions $\{z_k\}$ have been determined, the constants C_i, along with the approximate eigenvalues for the original eigenvalue problem (2.8b), are found by solving the matrix eigenvalue problem [1],

$$(P + \lambda I)C = 0 \tag{3.4}$$

where the *ijth* element of the matrix P is $E[G(z_i)z_j]$.

3.1 Orthonormal trial functions

To construct the required basis functions, the set $\{\phi_i(x)\}$ of functions orthonormal with respect to $p_x(x)$, and the set $\{H_j(y)\}$ orthonormal with respect to $P_y(y)$ must first be found. Since y is normally distributed, the $\{H_j(y)\}$ take the form of Hermite polynomials. The required orthonormal set is given by,

$$\{H_j(y)\} = \frac{h_j(y)}{\|h_j\|} \tag{3.5}$$

where
$$\{h_j(y)\} = \left(\tfrac{D}{2c}\right)^{-\frac{j}{2}} h_j^{[1]}\left(y\left(\tfrac{D}{2c}\right)^{-\frac{1}{2}}\right)$$

$$\|h_j\| = \sqrt{E[h_j^2]}$$

and $\{h_j^{[1]}(y)\}$ are the standard Hermite polynomials orthogonal with respect to the normal probability distribution with density $\frac{1}{\sqrt{2\pi}} e^{-\frac{y^2}{2}}$.

The construction of $\{\phi_i(x)\}$ is approached using the Gram-Schmidt method. This method takes a finite set of linearly independent functions $\{U_n\}$ and generates an orthogonal set $\{\psi_n\}$ that spans the same subspace. A set of orthonormalised functions $\{\phi_n\}$ can then be obtained from the orthogonal sequence. Here $\{U_n\} = \{1, x, x^2, ..., x^n\}$.

3.2 Second and higher- order moments of x and y

The constructed orthonormal functions will involve the displacement variance $E[x^2]$, and higher-order moments such as $E[x^4]$. While the stationary probability density can be used to find the second order moments $E[x^2]$ and $E[y^2]$, recursion relations must be used to express any higher-order moments. If one considers $E[x^m]$ first, one observes that,

$$E[x^m] = \int_{-\infty}^{\infty}\int_{-\infty}^{\infty} x^m p_s(x,y)dxdy = \int_{-\infty}^{\infty}\int_{-\infty}^{\infty} x^m A e^{-\frac{c}{D}(y^2+x^2+\frac{\varepsilon}{2}x^4)} dxdy \tag{3.6}$$

Integrating by parts, one establishes the following recursion relation,

$$E[x^{m-2}] + \varepsilon E[x^m] = \frac{(m-3)DE[x^{m-4}]}{2c} \tag{3.7}$$

Using this relation one can find all the higher-order moments providing the displacement variance $E[x^2]$ is known. Directly from the work of To [8], one finds the displacement variance to be

$$E[x^2] = \sqrt{\frac{D}{4c\varepsilon}} U\left[1, \sqrt{\frac{c}{D\varepsilon}}\right] U\left[0, \sqrt{\frac{c}{D\varepsilon}}\right]^{-1} \tag{3.8}$$

where *U[a,b]* is again the parabolic cylindrical function.

It remains now to find an expression for $E[y^2]$, and the higher-order moments involving y. As before, one begins with,

$$E[y^m] = \int_{-\infty}^{\infty}\int_{-\infty}^{\infty} y^m p_s(x,y)dxdy = \int_{-\infty}^{\infty}\int_{-\infty}^{\infty} y^m A e^{-\frac{c}{D}(y^2+x^2+\frac{\varepsilon}{2}x^4)} dxdy \tag{3.9}$$

and integrates by parts; the following recursion relation is found,

$$E[y^m] = \frac{(m-1)DE[y^{m-2}]}{2c} \tag{3.10}$$

Furthermore, it follows trivially from this recurrence that,

$$E[y^2] = \frac{D}{2c}$$

Everything is now in place to use (3.1) to approximate the eigenfunctions of the system.

3.3 First order approximation, N=1.

Yar and Hammond [2] used the approximation in (3.1) with N=2, which amounts to a first-order approximation in x and y. In this case the basis functions are $z_1(x,y) = \phi_1(x), z_2(x,y) = H_1(y)$, where,

$$\phi_1 = \frac{x}{\|\psi_1\|} \quad \text{with} \quad \|\psi_1\| = \sqrt{E[x^2]} = \sigma_x$$

$$H_1 = \frac{y}{\|h_1\|} \quad \text{with} \quad \|h_1\| = \sqrt{E[y^2]} = \sigma_y = \frac{D}{2c}$$

and the matrix P of equation (3.4) takes the form,

$$P = \begin{bmatrix} E[G(\phi_1)\phi_1] & E[G(\phi_1)H_1] \\ E[G(H_1)\phi_1] & E[G(H_1)H_1] \end{bmatrix} = \begin{bmatrix} 0 & \frac{-\sigma_y}{\sigma_x} \\ \frac{\sigma_y}{\sigma_x} & -2c \end{bmatrix}$$

which yields eigenvalues completely consistent with a statistical linearization approach, as observed by Yar and Hammond.

A Maple routine [10] was used at this point to solve the matrix eigenvalue problem for a low level of excitation, where one would expect to observe the FRF of the underlying linear system; the parameters: D=0.005, c=0.05 and ε=0.05 were adopted (in order to make contact with the results obtained in [1]). With these choices, the approximate eigenvalues of the system were found to be,

-0.05±1.002*i*.

Looking at the eigenvalues gives a quick and simple validation as to whether the approximation is working, as the imaginary parts of the eigenvalues should represent the resonances of the system. In this case the imaginary parts of the eigenvalues are indicating that the primary resonance occurs at around unit frequency, which is expected.

From this approximation the autocorrelation function and FRF magnitude were computed using equations (2.14) and (2.16) and are shown below. The usual forms associated with the single-degree-of-freedom linear oscillator are observed.

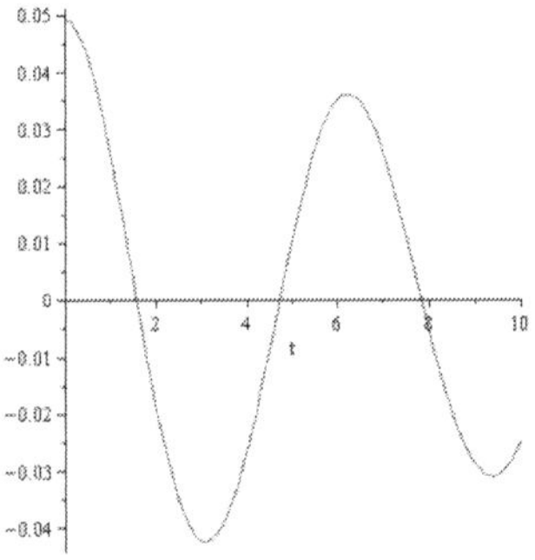

Figure 1. Autocorrelation function

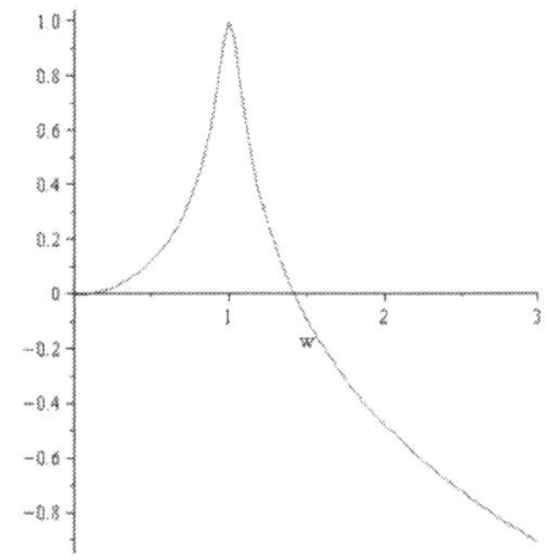

Figure 2. FRF magnitude.

3.3 Third order approximation, N=6.

Equation (3.1) was then used with N=6 to compute the third-order approximation of the FPK solution. The required basis functions in this case are $\{\phi_1(x), H_1(y), \phi_1(x)H_2(y), \phi_2(x)H_1(y), \phi_3(x), H_3(x)\}$, and P therefore becomes a 6x6 matrix.

$$P = \begin{bmatrix} E[G(\phi_1)\phi_1] & E[G(\phi_1)H_1] & E[G(\phi_1)\phi_1H_2] & E[G(\phi_1)\phi_2H_1] & E[G(\phi_1)\phi_3] & E[G(\phi_1)H_3] \\ E[G(H_1)\phi_1] & E[G(H_1)H_1] & E[G(H_1)\phi_1H_2] & E[G(H_1)\phi_2H_1] & E[G(H_1)\phi_3] & E[G(H_1)H_3] \\ E[G(\phi_1H_2)\phi_1] & E[G(\phi_1H_2)H_1] & E[G(\phi_1H_2)\phi_1H_2] & E[G(\phi_1H_2)\phi_2H_1] & E[G(\phi_1H_2)\phi_3] & E[G(\phi_1H_2)H_3] \\ E[G(\phi_2H_1)\phi_1] & E[G(\phi_2H_1)H_1] & E[G(\phi_2H_1)\phi_1H_2] & E[G(\phi_2H_1)\phi_2H_1] & E[G(\phi_2H_1)\phi_3] & E[G(\phi_2H_1)H_3] \\ E[G(\phi_3)\phi_1] & E[G(\phi_3)H_1] & E[G(\phi_3)\phi_1H_2] & E[G(\phi_3)\phi_2H_1] & E[G(\phi_3)\phi_3] & E[G(\phi_3)H_3] \\ E[G(H_3)\phi_1] & E[G(H_3)H_1] & E[G(H_3)\phi_1H_2] & E[G(H_3)\phi_2H_1] & E[G(H_3)\phi_3] & E[G(H_3)H_3] \end{bmatrix}$$

A Maple routine was once again used to solve the 6x6 matrix eigenvalue problem (3.4). FRFs and spectral density functions are yet to be computed for the third degree approximation; instead, for a quick validation of the approximation one can once again consider the approximate eigenvalues. For a higher level of excitation which is expected to force marked nonlinear behaviour: for D=1, c=0.05 and ε=0.001 (again, following [1]), the approximate eigenvalues of the system are found to be,

-0.016+1.027i	-0.016-1.027i
-0.070+3.081i	-0.070-3.081i
-0.048+1.012i	-0.048-1.012i

As expected, one pair of the eigenvalues have imaginary part close unity, representing the frequency corresponding to the primary resonance. The second pair have imaginary parts close to three, which confirms that the approximation is recreating the secondary resonance at three times the natural frequency. The third pair of eigenvalues suggest the presence of a double pole at around unit frequency, which is a conclusion consistent with a parallel study carried out using the Volterra series [9].

Journal of Physics: Conference Series **181** (2009) 012085 doi:10.1088/1742-6596/181/1/012085

4. Conclusions

Although the autocorrelation and spectral density functions of the third-order approximation have not yet been computed, one can conclude from the approximate eigenvalues that the third-order approximation is sufficient to indicate the presence of the secondary resonance at three times the natural frequency. It remains now to compute these functions for the third-order approximation and assess their accuracy by comparison with numerical simulation. Further study could naturally extend to a higher-order approximation to see if further harmonics at odd multiples of the natural frequency can be recreated.

References

[1] Yar M and Hammond J K 1986 Spectral analysis of a randomly excited Duffing system. *Proc. 4th Int. Modal Analysis Conf.* (Los Angeles).
[2] Yar M and Hammond J K 1986 Approximate eigenfunction analysis of first order non-linear systems with application to a cubic system *J. Sound Vib.* **111**(3) 457-66
[3] Nayfeh A H *Nonlinear Oscillations* (New York: Wiley)
[4] Johnson J P and Scott R A 1980 Extensions of eigenfunction-expansion solutions of a Fokker-Planck equation- II. Second order system *Int. J. Non-Lin. Mech.* **15** 41-56
[5] Caughey T K 1963 Nonlinear theory of random vibrations *Adv. in App. Mech.* **11** 209 53
[6] Risken H 1989 *The Fokker Planck Equation; Methods of solution and Application* (Springer)
[7] Atkinson J D 1973 Eigenfunction expansions for randomly excited non-linear systems *J. Sound Vib.* **30**(2) 153-72
[8] To C W S 2000 *Nonlinear Random Vibration* (Swets & Zeitlinger)
[9] Worden K and Manson G 1998 Random vibrations of a Duffing oscillator using the Volterra series *J. Sound Vib.* **217**(4) 781-9
[10] Maple 12 - software for algebraic computing. Maplesoft Ltd. 2009.

Journal of Physics: Conference Series **181** (2009) 012086 doi:10.1088/1742-6596/181/1/012086

Damage detection in turbine wind blades by vibration based methods

Ł Doliński [1], M Krawczuk [1,2]

[1] Gdansk University of Technology, Faculty of Electrical and Control Engineering, Narutowicza 11/12, 80-233 Gdansk, Poland, tel. +48 58 347 24 73,

[2] Polish Academy of Sciences, IFFM, Fiszera 14, 80-231 Gdansk, Poland, tel. +48 58 tel. +48 58 6995 206, fax. + 48 58 341 61 44,

E-mail: [1] l.dolinski@ely.pg.gda.gda.pl; [2] mk@imp.gda.pl

Abstract. The paper describes results of numerical simulation for damage localization in the composite coat of a wind turbine blade using modal parameters and a modern damage detection method (wavelet transform). The presented results were obtained in the first period of research on the diagnostic method, which is aimed at detecting damage in the blades of large wind turbines during normal operation. A blade-modelling process including the geometry, loads and failures has been introduced in the paper. A series of simulations has been carried out for different localizations and size of damage for finding the method's limits. To verify the results of numeric simulations a subscale blade has been built which has geometric features and mechanical properties similar to the computer model.

1. Introduction

The object of this research was the blade of a Horizontal Axial Wind Turbine, which is the most important sub-assembly of such a power plant. Nowadays, the blades are made of glass- or carbon fibres composite that are characterized by relatively small weight and higher strength compared with metallic structures. Employment of that new material technology makes possible building larger and more effective turbines. But such constructions are more exposed to the different kind of breakdowns and the continuous character of their operation causes that the blades are unceasingly loaded in different directions. Owing to real-time condition monitoring of all turbine elements it is possible to detect early an appearing damage. In modern constructions the diagnostic system is an the integral part of the structure. That makes it possible to create favourable conditions for failure-free operation in a set-up time. The most dangerous of composites failures are those which destroy the structure of the material. There are no external symptoms on the element surface, which makes damages very difficult to detect. Because of that, intensive work is conducted to create efficient methods of fault detection and localization but most of them require stopping the rotor.

Paper [1] presents results of fatigue tests of full-scale blades. The Authors have noticed connections between characteristic features of the emitted acoustic signal and mechanical properties (vibrations, modulus of elasticity and tip bending), changes of which can indicate the appearance of damage. Also paper [2] describes the application of acoustic emission for damage detection. Analysis of signals from sensors located on a blade surface allowed detection of existing defects and active zones where risk of damage was apparent. Descriptions of acoustic signal analysis of blades for diagnostics are also given in [3][4][5][6].

Ultrasonic computer scanning is utilized in [7] for damage detection. The method made use of a movable transmitter and a stationary receiver for creating a defects map of the tested surface. Also paper [8] describes the application of ultrasonic method for non-destructive testing. However, this technique is very time-consuming.

The presented approaches require specified conditions of measurement (restraint the blade, install complicated measuring equipment, apply a static load). The techniques which are free from those disadvantages are proposed in [9][10][11]. An infrared digital camera was used to find temperature changes of the blade surface. It has been shown that the method effectively indicates cracks as well as places threatened by damage.

The basic assumption of vibration damage detection is that vibration parameters such as natural frequencies, displacement, mode shapes or damping depend on the physical properties of the structure. For example, a method based on analysis of the transmittance function is presented in [12]. Results of the investigation presented in [13][14] determine that composite damage changed the vibration spectrum of the structure, which allowed creating a non-destructive testing method of the blade during normal running. Damages may significantly reduce the stiffness of the structure. In paper [15] global modal methods were used (mainly at low frequency). Signals from sensors of strain and acceleration were analyzed to determine a stiffness change of the tested element. Paper [16] shows that simple measurements and appropriate analysis techniques of mode shape make it possible to detect the damage. Wavelet transforms are used to analyze the mode shapes of the damaged plate and beam. That is why this technique was applied in this experiment.

2. Theoretical background

Analysing a signal in different frequency bands with different resolution by decomposing it into a coarse approximation and detailed information we call a wavelet transform [17]. The wavelet (means a small wave) is a kind of mathematical function used to divide a given function into different frequency components and study each component with a resolution that matches its scale. The wavelets are scaled and translated copies of a finite-length or fast-decaying oscillating waveform ("mother wavelet"). The shapes of exemplary mother wavelets are presented in figure 1.

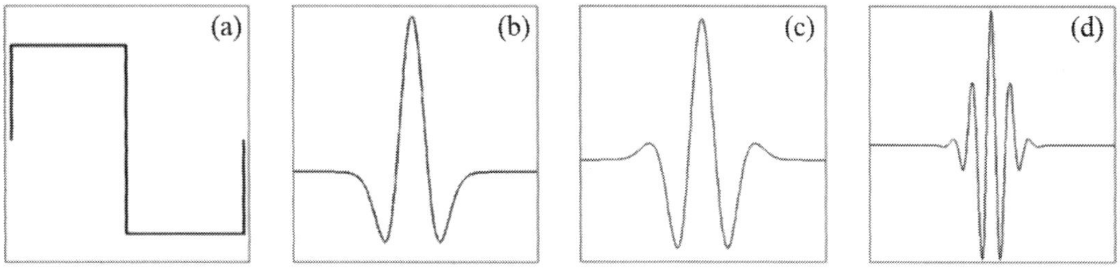

Figure 1. Shape of exemplary mother wavelet: a) haar; b) mexican hat; c) gaus4; d) morlet.

The width of the window is changed as the transform is computed for every single spectral component, which is probably the most significant characteristic of the wavelet transform. Wavelets are well-suited for representing functions that have discontinuities and sharp peaks, and for accurately deconstructing and reconstructing finite, non-periodic and/or non-stationary signals. Wavelet transforms can be used to represent continuous-time (analog) signals. We classified them into discrete wavelet transforms (DWTs) and continuous wavelet transforms (CWTs). CWTs operate over every possible scale and translation. The DWTs use a specific subset of scale and translation values or representation grid. For more detailed information please follow the literature [17][18].

Journal of Physics: Conference Series **181** (2009) 012086 doi:10.1088/1742-6596/181/1/012086

3. Numerical calculation

The tested blade has 1.74 meter length, without the twist of profile and one longitudinal spar. Glass fibre and epoxy resin have been used as the laminate components on the structure material. Reinforced fibres have been symmetrically arranged (45°/-45°). A changing thickness of the blade coating has been assumed. The blade was divided into three sections which had a different number of laminate layers. The total mass of the blade was about 2kg. Mechanical properties of the composite are presented in table 1. The geometrical characteristics of the blade are presented in table 2.

Table 1: Properties of composite material components.

	Epoxy resin	Glass fiber
Young modulus, E	3.43 GPa	66.5 GPa
Poisson ratio, ν	0.35	0.23
Kirchoff modulus, G	1.27 GPa	27 GPa
density, ρ	1250 kg/m3	2250 kg/m3

Table 2: Geometrical features of the blade.

	Dimension [mm]
Total length	1740
Max width	160
Tip width	32
Thickness of spar	2.0
Thickness of section 1 (near blade root)	2.0
Thickness of section 2 (middle part of blade)	1.5
Thickness of section 3 (near blade tip)	1.0

In the FEM modelling process the isoparametric shell element has been applied, which has eight nodes and six degrees of freedom at each node. A series of simulations was conducted. At first, the Block Lanchos method has been used to calculate the natural frequencies of the undamaged blade. In the next step, the damage has been applied to the composite coat. The crack has 2 mm width and four values of length (from 5% to 30% of the blade width) and for three localizations of the damage. All simulations and calculations have been conducted using applications of ANSYS and MATLAB® Software. The blade model and the place with damage is presented in figure 2.

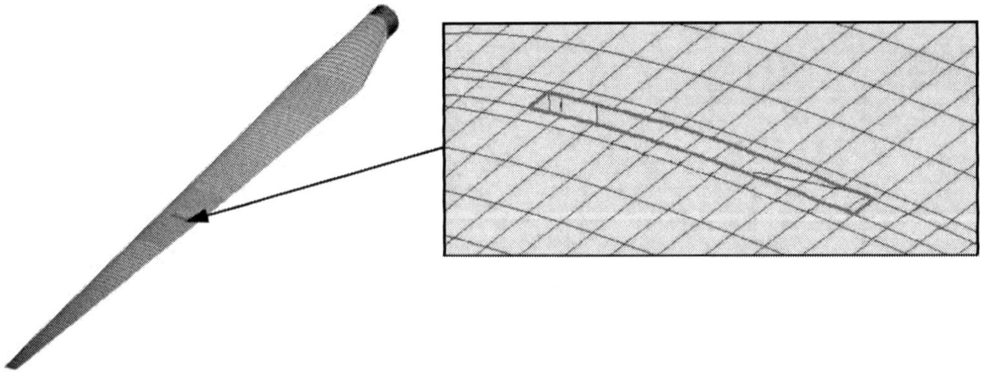

Figure.2. Blade model with the damage (red line).

4. Experimental measurements

A real blade (similar to the computer model) has been built to verify the results of simulation. The object was fixed on a special steel table which guarantees high stiffness and isolates from external sources of vibrations. An electromechanical shaker (GWV B100) has been used for vibration excitation (figure 3a). Measurements of vibrations have been carried out using a PSV- 400 Laser Scanning Vibrometer (figure 3b). The main advantages of a Laser Doppler vibrometer are automation, high sensitivity and non-contact capabilities [19]. Thanks to the sensor head with a laser Doppler vibrometer which enabled fast, accurate and non-contact vibration measurements of the whole surface, the influence of undesirable effects which could distort the results has been eliminated.

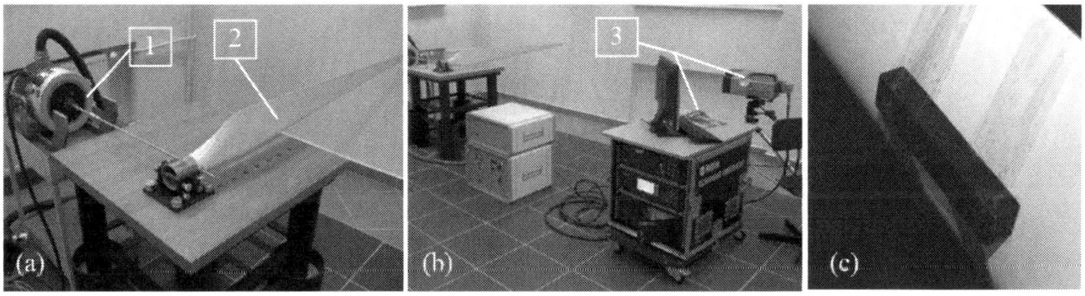

Figure 3. Experimental setup: a) shaker – 1; blade – 2; b) laser vibrometer – 3; c) model of damage.

The mode shapes of the object have been determined by velocity measurement for the undamaged and damaged blade. Damage has been applied in the form of an additional mass fixed to the surface of the object (figure 3.c). Measurements have been carried out for three values of additional mass and several localizations. All damage cases are gathered in table 3. The scan grid of measurement points is presented in figure 4. There are 290 scan points (5 lines, 58 points per line) for the first and second series of measurements. The last series has been carried out with one line (211 scan points). Each point has been excited with frequencies sweeping from 4 Hz to 250 Hz with a resolution of 0.1 Hz (7 mode shapes).

Table 3. Measurement cases.

Value of additional mass (percent of total mass)	Number of scan points (per line)	Number of damage localization
12%	58	3
6%	58	3
2%	211	7

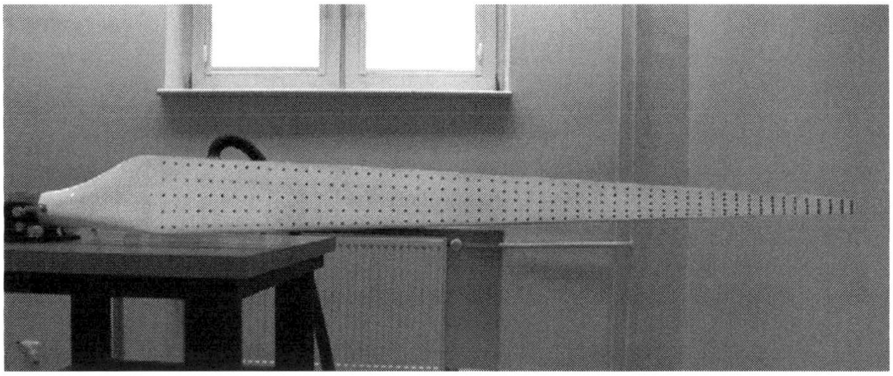

Figure 4. Research object and measurement grid.

5. Results

The first and seventh bending mode shape form experiments are presented in figure 5. As can be noticed, the signals for an undamaged and damaged blade differ from each other. Wavelet analysis was carried for one line localized at half of the blade width. Only bending of the form of mode shape was taken into account. Continuous wavelet transformation has been performed for each measured mode shape line. The signal from measurements has been denoised additionally. The difference between the undamaged state and the mode shape line for damaged blade has been calculated and transformed with the 'gaus4' mother wavelet. Figure 6.a.b.c presents three exemplary results obtained for different localization of damage (12% of total mass).

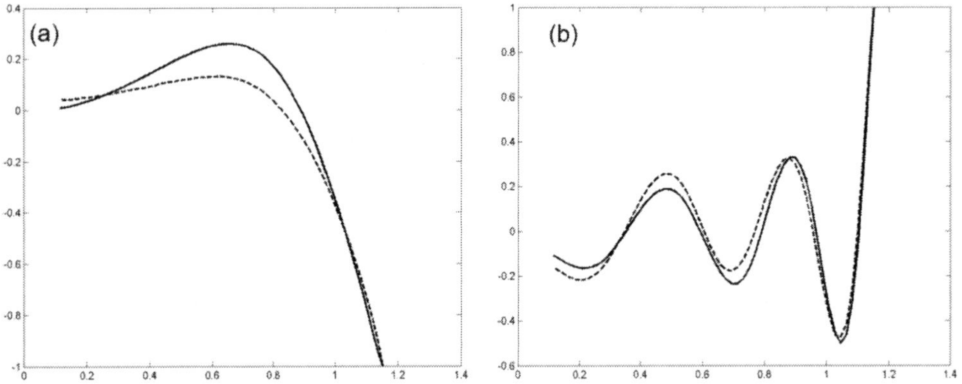

Figure. 5. Example of measurements results. Comparison of measurement lines for undamaged (solid line) and damaged (dashed line): (a) mode II, (b) mode VII.

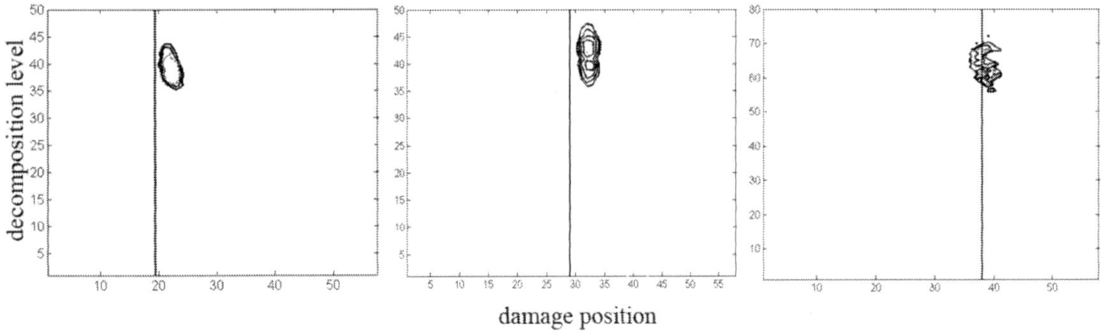

Figure 6. a), b), c) Results of CWT for three damage localizations;

The graphs show the maximal values of decomposition coefficients for growing decomposition level. The vertical line on the graph shows the real positions of damage. For all cases it can be noticed that the location of damage has been acceptably indicated.

Figure 7a.b shows the first and seventh bending mode shape from computer simulation. The CWT analysis was carried out for one line localized along the spar (136 points).

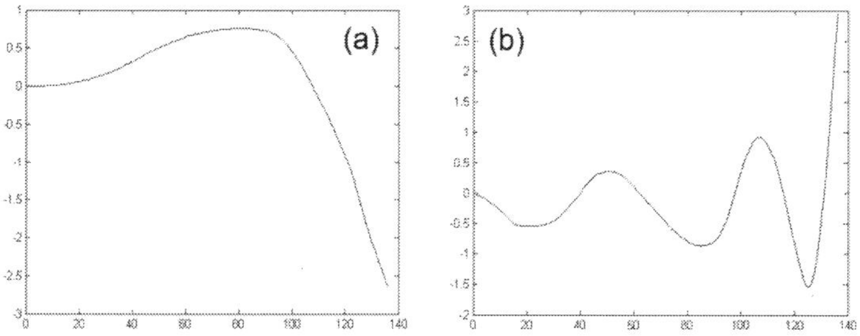

Figure. 7. Example of simulation results: (a) mode II, (b) mode VII.

Analysis of simulation results has shown that it is possible to define places of local rigidity changes. Unfortunately, in the case when those changes come/follow from the structure (e.g. change of element thickness – figure 8) it is impossible to detect and localize the damage without acquaintance of the initial/base state. Exemplary results of CWT analyses for the first mode shape and three cases of damage location are presented in figure 9.

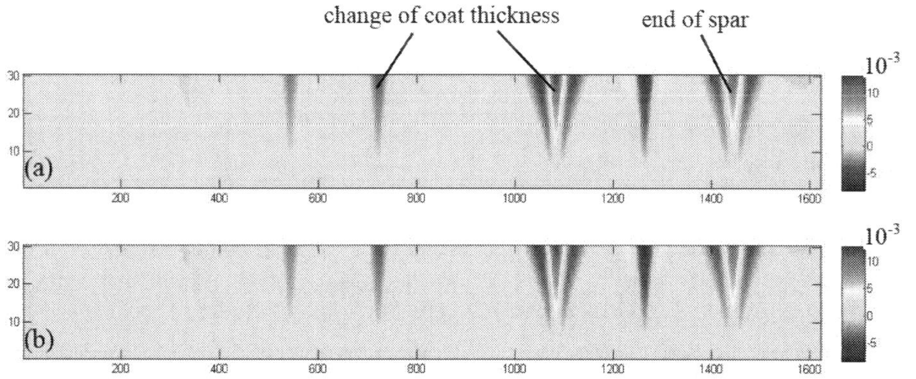

Figure. 8: Results of CWT for first mode shape: a) undamaged blade; b) damaged blade (30% width of blade).

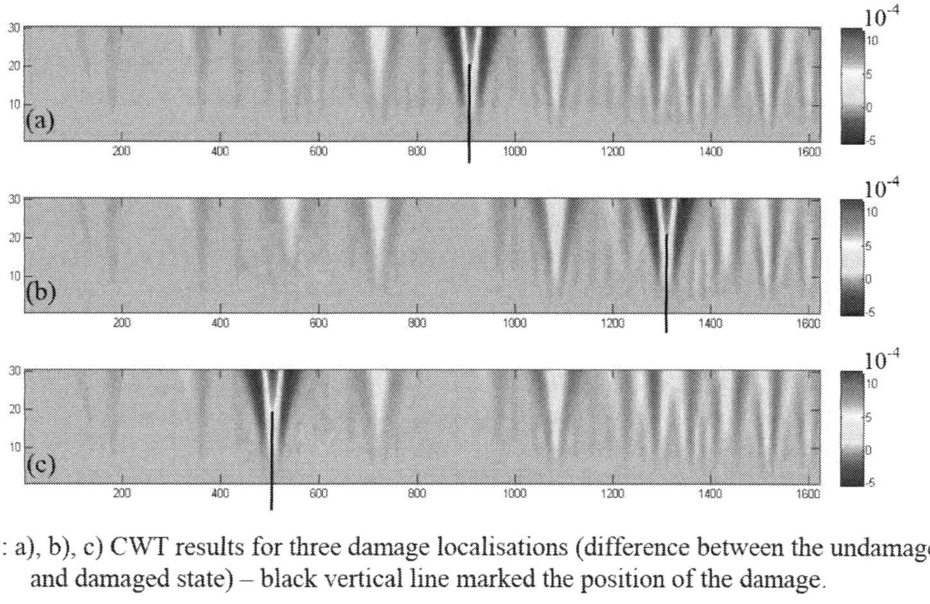

Figure. 9: a), b), c) CWT results for three damage localisations (difference between the undamaged and damaged state) – black vertical line marked the position of the damage.

Journal of Physics: Conference Series **181** (2009) 012086 doi:10.1088/1742-6596/181/1/012086

6. Conclusions

A method for damage localization in the composite blade of a wind turbine based on one-dimensional continuous wavelet analysis applied to mode shapes is presented. The mode shapes are determined by experiment and simulation. The experiments were conducted using a laser vibrometer. The Gaussian wavelet with different decomposition levels has been used as mother wavelet. The analysis of all damage cases has shown that localization of a relatively small damage is possible. The results obtained will allow to build a numerical model of a full-scale blade taking the advanced composite structure and service loads into consideration. A method of damage detection (cracks and delamination) will be created for this blade.

The trend of further research on this method will be concentrated on:

- improving the blade model for elimination of violent jump of coat thickness with simultaneous preservation of the character of decreasing thickness;
- increasing the sensitivity (analysis of higher frequency mode shapes and forced vibration);
- computer simulation of full-scale blade under service load;
- experimental measurements of blade with damage (crack);
- improving the measurement methodology (choice of measurement place and density of grid).

References

[1] Wei J., McCarty J. 1993. "Acoustic emission evaluation of composite wind turbine blades during fatigue testing", Wind Engineering Vol. 17, No. 6, pp. 266- 274

[2] Sutherland H., Beattie A., Hansche B., et al. 1993. "The Application of Non- Destructive Techniques to the Testing of a Wind Turbine Blade", SAND93-1380, Sandia National Laboratories.

[3] Blanch M. J., Dutton A. G. 2003. "Acoustic Emission Monitoring of Field Tests of an operating Wind Turbine", Key Engineering Materials Vols. 245-246 pp 475-482

[4] Joosse P. A., Blanch M. J., Dutton A. G., Kouroussis D. A., Philippidis T. P., Vionis P. S. 2002. "Acoustic emission monitoring of small wind turbine blades", Proc. 21st ASME Wind Energy Symp. in conjunction with 40th AIAA Aerospace Sciences Meeting pp 1–11 AIAA-2002-0063

[5] Dutton A. G. 2004. "Thermoelastic stress measurement and acoustic emission monitoring in wind turbine blade testing" European Wind Energy Conf. EWEC

[6] Kirikera G. R., Shinde V., Schulz M. J., Ghoshal A., Sundaresan M., Allemang R. 2007. "Damage localisation in composite and metallic structures using a structural neural system and simulated acoustic emissions", Mech. Syst. Signal Process. 21 280–97

[7] Gieske J. H., Rumsey M. A. 1997. "Non-destructive Evaluation (NDE) Of Composite/Metal Bond Interface Of A Wind Turbine Blade Using An Acousto-ultrasonic Technique", AIAA Aerospace Sciences Meeting, 249-254.

[8] Lading L., McGugan M., Sendrup P., Rheinlander J., Rusborg J. 2002. "Fundamentals for remote structural health monitoring of wind turbine blades" a preproject ANNEX B Risø-R-1341(EN) Report Risø National Laboratory, Denmark

[9] Beattie, A. G., and Rumsey, M. 1999. "Non-Destructive Evaluation of Wind Turbine Blades Using an Infrared Camera", ASME Wind Energy Symposium, 18th, Aerospace Sciences Meeting and Exhibit, 37th, Reno, NV, Jan. 11-14.

[10] Smith G. M., Clayton B. R., Dutton A. G., Irving A. D. 1993. "Infra-red thermography for condition monitoring of composite wind turbine blades feasibility studies using cyclic loading tests", 15th British Wind Energy Association Conference (BWEA15), York, UK, October

[11] Avdelidis N. P., Almond D. P., Ibarra-Castanedo C., Bendada A., Kenny S., Maldague X. 2006. "Structural integrity assessment of materials by thermography", Conf. Damage in Composite Materials CDCM, Stuttgart, Germany

[12] Ghoshal A., Sundaresan, M. J., Schulz, M. J., Pai P. F. 2000. "Structural Health Monitoring Techniques for Wind Turbine Blades", Journal of Wind Engineering and Industrial Aerodynamics, 85, 309-324.
[13] Krawczuk M., Ostachowicz W., Żak A. 1997. "Dynamics of cracked composite material structures", Computational Mechanics 20, Springer-Verlag, pp. 79-83
[14] Ge M., Lui E. M. 2005 "Structural damage identification using system dynamic properties" CS 83:2185 2196.
[15] Rolfes R., Gerasch W., Haake G., et al. 2006. "Early Damage Detection System for Towers and Rotor Blades of Offshore Wind Turbines", Third European Workshop on Structural Health Monitoring, Granada, Spain, pp. 455-462.
[16] Rucka M., Wilde K. 2006. "Application of continuous wavelet transform in vibration based damage detection method for beams and plates", 297(3-5):536-550.
[17] Mitsiti M. 1996. Wavelet Toolbox, For Use with MALAB. The Math Works, Inc., pp. 111-117.
[18] http://users.rowan.edu/~polikar/WAVELETS/WTtutorial.html; Polikar R.. The Wavelet Tutorial
[19] http://www.polytec.com/

Journal of Physics: Conference Series **181** (2009) 012087 doi:10.1088/1742-6596/181/1/012087

Investigation of the Nonlinear Dynamics of a Partially Cracked Plate

A Israr[1], L Atepor[2]

[1]School of Engineering and Physical Sciences, Heriot Watt University – Dubai Campus, Block 2, Dubai International Academic City, P O Box 294345, Dubai, UAE

E-mail: a.israr@hw.ac.uk

[2]Department of Mechanical Engineering, James Watt South Building, University of Glasgow, Glasgow, G12 8QQ

E-mail: katepor@yahoo.com

Abstract: In this paper the nonlinear vibration of an aircraft panel structure modelled as an isotropic cracked plate and subjected to transverse harmonic excitation is considered for studying the dynamic response, both analytically and experimentally. A crack is arbitrarily located at the centre of the plate, consisting of a continuous line. This mathematical model is in the form of Duffing equation with a cubic nonlinear term. The perturbation method of multiple scales is used to solve the algebraic equation, and then investigated with the results of the direct integration within *Mathematica*™ and finite element analysis in ABAQUS for the first mode only. In addition, experimental measurements are also carried out to verify the dependence of the cracked plate's fundamental mode shape and resonance frequency on the vibration displacement amplitude. An extermely close agreement between these results is observed.

1. Introduction

Metal plate structures are used in a variety of applications from aircrafts to ships and offshore platforms to bridges, power plants and cranes. Plate structures can be used as a basis for the analysis of vibration based damage detection in the field of dynamics. An important concern in the use of metal plate structures in the aircraft industry is their stability under high pressure loadings and stresses on aircraft wings and panels. Under these circumstances if there is even a small crack or minor irregularity on the surface of the plate it collapses the equilibrium completely and could cause failure. The dynamic responses of rectangular plates with cracks, or minor irregularities under different loading conditions, have been investigated in the past by many researchers for different boundaries conditions, and various methods have been proposed to deal with the problem [1-8]. The literature does not appear to contain any substantial references to analytical models for cracked plates undergoing forced vibration.

Israr *et al.* [9] developed an approximate analytical solution for damage detection in an aircraft panel structure modelled as a cracked isotropic plate without the application of a load, essentially for free vibration. Subsequently, Israr *et al.* [10] and [11] addressed a method for finding the vibration characteristics of plates that contain a crack that extends partially through the thickness by the application of periodic loading. Initially, certain simplifying assumptions were made in order to get an initial tractable solution to the vibration problem. Berger's formulation [12] was used to generate the nonlinear term within the model differential equation of motion. The authors presented results by the use of the method of multiple scales that showed the influence of crack length and boundary condition on the vibration characteristics of the plate.

In this study the results obtained by the perturbation method of multiple scales, direct integration with *Mathematica*™, finite element analysis in ABAQUS and experimental work for the model of the cracked plate are investigated in detail to see the behaviour of the cracked plate under critical circumstances.

2. Analytical Cracked Plate Model

Consider the rectangular plate of length l_1 in the x-direction and l_2 in the y-direction containing a crack which consists of a continuous line of length $2a$ arbitrarily located at the centre and parallel to the x-direction of the plate element for a given boundary conditions. A point load $\bar{P}_z$ based on the application of the appropriate delta function is introduced at the arbitrary location of (x_o, y_o) [10,11].

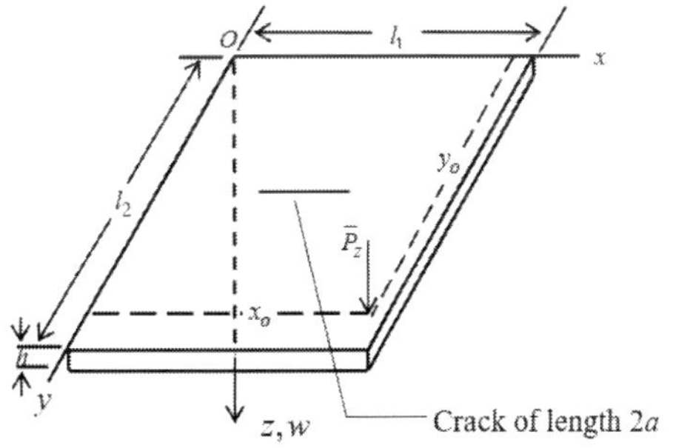

Figure 1. Isotropic plate loaded by arbitrary located concentrated force and small crack at the centre, and parallel to the x-axis

$$D\left(\frac{\partial^4 w}{\partial x^4}+2\frac{\partial^4 w}{\partial x^2 \partial y^2}+\frac{\partial^4 w}{\partial y^4}\right)=-\rho h\frac{\partial^2 w}{\partial t^2}+n_x\frac{\partial^2 w}{\partial x^2}+P_z+\frac{2aD\left(\frac{\partial^4 w}{\partial y^4}+\upsilon\frac{\partial^4 w}{\partial y^2 \partial x^2}\right)}{3\left(\frac{\alpha_{bt}^o}{6}+\alpha_{bb}^o\right)(3+\upsilon)(1-\upsilon)h+2a}-\frac{2a}{\left(6\alpha_{tb}^o+\alpha_{tt}^o\right)(1-\upsilon^2)h+2a}n_\infty\frac{\partial^2 w}{\partial y^2}. \quad (1)$$

where $D = Eh^3/12(1-\upsilon^2)$, P_z is the force acting at the surface of the plate, n_x, n_∞ are the membrane forces in the x & y-direction respectively. ρ, h and $2a$ are the density, thickness and crack length of the plate respectively and $\alpha_{bb}^o, \alpha_{tt}^o, \alpha_{tb}^o, \alpha_{bt}^o$ are the non-dimensional bending and stretching compliance coefficients at the crack centre respectively. Later, this P_z is replaced by a point load $\bar{P}_z$ based on the application of the appropriate delta function in equation (6) is introduced at the arbitrary location of (x_o, y_o) as shown in figure 1 to make it compatible with the experimental configuration.

Galerkin's Method is applied to reformulate the governing equation of the cracked plate into time dependent modal coordinates by the application of given boundary conditions, and Berger's formulation is used to express formally the in-plane forces, which can then be used to transform the governing equation of the cracked plate into a nonlinear system. The governing differential equation of the cracked plate subjected to transverse loading with cubic nonlinear term β_{mn}, and a weak linear viscous damping, μ in time domain form is [10,11]

$$\ddot{\psi}_{mn}(t)+2\mu\dot{\psi}_{mn}(t)+\omega_{mn}^2\psi_{mn}(t)+\beta_{mn}\psi_{mn}^3(t)=P_{mn}. \quad (2)$$

Journal of Physics: Conference Series **181** (2009) 012087 doi:10.1088/1742-6596/181/1/012087

where

$$\omega_{mn}^2 = K_{mn}/M_{mn}, \; \beta_{mn} = G_{mn}/M_{mn}, \; P_{mn} = \bar{P}_z D/M_{mn}, \tag{3),(4),(5}$$

$$\bar{P}_z = P_o(t)\,\delta(x - x_o)\,\delta(y - y_o), \; P_o(t) = p\cos\Omega_{mn} t, \tag{6),(7}$$

and

$$M_{mn} = \rho h/D \sum_{n=1}^{\infty}\sum_{m=1}^{\infty} A_{mn} \int_0^{l_1}\int_0^{l_2} X_m^2 Y_n^2 \, dxdy, \tag{8}$$

$$K_{mn} = \sum_{n=1}^{\infty}\sum_{m=1}^{\infty} A_{mn} \int_0^{l_1}\int_0^{l_2} \left(X_m^{iv} Y_n + 2X_m'' Y_n'' + Y_n^{iv} X_m - \frac{2a\left(\upsilon X_m'' Y_n'' + Y_n^{iv} X_m\right)}{3\left(\frac{\alpha_{bt}^o}{6} + \alpha_{bb}^o\right)(3+\upsilon)(1-\upsilon)h + 2a} \right) X_m Y_n \, dxdy, \tag{9}$$

$$G_{mn} = \sum_{n=1}^{\infty}\sum_{m=1}^{\infty} A_{mn}^3 \int_0^{l_1}\int_0^{l_2} \left(-F_{1mn} X_m X_m'' Y_n^2 + \frac{2aF_{2mn} X_m^2 Y_n Y_n''}{\left(6\alpha_{tb}^o + \alpha_{tt}^o\right)(1-\upsilon^2)h + 2a} \right) dxdy, \tag{10}$$

and

$$F_{1mn} = 6/h^2 l_1 l_2 \sum_{n=1}^{\infty}\sum_{m=1}^{\infty} \int_0^{l_1}\int_0^{l_2} \left[\left(\partial X_m/\partial x \right)^2 Y_n^2 + \upsilon \left(\partial Y_n/\partial y \right)^2 X_m^2 \right] dxdy, \tag{11}$$

$$F_{2mn} = 6/h^2 l_1 l_2 \sum_{n=1}^{\infty}\sum_{m=1}^{\infty} \int_0^{l_1}\int_0^{l_2} \left[\left(\partial Y_n/\partial y \right)^2 X_m^2 + \upsilon \left(\partial X_m/\partial x \right)^2 Y_n^2 \right] dxdy. \tag{12}$$

where *m, n* are plate mode designators, the quantity A_{mn} is a modal peak amplitude function, normalised in this case to unity, X_m and Y_m are the characteristic or modal functions of the cracked rectangular plate, and $\psi_{mn}(t)$ is the time dependent modal coordinate.

The solution for the governing differential equation of the cracked plate subjected to transverse loading is obtained by defining the characteristic functions depending upon the boundary conditions of the plate and then applied perturbation method of multiple scales will take the following form:

$$\sigma_{mn} = 3\beta_{mn} b^2/8\omega_{mn} \pm \sqrt{\lambda_{mn}^2 p^2/4\omega_{mn}^2 b^2 D^2 - \mu^2}. \tag{13}$$

This is a frequency response equation which gives the amplitude *b* as a function of the detuning parameter, σ_{mn}, this being a measure of deviation from the perfect forced resonance condition.

and

$$\lambda_{mn} = X_m(x_0)\,Y_n(y_0)/M_{mn}. \tag{14}$$

3. Numerical Analysis

Consider a test plate made of aluminium alloy 5083. It is used for various applications such as shipbuilding, aircrafts, rail cars, vehicle bodies, pressure vessels etc. It has low density and excellent thermal conductivity common to all aluminium alloys and has the material properties such as modulus of elasticity, E = 7.03 x 10^{10} N/m^2, density, ρ = 2660 kg/m^3 and Poisson's ratio, υ = 0.33. While the geometric properties of the plate are; length along *x*-direction, l_1=500 m, length along y-direction, l_2=1000 mm, half-crack length, a=10-25 mm and thickness of the plate, h = 10 mm. A load, p=10 N is acting upon the surface of the plate at some arbitrary specified point given here by x_o=375 mm and y_o=750 mm, and with a damping factor of μ=0.08.

The natural frequencies for cracked and un-cracked plate with three sets of boundary conditions i.e. Clamped-Clamped-Free-Free(CCFF), Clamped-Clamped-SimplySupported-SimplySupported (CCSS), and all sides simply supported (SSSS) and different aspect ratios have been calculated and tabulated in Table 1. It shows that the presence of the crack at the centre of the plate significantly influences the natural frequency of the first mode of the plate, in all three cases.

Table 1. Natural frequencies of the cracked plate model for diff. boundary conditions and aspect ratios

First mode natural frequency, ω_{mn} [rad/s] for a half-crack length, a = 10 and 25 [mm]										
Lengths of the sides of the plate		Two adjacent edges clamped, the other two free (CCFF)			Two adjacent edges clamped, the other two simply supported (CCSS)			All edges simply supported (SSSS)		
l_1 [mm]	l_2 [mm]	un-cracked	cracked 10 mm	cracked 25 mm	un-cracked	cracked 10 mm	cracked 25 mm	un-cracked	cracked 10 mm	cracked 25 mm
1000	1000	80.46	77.39	74.10	445.67	432.51	418.58	77.58	75.54	73.39
500	1000	231.06	229.95	228.80	1161.77	1154.27	1146.53	193.95	192.54	191.09
1000	500	231.06	213.85	194.61	1161.77	1089.98	1011.04	193.95	183.18	171.42
500	500	321.85	309.54	296.38	1782.66	1730.04	1674.31	310.32	302.17	293.57

3.1. Linear and Nonlinear Response Curves

Linear and nonlinear response curves are plotted for analysing the behaviour of a model of the cracked rectangular plate. To construct such a curve, one solves for, σ_{mn} in terms of b. Figure 2 shows the plots of amplitude, b as a function of σ_{mn} for given μ and p in the form of a frequency-response curve for three sets of boundary conditions CCFF, CCSS, and SSSS. Each point on this curve corresponds to a singular point.

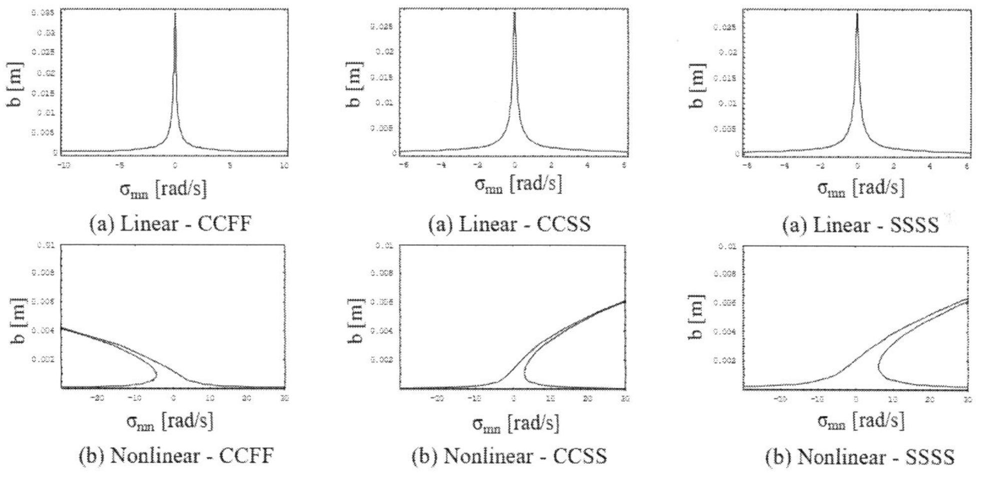

Figure 2. Response curves for three sets of boundary conditions and half-crack length of 10 mm

As shown, the nonlinearity bends the frequency-response curve away from the linear curve ($\beta_{mn} = 0$), to the right for hard springs i.e. $\beta_{mn} > 0$ for the cases of CCSS and SSSS, and bends to the left for soft springs i.e. $\beta_{mn} < 0$ for the case of CCFF. These are shown in figure 2(b), with some attendant change in the modal natural frequency, or in other words, as the amplitude of the excitation increases, the frequency response curves bend away from the $\sigma_{mn} = 0$ axis. In nonlinear problems the significant effect of including the cubic nonlinearity, β_{mn} is apparent from the numerical results depicted in figure 3 for the two sets of the plate aspect ratios 0.5/1 and 1/1. Figure 3 shows that the ratio of the linear 'b_L' and nonlinear 'b_{NL}' solution amplitude is very large for negative detuning. This exactly emulates the softening nonlinear characteristic for the case of CCFF boundary conditions as shown in figure 2(b). It can also be seen that this ratio reduces close to unity for zero and positive detuning, again fully in-line with the softening characteristic observable in figure 2(b) for the case of CCFF boundary condition. In addition, it can also be seen that there is a very small difference in amplitude ratio between the two aspect ratios of the plate in figure 3.

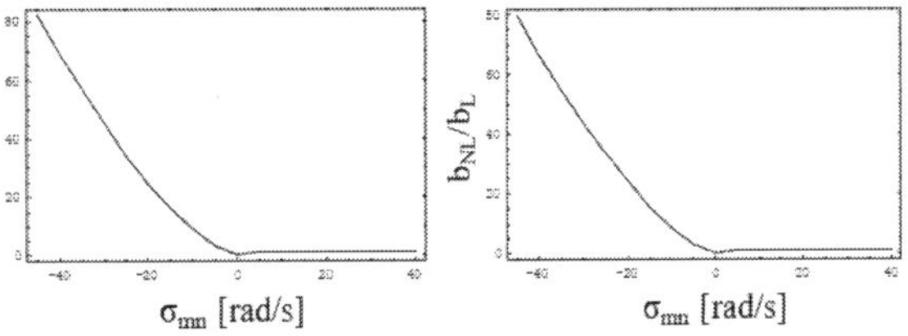

(a) Plate aspect ratio 0.5/1 (b) Plate aspect ratio 1/1

Figure 3. Comparison between linear and nonlinear response curves of the cracked rectangular plate and the boundary condition CCFF

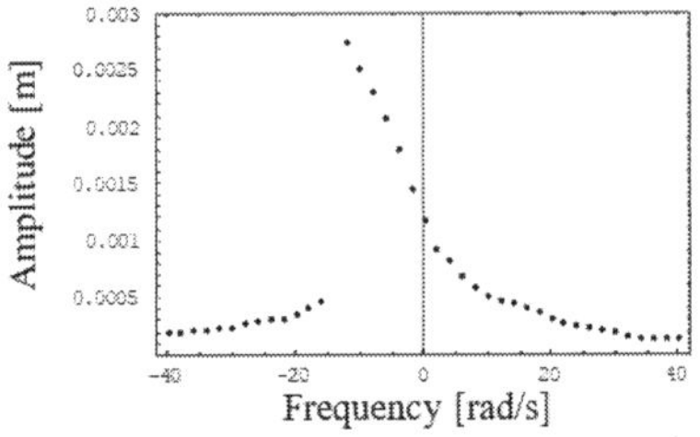

Figure 4. Nonlinear overhang in the form of the softening spring characteristic for an aspect ratio of 0.5/1 with zero initial conditions – *Mathematica*™ results

3.2. Direct Integration – NDSolve within *Mathematica*™

A program code is developed for integrating the differential equation (2) for given initial condition by the use of NDSolve integrator within *Mathematica*™ [13]. The results obtained are in the time domain and can be transformed into the frequency domain by running this program code several times for a range of frequency values from -40 rad/s to 40 rad/s to obtain a list of amplitude values. While going through all the data individually, one selects that portion of the amplitude values where the values are in steady-state condition, then one averages them for the plot of amplitude and frequency, and is shown in figure 4. These provide a basis for comparison with the approximate analytical solutions obtained from equation (13). Reasonably close agreement can be observed, however the numerical solution does not predict the same degree of nonlinear overhang in the form of the softening spring characteristic as the approximate analytical solution. The over-prediction of the softening overhang by the multiple scales solution is undoubtedly due to an over-correction to the solution from the first order perturbation contribution, and the assumptions that necessarily went with that in order to obtain a closed-form solution.

3.3. Finite Element (FE) Technique – ABAQUS/CAE 6.7-1

In this analytical framework, ABAQUS/CAE is used for evaluating the frequencies, displacements and mode shapes of the model of a cracked plate to investigate the theoretical predictions.

Figure 5. Plate without crack

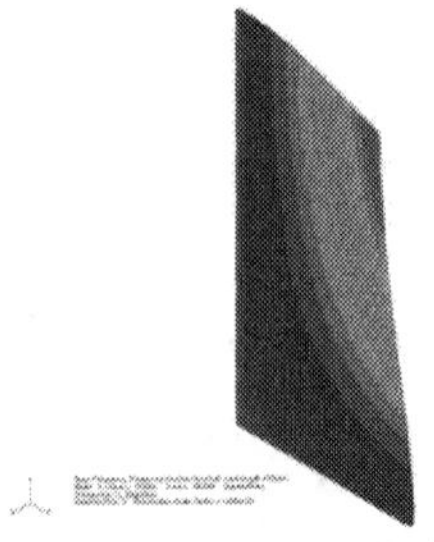

Figure 6. Plate with a half-crack length of 10 mm

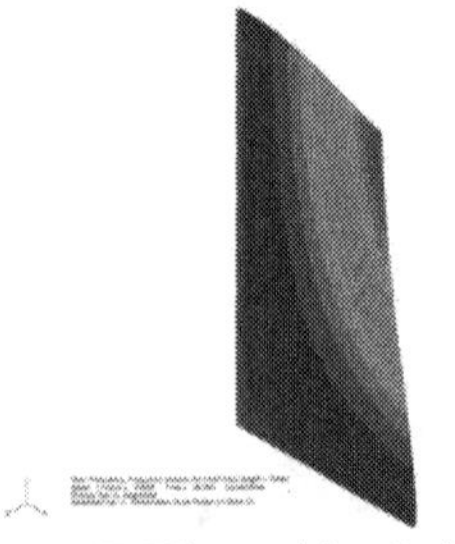

Figure 7. Plate with a half-crack length of 25 mm

The required inputs for these analyses consist of model geometry, material properties, loading, boundary conditions, and an initial crack configuration. The results obtained from ABAQUS/CAE are shown in figures 5, 6 and 7 for the nonlinear FE analysis of the plate with a part-through crack at the centre of the plate for a boundary condition of CCFF, and the aspect ratio of the plate is 0.5/1. Equally, results can readily be obtained by employing other boundary conditions, such as CCSS and SSSS. A decreasing trend in frequencies can be seen in figures 6 and 7, when damage is included in the form of crack at the centre of the plate. As expected, the amplitudes values also increase due to the insertion of a small crack in the plate. These frequency and amplitudes values are tabulated in Table 2, and the comparison is made between the theoretical model and the FE results.

Table 2. Finite element analysis results of a CCFF Plate for first mode of vibration

Dimension of the plate = 500x1000 mm		Theoretical Results		FE Analysis Results	
		Frequency [Hz]	Amplitude [mm]	Frequency [Hz]	Amplitude [mm]
Un-cracked Plate		36.77	1.1913	36.819	1.016
Cracked Plate	10 mm	36.60	1.1974	36.240	1.222
	25 mm	36.41	1.2045	36.232	1.386

4. Experimental Analysis

A heavy steel rig was specifically designed for the support of the plate's ends in such a way that it could be clamped in between the clamps on the two adjacent edges of the plate, and then these clamps are tightened by a series of screws. The remaining two edges of the plate are unconstrained. A rigid guided support made of a steel frame carrying the laser vibrometer is constructed, enabling displacement measurements to be made at any point on the surface of the plate. The plate is harmonically excited by an electro-dynamic exciter which imposes no added stiffness and no additional mass on the plate. The excitation of the exciter is monitored by means of a force transducer fitted into the exciter clamping fixture. This also provides a means for the control of the periodic load, as per the requirement. The excitation point is chosen to be at a distance of (37.5x75) mm from the fixed ends of the plate in the x and y directions respectively. The signal from the force transducer is monitored by an oscilloscope, and the amplitude signals in terms of voltages, are monitored by the spectrum analyzer. At different points of interest over the plate's surface, the signals are recorded for further analysis. These signals are ultimately converted into mechanical amplitudes for the generation of the first mode shape of the plate at the resonance frequency. Tests are carried out for an un-cracked and cracked plate of crack length 50 mm. This crack is at the centre of the plate and is parallel to the x-direction of the plate, so that comparisons can be made between the theoretical results and the experimental data. It is necessary initially to avoid having the excitation frequency too near to the unstable region where the jump effect could occur in the first mode. Curves show similar trends of decreasing resonance frequency with the increase of dynamic displacement amplitude for a given crack length.

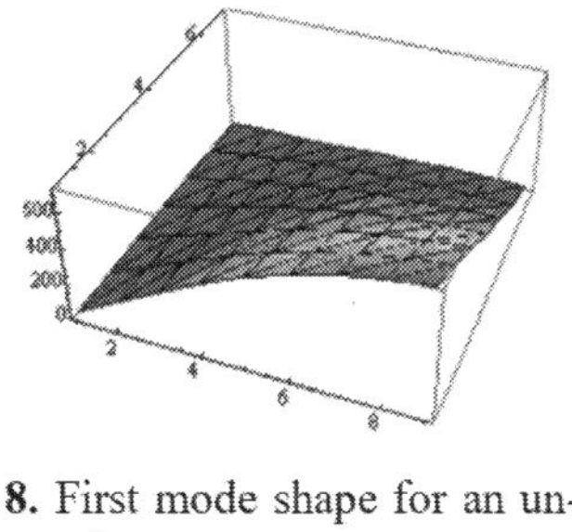

Figure 8. First mode shape for an un-cracked aluminium plate

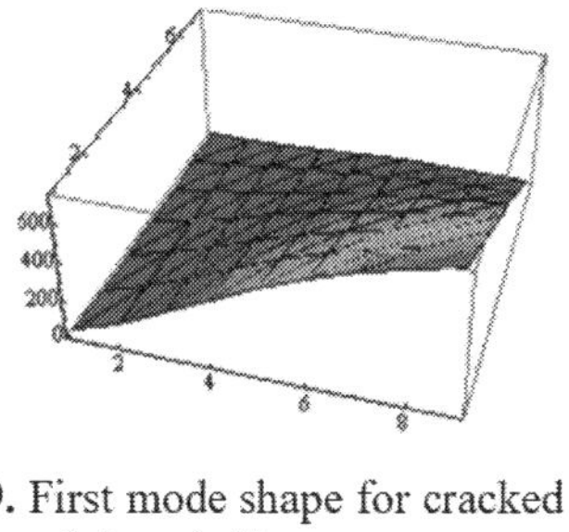

Figure 9. First mode shape for cracked aluminium plate of crack length 50 mm

Furthermore, while developing the analytical modal, shaker-structure interaction was ignored, as was possible nonlinear damping, and any initial geometrical distortion of the plate. It is possible that one or more of these factors could have played a significant role, and this could be further investigated. The measured responses of the plate (aspect ratio is 0.5/1) from repeated tests are shown in figures 8 and 9 for an un-cracked and cracked plate of crack length 50 mm respectively. Experimental investigation shows the same trend as explained earlier that the crack influences the excitation frequency and amplitude, this being summarised in Table 3.

Table 3. Experimental results of first mode of vibration for an un-cracked and cracked plates

Aluminium Plates	Frequency of excitation [Hz]	Amplitude at the tip of the free end of the aluminium plate [mm]
Un-cracked	24.0	0.90880
Cracked	23.6	0.93696

5. Results and Discussion

In this paper, the author studied, both experimentally and theoretically, the nonlinear behaviour, the frequency of excitation, amplitudes, and mode shapes. Three cases of boundary conditions - CCFF, CCSS and SSSS and two sets of crack length - were considered. CCFF showed negative cubic stiffness effect whereas CCSS and SSSS showed positive cubic characteristics.

Theoretically, the nonlinear vibration solutions of the cracked plate were investigated when it was subjected to transverse harmonic excitations, using the method of multiple scales, and the equation of motion was also numerically integrated by the use of NDsolve within *Mathematica*™. Interesting nonlinear behaviour was observed for the primary resonance condition. In this study it was shown conclusively, by using a first order multiple scales approximation, that the nonlinear characteristics of the steady-state responses are encoded within the non-autonomous modulation equations. A rectangular cross-section plate made of aluminium with a crack at the centre of the plate was considered for the numerical analysis. It was found that depending on increased crack length, the vibration frequency decreases and the amplitude increases. Furthermore, it was found that for a square CCFF plate there is an approximately 7.9% reduction in natural frequency in the presence of a centrally located crack of length 50 mm. However, the reduction in the value of natural frequency is lower for other plate aspect ratios (Table 1). There are limited references to such systems in the literature as there have not been many reported phenomena relating to nonlinear forced vibration of cracked plates to date. Finite element analyses were also performed using ABAQUS/CAE for a nonlinear vibrating cracked CCFF plate. Three dimensional profiles for various modes, deflection, and frequency of vibration were obtained (figures 5, 6, and 7). It was then illustrated that these FE analyses were fully in-line with the nonlinear dynamic regimes by means of comparison with the approximate analytical solution via the method of multiple scales and numerical integration.

Experimentally, the response of a rectangular aluminium cracked plate - CCFF to transverse harmonic excitations was observed. It was found that the amplitude of vibration is a function of the crack length of the plate. The modulation frequency, which depends on various parameters such as the amplitude and frequency of excitation, damping factors, etc., has to be near the natural frequency of the first mode. In such a case, we see a marginal increase in amplitude of the plate due to presence of crack of the first-mode component in the plate response. The analytical model was able to predict results qualitatively similar to the experimentally observed motions, however results differed quantitatively. Issues, such as the nonlinear damping, internal discontinuities within the plate, shaker-structure interaction, and initial curvature of the plate were ignored, when developing the analytical model. It is, therefore, possible that one or more of these factors might be playing a bigger role than expected, and this could be further investigated.

6. Conclusions

Based on the present observations, it can be concluded that conventional methods used for the reduction in frequency response in the cracked plate, might in fact lead to an increase in the amplitude of excitation for the first mode. A loss of local stability of plates with a small crack is possible under periodic loading. As shown, a simple cracked plate under harmonic excitation could display many different nonlinear dynamic phenomena. These types of nonlinear resonances have very high excitation level to be observed. Such unusual phenomena have been observed in real engineering structures and mechanical systems, therefore there is a potential need to study such phenomena in more detail in other cracked structures.

Finally, this research provides some basic theory and understanding of how nonlinear plate systems can be made to be more efficient. It has also initiated an identification of the nonlinear characteristics of cracked plate structures. This new approach can be used to configure an ideal and robust system, and hence more reliable and efficient industrial systems can be constructed for a vibration based analysis methodologies, and for further development of vibration based health monitoring in cracked plate structures.

7. Acknowledgement

We gratefully acknowledge the support and guidance provided by Matthew P Cartmell, James Watt Professor of Mechanical Engineering, University of Glasgow, Glasgow, UK in this research.

References

[1] Lynn E E, and Kumbasar N 1967 *In:10th Midwestern Mech. Conf., Fort Colins*, Free Vibration of Thin Rectangular Plates having Narrow Cracks with Simply Supported Edges 911-928.

[2] Aggarwala B D, and Ariel P D 1981 *Engineering Transactions*, Vibration and Bending of a Cracked Plate **29** 295-310.

[3] Hirano Y, Okazaki K 1980 *Bulletin JSME* Vibration of Cracked Rectangular Plates **23** 732-740.

[4] Nezu K 1982 *Bulletin JSME*, Free Vibrations of a Simple Supported Rectangular Plate with a Straight Through-notch **25** 16-23.

[5] Liew K M, Hung K C, and Lim M K 1994 *Engineering Fracture Mechanics*, A Solution Method for Analysis of Cracked Plates under Vibration **48**(3) 393-404.

[6] Ramamurti V, and Neogy S 1998 *Mechanics Based Design of Structures and Machines: an International Journal*, Effect of Crack on the Natural Frequency of Cantilever Plates - A Rayleigh–Ritz Solution **26**(2) 131-143.

[7] Khadem S E, and Rezaee M 2000(a), *Journal of Sound and Vibration*, Introduction of Modified Comparison Functions for Vibration Analysis of a Rectangular Cracked Plate **236**(2) 245-258.

[8] Khadem S E, and Rezaee M 2000(b) *Journal of Sound and Vibration*, An Analytical Approach for Obtaining the Location and Depth of an all-over Part-through Crack on Externally In-plane loaded Rectangular Plate using Vibration Analysis **230**(2) 291-308.

[9] Israr A, Cartmell M P, Krawczuk M, Ostachowicz W M, Manoach E, Tremdafolova I, Shishkina E V, and Palacz M 2006 *Modern Practice in Stress & Vibration Analysis VI*, P.Keogh, eds., Trans Tech Publications Ltd, Switzerland, On Approximate Analytical Solutions for Vibrations in Cracked Plates **5-6** 315-322.

[10] Israr A, Cartmell M P, Manoach E, Tremdafolova I, Ostachowicz W M, Krawczuk M, and Żak A 2009, *ASME Journal of Applied Mechanics,* Analytical Modelling and Vibration Analysis of Partially Cracked Rectangular Plate with Boundary Conditions and Loading **76**(1).

[11] Israr A 2008 *PhD Thesis*, University of Glasgow, Glasgow, UK, Vibration Analysis of Cracked Aluminium Plates.

[12] Berger H M 1955 *Journal of Applied Mechanics*, A New Approach to the Analysis of Large Deflections of Plates **22** 465-472.

[13] Wolfram S 2005 *Cambridge University Press* UK, The *Mathematica*™ Book.

Journal of Physics: Conference Series **181** (2009) 012088 doi:10.1088/1742-6596/181/1/012088

Damage localization from vibration data using hierarchical a priori assumptions

Antti Huhtala[1] **and Sven Bossuyt**[1,2]

[1] Department of Mathematics and Systems Analysis, Helsinki University of Technology, PO BOX 1100, FIN-02015 TKK, Finland

[2] Department of Mechanics of Materials and Constructions, Vrije Universiteit Brussel, Pleinlaan 2, B-1050 Brussel, Belgium

E-mail: ahhuhtal@math.tkk.fi, sven.bossuyt@vub.ac.be

Abstract. An inverse method for damage identification is described. The method assumes that a forward model is available to predict hypothetical vibration data, for an arbitrarily varying profile of the local stiffness of the structure. In general, the inverse problem of identifying this stiffness profile from the observed vibration data is an ill-posed problem, therefore some form of regularization is required. In the case of this method, the regularization results from a Bayesian formulation of the a priori assumption that the stiffness change is smooth almost everywhere, but could have an unknown number of discontinuities at unknown locations.

1. Introduction

Damage is assumed to be associated with localized loss of stiffness in a structure. This loss of stiffness causes changes in modal parameters in the structure, namely the vibration frequencies and mode shapes. Reconstructing the changes in stiffness that resulted in the measured vibration characteristics is generally an ill-posed problem and cannot be solved without some sort of regularization.

In this paper we apply Bayesian inference to the problem and use a hierarchical model to express our prior knowledge about the damage. Using statistical methods allows us to model the measurement errors and thus to better retrieve meaningful information from the data. This framework is applied to a cantilever beam, with the prior assumption that the change in stiffness is piecewise smooth.

Sensitivity based methods that use statistical measure variables, which our method is an example of, have also been proposed in the past, for example in [1]. These methods use a parametrized model of the structure. The model parameters are changed so that a chosen statistical variable, that measures how well the model explains the measurements, is optimized.

A completely different way to look at the problem is proposed in [2, 3]. Flexibility-based methods are based on the fact, that the flexibility matrix can be quite accurately constructed by just using a few of the lowest modes. The flexibility matrix is calculated for both the undamaged and the damaged structure. Damage localization is then based on the principle that vectors which are in the null space of the change in flexibility, correspond to zero stress in the damaged areas.

A comprehensive study of various inverse methods used in damage identification is presented in [4]. An overview is given in [5].

2. Forward model

We assume that the undamaged structure under free vibration can be modelled as

$$M\ddot{x} + Kx = 0, \tag{1}$$

where $M \in \mathbb{R}^{n \times n}$ is the mass matrix and $K \in \mathbb{R}^{n \times n}$ the stiffness matrix. The matrices M and K are assumed to be known for the structure. The vibration modes for such a structure are the solutions to the generalized eigenvalue problem

$$Kx = \omega^2 Mx, \tag{2}$$

where ω is the angular frequency of the vibration mode and x is the mode shape. The measurement of modal parameters from actual structures is less trivial and is discussed in [6].

We assume that damage to the structure affects only its stiffness and not its mass. Hence, the damaged structure under free vibration can be modelled as

$$M\ddot{x} + (K + \Delta K)\, x = 0. \tag{3}$$

As the matrices M and K are commonly built using the finite element method, we are motivated to write the matrix K as

$$K = \sum_{i=1}^{N} K_i^e, \tag{4}$$

where K_i^e are the element stiffness matrices. The change in stiffness caused by the damage is then written as

$$\Delta K = -\sum_{i=1}^{N} d_i K_i^e, \tag{5}$$

where d_i is a "damage index" corresponding to the loss of stiffness in element i. The d_i get values in the interval $[0, 1)$, where 0 means no damage and 1 would be complete loss of stiffness.

It is assumed that we can measure the frequencies for a few of the lowest modes, as well as some chosen components of the mode shapes for the same modes. The forward model is then a function f, such that

$$y = f(d), \tag{6}$$

where d is a vector of the damage indicies and y is a vector of the individual measurements in some order.

3. Hyperprior inverse method

In the Bayesian framework all variables are assumed to be random variables, which we only get realizations of. In this paper we refer to the random variables themselves with capital letters and to the realizations with lower case letters. Statistical inverse problems are discussed with more detail in [7].

It is assumed that measured data is produced by the forward model, but corrupted by random noise. Therefore we have the relation

$$Y = f(D) + \epsilon, \tag{7}$$

where ϵ is the noise, which we assume to be zero mean Gaussian with covariance $S^T S$. That is, ϵ has the probability density function

$$p_{noise}(\epsilon) = C \exp\left(-\frac{1}{2}\left\|S^{-1}\epsilon\right\|^2\right), \tag{8}$$

where C is a normalization constant.

We can then write

$$\epsilon = Y - f(D). \tag{9}$$

For a given D, the conditional probability density function for Y is

$$p_{Y|D}(y|d) = p_{noise}(y - f(d)) = C \exp\left(-\frac{1}{2}\left\|S^{-1}(y - f(d))\right\|^2\right). \tag{10}$$

By using Bayes' formula, we get the conditional probability for D

$$p_{D|Y}(d|y) = \frac{p_{Y|D}(y|d)p_D(d)}{p_Y(y)} \tag{11}$$

and assuming that we have some prior knowledge of the damage, we have $p_D(d) = p_{prior}(d)$. From these we get the posterior probability density for a given measurement of y,

$$p_{posterior}(d|y) = \frac{p_{Y|D}(y|d)p_{prior}(d)}{p_Y(y)}, \quad y = y_{measured}. \tag{12}$$

In our case, the prior knowledge itself depends on unknown variables, called hyperparameters, which are also estimated from the measured data. We assume that the hyperparameters are distributed according to a hyperprior distribution $p_{hyper}(\lambda)$. Thus we have the joint prior distribution

$$p_{prior}(d, \lambda) = p_{prior}(d|\lambda)p_{hyper}(\lambda). \tag{13}$$

Using this prior in Bayes' formula we get the joint posterior density

$$p_{posterior}(d, \lambda|y) = \frac{p_{Y|D}(y|d)p_{prior}(d|\lambda)p_{hyper}(\lambda)}{p_Y(y)}, \quad y = y_{measured}. \tag{14}$$

The posterior density is often considered the solution to an inverse problem. For single point estimates it is common to search for the mode of the posterior density, that is

$$(d, \lambda)_{MAP} = \arg\max p_{posterior}(d, \lambda|y), \quad y = y_{measured}, \tag{15}$$

which is referred to as the maximum a posteriori estimate or MAP estimate.

4. Application to a one-dimensional beam model

An Euler-Bernoulli beam of length L is discretized into N elements, each of length $h = \frac{L}{N}$. In each element we assume that the stiffness is constant. This produces a vibration model in the form of (1).

Let d be a vector of the damages. We know a priori that the stiffness along a cracked beam is piecewise smooth, by which we mean that the stiffness is smooth apart from an unknown number of jumps. The number and location of the jumps are to be estimated from the data.

We take

$$p_{prior}(d|\lambda) = C(\lambda) \exp\left(-\frac{1}{2}\sum_{j=1}^{N} \lambda_j \frac{1}{h^2}(d_j - d_{j-1})^2\right) = C(\lambda) \exp\left(-\frac{1}{2}\|\Lambda^{1/2} D d\|^2\right) \tag{16}$$

as the prior distribution, where

$$D = \frac{1}{h}\begin{pmatrix} 1 & & & \\ -1 & 1 & & \\ & \ddots & \ddots & \\ & & -1 & 1 \end{pmatrix} \in \mathbb{R}^{N\times N}, \tag{17}$$

and

$$\Lambda = diag(\lambda_1, \ldots, \lambda_N). \tag{18}$$

The factor $C(\lambda)$ is for normalization and it is known that

$$C(\lambda) = \tilde{C}\det(D^T\Lambda D)^{1/2} = \tilde{C}\det(D)\det(\Lambda)^{1/2}. \tag{19}$$

If the hyperparameters λ_i are all equal, this prior introduces a penalty for value changes in d, i.e. it favors smoothness of d. If, however, one λ_i is smaller than the rest, d can jump at that location. That is why we choose a hyperprior distribution that can allow for some extreme outliers. In this case, we have chosen

$$p_{hyper}(\lambda) = Cp_+(\lambda)\exp(-\frac{\gamma}{2}\sum_{i=1}^{N}\lambda_i), \tag{20}$$

where $p_+(\lambda)$ is 1 when all of the λ_i are non-negative and zero otherwise. The parameter γ specifies the width of the distribution and is a tuning parameter.

Now we have

$$\begin{aligned} &p_{posterior}(d,\lambda|y) \\ &= C\exp\left(-\frac{1}{2}\|S^{-1}(y-f(d))\|^2\right)\det(\Lambda)^{1/2}\exp\left(-\frac{1}{2}\|\Lambda^{1/2}Dd\|^2\right)\exp\left(-\frac{\gamma}{2}\sum_{i=1}^{N}\lambda_i\right) \\ &= \prod_{i=1}^{N}\lambda_i^{1/2}C\exp\left(-\frac{1}{2}\|S^{-1}(y-f(d))\|^2 - \frac{1}{2}\|\Lambda^{1/2}Dd\|^2 - \frac{\gamma}{2}\sum_{i=1}^{N}\lambda_i\right) \\ &= \exp\left(-\frac{1}{2}\|S^{-1}(y-f(d))\|^2 - \frac{1}{2}\|\Lambda^{1/2}Dd\|^2 - \frac{\gamma}{2}\sum_{i=1}^{N}\lambda_i + \frac{1}{2}\sum_{i=1}^{N}\log\lambda_i + \log C\right). \end{aligned} \tag{21}$$

We are interested in the MAP estimate, which we get by maximizing the posterior density.

$$\begin{aligned} (d,\lambda)_{MAP} &= \arg\max p_{posterior}(d,\lambda|y) = \arg\min -\log p_{posterior}(d,\lambda|y) \\ &= \arg\min\left(\frac{1}{2}\|S^{-1}(y-f(d))\|^2 + \frac{1}{2}\|\Lambda^{1/2}Dd\|^2 + \frac{\gamma}{2}\sum_{i=1}^{N}\lambda_i - \frac{1}{2}\sum_{i=1}^{N}\log\lambda_i - \log C\right) \\ &= \arg\min\left(\frac{1}{2}\|S^{-1}(y-f(d))\|^2 + \frac{1}{2}\|\Lambda^{1/2}Dd\|^2 + \frac{\gamma}{2}\sum_{i=1}^{N}\lambda_i - \frac{1}{2}\sum_{i=1}^{N}\log\lambda_i\right). \end{aligned} \tag{22}$$

We split the optimization problem into two smaller problems that we iteratively solve. This method, proposed in [8], simplifies the optimization by avoiding the need to calculate Hessians.

1. Set $d^k = 0$, where $k = 1$.
2. Set λ^k as the maximizer of $p(\lambda|d^k, y)$, which can be obtained analytically as

$$\lambda_i^k = \frac{1}{(Dd^k)_i^2 + \gamma}. \tag{23}$$

Journal of Physics: Conference Series 181 (2009) 012088 doi:10.1088/1742-6596/181/1/012088

3. Set d^{k+1} as the maximizer of $p(d|\lambda^k, y)$, which is obtained from the non-linear least squares problem

$$d^{k+1} = \arg\min \left(\frac{1}{2} \|S^{-1}(y - f(d))\|^2 + \frac{1}{2} \|\Lambda^{1/2} D d\|^2 \right). \tag{24}$$

4. Set $k = k + 1$ and repeat from 2. if not converged.

5. Results using simulated data

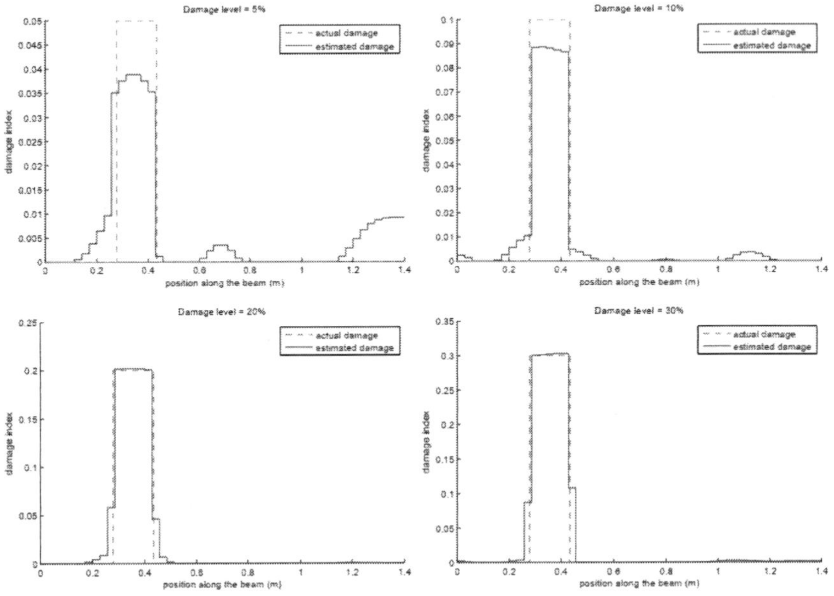

Figure 1. Reconstructions of simulated damage in a cantilever beam with a varying level of damage.

To avoid committing the inverse crime (which means using exactly the same model in simulating the data and as the forward model), the simulated measurements were taken from a finer finite element model than the one used in the forward model of the inverse method.

The forward model of the inverse method and the model used to simulate measurements both return frequencies and coarse shapes for 6 of the lowest modes. The coarse shapes are the mode shapes evaluated at 7 locations along the beam. This mimics measurement of the mode shapes by accelerometers or strain gauges. Gaussian noise is also added to the values when simulating measurements. The standard deviation of the noise is referred as the level of noise.

The simulated beam is 1.4 meters long, 60 millimeters wide and 5 millimeters thick (this is the direction of vibration) and is composed of steel. It is rigidly clamped on the left and is free on the right. The simulated accelerometers are located at 0.2, 0.4, 0.6, 0.8, 1.0, 1.2 and 1.4 meters from the left.

Figure 1 shows estimations for four cases. A different damage is induced to the same portion of the beam in each case. In each of the cases a constant damage index is set to the portion. The reconstructions have been calculated with $\gamma = 0.1$ and $N = 50$. Level of noise added to the frequencies is 0.01 Hz plus 0.1% of frequency magnitude and the level of noise added to each mode shape component is 1% of component magnitude.

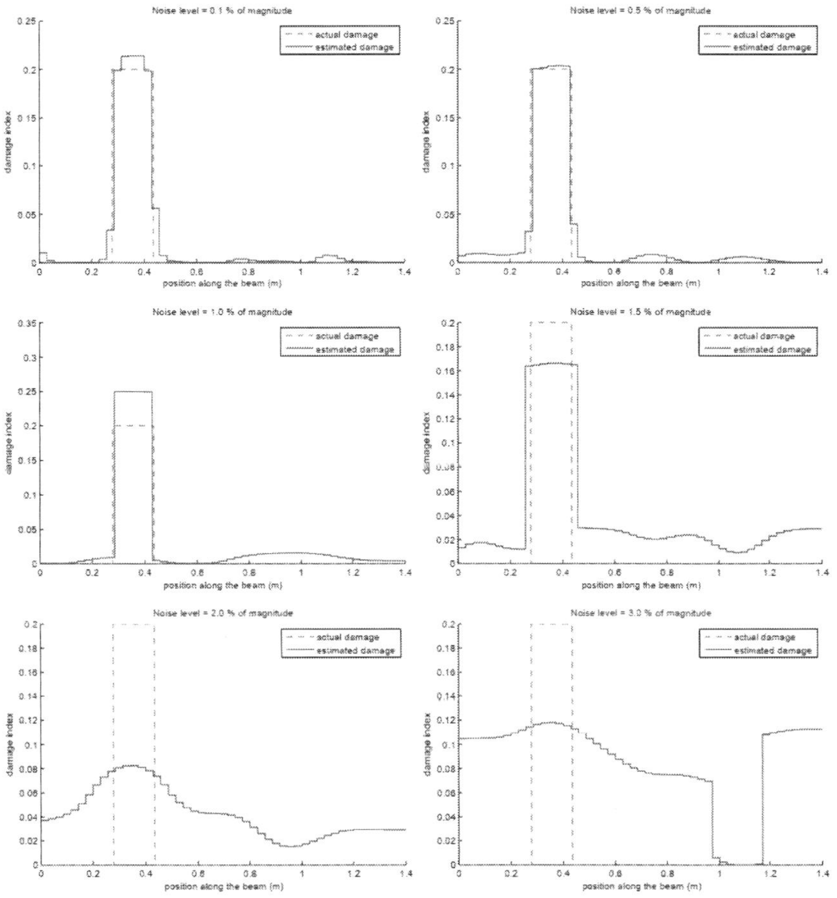

Figure 2. Reconstructions of simulated damage in a cantilever beam with a varying level of noise.

The reconstruction responds to the damage in a highly nonlinear way. The low level of damage (damage indices at 0.05) is hardly reconstructed, while for the higher levels of damage the reconstruction get progressively better.

Figure 2 shows the effect of noise on the reconstruction. In each case a constant damage index of 0.2 is induced to the same portion of the beam. Only the level of simulated measurement noise is changed between the reconstructions. The reconstructions have been calculated with $\gamma = 0.1$ and $N = 50$. In the cases, the level of noise added to the frequencies is proportional to the magnitude of the frequencies and the level of noise added to the mode shape components is proportional to the component magnitudes. The proportionality constants are 0.1%, 0.5%, 1.0%, 1.5%, 2.0% and 3.0% for each case respectively.

As with all inverse problems, the reconstruction is found to be highly sensitive to the amount of noise. Higher levels of noise make the reconstruction progressively more difficult and ultimately impossible.

Figure 3 shows the effect of a small undamaged area between two damaged areas. Both of the damaged areas are 10% of the total length of the beam. The damage levels in the areas are kept constant between each case. Only the length of the undamaged gap is changed. Level of noise added to the frequencies was 0.01 Hz plus 0.1% of frequency magnitude. Level of noise

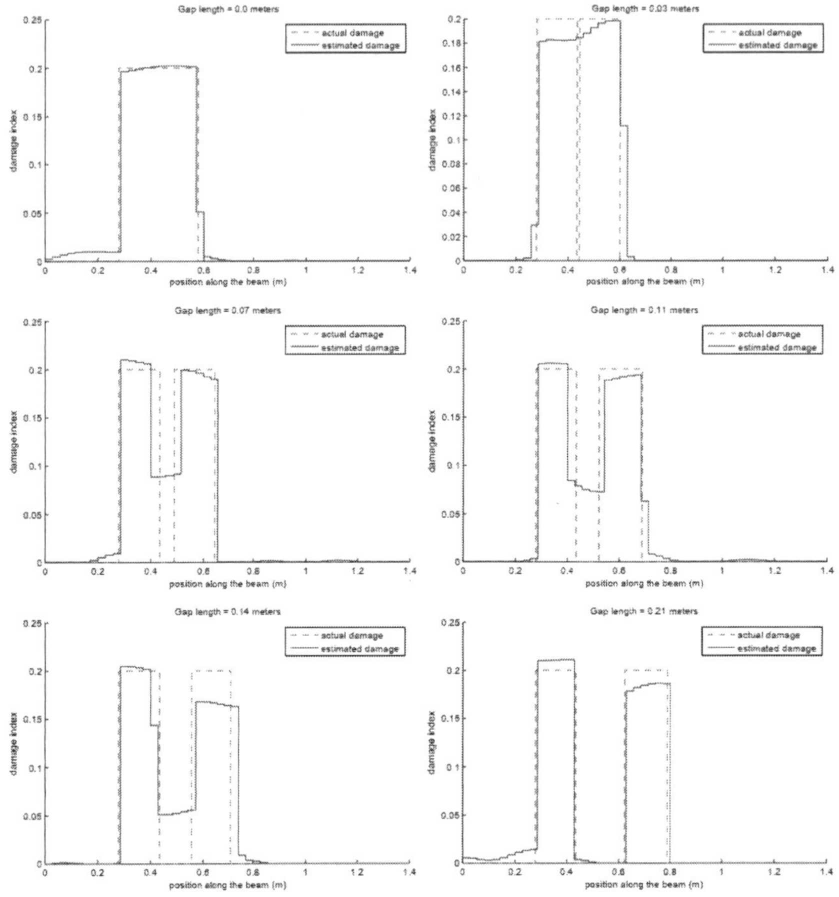

Figure 3. Reconstructions of simulated damage with a small-scale feature of varying size.

added to each mode shape component was 1% of component magnitude. The reconstructions have been calculated with $\gamma = 0.1$ and $N = 50$. The method starts to notice the gap even when it is fairly short. However, the gap length has to be quite long for the method to detect the damaged areas as completely separate. Nevertheless, the sensitivity is suprisingly high.

6. Conclusions and future work

We conclude that it is possible to solve the ill-posed problem of damage identification using the regularization method described, given enough meaningful data and that the damage is consistent with the prior information.

The method is highly sensitive to the amount of noise in the measurements, as are all inverse problems. This property can be explained by looking at the method as maximizing the likelihood that the difference in measurement and the estimate was produced by noise alone, given some prior information of the result. If the measurement noise is too high, the data might not be statistically meaningful enough, thus resulting in bad estimates.

Future research on the topic includes the measurement of damage in an actual beam. First an initial model is to be created for an undamaged beam. That model is then updated with the modal information obtained from measurements of the undamaged actual beam. Damage is then to be progressively inflicted on the beam and measurements are taken from several levels of damage. The described method is then to be used on damage identification for each of the damage cases and the results compared to those of other methods.

Also, a method for selecting values for the tuning parameter γ is required and is a topic for further research.

7. Acknowledgements

This research is supported by "ISMO", the Intelligent Structural Health Monitoring project of the Multidisciplinary Institute for Digitalisation and Energy (MIDE) of the Helsinki University of Technology. Additionally, S. Bossuyt is a researcher of the Research Foundation Flanders (FWO - Vlaanderen).

References

[1] Messina A, J W E and T C 1998 *Journal of Sound and Vibration* **216** 791–808
[2] Bernal D 2006 *Journal of Engineering Mechanics* **132** 651–658
[3] Pandey A K and Biswas M 1994 *Journal of Sound and Vibration* **169** 3–17
[4] Teughels A 2003 *Inverse Modelling of Civil Engineering Structures Based On Operational Modal Data* Ph.D. thesis Katholieke Universiteit Leuven URL `http://www.kuleuven.be/bwm/papers/teug03a.pdf`
[5] Friswell M I 2007 *Philosophical Transactions of The Royal Society A* **365** 393–410
[6] Mevel L, Beneviste A, Basseville M and Goursat M 2002 *IEEE Transactions on Signal Processing* **50** 41–48
[7] Kaipio J and Somersalo E 2004 *Statistical and Computational Inverse Problems* (Springer)
[8] Calvetti D and Somersalo E 2007 *Inverse problems* **23** 733–754

Gearbox Diagnostics Using Wavelet-Based Windowing Technique

Farag K. Omar[1] and A.M. Gaouda[2]

[1]Mechanical Engineering Department, UAE University, Al Ain, UAE
[2]Electrical Engineering Department, UAE University, Al Ain, UAE

E-mail: fomar@uaeu.ac.ae

Abstract. In extracting gear box acoustic signals embedded in excessive noise, the need for an online and automated tool becomes a crucial necessity. One of the recent approaches that have gained some acceptance within the research arena is the Wavelet multi-resolution analysis (WMRA). However selecting an accurate mother wavelet, defining dynamic threshold values and identifying the resolution levels to be considered in gearboxes fault detection and diagnosis are still challenging tasks. This paper proposes a novel wavelet-based technique for detecting, locating and estimating the severity of defects in gear tooth fracture. The proposed technique enhances the WMRA by decomposing the noisy data into different resolution levels while data sliding it into Kaiser's window. Only the maximum expansion coefficients at each resolution level are used in de-noising, detecting and measuring the severity of the defects. A small set of coefficients is used in the monitoring process without assigning threshold values or performing signal reconstruction. The proposed monitoring technique has been applied to a laboratory data corrupted with high noise level.

1. Introduction

Early detection of a machine element failure is of interest and importance to industry. Gearboxes are considered the most widely used power transfer machine element. They are almost present in any machine. Depending on the place and the importance of the machine, an early detection of the onset of damage in a gear set and the evaluation of the remaining life could have crucial impact on productivity and safety. Vibration analysis is a broadly used condition monitoring technique for gears as well as for other mechanical system components such as bearings. A vibration transducer, usually an accelerometer, is fastened to the gearbox casing. The vibration signals obtained from the transducer can be integrated to produce velocity or even displacement if needed. Vibration signals are then sampled and processed in various ways to highlight specific aspects of the signal which can be used in the detection and diagnostic of the gearbox condition.

Many approaches and techniques were developed in the past for the gear diagnostics. Those techniques are either based on time domain analysis [1-5], frequency domain analysis [6-8], or time-frequency domain analysis that highlights the time and frequency components strength instantaneously. This advantage of the time-frequency technique has attracted many researchers working in gear diagnostics. Gear vibration varies throughout one gear cycle due to variations in gear meshing stiffness, transmission error, and gear damage. Wang and Mcfadden used the spectrogram for early failure detection in a gearbox [9]. The spectrogram represents the signal energy distribution over the frequency domain and at every instant of time. For a typical spectrum analysis the abnormal disturbance in the signal energy at a single moment, e.g., damage at a single tooth, would be distributed over the time period of the gear rotation. However, the spectrogram would provide a higher sensitivity to changes of the short duration in the signal. Wang and Mcfadden found that the sensitivity of the spectrogram would be enhanced by removing the harmonics of the tooth meshing frequency before calculating the spectrogram. Later researchers directed their attention to the wavelet transform as it is powerful of detecting transients in non-stationary signals. The first attempt of using wavelet transform is reported by Wang *et al.* in 1995. They concluded that the wavelets are more suitable for gear damage detection than the spectrogram and the Gaussian enveloped oscillating wavelet is suitable for detecting various sizes of gear faults [10, 11]. Luo *et al.* used the wavelet transform as a band-pass filter around a number of natural frequencies in gear vibration power spectrum [12]. The technique presented enhanced the signal to noise ratio. The component peak values of the enhanced power spectrum components are used as indicators of condition and defect. However, the sensitivity of the approach to the degree of the tooth damage is not reported. Dalpiaz *et al.* compared the wavelet transform with the cepstrum and spectrum correlation density as being applied to vibration signals representing two damage severity levels of gear failure [13]. They found that the mean value of the wavelet cross-section is very sensitive to the presence and size of tooth crack. Lin and Qu introduced a de-noising method that is based on wavelet analysis to extract gear vibration features however, more work still needed to apply the technique on signals of faulty gears [14]. Baydar and Ball analyzed gear vibration and acoustics using a Gabor based wavelet [15]. They found that changes in the wavelet plots are clearly observed for cracks of 30% and more. More work is still needed to produce features from the wavelet coefficient images that are related to failure prognoses.

More recently, Saravanan *et al.* used Morlet wavelets to extract features from vibration signals of bevel gear. Then they used support vector machine, a classification technique based on statistical learning theory, to classify the extract features according to the gear tooth damage type, e.g., crack, wear and breakage [16]. However, the sensitivity of the technique for the damage type was not reported. A similar work was published by Rafiee *et al.* but using Daubechie 11 wavelet for feature extraction and the neural net technique for fault classification [17]. Jafarizadeh proposed a pre-processing time averaging technique before applying the technique [18]. The work presented in this paper considers using the Kasier's window as a pre-processing step before applying the wavelet technique. The details of the technique are presented in section 5.

2. Experimental Setup and Laboratory Data

The experimental setup of the gear testing consists of one stage gearbox, two electric motors, a DC motor controller, and high power resistors, as illustrated in figure 1. The gearbox is constructed from a 127 mm steel square section welded to a 6.3 mm plate. It includes two identical mild steel gears of 10.2 mm width, 15 teeth, 14.5° pressure angle, and 15.87 mm pitch diameter. The two gear shafts are supported with bronze bushings. The gears and bushings are sump lubricated with Mobil SHC 634 synthetic lubrication oil.

The electric motors are 1 HP permanent magnet DC motors used for driving and loading the gearbox. The DC motor controller can drive the gearbox between 200 and 1400 rpm with a speed accuracy of 2%. The load motor applies the breaking torque, dissipating the energy through the resistors. The resistors can be arranged in various ways to provide a wider load torque range. A shunt resistor is connected to the high power resistors in series to measure the current through the load motor.

The vibration signal is first collected by a PCB-307A piezoelectric accelerometer that is mounted on the gearbox plat inline with the gear meshing force. The accelerometer signal is amplified by a PCB-482A power supply unit. The signal is then passed through a data acquisition card (Metrabyte Dash-16F) which is controlled by software. A photo interrupter device is used in conjunction with a slotted disk to produce a pulse signal to trigger the A/D board and collect the data. This is required to facilitate time synchronous averaging. The data are then analyzed using Matlab software. The gear tooth crack is simulated using a thin saw cut at the tooth root. For the purpose of evaluating the efficiency of the proposed monitoring tool, different experiment runs were performed for various sizes of gear tooth cracks at the same speed of 1221 RPM. The sampling frequency was set at 83.33 KHz that generates 4096 samples for each single shaft rotation. The data signal to noise ratio was enhanced using the time averaging technique and with the help of the optical sensors.

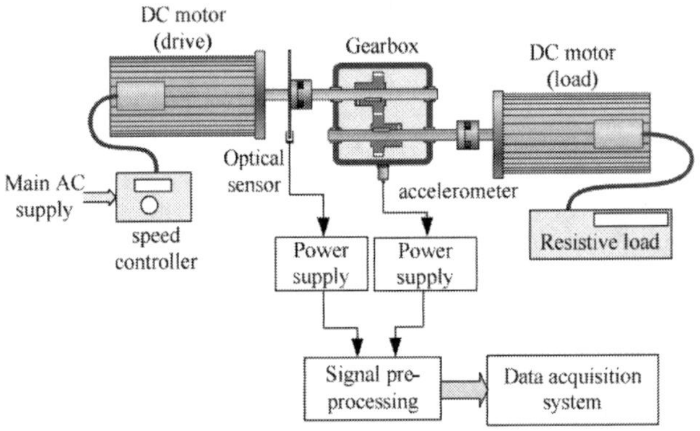

Figure 1. Experimental setup of gear vibration test.

3. Introduction to Windowing Wavelet Transform

The discrete wavelet transform (DWT) decomposes a signal $f(t)$ at different resolution levels and represents the signal as a series of approximate $c_j(k)$ and detail $d_j(k)$ expansion [19], [20]:

$$f(t)=\sum_k c_0(k)\phi(t-k)+\sum_k\sum_{j=0}^{J-1} d_j(k)2^{j/2}\psi(2^j t-k) \tag{1}$$

The discrete wavelet coefficients measure the similarity between the signal and the selected wavelet $\psi(t)$; hence provide a time-frequency localization of the signal. If the scaling function and wavelets form an orthonormal basis, then according to Parseval's theorem, the energy of the signal can be partitioned in terms of the expansion coefficient.

For any proposed new signal monitoring tool, the following characteristics represent important features for evaluating the proposed tool:

- *Sparsity*: The expansion coefficients of any event should have most of the important information in the smallest number of coefficients so that the rest of coefficients are small enough to be neglected or set equal to zero. This is important for data management, compression and denoising.
- *Separation*: During abnormal conditions, defects can be modeled as a linear combination of signals with different characteristics; the expansion coefficients should clearly separate those signals. Features of interest that classify each defect should be separated and localized at different resolutions. This is important for detection and classification.
- *Super-resolution*: A defect free sample and its characteristic set of expansion coefficients should have much better resolution than that with a traditional basis system. This is likewise important for detection, classification and estimation.
- *Stability*: The expansion coefficients that are extracted from a reference signal should not be significantly changed by defects or noise. This is important in auto-monitoring application and data measurement.
- *Speed*: The numerical calculation of the expansion coefficients in the new system should be of order $O(N)$ or $O(N\log(N))$. This is important for real-time application.

Significant improvement in the efficiency of any proposed monitoring tool can be achieved by increasing the similarity of the expansion system bases and different classes of gear tooth failures. The proposed technique uses a window function $w_k(t)$ in order to enhance the resemblance of the selected

Journal of Physics: Conference Series **181** (2009) 012089 doi:10.1088/1742-6596/181/1/012089

mother wavelet and processed signal. Hence, strengthen the important characteristics mentioned before. In this work, Kaiser's window of length w_N is selected in the windowing process which is presented as:

$$w[n] = \frac{I_o\{\beta[1-(n-u)^2/u^2]^{0.5}\}}{I_o(\beta(0,\alpha))} \qquad (2,$$

$$for\ n = 0,1,....,w_N - 1$$

where $I_o(x)$ is the modified Bessel function and u is the midpoint of the window function. The advantage of selecting Kaiser's window comes from the ability to adjust the window shape by changing α parameter [21], [22].

Using Mallat's algorithm, the set of expansion coefficients resulted from a windowing version of the signal at certain resolution can be expressed as:

$$wd_j(k) = \sum_m h_l(m-2k)\ wc_{j+1}(m) \qquad (3)$$

Figure 2 highlights the advantage of using the Kaiser's window with the wavelet analysis as applied to a test signal of 4.0 kHz sine wave. Figure 2a presents a direct application of the wavelet multi-resolution analysis where the data is processed through a rectangular window. Although large part of the test signal energy, as expected, is localized at the 4th resolution, a large number of coefficients are localized at adjacent resolutions and some of which have considerable large magnitudes. Simulation shows that maximum coefficient magnitude is close to 10 at the 3rd resolution and close to 20 at the 5th resolution. However, the proposed technique localizes all the coefficients in their true resolution level, as clearly shown in figure 2b, hence satisfies the *sparsity*, *supper-resolution* and *separation* requirements of the proposed coefficients-dependent monitoring tool.

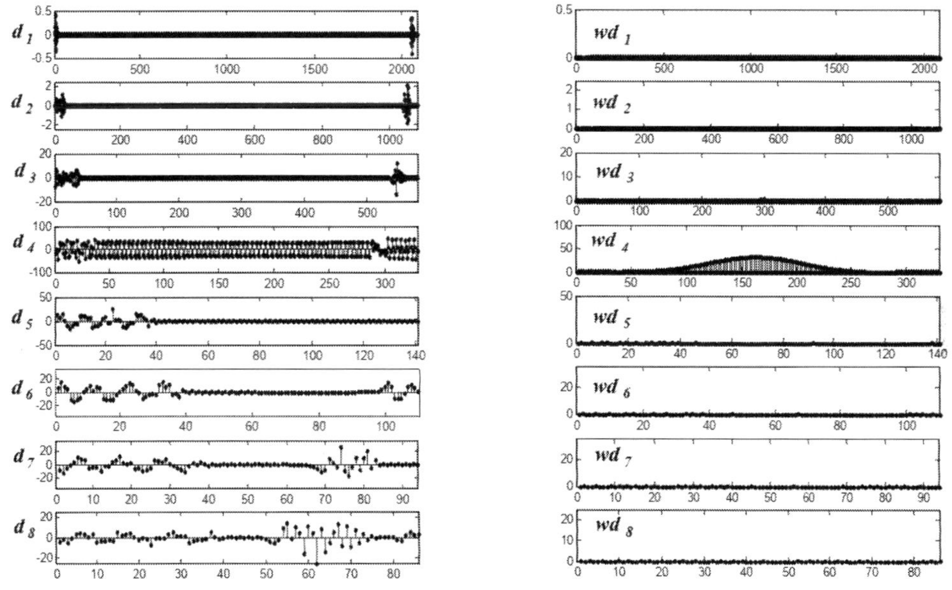

(a) (b)

Figure 2. Wavelet expansion coefficients localized at 8 resolutions of a virtual reference signal, (a) using rectangular window, (b) using Kaiser's window

Figure 2b shows the wavelet expansion coefficients of local maxima extracted using Kaiser's window $wd_j(k)$. At the super-resolution, where the test signal component is reside, the number of coefficients is reduced to a very small set and the magnitude of the maximum coefficient is almost the same as compared with the one shown in figure 2a at the same resolution level. Furthermore, for de-noising purposes a threshold value can be easily selected, if required, to eliminate the noise related coefficients. Also, the difficulty of selecting the right mother wavelet is reduced due to utilizing Kaiser's window.

Figure 3 compares the results of monitoring the absolute value of the maximum coefficients extracted directly and with that use of the Kaiser's window. The comparison is applied on a simulated noise free non-stationary signal and a similar signal corrupted with high noise levels, as shown in figure 3. Direct application of the WMRA produces coefficients ($|d_5|_{Max}^{Mag}$) that can detect some variations in the signal magnitude; however they are not able to trace these variations accurately and localize them on time. On the other hand, the proposed technique shows a stable detailed coefficients ($|wd_5|_{Max}^{Mag}$) that can detect accurately these variations and localize them on time. Hence reflect their magnitudes even with the presence of high noise levels.

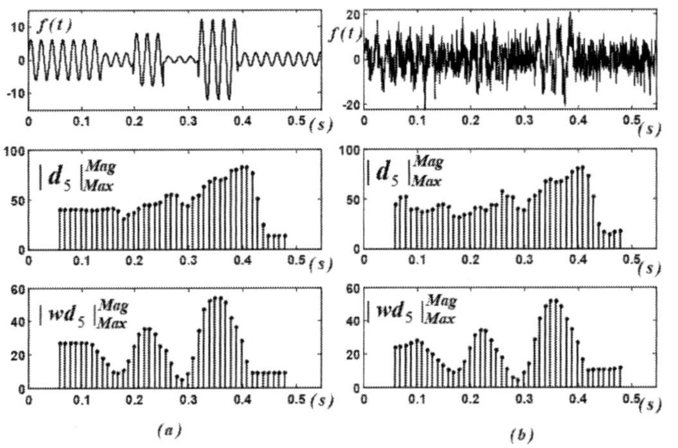

Figure 3. The variation in the maximum coefficients at the 5th resolution of a signal using rectangular window and Kaiser's window, a- noise-free signal and b- signal corrupted with high noise level.

4. Gearbox Condition Monitoring Technique

Any defect in a gearbox can be considered as a linear combination of time variant signals with different characteristics. The coefficients-based monitoring technique should provide a small set of expansion coefficients that clearly detect and measure the severity of the defect.

As data slides into Kaiser's window, the magnitudes of the maximum coefficients, $Max(wd_j)_{sig}$, detected at different resolution levels represent important features to monitor and localize any deviation in the gear condition. The sampling rate, f_{sam}, should be selected such that the healthy gear signal component is centered at the reference resolution level. The number of sampling points should be large enough to decompose the signal into resolution levels that cover the expected band where any defect spectra might be localized. The size of Kaiser's window, W_N, should not exceed the period of one gear tooth mesh and the data-sliding rate, N_S, should be kept very small compared to Kaiser's window size, W_N to enhance the tool monitoring accuracy.

Figure 4 is a pictorial representation of the proposed technique. When Kaiser's window is centered at w_{c1} the data processed represents a defect free signal. The information extracted will be similar to those extracted in the preprocessing stage from the healthy sample. As a new data set slides through Kaiser's window at w_{c2}, the proposed monitoring technique will generate expansion coefficients at certain resolutions that reflect the tooth failure event. The magnitude of the maximum coefficient, $Max(wd_4)_{sig}$, will change according to the severity of gear tooth fracture. The new magnitude of the maximum coefficient at each resolution $Max(wd_j)_{sig}$ can be easily detected and evaluated. These coefficients are localized in time at the center of Kaiser's window, w_c, since they were extracted from a windowing version of the data. Then the dynamic behavior of the magnitude variation of the extracted signal and the location of the defected tooth can be estimated using the proposed technique.

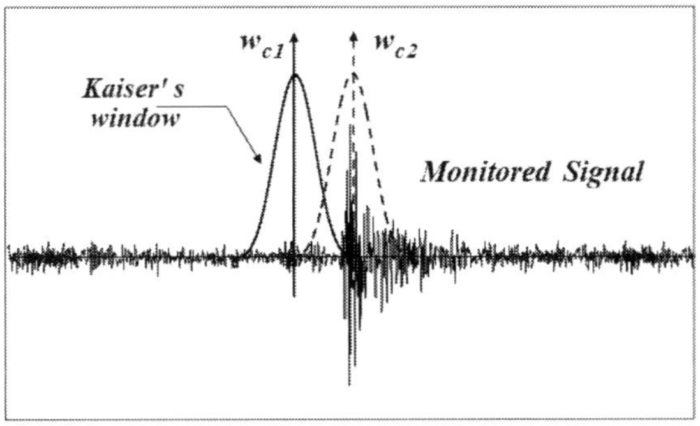

Figure 4. A pictorial representation of the proposed technique as data sliding through Kaiser's window.

5. Application and Results

A preprocessing stage is used to define the main features extracted from a healthy sample which will be used as a reference to monitor and diagnose any changes in the gear conditions. Figure 5 shows the results of the monitoring process under normal operation conditions, NOC, of a healthy gearbox. The vibration signal collected in laboratory of a healthy gear is shown in figure 5a. The normalized norm of the coefficients ($\| dj \|_2$) at different resolutions is used to define the distribution of signal's energy at 8th resolution levels during NOC. This feature vector will be used as a reference to detect any abnormal conditions in the gearbox. The maximum coefficients at the 4th resolution as data slides into Kaiser's window, as shown in figure 5c, are utilized to estimate the magnitude of defect in the gearbox and identify the tooth location. Monitoring abnormal operation conditions of a gearbox is shown in figure 6. The vibration signal collected of an unhealthy gear is shown in figure 6a. Any abnormal condition is detected as a change in the norm of the coefficients localized at different resolution in comparison with that representing NOC as shown in figure 6b. The variation in the magnitude of the maximum coefficients at the 4th resolution, as shown in figure 6c, will be used to estimate the magnitude of the tooth defect and to identify the tooth place.

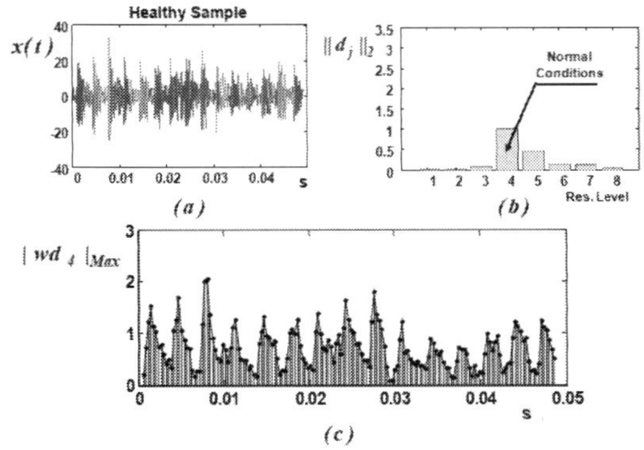

Figure 5. Monitoring normal operation conditions of a healthy gearbox, a- vibration signal of a healthy gear, b- norm of the coefficients at different resolutions and c- maximum coefficients at the 4th resolution as data sliding through Kaiser's window.

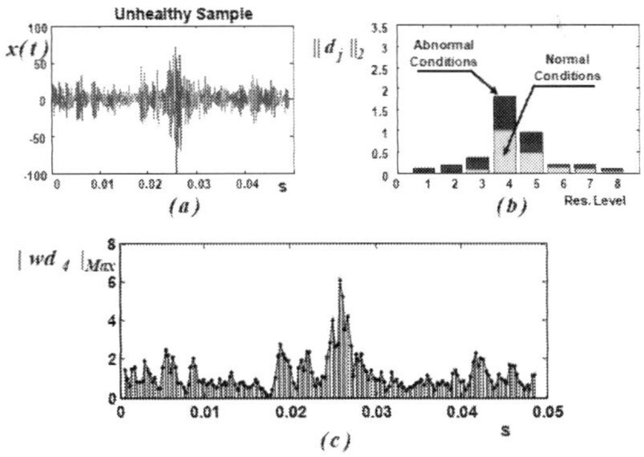

Figure 6. Monitoring abnormal operation conditions of a 50% crack in gear tooth, a- vibration signal of unhealthy gear, b- norm of the coefficients at different resolutions and c- maximum coefficients at the 4th resolution as data sliding through Kaiser's window.

Comparing results extracted under NOC, as shown in figure 5, and that of defected gear, as shown in figure 6, one can notice a change in the magnitude of the normalized coefficients' norm by 100% at the 4th and 5th and 200% at the 3rd resolutions. The magnitude of the coefficients related to gear tooth number 9 are increased from 2.0 under NOC to 6.0 at defected tooth location while other teeth show no change in their magnitudes.

The proposed technique is applied to monitor other sets of real gear vibration signals. Four data sets that represent healthy gear and three cracked tooth gears of 40%, 50%, and 70% crack sizes are used to evaluate the efficiency of the monitoring tool. Accurate results were reached that reflect the gear health condition, as shown in figure 7.

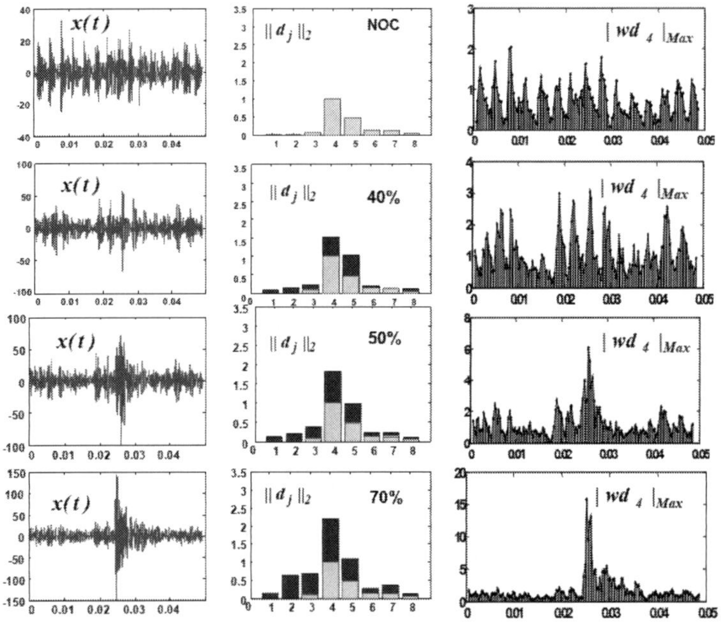

Figure 7. Monitoring and tracing fracture intensity increase in a gear tooth.

6. Conclusion

The proposed technique resolves the problem of selecting a suitable mother wavelet and defines a threshold for monitoring and evaluating gear tooth fracture. Processing vibration signals through Kaiser's window increases the similarity between the signal under processing and the selected mother wavelet. The norm of the wavelet expansion coefficients can be used to detect any abnormal conditions in the gear teeth. The maximum coefficients localized as processing data through Kaiser's window show clear features that can be used to measure and localize any damage in the gear tooth. The proposed tool is very sensitive to any variation in the signal and can be used in detecting early signs of any abnormal condition in the gearbox. Research is continuing to investigate the affect of the gearbox structure and gear types on the developed technique.

References

[1] Dalpiaz G Early 1990 Detection of Fatigue Cracks in Gears by Vibration Analysis Technique, *Osterreicchische Ingenieur-und Architekten-Zeitschrift (OIAZ)* Vol 135 (jg Heft).

[2] Ismail F, Martin H R and Omar F 1995 Statistical index for monitoring tooth cracks in a gearbox *15th Biennial Conference on Mechanical Vibration and Noise, ASME Design Engineering Division (Pub.) DE* Vol 84 No 3 Pt A/2 (Boston, MA, USA) pp 1413-1418.

[3] Andrede F A, Isat I and Badi M N M 2001 A new approach to time-domain vibration condition monitoring: gear tooth fatigue crack detection and identification by the Kolmogorov–

Smirnov test *Journal of Sound and Vibration* Vol 240 No 5 pp 909-919.

[4] Andrede F A, Isat I and Badi M N M 2002 Gear condition monitoring by a new application of the Kolmogorov–Smirnov test *Proceeding of the Institute of Mechanical Engineers–Part C* Vol 215 pp 793-800.

[5] Kar C. and Mohanty A R 2006 Multistage gearbox condition monitoring using motor current signature analysis and Kolmogorov–Smirnov test *Journal of Sound and Vibration* Vol 290 pp 337-368.

[6] Mathew J 1989 Monitoring the Vibration of Rotating Machine Elements – an overview *Diagnostics, Vehicle Dynamics and Special Topics ASME De*Vol 18-5 pp 15-22.

[7] Randall R B 1981 Cepstrum Analysis *Bruel Kjaer Technical Review* No 3.

[8] Tang H, Cha J, Wang Y and Zhang C 1991 The Principle od Cepstrum and Its Application in Quantitative Fault Diagnostics of Gears *Modal Analysis, Modeling, Diagnostics, and Control – Analytical and Experimental, ASME* De-Vol 38 pp 141-144.

[9] Wang W J and McFadden P D 1993 Early Detection of Gear Failure by Vibration Analysis – I. Calculation of Time-Frequency Distribution *Mechanical Systems and Signal Processing* Vol 7 No 3 pp 193-203.

[10] Wang W J and McFadden P D 1995 Application of orthogonal wavelets to early gear damage detection *Mechanical Systems and Signal Processing* Vol 9 pp 497-507.

[11] Wang W J and McFadden P D 1996 Application of wavelets to gearbox vibration signals for fault detection *Journal of Sound and vibration* Vol 192 No 5 pp 927-939.

[12] Luo G Y, Osypiw D and Irle M 2000 Real-Time Condition Monitoring by Significant and Natural Frequencies Analysis of Vibration Signal with Wavelet Filter and Autocorrelation Enhancement *Journal of Sound and vibration* Vol 236 No 3 pp 413-430.

[13] Dalpiaz G, Rivola A and Rubini R 2000 Effectiveness and Sensitivity of Vibration Processing Techniques for Local Fault Detection in Gears *Mechanical Systems and Signal Processing* Vol 14 No 3 pp 387-412.

[14] Lin J and Qu L 2000 Feature Extraction Based on Morlet Wavelet and its application to Mechanical Fault Diagnosis *Journal of Sound and Vibration* Vol 234 No 1 pp 135-148.

[15] Baydar N and Ball A 2003 Detection of Gear Failures Via Vibration and Acoustic Signals using wavelet Transform *Mechanical Engineering Signal Processing* Vol17 No 4 pp 787-804.

[16] Saravanan N, Siddabattuni V N S K and Ramachandran K I 2008 A comparative study on classification of features by SVM and PSVM extracted using Morlet wavelet for fault diagnosis of spur bevel gear box *Expert Systems with Applications* Vol 35 pp 1351–1366.

[17] Rafiee J, Tse P W, Harifi A and Sadeghi M H 2008 A novel technique for selecting mother wavelet function using an intelligent fault diagnosis system *Expert Systems with Applications Preprint* doi:10.1016/j.eswa.2008.05.052.

[18] Jafarizadeh M A, Hassannejad R, Ettefagh M M and Chitsaz S 2008 Asynchronous input gear damage diagnosis using time averaging and wavelet filtering *Mechanical Systems and Signal Processing* Vol 22 pp 172–201.

[19] Gaouda A M, Kanoun S H, Salama M M A and Chikhani A Y 2002 Wavelet-Based Signal Processing Techniques for Disturbance Classification and Measurement *IEE Proceedings – Generation, Transmission and Distribution* Vol 149 No 3 pp 310-318.

[20] Gaouda A M, El-Hag A, Abdel-Galil T K, Salama M M A and Bartnikas R 2008 On-Line Detection and Measurement of Partial Discharge Signals in a Noisy Environment *IEEE Transaction on Dielectrics and Electrical Insulation* Vol 15 No 4 pp 1162-1173.

[21] Sanjit K, Mitra and James F 1993 *Kaiser Handbook for Digital Signal Processing* (John Wiley & Sons).

[22] Burrus C S, McClellan J H, Oppenheim A V, Parks T W, Schafer R W and Schuessler H W 1994 *Computer-Based Exercises for Signal Processing using Matlab Matlab Curriculum Series* (Prentice hall).

Journal of Physics: Conference Series **181** (2009) 012090 doi:10.1088/1742-6596/181/1/012090

Methods for detecting fatigue cracks in gears

Aleš Belšak, Jože Flašker

University of Maribor, Faculty of Mechanical Engineering,
Smetanova ulica 17, 2000 Maribor, Slovenia

ales.belsak@uni-mb.si

Abstract. A crack in the tooth root is the least desirable damage caused to gear units and it often causes failure of gear unit operation. For fault analyses presented in this article, gear units with real damages or faults produced on the basis of numerical simulations of real operating conditions are used; tests were carried out in a laboratory test plant. Damages in gear units can be identified by monitoring vibrations. The influences of a crack in a single-stage gear unit on produced vibrations are presented. Significant changes in tooth stiffness are caused by a fatigue crack in the tooth root whereas, in relation to other faults, changes of other dynamic parameters are more expressed. Non-stationary signals are analysed, using the family of Time Frequency Analysis tools, which include Wavelets and Joint Time Frequency Analyses.

1. Introduction

The aim of maintenance is to keep a technical system (e.g. a gear unit) in the most suitable working condition, whereas its purpose is to discover, to diagnose, to foresee, to prevent and to eliminate damages. On the other hand, the purpose of modern maintenance is not only to eliminate failures but also to identify the stage before a sudden failure of system operation. The objectives of diagnostics are to define the current condition of the system and the location, shape and reason of damage formation.

A gear unit is a complex dynamic model. Its elements enable transmission of rotating movement. The movement is usually periodical, faults and damages represent a disturbing quantity or impulse. Disturbances are denoted by local and time changes in vibration signals. Consequently, time-frequency changes can be expected [1,2]. This idea is based on kinematics and operating features.

Fourier, adaptive and wavelet transforms, and Gabor expansion represent various time-frequency algorithms [3]. The basic idea of all linear transforms is to perform, in advance, the comparison with the elementary function [4]. Different signal presentations can be obtained by means of different elementary functions.

2. Introduction

Many authors had been developing algorithms without interference parts that reduce usability of individual transforms. Qian [4] significantly enhanced the adaptive transform of a signal, which represented a difference to Cohen's class.

The adaptive transform of a signal *x(t)* is expressed in the following way:

$$x(t) = \sum_{p} B_p \cdot h_p(t) \tag{1}$$

Journal of Physics: Conference Series **181** (2009) 012090 doi:10.1088/1742-6596/181/1/012090

where the following equations were used to analyse coefficients

$$B_p = \langle x, h_p \rangle \tag{2}$$

In this way, a similarity between the measured signal $x(t)$ and elementary functions $h_p(t)$ of transform is expressed.

It is possible to determine the time-dependent adaptive spectrum as follows

$$P_{ADT}(t,\omega) = \sum_p |B_p|^2 \cdot P_{WV} h_p(t,\omega) \tag{3}$$

As a rule, the choice of elementary functions has no impact upon the adaptive transform since it allows arbitrary elementary functions. Elementary functions, applied for adaptive representation of a signal with equation (1), are usually very general. In practice, however, this is not always so. To stress the time dependence of a signal, the elementary functions should be localised in regard to time and frequency. Also, it is desirable that they can use the presented algorithm in a relatively simple manner. A Gauss type signal has very favourable characteristics, and it is considered a basic choice when it comes to an adaptive representation.

3. Wavelet Analysis

This is how the continuous wavelet transform of function $x(t) \in L^2(\Re)$ at the time and scale is expressed [5]:

$$W\, x(u,s) = \langle x, \psi_{u,s} \rangle = \int_{-\infty}^{+\infty} x(t) \cdot \frac{1}{\sqrt{s}} \cdot \psi^*\left(\frac{t-u}{s}\right) \cdot dt = x(t) \otimes \overline{\psi}_s(t) \tag{4}$$

When it comes to the continuous wavelet transform, the observed function $x(t)$ is multiplied by a group of shifted and scaled wavelet functions. Time and frequency dissemination of the continuous wavelet transform changes simultaneously. Locally limited functions (wavelets) are used to analyse the observed function $x(t)$; therefore, the continuous wavelet transform is very sensitive to local non-stationarities.

Gabor wavelet function, which is acquired using a frequency modulation of the Gauss window function, represents an approximately analytical wavelet function [5]:

$$\psi_{Gabor}(t,\sigma,\eta) = \frac{1}{\sqrt[4]{\sigma^2 \cdot \pi}} \cdot e^{-\frac{t^2}{2\cdot\sigma^2}} \cdot e^{i\cdot\eta\cdot t} \tag{5}$$

4. Practical Example

The measurements were carried out in the test plant (presented in Fig. 1) of the Computer Aided Design Laboratory of the Faculty of Mechanical Engineering, University of Maribor. A one-stage helical gear unit is at the spot where vibrations were measured [6].

A single stage gear unit was used. A helical gear unit with straight teeth was integrated into the gear unit. A carburised spur gear pair of module 4 mm was in each gear unit, the pinion had 19 and the wheel 34 teeth. The presented results are relating to a nominal pinion torque of 20 Nm and nominal pinion speed of 1200 rpm (20 Hz), which is, in industry, a load condition very much characteristic of this type of gear units. Tests were performed under constant loads. Accelometers, fixed on the housings, were used to measure vibrations.

Journal of Physics: Conference Series **181** (2009) 012090 doi:10.1088/1742-6596/181/1/012090

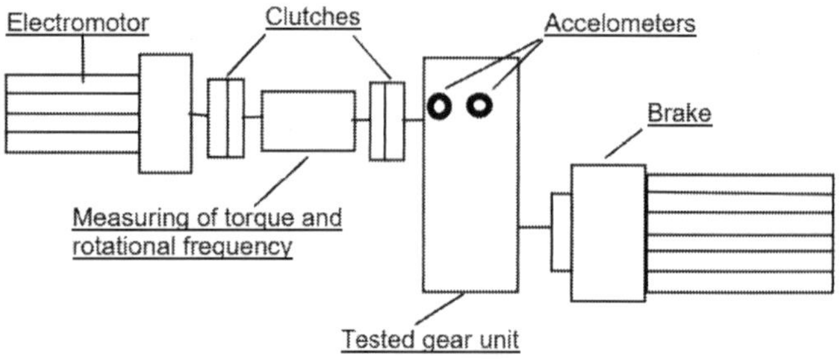

Figure 1: Test plant and a part of measuring equipment

A gear unit with a fatigue crack in the tooth root of a pinion was measured; the operating conditions were such as they are normally associated with this type of a gear unit. A standard gear pair, teeth quality 6, that was used had a crack in the tooth root of a pinion.

The characteristics of elementary functions are restricted. Consequently, adaptive spectrogram has a fine adaptive time-frequency resolution. Time-frequency resolution of the transform is adapted to signal characteristics. Gauss function (impulse) and linear chirp with Gauss window can be used as an elementary function.

The signal was 1 s long; on average, it had 14000 measuring points. It is possible to note some pulsation sources but, in relation to adaptive spectrogram, they are not very expressed (Fig. 2). This indicates a higher level of energy accumulation in the origins. Monitoring the increase or decrease (complete disappearance) in appropriate frequency components with rotational frequency of 20 Hz is of particular interest. This is typical of the 3rd harmonic of mesh frequency. 1530 Hz is expressed only in association to the presence of a crack. This phenomenon can be noted in the adaptive spectrogram (Fig. 3). In relation to a single engagement of a gear pair with a crack in one rotation of a shaft, the pulsation (the area marked with a continuous line) is expressed. Similarly, between the 6th and the 9th harmonics (the area marked with a dashed line), sources indicating pulsating portions of individual components, with the frequency of 20 Hz, can be observed.

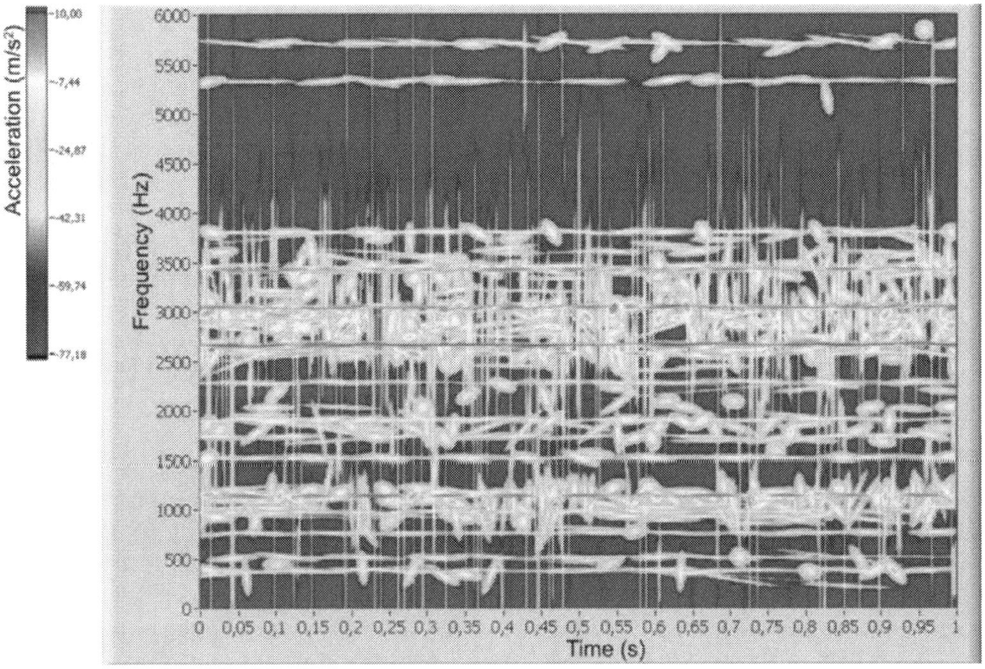

Figure 2: Adaptive spectrogram of a faultless gear unit

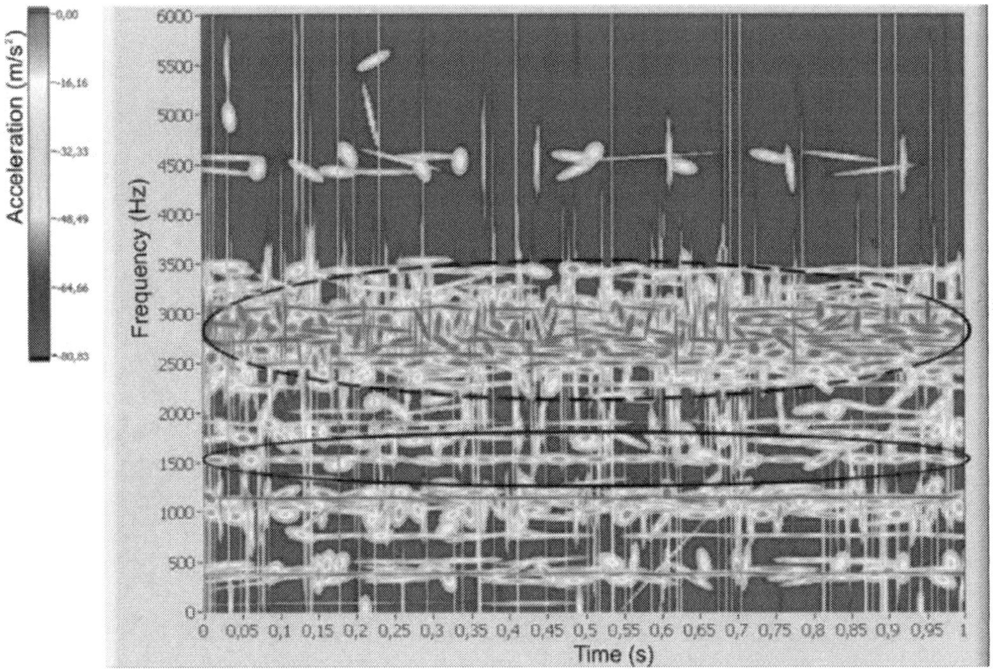

Figure 3: Adaptive spectrogram of a gear unit with a pinion with a crack

A scalogram of analytical wavelet transform with Gabor wavelet function represents square values of amplitudes of wavelet coefficients. On the grounds of the connection between the scale and frequency, the representation is performed in a time-frequency domain. This is favourable when it

Journal of Physics: Conference Series **181** (2009) 012090 doi:10.1088/1742-6596/181/1/012090

comes to technical diagnostics as it is much simpler to determine adequate characteristics in time-frequency domain (frequency scalogram) than in time-scale domain (scalogram). The transform matches the Parseval characterstic of energy preservation on the basis of normalization, meaning that the energy of wavelet transform is equal to the energy of the original signal in time domain.

Wavelet analysis is primarily appropriate in relation to non-stationary phenomena with local changes. Therefore, the analysis was performed to determine the condition associated with the presence of a crack in a tooth root. More specifically, the purpose of the analysis was to establish the location of the crack.

The analytical continuous wavelet transform (parameters: η = 8 and σ = 2) was used for the analysis. Nyquist frequency and the frequency of sampling the measured time signal were used to obtain the highest frequency in the signal (7000 Hz). After that, based on the known connection between scale and frequency, the scale type for constant frequency distribution was determined. The representation of the frequency scalogram is given in the form of wavelet coefficients or their square values.

In relation to the faultless gear, the figures in the frequency scalogram indicate no particularities in expressed components that would denote local changes. This is so in relation to a square representation (Fig. 4) of wavelet coefficients. In relation to the analysis of the signal produced by a gear with a crack, a local change in wavelet coefficients, in time at the value of 12 ms, is very clearly indicated in frequency scalograms (Fig. 5). In relation to the tooth with the crack in its root, a local change, i.e. the presence of transients, can be observed. If the wavelet length is 50 ms, which represents one rotation of the pinion, and if there are 19 teeth along the circumference, the increased amplitude is located at 12 ms. It belongs to the fourth tooth in the direction of rotation from the reference positional point of the gear unit.

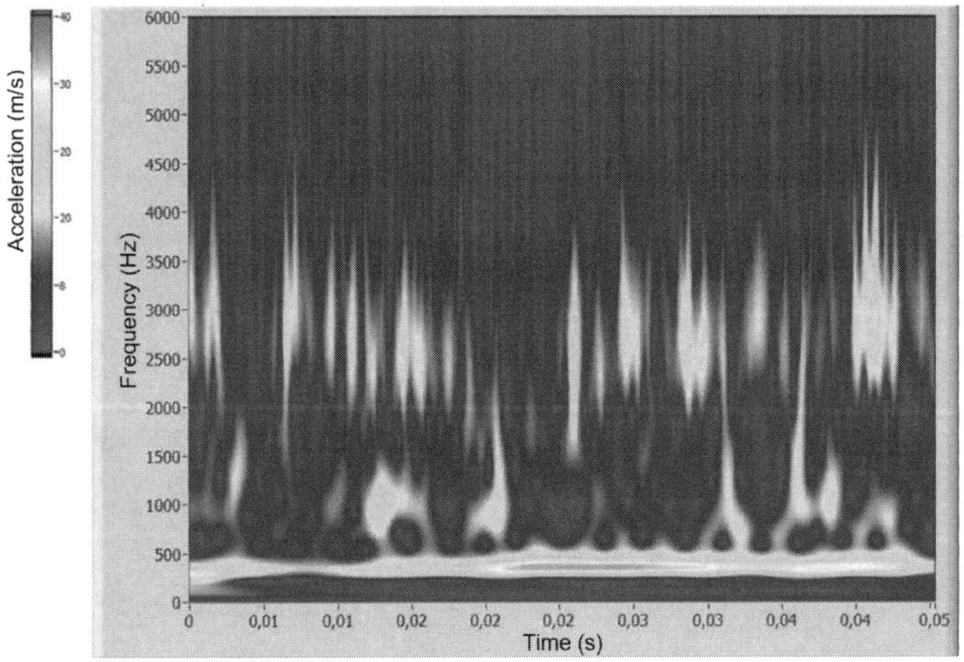

Figure 4: Frequency scalogram of square wavelet coefficient of the reference gear unit

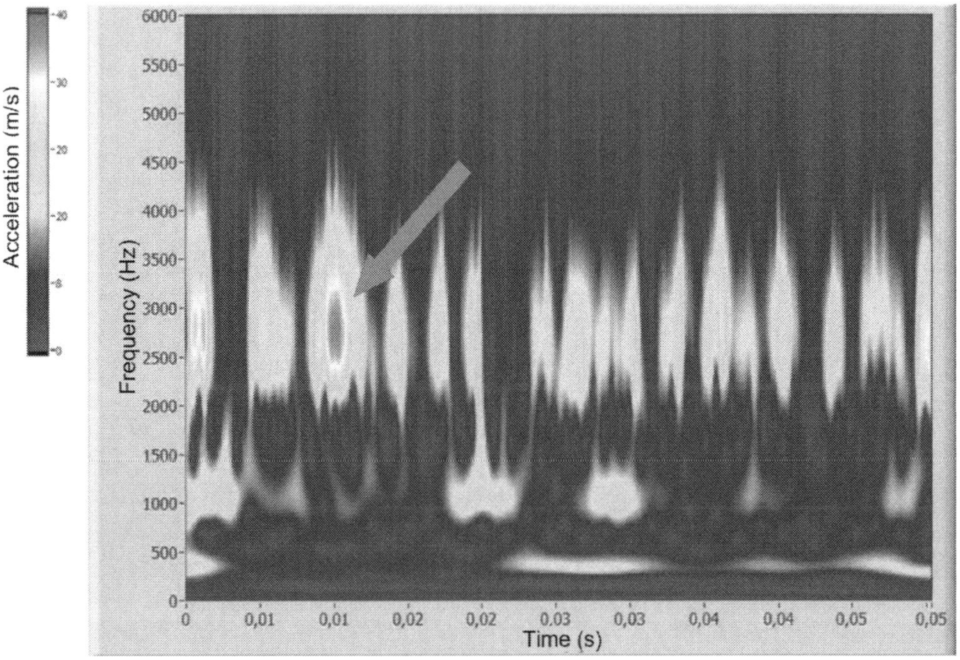

Figure 5: Frequency scalogram of square wavelet coefficient of the gear unit with a gear with a crack in a tooth root

5. Conclusions

The purpose of vibration analysis is to detect a fault in industrial gear units. On the basis of the methods presented, the safety of operation and, as a result, the reliability of monitoring operational capabilities can be improved. The reliability of monitoring life cycle of a gear unit can be increased by means of suitable spectrogram samples and a clear presentation of the pulsation of individual frequency components. By using wavelet transform, changes can be identified very quickly and the presence of a damage can be determined at the level of an individual tooth.

References

[1] P. Fladrin, Time-Frequency/Time-Scale Analysis, Academic Press, San Diego, (1999)
[2] A. Mertins, Signal Analysis, John Wiley & Sons Inc., New York, (1999).
[3] V. C. Chen, H. Ling, Time-Frequency Transforms, Artech Hause Publishers, Boston, (2002)
[4] S. Qian, D. Chen, Joint Time-Frequency Analysis, Prentice Hall, Upper Saddle River, (1996)
[5] S. Mallat, A Wavelet Tour of Signal Processing, Academic Press, San Diego, (1999)
[6] A. Belsak, Time-Frequency Analysis of the Condition of Gear Units (Abstract in English), Master Thesis, University of Maribor, Faculty of Mechanical Engineringg, (2004).

3D time-domain spectral elements for stress waves modelling

Pawel Kudela, Wieslaw Ostachowicz

Polish Academy of Sciences, Institute of Fluid-Flow Machinery, Fiszera 14 St., Gdansk, Poland

E-mail: pk@imp.gda.pl

Abstract. Elastic stress waves induced by piezoelectric transducers are extensively used for damage detection purposes. Induced high frequency impulse signals cause that stress wave modelling by the finite element method is inefficient. Instead, numerical model based on the time-domain spectral element method has been developed to simulate stress wave propagation in metallic structures induced by the piezoelectric transducers. The model solves the coupled electromechanical field equations simultaneously in three-dimensional case. Visualisation of the propagating elastic waves generated by the actuator of different shapes and properties has been performed.

1. Introduction

The research area of smart materials has attracted many researchers, because of promising potential in applications such as: morphing, energy harvesting, vibration control and damage detection. Taking advantages of piezoelectric effect in devices such as piezotransducers or recently Macro Fibre Composites (MFC) embedded into structure it is possible to built structures which merge superb mechanical properties of composites with tunning opportunity of static and dynamic response. Moreover, piezo-actuators and piezo-sensors can be used for Structural Health Monitoring. The analysis of piezoelectric composite structures such as piezoelectric laminated plates or structural elements activated by piezo-actuators requires efficient and accurate electromechanical modelling of both the mechanical and electric responses such as mechanical displacements and electric potentials. Exact 3D analytical solutions have been presented for the piezoelectric response of simply supported flat panels and rectangular plates in [1–4]. Since the exact 3D analytical solutions are available only for some regular shapes with specified simple boundary conditions, it is desirable to use approximate method for more complex structures (i. e. the finite element method, the finite difference method, the boundary element method).

Modelling of mechanical displacement field by the full 3D finite elements [5; 6] typically results in huge number of degrees of freedom and high computational cost. For this reason many 2D models with different assumptions made on the through-the-thickness mechanical displacement field displacement have been proposed including layerwise approximation [7–9] or multi-layer modelling [10]. More recently method of sublayers has been proposed [11] which combines a 2D single-layer representation model for the mechanical displacement field with a 3D layerwise-like approximation for the electric potential field. However, these models are rarely used in stress waves problems.

Journal of Physics: Conference Series **181** (2009) 012091 doi:10.1088/1742-6596/181/1/012091

Furthermore, the through-the-thickness electric potential field distribution can be approximated by various models: linear [12] for thin actuators or nonlinear [10; 13] or trigonometric [7]. Kapuria [14] even proposed an electromechanical model that combines the displacement field approximations of the third order zigzag theory with a layerwise approximation for the electric potential. Each electric potential representation model has its own advantages and disadvantages in terms of accuracy and ease of use and computational cost.

Moreover, in the case of elastic wave propagation, the major numerical challenge is to minimize numerical dispersion error [15] (error of phase and group velocities). Minimisation of the error requires sufficient discrete spatial resolution per minimum wavelength. In order to obtain solution with a less than 1% dispersive error, explicit finite element method requires more than 20 elements per minimum wavelength [15]. Thus, application of 3D finite elemtnts for wave propagation modelling is impractical.

The spectral element method (SEM) in the time domain proposed by Patera [16] is an alternative to the existing numerical methods in the field of stress waves modelling. This method originates from the use of spectral series for the solution of partial differential equations [17]. The idea of SEM is very similar to FEM except for the specific approximation functions it uses. Elemental interpolation nodes are located at points corresponding to zeros of an appropriate family of orthogonal polynomials (Legendre or Chebyshev). A set of local shape functions consisting of Lagrange polynomials, which are spanned on these points, are built and used. As a consequence of this, as well as the use of the Gauss-Lobatto- Legendre integration rule, a diagonal form of the mass matrix is obtained. In this way the cost of numerical calculations is much less expensive than in the case of any classic FE approach. Moreover, the numerical errors decrease faster than any power of $1/p$ (so called spectral convergence), where p is the order of the applied polynomial.

The main fields of application of SEM nowadays include: fluid dynamics [18], heat transfer [19], acoustics [20; 21], seismology [22; 23], etc. The application of SEM for problems of propagating waves in anisotropic crystals has been shown in [24].

The first attempt to the use of SEM for problems of propagation of elastic waves in 2D structural elements with crack has been done by Zak et al. [25]. 36-node spectral membrane element with two degrees of freedom per node has been developed. The crack has been modelled by simple splitting of the nodes between appropriate spectral elements. This approach has been extended to isotropic and composite plates [25–27].

The simplified actuator-induced Lamb wave propagation analysis using 3D SEM has been performed in [28]

The aim of this paper is to develop higher order 3D spectral element, which assures fast convergence of mechanical displacement field as well as electric potential field. Both actuation and sensing is taken into account. Moreover, presented special time integration algorithm based on the central difference scheme and element by element approach overcomes problem of cost of memory resources.

2. Spectral element formulation

Spectral element is derived in the same manner as finite element based on weight residual method. The SEM approximates the field variables in elements using higher-order one-dimensional Lagrange polynomial and its tensor product. The displacement field can be approximated as follows:

$$\begin{bmatrix} u^e(\xi,\eta,\zeta) \\ v^e(\xi,\eta,\zeta) \\ w^e(\xi,\eta,\zeta) \end{bmatrix} = \mathbf{N}^e\hat{\mathbf{u}}^e = \sum_{k=0}^{n}\sum_{j=0}^{n}\sum_{i=0}^{n} N_i^e(\xi)N_j^e(\eta)N_k^e(\zeta)\,\mathbf{I}_3 \begin{bmatrix} \hat{u}^e(\xi_i,\eta_j,\zeta_k) \\ \hat{v}^e(\xi_i,\eta_j,\zeta_k) \\ \hat{w}^e(\xi_i,\eta_j,\zeta_k) \end{bmatrix} \quad (1)$$

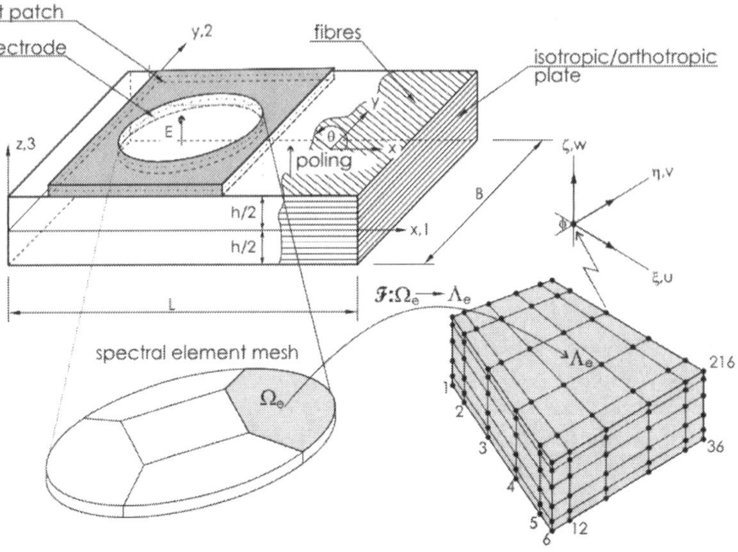

Figure 1. Modelling of the plate with pzt bonded to the surface

where N_i^e is the one-dimensional shape function (nth-order 1D Lagrange interpolation function at $n+1$ Gauss-Lobatto-Legendre points), $\hat{\mathbf{u}}^e$ – nodal degrees of freedom, $\mathbf{I}_3$ – unit matrix of the size 3x3. Approximation of potential and geometry of the element can be assumed similarly. Distribution of the nodes in the fifth-order 3D spectral element is shown in Fig. 1. The element has 216 nodes with three mechanical degrees of freedom ($\hat{u}$, $\hat{v}$, $\hat{w}$) and one electrical degree of freedom ($\hat{\phi}$) per node.

In this paper spectral elements are used for modelling propagating waves in isotropic or orthotropic plate excited by piezotransducer bonded to the upper surface of the plate (see Fig. 1). However, laminates can be considered too by using layers of spectral elements of various properties in each lamina.

2.1. Linear piezoelectric constitutive equations

The linear piezoelectric constitutive equations in Voigt notation can be expressed as:

$$\sigma = \mathbf{c}^E \varepsilon - \mathbf{e}^T \mathbf{E} \tag{2}$$

$$\mathbf{D} = \mathbf{e}\varepsilon + \mathbf{g}\mathbf{E} \tag{3}$$

where ε is the strain vector, $\mathbf{D}$ the electric displacement vector, σ the stress vector, $\mathbf{c}^E$ the elasticity matrix under constant electric field, $\mathbf{e}$ the piezoelectric constant matrix, $\mathbf{E}$ the electric field vector, $\mathbf{g}$ the dielectric constant matrix.

The electric field vector $\mathbf{E}$ is related to the electric potential field ϕ by using a gradient vector ∇ [5]:

$$\mathbf{E} = -\nabla\phi \tag{4}$$

2.2. Approximations of the field variables

Taking into account small deformations of any point in the solid linear strains can be approximated as:

$$\varepsilon^e(\xi, \eta, \zeta) = \mathbf{B}_u^e \hat{\mathbf{u}}^e \tag{5}$$

where $\mathbf{B}_u^e$ is strain–nodal displacement matrix calculated as:

$$\mathbf{B}_u^e = \mathbf{L}\,\mathbf{N}^e(\xi,\eta,\zeta)$$

$$\mathbf{L} = \begin{bmatrix} \frac{\partial}{\partial x} & 0 & 0 \\ 0 & \frac{\partial}{\partial y} & 0 \\ 0 & 0 & \frac{\partial}{\partial z} \\ \frac{\partial}{\partial y} & \frac{\partial}{\partial x} & 0 \\ 0 & \frac{\partial}{\partial z} & \frac{\partial}{\partial y} \\ \frac{\partial}{\partial z} & 0 & \frac{\partial}{\partial x} \end{bmatrix}, \quad \begin{bmatrix} \frac{\partial}{\partial x} \\ \frac{\partial}{\partial y} \\ \frac{\partial}{\partial z} \end{bmatrix} = \mathbf{J}^{-1} \begin{bmatrix} \frac{\partial}{\partial \xi} \\ \frac{\partial}{\partial \eta} \\ \frac{\partial}{\partial \zeta} \end{bmatrix} \tag{6}$$

where $\mathbf{J}$ is the Jacobi matrix. Using Eq. (4) and electric field approximation similar to Eq. (1) electric field vector is given by equation:

$$\mathbf{E} = -\mathbf{B}_\phi \hat{\phi} \tag{7}$$

where $\mathbf{B}_\phi$ is electric field–nodal potential matrix calculated as:

$$\mathbf{B}_\phi = \begin{bmatrix} \frac{\partial}{\partial x} \\ \frac{\partial}{\partial y} \\ \frac{\partial}{\partial z} \end{bmatrix} \mathbf{N}^e(\xi,\eta,\zeta) \tag{8}$$

2.3. Elementary governing equations of motion

The governing equations of motion of piezoelectric solid element can be derived by using Hamilton's variational principle in which the total work done by the external mechanical and electrical forces is taken into account. Finally equations of motion of piezoelectric element can be written in the matrix form:

$$\mathbf{M}^e\ddot{\hat{\mathbf{u}}}^e + \mathbf{K}_{uu}^e\hat{\mathbf{u}}^e - \mathbf{K}_{u\phi}^e\hat{\phi}^e = \mathbf{f}^e \tag{9}$$

$$\mathbf{K}_{\phi u}^e\hat{\mathbf{u}}^e + \mathbf{K}_{\phi\phi}^e\hat{\phi}^e = \mathbf{q}^e \tag{10}$$

where $\mathbf{M}^e$ denotes the elementary mass matrix, $\mathbf{K}_{uu}^e$ the elementary stiffness matrix, $\mathbf{K}_{u\phi}^e$ and $\mathbf{K}_{\phi u}^e$ the piezoelectric coupling matrices, $\mathbf{K}_{\phi\phi}^e$ the dielectric permittivity matrix, $\hat{\phi}^e$ the elementary electric potential vector, $\mathbf{f}^e$ the nodal external force vector, $\mathbf{q}^e$ the nodal externally applied charge vector. Particular matrices and vectors from Eq. (10) are defined as:

$$\mathbf{M}^e = \int_{V_e} \mathbf{N}^T \rho \mathbf{N} \mathrm{d}V_e \tag{11}$$

$$\mathbf{K}^e_{uu} = \int_{V_e} \mathbf{B}^T_u \mathbf{c} \mathbf{B}_u \mathrm{d}V_e \tag{12}$$

$$\mathbf{K}^e_{u\phi} = -\int_{V_e} \mathbf{B}^T_u \mathbf{e}^T \mathbf{B}_\phi \mathrm{d}V_e \tag{13}$$

$$\mathbf{K}^e_{\phi u} = (\mathbf{K}^e_{u\phi})^T \tag{14}$$

$$\mathbf{K}^e_{\phi\phi} = \int_{V_e} \mathbf{B}^T_\phi \mathbf{g} \mathbf{B}_\phi \mathrm{d}V_e \tag{15}$$

$$\mathbf{f}^e = \int_{V_e} \mathbf{N}^T \mathbf{P}_b \mathrm{d}V_e + \int_{\Gamma_S} \mathbf{N}^T \mathbf{P}_S \mathrm{d}\Gamma_S + \mathbf{P}^e \tag{16}$$

$$\mathbf{q}^e = \int_{\Gamma_\phi} \mathbf{N}^T_\phi \mathbf{q}_0 \mathrm{d}\Gamma_\phi \tag{17}$$

The governing equation of motions (9–10) can be further condensed in such a way that the unknow potentials are sacrificed in favor of the unknown displacements:

$$\mathbf{M}\ddot{\mathbf{u}} + (\mathbf{K}_{uu} + \mathbf{K}^S)\,\hat{\mathbf{u}} = \mathbf{f} + \mathbf{f}^{\mathbf{A}} \tag{18}$$

where $\mathbf{K}^S$ denotes stiffness matrix induced by electromechanical coupling which depends on electric boundary condition (open circuit, closed circuit, actuator) and is obtained as:

$$\mathbf{K}^S = \mathbf{K}^T_{\phi u} \mathbf{K}^{-1}_{\phi\phi} \mathbf{K}_{\phi u} \tag{19}$$

It should be noticed that the elementary dielectric permittivity matrix is not positive definite. To make static condensation (18) possible, electric boundary conditions must be applied. Imposed electrical boundary conditions are applied to matrices $\mathbf{K}_{\phi u}$ and $\mathbf{K}_{\phi\phi}$.

2.4. Actuation

Actuation is accomplished by the equivalent mechanical force vector $\mathbf{f}^A$ of the applied voltage of the piezoelectric actuator which can be expressed as follows:

$$\mathbf{f}^A = \mathbf{K}^A_{u\phi} \hat{\phi}^{(A)} \tag{20}$$

In case of excitation V_1 applied at the top surface of the piezoelectric electrode and assuming that the electric potential is zero at the bottom surface of the actuator, the induced potential distribution can be calculated using formula:

$$\phi^A = -(\mathbf{K}^A_{\phi\phi})^{-1} \mathbf{K}_{\phi\phi,1} V_1 \tag{21}$$

in which $\mathbf{K}^A_{\phi\phi}$ and $\mathbf{K}_{\phi\phi,1}$ are the submatrices of the matrix $\mathbf{K}_{\phi\phi}$ corresponding to electrical boundary conditions. For example in case of nth-order 3D spectral element $\mathbf{K}^A_{\phi\phi}$ matrix correspond to set of node numbers (I, I), $I \in [n \cdot n + 1 : n \cdot n \cdot n - n \cdot n]$ whereas $\mathbf{K}_{\phi\phi,1}$ matrix correspond to set of node numbers (I, I_1), $I_1 \in [1 : n \cdot n, n \cdot n \cdot n - n \cdot n + 1 : n \cdot n \cdot n]$, where colon denotes range of natural numbers.

2.5. Sensing

Sensing is performed according to the equation:

$$\hat{\phi}^{(S)} = -(\mathbf{K}^S_{\phi\phi})^{-1}\mathbf{K}^S_{\phi u}\,\hat{\mathbf{u}} \tag{22}$$

where $\mathbf{K}^S_{\phi\phi}$ and $\mathbf{K}^S_{\phi u}$ are the corresponding submatrices of $\mathbf{K}_{\phi\phi}$ and $\mathbf{K}_{\phi u}$, respectively (open circuit is considered here).

2.6. Solving equation of motion

Assuming notation $\mathbf{F}_t = \mathbf{f} + \mathbf{f}^{\mathbf{A}}$, $\mathbf{K} = \mathbf{K}_{uu} + \mathbf{K}^S$ and using central difference scheme, the Eq. (18) can be rearranged into the form:

$$\underbrace{\left(\frac{1}{\Delta t^2}\mathbf{M}\right)}_{\mathbf{M}_0}\mathbf{u}_{t+\Delta t} = \mathbf{F}_t - \underbrace{\mathbf{K}\mathbf{u}_t}_{\hat{\mathbf{F}}} + \underbrace{\left(\frac{1}{\Delta t^2}\mathbf{M}\right)}_{\mathbf{M}_0}\mathbf{u}_t - \underbrace{\left(\frac{2}{\Delta t^2}\mathbf{M}\right)}_{\mathbf{M}_2}\mathbf{u}_{t-\Delta t} \tag{23}$$

In Eq. (23) vector $\hat{\mathbf{F}}$ can be calculated element by element without necessity of global stiffness matrix assembly. Proposed algorithm for explicit integration of equation of motion is as follows:

- Loop over elements e, $e \in [1, n_{el}]$
  - Calculation of characteristic mass matrix $\mathbf{m}^e$
  - Assembly of the diagonal mass matrix into the global vector form: $\mathbf{M} = \overset{n_{el}}{\underset{e=1}{\mathsf{A}}}\,\mathrm{diag}\,(\mathbf{m}^e)$, where $\overset{n_{el}}{\underset{e=1}{\mathsf{A}}}$ denotes assembly function
- End of the loop over elements e
- Calculation of constants $a_0 = 1/\Delta t^2$, $a_2 = 2a_0$ and auxiliary vectors $\mathbf{M}_0 = a_0\,\mathbf{M}$, $\mathbf{M}_2 = a_2\,\mathbf{M}$, $\tilde{\mathbf{M}}^\alpha = 1/\mathbf{M}_0^\alpha$
- Calculation of displacement vector $\mathbf{u}_{t-\Delta t}$ or setting equal zero
- Initialisation $t = t_0$
- Loop over time steps t
  - Loop over elements e
    - Calculation of characteristic stiffness matrix $\mathbf{k}^e$ for each element
    - Multiplication $\hat{\mathbf{f}} = \mathbf{k}^e\mathbf{u}^I_t$, where I denotes degrees of freedom corresponding to element e
    - Assembly of global vector $\hat{\mathbf{F}} = \overset{n_{el}}{\underset{e=1}{\mathsf{A}}}\left(\hat{\mathbf{f}}^e\right)$
  - End of loop over elements e
  - Calculation of the effective force vector $\tilde{\mathbf{R}} = \mathbf{F}_t - \hat{\mathbf{F}} + \mathbf{M}_0^\alpha\mathbf{u}_t^\alpha - \mathbf{M}_2^\alpha\mathbf{u}_{t-\Delta t}^\alpha$, where indices α denotes multiplication without summation (vectorisation)
  - Solution at the time $t + \Delta t$ using vectorisation $\mathbf{u}^\alpha_{t+\Delta t} = \tilde{\mathbf{M}}^\alpha\,\tilde{\mathbf{R}}^\alpha$
  - $\mathbf{u}_{t-\Delta t} = \mathbf{u}_t$, $\mathbf{u}_t = \mathbf{u}_{t+\Delta t}$, $t = t + \Delta t$
- End of the loop over time steps t

3. Numerical calculations

Numerical calculations have been carried out for two different problems. The first is static analysis of a piezoelectric bimorph which was performed in order to validate modell. Numerical results of the bimorph consisted of 48 spectral elements have been compared with various

Journal of Physics: Conference Series **181** (2009) 012091 doi:10.1088/1742-6596/181/1/012091

approaches given in [11]. The error was less than 2% for maximum deflection ($L/h = 10$) and 0% for electric potential in comparison to full 3D finite element modell (ANSYS SOLID5 9600 elements).

The second problem is actuation and reception of elastic waves in aluminium plate of dimensions 315x315x1 mm^3 by circular and rectangular PZT-4 piezoelectric ceramics perfectly bonded to the surface of the plate. Piezoelectric material constants can be found in [11], whereas for aluminium it has been assumed: Young modulus $E = 71GPa$, Poisson ratio 0.33, mass density $\rho = 2700$ kg/m^3. It has been assumed that excitation signal is five tone burst sine modulated by Hanning window with the carrier frequency of 100 kHz. According to Lamb's wave dispersion equation [29], at this frequency-thickness value, there are only fundamental symmetric and antisymmetric modes. The more dispersive antisymmetric wave mode has a wavelength of 9.6 mm, and the symmetric wave mode has a wavelength of 54.3 mm. In order to assure accurate solution about ten nodes of spectral elements per minimum wavelength should be used [28].

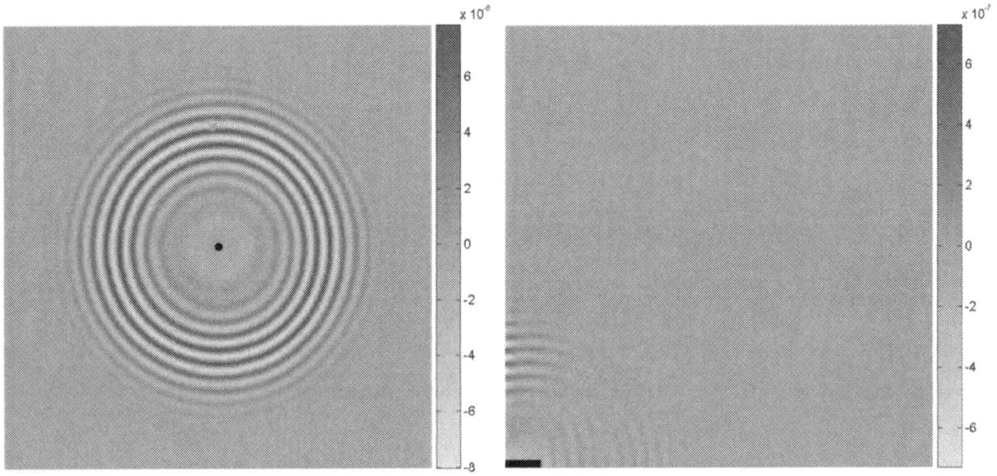

(a) circular piezotransducer (b) rectangular piezotransducer

Figure 2. Snapshot of guided elastic wave propagation excited by piezotransducer (transverse displacement after the time 70μs)

Exemplary results of numerical simulations for circular piezotransducer of diameter 5 mm and rectangular piezotransducer of the width 5 mm and length 25 mm are shown in Fig. 2a and Fig. 2b, respectively.

4. Conclusions

In spite of long calculation time, 3D spectral elements are excellent for modelling of wave propagation phenomena including electromechanical coupling. Both mechanical and electrical displacements converge very fast with increasing mesh density. Hence, accuracy the same or better as in case of finite elements can be obtained in much less time.

References

[1] Ray M, Bhattacharya R and Samanta B 1993 *AIAA Journal* **31** 1684–1691
[2] Tzou H and Tiersten H 1994 *Smart Mater. Struct.* **3** 255–265
[3] Bisegna P and Maceri F 1996 *ASME J. Appl. Mech.* **63** 628–638

[4] Ray M, Bhattacharya R and Samanta B 1998 *Comput. Struct.* **66** 737–743
[5] Tzou H and Tseng C 1990 *J. Sound Vib.* **138** 17–34
[6] Yao L and Lu L 2003 *Int. J. Numer. Methods Eng.* **58** 1499–1522
[7] Fernandes A and Pouget J 2002 *Eur. J. Mech. Solids* **21** 629–651
[8] Saravanos D, Heyliger P and Hopkins D 1997 *Int. J. Solids Struct.* **34** 359–378
[9] Reddy J 1997 *Mechanics of Laminated Composite Plates: Theory and Analysis* (CRC Press)
[10] Bisegna P and Caruso G 2001 *J. Solids Struct* **38** 8805–8830
[11] Wang S 2004 *Int. J. Solids Struct.* **41** 4075–4096
[12] Mindlin R 1972 *Int. J. Solids Struct.* **8** 895–906
[13] Yang J 1999 *Smart Mater. Struct.* **8** 73–82
[14] Kapuria S 2001 *Int. J. Solids Struct.* **38** 9179–9199
[15] Mullen R and Belytschko T 1982 *Int. J. Numer. Methods Eng.* **1** 11 29
[16] Patera A 1984 *J. Comput. Phys.* **54** 468–488
[17] Boyd J 1989 *Chebyshev and Fourier Spectral Methods* (Springer)
[18] Canuto C, Hussaini M, Quarteroni A and Zang T 1991 *Spectral methods in fluid dynamics* 3rd ed (Springer)
[19] Spall R 1995 *Int. J. Heat Mass Transfer* **15** 2743–2748
[20] Dauksher W and Emery A 1997 *Finite Elem. Anal. Des.* **26** 115–128
[21] Seriani G 1997 *J. Comput. Acoust.* **5** 53–69
[22] Komatitsch D and Vilotte J 1998 *Bull. Seismol. Soc. Amer.* **88** 368–392
[23] Seriani G 1999 *Phys. Chem. Earth.* **24** 241–249
[24] Komatitsch D, Barnes C and Tromp J 2000 *Geophisics* **65** 1251–1260
[25] Żak A, Krawczuk M, Ostachowicz W, Kudela P and Palacz M 2006 *Proceedings of the Third European Workshop Structural Health Monitoring 2006* pp 316–323
[26] Kudela P, Żak A, Krawczuk M and Ostachowicz W 2007 *J. Sound Vib.* **302** 728745
[27] Kudela P, Ostachowicz W and Żak A 2007 *Key Engineering Materials* vol 347 (Switzerland: Trans Tech Publications) pp 537–542
[28] Kim Y, Ha S and Chang F K 2008 *AIAA J.* **46** 591–600
[29] Rose J 1999 *Ultrasonic waves in solid media* (Cambridge: Cambridge University Press)

Journal of Physics: Conference Series **181** (2009) 012092 doi:10.1088/1742-6596/181/1/012092

Quadratic Partial Eigenvalue Assignment Problem with Time Delay for Active Vibration Control

Jesse M. Pratt[1], Kumar V. Singh[1] and Biswa N. Datta[2]

[1]Mechanical and Manufacturing Engineering Department,
Miami University, Oxford, OH 45056, USA
[2]Department of Mathematical Sciences,
Northern Illinois University, De Kalb, Illinois 60115, USA

singhkv@muohio.edu

Abstract. Partial pole assignment in active vibration control refers to reassigning a small set of unwanted eigenvalues of the quadratic eigenvalue problem (QEP) associated with the second order system of a vibrating structure, by using feedback control force, to suitably chosen location without altering the remaining large number of eigenvalues and eigenvectors. There are several challenges of solving this quadratic partial eigenvalue assignment problem (QPEVAP) in a computational setting which the traditional pole-placement problems for first-order control systems do not have to deal with. In order to these challenges, there has been some work in recent years to solve QPEVAP in a computationally viable way. However, these works do not take into account of the practical phenomenon of the time-delay effect in the system. In this paper, a new "direct and partial modal" approach of the quadratic partial eigenvalue assignment problem with time-delay is proposed. The approach works directly in the quadratic system without requiring transformation to a standard state-space system and requires the knowledge of only a small number of eigenvalues and eigenvectors that can be computed or measured in practice. Two illustrative examples are presented in the context of active vibration control with constant time-delay to illustrate the success of our proposed approach. Future work includes generalization of this approach to a more practical complex time-delay system and extension of this work to the multi-input problem.

1. Introduction and Background

Control of undesirable vibrations to avoid failure in structure can be achieved by passive and active means. Passive control devices based upon the classical Frahm's dynamic vibration absorber [1] are being utilized in automobile, aerospace and civil engineering applications [2-5]. These passive devices have several advantages due to simple design, reliability, low cost, no power requirements, good stability and larger broadband response. However, they are limited in their application to those systems only in which the dynamics of the structure do not change considerably or are limited within a small range of vibration amplitude and frequencies. Although passive devices are widely being used they are restricted in their application due to lack of adaptability with respect to the changes in dynamic environment, limited utilization for either storing or dissipating energy, its capability of producing only local controlling force, and its characteristics of altering the global dynamics of the system by introducing additional natural frequencies in the case of dynamic absorber. To overcome these limitations of the passive devices, active and/or semi-active control strategies are being developed and studied extensively for wide range of applications for vibration suppression [6-11]. Active Vibration Control (AVC) techniques are capable of supplying the desirable controlling forces in a dynamic environment (real time) by combining sensors (to sense the states), computers (to compute the controlling force in real time) and actuators (to supply the controlling force) in order to suppress the vibration corresponding to a wide range of dynamic operating conditions.

In vibration analysis the dynamics of the open loop multi-degree-of-freedom discrete system is governed by the following second order matrix differential equation,

$$\mathbf{M}\ddot{\mathbf{v}}(t)+\mathbf{C}\dot{\mathbf{v}}(t)+\mathbf{K}\mathbf{v}(t)=\mathbf{0}, \tag{1}$$

where $\mathbf{M},\mathbf{C}$ and $\mathbf{K}$ are the mass, damping and stiffness matrices respectively of dimension $(n\times n)$. By using separation of variables of the form

$$\mathbf{v}(t)=\mathbf{x}e^{\lambda t}, \tag{2}$$

where $\mathbf{x}$ a constant vector, in (1) leads to the problem of finding the eigenvalues and eigenvectors of the quadratic pencil

$$P(\lambda)=\lambda^2\mathbf{M}+\lambda\mathbf{C}+\mathbf{K}. \tag{3}$$

The characteristic roots of the quadratic polynomial (3) λ_k are known as eigenvalues, and the corresponding vectors $\mathbf{x}_k\neq 0$ are corresponding eigenvectors which satisfy

$$P(\lambda_k)\mathbf{x}_k=0, \quad \text{for } k=1,2,\cdots,n. \tag{4}$$

For AVC application the system modeled by (1) is modified by applying a controlling force $\mathbf{b}u(t)$,

$$\mathbf{M}\ddot{\mathbf{v}}(t)+\mathbf{C}\dot{\mathbf{v}}(t)+\mathbf{K}\mathbf{v}(t)=\mathbf{b}u(t), \tag{5}$$

with a constant vector $\mathbf{b}\in\mathbb{R}^{n\times 1}$, defining the location of control and associated control force $u(t)$ as,

$$u(t)=\mathbf{f}^T\dot{\mathbf{x}}(t)+\mathbf{g}^T\mathbf{x}(t), \tag{6}$$

where, $\mathbf{f}\in\mathbb{R}^{n\times 1}$ and $\mathbf{g}\in\mathbb{R}^{n\times 1}$ are velocity and displacement feedback control gain vectors respectively. By substituting (2) in (5), we obtain the quadratic matrix pencil associated with the controlled system

$$P_c(\lambda)=\lambda^2\mathbf{M}+\lambda\left(\mathbf{C}-\mathbf{b}\mathbf{f}^T\right)+\left(\mathbf{K}-\mathbf{b}\mathbf{g}^T\right). \tag{7}$$

Since vibration in a structure is caused by a few resonant or unstable eigenvalues, of practical importance here is to find the feedback vectors $\mathbf{f}$ and $\mathbf{g}$ such that these small numbers of unwanted eigenvalues are reassigned to suitably chosen numbers without altering the remaining large number of eigenvalues and eigenvectors of the pencil $P(\lambda)$. The last phenomeon is known as *no spill-over* and this problem is known as *Quadratic Partial Eigenvalue Assignemt Problem* (QPEVAP). An obvious approach to solve this problem is to transform it to a standard first-order control problem and then use any of the several excellent methods available in the literature [12]. However, there are several practical computational challenges in solving QEVAP which the traditional pole placement problem does not have to deal with. These include (i) the problem must be solved using the knowledge of a small number of eigenvalues and eigenvectors of the quadratic pencil $P(\lambda)$ which are all that can be computed with the state-of-the-art computational technique, such as Jacobi-Davidson methods [12], (ii) transformation to a first-order control system or reduction of the order of the model must be avoided, since these processes may give rise to instability in computation and some practically inherent properties of the physical system, such as the symmetry, sparsity, etc., may be completely destroyed, and above all (iii) the no spill-over must be guranteed mathematically, since in practice it is impossible to verify them by recomputing them after application of the feedback. Meeting these challenges, a new approach has been developed in [13] for the QPEVAP, but the system considered there does not take into account of the time-delay. Implementation of AVC strategy, in real engineering applications using sensors and actuators in the feedback loop, is subjected to short time delays that deteriorates the control performance and may lead to instability. Suppose the controlled system (5) is modified by considering a constant time delay τ in the feedback loop as,

$$\mathbf{M}\ddot{\mathbf{v}}(t)+\mathbf{C}\dot{\mathbf{v}}(t)+\mathbf{K}\mathbf{v}(t)=\mathbf{b}u(t-\tau), \tag{8}$$

with the choice of the controlling force,

$$u(t-\tau)=\mathbf{f}^T\dot{\mathbf{x}}(t-\tau)+\mathbf{g}^T\mathbf{x}(t-\tau), \tag{9}$$

where, the control gain vectors $\mathbf{f}$ and $\mathbf{g}$ depend upon the inherent time delay τ in the feedback loop. By substituting (2) in (8) we obtain the following characteristic quasi-polynomial,

$$P_c(\lambda)=\mathbf{M}\lambda^2+\mathbf{C}\lambda+\mathbf{K}-(\mathbf{b}\mathbf{f}^T\lambda+\mathbf{b}\mathbf{g}^T)e^{-\lambda\tau}, \tag{10}$$

where the dynamics of the controlled system are characterized by the eigenvalues of the closed loop pencil (10). Numerous efforts have been made in the past to study the existence of the time delayed AVC strategies and its effect on system's stability. The problem of delayed feedback control is summarized in the monographs [14-17] and the recent review papers [18, 19]. However, solution of time delayed AVC in a quadratic setting and effect of the time delay in partial pole assignment is almost non-existent in the literature. The partial pole assignment problem for the time delayed system (8) is to find the control gain vectors $\mathbf{f}$ and $\mathbf{g}$ such that

$$\det\left(\mathbf{M}\hat{\lambda}_k^2+\mathbf{C}\hat{\lambda}_k+\mathbf{K}-(\mathbf{b}\mathbf{f}^T\hat{\lambda}_k+\mathbf{b}\mathbf{g}^T)e^{-\hat{\lambda}_k\tau}\right)=0, \quad \text{for } k=1,2,\cdots,2n, \tag{11}$$

has the desired poles $\{\hat{\lambda}\}_1^{2n}=(\mu_1,\cdots,\mu_m,\lambda_{m+1},\cdots,\lambda_{2n})$. That it contains m numbers of newly assigned poles $(\mu_1,\cdots,\mu_m)$ without affecting the remaining poles $(\lambda_{m+1},\cdots,\lambda_{2n})$ of the open loop system (1).

In section 2, by using an orthogonality relation from [13], we have derived an implicit solution to the QPEVAP with time delay. This method ensures ***no spillover*** will occur. We emphasize that our results are applicable in the general case and do not assume that $\mathbf{M},\mathbf{C},\mathbf{K}$ are simultaneously diagonalizable. The method will be most advantageous in practical applications where n is large and when it is required to assign only the eigenvalues $\lambda_1,\ldots,\lambda_m$, where m is much smaller than $2n$. This situation is typical in the vibration control and stabilization of flexible, large space structures [20, 21]. The method uses only the eigenvalues $\lambda_1,\ldots,\lambda_m$ and their associated eigenvectors. These can be computed by Jacobi Davidson method (refer to the 2nd ed. of [12] to be published by SIAM, 2009), or by modal analysis measurements. Two illustrative examples are given in Section 3 validating the results for pole assignment without delay and analyzing the effect of time delay on the control gain vectors. The stability of the controlled system and its dependency on the time delay is shown in Section 4. Brief concluding remarks are presented in the final section.

2. Formulation for Partial Pole Assignment with Delayed Feedback Control Gains

Consider the following orthogonality relation proved in [13]. Suppose $\mathbf{M},\mathbf{C},\mathbf{K}\in\mathbb{R}^{n\times n}$ are symmetric and $\mathbf{M}$ is positive definite. Let the $n\times 2n$ system of equations

$$\mathbf{M}\mathbf{X}\mathbf{\Lambda}^2+\mathbf{C}\mathbf{X}\mathbf{\Lambda}+\mathbf{K}\mathbf{X}=\mathbf{0}, \tag{12}$$

where $\mathbf{X} \in \mathbb{C}^{n\times 2n}$ and $\mathbf{\Lambda} = diag\{\lambda_1,...,\lambda_{2n}\} \in \mathbb{C}^{2n\times 2n}$ with distinct λ_i, be an eigen-decomposition of the quadratic open loop pencil (3). Then

$$\mathbf{\Lambda X}^T\mathbf{MX\Lambda} - \mathbf{X}^T\mathbf{KX} = \mathbf{D}_1 \tag{13}$$

where $\mathbf{D}_1$ is a diagonal matrix. Using this orthogonality relation, we now present a solution to the following partial eigenvalue assignment problem for the pencil(3).

Given m complex numbers $\mu_1,...,\mu_m$, $m \le n$, and a vector $\mathbf{b} \in \mathbb{R}^{n\times 1}$, we are required to find $\mathbf{f},\mathbf{g} \in \mathbb{R}^{n\times 1}$ which are such that the closed loop pencil (10) has the spectrum $\{\mu_1,...,\mu_m,\lambda_{m+1},...,\lambda_{2n}\}$. We will now use the following partitions of eigenvalue and eigenvector matrices,

$$\mathbf{X} = \begin{pmatrix} \underset{m}{\mathbf{X}_1} & \underset{2n-m}{\mathbf{X}_2} \end{pmatrix}, \text{ and, } \mathbf{\Lambda} = \begin{pmatrix} \underset{m}{\mathbf{\Lambda}_1} & \\ & \underset{2n-m}{\mathbf{\Lambda}_2} \end{pmatrix}, \tag{14}$$

where $\mathbf{X}_1$ and $\mathbf{\Lambda}_1$ are pair wise self-conjugate. The eigenvalue matrix $\mathbf{\Lambda}_1$ is consist of m eigenvalues $\mathbf{X}_1$ is the eigenvectors corresponding to these eigenvalues.

THEOREM 2.1: *Let*

$$\mathbf{f} = \mathbf{MX}_1\mathbf{\Lambda}_1\boldsymbol{\beta}, \qquad \mathbf{g} = -\mathbf{KX}_1\boldsymbol{\beta}, \qquad \boldsymbol{\beta} \in \mathbb{C}^{m\times 1}, \tag{15}$$

then for any choice of $\boldsymbol{\beta}$

$$\mathbf{MX}_2\mathbf{\Lambda}_2^2 + \mathbf{CX}_2\mathbf{\Lambda}_2 + \mathbf{KX}_2 - (\mathbf{bf}^T\mathbf{X}_2\mathbf{\Lambda}_2 + \mathbf{bg}^T\mathbf{X}_2)e^{-\mathbf{\Lambda}_2\tau} = \mathbf{0}. \qquad \textbf{(no spillover)} \tag{16}$$

Proof: Note that the left hand side of (16) equals

$$-(\mathbf{bf}^T\mathbf{X}_2\mathbf{\Lambda}_2 + \mathbf{bg}^T\mathbf{X}_2)e^{-\mathbf{\Lambda}_2\tau}, \tag{17}$$

since $\mathbf{\Lambda}_2$, $\mathbf{X}_2$ are eigenmatrix pairs of the open loop pencil. Furthermore, by substituting (15) in (17) and rearranging we have,

$$-\mathbf{b}(\boldsymbol{\beta}^T e^{-\mathbf{\Lambda}_2\tau}\mathbf{\Lambda}_1\mathbf{X}_1^T\mathbf{MX}_2\mathbf{\Lambda}_2 - \boldsymbol{\beta}^T e^{\mathbf{\Lambda}_2\tau}\mathbf{X}_1^T\mathbf{KX}_2) = -\mathbf{b}\boldsymbol{\beta}^T e^{\mathbf{\Lambda}_2\tau}(\mathbf{\Lambda}_1\mathbf{X}_1^T\mathbf{MX}_2\mathbf{\Lambda}_2 - \mathbf{X}_1^T\mathbf{KX}_2) = 0, \tag{18}$$

since from [22],

$$\mathbf{\Lambda}_1\mathbf{X}_1^T\mathbf{MX}_2\mathbf{\Lambda}_2 - \mathbf{X}_1^T\mathbf{KX}_2 = \mathbf{0}, \tag{19}$$

and hence from (12) the relationship (16) can be proved.

In order to use this theorem to solve the partial pole assignment problem, we need to choose $\boldsymbol{\beta}$ such that it will move $\{\lambda_j\}_{j=1}^m$ of the pencil (3) to $\{\mu_j\}_{j=1}^m$ in (10). If there is such a vector $\boldsymbol{\beta}$ then there exists an eigenvector matrix $\mathbf{Y} \in \mathbb{C}^{n\times m}$,

$$\mathbf{Y} = (\mathbf{y}_1,...,\mathbf{y}_m), \mathbf{y}_j \ne 0, j = 1,...,m, \tag{20}$$

such that

$$\mathbf{MYD}^2 + \mathbf{CYD} + \mathbf{KY} - (\mathbf{bf}^T\mathbf{YD} + \mathbf{bg}^T\mathbf{Y})e^{-\mathbf{D}\tau} = \mathbf{0}. \tag{21}$$

where, $\mathbf{D} = diag\{\mu_1,...,\mu_m\}$, and $e^{-\mathbf{D}\tau} = diag\{e^{-\mu_1\tau},...,e^{-\mu_m\tau}\}$. Now by substituting for $\mathbf{f}$ and $\mathbf{g}$ from (15) in (21) and after rearranging we obtain,

$$\mathbf{MYD}^2 + \mathbf{CYD} + \mathbf{KY} = \mathbf{b}(\mathbf{f}^T\mathbf{YD} + \mathbf{g}^T\mathbf{Y})e^{-\mathbf{D}\tau} = \mathbf{b}\boldsymbol{\beta}^T(\boldsymbol{\Lambda}_1\mathbf{X}_1^T\mathbf{MYD} - \mathbf{X}_1^T\mathbf{KY})e^{-\mathbf{D}\tau} = \mathbf{b}\boldsymbol{\beta}^T\mathbf{Z}^T = \mathbf{bc}^T, \quad (22)$$

where,

$$\mathbf{Z} = e^{-\mathbf{D}\tau}(\mathbf{DY}^T\mathbf{MX}_1\boldsymbol{\Lambda}_1 - \mathbf{Y}^T\mathbf{KX}_1), \quad (23)$$

and,

$$\mathbf{c} = \mathbf{Z}\boldsymbol{\beta}, \quad (24)$$

is a vector that will depend on the scaling chosen for the eigenvectors in $\mathbf{Y}$. In order to obtain $\mathbf{Y}$, we can solve for each of the eigenvectors $\mathbf{y}_i$ using the equations

$$(\mu_j^2\mathbf{M} + \mu_j\mathbf{C} + \mathbf{K})\mathbf{y}_j = \mathbf{b}, j = 1,...,m. \quad (25)$$

This corresponds to choosing the vector $\mathbf{c} = (1,...,1)^T$, so, having computed the eigenvectors, we could solve the m-square system

$$\mathbf{Z}\boldsymbol{\beta} = (1,...,1)^T. \quad (26)$$

for $\boldsymbol{\beta}$, and hence determine the control gain vectors $\mathbf{f}$ and $\mathbf{g}$.

3. Examples

In this section we present two illustrative examples associated with single degree and two degree of freedom non-conservative system for which control gains assuring the partial pole assignment with time delay is considered.

3.1 Example 1 (Pole assignment in Single Degree of Freedom System)

Let, $\mathbf{M} = 1$, $\mathbf{C} = .01$, $\mathbf{K} = 5$, $\mathbf{b} = 1$, $\tau = 0.1$ and with these values the open loop system eigenvalues and eigenvectors satisfying (4) are,

$$\lambda_{1,2} = -0.005 \pm 2.2361i \text{ and } \mathbf{x}_1 = -0.0009{+}0.4082i \quad \mathbf{x}_2 = -0.0009 - 0.4082i. \quad (27)$$

We want to calculate the control gains such that the closed loop system will replace the poles (27) to their new values: $\mu_1 = -1$, $\mu_2 = -3$. First we find $\mathbf{Y}$ from (25) by using the assigned eigenvalues $\mu_{1,2}$ as: $\mathbf{Y} = \begin{pmatrix}0.1669, & 0.0005\end{pmatrix}$. By denoting,

$$D = \begin{pmatrix} -1 & 0 \\ 0 & -3 \end{pmatrix}, \quad \text{and} \quad e^{-D\tau} = \begin{pmatrix} e^{0.1} & 0 \\ 0 & e^{0.3} \end{pmatrix}, \quad (28)$$

we compute from (23) and (24),

$$\mathbf{Z} = \begin{pmatrix} -0.1676 + 0.3766i & -0.1676 - 0.3766i \\ -2.1309 + 0.1014i & -2.1309 - 0.1014 \end{pmatrix}, \text{ and } \boldsymbol{\beta} = \begin{pmatrix} -1.3499 - 0.7269i \\ -1.3499{+}0.7269i \end{pmatrix} \quad (29)$$

with $\mathbf{c} = \begin{pmatrix}1 & 1\end{pmatrix}^T$. The values of control gain now can be estimated from (15) as,

$$\mathbf{f} = -2.4646 \quad \text{and} \quad \mathbf{g} = 2.9553, \quad (30)$$

Journal of Physics: Conference Series **181** (2009) 012092 doi:10.1088/1742-6596/181/1/012092

which is the same value as obtained in [23] using the receptance method. The control gains obtained in (30) satisfy (11), ensuring the determinant to be smaller than 10^{-13} with $\{\hat{\lambda}\}_1^2 = (\mu_1, \mu_2)$. Now by varying the constant time delay from $\tau = 0$ (case of no delay in the feedback loop) to $\tau = 2$ seconds, the steps (27) to (29) are repeated and control gains for each τ are computed and plotted in the Figure 1. It is clear from Figure 1 that the requirements on the control gain vectors changes substantially as the time delay in the feedback loop changes.

3.2 Example 2 (Partial Pole assignment in Two Degree of Freedom System)

Let, $$\mathbf{M} = \begin{pmatrix} 1 & 0 \\ 0 & 1 \end{pmatrix}, \mathbf{C} = \begin{pmatrix} 0.1 & -0.1 \\ -0.1 & 0.1 \end{pmatrix}, \mathbf{K} = \begin{pmatrix} 2 & -1 \\ -1 & 1 \end{pmatrix}, \mathbf{b} = \begin{pmatrix} 0 \\ 1 \end{pmatrix}, \quad \tau = 0.01$$

For the above system, the open loop eigenvalues are:

$$\lambda_{1,2} = -0.0053 \pm 0.6183i, \; \lambda_{3,4} = -0.0947 \pm 1.6145i. \tag{31}$$

We wish to replace the eigenvalues $\lambda_{1,2}$ with $\mu_{1,2} = -1 \pm i$ without altering $\lambda_{3,4} = -0.0947 \pm 1.6145i$. From the definition in (14) we have,

$$\mathbf{X}_1 = \begin{pmatrix} -0.4474+0.0123i & -0.4474-0.0123i \\ -0.7233 & -0.7233 \end{pmatrix}, \; \mathbf{\Lambda}_1 = \begin{pmatrix} -0.0053-0.6183i & 0 \\ 0 & -0.0053+0.6183i \end{pmatrix}. \tag{32}$$

By following the procedure in the previous example we compute

$$\mathbf{Z} = \begin{pmatrix} 0.2002-0.1283i & -0.1513+0.3960i \\ -0.1513-0.3960i & 0.2002+0.1283i \end{pmatrix}, \text{ and } \quad \boldsymbol{\beta} = \begin{pmatrix} -2.8539+2.1735i \\ -2.8539-2.1735i \end{pmatrix}, \tag{33}$$

with $\mathbf{c} = \begin{pmatrix} 1 & 1 \end{pmatrix}^T$. Now the values of control gain vectors now can be estimated from (15) as,

$$\mathbf{f} = \begin{pmatrix} -1.2594 & -1.9659 \end{pmatrix}^T, \quad \mathbf{g} = \begin{pmatrix} -0.8718 & -1.6281 \end{pmatrix}^T \tag{34}$$

Once the control gain vectors are obtained, by computing

$$D_k = \det\left(\hat{\lambda}_k^2 \mathbf{M} + \hat{\lambda}_k \mathbf{C} + \mathbf{K} - \mathbf{b}\left(\hat{\lambda}_k \mathbf{f}^T + \mathbf{g}^T\right) e^{-\hat{\lambda}_k \tau}\right), \text{ for } k = 1,2,3,4 \tag{35}$$

where $\hat{\lambda}_{1,2} = -1 \pm i$, $\hat{\lambda}_{3,4} = -0.0947 \pm 1.6145i$, it is verified that $|D_k| < 10^{-13}$ for $k = 1,2,3,4$, validating the desired partial pole assignment.

4. A Note on Stability

It is important to note that the characteristic quadratic pencil associated with the controlled system without time delay in the feedback loop (7) is a polynomial; hence the n dimensional system will have $2n$ roots. The location of these roots in the complex plane defines the stability of the system. Hence if the assigned poles have negative real part then the controlled system must be stable. However the characteristic equation of a controlled system with delay (10) is a *quasi-polynomial* and has an infinite number of roots in the complex plane satisfying (35). Hence assigning the poles with negative real part may not necessarily make the system stable because there may be one or more unstable roots satisfying(35). This point is highlighted in the recent publication by Ram et. al. [23] and he has shown that for pole assignment in delayed system a posteriori analysis has to be done to identify and assign the primary eigenvalues ensuring that such assignment will avoid instability.

For example, reconsider example 2 in the previous section and assign partial poles $\lambda_{1,2}$ to new stable real and negative poles, which are further from imaginary axis, $\mu_1 = -1$ and $\mu_2 = -50$. The pole assignment is achieved and the associated control gains satisfy(35). However when plotting determinant (35) for a range of real valued $\hat{\lambda}$, as shown in figure (2), it is clear that there exist at least one positive real root which can bring the system to instability even if the assigned poles are stable. Hence by following the posteriori analysis based on second order Taylor series expansion in [23] we obtain the following eigensystem,

$$\left(\begin{bmatrix} \mathbf{0} & \mathbf{I} & \mathbf{0} \\ \mathbf{0} & \mathbf{0} & \mathbf{I} \\ \left(\mathbf{K} - \mathbf{b}\mathbf{g}^T\right) & \left(\mathbf{C} - \mathbf{b}\mathbf{f}^T + \tau\mathbf{b}\mathbf{g}^T\right) & \left(\mathbf{M} + \tau\mathbf{b}\mathbf{f}^T - \tau^2\mathbf{b}\mathbf{f}^T/2\right) \end{bmatrix} - \lambda \begin{bmatrix} \mathbf{I} & \mathbf{0} & \mathbf{0} \\ \mathbf{0} & \mathbf{I} & \mathbf{0} \\ \mathbf{0} & \mathbf{0} & \left(\tau^2\mathbf{b}\mathbf{f}^T/2\right) \end{bmatrix} \right) \begin{pmatrix} \mathbf{z} \\ \lambda\mathbf{z} \\ \lambda^2\mathbf{z} \end{pmatrix} = \begin{pmatrix} \mathbf{0} \\ \mathbf{0} \\ \mathbf{0} \end{pmatrix}. \quad (36)$$

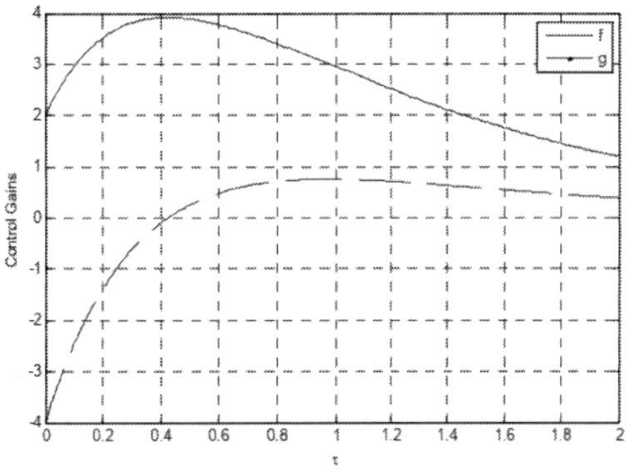

Figure 1: The effect of time delay on the control gains towards the assignment of poles

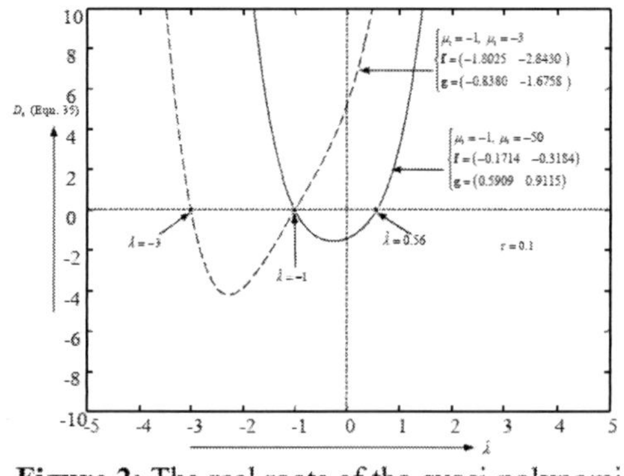

Figure 2: The real roots of the quasi-polynomial $\hat{\lambda}^2\mathbf{M} + \hat{\lambda}\mathbf{C} + \mathbf{K} - \mathbf{b}\left(\hat{\lambda}\mathbf{f}^T + \mathbf{g}^T\right)e^{-\hat{\lambda}\tau}$

The primary eigenvalues of the system (36) are $\hat{\lambda}_1 = 0.5614$, and $\hat{\lambda}_2 = -0.9999$, which is consistent with the real roots shown in figure 1. Consider another case for the same example 2 and now we assign partial poles $\lambda_{1,2}$ to $\mu_1 = -1$ and $\mu_2 = -3$. The characteristic real roots for this case are negative and therefore the system is stable. Moreover the primary eigenvalues for such assignment from (36) are also found to be stable: $\hat{\lambda}_1 = -2.9731$, and $\hat{\lambda}_2 = -1.0002$ .Hence if the primary poles are assigned during partial pole assignment the system will remain stable.

5. Conclusions

This paper proposes a new approach for partial eigenvalue assignment in active vibration control for a second-order model with time-delay. This new approach has several desirable practical features, most important of which is that the approach is implementable with the knowledge of only a few eigenvalues and eigenvectors of the associate quadratic eigenvalue problem which are all that are computable using the state-of-the-art computational technique or can be measured in vibration laboratory. Furthermore, the no spill-over property of the closed-loop system is guaranteed with a mathematically proved result. Thus the approach is applicable to control dangerous vibration and other forms of instability in large real-life structures. Such practically implementable methods for vibration control of large real-life structures are almost nonexistent. The research on extending the approach to partial eigenstructure assignment which involves reassigning not only a few unwanted eigenvalues but the associated eigenvectors as well, and to the multi-input time-delay feedback design problem, are currently underway. It is expected that by the time of the conference, validation studies of this technique with the lab scale experiment will be available.

6. Acknowledgements
This research study is supported in part by the Committee of Faculty Research (CFR) grant at Miami University, 2008-09. B. N. Datta's work is partially supported by NSF grant DMS-0505784 and vibration and acoustic grant from the College of Engineering at Northern Illinois University.

References

[1] Frahm, H., "Devices for Damping Vibration of Bodies", US Patent No. 989958, 1909.
[2] Harris, C. M., Shock and Vibration Handbook, 4th Ed., McGraw-Hill, New York, 1997.
[3] Nakra, B. C., "Vibration Control in Machines and Structures Using Viscoelastic Damping," Journal of Sound and Vibration, 211(3), 449–465, 1998.
[4] Soong, T., Dargush, G., Passive Energy Dissipation in Structural Engineering, Wiley, NY, 1997.
[5] Pearson, J. T., Goodall, R. M., and Lyndon, I., "Active control of helicopter vibration", Computing & Control Engineering Journal, 277-284, 1994.
[6] Fuller, C., Elliot, S., Nelson, P., "Active Control of Vibration", Academic Press, San Diego, 1996.
[7] Housner, G. W., Bergman, L.A., Caughey, T. K., Chassiakos, A. G., Claus, R. O., Masri, S.F., Skelton, R. E., Soong, T. T., Spencer, B.F. and Yao, J. T. P., "Structural control: past, present and future", ASCE Journal of Engineering Mechanics, 123(9), 897-971, 1997.
[8] Alkhatib, R., and Golnaraghi, M. F., "Active Structural Vibration Control: A Review", The Shock and Vibration Digest, 35(5), 367–383, 2003.
[9] Soong, T., Active Structural Control: Theory and Practice, Longman Scientific & Technical, 1990.
[10] Inman D., Vibration with control, measurements and stability, Prentice-Hall, New Jersey, 1989.
[11] Achkire, Y. and Preumont, A., "Active tendon control of cable stayed bridges", Earthquake Engineering and Structural Dynamics, 25(6), 585-597, 1996.
[12] Datta, B. N., Numerical Methods for Linear Control Systems Design and Analysis. Elsevier Academic Press, Boston, 2003.
[13] Datta, B.N., Elhay, S. and Ram, Y.M., "Orthogonality and partial pole assignment for the symmetric definite quadratic pencil", Linear Algebra and its Applications 257:29-48 1997.
[14] Stepan G., "Retarded dynamical systems: stability and characteristic functions", Essex Longman Scientific and Technical, 1989.
[15] Diekmann, O., Gils, S.A.V., Lunel, S.M.V. and Walther, H.O., Delay equations, functional, complex and nonlinear analysis, Springer-Verlag, New York, 1995.
[16] Gu K., Kharitonov V.L. and Chen J., Stability of time-delay systems, Birkhauser, Boston, 2003.
[17] Hu H.Y. and Wang, Z.H., Dynamics of controlled mechanical systems with delayed feedback, Springer, New York, 2002.
[18] Richards, J. P., "Time-delay systems: an overview of some recent advances and open problems", Automatica, in press.
[19] Watanabe, K., Nobuyama, E., and Kojima, K., "Recent advances in control of time-delay systems a tutorial review", *35th IEEE CDC'96* (Conference on decision and control), 2083–2089, 1996.
[20] Balas, M.J., "Trends in large space structure control theory: Fondest dreams, wildest hopes", IEEE Trans. Automat. Control AC-22:522-535 1982.
[21] Datta, B.N., "Linear and numerical linear algebra in control theory: Some research problems", Linear Algebra Appl. 197,198: 755-790 1994.
[22] Datta, B. N., and Sarkissian, D. R.. Theory and computations of some inverse eigenvalue problems for the quadratic pencil. StructuredMatrices in Operator Theory, Control, and Signal and Image Processing, vol. 280 of Contemporary Mathematics, 221–240. AMS, Providence, RI, 2001
[23] Ram, Y.M., Singh, A. and Mottershead, J., "State feedback control with time delay", Mechanical Systems and Signal, Processing, 2008, doi:10.1016/j.ymssp.2008.04.012.

Hybrid fuzzy and sliding-mode control for motorised space tether spin-up when coupled with axial oscillation

Yi Chen , Matthew P. Cartmell

Department of Mechanical Engineering, University of Glasgow, Glasgow, G12 8QQ, UK

E-mail: yichen@mech.gla.ac.uk, matthewc@mech.gla.ac.uk

Abstract. A hybrid fuzzy sliding mode controller is applied to the control of motorised tether spin-up coupled with an axial oscillation phenomenon. A six degree of freedom dynamic model of a motorised momentum exchange tether is used as a basis for interplanetary payload exchange. The tether comprises a symmetrical double payload configuration, with an outrigger counter inertia and massive central facility. It is shown that including axial elasticity permits an enhanced level of performance prediction accuracy and a useful departure from the usual rigid body representations, particularly for accurate payload positioning at strategic points. A special simulation program has been devised in *MATLAB* and *MATHEMATICA* for a given initial condition data case.

1. Introduction

Space tethers can be used for orbit raising, lowering, and maintenance, and in principle can also be used for interplanetary propulsion of appropriate payloads. There are two distinct types of space tether; the mechanical 'momentum exchange' system in which orbital energy is used to manipulate payload mass advantageously, and the electrodynamic 'ED' tether which uses the geomagnetic field in conjunction with a conductive tether core and an on-board plasma generator. The ED tether can perform either as a gigantic motor or as a dynamo in order either to raise or lower payload altitude. In this case the momentum exchange variant is considered, albeit with additional energy provided by an exciting motor. The application of momentum exchange tethers to space vehicle propulsion from Low Earth Orbit (LEO) has been explored in considerable depth during the last a few decades, with many ingenious designs exploiting the principle of momentum transfer. The motorised momentum exchange tether (MMET) was first proposed by Cartmell [1]. The modelling and conceptual design were developed further, in particular MMET modelling as a rigid body, Ziegler and Cartmell [2]; MMET modelling with axial elasticity, Chen and Cartmell [3]. The conceptual schematic of the MMET system is shown in Figure 1 (A-braided propulsion tether tube; B-braided outrigger tether tube; C-launcher mass, rotor; D-launcher mass, stator; E/F-outrigger mass; G/H-payload mass), in which it is excited by a motor and uses angular generalised co-ordinates to represent spin and tilt, together with an angular co-ordinate for circular orbital motion. Another angular co-ordinate defines backspin of the propulsion motor's stator components. The payload masses (H and G in Figure 1) are fitted to each end of the tether sub-span (A in Figure 1), and the system orbits a source of gravity in space, in this case the Earth. The use of a tether means that all parts of the system have the same angular velocity as the overall centre of mass (COM, C and D in Figure 1). As implied in Figures 1 and 2, the symmetrical double-ended motorized spinning tether can be applied as an orbital transfer system in order to exploit momentum exchange for propelling and transferring payloads.

Variable structure control (VSC) with sliding mode control was introduced in the early 1950s by Emelyanov and was published in the 1960s [4], and further developments were proposed by several researchers [5][6]. Sliding mode control (SMC) has been recognized as a robust and efficient control method for complex high order nonlinear dynamical systems. The major advantage of sliding mode control is its low sensitivity to a system's parameter changes under various uncertainty conditions, and the fact that it can be used to decouple system motion into independent partial components of lower dimension, which reduces the complexity of the system control and feedback design. A major drawback of traditional SMC is chattering, which is generally disadvantageous within control systems.

It has been well recognized that fuzzy logic controllers (FLC) can be effective and robust for various applications, and conventionally the FLC rule-base is based on practical human experience, however the linguistic expression of the FLC rule-base can sometimes make it difficult to guarantee the stability and robustness of the control system. In recent years, a lot of literature has been generated in the area of fuzzy sliding mode control (FSMC) and has covered the chattering phenomenon. When FLC is involved in designing a FSMC based controller, this can be harnessed to help to avoid the chattering problem. The feature of the smooth control action of fuzzy logic can be helpful for overcoming the disadvantages of chattering. This is why it can be useful to merge FLC with SMC to create the FSMC [7][8][9][10][11][12]. A hybrid fuzzy sliding mode control method is applied here to the tether for sub-span length changing in order to provide spin-up control of the MMET system.

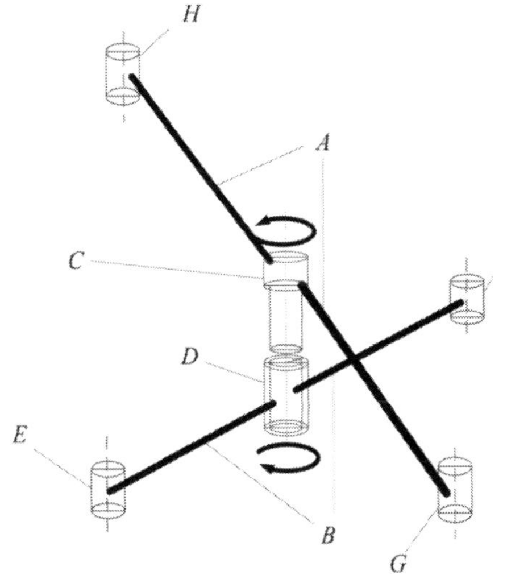

Figure 1. Conceptual Schematic of the Motorised Momentum Exchange Tether

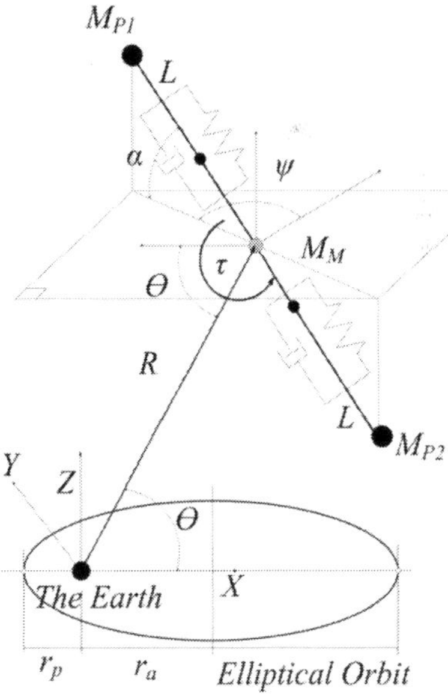

Figure 2. Modelling of the axially elastic MMET

2. Six degree of freedom MMET Model

A six degree of freedom non-planar tether model, which includes an axial elasticity coordinate and a solid rolling coordinate, is proposed for the MMET system in Figure 2. This discretised MMET system comprises a symmetrical and cylindrical double payload configuration, a cylindrical motor facility, and two axially flexible tubular tether sub-spans. In the discretised non-planar tether model, environmental effects such as solar radiation, residual aerodynamic drag in low Earth orbit and electrodynamic forces, that may also influence the modelling, are reasonally assumed to be negligible in this context. The motor consists of a central rotor, which is attached to the propulsion tethers, and a stator which locates the rotor by means of a suitable bearing. The power supplies, control systems, and communication equipment are assumed to be fitted within and surrounding the stator assembly in a practical installation. The stator also provides the necessary reaction that is required for the rotor to spin-up in a friction free environment. The motor torque acts about the motor drive axis, and it is assumed that the motor drive axis will stay normal to the spin plane of the propulsive tethers and payloads.

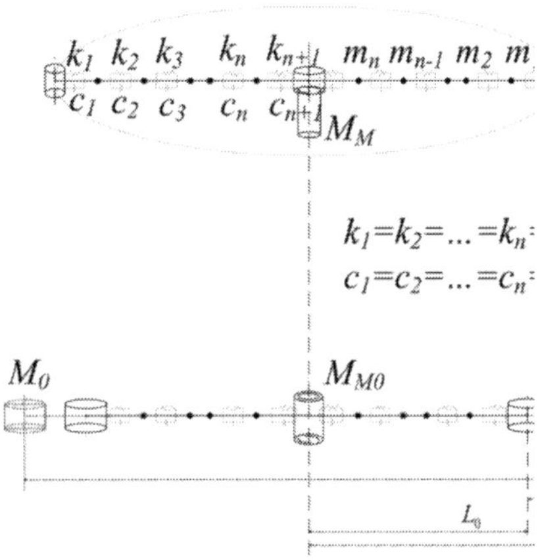

Figure 3. Modelling of the axial elastic for motorised momentum exchange tether

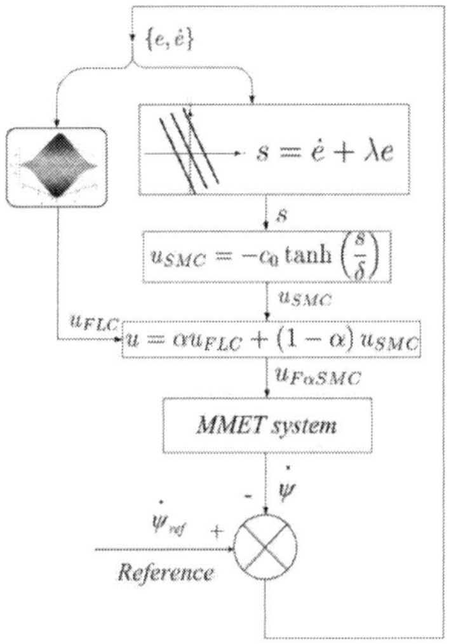

Figure 4. Hybrid fuzzy and sliding-mode control flow diagram

The elasticity of the tether system is considered to be distributed symmetrically along each tether sub-span. The tether and the motor are connected by a series of discrete spring-damper groups as shown in Figure 3. When the tether moves out of the orbital plane, the motor drive axis remains orthogonal to the spin plane, meanwhile the motor torque will act about the principal axis through its centre of mass. The length of the discretised MMET from payload to COM is given by $L(t) = L_0 + L_x(t)$, where the time variant length $L(t)$ of the tether is the sum of the static length, L_0, and the elastic length, $L_x(t)$, of the discretised tether. There are six generalised coordinates in this model, in the form of four rotational coordinates (ψ, θ, α, γ) and two translational coordinates (L_x, R). Coordinate ψ defines the spin-up performance of the MMET system and is the in-plane pitch angle. This denotes the angle from the x_0 axis to the projection of the tether onto the orbit plane. θ is the circular orbit angular position, effectively the true anomaly. α is the out-of-plane angle, from the projection of the tether onto the orbit plane to the tether, and is always within a plane normal to the orbit plane. γ is the rolling angle, which lies between the torque-plane and the tether-spin-plane. R is the distance from the Earth to the MMET COM, and L_x is the axial elastic length. Lagrange's equations are used to obtain the dynamical equations of motion based on the six generalised coordinates [3]. Lateral vibrations are included in this study because terrestrial and in-space deployment tests have shown that centripetal stiffening and gravity gradient effects in oscillating (librating) and spinning tethers, and just gravity gradient stabilisation in hanging tethers, provide very useful lateral rigidisation of otherwise flexible tethers, irrespective of whether they are single line designs or highly redundant multi-line structures. On that basis the predominant elastic effect is in axial stretch, and to some extent in torsion along the length. In this paper we solely consider the axially elastic effect and neglect any relatively small lateral oscillations, or torsional effects. Axial stretch is particularly important as even an extension of 0.1% will generate an axial displacement of the payload of 50 m when using a tether with a typically pragmatic sub-span length of 50 km. This would obviously have a very serious consequence for payload collection and delivery accuracy.

3. Hybrid Control - FαSMC

The chattering phenomenon is an acknowledged drawback of sliding mode control, and is usually caused by unmodelled system dynamics. In traditional SMC design, a sign function is conventionally applied but this can also lead to chattering in practice. Therefore a special boundary layer around the sliding surface is introduced to solve the chattering problem [5][6]. To make the necessary enhancement to the FSMC method, a hybrid control law is introduced here to combine fuzzy logic control with a soft continuous switching hyperbolic tangent control law, which is named FαSMC and given in equation 1. The flow diagram of the FαSMC approach is given in Figure 4. The hybrid control effects of FLC and SMC are combined in equation 1, where α is a proportionality factor, included to balance the weight of the fuzzy logic control to that of the hyperbolic tangent control. Clearly, $\alpha = 0$ represents hyperbolic tangent control, and $\alpha = 1$ represents fuzzy logic control, $\alpha \in [0,1]$.

$$u_{F\alpha SMC} = \alpha u_{FLC} + (1-\alpha)\, u_{SMC} \tag{1}$$

3.1. Fuzzy Logic Controller Design

Fuzzy control provides a convenient method for constructing nonlinear controllers via the use of heuristic information. An operator can act as a human-in-the-loop controller so that an appropriate fuzzy control implementation can provide a human-experience based representation, and an implementation of the ideas that a human could have about how to achieve high-performance control [13][14]. The structure of the FLC for the MMET system is shown in Figure 5, in which the FLC flow diagram is characterized by a set of linguistic description rules based on conceptual expertise which arises from typical human situational experience. The 'If-Then' rule base is then applied to describe the experts' knowledge, the 2-in-1-out FLC rule-base for the spin-up of the 6DOF MMET system is given in Figure 6. Figure 6 is the FLC rule-base cloud for the 2-in-1-out FLC, which came from previous experience gained from observing for the tether length changes during angular velocity control. Briefly, the main linguistic control rules are: (1) when the angular velocity decreases, the length tether increases; Conversely, when the angular velocity increases,the tether length decreases. (2) When the angular acceleration increases, the tether length increase can reduce the error between the velocity and the reference velocity; otherwise, when the angular acceleration decreases, the tether length decreases as well. The full rule-base is given in Figure 6 as a rule-base 3D plot, which defines the relationship between the two inputs of Error (E) and Change in Error (EC), with one output of the Fuzzified Length (FL).

Fuzzification is the process of decomposing the system inputs into fuzzy sets, that is, it is used to map variables from practical space to fuzzy space. The process of fuzzification allows the system inputs and outputs to be expressed in linguistic terms so that rules can be applied in

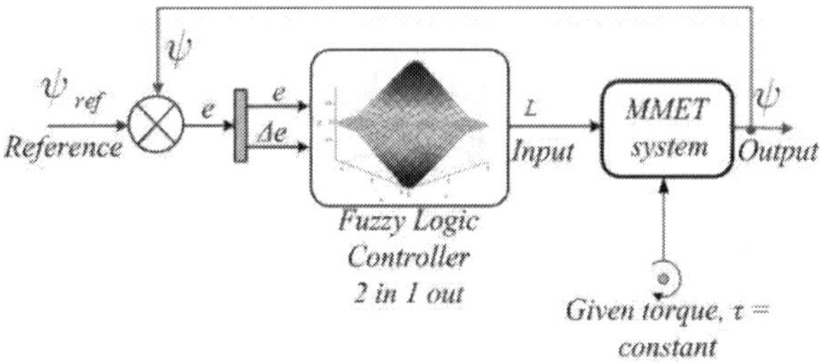

Figure 5. FLC flow diagram

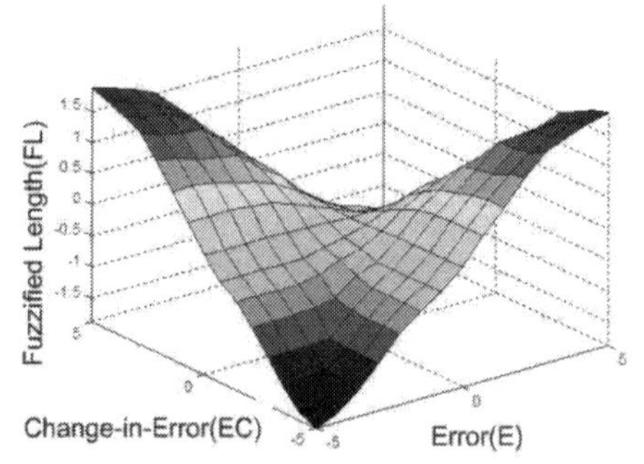

Figure 6. 2-in-1-out FLC rule-base 3D plot

a simple manner to express the complicated system. In the FLC for the MMET system, there are nine elements in the fuzzy sets for two inputs of E and EC, and one output of FL. These are: $< NB, NM, NS, NZS, ZE, PZS, PS, PM, PB >$. The Fuzzy Inference System (FIS) uses Mamdani-type inference [15] for the 2-in-1-out FLC. Defuzzification is the opposite process of fuzzification and is used to map variables from fuzzy space to practical space. A membership function (MF) is a curve that defines how each point in the input space is mapped to a membership value between 0 and 1. The MF for the MMET system is the symmetric Gaussian function. The inputs of E and EC are interpreted from this fuzzy set, and the degree of membership is interpreted.

3.2. Soft Switching Sliding Mode Control Design

The objective is to consider the non-linear tether system as the controlled plant, and therefore defined by the general state-space equation $\dot{x} = f(x, u, t)$, where $x \in R^n$ is the state vector, n is the order of the non-linear system, $u \in R^m$ is the input vector, and m is the number of inputs. $s(e, t)$ is the sliding surface of the hyperplane, which is given below and shown in Figure 7, where λ is a positive constant that defines the slope of the sliding surface. In the MMET system, we let n = 2 given that as it is a second-order system in which s defines the position and velocity errors.

$$s(e, t) = \left(\frac{d}{dt} + \lambda\right)^{n-1} e \tag{2}$$

$$s = \dot{e} + \lambda e \tag{3}$$

From equations 2 and 3, the second-order tracking problem is now replaced by a first-order stabilization problem in which the scalar s is kept at zero by means of a governing condition [8]. This is obtained from use of the Lyapunov stability theorem, given in equation 4, and stating that the origin is a globally asymptotically stable equilibrium point for the control system. This equation is positive definite and its time derivative is given in inequality 5 , to satisfy the negative definite condition, that the system should satisfy the inequality 5.

$$V(s) = \frac{1}{2}s^2 \tag{4}$$

$$\dot{V}(s) = s\dot{s} < 0 \tag{5}$$

A soft switching control law is introduced [9] for the major sliding surface switching activity in equation 6, where c_0 is an assumed positive damping ratio for the switching control law. This law needs to be chosen in such a way that the existence and the reachability of the sliding-mode are both guaranteed, noting that δ is an asssumed positive constant which defines the thickness of the sliding mode boundary layer [10].

$$u_{SMC} = -c_0 \tanh\left(\frac{s}{\delta}\right) \tag{6}$$

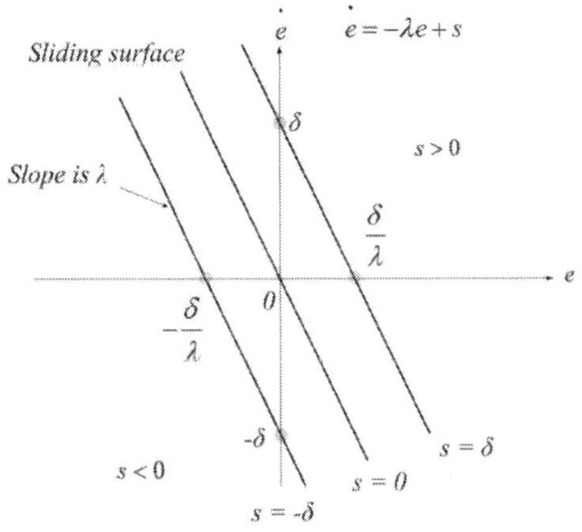

Figure 7. Sliding surface

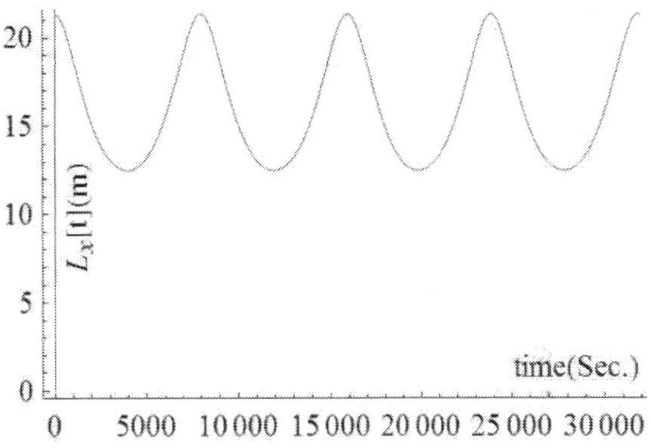

Figure 8. Response of axial displacement of tether sub-span

4. Simulation and Conclusion

Numerical results are obtained using a specially devised co-simulation toolkit of $MATLAB$ and $MATHEMATICA$ functions in an integrated program to provide a new toolbox, known henceforth as $SMATLINK$, which integrates control in $MATLAB/SIMULINK$ with MMET modelling in $MATHEMATICA$. The velocity and acceleration of ψ are selected as error (e) and change-in-error (ec) feedback signals for the the 6DOF MMET spin-up control.

Unless stated otherwise all the results are generated using the following MMET system parameters [3]: the number of mass points $N = 20$, gravitational constant $\mu = 3.9877848 \times 10^{14}$ m^3s^{-2}, propulsion tether payload mass $M_P = 1000\ kg$, mass of motor facility $M_M = 5000\ kg$, static length of propulsion tether $L_0 = 50000\ m$, orbit eccentricity $e = 0.2$, periapsis distance $r_{per} = 6.890 \times 10^6\ m$, apoapsis distance $r_{apo} = 1.0335 \times 10^7\ m$, radius of tether outer and inner tube $r_{Touter} = 0.1\ m$ and $r_{Tinner} = 0.08\ m$, radius of motor facility $r_M = 0.5\ m$, radius of payload $r_P = 0.5\ m$, the undeformed tether tube cross-sectional area $A = 1.13097 \times 10^{-2}$ m^2 , the tether density $\rho = 970\ kg/m^3$, initial angular $\psi_0 = 0\ rad$, initial angular velocity $\dot{\psi}_0 = 0\ rad/s$, motor torque $\tau = 2.5 \times 10^6\ Nm$, damping coefficient $c_i = 2 \times 10^6\ Ns/m$, and stiffness $k_i = 2 \times 10^9\ N/m$.

The hybrid fuzzy sliding mode control system parameters require a judicious choice of the FLC scaling gains, which in this case are taken to be $\{Ke, Kec, Ku\} = \{1.0, 1.0, 21000\}$. The $SMATLINK$ program deals with normalisation of the inputs, so the fuzzification gain factors

Journal of Physics: Conference Series **181** (2009) 012093 doi:10.1088/1742-6596/181/1/012093

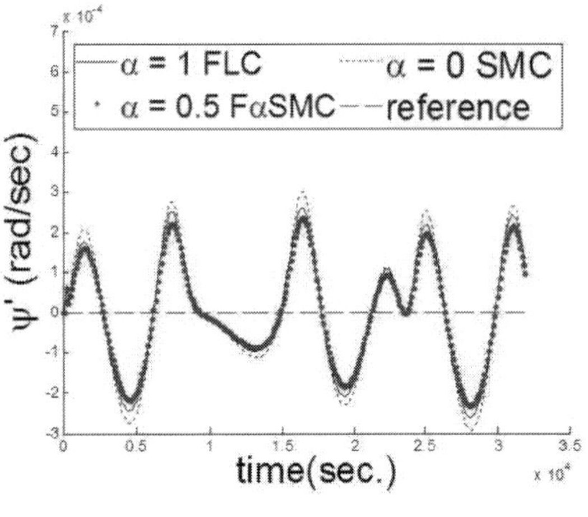

Figure 9. The angular velocity response 6DOF MMET spin-up

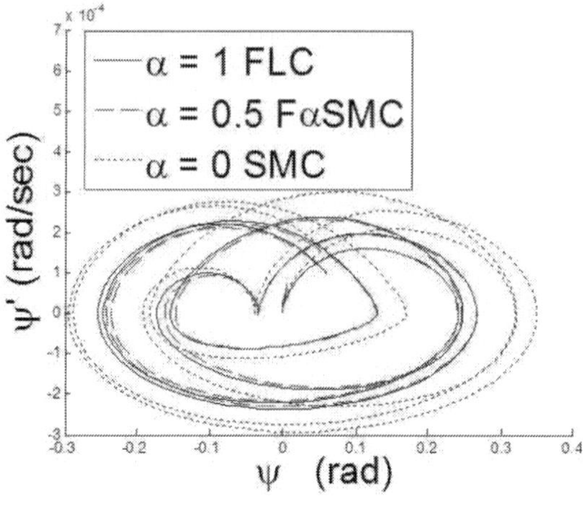

Figure 10. Phase plot for angular response

of Ke, Kec are set to 1.0, and Ku is the defuzzification gain factor which is used to map the FL to the practical tether length range. Similarly, the SMC damping coefficient $c_0 = -3000$ is required to expand the normalised output to a practical tether length. The thickness of the sliding mode boundary layer is given by $\delta = 0.8$, and the slope of the sliding surface $\lambda = 0.0014$. Both data came from previous 6DOF MMET system spin-up simulation results, but without control. In this simulation using the FαSMC, $\alpha = 0.5$ is required to balance the control weight between FLC and SMC. It is easy to switch the controller between SMC and FLC when a proper value of α is selected $(0 < \alpha < 1)$, the hybrid fuzzy sliding-mode controller is generated combining fuzzy logic control with a soft continuous switching hyperbolic tangent control law based on equation (2). All the control methods have an effect on the spin-up of the MMET system from the given initial conditions.

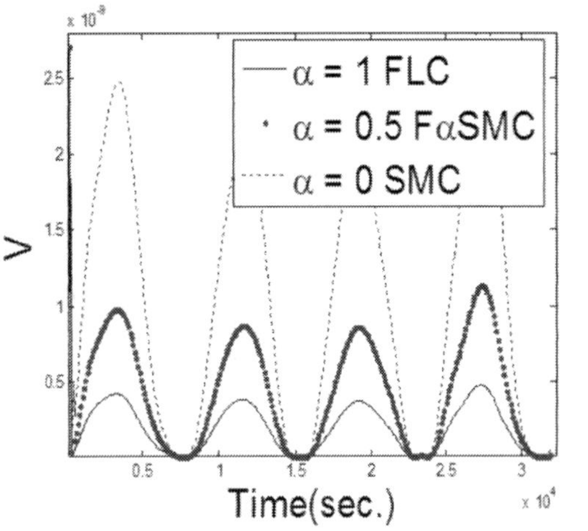

Figure 11. Lyapunov function for the 6DOF MMET system

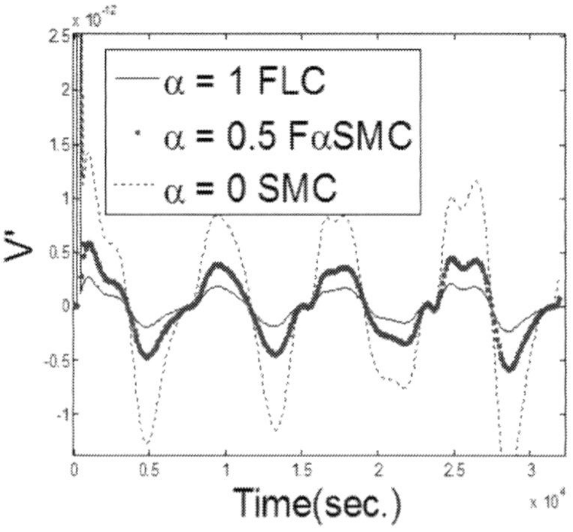

Figure 12. Derivative of Lyapunov function for the 6DOF MMET system

Figure 8 gives the axial elastic behaviour of the MMET in the simulation with the appearance of stable axial oscillation. Figure 9 shows the time responses for the spin-up velocity $\dot{\psi}$ with different values of $\alpha = \{0.0, 0.5, 1.0\}$ for $\{SMC,\ F\alpha SMC,\ FLC\}$ control respectively of the spin-up. The phase plots are shown in Figure 10 as limit cycles whose behaviour for the spin-up coordinate ψ, clearly corroborates interpretations of steady-state. Figures 11 and 12 show the plots for the Lyapunov function and its derivative, which showing that the FαSMC control with different value of α (same as in Figures 9 and 10) for the MMET system satisfies the Lyapunov stability condition in inequality 5.

5. Future work

The work in this paper has shown that including axial elasticity within an MMET model has a significant bearing on overall performance, the parameter settings for the FαSMC need further consideration because the current simulation results come from manual parameter tests. A MMET system with axial and lateral elastic effects could be considered for the further MMET modelling studies. In order to enhance the parameter selection process and validation, some computational intelligence (CI) optimisation tools, such as Genetic Algorithms (GA) and Artificial Neural Networks (ANN), could be applied for parameter selection for the FLC, SMC and FαSMC. This can hopefully give some reference data for the parameter settings. A GA has been used as an optimisation tool for parameter selection of the MMET system payload transfer from LEO to geostationary Earth orbit (GEO), and the GA's optimisation ability has therefore been reasonably demonstrated [16].

References

[1] M. P. Cartmell, (1998), "Generating Velocity Increments by means of a Spinning Motorised Tether", 34th AIAA/ASME/SAE/ASEE Joint propulsion Conference and Exhibit, Cleveland Conference Centre, Cleveland, Ohio, USA, AIAA-98-3739.

[2] S.W. Ziegler,and M. P. Cartmell, (2001), "Using Motorised Tethers for Payload Orbital Transfer", Journal of Spacecraft and Rockets, 38 (6), November / December, pp 904-913.

[3] Y. Chen, M.P. Cartmell, (2007) "Dynamical Modelling of The Motorised Momentum Exchange Tether Incorporating Axial Elastic Effects", Advanced Problems in Mechanics , 20-28 June, Russian Academy of Sciences, St Petersburg, Russia.

[4] S. V. Emelyanov, (1967), "Variable Structure Control Systems(in Russian)", Moscow: Nauka.

[5] Y. Itkis, (1976), "Control Systems of Variable Structure", New York: Wiley

[6] V. A. Utkin, (1978), "Sliding Modes and Their Application in Variable Structure Systems", Moscow: Nauka (in Russian) (also Moscow: Mir, 1978, in English)

[7] John Y. Hung, Weibing Gao, James C. Hung, (1993), "Variable structure control: a survey", IEEE Transactions on Industrial Electronics,pp 2-22.

[8] J.J.E. Slotine, and W.P. Li, (1991), "Applied Nonlinear Control", Prentice-Hall International.

[9] K. C. Ng, Y. Li, D. J. Murray-smith, K. C. Sharman, (1995), "Genetic Algorithm applied to Fuzzy Sliding Mode Controller design", Genetic Algorithms in Engineering Systems: Innovations and Applications, GALESIA. First International Conference on (Conf. Publ. No. 414) 12-14 Sep 1995 Page(s):220 - 225.

[10] J.J. E. Slotine, (1982), "Tracking control of non-linear systems using sliding surfaces with application to robot manipulations", PhD Dissertation, Laboratory for Information and Decision Systems, Massachusetts Institute of Technology.

[11] Brian O'Dell,(1997), "Fuzzy Sliding Mode Control: A Critical Review", Oklahoma State University, Advanced Control Laboratory, Technical Report ACL-97-001.

[12] S. G. Tzafestas1 ,G. G. Rigatos1, (1999), "A Simple Robust Sliding-Mode Fuzzy-Logic Controller of the Diagonal Type", Journal of Intelligent and Robotic Systems, Volume 26, Numbers 3-4 , pp 353-388.

[13] Kevin M. Passino , Stephen Yurkovich, (1998), "Fuzzy Control", Addison Wesley Longman, Menlo Park, CA.

[14] L.A. Zadeh, (1965), "Fuzzy Sets", Information and Control, Volume 8, Number 3, pp. 338–353.

[15] E.H. Mamdani,(1977), "Applications of fuzzy logic to approximate reasoning using linguistic synthesis", IEEE Transactions on Computers, Vol.26, No. 12, pp. 1182-1191.

[16] Y. Chen, M.P. Cartmell,(2007), "Multi-objective optimisation on motorised momentum exchange tether for payload orbital transfer", IEEE Congress on Evolutionary Computation, Publication Date: 25-28 Sept. 2007, pp. 987-993.

Journal of Physics: Conference Series **181** (2009) 012094 doi:10.1088/1742-6596/181/1/012094

Nonlinear vibrations of an inclined beam subjected to a moving load

Ahmad Mamandi[1], Mohammad H. Kargarnovin[2] and Davood Younesian*[3]

[1]PhD Student, Mechanical and Aerospace Engineering Department, Science and Research Branch, Islamic Azad University, Tehran, Iran

[2]Professor, Mechanical and Aerospace Engineering Department, Science and Research Branch, Islamic Azad University, Tehran, Iran and Mechanical Engineering Department, Sharif University of Technology, Tehran, Iran

[3]Assistant Professor, School of Railway Engineering, Iran University of Science and Technology, Tehran, Iran

*Corresponding Author. E-mail: Younesian@iust.ac.ir

Abstract. In this paper, the nonlinear dynamic responses of an inclined pinned-pinned Euler-Bernoulli beam with a constant cross section and finite length subjected to a concentrated vertical force traveling with constant velocity is investigated by using the mode summation method. Frequency analysis of the PDE's governing equations of motion for steady-state response is studied by applying multiple scales method. The nonlinear dynamic deflections of the beam are obtained by solving two coupled nonlinear PDE's governing equations of planar motion for both longitudinal and transverse oscillations of the beam. The dynamic magnification factor and normalized time histories of mid-point of the beam are obtained for various load velocity ratios and the numerical results are compared with those obtained from traditional linear solution. It is found that quadratic nonlinearity renders the softening effect on the dynamic response of the beam under the act of traveling load. Also stability analysis of the steady-state response for the modes equations having quadratic nonlinearity is carried out and it is observed from the amplitude response curves that for the case of internal-external primary resonance, both saturation phenomenon and jump phenomenon are predicted for the longitudinal excitation.

1. Introduction

The linear and nonlinear analysis for forced vibrations of structural elements, such as beams and plates under the act of traveling loads is of considerable practical importance and many analytical and numerical methods have been proposed in the past decades to investigate extensively the dynamic behavior of all customary and non customary shapes of isotropic, composite and FGM[1] made engineering structural elements in the academic literatures. However, until now little attention has been paid to study for dynamic response of the coupled longitudinal and transverse oscillations of an inclined Euler-Bernoulli beam subjected to a vertical concentrated moving load. Bridges on which vehicles or trains travel, and trolleys of overhead traveling cranes that move on their girders may be modeled as moving loads on simply supported beams. Also applications in aerospace industries include the rocket launcher systems and any kinds of firearms in which the inclined beams play an

[1] Functionally Graded Material

important part in the system. A comprehensive treatment of the subject of vibrations of structures due to moving loads, which contains a large number of related cases, is that of which studied in [1]. A common analytical method to tackle with issues of moving force traveling on infinite length beams is the method of integral transforms. On the other hand, in order to deal with problems including finite length beams, the modal analysis is commonly used. Many of researchers have studied the linear dynamic analysis of elastic single-span or multi-span beams traveled by both constant and variable magnitude moving load with various boundary conditions. They usually focus on the effects of the acceleration, deceleration or the dynamical behavior of beams and under one-axle or two-axle load draw some conclusions from structural design point of view [2-8].

The main objective in this study is to focus on the influence of a moving load excitation on the nonlinear coupled PDE's governing equations of motion for an inclined Euler-Bernoulli beam considering longitudinal and transversal oscillation interactions. Frequency response of the system having quadratic non-linearity and stability analysis of steady-state response in the case of internal-external primary resonance are carried out in this paper.

2. Mathematical modeling and definitions

A one-dimensional inclined Euler-Bernoulli beam of length l traveled by a concentrated vertical force F is considered (see Fig. 1). In our up-coming analysis, it is assumed that when the force enters the left end of the beam, all initial conditions are zero, i.e. the beam is at rest at time $t = 0$. It is further assumed that the force will be in full contact with the beam during its motion, i.e. no separation occurs. The nonlinear dynamic behavior of the coupled longitudinal and transverse vibrations of a uniform beam under simultaneous axial force and the effect of a moving force is sought. Additionally, it is assumed that the beam's supports are restricted against any translational movement hence, the lateral deflection produces some kind of mid-plane stretching and this one in turn introduces some order of nonlinearity in governing equations of motion. Under these conditions following coupled nonlinear partial differential system of equations are initiated [9]:

$$\rho A u_{tt} + c_1 u_t - EAu_{xx} = \frac{1}{2}(EA-P)\left[\left(1-2u_x\right)w_x^2\right]_x - F\sin\theta\delta(x-\varsigma) \tag{1}$$

$$\rho A w_{tt} + c_2 u_t - Pw_{xx} + EIw_{xxxx} = (EA-P)\left(ew_x\right)_x - F\cos\theta\delta(x-\varsigma) \tag{2}$$

where u, w and F represent the longitudinal time dependent displacement of a point in the mid-plane, lateral deflection and lateral concentrated moving force, respectively. Moreover, subscribes nt and nx stand for the nth order derivative with respect to time (t) and space coordinate (x), respectively, and $e = u_x - u_x^2 + \frac{1}{2}w_x^2$. Furthermore, ρ, A, EI and δ are the beam's mass density, sectional area, beam's flexural rigidity and Kronecker delta operator, respectively. In addition, c_1 and c_2 are two constant coefficients corresponding to the structural modal damping of the beam also P is the prescribed either axial tensile or compressive load. Note that the $\delta(x-\zeta)$ is the Dirac's delta function in which ζ is the instantaneous position of the moving force with velocity of v on the beam where $\zeta = vt$.

In this study we restrict our attention to the non-rotating planar oscillation and seek only for the first correction which accounts for the beam stretching. This means that to neglect the cubic derivative term $2u_x w_x^2$ in Eq. (1) and the also u_x^2 and $\frac{1}{2}w_x^2$ terms in e [9]. Therefore, it is appropriate to consider $e = u_x$, which is in accordance with the case of thin beam with small radius of gyration. After employing above restrictions on Eqs. (1) and (2), the general governing coupled partial differential equations of planar vibration for an inclined Euler-Bernoulli beam considering linear viscous damping are obtained as:

$$u_{tt} + \frac{c_1}{\rho A}u_t - \frac{E}{\rho}u_{xx} = \frac{(EA-P)}{\rho A}[w_x w_{xx}] - \frac{F\sin\theta}{\rho A}\delta(x-\varsigma) \tag{3}$$

$$w_{tt} + \frac{c_2}{\rho A}w_t - \frac{P}{\rho A}w_{xx} + \frac{EI}{\rho A}w_{xxxx} = \frac{(EA-P)}{\rho A}[u_{xx}w_x + u_x w_{xx}] - \frac{F\cos\theta}{\rho A}\delta(x-\varsigma) \tag{4}$$

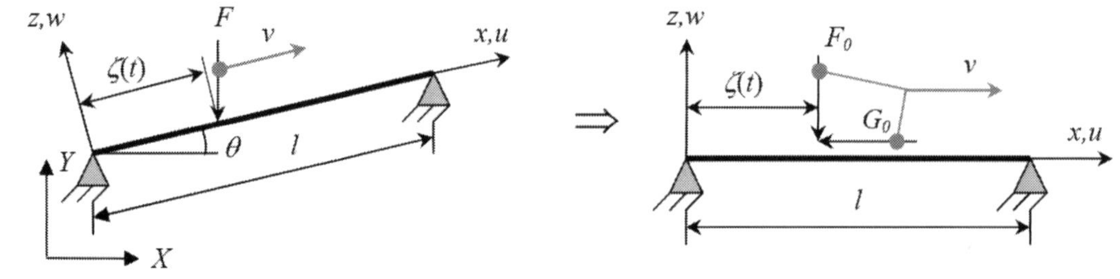

(a) (b)

Fig. 1 (a)- An inclined pinned-pinned beam subjected to a moving lateral load
(b)- Equivalent model

3. Solution method

Since response functions and their associated frequencies of a vibrating system are intrinsic properties of that system, one of the routines in the traditional dynamic analysis of a continuum is the modal analysis and expanding the time-spatial response function in terms of linear undamped natural modes of system obtained by neglecting the effect of foundation and damping terms. Moreover, it is proved that the infinitesimal motions are well described by such technique. In what follows, in order to clarify the problem, the conventional series expansion solution based on modal analysis approach or mode summation method for a beam with pinned-pinned ends due to a moving load would be presented. According to the separation of variable approach, the response of our continuous system in terms of the linear free-oscillation modes can be assumed as follows:

$$u(x,t)=\sum_{n=1}^{\infty}\xi_n(t)\sin\frac{n\pi x}{l} \text{ and } w(x,t)=\sum_{n=1}^{\infty}\eta_n(t)\sin\frac{n\pi x}{l} \tag{5}$$

The boundary conditions for a pinned-pinned beam with immovable ends are expressed as: $u(0,t)=u(l,t)=0, w(0,t)=w(l,t)=\frac{\partial^2 w(0,t)}{\partial x^2}=\frac{\partial^2 w(l,t)}{\partial x^2}=0,$ and initial conditions are: $w(x,0)=\frac{\partial w(x,0)}{\partial t}=0$.

Substituting Eq. (5) into Eqs (3) and (4) leads to:

$$\sum_{n=1}^{\infty}[(\xi_m)_{tt}+2\mu\lambda_m(\xi_m)_t+\lambda_m^2\xi_m]\sin\frac{n\pi x}{L}=-\pi^3\kappa\sum_{m=1}^{\infty}\sum_{k=1}^{\infty}k^2 m\eta_m\eta_k.\cos\frac{m\pi x}{l}\sin\frac{k\pi x}{l}-\frac{F\sin\theta}{\rho A}\delta(x-\varsigma) \tag{6}$$

$$\sum_{n=1}^{\infty}[(\eta_m)_{tt}+2\mu\omega_m(\eta_m)_t+\omega_m^2\eta_m]\sin\frac{m\pi x}{l}=-\pi^3\kappa\sum_{m=1}^{\infty}\sum_{k=1}^{\infty}mk\eta_k\xi_m.(k\cos\frac{m\pi x}{l}\sin\frac{k\pi x}{l}+m\sin\frac{m\pi x}{l}\cos\frac{k\pi x}{l}) -\frac{F\cos\theta}{\rho A}\delta(x-\varsigma) \tag{7}$$

in which $\kappa=\frac{EA-P}{\rho A l^3}$, $\lambda_m=\frac{m\pi}{l(\rho A)^{1/2}}$ and $\omega_m=\frac{m\pi}{l}(\frac{EI}{\rho A}\frac{m^2\pi^2}{l^2}+\frac{P}{\rho A})^{1/2}$. Multiplying both sides of Eqs. (6) and (7) by $\sin\frac{n\pi x}{l}$ and integrating over the interval $[0,l]$ leads to:

$$\begin{cases}(\xi_n)_{tt}+2\mu\lambda_n(\xi_n)_t+\lambda_n^2\xi_n=\frac{-2F\sin\theta}{\rho A l}\sin\frac{n\pi vt}{l}-\frac{n\pi^3}{4}\kappa\sum_{m=1}^{\infty}m\eta_m\left[|n-m|\eta_{|n-m|}+(m+n)\eta_{m+n}\right] & (8)\\ (\eta_n)_{tt}+2\mu\omega_m(\eta_n)_t+\omega_m^2\eta_n=\frac{-2F\cos\theta}{\rho A l}\sin\frac{n\pi vt}{l}-\frac{n\pi^3}{4}\kappa\sum_{m=1}^{\infty}m\xi_m\left[|n-m|\eta_{|n-m|}+(m+n)\eta_{m+n}\right] & (9)\end{cases}$$

Eqs. (8) and (9) are numerically solved to determine the values ξ_n and η_n then by substituting them in Eq. (5) one obtains the longitudinal and transverse modal co-ordinates for the beam, respectively.

4. Numerical results

Journal of Physics: Conference Series **181** (2009) 012094 doi:10.1088/1742-6596/181/1/012094

In order to obtain numerical results, the material properties of the beam are as follows: $E = 200 \times 10^9$ N/m^2, $\rho = 7850$ kg/m^3 (steel), $l = 8$ m, $\mu = 0.033$; $A = 196 \times 10^{-4}$ m^2. It is assumed that the beam has a square cross-section. Results for the dynamic magnification factor D_d, which is the ratio of the maximum magnitude of the dynamic deflection at the mid-span of the beam to the maximum static deflection at the same location, are shown in Fig. 2 for different values of velocity ratio α, in which $\alpha = \frac{T_f}{T} = \frac{v}{v_f}$, where T_f denotes the first natural period (fundamental period) of the beam while T denotes the total time taken by the moving load to cross from one end to the other and also the velocity of a reference load that would take the time of T_f to traverse the beam of length l is represented by v_f. Fig. 2 shows the effect of the moving load size on the dynamic magnification factor D_d of a horizontal steel beam for various values of load velocity. It can be seen that by increasing the magnitude of the moving load the dynamic displacement grows accordingly in such a way that it does not follow the well-known linear force-deflection relation as in the liner systems. It can be observed that displacement results in the transverse direction of the linear and nonlinear theories are very close to each other up to the $F = 6000$ N and maximum D_d for linear and nonlinear solutions are almost the same and equals to 1.73 at $\alpha = 1.2$. However, this figure further reveals that after this value of F, deviation between the linear and the nonlinear theories increases for the considered parameters and the dynamic deflections of the nonlinear cases are larger than those in the linear case. One of the reasons for this difference is mainly due to the effect of larger deflection on the higher values of the transverse traveling force. Furthermore, as seen from this figure, the velocity of the moving load plays an important role on the load-displacement curves and for the considered velocity ranges the difference between the two theories turns to be minimum. It can be seen that the maximum value of dynamic magnification factor D_d is about 1.85 at $F = 20000$ N for a steel beam under the act of moving force and occurs locally at $\alpha = 1.2$. Also, the maximum and minimum D_d occur at $\alpha = 1.2$ and $\alpha = 0.4$, respectively for steel beams under different values of moving loads corresponding to the velocity ratios up to $\alpha = 2$. In most practical applications in the mechanical and civil engineering fields these values play an important role from design point of view. Moreover, as it was seen from the trend of curves in this figure, the dynamic magnification factor for a pinned-pinned beam consists of two regions: under critical region and an overcritical region. In the under critical region the dynamic magnification factor, D_d increases and decreases by increasing α. In this region, D_d increases up to $\alpha = 0.3$ and then its variation reverses and it reaches to a minimum value at $\alpha = 0.4$. The main increase in the variation of D_d occurs in the intervals $0.4 \leq \alpha \leq 1.2$. In the overcritical region D_d decreases as α increases. In other words, in the under critical region the beam's dynamic deflection increases by increasing the velocity of the load and in the overcritical region, the beam's dynamic deflection decreases by increasing the velocity of the load, v.

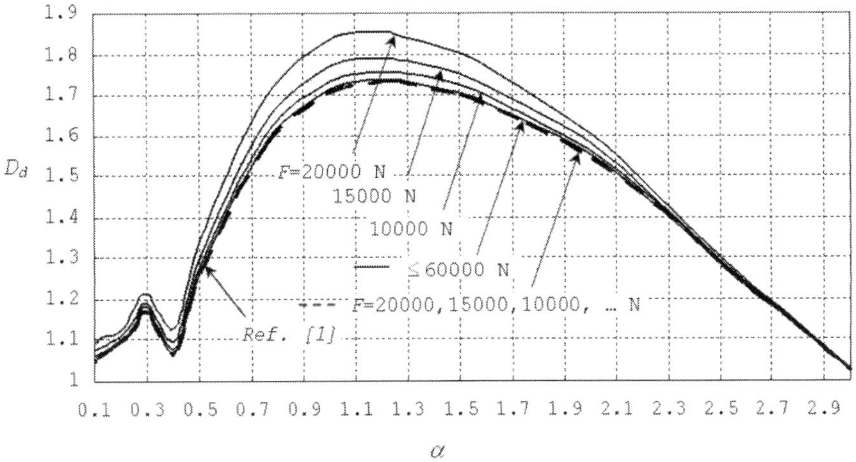

Fig. 2 Dynamic magnification factor (D_d) of an isotropic horizontal steel beam versus α for $l = 8\ m$, $P = 0$, $\mu = 0$, $n = 6$; (-----) linear solution [1], (——) nonlinear solution

Journal of Physics: Conference Series **181** (2009) 012094 doi:10.1088/1742-6596/181/1/012094

In Fig. 3, the maximum dynamic displacements of the beam under the moving load are obtained for different values of the moving load, which varies from $F = 2000$ N to $F = 30000$ N with an increment of 2000 N for $\alpha = 1$, considering both linear and nonlinear solutions. It can be seen that the dynamic deflection of the nonlinear analysis are higher than those obtained from the linear solution representing a softening behavior because of the existence of quadratic nonlinearity characteristic in the mode equations of motion of the beam. Thus this system is equivalent to a nonlinear soft spring [9]. Also, it is seen from this figure that displacements of the linear and the nonlinear solutions are very close to each other until the value of $F = 10000$ N; however, after this point, displacements shows a sizeable difference between the linear and the nonlinear solutions for the considered parameters.

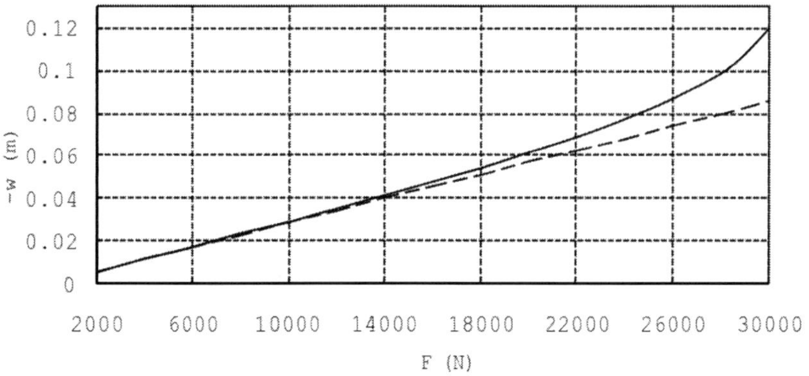

Fig. 3 Variation of the beam's mid-point dynamic displacement versus magnitude of the moving load $l = 8$ m, $A = 0.0196$ m^2, $\mu = 0$, $\theta = 0°$, $\alpha = 1$, $n = 6$; (-----) linear solution, (——) nonlinear solution

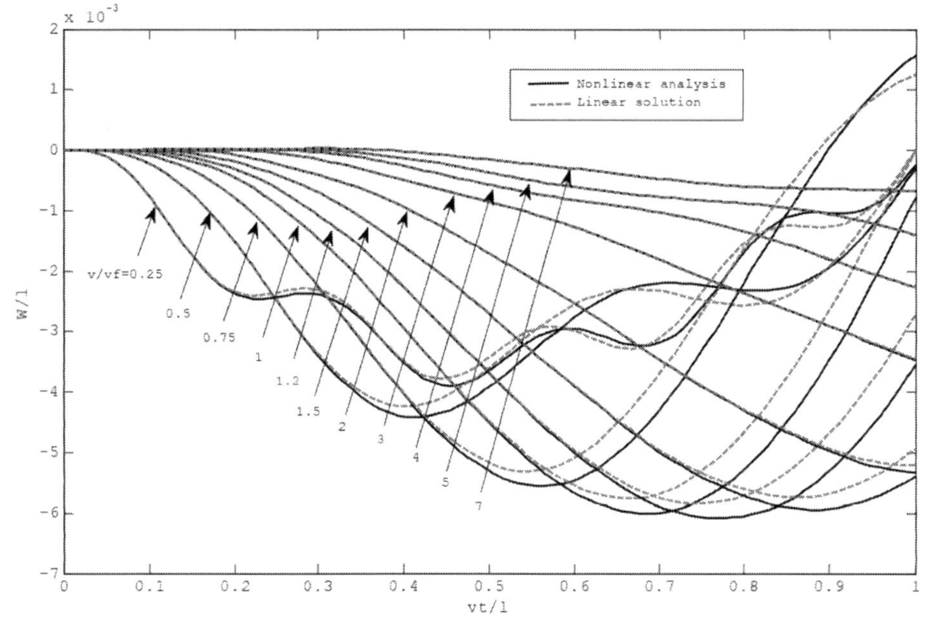

Fig. 4 Dimensionless dynamic deflection of mid-point of the beam versus *vt/l* for $l = 8$ m, $F = 20000$ N, $P = 0$, $\mu = 0$, $\theta = 36°$, $n = 6$

Fig. 4 shows the results for the normalized time histories of transverse deflection of the center of the beam versus *vt/l*, for different velocity ratios.

To validate the numerical results obtained from presented nonlinear dynamic analysis the outcome answers from nonlinear solution are compared in Fig. 4 with those from the linear solution. As it was seen, the nonlinear solutions give larger deflections compared to those of linear analysis. Also, it can be seen that the peak values do not almost occur simultaneously. In addition, it can be observed that the maximum dynamic displacement initially increases as *vt/l* increases. However, maximum dynamic displacement decreases when vt/l increases beyond the value of $\alpha = 1.2$. Also, it is seen that for higher velocity ratio of v/v_f, i.e. $\alpha \geq 3$, the time history curves of the beam from both theories overlays. It can be observed that the deflection at the mid-span actually becomes positive (upward) for the lower velocity ratio cases, i.e. $\alpha = 0.75$ when the traveling load is about to leave the beam. In addition, the maximum dynamic displacement occurs while the force leaves the beam at $\alpha = 1.5$. It can be easily seen that the time which the maximum mid-span displacement occurs moves to the right side as α increases.

5. Frequency analyses of forced oscillations of an inclined beam having quadratic nonlinearities

In this section we study the internal-external primary resonances for the forced response considering coupled longitudinal and transverse oscillations of an inclined Euler-Bernoulli beam with linear viscous damping under the act of traveling load. At this point, it is convenient to introduce the following dimensionless variables: $X=\frac{x}{l}$, $\tau=\frac{r_g}{l^2}\sqrt{\frac{E}{\rho}}t$, $N=\frac{Pl^2}{r^2EA}$, $W=\frac{w}{l}$, $U=\frac{u}{l}$, $r=\frac{r_g}{l}$, where r_g is the radius of gyration of the beam. The dimensionless form of Eqs. (1) and (2) will be [9]:

$$r^2U_{\tau\tau}-U_{XX}=(1-r^2N)W_XW_{XX}-2vU_\tau-\frac{Fl\sin\theta}{EA}\delta[l(X-V\tau)] \tag{10}$$

$$r^2(W_{\tau\tau}-NW_{XX}+W_{XXXX})=(1-r^2N)[U_{XX}W_X+U_XW_{XX}]-2r^2\mu W_\tau-\frac{Fl\cos\theta}{EA}\delta[l(X-V\tau)] \tag{11}$$

in which the dimensionless function of $U(X,\tau)$ and $W(X,\tau)$ are taken in following form:

$$U(X,\tau)=\sum_{n=1}^{\infty}\xi_n(\tau)\sin(n\pi X) \text{ and } W(X,\tau)=\sum_{n=1}^{\infty}\eta_n(\tau)\sin(n\pi X) \tag{12}$$

Substituting Eq. (12) into Eqs. (10) and (11) and multiplying both sides by $\sin(n\pi X)$ and integrating over the interval [0,1] and considering the linear modal damping v_n and μ_n, leads to the mode equations as:

$$(\xi_n)_{\tau\tau}+\frac{2v_n}{r^2}(\xi_n)_\tau+\lambda_n^2\xi_n=\frac{g\sin(n\pi V\tau)}{r^2}-\frac{n\pi^3\kappa}{4r^2}\sum_{m=1}^{\infty}m\eta_m\left[|n-m|\eta_{|n-m|}+(m+n)\eta_{m+n}\right] \tag{13}$$

$$(\eta_n)_{\tau\tau}+2\mu_n(\eta_n)_\tau+\omega_n^2\eta_n=\frac{f\sin(n\pi V\tau)}{r^2}-\frac{n\pi^3\kappa}{4r^2}\sum_{m=1}^{\infty}m\xi_m\left[|n-m|\eta_{|n-m|}+(m+n)\eta_{m+n}\right] \tag{14}$$

where $f=\frac{-2F\cos\theta}{EA}$, $g=\frac{-2F\sin\theta}{EA}$, $\omega_n=n\pi(n^2\pi^2+N)^{\frac{1}{2}}$ and $\lambda_n=n\pi/r$. Let ε be a small but finite dimensionless parameter in the order as the amplitudes of oscillation, thus we have:

$$\hat{\xi}_n=\frac{\xi_n}{\varepsilon},\ \hat{\eta}_n=\frac{\eta_n}{\varepsilon},\ \varepsilon\hat{v}_n=r^{-2}v_n,\ \varepsilon\hat{\mu}_n=\mu_n,\ \varepsilon\hat{g}_n=r^{-2}g_n,\ \varepsilon\hat{f}_n=r^{-2}f_n \text{ and } \hat{\kappa}=\pi^3\kappa(4r^2)^{-1} \tag{15}$$

in which $f_n = f\,sin(n\pi V\tau)$ and $g_n = g\,sin(n\pi V\tau)$. Substituting Eq. (15) into Eqs. (13) and (14) and dropping the hats in the results, one would get the quadratic nonlinear mode equations:

$$(\xi_n)_{\tau\tau}+\lambda_n^2\xi_n=\varepsilon\left[-2v_n(\xi_n)_\tau-n\kappa\sum_{m=1}^{\infty}m\eta_m(p\eta_p+q\eta_q)\right]-2\varepsilon g\cos(n\pi V\tau+\pi/2) \tag{16}$$

$$(\eta_n)_{\tau\tau}+\omega_n^2\eta_n=\varepsilon\left[-2\mu_n(\eta_n)_\tau-n\kappa\sum_{m=1}^{\infty}m\xi_m(p\eta_p+q\eta_q)\right]-2\varepsilon f\cos(n\pi V\tau+\pi/2) \tag{17}$$

where $p=|n-m|$ and $q=n+m$. We use the method of multiple scales and seek an approximate solution for Eqs. (16) and (17) for small but finite amplitudes in the form of:

$$\xi_n(\tau,\varepsilon)=\xi_{n0}(\tau_0,\tau_1)+\varepsilon\xi_{n1}(\tau_0,\tau_1)+O(\varepsilon^2) \text{ and } \eta_n(\tau,\varepsilon)=\eta_{n0}(\tau_0,\tau_1)+\varepsilon\eta_{n1}(\tau_0,\tau_1)+O(\varepsilon^2) \quad (18)$$

where $\tau_n\big|_{n=0,1,2,...}=\varepsilon^n\tau$, i.e., $\tau_0=\tau$ and $\tau_1=\varepsilon\tau$, represent different independent time scales. Substituting Eq. (18) into Eqs. (16) and (17) and noting that $d/d\tau=D_0+\varepsilon D_1+...$, $d^2/d\tau^2=D_0^2+2\varepsilon D_0D_1+...$, where $D_0=d/d\tau$, $D_1=d/d\tau_1$ and equating coefficients of like powers of ε, and considering $\xi_{n0}=A_n(\tau_1)\exp(i\lambda_n\tau_0)+c.c.$ and $\eta_{n0}=B_n(\tau_1)\exp(i\omega_n\tau_0)+c.c.$, we obtain following set of second order ordinary differential equations:

$$O(\varepsilon):\quad D_0^2\xi_{n1}+\lambda_n^2\xi_{n1}=-2i\lambda_n(D_1A_n+v_nA_n)\exp(i\lambda_n\tau_0)-n\kappa\sum_{m=1}^{\infty}mB_m\{pB_p\exp[i(\omega_m+\omega_p)\tau_0]$$
$$+p\bar{B}_p\exp[i(\omega_m-\omega_p)\tau_0]+qB_q\exp[i(\omega_q+\omega_m)\tau_0]+q\bar{B}_q\exp[i(\omega_m-\omega_q)\tau_0]-\frac{g}{2}\exp[i(\Omega_n\tau+\pi/2)]+c.c. \quad (19)$$

$$D_0^2\eta_{n1}+\omega_n^2\eta_{n1}=-2i\omega_n(D_1B_n+\mu_nB_n)\exp(i\omega_n\tau_0)-n\kappa\sum_{m=1}^{\infty}mA_m\{pB_p\exp[i(\lambda_m+\omega_p)\tau_0]$$
$$+p\bar{B}_p\exp[i(\lambda_m-\omega_p)\tau_0]+qB_q\exp[i(\lambda_m+\omega_q)\tau_0]+q\bar{B}_q\exp[i(\lambda_m-\omega_q)\tau_0]-\frac{f}{2}\exp[i(\Omega_n\tau+\pi/2)]+c.c. \quad (20)$$

where $\Omega_n=n\pi V$, (n = 1, 2, 3, ...). The bar sign over any parameters and *c.c.* stand for the complex conjugates terms. Inspection of Eqs. (19) and (20) shows that when $\lambda_n\cong\omega_m\pm\omega_p$ or $\lambda_n\cong\omega_m\pm\omega_q$ there is an extra link, or term, connecting ξ_{n1} and η_{n1}. This is referred to as an internal resonance.

5.1. Internal-external resonance in longitudinal excitation

Internal resonance could be described by these conditions: $\lambda_n\cong\omega_m\pm\omega_p$ and or $\lambda_n\cong\omega_m\pm\omega_q$ and longitudinal primary external resonance condition would be defined when $\Omega_n\cong\lambda_n$. As a case study for the internal resonance portion we consider the case in which λ_n is near $\omega_m+\omega_p$, other cases could be treated as following. The solvability conditions of Eqs. (19) and (20) are:

$$-2i\lambda_n(D_1A_n+v_nA_n)-n\kappa\sum_{m=1}^{\infty}mpB_mB_p\exp(-i\sigma_I\tau_1)-g\exp(\sigma_1\tau_1)=0$$
$$-2i\omega_n(D_1B_n+\mu_nB_n)-n\kappa\sum_{m=1}^{\infty}mpA_m\bar{B}_p\exp(i\sigma_I\tau_1)=0 \quad (21)$$

in which $\lambda_n=\omega_m+\omega_p+\varepsilon\sigma_I$ and hence $\Omega_n=\lambda_n+\varepsilon\sigma_1$. We introduce the polar notation for amplitudes $A_n=\frac{1}{2}a_n\exp(i\alpha_n)$ and $B_n=\frac{1}{2}b_n\exp(i\beta_n)$, where a_n, b_n, α_n and β_n are all real function of τ_1. Also to obtain numerical solutions, we consider n = 1, m = 2 and p = 3, thus for the steady-state response we need to solve the following differential algebraic equations (DAE's) simultaneously

$$a_1'=-v_1a_1+\frac{3\kappa}{2\lambda_1}b_2b_3\sin\gamma_I+\frac{g}{\lambda_1}\sin\gamma_1,\ a_1\gamma_1'=a_1\sigma_1-\frac{3\kappa}{2\lambda_1}b_2b_3\cos\gamma_I+\frac{g}{\lambda_1}\cos\gamma_1,\ b_2'=-\mu_2b_2-\frac{3\kappa}{2\omega_2}a_1b_3\sin\gamma_I$$
$$b_2\beta_2'=\frac{3\kappa}{2\omega_2}a_1b_3\cos\gamma_I,\ b_3'=-\mu_3b_3+\frac{3\kappa}{2\omega_3}a_1b_2\sin\gamma_I,\ b_3\beta_3'=\frac{3\kappa}{2\omega_3}a_1b_2\cos\gamma_I \quad (22)$$

where $\gamma_I=\sigma_I\tau_1+\alpha_1-\beta_2-\beta_3$ and $\gamma_1=\sigma_1\tau_1-\alpha_1+\pi/2$. For the steady-state response we will note that $a_1'=b_2'=b_3'=\gamma_I'=\gamma_1'=0$, where the prime over parameters denotes the first derivative with respect to τ_1. In Figs. 5, the amplitudes a_1, b_2 and b_3 have been plotted as functions of g, the amplitude of longitudinal excitation. The so called saturation phenomenon and jump phenomenon are illustrated in this figure. In Fig. 5-a there is a small detuning of the external resonance corresponded to the moving load velocity, while the internal resonance is perfectly tuned. Fig. 5-b identifies the situation in which the external resonance and internal resonance are perfectly tuned. As illustrated in Fig. 5-a in addition to the saturation phenomenon, there is a jump phenomenon associated with the variation of the

Journal of Physics: Conference Series **181** (2009) 012094 doi:10.1088/1742-6596/181/1/012094

amplitude of the excitation of g. When g increases slowly from zero, a_1 follows increases linearly and b_2 and b_3 are zero and more increases in the value of g renders a_1 to have the saturated value of a_{1s} and b_2 and b_3 suddenly jump from zero to their upper branch of corresponded curves, accordingly. Further increases in g, b_2 and b_3 follow to the right direction along their corresponded curves. When g decreases slowly from a large value, a_1 follows along straight line which has a constant value and b_2 and b_3 follow along their upper branch curves. Further decrease in the value of g, causes b_2 and b_3 jump down from their bend points and follow the linear solution back to the origin. These processes are shown graphically by the bilateral arrows in the Fig. 5-a and b.

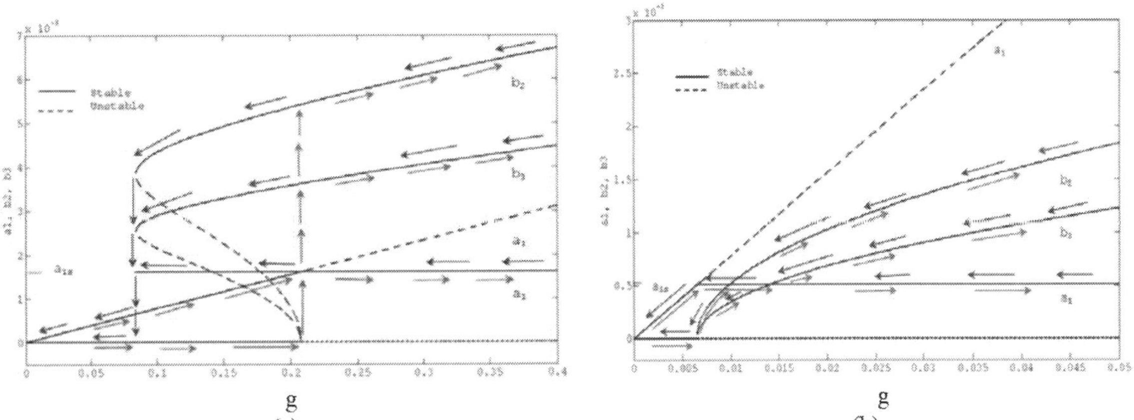

(a) (b)

Fig. 5 Variation of a_1, b_2 and b_3 (amplitude response) vs. g for an inclined pinned-pinned beam longitudinal primary resonant case: $\mu_2 = \mu_3 = 0.165$, $\nu_1 = 0.1$, $r = 0.0245$, $\varepsilon = 0.2$, $\lambda_1 \simeq \pi V$

(a) $\sigma_I = 0, \sigma_1 = 1$, (b) $\sigma_I = \sigma_1 = 0$

6. Conclusions

Nonlinear dynamic analysis of an inclined Euler-Bernoulli beam under the action of a concentrated moving load was carried out. Numerical results were compared with the linear solution and frequency analysis was performed by using multiple scales method. It was found that because of having cubic nonlinearity in mode equations, the system behaves like a nonlinear soft spring; it means that by increasing of magnitude of moving load, the dynamic deflections become larger than those obtained by linear solution. Steady-state solutions were obtained and it was found that in the case of internal-external longitudinal primary resonance, there are two possible solutions: either longitudinal amplitude is nonzero and transversal amplitudes are zero, or both longitudinal and transversal amplitudes are all nonzero. That is we have a jump phenomenon associated with this motion. In addition to the jump phenomenon, the saturation phenomenon is also associated with this motion. This means that by increasing either of external detuning or internal detuning parameters the stable region corresponded to the zero solutions for transverse amplitudes will be extended accordingly.

References

[1] Fryba, L.: Vibration of Solids and Structures under Moving Loads. Thomas Telford Publishing, London (1999)

[2] Lou, P., Dai, G.L., Zeng, Q.Y.: Finite-element analysis for a Timoshenko beam subjected to a moving mass. Journal of Mechanical Engineering Science, IMechE 220(C), 669-678 (2006)

[3] Michaltsos, G.T.: Dynamic Behavior of a single-span beam subjected to loads moving with variable speeds. Journal of Sound and Vibration, 258(2), 359-372 (2002)

[4] Olsson, M.: On the fundamental moving load problem. Journal of Sound and Vibration 145(2), 299-307 (1991)

[5] Kargarnovin, M.H., Younesian, D.: Dynamics of Timoshenko beams on Pasternak foundation under moving load. Mechanics research communications 31,713-723 (2004)

[6] Kargarnovin, M.H., Younesian, D., Thompson, D.J., Jones, C.J.C.: Response of beams on nonlinear viscoelastic foundations to harmonic moving loads. Commputers & Structures 83,

Journal of Physics: Conference Series **181** (2009) 012094 doi:10.1088/1742-6596/181/1/012094
1865-1877 (2005)
[7] Kargarnovin, M.H., Younesian, D., Thompson, D.J., Jones, C.J.C.: Parametrically excited vibration of a Timosheno beam on viscoelastic foundation subjected to a harmonic moving load. Nonlinear Dynamics. 45, 75-79 (2006)
[8] Wu, J.J.: Dynamic analysis of an inclined beam due to moving loads. Journal of Sound and Vibration. 288, 107-133 (2005)
[9] Nayfeh, A.H., Mook, D.T.: Non Linear Oscillations. Wiley-Interscience, New York (1995)

Thermo-Mechanical Analysis Using a Multiphysics Approach

C. Delprete[1] **F. Freschi**[2] **M. Repetto**[2] **and C. Rosso**[1]

[1] Dipartimento di Meccanica, Politecnico di Torino, Corso Duca degli Abruzzi, 24, Torino
[2] Dipartimento di Ingegneria Elettrica, Politecnico di Torino, Corso Duca degli Abruzzi, 24, Torino

E-mail: cristiana.delprete@polito.it

Abstract. In the paper the Cell Method, a discrete method for solving partial differential equations, is applied to a time dependent thermo-mechanical problem. The basic equations are developed with a multiphysics approach and results, both in two-dimensional and tree-dimensional models, are presented. By means of the comparison with finite element approach, some advantages of the proposed methodology are highlighted, in particular quick model construction, capability to separate thermal strain from the mechanical one and, as a consequence, the capability to model the strain and stress evolution in a time dependent problem, considering possible mutual effects between thermal and mechanical fields.

1. Introduction

Nowadays, more and more applications have to be studied by means of multiphysics approaches, because they are characterised by the interaction of different physical phenomena. An example is the Thermo-Mechanical Fatigue (TMF) where thermal field effects add up to the effects due to mechanical actions given by gravitational or surface forces. Usually, the two problems are separately studied, using two different models, and then combined assuming the superposition principle. This methodology is effective if the non-linearities of the material are not considered, but the residual life assessment of a component undergone TMF is strictly linked to the evaluation of the plastic strain accumulation in the *hot points* of the component. As a consequence, the principle of effects superposition shows its limits, it is then necessary to follow step by step the time evolution of the mechanical and thermal effects and their interactions.

A collaboration between the Department of Electrical Engineering (Dipartimento di Ingegneria Elettrica) and the Department of Mechanics (Dipartimento di Meccanica) of Politecnico di Torino is developing a Matlab based code (named DualLab) that allows the study of several multiphysics problems using the discrete formulation of the physical laws given by Cell Method (CM) approach. A portion of the DualLab code is dedicated to the solution of thermo-mechanical problems, where it is possible to apply in the same time and on the same model the mechanical and thermal boundary conditions.

2. Integrated approach to the thermo-mechanical problem

2.1. Discrete approach to the Physical Laws

The most important contribution provided by the theoretical analysis made by Tonti [1, 2, 3] is the rigorous classification of physical laws. This approach allows the underlining of analogies and common structures of the different physical laws. At the basis of Tonti's analysis there is the classification of the temporal and geometric structures on which problem variables are defined. Such variables do not present a variation on their own domain, as in the usual differential approach, but they are associated to n-dimensional elements of the space and time, according to the physical nature of the variable. For example, in the mechanical problem, Tonti's classification makes reference to the surface forces (expressed in newtons) instead of pressures (expressed in pascals). Variables of this kind are called *global* in order to distinguish them from the variables linked to density of field quantities that present a punctual variation. The definition of surface forces as a global variable allows its direct use in the equilibrium equation (topological law), without using integral operator acting on the surface. This is the fundamental consideration that is the basis of the *Algebraic Formulation* of the field problem, because the problem domain is geometrically discretized. As a consequence of the *Algebraic Formulation*, the applicability of this approach to the solution of a physical problem is easy to implement in numerical solution code.

Referring to [1] for a detailed presentation of the technique, here only its main characteristics are outlined:

- primal and dual meshes, characterised by a one-to-one correspondence between primal and dual geometric entities is used and a simplicial mesh (1-d, 2-d or 3-d with respect to the problem dimension) is taken as primal complex K while a barycentric one is used as dual complex $\tilde{K}$;
- standard differential operators are translated in finite form by means of incidence matrices: $\mathbf{G}$ is the discrete equivalent of gradient, $\mathbf{C}$ of curl and $\mathbf{D}$ of divergence operators defined on primal complex K. Similar matrices are defined on dual complex $\tilde{K}$. Using the duality relationship, it is possible to write: $\tilde{\mathbf{D}} = -\mathbf{G}^T$, $\tilde{\mathbf{C}} = \mathbf{C}$, $\tilde{\mathbf{G}} = \mathbf{D}^T$.
- considering for example the thermal problem, the temperature difference Γ, configuration variable, is related to a *primal* edge l and the source variable, the thermal flux Φ, acts on the *dual* surface $\tilde{S}$ associated to the *primal* edge;
- the constitutive equation, for instance on an orthogonal grid, $\Phi = -\lambda\frac{\tilde{S}}{l}\Gamma$, relates the configuration and source variables and considers the size of the *primal* edge, the size of the *dual* surface and the thermal conductivity λ of the material.

In the present paper the above described approach is applied to a thermo-mechanical problem.

2.2. Algebraic formulation of thermal problems

Using Tonti diagram for the thermal problem, depicted in Figure 1, it is possible to identify four global variables: temperature and temperature differences (as configuration variables), thermal flux and heat generation (as source variables).

Primal cell complex is obtained using a triangular mesh. Temperature difference is defined on *primal* edges. By making reference to Figure 3, temperature gradient along primal edge α is obtained as difference between nodal temperatures $\theta_i - \theta_k$. The whole array of temperature differences can thus be obtained from the vector of temperatures, by using edge-to-node incidence matrix $\mathbf{G}$. In this matrix each row corresponds to an edge and is full of 0s but for the positions of the nodes on which edge insists: $+1$ if edge i comes out of node j and -1 if edge i enters into node j. Construction of matrix $\mathbf{G}$, for the geometrical layout of Figure 4, is highlighted in Figure 5. $\mathbf{G}$ is the discrete counterpart of gradient differential operator and

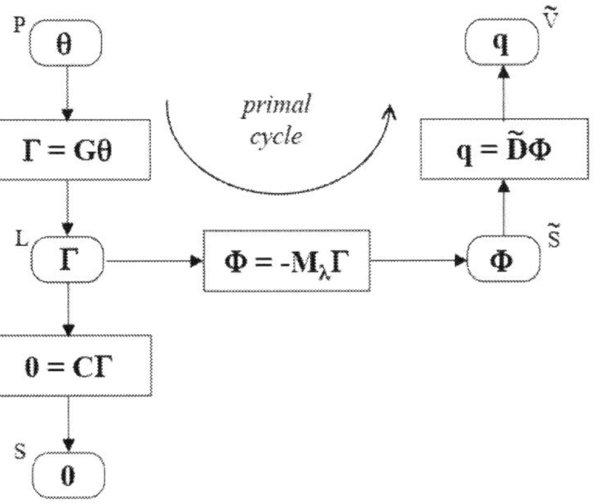

Figure 1. Tonti diagram for steady state thermal conduction.

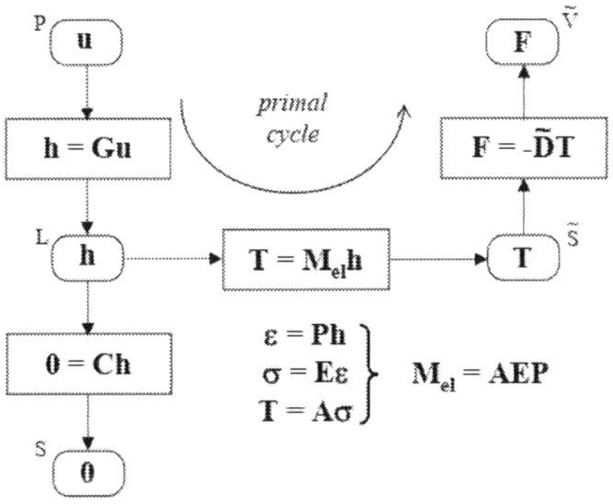

Figure 2. Tonti diagram for steady state mechanical problem.

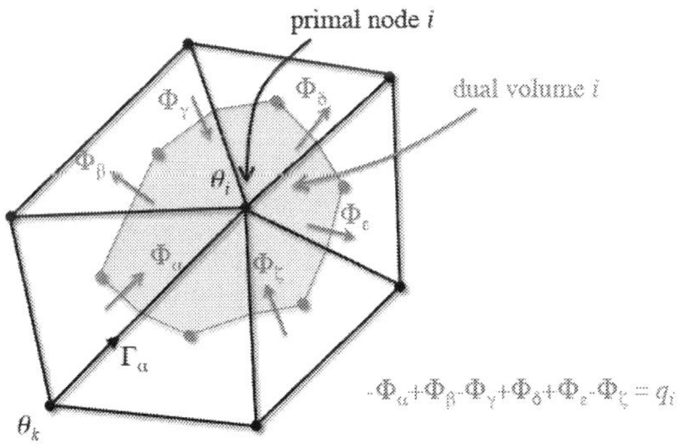

Figure 3. 2-d example of primal (black) and dual (red) complexes: θ are the temperatures defined on primal nodes, Φ are thermal fluxes on dual faces and q are the heat produced in dual volumes.

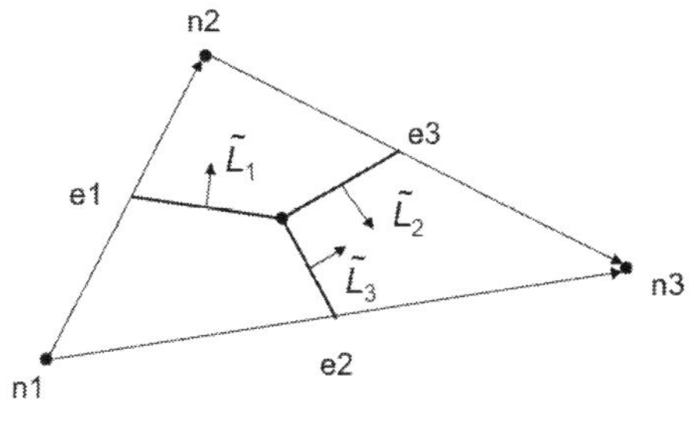

Figure 4. Primal 2-d cell with primal nodes (n_i), primal edges (e_i) and dual surfaces $\tilde{L}_i$

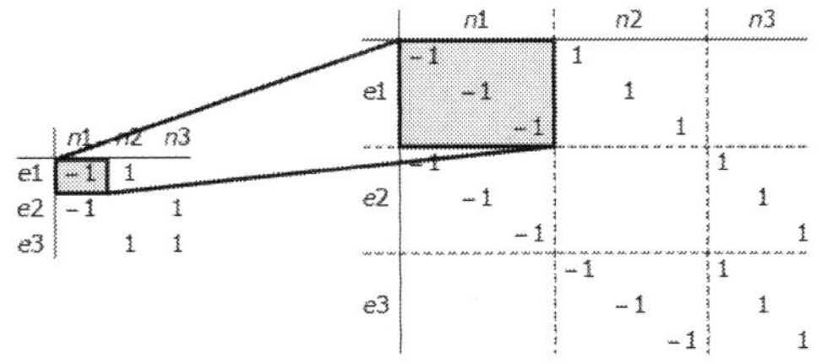

Figure 5. Definition of incidence matrices

since it only depends on the discretization, so it is the same for all physical problems. Thus, temperature difference can be computed as:

$$\Gamma = \mathbf{G}\theta, \tag{1}$$

where θ are node temperatures, Γ are temperature differences along the *primal* edges.

Again referring to Figure 3, balance equation on dual volume i is obtained by summing up all thermal fluxes on surfaces which bound it. Orientation of surfaces is translated in a $\pm$ sign to be applied to each flux. All these ± 1 coefficients can be collected in a $\tilde{\mathbf{D}}$ which is the incidence matrix that represents the relationship between *dual* volumes and surfaces and that expresses in algebraic form the divergence applied to the *dual* complex.

In this way, balance equation (topological equation written on the *dual* complex) is defined applying a thermal balance operation to the *dual* volumes:

$$\tilde{\mathbf{D}}\Phi = q, \tag{2}$$

where Φ are thermal fluxes through *dual* surfaces, q are heat productions defined in *dual* volumes.

The relationship between thermal flux and temperature gradient is the constitutive equation of the problem that represents the discrete Fourier equation:

$$\Phi = -\mathbf{M}_\lambda \Gamma, \tag{3}$$

where $\mathbf{M}_\lambda$ is the constitutive matrix that considers the "size" of the problem and the material features. The constitutive matrix aspect depends on the mesh used to discretize the domain, if a structured mesh with orthogonal lines and surfaces is used, $\mathbf{M}_\lambda$ is a diagonal matrix. In general, if the mesh is not structured, the constitutive matrix is computed by means of local interpolation of the field variables and it becomes a sparse matrix [6], [7].

By operating on Equations (1), (2) and (3), deleting Φ e Γ and applying duality principle ($\tilde{\mathbf{D}} = -\mathbf{G}^T$), the solutive equation of the thermal problem assumes the form:

$$\mathbf{G}^T \mathbf{M}_\lambda \mathbf{G}\theta = q, \tag{4}$$

that can be easily assembled by means of sparse linear algebra routines and solved applying the thermal problem boundary conditions.

2.3. Algebraic formulation of mechanical problems

The mechanical problem can be divided in two aspects: elastostatic and elastodynamics problems. The first step is to study the elastostatic problem in order to express the stiffness matrix and to do this the same approach of section 2.2 can be used. The elastostatic problem can be analysed using the Tonti diagram of Figure 2. In analogy with the thermal problem the global variables have to be defined. The configuration variables are displacements u e their differences h, respectively related to *primal* nodes and to *primal* edges. The source variables are surface forces σ, linked to *dual* surfaces, and external forces F_e acting on the system and related to *dual* volumes.

On the basis of Figure 4, the displacement topological equation can be expressed on the *primal* complex as:

$$\mathbf{G_3} u = h, \tag{5}$$

The matrix $\mathbf{G_3}$ has the same form of the incidence matrix $\mathbf{G}$ of the thermal problem, because the discretization is the same, ad it is an expansion of $\mathbf{G}$ for each spatial direction, as shown in Figure 5. Using a different approach from [1, 5], the displacement of a generic point inside a primal cell is assumed to be *affine*. As described in Figure 4 for a 2-d problem, the only reported for briefness, the displacement components u, v have the form

$$u(x,y) = H_{xx}x + H_{xy}y + c_u \tag{6}$$
$$v(x,y) = H_{yx}x + H_{yy}y + c_v \tag{7}$$

hence, the relative horizontal displacements are:

$$\Delta u_1 \;= u_2 - u_1 = H_{xx}(x_2 - x_1) + H_{xy}(y_2 - y_1) \tag{8}$$
$$\Delta u_2 \;= u_3 - u_1 = H_{xx}(x_3 - x_1) + H_{xy}(y_3 - y_1) \tag{9}$$
$$\Delta u_3 \;= \Delta u_2 - \Delta u_1 \tag{10}$$

Since Δu_3 is a linear combination of Δu_2 and Δu_1, no information is due to this term, that can be neglected. It is useful to rewrite (8) in matrix form:

$$\begin{bmatrix} \Delta u_1 \\ \Delta u_2 \end{bmatrix} = \begin{bmatrix} L_{1x} & L_{1y} \\ L_{2x} & L_{2y} \end{bmatrix} \begin{bmatrix} H_{xx} \\ H_{xy} \end{bmatrix} \tag{11}$$

Then, operating on the matrices written for each considered direction,

$$\begin{bmatrix} H_{xx} & H_{yx} \\ H_{xy} & H_{yy} \end{bmatrix} = \mathbf{H}^T = \begin{bmatrix} P_{x1} & P_{x2} \\ P_{y1} & P_{y2} \end{bmatrix} \begin{bmatrix} \Delta u_1 & \Delta v_1 \\ \Delta u_2 & \Delta v_2 \end{bmatrix} \tag{12}$$

Using the classical strain expression coming from continuous mechanics, the (12) becomes:

$$\begin{bmatrix} \varepsilon_x \\ \varepsilon_y \\ \gamma \end{bmatrix} = \begin{bmatrix} P_{x1} & 0 & P_{x2} & 0 & 0 & 0 \\ 0 & P_{y1} & 0 & P_{y2} & 0 & 0 \\ P_{y1} & P_{x1} & P_{y2} & P_{x2} & 0 & 0 \end{bmatrix} \begin{bmatrix} \Delta u_1 \\ \Delta v_1 \\ \Delta u_2 \\ \Delta v_2 \\ \Delta u_3 \\ \Delta v_3 \end{bmatrix} \tag{13}$$

and in compact form

$$\varepsilon = \mathbf{P}h \tag{14}$$

The same reasoning allows the formulation of the 3-d problem.
The balance topological equation related to the forces acting on the *dual* volume is:

$$\tilde{\mathbf{D}}_{\mathbf{3}}T = F_e \tag{15}$$

Surface forces are defined on dual faces, see Figure 4. They are linear combination stress:

$$\begin{aligned} T_x &= \sigma_{xx}\tilde{a}_x + \sigma_{xy}\tilde{a}_y \\ T_y &= \sigma_{yx}\tilde{a}_x + \sigma_{yy}\tilde{a}_y \end{aligned} \tag{16}$$

and, writing (16) for each dual face inside a triangle, the relationship between surface force and stress can be written as:

$$\begin{bmatrix} T_{1x} \\ T_{1y} \\ T_{2x} \\ T_{2y} \\ T_{3x} \\ T_{3y} \end{bmatrix} = \delta \begin{bmatrix} \tilde{L}_{1x} & 0 & \tilde{L}_{1y} \\ 0 & \tilde{L}_{1y} & \tilde{L}_{1x} \\ \tilde{L}_{2x} & 0 & \tilde{L}_{2y} \\ 0 & \tilde{L}_{2y} & \tilde{L}_{2x} \\ \tilde{L}_{3x} & 0 & \tilde{L}_{3y} \\ 0 & \tilde{L}_{3y} & \tilde{L}_{3x} \end{bmatrix} \begin{bmatrix} \sigma_x \\ \sigma_y \\ \tau \end{bmatrix} \tag{17}$$

where δ is the problem depth. In compact form:

$$T = \mathbf{A}\sigma \tag{18}$$

The constitutive law in the elastostatic problem is usually expressed as a relationship between stress σ and strain ε and, in formula, it can be written as:

$$\sigma = \mathbf{E}\varepsilon, \tag{19}$$

assuming that stress and strain fields are uniform in the cell volume and the material (matrix $\mathbf{E}$) is defined as isotropic linear elastic.

As in the thermal problem and in the most physical problems, using the duality principle, the matrix $\mathbf{G}$ can be expressed as the transpose of $\tilde{\mathbf{D}}$. By assembly together, it is then possible to write the solution equation of the elastostatic problem, in formulae:

$$\mathbf{G_3}^T\mathbf{AEPG_3}u = F_e, \tag{20}$$

and in compact form:

$$\mathbf{G_3}^T\mathbf{M}_E\mathbf{G_3}u = F_e, \tag{21}$$

where the product of the three matrices, the two topological and the constitutive, gives the stiffness matrix of the analysed component. As demonstrated in [1, 5], the stiffness matrix obtained using (*Algebraic Formulation*) is the same deduced using the well-known Finite Element Method, with respect to the same model. By imposing the actual boundary conditions, (21) can be solved and the elastostatic quantities evaluated.

The elastodynamic problem involves the definition of the mass matrix. Using the *Algebraic Formulation*, the inertial properties of the single *dual* cell are related to the *primal*, this consideration allows the expression of the mass matrix as a diagonal matrix, where each element of the diagonal is the inertial property portion that is referred to the related *primal* node. In the 2-d case, mass matrix can be expressed as:

$$\mathbf{I} = \begin{bmatrix} m_i & 0 & 0 & 0 & 0 & 0 \\ 0 & m_i & 0 & 0 & 0 & 0 \\ 0 & 0 & m_j & 0 & 0 & 0 \\ 0 & 0 & 0 & m_j & 0 & 0 \\ 0 & 0 & 0 & 0 & m_k & 0 \\ 0 & 0 & 0 & 0 & 0 & m_k \end{bmatrix} \tag{22}$$

where $m_{i,j,k}$ are the masses of the *dual* cells related to i, j, k *primal* nodes. As a consequence, the elastodynamic problem can be solved using the well-known equation of motion.

2.4. Algebraic formulation of the thermo-mechanical problem

By comparing the (4) and (21), it is possible to notice that the unique difference between the two expressions is the constitutive matrix, whereas the model (the topological equations) is the same. In such a way the benefit of the Cell Method in multiphysics approach is clearly evident, it is sufficient to define a unique model. By combining (4) and (21) it is possible to solve a thermo-mechanical problem. In order to due that, it is important to define the mutual effects of thermal and mechanical field. In a first step, the authors only consider the effect of thermal field on the mechanical properties and responses of the studied system. Those effects are related to the variation of the Young modulus with the temperature and the thermal strains of the analysed component due to the nodal temperatures variation. For considering the first effect, it is possible to evaluate the temperature of each node (4), compute a mean temperature of the *dual* cell and, by means of a look-up table, evaluate the material feature of this cell; then it is possible to write the elastic constitutive matrix. The second effect can be evaluated considering the actions on the component due to the additional thermal strains. In other words, the thermal expansion effect is transformed in a set of additional nodal forces. In formulae, the thermal strains can be expressed as:

$$\left\{ \begin{array}{c} \varepsilon_x \\ \varepsilon_y \\ \gamma_{xy} \end{array} \right\} = \alpha \left\{ \begin{array}{c} \bar{\theta} - \theta_0 \\ \bar{\theta} - \theta_0 \\ 0 \end{array} \right\}, \text{ in general way } \varepsilon_{th} = \alpha \Delta \bar{\theta} \tag{23}$$

where $\bar{\theta}$ is the mean temperature of the cell, θ_0 is the reference temperature and α is the expansion coefficient of the material.

Then, using the same approach of the elastostatic problem, the additional thermal forces can be written as:

$$F_{th} = \mathbf{G_3}^T \mathbf{A} \sigma_{th} = \mathbf{G_3}^T \mathbf{A} \mathbf{E} \varepsilon_{th} = \alpha \mathbf{G_3}^T \mathbf{A} \mathbf{E} \Delta \bar{\theta}. \tag{24}$$

The thermo-mechanical problem is then formulated as:

$$\mathbf{I} \ddot{u}(t) + \mathbf{G_3}^T \mathbf{AEPG_3} u(t) - \alpha \mathbf{G_3}^T \mathbf{AE} \Delta \bar{\theta} = F_e(t). \tag{25}$$

2.5. Algebraic Formulation *of the time evolution*

Algebraic Formulation can be applied also to time evolution. The solution of (25) can be directly written in a discrete form. In fact, if the displacement u is *primal instants*, whereas velocity $v = \dot{u}$ to the *primal intervals*. It is possible to write that:

$$\tau_k v_k = u_{k+1} - u_k \tag{26}$$

$$\tau_k a_k = v_{k+1} - v_k \tag{27}$$

where k is the considered time increment and a_k is the acceleration.

The motion equation can be then expressed as:

$$\mathbf{I} v_{k+1} = (F_{k+1} + \alpha \mathbf{G_3}^T \mathbf{AE} \Delta \tilde{\theta}) \tau_k + \mathbf{I} v_k - \mathbf{G_3}^T \mathbf{AEPG_3} u_{k+1} \tau_k. \tag{28}$$

By knowing that $u_{k+1} = u_k + \tau_k v_k$, imposing initial condition, the thermo-mechanical problem can be evaluated in its time variation.

3. Results

In the present paper, authors do not consider the non-linearity in the constitutive matrices. So the analysed problems have material features dependent only on the temperature and not on the strain. In order to evaluate the proposed methodology, a 2-d and a 3-d (Figures 6 and 7, respectively) case studies are analysed and compared with numerical solution obtained by a commercial FEM software. As far as the case study mesh is concerned, both CM and FEM use the same discretization, with the same number of elements and nodes. It is worth noting that for FEM models element first order shape functions are chosen and different solution runs are performed for solving the conduction problem and the mechanical problem. In fact the output of the thermal problem solved with FEM are inserted, using a tailored data recovery procedure, as boundary conditions (nodal temperatures) in the mechanical problem in association with mechanical actions. In the CM procedure, the solution run is unique. In addition, CM procedure has the benefit of a quicker creation of the model, in fact, only a model is defined and the boundary conditions, both thermal and mechanical, are at the same time imposed. The case studies prove that results obtained with CM procedure are comparable to those obtained using FEM, confirming that the linear elastic stiffness matrices computed with the two methods are the same. Authors expect relevant differences when the non linearities of the material will be taken into account.

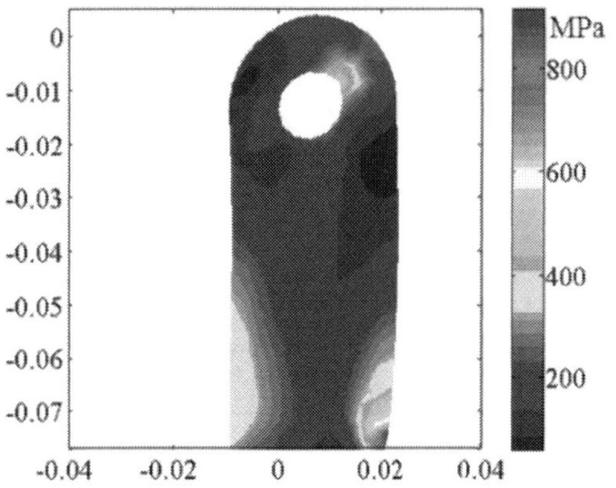

Figure 6. Thermo-mechanical model of a hook, stress distribution.

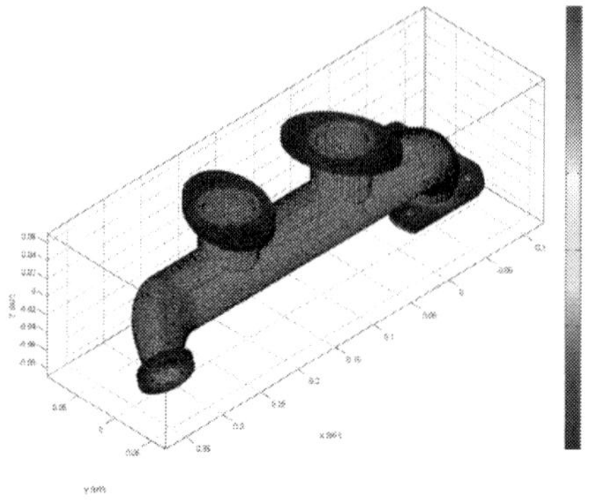

Figure 7. Thermo-mechanical model of an exhaust manifold, stress distribution.

4. Conclusion

In the present paper, thermo-mechanical problem is studied using Cell Method and the solution formulation is proposed. A 2-d and a 3-d case studies are analysed and compared with FEM solution. Some remarks can be done:

- Cell Method proves to be useful in Multiphysics approach, giving coherent results;
- the model construction is quicker than traditional FEM method, only one discretization is needed because it is congruent among the different Physics disciplines;
- Cell Method also provides a low-level coupling of the physical laws, making easy to solve multiphysics problem, also in time dependent problem;
- with the proposed technique, it is possible to evaluate the thermal and mechanical strains for each step of time evolution and a more accurate life assessment can be done.

The present paper is the starting work of a series of technique evaluations with respect to different experimental case studies and different coupling problems, such as magneto-inductive-thermo-mechanical problem and the integration of the stress and strain evaluation in a TMF life assessment postprocessor [8, 9].

References

[1] Tonti E, 2001 *CMES - Computer Modeling in Engineering and Sciences*, vol 2 (2), p 237-258
[2] Tonti E, 2001 *Journal of Computational Acoustics*, vol 9 (4), p 1355-1382
[3] Tonti E, 2001 *Progress in Electromagnetics Research*, (PIER 32 Special Volume on Geometrical Methods for Comp. Electromagnetics), p 1-44
[4] Freschi F, Giaccone L, Repetto M, 2008 *COMPEL - The International Journal for Computation and Mathematics in Electrical and Electronic Engineering*, vol 27 (6), p 1343-1357
[5] Cosmi F, 2005 *CMES - Computer Modeling in Engineering and Sciences*, vol 8 (3), p 191-200
[6] Repetto M, Trevisan F, 2004 *International Journal for Numerical Methods in Engineering*, vol 60 (2), p 755-772
[7] Repetto M, Trevisan F, 2003 *IEEE Transactions on Magnetics*, vol 39 (3), p 1135-1138
[8] Delprete C, Rosso C, Sesana R, 2008 *International Journal of Mechanics and Control*, vol 9, p 17-25
[9] Delprete C, Rosso C, Sesana R, 2006 *Proc. Int. Conf. 8th Biennial ASME Conference ESDA2006* (Torino, I) Paper ESDA2006-95470.

Differential operators associated with the equations of motion and nondissipative heat conduction in the Green–Naghdi theory of thermoelasticity

S. A. Lychev[1] **and A. V. Manzhirov**[1]

[1]Ishlinsky Institute for Problems in Mechanics of the Russian Academy of Sciences, prosp. Vernadskogo 101-1, Moscow, 119526 Russian Federation

E-mail: lychev@ipmnet.ru

Abstract. We consider a wave (hyperbolic) heat conduction theory of the Green–Naghdi type. In the framework of the continual approach, such a theory permits describing low-temperature phenomena of quantum nature (for example, the second sound effect) that are not within reach of the classical parabolic thermoelasticity. At low temperatures, the medium heat conductivity experiences a substantial increase, which should be reckoned with in the design and analysis of cryogenic devices. This change in heat conduction properties results from the wave character of heat propagation, which cannot be taken into account in classical heat conductivity models of the diffusion type but can be described by hyperbolic models of the Green–Naghdi type. That is why considerable attention has been paid to the development of hyperbolic heat conduction models. The present paper deals with analytic solution methods for the corresponding nonself-adjoint initial–boundary value problems.

1. Introduction

Thermal disturbances in some materials (in particular, liquid and crystalline helium) are known to propagate at low temperatures as progressive waves (Peshkov [1]). This phenomenon is usually referred to as "second sound." The wave nature of heat propagation substantially changes the medium heat conductivity, and this change should be taken into account when designing cryogenic devices. Various theories have been devised to explain this phenomenon (Tisza–Landau [2], Atkin et al. [3], etc.). One such theory is the Green–Naghdi theory of nondissipative hyperbolic thermoelasticity [4], in which the time primitive of the temperature is chosen as an independent constitutive variable. This theory results in very special nonself-adjoint boundary value problems, for which classical approaches apparently fail. The aim of this paper is to obtain closed solutions of related linear problems in the form of spectral expansions in complete biorthogonal systems of root functions (eigenfunctions and associated functions) of the corresponding nonself–adjoint operator pencils.

2. Preliminaries

Consider the motion of a body whose elements are identified with points $\boldsymbol{X}$ of a reference configuration. The position $\boldsymbol{x}$ occupied by a point $\boldsymbol{X}$ in the current configuration at time t is given by a sufficiently smooth one-to-one mapping $\boldsymbol{x} = \boldsymbol{\chi}(\boldsymbol{X}, t)$. In the current configuration, an arbitrary material volume $\mathscr{B}$ of the body fills some part $\mathscr{P}$ of the spatial region occupied by the entire body. This part is bounded by a closed surface $\partial\mathscr{P}$. The deformation gradient tensor $\boldsymbol{F} = (\boldsymbol{\nabla}_R\boldsymbol{\chi})^*$ is defined in a standard manner. (Here the asterisk stands for the transpose.) Further, assume that $J = \det \boldsymbol{F} > 0$. By $\boldsymbol{\nabla}_R$ and $\boldsymbol{\nabla}$ we denote the material and spatial gradients, respectively.

In the framework of the classical field theory [5], the field equations for mass conservation, linear and angular momentum, and energy balance can be postulated as

$$\frac{d}{dt}\int_{\mathscr{P}} \rho\, dV = 0, \quad \frac{d}{dt}\int_{\mathscr{P}} \rho\boldsymbol{v}\, dV = \oint_{\partial\mathscr{P}} \boldsymbol{t}\, dA + \int_{\mathscr{P}} \rho\boldsymbol{b}\, dV,$$
$$\frac{d}{dt}\int_{\mathscr{P}} \rho\boldsymbol{\chi}\times\boldsymbol{v}\, dV = \oint_{\partial\mathscr{P}} \boldsymbol{\chi}\times\boldsymbol{t}\, dA + \int_{\mathscr{P}} \boldsymbol{\chi}\times\rho\boldsymbol{b}\, dV,$$
$$\frac{d}{dt}\int_{\mathscr{P}} \rho\left(\frac{\boldsymbol{v}^2}{2}+\varepsilon\right) dV = \int_{\mathscr{P}} \rho\,(\boldsymbol{b}\cdot\boldsymbol{v}+r)\, dV + \oint_{\partial\mathscr{P}} (\boldsymbol{t}\cdot\boldsymbol{v}-q)\, dA.$$

Here ρ is the mass density in the current configuration, ε is the specific internal energy, q is the surface density of heat flux into $\mathscr{P}$, r is the external heat supply rate per unit mass, $\boldsymbol{t}$ is the surface traction, and $\boldsymbol{b}$ represents the external body forces per unit mass.

Taking into account the Euler–Cauchy hypothesis [5], we obtain linear relations between $\boldsymbol{t}$, q, and the outward unit normal $\boldsymbol{n}$ to $\partial\mathscr{P}$; namely, $\boldsymbol{t} = \boldsymbol{n}\cdot\boldsymbol{T}$ and $q = \boldsymbol{n}\cdot\boldsymbol{h}$. Let η be the entropy density per unit mass. Suppose that $q = \eta\theta$ and $\eta = \boldsymbol{n}\cdot\boldsymbol{\mathcal{S}}$, where $\boldsymbol{\mathcal{S}}$ is the entropy flux vector. Since the surface $\partial\mathscr{P}$ is arbitrary, it follows that $\boldsymbol{h} = \boldsymbol{\mathcal{S}}\theta$. The field entropy balance equation reads

$$\frac{d}{dt}\int_{\mathscr{P}} \rho\eta\, dV = -\oint_{\partial\mathscr{P}} \boldsymbol{n}\cdot\boldsymbol{\mathcal{S}}\, dA + \int_{\mathscr{P}} \rho s\, dV, \quad s = s_i + s_e.$$

Here s is the entropy production rate per unit mass. We suppose that it splits additively into the internal entropy production rate s_i and the external entropy production rate s_e. We assume that these terms are independent.

Let ρ_0 be the mass density in the reference configuration, let $\boldsymbol{P}$ be the first Piola–Kirchhoff stress, let $\boldsymbol{h}_R$ be the material heat flux vector, let $\boldsymbol{\mathcal{S}}_{\boldsymbol{R}}$ be the material entropy flux vector, and let ψ be the specific Helmholtz free energy; i.e.,

$$\rho_0 = J\rho, \quad \boldsymbol{P} = J\boldsymbol{F}^{-1}\boldsymbol{T}, \quad \boldsymbol{h}_R = J\boldsymbol{F}^{-1}\boldsymbol{h}, \quad \boldsymbol{\mathcal{S}}_{\boldsymbol{R}} = J\boldsymbol{F}^{-1}\boldsymbol{\mathcal{S}}, \quad \psi = \varepsilon - \theta\eta.$$

An application of the localization theorem yields the local balance equations

$$\dot{\rho}_0 = 0, \quad \boldsymbol{\nabla}_R\cdot\boldsymbol{P} - \rho_0\dot{\boldsymbol{v}} + \rho_0\boldsymbol{b} = \boldsymbol{0}, \quad \boldsymbol{P}^*\cdot\boldsymbol{F}^* = \boldsymbol{F}\cdot\boldsymbol{P}, \tag{1}$$
$$\boldsymbol{P} : \dot{\boldsymbol{F}} - \boldsymbol{\mathcal{S}}_{\boldsymbol{R}}\cdot\boldsymbol{\nabla}_R\theta - \rho_0\dot{\psi} - \rho_0\theta\,(s_i+s_e) - \rho_0\eta\dot{\theta} + \rho_0 r = 0, \tag{2}$$
$$\boldsymbol{\nabla}_R\cdot\boldsymbol{\mathcal{S}}_{\boldsymbol{R}} - \rho_0\,(s_i+s_e) + \rho_0\dot{\eta} = 0. \tag{3}$$

In conventional thermoelasticity, the temperature gradient $\boldsymbol{\nabla}_R\theta$ is chosen as an independent constitutive variable along with the temperature θ and the deformation gradient $\boldsymbol{F}$. In dissipationless thermoelasticity, $\boldsymbol{\nabla}_R\underset{.}{\theta}$ is is taken instead of $\boldsymbol{\nabla}_R\theta$, where $\underset{.}{\theta}$ is the so-called thermal displacement field defined as the time primitive

$$\underset{.}{\theta}(\boldsymbol{X},t) = \int_{t_0}^{t} \theta(\boldsymbol{X},\tau)d\tau + \underset{.}{\theta}^0, \quad \underset{.}{\theta}^0 = \text{const}, \tag{4}$$

of the temperature. Thus, the free energy ψ depends on the following independent variables:

$$\psi = \psi\left(\boldsymbol{X}, \underset{\cdot}{\theta}, \theta, \boldsymbol{\nabla}_R \underset{\cdot}{\theta}, \boldsymbol{F}\right). \tag{5}$$

We point out that the variable (4) is a rather old notion; it introduced as early as in 1921 by van Dantzig as "thermacy."

The constitutive equation can be obtained by the Noll–Coleman technique. The time derivative of the free energy is given by

$$\dot{\psi} = \frac{\partial \psi}{\partial \boldsymbol{\nabla}_R \underset{\cdot}{\theta}} \cdot \boldsymbol{\nabla}_R \dot{\underset{\cdot}{\theta}} + \frac{\partial \psi}{\partial \dot{\underset{\cdot}{\theta}}} \ddot{\underset{\cdot}{\theta}} + \frac{\partial \psi}{\partial \boldsymbol{F}} : \dot{\boldsymbol{F}} + \frac{\partial \psi}{\partial \underset{\cdot}{\theta}} \dot{\underset{\cdot}{\theta}} = \frac{\partial \psi}{\partial \boldsymbol{\nabla}_R \underset{\cdot}{\theta}} \cdot \boldsymbol{\nabla}_R \theta + \frac{\partial \psi}{\partial \theta} \dot{\theta} + \frac{\partial \psi}{\partial \boldsymbol{F}} : \dot{\boldsymbol{F}} + \frac{\partial \psi}{\partial \underset{\cdot}{\theta}} \theta. \tag{6}$$

By substituting (6) into the energy balance equation (2), we obtain

$$\boldsymbol{P} : \dot{\boldsymbol{F}} - \boldsymbol{\mathcal{S}}_{\boldsymbol{R}} \cdot \boldsymbol{\nabla}_R \theta - \rho_0 \left(\frac{\partial \psi}{\partial \boldsymbol{\nabla}_R \underset{\cdot}{\theta}} \cdot \boldsymbol{\nabla}_R \theta + \frac{\partial \psi}{\partial \theta} \dot{\theta} + \frac{\partial \psi}{\partial \boldsymbol{F}} : \dot{\boldsymbol{F}} + \frac{\partial \psi}{\partial \underset{\cdot}{\theta}} \theta + \eta \dot{\theta} \right) + \rho_0 r = \theta \rho_0 s_i + \theta \rho_0 s_e.$$

This equation holds for all thermodynamic processes. Since the internal and external entropy production are independent, we obtain the constitutive equations

$$\eta = -\frac{\partial \psi}{\partial \theta}, \quad \boldsymbol{P} = \rho_0 \frac{\partial \psi}{\partial \boldsymbol{F}}, \quad \boldsymbol{\mathcal{S}}_{\boldsymbol{R}} = -\rho_0 \frac{\partial \psi}{\partial \boldsymbol{\nabla}_R \underset{\cdot}{\theta}}, \quad s_i = \frac{\partial \psi}{\partial \underset{\cdot}{\theta}}, \quad s_e = \theta^{-1} r. \tag{7}$$

The constitutive equation

$$\boldsymbol{h}_R = -\rho_0 \theta \frac{\partial \psi}{\partial \boldsymbol{\nabla}_R \underset{\cdot}{\theta}}$$

for the material heat flux follows from the third equation in (7). If the internal energy does not explicitly depend on the temperature displacement, then the internal entropy production is zero.[1] This case was considered by Green and Naghdi [4].

3. Linear equations

In this section, we consider three states of the body, namely, the initial state, the first deformed state, and the final deformed state. We assume that the difference between the first and final states is infinitesimal. The thermodynamic variables and forces associated with the final deformed state are marked by an asterisk.

The position $\boldsymbol{\chi}^\star$ and temperature $\theta^\star$ in the final deformed state can be represented in the incremental form

$$\boldsymbol{\chi}^\star = \boldsymbol{\chi} + \varsigma \boldsymbol{\chi}', \quad \theta^\star = \theta + \varsigma \theta'. \tag{8}$$

Here ς is a real parameter small enough that its square and higher powers can be neglected, $\varsigma\boldsymbol{\chi}'$ and $\varsigma\theta'$ are the position and temperature increments, and $\boldsymbol{\chi}$ and θ are the position and temperature in the first deformed state.

The incremental form

$$\underset{\cdot}{\theta}^\star = \int_{t_0}^{t} \left[\theta(\tau) + \varsigma \theta'(\tau)\right] d\tau = \underset{\cdot}{\theta} + \varsigma \underset{\cdot}{\theta}', \quad \underset{\cdot}{\theta}' = \int_{t_0}^{t} \theta'(\tau)\, d\tau,$$

[1] The lack of dissipation makes it possible to give a variational statement, to which the Noether theorem applies. Thus, the governing equations of the Green–Naghdi theory can be obtained from the variational Lagrangian formalism [6, 7].

for the temperature displacement follows from (4) and $(8)_2$. The incremental forms for the deformation gradient and the temperature displacement gradient read

$$\boldsymbol{F}^\star = (\boldsymbol{\nabla}_R\cdot\boldsymbol{\chi}^\star)^* = \left(\boldsymbol{\nabla}_R\cdot(\boldsymbol{\chi}+\varsigma\boldsymbol{\chi}')\right)^* = \boldsymbol{F}+\varsigma\boldsymbol{F}', \quad \boldsymbol{F}' = \left(\boldsymbol{\nabla}_R\cdot\boldsymbol{\chi}'\right)^*, \quad \boldsymbol{\nabla}_R\theta^\star = \boldsymbol{\nabla}_R\theta + \varsigma\boldsymbol{\nabla}_R\theta'.$$

By combining this with (5), we obtain the expansion

$$\begin{aligned}\psi^\star &= \psi + \varsigma\left(\frac{d\psi^\star}{d\varsigma}\right)_{\varsigma=0} + \frac{\varsigma^2}{2}\left(\frac{d^2\psi^\star}{d\varsigma^2}\right)_{\varsigma=0} + o(\varsigma^2)\\ &= \psi + \varsigma\left(\frac{\partial\psi}{\partial\boldsymbol{F}}:\boldsymbol{F}' + \frac{\partial\psi}{\partial(\boldsymbol{\nabla}_R\theta)}\cdot\boldsymbol{\nabla}_R\theta' + \frac{\partial\psi}{\partial\theta}\theta'\right) + \frac{\varsigma^2}{2}\left[\boldsymbol{F}':\frac{\partial^2\psi}{\partial\boldsymbol{F}^2}:\boldsymbol{F}' + \boldsymbol{\nabla}_R\theta'\cdot\frac{\partial^2\psi}{\partial(\boldsymbol{\nabla}_R\theta)^2}\cdot\boldsymbol{\nabla}_R\theta'\right.\\ &\quad\left. + \frac{\partial^2\psi}{\partial\theta^2}(\theta')^2 + \boldsymbol{\nabla}_R\theta'\cdot\frac{\partial^2\psi}{\partial(\boldsymbol{\nabla}_R\theta)\partial\boldsymbol{F}}:\boldsymbol{F}' + \theta'\frac{\partial^2\psi}{\partial\theta\partial\boldsymbol{F}}:\boldsymbol{F}' + \boldsymbol{\nabla}_R\theta'\cdot\frac{\partial^2\psi}{\partial(\boldsymbol{\nabla}_R\theta)\partial\theta}\theta'\right] + o(\varsigma^2)\end{aligned}$$

of free energy in powers of ς near the first deformed state. Let us introduce some notation:

$$\mathfrak{A} = \frac{\partial\psi}{\partial\boldsymbol{F}}, \quad \mathfrak{B} = \frac{\partial\psi}{\partial(\boldsymbol{\nabla}_R\theta)}, \quad \mathfrak{C} = \frac{\partial\psi}{\partial\theta}, \quad \mathfrak{E} = \frac{\partial^2\psi}{\partial\boldsymbol{F}^2}, \quad \mathfrak{D} = \frac{\partial^2\psi}{\partial(\boldsymbol{\nabla}_R\theta)^2}, \quad \mathfrak{F} = \frac{\partial^2\psi}{\partial\theta^2},$$
$$\mathfrak{H} = \frac{\partial^2\psi}{\partial(\boldsymbol{\nabla}_R\theta)\partial\boldsymbol{F}}, \quad \mathfrak{K} = \frac{\partial^2\psi}{\partial(\boldsymbol{\nabla}_R\theta)\partial\theta}, \quad \mathfrak{L} = \frac{\partial^2\psi}{\partial\theta\partial\boldsymbol{F}}.$$

It follows from the commutativity of second derivatives that

$$\mathfrak{E} = \mathfrak{E}^{(3412)}, \quad \mathfrak{D} = \mathfrak{D}^*, \quad \mathfrak{H} = \mathfrak{H}^{(231)}.$$

Now we have

$$\frac{\partial\psi^\star}{\partial\boldsymbol{F}^\star} = \mathfrak{A} + \varsigma\left(\mathfrak{E}:\boldsymbol{F}' + \mathfrak{H}\cdot\boldsymbol{\nabla}_R\theta' + \mathfrak{L}\theta'\right),$$
$$\frac{\partial\psi^\star}{\partial(\boldsymbol{\nabla}_R\theta)^\star} = \mathfrak{B} + \varsigma\left(\mathfrak{D}\cdot\boldsymbol{\nabla}_R\theta' + \mathfrak{H}:\boldsymbol{F}' + \mathfrak{K}\theta'\right), \quad \frac{\partial\psi^\star}{\partial\theta^\star} = \mathfrak{C} + \varsigma\left(\mathfrak{F}\theta' + \mathfrak{L}:\boldsymbol{F}' + \boldsymbol{\nabla}_R\theta'\mathfrak{K}\right).$$

Since the tensor coefficients $\mathfrak{A}$, $\mathfrak{B}$, and $\mathfrak{C}$ determine the stress and entropy fields in the first deformed state,

$$\mathfrak{A} = \rho_0\boldsymbol{P}, \quad \mathfrak{B} = -\rho_0\boldsymbol{S_R}, \quad \mathfrak{C} = -\eta,$$

we obtain the expansions

$$\begin{aligned}\boldsymbol{P}^\star &= \rho_0\frac{\partial\psi^\star}{\partial\boldsymbol{F}^\star} = \boldsymbol{P}+\varsigma\boldsymbol{P}', & \boldsymbol{P}' &= \rho_0\left(\mathfrak{E}:\boldsymbol{F}' + \mathfrak{H}\cdot\boldsymbol{\nabla}_R\theta' + \mathfrak{L}\theta'\right),\\ \boldsymbol{S_R}^\star &= -\rho_0\frac{\partial\psi^\star}{\partial(\boldsymbol{\nabla}_R\theta)^\star} = \boldsymbol{S_R}+\varsigma\boldsymbol{S_R}', & \boldsymbol{S_R}' &= -\rho_0\left(\mathfrak{D}\cdot\boldsymbol{\nabla}_R\theta' + \mathfrak{H}:\boldsymbol{F}' + \mathfrak{K}\theta'\right),\\ \eta^\star &= -\frac{\partial\psi^\star}{\partial\theta^\star} = \eta + \varsigma\eta', & \eta' &= -\left(\mathfrak{F}\theta' + \mathfrak{L}:\boldsymbol{F}' + \boldsymbol{\nabla}_R\theta'\mathfrak{K}\right).\end{aligned} \tag{9}$$

The entropy flux vanishes under the conditions $\boldsymbol{\nabla}_R\theta' = \boldsymbol{0}$ and $\theta' = 0$. It follows that $\mathfrak{H} = \boldsymbol{0}$. By substituting (9) into $(1)_2$ and (3), we obtain the incremental equations

$$\boldsymbol{\nabla}_R\cdot\boldsymbol{P}' - \rho_0\ddot{\boldsymbol{\chi}}' + \rho_0\boldsymbol{b}' = \boldsymbol{0}, \quad \boldsymbol{\nabla}_R\cdot\boldsymbol{S_R}' + \rho_0\dot{\eta}' - \rho_0 s' = 0.$$

Here $\boldsymbol{b}' = \boldsymbol{b}^{\star} - \boldsymbol{b}$ is the body force increment and $s' = s^{\star} - s$ is the entropy production increment induced by the change in the current configuration.

By taking into account relations (9), one can state the equations of motion and heat conduction in the form

$$\nabla_R \cdot \left(\rho_0 \mathfrak{C} : (\nabla_R \boldsymbol{\chi}')^* + \rho_0 \mathfrak{L} \theta'\right) - \rho_0 \ddot{\boldsymbol{\chi}}' + \rho_0 \boldsymbol{b}' = \mathbf{0},$$
$$\nabla_R \cdot \left(\rho_0 \mathfrak{D} \cdot \nabla_R \theta'\right) + \left(\nabla_R \cdot \rho_0 \mathfrak{K}\right) \dot{\theta}' - \rho_0 \left[\mathfrak{F} \ddot{\theta}' + \mathfrak{L} : (\nabla_R \ddot{\boldsymbol{\chi}}')^*\right] + \rho_0 s' = 0.$$

These equations, together with the prescribed boundary conditions and the initial conditions

$$\boldsymbol{\chi}'\Big|_{t=0} = \boldsymbol{\chi}_0, \quad \theta'\Big|_{t=0} = \theta_0, \quad \dot{\boldsymbol{\chi}}'\Big|_{t=0} = \dot{\boldsymbol{\chi}}_0, \quad \dot{\theta}'\Big|_{t=0} = \dot{\theta}_0,$$

form a well-posed statement of the initial–boundary value problem.

4. Differential operators

Consider the space of square integrable four-component complex-valued vector functions $(\boldsymbol{\chi}, \theta)$ with the inner product

$$\langle (\boldsymbol{\chi}_1, \theta_1), (\boldsymbol{\chi}_2, \theta_2) \rangle = \int_{\mathscr{B}} (\boldsymbol{\chi}_1 \cdot \boldsymbol{\chi}_2 + \theta_1 \theta_2)\, dV.$$

The initial–boundary value problem can be stated as the following Cauchy problem with operator coefficients:

$$\mathcal{L}(\boldsymbol{\chi}, \theta) = \boldsymbol{f}, \quad \mathcal{L}(\boldsymbol{\chi}', \theta') = \mathcal{A}_0 (\boldsymbol{\chi}', \theta') + \mathcal{A}_1 \frac{\partial}{\partial t} (\boldsymbol{\chi}', \theta') + \mathcal{A}_2 \frac{\partial^2}{\partial t^2} (\boldsymbol{\chi}', \theta'), \quad \boldsymbol{f} = -(\rho_0 \boldsymbol{b}, \rho_0 s'),$$
$$\mathcal{A}_0 = \begin{pmatrix} \mathbb{A} \vdots \nabla_R \nabla_R + \mathbb{B} : \nabla_R & \mathbb{C} \cdot \nabla_R + \mathbb{D} \\ 0 & \mathbb{E} : \nabla_R \nabla_R + \mathbb{F} \cdot \nabla_R \end{pmatrix}, \quad \mathcal{A}_1 = \begin{pmatrix} 0 & 0 \\ 0 & \mathbb{G} \end{pmatrix}, \quad \mathcal{A}_2 = -\begin{pmatrix} \rho_0 & 0 \\ \mathbb{K} : \nabla_R & \mathbb{H} \end{pmatrix}.$$

Here we have used the notation

$$\mathbb{A} = \rho_0 \mathfrak{C}^{(2143)}, \quad \mathbb{B} = \nabla_R \cdot \rho_0 \mathfrak{C}^{(1243)}, \quad \mathbb{C} = \rho_0 \mathfrak{L}, \quad \mathbb{D} = \nabla_R \cdot \rho_0 \mathfrak{L},$$
$$\mathbb{E} = \rho_0 \mathfrak{D}, \quad \mathbb{F} = \nabla_R \cdot \rho_0 \mathfrak{D}, \quad \mathbb{G} = \nabla_R \cdot \rho_0 \mathfrak{K}, \quad \mathbb{H} = \rho_0 \mathfrak{F}, \quad \mathbb{K} = \rho_0 \mathfrak{L}^*.$$

The problem in question gives rise to the polynomial operator pencil

$$\mathcal{L}_\nu = \mathcal{A}_0 + \nu \mathcal{A}_1 + \nu^2 \mathcal{A}_2,$$

The domain $\mathcal{D}$ of the pencil $\mathcal{L}_\nu$ is determined by the boundary operator $\mathcal{B}$ as follows:

$$\mathcal{D} = \left\{ (\boldsymbol{\chi}, \theta) \middle| \mathcal{B}(\boldsymbol{\chi}, \theta)|_{\partial\mathscr{P}} = \mathbf{0} \right\}.$$

The adjoint pencil $\mathcal{L}_\nu^*$ can be obtained directly from the definition:

$$\forall (\boldsymbol{\chi}_1, \theta_1) \in \mathcal{D}, \ (\boldsymbol{\chi}_2, \theta_2) \in \mathcal{D}^* \quad \langle \mathcal{L}_\nu (\boldsymbol{\chi}_1, \theta_1), (\boldsymbol{\chi}_2, \theta_2) \rangle = \langle (\boldsymbol{\chi}_1, \theta_1), \mathcal{L}_\nu^* (\boldsymbol{\chi}_2, \theta_2) \rangle.$$

Indeed, a standard argument based on the Gauss divergence theorem shows that

$$\langle \mathcal{L}_\nu (\boldsymbol{\chi}, \theta), (\boldsymbol{\chi}^*, \theta^*) \rangle - \langle (\boldsymbol{\chi}, \theta), \mathcal{L}_\nu^* (\boldsymbol{\chi}^*, \theta^*) \rangle = [(\boldsymbol{\chi}, \theta), (\boldsymbol{\chi}^*, \theta^*)]_{\partial\mathscr{B}}$$
$$= \oint_{\partial\mathscr{B}} \rho_0 \boldsymbol{n}_0 \cdot \Big[(\mathfrak{C} : (\nabla_R \boldsymbol{\chi})^*) \cdot \boldsymbol{\chi}^* - \boldsymbol{\chi} \cdot \mathfrak{C}^{(4321)} : (\nabla_R \boldsymbol{\chi}^*)^* + \theta \mathfrak{L} \cdot \boldsymbol{\chi}^* + \mathfrak{D} \cdot (\nabla_R \theta) \theta^* - \theta \mathfrak{D}^* \cdot \nabla_R \theta^* + \mathfrak{L}^* \cdot \boldsymbol{\chi} \theta^* \Big]\, dA \tag{10}$$

Journal of Physics: Conference Series **181** (2009) 012096 doi:10.1088/1742-6596/181/1/012096

provided that

$$\mathcal{L}_\nu^* = \mathcal{A}_0^* + \overline{\nu}\mathcal{A}_1 + \overline{\nu}^2\mathcal{A}_2^*,$$

$$\mathcal{A}_0^* = \begin{pmatrix} \mathbb{A}^* : \nabla_R\nabla_R + \mathbb{B}^* : \nabla_R & 0 \\ \mathbb{K} : \nabla_R & \mathbb{E}^* : \nabla_R\nabla_R + \mathbb{F}\cdot\nabla_R \end{pmatrix}, \quad \mathcal{A}_2^* = -\begin{pmatrix} \rho_0 & \mathbb{C}^*\cdot\nabla_R + \mathbb{D} \\ 0 & \mathbb{H} \end{pmatrix},$$

$$\mathbb{A}^* = \rho_0\mathfrak{E}^{(3412)}, \quad \mathbb{B}^* = \nabla_R\cdot\rho_0\mathfrak{E}^{(3421)}, \quad \mathbb{C}^* = \rho_0\mathfrak{L}^*, \quad \mathbb{E}^* = \rho_0\mathfrak{D}^*.$$

We define the domain $\mathcal{D}^*$ of the adjoint pencil $\mathcal{L}_\nu^*$ as the set of vector functions $\boldsymbol{k}^* = (\boldsymbol{\chi}^*, \theta^*)$ such that the bilinear form (10) is zero for arbitrary $(\boldsymbol{\chi}, \theta) \in \mathcal{D}$. Hence we obtain the adjoint boundary operator $\mathcal{B}^*$ in closed form. It is well known that the system of root functions of the pencil $\mathcal{L}_\nu$ is biorthogonal to that of the adjoint pencil. Note that if $\mathcal{D} = \mathcal{D}^*$, then this system is a basis. This provides a theoretical background for the representation of the solution in the form of a spectral expansion.

5. Solution in the general case

The obtained solutions have the form of expansions based on the systems of eigenfunctions $\boldsymbol{k}_i$ and $\boldsymbol{k}_j^*$ that can be found as the solutions of the mutually adjoint boundary value problems

$$\mathcal{L}_\nu\boldsymbol{k}_i = \boldsymbol{0}, \quad \mathcal{L}_\nu^*\boldsymbol{k}_j = \boldsymbol{0}, \quad \boldsymbol{k}_i = (\boldsymbol{\chi}_i, \theta_i), \quad \boldsymbol{k}_j^* = (\boldsymbol{\chi}_j^*, \theta_j^*); \quad i = 1, \ldots\infty; \quad j = 1, \ldots\infty. \tag{11}$$

We assume that the body $\mathscr{B}$ is compact. Hence the set of eigenfunctions and the corresponding set of eigenvalues are countable. According to [8], the following biorthogonal relations hold for $\boldsymbol{k}_i$ and $\boldsymbol{k}_j^*$:

$$\langle\mathcal{A}_1\boldsymbol{k}_i, \boldsymbol{k}_j^*\rangle + (\nu_i + \nu_j)\langle\mathcal{A}_2\boldsymbol{k}_i, \boldsymbol{k}_j^*\rangle = 0, \quad \langle\mathcal{A}_0\boldsymbol{k}_i, \boldsymbol{k}_j^*\rangle - \nu_i\nu_j\langle\mathcal{A}_2\boldsymbol{k}_i, \boldsymbol{k}_j^*\rangle = 0.$$

These properties permit one to represent the solution in the form [8]

$$\boldsymbol{y} = \sum_{i=1}^{\infty}\left[\left(\langle\mathcal{A}_1^*\boldsymbol{k}_i^* + \nu_i\mathcal{A}_2^*\boldsymbol{k}_i^*, \boldsymbol{y}_0\rangle + \langle\mathcal{A}_2^*\boldsymbol{k}_i^*, \dot{\boldsymbol{y}}_0\rangle\right)\mathbf{exp}(\nu_i t) - \int_0^t \langle\boldsymbol{f}(\tau), \boldsymbol{k}_i^*\rangle\,\mathbf{exp}\big(\nu_i(t-\tau)\big)\,d\tau\right]\boldsymbol{k}_i\mathcal{Q}_i^{-1},$$

where

$$\boldsymbol{y} = (\boldsymbol{\chi}, \theta), \quad \boldsymbol{y}_0 = (\boldsymbol{\chi}_0, \theta_0), \quad \dot{\boldsymbol{y}}_0 = \left(\dot{\boldsymbol{\chi}}_0, \dot{\theta}_0\right), \quad \boldsymbol{f} = (\rho_0\boldsymbol{b}, \rho_0 s), \quad Q_i = \langle\mathcal{A}_1\boldsymbol{k}_i, \boldsymbol{k}_i^*\rangle + 2\nu_i\langle\mathcal{A}_2\boldsymbol{k}_i, \boldsymbol{k}_i^*\rangle.$$

The spectral expansions in complete biorthogonal systems of root functions (eigenfunctions and associated functions) are of the same form and discussed in details in [8].

In the special case $\mathcal{A}_1 = 0$ and $\operatorname{Re}\nu_i = 0$, this representation can be written as follows:

$$\boldsymbol{y} = \sum_{i=1}^{\infty}\left[\left(\langle\mathcal{A}_2^*\boldsymbol{k}_i^*, \boldsymbol{y}_0\rangle\cos(\lambda_i t) + \frac{1}{\lambda_i}\langle\mathcal{A}_2^*\boldsymbol{k}_i^*, \dot{\boldsymbol{y}}_0\rangle\right)\sin(\lambda_i t) - \frac{1}{\lambda_i}\int_0^t \langle\boldsymbol{f}(\tau), \boldsymbol{k}_i^*\rangle\sin\big(\lambda_i(t-\tau)\big)\,d\tau\right]\boldsymbol{k}_i\mathcal{N}_i^{-1}.$$

Here $N_i = 2\langle\mathcal{A}_2\boldsymbol{k}_i, \boldsymbol{k}_i^*\rangle$ and $\lambda_i = \operatorname{Im}\nu_i$.

Journal of Physics: Conference Series **181** (2009) 012096 doi:10.1088/1742-6596/181/1/012096

6. Solution in a special case

To find a simple example of the proposed representation, consider an isotropic homogenous body and assume that the first deformed state is natural. (That is, the initial stress and entropy flux are zero.) In this case, without loss of generality, it can be assumed that $\boldsymbol{\chi}$ represents the displacements and

$$\rho = \text{const}, \quad \boldsymbol{T} = \boldsymbol{0}, \quad \boldsymbol{S} = \boldsymbol{0}, \quad \boldsymbol{F} \approx \boldsymbol{I}, \quad \nabla_R \approx \nabla, \quad \mathfrak{C} = \frac{\lambda}{\rho}\boldsymbol{I}\otimes\boldsymbol{I} + \frac{2\mu}{\rho}(\boldsymbol{I}\otimes\boldsymbol{I})^{(1324)},$$
$$\mathfrak{L} = \frac{\alpha}{\rho}\boldsymbol{I}, \quad \mathfrak{D} = \frac{\Lambda}{\rho}\boldsymbol{I}, \quad \mathfrak{F} = \frac{\varkappa}{\rho}\boldsymbol{I}, \quad \tilde{\mathfrak{K}} \approx \mathfrak{K} = \text{const},$$

where λ and μ are the Lamé elastic moduli and Λ, $\varkappa$, and α are the thermomechanical moduli. Thus, the mutually adjoint systems (11) are reduced to

$$\mu\nabla^2\boldsymbol{\chi}_i + (\lambda+\mu)\nabla\nabla\cdot\boldsymbol{\chi}_i - \alpha\nabla\theta_i - \nu_i^2\rho\boldsymbol{\chi}_i = \boldsymbol{0}, \quad \nabla^2\theta_i - \nu_i^2\beta\theta_i - \nu_i^2\gamma\nabla\cdot\boldsymbol{\chi}_i = 0,$$
$$\mu\nabla^2\boldsymbol{\chi}_i^* + (\lambda+\mu)\nabla\nabla\cdot\boldsymbol{\chi}_i^* + \overline{\nu}_i^2\alpha\nabla\theta_i^* - \overline{\nu}_i^2\rho\boldsymbol{\chi}_i^* = \boldsymbol{0}, \quad \nabla^2\theta_i^* - \overline{\nu}_i^2\beta\theta_i^* + \gamma\nabla\cdot\boldsymbol{\chi}_i^* = 0. \qquad (12)$$

Here $\beta = \varkappa/\Lambda$ and $\gamma = \alpha/\Lambda$. Now let us apply the Stokes–Helmholtz resolution

$$\boldsymbol{\chi}_i = \nabla\phi_i + \nabla\times\boldsymbol{\psi}_i, \quad \nabla\cdot\boldsymbol{\psi}_i = 0, \quad \boldsymbol{\chi}_i^* = \nabla\phi_i^* + \nabla\times\boldsymbol{\psi}_i^*, \quad \nabla\cdot\boldsymbol{\psi}_i^* = 0. \qquad (13)$$

where ϕ_i, ϕ_i^* and $\boldsymbol{\psi}_i$, $\boldsymbol{\psi}_i^*$ are scalar and vector functions, respectively. By substituting (13) into (12), we obtain

$$(\lambda+2\mu)\nabla^2\phi_i - \alpha\theta_i - \nu_i^2\rho\phi_i = 0, \quad \nabla^2\theta_i - \nu_i^2\beta\theta_i - \nu_i^2\gamma\nabla^2\phi_i = 0, \quad \mu\nabla\boldsymbol{\psi}_i - \nu_i^2\rho\boldsymbol{\psi}_i = \boldsymbol{0},$$
$$(\lambda+2\mu)\nabla^2\phi_i^* + \overline{\nu}_i^2\alpha\theta_i^* - \overline{\nu}_i^2\rho\phi_i^* = 0, \quad \nabla^2\theta_i^* - \overline{\nu_i}^2\beta\theta_i^* + \gamma\nabla^2\phi_i^* = 0, \quad \mu\nabla\boldsymbol{\psi}_i^* - \overline{\nu}_i^2\rho\boldsymbol{\psi}_i^* = \boldsymbol{0}.$$

In this section, we assume that the desired solutions are generated by scalar potentials alone[2] ($\boldsymbol{\psi}_i = \boldsymbol{\psi}_i^* = 0$). It follows in a standard way that

$$\begin{pmatrix}\phi_i\\ \theta_i\end{pmatrix} = \begin{pmatrix}\varrho\\ \varsigma\end{pmatrix}\Xi_i.$$

Here ϱ and ς are constants, Ξ_i is a solution of the auxiliary equation

$$\nabla^2\Xi_i = \zeta_i\Xi_i$$

of the Helmholtz type, and ζ_i satisfies the algebraic equation

$$\begin{vmatrix}(\lambda+2\mu)\zeta_i - \rho\nu_i^2 & -\alpha\\ -\gamma\zeta_i\nu_i^2 & \zeta_i - \beta\nu_i^2\end{vmatrix} = 0.$$

By solving the latter for ν_i, we obtain

$$\nu_{i1,i2}^2 = \frac{\zeta_i}{2\rho\beta}\xi_{1,2}, \ \xi_{1,2} = \rho + \alpha\gamma + \beta(\lambda+2\mu) \pm \omega, \ \omega = \sqrt{[\rho+\alpha\gamma-\beta(\lambda+2\mu)]^2 + 4\alpha\beta\gamma(\lambda+2\mu)}.$$

[2] This assumption, of course, restricts the problem under consideration to potential displacement fields, body forces, and initial data. However, the rotational components of the above-mentioned fields do not interact with the temperature field. Thus, the rotational components can be omitted in the case of purely thermal excitation.

We point out that the coincidence of the roots $\nu_{i1}^2 = \nu_{i2}^2$ corresponds to the case of multiple eigenvalues, where the corresponding root space has nonempty chains of associated functions. Further, in the case of simple spectrum the eigenfunctions can be obtained as

$$\boldsymbol{k}_{i1} = (\boldsymbol{\chi}_{i1}, \theta_{i1}) = \left(\left[1 - \frac{\xi_1}{2\rho}\right] \boldsymbol{\nabla}\Xi_i, \gamma\nu_{i1}^2\Xi_i\right), \quad \boldsymbol{k}_{i2} = (\boldsymbol{\chi}_{i2}, \theta_{i2}) = \left(\left[1 - \frac{\xi_2}{2\rho}\right] \boldsymbol{\nabla}\Xi_i, \gamma\nu_{i2}^2\Xi_i\right),$$

$$\boldsymbol{k}_{i1}^* = (\boldsymbol{\chi}_{i1}^*, \theta_{i1}^*) = \left(\left[1 - \frac{\xi_1}{2\rho}\right] \boldsymbol{\nabla}\Xi_i, -\alpha\Xi_i\right), \quad \boldsymbol{k}_{i2}^* = (\boldsymbol{\chi}_{i2}^*, \theta_{i2}^*) = \left(\left[1 - \frac{\xi_2}{2\rho}\right] \boldsymbol{\nabla}\Xi_i, -\alpha\Xi_i\right),$$

It is clear that the solution thus obtained depends on the boundary conditions for the auxiliary scalar function Ξ. Consider the Neumann type boundary conditions $\boldsymbol{n}\cdot\boldsymbol{\nabla}\Xi\Big|_{\partial\mathscr{B}} = 0$. This corresponds to the following boundary conditions in terms of the variables $\boldsymbol{\chi}$ and θ (where $\boldsymbol{I}$ is the unit tensor)

$$\boldsymbol{n}\cdot\boldsymbol{\chi} = 0, \quad \boldsymbol{n}\cdot\boldsymbol{T}\cdot(\boldsymbol{I} - \boldsymbol{n}\otimes\boldsymbol{n}) = \boldsymbol{0}, \quad \boldsymbol{n}\cdot\boldsymbol{\nabla}\theta = 0.$$

Now let $\mathscr{B}$ be a cube with edge length π, and let x, y, z be Cartesian coordinates. Assume that the body force and the entropy production are zero, and so are the initial displacements, velocities, and temperature rate; i.e.,

$$\boldsymbol{b} = \boldsymbol{0}, \quad s = 0, \quad \boldsymbol{\chi}\Big|_{t=0} = \boldsymbol{0}, \quad \dot{\boldsymbol{\chi}}\Big|_{t=0} = \boldsymbol{0}, \quad \dot{\theta}\Big|_{t=0} = 0.$$

Let the initial temperature distribution be given by $\theta_0 = H(z) - H(z-h)$, $0 > h > \pi$, where $H(z)$ is the Heaviside step function. The solution of the auxiliary problem is

$$\Xi_k = \cos nx \cos my \cos lz, \quad \zeta_k = -n^2 - m^2 - l^2.$$

Finally, we obtain the desired solution in the form

$$\boldsymbol{\chi} = w\boldsymbol{e}_z, \quad w = \sum_{i=1}^{\infty} \frac{2\alpha\beta}{\pi i^2 \omega} \sin(ih)\sin(iz)\left(\cos(|\nu_{2i}|t) - \cos(|\nu_{1i}|t)\right),$$

$$\theta = \frac{h}{\pi} + \sum_{i=1}^{\infty} \frac{4\alpha\gamma}{\pi i} \sin(ih)\cos(iz)\left[\frac{\xi_1\cos(|\nu_{1i}|t)}{(\xi_1 - 2\rho)^2 + 4\alpha\gamma\rho} + \frac{\xi_2\cos(|\nu_{2i}|t)}{(\xi_2 - 2\rho)^2 + 4\alpha\gamma\rho}\right].$$

The results can be summarized as follows. The differential operator pencil associated with the equations of motion and nondissipative heat conduction is not self-adjoint. The spectrum of the above-mentioned operators in the isotropic homogeneous case is real. The solution of the corresponding initial–boundary value problem can be represented in the form of a spectral expansion.

Acknowledgments

The research was financially supported by the Russian Foundation for Basic Research (under grants Nos. 08-01-00553-a, 08-01-91302-IND a, and 09-08-01194-a) and by the Branch of Power Engineering, Mechanical Engineering, Mechanics, and Control Processes of the Russian Academy of Sciences (Program No. 13 OE).

References

[1] Pitaevskii L. P. 1968 *Phys. Uspekhi* **95** 139–44.
[2] Landau L. D. 1967 *Phys. Uspekhi* **11** 495–520.
[3] Atkin R. J., Fox N. and Vasey M. W. 1975 *J. Elasticity* **5** 237–48.
[4] Green A. E. and Naghdi P. M. 1993 *J. Elasticity* **61** 189–208.
[5] Truesdell C. and Toupin R. A. 1960 *Handbuch der Physik. The Classical Field Theories.* Band III/1 pp 226–858.
[6] Kalpakides V. K. and Maugin G. A. 2004 *Rep. Math. Phys.* **53** 371–91.
[7] Lychev S. A. and Semenov D. A. 2008 *Vestnik SSU* **2** 183–217 (in Russian).
[8] Lychev S. A. 2008 *Mech. Solids* **43** 769-784

Innovative Unmanned Airship Structural Analysis: Dual-Hull and Exoskeletal Configurations

Alessio Cappadona[1], Roberto Lecca[1], Matteo Vazzola[1], Piero Gili[1], Pierangelo Farina[2], Cecilia Surace[3]

[1]Politecnico di Torino, Dept. of Aeronautical and Space Engineering, corso Duca degli Abruzzi 24, Torino, Italy

[2]BLUE Engineering, Via Albenga, 98, Rivoli (TO), Italy

[3]Politecnico di Torino, Dept. of Structural and Geotechnical Engineering, corso Duca degli Abruzzi 24, Torino, Italy

cecilia.surace@polito.it

Abstract. The research study described concerns the design and development of an innovative airship concept which is remotely-controlled and intended to be used for monitoring, surveillance, exploration and reconnaissance missions. Two potential structural configurations have been analysed: the first consists of a dual-hull configuration, characterised by the presence of a primary support structure flanked by the two inflated sections held to the structure by appropriated bindings. The second is an exoskeletal configuration which features a single inflated section incorporating two separate elements held internally by a system of ribs. The final aim of this study is to analyse both configurations to determine the most appropriate solution in terms of performance and cost.

1. Introduction

In recent years there has been a considerable surge of interest in the development of unmanned aerial vehicles for applications both in the civilian and military sectors. Radio-controlled airships represent the most suitable solution for missions of surveillance and exploration at low altitudes and low velocities. Furthermore, airships are also being considered for radio or television platforms, radio relays and specialist scientific applications. Due to the nature of their propulsion they are practically silent, non-polluting, ecological and hence suited for employment in environmental and oceanographic fields, as well as for use in atmospheric analysis and nocturnal surveillance.

Such "lighter-than-air" platforms combine the advantages of the capability of vertical takeoff with the feasibility of extensive flight times over a range of velocities and altitudes. Moreover, in comparison with helicopters, for example, the absence of rotors reduces the cost of structural design and the effects of the resulting substantial vibration on the payload (likely to be television cameras or surveillance equipment). Thus overall, due to their inherent characteristics regarding performance and safety, airships can be used effectively in commercial, civil and military applications.

For the applications described above it is conceivable to employ a classical airship design, however such designs have intrinsic weaknesses – such as problems with control at low velocities, in particular during landing and takeoff, and in adverse weather conditions. Hence it was proposed to design a new type of radio-controlled lighter-than-air platform which is novel in terms of its unconventional shape and control system.

Conventional airships are manoeuvred (in addition to a possible longitudinal orientation of thrusting propellers) by aerodynamic surfaces that have one part fixed and one part mobile. The construction of this new design is distinctly different to that of a classical configuration, in that it is based on a structure of two adjacent hulls separated by a central plane, which has both a structural function and serves to contain or support the payload and power unit. The dual-hull design (named "Elettra Twin Flyers (ETF)") has the important distinction of a reduced lateral cross-sectional area (hence reduced sensitivity to cross winds) for a similar volume of gas, thus still retaining comparative buoyancy. The airship features ballonets inside each envelope which can be inflated and deflated with air during descent and ascent respectively. The landing gear is comprised of a system of legs. The features described are represented in Figure 1.

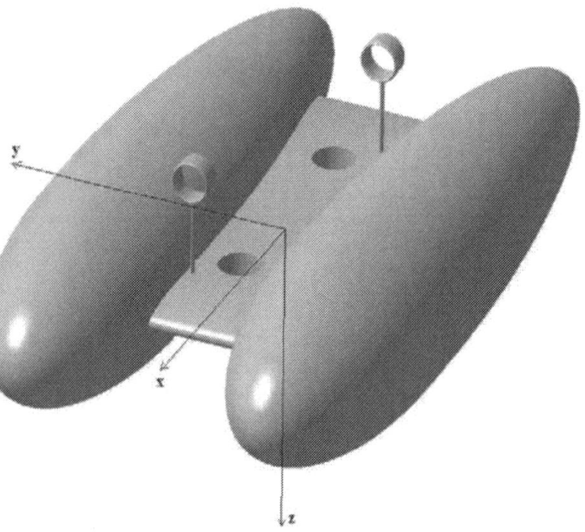

Figure 1: The design of *Elettra Twin Flyers.*

With appropriate rotation and variation of the thruster forces generated automatically through the control system, this configuration enables rotation around three axes and translation in three directions. Furthermore, it is possible to remain hovering with the prow (the front of the ship, corresponding to the positive x-direction, as illustrated in figure 1) orientated in any direction with winds incident at any angle. During hovering, altitude is maintained by the helium and the propellers orientated along the vertical axis, while during forward movement additional aerodynamic lift is also generated as a result of air-flow over the hulls with suitable angle of attach. As such, the complete elimination of aerodynamic control surfaces not only increases manoeuvrability at low velocities, but eliminates a source of disturbance during operation in adverse weather conditions. An obvious consequence of this approach is the need for higher power required for manoeuvres, and a reduction in manoeuvrability with increasing velocity, when compared to airships of conventional design. However, for aircraft used primarily for surveillance applications, this is not thought to represent a particular problem.

The current design is an evolution of previous designs whereas the concept behind the control system is identical comprising a set of 8 or 10 propellers, of which 2 or 4 are fixed and 4 or 8 are manoeuvrable. Similar to previous designs, with the new ETF06 there are no movable aerodynamic control surfaces.

Aside from the use of electric motors, low environmental impact is guaranteed by the chosen power system – hydrogen fuel cells – and auxiliary batteries or supercapacitors to cope with energy peaks resulting from abrupt manoeuvres.

After construction of the flight simulator and a working (albeit simplified) "demonstrator" airship, two more advanced designs were proposed regarding the structure of the prototype. They are the result of a natural evolution of the demonstrator and will use the same control strategy. The two alternative solutions are as follows:

- The first is a direct derivation of the demonstrator – the "dual-hull" solution. This is characterised by a main load-bearing structure, to which are attached on either side the two gas envelopes, attached by arms of an elongated S-shape.
- The second solution is distinctly different from the demonstrator, above all because, unlike the latter, it is a rigid structure – this is known as the "exoskeletal" design. It features a single hull formed from the union of two parallel hulls, supported internally by a system of structural ribs.

The structural development of both designs will be driven by a series of structural analyses, the object being to arrive at a solution with minimum weight, minimum volume, maximum safety and minimum cost.

2. Dual-Hull Configuration

The dual-hull configuration (Figure 2) is the classic design of two hulls separated by a central structure: The *nacelle* (the compartment in which the payload, power supply and suchlike are located) is positioned beneath the line of contact between the two hulls, enabling television cameras, if installed, to be angled and point downwards without any visual obstructions; additionally, it reduces the volume of the structure. Two large vertical propellers are positioned both at the prow and stern. Eight further propellers are used for orientation; however these could be reduced to four, the minimum number required for full manoeuvrability. The connecting structure is a central tube running in the longitudinal direction between one supporting arm and the other – these arms are used to hold the propellers in position and are hence rigid.

The design of the various solutions was performed using the CAD package CATIA V5. In total, three solutions were investigated; however the relative dimensions of the structure remained unchanged – only the total length was varied between 25, 34 and 36 metres. Throughout the design process the implications of construction were considered, in such a way so as not to render the structure prohibitively expensive. Hence components were designed to have symmetry – the substantial size of the various components does not permit entire pieces to be manufactured at once, given the limitations of the dimensions of the available moulds. An additional advantage of symmetry is that the number of moulds required is reduced, with a subsequent reduction in production cost.

The structure (Figure 3) is comprised of a central longitudinal member, to which are connected four supporting strips or arms of an elongated S-shape running along the surface of the hulls, and three *septa* (dividing structures) which support the nacelle. Inside the latter the payload is contained, which may include a range of video and photographic devices, equipment relating to mapping of terrain, and the power unit. Both the septa and the supporting arms are attached to belts encompassing the hulls to prevent the membranes from assuming the typical banana-shape which would otherwise occur.

The central tube is the part to which all the perpendicular components are attached – the supporting arms and the septa. Its length is proportional to the sum of the two semi-axes of the ellipses that constitute the hulls (around 0.6 times the total length); it must be sufficiently long to allow the supporting arms to encompass the hulls. Additionally, the tube supports the two vertical propellers at either extremity.

The supporting arms represent the most critical part of the design since due to their long and slender shape, are likely to experience the most severe deflections of the entire structure. Their function is to impede the hulls from changing shape, as well as being points of attachment for the eight longitudinal manoeuvring propellers. It is critical that these components be resistant to excessive deformation: the control of the airship depends completely on the positioning of the propellers – if these are displaced significantly there would be problems in manoeuvring.

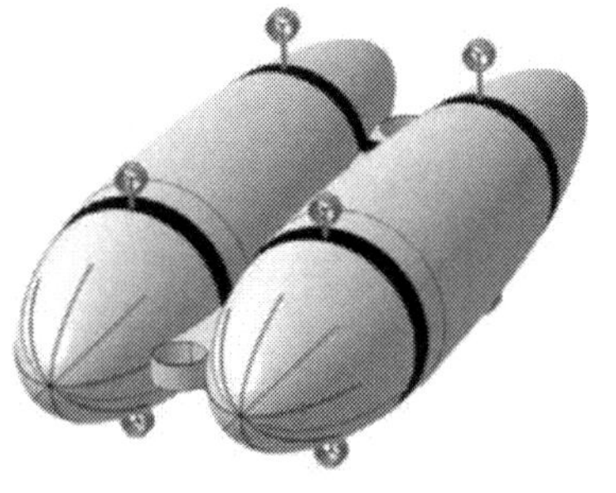

Figure 2 : Dual-hull configuration

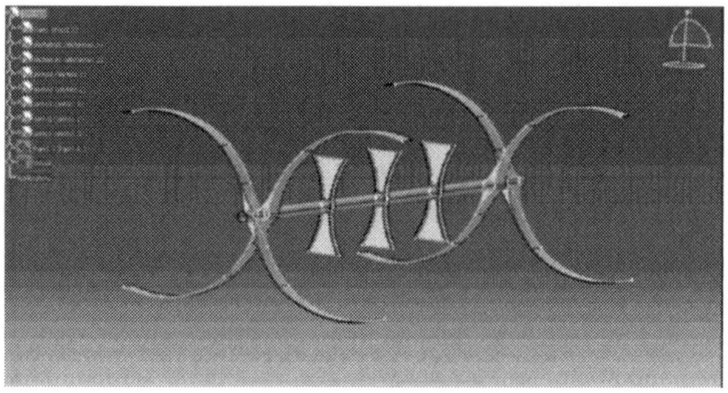

Figure 3: Illustration of structural assembly.

The central septa has two functions. The first regards the hulls – together with the supporting arms they keep the membranes in position with the aid of straps encompassing the membranes. The second is to support the nacelle below. In the model created for the simulation in MSC.NASTRAN, the nacelle was assumed to be positioned hanging from the septa in the space between the two hulls: the connection was simulated with rigid elements. As with the supporting arms, the septa were intentionally designed to have planes of symmetry – in fact they exhibit symmetry in every plane. As such, due to their symmetrical nature it may no longer be necessary to suspend the nacelle, rather, simply attach it to the septa.

The dimensions of the hulls were subjected to careful aerodynamic investigation, such that the air resistance due to its shape was as low as possible, even at low velocities. The slenderness ratio of the hulls (ratio of length to radius) is the most important geometric parameter in determining the hull drag. The shape of the hull is constructed of two half-ellipses, one having its major axis equal to a third of the total length of the airship; the second having a major axis of the remaining two-thirds of the length. The minor axes of both, clearly, are the same and equal to approximately one-eighth of the total length. Hence it is a solid of revolution whose dimensions are fundamental to subsequent calculations – the volume of helium that can be contained (and hence the aerostatic lift); and the surface area of the hulls, which is necessary to calculate the mass.

3. Exoskeletal Configuration

The second configuration is that of the exoskeletal design (Figure 4), formed by drawing together and uniting the two hulls. The principal advantage is that for an equivalent size there is an increase of around 20% in the volume available for helium, thus in theory it is possible to increase the payload or decrease the length of the structure. In reality it is necessary to evaluate the mass of the structure of this second solution, which is found to be greater than the dual-hull design of the same length. There are four vertical propellers, positioned at the prow and the stern, or along the sides of the airship.

The structure of the exoskeletal design is geodesic and constructed of eight "ribs" of a T-section (Figure 5). Thus the propellers can be attached in more different positions than compared to the dual-hull. Despite this, the number of propellers was reduced for eight to four – two positioned on top of the airship and two on the underside, all aligned with the longitudinal axis. In this case the nacelle is also positioned on the underside.

In this design the envelope is no longer a single entity, instead it is divided into an internal membrane containing the helium, and an outer layer purely to retain the shape of the structure.

There are two main causes of deformation in the exoskeletal design: the buoyancy that is acting upwards on the upper "spine" of the airship, and the weight of the nacelle that acts downwards on the belly. The result is a dilation of the structure along the vertical direction, resulting in deformations greater than one metre. The solution to this problem was the application of a series of restraining cables in the vertical-longitudinal plane, with a consequent, albeit small, increase in mass (around 10kg). A small complication then arises, from a manufacturing point of view, concerning the attachment of these cables to the inside of the membrane, given that any attachments must be impermeable to helium.

An additional increase of weight is caused by the necessity of having a structural network of a sufficient density so as not to allow the membrane to dilate between bracing cables – this would cause problems from an aesthetic and aerodynamic perspective.

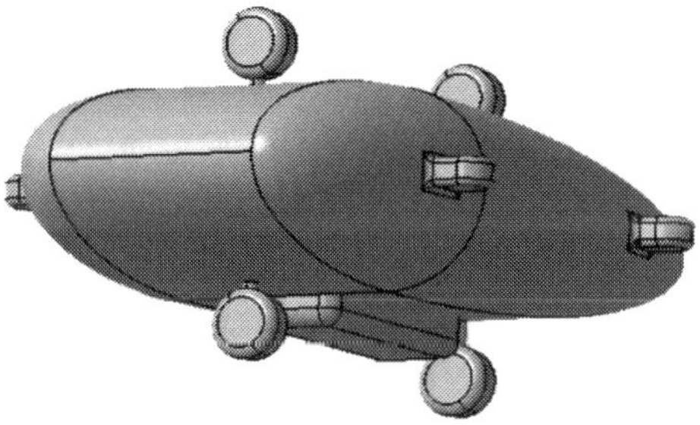

Figure 4 : Exoskeletal configuration

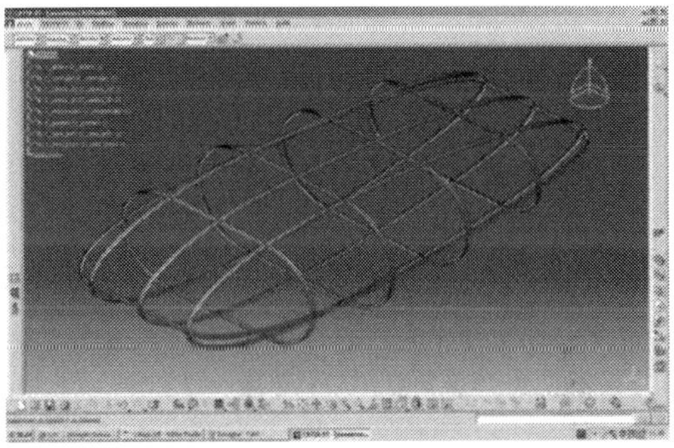

Figure 5: Illustration of structural assembly

4. Cost Function Definition

To select the best solution, a cost function was introduced that accounted for a number of design criteria, such that a more desirable design would produce a lower cost index. Each criterion was weighted according to its perceived importance to give an objective view of the suitability of the design.

The following factors were considered:

- The dimensions: the cost of this parameter is considered to be linearly dependent and given a medium-high importance.
- The weight: evaluated in a similar manner to the dimensions, but with a medium-low importance.
- Sensitivity to wind: this is an aerodynamic consideration, evaluated in terms of aerodynamic derivatives of stability. The cost is given a quadratic relationship and a high importance.
- The handling in mid-air and the positioning of the payload: since in both cases it is difficult to define a parameter that accurately represents these criteria, they were judged quantitatively and assigned a medium importance.
- Cost of purchase and running costs: a linear parameter with a medium-high importance.
- Reliability, service life and maintenance (cost and duration of maintenance): these are parameters that can be considered similar if not equal for both solutions (or at least are subject to discretion and the design objectives of the project, having clear cost implications) They have linearly-increasing costs and have been attributed a medium-high importance.

5. Material Selection

In both cases the material was chosen to be a multilayered composite of carbon/epoxy arranged in four layers [0/45/-45/90]. The mechanical properties are reported in [5], [6].

A preliminary series of analyses were performed using simply a thickness of 2 or 2.5mm on all the surfaces designed with the software CATIA. Given the unsatisfactory results obtained, it was subsequently decided to increase the moment of inertia of the section by substituting the material with one of a sandwich structure. The sandwich structure comprised two 1.6mm sections of the four layers of the composite described above, in the same arrangement, either side of a central 25mm lightweight orthotropic honeycomb core. The nature of the layup resulted in macroscopic quasi-isotropic material properties. All materials in the model were assumed to be linear elastic and homogeneous. The honeycomb material had elastic moduli of 121MPa and 100MPa in the x and y-directions respectively, with a density of 50g/dm^3. In comparison, the composite had elastic moduli of 157GPa in the direction of the fibres and 8GPa orthogonal to them.

6. External Load Definition

The loads considered acting on the structure are from two sources – those resulting from the mass of the structure, and the forces from the propellers. The propeller forces were simulated as point forces along the axis of rotation of the motors. The magnitude and direction of these forces was varied to simulate the gamut of possible manoeuvres, and in particular the loading cases which resulted in the most severe deflections of the structure.

The buoyancy was applied to the upper part of the structure in the case of the exoskeletal design and to the septa on in the simulation of the dual-hull design. The value of the buoyancy was calculated based on the volume of helium and applied in discrete sections across the structure. This approach was necessary at this stage in the analysis since the interactions between the helium envelopes and the structure were not explicitly modelled.

The payload and the avionics were modelled as added masses that were distributed over the lower part of the cargo bay. The landing gear was considered in a similar manner. The mass of the envelopes was distributed over the entire structure. The aerodynamic forces were not considered in the analysis – they would require a distributed pressure over the surfaces of the model. However, they are not thought to contribute significantly to the loading. In order to properly consider these effects, a coupled fluid-structure interaction model would be required.

The inertial loads on the structure are accounted for using the option of *inertia relief* that is available in Patran – this is detailed in the following paragraph.. All loads on the structure were considered to be static – that is, inertial effects during loading were not accounted for. This is justified by the slow application of pressure and motor forces in reality.

7. Structural Analysis

The structural analysis was performed using two software packages – firstly the pre-processing in MSC.Patran and then the structural solution using MSC.Nastran. A major problem in the simulation of both structures was that the software required that both models were constrained in some manner, otherwise the stiffness matrix would become singular on inversion. To avoid this problem it was decided to use the method of *inertia relief*.

Inertia relief is a command in MSC.Patran that allows the simulation of an unconstrained structure under linear static conditions. This approach avoids the problem of unrealistic stress concentrations which would arise in the use of conventional constraints. Typical applications of this method include the simulation of aircraft in flight, satellites in space or, as in this situation, an airship under various loading conditions. Under inertia relief the structure is unconstrained, but the inertia of the structure resists the loading in such a way that the entire structure is in a state of equilibrium.

In a simulation using inertia relief, the analyst selects a point in the structure to be a "support". The FEA solver then applies a distribution of uniform acceleration such that the induced inertial forces and the design loads (pressure and forces) produce a reaction force of zero magnitude on the same support, thus resulting in a system of forces in equilibrium.

When the "inertia relief" option is invoked in a static analysis, MSC.NASTRAN calculates the resultant force in all directions, and thus the field of acceleration that must be applied to the entire structure such that it will be in equilibrium.

8. Results and Comparison of the Two Models

The structural design in this project was assisted by a number of structural analyses, which have allowed the calculation of member thicknesses, the geometry and hence the total mass of the proposed structure. These values, combined with information regarding the loads on the structure, were compared with the aerostatic lift generated from the buoyancy of the helium. The objective was to ascertain the necessary length of the airship – it was found that for the exoskeletal design required a minimum length of 32m, while the dual-hull could be no less than 36m. Once the required length was established such that the lift balanced the weight, a structural comparison was performed between the two solutions.

8.1 Displacements

Table 1 shows the results from the two models under a variety of loading situations.

	Exoskeleton	Double-hull
LOAD CASES	Displacements[m]	Displacements[m]
Hovering	0,132	0,336
Mot. Long. 1	0,147	0,354
Mot. Long. 2	0,149	0,340
Mot. Long. 3	0,147	0,971
Mot. Long. 4	0,144	0,752
Mot. Vert. 1	0,182	0,336
Mot. Vert. 2	1,150	0,344
Mot. Vert. 3	0,147	0,384
Mot. Vert. 4	0,144	//
Parking 1	0,533	//
Parking 2	0,773	0,307

Table 1

As can be seen in the above table, both models exhibit displacements that are within acceptable limits of 500mm (this value was verified using the ETF simulator by updating the propellers positions in the situation with the structure deformed in order to check that the airship maintains complete controllability) for the majority of load cases. However, to arrive at these results it was necessary to modify the designs to some extent. For the exoskeletal model, bracing cables were used in the vertical direction in such a way as to attach the upper surface of the airship to its equatorial rib, thus impeding the structure from deforming in the vertical sense: this alteration significantly reduced displacement.

Regarding the dual-hull model, restraining straps were simulated that joined the top and bottom ends of the supporting arms, to observe whether the displacements would still be as substantial. The results showed that displacements relative to the two S-shaped arms were notably decreased. For the dual-hull model, the simulation was not entirely realistic since the hulls themselves were not included, thus it was not possible to investigate the manner in which the restraining straps interacted with the membrane and the resultant stresses on both bodies.

8.2 Stress Analysis

In comparison, it was found that the results from the two models were quite similar. The dual-hull design performed particularly well due to its optimum distribution of stress, which did not surpass 300MPa at any point in the structure. Conversely the exoskeletal solution yielded critical stresses under forces from propellers in the vertical axis, when the structure was completely supported by the ground (the takeoff situation). In reality these comparisons were drawn without considering the alleviating effect of the buoyancy, which reduces the load by 5% in operational conditions. During construction, however, forces from propellers and buoyancy are not present, so the structure could indeed experience the magnitudes of stress calculated.

Table 2 shows the results from the inertia relief analysis for various loading situations.

LOAD CASES	Exoskeleton Stress[MPa]	Double-hull Stress[MPa]
Hovering	353	144
Mot. Long. 1	367	193
Mot. Long. 2	346	269
Mot. Long. 3	354	280
Mot. Long. 4	364	264
Mot. Vert. 1	375	133
Mot. Vert. 2	354	155
Mot. Vert. 3	369	163
Mot. Vert. 4	1120	//
Parking 1	435	//
Parking 2	810	137

Table 2

The problem of the magnitude of the stress can be resolved by increasing the thickness of the material, however this will of course increase the weight of the structure, and cause further problems with the production, transportation and construction.

8.3 Aerodynamics

Numerical analyses were performed of turbulent incompressible flow on both models [10], which revealed that there was a distinct advantage from an aerodynamic point of view for the exoskeletal design, in particular in the "cross-wind" situation (hovering with air flow perpendicular to the axis of the ship, as shown in Figure 6). Conversely, the presence of the two hulls actually accentuated the strength of vortices since there is additional flow separation between the two hulls as well as the wake behind the entire structure. Aside from the increase in drag, this could be dangerous if the frequency of aerodynamic excitation is coincident with the natural modes of the structure. Such a coupling could not only cause damage to the structure but also, in extreme situations, may result in structural failure. Due to this fact and additional aerodynamic considerations, the exoskeletal design is considered to be superior.

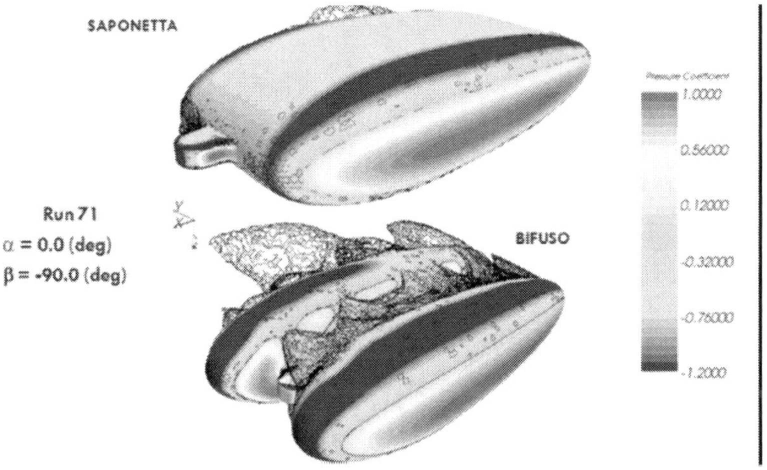

Figure 6: Air flow perpendicular to the axis of the airship

8.4 Costing

A further comparison can be drawn from a construction point of view – it should be noted that for the dual-hull solution only the gas envelopes are 36m long; the entire structure does not surpass 22m in length. This is surely a point in favour of this design – the number of pieces to assemble and disassemble is considerably less than the alternative solution, which would also make transportation less expensive than the exoskeletal design. Cost coefficients were evaluated for each design, taking into account a number of factors: the number of components, the number of moulds used, the number of connections and fixings, the difficulty of construction and the quantity of material necessary. A comparison between the two revealed that the dual-hull design offered a saving of around 15-20%.

8.5 Cost function consideration

Of the other parameters in the cost function, values are given to the greatest accuracy that can be permitted in the early stages of the design, or neglected from this first analysis due to the difficulty of determining them at this point. The parameters that have been examined above are thought in any case to be most significant, and are hence the basis for the following conclusions.

9. Conclusions

In summary, the cost function reveals that the structural and financial advantages of the dual-hull design are roughly balanced by the aerodynamic benefits of the exoskeletal design. As such, a third design (shown in Figure 7), is in the development stage, which will unite the aerodynamic advantages of the exoskeletal design with the construction benefits of the dual-hull: Illustrated are the internal cables (for structural loads) and supports (to maintain the position and form of the membrane) together with the load-bay (for the avionics, power systems, payload etc.). A robust longitudinal keel has been introduced, this being the sole rigid structural element, on which act all the loads in the system, including propulsive and aerodynamic forces and buoyancy. Instead the command and control systems are to remain unchanged.

Figure 7: The new structural solution

Acknowledgements
The authors would like to thank Will Backer for his assistance in the writing of this article. The activity presented in the article is part of the project entitled *Innovative solutions for control systems, electric plant, materials and technologies for a non-conventional remotely-piloted aircraft* funded by the Piedmont Region.

References

[1] Elfes A., Bueno S.S., Bergerman M., Ramos J.Jr.G., "A Semi-Autonomous Robotic Airship for Environmental Monitoring Missions", *Proceedings of the 1998 IEEE International Conference on Robotics & Automation*, Leuven, Belgium, May 1998, pp. 3449-3455.
[2] Zaugg D.A., Arnold D.V., Jensen M.A., "Ocean Surface and Landside Probing with a Scanning Radar Altimeter", *Proceedings of the 2000 International Geoscience and Remote Sensing Symposium,* vol. 1, 2000, pp. 120-120.
[3] Hain J.H.W., "Lighter-than-air Platforms (Blims and Aerostats) for Oceanographic and Atmospheric Research and Monitoring", *Proceedings of the Oceans 2000 MTS IEEE Conference and Exhibition*, vol. 3, 2000, pp. 1933-1936.
[4] Cappadona A., "Analisi Strutturale su Dirigibile non Convenzionale a Controllo Remoto: Configurazione Esoscheletro", MSc Thesis, October 2008 .
[5] Lecca R., "Analisi Strutturale su Dirigibile non Convenzionale a Controllo Remoto: Configurazione Bifuso", MSc Thesis, October 2008.
[6] Battipede M., Lando M., Gili P.A. and Vercesi P., "Peculiar Performance of a New Lighter-Than-Air Platform for Monitoring", *AIAA, 4th ATIO Forum,* September 2004, Chicago, Illinois.
[7] CATIA V5R18, Computer Aided Design (CAD) software, Dassault Systems, France.
[8] MSC Nastran 2005, Finite element method (FEM) software MSC Software, USA.
[9] MSC Patran 2005, FEM pre and post-processing software, MSC Software, USA.
[10] Visone M., (06/2008), ETF06 Airship Aerodynamic Investigations.

AUTHOR INDEX

AUTHOR INDEX

AUTHOR INDEX

Institute of Physics
Dirac House, Temple Back
Bristol BS1 6BE UK

CURRAN ASSOCIATES INC
proceedings
.com